The Hive and the Honey Bee

A new book on beekeeping which continues the tradition of "Langstroth on the Hive and the Honeybee"

EDITED BY JOE M. GRAHAM

*With Individual Chapters Written
by a Staff of Specialists*

J. T. Ambrose, E. L. Atkins, A. Avitabile, G. S. Ayers, M. S. Blum, S. L. Buchmann, D. M. Caron, E. Crane, C. C. Dadant, A. Dietz, E. H. Erickson, B. Furgala, N. E. Gary, J. R. Harman, E. W. Herbert, Jr., R. A. Hoopingarner, E. E. Killion, D. A. Knox, H. H. Laidlaw, Jr., D. M. McCutcheon, R. E. Page, Jr., H. Rodenberg, M. T. Sanford, J. O. Schmidt, H. Shimanuki, R. W. Shuel, R. E. Snodgrass, E. E. Southwick, J. E. Tew, G. D. Waller, J. W. White, Jr., J. L. Williams, and M. L. Winston

REVISED EDITION

DADANT & SONS • HAMILTON, ILLINOIS
Publishers of the American Bee Journal

Revised 1949
Second Printing 1954
Third Printing 1960
Fourth Printing 1961
Extensively Revised 1963
Second Printing 1966
Third Printing 1970
Fourth Printing 1973
Extensively Revised 1975
Second Printing 1975
Third Printing 1976
Fourth Printing 1978
Fifth Printing 1979
Sixth Printing 1982
Seventh Printing 1984
Eighth Printing 1986
Ninth Printing 1987
Extensively Revised 1992
Second Printing 1993
Third Printing 1997
Fourth Printing 1999
Fifth Printing 2000
Sixth Printing 2003
Seventh Printing 2005
Eighth Printing 2008
Ninth Printing 2010

Library of Congress Catalog Card Number: 92-81904

ISBN 0-915698-09-9

Use permitted with acknowledgment

PRINTED IN THE U.S.A. BY BOOKMASTERS, INC.
ASHLAND, OHIO

TYPESETTING AND PRE-PRESS WORK BY M & W GRAPHICS, INC.
CARTHAGE, ILLINOIS

"I have determined, in writing this book, to give facts, however wonderful, just as they are; confident that in due time they will be universally received; and hoping that the many wonders of the economy of the honey-bee will not only excite a wider interest in its culture, but lead those who observe them to adore the wisdom of Him who gave them such admirable instincts."

L. L. Langstroth

FOREWORD

LANGSTROTH *on the Hive and the Honey-Bee, a Beekeeper's Manual* was the title of the original book by the Reverend L. L. Langstroth in 1853. This practical treatise made available to the world his fundamental discovery of the bee space and his invention of the top-opening, movable-frame hive which made modern beekeeping possible. More than a century has passed since Langstroth's discovery, but our system of beekeeping still is based on his hive and methods.

Langstroth's book was well received: a second edition appeared in 1857, a third edition just 2 years later in 1859; and his last revision in 1875. Not wishing to attempt another edition alone, because of ill health, he began to correspond with Charles Dadant and his son, Camille Pierre Dadant, in 1881. This correspondence was to be interrupted often by Langstroth's nervous trouble until it became necessary for the Dadants to revise and publish the book. Langstroth entrusted the rewriting of the book to them in 1885.

The Dadant revision bore the title *Langstroth on the Hive and the Honey Bee, revised, enlarged, and completed by Chas. Dadant and Son.* It was published in 1889. A second edition appeared in 1893, a third in 1896, and the fourth in 1899. Charles Dadant translated it into French in 1891; a Russian edition appeared in 1892. It was later published in Italian, Spanish, and Polish.

Charles Dadant died in 1902, and the four succeeding revisions were the work of C.P. Dadant, retaining the title *Langstroth on the Hive and the Honey Bee.* His first revision was published in 1907 and was called the 20th Century edition. Consequently, succeeding revisions were called the 21st edition (1922), the 22nd edition (1923), and the 23rd edition (1927). C. P. Dadant passed away in 1938.

James C. Dadant, a grandson, undertook the task of a new revision about the time of World War II. It was his proposal that a completely new book should be written by a group of authors, each dealing with the subject with which he was most familiar. But the war intervened and James was called to service. The task of carrying out the complete revision then evolved to Roy A. Grout, *American Bee Journal* editor. The new book was given the title *The Hive and the Honey Bee.*

The first multi-author edition was published in 1946 and a somewhat more revised edition appeared in 1949. An extensive revision was released in 1963. In 1975 the book was again extensively revised and enlarged by Vern Sisson, *American Bee Journal* editor, who had started the book before leaving Dadant & Sons, Inc., in 1974, and completed by Howard Veatch, publications director. The 1992 edition continues this tradition by further updating and enlarging upon this classic beekeeping book.

Previous authors of portions of the book include: Robert Banker, Colin G.

[v]

Butler, Gladstone H. Cale, Sr., Gladstone H. Cale, Jr., Walter J. Diehnelt, Thomas A. Gochnauer, Carl E. Killion, Sr., E. C. Martin, Everett Oertel, James Powers, Frank A. Robinson, Walter C. Rothenbuhler, Friedrich Ruttner, William A. Stephen, Gordon F. Townsend, Peter C. Witherell and Harvey F. York, Jr. Their previous chapters and contributions are deeply appreciated.

Special mention is made here of the death of one of the present authors of this revision, Dr. Elton W. Herbert, Jr., research entomologist with the USDA, ARS Bee Research Laboratory in Beltsville, Md. His chapter, *Honey Bee Nutrition*, was largely completed before his death, but his colleagues at the Bee Laboratory, notably David A. Knox and Hachiro Shimanuki, were kind enough to finish the final preparation of Dr. Herbert's chapter. His contribution and the help of his colleagues are greatly appreciated.

Special mention is also made here of Chapter 4 by the late Dr. R. E. Snodgrass, a world authority on the anatomy and physiology of insects. His work on honey bee anatomy still stands as the recognized reference standard and therefore is reproduced here largely as it appeared in the previous edition of *The Hive and the Honey Bee*. However, Dr. Eric H. Erickson has incorporated many fascinating scanning electron micrographs of much of the honey bee's anatomy to complement Dr. Snodgrass' drawings. He also has reviewed and updated Dr. Snodgrass' chapter where necessary.

In this extensively revised 1992 edition of *The Hive and the Honey Bee,* new authors have been selected and new chapters added, in an attempt to present an up-to-date worldwide picture of beekeeping, suitable for practical as well as classroom reference.

Charles Dadant emigrated to America in 1863 with his son, Camille Pierre, and, together, they dedicated their lives to beekeeping throughout the world. The present Dadant organization, which represents the fourth and fifth generations, is privileged to carry on that tradition of devotion to the knowledge of bees and beekeeping and celebrates one hundred and twenty-nine years of service to the industry with the publication of this book.

ACKNOWLEDGMENTS

DADANT & SONS, INC., publishers of the *American Bee Journal*, especially desire to express their thanks and appreciation to the co-authors of this revision and for illustrations furnished by them. Fully recognizing that these persons have contributed generously of their time and knowledge, due acknowledgment is given here to:

John T. Ambrose	Eugene E. Killion
E. Laurence Atkins	David A. Knox
Alphonse Avitabile	Harry H. Laidlaw, Jr.
George S. Ayers	Doug M. McCutcheon
Murray S. Blum	Robert E. Page, Jr.
Stephen L. Buchmann	Harry Rodenberg
Dewey M. Caron	Malcolm T. Sanford
Eva Crane	Justin O. Schmidt
Charles C. Dadant	Hachiro Shimanuki
Alfred Dietz	Reginald W. Shuel
Eric H. Erickson	Robert E. Snodgrass
Basil Furgala	Edward E. Southwick
Norman E. Gary	James E. Tew
Jay R. Harman	Gordon D. Waller
Elton W. Herbert, Jr.	Jonathan W. White, Jr.
Roger A. Hoopingarner	Jon L. Williams

Mark L. Winston

Appreciation and acknowledgment likewise are given to the British Columbia Ministry of Agriculture for permission to publish the manuscript and photos of Doug M. McCutheon; to the University of California at Davis for permission to publish the manuscripts and photos of Dr. Norman E. Gary, Dr. Harry H. Laidlaw, Jr., and Dr. Robert E. Page, Jr.; to the University of California at Riverside for permission to publish the manuscript and photos of Dr. E. Laurence Atkins; to the University of Connecticut for permission to publish the manuscript of Dr. Alphonse Avitabile; to the University of Delaware for permission to publish the manuscript and photos of Dr. Dewey M. Caron; to the University of Florida for permission to publish the manuscript and photos of Dr. Malcolm T. Sanford; to the University of Georgia for permission to publish the manuscripts and photos of Dr. Murray S. Blum and Dr. Alfred Dietz; to the University of Guelph, Canada, for permission to publish the manuscript and photos of Dr. R. Shuel; to the International Bee Research Association for permission to publish the manuscript and photos of Dr. Eva Crane; to Michigan State University for permission to publish the manuscripts and photos of Dr. George S. Ayers, Dr. Jay R. Harman and Dr. Roger A. Hoopingarner; to the University of Minnesota for permission to publish the manuscripts and photos of Dr. Basil Furgala; to the National Honey Board for permission to publish the manuscript and photos of Harry Rodenberg; to New York State University College at Brockport for permission to publish the manuscript and photos of Dr. Edward E. Southwick; to North Carolina State University for permission to publish the manuscript and photos of Dr. John T. Ambrose; to the Ohio State University for permission to publish the manuscript and photos of Dr. James E. Tew; to Simon Fraser University, Canada, for permission to publish the manuscript and photos of Dr. Mark L. Winston; and to the United States Department of Agriculture for permission to publish the manuscripts and photos of Stephen L. Buchmann, Dr. Eric H. Erickson, David A. Knox, the late Dr. Elton W. Herbert, Dr. Justin O. Schmidt, Dr. H. Shimanuki, the late Dr. R. E. Snodgrass, Dr. Gordon D. Waller and Jon L. Williams.

The editor desires to express his special appreciation to Mildred and William Marshall of M & W Graphics, Carthage, Ill., for their expert counsel and help during the nearly five years of work involved in completing this new edition. Special thanks is also extended to Lenore Kimbrough of Hamilton, Ill., for doing the thorough job of assembling the book index.

Appreciation is also expressed to photographer Kenneth Lorenzen, University of California, Davis, for providing the flyleaf photographs and to artist Kathryn Hyndman, Hamilton, Ill., who redrew the spine illustration from the 1857 edition of L. L. Langstroth's *The Hive and the Honey-Bee* which is used on this edition's spine and front cover.

The following Dadant & Sons, Inc., employees are also thanked for their help with proofreading and other work: Terry Avise, David Coovert, Nick Dadant, Tim Dadant, Gerald Hayes, Steve Martin, Shirley Moeller and Tom Ross.

Various portions of previous editions of this book have been freely incorporated with or without individual citation. For the use of this material, due acknowledgment is gladly given to the authors.

TABLE OF CONTENTS

[ix]

Table of Contents   *xi*

Chapter   *Page*

*The Hive
and the Honey Bee*

CHAPTER 1

THE WORLD'S BEEKEEPING — PAST AND PRESENT

by EVA CRANE*

The present book is the direct successor to one written by Lorenzo Lorraine Langstroth, published in 1853 under the title *Langstroth on the Hive and the Honey-bee* (1853). This made known Langstroth's practical application of the concept of the bee-space in 1851, and laid the foundation of modern beekeeping. The first part of this chapter gives a brief account of the history of beekeeping with the honey bee *Apis mellifera,* and touches on some of the archeological discoveries made since the last edition of the book was published. Other bees of economic importance are also mentioned. Finally a summary is given of the extent of beekeeping in various parts of the world today.

BEEKEEPING UP TO 1600

Honey bees now live in all parts of the world except the extreme polar regions, but this was not always so. Until the 17th century they were confined to the Old World, where they had evolved and were widely distributed long before man appeared on earth. Primitive man harvested honey comb from bees' nests in hollow trees and rock crevices. The operation with *Apis mellifera* is shown in many rock paintings found in Africa, and some in Spain (Fig. 1), and with other species of bees in India and Australia. It is still carried out in various parts of the world, and honey can be a lifesaving food for primitive peoples in times of severe famine.

Beekeeping proper started when man learned to safeguard the future of swarms and colonies established from them, by a certain amount of care and supervision, keeping them in separate purpose-made hives which substituted for natural dwellings of bees. For convenience and safety, a number of hives were often sited together in an apiary. Hive construction depended on what materials were at hand, and on the local skills of various communities. It is almost certain that the beehive had no single origin: it was a likely development in any region populated by honey bees, when man changed from hunting and collecting food to producing it, and started a settled existence.

The earliest centers of culture were in the Middle East, in hot, dry open country. Pottery vessels were made during most of the Neolithic period, from perhaps 5000 B. C. onwards, and the first hives there may have been pottery

---

*Eva Crane, OBE, DSc, Honorary Life President, International Bee Research Association, Past Editor of *Bee World, Apicultural Abstracts,* and *Journal of Apicultural Research.* World authority on bees and beekeeping.

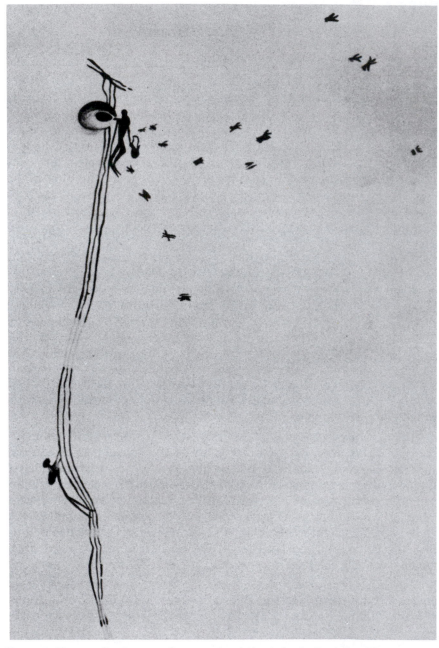

FIGURE 1. Honey-collecting scene from a rock painting in La Araña shelter, Bicorp, eastern Spain, made around 6000 B.C. and found in 1924 (copy by E. Hernández-Pacheco).

vessels in which swarms happened to settle. In ancient Egypt and other Mediterranean regions, long cylindrical hives were made of mud, clay or other materials, and used in a horizontal position. Figure 2 shows honey harvesting about 1450 B.C. The beekeepers used smoke to drive the bees from the back of the hive to the front, and honey combs could be removed without killing bees; the brood combs were recognized and could be left intact. Hives in Ancient Greece followed a somewhat similar style, and fragments of pottery hives have been excavated from 26 sites; Figure 3 shows a complete hive. In hot, dry regions the hives were often stacked together, and sometimes embedded in a wall, to provide insulation against the heat.

FIGURE 2. Honey harvesting from hives, shown in a wall painting *c.* 1450 B.C. in Rekhmire's tomb, West Bank, Luxor, Upper Egypt (Davies, 1944)

In the great forests of northern Europe, wild colonies in their nests in trees were tended from 2000 B.C. onwards. The earliest hives were probably logs from fallen trees, separated by chipping away the rest of the tree with axe and adz (a technique used throughout the Stone Age), and stood upright. Such hives can still be found in wooded areas of the north temperate zone (Fig. 4). Cork and other types of bark were also made into hives and, later, planks cut from tree trunks.

FIGURE 3. Two pottery hives in Greece; in front, hive from about 300 B.C. excavated near Mount Hymettus; behind, hive in use in 1973 on the island of Antiparos in the Cyclades (photo: M. I. Geroulanos).

In agricultural communities, techniques were developed for making containers of basket work as well as pottery. Baskets were also used to house bees, and they have changed little through the ages; those of coiled straw are made today in the same way as before 5000 B.C. Wicker baskets were woven from materials such as pliable hazel twigs. Two wicker hives are shown in Figure 5, which also provides one of the earliest illustrations of protective clothing for beekeepers.

All these primitive hives fulfilled certain necessary functions: they protected the bees and their combs from wind, rain, and extremes of heat or cold; their flight entrances were made small enough for the bees to guard; and there was some other opening through which the beekeeper could get at the honey and wax which constituted his harvest. Wood, bark, and clay were themselves weatherproof; straw and wicker hives were generally protected with an additional cover, and wicker hives were often plastered with mud and cow dung.

In northern Europe the upright log hives were often large, but with basket hives (skeps) it was easier to vary the size, and quite small ones were used where beekeepers wanted their colonies to swarm early to populate empty

FIGURE 4. Apiary of log hives, with two made from planks in the center, in North Carolina, U.S.A., *c.* 1960 (photo: W. A. Stephen).

hives. Primitive beekeeping then consisted of little more than catching and hiving the swarms that issued in early summer and, at the end of the season, killing the bees (for instance with burning sulfur, or by plunging the hive into boiling water) to get the honey and wax. More advanced beekeepers cut out the honey combs and left the bees and brood; if necessary they provided food in hives for winter.

The horizontal hives used farther south usually had a closure at the back, and the flight entrance at the front end; an Egyptian tomb later than Figure 2 seems to show this (Crane, 1983). Sometimes the front end also had a movable closure. The beekeeper gained access to honey combs from the back and inserted food there. From studies of traditional beekeeping now practised in the Mediterranean region, it seems likely that beekeepers in some Ancient countries — for instance Egypt, Arabia, Crete and Sicily — carried out

FIGURE 5. A well protected beekeeper with two wicker skeps (from Sebastian Münster's *Cosmographia*, Bern, Switzerland, 1535).

various colony manipulations, including the division of colonies in summer (see *e.g.* Crane, 1983).

In the New World there were no native honey bees, but species of stingless bees — social honey-storing bees in the family Meliponinae — are native to the tropics of America and Australasia. In Central America the Maya still keep some species in horizontal log hives (Schwarz, 1948), which are closed at the ends with wood or stone discs, each made to the shape and size required. Wood is perishable, and no hives are known to survive from earlier centuries. However, stone closures have been excavated at 12 sites, 6 in Mexico, 4 in Belize and 2 in Guatemala (see Crane & Graham, 1985). Figure 6 shows locations of limestone discs found at one site, and dotted lines have been drawn to join together 12 discs in pairs, which are likely to have been the two ends of the same hive. This site is on Cozumel, an island off the Yucatán peninsula where 255 of the discs have been found. Also on Cozumel, a pottery incense burner in the form of the Maya bee god Ah Mucan Cab was found, dated to about A.D. 1400, which incorporates models of four hives that appear similar to those used today. It is now in the museum at Mérida. Figure 7 shows some of the end closures from a site in Belize which may date from the Late Preclassic period (300 B.C. to A.D. 300); they provide the earliest remains so far found of beekeeping in the New World.

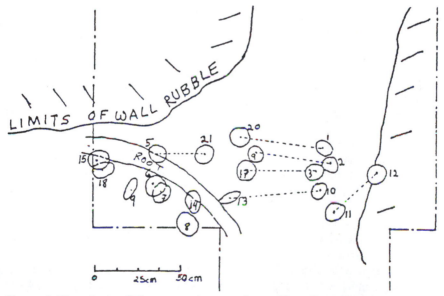

FIGURE 6. Plan of part of the excavated area at Buena Vista, Cozumel, Mexico, showing locations at which stone hive closures were found; those to the left of the root had been disturbed (Wallace, 1978). See text for explanation.

In northern Australia, evidence has been found that aborigines tended nests of stingless bees in trees (Dollin & Dollin, 1986). But unlike tree beekeepers in Europe, these people do not seem to have advanced to beekeeping with hives.

BEEKEEPING 1600 to 1851

Three separate streams of events, each of great significance in the history of bees and beekeeping, were set in motion from Europe in the 17th century. The honey bees, themselves, were taken to the New World, where previously honey and wax had been harvested only in the tropics from species of stingless bees. Beekeepers started to appreciate the fundamental facts of the life cycle and biology of bees. And developments in beekeeping methods gave beekeepers slightly more control over their bees, as well as greater opportunities for observing them inside the hive.

The Spread of Honey Bees over the World

Honey bees evolved in the Old World — Europe, Africa, and Asia. Prior to 1600 there were none in the New World — the Americas, Australia, and New Zealand. But early settlers in many parts of the New World managed to get hives of bees across the oceans to their new lands. European bees probably reached North America from England in 1622 (Smith, 1977), in the *Bona Nova*, the *Hopewell* or the *Discovery**. In 1985, with help from John Adams, I found the probable landing place of these ships, the site of the Old City Point Wharf on the James River, just below its confluence with the Appomattox. In 1688/89 bees were taken from France to St. Kitts and Guadeloupe in the Caribbean. The first honey bees were landed in Australia probably in 1822 and in New Zealand (from England) in 1839. Honey bees were not introduced to the west coast of North America until the early 1850s, when they were landed in California; from there they were taken to Oregon, and thence to British Columbia.

Discovery of Fundamental Facts about Honey Bees

The fact that honey bees could raise a queen from eggs or very young larvae was published in Germany by Nickel Jakob (1568). The first description of the queen bee as a female, which laid eggs and was the mother of all bees in the colony, was published by Luis Méndez de Torres in Spain (1586). Then in England Charles Butler showed (1609) that the drones were male bees, and Richard Remnant (1637) that the workers were females; Remnant had observed that they possessed "a neat place for the receipt of generation." Meanwhile in Italy, Prince Cesi (1625) had published the first drawings of bees made under a microscope. The primary facts about the mating of the

---

*In 1616, a ship also bound for Virginia took refuge in Bermuda during a hurricane. It carried hives of bees which were landed there, and the bees prospered (Hilburn, 1989).

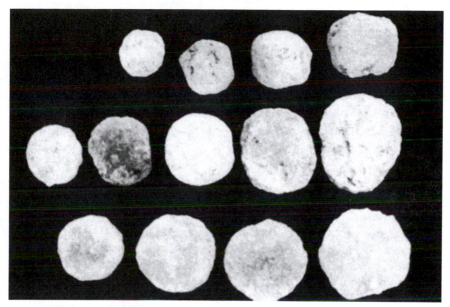

FIGURE 7. Thirteen of 37 hive closures excavated at Chan Chen, North Belize (photo: J. R. Andresen).

queen with a drone were not known until Anton Janscha in Slovenia described them (1771; see Fraser, 1951).

The nature of the materials collected by bees was also established only in the 18th century. Until Hornbostel in Germany described the true origin of beeswax (1744), beeswax was confused with pale yellow pollen carried to the hives, and thought to be collected from flowers. Nectar, on the other hand, was assumed to fall from the skies until 1717 when Vaillant in France showed that it is produced by flowers, in specific parts he called *mielliers*. The fact that the pollen collected by bees is the "male seed" of the flower, which fertilizes the ovum, was reported (1750) by Arthur Dobbs of Castle Dobbs in Ireland. Dobbs also observed that bees gathered pollen from only one kind of flower on each flight, and he suggested that disastrous cross-fertilization would result if this were not so. Then Sprengel (1793) established clearly the part played by bees in fertilizing flowers.

François Huber, a blind Swiss beekeeper, published his *Observations* (1792) which properly laid the foundations of modern bee science.

Developments in Beekeeping Techniques with Upright Hives

Between 1500 and 1851 certain skep beekeepers devised ways of taking the honey from hives without killing the bees. For instance, instead of killing a proportion of their colonies before winter, they united several colonies in a single skep, by driving the bees from a series of hives (each inverted) into another placed mouth down above it. Or an extension was added above a straw skep, over a hole left in the top; it might be a smaller skep (a cap) or a glass jar (a bell). The bees stored honey, but they did not rear brood there; it was a honey super, which could be removed complete with honey combs, without disturbing the brood nest. Alternatively an "eke" was placed under the skep when the flow started; it was a straw cylinder a few inches high that formed an extension of the skep downwards.

The use of hives made of wooden boards allowed new scope for inventiveness. Some hives consisted of a tier of boxes, whereas others — collateral hives — had boxes at the sides for honey storage. Fraser (1958) gives some details.

Throughout these centuries the minds of beekeepers in the most progressive areas were constantly occupied with the problem of getting more control over the bees and their activities, and of monitoring what was going on inside the hive, including the brood nest. Observation windows in hive walls were easy enough to make, but they did not enable the beekeeper to see much of what was going on inside. The Italian astronomer M. Maraldi found single-comb observation hives in the garden of the French Royal Observatory in Paris in 1687 (Réaumur, 1740). Huber's leaf hive (1792) came more than a century later; it consisted of a number of frames hinged together at one side like the leaves of a book, and the bees built combs in the frames. It was invaluable for his observations, but was an observation hive only and quite unsuited to practical beekeeping.

Meanwhile in England Sir George Wheler (1682) described woven wicker hives that he had seen used in Greece. The open end was uppermost, and was covered with wooden 'bars' about 1½ inches wide (Fig. 8). Each bar was made slightly convex on its underside, and the bees attached their combs along the ridges so formed, i.e. one along the underside of each bar. What distinguished this hive from all the other "bar hives" then known was the fact that the bees did not attach their combs to the hive walls. (The hive was wider at the top than at the bottom, so the hive walls sloped inward.) In spring the beekeeper removed half the combs from each hive and placed them in an empty one, thus doubling the number of occupied hives. The Greek beekeepers had in fact produced a workable movable-comb hive. Somewhat similar top-bar hives have been found in several parts of Vietnam (Fig. 9), where I saw some still in use in 1989.

Wheler's report had a considerable influence on hive development in European countries, and between 1650 and 1850 box hives with top-bars and frames were invented (Walker, 1928), but after two centuries of effort there was still failure on a fundamental point: whatever bars or frames were used, the bees attached their comb also to the walls of the hive, and the combs could therefore be removed from the hive only by cutting them out. For instance about 1806 a Ukrainian beekeeper Petr Prokopovich produced the first frame hive to be used on a commercial scale (about 10,000 of the hives were made; see Haydak, 1957). This hive had three vertical compartments in a single box, the top one having wooden frames with notched bee passages in the end-bars; the frames were removed from the back of the hive, but as the bees attached the frames to the hive walls with comb or propolis, this was not at all easy.

It is difficult for us now, with the problem solved and therefore no longer a problem, to enter into the minds of the experimenting beekeepers, who struggled to obtain a convenient workable hive, with combs they could easily remove from it. I do not think these struggles would have been pursued so

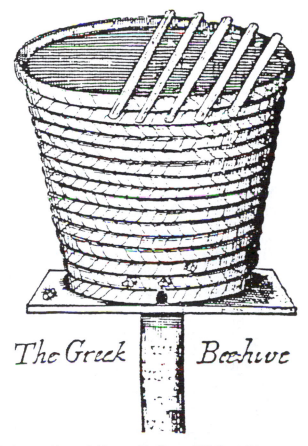

FIGURE 8. Early movable-comb hive used in Greece (Wheler, 1682).

assiduously if the same beekeepers had been using horizontal hives, with their greater scope for bee management; on the other hand it is possible that the honey bees of northern Europe would not have prospered in such hives.

The crucial step which gave the desired movable-comb wooden hive was not taken until 1851, at which time — as we have seen — European honey bees had completed their colonization of almost the whole world.

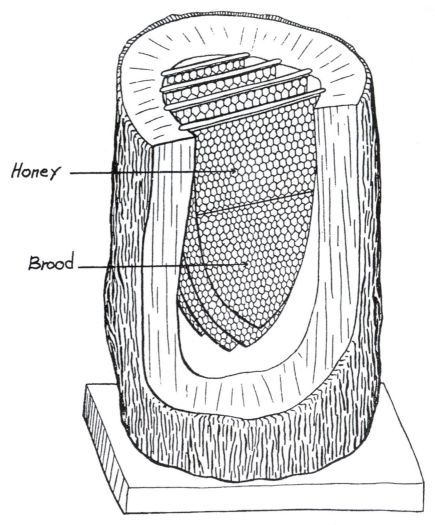

FIGURE 9. Vietnamese movable-comb hive (Toumanoff & Nanta, 1933).

BEEKEEPING 1851 AND AFTER

Lorenzo Lorraine Langstroth (Fig. 10) was born in Philadelphia. As a child he had shown a rather unusual interest in insects, and this was revived when, as a young pastor in Andover, Massachusetts, he visited a friend who kept bees, and saw a glass globe filled with combs of honey. Before he returned home he bought two colonies of bees in box hives. He soon also acquired a Huber leaf hive and in due course obtained various books on bees, including Huber's (1792), Bevan's (1838) and Munn's *Description of the bar-and-frame hive* . . . (1844 and 1851 editions). Johansson and Johansson (1967) found that Langstroth had marked the passage in Munn's book: "The frames with their contents can be lifted out [because of the proper space left between them]." Langstroth used a bar hive described by Bevan and put in it complete frames with "the distance of the bars from each other . . . nicely adjusted." He deepened the grooves on which the bars rested, leaving about 3/8 inch between the cover and the bars, because this facilitated the removal of the cover board on which the glasses rested. (This is the origin of our present top bee-space.) Then came the key development, which cuts the history of beekeeping into two halves. In the fall of 1851, Langstroth wrote in his diary:

> Pondering, as I had so often done before, how I could get rid of the disagreeable necessity of cutting the attachments of the combs from the walls of the hives, and rejecting, for obvious reasons the plan of uprights, close fitting (or nearly so) to these walls, the almost self-evident idea of using the same bee space as in the shallow chambers came into my mind, and in a moment the suspended movable frames, kept at a suitable distance from each other and the case containing them, came into being. Seeing by intuition, as it were, the end from the beginning, I could scarcely refrain from shouting out my "Eureka!" in the open streets.

Langstroth's intuition was justified. The bees did, in fact, "respect" *the bee-space left between the hive and the frames* in which the combs were built. They did not build comb across the space, and the frames could, therefore, easily be lifted out from the top of the hive. Being rectangular, the frames were economical in construction. All had the same length, so any one could be replaced by any other, in the same or in another hive. Also, unlike some other bar or frame hives, Langstroth's could be tiered — built up from a number of hive boxes one placed above another — and this is of great practical utility when working a heavy honey flow.

Langstroth's movable-frame hive was in common use in the United States by 1861. T. W. Woodbury introduced movable-frame hives into England in 1862, and from 1869 onwards the writings of Charles Dadant in the French and Italian journals spread its use in continental Europe; it was soon introduced in many other countries.

Figure 10. Lorenzo Lorraine Langstroth, 1810-1895.

With the use of this hive, modern beekeeping began, and development in the next half-century was in the nature of an explosion compared with the slow and halting progress of the previous centuries. In 1857, Johannes Mehring in Germany produced the first matrix for making beeswax comb foundation; the resulting sheets of foundation saved beeswax, and encouraged the bees to build regular worker comb in the frames. A superimposed honey chamber (super) furnished with such framed combs built on foundation was a valuable appliance. If means could be found to extract the honey without destroying the combs, the whole could be re-used in the hive. In 1865 Franz von Hruschka, an Austrian, produced a contrivance in which honey combs could be spun round and the honey extracted by centrifugal force (see Büdel, 1963).

Slot queen excluders had been used by Prokopovich and others in the early 1800s, to prevent a queen from getting into the honey chamber and thus to keep it free from brood; these were now improved. Then by using a bee escape such as that made in 1891 by E. C. Porter in the United States, it became possible to clear bees out of the honey chamber before removing the frames of honey.

The pattern of modern beekeeping was thus established in the half-century between 1850 and 1900. Equipment invented in that period, or based on principles appreciated then, has since undergone important developments, and also greater mechanization. Beekeeping is now much more efficient and productive than in Langstroth's time, but in principle the hive we use today is based on his design in 1851.

OTHER BEES OF ECONOMIC IMPORTANCE TO MAN

The honey bees in most of the New World, and in Europe whence they came, are various races of the European honey bee *Apis mellifera* —Italian, Caucasian, Carniolan, and so on. There are other races in the Mediterranean region, and in Africa south of the Sahara are tropical races of *Apis mellifera*. Ruttner (1988) gives information on individual races.

In Asia there is a greater variety of honey bees, three representative species all being widely distributed: *Apis cerana, Apis dorsata* and *Apis florea*. *Apis cerana*, the native hive bee of Asia, is very similar to *Apis mellifera* but slightly smaller. It is kept in hives similar to those used for *Apis mellifera* but smaller, and with smaller combs containing smaller cells. There are several races of *Apis cerana*, some of which extend through much of tropical Asia, and others along the eastern parts of Asia as far north as Korea, the U.S.S.R., and Japan. In areas where introduced European honey bees thrive, these surpass the native bees in honey yields, and in certain other characteristics desirable in hive bees. Beekeepers in agricultural areas of China, Japan, Taiwan, and elsewhere have therefore largely replaced their native hive bee by the European bee. In mountains and other uncultivated regions where *Apis mellifera* has not succeeded, beekeeping is still done with *Apis cerana*.

The two other *Apis* species both live in tropical Asia, and both build a nest in the open, consisting of only one comb. *Apis dorsata*, the giant bee or rock bee, gets its names from its large size and its habit of nesting on rock faces. It produces much honey, and in many countries of tropical Asia more honey is harvested from wild nests of *Apis dorsata* than from hives of *Apis cerana* or *Apis mellifera*. The comb of the little bee *Apis florea* yields only a few pounds of honey or less.

In the 1980s evidence was accumulated in parts of Asia for the existence of certain other honey bee ecotypes and species:
— a species *Apis koschevnikovi*, similar to *Apis cerana*, in Sabah (northeast Borneo)

— a possible species *Apis andreniformis*, similar to *Apis florea*, in part of Yunnan, China
— ecotypes of *Apis dorsata* or possibly different species,
laboriosa in the high Himalayas
breviligula in the Philippines, except Palawan
binghami in Sulawesi

Different species of stingless bees are native to tropical Asia, Africa, Australia, and America. In America and Africa they are kept in hives for their honey (Nogueira-Neto, 1970). Archeological evidence for early beekeeping with these bees in Central America was referred to earlier; see Figures 6 and 7.

In the present century, agricultural practices have become more intensive and mechanized, and attention has been directed towards keeping bees for pollination, as well as for producing honey and wax. Honey bees are unsurpassed as pollinators in that large numbers of bees can easily be brought to a crop, and the bees are good general pollinators. Nevertheless, certain other bees are individually far more efficient pollinators of certain crop plants. In parts of the temperate zones, certain bumble bees (*Bombus*) are reared for pollinating red clover and other crops whose florets have long corolla tubes, and also for greenhouse crops. A soil-nesting alkali bee *Nomia melanderi*, native to parts of the northwest of the United States, is a very efficient pollinator of alfalfa, and is reared commercially for the purpose. *Megachile rotundata*, an efficient pollinator from Europe, also pollinates alfalfa well and is more adaptable in its nesting habits; it nests in stems and can be mass-reared in "banks" of man-made tubes of the appropriate diameter.

THE PATTERN OF BEEKEEPING TODAY*

Beekeeping is now spread over almost all the habitable parts of the world, and according to official statistics produces over 2 billion pounds of honey a year from more than 50 million hives. The estimated human population of the world is about 5 billion, so there are nearly a hundred times as many people as hives of bees. Beekeeping is practised over a greater area of the earth's surface than perhaps any other single branch of agriculture, and the success of certain other branches of agriculture depends on it, because of the need for crop pollination.

The patterns of beekeeping tend to be different in the Old and New Worlds. In general the Old World is more densely populated with honey bees, as it is with people, but the New World gives richer honey harvests. The average annual honey yield per hive is now about 48 pounds in the New

---

*Figures in this section are quoted from *Bees and beekeeping: science, practice and world resources* (Crane, 1990), which gives detailed statistics and fuller information on beekeeping in the various regions.

World and 31 pounds in the Old World. There is a net "flow" of exported honey from the New World to the Old World, China being the only large exporter in the Old World, contributing 100 million pounds.

In the New World beekeeping equipment tends to be simple, uniform, and — because labor is expensive — mechanized as far as possible. Langstroth hives are commonly used, whereas some European countries have a wide variety of "standard" hives which take frames, and hence foundation, of different shapes and sizes. Some of the hives are complicated, and in ways that decrease rather than increase honey production. The greater mechanization in the New World enables an individual beekeeper to manage more hives — up to 1000, or even 2000 in the USA. In European countries one man can look after 100-300 hives, although this number is steadily increasing.

In the New World there are more commercial beekeepers, although they still number less than 10 per cent of the total. Average annual honey yields in most individual countries vary from 40 to 80 pounds per hive, and in the best beekeeping districts they are 200 or even 300 pounds. The Old World has a higher proportion of hobby beekeepers, who may get an annual harvest of 10 to 40 pounds of honey per hive. Sideline beekeepers — those who look after a substantial number of hives in their spare time from regular employment —are sure of an income even in a bad season.

In good areas beekeeping can be done (although not with maximum efficiency) without knowing a great deal about the bees themselves, and the beekeepers are not necessarily very interested in them — or in beekeepers' organizations. At the opposite end of the scale, the strongest beekeepers' organizations are in the European countries with a high density of hives, each giving a small return — for instance Austria, Czechoslovakia, the Netherlands, and Switzerland. Such beekeepers often need a good knowledge of bees in order to maintain their colonies at all.

Since the end of World War II, a factor that has affected the whole beekeeping industry — largely adversely — is the ever-increasing speed and availability of air transport. This enabled honey bees to be taken relatively easily from one country or continent to another, and unfortunately diseases and parasitic mites sometimes went with the bees. Starting in 1951, the tracheal mite *Acarapis woodi* was transported with bees to South America, then to Asia, Africa and North America. The damaging external mite *Varroa jacobsoni* was transported from Asia to Europe before 1970, and thence (or directly) to Africa and to South, Central and North America. Another damaging Asiatic external mite, *Tropilaelaps clareae*, is now also spreading around the world. In 1956 queens of tropical African honey bees were introduced to Brazil, and as a result of hybridization, "Africanized" bees have spread through South and Central America and into Mexico and the United

States. Crane (1988) provides some details of these early dispersals.

Individual Continents

Europe (excluding the U.S.S.R.) has about 15 million hives of bees, giving a density of 8.3 per square mile, compared with the world average of 1.3. The average honey yield is about 24 pounds per hive. In spite of the great density of bees, the honey produced does not satisfy the demand, and Western Europe is the world's great honey-importing region. Europeans are in general familiar with honey because it is part of their heritage from past centuries, when beekeeping belonged to the pattern of life of every rural community, and the towns were small enough to keep in touch with the country around them. Just as tradition has helped to maintain honey consumption and an interest in bees among the general public, so it has played a material part in sustaining centers of learning where research on bees was carried out, even where honey production was not economically important. Such centers were very active between 1950 and 1980, but in the 1980's there have been cuts in funding, and some centers have been closed or absorbed into more general ones.

The honey production of Europe (excluding the U.S.S.R.) is probably about 360 million pounds a year, and around 340 million pounds are imported into Western Europe from New World countries and China, of which about a third is re-exported. West Germany alone imports 163 million pounds. Countries of Eastern Europe import much less honey, and export more: for instance Hungary about 35 million pounds, and Bulgaria and Romania about 11 million pounds each.

U.S.S.R. Bees and honey have been important in Russia throughout historical times. The Soviet State gives much encouragement to beekeeping in both public and private sectors. Activities are organized by means of a centrally planned and controlled network of organizations throughout each of the 15 Republics of the Union, which includes large territories in both Europe and Asia. More than half the colonies of bees perished in World War II, but the number has since reached the pre-war figure of 10 million, and is now about 8 million; the annual honey production per hive has doubled within the last 10 years or so, to about 50 pounds. The total honey production is about 400 million pounds.

Asia. Statistics are still difficult to establish for many countries of Asia. There are possibly about 13 million hives in Asia excluding the U.S.S.R., many of them traditional types without movable frames. A considerable part of the honey produced in the tropical south is collected from wild nests of *Apis dorsata*. Beekeeping gives an average annual honey yield per hive of up to 40 pounds; the total production is recorded as around 550 million pounds. Production is increasing with the use of the more productive European bees. After 1950, China developed beekeeping greatly in suitable areas, and in

1980-1982 this country was the world's largest exporter of honey. At present it has about 6 million hives of European *Apis mellifera* and 1 million of the native *Apis cerana*; they produce 350 million pounds of honey a year of which 100 million pounds is exported. Japanese imports are at the present about 70 million pounds a year, higher than for any country except West Germany.

Greater quantities of royal jelly are produced and consumed in Eastern Asia than anywhere else. During the mid-1980's, China produced 880,000 pounds a year, Taiwan 510,000 pounds and Thailand 26,000 pounds; Japan imported 441,000 pounds.

Africa. Beekeeping conditions on the north coast of Africa are not dissimilar from those in other Mediterranean countries, but many beekeepers still use traditional hives. Africa has some 14 million hives of bees in all, and many in tropical countries south of the Sahara are also traditional hives, from which beeswax as well as honey is harvested; they give a wax yield that is about 8 to 10% of the honey yield (compared with 1½ to 2% with modern hives). In tropical Africa the native honey bees (*Apis mellifera*) swarm freely and build their nests in tree or rock cavities, and also in hives hung in trees. There are vast areas of wooded country that provide bee forage, and shade for hives. Some tribes developed a tradition of beekeeping and harvesting from wild colonies, and to these peoples the production and sale of beeswax has been an accepted form of livelihood. The total beeswax production of Africa is several thousand metric tons a year, and has represented most of the beeswax offered on the world market. The total amount of honey collected from hives in Africa is 200 million pounds; some of it is made into honey beer.

Farther south, productive beekeeping is carried out with modern hives, mostly by descendants of settlers from Europe (Anderson *et al.*, 1983). In the extreme south of Africa the climate is again more Mediterranean, and there is even a different native race of honey bees — the Cape bee *Apis mellifera capensis*.

The Americas. We have seen that the continents of the New World —North and South America, and Australasia — give higher harvests from bees than the Old World. Figures quoted for the 1970's (Crane, 1975) showed a higher differential than today; in the Old World increasing beekeeping efficiency has raised yields, and in parts of the Americas yields have been reduced by intensive agricultural practices, and variously affected by the spread of Africanized honey bees.

The United States and Canada together have about 5 million hives, with an average annual honey yield per hive of about 40 pounds in the United States and 140 pounds in Canada — the highest national average in the world. Beekeeping is still mostly a hobbyist pursuit, but there are several thousand full-time or part-time professionals, with holdings of up to several thousand hives. In areas in the East the climate is more like that of Europe, and honey yields are lower; most holdings are comparatively small, and — as in Europe — there tends to be a greater interest in the bees themselves.

There are nearly 8 million hives in South and Central America including Mexico. Tropical regions are the home of many species of native stingless bees (Meliponinae), and also of the Africanized honey bees that spread through and beyond Brazil from 1957 onwards. The average annual yield per hive in Central and South America as a whole is about 42 pounds, and the total annual production about 320 million pounds, 230 million pounds of this are exported — 120 million pounds from Mexico alone, which is at present the world's largest honey exporter.

Australasia. There are about 800,000 hives in Australasia and the Pacific islands together. Migratory beekeeping to a succession of honey flows in Australia makes it possible to obtain some of the highest annual honey yields per hive in the world; the average for the whole country is about 100 pounds. Most of the honey comes from different eucalypts, some of which give copious yields during their rather infrequent flowering periods, which may occur only once every 2, 4 or more years. The honey yield declines as the indigenous forest is cleared in successive areas, even where the stands of trees may be replaced by nectar-bearing agricultural crops.

Australia produces about 55 million pounds of honey a year and exports 22 million pounds or more. New Zealand produces about 13 million pounds and also exports. The eucalypts are not native to New Zealand, and beekeeping there is more similar to that in parts of America. Especially during the past few decades, beekeepers in New Zealand and Australia have contributed a number of developments in beekeeping equipment and techniques.

Pacific islands can be safe havens from diseases and pests, and queen rearing has been developed on some of them, notably Hawaii.

CONCLUSION

With all the changes noted in this chapter, two factors in beekeeping are, so far, beyond the power of man to change materially: the climate which determines what bee forage will flourish, and the habits of the bees themselves. Perhaps the challenge presented by these factors helps to mold beekeepers into the class of people that they are.

Beekeeping thus follows a varied and interesting pattern in different parts of the world. It is a pattern which has changed through the centuries with man's colonization of new regions, and which now changes every decade with changing agricultural practices, for these affect the forage which gives the bees — and the beekeepers — their harvest. As new areas are brought into cultivation, new crops grown, and new agricultural methods used, the pattern of beekeeping inevitably changes. New bee forage may be provided by new crops, but the promotion of a clean agriculture by killing weeds before they flower, and the rapid harvesting of fodder crops, both reduce the bees' forage. In some areas the control of insects which damage agricultural crops has destroyed many wild bees and other beneficial insects, whose nesting places may also be endangered by the reduction of waste land. This has left the so-called domesticated honey bee as the only pollinator available in large enough numbers and has brought a new form of return to the beekeeper, in pollination rental.

REFERENCES

Anderson, R. H.; Buys, B.; Johannsmeier, M. F. (1983) Beekeeping in South Africa. *Bull. Dep. Agric. Tech. Serv.* No. 394, 2nd ed.

Bevan, E. (1838) *The honey-bee; its natural history, physiology, and management* (London: Van Voorst) 2nd ed.

Büdel, A. (1963) Franz von Hruschka. Zu seinem 75. Todestag (9 May 1888). *Dtsch. Bienenw.* 14(5): 139-142.

Butler, C. (1609) *The feminine monarchie.* (Oxford: printed by Joseph Barnes) Facsimile reprint 1969 (Amsterdam: Theatrum Orbis), English Experience No. 81; also 1985 (Mytholmroyd, UK: Northern Bee Books).

Cesi, F. (1625) *Descrizzione dell' api* (broadsheet dedicated to Pope Urban VIII).

Crane, E., ed. (1975) *Honey: a comprehensive survey* (London: Heinemann).

Crane, E. (1983) *The archaeology of beekeeping* (London: Duckworth).

Crane, E.; Graham, A. J. (1985) Bee hives of the Ancient World. *Bee Wld* 66:23-41, 148-170. IBRA Reprint M117.

Crane, E. (1988) Africanized bees, and mites parasitic on bees, in relation to world beekeeping. Pp. 1-9 from *Africanized honey bees and bee mites* ed. G. R. Needham *et al.* (Chichester, UK: Ellis Horwood).

Crane, E. (1990) *Bees and beekeeping: science, practice and world resources* (Oxford: Heinemann Newnes).

Davies, N. de G. (1944) *The tomb of Rekhmire at Thebes* (Salem, NH, USA: Ayer Co.).

Dobbs, A. (1750) On bees and their method of gathering wax and honey. *Phil. Trans.* 46: 536-549.

Dollin, A.; Dollin, L. (1986) Tracing aboriginal apiculture of Australian native bees in the far north-west. *Australas. Beekpr* 88(6): 118-122.

Fraser, H. M. (1951) *Anton Janscha on the swarming of bees* (Foxton, UK: Apis Club).

Fraser, H. M. (1958) *History of beekeeping in Britain* (London: Bee Research Association).

Haydak, M. H. (1957) Petro Prokopovich - Ukrainian beekeeper, teacher and scientist. *Am. Bee J.* 97(12): 474-475.

Hilburn, D.J. (1989) Beekeeping in Bermuda: A short history. *Glean. Bee Cult.* 117 (10): 569-570.

Hornbostel, H. C. (1744) *Neue Entdeckung, wie das Wachs von den Bienen kömmt* (Hamburg: Vermis Bibliothek).

Huber, F. (1792) *Nouvelles observations sur les abeilles* (Geneva: Barde, Manget & Co.).

Jacob, N. (1568) Gründlicher Unterricht von der Biene und ihrer Nahrung (Görlitz, Germany).

Janscha, A. (1771) *Abhandlung von Schwärmen der Bienen* (Vienna: Joseph Kurzböck).

Johansson, T. S. K., Johansson, M. P. (1967) Lorenzo L. Langstroth and the bee space. *Bee Wld* 48(4): 133-143.

Langstroth, L. L. (1853) *Langstroth on the hive and the honey-bee, a bee keeper's manual* (Northampton, MA, USA: Hopkins, Bridgman & Co.).

Méndez de Torres, L. (1586) *Tractado breve de la cultivación y cura de las colmenas* (Alcalá de Henares: author) Reprinted 1983 (Torrejón del Rey, Spain: AMUVARI).

Munn, A. (1844, 1851) *A description of the bar-and-frame hive* . . . (London: Van Voorst).

Nogueira-Neto, P. (1970) *A criação de abelhas indigenas sem ferrão* (Meliponinae) (São Paulo: Chácaras & Quintais).

Réaumur, R. A. F. de (1740) *Mémoires pour servir à l'histoire des insectes* Vol. 5 (Paris: Imprimerie Royale).

Remnant, R. (1637) *A discourse or history of bees* . . . (London: R. Young for T. Slater).

Ruttner, F. (1988) *Biogeography and taxonomy of honeybees* (Berlin: Springer-Verlag).

Schwarz, H. F. (1948) Stingless bees (Meliponidae) of Western Hemisphere. *Bull. Am. Mus. nat. Hist.* No. 90.

Smith, D. A. (1977) The first honeybees in America. *Bee Wld* 58(2): 56.

Sprengel, C. K. (1973) *Das entdeckte Geheimniss der Natur* . . . (Berlin: Friedrich Vieweg).

Toumanoff, C.; Nanta, J. (1933) Enquète sur l'apiculture au Tonkin. *Bull. écon. Indochin.* 1015-1048; *see also* Toumanoff, C. (1933) Documentation sur l'apiculture annamite. *Bull. écon. Indochin.*: 169-175.

Walker, H. J. O. (1928) The hanging bar frame hive. *Brit. Bee J.* 56: 441-444, 453-457.

Wallace, H. (1978) The strange case of the panucho plugs: evidence of pre-Columbian apiculture on Cozumel. *Paper released for use in 1985.*

Wheler, G. (1682) *A journey into Greece* (London: T. Cademan), pp. 411-413.

CHAPTER 2

HONEY BEES OF THE WORLD

by ALFRED DIETZ*

INTRODUCTION

E xcept for the extreme polar regions, honey bees (*Apis mellifera* Linnaeus 1758) are found in all parts of the world. No other insect has received as much attention as the honey bee not only from a practical point of view in terms of honey and wax production, and their value as pollinators (Robinson *et al.*, 1989; Southwick and Southwick, 1989), but also from the scientific perspective. The literature about honey bees falls generally into two groups: (1) apiculture or beekeeping and (2) honey bee biology. The best known and most widely circulated books on beekeeping are those edited by Dadant and Sons (1975), and the A.I. Root Company (1978). The various aspects of honey bee biology, behavior, physiology and taxonomy are elaborated in volumes by Buttel-Reepen (1915), Nelson (1915), Snodgrass (1925) Ribbands (1953), Butler (1954), Büdel and Herold (1960), Zander and Weiss (1964), von Frisch (1965), Chauvin (1968), Lindauer (1975), Seeley (1985) and Winston (1987). The natural history of the genus *Apis* was recently described by Ruttner (1988a). Dictionaries of beekeeping terms have also been tabulated by Crane (1977), Morse (1985), and Herold and Pieterek (1985). Books on honey bee pests, predators and diseases have been compiled by Morse (1978), and Weiss (1984) among others.

GEOLOGICAL HISTORY OF HONEY BEES

The geological record of the earliest insects is still uncertain. The oldest insects were uncovered in rocks extending back to the Upper Carboniferous Period (Carpenter and Hermann, 1979). Their age dates back to about 350 million years. Orders of insects which are still present today already existed at the end of the Paleozoic Era.

Flowering plants had become established by the beginning of the next period, the Cretaceous, and with it the types of insects associated with these plants. Insects found in Baltic amber, now considered as of early Tertiary age, are an important link to extant genera and species. Ant fauna of the Baltic amber include 43 genera of which 24 are still present, whereas all but one of the genera of bees in such amber are extinct (Carpenter, 1952).

The superfamily Apoidea includes all bees which on morphological grounds are similar to the sphecoid wasps, even though the absence of an

*Alfred Dietz, Ph.D., Department of Entomology, The University of Georgia, Athens, GA 30602

adequate fossil record has made it impossible to determine the exact ancestral phyletic line (Wilson, 1971). However, Evans (1969), in his description of sphecoid wasps from the Cretaceous Period, reported on one specimen which appeared to be generalized enough to be considered as a possible ancestor of the Apoidea. This find could indicate that the present bee fauna probably originated more than 70 million years ago. The first appearance of bees is thus closely tied in with a change in food from insect prey to pollen and nectar obtained from the flowers of angiosperms. In some eusocial species, including honey bees, the larval food is derived from glandular secretion which ultimately is obtained from the pollen and nectar of flowering plants. Ribbands (1953) considers the development of the brood food glands in honey bees as a discriminating characteristic from their non-social relatives. Based on this assumption, he indicates that food has played a major role in the evolution of honey bees.

The geologically oldest and most completely preserved honey bees were found in Baltic amber in East Prussia and date to the upper Eocene period, or roughly 50 million years ago (Zander and Weiss, 1964). These bees have morphological characters which partially point to the present day Meliponini and partially to the Apini. Haskins (1970) reported on a find of a worker bee apparently of Meliponid affinities. This eusocial bee, complete with well developed pollen baskets, was discovered in the Mexican amber of Chiapas. It is believed to date from the Oligocene period and thus is roughly 30 to 40 million years old. Other bees found in Baltic amber (lower Oligocene) include representatives of the families Andrenidae, Apidae and Megachilidae (Carpenter and Hermann, 1979).

Fossil bees which are morphologically very similar to the present day Apini have been discovered in the lower Oligocene beds at Rott (Siebengebirge), Germany (Zander and Weiss, 1964). Some of these bees, such as *Synapsis henshawi* (Cockerell, 1907), originally described under the genus *Apis*, subgenus *Synapsis*, have been estimated as 30 million years old and are clearly species belonging to the present-day genus *Apis*. *Apis oligocenia* (Meunier) is its synonym (Maa, 1953). Other finds, dating to the upper Miocene period (about 15 million years ago), came from the Randecker Mar near Göttingen, Germany. Based on their wing venation, Armbruster (1938), and Zander and Weiss (1964), considers them to be very close to the present-day *Apis*. Their pollen-collecting structures, however, are morphologically more primitive. Another find, consisting of 17 individuals tightly enclosed in a piece of red marble sinter (stalactite), dates to the same period, and was made in Böttingen (Schwagen), Germany. These bees, described by Zeuner (1931, cited by Bischoff, 1960) as *Apis armbrusteri*, are also very similar to present-day honey bees. A thorough review of all fossil honey-bee species is presented by Maa (1953). A honey bee find in East Africa, from the

Upper Pleistocene period (100,000 years ago), has been reported by Bishoff (1960). This bee cannot be differentiated from the contemporary African honey bee subspecies.

Although fossil records will be useful in interpreting the social evolution of bees on a morphological basis, they provide no evidence about the evolution of social behavior. Nevertheless, the present-day assemblage of ancient forms of ants and bees provides a living record of evolution, including near-facsimiles of types that were dominant on our planet 50 million or more years ago. In such "living fossils" we can observe in detail the specific evolution of many of the behavioral and physiological patterns that have perfected the sociality of honey bees over an extremely long period of time.

ORIGIN OF HONEY BEE SPECIES

Honey bees have been separated by Maa (1953) into three genera and about 28 species on the basis of morphological differences. These morphological differences, however, are currently considered to be deviations of various honey bee races, and thus Maa's system of classification is not widely followed. According to Ruttner (1988a, 1988b), the subfamily Apinae with only one tribe (Apini) encompass four species and 24 races of *Apis mellifera*.

The genus *Apis* is apparently tropical in origin, most likely India and southeast Asia, and until recently included the following four traditional species: (1) the giant honey bee (*Apis dorsata*), (2) the little honey bee (*Apis florea*), (3) the eastern honey bee (*Apis cerana*), and (4) *Apis mellifera*, the western honey bee (Butler, 1975; Lindauer, 1975). These honey bee species are characterized by a remarkable conformity in a number of morphological and physiological characters (Koeniger, 1976).

The giant honey bee, *Apis dorsata* Fabricius (1798), and the little honey bee, *Apis florea* Fabricius (1787), are two traditional species of honey bees that live in the open on a single honey comb hung from the thick branch of a tree, a rocky cliff or the eaves of a building. Although the upper part of the comb has a diameter of up to 15-20cm, the brood comb has a thickness of 3.5 cm and pollen is stored between these two regions. The size of the single-comb nest of the giant honey bee varies from about .1 m to well above 1 m². The comb size of 12 measured colonies in the Philippines varied from 110 to 8.120 cm² and had an average of 23,300 cells. The depth of the brood cells is 16 mm and the cell diameter is 5.35-5.64 (Muttoo, 1956; Morse and Laigo, 1969; Seeley *et al.*, 1982; Ruttner, 1988). Queens, workers and drones are all produced in cells which are similar in size and shape. The size of the adult worker of this species is about the same as that of a queen of the Italian race of the western honey bee (Richards, 1953). Many of these colonies live in close proximity to each other. Lindauer (1975) reported that as many as 96 colonies of giant honey bees may be found in a single "bee tree."

FIGURE 1. a. Several single-comb nests of *Apis florea* offered for sale by "bee hunters" in Thailand.

b. Nest with capped honey storage cells on the upper part of the comb forming a forager dance platform. The brood cells are located below the honey storage area.

The nest of the little honey bee (*Apis florea*) is considerably smaller and is most often attached to a twig of a small tree. The single comb nest contains cells of four sizes. The large storage cells for honey are very deep and constructed in such a manner that the comb bulges out on either side and at the top. This extension of the comb results in a small horizontal surface upon which foraging bees dance when recruiting other workers (Fig. 1).

The size of the comb varies considerably and ranges from 130x130 mm to 350x740 mm. The small worker cells (2.7-3.1 mm) are located below the honey storage cells (Table 1). Corresponding to the decreasing body size of the

Table 1. Approximate Number of Worker Cells on Both Sides of the Comb of Honey bees.

Species and Races of Honey bees	Cells/decimeter^2	Cells/inch^2
Apis dorsata	787	50.8
Apis cerana	1243	80.0
Apis florea	1375	88.5
Apis mellifera scutellata	1000	64.4
Italian *(A. m. ligustica)*	857	55.3
Carniolan *(A. m. carnica)*	857	55.3
Caucasian *(A. m. caucasica)*	857	55.3
Dark bee *(A. m. mellifera)*	897	57.9

[A]From Goetze, 1964; Dadant, 1975; Dietz, 1982; Ruttner, 1988a.

bee from north to south, the diameter of worker brood cells shows a smiliar geographic variability. In northern India, the number of cells counted per dm² was 1190, while in southern India the count was 1560. Although the considerably larger drone cells (4.2-4.8 mm) are most found in the lower part of the comb, the pear-shaped queen cells, which are the largest of all cells, are located near the bottom (Richards, 1953; Muttoo, 1956; Lindauer, 1975; Free and Williams, 1979; Dietz, 1982; Ruttner, 1988).

The eastern honey bee (*Apis cerana*) and the western honey bee (*Apis mellifera*) are two species of bees which are closely related to each other. It has been suggested by Deodikar *et al.* (1959) that *Apis indica* (now *A. cerana*) gradually evolved into *Apis mellifera* which further differentiated into a number of African, Eurasian and Sino-Japanese races during its passage through the three main Himalayan migratory routes. Since the variety of forms of *Apis mellifera* is unusually large in the regions between the eastern Mediterranean and the Caucasus, one may look for its center of origin somewhere in the Near East.

Both honey bees have nests that consist of several parallel, vertical combs. Muttoo (1956) counted 6 to 8 combs in feral colonies in India. Seeley *et al.* (1982) reported an average of 5.6 combs with an area of 2825cm² on one side for 15 colonies in Thailand. Dark cavities are usually used as nesting sites.

The combs built by western honey bees are separated from each other by about 1/2 inch. Those of the eastern honey bee are slightly closer together. *Apis cerana*, which has just under seven worker cells per linear inch of its comb, however, can be kept in *Apis mellifera* hives, even though there are only five worker cells in the case of western honey bees (Table 1). The difference in size between drone and worker cells is less pronounced in *Apis cerana* as compared to *Apis mellifera* (Butler, 1975; Dietz, 1982; Ruttner, 1988).

Since *Apis mellifera* and *Apis cerana* are very similar, it was not clear for a long time whether there are several geographical races or variants of *A. cerana*. Buttel-Reepen (1906) classified *Apis indica* (now *cerana*) as a subspecies of *Apis mellifera* with six geographic varieties. Kerr and Laidlaw (1956) listed *Apis cerana* as *Apis mellifera cerana* because Fabricius, in 1793, described a bee in China as *A. cerana*. However, five years later (1798), Fabricius named the honey bees of India, *Apis indica*. Since there is general agreement that both geographic types belong to one and the same species, the term *Apis cerana* has priority (Lindauer and Kerr, 1960). According to Kerr and Laidlaw (1956), *Apis cerana* has at least the following varieties: *cerana, indica, javana,* and *phillipina*. Ruttner (1988) lists the following four subspecies: *A. cerana cerana, A. cerana indica, A. cerana japonica,* and *A. cerana Himalaya*.

In a study of *Apis cerana*, Tokudo (1935) found that there are several anatomical differences between it and the European honey bee. He also pointed out that they do not interbreed with each other. Similarly, Ruttner and Maul (1983) reported that it is genetically separated from *Apis mellifera*.

Apis cerana is apparently better adapted to hot climates and able to avoid the oriental hornet, a serious predator. *Apis mellifera*, on the other hand, is about one-third larger in size and is better adapted to extended periods of cold weather (Butler, 1975; Dietz, *et al.*, 1986). However, the Japanese honey bee, *A. cerana japonica* Radeszkowsky (1887), is a honey bee that is active at low temperatures (Goetze, 1964). Ruttner (1986) reported that the ecological requirements of *A. cerana* are similar to those of *A. mellifera*. Additionally, this species succeeded in colonizing wooded areas in the cool temperate zone, such as northern China to Ussuria in East Siberia. Thus, *A. cerana* has a rather large area of distribution, ranging from West Afghanistan to Japan.

All four traditional species of honey bees use the round and wag-tail dance to indicate a productive food source. The transition from the round dance to the wag-tail dance, however, occurs much sooner in *A. cerana*, *A. florea* and *A. dorsata* as compared to *A. mellifera*. Lindauer (1975) found that Indian honey bees use the wag-tail dance at a distance of 2 m from the food source. This is a clear indication that these honey bees are much more accurate in communicating the position of a food source, or a potential nesting site. Ruttner (1988) pointed out that it is biologically important for honey bee populations which live in marginal zones of existence to have a precise intra-colony communication in regard to available food sources. Especially since a few days, or even hours, fully exploited for food collection could contribute significantly to the survival of a colony.

Food source distance is indicated in all four traditional species of honey bees by the number of turns in the wag-tail dance that are made within a given time. The slowest dance rhythm is employed by *Apis florea* which also uses the wag-tail dance with great accuracy for food sources very near to the nest. The foraging area is consequently restricted to a radius of about 350 m. *Apis cerana*, which employs a slightly faster dance rhythm, extends its foraging area roughly to 750 m. The distance of the foraging area and the dance rhythm of *Apis dorsata* approaches that of *Apis mellifera*.

Apis florea deviates somewhat from the other species of honey bees in the communication of direction (Lindauer, 1975). Since this bee always builds its nest in the open and dances exclusively under the open sky and always on a horizontal surface of the nest, it is unable to communicate the angle between the line of flight and the food source onto a vertical area. The straight run portions of the wag-tail dance consequently point directly to the food source (Lindauer, 1975).

The other three traditional species of honey bees indicate direction to a target by transposing the angle between the food source and the sun to the angle between the straight run portion of the wag-tail dance and the vertical line of gravity. *Apis dorsata* seems to occupy an intermediate position between *Apis florea*, on the one hand, and *Apis cerana* and *Apis mellifera*, on the other. The worker communicates the direction to a food source on the vertical area of the comb in a way similar to the method used by *Apis cerana* and *Apis mellifera*, but it uses only the side of the comb with an unobstructed view of the sky (Lindauer, 1975).

All traditional four *Apis* species have the same number of chromosomes (n=16; Fahrenhorst, 1975) and use their sting apparatus and the alarm signal, or sting pheromone, isopentyl acetate (Boch *et al.*, 1962), in the common defense of their colony against intruders. Morse *et al.* (1967) reported the presence of isopentyl acetate in the sting extracts of all four traditional species of *Apis*. Even though there were quantitative differences in the alarm phero-mone, it was assumed that the alarm system of the Asiatic honey bees is similar to that of *Apis mellifera*. However, Koeniger *et al.* (1979), in a comparative study of the alarm function of sting extracts within the genus *Apis*, concluded that *Apis dorsata* and *Apis florea* possess an additional pheromone besides isopentyl acetate. This pheromone was identified as 2-decen-1-yl-acetate and appears to extend the alarm reaction as compared to the reaction resulting from pure isopentyl acetate. In general, the behavioral response to sting extracts showed only slight differences among the four species. However, the duration of the reaction to the sting extract was different and lasted about 3 minutes for *Apis cerana* and *Apis mellifera*, and 6-15 minutes for *Apis dorsata* and *Apis florea*.

Apis cerana does not produce as large a colony as *Apis mellifera*. It is thus kept in smaller hives by beekeepers that use modern methods (Kapil, 1971; Wongsiri, 1986). Even though these two species of bees are closely related, crossbreeding between *Apis cerana* and *Apis mellifera* does not produce viable offspring (Maul and Ruttner, 1969). Although both species share the same sex attractant, it is difficult to maintain them together in the same area (Ruttner, 1975a).

The possible existence of additional species was mentioned by Ruttner (1988), although he is of the opinion that important data are still missing for a definitive classification of these species. Since work on these bees is in progress and additional information is accumulating on the status of these new species, I consider it appropriate to list the available evidence which shows that there are three distinctly new species of honey bees. Perhaps the best known example is *Apis andreniformis*, the small dwarf honey bee of Southeast Asia. This honey bee was recently recognized as a new species by Wongsiri *et al.*

(1990). The bee is also found in the Southern China peninsula, Malaysia and Borneo, however, its exact distribution and sympatry with *Apis florea* are unknown. Also unknown are many comparative aspects of *Apis andreniformis* biology. The recognition of this species is based on distinctive endophallus characteristics in comparison with sympatric *Apis florea*. Additionally, Wongsiri *et al.* (1990) presented scanning electron micrographs of drone basitarsi and preliminary comparisons of wing venation. Jayasvasti (1989) showed that there are distinct differences in the number of barbs present on the stylet of *A. andreniformis* and *A. florea* (Fig. 2). Since the esterase isozymes between *A. andreniformis* and *A. florea* differ (Shao-Wen *et al.*, 1989), *Apis andreniformis* can be considered to be a distinct new species of honey bees.

Apis laboriosa, the world's largest honey bee, and closely related to *Apis dorsata*, was reported by Sakagami *et al.* (1980) as a new species. According to Ruttner (1988), the proposition to recognize this taxon as another species is based on the following three major arguments: (1) considerable quantitative differences supposed to be surpassing those found within one of the known *Apis* species, (2) presumably sympatric occurrence, and (3) ecological divergence (Roubik *et al.*, 1985).

FIGURE 2A. *A. florea* lancet

FIGURE 2B. *A. florea* stylet

FIGURE 2C. *A. andreniformis* lancet

FIGURE 2D. *A. andreniformis* stylet

FIGURE 2. Scanning electron micrographs showing the differences in the number of barbs on the lancets and stylets of the sting of the honey bee species *Apis florea* and *Apis andreniformis*.

A limited number of specimens of *Apis laboriosa*, a specialized mountain bee, were collected so far in Nepal, Butan Sikkim and Yunnan at altitudes between 1,300 and 4,100 m with the majority coming from the oak zone with cool temperatures at altitudes of 2,600-3,500 m (Sakagami *et al.,* 1980; Roubik *et al.,* 1985). In the oak zone this honey bee encounters ambient temperature regimes between 10° and -5°C, or colder during most of the year. There is no definitive evidence whether or not colonies of *Apis laboriosa* migrate to lower altitudes during the cool season. However, the capacity of microclimate regulation at cool temperature of an open-air-nesting colony is a truly remarkable achievement. Since a clear ecological isolation from *Apis dorsata* appears to exist, even though many questions about the physiology of *Apis laboriosa* remain unsettled (Ruttner, 1988), this honey bee is also a new species.

Apis koschevnikovi Buttel-Reepen (1906) is the third new species, based on a morphological comparison of this cavity dwelling honey bee and with that of *Apis cerana* which is sympatric with *A. koschevnikovi* in northern Borneo (Rinderer, 1988; Rinderer *et al.,* 1989). This bee, which also occupies multiple comb nests, has a rufus body color. Morphologically, *Apis koschevnikovi* is distinguished by (1) a much larger cubital index (see Fig. 5) as compared to the cubital index of *A. cerana*, (2) a hairy fringe on the hind tibia of the drone leg, and (3) most importantly, a unique endophallus (Tingek *et al.,* 1988). Variations in enzymes exists in the species of *A. koschevnikovi, A. cerana* and *A. dorsata*. Behaviorally, the drone flight time is different from that of other honey bee species (Koeniger *et al.,* 1988).

The defensive behavior of *Apis koschevnikovi* is not very strong and the bees can be manipulated without protective clothing or smoke. Returning foragers are inspected by the guard bees and escorted into the nest.

It is indeed interesting to note that since the last publication of a chapter on "Races of Bees" (Ruttner, 1975a) in *The Hive and the Honey Bee* (Dadant and Sons, 1975), three new species have been added to the list of honey bee species. Not only is this a truly remarkable occurrence, but it is also a clear illustration of the phenomenal increase in honey bees research on a worldwide basis.

RACIAL CHARACTERISTICS OF HONEY BEES

Geographic races and varieties of honey bees have evolved to such an extent that they are presently found in many distinct geographical areas. These geographic races, or subspecies, are relatively uniform in their native habitat, but they may differ from other races by tenuous or striking morphological characters. Also there are some admixtures of honey bee races where the perimeter of racial areas meet (zones of sympatry). Additionally, in some parts of the world where various geographic races of honey bees have been imported extensively, it is likely that many of these honey bees are racial mixtures. Overall uniformity of anatomical characters is uncommon among geographic races, even though uniformity in one or several characteristics is, at times, achieved. These honey bees are generally called by the name of the geographic race that they most closely resemble by comparing several morphological characteristics (Kerr and Laidlaw, 1956), and most geographic races of economic value are best known by the name of their country of origin, such as the Carniolan, Caucasian, or Italian (Dietz, 1982).

In bee breeding, "race" has not the same meaning as in the breeding of other animals. "Race" in the breeding of dogs, cattle, or chickens means a result of long planned breeding. The geographic races of bees are the results of natural selection in their homeland. That is, the bees became adjusted to their original environment, but not always to the economic requirements of beekeepers. Therefore, they are not the result, but the raw material for breeding.

Geographic races of honey bees can be readily distinguished not only by biometrical methods (Goetze, G. 1930, 1940; DuPraw, 1964, 1965a,b; Daly and Balling, 1978; Dietz, 1978; Sylvester and Rinderer, 1987; Ruttner, 1988), but also various newly developed techniques (Hall, 1988; Smith, 1988; Sheppard and Huettel, 1988). Biometrical honey bee researchers who studied the various races in Russia, China, India and Central Europe used size, hair coverage, color, veins of wings, length of tongue, number of hamuli on the wings, metatarsal width and other characteristics to distinguish races (Ruttner, 1975a). However, even in "pure races" such as the Carniolan, there are various characteristics which show considerable variations between individual colonies (Ruttner, 1984).

Size differences between the geographic races of bees can often be readily determined visually. Smaller differences in size can be objectively established by measuring individual parts of the body, such as tongue length, thorax and abdominal segment width, legs and wings. The northern Dark bee of Europe, *Apis mellifera mellifera* Linnaeus (1758) is larger than the southern races, such as the Carniolans, *Apis mellifera carnica* Pollman (1879), Cyprians, *Apis mellifera cypria* Pollmann (1879) and Italians, *Apis mellifera ligustica* Spinola (1808). African honey bee races are smaller in size. Also, smaller European bees, as a general rule, have longer legs, tongues and wings in proportion to the size of their bodies.

Differences in body size may affect comb cell size, since smaller geographic races of bees construct smaller cells (Ruttner, 1975a). Bee body size may also be influenced by the nutritional condition of the colony. Worker bees that are produced during the early part of the year, when the pollen supply is adequate, are generally smaller than those reared under nutritionally more rewarding conditions (Anderson and Dietz, 1976).

Grout (1937) studied the influence of brood cell size upon the size and variability of worker bees by using sheets of foundation with 857, 763 and 706 cells/cm². He found that worker bee size is influenced by cell size and that significantly larger bees are obtained with larger cell size. The maximum increase in tongue length in large bees was 2.07%.

Brood cells also undergo size reduction with use and age. The accumulation of old larval pupal cuticle and cocoon remnants, and the modification of cells in preparation for a future brood cycle may also contribute to cell size reduction and the subsequent reduction in the size of bees emerging from these smaller cells (Dadant, 1975). However, bees do reduce the size of the cell wall and extend it in length to compensate for any accumulation of materials in the cell, especially if these materials are in the base of the cell (Dietz, 1982).

In Caucasian, Carniolan and Italian races, the number of cells per square decimeter of brood comb is identical. However, there are definite differences in other races and in the other three species of honey bees (Table 1).

The number of worker cells produced by colonies of *Apis mellifera ligustica* allowed to establish their nests in unrestricted cavities was only 814 cells/dm² (Taber and Owens, 1970). The decrease of roughly 5% in the number of cells produced by the American strain of the Italian bee is possibly due to scientific queen rearing. The result of this selection process points to a larger bee and consequently an equivalently larger cell size.

The average cell diameter measurements of the various species and geographic races of bees are as follows: 2.87-3.00 mm for *Apis florea*; 4.02-4.20 mm for *Apis cerana*; 5.2-5.8 mm for *Apis mellifera*; and 5.42-6.35 mm for *Apis dorsata* (Zander and Weiss, 1964; Dadant, 1975; Seeley and Morse, 1976).

Cell depth is influenced by its use, such as brood rearing or food storage. Storage cells for honey are usually larger than cells used for brood rearing. Length of the queen's abdomen is the limiting factor in this respect. She is unable to attach the egg to the cell base if cell depth exceeds the length of her abdomen. Worker cell depth ranges from 7.53-8.50 mm for *Apis florea* to 9 mm for *Apis cerana* to 11 mm for *Apis mellifera* to 16.87-17.00 mm for *Apis dorsata* (Zander and Weiss, 1964; Seeley and Morse, 1976). The dimensions of *Apis florea* drone cells are 4.45 x 10.51 mm; those of *Apis mellifera* are 6.2 x 12.5 mm. *Apis dorsata* apparently lacks clearly defined drone cells (Seeley and Morse, 1976).

Measurement of tongue length of various geographic races of honey bees has shown considerable variations. Ruttner (1975a) reported a difference of 1.7 mm between the tongue length of the two extremes in honey bee races

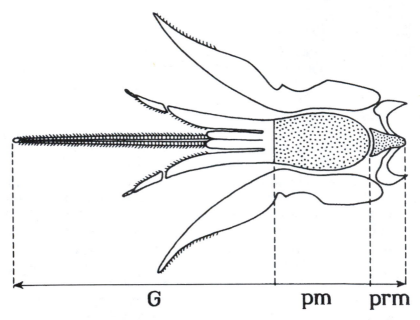

FIGURE 3. The tongue (proboscis) of a worker bee as it is measured to determine its length. G glossa, pm postmentum, prm prementum.

(Fig. 3), the Caucasian, *Apis mellifera caucasica* Gorbachev (1916), and the Egyptian *Apis mellifera lamarckii* Cockerell (1906; Syn. *A. m. fasciata*). Tongue length is of extreme importance in honey production. Bees with a long proboscis, such as the Carniolan, Caucasian and Italian, can work red clover, while races with short tongues cannot (Goetze, 1940).

The body color of honey bees is essentially black, but because of yellow pigments in the exoskeleton, there may be yellow areas (Kerr and Laidlaw, 1956). The yellow areas are apparently completely free of black pigment, or black pigment may appear to be sporadically present and thus darken the yellow. Superimposed on this is the color of the hairs which are black in some places, and in other areas may range from white to yellow. The intensity of the yellow color and the pattern of the yellow areas are hereditary, as is the hair color.

Roberts and Mackensen (1951) studied the inheritance of abdominal coloration in two geographic races of honey bees. They found that if an inbred yellow queen is instrumentally inseminated to a black drone, the resultant F_1 worker progeny are like neither parent but are banded intermediates in color. When an F_1 hybrid queen intermediate in color is mated to a black drone, the progeny produced ranged in color from intermediate to completely black. An F_1 hybrid queen intermediate in color mated to a yellow drone, produces progeny that range in color from intermediate to parental yellow. Drones produced by F_1 hybrid queens range from yellow to completely black and show gametic segregation for color. Roberts and Mackensen (1951) concluded that there are at least seven different loci for genes that affect the yellow coloration of the abdomen. Thus, many abdominal coloration patterns are possible, because 7 pairs of genes can produce 2181 different genotypes. Workers of a mixed population, if arranged according to color from the lightest to the darkest abdomen, will represent a continuous series.

Honey bees in some geographic races have wide and dense tomenta (Caucasian and "gray" Carniolan). Other races have only narrow and dispersed bands of hair or almost none at all. The Dark bees (*A. m. mellifera*) have overhairs on the abdomen that are 0.5 mm long, as compared to only 0.3 mm in most of the other races (Fig. 4). The color of the hairs on drones shows considerable variation. In general, it is black in the Caucasians, dark brown to black in the Dark bees of northern Europe, gray to grayish brown in the Carniolan, and yellow in the Italian (Ruttner, 1975a).

The abdomen of African drones, *Apis mellifera scutellata* Lepeletier (1836), have a brownish color and thus differ from Italian males. The color is genetically determined. Consequently, the genes which determine the yellow abdominal color are sex limited. Coloration, nevertheless, should not be considered as a good character to identify worker honey bees because of its large variability (Gonclaves, 1975). The color in honey bees is highly variable and is often unreliable as a taxonomic character (Goetze, 1940).

A morphological character of major importance in the taxonomy of geographic honey bee races is not only the venation pattern of the worker forewing, but also the angles, length, and width of the wing. The non-Linnean classification of honey bees by DuPraw (1965a,b) is based on 15 characters of the worker forewing. Goetze (1930, 1940) introduced the cubital index (Fig. 5) of the forewing as a reliable character to distinguish between similar populations of honey bees, such as *Apis mellifera mellifera* and *Apis mellifera*

FIGURE 4. The hairs on the abdomens of worker bees: Tomentum (band of hairs in the middle of three abdominal tergites of workers) and "overhairs" shown in profile. Left *mellifera*, right *carnica.*

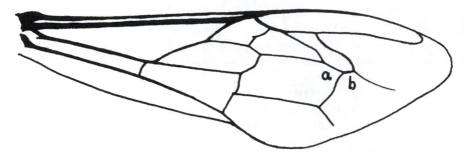

FIGURE 5. Venation of a worker fore wing. Ratio a:b = **cubital index.**

carnica. Many geographic races have been identified by means of morphological analysis, however, it must be remembered that behavioral and biological
characteristics must also be considered in order to delineate closely related
populations. DuPraw (1965a), according to Ruttner (1975c), was unable to
delimit *Apis mellifera adansonii* Latreille (1804) from *Apis mellifera capensis*
Escholtz (1822) on the basis of wing characters. However, if the fundamental
differences in reproduction and behavior of these two geographic races is
taken into consideration, they can be readily separated. The geographic races
of honey bees, consequently, may be very similar morphologically, but
strikingly different in a variety of behavioral and physiological characteristics
(Lauer and Lindauer, 1973; Menzel *et al.*, 1973; Kloft *et al.*, 1976; Velthuis,
1976; Dietz, 1982).

GEOGRAPHICAL DISTRIBUTION

In honey bees, the geographical races are the product of natural selection
in their respective homeland. They represent very different genotypes adapted
to dissimilar environments (Ruttner, 1975a). Inside the wide geographic races
of honey bees, there are also ecotypes characterized by their adaptation to the
environment in which they live (Louveaux, 1969). This genetic variation
needs to be taken into consideration in the selection of honey bees for
apicultural purposes. Much care and manipulation may not be required of
honey bees adapted to a given environment as compared to those which are
nonadapted. The nonadapted bees, however, may be much more prolific and
suitable for intensive beekeeping practices (Dietz, 1982).

The Western honey bee, *Apis melliera*, has been colonized in both
temperate and tropical zones of western Asia, Africa, and Europe (Fig. 6).
This colonization has resulted in about two dozen geographic races (Rothenbuhler, 1979, Ruttner, 1988). These races, or geographical variants have been
subdivided by Rothenbuhler *et al.* (1968) into the following four groups: 1.

African races, 2. Southeast European races, 3. Near East races, and 4. Northwest European races.

Based on integrated biometrics, Ruttner (1986; 1988) suggested a preliminary list of 24 subspecies in *Apis mellifera* grouped into either 3 or 4 categories. Table 2 shows his subspecies listing.

Table 2. List of Geographic Races of the Western Honey Bee, *Apis mellifera* (from Ruttner, 1986, 1988).

AFRICAN GROUP	NEAR EAST GROUP
A. m. scutellata Lepeletier (1836)	*A. m. anatoliaca* Maa (1953)
A. m. adansonii Latreille (1804)	*A. m. adami* Ruttner (1975d)
A. m. litorea Smith (1961)	*A. m. cypria* Pollman (1879)
A. m. monticola Smith (1961	*A. m. syriaca* Buttel-Reepen (1906)
A. m. lamarckii Cockerell (1906)	*A. m. caucasica* Gorbachev (1916)
A. m. capensis Escholtz (1821)	*A. m. meda* Sorikov (1929)
A. m. unicolor Latreille (1804)	*A. m. armeniaca* Sorikov (1929)
A. m. yemenitica Ruttner (1975b)	
CENTRAL MEDITERRANEAN AND SOUTHEASTERN EUROPEAN GROUP	**WESTERN MEDITERRANEAN AND NORTHWESTERN EUROPEAN GROUP**
A. m. sicula Montagano (1911)	*A. m. sahariensis* Baldensperger (1922)
A. m. ligustica Spinola (1806)	*A. m. intermissa* Buttel-Reepen (1906)
A. m. carnica Pollmann (1879)	*A. m. iberica* Goetze (1964)
A. m. macedonica Ruttner (1988a)	*A. m. mellifera* Linnaeus (1758)
A. m. cecropia Kiesenwetter (1860)	

TROPICAL AFRICAN GROUP

Kerr and Araujo (1958) and Rothenbuhler *et al.* (1968) placed all the bees north and south of the Sahara and *Apis mellifera lamarckii* from the Egyptian region into the African group. Smith (1961) distinguished five distinct races. Ruttner (1975b) recognized three major groups with 12 African races of honey bees. The Northwest African group includes *Apis mellifera intermissa* Buttel-Reepen (1906), *Apis mellifera sahariensis* Baldensperger (1922) and a new race, *Apis mellifera major* Ruttner (1975b). The Egypt and the Sudan group included *Apis mellifera lamarckii* and another new geographical variant, *Apis mellifera nubica* Ruttner (1975b).

The African group includes not only the largest number of geographic races, but also some of the best known, such as the notorious *Apis mellifera scutellata*, formerly *Apis mellifera adansonii* (Ruttner, 1981). In addition to *Apis mellifera adansonii* and *Apis mellifera scutellata* this group also includes *Apis mellifera litorea* Smith (1961), *Apis mellifera monticola* Smith (1961),

Apis mellifera unicolor Latreille (1804), a new subspecies, *Apis mellifera yemenitica* Ruttner (1975b), and *Apis mellifera capensis*. Fletcher (1978) lists an additional African race, *Apis mellifera nigritarum* Lepeletier (1836).

Since Ruttner (1988) has withdrawn *Apis mellifera major* as an African subspecies because of lack of information on biology, etc., and since *A. mellifera nubica* is now taken as a synonym for *A. mellifera yemenitica*, his former listing of honey bees of Tropical Africa is now reduced to 8 subspecies.

FIGURE 6. Shows the geographic distribution of the most important races of *Apis mellifera*. It is not complete inasmuch as some territories are not explored enough and exact judgment is not possible at the present time. This is true, for instance, of the bees of Iran and Anatolia, which probably will play a specific role in future breeding of bees.

These eight subspecies can be readily distinguished morphometrically. Furthermore, he has transferred *Apis mellifera sahariensis* and *Apis mellifera intermissa* from this group into the Western Mediterranean and Northwestern European category.

In addition to the parental stock (*Apis mellifera scutellata*) of the notorious Africanized honey bee, this group also includes some other interesting and important geographic races of honey bees, *Apis mellifera capensis*, the Cape Bee, is found in a very restricted area on the southwest coast of the Republic of South Africa (region of Cape Town). This bee is distinctly different from all other geographic races of honey bees because of the presence of large numbers of ovarioles in the ovaries of worker bees and well developed spermatheca. However, a spermatheca of a worker bee was never found to be filled with spermatozoa (Anderson, 1961). In queenless colonies of *A. m. capensis*, the workers soon start laying eggs which, in a high percentage, develop without fertilization into females. Thus, from these female eggs, queens may be raised. According to Mackensen (1943), the same phenomenon occurs in other races, but only in a very low frequency.

The African Bee, *Apis mellifera scutellata*, covers the central and eastern part of the continent, from Ethiopia to the Great Karoo in South Africa (Fig. 6). Originally, all bees south of the Sahara were called *A. m. adansonii*. With growing knowledge, however, it was found that this is no uniform type, but a group of several distinct geographic races. As the designation "*adansonii*" was given to the bees of Senegal by Latreille in 1804, this name has to be restricted to the bees of the African west coast (Ruttner, 1975a,b). The name *scutellata* was given to bees of South Africa by Lepeletier (1836), evidently referring to the conspicuous yellow scutellum. *A. m. scutellata* is a small bee with scarce pilosity, variable pigmentation on the abdomen (one or more yellow bands), mostly a bright yellow scutellum on the thorax and a characteristic wing venation. It has common features with *Apis mellifera capensis*, but is easily distinguished from all the other races (DuPraw, 1965a,b).

Apis mellifera monticola has been assigned a special position among honey bee races because it is the first taxonomic unit demonstrating isolation entirely by ecological factors showing a unique and distinct area of distribution (Ruttner, 1988). *A. m. monticola*, a "black bee" of the rain forests of the East Africa mountains, was collected by us (Dietz and Krell, 1986) on Mt. Meru (Kenya) at altitudes between 6,000-12,000 ft. Interestingly enough, we found a dead colony of black bees inside a log hive (Fig. 7) at an isolated, and by vehicle inaccessible, location at an estimated altitude of 12,000 ft. Although we do not know the cause of death, the appearance of the honey bees on the combs was similar to those which perish due to cold starvation (Dietz, 1968). This discovery is unique because we did not encounter another

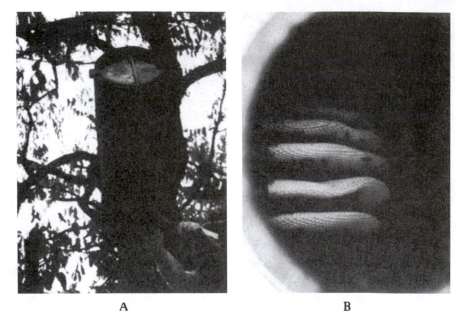

A   B

FIGURE 7. a. Customary log hive used by beekeepers in Kenya.
        b. Log hive showing several newly constructed combs of the East African mountain
bee *Apis mellifera monticola.*

dead African colony during our 30-day survey and sampling tour of Kenya in
1986. Equally important is the fact that we did not encounter another single
dead feral colony of Africanized honey bees either during our 3-year research
effort in Argentina, or our more recent research and monitoring effort in
Mexico (Dietz *et al.*, 1985, 1986, 1988a,b, 1989a,b, 1991; Dietz and Vergara,
1991; Eischen *et al.*, 1986; Krell *et al.*, 1985; Vergara *et al.*, 1989, 1991).

   A. m. monticola is a very gentle bee, able to maintain its racial integrity
even though perennial hybridization takes place with *A. m. scutellata* in a
fluctuating transitory zone (Smith, 1961). It occurs at altitudes similar to
where Africanized honey bees are found in South America (*e.g.* Santo
Domingo, Venezuela; San Juan, Argentina), and could be a useful stock in our
attempt to control Africanization (Dietz and Krell, 1986; Rinderer, 1986).

   Some 3 million years ago, or at the beginning of the Pleistocene epoch,
honey bees appeared in Africa (Ruttner, 1988). It is generally assumed that
with the beginning of this time period, honey bees have coexisted with early
man and honey has always been an essential food item. The type of beekeeping practiced in many African countries essentially has been destructive in
nature. That is, if a nest was discovered by honey hunters, the colony products
were collected without regard to the surviving bees and queen. However, in

some countries, especially Kenya, there are governmental programs now in place to teach not only beekeeping to interested persons, but also to provide various types of hives with removable top bars, or frames, to allow for colony manipulation and to prevent colony destruction (Dietz and Krell, 1986). Nevertheless, the majority of the feral swarm captures in Kenya, and probably most other African countries, is still in homemade log hives hanging in trees. Interestingly enough, we found that many of these log hives were not only well concealed, but also not readily accessible. Since feral swarms of African honey bees are not only exposed to predation by ants, wasps and other pests, but also to man, perhaps the most efficient predator, these bees display an intense defensive behavior. Although this type of behavior has been of little help in protecting their nests from the skilled African honey hunters, who also use smoke to disorient the bees, the defensiveness of *A. m. scutellata* has had a most serious impact on beekeeping in many parts of South America.

AFRICANIZED HONEY BEES IN THE AMERICAS

The geographic races of *Apis mellifera* in Africa show considerable variations. Here we find not only the largest and the smallest, but also the darkest and lightest honey bee known. Although most of the colonies exist in feral condition, Ruttner (1988) stated that there are probably more colonies found in Africa south of the Sahara than in the remaining part of the original *A. mellifera* area in Eurasia.

The more than 20 geographic races of *Apis mellifera* are found in many different climates, ranging from the tropics to temperate zones, from semi-desert to rain forests (Fig. 6). These honey bees not only display brood patterns that are genetically adapted to different floral seasons, but they also have diverse rates of reproduction, ranging from short-living, frequently reproducing colonies to long-living colonies with low reproductivity (Ruttner, 1988). In spite of this differentiation within the group, the African races as a whole show clear differences to all the other races of *Apis mellifera*, south of the Sahara more distinctly than north of it. Therefore, it is not too difficult to discriminate African honey bees and even their hybrids from European races by biometric methods (Daly and Balling, 1978; Rinderer *et al.*, 1990).

Smith (1961) was the first person to apply morphometric methods on honey bees of Africa. In addition, he introduced behavioral and ecological characteristics of the colony into the description of honey bee varieties. This was a major advance over the approach of earlier taxonomists who studied only a small number of museum specimens, without reference to live honey bee colonies (Ruttner and Kauhausen, 1985).

Although *Apis mellifera scutellata* has gained much publicity in recent years because of its rapid spread on the American continent, *A. m. adansonii*,

A. m. capensis and the gentle, black mountain bee, *A. m. monticola* are equally important and well known geographic races of African honey bees. A great deal of *scutellata*'s territory in Africa, extending from Sudan to South Africa, an open woodland called Miombo, is one of the most favorable beekeeping areas of the world (Guy, 1972). Most of the beeswax on the world market comes from this region, produced mostly by feral, or wild, colonies. This region is characterized by a generally hot climate with a long dry season, an abundant nectar and pollen flow and many enemies, including birds and man (Ruttner, 1975a).

Apis mellifera scutellata is a honey bee that is excellently adapted to such an environment. If required to escape unfavorable conditions, it not only produces migratory swarms as do other honey bees in Asia (*A. cerana, A. florea,* and *A. dorsata*), but the whole colony deserts the nest and may migrate over long distances. Even small swarms have in general a good chance to survive, because they migrate to areas with adequate food resources. Since losses due to enemies and climate are heavy, the colony divides or swarms extensively to make up for these losses. Although the climate is often dry in many parts of Africa, colonies can survive in unsheltered places such as branches of a tree, holes in the ground, or even cold temperatures. Fletcher (1978) reported that subzero temperatures are experienced seasonally by *Apis mellifera scutellata* over much of its range in Africa. He observed perennial colonies at altitudes of almost 2000 m where snow lasts up to a week at a time and absolute minimum temperatures of less than 0°C are found 6 months of the year.

Although extensively exploited by indigenous honey hunters, *Apis mellifera scutellata* has been able to survive and thrive in not only Africa, but also in all the other places where this honey bee has been introduced (Dietz *et al.*, 1985, 1986, 1989b). The best example has been the introduction in 1956 of a few queens of *A. m. scutellata* from South Africa into Brazil to improve the local stock (Kerr, 1957). Upon their release, these colonies, possessing African queens, multiplied, migrated and hybridized, soon occupying the whole State of São Paulo where they were introduced.

Ultimately, the Africanized honey bee began to disperse from the original point of introduction and migrated a rate of 100 to 300 miles a year (Michener, 1975; Taylor, 1977, 1985). By 1965, the bee had spread to Argentina (Dietz *et al.*, 1985). It reached Venezuela in 1973 and Mexico in 1986 (Elba Quintero, personal communication; Kaplan, 1990).

The implications of the invasion of Latin America by *A. m. scutellata* in general, and on beekeeping in particular, because of the highly defensive behavior of this bee (Collins *et al.*, 1982), the reduction in honey yields (Rinderer, 1988), among other undesirable traits, have been dealt with in numerous summary publications (Michener, 1972, 1975; American Farm Bur. Symp. 1986, Needham *et al.*, 1988, and others). In this connection, the question arises as to why the undesirable characters of a few queens whose progeny hybridized with millions of European queens and drones remained unchanged. The answer, according to Ruttner (1975a), is given by a simple genetic consideration: If the new stock gives many swarms a year and if each swarm, finding enough food and being able to live even in poorly sheltered places, has a fairly high chance to survive, the selective value of this type against a "European" type easily may be 10:1 or even more. Thus the "European" type is permanently out-numbered to an extreme degree. Since the "African" honey bees of Latin America are hybrids, Gonalves (1982) suggested that it is more appropriate to refer to them as "Africanized" honey bees.

Potentially, Africanized honey bees could be the single most severe insect pest in the United States. Its presence may hurt the beekeeping industry and could harm the pursuit of beekeeping as a recreation. The major economic impact will be on agriculture and public health. The financial rewards of the beekeeping industry in the United States are about $200 to $250 million annually, but the inclusion of all agricultural production linked to beekeeping results in a value of perhaps $10 billion, or more (Robinson *et al.*, 1989; Southwick and Southwick, 1989) on a national level. Aside from honey and beeswax, 200 crops are dependent or benefit from honey bees pollination (Rinderer, 1986). The value of honey bees in the pollination of non-crop plants, wild flowers and trees is not known and almost impossible to estimate (Dietz *et al.*, 1983). The public health issue deserves equal consideration. Based on extrapolations from Central American data, perhaps 200 or more persons are expected to die annually in the United States from massive stinging, and 2,000 to 3,000 may be hospitalized for the same reason (Rinderer, personal communication).

The behavioral differences between Africanized and European honey bees as observed by us and others (Collins *et al.*, 1982; Krell *et al.*, 1985; Dietz *et al.*, 1986, 1989a,b,c; Rinderer, 1988; Rinderer *et al.*, 1985; Taylor, 1985; Winston *et al.*, 1983) are a clear indication that major apicultural changes will occur when this insect becomes established in the United States. There does not seem to exist means to eradicate the African genes in regions where wild colonies can live uncontrolled in the open. A promising method to get better stock in the apiary is to requeen colonies with daughters of pure bred Carniolans or Italians (Martin, 1973). The offspring of these hybrids may give higher yields and are less ill-tempered than the uncontrolled Africanized bees. Of course, this means permanent importation or artificial insemination of European stock, and control of the queens in the colonies (Ruttner, 1975a).

During the last decade, researchers from the University of Georgia have worked continuously on Africanized honey bees and parasitic mites in Argentina, Brazil, Mexico and Venezuela (Dietz, 1986; Dietz *et al.*, 1986, 1987, 1989a,b,c, 1991; Krell *et al.*, 1985; Eischen and Dietz, 1986; Eischen *et al.*, 1986a,b, 1988, 1989; Mejia, 1989; Pettis *et al.*, 1989; Vergara *et al.*, 1989a,b, 1990a,b; Vergara, 1990). Additionally, we studied Asian honey bees in Thailand (Wongsiri *et al.*, 1990) and conducted a survey for gentle African bees in Kenya (Dietz and Krell, 1987). Our survey in Kenya included the collection of honey bee samples from feral and managed colonies, as well as composite bee samples from roadside flowers. The locations surveyed ranged from 3,500 feet to almost 12,000 feet in altitude and from dry acacia savannah to mountain rain forest. We observed a wide variation of behavior in response to colony manipulations and disturbances of the very gentle, black African mountain bee *A. m. monticola* and the highly variable, but often extremely defensive *A. m. scutellata*. It is interesting to note that *A. m. monticola* is able to maintain its subspecies purity in the presence of *A. m. scutellata*. In Africa, both subspecies are only separated by altitude of roughly 2,000 m, although Africanized honey bees were found by us at an altitude of 3,300 m in the province of San Juan, Argentina. (Dietz *et al.*, 1991).

During our work in Argentina, we concentrated in surveying most of Argentina north of the 40°S latitude (Dietz *et al.*, 1985). We collected bee samples from feral and managed honey bee colonies for morphometric identification and conducted field tests of colony defensive behavior with a procedure developed by USDA scientists (Collins and Kubasek, 1982; Collins, *et al.*, 1982). In addition to our effort in determining the distribution of managed and feral Africanized honey bee populations, we also attempted to identify some of the factors which influence this distribution. Along with other researchers (Taylor, 1977; Taylor and Spivak, 1984; Kerr *et al.*, 1982), we suspected colder temperatures to limit the natural spread of Africanized honey bees as they approached the more temperate regions of Argentina.

However, our initial overwintering study of different sized honey bee colonies at higher elevations (1,400 m) in Cordoba, Argentina (Krell *et al.*, 1985) showed that under the prevailing temperature conditions there were no significant differences in survival, food consumption or subsequent weight gain between Africanized and European colonies of either large or small size.

Further evidence that the survival of Africanized honey bees is not restricted to tropical or subtropical area was produced in two comparative studies with colonies of Africanized and European honey bees exposed to cold conditions and long-term confinement in the province of entre Rios, Argentina in 1984 (Dietz *et al.*, 1988) and in 1985 (Dietz *et al.*, 1989a). In both studies, six Africanized and six European honey bee colonies were confined in a refrigeration chamber at a temperature of 0°C - 2°C for 78 days and 91 days, respectively. Although Africanized honey bees colonies died sooner than European colonies in our 1984 study, there was no difference in survival in the 1985 study.

We also observed that European honey bees clustered more tightly than Africanized honey bees. Similar observations were made in Poland (Dietz and Nowakowski, 1981). Africanized honey bees also responded to colony disturbance within 1-3 sec., whereas European honey bees required between 5-10 sec. to respond. These distinct behavioral differences, also reported by Collins *et al.* (1982), may be the reasons why winter clusters in Africanized honey bees are not as tight as those of European honey bees. Although more Africanized honey bee colonies in our 1984 study died sooner than the European colonies, we do not know whether reduced longevity of honey bees played a role in our experiments. It has been suggested that the longevity of Africanized honey bees is shorter as compared to European honey bees. Woyke (1973) reported that Africanized honey bee colonies studied at cold temperatures in Poland wintered well at the beginning, but the bees began to die after about 3 months. However, only a single colony was wintered out-of-doors and this colony survived the winter and developed well in the spring and autumn. The same colony and the original queen also survived the following winter.

A fourth overwintering study was conducted in San Juan, Argentina under more severe temperature conditions than our study in Cordoba (Krell *et al.*, 1985), and for a longer time period (Dietz *et al.*, 1986, 1991). Average monthly temperatures reached a minimum of approximately -2°C at the study site. Colonies of Africanized and European honey bees showed no significant differences in the rate of winter survival. Except for one colony, all colony losses of Africanized honey bees occurred during the first 2 months, whereas European honey bee colonies continued to die during the whole overwintering period. This is very similar to our findings in the refrigeration chamber studies (Dietz *et al.*, 1988, 1989a) and suggests that those Africanized honey bee colonies which do not have the necessary attributes for cold temperature tolerance, or survival, are eliminated early during the cold temperature period. Thus overwintering for 3 to 4 months is less a question of worker longevity than other factors such as cluster formation and maintenance, cluster temperature control, etc. Based on these results, it was concluded that the ability to survive low temperatures for extended periods of time is equal for the Argentinean ecotypes of Africanized and European honey bees. Winter survival may therefore not be the main factor in limiting the distribution of Africanized honey bees in Argentina, or for that matter in the United States (Fig., 8). Environmental conditions such as nest site availability,

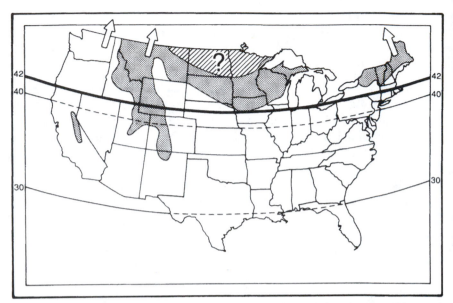

FIGURE 8. Potential distribution of Africanized honey bees in North America (from Dietz *et al.,* 1986).

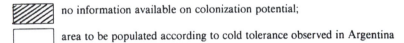 no information available on colonization potential;

☐ area to be populated according to cold tolerance observed in Argentina

water and food sources, which ultimately are influenced by weather conditions, appear much more limiting than low temperature conditions *per se* during the winter.

Floral preferences and pollination effectiveness of Africanized and Italian honey bees were the subject of another study in Argentina. In these investigations, we compared seed set of alfalfa, onions and carrots exposed to different pollination treatments. Although there were some differences in floral visitations between both groups of honey bees, there were no major differences in pollination efficiency between Africanized and European honey bees (Vergara *et al.,* 1991b). Rinderer (1988) reported that pollen collection of Africanized honey bees is different from that of European honey bees. That is, Africanized honey bees collect more pollen because a greater proportion of the foragers are concentrating on pollen foraging.

One of the major issues of the Africanized honey bee problem is the process of Africanization in areas with large populations of managed European honey bee colonies. Although the availability of resources, such as food and nest sites, are major factors in the successful colonization of new territory, migrant honey bee populations utilize several strategies in their conquest of new areas. Rinderer (1986) lists the following four conditions, or primary causes, of Africanization: 1. Parasitism, or usurpation, by Africanized queens of Europeans colonies (Michener, 1972, Dietz, 1982; Dietz *et al.*, 1989c; Danka and Rinderer, 1988; Vergara *et al.*, 1989a,b); 2. Parasitism by Africanized drones of European colonies (Rinderer *et al.*, 1985). That is, Africanized drones migrate into European honey bee colonies in large numbers, while Africanized honey bee colonies rarely allow European drones to enter. The result is an increase in Africanized and a decrease in European drone production. This condition gives rise to the third factor (3), which is differential drone production. The fourth (4) condition is the possibility that the mating flight distances of Africanized honey bee reproductives are smaller than those of European honey bee reproductives. Rinderer suggested that natural selection would tend to select for longer mating flights in areas where feral, or wild, colonies were more scattered, but mating flight with shorter distances where feral colonies were present in larger numbers.

The conditions outlined by Rinderer (1986) in his model play a major role in the process of Africanization of new territory. Not only are they consistent with our knowledge of honey bee biology, but they can also be used to explain how a few Africanized honey bee colonies can initiate the replacement on an established population of European honey bees in a short period of time. Since several aspects of these conditions are hypothetical, it may be appropriate to review some of the research findings in this area.

It has been known for a long time that rival honey bee queens which meet usually fight and one or the other is quickly stung and killed (Ribbands,1953). However, usurpation, or colony take-overs, by queens of European honey bees does not appear to be a common event. Over a period of perhaps 30 years, this phenomenon has been observed by me on only two occasions. Gonalves *et al.* (1982) also stated that in 18 years of working with honey bees in Brazil they have never observed this phenomenon. Although the occurrence of this phenomenon has been reported on several occasions (Michener, 1975; Dietz, 1982; Danka and Rinderer 1988), most of these reports were not supported research data. Nevertheless, usurpation of European honey bee colonies appears to be one of the strategies employed by Africanized honey bees and queens to colonize new areas. Moreover, during our research in Argentina and Brazil (Dietz *et al.*, 1985, 1986, 1988; 1989a,b,c; Dietz and Krell, 1991; Dietz and Vergara, 1991; Krell *et al.*, 1985, Mejia, 1989), we encountered numerous beekeepers who reported that they had experienced queen usurpation in European honey bee colonies by swarms and queens of Africanized honey bees.

As a result of these discussions with beekeepers, we conducted a series of studies on usurpation in honey bees to verify these anecdotal reports. Under the experimental conditions employed by us in Argentina, usurpation of resident European honey bee queens by small artificial swarms of Africanized honey bees and queens was significantly higher and more successful than the take-overs of Africanized honey bee colonies by European honey bees and queens (Dietz *et al.*, 1989c).

Our research on usurpation by Africanized queens in Argentina has been the impetus for studying this phenomenon in Mexico. Results from our studies showed that pioneer populations of Africanized honey bees successfully invaded managed European honey bee colonies. The number of colony invasions observed was considerably lower than the numbers of swarms captured during this period. However, usurpation of honey bee colonies is a reality. Colony invasions are probably not contingent on the presence of large populations of migratory honey bees (Vergara *et al.*, 1989a, 1991a).

Our studies of the process of Africanization of honey bees in Mexico has yielded unexpected results (Dietz *et al.*, 1989b). Although morphometric identification of Africanized honey bees has indicated their presence in the fall of 1986, instances of intense defensive behavior have not occurred until about 1-1/2 years after the initial discovery. So far Africanized honey bees have not shown extremely strong defensive behavior often encountered by us during our recent efforts in Argentina (Dietz, 1982; Dietz *et al.*, 1985, 1986, 1988, 1989 a,b,c,1991; Krell *et al.*, 1985; Vergara *et al.*, 1991b).

Results from preliminary studies on bait hive preferences demonstrated significant differences in the number of swarms captured at high and low nest site locations and in bait hives treated with and without pheromones. Colony invasions occurred intermittently during our study period. Significantly greater numbers of invasions occurred in queenless and strong, queen-right colonies as compared to weak and medium-sized colonies.

It is known that migratory Africanized swarms not only prefer to occupy cavities of a certain volume (Rinderer *et al.*, 1982), but they are also attracted to such nesting sites by honey bee pheromones (Witherell, 1985). In addition, Africanized honey bee swarms tend to occupy not only larger nest cavities, but also wider cavity sizes as compared to European swarms (Rinderer *et al.*, 1982). Perez de Leon *et al.* (1991) was able to show that if migratory swarms in Mexico are offered three types of bait hives, ranging from 16 to 46 liters, they overwhelmingly selected the largest cavity. This study also showed that in areas with heavy concentrations of swarms, there was no statistical difference in the selection of trips with or without pheromone lures. Under those conditions, nest site cavities appear to play a greater role than pheromone lures.

In a recent study on bait hive technology by researchers from the University of Georgia (Dietz *et al.,* 1989; Vergara, 1990) it was shown that the number of swarms captured in bait hives placed away from the bee yards (2.5 km) was greater than those captured near apiaries. Also the number of swarms captured at "high" nest site locations (about 3.5 m) was greater than the number captured at "low" bait hives sites (0.4-0.6 m). Additionally, more swarms were captured in bait hives provisioned with pheromones as compared to the capture rate of unlured bait hives. Based on these findings, we have concluded that bait hives placed away from apiaries will be more effective in capturing swarms than those located near bee yards. Also lured bait hives at "high" nest site locations were more attractive to swarms than "low" nest site locations.

Based on the results of our monitoring research on Africanized honey bees in the States of Tabasco and Veracruz, Mexico (Dietz *et al.,* 1989b); Vergara *et al.,* 1989a), it appeared that the initial, or pioneer, swarms of this population could reach the United States border in Texas not before 1991. However, the first migratory swarm of Africanized honey bees (AHB) to invade the United States was discovered in a trap or bait hive on October 15, 1990 near Hidalgo, Texas by USDA, ARS personnel (Kaplan, 1990). Although the Hidalgo, Texas find appears to be the first occasion where AHB has immigrated into the United States by natural range expansion (Dietz, 1990, see Kaplan, 1990), it should be remembered that since 1983 personnel from APHIS (Animal and Plant Health Inspection Service, USDA) have destroyed AHB swarms on 21 different occurrences. As this was written, Africanized honey bees were found in the Southwest and California. Part of this retardation in the northward movement of these honey bees must be credited to the development and initiation of the Integrated Plan to Control Africanized honey bees by the Mexican government and the United States Department of Agriculture (APHIS and ARS). A major effort by these agencies resulted in the placement of over 200,000 pheromone treated bait hives along projected invasion routes to capture and destroy invading swarms. At present, over 80,000 swarms have been captured and identified along the invasion routes of the Pacific in the states of Chiapas and Oaxaca and the Atlantic in the states of Campeche, Yucatan, Tabasco, Veracruz and Tamaulipas. Based on reports by Mexican and USDA officials (Elba Quintero, personal communication), more than 80,000 swarms have been captured and destroyed in a period of two years. About 60,000 of these swarms were Africanized honey bees.

We assume that destruction of the captured swarms has contributed to the partial suppression of migrating honey bee populations. Although swarms continue to enter from the south, and the influx of normal reproductive swarms from feral and managed colonies occurs, swarm captures and northward migration of honey bees subsided during 1990.

The presence of Africanized honey bees in Mexico has provided us with the opportunity to examine the process of Africanization in areas with large populations of managed European honey bees colonies for the first time. Although the date of entry of Africanized honey bees into the United States is of historical importance, the more critical questions to be addressed here from a biological standpoint are: What factors contribute to the observed time difference between the discovery of the first Africanized, or pioneer swarms, and the measurable evidence of defensive behavior in managed colonies? Are "pioneer" swarms distinct ecotypes of the subsequent "front" of the migrant Africanized honey bee populations? Pioneer swarms can be defined as the first swarms discovered in a geographical area indicating the early stages of range expansion of Africanized honey bees (AHB). These initial isolated swarms generally precede the advancing "front" of AHB by a period ranging from several months up to 2 years. The "front" of the AHB has reached a specific geographical area when a rapid increase, or surge, in numbers of honey bee swarms occur and 50 percent or more of the total captured are identified as Africanized honey bees.

Although resource availability is a major factor in the successful colonization of new territory, migrant honey bee populations utilize several strategies in their conquest of new areas. An examination of some of the strategies employed by pioneer swarms should contribute to an understanding of the process of Africanization.

The absence of intense colony defense by pioneer swarms in Mexico has been a most remarkable phenomenon. The extreme defensive behavior expected by all honey bee specialists, the apicultural community, and the general public in Mexico, was not observed until February, 1988, or 1-1/2 years after the discovery of the first swarm of Africanized honey bees (Quintero pers. com.). So far, we have not encountered intense colony defensive behavior in our work with Africanized honey bees during the period of 1987 - 1990 in Tabasco and Veracruz, Mexico (Dietz *et al.*, 1989b; Vergara *et al.*, 1991a). Thus, the major problem allied with these bees may not be evident until 1-1/2 to 2-1/2 years after Africanized honey bees have entered the United States.

Africanized honey bees normally display a significantly stronger defensive behavior. They respond more intensely, in greater numbers and for a longer time period (Collins, 1988). Since extreme defensive behavior is the major problem associated with Africanized honey bees, a critical question is whether or not there will be such a time difference between the discovery of pioneer swarms of Africanized honey bees and measurable evidence of massive defensive responses when they become established in the United States. This time difference may be important in devising control strategies.

The migration of Africanized honey bees from Brazil to Mexico would not have been possible if they were unable to compete for food sources with native bees, and resident European honey bees. The ability of honey bees to shift from one foraging site to another is indicative of their utilization of large foraging areas. Africanized honey bees have a highly evolved capacity to exploit varied and unpredictable resource conditions and are thus able to survive dearth periods which may be detrimental to European honey bees. Absconding, either disturbance induced or resource related (Rinderer, 1988), is an effective strategy of pioneer swarms to insure reproductive success.

Pioneer swarms entering new habitats have the opportunity to select the most appropriate nest sites available. What constitutes an ideal nest site for Africanized honey bees is not known (Seeley, 1988). However, it is known that these bees prefer larger over smaller cavities, *i.e.* 40 to 80 liter versus 10 to 20 liters (Rinderer *et al.*, 1982). Our studies in Mexico have shown significant differences in nest site preferences of pioneer swarms when offered bait hives with and without pheromones and high or low nest site locations. Significantly more swarms preferred bait hives treated with pheromones over those without pheromones. Also, over 90 percent of captured swarms choose high (4 m) over low 1 m) nest site locations (Vergara *et al.*, 1991a).

Swarm production in Africanized honey bees appears to depend on the continuing availability of food sources in the field (Rinderer, 1988). The swarming rate of feral tropical honey bees in French Guiana during the swarming season was 3 to 4 cycles in an 8-month period, resulting in one prime swarm and an average of 2 after-swarms per cycle (Winston, 1988). It is unlikely that similar swarming rates can be expected in the near future in Mexico and the United States, based on the present capture rate of pioneer swarms. Our studies in Mexico (Dietz *et al.*, 1989b; Vergara *et al.*, 1989a,g; 1991a; Vergara, 1990) have shown a major swarming cycle in March and a considerably smaller cycle in August. Although swarming occurs throughout the year, the possibility of other major yearly cycles exists, depending on seasonality of food sources and feral bee populations. In Panama, Boreham and Roubik (1987) found that peak swarming cycles coincided with peak flowering.

The process of Africanization of new areas, especially the temperate zones of the Americas, can be described as an incomplete puzzle of honey bee behavior. I agree with Laidlaw (1988) in his assessment that the Africanized honey bee problem is essentially one of bee behavior and that the resolution of this problem rests on the genetic change of behavior by dilution and breeding. During the last three decades many pieces of this behavioral puzzle have been identified, and progress has been made to shed light on the question addressed here: Are pioneer swarms distinct ecotypes of an advancing "front" of Africanized bees populations?

Based on our studies and the above discussions on defensive behavior, competition, nest site selection, swarming, and colony invasion, we can state with some certainty that the pioneer populations of Africanized honey bees in Mexico displayed different behavioral patterns than the honey bees we studied in Argentina (Dietz *et al.*, 1985; Krell *et al.*, 1985; Dietz *et al.*, 1986; Dietz *et al.*, 1989a,c). It is important to indicate that with the passage of time, some of the present pioneer populations of Africanized honey bees will change because of the constant pressure exerted upon them by other ecotypes of advancing Africanized honey bees. The concept of different ecotypes of Africanized honey bees migrating northward finds support not only in our studies in Argentina (Dietz, *et al.*, 1986, 1989a,b,c), but in the work by Boreham and Roubik (1987), who reported a decrease in the size of hybrid worker bees between 1982 and 1985 in Panama.

Additionally, moving pioneer swarms of Africanized honey bees may be comprised of colonies more sensitive to predation, or changes in other environmental factors. Such colonies appear to respond to these various stimuli by absconding instead of increasing the level of defensive response as seems to be the case with colonies that become established in new territory.

The various ecotypes of Africanized bees entering and passing through Mexico into the United States probably cannot be stopped. Laidlaw (1988) suggested that some of the migrants will mix with the resident European honey bees, resulting in a change of their behavior, while others will escape extensive hybridization and enter Texas essentially unchanged from the African honey bees released in Brazil. Hall and Muralidharan (1988) have recently reported the existence of some of these relatively unchanged populations of African honey bees in Central America and Mexico. Ultimately, the majority of the honey bee populations entering the United States will have different behavioral and genetic repertoires than their African ancestors.

NEAR EAST GROUP

This group is considered by Rothenbuhler *et al.* (1968) to be a transition between the African and Southeast European groups. Included in this category are the following six geographical variants: *Apis mellifera caucasica, Apis mellifera cypria, Apis mellifera syriaca* Buttel-Reepen (1906), *Apis mellifera anatoliaca* Bodenheimer (1941), *Apis mellifera persica* (Khabel Jasim, 1955), and *Apis mellifera armenica* Skorikov (1929).

Ruttner (1988a) lists seven subspecies and excludes *A. m. persica*. Included in his classification is a well characterized island race collected by Brother Adam in 1952 on the island of Crete (see Ruttner, 1975d; 1980). This bee, known as *Apis mellifera adami* Ruttner (1975), is interestingly enough similar to *A. m. syriaca* (Mediterranean east coast), but surprisingly not similar to bees of Greece (Ruttner, 1988a). Although Br. Adam (1983) described this honey bee as highly defensive, especially when kept in the cool climate of England, it is also able to winter without difficulty in a cold-temperate climate. Additionally, the brood rearing activity of *Apis mellifera adami* is not only exceptional because it continues during winter due to the island's favorable climate, but it also peaks in May. Almost every colony prepares to swarm and as many as 200 swarm cells are constructed for this purpose.

Another bee listed by Ruttner (1988), but not Rothenbuhler *et al.* (1968), is *Apis mellifera meda* Skorikov (1929). This bee is found in the Iranian highland west of the big deserts Kavir and Lut, extending with a relatively narrow zone as far west as the Mediterranean (Ruttner, 1988a).

According to Ruttner (1988a), Bodenheimer (1941) was the person who attempted a taxonomic classification of the honey bees of Anatolia. However, *Apis mellifera anatoliaca* was described and published in a formal taxonomic classification by Maa (1953). The central Anatolian bee, according to Brother Adam (1964), is small in size with a smudgy orange color which becomes brown on the posterior dorsal and ventral segments. The scutellum is generally dark orange in color. The two major defects displayed by this bee are her disposition to build excessive brace-combs and her excessive use of propolis.

Brother Adam (1977, 1983) was never compelled to use special protection, but even the best tempered Anatolian strains are highly susceptible to low temperatures and cannot be manipulated with impunity either early in the morning or late in the evening. In terms of good qualities, Brother Adam (1964) considers her without comparison in regard to foraging powers, thrift and wintering abilities. The pure Anatolian bee is unreliable in terms of maximum performance. It is only when this race is crossed with a good strain of Italian or Carniolan bees that the full economic potentialities of the Anatolian race comes to the foreground.

The western Anatolian bee does not possess the hardiness and thrift of the Central Anatolian variety (Brother Adam, 1977, 1983). The subtropical temperature of southwestern Turkey versus the severe winters, high summer temperatures and the absence of rain from mid-June to mid-October in central Anatolia, are undoubtedly responsible for this difference. Nevertheless, it is appropriate to compare this bee with the Italian bee, *Apis mellifera ligustica*, because both of them are similar in color and size. Additionally, Brother Adam (1983) who has studied the performance of this bee for three decades, reports not only energetic food-gathering activities, but also a wintering ability which is superior to most other races.

The gray Caucasian bee, *Apis mellifera caucasica*, is along with the Italian, the Carniolan and the Dark honey bee one of the four major races of

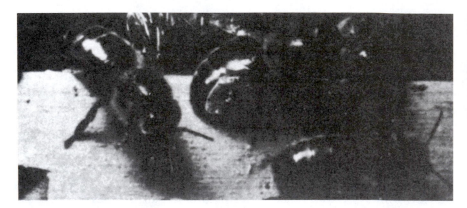

Mellifera

Ligustica

Carnica   Caucasica

Figure 9. The four races of well-known economic value. The individual bees are approximately three times life size. *(Photos courtesy of F. Ruttner except the one of Ligustica which is by G. P. Piana, Bologna, Italy).*

economic importance (Fig. 9). This honey bee has her homeland in the high valleys of the Central Caucasus (Ruttner, 1960; 1975a). In general, these bees are dark to black with brown spots in many cases on the first bands of the abdomen. The hairs of the worker are short and grey. The thoracic hairs of drones are black. In their native homeland, the Caucasian is a composite of many local strains and thus does not display a uniform color (Ruttner, 1975a).

The gray Caucasian bee has the longest tongue of any economically important race of honey bees. This bee is very similar in appearance to the Carniolan bee, except in wing venation and wax mirror (Ruttner, 1960, 1988). The Caucasian bees are rather gentle and calm on the comb. They are not inclined to excessive swarming and develop into strong colonies, but normally not before the early part of the summer. This bee will forage at somewhat lower temperatures and less favorable climatic conditions than the Italian bee (Dietz, 1968). According to Phillip (1928), these bees cap their honey-cells with white wax, do not drift to other hives and winter well. However, Ruttner (1960, 1975a) pointed out that they display a tendency to drifting and robbing. Also, their cappings of the honey cells often have a "wet" appearance because there is no air between the honey and the wax cappings. Additionally, *Apis mellifera caucasica* does not winter well in northern regions due to their susceptibility to Nosema disease (Alpatov, 1932; Avetisyan, 1978).

One of the major reasons why the Caucasian bee has been abandoned by most beekeepers in the United States has been its use of large amounts of propolis to fasten down combs and to reduce the size of the colony entrance, as well as the construction of brace-comb (Brother Adam, 1954; Dietz, 1968). The Caucasian bee is a better honey producer in Russia as compared to the Dark bee, *A. m. mellifera*. However, in western Europe, the honey production was lower than that of the Carniolan bee. The importance of this bee in bee breeding must be emphasized because it possesses a number of valuable qualities (Ruttner, 1975a).

CENTRAL MEDITERRANEAN AND SOUTHEAST EUROPEAN GROUP

Members of this assembly include the Italian honey bee, *Apis mellifera ligustica*, and the Carniolan honey bee, *Apis mellifera carnica*. With the exception of the Africanized honey bees, these two races of bees are not only the best known but also the economically most important. Rothenbuhler *et al.* (1968) includes the Greek bee, *Apis mellifera cecropia* Kiesenwetter (1860), as the only other member of this group. Ruttner (1988), added a new subspecies, *Apis mellifera macedonica* Ruttner (1988), the Macedonian bee and the Sicilian honey bee, *Apis mellifera sicula* Montagano (1911) to this category. The behavior of *A. m. macedonica* has been described as very gentle, not inclined to swarm and strong winter population. Good wintering results were also reported from England; however, the enthusiasm for the Macedonian bee quickly faded when losses due to nosematosis occurred (Ruttner, 1988).

The Sicilian honey bee, *A. m. sicula*, has a dark color. The hair of the thorax of workers and drones is not grey or brown as in other races or brown as in other dark races, but yellowish (Ruttner, 1988). This bee is rather gentle and quiet during manipulations. Brood rearing, including drones, occurs almost year around. The most surprising observation, according to Ruttner (1988), is that no swarm departs from the nest until a number of virgin queens have hatched. The resident queen and daughters live together peacefully in the nest prior to departure of the swarm. Since this behavior is similar to that of the Egyptian and the Syrian bee, *A. m. sicula* is totally isolated among European geographic races.

The Greek honey bee, *A. m. cecropia*, is not commercially used to a large extent, except in Greece. Aristotle reported that the bees of Greece are not uniform in color and those with yellow markings were considered the best (Brother Adam, 1954). The reason this bee has perhaps not attracted attention is due to its lack of bright color and uniformity of appearance, even though it is an excellent forager. The most outstanding qualities of the Greek bee are its numerical strength, good temperament and a disinclination to swarm. These bees are similar to the average Carniolan in regard to their gentleness and their quiet disposition during examination. They resemble the Caucasian bee in many of their characteristics, such as the tendency to propolize and the construction of brace-comb. Unlike the Italian or Eastern races of honey bees, brood production is often severely restricted at the end of July because of large accumulations of stores in the brood nest area. The cappings of the honey comb are rather watery in appearance. Brother Adam (1954) considers the Greek bee to be superior to the Caucasian and is of the opinion that this race may well prove to be of great value. However, the two remaining races of this group are the best known bees in apicultural circles throughout the world.

The Carniolan bee (*A. m. carnica*) and the Italian bee (*A. m. ligustica*) are probably the most widely distributed races of bees in the world. The most typical forms of *A. m. carnica* are found in upper Carniola and the two adjoining provinces of Carinthia and Styria. The geographical distribution of the Carniolan bee, however, extends far beyond these three regions and not only includes Yugoslavia, Rumania, Bulgaria and Hungary, but also the greater part of Austria (Brother Adam, 1964, 1977). Since the first importations of this bee came from Upper Carniola into English-speaking countries, the name Carniolan has been used by beekeepers throughout the world (Dietz, 1982).

Described as a grayish-black version of the Italian bee, and with the exception of color and gray pubescence, *A. m. carnica* approximates the Italian bee more closely than any other race of bee. The range of variation between the various strains of *A. m. carnica* is probably greater than any other known race (Ruttner, 1984). The Carniolan bee is without question a distinct variety of *Apis mellifera ligustica*. Aside from morphological differences, the variability in physiological distinctions between the various strains is very pronounced. (Brother Adam, 1951, 1977).

It is a slender bee with a fairly long tongue measuring 6.4-6.8 mm. Even though the Carniolan is mainly dark, there are often brown spots on the second and third terga of the abdomen and occasionally a leather-brown band. The hairs of the drones are gray to grayish brown (Ruttner, 1975a).

Perhaps the most outstanding trait of the Carniolan is its exceptional docility. Other important beneficial characteristics include longevity, hardiness, foraging ability and wintering ability. The Carniolan bee forms small colonies in the fall and thus is able to winter on a minimum of stores, which is distinctly different from the Italian bee which requires large stores during the winter period. Brood rearing is closely tied together with pollen availability. They maintain a large brood nest during the summer if pollen is abundantly available; however, a poor pollen supply will result in a reduction in brood rearing (Brother Adam, 1951; Ruttner, 1975a).

The Carniolan bee is not only an excellent honey producer and comb builder, but also caps the cells of the honey with paper-white wax cappings. The bee has a very good sense of orientation and is not inclined to robbing. The Carniolan bee collects less propolis than any other European race. The major deficiency of this bee is its strong disposition to swarming because of its great vitality and the fast development of its colonies. However, the excessive swarming tendency can be reduced within tolerable limits by selection (Brother Adam, 1951; Ruttner, 1975a).

The Carniolan is the most popular bee in Germany where it has replaced most of the other geographic races of honey bees (Ruttner, 1980). The Carniolan bee, also known as the Carnica bee in non-English speaking countries (Brother Adam, 1951, 1977), has finally also been accepted by beekeepers in the United States. Nevertheless, the dark color of these bees is still a factor in limiting acceptance by some American beekeepers. The queens of dark bees are also more difficult to locate in the hive as compared to yellow races of bees. Brother Adam (1951) considers the Carniolan bee as a mystery race in many respects because many of its hereditary potentials lie dormant and only become apparent in cross-breeding.

The continental strains which are popular in Central Europe originated in

Austria and are in some respects not identical to the Carniolan bee from Upper Carnica which was imported into England about a hundred years ago (Brother Adam, 1977). The Austrian strains of Carniolan bees use much more propolis as compared to the original Carniolan bee.

On a worldwide basis, the Italian bee, *A. m. ligustica*, is perhaps the most popular bee. Commercial beekeeping, as it is practiced today, would probably not have been possible without this strain (Brother Adam, 1951). The first "yellow bees" were brought from Venice, Italy, to Germany by Dzierzon in 1853, and based on his enthusiastic report, the first living Italian queens landed in the United States in 1859 (Ruttner, 1975a). Other imports into the United States occurred in 1860, and the sale of Italian queens in the United States began in 1861 (Phillips, 1928).

In general, each geographic race of honey bees consists of a great number of strains, and according to Brother Adam (1951), there are three distinct varieties of Italian bees: (1) the leather-colored kind; (2) the bright yellow type; and (3) the very pale lemon-colored variety. However, Goetze (1964) considers coloration of the *A. m. ligustica* much more complex than is generally assumed. In his opinion, the true *A. m. ligustica* is a yellow copy of the Carniolan bee since both are very similar in most morphological characteristics. In the original homeland, the extensions of shades of yellow vary considerably between different strains of bees. A fairly bright-colored form of *A. m. ligustica*, known as the Aurea or "golden bee" (Ruttner, 1975a), has only a small black spot on the tip of the abdomen. This type of bee is probably the result of selective breeding and may occasionally appear in nature as a mutation (Goetze, 1964). Many beekeepers are very fond of the "golden bee" not only because of her beauty, but also the ease with which the queen can be located during colony examination.

During colony examination, the Italian bee is relatively gentle and rather calm. The strong disposition to brood rearing results in large populations which are able to collect a considerable amount of nectar in a relatively short period of time. *A. m. ligustica* is not only a good housekeeper, and use very little propolis, but it is also less susceptible to European foulbrood. Italian bees are excellent foragers, good comb builders, and cover the honey with brilliant white wax cappings (Phillips, 1928; Park, 1938; Goetze, 1964; Dietz, 1968; 1982).

The Italian honey bee also has a number of serious defects which have made it difficult for her to become established in Europe north of the Alps. *A. m. ligustica* tends to breed excessively at the end of the main nectar flow, resulting not only in the overwintering with strong colonies but also a high food consumption. This bee does not overwinter well in areas with long and cold winters. In northern latitudes there is often a loss of worker bees due to early brood rearing resulting in a condition known as spring dwindling and thus slow colony development in the spring. However, Ruttner (1988) said that selected strains with adapted management methods are overwintered with good results in Finland.

Her poor sense of orientation causes her to drift from one colony to another and usually from a weak to a strong colony. The excellent ability of this bee to locate available food sources often results in robbing in an apiary if care is not exercised in colony examination during periods of nectar shortage. If robbing starts among different geographic races maintained in an apiary, *A. m. ligustica* is generally the instigator. The Italian bees are often very annoying because they angrily fly about the head of the beekeeper during colony manipulations in the apiary. This behavior pattern is not only evident during periods of nectar shortages, but also during the nectar flow (Brother Adam, 1951; Goetze, 1964; Dietz, 1968; 1982).

WESTERN MEDITERRANEAN AND NORTHWEST EUROPEAN GROUP

Related to the African geographic races through *Apis mellifera intermissa* and *Apis mellifera sahariensis* is the Northwest European group of Rothenbuhler *et al.*, 1968. Both geographical variants, however, had been removed from the grouping proposed by Rothenbuhler *et al.* (1968), placed in the African group by Ruttner (1975b) and in the Western Mediterranean category by Ruttner in 1988. The other honey bee races in this category are *Apis mellifera mellifera* and *Apis mellifera iberica* Goetze (1964).

The bees in this category are generally larger in size than most of the other geographic races (Goetze, 1940). *Apis mellifera iberica*, the dark bee of the Iberian peninsula, is the link between the African and the northwest European group (Goetze, 1964). Its appearance is very close to *A. m. mellifera* L., the Dark bees. Ruttner (1988) reports that in behavior the Iberian bee could almost be taken to be a variety of *A. m. intermissa* because of quick defense reaction, nervousness on comb, heavy use of propolis and propensity to swarm. Extremely defensive honey bees, identified as European honey bees by H. Daly (personal communication), were discovered by us near the 38th parallel in the Buenos Aires province of Argentina. Since this is a temperate zone, these defensive bees were tentatively identified by us as *Apis mellifera iberica*, or their hybrids (Dietz *et al.*, 1985).

The Dark bee, *Apis mellifera mellifera*, is one of the largest honey bees. This insect, however, has a relatively short tongue (5.7-64 mm). The color is uniformly dark with small yellow spots on the second and third abdominal tergites. Yellow bands are absent (Ruttner, 1975a). They develop somewhat slowly in the spring, but have strong colonies in late summer and during the winter. They overwinter well in severe climates. This bee is also susceptible to brood diseases and wax moth attacks (Phillips, 1928). Those deficiencies are mainly responsible for this bee being replaced by beekeepers with other races in the United States as well as in other countries (Dietz, 1982).

According to Ruttner (1988), the original homeland of *A. m. mellifera* covered all of France, the British Islands up to Scotland and Ireland, Central Europe north of the Alps and the plains of North Poland, and the USSR east of the Ural Mountains (Fig. 6). At present, these bees are kept in Norway up to 65° latitude (250 km north of Drontheim). Although there is a rich supply of heather nectar available in this region, it is not the length of winter, but rather the lack of suitable nesting sites which is the limiting factor in the distribution of this honey bee.

THE UTILIZATION OF NATURAL DIFFERENCES AMONG BEES

A successful bee economy needs, in addition to good flows and ambitious beekeepers, first of all a capable bee. Vigor and ability to develop the colony, gentleness and quietness on the combs, and the capability of gathering large amounts of stores are the qualities primarily required of the bee. The environment may make further claims as winter hardiness, resistance to specific diseases, response of brood rearing to seasonal variations, and orientation. One cannot expect from natural selection, the development of a race perfectly fulfilling the requirements of the modern beekeeper. But the comparison of the typical characteristics of the different geographic races shows that one race approaches the ideal more closely than another one. Sometimes a natural race can simply be replaced by another one in order to increase production. As has already been mentioned, this has been done many times. Certainly, careful consideration should be given to the climatic conditions of the original homeland in the case of transplantation of a race. Bees from a mild climate can hardly adjust themselves to a rugged environment.

Further increase of production can be achieved by selection within a race. To a certain extent, the results of this method have proved to be satisfactory, if the breeder was able to avoid intensive inbreeding. Inbreeding generally decreases the fertility and vigor of the colony, in extreme cases so far that the colony becomes incapable of living without regular addition of combs of brood from other colonies (Mackensen, 1956).

In other fields of animal breeding, synthetic races originated mostly by crossbreeding different geographic races, or even species, and by long selection and inbreeding a new stable combination developed. In bee breeding this way has not been used frequently so far, because of difficulties in controlled matings, although great possibilities exist in this field (Brother Adam, 1966).

In the cultivation of plants, one uses the heterosis effect (hybrid vigor) to a great extent today. Hybrid corn is the best known example of what can be achieved in this manner. Only a few kinds of animals are adaptable to this method; the honey bee is one of them because of the great number of its offspring. Hybrid vigor, with the performance of the first hybrid crossing being better than that of the better parent, can be achieved in the bee in two different ways. Hybrids or double hybrids from inbred lines have been used in breeding honey bees. The classical methods used in corn production show very promising results in bee breeding also (Cale, 1957, Cale and Gowen, 1956). In egg laying, as well as in honey production, the hybrids surpassed the control strains. The increase in production amounted to 34 percent.

CROSSBREEDING AMONG GEOGRAPHIC RACES

Hybrid vigor also occurs in many cases when two geographic races are crossed without previous inbreeding. Since there are mostly large genetical differences between the races, the results are greater than of any other method, but are not always repeatable. However, it must be pointed out that in a considerable number of experiments, hybrids did not show any hybrid vigor (Ruttner, 1968; Bornus, 1972).

The combining ability within a geographic race shows some differences as well as many other characteristics. In crossbreeding not all the strains of two races create the same beneficial heterosis effect. The specific combining ability has to be tested in every single case. Another difficulty in crossbreeding lies in the development of unfavorable characteristics in numerous cases—that is some hybrids may be very aggressive; while others may be very suceptible to diseases. Although crosses between different ecotypes of the same race could result in considerably higher yields than noncrossed controls, it is evident that well-planned crossbreeding will play an extraordinary role in beekeeping of the future. Selecting honey bees that are resistant to parasitic mites is presently under investigations by several researchers (Kulincevic and Rinderer, 1987; Peng *et al.,* 1987, and others). This type of research is very promising, especially since several resistance mechanisms have already been identified (Dietz and Hermann, 1988).

REFERENCES

Abushady, A.Z. (1949). Races of bees. *In* "The Hive and the Honey Bee" (R.A. Grout, ed.), pp. 11-20, Dadant & Sons, Inc., Hamilton, IL.

Alber, M. (1956). The size of comb cells as a racial characteristic. XVI. Intern. Beekeep. Congr., Vienna Apic. Abstr. 241/56.

Alley, H. (1883). The Bee-Keeper's Handy Book: or Twenty-two Years' Experience in Queen-Rearing. Salem Press, Salem, Mass.

Alpatov, W.W. (1929). Biometrical studies on variation and races of the honey bee *Apis mellifera* L.. *Quart. Rev. Biol.* 4:1-57.

Alpatov, W.W. (1932). Some data on the comparative biology of different bee races. *Bee World.* *13*:138-139.

Alpatov, W.W. (1948). The races of honeybees and their use in agriculture. *Sredi Prirodi* *4*:1-65. (In Russian).

American Farm Bureau Res. Foundat. (1986). Proc. of the Africanized Honey Bee Symp. Feb. 11-12, 1986. Atlanta, GA p 129.

Anderson, R.H. (1961). The development of egg laying workers in the Cape honey bee. XVIII. Intern. Beekeep. Congr. Madrid.

Anderson, L.M. and A. Dietz. (1976). Pyridoxine requirement of the honey bee (*Apis mellifera*) for brood rearing. *Apidologie, 7*:67-84.

Armbruster, L. (1938). Versteinerte Honigbienen aus dem obermiocanen Randecker Maar. *Arch. Bienenkd. 19*:(1)1-48, (2)73-93, (3/4)97-133.

Avetisyan, G.A. (1973). Breeds of Soviet Union bees: their selection and protection. 24th Intern. Beekeep. Congr., Buenos Aires.

Baldensperger, P.J. (1922). Sur l'apiculture en orient. *Proc. Intern. Apicult. Congr. 6*:59-64.

Bischoff, H. (1960). Stammesgeschichte der Biene. *In* "Biene und Bienenzucht" (Büdel, A. and E. Herold, *eds.*), pp. 1-4. Ehrenwirth Verlag, Munich.

Blum, R. (1951). Beekeeping in Israel. *Am. Bee J. 91*:378-379.

Boch, R., Shearer, D.A. and B.C. Stone. (1962). Identification of isoamyl-acetate as an active component in the sting pheromone of the honey bee. *Nature,* (London) *196*:1018-1020.

Bodenheimer, F.S. (1941). Studies on the honey bee and beekeeping in Turkey. Merkez Ziraat Macadela Enstitüsü, Ankara.

Boreham, M.M. and D.W. Roubik. (1987). Population change and control of Africanized honey bees (Hymenoptera:Apidae) in the Panama Canal area. *Bull. Ent. Soc. Am. 32*:34-89.

Brother Adam. (1951). In search of the best strains of bees. *Bee World, 32*:49-52, 57-62.

Brother Adam. (1954). In search of the best strains of bees: Second journey. *Bee World, 35*:193-203, 233-244.

Brother Adam. (1964). In search of the best strains of bees: Concluding journey. *Bee World, 45*:70-83, 104-118.

Brother Adam. (1977). In search of the best strains of bees: Supplementary journey to Asia Minor, 1973. *Bee World, 58*:57-66.

Brother Adam. (1983). In search of the best strains of bees. Northern Bee Books, Hebden Bridge, UK. 206 pp.

Butler, C.G. (1954). "The World of the Honey Bee." Collins, London.

Butler, C.G. (1975). The honey bee colony-life history. *In* "The Hive and the Honey Bee," pp. 39-74, Dadant and Sons, Hamilton, IL.

Buttel-Reepen. H. (1906). Apistica. Beiträge zur Systematik, Biologie, sowie zur geschichtlichen und geographischen Verbreitung der Honigbiene (*Apis mellifera* L), ihren Varietäten und der übrigen *Apis*-Arten. Veröff. Zool. Museum Berlin.

Buttel-Reepen, H. (1915). "Leben und Wesen der Bienen." Friedrich Vierweg & Sohn, Braunschweig.

Cale, G.H., Jr., and J.W. Gowen. (1956). Heterosis in the honey bee *Apis mellifera* L. *Genetics 41*:292-303. Apic. Abstr. 245/56.

Cale, G.H., Sr. (1957). How the new hybrids affects management. *Am. Bee J. 97*:48.

Carpenter, F.M. (1952). Fossil Insects. *In* "Insect" (A. Stefferud, ed.), pp. 14-19, The Yearbook of Agriculture, USDA, U.S. Government Printing Office, Washington, D.C.

Carpenter, F.M. and H.R. Hermann. (1979). Antiquity of sociality in insects. *In* "Social Insects" (H.R. Hermann, *ed.*) Vol. 1, pp. 81-89, Academic Press, N.Y.

Cockerell, T.D.A. (1906). New Rocky Mountain bees and other notes. *Canad. Entomol. 38*:160-166.

Cockerell, T.D.A. (1907). A fossil honey bee. *Entomologist, 40*:227-229.

Collins, A.M. and K.J. Kubasek. (1982). Field test of honey bee (Hymenoptera, Apidae) colony defensive behavior. *Ann. Entomol. Soc. Amer. 75*:385-387.

Collins, A.M. T.E. Rinderer, J.R. Harbo and A.B. Bolten. (1982). Colony defense by Africanized and European Honey Bees. *Science, 218*:72-74.

Collins, A.M., Rinderer, T.E., J.R. Harbo and M.A. Brown. (1984). Heritabilities and correlations for several characters in the honey bee. *J. Hered. 75*:135-140.

Collins, A.M. (1988). Genetics of honey-bee colony defense. pp. 110-115. *In* "Africanized Honey Bees and Bee Mites" (G.L. Needham, R.E. Page, Jr., M. Delfinado-Baker and C.E. Bowman, *eds.*), John Wiley and Sons, New York.

Crane, E. (1977). Dictionary of beekeeping terms. Apimondia Publishing House, Bucharest, Rumania. 206 pp.

Dadant, C.C. (1975). Beekeeping equipment. *In* "The Hive and the Honey Bee," pp. 303-328, Dadant & Sons, Hamilton, IL.

Daly, U.V. and S.S. Balling. (1978). Identification of Africanized honey bees in the Western Hemisphere by discriminant analysis. *J. Kans. Entomol. Soc. 51*:857-869.

Danka, R.G. and T.E. Rinderer. (1988). Social reproductive parasitism by Africanized honey bees. pp. 214-222. *In* "Africanized Honey bees and Bee Mites" (G.L. Needham, R.E. Page, Jr., M. Delfinado-Baker and C.E. Bowman, *eds.*), John Wiley and Sons, New York.

Darchen, R. and J. Lensky. (1963). Quelques problèmes soulevés par la création de socétés polygynes d'abeilles. *Insectes Soc. 10*:337-357.

Darwin, C.R. (1859). "On the Origin of Species by Means of Natural Selection, or the Preservation of Favoured Races in the Struggle for Life." 1st Ed. John Murray, London. 502 pp.

Deodikar, G.B., C.V. Thakar and Pushipa N. Shaw. (1959). Cytogenetic studies in Indian honey bees. 1. Somatic chromosome complement in *Apis indica* and its bearing on evolution and phylogeny. *Proc. Indian Acad. Aci. 49*:194-206.

Dietz, A. (1962). A short natural history of the honey bee family Apidae (Leach, 1817). *Australas. Beekeep. 63*:187-188.

Dietz, A. (1968). "Beekeeping in Maryland," Ext. Bull. 223. University of Maryland, College Park. 41 pp.

Dietz, A. (1972). The nutritional basis of caste determination in honey bees. *In* "Insect and Mite Nutrition" (J.E. Rodriguez, *ed.*), pp. 271-279, North-Holland Publ., Amsterdam.

Dietz, A. (1975). Nutrition of the adult honey bee. *In* "The Hive and the Honey Bee," pp. 125-156, Dadant and Sons, Hamilton, IL.

Dietz, A. (1978). An anatomical character suitable for separating drone honey bees of *Apis mellifera ligustica* from *Apis mellifera adansonii*. Proc. Apimondia Int. Symp. Apic. Hot Climate, 1978, pp. 102-106.

Dietz, A. and J. Nowakowski. (1981). Unpublished observations.

Dietz, A. (1982). Honey Bees. *In* "Social Insects" (H.R. Hermann, *ed.*). Vol. 3, pp. 323-360. Academic Press, N.Y.

Dietz, A. (1984). Africanized honey bees in Argentina. Program Symposium: The future of Africanized Bees in the Americas. Entomological Society of America meeting, San Antonio, TX. December 1984.

Dietz, A. (1986). The geographical distribution and levels of infestation of the mite *Varroa jacobsoni* Oudemans (Parasitiformes: Varroidae) in honey bee colonies in Argentina. *Am. Bee J. 126*:49-51.

Dietz, A. (1985a). Problems and prospects of maintaining a two queen colony system in honey bees throughout the year. *Am. Bee J. 125*:451-453.

Dietz, A. (1985b). Monogyny and induced polygyny in honey bee colonies. (Manuscript in preparation).

Dietz, A. (1986). Evolution. pp. 23-56. *In* "Bee Genetics and Breeding" (T.E. Rinderer, *ed.*). Academic Press, N.Y.

Dietz, A. (1990). Nutrition of larvae and adult honey bees. *In* "Honey bees and Beekeeping in Israel." (In press).

Dietz, A., R. Krell and M.S. Brower. (1983). Pollination of our seashores. Proc. 10th Pollination Conf. South. Ill. Univ., Carbondale, IL. pp. 57-66.

Dietz, A., R. Krell and F.A. Eischen. (1985). Preliminary investigation on the distribution of Africanized honey bees in Argentina. *Apidologie, 16*:99-108.

Dietz, A. and R. Krell. (1986). Survey for honey bees at different altitudes in Kenya. *Am. Bee J. 126*:829-830.

Dietz, A., R. Krell and J. Pettis. (1986). The potential limit of survival for Africanized bees in the United States. pp 87-100. Amer. Farm Bureau Res. Found. Africanized Honey Bee Symposium, Atlanta, GA, Feb. 11-12, 1986.

Dietz, A., D.S. Hurley, J.S. Pettis and F.A. Eischen. (1987). Controlling *Acarapis woodi* (Rennie) in package bee colonies with Apitol. *Amer. Bee J. 127*:843.

Dietz, A. and R. Krell. (1991). Overwintering of Africanized Honey Bees in Argentina. (Submitted for publication).

Dietz, A. and H.H. Hermann. (1988). Biology, Detection and Control of *Varroa jacobsoni:* A parasitic mite on honey bees. Lei-Act Publishers, Commerce, GA. 80 pp.

Dietz, A. and Carlos Vergara. (1990). Beekeeping in Argentina. (Submitted for publication).

Dietz, A., R. Krell and J. Pettis. (1988). Survival of Africanized and European honey bee colonies confined in a refrigeration chamber. pp. 237-242. *In* "Africanized Honey Bees and Bee Mites" (G.L. Needham, E. Page, Jr., M. Delfinado-Baker and C.E. Bowman, *eds.*). John Wiley & Sons, New York.

Dietz, A., J.F. Leitner, Vergara, C. and M. Mejia. (1989a). Effect of prolonged confinement in a refrigeration chamber on the survival of Africanized and European honey bee colonies. *Am. Bee J. 129*:815.

Dietz, A., C. Vergara, A. Perez de Leon and V.M. Butz. (1989b). Africanized honey bees in the Americas. Proc. 4th Intern. Conf. on Apiculture in Tropical Climates, Cairo, Egypt. Intern. Bee Res. Assoc., Cardiff, England, pp. 471-477.

Dietz, A., C. Vergara, M. Mejia and R. Krell. (1989c). Forced queen usurpation in colonies of Africanized and European honey bees in San Juan, Argentina. Proc. 32nd Intern. Apicult. Congr., Rio de Janeiro, Brazil. (In press).

Dietz, A., R. Krell, C. Vergara, M. Mejia and J. Pettis. (1991). Study on winter survival of Africanized and European honey bees in San Juan, Argentina. Apidologie, (In preparation).

Dulkin, A.L. and G.F. Treskova. (1953). Über die Bienen des Urals und der Taigo-Region. *Pchelovodstvo, 30*:26-29.

DuPraw, E.J. (1964). Non-Linnean taxonomy. *Nature* (London) *202*:849-852.

DuPraw, E.J. (1965a). The recognition and handling of honey bee specimens in non-Linnean taxonomy. *J. Apic. Res. 4*:71-84.

DuPraw, E.J. (1965b). Non-Linnean taxonomy and the systematics of honey bees. *Syst. Zool. 14*:1-24.

Dzierzon, J. (1845). Gutachten über die von Hrn. Direktor Stöhr im ersten und zweiten Kapitel des Generalgutachtens aufgestellten Fragen. Eichstadt. *Bienenzeitung 1 (11)*:109-113, *(12)* 119-121.

Eischen, F.A., J.S. Pettis and A. Dietz. (1986). Prevention of *Acarapis woodi* infestation in queen honey bees with Amitraz. *Am. Bee J. 126*:498-500.

Eischen, F.A., C. Vergara and A. Dietz. (1986b). Apitol, a new systemic acaricide for the control of *Acarapis woodi. Am. Bee Jour. 126*:830.

Eischen, F.A. and A. Dietz. (1986). *Acarapis woodi* studies in northeastern Mexico. Proc. Honey Bee Tracheal Mite (*Acarapis woodi*, R.) Scientific Symposium. St. Paul, Minn., July 8-9, 1986. pp. 52-53.

Eischen, F.A., T.E. Rinderer, and A. Dietz. (1986). Nocturnal defensive responses of Africanized and European honey bees to the greater wax moth (*Galleria mellonella* L.). *Animal Behav. 34*:1070-1077.

Eischen, F.A., J.S. Pettis and A. Dietz. (1987). A rapid method of evaluating compounds for the control of *Acarapis woodi* (Rennie). *Amer. Bee J. 127*:99-101.

Eischen, F.A., C. Vergara, A. Dietz and D. Cordoso-Tamez. (1988). Cymiazole, a systemic acaricide that control *Acarapis woodi* (Rennie) infesting honey bees. I. Laboratory tests. *Apidologie, 19*:367-376.

Eischen, F.A., D.D. Cordoso-Tamez, A. Dietz and G.O. Ware. (1989). Cymiazole, a systemic acaricide that controls *Acarapis woodi* (Rennie) infesting honey bees. II. An apiary test. *Apidologie, 20*:41-51.

Escholtz, J.F. (1822). Entomographien. Vol. 1. Reimer, Berlin.

Fletcher, D.J.C. (1978). The African Bee, *Apis mellifera adansonii*, in Africa. *Ann. Rev. Entomol. 23*:151-171.

Fletcher, D.J.C. and K.G. Ross (1985). Regulation of reproduction in eusocial Hymenoptera. *Ann. Rev. Entomol. 30*:319-43.

Free, J.B., and I.H. Williams. (1979). Communication by pheromones and other means in *Apis florea* colonies. *J. Apic. Res. 18*:17-25.

Georgandas, P. (1957). The Greek Bee. *Am. Bee J. 97*:314.

Goetze, G. (1930). Variabilitäts und Züchtungstudien an der Honigbiene mit besonderer Berücksichtigung der Langrüsseligkeit. *Arch. Bienenkd. 11*:185-236.

Goetze, G. (1940). "Die Beste Biene." Liefloff, Loth & Michaelis, Leipzig. 200 pp.

Goetze, G. (1956). Methodik der Selektion der Honigbiene auf Langrüsseligkeit. *Insectes. Soc. 3*:335-346.

Goetze, G. (1964). "Die Honigbiene in natürlicher und künstlicher Zuchtauslese. I. Systematik, Zeugung und Vererbung." Beih. Z. angew. *Entomol. 19*:1-120.

Gorbachev, A.N. (1916). The grey mountain Caucasian bee (*Apis mellifera caucasica*) and its place among other bees. Tiflis (In Russian with English summary).

Goncalves, L.S. (1975). Preliminary studies on the identification of the Africanized bees from Brazil. *Proc. Intern. Beekeep. Congr. 25*:345:347.

Goncalves, L.S. (1982) The economic impact of the Africanized honey bee in South America. Proc. 9th Congr. of IUSSI, Boulder, CO. "The biology of social insects." M.D. Breed and H.E. Evans, *eds*. pp. 134-137.

Grout, R.A. (1937). The influence of size of brood cell upon the size and variability of the honey bee (*Apis mellifera* L.). Iowa Agric. Expt. Sta. Bull. 218.

Guy, R.D. (1972). Commercial beekeeping with African bees. *Bee World, 53*:14-22, 159-166.

Haccour, P. (1960). Investigations on the Sahara bee in Morocco. C.R. *Soc. Sci. Nat. Moroc.* 6:96-98.

Hall, H.G. (1988). Characterization of the African honey-bee genotype by DNA restriction fragments. pp. 287-302. *In* "Africanized Honey Bees and Bee Mites" (G.L. Needham, E. Page, Jr., M. Delfinado-Baker and C.E. Bowman, *eds*.). John Wiley & Sons, New York.

Hall, H.G. and K. Muralidharan. (1988). African honey bee migration into Mexico followed with DNA markers. *Am. Bee J. 128*:803.

Herold, J. and H. Pieterek. (1985). "Das Kleine Imker-ABC." Ehrenwirth Verlag, Munich. p. 181.

Henriksen, C. and O. Hammer. (1957). An experiment in breeding long-tongued bees. Nord. Bitidskr. 9:11-19, *Apic. Abstr.323*:57.

Jayasvasti, S. (1889). Scanning electron microscopy analysis of honey bees (*Apis florea* F., *Apis dorsata* F., *Apis cerana* F., *Apis mellifera* L. and *Apis andreniformis* S.) stings in Thailand. First Asia-Pacific Conf. of Entomol. (APCE), Chiangmai, Thailand (Abstract).

Kapil, R.P. (1971). A hive for the Indian honey bee. *Apiacta 6*:107-109.

Kaplan, K.J. (1990). The Africanized Honey Bees. Agricultural Research, ARS, USDA, Washington, D.C. *38* (12):4-11.

Kempff Mercado, N. (1973). African (Brazilian) bee report published in Bolivia. *Am. Bee J. 113*:344.

Kerr, W.E. (1957). Introduction of African bees to Brazil. *Brazil Apic. 3*:211-213. Apic. Abstr. 184/589.

Kerr, W.E. and H.H. Laidlaw, Jr. (1956). General genetics of bees. *Adv. Genet. 8*:109-153.

Kerr, W.E. and V.P. Araujo. (1958). Racas de abelhas de Africa. *Garcia de orta. 6*:53-59.

Kerr, W.E., S. de Leon Del Rio and M.D. Barrionuevo. (1982). The southern limits of the distribution of the Africanized honey bee in South America. *Am. Bee J. 122*:196-198.

Kiesenwetter, E.A.H. (1860). Uber die Bienen des Hymettos. Berlin Entomol. *Nachr. 1860*:315-317.

Kleine, G. (1960). The characteristics of three races of bees: the Italian. *Am. Bee J. 100*:177.

Kloft, W., Djalal, A.S. and W. Drescher. (1976). Untersuchung der unterschiedlichen Futterverteilung in Arbeiterinnengruppen verschiedener Rassen von *Apis mellifica* L. mit Hilfe von ^{32}P als Tracer. *Apidologie, 7*:49-60.

Koeniger, N. (1976). Neue Aspekte der Phylogenie innerhalb der Gattung *Apis. Apidologie, 7*:357-366.

Koeniger, N., J. Weiss and U. Maschwitz. (1979). Alarm pheromones of the sting in the genus *Apis. J. Insect Physiol. 25*:467-476.

Koeniger, N., G. Koeniger, S. Tinger, M. Madran and T.E. Rinderer. (1988). Reproductive isolation by different time of drone flight between *Apis cerana* Fabricius, 1973 and *Apis vechti* (Maa, 1953). *Apidologie, 19*:103-106.

Krell, R., A. Dietz and F.A. Eischen. (1985). A preliminary study on winter survival of Africanized and European honey bees in Cordoba, Argentina. *Apidologie, 16*:109-118.

Kulincevic, J.M. and T.E. Rinderer. (1988). Breeding honey bees for resistance to *Varroa jacobsoni*: analysis of mite population dynamics. pp. 434-443. *In* "Africanized Honey Bees and Bee Mites" (G.L. Needham, E. Page, Jr., M. Delfinado-Baker and C.E. Bowman, eds.). John Wiley & Sons, New York.

Laidlaw, H.H. (1988). Thoughts on countering the Africanized-bee threat. pp. 209-213. *In* "Africanized Honey Bees and Bee Mites" (G.L. Needham, R.E. Page, Jr., M. Delfinado-Baker and C.E. Bowman, eds.). John Wiley & Sons, New York.

Lauer, J. and M. Lindauer. (1973). Die Beteiligung von Lernprozessen bei der Orientierung. *Fortschr. Zool.* 21:349-370.

Latreille, P.A. (1804). Notice des especes d'abeilles vivant en grande societe, et formant de cellules hexagonales, ou des abeilles proprement dites. *Ann. Mus. Nat. Hist. Natur, Paris* 5:161-178.

Lepeletier, A. (1836). Histoire naturelle des insectes. *Hymenopteres 1*:400-407. Roret, Paris.

Lindauer, M. (1957). Communication among the honeybees and stingless bees of India. *Bee World,* 38:3-14, 34-39.

Lindauer, M. (1975). "Verständigung im Bienenstaat." Fischer-Verlag, Stuttgart. 163 pp.

Lindauer, M. and W.E. Kerr. (1960). Communication between the workers of stingless bees. *Bee World,* 41:29-41, 65-71.

Linnaeus, C. (1758). Systema Naturae. 10th edit. Holmiae, Laur Salvii.

Louveaux, J. (1969). Ecotype in honey bees. *Proc. Intern. Beekeep. Congr. Munich,* 22:499-501.

Lunder, R. (1953). *Nord. Bitidskr.* 5:71-83. *Apic. Abstr.* 49/55.

Maa, T.C. (1953). An inquiry into the systematics of the Tribus Apidini or honeybees (Hymenoptera). *Treubia, 21*:525-640.

Mackensen, O. (1943). The occurrence of parthenogenetic females in some strains of honeybees. *J. Econ. Entomol.* 36:465-467.

Mackensen, O. (1956). Some effects of inbreeding in the honeybee. XVI. Intern. Beekeep. Congr. Vienna, Apic. Abstr. 247/56.

Martin, E.C. (1973). Can the African bee be stopped in Brazil? *Am. Bee J. 113*:291.

Maul, V. and F. Ruttner. (1969). The cause of the hybridization barriere between *Apis mellifera* L. and *Apis cerana* Fabr. (=Syn. A. *indica* Fabr.). *Proc. Intern. Beekeep. Congr. Munich,* 23:510, 561.

Mejia, M. (1989). The status of beekeeping and the impact of Africanized honey bees in Argentina. M.S. Thesis, University of Georgia, Athens, GA. pp. 113.

Menzel, R., Freudel, H. and U. Ruhl. (1973). Art-und rassenspezifische Unterschiede im Lernverhalten der Honigbiene (*Apis mellifica* L.). *Apidologie,* 4:1-24.

Michener, C.D. (Chairman) (1972). Final Report, Committee on the African Honey Bee. Div. of Biol. and Agr. N.R. National Acad. Sci., Washington, D.C.

Michener, C.D. (1975). The Brazilian bee problem. *Ann. Rev. Entomol.* 20:399-416.

Michener, C.D. (1974). "The Social Behavior of the Bees: A Comparative Study." Harvard Univ. Press. Cambridge, Mass. 404 pp.

Michener, C.D. and D.J. Brothers. (1974). Were workers of eusocial Hymenoptera initially altruistic or oppressed? *Proc. Nat. Acad. Sci., USA,* 71:671-674.

Morse, R., Shearer, D., Boch, R. and A.W. Benton. (1967). Observations on alarm substances in the genus *Apis. J. Apic. Res.* 6:113-118.

Muttoo, R.N. (1956). Facts about beekeeping in India. *Bee World,* 37:125-133; 154-157.

Montagano, J. (1911). Relation sur l'*Apis sicula. Proc. Int. Beekeep. Congr.* 5:26-29.

Nedel, O. (1960). Morphologie und Physiologie der Mandibeldrüse einiger Bienen-Arten. *Z. Morphol. u. Okol. Tiere,* 49:139-183.

Needham, G.L., E. Page, Jr., Delfinado-Baker, M. and C.E. Bowman. (1988). "Africanized Honey Bees and Bee Mites." John Wiley & Sons, New York. 572 pp.

Nunez, J.A. (1973). Quantitative investigation of the behaviour of *Apis mellifera ligustica* Spinola and *Apis mellifera adansonii* Latreille: conditioning food and informational factors, and foraging activity. Proc. XXVI. Intern. Beekeep. Congro. Bueños Aires.

Park, O.W. (1938). Is there a best race of bees? *Am. Bee J.* 78:366-368, 377, 414-417.

Peng, Y.S., Y. Fang, S. Xu and L. Ge. (1987). The resistance mechanism of the Asian honey bee, *Apis cerana*, to an ectoparasitic mite, *Varroa jacobsoni. J. Invert. Path.* 49:54-60.

Perez de Leon, A.A., A. Dietz and C. Vergara (1991). Nest cavity size preference of swarms of honey bees challenged with 3 types of bait hives. Manuscript in preparation.

Pettis, J.S., A. Dietz and F.A. Eischen. (1989). Incidence rates of *Acarapis woodi* (Rennie) in queen honey bees of various ages. *Apidologie, 20*:69-75.

Phillips, E.F. (1928). "Beekeeping." Macmillan, N.Y. 490 pp.

Pollmann, A. (1889). Wert der verschiedenen Bienenrassen und deren Varietäten. 2nd edit. Voigt, Berlin, Leipzig (the description of *Apis mellifera carnica* 1879 is listed in the 1st edition).

Radoev, L. (1969). Results of investigations of F_2 crossbreds between native Bulgarian and grey Caucasian mountain bees. *Proc. Intern. Beekeep. Congr. Munich, 22*:551.

Ribbands, R. (1953). "The Behavior and Social Life of Honey Bees." Bee Res. Association, Ltd., London. 352 pp.

Richards, O.W. (1953). "The Social Insects." MacDonald, London. 219 pp.

Rinderer, T.E. (1983). News Report, Information Division, ARS, USDA, Washington, DC.

Rinderer, T.E. (1986a). Africanized bees: The Africanization process and potential range in the United States. *Bull. Ent. Soc. Am. 32*:222, 224, 226-227.

Rinderer, T.E. (1986b). Africanized bees: An overview. *Am. Bee J. 126*:98-100, 128-129.

Rinderer, T.E. (1988). Evolutionary aspects of the Africanization of honey-bee populations in the Americas. pp. 13-28. *In* "Africanized Honey Bees and Bee Mites" (G.L. Needham, E. Page, Jr., M. Delfinado-Baker and C.E. Bowman, *eds.*). John Wiley & Sons, New York.

Rinderer, T.E. and H.A. Sylvester. (1981). Identification of Africanized bees. *Am. Bee J. 121*:512-516.

Rinderer, T.E., H.V. Daly, H.A. Sylvester, A.M. Collins, S.M. Buco, R.L. Hellmich and R.G. Danka. (1990). Morphometric differences among Africanized and European honey bees and their F_1 hybrids (Hymenoptera: Apidea). *Ann. Entomol. Soc. Am. 83*:366-351.

Rinderer, T.E., R.L. Hellmich, R.G. Danka and A.M. Collins. (1985). Male reproductive parasitism: A factor in the Africanization of European honey-bee populations. *Science, 228*:1119-1121.

Rinderer, T.E., Tucker, K.W. and A.M. Collins. (1982). Nest cavity selection by swarms of European and Africanized honey bees. *J. Apic. Res. 21*:98-103.

Rinderer, T.E., H.A. Sylvester, M.A. Brown, J.D. Villa, D. Pesante, and A. Collins. (1986). Field and simplified techniques for identifying Africanized and European honey bees. *Apidologie, 17*:33-48.

Rinderer, T.E., N. Koeniger, S. Tingek, M. Mardan and G. Koeniger. (1989). A morphological comparison of the cavity dwelling honey bees of Borneo *Apis koschevnikovi* (Buttel-Reepen 1906) and *Apis cerana* (Fabricus 1793). *Apidologie 20:*405-411.

Robinson, W.S.R., Nowogrodzki, R. and R.A. Morse. (1989). The value of honey bees as pollinators of U.S. crops. *Am. Bee J. 129:*411-423; *129:*477-487.

Roberts, W.C. and O. Mackensen. (1951). Breeding improved honey bees. II. Heredity and variation. *Am. Bee J. 92:*328-330.

Root, A.I. (1978). "The ABC and XYZ of Beekeeping." A.I. Root Co., Medina, Ohio.

Rothenbuhler, W.C. (1979). Semidomesticated Insects: Honey Bee Breeding. *In* "Genetics in Relation to Insect Management." (Hoy, M.A. and J.J. McKelvey, Jr., *eds.*), pp. 84-92. Rockefeller Foundation, Bellagio, Italy.

Rothenbuhler, W.C., Kulincevic, J.M. and W.E. Kerr. (1968). *Bee genetic. Ann. Rev. Genet. 2:*413-438.

Ruttner, F. (1960). Fortpflanzung und Vererbung. *In* "Biene und Bienenzucht." (A. Büdel and E. Herzog, *eds.*), pp. 5-22. Ehrenwirth Verlag, Munich.

Ruttner, F. (1968). Intraracial selection or race-hybrid breeding of honey bees. *Am. Bee J. 108:*394-396.

Ruttner, F. (1975a). Races of bees. *In* "The Hive and the Honey Bee," pp. 19-38, Dadant & Sons, Hamilton IL.

Ruttner, F. (1975b). The African races of honey bees. *Proc. Intern. Apicult. Congr. 25:*325-344.

Ruttner, F. (1975c). Biological and statistical methods as a means of an ingenious intraspecific classification in honey bee. *Proc. Intern. Beekeep. Congr. 25:*360-362.

Ruttner, F. (1975d). Die Kretische Biene, *Apis mellifera adami. Allg. dtsch. Inmkerztg. 9:*271-272.

Ruttner, F. (1976). Beekeeping in Crete. *Apiacta, 11:*187-191.

Ruttner, F. (1980). *Apis mellifera adami* (n ssp) Die Kretische Biene. *Apidologie, 11:*385-400.

Ruttner, F. (1981). On the taxonomy of honey bees of tropical Africa. *Proc. Intern. Beekeep. Congr. 28:*278-287.

Ruttner, F. (1984). "Zuchttechnik und Zuchtauslese bei der Biene." Ehrenwirth Verlag, Munich. 141 pp.

Ruttner, F. (1986). Geographical variability and classification. pp. 23-56. *In* "Bee Genetics and Breeding." (T.E. Rinderer, *ed.*). Academic Press, New York.

Ruttner, F. (1988). "Biogeography and Taxonomy of Honeybees." Springer Verlag, Berlin. 284 pp.

Ruttner, F. and D. Kauhausen. (1985). Honeybees of tropical Africa: ecological diversification and isolation. Proc. 3rd. Int. Conf. Apic. trop. Climates, Nairobi, Kenya, pp. 45-51.

Ruttner, F. and V. Maul. (1983). Experimental analysis of the reproductive interspecies isolation of *Apis mellifera* L. and *Apis cerana* Fabr. *Apidologie, 14:*305-327.

Sakagami, S.F. (1971). Ethosoziologischer Vergleich zwischen Honigbiene und stachellosen Bienen. *Z. Tierpsychol. 28:*337-350.

Sakagami, S.R. (1982). Stingless bees. *In* "Social Insects" (H.R. Hermann, *ed.*), Vol. 3, pp. 361-423. Academic Press, N.Y.

Sakagami, S.F., Matsumura T. and K. Ito. (1980). *Apis laboriosa* in Himalaya, the little known world largest honeybee (Hymenoptera, Apidae). *Insecta Matsumura New Series 19:*47-77.

Seeley, T.D. and R.A. Morse. (1976). The nest of the honey bee (*Apis mellifera* L.), *Insects Soc. 23:*495-512.

Seeley, T.D., Seeley, R.H. and P. Akratanakul. (1982). Colony defense strategies of the honey bee in Thailand. *Ecol. Monogr. 52*:43-63.

Seeley, T.D. (1985). "Honey bee ecology." Princeton University Press. New Jersey.

Seeley, T.D. (1988). What do we know about nest-site preferences of African honey bees. pp. 87-90. *In* "Africanized Honey Bees and Bee Mites" (G.L. Needham, R.E. Page, Jr., M. Delfinado-Baker and C.E. Bowman, *eds.*), John Wiley and Sons. New York.

Shao-Wen, L., M. Yu-Pin, J.T. Chang, L. Ju-Huai, Shao-Yu, H. and K. Bang-Yu (1987). A comparative study of esterase isozymes in 6 species of Apis and a genera of Apoidea. *J. Apic. Res. 25*:129-133.

Skorikov, A.S. (1929). Eine neue Basis für eine Revision der Gattung *Apis* L.. *Rep. Appl. Entomol. 4*:249-264 (In Russian with German summary).

Smith, F.G. 1961. The races of honeybees in Africa. *Bee World, 42*:255-260.

Smith, R.K. (1988). Identification of Africanization in honey bees based on extracted hydro-carbon. pp. 275-280. *In* "Africanized Honey Bees and Bee Mites" (G.L. Needham, E. Page, Jr., M. Delfinado-Baker and C.E. Bowman, *eds.*). John Wiley & Sons, New York.

Solodkova, N. and F. Guba. (1960). Testing hybrids in Ukrainian collective and state farms. *Pchelovodstvo 6*:28-31. Apic Abstr. 792/63.

Spinola, M. (1860). Insectorum Liguriae species novae aut rariores, etc. *Genuae I*:159.

Southwick, L., Jr,. and E.E. Southwick. (1989). A comment on 'value of honey bees as pollinators of U.S. Crops.' *Am. Bee J. 129*:805-807.

Sylvester, H.A. and T.E. Rinderer. (1987). Fast Africanized bees identification system (FABIS) manual. *Am. Bee J. 127*:511-516.

Taber, S. (1954). The frequency of multiple mating of queen honey bees. *J. Econ. Entomol. 47*:995-998.

Taber, S. and J. Wendel. (1958). Concerning the number of times queen bees mate. *J. Econ. Entomol. 51*:786-789.

Taber, S. and C.D. Owens. (1970). Colony founding and initial nest design of honey bees (*Apis mellifera* L.). *Anim. Behav. 18*:625-632.

Taylor, O.R. (1977). Past and possible future spread of Africanized honey bees in the Americas. *Bee World, 58*:19-30.

Taylor, O.R. (1985). African Bees: Potential impact in the United States. *Bull. Ent. Soc. Am. 31*:14-24.

Taylor, O.R. and M. Spivak. (1984). Climatic limits of tropical African honey bees in the Americas. *Bee World. 65*:38-47.

Taranov, G.F. (1956). Characteristics of the grey Caucasian mountain bee and its use for breeding purposes. XVI. Intern. Beekeep. Congr. Vienna. Apic. Abstr. 240/56.

Tokudo, Y. (1935). Studies on the honey bee, with special reference to the Japanese honey bee. *Imp. Zootech. Exp. Stn. Bull. 15*, 1-27.

Velthuis, H.H.W. (1976). Egg laying, aggression and dominance in bees. *Proc. Intern. Congr. Entomol. 15*:436-449.

Vergara, C. (1990). Africanized bee swarms preference for different bait hives in Mexico. *Am. Bee J. 130*:817-818.

Vergara, C., A. Dietz and A. Perez. (1989a). Usurpation of managed honey bee colonies by migratory swarms in Tabasco, Mexico. *Am. Bee J. 129*:824-825.

Vergara, C., A. Dietz and A. Perez de Leon. (1989b). Colony usurpation by migratory swarms in Mexico. Proc. XXXIInd Int. Apicult. Congr., Rio de Janeiro, Brazil. (In press).

Vergara, C., A. Dietz and A. Perez de Leon. (1991a). Studies on the process of Africanization in Mexico. Manuscript in preparation.

Vergara, C., A. Dietz, R. Krell, and M. Mejia. (1991b). Comparative pollination efficiency of Africanized and European honey bees. I: Alfalfa (*Medicago sativa*); II: Onion (*Allium cepa*); III: Carrot (*Daucus carota*). (In preparation).

Villa, J.D. (1986). Performance of Africanized colonies at high elevations in Colombia. *Am. Bee J. 126*:835.

von Frisch, K. (1967). "The dance language and orientation of bees." Harvard University Press, Cambridge, MA.

Weiss, K. (1984). "Bienen-Pathologie." Ehrenwirth Verlag, Munich. 252 pp.

Winston, M.L., O.R. Taylor and G.W. Otis. (1983). Some differences between temperate European and tropical African and South American honeybees. *Bee World, 64*:12-21.

Winston, M.L. (1987). "The biology of the honey bee." Harvard University Press, Cambridge, MA.

Winston, M.L. (1988). The impact of a tropical-evolved honey bee in temperate climates of North America. pp. 135-140. *In* "Africanized Honey Bees and Bee Mites." (G.L. Needham, R.E. Page, Jr., M. Delfinado-Baker and C.E. Bowman, *eds.*), John Wiley and Sons. New York.

Witherell, P.C. (1985). A review of the scientific literature relating to honey bee bait hives and swarm attractants. *Am. Bee J. 125*:823-829.

Wongsiri, S., You-Sheng, L and L. Zhi-Song. (1986). Beekeeping in the Guangdong province of China and some observations on the Chinese honey bee *Apis cerana cerana* and the European honey bee *Apis mellifera ligustica*. *Am. Bee J. 126*:748-752.

Wongsiri, S., A. Dietz and S. Pothichot. (1989). A comparison of instrumentally inseminated and naturally mated queens of *Apis cerana* Fabricius. Proc. XXXIInd Int. Apicult. Congr., Rio de Janeiro, Brazil. (In press).

Wongsiri, S., K. Limbipichai, P. Tangkanasing, M. Mardan, T. Rinderer, H.A. Sylvester, Koeniger G. and G. Otis. (1990). Evidence of reproductive isoltion confirms that *Apis andreniformis* (Smith, 1858) is a separate species from sympatric *Apis florea* (Fabricius, 1787). *Apidologie, 21*:47-52.

Woyke, T. (1973). Experiences with *Apis mellifera adansonii* in Brazil and Poland. *Apiacta, 8*:115-116.

Zander, E. and K. Weiss (1964). "Das Leben der Biene." Ulmer, Stuttgart. 189 pp.

Zeuner, F. (1931). Die Insektenfauna des Böttinger Mamors. *Fortschr. Geol. u. Paläontol. 9*:292.

CHAPTER 3

THE HONEY BEE COLONY: LIFE HISTORY

by MARK L. WINSTON*

For many of us involved in beekeeping, it is easy to forget that honey bees are feral organisms. In their domesticated state, they are provided with many-leveled domiciles, fully furnished with comb, and given extensive feeding and management to maximize their productivity. While this pampering of our bees is highly beneficial in the enhancement of honey production and pollination, we can easily lose sight of the natural origins of this well-domesticated insect. Honey bees, like any species, evolved in a variable and competitive world, where only those with flexible and adaptive traits survive.

Among the traits most important in the survival and evolution of honey bees is their life history, including nesting biology, development, caste-specific behaviors, and the timing of annual colony cycles. The social interactions in colonies have allowed honey bees to evolve a complex life history which can respond well to the rigors of survival in both tropical and temperate habitats, and can best be understood by weaving together the intricate architecture of the nest with the biology and functions of its inhabitants, the queen, workers, and drones. The study of honey bees in their natural setting is not only of academic interest, as fascinating as it can be, but is also of great importance to beekeeping, because only by understanding the biology and life history of this insect can we manage them most effectively. Indeed, studies of honey bee life history have contributed to almost all aspects of beekeeping, from hive design to such important management techniques as swarm prevention, wintering, nutrition, and disease prevention and control.

In this review I will examine some of the aspects of honey bee life history which are particularly important in understanding the success and function of colonies in their natural environment. I will concentrate here on honey bees in temperate climates, except for the last section; similar information on tropically-evolved honey bee races can be found in Seeley (1985), Winston (1987), and Winston, Otis, and Taylor (1983). Also, more extensive references and discussion on all of the topics presented here can be found in Seeley (1985) and Winston (1987).

THE STAGE: NEST ARCHITECTURE

Home for a honey bee colony in temperate climates is usually a tree cavity, or a hollow space between the walls, ceilings, or floors of a building (Fig. 1).

---

*Prof. Mark L. Winston, Ph.D., Professor, Biological Sciences, Simon Fraser University, Burnaby, British Columbia. Author of the book *The Biology of the Honey Bee* and numerous publications on honey bees.

The furnishings are both ultramodern and timeless: the stark, highly regular cells of the worker-constructed comb. The nest provides shelter from the weather and from predators, while the comb serves as a nursery, storage area, message center, and site of myriad social interactions.

The paramount importance of the nest to a honey bee colony is evident in the care with which a new nesting site is selected by a swarm. New sites are carefully scouted, since the characteristics of the new nest will determine in part the likelihood of the colony surviving to reproduce the following season, and for many years beyond. Scout bees leave the swarm and look for possible sites, extensively inspecting both the inside and outside of potential cavities to determine whether the site meets numerous criteria for a good nest. If so, this information is communicated to other workers back at the swarm cluster, and, when a consensus is reached by the swarm, the workers and queen take to the air and move en masse to colonize the new nest (von Frisch, 1967; Lindauer, 1951, 1955; Seeley, 1977)

Bees will accept a great diversity of characteristics in their nest sites, but exhibit some preferences for certain nest qualities which considerably enhance survival potential. One important nest site characteristic is cavity volume. In temperate climates, a nest which is too small is undesirable because there would not be sufficient space to store the honey which is critical to winter survival, and a small nest might overheat during the summers. In contrast, a colony which establishes itself in an overly large cavity would have difficulty maintaining temperature during the winters. For these reasons, most feral colonies are found in nests between 20 and 100 L in volume, with an average nest cavity volume of about 40 L, which is approximately the volume of one standard Langstroth hive body (Seeley and Morse, 1976: Wadey, 1948).

Many other nest site characteristics also contribute to a colony's future survival (Avitabile, Stafstrom, and Donovan, 1978; Morse and Seeley, 1978; Seeley and Morse, 1976). For example, sites which are heavily exposed to wind, sun, and rain are not generally accepted by workers, and southern exposures seem to be preferred. Also, colonies are usually found in cavities with few and small entrances, which would improve a colony's thermoregulatory capabilities as well as ease of defense. Excessively wet cavities also are avoided, although workers can plug drafty or damp cavities with plant resins to improve their habitability.

The comb itself is entirely constructed from wax secreted by worker glands and shaped into cells using the workers' mandibles and legs (recently reviewed by Hepburn, 1986). Comb consists of a back-to-back array of hexagonal cells, arranged in a mostly parallel series, with each comb a precise distance from its neighbor (Fig. 1, 1a, 1b). The well-regulated distance between each comb (3/8 of an inch), referred to as "bee space," allows for

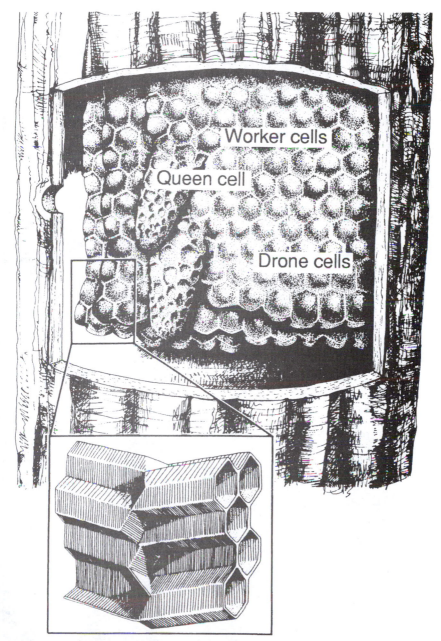

Figure 1. A typical, temperate-climate honey bee nest inside of a tree cavity, showing worker, drone, and queen cells. The insert shows the back-to-back array of hexagonal cells characteristic of honey bee comb.

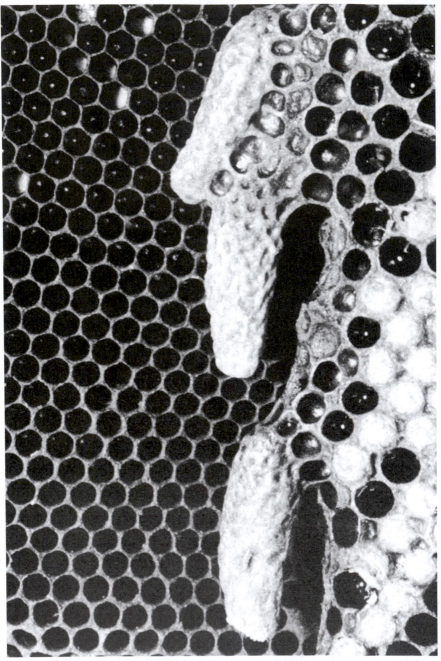

FIGURE 1a. Worker, queen, and drone cells. Photo by Kenneth Lorenzen.

FIGURE 1b. A worker, drone, and queen honey bee.

easy bee movement between combs and provides a cozy space in which to cluster and warm the brood. The hexagonal cells come in two sizes, the smaller worker cells and the somewhat larger drone cells, while the queens are reared in elongate, conical cells which hang from the comb edges. Only 10 to 20 of these cells are usually built at any one time, and they are torn down by the workers after queen emergence.

The regular pattern of cells gives an impression of organization, an impression which is reinforced by the way in which a colony localizes its nest functions. Generally, honey is stored in the upper and peripheral sections of the nest, while the brood is centrally located so that temperature can be easily maintained and nursing activities concentrated in one area; pollen is stored around the brood area for easy access (Fig. 2, 2a). The worker and drone cells are also reused, since adult workers fastidiously clean cells after brood emerges or when stored honey or pollen is removed. While workers are usually reared in worker-sized cells and drones in drone cells, both types of hexagonal cells are also used for honey or pollen storage. The cell diameters diminish over time because of accumulated cocoons, cast-off larval and pupal skins, and treatment given cells to prepare for brood rearing. Such old comb is brittle and dark, but nevertheless can be used by colonies for many years.

Many of the characteristics of feral nests have been incorporated into artificial hives for management. For example, the concept of bee space was a critical one in the development of movable frame beekeeping, since the proper between-frame distance not only is important for proper colony functioning, but also in preventing workers from constructing brace and bridge wax which would prevent easy frame removal. Proper nest organization also is an objective of good management, since colony productivity is enhanced by

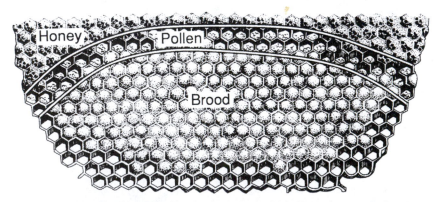

FIGURE 2. The organization of comb utilization by honey bees. Honey is generally found at the top, peripheral sections of comb, with pollen stored surrounding the central brood area.

FIGURE 2a. Beekeeper examining a frame organized as in Fig. 2. Photo by Kenneth Lorenzen.

maintaining a brood area in the bottom boxes of bee hives and allowing honey storage in the upper supers. Finally, aspects of hive structure such as one or two small entrances, and colony management recommendations such as locating hives in warm, south-facing sites, are all indicated by studies of feral nests.

THE PLAYERS: WORKERS, DRONES, AND THE QUEEN
Caste Development and Differentiation

There are three types of individuals or castes in a honey bee colony, the female queen and workers and the male drones. There is generally only one queen, who lays the eggs in the colony and controls many colony activities by her production of pheromones, the chemicals which strongly influence worker behavior and physiology. The workers, as their name implies, do almost all of the colony's work, including comb building, brood tending, defense, foraging, and many other tasks. There can be anywhere from a few thousand to 50 or 60 thousand workers in a colony, depending on the season and timing of the annual colony cycle. For drones, there are usually a few hundred or thousand present in the nest during the spring and summer, and their only function is to mate. They do no work in the nest, and die in the fall after the workers drive them from the nest as colonies prepare for winter.

An egg laid by the queen has the potential to develop into any one of the three castes, and its developmental direction is determined by both genetic and nutritional factors (Fig. 3). Usually, unfertilized eggs develop into drones, while a fertilized egg can develop either into a worker or a queen; female differentiation depends on the type of food the developing larva is fed. When the queen lays an egg, she determines whether that egg is being laid in a worker or a drone cell, likely by inspection of the cell with her forelegs prior to egg laying (Koeniger, 1969, 1970), and/or by the angle of her abdomen during oviposition. If the egg is going into a drone cell, the queen does not release any sperm as the egg travels down the oviduct; this unfertilized egg is haploid,

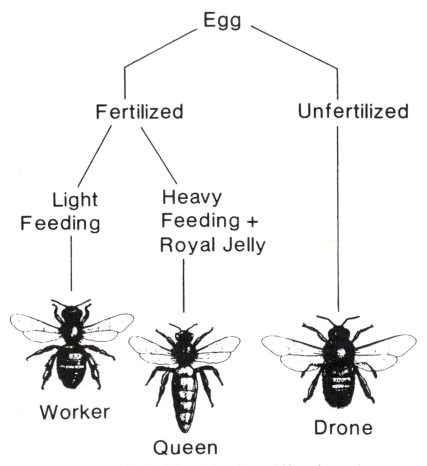

Figure 3. Factors determining the differentiation of queen-laid eggs into workers, queens, or drones. Reprinted by permission of the publishers from *The Biology of the Honey Bee* by Mark L. Winston, Cambridge, Massachusetts: Harvard University Press, Copyright © 1987 by Mark L. Winston.

having only one set of chromosomes, and will develop into a drone. In contrast, sperm are released and eggs fertilized when the queen lays in a worker or queen cell, resulting in a female diploid individual having both sets of chromosomes.

As with most aspects of honey bee biology, there are exceptions to these developmental pathways. Queens occasionally lay fertilized eggs in worker cells which can potentially develop into males called diploid drones. This phenomenon was first explained by Mackensen (1951), who determined that individuals homozygous at a particular gene locus will develop into diploid males. This condition is invariably lethal, however, since diploid male brood are quickly eaten by workers (Woyke 1963, 1969, 1973). Also, workers may lay eggs under certain conditions, which generally develop into drones since workers never mate and thus have no sperm to fertilize eggs. However, eggs laid by workers and queens sometimes develop parthenogenetically into females, although this developmental pathway is relatively rare (Anderson, 1963; Mackensen, 1943).

Usually, workers and queens develop from fertilized, queen-laid eggs; once fertilized, an egg has the potential to develop into either a queen or a worker. Which developmental pathway is followed depends on the quantity and quality of food which the growing larvae are fed. The food which is fed to developing bees is largely produced in the hypopharyngeal and mandibular glands of adult workers, and the brood food from these two glands has somewhat different compositions. The food fed to potential queens is called royal jelly, and is particularly rich in mandibular gland products and sugars which act as phagostimulants. Worker larvae are fed more hypopharyngeal gland secretions, and also more honey and pollen during the last few days of larval development. The developing larvae have the potential to mature into either workers or queens for the first three days of the larval growth, but by day four the larvae are committed to one pathway or the other. The direction of development seems to be determined by the levels of a hormone, juvenile hormone, on day three, and the quantity and quality of food fed to the larvae appears to determine the levels of this hormone (for recent reviews of caste determination, see Beetsma 1979, Brouwers, Ebert, and Beetsma 1987, Goewie 1978, de Wilde and Beetsma 1982, and Winston 1987).

All of the honey bee castes pass through the same four stages during their development: egg, larva, pupa, and adult (Fig. 4, 4a; Jay 1963). When the queen first lays an egg, she glues it to the floor of the cell so that it is standing at the bottom; the egg sags during the approximately 3-day period before hatching into the larval stage. The larvae are essentially feeding machines, designed for rapid growth, and consisting mainly of a digestive system, with no eyes, legs, antennae, wings, or sting. The larvae remain curled up in the bottom

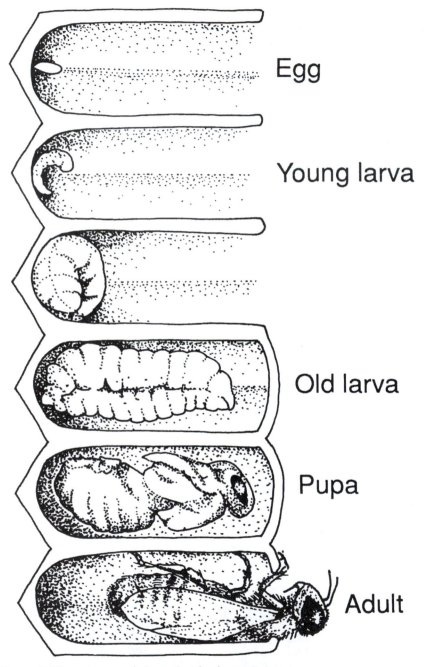

Egg

Young larva

Old larva

Pupa

Adult

FIGURE 4. The major stages in honey bee development.

of their cells for four to six days, feeding on secretions deposited there by the nurse bees. After this feeding period, the adult workers cap the cell with wax, and the larvae uncurl, stretch out fully in the cells, and spin a cocoon. This last larval stage is referred to as the prepupal stage, since the final larval molt includes a metamorphosis into the pupal stage. The pupa is the last stage before the final molt to the adult, during which the developing bee completes its metamorphosis; when development into the adult stage is mostly completed, the young bee chews her way through the cell capping and emerges.

The timing of bee development depends on the caste of the developing individual as well as environmental and genetic factors. While all three castes have egg stages which average 3 days, the remainder of their development shows different timing. For example, the mean duration of the uncapped larval period is about 4.5 days for queens, 5.5 days for workers, and 6.3 days for drones, and total development times average 16, 21, and 24 days for queens, workers, and drones, respectively. These times can be affected by colony conditions, such as temperature, humidity, and the number of nurse bees, and also by racial differences. Tropical bees tend to have shorter worker development times, often as low as 18 or 19 days to complete development (Tribe and Fletcher 1977; Winston 1979).

After emerging, the young bee completes its development during the next few days. The cuticle takes 12 to 24 hr to harden, and internal development requires additional feeding for proper glandular and fat body development.

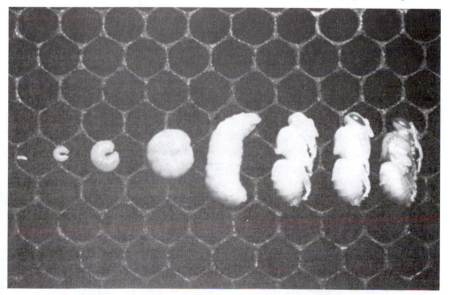

FIGURE 4a. Egg, larva, prepupa, and pupa. Photo by Kenneth Lorenzen.

The life span of emerged workers can vary from only a few days to almost a year, with "summer" workers living for average times of 15 to 38 days while workers can survive for 140 days and longer during the winter (Fukuda and Sekiguchi 1966; Ribbands 1953; Winston, Taylor, and Otis 1983). Adult drones generally survive for average times of 21 to 32 days during the summer, and usually do not survive the winter since workers expel them from the nest in the fall (Drescher 1969; Fukuda and Ohtani 1977; Jaycox 1956; Witherell 1972). Queens are the most long-lived of the three castes, generally surviving for one to three years, although queens can live for eight years or longer (Bozina 1961; Seeley 1978).

The Activities of Workers, Queens, and Drones

The three physically distinct castes, the workers, queen and drones, each have their own distinctive role to perform in the honey bee society. The drones are the simplest bees in terms of job performance, having only one task, which is mating with queens. When a young drone emerges from his cell, he is fed by workers for the first few days of his life, and then begins feeding himself from honey cells. After about two weeks, when drone semen has matured sufficiently for mating, drones begin their mating flights (Ruttner 1966). These flights generally take place in the afternoon, most frequently between 1400 and 1600, and each flight lasts an average of about 25 to 32 min; a drone may take several flights a day (Howell and Usinger 1933; Lensky *et al.* 1985; Oertel 1956; Otis 1986; Ruttner 1966; Witherell 1971). Most drones die before mating, since there are many more drones produced by colonies than queens. A successful drone can only mate once, however, since the drone's abdomen bursts during copulation. Drones have no other known function besides mating.

Queens also perform a limited although critically important repertoire of activities for the colony. Their major task is egg laying, and a queen will lay between 175,000 and 200,000 eggs annually. Workers feed adult queens brood food and possibly some honey to provide the raw materials for egg production and other queen functions, and increased feeding levels are associated with periods of higher egg production (Allen 1955, 1960; Haydak 1970). The other important function of the queen is to produce pheromones, the chemicals which control and organize many colony functions (reviewed by Free 1987 and Winston 1987). She is thus the mother and central authority figure in the nest, but otherwise performs little or no work.

It is the workers which, as their name implies, perform almost all of the work in the colony, including brood rearing, construction, defense, foraging, thermoregulation, cleaning, and many other tasks. In general, younger workers tend to perform within-colony tasks and older workers do outside jobs like guarding and foraging, but there is great variability in this temporally-based caste structure (Fig. 5). For example, the average age for the onset of foraging has been recorded as low as 18.3 days (Sakagami 1953) and as high as 37.9 days of the age (Winston and Fergusson 1985). Nevertheless, worker activities seem to be organized in loose groupings, with cell cleaning and capping performed by the youngest bees, brood and queen tending by slightly older workers, comb building, cleaning, and food handling by middle-aged workers, and ventilating, guarding, and finally foraging by the oldest workers. The demarcation between the first three categories of within-colony tasks and the final category of outside tasks is the most clearly defined boundary between these loose caste distinctions.

The ontogeny of worker tasks is also characterized by the performance of multiple tasks at any age, and considerable overlap between the ages when a worker performs various jobs (Lindauer 1952). On a given day, a worker might tend brood, feed the queen, pack pollen into a cell, and patrol the nest. While the daily performance of multiple tasks is the norm, occasionally a bee might specialize for much of her life on one task, such as water collection (Robinson, Underwood, and Henderson 1984) or grooming (Frumhoff and Schneider 1987; Winston and Punnett 1982). The transitions between tasks and task groupings occur gradually as workers age, with considerable overlap between days in the jobs which they perform.

Temporal caste structure in bees is closely linked to glandular develop-ment and resorption, particularly for the brood food and wax glands (King 1933; Simpson, Riedel, and Wilding 1968). The hypopharyngeal, mandibu-lar, and wax glands begin to enlarge shortly after a young worker emerges, reaching and maintaining their maximum sizes at 5 to 15 days of age, coincidental with brood rearing and comb building tasks. All three glands diminish in size and output coincidental with the transition from these jobs to other tasks. When necessary, older workers whose glands have resorbed can regenerate both brood food and wax glands; this suggests that, in addition to age-based factors, colony requirements may also influence glandular devel-opment (Free 1961; Kolmes 1985 a,b,c; Lindauer 1961; Milojevic 1940; Moskovljevic-Filipovic 1956; Nolan 1924; Rösch 1930). Pheromone-producing glands show a similar, age-based relationship with task; alarm pheromone production is low in younger bees, and rises as workers age (Allan *et al.* 1987; Boch and Shearer 1966; Masson and Arnold 1984; Robinson 1985, 1987; Shearer and Boch 1965).

The last aspect of division of labor which has been noticed by almost all researchers is that workers spend most of their time resting and patrolling. Even these behaviors may have considerable significance for the colony, however. Patrolling workers are presumably gathering information about

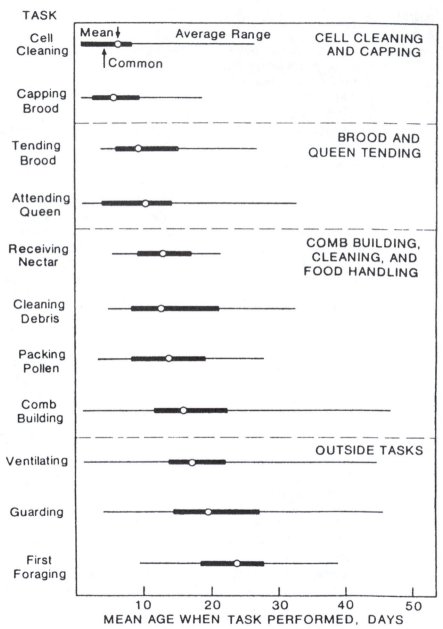

FIGURE 5. Overlap of age-related task performance by worker bees. Data are from the references cited in Winston 1987. Reprinted by permission of the publishers from *The Biology of the Honey Bee* by Mark L. Winston, Cambridge, Massachusetts: Harvard University Press, Copyright © 1987 by Mark L. Winston.

colony conditions which will determine what tasks they perform and the ontogeny of glandular development. Resting workers may be producing brood food or wax (Lindauer 1952), or possibly being held in reserve as a "quick-response team" for sudden catastrophes or opportunities.

The patterns of temporal caste ontogeny have been of great research interest over the years, since the ability to shift the ages at which workers perform tasks provides colonies with a flexible system for allocating work to those tasks which are most necessary. For example, workers begin foraging at younger ages, produce more wax, and build more comb when colonies are suddenly deprived of some of their comb (Fergusson and Winston 1988). The shift in foraging age presumably provides an additional work force for nectar foraging, which is necessary for wax production, while within the nest workers are increasing both their wax production and the jobs associated with comb building. Similar shifts in task performance have been found in colonies which suffer a sudden drop in adult worker population and/or a loss of many old or young workers, or a restriction on the entry of pollen (Free 1965; Fukuda 1960; Lindauer 1961; Milojevic 1940; Rösch 1930; Winston and Fergusson 1985). Also, workers may shift the frequencies at which various tasks are performed when colonies are mildly stressed (Kolmes 1985). The response of temporal caste structure seems to depend on the extent of disruption in normal colony activities; mildly stressed colonies may first respond by changing the frequency but not the age of task performance, while workers in more heavily stressed colonies respond by changing the ages of task performance.

There is also a strong genetic component to division of labor, with workers from different races, strains, or even patrilines within colonies showing differences in task ontogeny. This was first demonstrated by Winston and Katz (1982), who used a cross-fostering technique to demonstrate racially-based differences in temporal caste structure between European and Africanized honey bees. Calderone and Page (1987) have shown that workers from two genetically distinct strains of honey bees, selected high and low pollen hoarding lines, differed in the age at which they began foraging and in the relative frequency at which they foraged for pollen. Finally, workers from different patrilines within the same colonies showed differences in foraging age depending on their drone fathers, and these differences were most pronounced under conditions of stress, i.e. nectar dearth or comb deprivation (Kolmes, Winston, and Fergusson, 1989).

It is now apparent that workers are able to adjust their temporal division of labor schedules in response to a broad range of nest and external environmental conditions, and these adjustments provide a flexible caste system exquisitely tuned to colony requirements. Both genetic and environmental components are involved, and such factors as worker population, the amounts and type of brood and stored resources, comb building needs, spatial organization of the colony, weather, available forage, characteristics of worker physiology, and even the colony's racial origin all interact to determine what a worker will do at any time. Further, the flexibility inherent in this system allows colonies to shift their work schedules rapidly in response to colony needs, which has undoubtedly been a major factor in the survival and evolution of honey bee societies.

THE PLAY: SEASONAL CYCLE OF HONEY BEE COLONIES

The annual cycle of honey bee colonies in temperate climates is unusual for a social insect, since it not only involves individuals surviving the winter as a social unit, but also reproducing by swarming (colony fission) in the spring. Many of the characteristics of honey bees in temperate areas can best be understood as modifications of colony-level behaviors to survive in regions with cold winters and relatively short flowering seasons while maintaining a swarming mode of reproduction. These adaptations for colony survival in temperate climates include the use of protected cavities for nesting, storage of considerable quantities of honey, clustering for heat maintenance, mid-winter initiation of brood rearing, and spring swarming (Seeley 1985; Winston 1987).

Colonies survive the winter by using energy derived from consumption of stored honey to generate body heat and keep the nest at an adequate temperature for adult survival. Workers begin to cluster together when the temperature drops below 18°C; as the temperature continues to drop, the cluster contracts (Fig. 6). Individual workers periodically consume honey and move in and out of the cluster center; to generate heat, workers elevate their body temperatures by contracting the flight muscles in the thorax without moving the wings (Bastian and Esch 1970; Esch 1960, 1964; Esch and Bastian 1968; Roth 1965). Colonies can sometimes die under conditions of extreme cold, even with substantial honey reserves, if workers are unable to leave the cluster to get to honey located at the nest periphery. There is little or no brood rearing during the coldest parts of the winter, and cluster temperatures are maintained at a relatively cool 20°C (Corkins 1930; Gates 1914; Haydak 1958; Owens 1971; Phillips and Demuth 1914; Southwick and Mugaas 1971; Wilson and Milum 1927).

The new year for a honey bee colony in a cold temperate climate begins with the initiation of brood rearing late in the winter. When brood rearing commences, the thermoregulatory requirements for colonies become more rigid, since proper brood development is dependent on temperatures inside the cluster around the brood being maintained within one degree of 35°C (Himmer 1927; Simpson 1961). This early brood rearing is fueled by the vast quantities of honey and pollen stored in the nest during the previous summer, as well as by workers breaking up the metabolic reserves stored in their fat bodies. Colonies reach their minimum populations sometime in late winter, but increased brood rearing at this time gradually compensates for winter losses, and by early spring the worker population begins a dramatic rise which climaxes in swarming (Fig. 7; Allen and Jeffree 1956; Jeffree 1955; Lee and Winston 1987; Nolan 1925).

FIGURE 6. The diameter of a winter cluster in a honey bee colony at various temperatures.

Most swarming occurs in mid-spring, usually sometime in May or June (Fig. 7; Burgett and Morse 1974; Caron 1980; Fell *et al.* 1977; Jeffree 1951; Simpson 1959; Winston 1980; Lee and Winston 1987). However, swarms can issue as early as the first week of April and as late as September, although these late swarms have little chance of survival. Almost all feral colonies will swarm at least once in the spring, and those colonies rarely swarm again that season (Seeley 1978; Winston 1980).

Swarming begins with the rearing of new queens, and a colony generally will not swarm unless a replacement queen is being reared. Considerable research has been devoted to studying the factors which initiate queen rearing and subsequent swarming, since swarm prevention is one of the major management problems facing beekeepers worldwide. It is now apparent that there is no single factor which induces queen rearing; rather, a number of colony characteristics must be at their critical levels simultaneously before queen rearing will begin. These factors include colony size, brood nest congestion, worker age distribution, and, perhaps most significantly, a reduced transmission of queen substances in the colony. As colonies become more crowded prior to swarming, the distribution of the queen's pheromones, which normally prevent workers from rearing queens, may be partially inhibited, thereby allowing queen rearing to commence (For additional details of swarm preparation behavior, and the factors which influence swarming, see chapter 8 by Dr. Norman Gary, and Winston 1987).

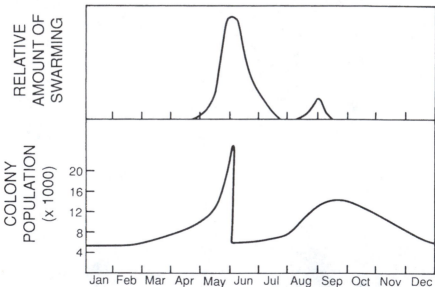

FIGURE 7. The relative amount of swarming and typical colony populations during one year in an area with a cold temperate climate.

Two types of swarms issue from colonies, the prime swarm and subsequent afterswarms. The prime swarm contains a majority of the workers in the nest and the old queen, and does not issue until the colony is well along in rearing replacement queens. Afterswarms, which issue about 10 to 15 days after prime swarms, contain one or more virgin queens and generally a smaller number of workers than the prime swarm. From zero to four afterswarms can issue from nests, although one or two afterswarms per colony are most commonly found for temperate-evolved bee races. The worker population in swarms generally decreases with swarm number; prime swarms average around 16,000 workers while first and second afterswarms have about 11,500 and 4000 workers, respectively (Lee and Winston 1987; Otis 1980; Winston 1979, 1980; Winston, Dropkin, and Taylor 1981). Swarms can be much larger or smaller than these values. However, as few as 1750 and as many as 50,750 workers were recorded in swarms from one study (Avitabile and Kasinkas 1977). Once swarming is completed, the remaining queens fight to the death, and the single survivor mates and begins to lay eggs. The swarms quickly find new nest sites and begin the business of constructing a new nest, rearing brood, and storing nectar and pollen for the winter.

The importance of swarms initiating their colonies rapidly is emphasized by the high rate of swarm mortality found in cold temperate climates. In three studies conducted in central New York state, Ontario, and British Columbia, only 24%, 8%, and 0% of colonies founded from swarms survived until the following season, respectively (Morales 1986; Seeley 1978; Lee and Winston 1987). In contrast, established colonies which had already survived their first winter in the New York and Ontario studies showed 78% and 45% survival until the next season. Once the New York colonies survived the first season, their average life span was almost 6 years, indicating that, while an individual swarm has a low probability of survival, those which do survive may persist for many years.

Many aspects of honey bee life history in temperate climates seem to reflect adaptive responses to the high risks associated with swarming. First, the use of stored honey and pollen for mid-winter brood rearing results in early population growth which permits spring swarming. Early swarming with relatively large swarms is important for honey bees, since there are strong positive correlations between swarm survival and both the date of swarming and worker population in swarms (Lee and Winston 1985a, b, 1987; Seeley and Visscher 1985). Second, colonies only produce as many afterswarms as their brood and adult populations can support; there is a positive relationship between the number of afterswarms produced and the brood and adult populations at swarming (Lee and Winston 1987; Winston 1979, 1980). Third, workers engorge heavily with honey prior to swarming, which provides colonies with an initial source of energy and wax until foragers can locate and recruit to flowers near the new nest (Otis, Winston, and Taylor 1981). Finally, many young workers issue with swarms, resulting in a young population in the new nest which can most vigorously accomplish the necessary comb building, brood rearing, and foraging tasks (Butler 1940; Meyer 1956; Winston and Otis 1978).

Once colonies have completed swarming and swarms have settled into their new nests, the task of gathering copious quantities of honey and pollen for the winter begins. A typical colony will require between 15 and 30 kg of pollen a year, and 60 to 80 kg of honey (reviewed in Winston 1987). The effort to collect this level of resources is enormous; a typical nectar or pollen load weighs only 10 to 40 mg, meaning that workers must make one million trips annually to collect pollen and almost four million trips to collect nectar. Furthermore, in temperate areas these trips must all be accomplished during the relatively short period when flowers are blooming. It is therefore not surprising that most feral colonies barely collect enough resources to survive, and many colonies starve to death during the winter (Lee and Winston 1985; Seeley 1978; Seeley and Visscher 1985).

Since food collection is so important for colonies, honey bees have evolved a recruitment system to increase foraging efficiency. The colony functions as an advanced information processor, taking in reports from scout bees concerning available resouces, and providing instructions to potential foragers based on both the incoming information and colony requirements. The foundation of this system is scouting, whereby individual workers find a resource, evaluate it, and communicate the location and quality of that resource to potential foragers back in the nest. Between 5 and 35% of the foragers in a colony will act as scouts at any one time; fewer workers scout during periods of resource abundance, while more workers scout when resources are scarce (Seeley 1983). Scouts may travel for many kilometers in search of flowers, although the median foraging radius of a colony is only a few hundred meters in agricultural areas, and about 2 km in forested regions (Gary, Witherell, and Marston 1972; Visscher and Seeley 1982). When scouts return to the nest, they use the well-known dance language to communicate the distance, direction, and quality of the flowers, thereby recruiting other workers to forage on the best available resources.

The nectar, pollen, and water which are collected by the foragers provide all of the colony's nutritional requirements. Both nectar and pollen are extensively processed by the workers for storage, so that food will be available during the many months when flowers are not available in the field. Nectar is the principal carbohydrate source for bees, and is carried back to the nest in the crop or honey stomach of the workers. Numerous enzymes are added to the nectar which invert the sugars to an easily digestible form and protect it from bacterial degradation. When a forager returns to the nest with a nectar load, she will regurgitate it to a receiving worker, who then evaporates some of the water from the nectar by repeatedly folding and unfolding her mouthparts. The partially evaporated nectar is then deposited in a cell, where the drying is continued by fanning bees until the nectar contains less than 18% water; this drying is important in preventing attack by yeasts. When the cell is full, other workers construct a wax capping over the cell. Once the nectar has been inverted and evaporated, had enzymes added to it, and been capped, it is honey, and in this form can be stored indefinitely; edible honey has been unearthed from the tombs of pharoahs after many thousands of years in clay pots!

Pollen is virtually the only protein source available to bees, and is required for proper growth and development. Workers pick up pollen at flowers on their body hairs and mouthparts, and transfer it to their hind legs for transport back to the nest as pellets. In the nest, a returning forager will moisten the pollen with regurgitated honey and enzymes before packing it into the bottom of cells by pushing the pellets in with her mandibles (Parker 1926). A phytocidal acid is added when pollen is packed into cells which prevents germination and bacterial degradation. Workers also add a thin layer of honey on top of the packed pollen to further protect it; pollen stored in this manner can last for many months.

As fall approaches, colonies reduce their brood rearing and foraging activities in preparation for winter. Drones are driven from colonies at this time, since colonies cannot afford the energy required for their winter maintenance (Free and Williams 1975; Morse, Strang, and Nowakowski 1967). As the temperature drops and days shorten, workers begin to form the winter cluster, awaiting whatever cues tell them to begin brood rearing in preparation for another season.

THE LIFE HISTORY OF TROPICAL HONEY BEES

Until recently, the vast majority of honey bee research has concentrated on bee races which evolved in temperate climates, probably because most bee-keepers and bee scientists worked in those regions. In recent years, however, there has been a dramatic increase in tropical bee research, due to the introduction of African honey bees into South America and heightened interest in developing beekeeping in tropical countries worldwide. Our increased awareness of honey bee biology in the tropics is fitting, since honey bees originated in tropical habitats and remain an important component of tropical ecosystems. When temperate- and tropical-evolved bees are compared, dramatic differences are evident in their life histories, the result of very different selective pressures in temperate and tropical habitats.

One major difference between temperate and tropical honey bees is that feral nests of tropical bees are much smaller than bees from temperate regions, about one-third to one-half their size, on average. Nest volumes in temperate regions average about 40 L, with most nest cavities in the range of 20 to 100 L (Seeley and Morse 1976; Wadey 1948); this is about one standard Langstroth deep super in size. In contrast, tropical honey bee nests average 20 to 25 L in volume, with nests as small as 7 L (Winston, Taylor, and Otis 1983). Also, tropical bees will frequently construct nests external to a cavity, suspended under branches or rock overhangs; this is uncommon in temperate-evolved bees.

A second major difference between temperate- and tropical-evolved honey bees is in their reproductive biology, especially swarming rates. A single tropical bee colony may produce between 6 and 12 swarms a year, and, when the swarm production of these offspring is included, almost 60 colonies can result from one original starting colony after a year (Otis 1980). In contrast, the highest reported swarming rate for temperate-evolved bees is only an average of 3.6 swarms annually, and most studies have shown swarm production rates of only 1.0 to 2.6 swarms per year (Lee and Winston 1987; Morales 1986; Winston 1980). The high reproductive rate characteristic of tropical honey bees is due to colonies swarming three or four times a year, often at intervals of two months or less. In addition, tropical bees produce more afterswarms than temperate bees each time they go through a swarming cycle (Winston 1979, 1980; Winston, Dropkin, and Taylor 1981).

Colonies of tropical-evolved honey bees also abscond much more than their temperate-evolved relatives. Absconding can be defined as the abandoning of a nest by a colony which forms a swarm and presumably reestablishes itself elsewhere, and is generally due to either disturbance or lack of resources. Both types of absconding are much more common for tropical honey bees. Among tropical-evolved bees, disturbance-induced absconding usually results from partial or total destruction of colonies by predators, destruction of comb by wax moths, fire in the nest proximity, heavy wasp or bird predation at the nest, difficulty in regulating temperature due to cold or excessive sunlight, and beekeeper's manipulations (Chandler 1976; Fletcher 1975/1976; 1978; Winston, Otis, and Taylor 1979; Woyke 1976). Resource-induced absconding seems to result from a scarcity of nectar, pollen, or water, and occurs primarily during the dearth season in tropical habitats; an average of 30% and up to 100% of colonies will abscond under those conditions. Presumably, the absconding swarms leave the vicinity of the nest and search for areas with better resources. Absconding swarms may travel as far as 160 km or more before constructing a new nest, migrating through areas of poor resources until they discover a better area with abundant but localized floral resources, such as Eucalyptus plantations in Africa (Cosenza 1972; Fletcher 1978; Nightingale 1976; Winston, Otis, and Taylor 1979).

A fourth difference between these two groups of bee races is in the extent of their defensive behavior, a subject of great interest to beekeepers. Generally, tropical-evolved honey bee colonies are much more aggressive than temperate-evolved bees. In various studies, colonies of tropical races have been shown to react more quickly to disturbances, to sting more, and to pursue attackers longer and farther than temperate-derived colonies (Collins and Kubasek 1982; Collins *et al.* 1982; Stort 1974 a,b, 1975 a,b,c). Ferocious attacks by tropical bees have certainly been well-documented by the press, earning the South American imports the nickname "killer bee."

There are clearly profound differences in life history between temperate and tropical honey bee races; tropical-evolved bees build smaller colonies, produce more swarms, abscond more frequently, and are feistier than temperate-evolved colonies. These differences can be largely explained by two major selective forces, climate and predation pressure. In temperate habitats, a long cold winter is a powerful factor selecting for large colonies with considerable investment in stored honey. Absconding is a particularly poor strategy, since absconding swarms would be unlikely to collect sufficient honey to survive the following winter. In contrast, large colonies with massive quantities of stored honey are not necessary in the warm tropics, and colonies can invest proportionately more energy in reproduction and less in honey production. Absconding is a viable strategy, particularly in areas with patchy resources where a good honeyflow may be occurring elsewhere. Colony defensive behavior can be explained by the significantly greater predation pressure in tropical habitats, due to a great variety of predators including ants, toads, armadillos, anteaters, honey badgers, and man. Also, frequent swarming may provide some escape from predation pressure by offsetting high colony death rates in the tropics. Whatever the causes, it is clear that understanding differences between honey bee races will become increasingly important in bee breeding and in devising management schemes for bees introduced deliberately or accidentally into habitats in which they did not evolve.

ACKNOWLEDGMENTS

I am grateful to Harvard University Press for permission to reprint some figures from *The Biology of the Honey Bee* (1987). I would also like to acknowledge funding from the U.S. Deptartment of Agriculture and National Science Foundation, the Natural Sciences and Engineering Research Council of Canada, the British Columbia Science Council, and Simon Fraser University, which have provided me with substantial research support.

REFERENCES

Allan, S.A., K.N. Slessor, M.L. Winston, G.G.S. King. (1987). The influence of age and task specialization on the production and perception of honey bee pheromones. *J. Insect Physiol,* *33*: 917-922.

Allen, M.D. (1955). Observations of honeybees attending their queen. *Brit. J. of Anim. Behav.* *3*:66-69.

Allen, M.D. (1960). The honey bee queen and her attendants. *Anim. Behav. 8*:201-208.

Allen, M.D., and E.P. Jeffree. (1956). The influence of stored pollen and of colony size on the brood rearing of honeybees. *Ann. Appl. Biol. 44*:649-656.

Anderson, R.H. (1963). The laying worker in the Cape honey bee, *Apis mellifera capensis. J. Apic. Res. 2*:85-92.

Avitabile, A., and J.P. Kasinskas. (1977). The drone population of natural honeybee swarms. *J. Apic. Res. 16*:145-149.

Avitabile, A., D. Stafstrom, and K.J. Donovan. (1978). Natural nest sites of honeybee colonies in trees in Connecticut USA. *J. Apis Res. 17*:222-226.

Bastain, J., and H. Esch. (1970). The nervous control of the indirect flight muscles of honey bee. *Z. vergl. Physiol. 67*:307-324.

Beetsma, J. (1979). The process of queen-worker differentiation in the honeybee. *Bee World 60*:24-39.

Boch, R., and D.A. Shearer. (1966). Iso-pentyl acetate in stings of honeybees of different ages. *J. Apic. Res. 5*:65-70.

Bozina, K.D. (1961). How long does the queen live? *Pchelovodstvo 38*:13.

Brouwers, E.V.M., R. Ebert, and J. Beetsma. (1987). Behavioural and physiological aspects of nurse bees in relation to the composition of larval food during caste differentiation in the honeybee. *J. Apicultural Research 26*:11-23.

Burgett, D.M., and R.A. Morse. (1974). The time of natural swarming in honey bees. *Ann. Entomol. Soc. Amer. 67*:719-720.

Butler, C.G. (1940). The ages of the bees in a swarm. *Bee World 21*:9-10.

Calderone, N.W. and R.E. Page, Jr. (1987). Genotypic variability in age polyethism and task specialization in the honey bee, *Apis mellifera* (Hymenoptera: Apidae). *Behav. Ecol Sociobiol. 22*:17-25.

Caron, D.M. (1980). Swarm emergence date and cluster location in honeybees. *Amer. Bee J. 119*:24-25.

Chandler, M.T. (1976). The African honey bee *Apis mellifera andansonii*: The biological basis of its management. In "Apiculture in Tropical Climates" (ed., E. Crane.) Internat. Bee Res. Assoc., Lond. pp. 61-68.

Collins, A.M., and K.J. Kubasek. (1982). Field test of honey bee (Hymenoptera: Apidae) colony defensive behavior. *Ann. Entomol. Soc. Amer. 75*:383-387.

Collins, A.M., T.E. Rinderer, J.R. Harbo, and A.B. Bolten. (1982). Colony defense by Africanized and European honey bees. *Science 218*:72-74.

Corkins, C.L. (1930). The winter activity in the honeybee cluster. *Rpt. Ia. St. Apiarist* pp. 44-49.

Cosenza, G.W. (1972). Estudo dos enxames de migracao de abelhas africans. *I Congr. bras. Apic.* Florianopolis, pp. 128-129.

Drescher, W. (1969). Die Flugaktivit.t von Drohnen der Rasse *Apis mellifica carnica* L. und. *A. mell. ligustica* L. in Abh.ngigkeit von Lebensalter und Witterung. *Zeitschr. Bienenfor-schung. 9*:390-409.

Esch, H. (1960). Üeber die Körpertemperaturen und den Wärmehaushalt von *Apis mellifica. Z. Vergl. Physiol. 43*:305-355.

Esch, H. (1964). Uber den Zusammenhang zwischen Temperatur, Aktionspotentiaten und Thoraxbewegungen bei der Honigbiene *(Apis mellifica). Z. vergl. Physiol. 48*:547-551.

Esch, H., and J.A. Bastain. (1968). Mechanical and electrical activity in the indirect flight muscles of the honey bee. *Z. vergl. Physiol. 58*:429-440.

Fell, R.D., J.T. Ambrose, D.M. Burgett, D. DeJong, R.A. Morse, and T. Seeley. (1977). The seasonal cycle of swarming in honeybees. *J. Apic. Res. 16*:170-173.

Fergusson, L.A. and M.L. Winston. (1988). The influence of wax deprivation on temporal polyethism in honey bee *(Apis mellifera* L.) colonies. *Can. J. Zoology 66*:1997-2001.

Fletcher, D.J.C. (1975/1976). New perspectives in the causes of absconding in the African bee Part I. *S. African Bee J. 47*:11, 13-14. (1975) *Part II 48*:6-9. (1976).

Fletcher, D.J.C. (1978). The African bee, *Apis mellifera adansonii*, in Africa. *Ann. Rev. Entomol. 23*:151-171.

Free, J.B. (1961). Hypopharyngeal gland development and division of labour in honey-bee *(Apis mellifera* L.) colonies. *Proc. Roy. Entomol. Soc. London (A)36*:5-8.

Free J.B. (1965). The allocation of duties among worker honeybees. *Symp. Zool. Soc. London 14*:39-59.

Free, J.B. (1987). Pheromones of Social Bees. Cornell University Press, Ithaca, N.Y.

Free, J.B., and I.H. Williams. (1975). Factors determining the rearing and rejection of drones by the honeybee colony. *Anim. Behav. 23*:650-675.

von Frisch, K. (1967). The dance language and orientation of bees. Harvard University Press, Cambridge, Mass.

Frumhoff, P.C. and S. Schneider. (1987). The social consequences of honey bee polyandry: the effects of kinship on worker interactions within colonies. *Anim. Behav. 35*:255-262.

Fukuda, H. (1960). Some observations on the pollen foraging activities of the honey bee, *Apis mellifera* L. (Preliminary Report). *J. Fac. Sci. Hokkaido Univ. Ser. VI Zool. 14*:381-386.

Fukuda, H., and T. Ohtani. (1977). Survival and life span of drone honeybees. *Res. Pop. Ecol. 19*:51-68.

Fukuda, H., and K. Sekiguchi. (1966). Seasonal change of the honeybee worker longevity in Sapporo, North Japan, with notes on some factors affecting the life span. *Jap. J. Ecol. 16*:206-212.

Gary, N.E., P.C. Witherell, and J.M. Marston. (1972). Foraging range and distribution of honey bees used for carrot and onion pollination. *Envir. Entomol. 1*:71-78.

Gates, B.N. (1914). The temperature of the bee colony. *Bull. U.S. Dep. Agric. No. 96*:1-29.

Goewie, E.A. (1978). Regulation of caste differentiation in the honey bee *(Apis mellifera* L.). *Meded. LandbHoogesch. Wageningen No. 78*-15, 1-75.

Haydak, M.H. (1958). Wintering of bees in Minnesota. *J. Econ. Entomol 51*:332-334.

Haydak, M.H. (1970). Honey bee nutrition. *Ann. Rev. Entomol. 15*:143-156.

Himmer, A. (1927). Der soziale Wärmehaushalt der Honigbiene, II, Die Wärme der Bienen-brut. Erlanger Jb. *Bienenk. 5*:1-32.

Howell, D.E., and R.L. Usinger. (1933). Observations on the flight and length of life of drone bees. *Ann. Entomol. Soc. Amer. 26*:239-246.

Jay, S.C. (1963). The development of honeybees in their cells. *J. Apic. Res. 2*:117-134.

Jaycox, E.R. (1956). Factors affecting the attainment of sexual maturity by the drone honeybee *(Apis mellifera* L.) Ph.D. Dissertation, Univ. of Calif., Davis.

Jeffree, E.P. (1951). The swarming period in Wiltshire. *Wiltshire Beekeepers Gazette 74*:2-3.

Jeffree, E.P. (1955). Observations on the decline and growth of honey bee colonies. *J. Econ. Entomol. 48*:723-726.

King, G.E. (1933). The larger glands in the worker honey-bee—A correlation of activity with age and with physiological functioning. Ph.D. Thesis Univ. Ill., Urbana, Ill.

Koeniger, N. (1969). Experiments concerning the ability of the queen (*Apis mellifica* L.) to distinguish between drone and worker cells. XXII Internat. Beekeeping Cong. Summ. p.138.

Koeniger, N. (1970). Factors determining the laying of drone and worker eggs by the queen honeybee. *Bee World 51*:166-169.

Kolmes, S.A. (1985a). An ergonomic study of *Apis mellifera* (Hymenoptera:Apidae). *J. Kansas Entomol. Soc. 58*:413-421.

Kolmes, S.A. (1985b). An information-theory analysis of task specialization among worker honey bees performing hive duties. *Anim. Behav. 33*:181-187.

Kolmes, S.A. (1985c). A quantitative study of the division of labour among worker honey bees. *Z. Tierpsychol. 68*:287-302.

Kolmes, S.A., M.L. Winston, and L.A. Fergusson. (1989). The division of labor among worker honey bees: The effects of multiple patrilines. *J. Kans. Entomol. Soc. 62*:80-95.

Lee, P.C. and M.L. Winston. (1985a). The influence of swarm population on brood production and emergent worker weight in newly founded honey bee colonies (*Apis mellifera*). *Insectes Sociaux 32*:96-103.

Lee P.C., and M.L. Winston. (1985b). The effect of swarm size and date of issue on comb construction in newly founded colonies of honey bees (*Apis mellifera* L.) *Can. J. Zool. 63*:524-527.

Lee, P.C. and M.L. Winston. (1987). Effects of reproductive timing and colony size on the survival, offspring colony size and drone production in the honey bee (*Apis mellifera*). *Ecological Entomology 12*: 187-195.

Lensky, Y., P. Cassier, M. Notkin, C. Delorme-Joulie and M. Levinsohn. (1985). Pheromonal activity and fine structure of the mandibular glands of honeybee drones (*Apis mellifera* L.) (Insecta, Hymenoptera, Apidae). *J. Insect Physiol. 31*:265-276.

Lindauer, M. (1951). Bee dances in the clustered swarm. *Naturwissenschaften 22*:509-513.

Lindauer, M. (1952). Ein Beitrag zur Frage der Arbeitsteilung im Bienenstaat. *Z. vergl. Physiol. 34*:299-345. (Translation, *Bee World 34*:63-73, 85-90).

Lindauer, M. (1955). Schwarmbienen auf Wohnungssuche. *Z. vergl. Physiol. 37*:263-324.

Lindauer, M. (1961). Communication among social bees. Harvard University Press, Cambridge Mass.

Mackensen, O. (1943). The occurrence of parthenogenetic females in some strains of honeybees. *J. Econ. Entomol. 36*:465-467.

Mackensen, O. (1951) Viability and sex determination in the honey bee (*Apis mellifera* L.) *Genetics 36*:500-569.

Masson, C., and G. Arnold. (1984). Ontogeny, maturation and plasticity of the olfactory system in the worker bee. *J. Insect Physiol. 30*:7-14.

Meyer, W. (1956). Arbeitsteilung im Bienenschwarm. *Insectes Sociaux 3*:303-324.

Milojevic, B.D. (1940). A new interpretation of the social life of the honeybee. *Bee World 21*:39-41.

Morales, G. (1986). Effects of cavity size on demography of unmanaged colonies of honey bees. (*Apis mellifera* L.). M.Sc. Thesis, Univ. of Guelph, Ontario.

Morse, R.A., and T.D. Seeley. (1978). Bait hives. *Glean. Bee Cult. 106*:218-220, 242.

Morse, R.A., G.E. Strang, and J. Nowakowski. (1967). Fall death rates of drone honey bees. *J. Econ. Entomol. 60*:1198-1202.

Moskovljevic-Filipovic, V.C. (1956.) [The influence of normal and modified social structures on the development of the Pharyngeal glands and the work of the honeybee (*Apis mellifera* L.).] *Monogr. Srpska. Akad. Nauka. 262*:101 pp.

Nightingale, J. (1976). Traditional beekeeping among Kenya tribes, and methods proposed for improvement and modernisation. Pp. 15-22 from *Apiculture in Tropical Climates.* Ed. E. Crane, London, UK: Intern. Bee Res. Assoc.

Nolan, W. J. (1924). The division of labor in the honeybee. *N.C. Beekeeper (Oct)*:10-15.

Nolan, W. J. (1925). The brood-rearing cycle of the honeybee. *Bull. U.S. Dept. Agric. No.1349*:1-56.

Oertel, E. (1956). Observations on the flight of drone honeybees. *Ann. Entomol. Soc. Amer. 49*:497-500

Otis, G.W. (1980). The swarming biology and population dynamics of the Africanized honey bee. Ph.D. Dissertation, Univ. of Kansas.

Otis, G.W., M.L. Winston, and O.R. Taylor, Jr. (1981). Engorgement and dispersal of Africanized honeybee swarms. *J. Apic. Res. 20*:3-12.

Owens, C.D. (1971). The thermology of wintering honey bee colonies. *Tech. Bull., U.S. Dep. Agric. 1429*:1-32.

Parker, R.L. (1926). The collection and utilization of pollen by the honeybee. *Mem. Cornell agric. Exp. Sta. No.98*:1-55.

Phillips, E.F., and G.S. Demuth. (1914). The temperature of the honeybee cluster in winter. *Bull. U.S. Dep. Agric. No. 93*:1-16.

Ribbands, C.R. (1953). The behaviour and social life of honeybees. London: Bee Res. Assn. Ltd. (Republished, 1964, by Dover Publ., Inc., New York.)

Robinson, G.E. (1985). Effects of a juvenile hormone analogue on honey bee foraging behaviour and alarm pheromone production. *J. Insect Physiol. 31*:277-282.

Robinson, G.E. (1987). Regulation of honey bee age polyethism by juvenile hormone. *Behav. Ecol. Sociobiol. 20*:329-338

Robinson, G.E., B.A. Underwood, and C.E. Henderson. (1984). A highly specialized water-collecting honey bee. *Apidologie 15*:355-358.

Rösch, G.A. (1930). Untersuchungen über die Arbeitsteilung im Bienenstaat. 2. Teil: Die Tätigkeiten der Arbeitsbienen unter experimentell veränderten bedingunen. *Z. f. vergl. Physiol. 12*: 1-71.

Roth, M. (1965). La production de chaleur chez *Apis mellifica. Ann. Abeille 8*:5-77.

Ruttner, F. (1966). The life and flight activity of drones. *Bee World 47*:93-100.

Sakagami, S.F. (1953). Untersuchungen .ber die Arbeitsteilung in einem Zwergvolk der Honigbiene. Beitr.ge zue Biologie des Bienenvolkes, *Apis mellifera* L. *I Jap. J. Zool. 11*:117-185.

Seeley, T.D. (1977). Measurement of nest cavity volume by the honey bee (*Apis mellifera*). *Behav. Ecol. Sociobiol. 2*:201-227.

Seeley, T.D. (1978). Life history strategy of the honey bee *Apis mellifera. Oecologia 32*:109-118.

Seeley, T.D. (1983). Division of labor between scouts and recruits in honeybee foraging. *Behav. Ecol. Sociobiol. 12*:253-259.

Seeley, T.D. (1985). Honeybee Ecology. Princeton University Press, Princeton, N.J.

Seeley, T.D., and R.A. Morse. (1976). The nest of the honey bee (*Apis mellifera*). *Insectes Sociaux 23*:495-512.

Seeley, T.D., and P.K. Visscher. (1985). Survival of honeybees in cold climates: the critical timing of colony growth and reproduction. *Ecol. Entomol. 10*:81-88.

Shearer, D.A., and R. Boch. (1965). 2-Heptanone in the mandibular gland secretion of the honey-bee. *Nature 206*:530.

Simpson, J. (1959). Variation in the incidence of swarming among colonies of *Apis mellifera* throughout the summer. *Insectes Sociaux 6*:85-99

Simpson, J. (1961). Nest climate regulation in honeybee colonies. *Science 133*:1327-1333.

Simpson, J., I.B.M. Riedel and N. Wilding. (1968). Invertase in the hypopharyngeal glands of the honeybee. *J. Apic. Res. 7*:29-36.

Southwick, E.E., and J.N. Mugaas. (1971). A hypothetical homeotherm: the honeybee hive. *Comp. Biochem. Physiolo. 40A*:935-944.

Stort, A.C. (1974a). Genetic study of aggressiveness in two subspecies of *Apis mellifera* in Brazil. I. Some tests to measure aggressiveness. *J. Apic. Res. 13*:33-38.

Stort, A.C. (1974b). Genetical study of aggressiveness of two subspecies of *Apis mellifera* in Brazil. IV. Number of stings in the gloves of the observer. *Behav. Genetics 5*:269-274.

Stort, A.C. (1975a). Genetic study of aggressiveness in two subspecies of *Apis mellifera* in Brazil. IV. Number of stings in the gloves of the observer. *Behav. Genet. 5*:269-274.

Stort, A.C. (1975b). Genetic study of aggressiveness of two subspecies of *Apis mellifera* in Brazil. II. Time at which the first sting reached a leather ball. *J. Apic. Res. 14*:171-175.

Stort, A.C. (1975c). Genetic study of the aggressiveness of two subspecies of *Apis mellifera* in Brazil. V. Number of stings in the leather ball. *J. Kansas Entomol. Soc. 48*:381-387.

Tribe, G.D., and D.J.C. Fletcher. (1977). Rate of development of the workers of *Apis mellifera adansonii* L. In "African bees: their taxonomy, biology, and economic use." (ed. D.J.C. Fletcher). Pretoria, South Africa: Apimondia. pp. 115-119.

Visscher, P.K., and T.D. Seeley. (1982). Foraging strategy of honeybee colonies in a temperate deciduous forest. *Ecology 63*:1790-1801.

Wadey, H.J. (1948). Section de chauffe? *Bee World 29*:11.

de Wilde, J. de., and J. Beetsma. 1982. The physiology of caste development in social insects. *Adv. Insect Physiol. 16*:167-246.

Wilson, H.F., and V.G. Milum. (1927). Winter protection for the honeybee colony. *Res. Bull. Wisc. Agric. Exper. Sta. 75*:1-47.

Winston, M.L. (1979). Intra-colony demography and reproductive rate of the Africanized honeybee in South America. *Behav. Ecol. Sociobiol. 4*:279-292.

Winston, M.L. (1980). Swarming, afterswarming, and reproductive rate of unmanaged honey-bee colonies (*Apis mellifera*). *Insectes Sociaux 27*:391-398.

Winston, M.L. (1987). The Biology of the Honey Bee. Harvard University Press, Cambridge, Mass.

Winston, M.L., J.A. Dropkin, and O.R. Taylor. (1981). Demography and life history characteristics of two honey bee races (*Apis mellifera*). *Oecologia 48*:407-413.

Winston, M.L., and L.A. Fergusson. (1985). The effect of worker loss on temporal caste structure in colonies of the honey bee (*Apis mellifera* L.). *Can. J. Zool. 63*:777-780.

Winston, M.L., and S.J. Katz. (1982). Foraging differences between cross-fostered honeybee workers (*Apis mellifera*) of European and Africanized races. *Behav. Ecol. Soc. 10*:125-129.

Winston, M.L., and G.W. Otis. (1978). Ages of bees in swarms and afterswarms of the Africanized honeybee. *J. Apic. Res. 17*:123-129.

Winston, M.L., G.W. Otis, and O.R. Taylor. (1979). Absconding behaviour of the Africanized honeybee in South America. *J. Apic. Res. 18*:85-94.

Winston, M.L., O.R. Taylor, and G.W. Otis. (1983). Some differences between temperate European and tropical African and South American honeybees. *Bee World 64*:12-21.

Winston, M.L., and E.N. Punnett. (1982). Factors determining temporal division of labor in bees. *Can. J. Zool. 60*:2947-2952.

Witherell, P.C. (1971). Duration of flight and of interflight time of drone honey bees, *Apis mellifera. Ann. Entomol. Soc. Amer. 64*:609-612.

Witherell, P.C. (1972). Flight activity and natural mortality of normal and mutant drone honeybees. *J. Apic Res. 11*:65-75.

Woyke, J. (1963). What happens to diploid drone larvae in a honeybee colony? *J. Apic. Res. 2*:73-75.

Woyke, J. (1963). What happens to diploid drone larvae in a honeybee colony? *J. Apic. Res. 2*:73-75.

Woyke, J. (1969). A method of rearing diploid drones in a honeybee colony. *J. Apic. Res. 8*:65-74.

Woyke, J. (1973). Reproductive organs of haploid and diploid drone honeybees. *J. Apic. Res. 12*:35-51.

CHAPTER 4

THE ANATOMY OF THE HONEY BEE

by R. E. SNODGRASS[1]
and E. H. ERICKSON[2]

Too often we forget that honey bees are simply insects. Of course, insects themselves are quite remarkable—EHE

PREFACE

This chapter was first written by the late R. E. Snodgrass in 1946 and then extensively revised by him for a later edition. It is based on Snodgrass' earlier works published first in 1910 and expanded in 1925. Since Snodgrass' work remains the subject authority, I have in this revision changed little of his text except to update it with new information as appropriate. I have added a number of Scanning Electron Micrographs (plates) to supplement his many fine drawings. Because those drawings are extensively discussed in the accompanying text, I have elected to insert the plates into the text using a separate numeric series. These micrographs are taken from *A Scanning Electron Microscope Atlas of the Honey Bee* by E. H. Erickson, Jr., S. A. Carlson and M. B. Garment © 1985 and reprinted with permission by Iowa State University Press, Ames.

—ERIC H. ERICKSON

The anatomy of an animal is the assemblage of structual parts that enables the animal to do the things necessary for the maintenance of its individual existence, and for the perpetuation of its kind. As an individual the animal must obtain and distribute to its tissues both food and oxygen, eliminate waste matter, and correlate the activities of its various organs with one another, and its own activities with changing conditions of the environment. Hence, the animal has a locomotor system, feeding and digesting organs together with a system of food distribution, a respiratory system, an excretory system, and a nervous system. To provide for the continuance of its species it has a reproductive system. In addition, nearly every animal has some specialty of its own—it may eat only a particular kind of food, it may inhabit a special kind of environment, it may adopt a particular method of lomotion, it may be individualistic or socialistic—and according to its habits or its way of life, it is equipped with special anatomical mechanisms.

---

[1]Robert E. Snodgrass, Ph.D. (deceased) Honorary Research Associate, Smithsonian Institute. Author of *Anatomy of the Honey Bee.* World authority on insect morphology.

[2]Eric H. Erickson, Ph.D. Director, Carl Hayden Bee Research Center. Coauthor of *A Scanning Election Microscope Atlas of the Honey Bee.* Honey Bee Biologist and Pollination Ecologist.

To understand why an animal is made as it is, we study its structure and functions; to understand how it comes to be what it is, we must know something of its development. The honey bee, as every other complex animal, begins life as a single cell, the egg, or *ovum*, but the development of the embryo produces first the *larva*. The adult bee is then formed in an intermediate state known as the *pupa*.

DEVELOPMENT

In the narrowed inner end of each tubule in the queen's ovary (Fig. 22,B) are the primary female germ cells *(GCls)* or *oogonia*. Further down the tube these cells multiply and differentiate into larger cells, or *oocytes*, which become the eggs *(E)*, and into smaller cells called *nurse cells*, or *trophocytes*, because they will be absorbed as nutriment by the growing egg cells. As new oocytes are formed, the egg tubes lengthen at their inner ends, and the older eggs, each accompanied by a mass of nurse cells, increase in size. An egg tubule thus becomes a succession of egg chambers *(EC)* alternating with nurse cell chambers *(NC)*. The fullgrown eggs *(E)* in the lower ends of the tubes have absorbed practically all their nurse cells, and the walls of the containing chambers, or follicles, now secrete over each egg a shell, known as the *chorion*.

In a similar manner the tubules of the male testes produce spermatozoa, but the spermatozoa remain as free, active, threadlike individuals (Plate 1).

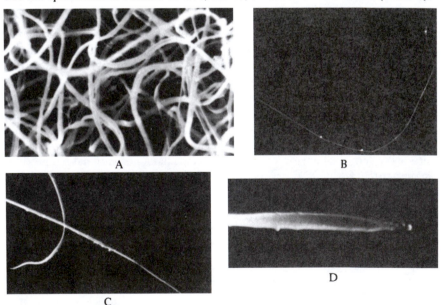

PLATE 1. A. Tangled mass of thread-like spermatozoa (X 2,053). B. A single sperm (X 350). C. Head and tail of a sperm (X 1,688). D. Close-up of the head of a sperm (X 9,490).

The mature egg is discharged through the oviduct into the genital exit passage, or *vagina* (Fig. 22, *A, Vag*), with which is connected a small sac, the *spermatheca (Spt)*, containing the spermatozoa received from drones at the time of mating. A minute aperture, the *micropyle*, at the anterior end of the egg admits the sperm, but by some regulatory mechanism the queen discharges sperm on some eggs and withholds it from others (Plate 2). The eggs that become fertilized develop into female bees; those unfertilized become drones. By some stimulus the egg develops whether fertilized or not.

The newly laid egg (Fig 1, A) is elongate, rounded at the ends and slightly convex on what will be the under surface of the embryo. The interior consists of the egg cytoplasm and the nutritive material *(Y)* derived from the nurse cells in the ovary (called *yolk* because it is yellow in a bird's egg). The egg within the chorion is invested in a delicate *vitelline membrane (Vit)*. Just within the membrane is a peripheral *cortical layer* of cytoplasm. The egg nucleus *(Nu)* is contained in a small cytoplasmic body near the anterior end of the egg.

Development begins with the division of the nucleus and the resulting nuclei. The *cleavage cells* (Fig. 1, B, *CCls*) thus formed migrate out into the cortical cytoplasm, where they form a layer of cells on the surface of the egg, which is the *blastoderm* (C, *Bld*). Soon the lower part of the blastoderm

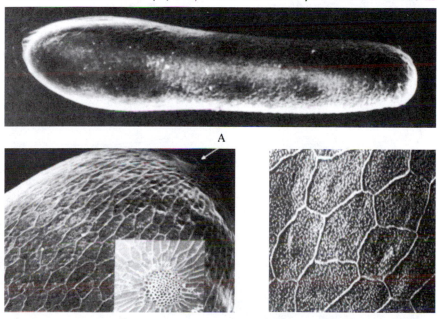

PLATE 2. A. Worker egg (X 85). B. Close-up of the large end of an egg (X 300) with the micropyle (inset X 300). C. Higher magnification of the egg shell (chorion) (X 575).

becomes thickened, forming what is known as the *germ band* (D, *GB*), while
the dorsal blastoderm *(dBl)* by contrast becomes very thin. The germ band is
the beginning of the embryo. Its edges grow upward on the sides and around
the ends of the egg *(E)* as the thin dorsal blastoderm contracts and finally
disappears, allowing the germ band to close over the back. The wall of the
embryo is now complete, but as yet there are no internal organs.

While the germ band is growing upward on the sides of the egg (Fig. 1, E),
it becomes differentiated into a pair of lateral plates and a median ventral
plate. The ventral plate sinks into the egg and the lower ends of the lateral
plates come together beneath it. The lateral plates thus become the body-wall
epidermis of the insect, which secretes the external cuticle. The ventral plate
becomes mesoderm, from which are derived the muscles, fat tissue, the heart,
and the internal reproductive organs. The stomach, or *ventriculus* (Fig. 2, B,
Vent), is formed of strands of endodermal cells proliferated from the two ends

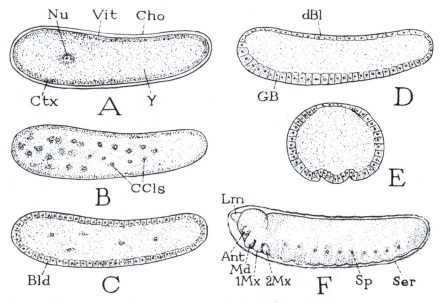

FIGURE 1. Development of the embryo in the egg (diagrammatic from Nelson. Embryology of
the Honeybee. (1915).

A, lengthwise section of egg in chorion. B, cleavage cells in the yolk, resulting from repeated
division of the nucleus and daughter nuclei, migrating to the cortex. C, blastoderm formed by
cleavage cells in the cortex. D, blastoderm differentiated into thick ventral germ band and thin
dorsal blastoderm. E, cross section of egg, germ band grown dorsally, differentiated into lateral
plates and ventral plate. F, the young embryo.

Ant, antenna; *Bld*, blastoderm; *CCls*, cleavage cells; *Cho*, chorion; *dBl*, dorsal blastoderm;
GB, germ band; *Lm*, labrum; *Md*, mandible; *1Mx*, first maxilla; *2Mx*, second maxilla; *Nu*,
nucleus; *Ser*, serosa; *Sp*, spiracle; *Vit*, vitelline membrane.

of the larva that come together and enclose the yolk, which will be the food of the growing embryo. At the mouth a tubular ingrowth of the ectoderm forms the *stomodeum (Stom)*, which opens into the anterior end of the stomach. Likewise an ingrowth from the anus forms the *proctodeum (Proc)*, which unites with the posterior end of the stomach. The insect alimentary canal always consists of these three parts.

The nervous system is of ectodermal origin, being formed from cells given off internally from the midventral line of the embryo. The nerve cells become aggregated first into paired segmental masses, or ganglia, connected crosswise by fiber commissures and lengthwise by connectives. The primary ganglia, however, unite to form a compound ganglion in each segment (Fig. 20), from which nerves grow outward to the muscles, glands, and other organs. Ganglia of the head form the brain. Nerves from the sense organs originate from cells of the epidermis and grow inward to the ganglia.

The tracheal respiratory system (Fig. 19) likewise is ectodermal, originating from tubular ingrowths along the sides of the body, the external openings of which are the spiracles.

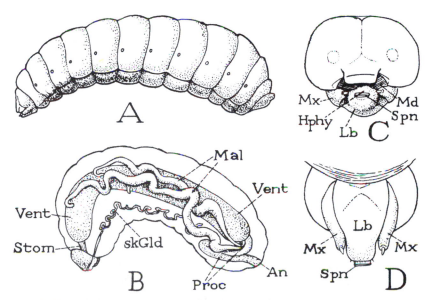

FIGURE 2. The larva.

A, a mature larva. B, same, alimentary canal, with Malpighian tubules, and silk glands of left side (from Nelson. Morphology of the Honeybee Larva. 1924). C, head, anterior. D, head, ventral.

An, anus; *Hphy,* hypopharynx; *Lb,* Labium; *Mal, Malpighian tubules; Md,* Mandible; *Mx,* maxilla; *Proc,* proctodeum; *skGld,* silk gland; *Spin,* spinneret; *Stom,* stomodeum; *Vent,* ventriculus.

Externally the young embryo (Fig. 1, F) becomes differentiated into a head region and a body. On the head, rudiments of the labrum *(Lm)*, antennae *(Ant)*, mandibles *(Md)*, and two pairs of maxillae *(1Mx, 2Mx)* are formed as small lobes. Wings and legs, however, are not apparent on the embryo or larva because their rudiments are sunken into shallow pockets of the epidermis beneath the cuticle. The full complement of 10 spiracles is present on each side of the embryo.

When the embryo is fully developed as a young larva, the latter hatches from the egg. During its life the larva goes through five stages of growth, or *instars*, moulting its cuticle after each instar. The bee larva (Fig. 2, A) is a very simple creature without external legs or wings as befits its life of inactivity in the comb cell. It has little to do except to eat the food the nurse bees give it. The larva has a small head and a body of 13 segments undifferentiated into thorax and abdomen. On the front of the head *(C)* are two small discs that mark the sites of the sunken antennae. The feeding organs include a pair of small mandibles *(Md)* and a pair of simple maxillae *(Mx)*. Between the maxillae is a median lobe on which the duct of the silk glands opens between raised lips, forming a spinneret *(Spn)*. The spinneret lobe is formed by a union of the hypopharynx *(Hphy)* with the end of the labium *(Lb)*, the latter being fully exposed on the underside of the head *(D)*. The silk glands of the honey bee larva are the salivary glands of most other insects, the terminal duct of which opens between the bases of the ununited hypopharynx and labium.

Since the principal function of the bee larva is eating, it is provided with an enormous stomach, or ventriculus, a cylindrical sac (Fig. 2, B, *Vent*) almost as long as the body. A short intake tube, the stomodeum *(Stom)*, goes from the mouth to the stomach, and a looped intestine, or proctodeum *(Proc)*, connects the rear end of the stomach with the terminal anus *(An)*. Arising from the inner end of the proctodeum are four Malpighian tubules, two on each side *(Mal)*, which are the excretory organs of the insect. In a young bee larva the tubules are slender, but in the mature larva (B) they become greatly distended by the accumulation of secretion within them. In order to preserve cleanliness in the larval cell, both the Malpighian tubules and the stomach are shut off from the intestine until the larva is mature at the time the cell is capped. Then the tubules and the stomach open into the intestine and their contents are discharged from the anus into the inner end of the cell. Now the larva spins its cocoon and moults for the last time, but it does not shed the cuticle. The body wall has already taken on the form of the young pupa (Fig. 3, A), which is enveloped in the larval cuticle *(lCt)*.

The antennae, legs, and wings of the pupa are now everted, compound eyes and the adult mouth parts are present (C). Otherwise, however, the young pupa still retains larval features. The three thoracic segments (1,2,3) are of about equal size, and there is no constriction between the thorax and the abdomen. On the posterior part of the abdomen of a female pupa (B) are the rudiments of the sting. At a later stage (D), still within the larval cuticle, the thorax has taken on more of the adult form by an expansion of the middle segment (2) at the expense of the first and third segments (1,3). There is as yet, however, no constriction separating the abdomen from the thorax. With the completion of development the mature pupa (E) distinctly resembles an adult bee. The thorax is now well separated from the abdomen, but it is to be noted

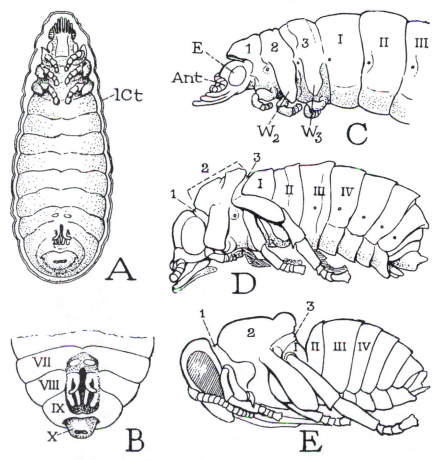

FIGURE 3. Development of the pupa.

A, the young pupa in unshed larval cuticle. B, same end of abdomen, with rudiments of sting on under surface. C, same as A, lateral view, more enlarged. D, later stage of pupa, still in larval cuticle. E, the mature pupa.

Ant, antenna; *E*, compound eye; *ICt*, larval cuticle; W_2, W_3, mesothoracic and metathoracic wings; 1, 2, 3, thoracic segments; *I-X*, abdominal segments.

that the constriction is between the primary first *(I)* and second *(II)* abdominal segments. The reduced first segment is intimately united with the thorax and becomes virtually a part of the thorax, known as the *propodeum*.

The fully formed pupa throws off the larval cuticle, and undergoes no further external change. Within it, however, the special larval tissues break down and go into dissolution as food for the growing adult tissues. Then, when the adult is completed within the pupa, it splits the pupal shell and emerges as an adult bee. The changeover from the larva to the adult is called the "metamorphosis" of the insect, but it is largely a replacement of the larva by the adult.

THE ADULT BEE

The adult honey bee is constructed on the general plan of an insect, but it leads a highly specialized kind of life and for this reason is provided with special mechanisms and gadgets that enable it to live in its particular way. Hence, in studying the bee, while we must give attention to its fundamental insect organization, special interest pertains to the structures and modifications of organs that adapt the bee to its manner of living and differentiate it from other insects.

In its general structure the bee resembles any other insect, though its form (Fig. 4) is obscured by the dense coating of hairs with which the body is covered (Plate 3). The bee's coat is particularly fluffy because many of the hairs are featherlike, the shaft of each having many short side branches.

The *head* of the insect (Fig. 4, *H*) carries the eyes, the antennae, and the organs of feeding. It is joined to the next body division, the *thorax (Th)*, by a

A   B

PLATE 3. A. Branched body hairs (X 85). B. Pollen grains in a tangled mass of hairs (X 750).

slender, flexible neck. The thorax and the third section of the trunk, or *abdomen (Ab)*, are composed of a succession of rings called *segments*. In most insects the thorax consists of only three segments, but in the bee and related insects it includes four segments, which are the *prothorax (1)*, the *mesothorax (2)*, the *metathorax (3)*, and the *propodeum (I)*. The propodeum of the bee is the first abdominal segment of most other insects. The prothorax carries the first pair of legs (L_1); the mesothorax and the metathorax, in addition to

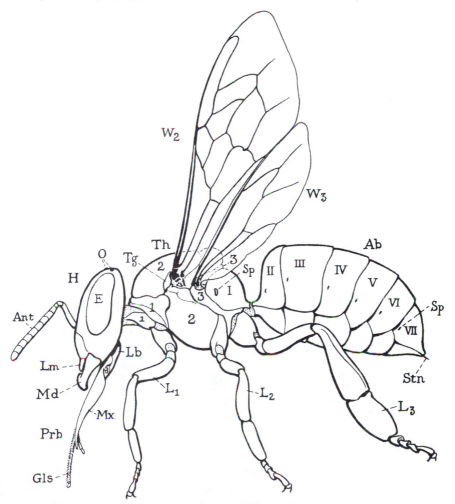

FIGURE 4. External structure of a worker bee as seen when the hairy covering is removed.

Ab, abdomen; *Ant*, antenna; *E*, compound eye; *H*, head; *I*, propodeum; *II-VII*, abdominal segments; L_1, L_2, L_3, legs; *Md*, mandible; *Prb*, proboscis; *Sp*, spiracle; *Th*, thorax; W_2, W_3, wings; *1*, prothorax, *2*, mesothorax; *3*, metathorax.

carrying each a pair of legs (L_2, L_3), support also the two pairs of wings (W_2, W_3). The thorax is clearly the locomotor center of the insect. A short stalk, the *peduncle*, attaches the thorax to the abdomen which contains the principal internal organs and bears the sting.

The feeding organs of the bee consist of the same parts as do those of a grasshopper or a cricket, but the parts are very different in form because in the bee they are adapted to the ingestion of pollen, as well as of liquid food to be obtained from the depths of flower corollas, while such insects as grasshoppers and crickets merely bite off, chew, and swallow particles of solid food.

The wings of the bee are adapted for swift flight and also for sustaining a load. The legs are modified in their structure for various uses besides that of locomotion. The sting of the bee represents the ovipositor of other female insects, sufficiently remodeled in structure to serve for piercing and for the injection of poison instead of eggs.

Most of the internal organs of the bee are much the same as in other insects, but the alimentary canal has a special adaptation for carrying nectar or honey. The respiratory system is greatly amplified. In the queen the ovaries are so highly developed as to be capable of producing a great number of eggs which can be discharged continually over long periods of time. Special glands of the head produce a rich food substance for the brood. Glands of the abdomen form wax for comb building. Near the end of the body is a gland that secretes a scent by which bees get information from one another.

THE HEAD, THE ANTENNAE, AND THE ORGANS OF FEEDING

Though the head of an insect is a cranium-like structure with continuous walls, its embryonic development shows that it is formed by the close union of several segments like those of the thorax and abdomen. The segmental structure of the head, moreover, is attested by the fact that the head carries four pair of appendages. These appendages are the *antennae*, the *mandibles* or jaws of the insect, the *maxillae*, and the *labium*, which last represents a second pair of maxillae united. In the bee the maxillae and the labium together form the *proboscis*, an organ for feeding on liquids. The head bears also the eyes, usually a pair of large lateral *compound eyes* and, between the latter, usually three small simple eyes called *ocelli*.

Structure of the Head

The head of the honey bee is triangular as seen from in front (Fig. 5, *A*), flattened from before backward, somewhat concave on the posterior surface *(B)*, and is set on the thorax by a narrow membranous neck. The lateral angles are capped by the compound eyes (A, *E*), and on the top of the head are three ocelli *(O)* (Plate 4). The antennae *(Ant)* arise close together near the center of the face. Below their bases a prominent arched groove *(es)* sets off a large area known as the *clypeus (Clp)*, from the lower margin of which is suspended a broad, movable flap, the *labrum (Lm)*. Attached laterally to the lower part of the head behind the labium are the two jawlike mandibles *(Md)*, and behind the mandibles, better seen from the back of the head (B) are suspended the two maxillae *(Mx)* and the median labium *(Lb)*. The long distal parts of the maxillae and the labium, shown spread out at A of Fig. 6, either project downward or are folded back below and behind the head (Fig. 5, *E*); but in their functional position they are brought together to form a tubular proboscis (Fig. 4, *Prb*) for feeding on liquids.

On the back of the head, as seen when it is detached from the body (Fig. 5, *B*), is a central opening, the *neck foramen (For)*, by which the cavity of the head communicates with that of the body, and which gives passage for the oesophagus, nerves, blood vessel, air tubes, and salivary duct. Below the foramen the hard wall of the head is cut out in a large horseshoe-shaped notch with a membranous floor in which are implanted the long bases of the maxillae and labium. The depression of the notch, therefore, is designated the *proboscis fossa (PF)*.

Internally the walls of the head are braced by two large tubes (Fig. 5, C,

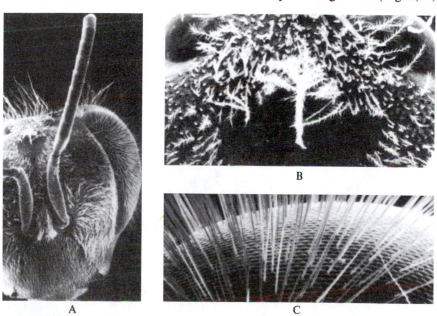

A   C

PLATE 4. A. Worker face showing ocelli at the top between the antennae, and a large compound eye. Small black arrow points toward one of the two anterior tentorial pits (X 15). B. Close-up of the three ocelli (X 80). C. Close-up of the surface of the compound eye with its numerous facets and hairs (X 120).

The Hive and the Honey Bee

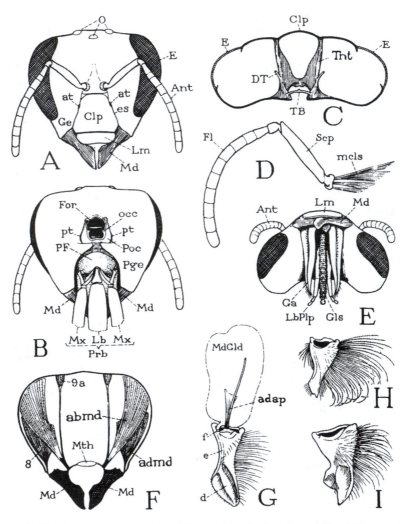

FIGURE 5. The head, antennae, and mandibles of a worker bee (except H and I).

A, facial view of head. B, rear view of head. C, horizontal section of head showing internal tentorium. D, antenna. E, under view of head and folded proboscis. F, transverse verticle section of head showing mandibles and their muscles. G, mandible and mandibular gland, mesal view. H, mandible of drone. I, mandible of queen.

Abmd, abductor muscle of mandible; *adap*, tendon of abductor muscle of mandible; *admd*, adductor muscle of mandible; *Ant*, antenna; *at*, anterior tentorial *p* & *Clp*, clypeus; *d*, channel of mandible; *e*, groove of mandible; *E*, compound eye; *es*, suture defining clypeus; *f*, orifice of mandibular gland; *Fl*, flagellum; *For*, neck foramen of head; *Ga*, galea; *Gls*, tongue (glossa); *Lb*, labium; *LbPlp*, labial palpus; *Lm*, Labrum; *Mcls*, muscles; *Md*, mandible; *MdGld*, mandibular gland; *Mth*, mouth; *Mx*, maxilla; *O*, ocelli; *PF*, proboscis fossa; *Prb*, proboscis; *pt*, posterior tentorial pit; *Scp*, scape; *TB*, tentorial bridge; *Tnt*, tentorium.

Tnt) that extend through the head cavity from the sides of the neck foramen to the grooves of the face at the sides of the clypeus *(Clp)* (Plate 5). The posterior ends of the tubes are bridged by a slender cross-tube *(TB)*, which may be seen from behind just within the neck foramen (B). The tubes and the connecting bridge constitute the *tentorium.*

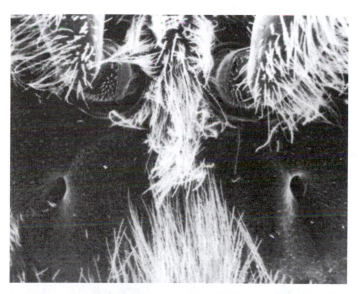

A

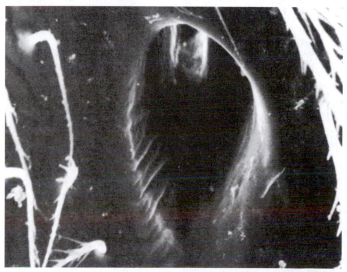

B

PLATE 5. A. Anterior tentorial pits below the antennal bases of the drone (X 50). B. View into a worker tentorial pit showing the tubular nature of the tentorium (X 567).

The Antennae

The antennae are freely movable appendages with their bases set into small socketlike membranous areas of the head wall (Fig. 5, *A*). Each antenna is pivoted on a single articular point of the socket rim, and is provided with four muscles (D, *Mcls*) arising on the tentorial tube of the same side of the head. Each appendage, moreover, has an elbow-like joint between its basal stalk, or *scape (Scp),* and the flexible distal part called the *flagellum (Fl).* The scape of the drone antenna is shorter than that of the worker, but the flagellum is much longer and consists of 12 short rings, while there are only 11 in the worker and the queen. The antennae are important sensory organs. Each appendage is penetrated by a large double nerve from the brain. The flagellum is covered with small innervated hairs and other minute sensory structures of several kinds (Plates 6 & 7). It is difficult to determine the function of each variety of sense organ, but the antennae are responsive particularly to stimuli of touch and odor. Hairs and pegs may be either touch or odor receptors depending upon their structure. Pit and plate organs are usually odor receptors but some may sense temperature or humidity.

The Mandibles

The mandibles (Fig 5, A, *Md*) are suspended from the head at the sides of the mouth (F, *Mth*), which lies immediately behind the base of the labrum. Each jaw has an anterior and a posterior point of articulation on the head, and is provided with only two muscles (F, *abmd, admd*) which are attached on opposite sides of the axis of movement. The mandibles, therefore, swing sideways; but, because the anterior articulations are higher than the posterior, the points of the jaws turn inward and backward when the mandibles close.

The mandible of the worker bee (Fig. 5, *G*) is thick at the base, narrowed through the middle, and widened again distally in an expansion with a concave inner surface traversed by a median channel *(d).* From the channel a groove *(e)* runs upward to an aperture *(f)* at the base of the mandible, which is the outlet of a large, saclike *mandibular gland (MdGld)* that lies in the head above the mandible. The gland secretes a clear liquid, the purpose of which is not definitely known, but the secretion is supposed to be used for softening wax (Plate 8, A,B,C). The mandibular glands are largest in the queen; in the drone they are reduced to small vesicles. The worker bee uses its mandibles for eating pollen, for working the wax in comb building, for holding the base of the outstretched proboscis, and for doing any of the chores about the hive that require a pair of grasping instruments. The mandibles of the queen (Fig. 5, *I*) are larger than those of the worker but they lack the special features of the worker mandibles and each has a broad innerlobe near the pointed apex. The drone mandibles *(H),* on the other hand, are smaller than those of the worker, and each is sharply notched at the base of the apical point.

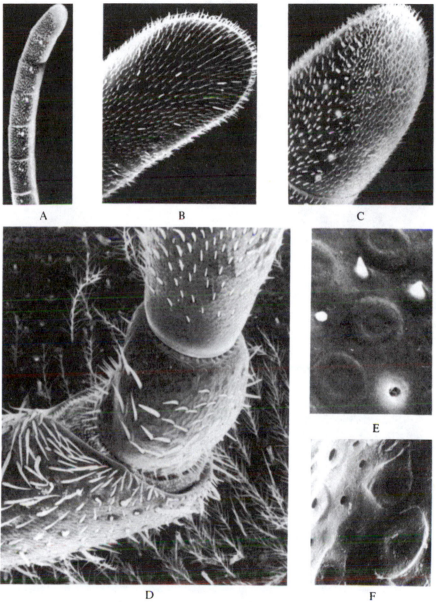

PLATE 6. A. Worker antenna (side view) (X 31). B. Worker antennal tip (top view) (X 149). C. Worker antennal tip (side view) (X 176). D. Close-up of basal segments of worker antenna showing in detail the socket articulation between segments (X 186). E. Close-up of plate (pore plate) and peg organs on the drone antennal surface (X 3,300). F. Close-up of a cluster of drone pit organs (X 2,062).

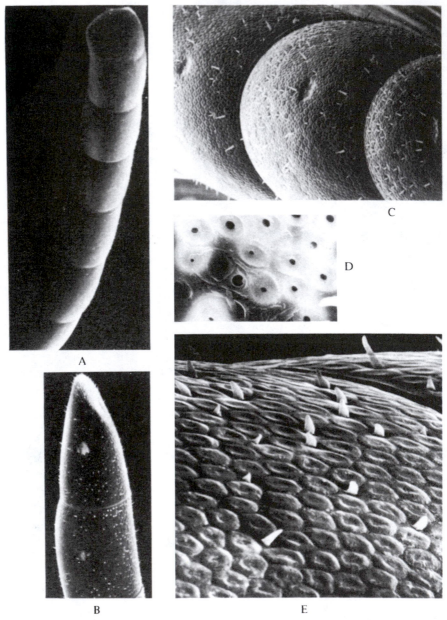

PLATE 7. A. Drone antenna (top view) (X 60). B. Tip of a drone antenna displaying several sensory organs including clusters of pit organs (X 70). C. Enlargement of three antennal segments (X 210). D. Close-up of a cluster of pit organs (X 789). E. Close-up of the high density of plate organs on the drone antenna (X 1,015).

The Proboscis

The proboscis of the bee is not a permanently functional organ as it is in most other sucking insects; it is temporarily improvised by bringing together the free parts of the maxillae and the labium to form a tube for ingesting liquids—nectar, honey, or water. The maxillary and labial components of the proboscis are closely associated at their bases, which are suspended in the ample membrane of the fossa on the back of the head (Fig. 6, *A*). The base of the median labium includes a long, cylindrical distal part termed the *prementum (Prmt)*, and a small, triangular proximal plate, the *postmentum (Pmt)*. The

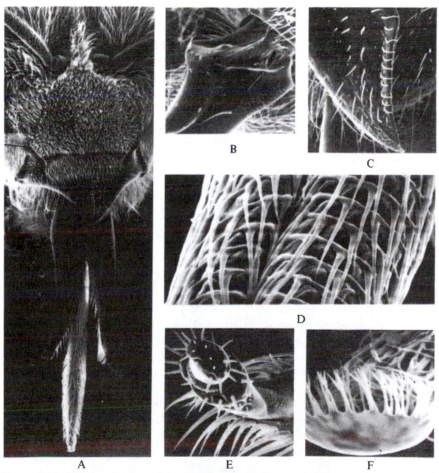

PLATE 8. A. Worker bee mouth parts (X 16). B. Open mandible showing opening of mandibular gland (black arrow and in A. at left) (X 48). C. Tip of a mandible in the closed position (X 53). D. Shaft of the tongue (X 340). E. Tip of labial palp with its numerous touch and taste receptors (X 174). F. Tip of the tongue (X 313).

prementum carries at its end the slender hairy *tongue (Gls)*; a pair of short lobes, the *paraglossae (Pgl)*, embracing the base of the tongue; and a pair of slender *labial palpi (LbPlp)*. Each palpus consists of two long basal segments and two short apical segments, and is individually movable by a muscle arising in the prementum. In each maxilla the principal basal plate (A, *St*) is the *stipes* (plural *stipites*), but the stipes is suspended by a slender rod *(Cd)*, the *cardo* (plural *cardines*), that articulates with a knob on the margin of the proboscis fossa. The distal ends of the two cardines are yoked to the postmentum of the labium by a V-shaped sclerite known as the *lorum (Lr)*. Each stipes carries a long, free, tapering, blade-like lobe, the *galea (Ga)*, and , arising laterally to it, a very small maxillary palpus *(MxPlp)*.

When the proboscis is not in use its basal parts are drawn up behind the head by swinging on the suspending cardines (Fig. 6, H) (Plate 9), while at the same time the distal parts are folded back against the prementum and stipites. When the bee would imbibe liquid, the proboscis is protracted by swinging downward on the cardines, and its distal parts are extended. The broad maxillary galeae and the labial palpi are brought together around the tongue (Fig. 6, B, G) in such a manner as to form a tube (D), closed anteriorly by the overlapping galeae *(Ga)* and posteriorly by the palpi *(LbPlp)*, with the tongue *(Gls)* occupying an axial position and projecting beyond the enclosing parts (G). The two small end-segments of the palpi diverge at the end of the tube and probably have a sensory function (Plate 8E). The tongue now begins a rapid back-and-forth movement, while its flexible tip is swung around with an agile lapping motion. Apparently by the action of the tongue the liquid food is drawn up into the canal of the proboscis (D, *fc*).

The long hairy tongue of the bee is an extension from the end of the labial prementum (Fig. 6, A, *Gls*). It has a closely cross-lined appearance owing to the presence in its wall of hard rings bearing the hairs, separated by narrow, smooth, membranous intervals (Plate 8D). Because of this structure the tongue can be shortened and lengthened. On its base is a bonnet-shaped plate (Fig. 7, A) supported on a pair of arms *(b)* which are extensions of two strap-like bars in the sidewalls of the prementum (D, *a*). The under or posterior side of the tongue is traversed by a deep groove (Fig. 6, A, *sc*) with thin membranous walls, through the middle of which runs a long rod-like thickening (D, *rd*) which is grooved on its free surface. At the tip of the tongue the rod ends in a small spoon-shaped lobe, the *flabellum (A, G, Fbl)*, that has a smooth, rounded under-surface but is armed on the margin and the upper surface by minute branched spines (Plate 8F). Basally the tongue rod curves backward to be firmly attached to the end of the posterior wall of the prementum (Fig. 7, A), and on this curved part are inserted two long muscles (20) arising in the prementum. It is the pull of these muscles on the basal

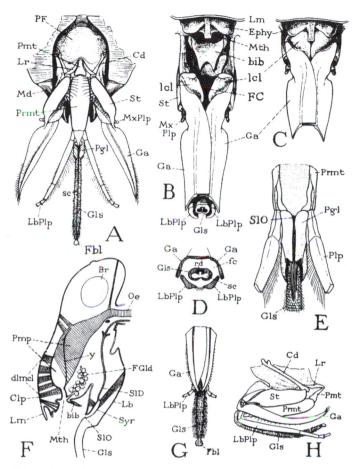

FIGURE 6. The proboscis and the sucking apparatus of a worker bee.

A, proboscis, seen from behind, with its parts artificially spread out. B, front view of base of proboscis pulled down from head, exposing mouth *(Mth)* and deep food channel *(FC)* leading up to mouth. C, same, food channel closed by base of proboscis brought up against the mouth. D, cross section through middle of proboscis. E, base of tongue, paraglossae, labial palpi, and distal end of labial prementum, anterior view. F, lengthwise verticle section of head showing sucking pump *(Pmp)*, salivary syringe *(Syr)*, and associated structures. G, distal part of proboscis with tongue protruded. H, side view of proboscis folded beneath head. *bib,* biblike fold from lower lip of mouth; *Br,* brain; *Cd,* cardo; *Clp,* clypeus; *dlmcl,* dilator muscles of sucking pump; *Ephy,* epipharynx; *Fbl,* flabellum; *fc,* food canal of proboscis; *FC,* food channel on base of proboscis; *FGld,* food gland; *Ga,* galea; *Gls,* tongue (glossa); *Lb,* labium; *LbPlp,* labial palpus; *lcl,* lacinial lobe of maxilla; *Lm,* Labrum; *Lr,* lorum; *Md,* mandible; *Mth,* mouth; *MxPlp,* maxillary palpus; *Oe,* oesophagus; *PF,* proboscis fossa on back of head; *Pgl,* paraglossa; *Plp,* labial palpus; *Pmp,* sucking pump; *Pmt,* postmentum; *Prmt,* prementum; *rd,* rod of tongue; *sc,* salivary canal of tongue; *SlD,* salivary duct; *SlO,* orifice of salivary duct; *St,* stipes; *Syr,* salivary syringe.

curvature of the rod that shortens the tongue (B); extension is due apparently to the elasticity of the rod which straightens again when the muscles relax. Thus are produced the movements of protraction and retraction of the tongue from the end of the proboscis; but since the rod lies close to the posterior margin of the tongue its retraction gives the tongue also a slight backward curvature (B). The lapping motion of the tongue tip evidently is produced by the rod muscles acting separately in opposition to each other.

The food canal of the proboscis leads up into a channel on the base of the proboscis between the bases of the two maxillae and the labium, which appears as an open trough-like cavity when the proboscis is lowered from the head (Fig. 6, B, *FC*). At the upper end of this channel is the *mouth (Mth)* partly hidden behind a large soft lobe, the *epipharynx (Ephy)*, projecting from beneath the labrum *(Lm)*. The lower lip of the mouth is extended in a broad,

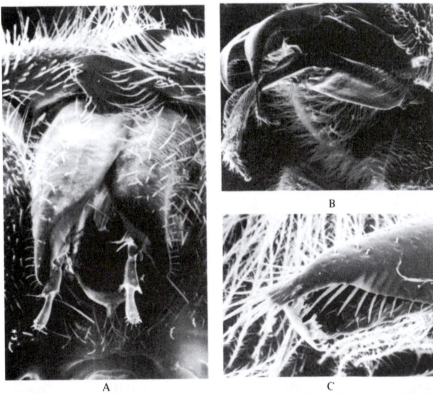

PLATE 9. A. Partially retracted proboscis of the drone as seen from beneath the insect (X 50). B. Partially retracted proboscis of the worker as seen from the side (X 30). C. Retracted tip of the proboscis of the worker. One galea has been removed to expose the remaining mouth parts. Note the lighter areas along the edge of the galea worn by the stiff spines of the opposing galea (X 120).

two-pointed, bib-like fold *(bib)* that hangs down against the labial floor of the food channel (Fig. 8, A, *bib*). In the functional feeding position, the base of the proboscis is drawn up against the mouth (Fig. 6, C), by which act the food channel (B, *FC*) is outwardly closed by the appression of two cushion-like lobes on the maxillae *(lcl)* against the epipharynx *(C, Ephy)*. There thus is established a continuously closed passageway from the tip of the proboscis to the mouth, through which the liquid food is drawn up to the latter. The sucking apparatus, however, is contained within the head.

When, after feeding, the proboscis is drawn up and folded behind the head

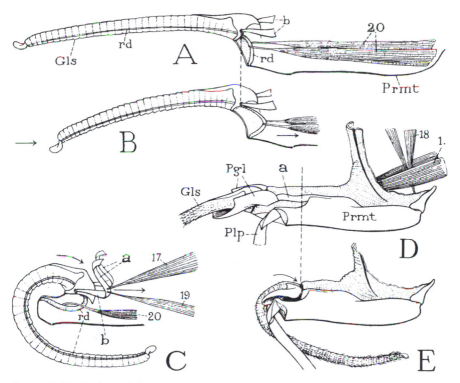

FIGURE 7. Mechanisms of the proboscis of a worker bee.

A, diagram of tongue extended from prementum, showing tongue rod and its muscles. B, tongue shortened by pull of muscles on base of tongue rod. C, tongue retracted and automatically curved backward by pull of muscles *(17, 19)* attached on supporting arms *(a)* of tongue base. D, base of labium, with tongue and paraglossae extended. E, same with tongue and paraglossae retracted and tongue curved back as at C.

a, supporting arms of tongue and paraglossae; *b*, pivotal supports of tongue; *Gls*, tongue; *Pgl*, paraglossae; *Plp*, labial palpus; *Prmt*, prementum; *rd*, flexible rod of tongue; *17, 19*, retractor muscles of tongue and paraglossae; *18*, adductor muscle of labium; *20*, muscles of tongue rod.

(Fig. 6, H) (Plate 9), the tongue appears to be much shorter than before. Its decrease in length is due partly to the contraction of its rings, but it may be noted (Fig. 7, E), by comparison with the extended position (D), that the bases of the tongue and paraglossae *(Pgl)* have been deeply retracted into the end of the prementum. The muscles that retract the proboscis are a pair of long labial muscles (D, *17*) arising on the top of the head and inserted on the distal ends of the premental bars (C, *a*) along with another pair of muscles *(19)* arising in the prementum. When these muscles contract they bend the premental bars sharply inward (C), and the bars bring with them the attached bases of the tongue and paraglossae. The tongue rod *(rd)*, however, is thus pulled so far out of the base of the tongue that its tension automatically curves the tongue backward beneath the prementum. Since there is no mechanism for extending the retracted parts, the reverse action probably results either from the elasticity of the flexed bars or from blood pressure caused by drawing the base of the proboscis close against the head. The flexing of the labial palpi and the maxillary galeae is produced by specific flexor muscles attached on the bases of these appendages.

The Sucking Pump

The sucking apparatus of the bee is a large, muscle-walled sac lying in the head (Fig. 6, F, *Pmp*) extending from the mouth *(Mth)* to the neck foramen, where its narrowed upper-end becomes continuous with the oesophagus *(Oe)*. Each lateral wall of the pump is traversed obliquely by a slender rod *(y)*, extending upward from a plate on the floor of the mouth (Fig. 8, B, *opl*). In other insects these rods and the oral plate pertain to a large tongue-like lobe known as the *hypopharynx* projecting between the upper ends of the rods and opens into the *pharynx,* which is the first part of the alimentary canal. The preoral food cavity before the hypopharynx is the *cibarium.* In the bee, therefore, the sucking pump is a combination of the preoral cibarium and the postoral pharynx. The hypopharynx of the bee is represented by the oral plate, the bib-like fold (Fig. 6, F, *bib*) hanging from the latter, and the infolding ending at the opening of the salivary duct *(SlD)* into the so-called salivary syringe *(Syr)*.

The functional mouth of the bee *(Mth)* is simply the opening into the cibarium between the labrum and the hypopharynx. The cibarium is the operative part of the pump, five pairs of thick bundles of dilator muscle fibers *(dlmcl)* from the clypeus being attached on its anterior wall. Between these muscles (not shown on the figure) are strong compressor fibers running obliquely crosswise on the cibarial wall. Liquids are sucked up from the canal of the proboscis by the action of the dilator muscles; contraction of the compressor muscles then closes the mouth and drives the liquid into the muscular pharynx, from which it is driven into the narrow oesophagus.

Inasmuch as regurgitation of nectar and honey is an important function of the bee's feeding apparatus, it is probable that the pump can serve for both ingestion and egestion.

The Salivary System

Between the root of the tongue and the distal end of the labial prementum anteriorly is a deep depression (Fig. 6, E, *SlO*), mostly concealed by the overlapping paraglossae *(Pgl)*. At the bottom of this depression is an opening that leads into a small pocket of the prementum (F, *Syr*). By exposing the pocket, it is seen that its walls are provided with dilator and compressor muscles, and that into its inner end opens the common duct of the salivary glands *(SlD)*. This apparatus is a pump for the ejection of the saliva and may be termed the *salivary syringe.*

The saliva is secreted by two pair of glands discharging finally into one median duct. The glands of one pair lie in the back of the head, those of the other pair in the ventral part of the thorax. The *thoracic glands* (Fig.8, C, *ThGld*) consist of masses of elongate or tubular saccules at the ends of branching ducts (E) that lead into a pair of reservoir sacs (C, *Res*). From the reservoirs two ducts go forward and unite just behind the head in the common median duct *(SlD)* that enters the neck foramen of the head and empties into the salivary syringe *(Syr)*. The *head glands* (C, *HGld*) are flat masses of small pear-shaped bodies (D) spread over the posterior head wall. Their ducts unite within the head with the common duct from the thoracic glands. The thoracic glands are developed from the silk glands of the larva, and correspond with the usual salivary glands of other insects; the head glands are developed in the pupa as outgrowths from the common salivary duct.

From the salivary syringe the saliva is ejected into the cavity on the labium at the root of the tongue, but it is here confined by the overlapping paraglossae (Fig. 6, E, *Pgl*), and within the latter is conveyed around the base of the tongue into the channel-like groove on the posterior or undersurface of the tongue (A, D, *sc*). Through this channel probably it is conducted to the tip of the tongue where it flows out over the smooth undersurface of the flabellum (A, *Fbl*) to mix with the nectar or honey being drawn into the proboscis, or is used as a solvent if the bee is feeding on sugar.

The Brood-Food Glands

The glands of the worker bee that produce the food material called "royal jelly," which is fed to the queen, drones, and larvae, are two long strings of small saccules closely packed in many loops and coils in the sides of the head (Fig. 8, A, *FGld*). The axial ducts open separately by two small pores on the lateral angles of the oral plate on the floor of the mouth (B, *opl*). Since this plate belongs to the hypopharynx, the food glands are *hypopharyngeal* and not "pharyngeal" glands as they have long been called. The rodlike arms *(y)* of the oral plate give attachment to muscles from the head wall, and probably have some function in discharging the royal jelly from the mouth. The food evidently must run down the bib-like flap that hangs from the edge of the oral plate (A, B, *bib*) and accumulate *(A, RJ)* in the food channel on the base of the proboscis. The open channel thus serves as a feeding trough for other adult bees, which obtain the royal jelly by thrusting the end of the proboscis over the base of the tongue of the feeder bee, the proboscis of the latter being turned back, the mandibles opened, and the labrum raised. When the nurse bees feed the larvae, however, the royal jelly is said to be discharged from between the partly opened mandibles.

THE THORAX, THE LEGS, AND THE WINGS

The thorax of an insect is the middle division of the trunk which carries the legs and wings. Its cavity is largely occupied by the muscles of the locomotor appendages and the muscles that move the head and the abdomen, the other internal organs being mostly contained in the head and the abdomen (Fig. 17, A). The nerve centers of the thorax, however, are particularly large (Fig. 20) because they control the activities of the thoracic muscles.

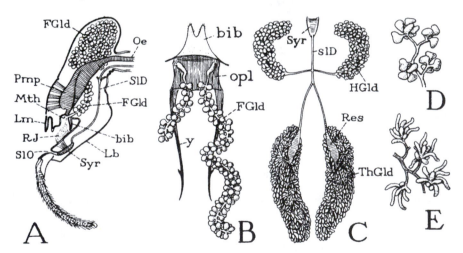

FIGURE 8. Glands of the head and thorax of a worker bee.

A, vertical section of head showing food gland *(FGld)* of right side. B, under surface of oral plate, showing openings of food glands. C, general view of the salivary system, including head glands *(HGld)*, thoracic glands *(ThGld)*, ducts, and salivary syringe *(Syr)*. D, detail of head gland. E, detail of thoracic gland.

bib, biblike fold from lower lip of mouth; *FGld,* food gland; *HGld,* head salivary gland; *Lb,* labium; *Lm,* labrum; *Mth,* mouth; *Oe,* oesophagus; *opl,* oral plate on floor of mouth; *Pmp,* sucking pump; *Res,* reservoir of thoracic gland; *RJ,* royal jelly; *SlD,* salivary duct; *SlO,* orifice of salivary duct; *Syr,* salivary syringe; *ThGld,* thoracic salivary gland; *y,* arm of oral plate.

Structure of the Bee's Thorax

The thorax of the bee and related Hymenoptera, as already noted, consists of four body segments which, beginning with the first, are designated the *prothorax* (Fig. 4, 1), the *mesothorax* (2), the *metathorax (3),* and the *propodeum* (I), but the several segments are so closely united that it is difficult to observe their limits. In studying a thoracic segment we distinguish a back plate or *notum,* a ventral plate or *sternum,* and a plate or group of plates on each side called the *pleuron.*

The prothorax of the bee is merged with the neck to form a slender support for the head, and carries the first pair of legs. Its back plate, the *pronotum* (Fig. 9, N_1), is like a collar on the front of the mesothorax, and is expanded on each side in a flat lobe *(spl)* that covers the first pair of breathing apertures. The pleural and sternal plates of the prothorax support the first legs *(L₁,),* and the head is pivoted on a pair of peg-like processes (a) projecting from the anterior ends of the pleura. The mesothorax is the largest part of the thorax. The mesonotum *(N₂)* lies above the wing bases *(W₂)* forming the uppermost bulge of the thoracic wall and sloping steeply downward to the pronotal collar. Below the wings the pleural and sternal walls of the segment *(Pl₂, S₂)* are continuous from one side to the other. The metathorax is a narrow band

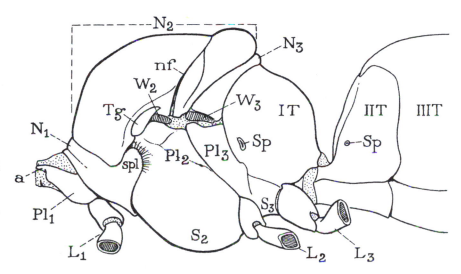

FIGURE 9. The thorax and base of the abdomen, left side, of a worker bee.

a. pivotal point supporting head; *I T,* back plate of propodeum; *II T, III T,* back plates (terga) of first and second abdominal segments; L_1, L_2, L_3, bases of legs; N_1, pronotum; N_2, mesonotum; N_3, metanotum; *nf,* notal fissure; Pl_1, pleuron of prothorax; Pl_2, pleuron of mesothorax; Pl_3, pleuron of metathorax; S_2, S_3, sternal areas of mesothorax and metathorax; *Sp,* spiracle; *spl,* lobe of pronotum covering first spiracle; *Tg,* tegula; W_2, W_3, bases of wings.

angularly bent forward on the sides, closely wedged between the mesothorax and the propodeum. The metanotum *(N₃)* widens somewhat toward the wing bases *(W₃)*; the pleural plates *(Pl₃)* are continuous with the sternum *(S₃)*, as in the mesothorax. The fourth thoracic segment, or propodeum, consists mostly of a large back plate *(IT)*, firmly united with the metathorax. It has no pleural elements and its sternum is a weak ventral plate behind the third legs. Posteriorly the propodeum is abruptly narrowed to give attachment to the abdominal petiole. Further details of the thoracic structure will be described in connection with the wings and their mechanism.

The Legs of the Bee

The three pair of legs of an insect are seldom alike in size or shape, but each is divided into six principal parts or *segments,* movable on each other at flexible *joints* (Fig. 10, A). The basal leg segment is the *coxa (Cx)*; the second segment is the *trochanter (Tr)*; the third, usually a long segment, is the *femur (Fm)*; the fourth is the *tibia (Tb)*; the fifth is the *tarsus (Tar)*; and the last is the *pretarsus (Ptar)*. The tarsus, however, is subdivided into several small parts or *tarsomeres.* The pretarsus is a very small segment but it carries a pair of lateral *claws (E, Cl)* and a median lobe termed the *arolium (Ar)*.

The joints between the leg segments are mostly hinges with motion limited to one plane; no part of an insect's leg having anything like the freedom of movement at the joints of a vertebrate limb as the human arm. As a consequence the insect has little choice as to what it can do with its legs, and hence, all individuals of a species do the same things in practically the same way. The limitation of action at the joints is partly compensated by the number of segments that move in different directions.

Each leg of the bee is hinged to the body on an obliquely transverse axis and, therefore, swings as a whole only forward and backward. At the first leg joint, that between the coxa and the trochanter (Fig. 10, A), the part of the leg beyond the coxa turns up and down in a plane at right angles to the plane of movement of the entire leg on the body. The muscles of the trochanter hence raise and lower the leg at the coxotrochanteral joint; but if the feet are against a support the contraction of the depressor muscles lifts the body on the legs. The trochanter has several small muscles arising in the coxa, but, to increase the lifting power, each trochanter is provided with a long muscle arising on an internal framework within the thorax (Fig. 10, C, *86*). The trochanter is joined to the femur by an oblique hinge in a vertical plane which thus does not interfere with the lifting power of the trochanteral muscles, though it gives only a slight backward bend to the femur. The articulation between the femur and the tibia is a typical knee joint, at which the tibia can be extended or flexed on the end of the femur by long extensor and flexor muscles arising in the femur.

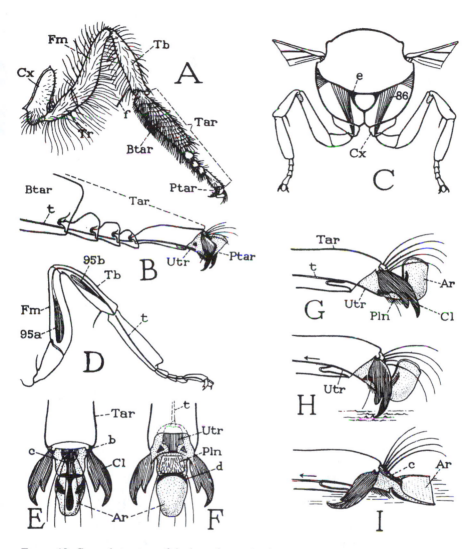

Figure 10. General structure of the legs of a worker bee.

A, middle leg of worker. B, tarsus and foot (pretarsus) more enlarged. C, cross section of mesothorax, with middle legs. D, outline of middle leg, showing muscles and tendon of pretarsus. E, pretarsus, upper surface. F, same lower surface. G, diagram of foot (pretarsus) with claws extended. H, same, with claws grasping a rough surface. I, same, arolium spread out on a smooth surface on which claws fail to hold.

Ar, arolium; *b,* articular knob of claw; *Btar, basitarsus; c,* handle-like bar of arolium braced on end of tarsus; *Cl,* claw; *Cx,* coxa; *d,* elastic band in under wall of arolium; *e,* internal framework of thorax; *f,* spine of tibia; *Fm,* femur; *Pln,* planta; *Ptar,* pretarsus (foot); *t,* tendon of pretarsal muscles, *Tar,* tarsus; *Tb,* tibia; *Tr,* trochanter; *Utr,* unguitractor plate.

The tarsus of the bee (Fig. 10, A, B, *Tar*) consists of five parts, or tarsomeres, of which the first is much longer and thicker than the others in all the legs, and is distinguished as the *basitarsus (Btar)*. The large, flat basal subsegment of the hind tarsus (Fig. 11, C, *Btar*) is commonly called the "planta" by writers on the honey bee, but the word *Planta* in Latin means "the sole of the foot," and the term is so used in general zoology. In insects, the planta is properly a small ventral sclerite of the pretarsus (Fig. 10, F, *Pln*). The tibiotarsal joint differs from the other leg joints in that it allows more freedom of movement to the tarsus, which has three muscles attached on the basitarsus that give three separate movements. The large basitarsus is followed by three very small tarsomeres, and the fifth longer tarsomere carries at its extremity the pretarsus *(Ptar)*. Between the subsegments of the tarsus (B) there are no muscles, the tarsomeres being flexible on each other but having no power of individual movement. The entire tarsus, however, is traversed by the tendon (B, *t*) of the pretarsal muscles arising in the femur and the tibia (D, *95a, 95b*).

The pretarsus, which might be termed the "foot" of the insect, is a very important part of the leg since it bears the organs by which the insect clings to supporting surfaces (Plate 10). The pretarsal claws (Fig. 10, E, F, *Cl*) maintain a hold on rough surfaces; the arolium *(Ar)* adheres to smooth surfaces. The claws of the bee are double pointed and their deep bases are implanted vertically in the lateral walls of the pretarsus, but each claw is articulated to a small knob (E, *b*) on the end of the tarsus. The arolium (Plate 10, B, C) projects from the end of the pretarsus between the claws. When not in use (G, H) it is turned upward and appears to be merely a soft, oval lobe, though closer inspection shows that it is deeply concave on its upper or anterior surface (E), that is, its sides are folded together upward. The basal lip of the aroliar cavity is braced against the end of the tarsus by a bottle-shaped sclerite *(s)* on the upper wall of the pretarsus bearing five or six long curved bristles. The arolium, thus, resembles a scoop with a long handle. Its convex outer wall contains a U-shaped elastic band (F, *d*).

In the lower wall of the pretarsus (Fig. 10, F) are two median plates; the stronger proximal one, which is partly concealed in a pocket at the end of the tarsus, is the *unguitractor plate (Utr)*; and the distal one, covered with strong spines, the *planta (Pln)*. The concealed proximal end of the unguitractor plate is connected with the end of a strong internal tendon *(t)* that runs through the entire tarsus (D) into the tibia where it divides into two branches; one branch gives attachment to a muscle in the tibia *(95b)*, and the other goes on into the femur and ends in a long muscle arising in the base of the femur *(95a)*. These muscles operate both the claws and the arolium by their pull on the tendon and the attached unguitractor plate.

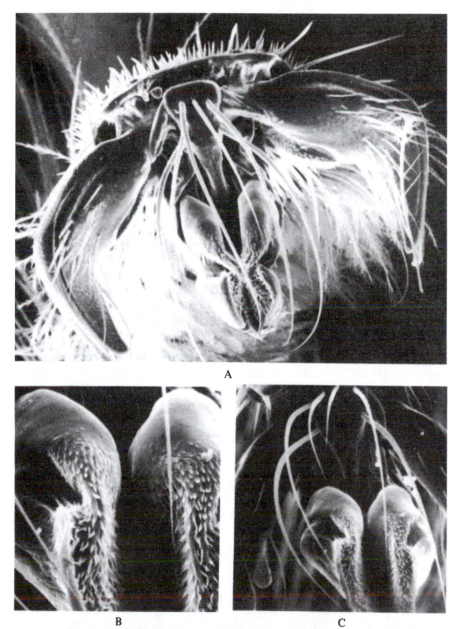

A

B   C

PLATE 10. A. Last tarsal segment (foot) of the hind leg of the drone with its numerous sensory hairs for taste and touch (X 134). B. Expanded arolium of the queen used for walking on smooth surfaces (X 284). C. Queen arolium in its normal position between the tarsal claws (X 107).

The mechanism of the foot structure is illustrated diagrammatically at G, H, and I of Fig. 10. At G the claws are extended and the arolium *(Ar)* turned upward in the usual position. At H the unguitractor plate *(Utr)* has been retracted by its muscles into the end of the tarsus, and, as the plate is closely associated with the bases of the claws, the claws are flexed until irregularities of the surface of contact restrain their points. In this way the bee is enabled to cling to rough surfaces. At I the bee is supposed to be resting on a smooth surface on which the claws have not been able to hold; the continued pull of the muscles, therefore, has turned the claws so far forward that they sprawl out helplessly with their points upward. But now the traction of the unguitractor plate is exerted on the planta and finally on the base of the arolium. The aroliar handle *(c)*, however, being braced against the tarsus, prevents a retraction of the arolium, but the tension and pressure on the base of the arolium, together with the pressure of the leg against the support, flatten the aroliar scoop to the form of a dustpan. The broad, soft undersurface of the spread-out arolium now adheres to the smooth surface which the claws failed to grasp. It has been said that the adhesive property of the arolium is due to a sticky liquid exuded from the spines of the planta. On release of the muscle pull on the unguitractor plate, the elastic band in the under wall of the arolium causes the latter to fold up again, and the claws, by the elasticity of their basal connections, are once more extended.

Though the legs are primarily organs of locomotion, various specialized parts of the legs of the honey bee serve for purposes other than that of walking or running. The brushes of stiff hairs on the inner surfaces of the long basal segments of the anterior tarsi (Fig. 11, I, g) are used for cleaning pollen or other particles from the head, the eyes, and the mouth parts. Similarly the bushy middle tarsi (Fig. 10, A) serve as brushes for cleaning the thorax. The long spines at the ends of the middle tibiae *(f)* are said to be used for loosening the pellets of pollen from the pollen baskets of the hind legs, and also for cleaning the wings and the spiracles (Plate 11). The wax scales are removed from the wax pockets of the abdomen by means of the legs, but there is some difference of opinion as to just how it is done. The special structures of chief interest on the legs, however, are the antenna cleaners on the forelegs of all castes, and the pollen-collecting and pollen-carrying apparatus on the hind legs of the worker bees.

The Antenna Cleaner

The structures used by the bee for cleaning its antennae are situated on the inner margins of the forelegs just beyond the tibiotarsal joints (Fig. 11, I, *h*). Each antenna cleaner consists of a deep semicircular notch on the basal part of the long basitarsus *(A, i)*, and of a small clasp-like lobe *(j)* that projects over the notch from the end of the tibia (Plate 12). The margin of the notch is fringed

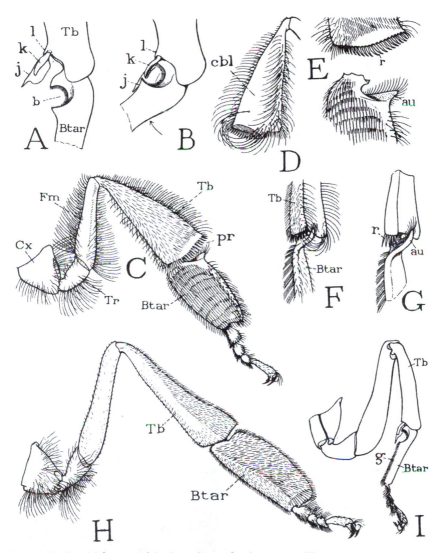

FIGURE 11. Special features of the legs of a worker bee (except H).

A, antenna cleaner of first leg, open. B, same, closed. C, hind leg of worker, inner surface, showing pollen-collecting brush on basitarsus *(Btar)* and pollen press *(pr)*. D, pollen basket (corbicula) on outer surface of hind tibia (C, *Tb*). E, end of hind tibia with pollen rake *(r)* and opposing end of basitarsus with auricle *(au)*. F, pollen press between tibia and basitarsus, dorsal view. G, same, better seen after removal of tibial hairs. H, hind leg of drone. I, first leg of worker, showing position of antenna cleaner.

au, auricle; *Btar,* basitarsus; *cbl,* pollen basket; *Cx,* coxa; *Fm,* femur; *g,* tarsal brush of first leg; *h,* antenna cleaner; *i,* notch of antenna cleaner; *j,* clasp of antenna cleaner; *k,* lobe of clasp; *l,* stop-point for clasp; *pr,* pollen press; *r,* pollen rake; *Tb,* tibia.

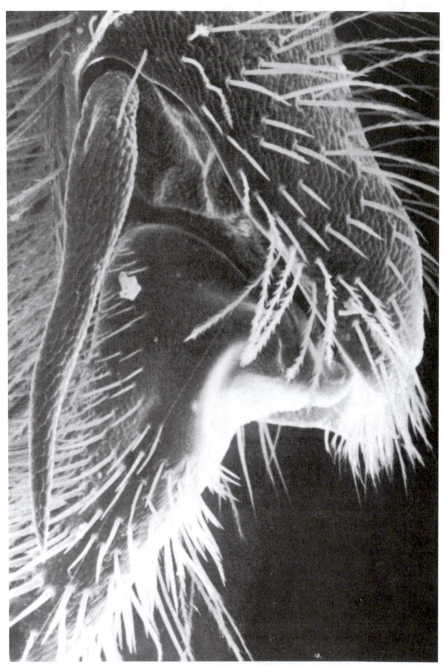

PLATE 11. Tibial spine on the middle leg of the drone (X 355).

with a comb-like row of small spines. The clasp is a flattened appendage, tapering to a point and provided with a small lobule *(k)* on its anterior surface; it is flexible at its base but has no muscles. As this gadget is used by the bee, the open tarsal notch is first placed against the antennal flagellum by appropriate movements of the leg; then by flexing the tarsus against the tibia (B), the flagellum is brought against the tibial clasp which resists the pressure because of the small stop-point *(l)* behind its base. The flagellum, thus held in the notch by the clasp, is now drawn upward between the comb of the notch and the scraping edge of the clasp. The antenna cleaner is present in the queen and the drone as well as in the worker.

The Pollen-Collecting Apparatus and the Pollen Baskets

The hind legs differ from the other legs in their large size and the broad, flattened form of the tibia and basitarsus (Fig. 11, C), which latter parts differ also as between the worker (C), the queen, and the drone (H). It is only in the worker that there is any evident reason for the special shape of the hind legs (Plate 13). The smooth, somewhat concave, outer surface of the hind tibia of the worker is fringed with long, curved hairs (D), and the space thus enclosed is the so-called *pollen basket* or *corbicula (cbl)* in which pollen (also propolis) is carried to the hive (Plate 13 A, B). The pollen stored in the baskets is first collected from the body by the fore and middle legs and deposited on the large flat brushes (combs) on the inner surfaces of the broad basal segments of the hind tarsi, each of which is covered with about ten transverse rows of stiff spines projecting posteriorly (C, *Btar*) (Plate 13 E). The apparatus for transfer of the pollen from the brushes to the baskets are the deep notches in the upper margins of the legs between the tibiae and the tarsi *(pr)*. The tibial margin of each notch is armed with a *rake* of short, stiff spines (E, *r*); the opposing tarsal margin is flattened transversely and extended laterally into a small triangular lip, or *auricle (au)*, fringed with hairs (Plate 13 C, D, E).

The transfer of the pollen from the collecting brushes to the baskets is accomplished as follows: When the basitarsal brushes are sufficiently loaded with pollen, the leg of one side is rubbed against the other in such a manner that the rake on the end of the tibia scrapes off a small mass of pollen from the tarsal brush of the opposite leg. The detached pollen grains fall on the flat surface of the auricle which is beveled upward and outward. Consequently, when the tarsus is closed against the tibia (Fig. 11, F, G), the pollen on the auricle is forced upward and pressed outward against the outer surface of the tibia where, being wet and sticky, it adheres to the floor of the pollen basket. A repetition of this process, first on one side, then on the other, successively packs more pollen into the lower ends of the baskets, until finally both are filled.

The pollen baskets are used also for the transport of propolis, but the pollen presses play no part in loading the baskets in this case. Propolis is a resinous gum collected by bees with their mandibles from trees or other plants. The resin particles, it is said, are gathered up with the fore and middle legs and placed directly in the baskets of the hind legs.

The Wings of the Bee

The wings of an insect are flat, thin, two-layered extensions of the body wall, strengthened by tubular thickenings called *veins*. They arise from the sides of the mesothorax and the metathorax between the notal and pleural plates of these segments (Plate 13 A). In the honey bee the fore wings (Fig. 12, A) are much larger than the hind wings (B) and their venation (Plate 13 D) is stronger, but the two wings of each side work together in flight. To insure unity of action the wings are provided with a coupling apparatus formed by a series of upturned hooks on the front margin of each hind wing (B, *h*), and a decurved fold on the rear margin of the fore wing (A, *f*). When the wings are extended preparatory to flight, the fore wings are drawn over the hind wings, and the hooks of the latter automatically catch in the marginal folds of the former (E) (Plate 13 B, C).

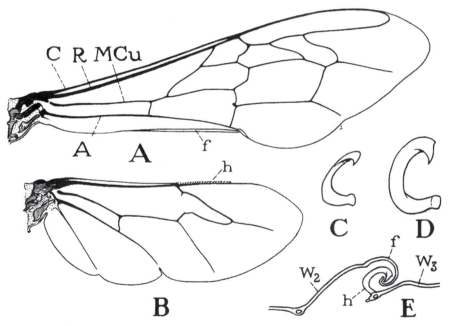

FIGURE 12. The wings of a worker bee.

A, right fore wing of worker. B, right hind wing. C, hind wing hook of worker. D, wing hook of drone. E, section of fore and hind wings showing interlocking by fold and hooks.

A, anal vein; *C,* costal vein; *f,* marginal fold of fore wing; *h,* marginal hooks of hind wing; *R,* radial vein; *MCu,* median and cubital veins united; W_2, fore wing; W_3, hind wing.

Each wing is hinged by its narrowed base to the margin of the back plate of its segment, and is supported from below on the upper edge of the corresponding pleuron. The wings are thus free to move up and down; but progressive flight requires other movements, including a forward and backward motion of each wing, and a twisting or partial rotation of the wing on its long axis. These latter components of the flight movement depend on details of structure in the wing bases. All the wing movements, except such as may result from air pressure during flight, are produced by thoracic muscles; but most of the muscles involved are attached not on the wings themselves, but on movable parts of the thorax that indirectly affect the wings (Plate 15). Hence, for an understanding of the flight mechanism, we shall have to make a further study of the thoracic structure. Moreover, the wings when not in active use are folded horizontally backward over the abdomen. From this position of rest they can quickly be extended to the position of activity. In addition to the flight mechanism, therefore, we must distinguish a mechanism of *flexion* and *extension* for each wing individually. It will be logical to give attention first to the structures that produce these horizontal movements.

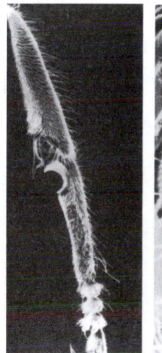
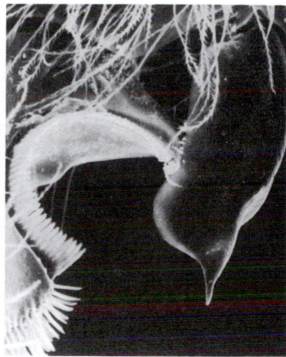

A   B

PLATE 12. A. Drone fore leg with antenna cleaner (comb) (X 20). B. Comb close-up (X 152).

If the base of one of the wings, a fore wing for example (Fig. 13, A), is spread out flat it will be seen that it contains several small plates or *axillaries*. Two of the axillaries, the *first (1Ax)* and the *fourth (4Ax)*, are the hinge plates by which the wing articulates on the edge of the notum. Another, the *second axillary (2Ax)* lying behind the first, rests on the upper edge of the pleuron (B, Pl_2) and constitutes the pivotal plate of the wing base. A third long sclerite (A, *3Ax*) extends outward along the thickened margin of the basal wing membrane, and this axillary is the skeletal element of the flexing mechanism. On its proximal part are attached three small muscles (F) arising on the pleuron.

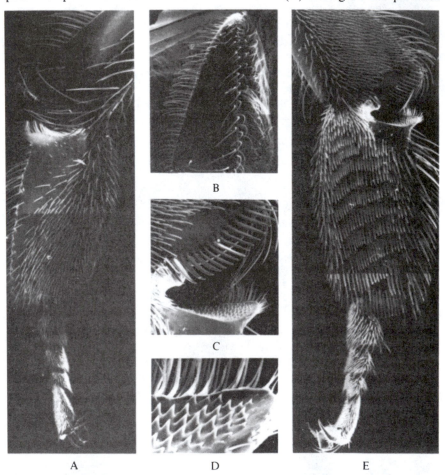

A   D   E

PLATE 13. A. Outer surface of the worker hind leg showing the pollen press and the pollen basket at the top (X 20). B. The pollen basket (X 15). C. The pollen press and rastellum (X 48). D. The floor of the pollen press (X 200). E. Inner surface of the worker hind leg showing the rastellum, pollen press and pollen comb (X 23).

These muscles in contraction lift the outer end of the third axillary and revolve it toward the back, producing necessarily a fold in the wing base which causes the extended wing to turn horizontally backward. The flexor action of the third axillary can be well illustrated with a piece of paper cut and creased as at D of Fig. 13. By lifting the point *d* at the outer end of the "axillary" *(3Ax)* and revolving it upward to the left *(E)*, the base of the "wing" turns with it along the line *bc*; and the distal part folds horizontally backward along the line *ab*, as the triangle abc turns over.

The extension of the flexed wing is caused by the movement of a small sclerite resting on the pleuron beneath the front part of the wing (Fig. 13, B, *Ba*) and connected with the latter by a tough membrane. On this sclerite, called the *basalare*, is attached a long muscle (G, I, 77) from the lower part of the pleuron, which by contracting turns the basalare inward on the pleuron (H, *Ba*), and thus pulls indirectly on the wing base before the pivotal second

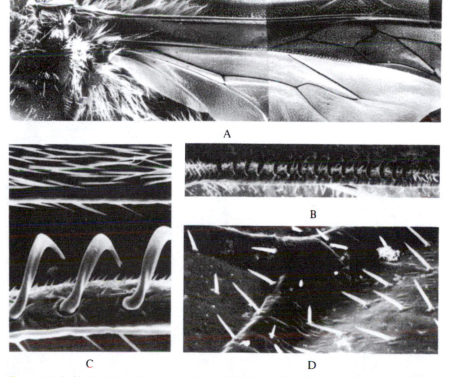

A

B

C   D

PLATE 14. A. Undersides of the worker fore and hind wing showing wing veination and hooks on the leading edge of the hind wing. (X 13). B. Enlargement of the zone with wing hooks (queen) (X 285). C. Close-up of three wing hooks (queen) with opposing trailing edge of fore wing (X 81). D. Close-up of the queen wing surface showing veins and surface spines (X 242).

axillary, with the result that the wing is swung forward on the latter. The action is easily demonstrated by pressing the point of a needle against the basalare.

The flight movements of the wing, as already stated, are compound. They include an up-and-down component, a forward-and-backward component, and a torsion, or partial rotary movement of the wing on its long axis. The up-and-down strokes are caused by vibrations of the back plate of the wing-bearing segment. Because the two wings of a segment are supported from below on the pleura, a depression of the back plate causes the wings, in the manner of a pump handle, to go up (Fig. 14, A), as can be demonstrated by pressing downward on the back of a bee. A reverse movement of the back (B) depresses the wings. The major part of the wing mechanism produces the vibrations of the back plates.

If the thorax of a bee is cut open (Fig. 14, D, E) it will be seen that it is almost filled with great masses of muscle fibers. In each side of the mesothorax is a thick column of vertical fibers (72) attached dorsally on the notum, and

PLATE 15. Complex of plates at the base of wings (wings removed, worker) (X 56).

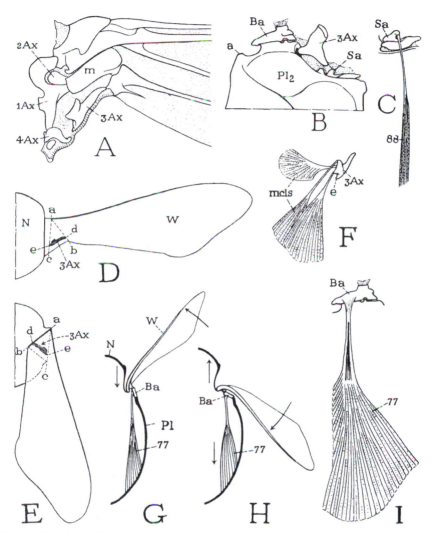

FIGURE 13. Details of the wing structure, and wing mechanisms.

A, base of fore wing, flattened, showing axillary sclerites and bases of veins. B, upper part of left mesopleuron supporting basalar sclerite *(Ba)*, second axillary *(2Ax)*, and subalar sclerite *(Sa)*. C, subalar sclerite and its muscle. D, diagram of extended wing and lines folding in base. E, diagram of wing turned horizontally over back. F, third axillary (flexor sclerite) of fore wing and its muscles. G, diagram of raised wing with front margin elevated. H, same, wing lowered, front margin depressed by contraction of basalar muscle *(77)*. I, basalar sclerite of fore wing and its muscle.

a-b, b-c, lines of folding in wing base; *1Ax, 2Ax, 3Ax, 4Ax*, first, second, third, and fourth axillary; *Ba*, basalar sclerite; *d*, outer end of third axillary; *m*, median plate of wing base; *Mcls*, muscles; *N*, notum; *Pl*, pleuron; *Sa*, subalar sclerite; *W*, wing.

another smaller muscle (E, *75*) is attached on the margin of the notum. These muscles are the depressors of the back and, therefore, the elevator muscles of the wings (A). Between the first pair are two flat bundles of fibers (D, E, *71*) running obliquely lengthwise from the median area of the mesonotum to a strong, internal U-shaped band, the *second phragma (D, 2Ph)*, extending from the mesonotum far back into the propodeum. These are the *depressor muscles of the wings* because their contraction compresses the mesonotum in a lengthwise direction, and hence elevates the back, causing the wings to turn downward (B). It will be recalled that the two wings on each side are hooked together during flight. The mesothorax has wing-elevator muscles of its own, but the downstroke of both pair of wings is produced by the mesothoracic muscles and depends on the coupling of the fore and hind wings with each other.

In most insects the back plates of the wing-bearing segments are sufficiently flexible to respond by vibratory movements to the alternating pull of the vertical and lengthwise muscles attached on them. The mesonotum of the bee, however, is a rigid and strongly convex plate. In order to perform its function in connection with the wings it is cut by a deep, crosswise groove (Fig. 14, D, F, *nf*) into a larger anterior plate, (F, *1N$_2$*) and a smaller posterior plate *(2N$_2$)*. The middle part of the groove on the top of the back acts as a hinge between the two plates; the lateral parts open out into actual clefts (G, *nf*) having the edges united by infolded membranes. The front angles of the anterior notal plate are firmly braced on the pleura (as indicated by the arrow at *a*); the posterior notal plate is supported on the metanotum *(N$_3$)*. Contraction of the vertical muscles (G, *72*), therefore, depresses the mesonotum at the hinge line *(d)* on the back and opens the lateral clefts *(nf)* between the two notal plates. Conversely, the contraction of the lengthwise muscles (F, *71*) restores the notum to its original shape by closing the lateral clefts. The opening and closing of the clefts, however, necessarily is accompanied, respectively, by a downward and upward movement of the adjoining tergal margins *(e, f)*, as is seen by comparing their positions in F and G relative to the line *g*. The action of the mesonotum may be exactly imitated by compressing half of a hollow rubber ball having a meridional cleft on each side (C). Each wing, being attached to the margin of the back just before and behind the notal cleft, is thus hinged at the points of greatest vertical movement in the notal margin, and it is this movement that causes the up-and-down wing strokes during flight.

A mere flapping of the wings cannot produce flight; the driving force results from a propeller-like twist given to each wing during the upstroke and the downstroke. As the wing descends it goes also somewhat forward and its anterior margin turns downward; during the upstroke the action is reversed.

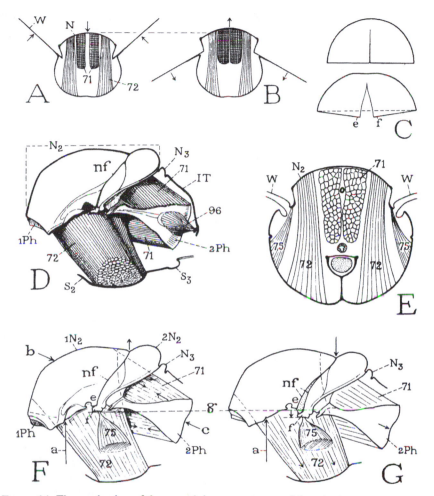

FIGURE 14. The mechanism of the up-and-down movement of the wings.

A, cross section of mesothorax with wings elevated. B, same, with wings depressed. C, diagrams of movements of mesonotum that depress and elevate the wings. D, thorax with left wall removed showing muscles. E, cross section of mesothorax through wing bases. F, diagram of position of back plates of mesothorax $(1N_2, 2N_2)$ when muscle 71 is contracted. G, same when muscles 72 and 75 contract, the notal fissure (nf) opened on the side, the marginal points e and f lowered. (compare F and G with C).

a, point of support of mesonotum on pleuron; b-c, direction of contraction of muscle 71; d, hinge line on back between plates of mesonotum; e, f, points of greatest movement on margin of mesonotum as notal fissure (nf) opens and closes; g, horizontal line; IT, back plate of propodeum; N, *notum*; N_2, mesonotum, divided by notal fissure (nf) into anterior plate $(1N_2)$ and posterior plate $(2N_2)$; N_3, mentanotum; nf, notal fissure; $1Ph$, first phragma; $2Ph$, second phragma; S_2, mesosternum; S_3, metasternum; W, wing; 71, lengthwise muscles of mesothorax (depressors of wings); 72, 75, vertical muscles of mesothorax (elevators of wings).

The tip of the vibrating wing, therefore, if the insect is held stationary, describes a figure 8. The mechanism that produces these components of the flight movement includes the basalar sclerite and its muscle (Fig. 13, B, I, *Ba*), already described as causing the extension of the flexed wing; and a similar musculated sclerite (C, *Sa*) designated the *subalare*, resting on the pleural margin beneath the posterior part of the wing base (B, *Sa*). The alternating pull of the basalar and subalar muscles on their respective sclerites not only turns the wings during flight forward and backward, but at the same time deflects their forward margins during the downstroke and reverses the movement during the upstroke. The action of the basalare is illustrated diagrammatically at G and H of Fig. 13. At G the wing is raised by the depression of the notum *(N)*, and its rear margin is turned downward by the pull of the subalare (not shown) on the rear part of its base. At H the wing is in the downstroke position caused by the elevation of the notum; but it is also turned forward, with its front margin deflected by the pull of the basalar muscle (77) on the basalar plate *(Ba)*, which latter in turn pulls downward and forward on the anterior part of the wing base.

The efficiency of the insect flying machine is most surprising, considering the simplicity of the flight mechanism. It is to be observed that the wings not only act as organs of propulsion, but also direct the course of flight as the insect has no other apparatus for steering. Many insects, moreover, can change abruptly the direction of flight without altering the position of the body, going forward, backward, or sideways with equal ease, while finally they can remain stationary, hovering at one point in the air. No airplane yet invented can perform in this manner.

THE ABDOMEN

The abdomen contains the principal viscera of the insect (Fig. 17, A), such as the stomach, intestine, and reproductive organs, and bears externally the structures concerned with mating and egg laying. Its general outer form is simple as compared with that of the thorax or head (Fig. 4, *Ab*) and its component segments are nearly always distinct.

The Abdomen of the Bee

The bee larva has ten abdominal segments; but in the adult bee and other related Hymenoptera the abdomen is reduced to nine segments by the transfer, during the pupal stage, of the first larval segment to the thorax. In order to keep the correspondence, or *homology*, of the segments in mind, it is customary to number the segments beginning with the transposed first segment, or propodeum, as segment *I*. The abdomen of the bee, however, is further shortened by a reduction and retraction of some of the posterior segments. Thus, in the worker and the queen the abdomen appears to have only six segments (Fig. 4, 15, A), which are segments *II* to *VII*, the tergal and sternal plates of the last

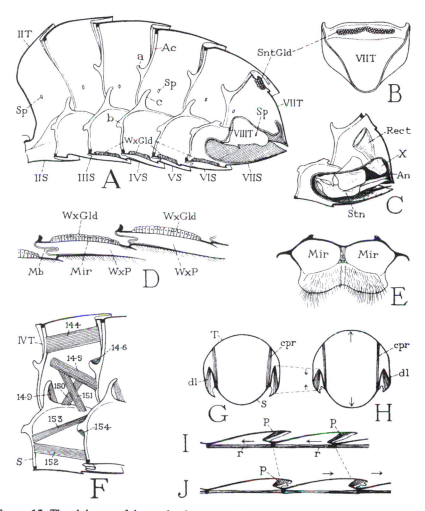

FIGURE 15. The abdomen of the worker bee.

A, internal view of right half of abdomen of worker. B, underside of back plate of abdominal segment *VII* showing scent gland. C, end of abdomen opened on left side, showing sting chamber. D, vertical lengthwise section of two consecutive sternal plates, showing wax glands and wax pockets (as seen also at A). E, Outer surface of a sternal plate (segment *V*) with smooth "mirrors" *(Mir)* beneath wax glands. F, diagram of muscles in right half of a typical abdominal segment. G, H, diagrams illustrating mechanism of vertical compression and expansion of an abdominal segment. I, J, diagrams of mechanism of lengthwise contraction and protraction of abdominal segments.

An, anus; *cpr,* compressor muscle; *dl,* dilator muscle; *Mb,* intersegmental membrane; *Mir,* mirror; *p,* protractor muscle; *r,* retractor muscle; *Rect,* rectum; *s,* sternum; *SntGld,* scent gland; *Sp,* spiracle; *Stn,* sting; *T,* tergum; *WxGld,* wax gland; *WxP,* wax pocket; *X,* ninth abdominal segment, concealed in sting chamber.

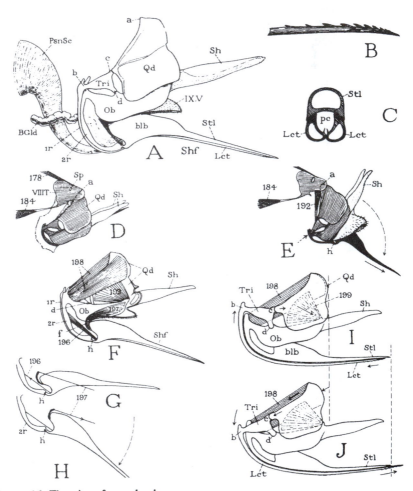

FIGURE 16. The sting of a worker bee.

A, entire stinging apparatus, left side. B, end of a lancet. C, cross section of shaft of sting. D, sting in position of repose, suspended from wall of sting chamber between spiracular plates *(VIII T)*. E, sting in position of protraction (arrows indicate the two essential movements). F, diagram of sting and its muscles. G, diagram of shaft of sting held in position of repose by muscle *196*. H, same, shaft turned down (as at E) by contraction of muscle *197*. I, J, mechanism of retraction (I) and protraction (J) of lancet.

a, attachment of quadrate plate with spiracular plate; *b,* apex of triangular plate continuous with lancet; *BGld,* "alkaline" gland of sting; *blb,* bulb of stylet; *c,* hinge of triangular plate on quadrate plate; *d,* hinge of triangular plate on oblong plate; *f,* forked rod (furcula) giving attachment to depressor muscles (197) of shaft; *h,* hinge of bulb with its basal arm *(2r)*; *Lct,* lancet; *Ob,* oblong plate; *pc,* poison canal; *PsnSc,* poison sac (see Fig. 22C); *Qd,* quadrate plate; *1r,* basal arm (ramus) of lancet; *2r,* basal arm of bulb and stylet; *Sh,* sheath lobes; *Shf,* shaft of sting; *Sp,* spiracle; *Stl,* stylet; *Tri,* triangular plate; *VIII T,* spiracular plate associated with sting base.

forming the conical apex of the body. Segments VIII, IX, and X are not only concealed within segment VII, but they are so reduced in size and altered in form as scarcely to be recognized as segments. In the drone the exposed part of the abdomen ends above with the back plate of segment VIII, and below with the sternal plate of segment IX (Fig. 21, C). In both sexes the tenth segment is reduced to a small conical lobe (Fig. 15, C, X) bearing the anus *(An)*; in the

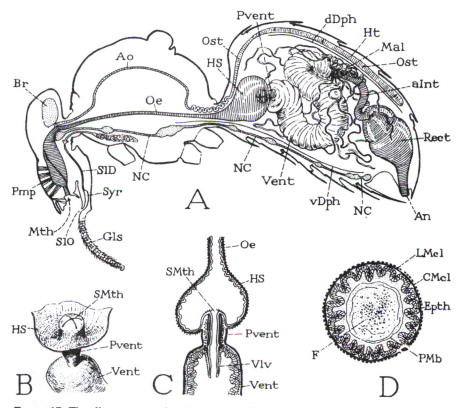

FIGURE 17. The alimentary canal and other internal organs of a worker bee.

A, lengthwise section of a worker bee, showing alimentary canal, dorsal blood vessel, diaphragms, brain, and ventral nerve cord. B, inner end of honey stomach cut open to show stomach mouth *(SMth)* at summit of proventriculus *(Pvent)*. C, lengthwise section of honey stomach, proventriculus and anterior end of ventriculus. D, cross section of stomach (ventriculus).

An, anus; *alnt,* anterior intestine; *Br,* brain; *CMcl,* circular muscles; *dDph,* dorsal diaphram; *Epth,* epithelium (cellular layer of stomach); *F,* food material; *Gls,* tongue; *HS,* honey stomach; *Ht,* heart; *LMcl,* longitudinal muscles; *Md,* mandible; *Mth,* mouth; *NC,* nerve cord; *Oe,* oesophagus; *Ost,* ostium; *Pmp,* sucking pump; *PMb,* peritrophic membrane; *Pvent,* proventriculus; *Rect,* rectum; *SlD,* salivary duct; *SlO,* salivary orifice; *SMth,* stomach mouth; *Syr,* salivary syringe; *vDph,* ventral diaphragm; *Vent,* ventriculus; *Vlv,* proventricular valve.

female it is entirely concealed in a chamber at the end of the abdomen containing the sting *(Stn)*.

Each of the exposed abdominal segments has a large back plate, or *tergum* (Fig. 15, A. G, *T,*), and a smaller ventral plate, or *sternum (S)* (Plate 16). The sucessive terga and sterna overlap from before backward, but are connected by infolded *intersegmental membranes.* Likewise the terga overlap the sides of the sterna (G), and the two plates are connected on each side by an infolded lateral membrane. Hence the abdomen is distensible and contractile in both a lengthwise direction (I, J) and a vertical direction (G, H), as may be observed when a bee is breathing strongly.

The mechanism of the abdominal movements is fairly simple. Between the consecutive tergal and sternal plates are stretched long *retractor muscles* (Fig. 15, F, 144, 145, 152, 153) that by contraction (I, *r*) pull the segments together. The opposite movement, or extension of the abdomen, is produced by short *protractor muscles* that arise on projecting lobes of the front margins of the terga and sterna (F, 146, 154), and are attached posteriorly on the overlapping rear margins of the preceding plates. These muscles by contraction (J, *p*) shorten the overlap of the plates and push the segments apart. The vertical movements are produced in the same way by lateral muscles between the terga and sterna. The *compressor muscles* are two crossed muscles in each side of each segment (F, 150, 151), which by contracting (G, *cpr*) draw the tergum and sternum together. The *dilator muscles* extend from the upper ends of long lateral sternal arms (F, 149; G, *dl*) to the lower edges of the terga, and hence expand the abdomen vertically by pulling the edges of the overlapping plates toward each other (H).

The abdomen is connected with the propodeum of the thorax by a short but narrow stalk, the *petiole,* and acquires thereby a high degree of mobility on the thorax (Plate 17). The principal muscles that move the abdomen as a

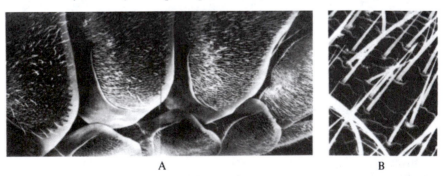

A   B

PLATE 16. A. Overlapping plates of the worker abdomen (X 15). B. Texture of the queen cuticular surface (X 259).

whole are those of the transposed propodeum, which are the intersegmental muscles between the primary first and second abdominal segments.

The external features of chief interest in the bee' s abdomen are the *wax glands* with their accompanying wax pockets, the *scent gland,* and the *sting.* In the drone there are two pair of small plates associated with the genital aperture, but they will be described in connection with the other parts of the reproductive system.

The Wax Glands

The abdominal sterna have long posterior extensions that widely underlap in each case the sternum of the segment behind (Fig. 15, A). The anterior

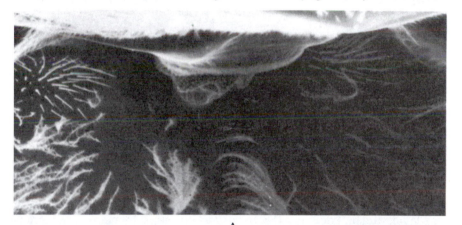

A

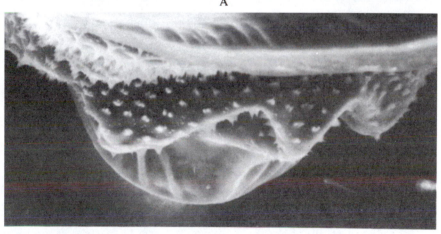

B

PLATE 17. A. Top view of the membraneous connectives of the worker petiole (X 104). B. Enlargement of the view in A. above (X 464).

underlapped parts of sterna *IV, V, VI,* and *VII* in the worker each present two large, smooth, glistening oval areas separated by a narrow, darker median band (E); the exposed part beyond is densely clothed with hairs. The polished oval spaces are known as the *mirrors (Mir);* they are the areas of the sterna covered internally by the wax-secreting glands (A, D, *WxGld*) (Plate 18). These glands are merely specialized parts of the body-wall epidermis, which, during the wax-forming period in the life of the worker, become greatly thickened and take on a glandular structure. The wax is discharged as a liquid through the mirrors and hardens to small flakes in the pockets (D, *WxP*) between the mirrors and the long underlapping parts of the preceding sterna (Plate 18). After the wax-forming period the glands degenerate and become a flat layer of cells.

The Scent Gland

The scent-producing gland of the worker bee lies internally against the back of the abdominal segment *VII* (Fig. 15, *A, SntGld*). Seen from the inner surface (B) the gland appears as a band of large cells extending crosswise near the anterior margin of the tergal plate (Plate 19). Outside the gland (A) is an elevation of the tergum with a smooth and slightly concave surface, suggestive of being an evaporating dish for the gland secretion; but the gland cells discharge their products by minute, individual ducts opening into the pocket at the base of the tergal plate (Plate 19).

The Sting

The sting of the bee is similar in its structure and mechanism to an egg-laying organ, known as the *ovipositor,* possessed by many other female

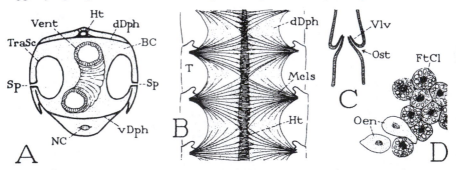

Figure 18. The body cavity, diaphragm, heart, fat cells, and oenocytes.

A, diagrammatic cross section of an abdominal segment. B, part of heart and dorsal diaphragm seen from below against the abdominal terga. C, lengthwise section of heart through a pair of ostia. D, group of fat cells and oenocytes.

BC, body cavity (filled with blood); *dDph,* dorsal diaphragm; *FtCl,* fat cells; *Ht,* heart; *Mcls,* muscles; *NC,* nerve cord; *Oen,* oenocytes; *Ost,* ostium; *Sp,* spiracle; *T,* tergum; *TraSc,* tracheal air sac; *vDph,* ventral diaphragm; *Vent,* ventriculus; *Vlv,* valve-like inner end of funnel-shaped ostial opening.

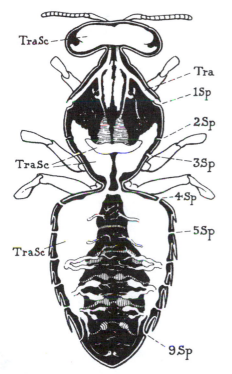

FIGURE 19. General view of the tracheal respiratory system of a worker bee, seen from above. *1Sp-9Sp,* spiracles; *Tra,* tracheal trunk from first spiracle; *TraSc,* tracheal sacs. Though not designated, the sixth, seventh, and eighth spiracles are the openings shown between the fifth and ninth spiracles.

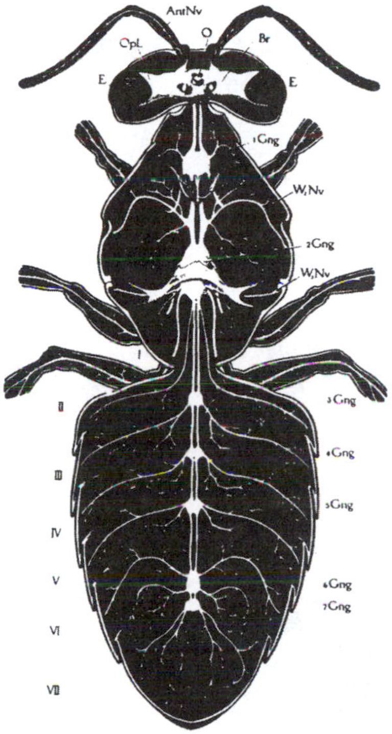

FIGURE 20. The nervous system of a worker bee seen from above. *AntNv,* antennal nerve; *Br,* brain; *E,* compound eye; *1Gng-7Gng,* ganglia of ventral nerve cord; *I,* propodeum; *II-VII,* abdominal segments; *O,* ocellus; *Opl,* optic lobe of brain; *W₂Nv,* nerve to first wing; *W₃Nv,* nerve to second wing.

insects, including most of the Hymenoptera. The ovipositor of some species is also a piercing organ capable of being inserted into the bodies of other insects, or of penetrating plant tissues, even hard wood; but in such cases its function is merely to form a hole in which the eggs may be deposited. The sting of the stinging Hymenoptera (Plate 20), therefore, is very evidently an ovipositor that has been remodeled in a few ways for the injection of poison instead of eggs

The sting is ordinarily contained within a chamber at the end of the abdomen (Fig. 15, C) from which only its effective part, the familiar, tapering, sharp-pointed *shaft* is protruded (Plate 20B). If the sting is removed from the body, or is examined in place within the abdomen, it is seen to include a large basal structure (Fig. 16, A) from which the shaft *(Shf)* is suspended by a pair of curved arms (only one arm visible on each side) (Plate 20A). The sting base in turn is suspended in the membranous wall of the sting chamber (Fig. 15, C).

Though the shaft of the sting appears to be a solid structure, it is composed of three separable pieces, one above termed the *stylet* (Fig. 16, A, C, *Stl*) and two below known as the *lancets (Lct).* The stylet tapers to a point, but swells proximally into a long bulb-like enlargement (A, *blb*) containing a deep cavity open below and continued as a shallow groove on the undersurface of the stylet (C). The lancets are long, slender, sharp-pointed rods lying side by side along the lower edges of the bulb and stylet. Grooves on their upper surfaces fit snugly over track-like ridges of the bulb and stylet (C) so that, while held firmly in place, the lancets can slide freely back and forth (Plate 21A&B). The lower edges of the lancets are in contact with each other; there is, thus, enclosed between the stylet and the lancets, a channel (C, *pc*) leading from the bulb to the tip of the shaft.

This channel is the *poison canal* of the sting. The poison liquid is poured into the base of the bulb from a large *poison sac* (A, *PsnSc*), which is the

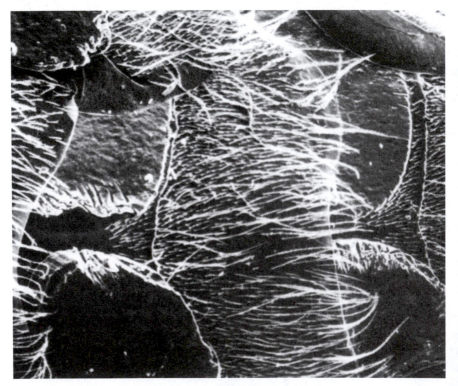

PLATE 18. Wax mirrors with wax scales in place (X 42).

reservoir of a long, slender poison gland (Fig. 22, C, *PsnGld*) that opens into its upper end. A second thick tubular gland associated with the sting (Fig. 16, A, *BGld*) opens externally below base of the bulb (Plate 20A). The terminal part of the stylet is armed with three pair of small lateral teeth; the lancets each have a series of nine or ten recurved barbs along the outer side near the end (Fig. 16, B).

The basal arms of the shaft are each composed of two closely appressed rods, one (Fig. 16, A, *2r*) narrowly attached to the base of the bulb and the other *(lr)* continuous with the lancet of the same side. The lancet arms slide on the bulb arms by ridge-and-groove connections continuous with those by which the lancets slide on a bulb and stylet.

The supporting basal structure of the sting is the motor apparatus of the stinging mechanism (Plate 20A). On each side it presents three plates. The largest and uppermost plate is a four-sided sclerite known as the *quadrate plate* (Fig. 16, A, *Qd*); another is a small *triangular plate (Tri)* lying before the quadrate plate and connected anteriorly by its apex with the upper end of the basal arm of the lancet of the same side *(lr)*; the third *(Ob)*, because of its shape called the *oblong plate,* lies below the other two and is connected anteriorly with the corresponding basal arm of the bulb *(2r)*. Posteriorly each oblong

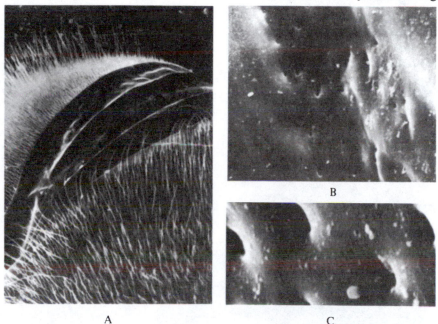

A   C

PLATE 19. Worker scent gland exposed (X 42). B. Pores through the cuticle at the site of the gland (X 940). C. Close-up of the pores (X 7,885).

plate bears a long, soft, finger-like lobe *(Sh)*. The triangular plate articulates by its upper basal angle *(c)* with the anterior angle of the quadrate plate, and by its lower angle *(d)* with the upper edge of the oblong plate. The two quadrate plates are closely associated with two other plates partly overlapping them (D, *VIII,T*), which contain a pair of spiracles and, hence, are known as the *spiracular plates*. The spiracular plates belong to abdominal segment *VIII*, the quadrate plates to segment *IX*.

When the sting is not in action it is entirely retracted within the sting chamber of the abdomen (Fig. 15, C). In the retracted position (Fig. 16, D),

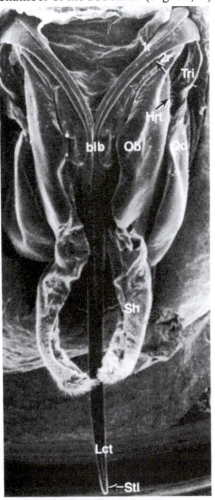

A

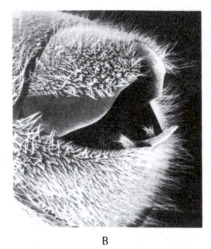

B

PLATE 20. A. Queen sting (X 50). B. Sting chamber (X 19).

the shaft is turned up so that its base is concealed between the oblong plates and its distal part ensheathed between the two projecting lobes of the latter *Sh)*. The protraction of the shaft from the abdomen (E) involves two simultaneous acts: first, a backward swing of the entire supporting apparatus; second, a downward swing of the shaft on its basal arms until it stands at right angles to the oblong plates. The backward swing of the whole structure gives the outward thrust to the extended shaft. This movement apparently is produced by a sharp upward tilt of sternum *VII* which lies immediately beneath the sting (Fig. 15, C). The deflection of the shaft is caused by the contraction of a pair of large muscles arising on the inner faces of the oblong plates (Fig. 16, F, 197). and inserted on a Y-shaped rod *(f)* attached by its stalk to the base of the bulb. The backward pull of these muscles on the bulb turns the entire shaft downward (H) on the flexible hinges *(h)* between the bulb and its basal arms *(2r)*. The retraction of the sting into the chamber probably results from the restoration of sternum *VII* to its usual position; the shaft is drawn up again between the ensheathing lobes by a pair of slender muscles stretched between the base of the bulb and its supporting arms (Fig. 16, F, G, 196).

When the worker bee stings, the end of the abdomen is abruptly bent downward and with a sudden jab the tip of the out-thrust shaft is inserted into the flesh of the victim. Now another part of the sting mechanism comes into play by which the lancets are alternately forced deeper and deeper into the wound, holding each gain by the recurved points along their sides. The movements of the lancets depend on an interaction between the quadrate plates and the triangular plates of the motor apparatus. Two muscles arising on each quadrate plate (Fig. 16, F, 198, 199) are attached, one (198) on the anterior end of the oblong plate *(Ob)*, the other (199) on the posterior end. The alternating pull of these two muscles vibrates the quadrate plate forward and backward (as shown at I and J). The forward movement of the quadrate plate (J) pushes on the upper basal angle *(c)* of the triangular plate *(Tri)*; the latter revolves on its fulcral support *(d)* on the oblong plate, and the depression of its apical angle *(b)* pushes down on the basal arm of the lancet *(Lct)*; and the lancet thus slides backward on the stylet and its tip protrudes from the end of the shaft. Conversely, as long as the lancet is free to move, the contraction of muscle *199* (I) pulls the quadrate plate backward, lifts the apical angle *(b)* of the triangular plate, and retracts the lancet.

If, however, the stinging bee has succeeded in inserting the point of the sting in an intended victim and the lancet first thrust out holds in the skin by means of its barbs, the force of the retractor muscle (I, 199) is now expended on the oblong plate which revolves downward anteriorly on its articulation *(d)* with the triangular plate and drives the attached stylet into the wound made by the lancet. The lancet of the opposite side is then thrust out in the same way as the first and takes a still deeper hold in the flesh. Thus, by repeated alternating thrusts of the lancets the point of the shaft sinks deeper and deeper, and the action of the sting apparatus continues even when the sting is separated from the body of the bee. Valves on the basal parts of the lancets drive the poison through the poison canal of the shaft and the liquid is expelled, not at the penetrating tip of the sting, but from a ventral cleft near the ends of the lancets.

The sting of the queen is longer than that of the worker and is more solidly attached within the sting chamber. Its shaft is strongly decurved beyond the bulb. The lancets have fewer and smaller barbs than those of the worker (Plate 21), but the poison glands are well developed and the poison sac is very large. A small sensillum which looks like a button is associated with each barb on both lancets and the stylet (Plate 21G). These sensors likely register the depth of sting insertion.

THE ALIMENTARY CANAL

The food tract begins at the *mouth* in the lower wall of the head (Fig. 17, A, *Mth*). The mouth opens into the cavity of the *sucking pump (Pmp)* which stands vertically in the head. At its upper end the pump narrows to the slender, tubular *oesophagus (Oe)*, which turns posteriorly through the neck and thorax, and in the anterior end of the abdomen enlarges into a thin-walled sac *(HS)*. This sac corresponds with the *crop* of other insects, but it is commonly known to students of the bee as the *honey stomach* because it is used by the bee for carrying nectar or honey. Following the honey stomach is a short narrow part of the food canal called the *proventriculus (Pvent)*. Next comes a long, thick, cylindrical sac *(Vent)* looped crosswise in the abdomen, usually in an S-shaped curve, which is the true stomach of the insect, or *ventriculus*. Following the stomach is the intestine, but the latter is distinctly divided into two parts: first, a narrow *anterior intestine, (alnt)* which is looped or coiled in various ways according to the position of the stomach; and second a large, pear-shaped *posterior intestine,* or *rectum (Rect)*, opening by its tapering extremity through the *anus (An)* into the cavity that contains the sting (Fig. 15, C).

The structure of the sucking pump has already been described in connection with the feeding mechanism. The oesophagus is a tube with muscular walls, which passes the food along its length by successive waves of constrictions in the manner by which most animals swallow their food. The honey stomach, being the crop of the insect, serves the bee not only as a nectar-carrier but also as a storage place for food material. It is greatly distensible because its inner wall is thrown into numerous folds (Fig. 17, C, *HS*). The proventriculus is a regulatory apparatus that controls the entrance of food into the stomach. Its anterior end projects like a thick plug into the honey stomach (B, C) and

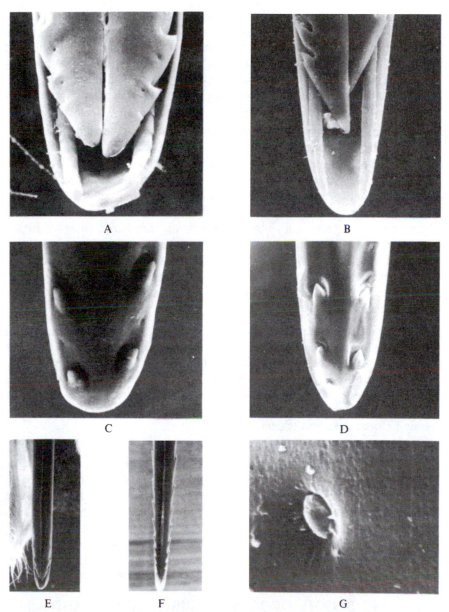

PLATE 21. Comparison of the stings of the queen and worker. A. Tips of the queen lancets (X 270). B. Tips of the worker lancets (X 270). C. Tip of the queen stylet (X 270). D. Tip of the worker stylet (X 270). E. Shaft of the queen sting (X 110). F. Shaft of the worker sting (X 110). G. Close-up of a sensillum for detecting pressure (X 7,615). Note that one of these sensors is associated with each barb on both the stylet and lancets.

contains an X-shaped opening between four thick, bristly, triangular, muscle-controlled lips *(B, SMth)*. This structure constitutes a *stomach mouth*; by its action nectar or honey can be retained in the honey stomach while pollen is taken out and delivered to the ventriculus. The posterior end of the proventriculus extends in a long funnel-like fold (C, *Vlv*) into the anterior end of the ventriculus *(Vent)*, and probably acts as a valve to prevent regurgitation from the stomach.

The ventriculus is the part of the insect alimentary canal in which digestion and absorption of food material take place. Its inner wall (Fig. 17, D, *Epth*) is a thick cellular layer (epithelium) thrown into numerous crosswise folds that not only greatly increase the extent of the digestive surface but also allow for expansion. Outside the cellular wall is first a layer of circular muscle fibers *(CMcl)*, and surrounding the latter a sheath of lengthwise fibers *(LMcl)*. Inside the stomach is a very thin, irregular *peritrophic membrane,* or several such membranes *(PMb)*, forming a delicate cylindrical covering around the food mass *(F)*. The cellular layer of the ventricular wall secretes the digestive juices and enzymes; the products of digestion go through the thin peritrophic membrane and are discharged through the stomach wall directly into the surrounding blood.

The intestinal tract usually is a relatively small part of the food canal. The narrow anterior intestine and the sac-like rectum (Fig. 17, A, *alnt, Rect*) serve principally for the discharge of waste matter and for the absorption of water, but the rectum is also a storage chamber for the retention of feces until the latter can be evacuated outside the hive. In overwintering bees, the rectum may become so greatly distended as to occupy a large part of the abdominal cavity before defecation occurs.

At the junction of the intestine with the ventriculus there open into the intestine a great number of long, thread-like tubes, probably a hundred or more of them. These tubes are known as the *Malpighian tubules* (Fig. 17, A, *Mal*); they are not digestive glands but excretory organs that remove waste products of metabolism from the blood, including both nitrogenous substances and salts. The tubules extend long distances in the body cavity, winding and twisting in numerous convolutions through the spaces about the other organs, where they are directly bathed by the blood. The products discharged into the intestine are eliminated along with the waste food matter.

THE BLOOD, ORGANS OF CIRCULATION, AND ASSOCIATED TISSUES

The spaces in the body of an insect not occupied by the organs or other tissues are filled with a liquid which is the *blood,* or *haemolymph.* Floating in the blood are numerous blood cells, or *haemocytes*, of several kinds, but the blood cells do not serve for the transport of oxygen; they resemble the white blood cells of vertebrate animals. The blood liquid also carries but little oxygen; its principal known functions are the distribution of digested food material absorbed from the alimentary canal, the reception of waste products of metabolism which are removed by the excretory organs, and the transport of carbon dioxide to be eliminated through the respiratory organs and the skin. The blood is kept in circulation through the body by a pulsating tubular blood vessel and by vibratory membranes. The blood of the honey bee is of a pale amber color.

The single blood vessel is a long slender tube (Fig. 17, A, *Ao, Ht*) extending forward along the midline of the back in the abdomen from abdominal segment *VI* through the thorax and into the head, where it opens beneath the brain *(Br)*. The abdominal part of the vessel is called the *heart (Ht)* and the thoracic part, the *aorta (Ao)*. The sides of the heart are perforated by five pair of slits, the *ostia (Ost)*, in abdominal segments *II* to *VI*, inclusive, through which the blood enters the heart. The heart drives the blood forward by rhythmic pulsations of its muslcular walls. The lips of the ostia project forward into the heart cavity (Fig. 18, C) and act as valves *(Vlv)* to prevent the escape or backward flow of the blood. The aorta has no ostia, is thrown into numerous small loops where it enters the thorax, and ends openly beneath the brain.

The heart is supported on a thin membrane, called the *dorsal diaphragm* (Fig. 18, A, *d,Dph*), stretched across the upper part of the abdominal cavity in segments *III* to *VII* (Fig. 17, A). The membrane contains five pair of fan-shaped bundles of fine muscle fibers attached laterally on the anterior margins of the tergal plates and spreading toward the heart where the fibers break up into numerous branching fibrils (Fig. 18, B, *Mcls*). The diaphragm shuts off above it in a *pericardial cavity* containing the heart, but its lateral margins are free between the muscle attachments, thus leaving openings by which the blood can enter the pericardial cavity from the general body cavity. A rhythmic contraction of the diaphragm muscles causes the dorsal diaphragm to pulsate in a forward direction.

In the ventral part of the body above the nerve cord is a similar *ventral diaphragm* (Fig. 18, A, *vDph*), but this diaphragm extends from the meta-thorax into segment *VII* of the abdomen. It is more strongly muscular than the dorsal diaphragm and beats in a backward direction.

The blood discharged into the head from the aorta bathes the organs in the head and flows backward through the thorax and the abdomen by fairly well-defined channels, circulating also through the antennae, the wings, and the legs. The backward flow of the blood in the abdomen is assisted by the pulsations of the ventral diaphragm; from the lower part of the abdomen it goes upward into the pericardial cavity where it is driven forward along the sides of the heart probably by the vibrations of the dorsal diaphragm, and finally again enters the heart through the ostia. Small pulsating membranes occur in the head between the antennal bases and in the upper part of the thorax, but the movements of these membranes appear to be produced by neighboring muscles.

Scattered all through the body cavity of the insect but especially in the abdomen are irregular masses of a soft, usually white tissue composed of large, loosely united cells (Fig. 18, D, *FtCl*). These cell masses are known collectively as the *fat body* because the cells contain, enmeshed in their cytoplasm, small droplets of oily fat. The fat tissue is particularly abundant in the larva; but the larval cells contain, besides large amounts of fat, also glycogen and, toward the end of the larval life, numerous minute protein granules. The fat body, in short, is a storage tissue for the conservation of elaborated food products not immediately needed by the individual. The larval fat cells, thus, carry over a large supply of food material into the pupal stage where it is thrown out into the blood by dissolution of the cells, and consumed by the pupal tissues developing into the adult organs. An underfed larva cannot properly transform into a mature bee.

Intermingled with the fat cells are other cells of larger size having a pale yellowish color, known as *oenocytes* (Fig. 18, D, *Oen*), and strands of special cells, called *nephrocytes,* lie along the sides of the heart, but nothing definite is known concerning the functions of these cells.

THE RESPIRATORY SYSTEM

The chemical changes that go on constantly inside the body cells of living things require oxygen for consumption, and produce carbon dioxide which must be eliminated. All multicellular animals, therefore, are confronted with the problem of how to supply their tissues with oxygen and how to remove the carbon dioxide. Some have solved the problem in one way, others in other ways, but the anatomical means adopted, whatever its nature, consititutes a *respiratory system.*

Certain small, soft-bodied insects and some insect larvae have the very simple system of exchanging respiratory gases by diffusion directly through the skin. With most insects, however, the integument is too hard and dense for skin breathing and the majority of species, even very minute insects, have long, tubular, many-branched, thin-walled ingrowths of the integument that conduct air from outside the body to all the living tissues inside the body. These air tubes are called *tracheae;* collectively the tracheae constitute a *tracheal respiratory system.* The fine terminal branches of the tracheae go to practically all the cells of the body, the tissues thus receiving their oxygen direct without transport by the blood. The blood absorbs only what oxygen it needs for its own use.

The tracheal system of the bee is highly elaborate (Fig. 19), but a large part of it consists of air sacs *(TraSc)* which are thin-walled expansions of tracheae. The tracheal tubes open from the exterior by small breathing pores, the *spiracles (Sp)*, situated along the sides of the body. Most insects have ten spiracles on each side, two on the thorax and eight on the abdomen. The adult honey bee, however, has three thoracic and seven abdominal spiracles because of the transfer of the first abdominal segment to the thorax. The first and largest spiracles of the bee *(1Sp)* lie between the prothorax and the meso-thorax, but each is concealed beneath the lateral lobe of the pronotum (Fig. 9, *spl)*, and is further protected by a dense fringe of long hairs on the covering lobe. Nevertheless, the first spiracle is entered by the parasitic mites, *Acarapis woodi,* that accumulate in the large tracheal trunk (Fig. 19, *Tra*) proceeding from this spiracle and cause a form of paralysis called acarine disease. The second spiracle *(2Sp)* is very small and lies between the upper angles of the pleural plates of the mesothorax and metathorax, but is normally concealed between these plates and, therefore, is not seen in Figure 9 of the thorax. The third spiracle *(3Sp)* is fully exposed on the side of the propodeum (Fig. 9). The next six *(4Sp-9Sp)* are in the lower parts of the first six tergal plates of the abdomen (Fig. 15, A), but the tenth is not visible externally since it is in the so-called spiracular plate associated with the base of the sting (Fig. 15, A; 16, D, *VIII T*). All the spiracles, except the minute second spiracle, have an apparatus for closing the spiracular orifice to prevent the escape of inhaled air or to control air movement through the tracheae.

Respiration in the bee is effected partly by dorsoventral and lengthwise contractions and expansions of the abdomen, produced as already explained by opposing sets of abdominal muscles (Fig. 15, G-J). For the inhalation of air by respiratory movements, however, it is necessary that some part of the air passages inside the body should be able to expand. The tracheal tubes are more or less rigid by reason of spiral thickening in their walls that serve to keep them open, and, hence, do not respond much to increase and decrease of pressure around them. In the bee, therefore, as in most insects that breath actively, parts of the tracheae are dilated into thin-walled *air sacs* that expand and collapse in the manner of lungs in response to the expansion and contraction of the rigid parts of the body wall. The honey bee is particularly well supplied with tracheal air sacs (Fig. 19, *TraSc*), two large sacs occupying much of the space in the sides of the abdomen while smaller ones are distributed all through the thorax, in the head, and even in the legs.

From the air sacs the tracheae continue as branching tubes that ramify to all the parts of the body, the appendages, and the internal organs. Finally the tracheae end in groups of minute tubes, called *tracheoles*, which terminate blindly against or within the tissue cells. The distal parts of the tracheoles are filled with liquid that absorbs oxygen from the air delivered through the tracheae. The physiological activity of the cells causes the oxygen-saturated liquid in the tracheoles to be drawn through the tracheole walls and the cells walls into the protoplasm of the cells; when the cell activity (metabolism) decreases or returns to a minimum, the liquid again accumulates in the tracheoles and absorbs a fresh supply of oxygen. The carbon dioxide produced in the cells cannot be directed into the tracheoles; it is mostly discharged directly into the surrounding blood, from which it is carried off by diffusion through the trachea or through the softer parts of the integument.

The metabolism of insects produces heat as in other animals, and the body temperature increases considerably during muscular activity, but the insect has no means of retaining the heat within its body; consequently, it is given off almost immediately by radiation. Beekeepers well know that bees raise the temperature of the hive in winter by rapid vibration of the wings, which involves sustained action of the huge masses of muscle fibers in the thorax.

THE SENSORY AND NERVOUS SYSTEM

A distinctive feature of animals is their ability to adjust their actions to conditions of the environment, particularly changing conditions. Animals are able to do this because they have cells or groups of cells close to the exterior of the body that are specifically sensitive to the common forms of energy in nature that are not destructive to them. These specialized cells and associated structures are known as *sense organs,* but the term does not necessarily imply conscious perception on the part of the animal.

From the receptive cells of the sense organs, *sensory nerves* extend inward to the central nervous system. Another set of fibers, called *motor nerves,* goes outward from cells in the central system to the body muscles and glands. A third set of intermediary *association fibers* connects the ends of the incoming sensory nerves with the roots of the outgoing motor nerves. In this way there is established a nerve circuit from the outlying sense organs through the central system to the muscles or glands, and the stimuli received from outside the body thus set up a nerve impulse that finally activates the motor system, or causes certain glands to produce a secretion. What the animal does in response to an external stimulus is called *reaction.*

The nature of the reaction depends on the external stimulus and on the internal nerve pathways that are affected. If the animal has conscious control of its actions, violation may determine the path of the outgoing nerve impulse, otherwise the action is a *reflex;* coordinated reflexes are *instincts.* The sensory and nervous systems of insects are well developed and organized, and the honey bee is in many respects one of the most highly endowed in its powers of sensory reception and motor reaction.

Reaction to touch, or external pressure, is probably the most primitive of all the senses. In adult insects, as compared with soft-skinned larvae, the general body surface has relatively little sensitiveness to pressure because of the hardness of its outer covering. Hence, most of the sensory nerves of the skin end in cells at the bases of hairs. The hairs being delicately poised are easily moved by contact with objects or currents of air. Therefore, an innervated hair and its associated sense cell constitute an *organ of touch*. It is not known how many of the hairs of the bee are sensory organs, but innervated hairs occur on various parts of the body and appendages, and are particularly numerous on the antennae. Some insects respond to sound by the vibration of sensory hairs. Bees readily detect substrate vibrations through sensors in their legs. Such vibrations may be induced by airborne sound. Some believe that bees hear airborne sound via antennal vibration. However, evidence to support this theory is limited.

Some very small, thin-walled hairs of the bee, or hairlike structures reduced to small pegs, are supposed to be capable of being stimulated by minute particles of matter in the air or liquids, and hence are regarded as *organs of smell or taste*. These organs are innervated each from a group of sense cells which sends a nerve strand into the hair or peg. Organs of this kind occur on the antennae and on parts near the mouth; some on the antennae are sunken into deep flask-shaped cavities. The most numerous organs on the antennae, however, appear on the surface as minute oval disks or plates. Each plate has a groove around the margin and covers a large group of sense cells. These structures are known as *plate organs;* they are the principal organs of smell in the bee. It has been estimated that there are 5 or 6 thousand plate organs on the antennal flagella in the worker, 2 or 3 thousand in the queen, and perhaps 30 thousand in the drone.

Still other organs, having the form of minute bells or inverted cups sunken into the body wall with nerve ending inside, are distributed in groups on various parts of the body and appendages. These organs have been described as "olfactory pores," but recent experiments appear to show that they respond to strains, stresses, or bending of the integument, and thus give the insect "information" concerning its own actions.

Imaginative people like to speculate on the possibility of insects having some sense "totally unknown to us," but, considering that the senses are internal reactions on the part of the animal to external conditions of the environment, a supposed "unknown sense" must be based on something real in nature. An insect's reactions to sensory stimuli can be determined experimentally and its sense organs can be studied under the microscope, but inasmuch as its sense organs are so different from ours, and because various kinds of them occur on the same parts of the body or appendages, it is a difficult matter in the case of most of them to determine specifically what stimulus is effective on each kind of organ.

The best known of the insect sense organs are the eyes, because an insect eye is like any other eye in that its essential parts are an external *lens* for focusing light, and a sub-lying, light-sensitive *retina* connected by nerves with the brain. The bee, as we have observed, has three small simple eyes, or *ocelli,* on top of the head (Fig. 5, A, *O*), and a pair of large *compound eyes (E)* on the sides of the head. A simple eye has one lens for the entire retina; a compound eye has many small lenses, and the retina is divided into parts corresponding with the lenses. It is supposed that the insect "sees" with a compound eye as many points of light as there are divisions of the eye, and, thus, gets a mosaic picture of the object or scene before it — and there may be several thousand separate light-receptive parts of the eye. However, it is impossible for us to know what the final effect on the insect's brain may be. It is certain that insects respond most quickly to movements of objects, but many insects, including the honey bees, perceive differences of color, shape, and position. Most insects "see" the higher colors of the spectrum visible to us (green, blue, violet), and even the ultraviolet which we do not see, but they are more or less insensitive to the lower red rays.

The ocelli were once thought to be organs for near vision, but a current idea of their function is that they keep the insect continually in a state of stimulation to light, and, thus, make the compound eyes more quickly responsive. Many larval insects and other arthropods such as spiders have ocelli but no compound eyes.

The central nervous system of the bee (Fig. 20), as in insects generally, is fairly simple in its general structure. It consists of a *brain (Br)* in the head above the pharynx, and of a *ventral nerve cord* in the lower part of the body extending from the head to the posterior part of the abdomen. The brain is principally a sensory center, as it receives the nerves from the eyes and the antennae. The ventral nerve cord consists of a series of small, segmental nerve masses, or *ganglia,* united by paired intervening *connectives.* The first three ganglia of the ventral system are always condensed into a large composite ganglion, termed the *suboesophageal ganglion,* lying in the power part of the head and supplying nerves to the feeding organs. The first body ganglion *(1Gng)* pertains to the prothorax. The second in the honey bee *(2Gng)* lies in the posterior part of the thorax, but it is composed of four primary ganglia belonging to the mesothorax, the metathorax, the propodeum, and the first abdominal segment, and supplies nerves to all these segments. In the abdomen are five ganglia. The first two *(3Gng, 4Gng)* are displaced forward so that each innervates the segment behind it; the third *(5Gng)* lies in its own segment *(V)* and gives its nerves to this segment; the fourth *(6Gng)* is in segment *VI* but innervates segments *VI* and *VII;* the fifth *(7Gng)*, in segment *VII*, supplies nerves to segments *VIII, IX,* and *X.*

The brain and the ventral ganglia are masses of nerve cells and nerve fibers. The cells of the central system give rise to the fibers of the motor and the association nerves, but the fibrous parts of the ganglia include also the branching ends of the incoming sensory nerves from the peripheral sensory nerve cells. The organization of the brain or of a ventral ganglion depends mostly on the intricate tracts of intercommunication formed by the association fibers, which connect all parts of the central system. The brain of an insect owes its importance to the fact that it receives the sensory nerves from the eyes and the antennae, and transmits the nervous impulses from these sensory organs to the motor centers of the ventral nerve cord. The environmental stimuli received through the eyes and the antennal sense organs, thus, direct many of the insect's natural activities. If the head is removed from the insect it is deprived of these stimuli, as well as the ability to eat, but still retains the power of acting through its body motor centers. A decapitated insect is, therefore, able to walk or even to fly, and a bee without a head is even able to sting.

THE REPRODUCTIVE SYSTEM

The reproductive organs of insects usually include external and internal structures, but in the honey bee the organs that subserve the reproductive function are almost entirely internal, the intromittent organ of the drone being a large sac within the abdomen that is everted only at the time of mating, and the instrument of the female used for egg laying by other insects being converted into a sting. Furthermore, the organs are fully developed only in the drone (Fig. 21, A) and the queen (Fig. 22, A). Female organs are present in the worker, but they are greatly reduced in size (Fig. 22, C), and only under certain conditions do they produce eggs. The mature reproductive cells of the male and the female, respectively, are the *spermatozoa* and the *eggs,* or *ova,* but they are developed from primary germ cells, set apart as such in the young embryo, that have little to distinguish them visibly from other cells of the body.

The Male Reproductive Organs

The organs of the male that contain the primary reproductive cells, and in which the latter develop into spermatozoa are the *testes.* In the drone bee the testes are a pair of small flattened bodies (Fig. 21, A, *Tes*) lying in the sides of the abdomen. From each testis there proceeds posteriorly a duct, the *vas deferens* (plural *vasa deferentia*), at first coiled *(Vd)*, but soon enlarging into a long slender sac, the *seminal vesicle (SV).* The narrowed posterior ends of the two vesicles enter the lower ends of a pair of huge *mucous glands (MuGlds)*, lying side by side, and the two glands open together into a single long outlet tube, which is the *ejaculatory duct (Dej)*. Finally, the ejaculatory duct opens into the anterior end of a large complex structure *(Pen)*, termed the *penis* because by inversion at the time of mating it serves to discharge the spermatozoa.

Three consecutive parts may be noted in the structure of the inverted penis. The inner half is a large pear-shaped swelling, or *bulb (Blb)*, which receives the ejaculatory duct at its anterior end, and has a pair of dark plates in its thick dorsal wall. The bulb passes into a narrowed, usually twisted neck, or *Cervix (Cer)*, having a series of dark crescent-shaped thickenings along its lower side and a fringed lobe *(fml)* projecting from its dorsal wall. The neck ends in a large thin-walled sac, or *bursa (Brs)*, from which project a pair of crumpled horn-like pouches, the *bursal cornua (bc)*. The bursa opens to the exterior by a wide orifice *(C, Phtr)* beneath the anus *(An)* and between a pair of small valve-like plates *(pv)*. These plates and a still smaller pair *(lp)* at their

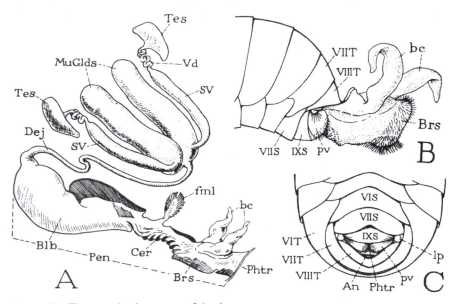

Figure 21. The reproductive organs of the drone.

A, general view of internal organs of reproduction as seen from left side. B, end of the abdomen of a drone with penis partly everted. C. end of drone abdomen, under surface, (penis not exerted).

An, anus; *bc,* bursal cornua; *Blb,* bulb of penis; *Brs,* bursa of penis; *Cer,* neck (Cervix) of penis; *Dej,* ejaculatory duct; *fml,* fimbriate lobe; *lp,* parameral plate; *MuGlds,* mucous glands; *Pen,* penis; *Phtr,* external opening of inverted penis; *pv,* penis valve; *SV,* seminal vesicle; *Tes,* testis; *Vd,* vas deferens; *VI S-IX S,* sternal plates of abdomen; *VI T-VIII T,* tergal plates of abdomen.

outer angles are the only representatives in the honey bee of a large, often complex, external copulatory organ present in most other male Hymenoptera.

The mature spermatozoa are minute bodies with long vibratile tails. From the testes they pass down the *vasa deferentia* into the sperm vesicles, where they are temporarily stored with their heads buried in the soft cellular walls of the vesicles. In the mating season, spermatozoa are sent on down through the ejaculatory duct in a secretion from the mucous glands, and a mass of sperm and mucus now fills the bulb of the penis.

The Female Reproductive Organs

In the female the primary germ cells are housed in the *ovaries,* and in these organs they undergo all but the last stage of development into eggs ready for fertilization. The ovaries of the queen bee (Fig. 22, A, *Ov*) are two huge, pear-shaped masses of slender, closely packed tubules, termed *ovarioles (B)*. At the posterior end of each ovary the ovarioles come together in a *lateral oviduct (A, Odl)* and these two ducts unite in a short *common oviduct (Odc)*. The last is continuous with a wide terminal sac, the *vagina (Vag),* which opens to the exterior by a median orifice (*VO)* in a depression of the body wall at the base of the sting. At the sides of the genital orifice are two other openings *(PO)* which are the mouths of two large pouches *(P)* embracing the sides of the vagina. Lying on the dorsal wall of the vagina is a spherical body *(Spt,* shown turned to one side in the figure), which is the female receptacle for the spermatozoa and is, hence, termed the *spermatheca*. The spermatheca is connected with the vagina by a short duct *(SptDct)*. A pair of tubular *spermathecal glands (SptGlds)* open into the distal part of the duct.

Insemination of the Queen and Fertilization of the Eggs

At the time of mating, the sperm mass in the penis bulb of the drone is discharged by eversion of the penis into the vaginal pouch of the female. It has been supposed that the male organ is first anchored in the female by the eversion of its lateral cornua (Fig. 21, A, *bc*) into the lateral genital pouches of the queen (Fig. 22, A, *P*), but apparently this plausible idea has not been verified by observation. Furthermore, some investigators say that the entire penis is turned inside out, while others claim that only the end pouch is everted.

Whatever may be the facts concerning the copulatory process and the extent of the penis eversion, when the mating queen finally separates herself from the drone, only the penis bulb remains in her genital tract, the male organ having been torn apart at its weakest point between the bulb and the penis neck. The spermatozoa discharged by the drone are first stored in the distended lateral oviducts. As soon as the remains of the male organ and the mucus are removed from the vagina, by muscular contractions the female forces the spermatozoa into the vagina. Here they are stopped by a valve-like fold of the vaginal wall, which causes them to be directed into the spermathecal duct, and finally into the spermatheca. The spermatozoa retain their vitality within the spermatheca throughout the productive life of the queen. The queen on her nuptial flights may mate with several drones in succession.

When an egg in the ovary is ready to be discharged, the lower end of its follicle opens and the egg passes down the oviduct into the vagina. The deserted follicle shrivels and is absorbed as the next one above takes its place, and the ovariole regains its length by growth at the upper end where the new eggs are forming. Considering the great number of ovarioles in the two ovaries of the queen bee, it is clear how eggs may be matured continuously and take their places consecutively in the vagina.

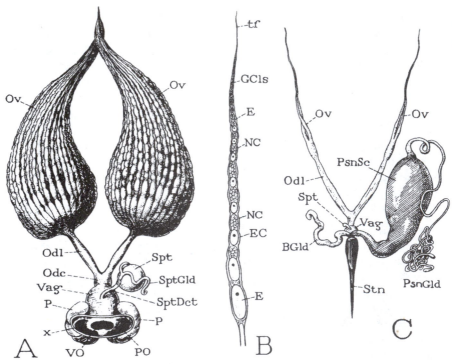

FIGURE 22. The female reproductive organs.

A, ovaries, genital ducts and genital pouches of the queen. B, single ovariole, diagrammatic, showing succession of egg cells and nurse cells. C, reproductive organs of a worker, together with shaft of sting, sting glands, and poison sac.

BGld, "alkaline" gland of sting; *E,* egg; *EC,* egg chamber; *GCls,* undifferentiated germ cells; *NC,* nurse chamber; *Odc,* common oviduct; *Odl,* lateral (paired) oviduct; *Ov,* ovary; *P,* lateral genital pouch; *PO,* opening of lateral pouch; *PsnGld,* poison gland of sting; *PsnSc,* poison sac; *Spt,* spermatheca; *SptDct,* spermathecal duct; *SptGld,* spermathecal gland; *Stn,* shaft of sting; *tf,* terminal filament; *Vag,* vagina; *VO,* opening of vagina; *x,* cut edge of body wall around genital openings.

As the eggs descend the oviducts and enter the vagina they undergo the last act of their maturing process. This consists of two consecutive divisions of the egg nucleus, one of the new nuclei becoming the definitive egg nucleus, the others being absorbed. The nuclei of all cells contain small bodies known as *chromosomes*. In the female honey bee the accepted number of nuclear chromosomes is 32, in the male 16. At the first division of the egg nucleus half the chromosomes go into each newly formed nucleus, the persisting egg nucleus being, thus, reduced to 16 chromosomes. At the second division the chromosomes split so that 16 are retained in each nucleus. The eggs are now ready for fertilization by the spermatozoa, which are discharged upon them from the spermatheca. But only eggs that are to produce female larvae are fertilized; eggs that are to produce males are not fertilized. When a spermatozoon, containing 16 chromosomes, enters an egg through the micropyle and its nucleus unites with the egg nucleus, containing also 16 chromosomes, the fertilized egg will contain 32 chromosomes, and consequently develops into a female bee. On the other hand, an unfertilized egg remains with only 16 chromosomes in its nucleus and develops only into a drone. The eggs finally are discharged from the vaginal orifice at the base of the sting. Whether the fertilized egg becomes a queen or a worker now depends on the feeding of the larva.

CHAPTER 5

PHYSIOLOGY AND SOCIAL PHYSIOLOGY OF THE HONEY BEE

by EDWARD E. SOUTHWICK*

To understand the physiology of honey bees is to understand how the insect carries on its various life processes. The physiology of the individual honey bee is not remarkably different from that of other bees or single insects of most other species (*e.g.*, see Blum 1985, Kerkut and Gilbert 1985, or Rockstein 1987, 1978). However because of its eusocial behavior (or true social behavior with offspring assisting parent in care and provision of the family), there are aspects that need to be examined in the context of the social group. There is actually a "group physiology" that is very different than that shown by isolated individuals. This chapter is, therefore, divided into first a discussion of the aspects of physiology of individual honey bees, and second, a look at the physiology of the social group.

I. PHYSIOLOGY OF THE INDIVIDUAL

A. The Integument

The exoskeleton or cuticle is the outer integument of the body of the bee. It separates the internal environment from the external one and aids in maintaining homeostasis within the body. It is not permeable to water or air and thus protects the bee from dessication. There are openings in the cuticle which permit exchange of gasses and nutrients. The hinged mandibles allow for grasping and chewing pollen as well as the manipulation of non-nutritive wax. The multipart proboscis provides for both lapping and sucking fluids such as water, honey, flower nectar, and plant resins and saps. The tracheae and the anterior and posterior ends of the digestive tract are invaginations of the body wall and are lined with cuticle. The cuticle is composed of chitin, proteins, and pigments. The proteins allow for the hardness and rigidity (sclerotin) of the exocuticle, and elasticity and extensibility (resilin) of the endocuticle (Porlier 1987, Wigglesworth 1984). Chitin, a nitrogenous polysaccharide $(C_8H_{13}NO_5)_n$, has a chemical structure like cellulose in plants. It is insoluble in water, alcohol, dilute acid, and alkalis, and it is not broken down by mammalian digestive enzymes. It can be broken down by certain bacteria and other insects, and some land snails. Its fiberlike nature is similar to fiberglass. Waxes are secreted on the outer surface of the cuticle and protect the bee from water loss by transpiration. Also because of this waxy covering, the bees are not easily wetted.

*Edward E. Southwick, Ph.D., Professor, Department of Biology, State University College, Brockport, N.Y.

Branched hairs are present over all the body (Figure 1) and can be used to distinguish bees phylogenetically (Lanham 1979, Ruttner 1988). These hairs are present on the body at emergence and are gradually worn off as the bee ages. The longer hairs function primarily in pollen collection and transport, and combined with shorter hairs of the thorax, also have heat retention properties when the bees are grouped together under low temperature conditions (Southwick 1985b).

B. Digestion and Metabolism

Bees, by nature, are vegetarians and consume only flower nectars and pollen. They derive their total nutrient requirements from these two resources. Although small quantities of amino acids, lipids, minerals, salts, and vitamins are available in nectar and detectable by bees (Baker and Baker 1975, Inouye and Waller 1984, Sadwick 1983), this is primarily a resource for high energy carbohydrates. Pollen is the only good source of protein. In the natural environment, flower nectar sugar concentrations are quite variable ranging from almost no sugar in flowers filled with dew or rain water, to over 80% (*i.e.*, 80g sugar dissolved in 100 g solution, or a 3.30 molar solution) in open flowers exposed to the drying air. Average nectar concentration found in the floral standing crop is about 40% (1.37 molar) (Beutler 1930, Southwick *et al.* 1981.).

The alimentary canal is comprised of the foregut and crop, midgut (where the digestive enzymes are secreted and absorption takes place) and the hindgut. The foregut conducts food into the crop (honey stomach). Here the food is stored and released through a muscular sphincter (proventriculus, *i.e.*, the "honey stopper") in small quantities at a time into the midgut. Through the action of the spiney proventriculus, pollen grains are filtered out and passed on to the ventriculus while clarified nectar or honey is retained in the honey

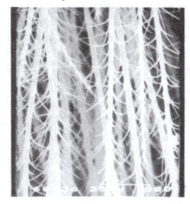

FIGURE 1. Branched hairs on the thorax of *Apis mellifera*. These scanning electron micrographs were taken after gold ion coating by R. Barkhausen, University of Frankfurt, FRG.

stomach. Salivary secretions of invertase (a hexose enzyme) can begin to break down the food somewhat (*i.e.*, complex sugars into simple sugars) in the honey stomach. But, the digestive proteolytic enzymes for pollen breakdown are produced and released beyond the honey stomach in the midgut. In the ventriculus of the midgut, digestion and absorption of nutrients takes place. These nutrients pass through the membranous stomach walls right into the hemolymph (blood). In the hindgut, absorption of water and inorganic ions takes place. The inorganics are important in maintenance of blood ionic composition. Feces are also formed here and released to the outside through the anus.

In careful studies on pollen travel in the gut of honey bees by Peng *et al.* (1986), pollen was found to take about 30 min. after feeding to reach the anterior region of the midgut. One half hour later it enters the middle region where the pollen wall begins to be broken down around the germination pore. Pollen cytoplasm extrudes gradually during the following two hours, and the nutrients can be absorbed. Empty and broken pollen grains are found in the rectum after 15 hours.

Carbohydrate reserves in the honey bee are present as glycogen (stored in tissue) and trehalose (in the hemolymph). These reserves can readily be transformed into glucose. Glycogen (a branched chain polysaccharide) is synthesized from glucose in a series of chemical reactions similar to those found in vertebrates (Chippendale 1978). The glycogen reserves are most abundant in the fat body, flight muscles and intestinal tissues. The catalyst for glycogen synthesis, glycogen synthase, is bound to glycogen particles in the cytosol. Deposits of glyogen fluctuate with activity level and age. The mobilization of glucose and trehalose from glycogen stores is under metabolic and hormonal regulation. Trehalose concentration in the hemolymph is an important reserve sugar because it can be readily hydrolyzed to glucose, which is the high energy food for active muscles (Storey 1985). Trehalose is synthesized primarily in the fat body and released into the hemolymph. The breakdown of these sugars provides the source for cellular activity. This source is ATP (adenosine triphosphate), the hydrolysis of which releases the energy required for muscle contraction. Oxygen is needed for efficient carbohydrate utilization (serving as the last electron acceptor in the electron transport chain) especially in flight and during shivering activity.

Aggregations of specialized cells in the body cavity comprise the "fat body." This "meshwork" (Wigglesworth 1984) of lobes and delicate tissue membranes has high relative surface exposed to the blood. Fat bodies have important functions as food-storage reservoirs (lipid, glycogen, protein), and have high concentrations of mitochondria and enzymes (*e.g.*, esterase or lipase, which frees fatty acids from fat stores; succinoxidase and other enzymes of the Krebs cycle that provide hydrogen for the cytochrome transport chain; transaminases that convert amino acids into other amino acids; and enzymes that convert glucose into trehalose, deamination, synthesis of uric acid, oxidation of purines, etc.). Fat bodies also contain RNA (ribonucleic acid) and they synthesize protein for the blood (and egg development in fertile queens). They are important sites of intermediary metabolism. Although in most insects fat bodies are best developed in the late larval instars, and are often depleted by the end of metamorphosis, honey bees show a seasonal change with fat bodies appearing in "winter bees" but absent in "summer bees."

C. Respiration

In the year 1669, the Italian physiologist, Marcello Malpighi, discovered that insects breathe through tracheal tubes which open through the integument via spiracles. These tubes are on the order of 2 μm in diameter, and branch into tracheoles measuring 0.6 to 0.8 μm in diameter for 200 to 400 μm in length when they then taper to blind ends of 0.2-0.5 μm diameter (Wigglesworth 1983).

The proper balance of oxygen and carbon dioxide is maintained by passive diffusion of the gases throughout the tracheolar system. Adequate oxygen is supplied through this mechanism as the tracheoles come to within a few μm of mitochondria in the cells. The tracheoles supply the flight muscle with as much as 4 ml oxygen/g muscle/min without the accumulation of an oxygen debt (Wigglesworth 1983). Carbon dioxide elimination is achieved through the same means (diffusion), but it also diffuses more readily through the tissues than does oxygen. Thus, carbon dioxide is lost also directly through the tissues and the cuticle. Carbonic anhydrase may be important in releasing carbon dioxide as well. This enzyme catalyses the formation and breakdown of carbonic acid ($H_2CO_3 \rightarrow H_2O + CO_2$).

Ventilatory exchange can be greatly increased by movement of the abdomen, especially accordian-like longitudinal telescoping movements. Such movements, which can be seen with the naked eye on honey bees sitting still, increase convective exchange of air in the trachael system network.

The respiratory centers in the nervous system seem to lie in the segmentel ganglia and thoracic segments (Wigglesworth 1984). A reduced concentration of oxygen or increased concentration of carbon dioxide may stimulate respiratory centers probably through changes in acidity at these sites (Seeley 1974).

D. Blood and Circulation

The blood, or hemolymph, fills the body cavity (hemocoele) in an "open" circulation system. One blood vessel moves blood along the midline of the back (dorsum). The posterior segment of vessel (heart) pulsates and has valved openings (ostia) through which the blood enters. The anterior segment (aorta) is a tube of contractile tissue passing through the brain in the head above the esophagus and then ending. The blood leaves the vessel and percolates backwards bathing all the tissues. Under high temperature stress the circulating blood is utilized in keeping a cool head (Heinrich 1980). At cool temperatures (5°C), there is apparently more blood circulated to the head (Crailsheim 1985). Under non-stress temperatures at rest, the blood pressure in the body cavity may be less than the surrounding atmospheric pressure. The rigid cuticle prevents collapse. The blood pressure increases during movement.

The heart action causes the blood to circulate. Like a mammalian heart, blood enters the organ at a pressure less than that in the general body cavity. A weak positive pressure is exerted by the heart contraction driving the blood forward through the aorta. In the thorax are accessory contractile tissues surrounding the vessel which help to move the blood from the thoracic cavity through the wing veins and return it to the body cavity again.

The heart beat is myogenic with the rhythmic contraction originating within the heart muscle tissue itself. It does not need nervous impulses from any other "higher" nerve center. The rate is affected by both nervous and hormonal controls. Acceleration of heart rate is usually brought about by acetyl choline. Epinephrine also causes acceleration of the rate.

Table 1 - Properties of honey bee hemolympth (after Wigglesworth 1984)

color	clear with a greenish tinge?
red blood cells	none
specific gravity	1.045
pH (acidity)	6.83 (becomes more acid in pupal stage; varies with age and sex)
chief simple sugar	glucose
chief carbohydrate	trehalose (a non reducing disaccharide; blood concentration as high as 5%)
buffers	bicarbonates, phosphates, amino acids (ca 15% of total nitrogen), proteins (up to 6%)
molecular concentration	1.5% (expressed in terms of [NaCl] of same osmotic pressure)
other organic metabolites	succinate, pyruvate, citrate, a-glycerophosphate, glucose-6-phosphate, etc.
plasmatocytes	phagocytic blood cells

Hemolymph proteins have been identified by electrophoresis and antigen-antibody precipitation, and at least 20 proteins have now been distinguished. The bee blood carries about 20 times the concentration of amino acids as found in human blood, and has about twice the molecular concentration of mammalian blood (ca.0.9%). The plasmatocytes either circulate in the blood or adhere to tissue surfaces where they ingest dead bacteria and dead tissue cells. The blood is unimportant in the transport of oxygen to the tissues. Both oxygen and carbon dioxide are in physical solution, although the transport of dissolved carbon dioxide may be more important as it dissolves more readily then oxygen. Blood transports nutrients and hormones to the bathed tissues, and waste products to the excretory organs, and serves as a reservoir of food materials. In larvae, there is no blood clotting mechanism, so the blood is unimportant in closing wounds as it is in adults (Wigglesworth 1984).

The blood has an important role in transmitting hydrostatic pressure from one region of the body to another. It also functions in the mass transfer of heat by movement of warm blood from one region to another.

E. Salt and Water Balance

Bees have Malpighian tubules (named after Malpighi) that function in the control of salt and water balance. These are analogous to the glomerulus structure of the mammalian kidney. They are elongated tubular glands bathed in the blood that open into the junction between the midgut and hindgut. They contain microvilli which greatly increase the surface area for passive and active transport of substances including mineral salts, water, end products and by products of metabolism (nitrogenous compounds, sulfur compounds, and phosphates), acids, bases and some ions. Uric acid is the end product of protein (pollen) catabolism. The primary urine is iso-osmotic with the hemolymph, however the composition of the urine voided finally from the anus is changed by reabsorption of water and specific compounds. The final fluid which the Malpighian tubules excrete is a "urine" with a yellowish pigment. This is combined with undigested food materials and eliminated from the body in the feces. If the bee has imbibed much water or is "water loaded" the fluid will be thin because little of the water is reabsorbed. If, on the other hand, the bee is dessicated, much water is reabsorbed and the fluid excreted is thicker.

Throughout larval life, excreted waste products accumulate in the gut. These are not eliminated and therefore do not contaminate the food on which the larvae are resting, but are eliminated at the bottom of the wax cell in the comb just prior to pupation. In the pupal stage, the small amount of waste excreted is stored in the gut until emergence at which time it is eliminated as a pastey meconium (Hagedorn personal communication).

Considerable water can be lost from a bee through respiration. Although the air taken in through the spiracles contains the same concentration of water vapor as environmental air, the air that escapes back out of the spiracles is saturated with water. This water can be seen as a condensate from honey bees kept in a glass container under cool conditions. Another route of water loss is by evaporation from the cuticular surface. The waxey cuticular covering, however, prevents much transpiration of water, and the primary loss is due to respiratory loss through the spiracles. As environmental temperature increases, the rate of water loss increases.

F. Sensory Physiology and Nerve Function

The central nervous system (CNS) in a honey bee consists of nerve ganglia comprised of neuronal cell bodies and processes in the body segments. Paired longitudinal nerves connect one ganglion with the next. The brain is several ganglia fused together (three pre-oral segmental ganglia). The neuronal cell bodies within the ganglia are not directly bathed by the hemolymph. Nutrients are conveyed via glial cells around the neurons (Mordue *et al.* 1980).

The nerve impulse or spike depends on changes in nerve membrane permeability to sodium and potassium ions. Sufficient stimuli alter the membrane permeability enough so that positively charged sodium ions rush in causing an impulse to occur. Once a spike has occurred, there follows a refractory period for a short time during which further stimulation does not result in another spike.

Sensory receptors in honey bees are sensitive to stimuli from a variety of sources. Mechanoreceptors responding to mechanical insults (touch, gravity) usually consist of socketed hairs on the bee's body which when deformed cause tactile stimulation. Chemoreceptors (taste, smell) found mostly on the antennae are sensitive to chemical compounds to which they are exposed via pores in neuronal plate endings. The threshold for perception of taste of sugars dissolved in nectar is between 1 and 2% sugar (i.e., 1-2 g sugar dissolved in 100 g of solution, also expressed as 0.03 - 0.06 molar, Frisch 1971). This is not to be confused with their threshold of acceptance. Honey bees may not accept nectars even with concentrations of sugar much higher than 2%, depending on conditions. If many plants are in bloom, the acceptance threshold is high, perhaps as high as 40% sugar. In the fall of the year in temperate geographical areas, when few plants are blooming, the threshold may be as low as 5% (Frisch 1971). These insects can also perceive differences among the various sugars such as fructose, glucose and sucrose (as well as many others) in addition to salts, and the tastes we equate to bitter and sour.

Chemosense has importance in collective colony coordination and other behaviors, particularly related with the queen and worker chemical pheromones (see chapter by Blum, Free 1987). Like flies (Diptera), honey bees show a variety of "footprint" chemicals. These secretions have been characterized by chromatographic techniques (Lensky *et al.* 1987). Included are alkanes, alkenes, alcohols, organic acids, ethers, esters and aldehydes. Tarsal glands secrete 12 compounds specific to queens, 11 specific for workers, and one specific to drones.

Honey bees also have receptor neurons sensitive to temperature, and to carbon dioxide and water vapor contents of the air. The one dorsal and two lateral ocelli contain light-sensitive receptor cells (*i.e.,* sensitive to light intensity) as do the ommatidia of the compound eyes. Each ommatidium registers an average light intensity from a small area of the visual field with the whole eye forming a mosaic pattern (apposition). These compound eyes also respond to polarized light and colors including ultra-violet light in a wave-length band of 0.30 to 0.65 μm (Figure 2, Frisch 1965, Rossel and Wehner 1986). The image formation is adequate for efficient discrimination of shapes and contrast, and detection of movement (Mordue *et al.* 1980).

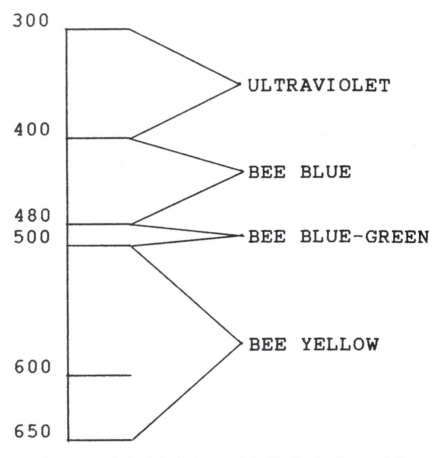

FIGURE 2. Spectrum of color detectable by honey bees. Wavelengths of light are indicated in nanometers (1nm = 10^{-9}m). Colors are given as seen by the bees. Humans cannot see in the uv (<400nm); bees cannot see in the red region (>650 nm). (modified from Frisch 1965).

Sound perception by honey bees is poorly documented. That they can detect vibration is shown by a complete and instantaneous slow-down of usual activities within an observation hive when a "C" tuning fork is placed against the glass. The bees do, of course, produce a variety of sounds. Wing beat frequency is tied in with most sound production, but there is also queen piping and high frequency sounds of workers within the hearing range of wax moths (*Galleria mellonella*) (Spangler, 1986).

Neuronal inputs are integrated for complex behaviors such as foraging and search, decision making, learning and memory (Menzel and Mercer 1987). For orientation and navigation, honey bees use landmarks and other visual cues including polarized and uv light, and sun and moon positions. Bees use the sun's position and the pattern of polarized light in the sky to keep track of the direction from hive to food sources or new homesites. They communicate this information to their colony mates with their waggle dances. Foragers compensate for the rotation of the earth relative to the sun's position using an internal "clock" (Dyer 1987). The physiological or biochemical mechanisms for time-compensation are poorly understood (Eckert & Randall 1983). If bees are placed on a day-night schedule in the laboratory with dawn and dusk shifted by several hours, they enter the incorrect time into their internal clocks and orient with a compass deviation corresponding to the artificial phase shift in the day-night cycle.

Bees have been observed by Roubik (pers comm) in Central America foraging on warm nights under a bright full moon. Further, they can utilize geomagnetic cues in navigation. They can detect and orient to the earth's magnetic field (Gould *et al.* 1980). Drones and queens may be able to orient to even small changes in the magnetic field in location of mating areas (drone congregation areas). Just how this magnetic field is detected is still unanswered. Honey bees and a few other animals including some birds, mollusks, and bacteria, contain magnetite, an iron containing area within the body.

Extensive behavioral experiments have shown conclusively that honey bees exhibit learning (*e.g.*, Dyer 1987, Frisch 1965, Menzel and Mercer 1987). They can learn to distinguish odors, learn and remember shapes, tell time of day and a large variety of other tasks. However, the neurophysiological basis for this learning eludes us (Gould 1985, 1986).

G. Reproduction, Development, Aging and Caste Structure

Like many insects, honey bees develop through complete metamorphosis from egg to larva to pupa to adult. This development occurs in the cells of the wax comb. In the pupal stage, the cell is sealed with a wax cap by house bees. The developmental steps include growth and morphogenesis. The details of how differentiation occurs are not fully understood. Certain hormones carried in the blood are known to play an important part in development and division of labor, especially the juvenile hormones (JH) (Robinson 1986, Sasagawa *et al.* 1986). JH are secreted from the copora allata and control metamorphosis and reproductive development. JH are not limited to juveniles, but are also found in adult honey bees. The titer of JH is usually high in larval stages (promoting juvenile growth and development, but not metamorphosis), but low during pupation. The variations in the titer of JH during development are often analyzed in an assay utilizing the wax moth, *Galleria* (see Mordue, *et al.* 1980). After completing development, the adult chews her way through the cap and emerges from the cell. Upon emergence, the young adult immediately fills the role of housekeeper by cleaning out her cell and surrounding cells, too. She cannot yet fly nor sting. Her physical size is determined by the size of the cell in which she developed, and she grows no larger as an adult. She soon begins to feed the older larvae with honey and pollen from the supplies stored elsewhere in the comb. After a couple of days, hypopharyngeal glands in her head mature enough to produce a special highly nutritious, protein-rich substance, royal jelly, which she feeds to the very young larvae (Takenaka 1988). Then, this worker assumes other roles in the social context of the colony.

JH, food and pheromones act in concert to control castes within the colony. For example, the undeveloped ovaries of workers are associated with low JH from inactive corpora allata brought about by exposure to queen contact pheromones. This is licked off the queen by her attendants and passed on to other workers by trophallaxis. Queen development is due in part to her exclusive feeding on royal jelly. Determinants of sex and other morphological differences between the sexuals are genetically determined. Drones develop from unfertilized eggs and are haploid (1n), while workers and the queen develop only from fertile eggs and are diploid (2n). The adult queen endocrine system has little effect on egg development, although it may influence her sexual behavior. A virgin queen may be induced to lay drone eggs artificially by anesthesia with carbon dioxide gas. Virgin queens normally mate with 8-19 drones in the air in a drone congregation area (DCA) (Koeniger 1988). Their specific activities during their mating flight have been clarified especially well by the study and excellent photographs of Koeniger (1986). Drones are attracted to the flying queen by her release of sex pheromones. The sperm from several drones that succeed in mating is not stored separately by drone, but is instead stored as mixed semen in the spermatheca (Moritz 1983). Once mated, the queen returns to her nest, and within a day or two begins her egg-laying activities.

Honey bees show aging characteristics. They show a deterioration in flight performance and a change in thermal preference to cooler temperatures. A number of changes in ultrastructure and biochemistry may be involved. These changes are likely to be genetically programmed.

H. Muscle Action

In vertebrates, a single nerve impulse usually evokes a single muscle contraction. This also holds for muscles involved in walking honey bees. The walking gait usually alternates between the first and third leg on one side and the second leg on the other side in contact with the flower or nest surface at one time, and the legs on the opposite sides of the thorax at the next step. This, of course, varies with the speed of walking. Central control is likely and has been found in other insect species. Sensory feedback is likely to be important at each step. The flight muscles of bees (and wasps, flies, beetles, and true bugs) however, are oscillating fibrillar muscles and show no relation between the timing of the individual contractions of the muscles and the timing of the motor nerve impulses (Eckert and Randall 1983). Bee flight muscles are asynchronous (because they do not contract in synchrony with each motor impulse). The muscle maintains an active state of fibrillation from a train of motor nerve impulses, but the contraction rate is much higher than the rate of nervous impulses. The thorax changes shape during flight muscle contractions in such a way as to favor two stable states, up or down (Figure 3). The elevator muscles elevate the wings by pulling down the roof of the thorax, snapping past the "click" point. This elevates the wings, stretches and activates the depressors, and slackens the elevator muscles suddenly, inactivating them. The depressor muscles lower the wings by shortening the thoracic exoskeleton (front to back) thereby expanding the thorax dorosoventrally. These depressors produce an upward deformation of the roof of the thorax until it "clicks" back to the raised position. The frequency of contraction depends on the mechanical resonance of the flight apparatus. Thus if wings are clipped short, the beat frequency increases, even with no change in nerve impulse activity. The wings assume an airfoil shape during their up and down beats, and the efficiency of flight and hovering is directly related to this (McMasters 1989). During shivering activity, these flight muscles vibrate at amplitudes so small that there is no apparent wing movement (Esch 1964).

The glycogen stored in flight muscle supports flight or shivering metabolism for only 10-20 min (Wigglesworth 1984). After this time, trehalose taken from the blood is the primary fuel, along with sugars from the honey stomach which can be used. Both potassium and calcium ions are important in the muscle contraction. When foraging honey bees are taken from flowers, the sugar dissolved in the hemolymph averages about 2.6% (0.08 molar). Blood sugar concentrations as high as 11.5% (0.35 molar) have been recorded. Should the concentration fall below 1% (0.03 molar), the forager is no longer able to fly. A bee will perch motionless with blood sugar concentration under 0.5% (0.02 molar). An hour of continuous flight may use up to 10 mg sugar to power the flight muscles. The sugar normally carried in the blood provides for about 15 min of flight (or about 5.5 km distance). When bees feed on "rich"

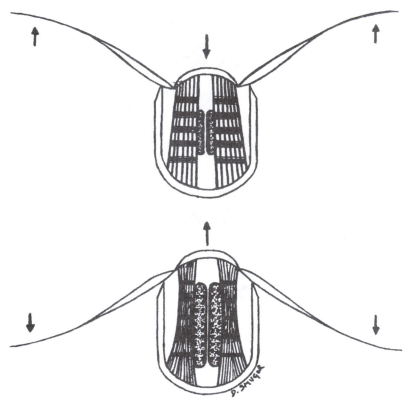

FIGURE 3. Asynchronous flight muscles in the thorax of a honey bee. This cross-sectional sketch shows contraction of the elevators pulling the roof of the thorax down raising the wings and stretching the longitudinal depressors, followed by contraction of the depressors causing the roof to move up and the wings to move down.

sugar solutions, they can fly as fast as 29 km/h, while a dilute solution results in slower flight speeds (4-7 km/h).

I. Body Temperature Control

Individual honey bees can achieve fairly high body temperatures during flight. The flight muscle activity incorporated during flight generates excess heat even at cool ambient temperatures. Heinrich (1980) recorded body temperatures as high as 47°C in flying honey bees. The high temperatures achieved in flight stimulate modes of reducing heat load including increased blood flow to the head and evaporative cooling. It is apparent also that the temperature may vary substantially in different regions of the active bee. The head temperature may be as much as 3°C cooler than the thorax. The honey bee under these conditions probably regulates head temperature, with thorax and abdomen temperatures passively following.

In flight, honey bees can maintain a fairly high thorax temperature over a wide range of air temperatures. Higher flight performance would be expected at higher thorax muscle temperatures. However, as isolated individuals, honey bees cannot maintain a high enough thorax temperature for flight (>27°C) at air temperatures of less than about 12°C (Heinrich 1979). When air (and body) temperatures decrease below 10°C, honey bees fall into a cold comatose condition (cold coma) from which they cannot voluntarily recover without external warming.

II. PHYSIOLOGY OF THE SOCIAL GROUP

The 20 to 80 thousand honey bees living together in a typical colony function as a social entity which shows consolidated group responses in food procurement, defense, recognition of kin, daily rhythms of metabolism and breathing, and temperature regulation. The stability and control exhibited at the colony-level are not seen in individual members isolated from the group nor are they just the sum of behaviors of the individuals comprising the group.

A. Physiology of Social Behavior

The colony has no central nervous control, of course, but rather is regulated chemically by volatile chemical pheromones released primarily from the queen. Pheromones are chemical substances that are secreted externally and elicit behavioral or physiological responses in other members of the species. A number of pheromones have been identified that are important for their social organization. These include several chemicals produced by the queen that identify her as the queen, help in calming the colony, and at the same time prevent development of the ovaries of the female workers as discussed above. Should a queen be removed from a colony, there is soon a commotion and the workers become easily agitated. The classical *queen substance,* a contact pheromone synthesized in the queen's mandibular glands, is known to be the key control chemical in the honey-bee colony. This *queen substance* is known to calm the colony. It has a composition of long chain fatty acids (8-hydroxy-octenoic acid, 9-oxo-2-decenoic acid, 10-hydroxy-2-decenoic-acid, 9-hydroxy-2-decenoic acid and 10-hydroxy-decenoic acid). The absence of these queen pheromones enables workers to detect the loss of the queen after about 30 min, so some must be highly volatile. Other components of the queen repertoire of pheromones are obtained by workers by licking or on their antennae and are likely of low volatility. The ODA and HDAs (above) stimulate brood rearing, comb building, hoarding, and foraging. These are also apparently important in the maintenance of her court of workers. 10-HDA is also a dominant component of the mandibular glands of worker bees. Other pheromones have been identified with the comb, mating, nest and colony, trail and scent marking of food sources, Nassanoff gland, alarm, drone (drone population in a colony is probably under pheromone control), and drone comb.

Workers also produce several important pheromones. One is released from an abdominal gland (Nassanoff gland) at the entrance to their nest or on the surface of a swarm to help returning foragers and scouts find home. A different chemical (isopentyl acetate) is released when there is danger. This brings on a group alarm behavior described below. Other chemicals are released which aid in the recognition of members of the particular social group (Moritz and Southwick 1987). Detailed analysis of pheromones is provided in the chapter by Blum.

The individual worker bees show polyethism, that is, they usually perform specific tasks dependent on their age since emergence. In their temporal division of labor, the young workers are most active near the center of the nest at first, and as they age, their labor moves outward from the nest center. This seems to be an adaptation that enhances the likelihood of survival of the worker as well as the colony (Seeley 1985). Colony-level homeostasis and adaptations help to explain the successful spread of the honey bee throughout the world.

One of the curious physiological controls resulting from social interaction is the drone "titer" in a colony. This titer, or concentration of drone number (*i.e.,* drone/worker ratio), is maintained at determined levels depending on the time of the year. That is, there are few drones in the winter, and then more and more as the summer season progresses. The queen is inhibited from laying drone (unfertilized) eggs when "plenty" of drones are present in the colony. Should the drone titer fall, the queen lays more eggs, thus balancing this titer at

Table 2: Pheromones identified in the honey bee. Over 30 pheromones have effects in honey-bee colony behavior. Several of the best known are given below (after Free 1987).

Name	Key Chemical	Source	Effect
queen substance	9-oxo-2-decenoic acid (9-ODA) 9-hydroxy-2-decenoic acid (9-HDA)	Queen's mandibular glands	*sex attractant *worker attraction
	10-hydroxy-2-decenoic acid (10-HDA)	also, dominant cmpd of mandib gland of worker	*colony & swarm stabilization *worker ovary inhibitor
alarm odor	iso-pentyl-acetate	lining of sting chamber	*defensive behavior
alarm odor	2-heptanone	worker's mandibular glands	*defensive behavior
scent gland secretion	geraniol, citral	worker's abdominal Nassanoff gland	*recruitment

the proper level. It was noticed by Rinderer *et al.* (1985, 1987) that this drone titer maintenance influences the population of European honey bees in geographical areas where Africanized honey bees are also present in the Americas. Africanized drones returning from afternoon flights to DCAs frequently return to and are accepted in a different colony than that from which they came. Their presence effectively raises the drone titer in the colony and inhibits the European queen from laying more drone eggs. At the same time, because of their absence from their home colony, the drone titer is reduced causing their original Africanized queen mother to lay even more drone eggs. Through this mechanism in warm geographical regions where Africanized races are present along with European races, there is considerable pressure for the population takeover by the Africanized strain.

B. Colony Food Procurement

Honey-bee workers forage for food not according to their own needs, but in response to the needs of the colony. Through careful behavioral studies, led by Karl von Frisch and Martin Lindauer of the University of Munich, Germany, the colony-level control mechanisms for food and water procurement have been elucidated (Frisch 1965, Lindauer 1967). Food and water foragers are usually the older experienced bees in the colony. They have become familiar with the landscape of the area in the vicinity of the nest site and have located flower patches containing good nectar sources, or places to obtain water. When nectar collecting is in progress, the foragers return to the nest with their nectar load and transmit it to younger house bees. Their acceptance by the house bees gauges the nectar demand. When delivery times are long, taking perhaps 2-3 min or more, the foraging bees' "eagerness" for collecting dramatically decreases. On the other hand, if they are quickly relieved of their nectar loads, they are stimulated to forage for more, or even to recruit other foragers to the same flower resource. If there is a shift in demand in the colony, for example to more dilute nectar or water because of a heat buildup, the delivery time for rich nectar will increase and that for dilute nectar or water will decrease resulting in more active foraging and recuitment to dilute sources. There is thus a colony-level consensus reached through the activities and communication of numerous individuals. This foraging activity has been described as an "information-center strategy" more recently by Seeley (1985) and his associates.

C. Circadian Rhythms

Honey-bee activities follow a daily cycle during the summer season (*e.g.,* Frisch and Aschoff 1987, Moore 1983). As honey bees are diurnal (day active), during the daylight hours they forage for nectar and pollen, and carry out many other activities. At night, when nearly all the foragers are back in the colony, their activities are greatly reduced. Metabolic measurements of groups of bees and of entire colonies made by monitoring respiratory oxygen consumption and production of carbon dioxide show a daily rhythm. At moderate temperatures (10-20°C) lowest metabolic rates are found to occur at about 4 o'clock in the morning and highest rates at about 3 o'clock in the afternoon. When temperature is held constant under laboratory conditions, the cyclic colony metabolism is keyed to photoperiod. The day-night change is not small, being as much as eight times higher in the day. Even under constant low light or dark conditions, the rhythmic metabolism persists, indicating a truly internal (endogenous) social clock. This social circadian rhythm of metabolism is apparent in colonies with or without brood. The evidence indicates that the individual clocks of the bees are set and coordinated by the group. Tactile interactions among bees are responsible for this social coordination, although the actual mechanism of setting the internal clock of the group remains a mystery (Southwick and Moritz 1987a).

D. Social "Breathing"

The high metabolic rate of the bees requires adequate exchange of respiratory gases between the colony and the environment. Circulation of the atmosphere in the nest cavity is maintained by fanning behavior of some of the bees. The ventilation activities of fanning bees may be observed at the entrances to natural nests or manmade hives on hot summer days. Their nest cavities usually have only one entrance so the fanning functions in a way that assures outflow of stale air and inflow of fresh air, sometimes in an in-out ventilatory air movement. The bees form legions of fanners in fixed positions about the bottom and sides of the cavity and even outside the entrance or on the landing platform. The inside fanners circulate the inside air, but the behavior of the fanners at the entrance is the crucial factor in determining the direction of air flow into and out of the nest cavity (Southwick and Moritz 1987b). When these few bees fan with their abdomens pointed outward at the entrance, stale warm air flows outward, and when they stop fanning fresh air flows in. Respiratory frequencies of small colonies have been measured in the laboratory.

The social control of respiratory activity of the bee colony shows a conspicuous daily pattern, like that of metabolism, even under constant dark conditions. Instead of a muscular thoracic pump such as utilized by intact mammals or birds, the honey-bee colony actively changes direction of air flow by the behavior of entrance fanners.

E. Alarm Response

Another group physiological activity involves the organized colonial defense demonstrated by a colony of bees. The pattern of defensive behavior commences with the alerting of the colony by the guard bees. This is followed by numerous bees taking flight in defense of the colony. Alerted or stinging workers release alarm pheromones (especially isopentyl acetate) that elicit colonial defense. The response to alarm pheromone has been quantified and shows that defensive behavior is highly dependent on the number of bees in the colony or group (Southwick and Moritz 1985). There is very little response from one or a few bees. The response *per bee* increases as the number of individuals increases in the group giving an ever larger group response. Individuals respond differently when they are a part of a group than they do when solitary. When one encounters solitary honey bees feeding in the field, there is rarely any defensive behavior exhibited. Here the bees demonstrate a colonial pattern of temperament not observable in individuals.

The intensity of defensive behavior of honey bees is highly dependent on external environmental factors. In studies of defensive response to the alarm pheromone as shown by numbers of stings in a leather target, Southwick and Moritz (1987c) showed that weather factors affected the results more than might be expected (Figure 4). For example, contrary to the idea that the best time to manipulate a hive of bees is when most of the foragers are out foraging on a nice sunny day, their results showed that bee stings increased with

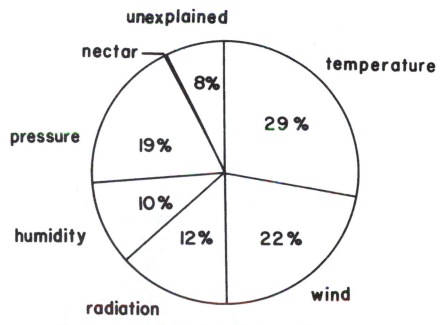

FIGURE 4. Climatic components of sting behavior in honey bees. Sting behavior is explained by the variables indicated, and can be predicted if these meteorological variables are known. More stinging results under conditions of high temperature, low wind, full sun, high humidity, and high pressure. Nectar flow had little effect on stinging (from data reported in Southwick and Moritz 1987).

temperature and solar radiation, and decreased with wind. There was a small increase in stings with increased barometric pressure as well. Practically no correlation is shown between nectar flow and stingyness. Of course, these results are only measured in terms of actual stings delivered. It may be that guard bees become "ornery" flying at the beekeeper bombarding the head and so forth without actually stinging, but, nevertheless, making it difficult to handle the colony.

Japanese honey bees show a balling defensive behavior against invading wasps (Ono, *et al.* 1988, 1987). Many workers cling to the invader and form a small group in the form of a ball. They apparently shiver and produce sufficient amounts of heat to raise the internal temperature of the ball, where the wasp is located, to lethal levels for the wasp. Temperatures in excess of 43°C kill the wasp, but are tolerated by the bees.

F. Recognizing Relatives

Members of a honey-bee colony are able to recognize their own nestmates and distinguish them from foreign visitors of other colonies. Behavioral studies suggest that genetically determined odors are used as recognition cues by individual workers. It has recently been demonstrated using quantitative physiological tests that worker bees use volatile odors in discerning related drones and workers, and thus probably odor cues are crucial for the identification of kin in honey bees (Moritz and Southwick 1987).

G. Social Temperature Regulation

One of the most interesting and most studied colony-level controls is that of temperature regulation. A thermal profile of the bee colony shows high stable temperatures in the central core area (where the brood is located) and concentric isotherms of lower temperatures at distances from the center (Owens 1971) (Figure 5). Through social homeostatic adjustments the colony maintains high relatively constant core temperatures (especially if there is brood present) under a variety of environmental conditions (Heinrich 1985, Southwick 1988, 1983). A bee hive placed on a lava field in full sun where the air temperature reached 70°C was able to maintain its normal hive temperature of 35°C (Lindauer 1954). Colonies tested at air temperatures as low as -80°C are able to maintain their brood nest temperature at 35°C (Southwick 1987, Southwick and Heldmaier 1987). This tremendous environmental temperature range is tolerated through the employment of a number of specific mechanisms of temperature control by the colony (Figure 6) (Southwick 1988, 1985a).

Wide spacing among individuals in the hive, fanning and evaporative cooling are used in hot weather. A critical factor in this cooling is the constant high temperature of 35°C the bees maintain within the hive. This high temperature is always above the dew point (the temperature of the air at which

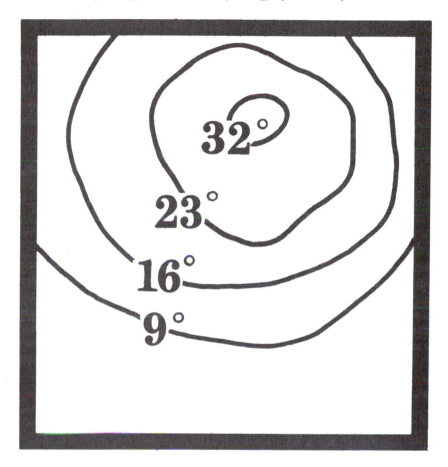

5 °C

FIGURE 5. Thermal profile of a colony of honey bees in a nest at an air temperature of about 5°C. The high core temperature is maintained and the bees at the cluster surface are kept at about 9°C.

the water vapor reaches saturation). The efficiency of the evaporative cooling (in which approximately 580 calories of heat are lost for each gram of water evaporated) is directly dependent on the difference between the hive temperature and the dew point. The bee colony has a tremendous advantage in cooling as it nearly always maintains a large difference between hive temperature and dew point. The actual water vapor pressure in the hive is, therefore, always lower than the saturation vapor pressure resulting in a large saturation deficit.

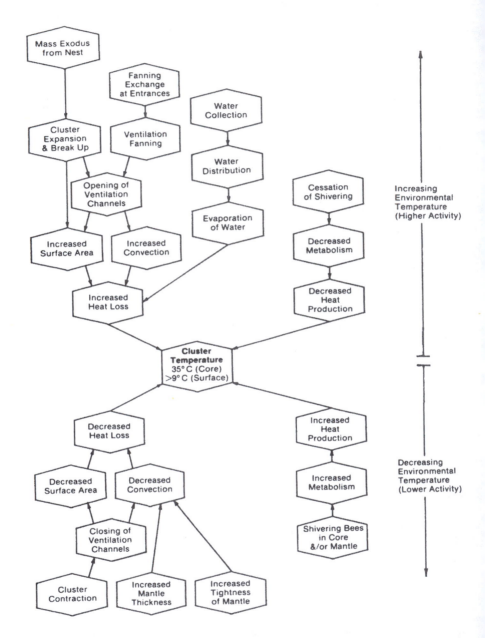

Figure 6. Flow diagram of temperature control in a honey bee colony in hot and cold environments (modified from Southwick 1988).

The bee can take advantage of this by holding a water droplet in a thin film on its mouth parts, and placing small droplets on surfaces throughout the hive, and then causing air currents by fanning behavior. So even if a colony of bees is in a very hot environment such as a desert, it has no problem cooling as long as there is a sufficient supply of water. The same saturation deficit that allows for effective cooling insures high rates of water loss from the stored nectar, resulting in honey of low water content (nectar has about 60% water while honey is only about 20% water). Even in cool rainy weather in the spring and fall of the year, the saturation deficit is still maintained because of the high hive temperature, and evaporation of water occurs.

As the weather cools, clustering reduces the exposed surface area of the bees where heat exchange takes place (Moebus 1978). As the temperatures drop further, the bees draw closer together forming an ever tighter cluster, conserving heat. This cluster expands and contracts in direct response to environmental temperatures at least around 15°C. Heat generated by the bees is further conserved by outer layers of bees forming an insulating shell around the core. The shell is composed of tightly compacted layers of bees with their heads pointed inward, and abdomens exposed to the cold. Their inward directed heads may be important in their precise thermal control since they have been shown to control head temperature as individuals under some circumstances (Heinrich 1980). The interlacing of thoracic hairs forms small dead air spaces creating an effective insulating coat for the colony; something they could not achieve as individuals. Heat retention by the hair is reflected in metabolic energy savings at low temperatures (Southwick 1985b).

A further heat savings is achieved in cold conditions since the bees cluster between the wax combs. The bees form a compact living layer over the comb effectively trapping air in the empty cells. As much as two-thirds of the surface area of the cluster is insulated by these combs in the nest cavity (Figure 7) (Southwick 1985a).

When all means of heat savings are employed and the temperature continues to fall, the honey bee cluster generates more heat metabolically by burning up the honey stores as a fuel (Ritter 1982). This metabolic heat production increases with decreasing air temperature (Figure 8). Esch (1964) of Notre Dame University found that the shivering flight muscles of the thorax are the heat producers in honey bees. The thoracic muscle is one of the most metabolically active tissues known. The shivering amplitude is too small to be seen, but Esch was able to measure the muscle electrical potential showing rapid low-amplitude contractions were taking place. Muscle tissue in honey bees utilizes predominantly carbohydrate as a fuel.

Honey-bee clusters of different sizes show different responses to falling temperature. Generally, bigger is better! The larger the population of bees, the less it costs each bee for temperature maintenance during the winter. To be a member of a small cluster is very costly. The high broodnest temperature probably increases the resistance to some diseases (Seeley 1985).

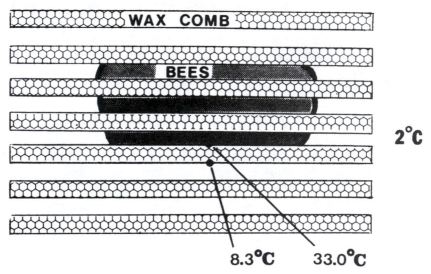

FIGURE 7. Effective insulation of empty honeycomb adjacent to a winter cluster of honey bees. Temperature difference on the cluster side of the comb and 2 cm through on the side away from the cluster is 25°C in a 2°C air temperature (from Southwick 1985a).

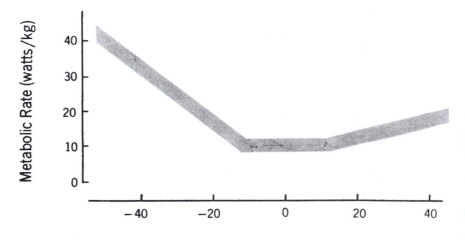

FIGURE 8. The pattern of metabolic heat production as affected by ambient temperature for colonies of honey bees. As the temperature falls below about -10°C, the bees must expend more and more metabolic energy (and consume more honey stores) to keep warm.

Precise thermal control, especially under low temperature conditions, is costly to the bees and requires cooperation among members of the colony. Some races of honey bees are less well adapted to the conditions encountered in long north temperate winters. Because of this, geographical limits for northern overwintering can be expected to be tied in with climatological factors and physiological constraints as well as behavioral attributes of the particular races (Southwick *et al.* 1990).

III. CONCLUSION

Overall, the honey-bee colony is a remarkably well organized social group showing regulatory functions that result in precise controls. Although individuals show normal physiological functions as other insects, they function differently within the perennial colony, serving as components of a larger entity to meet the demands placed upon them by stresses of the natural environment. Further study of this harmonious social system may shed light on our understanding of selective forces and adaptation in physiology, ecology and behavior.

REFERENCES

Baker, H.G. and I. Baker (1975). Nectar constitution and pollinator-plant coevolution. Pp. 100-140 In: L.E. Gilbert and P.H. Raven, (eds.). Co-evolution of animals and plants. Univ. Texas Press, Austin.

Beutler, R. (1930). Biologisch-chemische Untersuchungen am Nektar von Immenblumen. *Zeits. vergl. Physiol. 12*:72-176.

Blum, M.S. (1985). Fundamentals of insect physiology. John Wiley & Sons, New York, 598p.

Chippendale, G.M. (1978). The functions of carbohydrates in insect life processes. P.1-55 In: M. Rockstein (*ed.*). Biochemistry of insects. Academic Press, New York.

Crailsheim, K. (1985). Distribution of haemolymph in the honeybee (*Apis mellifica*) in relation to season, age and temperature. *J. Insect Physiol. 31*:707-713.

Dyer, F.C. (1987). Memory and sun compensation by honey bees. *J. Comp. Physiol. 160*:621-633.

Eckert and Randall (1983). Animal physiology: mechanisms and adaptations. 2nd ed. Freeman & Co., New York, p.333-380.

Esch, H. (1964). Ueber den Zusammenhang zwischen Temperatur, Aktionspotentialen und Toraxbewegungen bei der Honigbiene (*Apis mellifica* L.). *Z.vergl. Physiol. 48*:547-551.

Free, J.B. (1987). Pheromones of social bees. Chapman and Hall, UK, 218p.

Frisch, B. and J. Aschoff. (1987). Circadian rhythms in honeybees: entrainment by feeding cycles. *Physiol. Entomol. 12*:41-49.

Frisch, K. von. (1971). Bees, their vision, chemical senses, and language. Cornell University Press, Ithaca, New York.

Frisch, K. von. (1965). Tanzsprache und Orientierung der Bienen. Springer-Verlag, Berlin.

Gould, J.L. (1986). The locale map of honey bees: do insects have cognitive maps? *Science 232*:861-863.

Gould, J.L. (1985). How bees remember flower shapes. *Science 227*:1492-1494.

Gould, J.L., J.L. Kirschvink, K.S. Deffeyes, M.L. Brines. (1980). Orientation of demagnetized bees. *J. exp. Biol. 86*:1-8.

Heinrich, B. (1987). The social physiology of temperature regulation in honey bees. Pp.393-406 *In*: Experimental Behavioral Ecology, Fischer Verlag, Berlin.

Heinrich, B. (1980). Mechanisms of body-temperature regulation in honeybees, *Apis mellifera. J. exp. Biol. 85*:61-72.

Heinrich, B. (1979). Thermoregulation of African and European honeybees during foraging, attack, and hive exits and returns. *J. Exp. Biol. 80*:217-229.

Inouye, D.W. and G.D. Waller. (1984). Responses of honey bees (*Apis mellifera*) to amino acid solutions mimicking floral nectars. *Ecology 65*:618-625.

Kerkut, G.A. and L.I. Gilbert (eds.). (1985). Comprehensive insect physiology, biochemistry and pharmacology. 13vols. Pergamon Press, New York esp vols 3,4,7.

Koeniger, G. (1988). Mating behavior of honey bees. Pp. 167-172 *In*: G.R. Needham, R.E. Page, Jr., M. Delfinado-Baker, C.E. Bowman. Africanized honey bees and bee mites. John Wiley & Sons, New York.

Koeniger, G. (1986). Reproduction and mating behavior. Pp.255-280 *In*: T.E. Rinderer (*ed.*). Bee genetics and breeding. Academic Press, London, U.K.

Lanham, U.N. (1979). Possible phylogenetic significance of complex hairs in bees and ants. *J. New York ent. soc. 87*:91-94.

Lensky, Y., A. Finkel, P. Cassier, A. Teeshbee, R. Schlesinger. (1987). The tarsal glands of honeybee (*Apis mellifera* L.) queens, workers and drones—chemical characterization of footprint secretions. *Honeybee Sci. 8*:97-102. (in Japanese)

Lindauer, M. (1967). Recent advances in bee communication and orientation. *Ann. Review Entomol. 12*:439-470.

Lindauer, M. (1954). Temperaturregulierung und Wasserhaushalt im Bienenstaat. *Zeit. vergl. Physiol. 36*:391-432.

McMasters, J.H. (1989). The flight of the bumblebee and related myths of entomological engineering. *American Scientist 77*:164-169.

Menzel, R. and A. Mercer. (1987). Neurobiology and behavior of honeybees. Springer-Verlag, Berlin. 334pp.

Moebus, B. (1978). Um die Wintertraube bei den Bienen. *Bienenwelt 20*:170-182

Moore, D.J. (1983). The honeybee circadian clock: regulation of foraging and general locomotor behavior in *Apis mellifera*. Doc. Diss., Dept. Zool., Univ. Texas, Austin, TX, USA, 199pp.

Mordue, W., G.J. Goldsworthy, J.Brady, W.M. Blaney. (1980). Insect physiology. John Wiley & Sons, New York 108pp.

Moritz, R.F.A. (1983). Homogeneous mixing of honeybee semen by centrifugation. *J. Apic. Res. 22*:249-255.

Moritz, R.F.A. and E.E. Southwick. (1987). Metabolic test of volatile odor labels as kin recognition cues in honey bees. *Journal of Experimental Zoology 243*:503-507.

Ono, M., I. Okada, and M. Sasaki. (1987). Heat production by balling in the Japanese honey bees, *Apis cerana japonica* as a defensive behavior against the hornet. *Vespa simillima xanthoptera* (Hymenoptera: Vespidae). *Experientia 43*:1031-1032.

Owens, C.D. (1971). The thermology of wintering honey bees. *U.S.D.A. Tech. Bull. 1429*:1-42.

Peng, Y.-S., M.E. Nasr, and J.M. Marston. (1986). Release of alfalfa, *Medicago sativa*, pollen cytoplasm in the gut of the honey bee, *Apis mellifera* (Hymenoptera: Apidae). *Ann. Entomol. Soc. Am. 79*:804-807.

Porlier, B. (1987). La cuticule de l'abeille: un role indispensable dans l'organisme de l'insecte. *Abeilles et Fleurs 370*:19-21.

Rinderer, T.E., R.L. Hellmich II, R.G. Danka, A.M. Collins. (1985). Male reproductive parasitism: a factor in the Africanization of European honey-bee populations. *Science 228*:1119-1121.

Rinderer, T.E., A.M. Collins, R.L. Hellmich II, R.G. Danka. (1987). Differential drone production by Africanized and European honey-bee colonies. *Apidologie 18*:61-68.

Ritter, W. (1982). Experimenteller Beitrag zur Thermoregulation des Bienenvolkes (*Apis mellifera* L.). *Apidologie 13*:169-185.

Robinson, G.E. (1986). Hormonal regulation of division of labor in honey bee (*Apis mellifera* L.) colonies. Doct. Diss. Cornell Univ., Ithaca, NY, USA, 118pp.

Rockstein, M. (ed.). (1978). Biochemistry of insects. Academic Press, New York. 649p.

Rockstein, M. (ed.). (1964). The physiology of insecta. Academic Press, New York. 3 vols.

Rossel, S. and R. Wehner. (1986). Polarization vision in bees. *Nature 323*:128-131.

Ruttner, F. (1988). breeding techniques and selection for breeding of the honeybee. Ehrenwirth Verlag, Munich, W. Germany.

Sadwick, S.E. (1983). Aminco acids in *Asclepias* nectar. Masters Diss., State Univ New York, Brockport.

Sasagawa, H., M. Sasaki, and I. Okada. (1986). Experimental induction of the division of labour in worker *Apis mellifera* L. by juvenile hormone (JH) and its analog. Pp.140-143 In: Proc. XXXth Int. Congr. Apiculture, Nagoya, Japan. Apimondia.

Seeley, T.D. (1985). Honeybee ecology. Princeton University Press, Princeton, New Jersey 201pp.

Seeley, T.D. (1974). Atmospheric carbon dioxide regulation in honey-bee (*Apis mellifera*) colonies. *J. Insect Physiol. 20*:2301-2305.

Southwick, E.E. (1988). Thermoregulation in honey-bee colonies. Pp. 223-236 *In*: M. Delfinado-Baker, G. Needham, R. Page and C. Bowman (*eds.*). Africanized honey bees and bee mites. Ellis Horwood Ltd., Chichester, U.K. (Distr. by Wiley in U.S.)

Southwick, E.E. (1987). Cooperative metabolism in honey bees: an alternative to antifreeze and hibernation. *Journal of Thermal Biology 12*:155-158.

Southwick, E.E. (1985a). Allometric relations, metabolism and heat conductance in clusters of honey bees at cool temperatures. *Journal of Comparative Physiology 156B*:143-149.

Southwick, E.E. (1985b). Bee hair structure and effect of hair on metabolism at cool temperature. *Journal of Apicultural Research 24*:144-149.

Southwick, E.E. (1983). The honey bee cluster as a homeothermic superorganism. *Comparative Biochemistry and Physiology 75A*:641-645.

Southwick, E.E. and G. Heldmaier. (1987). Temperature control in honey bee colonies. *BioScience 37*:395-399.

Southwick, E.E. and R.F.A. Moritz (1987a). Social synchronization of circadian rhythms of metabolism in honey bees (*Apis mellifera* L.). *Physiological Entomology 12*:209-212.

Southwick, E.E. and R.F.A. Moritz. (1987b). Social control of air ventilation in colonies of honey bees. *Apis mellifera. Journal of Insect Physiology 33*:623-626.

Southwick, E.E. and R.F.A. Moritz. (1987c). Effects of meteorological factors on defensive behavior of honey bees. *International Journal of Biometeorology 31*(3):259-265.

Southwick, E.E. and R.F.A. Moritz. (1985). Metabolic response to alarm pheromone in honey bees. *Journal of Insect Physiology 31*:389-392.

Southwick, E.E., G.M. Loper and S.E. Sadwick. (1981). Nectar production, composition, energetics and pollinator attractiveness in spring flowers of western New York. *American Journal of Botany 68*:994-1002.

Southwick, E.E., D. Roubik, J. Williams, (1990) Comparative energy balance in groups of Africanized and European honey bees: Ecological implications. *Comparative Biochemistry and Physiology. In press.*

Spangler, H.G. (1986). High-frequency sound production by honeybees. *J. Apic. Res. 25*:213-219.

Storey, K.B. (1985). Metabolic biochemistry of insect flight. P.193-207 *In*: R. Gilles (*ed.*). Circulation, respiration and metabolism. Springer-Verlag, Berlin.

Takenaka, T. (1988). Protein synthesis by the hypopharyngeal glands of worker honeybees. *Honeybee Science 9*:13-18. (in Japanese)

Wigglesworth, V.B. (1984). Insect physiology. 8th ed. University Press, Cambridge, UK 191 pp.

Wigglesworth, V.B. (1983). The physiology of insect tracheoles. P.85-148 *In*: M.J. Berridge, J.E. Treherne, and V.B. Wigglesworth (*eds.*). Advances in insect physiology, Academic Press, New York.

CHAPTER 6

HONEY BEE NUTRITION

by ELTON W. HERBERT, Jr.*

Honey bees, like most animals, require proteins, carbohydrates, minerals, fats (lipids), vitamins, and water for normal growth and development. These nutritional needs are satisfied by the collection of nectar, pollen, and water. Nectar, which is collected by honey bees from either floral or extra floral nectaries of flowers, satisfies the carbohydrate requirement. Pollen, which is collected by honey bees from a wide range of flowering plants, normally satisfies the dietary requirement for proteins, minerals, lipids, and vitamins. Each class of nutrients will be briefly discussed beginning with the proteins. It should be stressed that these are the nutritional requirements of adult honey bees, not those of immature honey bees; these are different.

PROTEINS

The development of body tissue, muscles, and glands such as the hypopharyngeal glands, depends upon adequate amounts of protein in the honey bee's diet. During the early adult life of worker bees all nitrogen is derived from pollen protein. Consequently, young bees must consume a large quantity of pollen in the first two weeks of their adult life. Some worker bees begin consuming pollen within one to two hours after emergence. Within twelve hours after emergence, 50 percent or more of the workers have already consumed pollen in small amounts (Dietz 1969). However, mass consumption begins when the bees are 42 to 52 hours old (Hagedorn and Moeller 1967), and reaches a maximum when they are five days old. Within five days following emergence the nitrogen content of adult bees increases to 93 percent in the head, by 76 percent in the abdomen, and by 37 percent in the thorax (Haydak 1934). Simultaneously, their hypopharyngeal glands, fat bodies, and other internal organs develop (Maurizio 1954).

Nursing duties are normally finished and field duties are undertaken when bees are 10-14 days old. At that time the requirement for pollen decreases and the chief dietary constituent becomes carbohydrates which are obtained primarily from nectar and honey. Under abnormal conditions, such as when honey bees are required to rear brood continuously, proteinaceous food is consumed for a longer period and the hypopharyngeal glands of up to 70 percent of the bees 75 to 83 days old remain fully active (Haydak 1961, 1963).

Pollen is collected by honey bees from a great variety of plants, and the chemical composition and nutritive value vary according to the plant source.

*Elton W. Herbert Jr., Ph.D. (deceased), Research Entomologist, USDA, ARS Beneficial Insects Laboratory, Beltsville, Md. 20705

Many factors such as air temperature and soil moisture, pH, and fertility affect the nutritive value of pollen. Consequently, pollen from a single floral source may be different chemically from a similar pollen collected in a different area. The protein level of pollen collected from different plants ranges from eight to 40 percent. This large variability in chemical composition causes an equally large variability in food value for bees, and as a consequence, all pollen does not have the same physiological effects on honey bees (Todd and Bretherick 1942; Vivino and Palmer 1944; Maurizio 1954).

Pollen has been classified into four groups based upon the influence on bee longevity and development of the hypopharyngeal glands, ovaries, and fat bodies (Maurizio 1960). The first group of highly nutritious pollen includes fruit trees, willow, corn, and white clover. The second group of less nutritious pollen includes elm, cottonwood, and dandelion. The third group of pollen, with only a fair nutritional value, comes from alder and hazelnut. The fourth group of pollen, with the poorest nutritive value, includes various species of pine trees.

The annual pollen requirements for a colony of honey bees varies considerably depending on colony location, strength, and the floral sources. Haydak (1935) reported that honey bee colonies require 44 to 66 lbs. of pollen annually. Todd (1940) estimated the annual colony requirement to be 88 lbs. On an individual bee basis each larva requires more than 100 mg of pollen to complete its development (Haydak 1935). Schmidt and Buchmann (1985) reported that the average weight of nitrogen consumed per individual over a 28 day period was 3.07 mg, of which 0.60 mg was defecated in the form of 86.2% undigested food and 13.8% uric acid. Rashad and Parker (1958) reported that 66.5 mg of fresh pollen was needed to rear one larva. They further calculated that one cell contained 183 mg of bee bread which they stated was sufficient to rear 1.2 bees. Based on the above estimates, one pound of pollen would support the rearing of over 4,000 bees. Since a strong colony rears about 200,000 bees a year, a minimum of 44 pounds of pollen would be required (Nolan 1925).

DeGroot (1953) established that 10 amino acids must be present in the diet of honey bees for maximum development. By omitting each of 17 L-amino acids in turn, DeGroot found that for satisfactory development and survival the following 10 amino acids were essential: arginine, histidine, lysine, tryptophane, phenylalanine, methionine, threonine, leucine, isoleucine, and valine. DeGroot (1953) demonstrated that if a protein devoid of any one essential amino acid, but otherwise complete, was taken as the sole protein in the bees' diet, nitrogen equilibrium and development were impossible.

The amino acid composition of different pollens has been investigated by numerous researchers. Lunden (1954) investigated the pollen from five plant sources representing five families and also one mixed pollen, and found that the 10 essential amino acids are always present in about the same quantity. Bieberdorf *et al.* (1961) also reported that pollen extracts indicated the presence of essentially the same amino acids in all of the pollen analyzed, and the authors observed that in general honey bees more frequently visit plants with a larger number of amino acids than those having a lower number.

McCaughey *et al.* (1980) analyzed amino acids from various pollens in an attempt to show significant correlation between hypopharyngeal gland development and essential amino acid composition. All pollens contained detectable levels of the 17 amino acids. Glutamic acid, aspartic acid, and proline were the predominant amino acids, and showed the greatest variation in amount among species. Lysine and leucine were also found in high levels. Sarka and Wittner (1949) investigated 12 amino acids in *Zea mays* and they found that these amino acids accounted for about 50 percent of the total protein and 13 percent of the total dry weight. Nineteen amino acids were present in pollen collected by honey bees in Louisiana during one year (Kauffeld 1980). Proline, lysine, glycine, and phenylalanine showed considerable variation in amount among species throughout the study.

In mature pollen the predominant free amino acid is proline, with values often exceeding one percent of the fresh weight of pollen. Barker (1972) tested the effects of adding free proline to pollen substitutes to match those levels found in pollen. L-proline (1%) was mixed with soybean flour, honey, and water and fed to confined hives of honey bees. Honey bees fed natural pollen reared significantly more brood than those fed the pollen substitute, indicating that the superior nutritive value of pollen cannot be attributed to its unusually high content of free proline. Inouye and Waller (1984) measured the responses of honey bees to single amino acids dissolved in an aqueous solution of 30% sucrose at relatively high concentrations. Bees' responses to 24 L-amino acids, two DL mixtures, two phenolics, and a flavonoid were tested using artificial-flower feeders. There were no apparent correlations between characteristics of the amino acids and the bees' reaction to them, and no obvious difference between responses to essential and nonessential amino acids.

Auclair and Jamieson (1948) were unable to identify tryptophane in the free amino acids of protein hydrolysate of dandelion pollen or willow pollen. Kok (1952) analyzed pine pollen and found it deficient in cysteine, hydroxyproline, methionine, and tryptophane. Newly emerged honey bees held in cages were unable to rear brood when fed pure dandelion pollen diets (Herbert *et al.* 1970). Dandelion pollen analyzed chemically with an amino acid analyzer lacked tryptophane and phenylalanine; both are essential amino acids for bees. Dandelion pollen was also deficient in arginine, another essential amino acid. Dandelion pollen fortified with L-tryptophane and L-phenylalanine did not stimulate brood rearing; however, dandelion pollen diets fortified with L-arginine resulted in complete brood rearing. Loper and Berdel (1980) evaluated pollens with a bioassay that measured the relative nutritional efficiency of diets fed to young honey bees that were used to rear brood from eggs to the sealed stage. They found that no brood was raised by bees on a diet of dandelion pollen, or on dandelion pollen fortified with L-arginine. Knox *et al.* (1971) evaluated various pollens on the basis of honey bee longevity, development of the hypopharyngeal glands, and the amount of nitrogen in the pollens. The reduction in longevity of bees fed dandelion pollen was highly significant when compared with those fed either blueberry or cranberry pollen. Gilliam (1972) analyzed the electrophoretic patterns of hemolymph proteins of bees fed a complete diet, a protein deficient plus carbohydrate diet, or only a carbohydrate diet. Dandelion pollen was fed as the deficient diet. No electrophoretic differences among the three groups of honey bees were detected since the banding patterns were identical to those obtained from a separate group of control worker bees.

Two essential amino acids, L-histidine and L-methionine, were destroyed when pollen was treated with the ethylene oxide (EtO) for the destruction of *Bacillus larvae* White, the causative organism of American foulbrood disease (Herbert and Shimanuki 1971). No brood rearing was observed in any of the four cages of bees fed pollen fumigated with 450 or 850 mg/liter EtO. This inhibition was reversed when unfumigated pollen or fumigated pollen fortified with L-histidine plus L-methionine was fed. Boch *et al.* (1973) demonstrated that EtO fumigation of pollen resulted in a reduction of histidine. The addition of histidine to the fumigated pollen did not enable bees to rear brood to the sealed stage, indicating that nutrients other than histidine were also affected by fumigation.

Pollen stored in a freezer gradually loses its attractiveness and nutritive value for honey bees. This deterioration may be due, in part, to the loss of essential amino acids. Dietz and Haydak (1965) showed that the nutritional effects of stored pollen (3 years old) could be restored to the biological value level of fresh pollen by the addition of two amino acids, lysine and arginine. However, after 13 years of storage, the pollen was shown to have deteriorated to such an extent that the addition of amino acids could not restore it to its original nutritional value.

Honey bees exhibit preferences among pollen types when given a free choice which indicates that pollen odor may be crucial in pollen selection by foragers. Color may also be an important factor, but pH had little or no influence in the preferences of bees offered eight pollen types (Schmidt 1982).

Numerous chemicals have been identified as the sources of stimuli attracting honey bees to pollen. The source of stimuli in almond pollen was found in a benzene extract of the pollen (Doull 1974a). When a pollen supplement containing this fraction was compared with an untreated pollen supplement, it was preferred by more than 90 percent of the bees. Mixed pollen, gathered by bees, contained a free fatty acid which was shown to be an attractant for bees (Hopkins *et al.* 1969). Some of the attractive substances were soluble in chloroform, methanol, and/or water, but the whole pollen was more attractive than its chloroform-methanol extract and the chloroform-methanol extract of the same pollen was more attractive than the water extract (Doull and Standifer 1970). (See section on Lipids).

The location of pollen mixtures in the hive determines how quickly the diet will be consumed. Taber (1973) demonstrated that pollen mixtures placed in the brood nest of honey bee colonies near brood of known ages disappeared most rapidly when they were adjacent to unsealed brood. Pollen consumption was greater near older larvae than near sealed brood. Pollen supplement placed above honey combs adjacent to brood was also consumed, but more slowly than natural pollen. Doull (1974b) noted that the attractiveness of pollen in the hive declined as the distance from the brood area was increased. Honey bees' response to pollen declined significantly when the distance was increased from 1.0 to 1.5 cm, and even more at 1.5 and 2.0 cm. The response virtually disappeared at distances beyond 5 cm. This study demonstrated that pollen is unlikely to be used by the nurse bees unless it is very close to the brood.

The quantity of brood food that a worker larva receives is an important factor in its growth and development. Eischen *et al.* (1982) found a positive correlation between the number of workers, the adult life span, and dry weight of worker progeny they reared. When small numbers of workers cared for larger amounts of brood—resulting in inadequate larval nutrition—lighter, short-lived adults were reared. Workers in small colonies ate more pollen and reared more brood per individual than did workers in large colonies (Eischen *et al.* 1983). A positive correlation was found between pollen consumption by workers and the number of progeny reared per worker. Campana and Moeller (1977) reported that the number of honey bees reared depended more on the amount of food consumed (preference) than on the nutritive value (number of honey bees reared per gram of pollen consumed). Caged colonies of honey bees fed diets of bee-gathered pollen from each of five plant sources were able to rear brood for as long as 45 days.

The nutritive value of stored pollen for bees diminishes with time, depending on the drying conditions, the temperature and relative humidity, and the source of the pollen (Maurizio 1954; Haydak 1961; Dietz and Haydak 1965).

Fresh pollen, which was 100 percent effective in stimulating the development of hypopharyngeal gland development in worker bees, lost 76 percent of its effectiveness after being stored for one year. Two-year-old pollen did not stimulate gland development or support brood rearing (Haydak 1961). Brood rearing by bees fed 1- or 2-year-old pollen is often sporadic and spotty, lacking the biological potency of those fed fresh pollen.

Haydak (1961) showed that 8-year-old pollen was almost worthless for the development of young bees. Pollen stored at room temperature for 8 years, with or without sugar, was nutritionally inadequate to support brood rearing (Dietz and Stephenson 1975). Hagedorn and Moeller (1968) demonstrated that honey bees produced less brood when they were fed diets consisting of 75 percent soybean flour and 25 percent of 1-, 2-, or 4-year-old pollen. Pollen collected, dried, and stored frozen retains much of its nutritional value for honey bees; pollen frozen for 11 years was still adequate for brood rearing (Dietz and Stevenson 1980). Fresh pollen (2 parts) mixed with granulated sugar (1 part) and covered on top with a 5 cm layer of sugar to guard against surface mold was found to retain some of its nutritive value when stored at room temperature for up to two years (Townsend and Smith 1969). However, this pollen was unable to support as much brood as feeding fresh or frozen pollen pellets.

Pollen pellets collected by foragers are packed in cells by the honey bees, with a small cover of honey and glandular secretions added to the mass of stored pollen. This store of pollen is called "bee bread." Because different pollen loads are mixed in storing, the cells are stratified. Stored pollen becomes non-viable in 1 to 8 days depending on the plant species. Probably substances secreted by the honey bees to moisten the pollen while packing it in the comb are responsible for the rapid loss of germination capacity. The mandibular glands of honey bees also produce a germination-inhibiting factor. Stored pollen undergoes a number of biochemical changes which may be responsible for increased stabilization of the product or may lead to chemical changes that increase the digestibility and nutritive value for bees.

Beutler and Opfinger (1949) found that honey bees lived longer on pollen removed from the combs than on pollen collected in traps. The acidity of pollen increases (from 0.26 to 1.78 percent) and in water-soluble proteins (from 2.9 to 5.6 percent). A lactic acid-type metabolism occurs, indicating a lowering of oxygen tension; this probably contributes to the stability and conservation of pollen as a food source. Stored pollen generally has a specific bacterial flora; Pain and Maugenet (1966) found three genera in bee bread: *Pseudomonas, Lactobacillus,* and *Saccharomyces.* The microbiology of pollen and bee bread has been extensively studied by Gilliam (1979a, 1979b, 1989 *et al.*).

Many criteria have been used in studying the biological potency of various pollens (and other proteins) on the development and brood rearing of honey bees. These include 1) the longevity and amount of brood reared by bees offered the test diets; 2) the development of various glands such as the hypopharyngeal glands (Fig. 1); 3) growth, including changes in weight and nitrogen contents of bees reared on the test diets; and 4) colony weights.

The development of the hypopharyngeal glands in worker honey bees, being a consequence of their diet, offers an opportunity to study the nutritional value of various proteins. Maurizio (1954) divided hypopharyngeal gland development into four categories by size with stage 1 being totally undeveloped. The maximum development (stage 4) was evidenced by the main and side canals being completely concealed with no spaces visible between the gland lobi. Standifer *et al.* (1960) found that the rate of hypopharyngeal gland development was apparently related to the protein content of the diet; the best development was obtained at the highest levels of protein, but the lowest levels of protein promoted increased longevity. However, they also pointed out that the optimum level of one protein might not be the optimum level of another in promoting longevity of adults and optimal growth of larvae.

The complete morphological development of the hypopharyngeal glands did not necessarily mean that their secretions would provide food material suitable for larval nutrition. Brouwers (1982) measured the protein synthesis of isolated hypopharyngeal glands using a radio-labelled amino acid (leucine), and reported that there was no direct correlation between the rate of synthesis and the size of the lobes. Consequently, the development of the hypopharyngeal glands alone cannot serve as an indication of suitability of any food for brood-rearing, since well-developed glands may secrete a product that is deficient in a factor(s) essential for normal larval development.

CARBOHYDRATES

The carbohydrate requirement of adult honey bees is satisfied by nectar, from either floral or extra-floral nectaries of flowers. Nectar ranges in sugar content from 4% to 60% or higher depending upon the floral source and a number of environmental conditions such as temperature, humidity, and rainfall (Shuel 1975). Nectar and sugar solutions containing 30-50% sugar concentrations seem to elicit the maximum collection response by honey bees (Waller 1972). This is the reason nectar containing lesser amounts of sugar, such as in pear nectar, may not be attractive to bees. Sucrose, glucose, and fructose are the predominant sugars in nectar although many more have been identified, usually in trace amounts (White 1977). As the nectar is manipulated and enzymatically converted to honey by bees, most of the sucrose is inverted to glucose and fructose although the level of sucrose never reaches zero. Honey is essentially a carbohydrate material with 95-99.9% of the solids

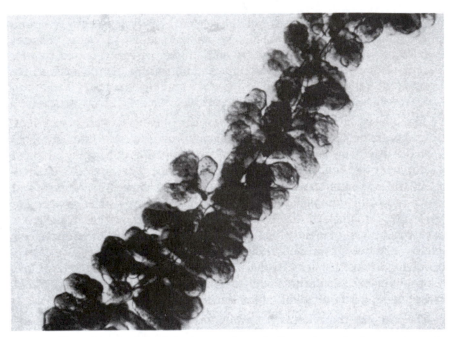

FIGURE 1. Hypopharyngeal gland development in worker honey bees: development in newly emerged adults above, and one week old nurse bees below.

being the sugars, glucose, and fructose, and with small amounts of at least 22 other more complex sugars (White and Doner 1980).

Although nectar and honey satisfy the carbohydrate requirement of adult honey bees, it is interesting that pollen and bee bread contain 30-35% sugar, which constitutes the major dry matter fraction. Polysaccharides, primarily starch and cell wall constituents, may comprise up to 50% of the dry weight. In bee-collected pollen the level of non-reducing sugar averages 2.71% and reducing sugars average between 18 and 41%. These values are reversed in hand-collected pollen.

The amount of honey a particular colony requires cannot be accurately determined as it depends upon a number of factors including the strength of the colony, the level of brood rearing, and the kinds and amounts of nectar available. Populous colonies in most northern regions of the U.S. require approximately 50-55 pounds of honey from the time brood-rearing stops in the fall until enough nectar can be collected in the spring to support the colony. Estimates of 60-80 pounds for winter and early spring survival have been reported. The amount of honey required for the entire season is much greater. The honey requirement for summer has been estimated at 95 pounds. This requirement, in addition to the honey needed for overwintering and that used in wax secretion, means a normal colony may require in excess of 150 pounds per year to survive in most northern states. Honey bees overwintered in less severe environments, especially in the southern and southwestern states, require less honey to maintain themselves during nectar dearths.

The sugars most readily accepted by honey bees are found naturally in either nectar (glucose, fructose, and sucrose) or in honey dew (melizitose and maltose). When von Frisch (1934) tested 34 carbohydrates and related compounds, he found that honey bees consider only seven of them sweet. Sugars which do not taste sweet to honey bees are of little or no nutritional value to them (v. Frisch 1965).

Honey bees show a definite preference for various sugars when offered a free choice. When concentrations of some sugars were offered in the field and laboratory, bees preferred—in descending order—sucrose, glucose, maltose, and fructose; mixtures of the sugars did not have an additive effect. Sucrose was the most preferred, and glucose the least to foraging bees given a choice of collecting syrup from an artificial flower (Waller 1972). Bees most often preferred sucrose to either glucose or fructose, but a mixture of equal parts of sucrose, glucose, and fructose was, like other mixtures, less attractive than sucrose or a mixture in which sucrose was dominant (Bachman and Waller 1977). Von Frisch (1934) found that glucose and fructose were only half as attractive as sucrose to foraging bees when they were allowed to choose among the three sugars.

Honey bees in cages preferred diets containing sucrose to those containing sugar combinations, honey, invert sugar, or isomerized corn syrup (Herbert and Shimanuki 1978a). Barker and Lehner (1974a) found that sucrose was superior to other sugars in both acceptance and nutritive value when they fed 13 sugars to honey bees and measured survival, water and sugar consumption.

Although honey bees can utilize many sugars in their diets, there are also several sugars that are either toxic or useless to honey bees because they lack the proper enzymes for digestion. Sugars which are either toxic to honey bees or reduce their longevity include mannose, lactose, galactose, and raffinose. Mannose is especially toxic and will kill honey bees within a few minutes of feeding. Both lactose and raffinose reduce the longevity of bees when offered to caged honey bees. Lactose is found in milk and milk products, and raffinose occurs naturally in soybeans. The toxicity of lactose is due to the effect of one of its components, galactose. Barker and Lehner (1974b) observed a toxic effect of lactose on honey bees; they also noted that newly emerged bees would starve rather than consume a 0.5M soloution of galactose or lactose. In addition, they found that these sugars were still toxic when fed in sucrose syrup, and did not sustain life any longer than water alone (Barker and Lehner 1974a).

LIPIDS

Bees require some dietary lipids (fatty acids, sterols, and phospholipids) in their diet as sources of energy, for the synthesis of reserve fat and glycogen, and as essential structural components of many cell membranes. Under normal conditions this lipid requirement is satisfied by the consumption of pollen. The lipid content in pollen varies, depending upon the plant species, from 1% to 20% (Lunden, 1954). On a dry weight basis, lipids are present in most pollen at a concentration of 4-6%. Todd and Bretherick (1942) reported a mean value of 5% for air-dried, bee-collected pollen. The maximum lipid values were obtained for dandelion (14.4%) and mustard (13.1%). Standifer (1966) reported a range of total lipid values of 1.5 to 18.9% in bee-collected pollen and 3.2 to 17.8% in hand-collected pollen.

Honey bees, like most insects, require sterols in their diet for normal growth development and reproduction. One class of sterols (cholesterol) is known to be essential for honey bees; and since bees, like most insects, are not able to manufacture these components, they must obtain them in their diet for normal development. Cholesterol or 24-methylene cholesterol is present in most pollen and both can be incorporated into the structural components of cells. Cholesterol is the precursor for the production of molting hormones in most insects. Barbier *et al.* (1960) isolated 24-methylene cholesterol from the pollen of apple, rockrose, and mixed pollen. This was the first isolation of a sterol in the plant kingdom. They also isololated 24-methylene cholesterol in equal proportions from queen bees and from worker bees with and without queens. Standifer *et al.* (1968) analyzed 15 plants from 11 families for sterols and found 24-methylene cholesterol to be the principal sterol, but it was not present in all species. Mass spectrographic analyses of 10 species showed that 24-methylene cholesterol was the major sterol in three, B-sitosterol in five, and stigmosterol and cholesterol in the two remaining species.

Sterol utilization and metabolism is one critical area of honey bee nutrition that has not been studied in depth, although some beneficial effects of adding cholesterol have been shown (Robinson and Nation 1966; Nation and Robinson 1968; Haydak and Dietz 1972). All insects, when examined in detail, have been found to require an exogenous source of sterol for normal growth, development, and reproduction; in all but two known cases (Heed and Kincher 1965; Chu *et al.* 1970) cholesterol satisfied this requirement. However, many phytophagous insects are capable of dealkylating and converting C_{28} and C_{29} phytosterols to cholesterol to satisfy their sterol requirements (Thompson *et al.* 1973; Svoboda *et al.* 1975).

The dietary sterols—cholesterol, campesterol, sitosterol, stigmasterol, and 24-methylene cholesterol—were tested for their ability to support brood-rearing in the honey bee by adding them singly to a chemically-defined worker bee diet (Herbert *et al.* 1980). It was found that a diet supplemented with cholesterol supported the greatest survival of worker honey bees, but diets supplemented with either 24-methylene cholesterol or cholesterol supported the production of nearly equivalent amounts of sealed brood with better results than any of the other three sterols tested (Table 1). Honey bees fed diets containing stigmasterol, sitosterol, campesterol, or no supplement produced less sealed brood in decreasing amounts.

The sterols of honey bee prepupae reared on diets containing various sterols were isolated, identified, and quantified (Svoboda *et al.* 1980). The major sterol present in each prepupae sample was 24-methylene cholesterol, but significant levels of sitosterol and isofucosterol were also present. The preponderance of 24-methylene cholesterol in all prepupae, regardless of the dietary sterol provided to the workers, suggested a unique system of utilization and metabolism of these dietary sterols by worker bees. Worker bees make available to the brood varying amounts of unchanged dietary sterols and 24-methylene cholesterol, sitosterol, and isofucosterol drawn from their own sterol pools. Although honey bees need sterols to satisfy their metabolic requirements, only cholesterol and 24-methylene cholesterol can be used (Svoboda *et al.* 1980). One of these must be available in the diet, apparently because honey bees cannot convert the C_{28} and C_{29} phytosterols to the C_{27} cholesterol.

Using labelled material, it was possible to determine precisely how much

Table 1. Effect of dietary sterols on honey bee sealed brood production (Herbert *et al.* 1980)

Dietary sterol	Sealed Brood					
	Control	Cholesterol	Campesterol	Stigmasterol	Sitosterol	24-methylene-cholesterol
Weeks 1 + 2	51.6*	254.1	70.9	6.5	70.9	38.7
Weeks 3 + 4	374.1	1,380.3	812.7	741.8	870.8	806.3
Weeks 5 + 6	593.4	690.2	374.1	715.9	574.1	799.8
Weeks 7 + 8	154.8	322.5	77.4	219.3	135.5	574.0
Weeks 9 + 10	. . . . .	116.1	. . . . . .	38.7	. . . . . .	380.6
Weeks 11 + 12	. . . . .	. . . . . .	. . . . . .	. . . .	. . . . . .	25.8
Total 12 Weeks	1,173.9	2,763.2	1,335.1	1,722.2	1,651.3	2,625.2
% production with cholesterol	42.5	100.0	48.3	65.6	62.9	95.0

*Total sealed brood (cm²) for four replicates during the two-week period.

sterol from the workers' diet was incorporated into the sterol pools of the worker honey bees (Svoboda *et al.* 1981). At various intervals the sterols of prepupae, newly emerged adults, and queens were analyzed and it was determined that there was insufficient radioactivity associated with cholesterol and/or demosterol in any of the samples to verify that any of the three C_{28} and C_{29} sterols was dealkylated and converted to cholesterol. There was no evidence for the conversion of campesterol or sitosterol to 24-methylene cholesterol. It was concluded that the major portion of the sterols incorporated into the tissues of the brood larvae originated from the worker bees used to establish the colony.

The selective transfer and utilization of sterols in honey bees that had been demonstrated in cage studies with artificial diets was also shown to occur under field conditions (Svoboda *et al.* 1983). Sterols from pollen collected by foraging honey bees at seven field sites were compared with the sterols of foraging adults and prepupae collected from colonies at each site. The major prepupae sterol was found to be 24-methylene cholesterol, even though some of the pollen samples contained little 24-methylene cholesterol.

Makisterone A, a 28-carbon moulting hormone, was identified as the major free pupae ecdysteroid in the honey bee (Feldlaufer *et al.* 1985). This was the first isolation and identification of a 28-carbon ecdysteroid in an insect species from the order Hymenoptera. No C_{27} ecdysteroids (20-hydroxyecdysone or ecdysone) were detected at this stage of development. Makisterone A was also isolated from the ovaries of queen bees (Feldlaufer *et al.* 1986a). The predominant neutral sterol present was 24-methylene cholesterol. To determine the sterol precursor of makisterone A, honey bee pupae (13 days post-oviposition) were injected with radio-labelled sterols and subsequently examined for labelled ecdysteroids (Feldlaufer *et al.* 1986b). Labelled campesterol was found to be converted to a compound that behaved identically to makisterone A when examined chromatographically.

Specific chemicals in pollen may serve as attractants for honey bees. Hügel (1962) reported that the material in pollen attractive to honey bees was a steroid or a mixture of steroids, especially 24-methylene cholesterol. Herbert *et al.* (1980) reported, however, that the presence of sterols in synthetic diets did not make any of the diets more attractive to honey bees than the sterol-deficient control diet. Other attractants included a free fatty acid, and an ester of the flavone pigment lutein which were isolated from mixed bee-collected pollen (Lepage and Boch 1968; Hopkins *et al.* 1969; Starratt and Boch 1971). The addition of either whole pollen lipids or the fraction soluble in cold acetone to bee diets resulted in a significant increase in the amount of diet consumed by caged bees. The addition of the fraction insoluble in cold acetone, or of an extract of the volatile substance in pollen, led to decreased food consumption (Robinson and Nation 1968). Taber (1963) reported that after pollen was extracted with hexane or ethyl ether, it was no longer attractive to honey bees, but honey bees were readily attracted to cellulose powder to which a small amount of this same pollen extract was added. Waller *et al.* (1970) added the oils anise, camomile, fennel, dark rum, and artificial honey essence to dry pollen substitutes to increase their attractiveness to honey bees. Anise oil, fennel oil, and artificial honey essence all appeared to be effective and economically feasible attractants when added to dry pollen substitutes.

VITAMINS

The role of vitamins in the growth and development of honey bees is unknown; we do know that they are essential for all living organisms. Pollens are generally rich in water-soluble vitamins and poor in the fat-soluble vitamins. In general, the vitamin requirements of honey bees are satisfied as long as pollen or supplementary protein foods are nutritionally adequate, abundant, and available in the hive. Normally, pollen contains the seven B-complex vitamins (thiamine, riboflavin, pyridoxin, pantothenic acid, niacin, folic acid, and biotin) which are essential for most insects (Dadd 1973).

Inositol and ascorbic acid, also water-soluble vitamins, are present in pollen (Nielsen 1955; Augustin and Nixon 1957).

Wahl and Back (1955) studied the effects of vitamins in pollen on honey bee longevity, hypopharyngeal gland development, and brood-rearing ability and found that longevity did not differ significantly between diets with and without vitamins. Pain (1956) demonstrated that vitamins had no effect on the longevity of honey bees, but that pantothenic acid had a great influence on the development of the hypopharyngeal glands. He concluded that vitamins exert a decisive influence on these glands in contrast to their minimal effect on longevity. Herbert and Shimanuki (1978d) showed that the hypopharyngeal glands of honey bees fed diets deficient in thiamine or riboflavin failed to develop beyond that stage found in newly emerged bees.

Haydak and Dietz (1965) studied the role of the B vitamins on the nutrition of honey bees, and concluded that for the growth of emerging bees and the development of their hypopharyngeal glands only an appropriate protein source is necessary, but for brood-rearing activities vitamins are indispensable. Haydak (1970) found that emerging bees' body nitrogen and hypopharyngeal glands developed normally when fed a diet of vitamin-free casein, minerals, and invert sugar, but that this diet was not adequate for brood rearing. Although the food in the brood cells was abundant and normal in color and consistency, the larvae did not develop beyond 2 to 3 days of age without the addition of B vitamins and cholesterol to the diet.

Haydak and Dietz (1972) tested the specific vitamin requirements of bees for brood rearing by feeding diets having a full vitamin supplement and lacking in turn pantothenic acid, pyridoxine, or thiamine. Colonies having a full vitamin supplement and those on a diet deficient only in pantothenic acid or thiamine reared brood until the end of the experiment. On all diets tested, except those lacking all vitamins and lacking only pyridoxine, normal adults were produced for at least three cycles of brood, although brood rearing diminished with time, demonstrating that pyridoxine is required for normal brood rearing. Anderson and Dietz (1976) confirmed the essential role of pyridoxine for brood rearing by caged bee studies. They reported that bees required 5.4 μg of pyridoxine to rear one larva to the sealed stage. Haydak and Dietz (1972) demonstrated that the bodies of newly emerged bees contain considerable amounts of both thiamine and pantothenic acid, and that bees may be able to synthesize pantothenic acid since royal jelly contains 20 times the amount found in bee-collected pollen. Barbier (1971) suggested that this may represent a biosynthesis of pantothenic acid by the mandibular glands of honey bees.

The effects of two other vitamins, gibberellic acid and inositol, on brood rearing have been reported. Complete larval and pupal development occurred in colonies of honey bees when adult bees were allowed to feed on an artificial diet containing gibberellic acid. In the absence of this vitamin, larvae died in the 3rd and 4th day of development. Worker bees were also able to rear brood when large amounts of inositol were substituted for gibberellic acid in the artificial diet (Nation and Robinson 1966, 1968).

The levels of vitamins in bee-collected pollen have been reported by several investigators including Vivino and Palmer (1944), and Hagedorn and Burger (1968). Herbert *et al.* (1987) measured the levels of thiamine (B$_1$) plus its monophosphate and diphosphate forms in fresh pollen and reported that the levels varied greatly depending on the floral source and time of year. All three forms of this vitamin were present in pollen but thiamine was the principal form, followed by the diphosphate and the monophosphate forms. Pollen apparently satisfies the honey bee's requirement for thiamine since there is no evidence that it can be synthesized by insects.

The precise functions and requirements of vitamin C in honey bees have not be demonstrated. Ascorbic acid undergoes rapid oxidation upon storage and changes from the active forms of the vitamins (ascorbic acid and dehydroascorbic acid) to nonactive products. Hagedorn and Burger (1968) found that the ascorbic acid content of pollen decreased with the age of pollen, as did the effectiveness of pollen for supporting brood rearing, hypopharyngeal gland development, and rate of growth. However, Standifer and Mills (1977) found no significant differences in the content of vitamin C in larval food produced by nurse bees fed alternately balanced and deficient diets. Herbert *et al.* (1985) showed that pollen is a rich source of vitamin C, but the levels varied considerably depending on the floral source of the pollen and the time of year (Fig. 2). The levels of vitamin C ranged from 136 μg/g pollen (collected in August) to 1943 μg/g pollen (collected in May). This study also demonstrated for the first time that bees are able to produce vitamin C; prepupae from colonies fed diets without vitamin C had equivalent levels of ascorbic acid to those fed the enriched diets.

Not much work has been done on the requirements of honey bees for the four fat-soluable vitamins A, D, E, and K. Goldsmith and Warner (1964) were the first to isolate a fat-soluble vitamin, and they demonstrated that vitamin A played a role in the vision of honey bees. They found this vitamin in the heads of bees but lacking in either thoraces or abdomens. Haydak and Palmer (1938) isolated vitamin E in pollen, but not in royal jelly. Vivino and Palmer (1944) identified small amounts of vitamin D and E in pollen, but both A and K were absent. They also found royal jelly, the protein-rich queen larvae food, to be completely devoid of any vitamin E activity. When Herbert and Shimanuki (1978e) fed caged honey bees a chemically-defined diet supplemented with either vitamin A, D, E, K, or with a complex of all four

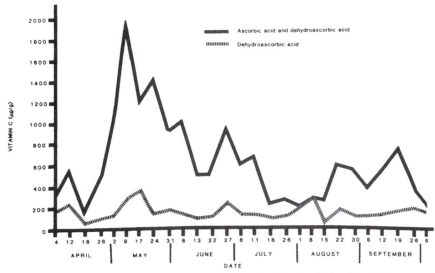

FIGURE 2. Vitamin C levels (μg/g) of fresh bee-collected pollen (April-October 1983) (Herbert *et al.* 1985).

vitamins, they found that honey bees fed diets containing A, K, or the complete complex reared twice as much brood as bees fed the control diet. The exact dietary requirements are difficult to determine, especially for fat-soluable vitamins, because the micro-organisms in the honey bee's digestive tract may produce many of the vitamins.

The precise functions and requirements of many vitamins have not been demonstrated for honey bees; often these nutritional requirements are implied by comparing the pollen chemistry with the nutritive value of pollen selected by bees. There is little information available on the capacity of bees to extract the chemical nutrients from the different pollen species.

The loss of vitamins may be one reason why some pollens and pollen substitutes are of little nutritional value to honey bees. Haydak (1945) reported that soybean flour was significantly deficient in riboflavin. Haydak (1949) also showed that the macin content of soybean is considerably below that of either pollen or dried brewers yeast. Vitamin deterioration also occurs in pollen. Pollen storage causes changes in ascorbic acid as well as pantothenic acid and probably other vitamins.

MINERALS

Less is known about the mineral requirement of honey bees than the other classes of nutrients. This requirement is usually met by the consumption of pollens, which generally contain between 2.5 and 6.5% ash on a dry weight basis (Todd and Bretherick 1942; Vivino and Palmer 1944; Lunden 1954;

Barbier 1971; Herbert and Shimanuki 1978b). Pollen contains the major minerals found in plant tissue and also trace quantities of several minor minerals. Although these inorganic elements are present in pollen in relatively small amounts, they are essential to such life processes as enzyme systems.

Mineral requirements of honey bees have not been established by nutritional studies because of the difficulty in preparing experimental diets deficient in trace minerals. Most often diets containing minerals have been based on the levels occurring in pollen; however, since pollen composition is highly variable, quantitative analyses of it may provide little information about the mineral requirement of honey bees. Often salt mixtures which were formulated for vertebrates have proved ineffective in honey bee diets (Nation and Robinson 1968; Herbert 1979). It was found that these salt mixtures often contain excessive amounts of calcium and sodium and insufficient amounts of potassium.

Studies of mineral composition have shown that potassium, phosphorus, calcium, magnesium, and iron are the most commonly occurring minerals in pollen (Todd and Bretherick 1942; Vivino and Palmer 1944). Nation and Robinson (1971) examined pollen from corn, citrus, long leaf and loblolly pines, and mixed bee-collected pollens for eight minerals. These eight minerals were found in the head, thorax, and abdomen of adult worker honey bees and also in royal jelly. The major elements in each body region were potassium, sodium, and magnesium, with the greatest concentration of the three ions in the head. Calcium was present in honey bees in much lower concentrations than in pollen. Dietz (1971) reported that phosphorus and potassium were the most abundant minerals in the body of the honey bee. Calcium, magnesium, sodium, and iron were present in considerably smaller quantities. Whereas the dry weight and nitrogen content of the body increases during the early life of a honey bee, the mineral constituents generally decline with age. The exception was iron, which increased considerably by the 6th day and then remained relatively constant during the remaining days.

Haydak and Dietz (1965) found that minerals were not essential in the diet of adult honey bees for the growth and development of their hypopharyngeal glands. However, Nation and Robinson (1968) found that the addition of a small amount of pollen ash to an artificial diet for adult honey bees improved their ability to rear brood. Nation and Robinson (1971) found potassium, sodium, calcium, magnesium, copper, manganese, and zinc in samples of various pollens, including mixed bee-collected pollen. Based on this study, the authors recommended that a salt mixture be included in diets of adult honey bees. They suggested that the ash content of an artificial diet should be approximately 3% of the dry weight. Herbert and Shimanuki (1978c) reported that when newly emerged bees were fed synthetic diets containing concentrations of pollen ash, a 1% concentration supported the greatest amount of brood rearing, although the ash content of pollen ranges from 2.4 to 3.5%. Wesson's salt mixture, designed for vertebrates, proved quite unsatisfactory for adult honey bees. Honey bees fed a synthetic diet containing Wesson's salt mixture consumed less diet and reared less brood than honey bees offered a similar diet fortified with pollen ash. The most striking feature was the difference between the level of sodium in Wesson's salts (3.3%) and in pollen ash (0.22%) (Table 2).

Excessive levels of minerals can be toxic to honey bees and even sucrose solutions normally attractive to honey bees are not consumed when the salt content exceeds one percent. Temnov (1958) found that the mineral salts present in honey dew are detrimental to honey bees and produce a reduction in the longevity. White and Doner (1980) noted that dark honey is higher in ash (minerals) than light honey and the amount of ash in honey dew honey (average 0.73%) is greater than the levels in floral honey (average 0.17%). This high ash level in honey dew honey may explain why honey bees overwinter poorly on this source.

WATER

Water is a vital element in the honey bee diet; as experiments have shown, bees without water will die within a few days. Water serves some very important functions in the bee, including carrying dissolved food materials to all parts of the body, assisting in the removal of waste products, and digesting and metabolizing food. Beekeepers should not overlook the water requirement of honey bees, which may be quite extensive, especially in the spring when large amounts of larval food must be secreted by nurse bees. The more brood a colony must feed, the greater the water requirement since larval food may contain as much as 66% water.

Honey bees also require water for the utilization of honey and crystallized

Table 2. Amounts of eight mineral elements found in pollen, pollen ash, and Wesson's salt mixture determined by atomic absorption spectrophotometry (ppm) (Herbert and Shimanuki 1978).

Element	Pollen (dry wt)	Pollen Ash 10 X	Wesson's salt 10 X
Copper	16	0.5	0.2
Iron	83	2.6	2.0
Manganese	35	1.1	0.1
Zinc	72	2.2	0.005
Magnesium	683	21.3	9.0
Calcium	1067	33.3	133.5
Sodium	70	2.2	33.0
Potassium	2933	91.7	41.7

sugar. Honey bees prefer 30-50% sucrose solutions in preference to higher or lower concentration (Waller 1972). Thick, granulated honey or dry granular sugar, which are sometimes fed to honey bees instead of sugar solutions, must first be liquified by honey bees before they can be utilized.

Water is also collected by honey bees to maintain the proper colony temperature in hot weather. The higher the temperature, the greater the water requirement. Water seldom is stored in the comb, but is collected as needed, most often from the nearest water source. Honey bees cool the inside of the hive by putting droplets of water on the comb, and increase evaporation by fanning their wings. They also assist cooling by evaporating nectar and water on their tongues.

Honey bees also use water to maintain the proper relative humidity in the brood nest to insure egg hatch and to prevent larval dessication. Doull (1976) demonstrated the importance of proper relative humidity for normal brood development. He found the optimum relative humidity for normal hatching to be between 90 and 95%. There was a significant decline in the number of normal larvae that emerged when eggs were incubated at 100% and 80% relative humidity. At 50% relative humidity only 29% of the eggs hatched into normal larvae, and no eggs hatched at humidities below 50%.

The water consumption of a colony varies with the strength of the colony and the colony location. Colonies of bees may require 200 grams or more a day during the brood-rearing period. The amount of water brought in annually, disregarding the water from the collected nectar, has been estimated at about 44 pounds. Outdoor-wintered colonies in cold climates can utilize water that condenses on the interior of the hive. This has been estimated to be as much as five liters per month (Nelson 1983).

Beekeepers should provide fresh water in the apiary if water is not naturally available nearby. This not only reduces the distance required for honey bees to fly to collect water, but it may prevent bees from collecting water from locations, such as swimming pools and bird baths, where they would create a nuisance. A novel watering device was developed by Moffett *et al.* (1977) for providing colonies of bees with water in the desert areas of Arizona where maximum daily temperatures frequently exceed 110°F. The super-like watering device, which holds about three gallons of water, was used in conjunction with a burlap covering of colonies to protect them during insecticide spraying.

PROTEIN SUPPLEMENTAL FEEDING

Beekeepers often feed colonies of honey bees supplemental diets during periods of pollen and nectar dearth. The brood-rearing activity and nutritional state of the colony, the quantity and quality of incoming pollen and nectar, and the food reserves in the hive will determine whether the bees benefit from supplemental foods (Standifer *et al.* 1977). Colonies are normally fed supplemental foods to produce strong colonies for package production, to develop colonies with optimum populations for pollination of crops, to build up colony population for autumn and spring divisions, for queen production, to help overcome pesticide damage, or to assist colonies to overcome diseases which may be associated with nutritional deficiencies.

When natural pollen is scarce, beekeepers can feed colonies of honey bees supplemental protein diets. These proteins are either fed as "pollen supplements" or "pollen substitutes." A pollen substitute is any material which, when fed to colonies of honey bees, replaces the pollen requirement of that colony for a short period of time. The same pollen substitute or any other proteins becomes a pollen supplement when pollen is added to the diet as an attractant or to increase its nutritive value.

Pollen substitutes can be fed outside the hive in trays where the substitute can be collected by foraging bees, or inside the hive as a moist patty over the brood frames (Fig. 3). Disadvantages of the outdoor feeding are that weak colonies often get less than they need, the substitute must be protected from

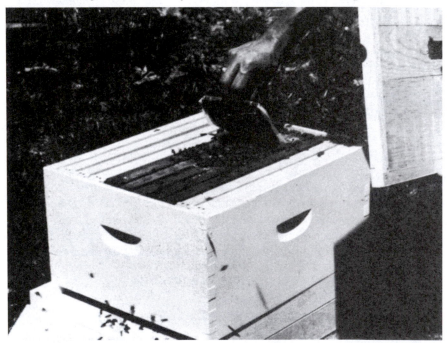

Figure 3. A pollen substitute, wrapped in waxed paper to prevent excessive moisture loss, is placed over the clustered honey bees.

inclement weather, and during bad weather bees are often unable to collect it. A moist patty placed inside the hive is, therefore, preferred, since its availability to the bees is not influenced by the weather. However, pollen substitute patties must maintain their moist consistency and high nutritional value over long periods. Ideally, pollen substitutes should fill completely all the roles of pollen, but a major difficulty is that they are less acceptable than natural pollen to honey bees, probably because they lack the specific attractants that occur in pollen. Most pollen substitutes offered to bees are nutritionally adequate, and some apparently surpass pollen in nutritive value, but when bees have a choice, they usually eat considerably more natural pollen than pollen substitute (Standifer *et al.* 1973a).

Spring supplemental feeding should begin 6 to 8 weeks in advance of the major nectar flow or the time when package bees or queens are to be produced. Fall supplemental feeding should be considered, especially if the hive contains inadequate stores of pollen for wintering. The feeding should begin early enough to allow the honey bees time to rear 1 to 2 additional cycles of brood. Fall feeding often results in colonies of honey bees going into winter with a larger population of younger bees, which may result in a more rapid spring buildup. Doull (1980) demonstrated the effects of supplemental feeding in a year-long test of a commercial pollen supplement. Colonies that received the supplement produced 38% more honey per colony and 28% more honey per bee than the colonies that did not receive any supplementary food.

Beekeepers and researchers have searched for years for protein sources that could be used in honey bee diets to effectively replace pollen. Protein sources tested include soybean flour, peanut meal, powdered skim milk, commercial casein, whole wheat flour, Wheast®, and various forms of yeast have been fed to colonies. Haydak (1936) fed 11 protein sources to honey bees and evaluated their value by measuring the dry weight and nitrogen content of bee thoraces. No difference was found in bees fed pollen, meat scraps, or commercial casein. In another study, Haydak (1937) found that colonies fed soybean flour or mixtures of skim milk powder with cottonseed or soybean meal reared brood normally, although the amount of brood was less than in a caged colony of bees fed pollen. Wahl (1954) found that confined colonies could rear brood when fed brewers yeast, Torula yeast, or soybean flour. Haydak (1959) recommended that the most effective and yet economical pollen substitute should consist of three parts soybean flour (expeller processed or solvent-extracted and heated afterwards with a fat content of 5 or 7%), one part dried brewers yeast, and one part dried skim milk.

Wahl (1963) compared the nutritive value of pollen, yeast, soybean flour, and dried milk. He found that natural pollen mixtures were able to initiate and maintain brood rearing in confined colonies of honey bees more effectively than the other tested proteins. The superior effectiveness of pollen was attributed to its high attractiveness to bees.

Haydak (1967) fed over 30 protein sources, either alone or in combinations, to newly emerged bees and evaluated the changes in dry weight and nitrogen content, adult mortality, quality and quantity of brood reared, and the population of the experimental colonies. On the basis of this study, Haydak substantiated his earlier studies and again recommended a mixture of soybean flour, dried brewers yeast, and dried skim milk as the most effective pollen substitute. Some formulations of pollen supplements and pollen substitutes that are being used today are shown in Table 3.

Standifer *et al.* (1970, 1973b) evaluated various pollen substitutes but concluded that their nutritional value was minor compared to that derived from fresh pollen. Standifer *et al.* (1970) measured the hypopharyngeal gland development of caged bees fed pollen from gum weed or saguaro and a pollen substitute containing soybean flour, dried brewers yeast, and dried skim milk.

Table 3. - Pollen supplement and pollen substitute diets. (Modified from Standifer *et al.*, 1977)

Protein source	Dry mix	Moist Patty
	Parts by weight	
Pollen Supplement Formula A		
Soybean flour:pollen (3:1 w/w)[a]	1	
Sucrose:water (2:1 w/w)	2	
Pollen Supplement Formula B		
Wheast or brewer's yeast:pollen (3:1 w/w)	1	
Sucrose:water (6:1 w/w)	2	
Pollen Substitute Formula A		
Brewer's yeast	2	3
Sucrose	3	3
Water[b]		2½
Pollen Substitute Formula B		
Soybean flour	2	3
Sucrose	3	3
Water[b]		2½
Pollen Substitute Formula C		
Wheast		3
Sucrose		3
Water[b]		4

[a] w/w × weight to weight.

[b] Add sufficient water to substitute diet mix to form a doughlike consistency.

They measured the degree of gland development after eight days and found that none of the diets produced maximum gland development.

Standifer *et al.* (1973a) measured the brood rearing, dry weight, and nitrogen content of emerging bees from colonies confined in outdoor flight cages. The bees were fed two rations containing various ratios of soybean flour, dry brewers yeast, meat scraps, and dried skim milk. A third ration tested contained all the above protein sources with the addition of 17% pollen. Colonies fed the pollen-fortified diet supported significantly more brood than the colonies fed the other two diets, but this diet still did not yield brood-rearing levels comparable to those of bees in uncaged control colonies.

Weaver (1964) formulated a mixture of nutrients on which a colony of confined bees continued to rear brood for six months. The solid portion of the diet contained cottonseed flour, cholesterol, corn oil (dissolved in chloroform and mixed with the cottonseed flour), and a Wesson's salt mixture. The liquid portion of the diet contained water, sucrose, enzymatic casein hydrolysate, and a vitamin mixture.

Nation and Robinson (1966) developed a semi-defined diet containing vitamin-free casein, gelatin, zein, and egg albumin. This supported excellent egg laying by the queen, but the larvae did not develop beyond 3 to 4 days of age. For larval development beyond four days of age and normal pupation, a minimum of 7.5% by weight of natural pollen or certain appropriate fractions of pollen need to be added to the artificial diets. Bees were able to rear brood through more than one generation when the pollen in the artificial diet was replaced with gibberellic acid. Additional studies were conducted in which diets were formulated either as liquid or solid preparations. The all-solid formulation contained gelatin, vitamin-free casein, zein, and egg albumin. The liquid-dry formulation was modified after Weaver (1964). The consumption of the artificial diets was poor when compared to control diets of honey and pollen, and the addition of an ether extract from 5 g of fresh pollen did not improve either the acceptance or the nutritive value of the diet. Reducing the amount of cholesterol from 1% to 0.5% and subsequently to 0.25% by weight of dry ingredients seemed to significantly increase the acceptance of the food. However, the colonies fed the test diets were still unable to rear larvae beyond 3 to 4 days of age. The addition of 10% or 7.5% by weight of whole pollen to the basic artificial diet resulted in complete brood development, indicating that the basic diet was deficient in some way that was critical for development of older larval stages. The addition of 10% or 20% of pollen lipids to the basic artificial diet did not promote brood development; one colony reared all stages of brood throughout the entire test period while feeding only on the residue remaining after the lipids had been extracted from the whole pollen. Other colonies reared brood when they were given the basic artificial diet fortified with 33%, by weight, of the residue remaining after the lipid and water-soluble materials were removed from whole pollen. These results indicate that lipids in pollen are not essential for brood rearing since one colony was able to maintain brood rearing for more than 50 days after lipids were extracted from their pollen.

Weaver and Chauthani (1967) succeeded in developing an all-liquid pollen substitute in which they substituted a yeast extract for the vitamin mixture. The authors also developed a candy diet containing sugar, casein hydrolysate, cellulose, cholesterol hydrogen succinate, crude phospholipid, sodium carbonate, Wesson's salts, and a vitamin mixture.

Nation and Robinson (1968) attempted to stimulate brood rearing in colonies fed an artificial diet containing proteins, carbohydrates, lipids, vitamins, and minerals but failed until inositol was added to the artificial food. Honey bees fed this diet continued to rear brood normally for 72 days. The authors found that the addition of 3.5 mg of pollen ash per gram of artificial diet without inositol enabled bees to rear brood for 44 days. The addition of both pollen ash and inositol to the artificial diet gave no better results than either alone. An all-liquid diet containing water, sucrose, casein, enzymatic hydrolysate, inositol, Wesson's salt mixture, peptone digest, corn oil, and vitamins was also formulated; but queens in colonies fed this diet failed to lay eggs during 27 days on the diet. The authors concluded that Wesson's salt mixture is evidently not a correct mixture, or not properly balanced for brood rearing; but this can be corrected by adding pollen ash to the basic diet.

Expeller-processed soybean flour has been used for many years by bee-keepers as a protein source for honey bees. The material has been fed either alone or mixed with other ingredients such as yeast, dried skim milk, or pollen and then fed as a pollen supplement. Unfortunately, all soybean products do not elicit the same biological response when offered to colonies of honey bees, probably because of the method used to remove the soybean oil from the raw bean. Early processing of the beans for oil and meal was by hydraulic pressing (expeller process). This process generated some heat and left from 4 to 6% oil (fat) in the meal. Another method now used extensively to remove the oil is washing the beans with organic solvents such as hexane. After this treatment, the resulting meal contains only 1% oil, but some oil may be added back to the final product. This meal is then ground and may be toasted for varying lengths of time to destroy naturally occurring proteolytic enzyme inhibitors (trypsin inhibitors) that interfere with the digestion of soybean protein in higher animals. Some undesirable flavor factors are also removed by the toasting process.

The amount of oil remaining in soybean meal after extraction may also affect the effectiveness of soybean flour as a full pollen substitute. Haydak

(1940) fed honey bees soybean diets with three levels of fat and concluded that lipids are important for brood rearing and that most were removed by extraction of oil with the solvent methods. Maurizio (1951) likewise reported differences in biological effectiveness depending on the method of processing. Haydak and Tanquary (1942) noted that the mortality of bees increased and that brood rearing decreased as the fat content of the food diminished.

Erickson and Herbert (1980) tested two new and inexpensive soybean products which proved to be satisfactory pollen substitutes. They concluded that the selection of a soy product to be a pollen substitute or supplement should be based on: 1) soyflour should be fully toasted to remove trypsin inhibitors; 2) soyflour should have a fat content of about 7% if fed without pollen; 3) soyflour should have a fat content of about 0.5%-1% if fed with 10-20% pollen, as in a pollen supplement; and 4) soyflour should have a high protein content (about 45 to 60%).

The optimum amount of protein to feed caged honey bees was tested by Herbert *et al.* (1977). Caged bees were fed diets containing either 5, 10, 23, 30, or 50% protein and the amount of sealed brood reared in each unit was measured. At low protein levels (5 and 10%) the queens laid eggs and brood rearing continued for short periods. At high protein levels (50%), the bees appeared to suffer from protein toxicity resulting in an inability of bees to defecate. Apparently excessive protein levels cause imbalances that affect normal biological processes, and these imbalances could also make the cost of supplemental feeding prohibitive. The optimum protein level to feed honey bees was found to be 23%. Lehner (1983) fed small colonies of honey bees for six weeks on diets of protein concentration increasing from 5% to 30%, using soya flours and a Torula yeast® product. Honey bees fed the 5% level of protein reared the smallest amount of brood. Colonies fed pollen raised more brood than those fed the test diets and on average produced populations about twice as large.

The effect of the pH of a pollen substitute on the resultant worker jelly produced by nurse bees was measured by offering caged honey bee colonies diets ranging in pH from 4.1 - 8.0 (Herbert and Shimanuki 1983). There were no significant differences in the pH of worker jelly produced and over 80% of all jelly sampled had a pH with the range of 4.0 - 4.3 (Table 4). The most brood was reared by honey bees offered diets with pH of 4.7 and 5.5. This study demonstrated that bees have the capability to buffer diets with extreme pH's and produce brood food within a narrow pH range.

CARBOHYDRATE SUPPLEMENTAL FEEDING

The amount of honey a colony requires depends upon a number of factors including population, brood rearing, and wax secretion. Strong colonies in most northern regions consume 50-55 pounds of honey from the time brood rearing stops in the fall until enough nectar can be collected in the spring to support the colony. Estimates of 60-80 pounds for winter and early spring's survival have been reported. The honey requirement for summer has been estimated at 95 pounds. Therefore, a normal colony may require in excess of 150 pounds a year to thrive in most northern states. Honey bees overwintered in less severe environments, especially in the southern and southwestern states, require less honey to maintain them during nectar dearths.

There are a number of feeding methods that can be used by beekeepers during periods when honey stores are depleted. The easiest method is to place comb(s) of sealed honey as close as possible to the cluster. If comb honey is not available, the bees can be provided with diluted liquid honey. Add about 1/3 gallon of warm water per gallon of honey to make a 60% sugar solution; 3/4 gallon of water per gallon of honey will produce a solution slightly less than 50%. Care should be taken to use only honey from disease-free colonies; honey purchased from unknown sources should never be used.

The ideal carbohydrate supplement is sucrose. For fall feeding honey bees should be fed a concentrated solution of 2 parts sugar to 1 part water (by weight) or by mixing 16 pounds of sugar with 1 gallon of warm water. In the spring a less concentrated solution of sugar containing 1 part sugar to 1 part water (weight by weight) can be prepared by mixing 8 pounds of sugar with 1 gallon of warm water. In the spring honey bees have more time to concentrate and ripen the solution so the sugar concentration is less critical than that for fall feeding.

There are sugar formulations that can be purchased for bee feeds. For example, drivert sugar (92% sucrose and 8% invert sugar), Type 50 sugar syrup (77% glucose-fructose and 23% water), and various concentrations of high fructose corn syrup (42%, 55%, and 92% fructose). The use of high fructose corn syrup is increasing because it is less expensive than mixing a sucrose

Table 4. The pH of worker jelly produced (1) by caged honey bees offered pollen or pollen substitute at various pH levels (Herbert and Shimanuki 1983).

Diet pH	16 June	2 July	15 July	10 August
4.1	4.13 ± .02	4.16 ± .02	4.21 ± .14	4.09 ± .03
4.7	4.09 ± .01	4.22 ± .03	4.45 ± .04	3.94 ± .12
5.5	4.06 ± .03	4.25 ± .01	4.04 ± .03	4.11 ± .03
6.6	4.08 ± .02	4.22 ± .02	4.06 ± .03	4.00 ± .03
8.0	4.07 ± .02	4.17 ± .02	4.08 ± .02	3.94 ± .04
Pollen (2)	4.18 ± .32	4.18 ± .01	4.37 ± .21	4.24 ± .02

(1) Mean ± S.E. of four replications, with four subsamples in each replication.

(2) The pH of the pollen diet was as follows: 16 June (pH 4.80), 2 July (4.86), 15 July (4.45), and 10 Aug. (4.02).

syrup. Researchers have shown that honey bees readily consume and store high fructose corn syrup even though its biological value may be somewhat less than sucrose. The major disadvantage of feeding high fructose corn syrup is the danger of adulterating surplus honey as the bees store this product and treat it as honey.

There are a number of methods for feeding syrup to a colony. Perhaps the easiest method is placing the syrup in a jar, pail, or can and inverting this over an inner cover placed over the cluster (Fig. 4). The number of holes in the lid should be small enough so the feeder will not leak, but numerous enough to allow bees to feed. The more holes in the lid, the faster the bees will take the syrup. The feeder should be placed inside an empty hive body and covered with a hive cover to conserve colony heat and prevent robbing. Another method of feeding is the division board feeder. This is an internal feeder that

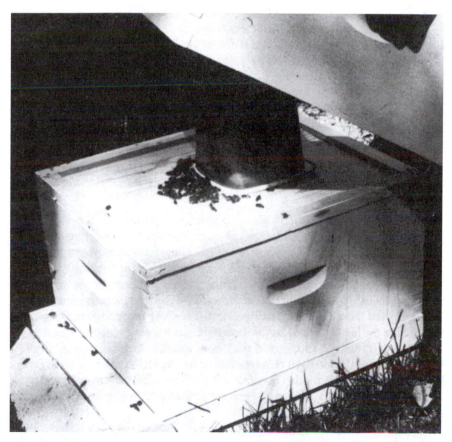

FIGURE 4. Feeding sugar syrup to a honey bee colony.

replaces a comb in the colony. The feeder is filled with syrup and a strip of wood or float is placed on top of the solution to allow bees to feed without drowning. These units can be refilled without disrupting the cluster.

Entrance feeders (Boardman feeders) are used primarily by hobby bee-keepers. The disadvantage of this method is that the solution is exposed to the weather and the syrup could leak, especially as the temperatures rise and fall. A leaking feeder could lead to robbing. Other disadvantages of the entrance feeder is that it is too far from the cluster to be of much benefit in cooler weather and the feeder jar is vulnerable to damage by vandals.

Honey bees can also be provided syrup by sprinkling it into the cells of an empty comb which is placed near the cluster. Some beekeepers offer dry sugar, usually on the inner cover, instead of feeding a solution, but this method requires more energy output by the bees since the dry sugar must first be liquefied before it can be utilized.

Regardless of the feeding method, the beekeeper should monitor his colonies closely and feed before the colonies are in danger of starving. The cost of the sugar will be offset by stronger colonies in the spring and greater honey yields.

LARVAL NUTRITION

The nutritional requirements of immature honey bees differ from those of adult bees. A comprehensive review of the literature is beyond the scope of this chapter, but a chapter on honey bee nutrition would not be complete without at least touching on this subject. It will become readily obvious that our knowledge in this area is fragmentary and that much research remains to be done.

Worker bee larvae are fed a glandular secretion produced by the mandibular and hypopharyngeal glands of adult nurse bees, normally during the first 2 weeks of their adult life. The secretions of the hypopharyngeal glands are clear, containing mostly protein, while those secretions from the mandibular glands are white, containing mostly lipid components. Worker larvae are fed a so-called "worker jelly" which consists of 20-40% white component for the first 2 days of larvae life. On the third day the white component ceases to be fed and for the last 2 days of larval development only the clear secretion from the hypopharyngeal gland, mixed with honey and pollen, is provided. Pollen supplies little of the nitrogen requirement of larvae and may be a contaminant from honey stomach fluids (Haydak 1970). This diet is often referred to as "modified worker jelly." Drone larvae are probably fed a similar diet, but they receive more food than worker larvae.

Queen larvae, throughout their larval development, are fed "royal jelly," which consists of equal amounts of secretions from the hypopharyngeal and mandibular glands. Queen larvae always have large amounts of royal jelly available and often much dried food is left uneaten when the larva pupates. Normally, little or none is left in worker cells after pupation (Free 1977).

The composition of royal jelly has been studied by a number of investigators (Haydak 1943; Rembold 1965; Beetsma 1979; Lercker *et al.* 1982; Howe *et al.* 1985; and others). Royal jelly is composed of 11.9% protein, 67.1% moisture, and 4.3% lipids. An amino acid analysis showed 17 standard protein amino acids and 5 unidentified ninhydrin-positive compounds (Howe *et al.* 1985). Royal jelly contains considerable lipid material and the acid, 10-hydroxy-2-decenoic. Pollen sucrose as well as nectar sucrose is metabolized by nurse bees to this acid and other nutritional substrates in royal jelly. Brown *et al.* (1962) uniformly labelled sucrose with carbon-14 and fed confined honey bees. Five hours after confinement, 13 grafted queen cells were given to the confined bees. Radioactivity was found in the jelly and the levels did not decrease for up to 7 days after feeding. Boch *et al.* (1979) found octanoic and other volatile acids in the mandibular glands of honey bees and in royal jelly. Pain *et al.* (1962) identified adipic, pimelic and suberic acid, and 24-methylene cholesterol in royal jelly and worker jelly as well. Lercker *et al.* (1982) found quantitative differences in the chemical composition of royal jelly samples collected in spring and summer. The main difference was that the free fatty acids increased markedly in 10-hydroxydecanoic acid in summer.

There is a considerable difference between the vitamin composition of royal jelly and the larval food of workers. For example, royal jelly contains 10-fold more pantothenic acid than in worker jelly. Other vitamins present in high levels in royal jelly include folic acid and biopterin.

The complete nutritional requirements of honey bee larvae have not been determined because to date no satisfactory diet has been developed to allow complete larval development in the laboratory. Hoffman (1960) raised 23% of the test larvae to pupation by transferring 2.5 mg larvae to queen cups in the laboratory and feeding them worker jelly. Dietz and Haydak (1967) found that survival of larvae depended on the moisture content of the food. Shuel and Dixon (1968) reared larvae from 3.5 days to pupation and showed the necessity of sugar in the larval diet. Shuel *et al.* (1978) reared larvae less than 1 day of age in royal jelly to adulthood. Rembold and Lackner (1981) formulated a basic diet composed of royal jelly sugars and water on which 75% of 1-day larvae reached adulthood. When Vandenberg and Shimanuki (1987) fed 1-day-old larvae a mixed diet consisting of royal jelly, water, glucose, fructose, and yeast extract—or royal jelly for 1 or 2 days followed by the mixed diet—88-96% survived to the defecation stage. Adult weights were higher when larvae were reared on the mixed diet alone than when reared on royal jelly for 1 day followed by the mixed diet.

Young female larvae, less than 3 days old, can develop into either a queen or a worker bee depending on the quantity and type of food they receive from the nurse bees. The diets of queen and worker larvae differ in several respects, with the most striking gross dietary difference in the concentration of sugar (Dixon and Shuel 1963). Brouwers (1984) reported that glucose was predominant during the early larval stages of workers and drones, but fructose became the main sugar component in the food of older larvae. Glucose remained the main sugar throughout the larval development of queens.

The sugar content of the larval food appears to be the crucial factor that triggers the rate of food intake, which is higher in queen larvae than in worker larvae. The rate of food intake probably regulates the activity of the corpora allata, and thereby the amount of juvenile hormone produced (Beetsma 1979). A high juvenile hormone level during the third day of larval development induces differentiation into a queen; a low level results in the development of a worker.

Rembold and Hansen (1964) suggested that the differentiating substances are synthesized by the mandibular glands of nurse bees while rearing queens. Peng and Jay (1977) removed the mandibular glands from 10-day-old nurse bees and found they were able to rear only a few larvae to the pupal stage while bees with mandibular glands reared many pupae or adults which were queens or queenlike intermediates. Another study (Peng and Jay 1979) determined that the mandibular gland secretion is less important as a larval food than that of the hypopharyngeal glands, and if a "queen determining substance" exists, the mandibular glands are not its only source.

This brief discussion of the effect of diet on caste determination does not attempt to assimilate the diverse literature on this subject. For a more complete review of theories of the effects of nutrition, hormonal deficiency, the moisture content of royal jelly, and food consumption during specific larval periods, see Dietz, 1972; Dietz *et al.* 1979.

REFERENCES

Anderson, L.M., and A. Dietz. (1976). Pyridoxine requirement of the honey bee (*Apis mellifera*) for brood rearing. *Apidologie* 7:67-84.

Auclair, J.L., and C.A. Jamieson. (1948). A qualitative analysis of amino acids in pollen collected by bees. *Science* 108:357-358.

Augustin, R., and D.A. Nixon. (1957). Grass pollen constituents; the meso-inositol content. *Nature* 179:530-531.

Bachman, W.W., and G.D. Waller. (1977). Honey bee responses to sugar solutions of different compositions. *J. Apic. Res.* 16:165-169.

Barbier, M. (1971). Chemistry and biochemistry of pollens. *Prog. Phytochem.* 2:1-28.

Barbier, M., M.F. Hügel, and E. Lederer. (1960). Isolation of 24-methylene-cholesterol from pollen of different plants. *Soc. de Chim. Biol.* 42:91-97.

Barker, R.J. (1972). Whether the superiority of pollen in diet of honey bees is attributable to its high content of free proline. *Ann. Entomol. Soc. Amer.* 65:270-271.

Barker, R.J., and Y. Lehner. (1974a). Acceptance and sustenance value of naturally occurring sugars fed to newly emerged adult workers of honey bees (*Apis mellifera* L.). *J. Exp. Zool.* 187:277-286.

Barker, R.J., and Y. Lehner. (1974b). Influence of diet on sugars found by thin-layer chromatography on thoraces of honey bees, (*Apis mellifera* L.). *J. Exp. Zool.* 188:157-163.

Bettsma, J. (1979). The process of queen-worker differentiation in the honey bee. *Bee World* 60:24-39.

Beutler, R., and E. Opfinger. (1949). Pollenernährung und Nosemabefall der honigbiene. *Z. Vergleich. Physiol.* 32:383-421.

Bieberdorf, F.W., A.L. Gross, and R. Weichlein. (1961). Free amino acid content of pollen. *Annals of Allergy* 19:867-876.

Boch, R.D., A. Shearer, and H. Shimanuki. (1973). Effect of ethylene oxide fumigation on amino acid composition of pollen. *Environ. Entomol.* 2:937-938.

Boch, R.D., A. Shearer, and R.W. Shuel. (1979). Octanoic and other volatile acids in the mandibular glands of the honey bee and in royal jelly. *J. Apic. Res.* 18:250-253.

Brouwers, E.V.M. (1982). Measurement of hypopharyngeal gland activity in the honey bee. *J. Apic. Res.* 21:193-198.

Brouwers, E.V.M. (1984). Glucose/Fructose ratio in the food of honey bee larvae during caste differentiation. *J. Apic. Res.* 23:94-101.

Brown, W.H., E.E. Felauer, and M.V. Smith. (1962). Biosynthesis of royal jelly from sucrose. *Nature* 195:75-76.

Campana, J., and F.E. Moeller. (1977). Honey bees: preference for and nutritive value of pollen from five plant sources. *J. Econ. Entomol.* 70:39-44.

Chu, H.M., D.M. Norris, and L.T. Kok. (1970). Pupation requirement of the beetle, *Xyleborus ferrugineus*: sterols other than cholesterol. *J. Insect Physiol.* 16:1379-1387.

Dadd, R.H. (1973). Insect nutrition: current developments and metabolic implications. *Ann. Rev. Entomol.* 18:381-420.

De Groot, A.P. (1953). Protein and amino acid requirements of the honey bee. *Physiol. Comp. Oecol.* 3:1-90.

Dietz, A. (1969). Initiation of pollen consumption and pollen movement through the alimentary canal of newly emerged honey bees. Ann. Entomol. *Soc. Amer.* 62:43-46.

Dietz, A. (1971). Changes with age in some mineral constituents of worker honey bees: I. Phosphorous, potassium, calcium, magnesium, sodium and iron. *J. Ga. Entomol. Soc.* 6:54-57.

Dietz, A. (1972). Nutritional basis of caste determination in honey bees. *In* Insects and mite nutrition. North-Holland Publishing Co., Amsterdam, pp. 271-279.

Dietz, A., and M.H. Haydak. (1965). Causes of nutrient deficiency in stored pollen for the development of newly emerged honey bees. Proc. Int. Beekeeping Cong. Bucharest. 20:222-225.

Dietz, A., and M.H. Haydak. (1967). Caste determination in honey bees. The significance of moisture in larval food. Proc. Int. Apicult. Congr., College Park, MD, 21:470 (Abst.).

Dietz, A., and H.R. Stevenson. (1975). The effect of long-term storage on the nutritive value of pollen for brood rearing of honey bees. *Amer. Bee J.* 115:476-477,482.

Dietz, A., H.R. Herman, and M.S. Blum. (1979). The role of exogenous JH I, JH III, and Anti-JH (Precocene II) on queen induction of 4½ day-old worker honey bee larvae. *Jour. Insect Physiol.* 25:503-512.

Dietz, A., and H.R. Stevenson. (1980). Influence of long-term storage on the nutritional value of frozen pollen for brood rearing of honey bees. *Apidologie* 11:143-151.

Dixon, S.E., and R.W. Shuel. (1963). Studies in the mode of action of royal jelly in honey bee development—III. The effect of experimental variation in diet on growth and metabolism of honey bee larvae. *Can. J. Zool.* 41:733-739.

Doull, K.M. (1974a). Effects of attractants and phagostimulants in pollen and pollen supplement on the feeding behavior of honey bees in the hive. *J. Apic. Res.* 13:47-54.

Doull, K.M. (1974b). Effect of distance on the attraction of pollen to honey bees in the hive. *J. Apic. Res.* 13:27-32.

Doull, K.M. 1976. The effects of different humidities on the hatching of the eggs of honey bees. *Apidologie* 7:61-65.

Doull, K.M. (1980). Relationships between consumption of a pollen supplement, honey production, and brood rearing in colonies of honey bees, *Apis mellifera* L. *Apidologie* 11:361-365.

Doull, K.M., and L.M. Standifer. (1970). Feeding responses of honey bees in the hive. *J. Apic. Res.* 9:129-132.

Eischen, F.A., W.C. Rothenbuhler, and J.M. Kulincevic. (1982). Length of life and dry weight of worker honey bees reared in colonies with different worker-larva ratios. *J. Apic. Res.* 21:19-25.

Eischen, F.A., W.C. Rothenbuhler, and J.M. Kulincevic. (1983). Brood rearing associated with a range of worker-larva ratios in the honey bee. *J. Apic. Res.* 22:163,168.

Erickson, E.H., and E.W. Herbert Jr. (1980). Soybean products replace expeller-processed soyflour for pollen supplements and substitutes. *Amer. Bee J.* 120:122-126.

Feldhaufer, M.F., E.W. Herbert Jr., J.A. Svoboda, M.J. Thompson, and W.R. Lusby. (1985). Makisterone A: the major ecdysteroid from the pupa of the honey bee, *Apis mellifera. Insect Biochem.* 15:597-600.

Feldlaufer, M.F., J.A. Svoboda, and E.W. Herbert Jr. (1986a). Makisterone A and 24-methylene cholesterol from the ovaries of the honey bee, *Apis mellifera* L. *Experientia* 42:200-201.

Feldlaufer, M.F., E.W. Herbert Jr., J.A. Svoboda, and M.J. Thompson. (1986b). Biosynthesis of makisterone A and 20-hydroxyecdysone from labeled sterols by the honey bee, *Apis mellifera* L. *Arch. Insect Biochem. Physiol* 3:415-421.

Free, J.B. (1977). The social organization of honey bees. Camelot Press Ltd., Southhampton, p. 66.

Frisch, K.v. (1934). Uber den Geschmacksinn der Bienen *Z. vergl. Physiol.* 21:1-156.

Frisch, K.v. (1965). Tanzsprache und Orientierung der Bienen. Springer-Verleg, Berlin, p. 578.

Gilliam, M. (1972). Lack of effect of deficient diets on hemolymph proteins of adult worker honey bees. *In* Insect and mite nutrition, North-Holland Publishing Co., Amsterdam, pp. 281-284.

Gilliam, M. (1979a). Microbiology of pollen and bee bread: the yeasts. *Apidologie* 10:43-53.

Gilliam, M. (1979b). Microbiology of pollen and bee bread; the genus *Bacillus. Apidologie* 10:269-274.

Gilliam, M., D.B. Prest, and B.J. Lorenz. (1989). Microbiology of pollen and bee bread; taxonomy and enzymology of molds. *Apidologie 20*:53-68.

Goldsmith, T.H. and L.T. Warner. (1964). Vitamin A in the vision of insects. *J. Gen. Physiol. 47*:443-441.

Hagedorn, H.H., and M. Burger. (1968). Effect of the age of pollen used in pollen supplement on the nutritive value for the honey bee. II. Effect of vitamin content on pollens *J. Apic. Res. 7*:97-101.

Hagedorn, H.H., and F.E. Moeller. (1967). The rate of pollen consumption by newly emerged honey bees. *J. Apic. Res. 6*:159-162.

Hagedorn, H.H., and F.E. Moeller. (1968). Effect of the age of pollen used in pollen supplement on the nutritive value for the honey bee. I. Effect on thoracic weight, development of hypopharyngeal glands, and brood rearing. *J. Apic. Res. 7*:89-95.

Haydak, M.H. (1934). Changes in total nitrogen content during the life of the imago of the worker honey bee. *J. Agric. Res. 49*:21-28.

Haydak, M.H. (1935). Brood rearing by honey bees confined to a pure carbohydrate diet. *J. Econ. Entomol. 28*:657-660.

Haydak, M.H. (1936). Value of foods other than pollen in nutrition of the honey bee. *J. Econ. Entomol. 29*:870-876.

Haydak, M.H. (1937). Further contribution to the study of pollen substitutes. *J. Econ. Entomol. 30*:637-642.

Haydak, M.H. (1940). Comparative value of pollen and pollen substitutes. II. Bee bread and soybean flour. *J. Econ. Entomol. 33*:397-399.

Haydak, M.H. (1943). Larval foods and development of castes. *J. Econ. Entomol. 36*:778-792.

Haydak, M.H. (1945). Value of pollen substitute for brood rearing of honey bees. *J. Econ. Entomol. 38*:484-487.

Haydak, M.H. (1949). Causes of deficiency of soybean flour as a pollen substitute for honey bees. *J. Econ. Entomol. 42*:573-579.

Haydak, M.H. (1959). Pollen substitutes—still a controversy? *Amer. Bee J. 99*:131-132.

Haydak, M.H. (1961). Influence of storage on the nutritive value of pollens for newly emerged honey bees. *Amer. Bee J. 101*:354-355.

Haydak, M.H. (1963). Influence of storage on the nutritive value of pollen for brood rearing by honey bees. *J. Apic. Res. 2*:105-107.

Haydak, M.H. (1967). Bee nutrition and pollen substitutes. *Apiacta 1967 (1)*:3-8.

Haydak, M.H. (1970). Honey bee nutrition. *Ann. Rev. Entomol. 15*:143-156.

Haydak, M.H., and A. Dietz. (1965). Influence of the diet on the development and brood rearing of honey bees. Proc. XV Intl. Beekeeping Cong. Bucharest, 1-6.

Haydak, M.H., and A. Dietz. (1972). Cholesterol, pantothenic acid, pyridoxine and thiamine requirements of honey bees for brood rearing. *J. Apic. Res. 11*:105-109.

Haydak, M.H., and L.S. Palmer. (1938). Vitamin E content of royal jelly and bee bread. *J. Econ. Entomol. 31*:576-577.

Haydak, M.H., and M.C. Tanquary. (1942). Various kinds of soybean flour as pollen substitutes. *J. Econ. Entomol. 35*:317-318.

Heed, W.B., and H.W. Kincher. (1965). Unique sterol in the ecology and nutrition of *Drosophila pachea. Science 149*:758-761.

Herbert, E.W. Jr. (1979). A new ash mixture for honey bees maintained on a synthetic diet. *J. Apic. Res. 18*:144-147.

Herbert, E.W. Jr., W.G. Bickley, and H. Shimanuki. (1970). The brood rearing capacity of caged honey bees fed dandelion and mixed pollen diets. *J. Econ. Entomol. 63*:215-218.

Herbert, E.W. Jr., and H. Shimanuki (1971). Fumigation of pollen with ethylene oxide and its effect on brood rearing of honey bees. *J. Econ. Entomol. 64*:877-879.

Herbert, E.W. Jr., and H. Shimanuki. (1978a). Consumption and brood rearing by caged honey bees fed pollen substitutes fortified with various sugars. *J. Apic. Res. 17*:27-31.

Herbert, E.W. Jr., and H. Shimanuki. (1978b). Chemical composition and nutritive value of bee-collected and bee-stored pollen. *Apidologie 9*:33-40.

Herbert, E.W. Jr., and H. Shimanuki. (1978c). Mineral requirements for brood rearing by honey bees fed a synthetic diet. *J. Apic. Res. 17*:118-122.

Herbert, E.W. Jr., and H. Shimanuki. (1978d). Effects of thiamine- or riboflavin-deficient diet fed to newly emerged honey bees, *Apis mellifera* L. *Apidologie 9*:341-348.

Herbert, E.W. Jr., and H. Shimanuki. (1978e). Effects of fat soluble vitamins on the brood rearing capabilities of honey bees fed a synthetic diet. *Ann. Entomol. Soc. Amer. 71*:689-691.

Herbert, E.W. Jr., and H. Shimanuki. (1979). A new ash mixture for honey bees maintained on a synthetic diet. *J. Apic. Res. 18*:144-147.

Herbert, E.W. Jr., and H. Shimanuki. (1983). Effect of diet pH on the consumption, brood rearing, and pH of worker jelly produced by caged honey bees. *Apidologie 14*:191-196.

Herbert, E.W. Jr., H. Shimanuki, and D. Caron. (1977). Optimum protein levels required by honey bees (Hymenoptera: Apidae) to initiate and maintain brood rearing. *Apidologie 8*:141-146.

Herbert, E.W. Jr., J.A. Svoboda, M.J. Thompson, and H. Shimanuki. (1980). Sterol utilization in honey bees fed a synthetic diet: effects on brood rearing. *J. Insect Physiol. 26*:287-289.

Herbert, E.W. Jr., J.T. Vanderslice, and D.J. Higgs. (1985). Vitamin C enhancement of brood rearing by caged honey bees fed a chemically defined diet. *Arch. Insect Biochem. Physiol. 2*:29-37.

Herbert, E.W. Jr., J.T. Vanderslice, M.H. Huang, and D.J. Higgs. (1987). Levels of thiamine and its esters in bee-collected pollen using liquid chromatography and robotics. *Apidologie 18*:129-136.

Hoffman, I. (1960). Rearing worker honey bee larvae in an incubator. *Bee World 41*:10-11.

Hopkins, C.Y., A.W. Jevans, and R. Boch. (1969). Occurrence of octadecatrans-2, cis, -12-trienoic acid in pollen attractive to the honey bee. *Can. J. Bio. 47*:433-436.

Howe, S.R., P.S. Dimick, and A.W. Benton. (1985). Composition of freshly harvested and commercial royal jelly. *J. Apic. res. 24*:52-61.

Hügel, M.F. (1962). Etude de qualques constituants du pollen. (A study of some components of pollen.) *Ann. de l'Abeille 5*:97-133.

Inouye, D.W., and G.D. Waller. (1984). Responses of honey bees (*Apis mellifera*) to amino acid solutions mimicking floral nectars. *Ecology 65*:618-625.

Kauffeld, N.M. (1980). Chemical analysis of Louisiana pollen and colony conditions during a year. *Apidologie 11*:47-55.

Knox, D.A., H. Shimanuki, and E.W. Herbert Jr. (1971). Diet and the longevity of adult honey bees. *J. Econ. Entomol. 64*:1415-1416.

Kok, J.C.N. (1952). Qualitatieve en semi-quantitaleve analyse von de aminozuren bij enkele pollensoorten; unpublished. Cited by De Groot, A.P. 1953.

Lehner, Y. (1983). Nutritional considerations in choosing protein and carbohydrate sources for use in pollen substitutes for honey bees. *J. Apic. Res.* 22:242-248.

Lepage, M., and R. Boch. (1968). Pollen lipids attractive to honey bees. *Lipids* 3:530-534.

Lercker, G., P. Capella, L.S. Conte, F. Ruini, and G. Giordani. (1982). Components of royal jelly. II. The lipid fraction, hydrocarbons and sterols. *J. Apic. Res.* 21:178-184.

Loper, G.M., and R.L. Berdel. (1980). The effects of nine pollen diets on brood rearing of honey bees. *Apidologie* 11:351-359.

Lunden, R. (1954). A short introduction to the literature on pollen chemistry. *Svensk. Kem. Tidskr.* 66:201-213.

McCaughey, W.F., M. Gilliam, and L.N. Standifer. (1980). Amino acids and protein adequacy for honey bees of pollens from desert plants and other floral sources. *Apidologie* 11:75-86.

Maurizio, A. (1951). Tests of pollen substitutes. *Schweiz. Bienenztg.* 74:111-118.

Maurizio, A. (1954). Pollen nutrition and life processes of the honey bee. *Landwirtschaftliches Jahrbuch der Schweiz.* 68:115-182.

Maurizio, A. (1960). Bienenbotanik. *In* Budel, A., and E. Herold, Biene and Bienenzucht. Ehrenwirth Verlag, Munchen, pp. 68-104.

Moffett, J.O., A. Stoner, and H.L. Wardecker. (1977). The Wardecker waterer. *Amer. Bee J.* 117:364-365,378.

Nation, J.L., and F.A. Robinson. (1966). Gibberellic acid: Effects of feeding in an artificial diet for honey bees. *Science* 152:1765-1766.

Nation, J.L., and F.A. Robinson. (1968). Brood rearing by caged honey bees in response to inositol and certain pollen fractions in their diet. *Ann. Entomol Soc. Amer.* 61:514-517.

Nation, J.L., and F.A. Robinson. (1971). Concentration of some major and trace elements in honey bees, royal jelly and pollens, determined by atomic absorption spectrophotometry. *J. Apic. Res.* 10:35-43.

Nelson, D.L. 1983. Honey bees and thermoregulation—a review. *Can. Beekeeping* 11:31-33.

Nielsen, N. (1955). Investigations on the chemical composition of pollen from some plants. *Acta. Chem. Scand.* 9:1100-1106.

Nolan, W.J. (1925). Colony population and honey crops. *Glean. Bee Cult.* 53:366-368.

Pain, J. (1956). Vitamins and ovarian development of worker bees. *Soc. de Biol., Paris* 145:1505-1507.

Pain, J., M. Barbier, D. Bogdanovsky, and E. Lederer. (1962). Chemistry and biological activity of the secretions of queen and worker honey bees (*Apis mellifera* L.). *Comp. Biochem. Physiol.* 6:233-241.

Pain, J., and J. Maugenet. (1966). Biochemical and physiological research on pollen stored by bees. *Ann. Abeille* 9:209-236.

Peng, Y.S., and S.C. Jay. (1977). Larval rearing by worker honey bees lacking their mandibular glands. I. Rearing by small numbers of worker bees. *Can. Ent.* 109:1175-1180.

Peng, Y.S., and S.C. Jay. (1979). Larval rearing by worker honey bees lacking their mandibular glands. II. Rearing by larger numbers of worker bees. *Can. Ent.* 111:101-104.

Rashad, S.E., and R.L. Parker. (1958). Pollen as a limiting factor in brood rearing and honey production. *Trans. Kansas Acad. Sci.* 61:237-248.

Rembold H. (1965). Biologically active substances in royal jelly. *In* Vitamins and Hormones, Academic Press, N.Y. 23:359-382.

Rembold, H., and G. Hansen. (1964). Uber den Weiselzellenfuttersaft der Honigbiene. VIII. Nachweis des determinierenden Prinzips im Futtersaft der Koniginnenlarven. *Hoppe-Seyler's Z. Physio. Chem.* *339*:251-254.

Rembold, H., and Lackner B. (1981). Rearing of honeybee larvae *in vitro*: effect of yeast extract on queen differentiation. *J. Apic. Res. 20*, 165-171.

Robinson, F.A., and J.L. Nation. (1966). Artificial diets for honey bees, *Apis mellifera. Fla. Entomol. 49*:175-184.

Robinson, F.A., and J.L. Nation. (1968). Substances that attract caged honey bee colonies to consume pollen supplements and substitutes. *J. Apic. Res.* *7*:83-88.

Sarka, B.C., and S.H. Wittner. (1949). Quantitative estimation of some amino acids in sweet corn pollen. *Arch. Biochem. 22*:353-356.

Schmidt, J.O. (1982). Pollen foraging preferences of honey bees. *Southwestern Entomol. 7*:255-259.

Schmidt, J.O., and S.L. Buchmann. (1985). Pollen digestion and nitrogen utilization by *Apis mellifera* L. (Hymenoptera: Apidae). *Comp. Biochem. Physiol. 82*:499-503.

Shuel, R.W. (1975). The production of nectar. *In* The Hive and the Honey Bee, Dadant & Sons, Inc., pp. 265-282.

Shuel, R.W., and S.E. Dixon. (1968). The importance of sugar for the pupation of the worker honey bee. *J. Apic. Res.* *7*:109-112.

Shuel, R.W., S.E. Dixon, and G.B. Kinoshita. (1978). Growth and development of honey bees in the laboratory on altered queen and worker diets. *J. Apic. Res.* *5*:93-98.

Standifer, L.N. (1966). Some lipid constituents of pollens collected by honey bees. *J. Apic. Res. 5*:93-98.

Standifer, L.N., M. Davys, and M. Barbier. (1968). Pollen sterols—a mass spectographic survey. *Phytochemistry 7*:1361-1365.

Standifer, L.N., M.H. Haydak, J.P. Mills, and M.D. Levin. (1973a). Influence of pollen in artificial diets on food consumption and brood production in honey bee colonies. *Amer. Bee J. 113*:94-95.

Standifer, L.N., M.H. Haydak, J.P. Mills, and M.D. Levin. (1973b). Value of three protein rations in maintaining honey bee colonies in outdoor flight cages. *J. Apic. Res. 12*:137-143.

Standifer, L.N., W.F. McCaughey, F.E. Todd, and A.R. Kemmerer. (1960). Relative availability of various proteins to the honey bee. *Ann. Entomol. Soc. Amer. 53*:618-625.

Standifer, L.N., R.H. MacDonald, and M.D. Levin. (1970). Influence of the quality of protein in pollens and of a pollen substitute on the development of hypopharyngeal glands of honey bees. *Ann. Entomol. Soc. Amer. 63*:909-910.

Standifer, L.N., and J.P. Mills. (1977). The effects of worker honey bee diet and age on the vitamin content of larval food. *Ann. Entomol. Soc. Amer.* 70:691-693.

Standifer, L.N., F.E. Moeller, N.M. Kauffeld, E.W. Herbert Jr., and H. Shimanuki. (1977). Supplemental feeding of honey bee colonies. USDA Agriculture Information Bulletin No. 413, 8 pages.

Starratt, A.N., and R. Boch. (1971). Synthesis of octadeca-trans-2, cis-9, cis-12-trienoic acid and its evaluation as a honey bee attractant. *Can. J. Biochem. 49*:251-254.

Svoboda, J.A., E.W. Herbert Jr., W.R. Lusby, and M.J. Thompson. (1983). Comparison of sterols of pollens, honey bee workers, and prepupae from field sites. *Arch. Insect Biochem. Physiol. 1*:25-31.

Svoboda, J.A., E.W. Herbert Jr., M.J. Thompson, and H. Shimanuki. (1981). The fate of radiolabelled C_{28} and C_{29} phytosterols in the honey bee. *J. Insect Physiol.* 27:183-188.

Svoboda, J.A., J.N. Kaplanis, W.E. Robbins, and M.J. Thompson. (1975). Recent developments in insect steroid metabolism. *Ann. Rev. Entomol.* 20:205-220.

Svoboda, J.A., M.J. Thompson, E.W. Herbert Jr., and H. Shimanuki. (1980). Sterol utilization in honey bees fed a synthetic diet; analysis of prepupal sterols. *J. Insect Physiol.* 26:291-294.

Taber, S. (1963). Why bees collect pollen. XIX Int. Apic. Congr.:675.

Taber, S. (1973). Influence of pollen location in the hive on its utilization by the honey bee colony. *J. Apic. Res.* 12:17-20.

Temnov, V.A. (1958). Composition and toxicity of honeydew. Proc. Int. Beekeeping Congr. Bologna-Rome 17:657-662.

Thompson, M.J., J.N. Kaplanis, W.E. Robbins, and J.A. Svoboda. (1973). Metabolism of steroids in insects. *Adv. Lipid Res.* 11:219-265.

Todd, F.E. (1940). Stimulation of brood rearing. Newsletter, *Bur. Ent. and Plant Quar.* 7:32-33.

Todd, F.E., and O. Bretherick. (1942). The composition of pollens. *J. Econ. Entomol.* 35:312-316.

Todd, F.E., and C.B. Reed. (1970). Brood measurement as a valid index to the value of honey bees as pollinators. *J. Econ. Entomol.* 63:148-149.

Townsend, G.F., and M.V. Smith. (1969). Pollen storage for bee feed. *Amer. Bee J.* 109:14-15.

Vandenberg J.D., and H. Shimanuki. (1987). Technique for rearing worker honey bees in the laboratory. *J. Apic. Res.* 26:90-97.

Vivino, E.A., and L.S. Palmer. (1944). Chemical composition and nutritional value of pollen. *Arch. of Biochem.* 4:129-136.

Wahl, O. (1954). Investigations on the food value of pollen substitutes for the honey bee. *Insects Sociaux* 1:285-292.

Wahl, O. (1963). Comparative investigations on the nutritive value of pollen, yeast, soy flour, and powdered milk for the honey bee. *Ztschr. f. Bienenforsch.* 6:209-280.

Wahl, O., and E. Back. (1955). Effect of vitamins in pollen on the length of life, pharyngeal gland development, and ability to rear brood in the honey bee. *Naturwissenschaften* 42:103-104.

Waller, G.D. (1972). Evaluating responses of honey bees to sugar solutions using an artificial-flower feeder. *Ann. Entomol. Soc. Amer.* 65:857-862.

Waller, G.D., M.H. Haydak, and M.D. Levin. (1970). Increasing the palatability of pollen substitutes. *Amer. Bee J.* 110:302-304.

Weaver, N. (1964). A pollen substitute for honey bee colonies. *Glean. Bee Cult.* 92:550-553.

Weaver, N., and A.R. Chauthane. (1967). An all-liquid pollen substitute for honey bee colonies. *Amer. Bee J.* 107:134-135.

White, J.W. (1975). Honey. *In* The Hive and the Honey Bee, Dadant and Sons, pp. 491-530.

White, J.W., and L.W. Doner. (1980). Honey composition and properties. *In* Beekeeping in the United States. USDA Agriculture Handbook 335, 82-91.

CHAPTER 7

HONEY BEE GENETICS AND BREEDING

by ROBERT E. PAGE, JR.[1]
and HARRY H. LAIDLAW, JR.[2]

INTRODUCTION

Honey bees are social insects. They live in family groups (*colonies*) consisting of a single egg-laying individual, the *queen,* up to a few thousand males (*drones*) and tens of thousands of normally non-reproductive *workers*. The intricacies of the resultant colony organization and individual bee activities have intrigued biologists and naturalists for centuries. Much of our knowledge of honey bee morphology and of colony and individual life history and behavior are legacies of these early naturalists and scientists. The desire of these early researchers to understand heredity in bees and to breed them, however, was largely unfulfilled. It is assumed that this was due partly to the difficulty of observing a colony in the dark cavity where it lived, and partly to the inability to control the matings of queens, which mate only in flight.

It is interesting to note that the abbot, Gregor Mendel, who founded modern genetics, endeavored to breed honey bees in his small abbey apiary (Fig. 1). His experiments failed, presumably because he did not know that queens mate more than once and that drones mate only once and die at mating, and because he could not control the queens' mating. Mendel tried to mate queens in an enclosure he built (Fig. 2), but like similar efforts of others these attempts failed (Iltis, 1932). We should be thankful that he decided to study the genetics of garden peas before honey bees, otherwise modern genetics may have suffered a major setback in development.

In this chapter, we present the elementary knowledge of honey bee genetics and breeding that we think is necessary to selectively improve honey bee stocks. We begin with reproductive cytology and parthenogenesis, the very genetic characteristics that make honey bee breeding unique. We discuss the mating behavior of queens and drones and how the mating behavior affects our ability to breed bees. Then, we present some particular breeding systems and discuss the advantages and difficulties of each.

[1]Robert E. Page, Jr., Ph.D., Associate Professor, Department of Entomology, University of California, Davis.

[2]Harry H. Laidlaw, Ph.D., Professor Emeritus, Department of Entomology, University of California, Davis.

Figure 1. Mendel's monastery apiary. (Extract taken from *Life of Mendel©* by Hugo Iltis, reproduced by kind permission of Unwin Hyman, Ltd.)

Cytology

The honey bee's *genome* is a single *chromosome set* of 16 individual and different chromosomes. Each chromosome has in specific order a series of *loci* with *genes* that are specific sequences of *nucleotides* that constitute a region of DNA (*deoxyribonucleic acid*) and represent the genetic information necessary for the development and functioning of the bee. Genes may exist in alternate forms (*alleles*), and may change form (*mutation*).

All *mature* reproductive cells are *gametes*. Gametes have only one genome (*haploid*). Female bees have two genomes (*diploid*) in their immature reproductive cells (newly formed eggs). The chromosomes and their genes of the two genomes of the egg are paired. This genetic arrangement is called a *genotype*. The genic composition of the genotype is the same as that of the female herself. Before the egg can develop into a larva it must mature, a process in which one member of each pair of chromosomes is eliminated (*meiosis*; Fig. 3). This is accomplished by two *nuclear* divisions. During the first division (*meiotic*) each chromosome is replicated and each replicated pair remains attached at a single point (the *centromere*). The replicated pairs of chromosomes (*dyads*) of each set then pair with the corresponding chromosome of the sister set forming a *tetrad*. The individual chromosomes of each genomic set exchange segments with the corresponding chromosome of the sister set, then the dyads segregate and recombine into two new genomes each of which consists of chromosome dyads derived from both premeiotic genomes. The two new sets separate so that the offspring receive only one member of each pair of genes. A second division (*mitotic*) separates the dyads to leave the egg with four chromosome sets, three of which are non-functional *polar bodies*, and one of which is the *pronucleus* of the egg which unites with

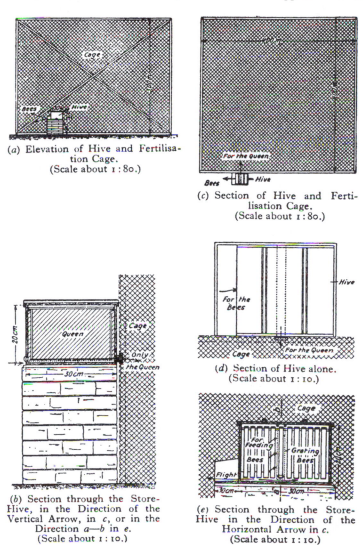

(a) Elevation of Hive and Fertilisation Cage.
(Scale about 1 : 80.)

(c) Section of Hive and Fertilisation Cage.
(Scale about 1 : 80.)

(b) Section through the Store-Hive, in the Direction of the Vertical Arrow, in c, or in the Direction a—b in e.
(Scale about 1 : 10.)

(d) Section of Hive alone.
(Scale about 1 : 10.)

(e) Section through the Store-Hive in the Direction of the Horizontal Arrow in c.
(Scale about 1 : 10.)

FIGURE 2. Mendel's queen bee mating cage. (Extract from *Life of Mendel*© by Hugo Iltis, reproduced by kind permission of Unwin Hyman, Ltd.)

one of the sperm that may enter the egg to form a *zygote* that will begin development of a bee, normally a female (Meves, 1907; Nachtsheim, 1913).

If the egg is not fertilized, the pronucleus itself begins development of a haploid drone which has only one genome. The reduction division of a drone matures a *spermatozoon* by a meiotic process that aborts the reduction division. Consequently, all sperm of a drone are genetically identical and are genetic replications of the genome formed during the maturation of the egg that gave rise to the drone.

Parthenogenesis

Parthenogenesis, the origin of an individual from an unfertilized egg, is the hallmark of the insect order Hymenoptera (bees, wasps and ants). Males (drones) are produced from unfertilized eggs and, except in special circumstances, have no father, a kind of parthenogenesis called *arrhenotoky*. The origin of drones from unfertilized eggs has the peculiar effect of making the drone's mother the genetic father of the drone's progeny. Drones only have daughters, no sons. This peculiarity necessitates adaptive modifications of usual genetic and breeding methods when they are used with the honey bee.

Arrhenotoky

In 1845, Johann Dzierzon published his theory that male honey bees are derived from unfertilized eggs. He proposed that a drone has a mother, but no father. Nachtsheim (1913) demonstrated that female-destined fertilized eggs have 32 chromosomes (16 pairs) while drones have just 16 composed of one member of each chromosome pair. Subsequently, it has been shown that this form of parthenogenesis is characteristic of the Hymenoptera. Males arise from unfertilized eggs laid by either queens or laying workers, while females are normally biparental. Eggs of honey bees, therefore, have a peculiar capability of undergoing development and hatching even without fertilization. Such eggs should not be called "infertile," but "unfertilized."

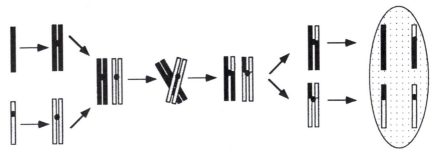

FIGURE 3. A diagram demonstrating the process of meiosis in an egg. A single chromosome pair (one chromosome from each parent) is shown going through the stages discussed in the text.

After an egg is laid, it goes through a critical period of time where it is receptive to the fusion of its pronucleus with any one of several pronuclei of sperm that penetrate the egg (von Siebold, 1856; Dupraw, 1967). If fusion does not occur, the pronucleus of the egg will begin division and establish cell nuclei with only a single set of chromosomes. The differentiated body tissues of drones, however, contain about the same amount of DNA as do those of diploid workers and queens, due to *endomitosis*, the replication of chromosomes within cells without cell division (Merriam & Ris, 1954). In fact, most differentiated cells of drones, workers, and queens are *endopolyploid*, that is, they contain several (some more than 16) copies of the chromosome sets (Painter, 1945).

There must be a genic mechanism of sex determination in order for haploid individuals to consistently develop into males while diploids are consistently female. In the honey bee, and indeed many other Hymenoptera, this determination is apparently under the control of a single gene. In the 1920's, P.W. Whiting and associates began unraveling the mysteries of sex determination in the parasitic wasp, *Habrobracon juglandis*, and first demonstrated the existence of biparental males (Whiting, 1943). These males were derived from fertilized eggs of inbred stocks. Whiting proposed, and then elegantly demonstrated over the next 20 years, that sex was determined by a single major gene that had many different forms, called alleles. Individuals that are *heterozygous* at this sex determining gene locus, that is, have two different alleles, develop into females; *homozygous* individuals, having two identical alleles, develop into diploid males.

The development of instrumental insemination of queen honey bees (see Laidlaw 1987 for history of the development of instrumental insemination) allowed breeders to use the inbred-hybrid method of breeding that had proven so effective in plant breeding. However, it soon became apparent that honey bees were not very amenable to inbreeding because it resulted in tremendous inbreeding depression. Queens that had been inseminated with the semen of their own sons produced spotty brood patterns as a consequence of the removal of the larvae by workers. Otto Mackensen (1951) proposed that the observed spotty brood pattern (about 50% of the cells were empty or contained brood of significantly younger age) was a consequence of lethal homozygosity at a sex locus as described by Whiting for *Habrobracon*. Mackensen (1955) went on to estimate that there were eleven different "lethal" alleles for sex in his North American honey bee population. Laidlaw *et al.* (1956) subsequently estimated twelve lethal alleles in a population in Piracicaba, Brazil; Adams *et al.* (1977) estimated a minimum of 18.9 lethal sex alleles in Rio Claro, Brazil; and Woyke (1976) estimated only 6 lethal sex alleles on the isolated honey bee sanctuary of Kangaroo Island off the coast of Australia.

The *Habrobracon* model of sex determination was supported by the studies of W.C. Rothenbuhler (1957). Rothenbuhler found wild type patches of male eye tissue in mosaic drones (see discussion of mosaics below) that were derived from queens that were homozygous for recessive eye mutation genes. The wild-type male eye facets must have been biparental in origin, and were diploid.

J. Woyke unraveled the mysteries of honey bee sex determination with an elegant scientific investigation (see Woyke, 1986 for complete review of this subject). First, he demonstrated that the empty cells in combs of "shot brood" in inbred colonies had contained eggs that were viable, but that the larvae were removed by the workers within about 6 hours of hatching. Woyke later demonstrated that the removed larvae were diploid drones and proposed that they were homozygous for the sex locus (Woyke, 1963a, 1963b, 1965). He also developed a technique for raising diploid males to maturity so that they could be studied. His method was to raise them on queen royal jelly in queen cells, either in the colony or in the laboratory, for the first 72 hours after hatching. After 72 hours they were placed back into the colony into drone-sized cells where the workers then accepted them and completed their development.

Woyke and Skowronek (1974) studied spermatogenesis in diploid males. They found it was similar to that in haploid males and, consequently, resulted in the production of diploid spermatozoa. Diploid males, however, have testes that are much smaller than haploid males and produce fewer sperm cells (Woyke, 1973). Chaud-Netto (1980a, 1980b) inseminated normal, diploid queens with sperm from diploid males and produced triploid workers.

Thelytoky

Thelytoky is the origin of females from unfertilized eggs. The Cape bee of South Africa is well known for the ability of its workers to parthenogenetically produce diploid workers and queens as well as haploid drones. This results from the regeneration of a diploid nucleus soon following meiosis of the egg. Not only is thelytoky of novel interest to geneticists, it provides a valuable tool for the study of honey bee behavior (Moritz & Hillesheim, 1985; Moritz & Klepsch, 1985).

The ability of workers of the Cape bee, *Apis mellifera capensis*, to produce females (workers and queens) by parthenogenesis became a topic of great interest and debate early in the twentieth century when Onions (1912), and Jack (1917) reported that females of this race developed from unfertilized eggs laid by workers of queenless colonies (cf. van Warmelo, 1912; Attridge, 1919). Mackensen (1943) demonstrated that this phenomenon is not unique to the Cape bee. About 23, 9, and 57 percent of virgin queens from his stocks of *A. m. caucasica, A. m. ligustica,* and a specially selected "golden" strain, respectively, produced some (less than 1%) workers after they were induced to lay eggs by treatment with carbon dioxide.

Tucker (1958) determined that thelytoky occurred most often among progeny of the first brood of a virgin queen. Thelytoky can be induced by causing cessation of oviposition by a queen for a period by confining her and then allowing her to lay again. The origin of the impaternate females was hypothesized by Tucker to be *automictic* where, following meiosis, the egg pronucleus fuses with one of the egg polar bodies and forms a diploid nucleus that undergoes development (see also Verma and Ruttner, 1983).

Mosaics and Gynandromorphs

One of the most interesting discoveries during a colony inspection is the occurrence of individuals that are part drone and part worker. These sex mosaics, called *gynandromorphs*, are not common but do occur occasionally in some colonies. Gynandromorphs can be bilateral with one side male and the other female, or segmental where one end is male and the other female (this is the one that surprises drone collectors!), or individuals can have mosaic patches of different sexes throughout the body. Some gynandromorphs have functional male sex organs and others may have functional female sex organs. These can have special uses in honey bee genetics studies.

Rothenbuhler and his associates (Rothenbuhler *et al.*, 1949) discovered gynandromorphic bees in research stocks and selected a strain of bees that produced these sex mosaics in much higher than normal frequencies. Using visible mutant marker genes, they found that the primary mechanism leading to gynandromorphs is the result of *polyspermy*—more than one sperm penetrates the egg—with cleavage and subsequent development of an accessory sperm nucleus (Rothenbuhler *et al.*, 1952). This nucleus leads to the production of the male tissue while the zygotic nucleus formed by the union of the egg and sperm pronuclei produces the female tissue.

This is not the only mechanism, however. Mackensen (1951) reported a case where the female parts were biparental and the male parts were maternal in origin, apparently arising from one of the haploid polar bodies. Tucker (1958) found gynandromorphs from unmated queens suggesting union of egg pronuclei to form biallelic female tissue and independent development of a haploid nucleus to generate male tissue. Laidlaw and Tucker (1964) reported the most bizarre mechanism of all: the fusion of two accessory sperm to form female tissue while the egg pronucleus fused with a sperm pronucleus (as is normal) and also produced the female tissue. Fusion of sperm to form female tissue was reported for the silk worm by Hasimoto in 1934.

Same-sex mosaics also occur. These are only detectable if the different tissues derived from different cleavage nuclei contain different visible genetic markers such as eye or integument color mutations. Tucker (1958) reported mosaic males that developed from binucleated eggs. Taber (1955) found mosaic females that developed from dizygotic eggs formed from the union of two sperm with two egg nuclei.

GENETIC ORGANIZATION OF COLONIES

A honey bee colony consists of a single long-lived queen, anywhere from zero to several thousand drones, depending on the time of year, and usually tens of thousands of workers. Honey bee queens mate with many drones resulting in a complex structure of genetic relationships within a colony that affect behavioral interactions of individuals and the progress of selective breeding programs (Laidlaw & Eckert, 1950; Polhemus *et al.*, 1950; Rothenbuhler, 1958, 1960).

Mating Behavior

Honey bee queens mate on the average with between about 7 and 17 different males (Adams *et al.,* 1977; Laidlaw *et al.,* 1956; see Page, 1986 for review). Different methods in different populations have resulted in different estimates. These matings take place while queens and drones are in flight and within 5-10 days after the virgin queen emerges as an adult. Each queen takes one or a series of mating flights; the series may extend over several days. The first flight may be either an orientation flight or a copulation flight. On any given flight she mates with from 6-17 males and may mate again with one or more males on a flight the following day (Nolan, 1932; Roberts, 1944; Taber, 1954; Triasko, 1951; Woyke, 1955, 1962). Males mate just once, depositing 6-10 million spermatozoa into the oviducts of the queen, then die. Pairings of drones and queens are probably random, that is, there is no apparent assorting of mates (Page & Metcalf, 1988). However, males do seem to "prefer" to mate with a queen that already is showing a mating sign, the evidence that she has already mated at least once on that particular flight (G. Koeniger, 1990).

Queens return to the hive after the mating flight where a total of about 4-7 million (Mackensen & Roberts, 1948; Woyke, 1964) of the sperm deposited by the drones in the *oviducts* of the queen migrate by primarily active processes (Ruttner & Koeniger, 1971) into the *spermatheca* (the sperm storage organ) over a period of about 40 hours (Woyke, 1983). The sperm that enter the spermatheca are fairly well mixed, representative of most of the drones, and last the egg-laying life of the queen (Laidlaw & Page, 1984).

Subfamily Structure of Colonies

A colony, therefore, consists of a large number of *subfamilies* of workers (Fig. 4). Members of the same subfamily share both a queen mother and a drone father and have on the average 75% of their genes in common by descent—they are called *super sisters*. Individuals with different drone fathers

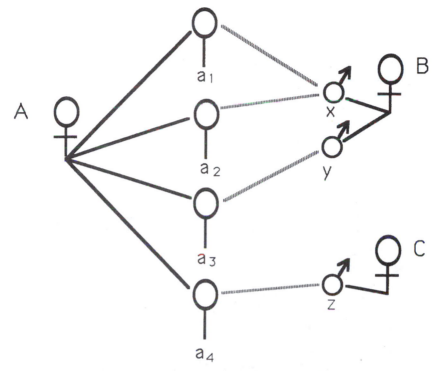

FIGURE 4. This hypothetical pedigree demonstrates the different relationships that are possible within and among families of honey bees. A - C represent queens. Uncrossed female symbols with lower case letters represent worker offspring. Male symbols designated with lower case letters x - z represent the haploid males. Solid lines represent the egg gametes while dashed lines represent sperm gametes. Individuals a_1 and a_2 are super sisters to each other and full sisters of a_3. Individuals a_1, a_2, and a_3 are half sisters of a_4.

belong to different subfamilies and, if the drone fathers are unrelated, the subfamilies share 25% of their genes in common, and the workers of a subfamily are *half sisters* with workers of a different subfamily (Laidlaw & Eckert, 1950; Polhemus *et al.,* 1950; Laidlaw, 1974).

Full sister relationship can also exist if "brother" drones (derived from different gametes of the same mother queen) inseminate a queen. All of the sperm that one drone produces contain identical genomes (with the exception of random mutations) that are derived from the egg gamete produced by the queen (female) because drones hatch from unfertilized eggs, have no father, and are haploid. The queen is, therefore, the originator of all honey bee genomes. Different eggs laid by a queen contain different genomes due to genetic recombination during maturation of the egg, and the sperm produced by males derived from different eggs of the same queen will differ accordingly. This results in a genetic relationship of 0.50 between individual progeny of (brother) drones, the equivalent of full sisters in diploid systems, and is equivalent to diploid full sisters with respect to the origin of the genomes (Page & Laidlaw, 1988). This "genetic pairing" terminology is accurate and appropriate for honey bee genetics and will be used later in the section on systems of mating.

More of the genetic variability of a population resides within colonies and less between them as the number of matings increases (Crow & Kimura, 1970). For selective breeding, it is usually best to have less genetic variability within colonies, that is, fewer matings of the colony queen, and more variability between colonies. This is because selection is usually based on the characteristics of individual colonies and the genetically based differences between them are more easily detected with greater genetic differences.

The exception is when within-family selection is applied. Selection within families is used when avoiding inbreeding is important for a breeding program. Daughter queens are produced from breeder queen mothers and tested. The superior performing daughter of each queen (usually based on the performance of her colony) is then selected to replace her mother as a parent in the next generation. With this selection scheme, more matings of queens results in a greater selection response (Falconer, 1981).

Another consequence of subfamily colony structure on breeding results from the complex interactions of workers. Interactions of individuals of different genotypes confound the ability of breeders to predict the response of their breeding population to selection (Moritz, 1986, 1987). For instance, individuals of a single subfamily may have a high genetic predisposition to defend the hive and, as a consequence, make the colony very defensive while the other 16 or so subfamilies may be very nondefensive. The behavior of the colony will not reflect its true genetic composition because of this single subfamily, and the bee breeder may mistakenly not select its queen as a breeder. Behavioral dominance of this type is very likely in honey bee societies as a consequence of expressed genetic variability among colony members for many behavioral traits (see discussion below).

SELECTIVE BREEDING
Successful selective breeding of honey bees has four essential components: 1. **Stock selection**—in order to select stock, there must be identifiable, selectable differences among individuals (or colonies) that constitute the potential parental population. 2. **Genetic variability**—these differences among individuals must be due at least in part to genetic differences of the reproductive individuals—queens and drones. 3. **Controlled matings**—matings must

be controlled when superior colonies (or individuals) are identified and reproductive progeny are raised. 4. **Stock maintenance**—superior stocks must be maintained over time or they have no value.

Stock Selection

Selection of superior performing stock is essential to a breeding program (Laidlaw, 1954, 1956, 1981). It requires accurate data collection and record keeping in order to identify those queens that are mothers of superior performing colonies so that they can be selected to be the queen and drone mothers for the next generation. Both the genotype of the queen and the genotypes of the workers of the colony, as well as the particular environment in which measurements are taken, contribute to the overall observable colony traits. Therefore, it is imperative to control the environmental variability as much as possible by developing repeatable, appropriate assays for evelution.

Selection of drones is direct selection of genomes generated in queens and is the most powerful method of selection. This type of selection may be useful when selecting directly for drone traits or for traits that drones and workers may have in common such as color and, perhaps, insecticide or disease resistance, development time or weight, length of life.

Genetic Variability

The Western honey bee, *Apis mellifera*, varies tremendously in appearance and behavior throughout its natural range in Europe, Asia, and Africa (Ruttner, 1988). These geographical differences are the result of millions of years of natural selection shaping the characteristics of the local bees to local environmental conditions. Variability also exists within local populations. Although this within population variability is usually less than that observed between geographically distant populations, or races, it is this material for natural selection to adapt populations to local conditions and for honey bee breeders to design programs to select superior stocks.

Western honey bees have been transported throughout the world. Different races have been tried in virtually every developed country where honey bees are kept. In some cases the native race has been more or less replaced by one introduced from a different location. This has occurred in the Nile delta of Egypt where the highly preferred Carniolan race has replaced the native Egyptian bee (Page *et al.*, 1981). This has also occurred to some extent in Germany where the native German bee has been replaced, through the action of importation and differential propagation with the Carniolan.

German bees were brought to North America by the first settlers. There are no native honey bees in the Americas; so these German bees spread slowly across the North American continent and were the dominant commercial and feral honey bee until the early 20th century. However, in the mid-19th century North American beekeepers began experimenting with other races of bees and found several to be superior to Germans. Doolittle (1915) discusses the attributes of different races of bees such as Syrians, Germans, and Italians kept in his apiary in New York State. Pellett (1929) documents the introduction of Syrian, Cyprian, Italian, Carniolan, Caucasian, Banat (a bee from Hungary that is very similar to Carniolans), Egyptian, and Punic (from North Africa) bees. In addition, Saharan bees (also from North Africa) were introduced into North America in the early 1920's and maintained by a queen breeder. It is likely that other races were also imported.

With the introduction of the Italian bee came the development of the queen-rearing industry. Many of the methods used in commercial queen rearing today were incorporated by G.M. Doolittle into a queen rearing system in response to his need to produce more Italian queens for sale to his beekeeping friends. Italians were favored because they were believed to be more resistant to European foulbrood disease, were more manageable than the dominant German bees, and because beekeepers liked the yellow color of the queens and workers. Other queen producers favored Carniolan or Caucasian bees and propagated them. With the growth of the queen production industry in North America came the replacement of the German bee with what is now truly a North American honey bee that represents a mix of the many races imported into this country from Europe, Asia, and Africa but is probably predominantly of Italian origin. These bees of broad genetic background constitute the foundation stock for selective breeding in North America.

German bees were also introduced into southern Brazil in 1839 and spread very slowly (Nogueira-Neto, 1964; W.E. Kerr, personal communication). The slow spread of these European bees was probably due to their poor adaptedness to tropical environments. In 1956, honey bees from South Africa were introduced into southern Brazil as part of a controlled honey bee breeding project aimed at improving commercial stocks. These African bees were adapted to Brazil's tropical environment and spread extensively but were, like most African bees, extremely defensive and prone to swarming and absconding. The hybrids resulting from interbreeding with the resident German, Italian, and Caucasian bees were superior honey producers and were manageable. However, by 1962 it was apparent that a *feral* (unmanaged) population of honey bees with the objectionable African characteristics had become well established and was displacing the resident European bees (Nogueira-Neto, 1964).

This "Africanized" feral population has spread throughout South and Central America since 1956, and in 1990 reached the United States. Along with the spread of the feral population has come a "conversion" of the bees maintained in commercial apiaries from European-type to Africanized. Bees in parts of Mexico, Costa Rica, Panama, Venezuela, and Brazil now seem to be mostly African in origin. This has been determined by analyzing the weight of workers and the size relationships (morphometrics) of different body parts (Daly & Balling, 1978; Boreham & Roubik, 1987; Buco *et al.,* 1987), specific proteins produced by worker bees (Lobo *et al.,* 1989; Smith *et al.,* 1989); DNA (Hall, 1988; Severson *et al.,* 1988; Hall & Muralidharan, 1989; Smith, *et al.,* 1989), the waxy layer (cuticular hydrocarbons) that bees secrete on the surface of their bodies (R.-K. Smith, 1988; Carlson, 1988), the chemical composition of wax combs, the sizes of cells of worker brood comb, and defensive behavior (Spivak *et al.,* 1988).

Feral Africanized bees in Mexico have been shown, by studies of DNA (the chemical genetic code) that resides in the *mitochondria* of cells of the body of honey bees, to be direct descendants of bees from South Africa (daughters and sons receive their mitochondria through eggs from queens but not sperm from drones). These results suggest a continual maternal lineage of bees in Mexico with those imported into Brazil in 1956. This does not mean that these bees are "pure" African but does suggest that the Africanized population is largely a consequence of the swarming and migration of colonies originating in southern Brazil. Bees can have maternally inherited African-type mito-chondria and still have both African and European nuclear genes. Nuclear genes, not mitochondria, are responsible for the observable characteristics of bees. European nuclear genes can mix with the Africanized population by mating success of European drones and European-Africanized hybrid drones with Africanized queens. Morphometric (Buco *et al.,* 1987), behavorial (Spivak *et al.,* 1988), hydrocarbon (R.-K. Smith, 1988; Carlson, 1988), and protein analyses (Lobo *et al.,* 1989) demonstrate that at least some mixing with European genes has occurred. This is most evident in the southern range of Africanized bees.

It is evident that the "collision" of two diverse populations of honey bees is occurring in North America. The consequences of this collision, however, are not yet evident. The blending of the more-or-less mixed Africanized and mixed-origin North American gene pools may provide new genetic variability upon which natural and artificial selection can act to produce an even better North American honey bee. Alternatively, Africanized honey bees, with their objectionable characteristics, may come to dominate the North American continent as they have South and Central America. North America, however, unlike South and Central America, has a strong queen rearing industry available that has already replaced one objectionable bee, the Germans, with the highly successful commercial bee used today. It is necessary that the queen rearing industry remain strong and supply the good honey bee stocks that will be needed following Africanization of North American populations.

Behavioral Variability

Behavioral differences among races of honey bees are a consequence of differences in the way that workers are allocated to particular tasks within the colony. A division of labor characterizes honey bee societies and it is this division of labor that determines the economic value of a colony. Below we present evidence for genetic variability within North American commercial honey bee populations for several economically important traits that are directly attributable to the behavior of individual adult workers.

Hygienic Behavior—The first conclusive evidence demonstrating how genetic variability accounts for variability in behavior came from the now classic studies of Walter C. Rothenbuhler. Rothenbuhler (1964a, 1964b) studied the mechanism of resistance to American foulbrood disease of a strain of honey bees that he developed from stocks obtained from Mr. Edward G. Brown of Sioux City, Iowa (Rothenbuhler, 1958). By careful study of these stocks he was able to demonstrate that a major component of disease resistance was a result of the hygienic behavior of the workers. He showed that hygienic behavior consists of two independent behavioral events: the uncapping of cells containing dead larvae or pupae, and the removal of the dead brood. Some of his colonies had workers that did neither task, others did one task but not the other, and some did both. Those that performed both tasks had relatively more resistance than those that did not. He also demonstrated that the mode of inheritance of these traits was simple and involved few genes and that these genes are independent (cf. Moritz, 1988).

Defensive Behavior—Studies of defensive behavior have also demonstrated genetic variability. Rothenbuhler (1964b) noted that not only were the workers of the Brown strain hygienic, they were also very defensive. Boch and Rothenbuhler (1974) studied defensive behavior in two strains: the very defensive Browns and the relatively indefensive Van Scoys. They demonstrated that the Brown strain was genetically more predisposed to respond defensively to hive disturbance, such as the hive being opened without the use of smoke, to human breath, and to the *alarm pheromone* (a chemical produced by glands associated with the sting of worker bees that stimulates other workers to sting) isopentyl acetate than were Van Scoy strain bees. In addition, workers of the Brown strain produced a larger quantity of alarm pheromone than did the Van Scoys. Collectively, these genetically determined differences made the Brown strain bees much more defensive.

Genetic variability for colony defensive behavior has also been demonstrated by comparing the defensive responses of Africanized honey bees with bees of European descent, and with their hybrids. Stort (1974, 1975a, 1975b, 1975c) demonstrated significant differences for the time it takes a colony to respond to a disturbance, the number of workers responding by stinging a moving target, and the distance that workers pursue the observer. Studies by Collins *et al.* (1982, 1984) have verified the earlier findings of Stort: Africanized bees are much more defensive. Collectively, these and other studies of the genetics of defensive behavior suggest that selective breeding programs should be able to significantly alter the defensive behavior of commercial honey bee stocks.

Pollen Collecting—Hellmich *et al.* (1985) reported the results of a breeding program designed to select strains of honey bees on the basis of the quantity of pollen that they stored in combs (hoarded). For four generations they selectively bred one strain for high and another for low pollen-hoarding. At the end of the fourth generation of selection, the high pollen-hoarding strain stored more than four times more pollen in combs than did the low strain. The criterion for selecting parents within strains was solely the amount of pollen stored in combs.

Calderone and Page (1988) studied the effects of this selection program on individual worker behavior of the high and low pollen-hoarding strains. Workers of the same age from the two strains were raised and studied together in the same hive environment. They demonstrated that differences between the high and low pollen-hoarding strains were due at least in a large part by differences in the individual foraging behavior of adult workers. Workers of the high pollen-hoarding strain were 3.6 to 16 times more likely to collect pollen than workers of the low strain. This study along with that of Hellmich *et al.* (1985) also demonstrated the large amount of genetic variability for individual foraging behavior available in populations of commercial honey bees in North America.

Genetically-based preferences for pollen from specific kinds of plants have been suggested by two-way selection for the percentage of alfalfa pollen loads collected by foraging workers (Nye & Mackensen, 1965, 1968, 1970; Mackensen & Nye, 1966, 1969). After seven generations of selection, an average of 87% of the pollen loads of workers from the high alfalfa pollen collecting strain were alfalfa, compared with only 8% of pollen loads collected by the low-strain workers.

Nectar Collecting Behavior—Calderone and Page (1988) also determined that bees of the low pollen-hoarding strain returned from foraging trips carrying pollen far less frequently than did workers of the high line. In a subsequent experiment Calderone (1988) determined that workers of the low strain returned to the hive carrying nectar loads with much greater frequency than did high-strain workers. In effect, selection for high pollen hoarding appears to have resulted in reduced individual nectar collection, while selection for low pollen hoarding increased the nectar collecting propensity of individuals. It, therefore, seems that these traits covary genetically, and in a selection program each may constrain the other.

Hoarding behavior of caged workers, that is, taking food from a feeder and storing it in a comb in the cage, has also been shown to have genetic variability. Rothenbuhler and his collaborators developed a laboratory assay that was designed to determine the tendency of workers of different genetic strains to store sugar syrup. They believed that a reliable laboratory assay would make selection for increased honey production easier than dealing with field colonies, provided that the laboratory assay correlated well with colony honey production. This was demonstrated with unselected stocks (Kulincevic & Rothenbuhler, 1973). A two-way selection for fast and slow syrup-storing behavior was conducted for five generations after which 50 caged workers of the fast line removed 20ml of sugar syrup from the feeder and stored it in the cage comb nearly three times as fast as workers of the slow line. Field tests, however, did not demonstrate close correlations between laboratory hoarding and honey production in the selected strains. Although these studies have failed to demonstrate the efficacy of the hoarding behavior laboratory study for selecting high honey producing stocks, they do demonstrate substantial genetic variability among workers for removing and storing sugar syrup, a trait that is perhaps important for honey production.

Behavioral Variability Among Subfamilies—Genetic variability has been demonstrated between members of different subfamilies within the same colony for several individual behavioral traits associated with colony division of labor. Differences in subfamily representation among different tasks have been observed (Robinson & Page, 1988, 1989) for the tasks of guarding the entrance, removing dead bees from the hive, dancing on the cluster of a swarm, foraging for nectar, and foraging for pollen. Additional studies have demonstrated genetic variability between subfamily members within colonies for caring for queen larvae (Page *et al.,* 1989), and grooming behavior (Frumhoff & Baker, 1988). Within queenless colonies, differences have been found between different subfamilies whose workers engaged in the exchange of food, egg-laying behavior, egg eating, and larval care (Moritz & Hillesheim, 1985; Hillesheim *et al.,* in press.)

Variability Among Components of Division of Labor—Worker honey bees change their behavior as they age, a phenomenon known as *age polyethism.* At different stages in their lives they perform different sets of tasks. In the section above we presented the results of some studies that have demonstrated differences among workers in the probability that they perform particular tasks as a consequence of their being genetically different. However, there are several components of division of labor, including: the probability that an individual performs a particular task within a "set" of tasks associated with a particular behavioral/age state (Seeley, 1982); the rate at which an individual changes from one behavioral/age state into another; and the sensitivity of an individual to environmental signals (stimuli) that result in behavioral state changes (Fig. 5).

Africanized honey bees become foragers at an earlier age than bees of European descent when they are raised in their own colonies (Winston & Katz, 1982). Calderone and Page (1988) showed that workers of the high pollen-hoarding strain made the transition to foraging earlier than workers of the low strain when they were co-fostered. These studies demonstrate genetic variability for age polyethism.

A division of labor that is less age dependent is established when the age composition of the workers of a colony is severely altered so that it is composed only of similarly-aged adults (Nelson, 1927). Some individuals begin foraging earlier in life (precociously) if the colony consists of all young

DIVISION OF LABOR

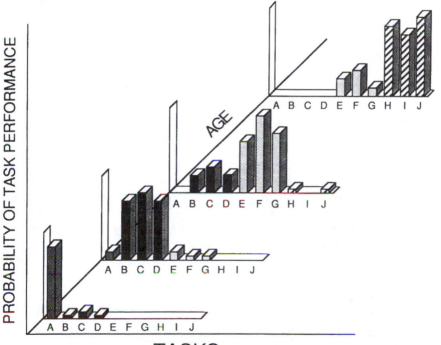

TASKS

FIGURE 5. A diagramatical representation of division of labor. As an individual worker honey bee ages the probabilities change that she will perform different tasks. From G.E. Robinson and R.E. Page. 1989. Used with the permission of Westview Press.

bees. If composed of all older aged bees, some continue to perform within-nest tasks beyond the time they normally make the transition to foraging, or some revert to within-nest tasks such as the care of larvae (Rösch, 1930; Lindauer, 1961). Genetic variability has been demonstrated with respect to the likelihood that individual workers undergo accelerated development to become precocious foragers, or retarded development to remain overaged nest bees.

Rothenbuhler (Rothenbuhler & Page, 1989) established two small colonies in observation hives. Each colony consisted initially of workers that had emerged as adults within the previous 48 hour period. The workers in each colony came from two different strains of bees that had been selected for high (the Brown strain) and low (the Van Scoy strain) hygienic behavior (discussed above). Three to five weeks later, similarly-aged but genetically different workers were performing different tasks. Workers of the Brown strain were engaged primarily in within-nest tasks including cell cleaning and defensive behavior while workers belonging to the Van Scoy strain and F_1, hybrids foraged. Most Van Scoy workers and F_1, hybrids developed to the forager behavioral state while most Brown strain workers remained in the nest.

Genetic differences in sensitivity of workers to environmentally induced developmental changes, such as precocious foraging in the absence of foraging-age bees, are reflected by differences in the amount of *juvenile hormone* (JH) circulating in the blood (*hemolymph*). Robinson *et al.* (1989) established colonies with 2000 workers that were less than three days old. Each colony consisted of three distinguishable subfamilies. The subfamily membership of "precocious" foragers and of "old" nurse bees (care for the larvae) were compared with "normal" and aged foragers and nurses. Members of some subfamilies were more likely to become precocious foragers while others were more likely to remain nurse bees. In addition, hemolymph samples of foragers and nurses showed that precocious foragers had higher concentrations of JH than did same-aged nurses. Old nurses had lower concentrations than did foragers of the same age. These results, along with others demonstrating the role of JH in regulating foraging behavior of workers (Jaycox *et al.*, 1974; Jaycox, 1976; Robinson, 1985, 1987), suggest a mechanism of developmental behavioral change that involves the genotypes of individuals, JH levels, and environment.

Disease Resistance

Disease resistance can result from behavioral and/or physiological mechanisms. The American foulbrood resistant Brown and the susceptible Van Scoy strains of Rothenbuhler differed with respect to several characteristics besides the hygienic behavior discussed above. Hygienic behavior confers resistance by removing the source of pathogens, the dead larvae, from the colony and may be effective against many disease pathogens including the causative agent of chalkbrood, *Ascosphaera apis* (Gilliam *et al.,* 1983). However, the larvae themselves also may have a physiological resistance. Larvae from the resistant Brown line were much more likely to survive after innoculation of their brood food with spores of *Bacillus larvae* than were larvae from susceptible colonies (Rothenbuhler & Thompson, 1956).

Adult workers can also confer a degree of resistance upon larvae. Fewer *B. larvae* spores germinated *in vitro* in an aqueous medium containing brood food from adults of the Brown than from the Van Scoy strain (Rose & Briggs, 1969). These results suggest that the brood food of the disease-resistant Brown strain contained some antibacterial properties. AFB innoculated larvae from the same source were more likely to survive in colonies containing AFB resistant workers than in colonies with susceptible adult workers (Thompson & Rothenbuhler, 1957).

Two-way selection was successful in producing strains of honey bees that were relatively susceptible or resistant to infestation with the tracheal mite, *Acarapis woodi*. Gary and Page (1987) developed a laboratory assay for studying the relative resistance among worker progeny of different queens. High and low resistant strains were then selected on the basis of this assay. After just two generations of selection within strains, workers from the susceptible strain had 2.4 times more adult tracheal mites than workers from the resistant strain (Page & Gary, 1990). Bees of each strain placed together in infested field colonies showed the same differences in levels of infestation that were shown in laboratory studies, demonstrating that selection for resistance based on the laboratory assay resulted in relative resistance in the field. The mechanism of this resistance and whether it confers colony-level resistance or tolerance to tracheal mites is unknown.

Biochemical Variability

Several biochemical variants have been identified in honey bee populations (see Sylvester, 1986 for a review). Each is presumed to be under the control of a single gene that has different forms (alleles). Most are *enzymes* involved in honey bee metabolism and are unlikely to be important in considerations of selective breeding, but have proven to be very valuable as markers in behavioral studies (see discussion above) because they allow behavioral data to be collected "blindly" and probably do not directly influence any behavioral trait.

The relative frequencies of different alleles of the enzymes *malate dehydrogenase* (Mdh) and *hexokinase* (Hk) have been shown to vary among adults of populations of North American commercial bees and Africanized bees in South and Central America and Mexico (Spivak *et al.,* 1988; Lobo *et al.,* 1989; Smith *et al.,* 1989).These variants may be useful for studying the process of Africanization in the United States.

Visible Mutants

Visible mutants are those directly observable differences among individuals that are heritable. While mutations are the basis of genetic change, and alternative alleles represent *mutations*, we usually think of mutations as genetic variants that are responsible for traits of individuals that differ from those that are common. There is a long catalog of visible mutants for honey bees, many of which have been shown to be controlled by a single gene (see Tucker, 1986 for review). They affect the color of the body, the hairs on the surface of the bee, eye color, wing size and shape, the shape, size, and occurrence of the compound eyes, and the configuration of the sting. Like the biochemical mutants, visible mutants are unlikely to be important in selective honey bee breeding but have proven to be very valuable for research.

Mating Designs

Honey bee stock improvement minimally requires two equally important elements. First is the identification of superior parental stock (queens) based on the particular characteristics of interest to the breeder and, second, mating the superior stock. The rest is often trial and error. Either the traits selected are heritable, that is, the variability observed for the trait has a sufficient amount of genetic determination, or they are not heritable in which case there will not be a response to selection. The particular mating design (Laidlaw & Page, 1986) used to develop stock is important and should fit the objectives of the breeding program.

It is easiest, and genetically most correct, to consider matings to be between queens (Laidlaw & Eckert, 1950; Polhemus *et al.,* 1950; Rothenbuhler, 1960). This is a consequence of the haplodiploid genetic system where males are derived directly from unfertilized eggs of queens, as discussed above. The drone, therefore, permits the queen to function both as mother of her colony and as father of other colonies or subfamilies of other colonies. Drones are little more than carriers and multipliers of their mother's mature reproductive cells.

Inbred-Hybrid Breeding—The objective of an inbred-hybrid breeding program is to take advantage of hybrid vigor that often occurs when two inbred lines are crossed. The inbred lines themselves display inbreeding depression— a loss of vigor—but when crossed to other inbred lines they often display superior characteristics. Inbred lines are selected on the basis of their combining ability with other lines, not solely on their own characteristics. Lines that combine well with some specific lines but poorly with others have *specific combining ability* while those that combine well with most or all lines display *general combining ability*. Inbred lines should be initiated from superior stock, and selection may continue within lines while they are developed if mating designs are used that result in relatively slow inbreeding. Selection for some characteristics is difficult or impossible once inbreeding depression is significant and, therefore, such lines may not continue to be selected while being inbred.

Self Fertilization—Virgin queens are stimulated to lay eggs by treatment with carbon dioxide (Mackensen, 1947). Two, two-minute treatments administered one or two days apart are usually sufficient. Drones produced from these virgin queens are then used to instrumentally inseminate their mother (Fig. 6). This results in the most rapid rate of inbreeding. Brood viability will be no better than 50 percent because there will be only two sex alleles within self-fertilized lines and, as a consequence, homozygosity for sex alleles in half the diploid progeny. These homozygous individuals are diploid males that are eaten as larvae by the workers. Inbreeding proceeds at the same rate regardless of the number of males used to inseminate queens.

Mother-Daughter Mating—This type of mating is effectively back-crossing to the mother (Laidlaw & Eckert, 1950; Polhemus *et al.,* 1950.) A virgin queen is mated to drones derived from her mother. One or many drones may be used for inseminating queens without affecting the rate of inbreeding. Mother-daughter matings are the most frequently used in breeding programs where rapid inbreeding is the objective (Fig. 7).

Sister Matings—Matings can be among super, full, or half sisters depending on the number and origin of drones used for each insemination. Super-sister matings result in the most rapid inbreeding. A queen is inseminated with the semen of a single drone which results in all worker and queen progeny being super sisters (Fig. 8). Virgin queen progeny are raised from the super sisters, some of which produce drones that are used to inseminate their sister queens. Super-sister matings result in a rate of inbreeding (per generation) greater than mother-daughter and only slightly slower than selfing. Queens inseminated by a single drone are necessary for super sister matings, but usually do not have enough sperm to lay enough fertilized eggs to maintain populous field colonies.

If more than one drone from the same mother are used for insemination, then daughter queens of each drone will be super sisters to each other but full sisters to daughters of other fathers, and under most circumstances it will not be apparent what the exact relationships are among the queens used for breeding (Fig. 9). Full-sister matings result in reduced rates of inbreeding compared to super-sister matings, but may be desirable when full-sized field colonies are needed to evaluate the lines.

Half-sister matings can result in inbreeding if the drone mothers are related. Drone mothers can be super, full, or half sibs depending on the number and origin of males used to inseminate queens. Inbreeding is slower with half than full sister matings.

Sister matings have the advantage over mother-daughter matings because the sister used as a drone source can be naturally mated and the resultant colony, therefore, can be free of most of the effects of inbreeding depression, and can be more accurately evaluated. Perhaps the greatest difficulty in inbreeding is producing a sufficient number of drones when they are needed

Figure 6. Self-fertilization.

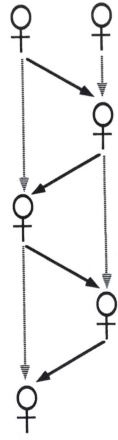

Figure 7. Mother-daughter mating.

from inbred colonies. Workers within colonies with naturally mated inbred queens are not inbred and, therefore, are vigorous and produce more and healthier drones. The disadvantage is that the generation time is longer, thereby slowing the overall rate of inbreeding in terms of actual, as opposed to generation, time.

There are many other possible mating schemes that can be used for inbreeding lines. Each has a different rate of inbreeding associated with it both in terms of generations and actual time. For a more complete set of designs and rates of inbreeding associated with them see Laidlaw and Page (1986).

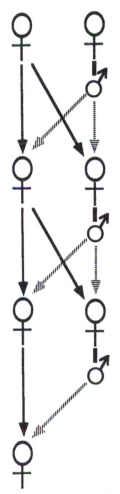 

FIGURE 8. Super-sister mating.   FIGURE 9. Full-sister mating.

Difficulties—Inbred lines are difficult to maintain and are easily lost. This is primarily a consequence of their ending up with only two sex alleles within the lines. With two sex alleles, brood viability is only 50% and colonies display a characteristic called "shot brood" where the sealed brood has many uncapped cells and the brood is of mixed age. This in combination with other effects of inbreeding depression results in a high likelihood of colony failure and loss of lines. Losses can be minimized by mating inbred queens with 1 or 2 drones of the virgin's inbred line and also with 5 or more drones not related to the virgin, and whose progeny are distinguishable from the inbreds. The brood of such queens is uniform and most of the worker bees show heterosis.

Drone sterility may also become a problem in inbred lines. Drones that yield small amounts of semen should not be used in instrumental insemination.

Breeding Successes—Inbred-hybrid breeding systems have been used commercially to produce some well known superior stocks. Starline and Midnite queens originated by the late Dr. G. H. Cale, Jr., of Dadant and Sons, Inc., were developed using this method. In addition, the United States Department of Agriculture honey bee laboratory in Madison, Wisconsin, in cooperation with the USDA laboratory in Baton Rouge, Louisiana, maintained a breeding and stock program for more than 40 years and produced some superior performing hybrids. Harry Laidlaw of the University of California, Davis, used mother-daughter matings and inbred queen selection to develop several lines of superior honey-producing bees over a period of about 10 years.

Closed Population Breeding—In contrast to inbred-hybrid breeding, the objective of a closed population breeding program is to progressively improve a breeding population by selection while maintaining high brood viability and genetic variability (Laidlaw & Eckert, 1950; Laidlaw, 1954, 1956, 1981; Page & Laidlaw, 1982a, 1982b, 1985; Page *et al.,* 1982, 1983, 1985). Matings are controlled either by instrumental insemination or by allowing queens to mate only in isolated areas that are free of drones from unselected colonies. Below we outline three possible mating systems that can be used with closed populations that are designed to minimize the loss of sex alleles and other genetic variability. Breeder queens are always used as both queen and drone mothers.

1. Queens of superior-performing colonies are identified and constitute the breeders for the initial generation. Several virgin queen daughters and many drones from all of the breeding queens are produced. Daughters from all breeder queens are each instrumentally inseminated with the semen of 10 drones that are selected at random from the entire closed population (or

queens are naturally mated in isolated areas) and placed in hives for evaluation. The more daughters evaluated, the greater the potential response to selection. Superior performing daughters are selected from among all the queen progeny of that generation. Using this system, about 35-50 breeder colonies must be selected and maintained each generation to have a 95% probability of maintaining enough sex alleles for at least 85% brood viability for 20 generations of selection (Page & Laidlaw, 1982a, 1982b).

2. Using the same replacement queen production method as above, each breeder queen is, instead, replaced by one of her own superior performing daughters each generation. With this system, about 25 breeder colonies are needed to maintain at least 85% viable brood for at least 20 generations. This system has the advantage that fewer colonies are needed to maintain brood viability, but has the disadvantage that selective progress will be slower (Page *et al.,* 1983; Moritz, 1986).

3. Semen from a large, equal number of drones from each breeder queen of a closed population can be pooled, homogenized, and used to inseminate all daughter queens (Laidlaw, 1981; Page & Laidlaw, 1982b; Moritz, 1983). This system should result in (a) all queens being effectively mated to the entire gene pool, (b) reduced variability among queens for brood viability, (c) the selection of replacement queens on the basis of the general combining ability of their own genotypes, and (d) the maintenance of more sex alleles in the population if queens are selected for high brood viability (Kubasek, 1980; Page *et al.,* 1983). Selection of replacement breeder queens can be done as in 1 or 2 above.

Breeding Successes—Closed population programs with different kinds of mating schemes have been responsible for producing high and low pollen hoarding strains (Hellmich *et al.,* 1985), a high alfalfa pollen collecting strain (Nye & Mackensen, 1965, 1968, 1970; Mackensen & Nye, 1966, 1969), improved honey production in Alberta, Canada (Szabo & Lefkovitch, 1987), and British Columbia, Canada (J. Corner, personal communication) and tracheal mite resistance (Page & Gary, 1990). In addition, programs are currently used by commercial queen breeders in California and both commercial and government sponsored breeding programs throughout Australia.

STOCK MAINTENANCE

After superior stocks are produced, they must be maintained in order to have any real value. The static maintenance of stock should never be a breeding objective. Instead, breeding efforts should be directed toward establishing long-term programs that continue to improve, maintain, and replace stock. Gains from a selective breeding program are easily lost when selection is relaxed.

Inbred-hybrid breeding systems have the greatest risk of losing stocks that have taken a lot of time and effort to develop. An inbred-hybrid program should be designed to continually produce new inbreds, replace lost lines, and cull those that do not perform well when combined. Closed population systems can be used to maintain particular stocks, such as European races of bees in an Africanized environment, but probably will not maintain selective improvements for very many generations if selection is not continued. The development and maintenance of closed populations, like the development and maintenance of inbred stocks, can require a lot of work (Severson *et al.*, 1986). The rewards, however, of well organized and operated bee breeding programs can far exceed the program costs through increased honey production and pollination efficiency.

CONCLUSIONS

There is evidence that some diseases can be controlled by producing genetically-resistant bees, and there is some reason to believe that most diseases can eventually be controlled through breeding. Undesirable behavioral traits can be minimized or eliminated by breeding, and good traits maximized or incorporated in a stock. The potential for selective breeding is great; but to be adequately employed a knowledge of basic bee genetics, methods of selection, mating behavior, and instrumental insemination is required. As a consequence of the spread of tracheal and *Varroa* mites across North America, and the inevitable spread of Africanized bees, future challenges for bee breeding will be formidable. But, we can now meet these challenges with a legacy of more than 200 years of scientific investigation into bee genetics and more than 100 years of queen-production technology.

REFERENCES

Adams, J., E.D. Rothman, W.E. Kerr, and Z.L. Paulino. (1977). Estimation of the number of sex alleles and queen matings from diploid male frequencies in a population of *Apis mellifera. Genetics 86*:583-596.

Attridge, A.J. (1919). Punics or African bees and parthenogenesis. *Am. Bee J. 59*:419-420.

Boch, R. and W.C. Rothenbuhler. (1974). Defensive behaviour and production of alarm pheromone in honeybees. *J. Apic. Res. 13*:217-221.

Boreham, M.M. and D.W. Roubik. (1987). Population change and control of Africanized honey bees (Hymenoptera:Apidae) in the Panama Canal area. *Bull. Entomol. Soc. Am. 33*:34-39.

Buco, S.M., T.E. Rinderer, H.A. Sylvester, A.M. Collins, V.A. Lancaster, and R.M. Crewe. (1987). Morphometric differences between South American Africanized and South African (*Apis mellifera scutellata*) honey bees. *Apidologie 18*:217-222.

Calderone, N.W. and R.E. Page. (1988). Genotypic variability in age polyethism and task specialization in the honey bee, *Apis mellifera* (Hymenoptera:Apidae). *Behavioral Ecology and Sociobiology 22*:17-25.

Calderone, N.W. (1988). The genetic basis for the evolution of the organization of work in colonies of the honey bee, *Apis mellifera* (Hymenoptera:Apidae). Ph.D. dissertation, The Ohio State University, Columbus, Ohio.

Carlson, D.A. (1988). Africanized and European honey-bee drones and comb waxes: analysis of hydrocarbon components for identification. *In*: G.R. Needham, R.E. Page, M. Delfinado-Baker, and C.E. Bowman (*eds.*), Africanized Honey Bees and Bee Mites, pp. 264-274, Ellis Horwood, Ltd., Chichester, United Kingdom.

Chaud-Netto, J. (1980a). Estudos biológicos com rainhas triplóides de *Apis mellifera* 1. Produção de ovos abortivos por rainhas virgens. *Ciên. Cult.* (São Paulo)*32*:483-486.

Chaud-Netto, J. (1980b). Comprovacão genética e citológica de triplóidia em descendentes de cruzamentos controlados entre rainhas e zangões diplóides de *Apis mellifera* (Hymenoptera, Apidae). *Ciên. Cult.* (São Paulo) *32*:351-355.

Collins, A.M., T.E. Rinderer, J.R. Harbo, and A.B. Bolten. (1982). Colony defense by European and Africanized honey bees. *Science 218*:72-74.

Collins, A.M., T.E. Rinderer, J.R. Harbo, and M.A. Brown. (1984). Heritabilities and correlations for several characters in the honey bee. *J. Hered 75*:135-140.

Crow, J.F. and M. Kimura. (1970). An Introduction to Population Genetics Theory. Burgess Publishing Co., Minneapolis, Minnesota.

Daly, H.V. and S.S. Balling. (1978). Identification of Africanized honeybees in the Western Hemisphere by discriminant analysis. *J. Kans. Entomol Soc. 51*:857-869.

Doolittle, G.M. (1915). Scientific Queen Rearing, 6th edition. American Bee Journal, Hamilton, Illinois.

DuPraw, E.J. (1967). The honeybee embryo. *In*: F.H. Wilt and N.K. Wessels (*eds.*), Methods in Developmental Biology, pp. 183-217, Crowell, New York.

Dzierzon, J. (1845). Gutachten über die von Herrn Direktor Stöhr in ersten und zweiten Kapitel des General-Gutachtens aufgestellten Fragen. *Bienenzeitung 1*:109-113, 119-121.

Falconer, D.S. (1981). Introduction to Quantitative Genetics, 2nd edition. Longman, London.

Frumhoff, P.C. and J. Baker. (1988). A genetic component to division of labour within honey bee colonies. *Nature 333*:358-361.

Gary, N.E. and R.E. Page. (1987). Phenotypic variability in susceptibility of honey bees, *Apis mellifera* L., to infestation by tracheal mites, *Acarapis woodi* Rennie. *Exp. Appl. Acarol. 3*:291-305.

Gilliam, M., S. Taber, and G.V. Richardson. (1983). Hygienic behavior of honey bees in relation to chalkbrood disease. *Apidologie 14*:29-39.

Hall, H.G. (1988). Characterization of the African honey-bee genotype by DNA restriction fragments. *In*: G.R. Needham, R.E. Page, M. Delfinado-Baker, and C.E. Bowman (*eds.*), Africanized Honey Bees and Bee Mites, pp. 287-293, Ellis Horwood Ltd., Chichester, United Kingdom.

Hall, H.G. and K. Muralidharan. (1989). Evidence from mitochondrial DNA that African honey bees spread as continuous maternal lineages. *Nature 339*:211-213.

Hasimoto, H. (1934). Formation of an individual by union of two sperm nuclei in the silkworm. *Bull. Imper. Sericul. Exper. Sta. 8*:463.

Hellmich, R.L., J.M. Kulincevic, and W.C. Rothenbuhler. (1985). Selection for high and low pollen-hoarding honey bees. *J. Hered 76*:155-158.

Hillesheim, E., N. Koeniger, and R.F.A. Moritz. (1989). Colony performance in honeybees (*Apis mellifera capensis* Esch.) depends on the proportion of subordinate and dominant workers. *Behav. Ecol Sociobiol. 24*:291-296.

Iltis, H. (1932). Life of Mendel. Translation by Eden and Cedar Paul, W.W. Norton and Co. Inc., New York.

Jack, R.W. (1917). Parthenogenesis amongst the workers of the Cape honeybee: G.W. Onion's experiments. *Trans. R. Entomol. Soc. London 64*:396-403.

Jaycox, E.R. (1976). Behavioral changes in worker honey bees (*Apis mellifera* L.) after injection with synthetic juvenile hormone (Hymenoptera:Apidae). *J. Kans. Entomol. Soc. 49*:165-170.

Jaycox, E.R., W. Skowronek, and G. Gwynn. (1974). Behavioral changes in worker honey bees (*Apis mellifera*) induced by injections of a juvenile hormone mimic. *Ann. Entomol. Soc. Am. 67*:529-534.

Koeniger, G. (1990). The role of the mating sign in honey bees, *Apis mellifera* L.: does it hinder or promote multiple mating? *Anim. Behav. 39*:444-449.

Kubasek, K.J. (1980). Selection for increased number of sex alleles in closed populations of the honey bee. An investigation via computer simulation. MS Thesis, Louisiana State University, Baton Rouge, Louisiana.

Kulincevic, J.M. and W.C. Rothenbuhler. (1973). Laboratory and field measurements of hoarding behaviour in the honeybee. *J. Apic. Res. 12*:179-182.

Laidlaw, H.H. (1954). Beekeeping management for the bee breeder. *Am. Bee J. 94*:92-95.

Laidlaw, H.H. (1956). Organization and operation of a bee breeding program. *In*: Proceedings of the Tenth International Congress of Entomology, Vol. 4, pp. 1067-1078.

Laidlaw, H.H. (1974). Relationships of bees within a colony. *Apiacta 9*:49-52.

Laidlaw, H.H. (1981). Honey bee genetics and its application to pollinator breeding. *Honeybee Sci. 2*:1-4. (In Japanese with English summary.)

Laidlaw, H.H. (1987). Instrumental insemination of honey bee queens: its origin and development. *Bee Wld. 68*:17-36,71-88.

Laidlaw, H.H. and J.E. Eckert. (1950). Queen Rearing. Dadant and Sons, Hamilton, Illinois.

Laidlaw, H.H., F.P. Gomes, and W.E. Kerr. (1956). Estimations of the number of lethal alleles in a panmictic population of *Apis mellifera*. *Genetics 41*:179-188.

Laidlaw, H.H. and R.E. Page. (1984). Polyandry in honey bees (*Apis mellifera* L.): sperm utilization and intracolony genetic relationships. *Genetics 108*:985-997.

Laidlaw, H.H. and R.E. Page. (1986). Mating designs. *In*: T.E. Rinderer (*ed.*), Bee Genetics and Breeding, pp.323-344, Academic Press, Orlando, Florida.

Laidlaw, H.H. and K.W. Tucker. (1964). Diploid tissue derived from accessory sperm in the honey bee. *Genetics 50*:1439-1442.

Lindauer, M. (1961). Communication among Social Bees. Harvard University Press, Cambridge, Massachusetts.

Lobo, J.A., M.A. Del Lama, and M.A. Mestriner. (1989). Population differentiation and racial admixture in the Africanized honeybee (*Apis mellifera* L.). *Evolution 43*:794-802.

Mackensen, O. (1943). The occurrence of parthenogenetic females in some strains of honeybees. *J. Econ. Entomol. 36*:465-467.

Mackensen, O. (1947). Effect of carbon dioxide on initial oviposition of artificially inseminated and virgin queen bees. *J. Econ. Entomol. 40*:344-349.

Mackensen, O. (1951). Viability and sex determination in the honey bee (*Apis mellifera* L.). *Genetics 36*:500-509.

Mackensen, O. (1955). Further studies on a lethal series in the honey bee. *J. Hered. 46*:72-74.

Mackensen, O. and W.P. Nye. (1966). Selecting and breeding honeybees for collecting alfalfa pollen. *J. Apic. Res.* 5:79-86.

Mackensen, O. and W.P. Nye. (1969). Selective breeding of honeybees for alfalfa pollen collection: sixth generation and outcrosses. *J. Apic. Res.* 8:9-12.

Mackensen, O. and W.C. Roberts. (1948). A Manual for the Artificial Insemination of Queen Bees. USDA Agricultural Research Administration Bureau of Entomology and Plant Quarantine. ET 250.

Meves, F. (1907). Die Spermatocytenteilungen bei der Honigbiene (Apis mellifica L.) nebst Bemerkungen über Chromatinreduktion. *Arch. Mikrosk. Anat. Entwicklungsmech.* 70:414-491.

Merriam, R.W. and H. Ris. (1954). Size and DNA content of nuclei in various tissues of male, female, and worker honeybees. *Chromosoma* 6:522-538.

Moritz, R.F.A. (1983). Homogenous mixing of honeybee semen by centrifugation. *J. Apic. Res.* 22:249-255.

Moritz, R.F.A. (1986a). Comparison of within-family and mass selection in honeybee populations. *J. Apic. Res.* 25:146-153.

Moritz, R.F.A. (1986b). Estimating the genetic variance of group characters: social behaviour of honeybees (*Apis mellifera* L.). *Theor. Appl. Genet.* 72:513-517.

Moritz, R.F.A. (1987). Phenotype interactions in group behavior of honey bee workers (*Apis mellifera* L.). *Behav. Ecol. Sociobiol.* 21:53-57.

Moritz, R.F.A. (1988). A reevaluation of the two-locus model for hygienic behavior in honeybees (*Apis mellifera* L.) *J. Hered.* 79:257-262.

Moritz, R.F.A. and E. Hillesheim. (1985). Inheritance of dominance in honeybees (*Apis mellifera capensis* Esch.). *Behav. Ecol. Sociobiol.* 17:87-89.

Moritz, R.F.A. and A. Klepsch. (1985). Estimating heritabilities of worker characters: a new approach using laying workers of the cape honeybee (*Apis mellifera capensis* Esch.) Apidologie 16:47-56.

Nachtsheim, H. (1913). Cytologische Studien über die Geschelechtbestimmung bei der Honigbiene (*Apis mellifica* L.). *Arch. fur Zellforsch.* 11:169-241.

Nelson, F.C. (1927). Adaptability of young bees under adverse conditions. *Am. Bee J.* 67:242-243

Nogueira-Neto, P. (1964). The spread of a fierce African bee in Brazil. *Bee Wld.* 45:119-121.

Nolan, W.J. (1932). Multiple matings of the queen bee. *Rep. Md. St. Beekeeprs. Assoc.* 23:20-34.

Nye, W.P. and O. Mackensen. (1965). Preliminary report on selection and breeding of honeybees for alfalfa pollen collection. *J. Apic. Res.* 4:43-48.

Nye, W.P. and O. Mackensen. (1968). Selective breeding of honeybees for alfalfa pollen: fifth generation and backcrosses. *J. Apic. Res.* 7:21-27.

Nye, W.P. and O. Mackensen. (1970). Selective breeding of honeybees for alfalfa pollen collection: with tests in high and low alfalfa pollen collection regions *J. Apic. Res.* 9:61-64.

Onions, G.W. (1912). South African "fertile worker bees." *Agric. J. Union of South Africa* 3:720-728.

Page, R.E. (1986). Sperm utilization in social insects. *Ann. Rev. Entomol.* 31:297-320.

Page, R.E. and N.E. Gary. (1990). Genotypic variation in susceptibility of honey bees, *Apis mellifera*, to infestation by tracheal mites, *Acarapis woodi. Exp. Appl. Acarol.* 8:275-283.

Page, R.E., M.M. Ibrahim, and H.H. Laidlaw. (1981). The history of modern beekeeping in Egypt. *Glean. Bee Cult. 109*:24-26.

Page, R.E. and H.H. Laidlaw. (1982a). Closed population honeybee breeding. 1. Population genetics of sex determination. *J. Apic. Res. 21*:30-37.

Page, R.E. and H.H. Laidlaw. (1982b). Closed population honeybee breeding. 2. Comparative methods of stock maintenance and selective breeding. *J. Apic. Res. 21*:38-44.

Page, R.E. and H.H. Laidlaw. (1985). Closed population honeybee breeding. *Bee Wld. 66*:63-72.

Page, R.E. and H.H. Laidlaw. (1988). Full sisters and super sisters: a terminological paradigm. *Anim. Behav. 36*:944-945.

Page, R.E., H.H. Laidlaw, and E.H. Erickson. (1982). A closed population breeding program for honey bees. *Am. Bee J. 122*:350,351,354,355.

Page, R.E., H.H. Laidlaw, and E.H. Erickson. (1983). Closed population honeybee breeding. 3. The distribution of sex alleles with gyne supersedure. *J. Apic. Res. 22*:184-190.

Page, R.E., H.H. Laidlaw, and E.H. Erickson. (1985). Closed population honeybee breeding. 4. The distribution of sex alleles with top crossing. *J. Apic. Res. 24*:38-42.

Page, R.E. and R.A. Metcalf. (1988). A population estimate of Mdh allozyme frequencies for the honey bee, *Apis mellifera* L. (Hymenoptera: Apidae). *Pan-Pac. Entomol. 64*:285-289.

Page, R.E., G.E. Robinson, and M.K. Fondrk. (1989). Genetic specialists, kin recognition, and nepotism in honey-bee colonies. *Nature 338*:576-579.

Painter, T.S. (1945). Nuclear phenomena associated with secretion in certain gland cells with especial reference to the origin of cytoplasmic nucleic acid. *J. Exp. Zool. 100*:523-541.

Pellett, F.C. (1929). Practical Queen Rearing, 5th edition. American Bee Journal, Hamilton, Illinois.

Polhemus, M.S., J.L. Lush, and W.C. Rothenbuhler. (1950). Mating systems in honey bees. *J. Hered. 41*:151-155.

Roberts, W.C. (1944). Multiple mating of queen bees proved by progeny and flight tests. *Glean. Bee Cult. 72*:255-259.

Robinson, G.E. (1985). Effects of a juvenile hormone analogue on honey bee foraging behaviour and alarm pheromone production. *J. Insect Physiol. 31*:277-282.

Robinson, G.E. (1987). Modulation of alarm pheromone perception in the honey bee: evidence for division of labor based on hormonally regulated response thresholds. *J. Comp. Physiol. A 160*:613-619.

Robinson, G.E. and R.E. Page. (1988). Genetic determination of guarding and undertaking in honey-bee colonies *Nature 333*:356-358.

Robinson, G.E. and R.E. Page (1989a). Genetic basis for division of labor in an insect society. *In:* M.D. Breed and R.E. Page (*eds.*), The Genetics of Social Evolution, pp. 61-80, Westview Press, Inc., Boulder, Colorado.

Robinson, G.E. and R.E. Page. (1989b). Genetic determination of nectar foraging, pollen foraging, and nest-site scouting in honey bee colonies. *Behav. Ecol. Sociobiol. 24*:317-323.

Robinson, G.E., R.E. Page, and M.K. Fondrk. (1990). Intracolonial variation in worker oviposition, oophagy, and larval care in queenless honey bee colonies. *Behav. Ecol. Sociobiol 26*:315-323.

Robinson, G.E., R.E. Page, C. Strambi, and A. Strambi. (1989). Hormonal and genetic control of behavioral integration in honey bee colonies. *Science 246*:109-112.

Rösch, G.A. (1930). Untersuchungen über die Arbeitsteilung im Bienenstaat, II. Die Tätigkeiten der Arbeitsbienen uter experimentell veranderten Bedingunen. *Z. Vergl. Physiol.* *12*:1-71.

Rose, R.I. and J.D. Briggs. (1969). Resistance to American foulbrood in honey bees, IX. Effects of honey-bee larval food on the growth and viability of *Bacillus larvae. J. Invert. Pathol. 13*:74-80.

Rothenbuhler, W.C. (1957). Diploid male tissue as new evidence on sex determination in honey bees. *J. Hered. 48*:160-168.

Rothenbuhler, W.C. (1958). Genetics and breeding of the honey bee. *Ann. Rev. Entomol. 3*:161-180.

Rothenbuhler, W.C. (1960). A technique for studying genetics of colony behavior in honey bees. *Am. Bee J. 100*:176,198.

Rothenbuhler, W.C. (1964a). Behaviour genetics of nest cleaning in honey bees. I. Responses of four inbred lines to disease-killed brood. *Anim. Behav. 12*:578-583.

Rothenbuhler, W.C. (1964b). Behavior genetics of nest cleaning in honey bees. IV. Responses of F_1, and backcross generations to disease-killed brood. *Am. Zoologist 4*:111-128.

Rothenbuhler, W.C., J.W. Gowen, and O.W. Park. (1952). Androgenesis with zygogenesis in gynandromorphic honeybees (*Apis mellifera* L.). *Science 115*:637-638.

Rothenbuhler, W.C. and R.E. Page. (1989). Genetic variability for temporal polyethism in colonies consisting of similarly-aged worker honey bees. *Apidologie 29*:433-437.

Rothenbuhler, W.C., M.S. Polhemus, J.W. Gowen, and O.W. Park. (1949). Gynandromorphic honey bees. *J. Hered. 40*:308-311.

Rothenbuhler, W.C. and V.C. Thompson. (1956). Resistance to American foulbrood in honey bees. I. Differential survival of larvae of different genetic lines. *J. Econ. Entomol. 49*:471-475.

Ruttner, F. (1988). *Biogeography and Taxonomy of Honeybees.* Springer-Verlag, New York.

Ruttner, F. and G. Koeniger. (1971). Die Füllung der Spermatheka der Bienenkönigin. Activ Wanderung oder passiver Transport der Spermatozoen? *Z. Vgl. Physiol. 72*:411-422.

Seeley, T.D. (1982). Adaptive significance of the age polyethism schedule in honeybee colonies. *Behav. Ecol. Sociobiol. 11*:287-293.

Severson, D.W., J.M. Aiken, and R.F. Marsh. (1988). Molecular analyses of North American and Africanized honey bees. *In:* G.R. Needham, R.E. Page, M. Delfinado-Baker, and C.E. Bowman (*eds.*), Africanized Honey Bees and Bee Mites, pp. 294-302, Ellis Horwood Ltd., Chichester, United Kingdom.

Severson, D.L., R.E. Page, and E.H. Erickson. (1986). Closed population breeding in honey bees: a report on its practical application. *Am. Bee J. 126*:93-94.

Smith, D.R., O.R. Taylor, and W.M. Brown. (1989). Neotropical Africanized honey bees have African mitochondrial DNA. *Nature 339*:213-215.

Smith, R.-K. (1988). Identification of Africanization in honey bees based on extracted hydrocarbons assay. *In:* G.R. Needham, R.E. Page, M. Delfinado-Baker, and C.E. Bowman (*eds.*) Africanized Honey Bees and Bee Mites, pp. 275-280, Ellis Horwood Ltd., Chichester, United Kingdom.

Spivak, M., T. Ranker, O.R. Taylor, W. Taylor and L. Davis. (1988). Discrimination of Africanized honey bees using behavior, cell size, morphometrics, and a newly discovered isozyme polymorphism. *In:* G.R. Needham, R.E. Page, M. Delfinado-Baker, and C.E. Bowman (*eds.*), Africanized Honey Bees and Bee Mites, pp. 313-324, Ellis Horwood Ltd., Chichester, United Kingdom.

Stort, A.C. (1974). Genetic study of the aggressiveness of two subspecies of *Apis mellifera* in Brazil. I. Some tests to measure aggressiveness. *J. Apic. Res. 13*:33-38.

Stort, A.C. (1975a). Genetic study of the aggressiveness of two subspecies of *Apis mellifera* in Brazil. II. Time at which the first sting reached the leather ball. *J. Apic. Res. 14*:171-175.

Stort, A.C. (1975b). Genetic study of the aggressiveness of two subspecies of *Apis mellifera* in Brazil. IV. Number of stings in the gloves of the observer. *Behav. Genet. 5*:269-274.

Stort, A.C. (1975c). Genetic study of the aggressiveness of two subspecies of *Apis mellifera* in Brazil. V. Number of stings in the leather ball. *J. Kans. Entomol. Soc. 48*:381-387.

Sylvester, H.A. (1986). Biochemical genetics. *In*: T.E. Rinderer (*ed.*), Bee Genetics and Breeding, pp. 177-203, Academic Press, Orlando, Florida.

Szabo, T.I. and L.P. Lefkovitch. (1987). Fourth generation of closed-population honeybee breeding. 1. Comparison of selected and control strains. *J. Apic. Res. 26*:170-180.

Taber, S. (1954). The frequency of multiple mating of queen honey bees. *J. Econ. Entomol. 4*:995-998.

Taber, S. (1955). Evidence of binucleated eggs in the honey bee. *J. Hered. 46*:156.

Thompson, V.C. and W.C. Rothenbuhler. (1957). Resistance to American foulbrood in honey bees. II. Differential protection of larvae by adults of different genetic lines. *J. Econ. Entomol. 50*:731-737.

Triasko, V.V. (1951). Sign indicating the mating of queens. *Pchelovodstvo 11*:25-31. (In Russian, abstracted in Apic. Abst., May, 1953).

Tucker, K.W. (1958). Automictic parthenogenesis in the honey bee. *Genetics 43*:299-316.

Tucker, K.W. (1986). Visible mutants. *In*: T.E. Rinderer (*ed.*), Bee Genetics and Breeding, pp. 57-90, Academic Press, Orlando, Florida.

van Warmelo, D.S. (1912). South African fertile-worker bees and parthenogenesis. *Agric. J. Union of South Africa 3*:786-789.

von Siebold, C.T. (1856). *Wahre Parthenogenesis bei Schmetterrlingen und Bienen. Ein Beitrag zur Fortpflanzungsgeschiehte der Tiere*, Leipzig.

Verma, S. and F. Ruttner. (1983). Cytological analysis of the thelytokous parthenogenesis in the Cape honeybee (*Apis mellifera capensis* Esch.). *Apidologie 14*:41-57.

Whiting, P.W. (1943). Multiple alleles in complementary sex determination of *Habrobracon*. *Genetics 28*:365-382.

Winston, M.L. and S.J. Katz. (1982). Foraging differences between cross-fostered honeybee workers (*Apis mellifera*) of European and Africanized races. *Behav. Ecol. Sociobiol. 10*:125-129.

Woyke, J. (1955). Multiple mating of the honeybee queen (*Apis mellifica* L.) in one nuptial flight. *Bull. Acad. Pol. Sci. CI:II*-Vol. III, No. 5, 175-180.

Woyke, J. (1962). Natural and artificial insemination of queen honeybees. *Bee Wld. 43*:21-25.

Woyke, J. (1963a). Drone larvae from fertilized eggs of the honeybee. *J. Apic. Res. 2*:19-24.

Woyke, J. (1963b). What happens to diploid drone larvae in a honeybee colony. *J. Apic. Res. 2*:73-75.

Woyke, J. (1964). Causes of repeated mating flights by queen honeybees. *J. Apic. Res. 3*:17-23.

Woyke, J. (1965). Genetic proof of the origin of drones from fertilized eggs of the honeybee. *J. Apic. Res. 4*:7-11.

Woyke, J. (1973). Reproductive organs of haploid and diploid drone honeybees. *J. Apic. Res.* *12*:35-51.

Woyke, J. (1976). Population genetic studies on sex alleles in the honeybee using the example of the Kangaroo Island Bee Sanctuary. *J. Apic. Res. 15*:105-123.

Woyke, J. (1983). Dynamics of entry of spermatozoa into the spermatheca of instrumentally inseminated queen honeybees. *J. Apic. Res. 22*:150-154.

Woyke, J. (1986). Sex determination. *In*: T.E. Rinderer (*ed.*), Bee Genetics and Breeding, pp. 91-119, Academic Press, Orlando, Florida.

Woyke, J. and W. Skowronek. (1974). Spermatogenesis in diploid drones of the honeybee. *J. Apic. Res. 13*:183-190.

CHAPTER 8

ACTIVITIES AND BEHAVIOR OF HONEY BEES

by NORMAN E. GARY*

INTRODUCTION

How much actually is known about honey bee activities and behavior, compared to what is still to be learned by man? No one can say for sure. Even though there are thousands of publications on this subject perhaps only 1% of the bee's behavioral repertoire has been discovered and documented by scientists. Only a naive person would expect to learn "all there is to know" about bees and their activities. Bee behavior is quite complex and has required millions of years to develop to its present stage. Obviously this chapter cannot contain all of the information from thousands of research documents, published in many languages during the last 200 or so years. However, we can present a wealth of basic information that should satisfy one's intellectual curiosity, and provide a firm informational foundation to support practical beekeeping activities. The available bits of information are similar to pieces of a jigsaw puzzle. Although the missing pieces frustrate us at times and tend to lead us down trails of misunderstanding, there is enough information to provide a rather satisfying profile of the typical behavioral life of the honey bee. This chapter is based on the behavior of European bees, the race of bees that is used commonly in the United States. Africanized bees are considerably different and are discussed at the end of this chapter.

Because bees normally nest inside dark cavities with small entrance holes, such as a hollow tree, observations of activities of the honey bee society inside the nest is an exquisite pleasure that was not possible until the development of the glass-walled observation hive around 200 years ago. Viewing bees on their combs in an observation hive is virtually the only way to observe *normal* behavior because smoking and opening a hive severely disrupts routine bee behavior. The express reason for using smoke is to cause a chaotic situation in which the defensive behavior is momentarily suspended as a means of preventing stings.

At this time mankind is on the threshold of astounding new developments in technology which may provide the means of documenting bee behavior that has been elusive heretofore because appropriate research tools were not available. For example, new electronic devices are being developed to follow

*Norman E. Gary, Ph.D., Professor of Entomology and Apiculture, Department of Entomology, University of California, Davis, California 95616. Research in honey bee behavior, especially mating, foraging, stinging, and communication. Author of over 100 scientific and popular publications.

and record the flight path of bees outside the hive. Where and how far do bees fly to various destinations associated with foraging, mating, or migrating to new nests? Radar was used successfully (Loper *et al.*, 1987) to locate and track congregations of flying drones. The prototype of a solar-powered microchip transmitter, developed for tracking individual bees by telemetry, was announced in 1988 (McManus, 1988). Who could have predicted, only several decades ago, that scientists would someday be able to track the aerial pathways of flying bees by means of a radio transmitter! Such novel developments may seem like frivolous fun for research scientists. However, developments of this nature represent milestones in our quest for new knowledge. The observation hive was just as novel when it was first developed. Without new tools to help us gather factual information, we are left with only our imaginations to fill in the missing pieces of the behavioral picture. This has generated a major problem in beekeeping literature in that oft-repeated speculation and theories gradually assume the status of fact, which then stifles further research. Why investigate something when everyone thinks that the answer is already known?

Although powerful research tools, such as the observation hive and electronic devices, have facilitated the resolution of many mysteries of bee behavior, it is ironic that, as more information is developed, the net result is an avalanche of tantalizing new questions that stimulate us to even greater efforts to learn more about bee behavior.

Activities of honey bees have always fascinated man. The expression "busy as a bee" adequately describes the non-stop activities inside the hive. It is intriguing to realize that behavior is slowly evolving even as you are reading.

Objectively Understanding Bee Behavior

One of the greatest pitfalls in one's quest to understand bee behavior is the tendency to be anthropomorphic when interpreting observations of bee activities. This simply means that, as humans, we are inherently biased. We tend to ascribe human attributes to bees, such as the ability to think, to have an overall awareness of events, to plan ahead, and to do all of those things that we might do if we were in their situation. . . with human brains! For example, a beekeeper might explain to a beginner that bees forage intensively during the summer to save enough honey to survive the long, cold winter season when there are no flowers. But think a moment! Is an insect capable of knowing or understanding what winter is, that cold weather is coming months after this bee will have died, and that a shortage of honey may lead to the death of the colony? I don't think so! No, it is man that understands the complex concepts of winter, nutritional requirements and the consequences of survival or death. Depending upon a person's training and orientation, our natural tendency to be anthropomorphic may elicit feelings of satisfaction, entertainment, disgust, alarm or some other emotion but, most certainly, **anthropomorphic thinking does not lead to an objective assessment of observed behavior.** Unfortunately anthropomorphic interpretations contribute no more to understanding the bee's behavior than the converse situation in which one attempts to interpret human behavior by mentally transferring the attributes of bees to humans!

A clear distinction must be made between two very different creatures. It is only natural that we as humans tend to ascribe certain human characteristics to anything we observe, particularly animals. Thus, bees sometimes are considered to be **angry** when they sting, **clever** when they build comb, or **ambitious** when they forage long hours. These words simply are not appropriate to describe an insect's behavior.

Anthropomorphic interpretation of behavior is especially likely when the observed animals have a high degree of social organization with functions that seem to parallel those of our society, as is the case with honey bees. One problem in interpreting bee behavior arises from our rather sketchy knowledge of precisely how the bee senses its world in terms of vision, taste, smell, hearing, and touch. How is information received by the sensory system actually "processed" within the nervous system? After sensing stimuli does the bee's brain permit thought processes or does it react mechanically and helplessly, such as when a moth is drawn to certain death by the powerful light stimulus of an open flame, to which it could respond in no other manner? In the absence of such knowledge, we are reduced to the level of speculating on such matters and, in doing so, we tend to apply our own values, usually at the expense of being objective. However, there is no excuse for such speculation in the light of present knowledge. Bees, like other insects, are reacting like tiny biological robots to signals (stimuli) in their immediate environment (external and internal) because their nervous systems are "hard-wired" or "programmed" genetically to react in this manner. In all probability bees are reacting without thought or conscious awareness of the mechanisms and consequences of their behavior. Such mechanisms have been honed to a sharp edge of adaptiveness by millions of years of evolution in which defective systems simply did not survive to reproduce.

The General Nature of Behavioral Events in the Colony

The notion that there must be some kind of plan for actually assigning "duties" to various worker bees has persisted for a long time. This idea has been perpetuated by many authors. Textbooks generally refer to bee activities under "division of labor," implying that bees consciously and actively divide "responsibilities," "tasks," or "duties" and also "perform labor." Obviously, these are anthropomorphic terms that, if accepted literally, would lead one to assume that there must be some kind of administrative hierarchy within the colony which dictates "who does what and when." The implication usually is that the queen bee somehow organizes the activities or supervises the "work" that goes on inside the hive. This idea was expressed by earlier writers when they mistakenly alluded to the king bee, the largest bee in the colony, which they assumed must be the leader. It must have been very embarrassing when the king bee was dissected later to reveal large ovaries, leaving no doubt that the sex of the "king" bee was indeed female.

Whether king or queen, both names imply a degree of dictatorship or ruling capacity. Although the activities of worker bees are affected in many ways by the queen, we know now that it is the pheromones from the queen, rather than her physical presence, that affect worker bee behavior and physiology (Free, 1987). The queen is not actively and consciously guiding or organizing worker bee activities. In fact, all activities within the hive can proceed, at least for a period of days, in the absence of a queen.

Because the bee is an insect, the vast quantity of information on the behavior of other insect species is helpful in understanding the root causes of behavior in bees. In general, behavior usually is caused by multiple factors, including those from within as well as external to the bee.

INTERNAL FACTORS AFFECTING BEHAVIOR

Some kinds of activities are limited by internal conditions, such as the stage of development. For example, wax secretion, and the behaviors associated with wax handling, are not possible until the wax glands are mature. Furthermore, very young bees less than one day old are unable to sting because the sting structure is not developed fully. Also, flight is impossible for very young bees, presumably because their nervous system and muscles are not fully developed. Another internal cause of behavior is the presence of various hormones that are secreted into the blood as various secretory cells develop with age.

Physical stimuli inside the body also affect behavior. For example, nerve cells sense the stretching of the honey stomach. The "messages" from these nerves either stimulate or inhibit feeding behavior.

The genetic composition of the bee is now known to exert a major effect on behavior. Each bee has a tendency to express different behaviors according to their genetic profile. For example, Rothenbuhler (1968) demonstrated that two specific genes control the housecleaning behavior of worker bees that remove the remains of dead larvae or pupae from brood cells. Workers endowed with one of the genes uncapped the cells, but did not remove the dead brood. Other workers that contained the other gene did not uncap cells but removed dead larvae or pupae from cells that had been uncapped. Enough workers of both genetic profiles must be present in the colony if there is to be effective housecleaning activity. The net effect is a degree of resistance to brood diseases, especially American foulbrood. In other research it has been demonstrated that genetic components strongly affect the tendency of foragers to collect pollen versus nectar. (Nye & Mackensen, 1968; Hellmich *et al.*, 1985). Another pioneer effort in demonstrating the effects of genetics on behavior was made by Calderone & Page (1988) who found that workers from two genetically distinct strains differed in the age at which they began foraging and in the relative frequency at which they foraged for pollen. Genetic manipulation offers the greatest opportunity for significantly changing the behavior of bees to achieve practical objectives, *e.g.*, reducing the tendency to sting. Perhaps the most vivid example of the impact of genetic makeup on behavior is found in the dramatic defensive behavior of African or Africanized bees.

The combination of these internal factors, which are invisible to the observer, tend to cause, or make more likely, specific activities of bees. Consequently, these activities caused by "invisible factors" are frequently mistaken as evidence for cognition. Thus, the illusion is created that bees have a significant degree of man's intelligence and awareness. Our intense and enthusiastic, sometimes emotional, involvement with bees often makes us forget that bees are, after all, only insects!

EXTERNAL CAUSES OF BEHAVIOR
External factors or stimuli also stimulate various responses. Sounds, chemicals (odors), touch, light, and magnetic fields are detected by thousands of specialized sensory cells. Nerve impulses from these cells speed along neural pathways of the nervous system and cause the bee to behave in a stereotyped manner when stimulated by the appropriate mixture or "configuration" of stimuli. However, even though certain behaviors are predictable and can be caused at will by the experimenter, bees nonetheless do not always respond the same way. Different levels of sensitivity of the sense organs to stimuli tends to "filter" or bias the information that goes into the nervous system. Furthermore, the very act of sensing stimuli momentarily alters the sensitivity of the sensory cells and other neural centers. Some cells momentarily become fatigued to the extent that they are temporarily insensitive. After they rest in the absence of the stimulus, their sensitivity is restored. Other cells become only partially fatigued so that the stimulus strength must be increased to achieve the initial level of stimulation. These checks and balances in the nervous system operate to produce a dynamic balance between the bee's current status relative to transient external stimuli. Once we understand the dynamics of the nervous system we'll not be surprised when individual bees

behave differently in response to identical stimuli, owing to their differences in sensitivity. Thus, a bee exposed to powerful chemical stimuli, such as repellents used in honey harvesting, may react initially, then cease to respond as the olfactory sensory cells temporarily become fatigued and insensitive to the odor. Our sense of smell operates in much the same manner.

TIME FACTORS IN BEHAVIOR

The speed with which bees behave is somewhat astounding. Man cannot, with the naked eye, observe the individual wing beats of a flying bee. Neither is it possible to watch the rapid leg movements as a bee walks along a surface. Numerous other examples show that many of the activities occur so rapidly that we cannot observe them accurately. These examples serve to illustrate that bees, and many other insects as well, behave with extreme speed. How is this possible? The explanation is very simple. The distances that nerve impulses have to move are **very short**. Thus, bees can receive and respond to stimuli in a matter of milliseconds, compared to seconds for man's response. That is why it is difficult to catch a bee with your hand. . . not that this is a smart thing to do!

Bee activities frequently are regulated by internal, physiological "time clocks" that trigger specific behaviors at specific times, especially on a circadian (an approximately 24 hour cycle) rhythm (Renner, 1960). Thus, bees remember the time of day and tend to arrive at nectar and pollen sources at the "correct" time, *i.e.*, the time of day when nectar or pollen becomes available at the respective plant species. Virtually all elements of behavior seem to be influenced by circadian rhythms, *e.g.*, walking movements (Moore & Rankin, 1985), visual sensitivity (Kaiser, 1979), and daily activity patterns (Spangler, 1972). Circadian behavior is not evidence of intelligence, but merely a "reflexive" response to signals within the nervous system, set in "motion" by certain stimuli, such as the onset of daylight or the taste of sugar at the time the bee initially visited a flower. Thereafter, the bee tends to respond at the same time each day. These "spontaneous" rhythms of behavior are rooted deeply inside the nervous system. In other insects it has been shown that the nerve impulses associated with "timed" behavior continue to function "on schedule" even after the nerve tissues are removed from the insect's body, at least until the tissue dies.

STEREOTYPED NATURE OF BEHAVIOR

The fact that the bee's nervous system is **relatively** simple in structure, compared to higher animals, profoundly affects its behavior. One result is that there are fewer response options in a given situation. Another consequence is a tendency for simple, stereotyped (repetitive) responses to given stimuli. Responses, such as extension of the proboscis when stimulated by sugar, tend to be very "reflexive." Also, the mechanical aspects of flexing the proboscis are essentially identical each time. Such simple responses are building blocks for more complex patterns of behavior. Simple responses are coupled together, like links in a chain, into complex, whole patterns of behavior that are repeated over and over.

Specific patterns of behavior are extremely similar for different individuals. For example, when a forager enters the hive with a load of pollen she typically goes through several minutes of behavior associated with locating a cell and unloading the pollen pellets. An observer is likely to marvel at the "searching" behavior by which bees, in the darkness of the hive, can locate a suitable cell. Actually, this is a whole pattern of behavior that is relatively fixed or stereotyped. After the first step in the pattern is triggered, then the subsequent stimuli within the pattern may lead to completion of the next step in the pattern in a chain reaction. An analogy is a row of dominoes in which each falling domino "drives" the next one, but always along a preset pathway! This behavioral pattern is shown in the honey bee by the following example: If the pollen pellets are gently removed from the hind legs of the forager as she is entering the hive (MacDonald, 1968), she will continue pollen unloading behavior, even to the point of going through the motions of unloading the non-existent pollen pellets into a cell! Another example of a common behavioral chain is oviposition behavior by the queen. She goes through the same pattern of "searching" for empty cells thousands of times during her life. Although the "search" for empty cells involves variable movements in different directions, actual egg-laying, beginning at the point when the queen "inspects" the cell, is one of the most stereotyped kinds of behavior in the hive.

Yet, there is some flexibility in patterns of behavior because the "fixed units of behavior," referred to earlier as building blocks, are similar to links in a chain in that they can be connected in various sequences. The actual order of coupling may be influenced by (1) genetic makeup, (2) learning as a consequence of experiences, or perhaps (3) other factors that are poorly understood at this time.

Influence of Learning on Bee Activities

Learning, by definition, merely means that a behavior has changed as a consequence of some previous behavioral experience. There are many examples of learning in the honey bee. For example, foraging bees quickly learn to associate the flower odor with the nectar food reward. Although bees learn to forage on certain crops at certain areas at certain times of day, intelligence per se is not implied by such a learning process. Again, the behavior is caused by relatively "reflexive" responses to transient stimuli. Learning does, however, introduce an element of great variability in bee behavior which is frequently mistaken as an indication of intelligence. For example, bees that forage on alfalfa eventually learn how to enter the flower without "tripping" it and getting hit by the flower structure. This is not unlike programming computers to "learn" certain tasks. However, regardless of the apparent sophistication of the equipment, intuitive thought and intelligence are not components of such systems. Although bees learn quickly, and apparently can remember certain events for the remainder of their lives, this does not justify considering them as clever, smart or intelligent!

GENERAL NATURE OF BEHAVIOR

An overview of behavior indicates that bee activity involves all of the preceding factors which *usually occur concurrently in combinations*. Unfortunately, the observer rarely can determine which factors are operating because behavioral events happen too fast to permit monitoring of all the conditions in the external and internal environment of the bee. Furthermore the bee's past learning experience combined with genotypic variability tend to confound one's interpretation of observations.

So we must conclude that intuitive thinking or cognition, akin to that of humans, is probably non-existent in bees, or at least inconsequential. There are strong indications that the potential for virtually all of insect behavior is genetically "programmed" into the system from the moment the egg is fertilized. For example, there is no opportunity for a worker larva to learn larval behavior, such as feeding movements, from other larvae or from adults. Also, once the larva is sealed in the cell, all of the complex cocoon-spinning behavior, and ultimately the emergence behavior of the adult, proceeds without benefit of a learning experience by association with adult bees. In another example, it can be demonstrated experimentally that a new colony of bees, when (1) composed entirely of newly emerged bees from a brood comb placed in an incubator, (2) given a queen cell to provide a queen, and (3) totally deprived of contact with experienced adult worker bees, will behave in the same manner as a normal colony after the bees become old enough to behave in various ways. Despite the lack of opportunity to associate with older bees, their behaviors are identical, including the complex dance communication system. This is proof that the basic building blocks of behavior are innate, and not developed by a cognitive thought process or learned through the communication of abstract information from other bees.

In summary, it is helpful in understanding bee activities if we assume that each bee is a tiny biological robot, "programmed" genetically to "perform" given activities that favor survival of the colony. Behavior is difficult to study, and sometimes frustrating when bees do something we don't want them to do, yet the challenge of observing and learning about bee activities is perhaps the most exciting aspect of keeping bees, whether you are a beekeeper, a scientist or just curious!

THE DIVISION OF LABOR CONCEPT

Many activities occur inside a hive at any given time, *e.g.*, comb building, fanning, nursing brood, house cleaning, food exchange, egg laying and grooming, to name a few. These activities seem to be performed in an organized manner. At least all of the needs of the colony tend to be met in a timely fashion. Such activities have been referred to as "tasks" or "labor." Extensive studies of intra-hive behavior reveal that individual bees tend to specialize in one or more of these activities (Free, 1965; Lindauer, 1953; Seeley, 1982, 1985). With such specialization comes a greater efficiency which favors the survival of honey bees as they compete with other species for food and nest space.

This entire process is popularly known as "division of labor," which may be defined as the variation in frequency at which individual bees are engaged in specific activities as a result of differences in age, environment and genotype. Division of labor is common for highly social insects in which there is a primary division of labor between reproductive and worker castes, and then a secondary division of labor among workers (also known as polyethism). Reproductive castes are engaged in those activities, such as mating, that eventually lead to oviposition (egg-laying). All of the other activities in the colony are performed by the worker caste.

The term "division of labor" implies that there is some sort of directed, conscious or active control over the behavior of each bee. The term "labor" is an unfortunate choice of words because it implies more than is intended. Perhaps the process of "dividing labor" can be understood more accurately by asking ourselves the question, "How does each bee 'know' what to do, how to do it and where and when the activity is appropriate?" Of course, bees don't have to "know" anything. We should actually compare a bee "at rest" to a computer that has been switched on, but is not actively processing data at the moment. All of the information is stored, the programming is complete and touching the appropriate keys can initiate a particular operation. Yet, the computer has no conscious awareness of the information it is processing or of the information it has stored. The same situation probably is true for a bee. Then one can ask the question, "If bees are simply little biological robots with computer-like brains, why do bees in the same colony behave in so many different ways at any given moment?" The answer is that the response of a bee is dependent upon its condition, and the condition of each individual bee in a colony is slightly different from all other bees at any given moment, in terms of some combination of the genotypic condition, internal and external stimuli and humoral factors (physiological state as influenced by hormones).

Another element is that the microenvironment for each bee is somewhat unique simply because it occupies a specific location that is not duplicated precisely anywhere else in the hive. For example, a nurse bee standing directly over a cell containing a larva, or perhaps pollen, receives a very different set of stimuli (especially odors) than a bee that is very near a queen and is being exposed to her powerful pheromones (odorous chemical "messengers").

Also, each bee has a different potential to behave, according to the influence of the immediately preceding behavior. For example, a bee that has just fed on honey will have a very different probability for feeding, as opposed to some other activity. Each bee, therefore, may be considered to have behavioral tendencies that change from moment to moment, even from second to second. In addition to the dynamic changes taking place inside the bee, the external stimuli are constantly changing. A stationary bee may suddenly be brought within range of the queen's highly stimulating phero-mones when the queen passes nearby. Similarly, a passing foraging bee may bear various stimulating flower odors.

Genotypic variability is a major factor in determining behavior. Our views have changed considerably since Ribbands (1952) erroneously concluded that division of labor is not genetically influenced. A number of studies indicate that major behavioral differences are attributable to genotypic varia-tions (Calderone & Page, 1988; Collins *et al.*, 1982; Frummhoff & Baker, 1988; Hellmich *et al.*, 1985; Moritz & Hillesheim, 1985; Robinson & Page, 1988, 1989; Winston & Katz, 1982). For example, Calderone & Page (1988) found major differences in genotypes concerning the age at first foraging flight, the location within the colony, and the frequency in gathering pollen. Robin-son & Page (1988) found a genetic component to guarding and undertaking behaviors. Robinson & Page (1989) also demonstrated a genetic basis for nectar and pollen foraging and nest-site scouting. Similarly, Frummhoff & Baker (1988) found striking patrilineal differences in the propensity of workers to groom nestmates.

If a given worker bee is to be of maximum utility to the colony, in terms of survival value, she must be exposed frequently to those stimuli that are associated with the most vital elements of the colony. Such exposure has been achieved through the evolutionary process by the development of so-called patrolling behavior, in which bees are "programmed" to walk and move around within the nest. By this simple expedient, each bee is exposed to all possible combinations of stimuli. During these movements individual bees tend to react to the most intensive stimulus situations, frequently referred to in a practical sense as the "needs" of the colony. Thus, it is not necessary or even possible that individual bees "know" the activities happening throughout the colony. By analogy, imagine one person working in a city of 40,000 people. The person merely responds to those situations with which he or she comes in contact, yet contributes services to the needs of the city. Performance of work is not contingent upon knowing the status of the entire city. By frequent movements within the hive, for example, hungry bees sense food location, nurse bees encounter larvae, and housecleaning bees detect debris. An important element is that the bees are reacting to transient conditions, and not anticipating future events.

Therefore, "division of labor" can be accounted for by these various mechanisms without necessarily implying the need for high levels of intelligence frequently attributed erroneously to honey bees. Thus, a worker bee that is feeding a larva merely happens to have the appropriate behavior in its genetic make-up, has reached the critical age of development (brood food glands that are enlarged and possibly stimulating locomotion) and has a nervous system that is highly adapted to sensing the odor of hungry larvae. The worker moves along in the darkness inside the hive, in typical "patrolling" behavior, until she encounters the stimulus of a hungry larva, at which point she reflexively feeds the larva. The act of feeding the larva changes her behavioral tendencies momentarily. This makes another kind of behavior more probable, such as food exchange or feeding on pollen.

In another example, much of normal worker behavior toward the queen, such as grooming and offering food, can be induced by exposing bees to queen bee odor that has been placed on some inanimate object, such as a cork. This behavior indicates that queen grooming behavior, sometimes considered as a somewhat intelligent act, is merely a sequence of reflexive responses to specific stimuli near the queen.

If you accept these interpretations, then it becomes meaningless or misleading to ask such questions as, "How do bees 'know' what to do and when to do it?" In all probability, bees do not actually "know" what they are doing or have any intuitive realization that they have done something when behavioral acts are completed. Even though bees function in a manner that appears to be highly organized, and involves some kind of conscious effort, all of their activities and behavior can be accounted for satisfactorily by reflexive or mechanistic responses, many of which are very simple. At least it makes sense that they are not actually "dividing their labor."

Effects of Worker Age on Activities Inside the Hive

As worker bees age, after emergence as adults, they are engaged in various activities that are correlated approximately with their age and physiological development. This may be thought of as "temporal division of labor" or "age polyethism." Younger workers tend to stay inside the hive during the first two to three weeks of adult life, then become active in foraging outside the hive during the remaining two to three weeks of their life.

Activities inside the hive have been studied by many researchers (*e.g.*, Calderone & Page, 1988; Lindauer, 1953; Rösch, 1925, 1930; Wenner,

1961a; Winston & Fergusson, 1986; Winston & Punnett, 1982; and many others). In general, these studies were performed with glass-walled observation hives that permit convenient viewing of normal intra-hive activities. Fortunately, bees are large enough that tiny numbered tags or distinctive paint marks can be placed on individuals for identification purposes. This permits the study of individual bees or groups of bees throughout their lives. Thus, the ages of observed bees are determined easily by placing combs containing mature brood in an incubator, tagging the newly emerged bees, and then introducing them into the experimental hive.

During the first three days after emergence, young worker bees typically clean the cells from which bees have recently emerged. Within a day of so after emergence they begin to feed nectar, diluted honey and pollen to larvae more than three days old. At approximately six to 12 days of age, after their brood food (hypopharyngeal) glands are mature enough to secrete royal jelly, they begin to feed young larvae less than three days old. During their third week as adults their activities become more varied and less related to age per se. Orientation flights, sometimes called "play flights," are taken near the hive entrance. Also, bees of this age may be seen cleaning debris and dead bees from the hive, packing pollen in cells, building comb, capping cells, ripening nectar, applying propolis and receiving nectar from foragers returning from the field. Near the end of the third week, some of the workers become guard bees.

It should be emphasized that there is great flexibility in the age-activity relationship. Experiments done by many investigators (*e.g.*, Lindauer, 1953; Ribbands, 1952; Sekiguchi & Sakagami, 1966; Winston & Punnett, 1982) indicate that there is a typical sequence of age-related activities in a normal colony.

However, bees of the same chronological age have the potential to change physiologically in response to changes in the colony. This phenomenon has been demonstrated experimentally: A population composed exclusively of young bees of the same age are placed into a small colony where they quickly adjust to the "needs" of the colony. The full gamut of activities are restored, many of which typically would not have been observed for the age group. For example, bees only one to two weeks old became active field foragers under such conditions. Conversely, very old bees can revert to the role of nurse bees if, for some reason, the nurse bees are removed from a colony population.

Another behavioral characteristic is that individual bees of a given age can and do engage in diverse activities within a short time, *e.g.*, a matter of minutes. The time spent in each activity may be very short. Typically there is "cooperation" by several bees in a common activity, such as capping a brood cell, even though it has been observed that it is possible for a single bee to cap an entire cell (Smith, 1959). Bees typically move around a great deal ("patrol-ling" behavior) and respond to the various conditions that stimulate behavior consistent with the physiological condition and experiences of each bee. Thus, the bee that is carrying a wax particle would be highly stimulated by a cell that is partially capped, or that contains the stimuli (whatever they are) of cells that "need" to be capped. In summary, there is a continuum of ages of bees, physiological conditions, stimulus situations, and genotypic constraints that collectively account for bee behavior.

It is important to realize that all bees do not necessarily engage in all types of activities. Some bees "age prematurely" and initiate field foraging, without having been a guard bee or engaging in some other kind of activity such as housecleaning.

Some of the most dedicated research concerning bee activities was done by Lindauer (1953), who continuously observed, day and night, individual bees throughout their lives. He found that one bee was inactive for a total of 68 hours, 53 minutes. A similar time (56 hours, 10 minutes) was spent in patrolling behavior, in which the bee simply moved around, "examined" cells and encountered other bees. This time period was interrupted frequently by definable activities. Cell cleaning behavior accounted for 11 hours, 44 min-utes. During a total of one hour, 50 minutes, the bee fed young larvae up to three days old. Feeding three- to six-day-old larvae occupied two hours, eight minutes. Other activities were comb building (six hours, 24 minutes), cell capping (12 hours, 27 minutes) guarding (34 minutes), "play" (orientation) flights (one hour, 15 minutes), and foraging in the field (nine hours, 59 minutes). In general, it was found that the ages of bees engaged in activities were:

Activity	Age of Bee
Cell cleaning:	1-25 days
Feeding larvae:	
Hatched to 1 day old:	7-25 days
1-2 days old:	2-24 days
2-3 days old:	1-30 days
3-4 days old:	1-28 days
5-5½ days old:	2-26 days
Comb building, including capping:	1-32 days

With the exception of activities relating directly to foraging, activities within the hive appear to be similar, day and night. Thus, bee colonies are never inactive. Periods of inactivity or "rest" of individual bees are common. Bees apparently don't sleep!

Communication

Bees are highly social animals. The basic requirement of social existence is effective communication. Without communication, an animal becomes semi-social or solitary because social interactions require the transfer of information, albeit simple signals. Communication as used here means simply the intra-specific transfer of stimuli that elicit behavioral and/or physiological responses in receptive individuals. It does not imply intelligence or awareness, or human-like understanding of the "messages" by the sender or receiver. Whether such communication should be considered as a "language" seems irrelevant to the pragmatic consideration of studying and understanding the significance and mechanisms of communication.

The basic modes of communication in bees are similar to those of man, that is, the use of various stimuli (such as light, chemical, and physical stimuli) that can be perceived by specific sensory organs. Each species of animal utilizes, to different degrees, the various possible stimulus modes during communication, according to the "life style" or habitat of the animal. During the evolution of man, the perception of visual stimuli, *i.e.*, light, must have been extremely advantageous. Thus, light perception (sight) became man's most sensitive means of perceiving the environment. Other senses are more significant to insects, especially the olfactory (smell) sense. The honey bee has adapted to a relatively closed nest environment that is dark. Consequently, senses relating to odor, touch, and sound stimuli, rather than vision, appear to be the primary stimuli involved in communication. Because visual and sound signals are so fundamental as means of communication in man, there is a tendency for us to underestimate the significance of chemical and tactual communication by bees. For example, it is easy to watch a dancing bee in an observation hive and mistakenly assume that the surrounding bees can see the dance pattern as we do. In considering honey bee communication, researchers tend to study one aspect of communication at a time, for example, sound, sometimes at the expense of understanding interactions with other elements. Communication more likely involves the interactions of multiple stimuli, such as the chemical, sound, and tactual signals.

COMMUNICATION ASSOCIATED WITH FOOD LOCATION

Spitzner (1788) apparently was one of the first to call attention to communication in honey bees when he described bee dances as a means by which bees communicate the intensity of a honeyflow and location of nectar sources. Spitzner's observations were largely forgotten or ignored until von Frisch (1920) published "Über die 'Sprache' der Bienen." Since that time many researchers have contributed a great quantity of information on bee communication, much of which is reviewed in a book by von Frisch (1967a), who shared a Nobel prize in 1973 for his research contributions in this area.

The primary experimental approach to studying the communication of food location has been to train bees from glass-walled observation hives to collect artificial nectar at artificial flowers or feeders (Fig. 1). Artificial nectar is made by dissolving ordinary table sugar (sucrose) in water to make a syrup. Several drops of a food flavoring or aromatic substance, such as peppermint, clove or lavender oils, are added to mimic the odor of flowers. Such artificial nectar is very stimulating to bees and apparently duplicates flower conditions adequately to enable valid experimentation.

FIGURE 1. Bees trained to forage at a small dish containing scented sugar syrup. A piece of colored paper beneath the dish aids in orientation. (Photo by Kenneth Lorenzen).

Bees are trained to collect sugar syrup from feeders placed at stations at known and strategic locations. Training bees is not difficult unless there is great competition from abundant nectar of high sugar concentration in the area. Methods of training have been described (Gary & Witherell, 1971; von Frisch, 1967b; Wenner, 1961b; and others). While the trained foragers are feeding at the station, they are captured and tagged or simply given distinctive paint marks in order to identify them when they are observed later inside the observation hive and at the feeder when they return on successive trips.

If the foraging bees are sufficiently stimulated by the food reward, they "perform" a dance on the vertical comb surface inside the brood nest. Von Frisch (1967a) described two types of dancing, the round dance and the wag-tail dance (Fig. 2). The round dance is "performed" by bees that forage less than approximately 10 meters from the hive. In the round dance, the bee, with quick short steps, runs around in narrow circles on the comb, often changing her direction so that she rushes once to the right, then to the left and again describes one or two circles in either direction. She may continue to "dance" for several seconds or even as long as a minute; then she may stop, move to a different location of the comb and begin dancing again. Finally, she moves rapidly toward the entrance and flies out again. The dance excites some of the nearby bees; they follow the dance movements of the "performer," with their antennae positioned on or near the dancing bee. Some of them leave the hive, presumably in "search" of the food source.

Newly recruited foraging bees that respond to the round dance appear to "search" in all directions from the hive. For many years it was thought that there is little or no information in the round dance regarding the compass direction of food from the hive location. A study by Kirchner *et al.* (1988) indicates that acoustic signals are emitted in the round dances and that there is directional information, and perhaps even some indication of distance. This is difficult to assess accurately because as yet no one has successfully traced the aerial flight paths of bees as they fly in the "search patterns."

As the distance from the hive to the food source increases to a range of approximately 10 to 100 meters the dance form changes into a crescent or sickle dance (Fig. 2). Beyond 100 meters the sickle form changes to the well-known "wag-tail" dance (Fig. 2), in which the dancing bee moves in a narrow half-circle to one side, then turns sharply and moves in a straight line over the imaginary radius of this circle to the starting point, then makes a

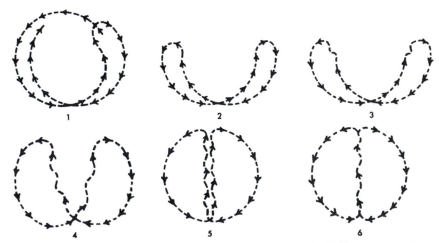

FIGURE 2. Honey bee dances (diagrammatic). (1) Round dance. (2) Sickle or crescent dance. (3-5) Transitions between the round dance and the wag-tail dance. (6) Wag-tail dance.

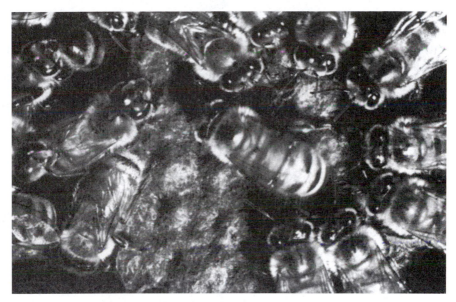

FIGURE 3. Worker bee (center) "performing" a wag-tail dance. She is closely attended by eight potential recruit bees. (Photo by Kenneth Lorenzen).

half-circle in the opposite direction, thus completing a full circle. Then the bee again runs in a straight line, retracing the initial straight-line path, until she reaches the initial starting point. While running in the straight line portion of the dance, vigorous, sideways wiggling motions are made with the body, especially the abdomen (Fig. 3).

Now that the basic mechanisms of communication are slowly being resolved, perhaps it is time to place less emphasis on the actual visual configuration of the dance. After all, bees can't see the dances! It seems appropriate to place more emphasis on those elements that are actually being sensed by the bees. For many years there was no conclusive evidence that bees can perceive airborne sound signals. However, Towne & Kirchner (1989) showed that bees in fact can perceive airborne sounds by detecting air-particle movements, rather than pressure oscillations as in human hearing. The most likely receptors of the air-particle oscillations appear to be the Johnston's organs in the antennae. Sounds produced by dancing bees are not perceived by following bees as vibrations in the comb (Michelsen *et al.*, 1986), as once believed.

Dancing bees produce sounds that seem to be an important component of communication (Esch, 1961, 1963, 1964; Wenner, 1962). During the wagging movements a train of pulsed sounds is made at the low frequency of 250-300 Hertz with a pulse duration of about 20 milliseconds and a repetition frequency of approximately 30 per second. These sounds are produced by the wings of the dancing bee (Michelsen *et al.*, 1987) and are inaudible to the human ear. Follower bees place their antennae very close to the dancer's abdomen in the zone where the air particle movements are most intense. The number of sound pulses is highly correlated with the distance to known experimental food sources (Wenner, 1962). Sound communication appears to be a common element for both round and wag-tail dances (Fig. 4). Some indication of distance to the food can be determined by observing the "straight runs" of dancing bees in a glass-walled hive. The number of straight runs per 15 seconds is correlated with those distances. For example, von Frisch (1967b) found the following approximate distances and times:

Distance (meters)	Straight runs per 15 seconds
100	9-10
600	7
1,000	4
6,000	2

Also, as distance increases the duration of each straight run increases, as found by von Frisch & Jander (1957).

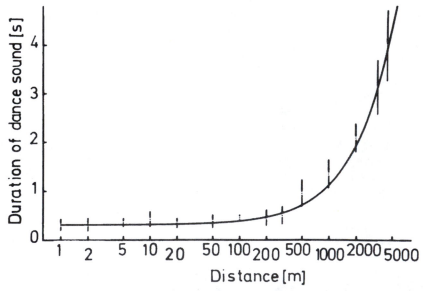

FIGURE 4. Indication of distance in round dances and wag-tail dances. There is a linear relationship between the distance to the feeding place and the duration of sound produced during the dance. Vertical bars indicate standard deviation. (From Kirchner & Lindauer, 1988).

The communication of direction is also an important component of honey bee dances. Such information obviously is advantageous to honey bees, particularly when food sources are great distances from the hive. Evidence for direction communication is found in the dance configuration. The direction that the dancing bee is heading during the straight run, **relative to gravity and not to compass direction**, corresponds to the compass direction of the food source outside the hive (Fig. 5). If the food source is directly toward the compass direction of the sun, the straight run portion of the dance is oriented straight up on the comb. If the food source is in the opposite compass direction from the sun, the dancing bee faces downward during the straight run. Food locations that are clockwise or counterclockwise from the sun's compass direction, are indicated by identical angular deviations of the straight run

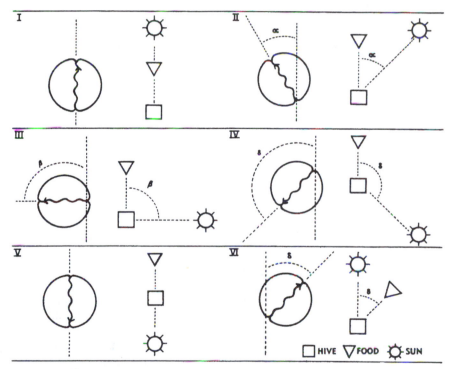

FIGURE 5. Relationship between the direction of food from the hive, relative to the compass direction of the sun, and the direction of dancing during the wag-tail dance. When the food source is in the same direction as the sun's compass direction, the bees "face" upward on the comb during the straight wag-tail run portion of the dance (I). The bees "face" downward when the food source is in the opposite direction from the sun's compass direction (V). When the source of food is to the left of the sun's compass direction, the bees dance at an angle counterclockwise from the top of the comb (II, III, IV). When the source of food is to the right of the sun's compass direction, the bees dance at an angle clockwise from the top of the comb (VI).

portion that are clockwise or counterclockwise from the vertically-oriented straight run that the bee danced when the food was in the direction of the sun.

To understand these relationships it is helpful to think of the location of the hive, food and the sun in one plane, as shown in Fig. 5. Imagine that the sun is always on the horizon in the appropriate compass direction indicated by the directional bias of the sun from the zenith (directly overhead). An easy method of determining the sun's compass direction at any given moment is as follows: The shadow of a perfectly vertical pole falls 180° from the compass direction of the sun. For example, if the shadow points due north from the base of the pole, then the compass direction of the sun is due south. As the sun moves across the sky, dancing bees change the orientation of the straight portion of the dance at the same rate, consequently always indicating the "correct" direction to the food source from the hive.

The above account of direction communication does not take into consideration the actual means whereby a bee, without visual reference inside the dark hive, can sense directional information in the dance. The most likely mechanism is that bees receiving the information can sense, directionally, air-particle oscillations with the antennae. Because the Johnston's organs are bilateral, they could yield information on the dance attender's spatial orientation relative to that of the dancer, which, in turn, would indicate the direction of the food source (Towne & Kirchner, 1989).

Another source of information during dance communication is the fragrance of the flower. There is evidence that flower odors cling to the waxy cuticle layer covering of the body. During the dance there is ample opportunity for the potential recruit bees to smell the fragrance, and then respond selectively to this odor later while "searching" in the field. Some scientists think that this means of communication may be as effective, or even more effective, than communication by dances. Others think that communication is by odor, and not by dances (*e.g.*, Wenner, 1971).

The flavor of nectar that is passed from the returning foragers to potential recruit foragers is also a potential source of information (von Frisch, 1946). This mechanism is probably very significant in alerting bees to rich food sources that contain high sugar concentration in the nectar. Follower bees vibrate the comb with 380 Hz "begging signals" which stimulate foragers to share a droplet of nectar (Michelsen *et al.*, 1986).

Bees also communicate the time of day that plants are yielding nectar and pollen. This is very important because many plant species yield nectar and/or pollen at rather precise, and sometimes brief, time periods during the day. The time of day when the bee is dancing indicates the time when food is present in the field, and bees have the ability to "remember" this time when they forage.

The quantity of nectar and pollen at the various sources probably is communicated by the number of dancing bees in the hive and frequency of dances. When nectar and pollen are abundant, the foragers can collect loads quickly. This situation is very stimulating to them. They become highly excited, return more frequently to the hive with full loads and a greater percentage of returning foragers "perform" more dances. When nectar and pollen sources are marginal, dance activity is diminished, and eventually stops if foraging conditions become too poor.

Foragers may encounter many hazards in the field, *e.g.*, predators, natural toxic materials present in the nectar and pollen of a small number of plant species, and pesticides. These hazards can prevent the return of affected foragers to the hive, or upset their normal behavior significantly, thus interfering with further communication that otherwise might have lead new foragers into hazardous foraging conditions. Such protection is also enhanced by the time delay mechanism in which experienced foragers typically make several round trips to food sources before they are stimulated sufficiently to dance in the hive. Thus, there is an opportunity for slow acting toxic materials or significant predation pressure to be expressed before communication takes place. By this mechanism hazards associated with foraging are communicated indirectly.

Other less obvious forms of communication are likely to be detected in future research. There may be other signals that are not known or recognized at present, such as electrostatic charges that build up on the bodies of foraging bees (Erickson, 1975). In any event, it seems clear that foraging-inexperienced bees in the hive are exposed to several kinds of information concerning the status and location of food resources in the field. The variety of behavior of bees following the dancer suggests that they may be receiving different components of the total available information. Even though all of the mechanisms are still not clear, many bees that follow the dancing bee eventually arrive at the food source where the experienced forager was known to be foraging. Many of the bees that follow dancing bees do not forage at the source indicated by the dance. This does not indicate that the message was not communicated, or that it was not received. Rather, it indicates that honey bee behavior is very complex and that bees can respond in various ways to the information they receive.

Apparently the communication mechanisms for pollen sources are the same as for nectar. Some plants yield pollen only. When pollen and nectar are available on the same plant species, most of the bees that collect pollen also collect nectar, a fact that has not been sufficiently emphasized in the literature.

Effective communication involves the activities of the "receiving" bees as much as the "sending" bees. Experimentally, communication is very difficult to evaluate precisely. How can a human observer know how much information is being transmitted and how much is being received? Must the receiver bee respond in an obvious activity as proof that message was actually received? If so, how soon after receiving the "message" must the response occur? These and many other difficult questions are discussed, but not always answered clearly, in greater detail elsewhere (Lindauer, 1971; von Frisch, 1967a; Wenner, 1971). It is known, however, that responses to dances are different in naive bees (those bees that have never foraged) as compared to experienced forager bees. Naive bees do not necessarily take their first foraging flight immediately after following a dancing bee, according to Lindauer (1953). Hours or even a day or more, may ensue before sufficient stimulation "accumulates" in the nervous system of the receiving bee to trigger the first foraging trip. Another complication is that an unknown number of foraging-inexperienced bees make their first foraging trip without actually using any information from dancing bees. However, most bees probably use at least some of the information contained in the dances. During the time between the first exposure to a dance, and leaving the hive to forage, potential recruits may be exposed to many dancing bees that are foraging at various locations on many different species, representing various levels of profitability (in an energetic or nutritional sense) to the colony. The most profitable food sources cause the greatest stimulation of the forager, and the differences in levels of stimulation are expressed by the number, vigor, and duration of dances. The potential exposure to many dances, as well as the time delay between exposure to dances and leaving the hive to forage, may be thought of as a mechanism that is parallel to the decision-making process by the human mind. In bees, then, the various alternative food sources are "reviewed and compared" during exposure to multiple dances of different bees. Ultimately the most stimulating food sources attract the most foraging activity. If all foraging-inexperienced bees responded immediately to the first dance to which they were exposed, a chaotic situation, parallel to thoughtless, irresponsible, impulsive and inefficient behavior in humans, could occur. In that kind of system, the colony could over-react to meager, erratic food sources, and could lose responsiveness to exploit richer sources that become available later during the day. Thus, the low frequency of response to any single dance should not be construed as evidence for the ineffectiveness of dance communication. The opposite may be true!

Many scientists have attempted, with variable success, to evaluate the communication system of bees. For example, in one study by Lindauer (1953) observations revealed that "of the 150 marked bees which departed after following the dancers, 91 brought home nectar or pollen, and 79 of these returned with the same kind of load as the dancers that they followed; moreover, 42 of them also indicated by their own dances that the direction and the approximate distance of their first source were the same as those given by the dancers." Wenner (1971) found that a very low percentage of bees that followed the dances actually arrived at the "correct" feeding station. Research by Gould *et al.* (1970) supports the existence of direction and distance communication. It is not surprising that different researchers have assigned different values to the efficiency of the various components of communication behavior of honey bees. Nevertheless, there seems to be a consensus that honey bees do, indeed, have a very effective communication system. It seems likely that multiple mechanisms of communication have evolved, and that the different stimuli (chemical, sound, tactual) have variable utility according to the everchanging variables in the foraging system, the colony, and the individual bees. For example, genetic variation probably predisposes individual bees to have differential sensitivity for the different kinds of stimuli. Thus, one bee may be able to smell with greater sensitivity than another bee and tend to utilize odor information to a greater extent. Perhaps the distance to food sources influences the mode of communication. Of what value is precise direction information if the food is only ten meters from the hive where it could be located by odor alone? However, if the food source is six kilometers from the hive, and downwind from the colony location, would not odor information alone be relatively inefficient, compared with distance and direction information? It seems very probable that multiple communication stimulus modes have advantages that would have constituted strong selective pressure during the evolution of bees.

Bees with previous foraging experience also respond to dances, but apparently only to the dances of bees that are foraging on the plant species on which the experienced bees had previously foraged. Such response probably is caused primarily by the odor stimulus. In fact, when bees have been trained to a feeding station using sugar syrup that has a distinctive odor, *e.g.*, peppermint, the liberation of this odor in the hive by any means is sufficient to stimulate many of the experienced foragers to visit the familiar source immediately (Wenner, 1971). This is a classical conditioned reflex response. Apparently many, perhaps most, of the experienced foragers do not leave the hive to forage until the first successful foragers each day return with the "correct" odor from the familiar food source. The odor stimulus is even more stimulating when the dances match the distance and direction corresponding to the foraging site to which the experienced bees are accustomed, according to Lindauer (1971). These older, experienced bees tend to congregate in different areas of the brood comb where dancing of "their group" occurs. These mechanisms favor very quick responses to the onset of nectar or pollen availability in the field, yet tend to avoid the expenditure of energy of "searching" by many foragers when the food source is not profitable at that particular time.

In summary, it seems clear that bees without any previous foraging experience are typically exposed to a great deal of information concerning the status of food resources in the field, *i.e.*, the direction, distance, odor, flavor, temporal availability, quality and abundance. The degree to which they actually utilize the potential information is yet to be determined in a precise manner. (Johnson, 1967; von Frisch, 1967b; Wenner, 1967).

Reference is frequently made to "scout" bees, those bees that presumably leave the hive and discover new food sources without utilizing information obtained from previous foragers. There is no question that such activity occurs. For example, experimental colonies containing only young bees that have had no contact with older experienced foragers, will initiate foraging activity. The frequency of scout bees in typical colonies is virtually impossible to determine. Yet scout bees probably are a very important segment of the population with respect to the "discovery" of new food sources. How does one prove that a bee has not received information from an experienced bee? The first experienced bees that visit a familiar source each day are sometimes called scout bees.

Some information associated with food sources apparently is not accounted for in the communication system. For example, there is evidence that bees do not communicate the altitude of food, relative to the ground level, *e.g.*, high in trees. However, this notion has been challenged by Wellington & Cmiralova (1979). Also, there is no evidence that the color of flowers is communicated inside the hive, although flower color is very important to foragers. This is very interesting because there would seem to be some advantage in doing so. Apparently, color is learned by each forager at the moment she approaches the flower as she associates the food stimuli with flower structure and color.

Many factors influence the state of excitement or level of stimulation of dancing bees, and probably affect the communication effectiveness. Sweetness of the nectar is a major factor. Also, various odors can stimulate dance activity, according to their concentration and quality. Apparently, even the similarity between feeding containers and the natural appearance of flowers is important (Kappel, 1952), there being more stimulation as the natural flower appearance is mimicked. Distance from the hive to food is extremely important. The numbers of new foragers at feeding locations farther from the hive is smaller than on those near the hive when all other factors are identical. With the same concentration of sugar solution 11% of those returning from distances of approximately 2,100 meters danced, but 68% danced when the food was 100 meters from the hive. When the concentration of the sugar solution at the more distant feeding station was increased, more bees began to dance (Boch,

1956). So the dance is the result of a certain level of excitement of the forager (Lopatina, 1956).

In colonies where the combs contained a small amount of honey, more foragers were induced to dance, and dances lasted longer than similar colonies containing full combs of honey. Environmental conditions in the hive can also directly affect the dance "readiness" of its foragers (Wittekindt, 1961).

Von Frisch (1952) was amazingly correct when he said, "What these animals can tell each other about feeding places they have discovered and how, in doing this, they deal with abnormally different situations is beyond anything we should have expected of insects. And their brains are not bigger than a grain of millet. Nothing could more clearly illustrate the marvelous structure of the nerve cells. It would be presumptuous to say that we can understand it."

THE DANCE LANGUAGE CONTROVERSY

A discussion of foraging communication in honey bees would be incomplete without addressing the so-called "dance language controversy," a long-standing debate that seems to have polarized many scientists into supporting the notion that bees use either the distance and direction information that is supposedly communicated in dances, or rely solely on their olfactory sense to locate forage sources. Some history of the evolution of our understanding of bee communication may be helpful. Initially Karl von Frisch generated the most comprehensive understanding of the foraging process, based upon his discovery of how to train bees to forage at feeding stations containing artificial nectar (sugar syrup) where he could manipulate the variables (*e.g.*, quantity, quality, distance, and temporal availability of food). At first he considered that communication was primarily by olfactory processes based upon the behavior of bees that were trained to forage at feeding stations located near the hive. He manipulated odor in such a way that bees revealed, by their behavior, that they truly were perceiving odors, and using this information in the foraging process. Later, as he worked with forage sources farther from the hive, he observed that components of the wag-tail dances were correlated with the distance and direction of forage sources. At that time he proposed the "dance language hypothesis," which implies that recruit bees use symbolic information in the dances as a means of finding the approximate location of the forage source. His findings were hailed as a great contribution to science. The discovery that bees could communicate with a "language" similar to man was exciting, and ultimately led to his sharing a Nobel prize in Medicine in 1973. However, Wenner & Johnson (1967) challenged the dance language hypothesis, based upon their own experiments which indicated, at least to them, that bees were using olfactory cues exclusively. Their challenge stimulated many other scientists to repeat von Frisch's experiments, and to devise still other experiments, all intended to resolve the debate. Noteworthy are the experiments of Gould (1975a), in which he successfully manipulated dancing bees so that their dances indicated, to potential recruits in an observation hive, the direction and distance to foraging sources that the dancing bees had never visited. A significant number of the recruits did indeed arrive at these stations, thereby persuading Gould, and many other scientists as well, that at least some of the bees truly were using information contained in the dances.

In attempts to prove that honey bees use dance information, several scientists have constructed mechanical dancing bees that mimicked real bees, in terms of sounds, movements, and odors. The object was to determine the extent to which naive bees could be directed to known food locations. After initial failures, Michelsen & Anderson (1989) claimed success (Fig. 6).

The dance language controversy is too complex and lengthy to be discussed in more detail here. Most of the pertinent information, plus a great deal of philosophical rhetoric, is presented in several papers (Gould, 1975b, 1976;

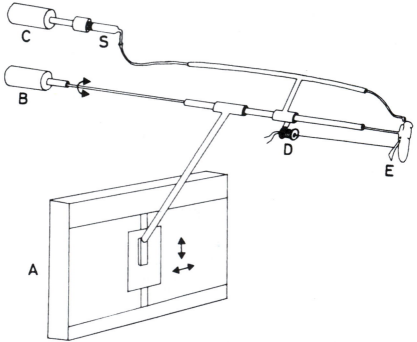

Figure 6. Diagram of mechanical bee (not to scale). The X-Y recorder (A) makes the figure-eight path. One step motor (B) turns the model (during the figure-eight and wagging). Another step motor (C) connected to a syringe (S) pumps scented sugar water through a soft tube to the front end of the model. An electromagnetic driver (D) moves the model's wing (E). (From Michelsen & Anderson, 1989).

Rosin, 1978, 1980a,b, 1988; Veldink, 1989; Wenner & Wells, 1987, 1990).

Some of the real problems that scientists have encountered in investigating honey bee communication are as follows:

(1) Inappropriate methodology—The best method to determine whether bees use the dance information is to document, with appropriate instrumentation, the complete aerial flight pathway of recruits that are known to have followed the dances, including the entire trip from the hive to the food and the return trip back to the hive. Because such sophisticated methods were not available, experimenters have been forced to use circumstantial evidence, *i.e.*, the arrival of recruits at the "correct" feeding stations where the dancing bees foraged, as the primary evidence that the dance information was used. Visual observations of bees just leaving the hive, or arriving at the feeding station, do not reliably indicate the overall flight path. Foragers arriving at a food station, and within the visual field of the observer, are undoubtedly using odor cues at that point. However, this observation does not suggest that these bees use only odor cues from the moment they depart from the hive, especially considering that food resources are sometimes farther than six kilometers away, and downwind from the hive.

(2) Inappropriate assumptions—Several assumptions seriously impair the experimental process:

(a) Information communicated by dances is not valid unless recruit bees arrive precisely at the feeding station where the dancing bee foraged. This assumption is not true. An analogy should suffice. Suppose a travel agent issued an invitation to visit a resort and distributed road maps to 100 people, yet no one arrived at the destination during a given observation time. This does not invalidate the map or the verbal invitation as legitimate means of communication. It merely means that, under the circumstances, other communications or conditions took priority.

(b) Bees must use either the olfactory senses or direction and distance information in the dances. Insect behaviorists know that complex behavior involves multiple stimuli, distributed in time and space, which activate multiple and complementary mechanisms. The honey bee communication system is complex, and includes many variables, *e.g.*, (1) the genotypic variation in individual bees as this affects their nervous system, (2) the dynamic changes in weather and plant conditions, and (3) the unpredictable levels of competition for limited forage resources by other insects and animals. The great success of honey bees in survival and competition obviously resides in the evolution of a flexible communication system that permits quick response to dynamic changes in the quality and distribution of foraging resources. No single mechanism or means of communication will operate as effectively as multiple complementary mechanisms. Even if as few as 1% of the recruits use the direction and distance information in a particular experiment, this result may validate the communication mechanism, providing that the experiment is properly controlled.

(c) Bees are "lower" animals and could not possibly have a true language, comparable to man. This is assumed by some people who equate the use of a language with intelligence and cognition. Does anyone dispute the existence of computer languages? There is no logical reason why bees cannot produce, receive, and act upon information within a given time frame in an adaptive manner without having the slightest degree of intelligence!

(d) The dance movements are some kind of artifact, without adaptive value to honey bees. Some supporters of the "olfactory hypothesis" of communication are content to ignore the dances completely, as if such a complex behavior, with so many positive correlates with forage resources, is a spurious event with no adaptive value. Actually, if one seeks to prove that either the "olfactory hypothesis" or the "dance language hypothesis" is true, then they must prove that the "opposing" hypothesis is false.

The best hope for resolution of the dance language controversy is further experimentation utilizing new technology. The controversy has been a positive experience in that it has stimulated extensive research on honey bee communication. Despite more than 50 years of intensive research a complete understanding of bee communication awaits even further research. Yet bees continue to communicate in the same way, oblivious to our arguments and controversies, which seem rather petty at times in relation to the "miraculous" behavior we witness in the honey bee. In any event, there does appear to be a consensus among scientists that bees do indeed utilize a complex and efficient communication system concerning foraging activities. And the bees could not "care" less about man's concern that this system does or does not fit the definition of a true language!

ORIENTATION AND NAVIGATION

Orientation during flight involves the use of landmarks, the position of the sun, and polarized light (Dyer & Gould, 1983; von Frisch & Lindauer, 1954). Dyer & Gould (1981) found that experienced bees are able to forage and communicate on cloudy days by determining the solar azimuth through a memory of the sun's course with respect to local landmarks. Gould (1986) reported that bees use their "knowledge" of the spatial relations between landmarks to navigate along novel routes, much the same as vertebrates.

Perception of objects in terms of estimating distance is accomplished by image speed on the retina, a phenomenon demonstrated first in the invertebrate world on honey bees (Kirchner, 1989). Furthermore, bees can calculate the sun's position by a biological clock (Gould, 1980), which is especially adaptive when foraging on heavily overcast days (Goodwin & Lewis, 1987).

If bees have to fly around an obstacle en route to the source of food, *e.g.*, buildings or high mountains, they will indicate by their dance the direction representing the most direct line between the hive and the feeding place, a path which they have never flown! Bees can perceive the position of the sun even when the sky is completely covered with clouds. This ability is due to the extreme sensitivity of a bee's eye to ultraviolet light, which penetrates through the clouds directly in front of the sun (von Frisch, 1958).

In the region of the equator, when the sun is in the zenith position (directly overhead), the bees stop dancing, and foraging flights cease at that time. If the bees are retained at the feeding place and allowed to return at noon, their dances are disoriented and remain so as long as the sun is closer than $2.5°$ from the zenith (Lindauer, 1957). New *et al.* (1961) calculated that when the distance of the sun from the zenith was less than $5°$, the frequency of the arrival of recruited bees was about half that at larger angles from the zenith. However, some recruited bees were still arriving, even when the zenith distance was less than $1°$. The observers concluded that either the perception of the relative position of the sun by some bees is extraordinarily acute, or that some method of communication is involved which is independent of the position of the sun.

While studying bees in swarms, Lindauer (1954a) found that bees "searching" for a new nest site danced even during the night. What is astonishing is that those bees correctly indicated the position of the sun during any time of night without seeing it. Experiments of Kalmus (1957) and Lindauer (1959) showed that orientation to the sun is innate, but that "calculation" of the path of the sun must be learned. It took 42 days for the bees from the southern hemisphere to learn to compensate correctly for the northern "reverse" movement of the sun (Lindauer, 1959).

When bees are about one week old, they take brief orientation flights in front of the hive and in the nearby vicinity. By this means they learn the appearance of the hive and surrounding landmarks. Successive flights are increased in duration and scope. Numerous young bees of a given colony commonly take their first flight at the same time. This activity has been referred to as "play flight." Such flights are especially noticeable on warm, sunny afternoons, especially after several days of inclement weather. The intensity of flight activity in the air in front of hive entrances is sometimes mistaken for robbing activity. However, closer scrutiny reveals no fighting.

Moreover, the bees do not fly away but join the group that hovers before the hive. These so-called play flights are of short duration, seldom lasting more than a few minutes. They end abruptly and activity at the front of the hive returns to a normal tempo. Each day, young workers, started approximately four weeks earlier as eggs, make their first orientation flight.

Some bees may transfer to other colonies, especially during a strong wind. Such transfer is termed "drifting." Rauschmeyer (1928) studied drifting of bees in German bee houses and found that, when there are no orientation landmarks, bees drift in extraordinarily large numbers (up to 50%). If there are enough orientation marks, few bees drift. Painting hives with different colors, particularly close to the entrance, is helpful. But when all hives are the same color, the tendency to drift is greater. According to Free (1959a), drifting occurs during the first and, to a lesser extent, the second week of life. Drifting is especially noticeable when bees have been confined for long periods or when colonies are moved to a new site. The amount of drifting varies considerably depending on the arrangement of hives. Straight rows of hives increase drifting. Orienting hive entrances in different directions considerably reduces drifting (Jay, 1969). Bees were more likely to drift from a weak to a strong colony, or from a queenless to a queenright colony (Free & Spencer-Booth, 1961), than vice versa.

OTHER DANCES

Bees "perform" dances other than those that are associated with foraging activities. Most of these dances have been described carefully but are poorly understood, even though their names indicate an association with various functions. Haydak (1929, 1945) described the following dances: "cleaning" dance, "joy" dance, and "massage" dance.

"CLEANING" DANCES

Particles of dust, hairs, or other foreign materials on the worker bee body may stimulate the "cleaning" dance, consisting of a rapid stamping of the legs and a rhythmic swinging of the body to the sides. At the same time the bee rapidly raises and lowers the body and cleans around the bases of the wings with the middle pair of legs. Such a "shaking" dance may be observed any time during the year, even during winter. Usually the bee nearest the dancer touches the latter with its antennae and begins to clean the dancer. With the mandibles spread widely the "cleaner" touches the thorax of the dancer just under the base of the wings. As soon as the dancer feels the touch of the cleaner, it stops dancing, slowly spreads out the wings of one side, bends the abdomen and curves the body to the side and somewhat upward as if accommodating the cleaner. The latter works energetically; its antennae are held close to the mandibles. With shear-like motions of the mandibles it cleans around the base of the wings. From time to time it stops, standing on the last two pairs of legs, the front pair being held in the air, and works with the mandibles as if chewing something it found while cleaning, holding the antennae close to the tips of the mandibles. Then the cleaner continues "clipping" with the mandibles over the scutum from the rear to the front, sometimes over the head, and in the grooves of the thorax; sometimes she climbs on the dancer, crawls to the other side and cleans under the opposite pair of wings and then ceases activity. The dancer may then clean her tongue, antennae, and body in general. She may continue to dance, and either the same cleaner or another bee starts the process of cleaning all over again.

Sometimes there are several "general cleaner bees" present at one time on a single comb. They clean various bees in succession, even if there is no dancing. Beecken (1934) also described the activity of general cleaners. Milum (1947, 1956) reported that such a bee "groomed" 26 bees in 25 minutes. Even drones may be groomed if they happen to be encountered. Milum observed that this is the activity of bees of non-foraging age.

"DVAV" ("JOY") DANCE

Haydak (1929, 1945) originally gave the name "joy" dance to this activity because he observed it only when the conditions in the hive were "optimum." The DVAV (Dorso-Ventral-Abdominal-Vibration) terminology proposed by Milum (1955), is less anthropomorphic, and is suggested as a more appropriate name. In this dance the front legs are placed on some part of the body of another bee and then five or six shaking movements are made up and down with the abdomen, with a simultaneous slight swinging forward and backward. Then it walks farther and performs such a dance while touching a series of bees in various locations. Sometimes one can see such a dance on a sealed queen cell. When a virgin queen had just emerged from a cell in a queenless colony, a large number of bees started "performing" such a dance.

This dance is "performed" by bees of predominantly field age, or at least old enough to perform field duties (Milum, 1955), ranging in age from nine to 151 days. It occurs at all hours of the day and night, during all seasons, with or without flight in progress, in queenless colonies and even in colonies near starvation. DVAV dances may be intermixed with wag-tail dances and they may be "performed" on the queen or queen cells, before swarming, or on virgin queens before they fly out. Bees that are shaken remain very still while being shaken. Allen (1959a,b) found that the age of the participating bees varied widely; most were older bees.

Fletcher (1975) concluded that DVAV dances inhibited certain activities of workers and queens. Gahl (1975) found that approximately 96% of DVAV dances were performed on bees that were older than the shaker and that the mean number of vibrations per shaking dance was 19.7 during a mean duration of 1.2 seconds. Schneider *et al.* (1986a,b) gathered evidence indicating that DVAV dances exerted a short- and long-term "priming" effect on foraging behavior. Schneider (1987) also reported that DVAV dances appeared to increase the number of different tasks, and the time spent on these tasks, by bees of non-foraging age.

Perhaps there is still much that is not known about DVAV dances.

"MASSAGE" DANCE

This dance begins when one of the bees on the comb bends its head in a particular way. One or more neighboring bees become excited and immediately begin to "investigate" her, using their antennae and front legs. They climb over and under her, pull the joints of the hind and middle legs, but mostly touch her sides from below with their antennae, mandibles, and front legs, cleaning their antennae periodically. The bee which aroused so much attention has the mandibles wide open and the upper part of the tongue protruded as with the nurse bees, but the protruded part is entirely dry. The bee turns her head toward the "examining" bee as soon as the latter approaches the front part of her body. Then she unfolds the entire tongue, extends the second pair of legs as if sitting on the third pair and constantly cleans the tongue with her front legs, stroking it downward.

"ALARM" DANCES

Schneider (1949) observed that, when foragers returned to the hive carrying a sugar solution contaminated with dinitrocresol, the bees became excited a few minutes after their arrival. Many field and hive bees started to "perform" an "alarm" dance. The bees ran in spirals or irregular zig-zags and vigorously shook their abdomens sideways. The flight activity of the bees stopped completely and neighboring bees began to respond to the dancers. As the poison spread, many more bees began to dance. The highest mortality occurred one to two hours later. In another two to three hours the colony returned to normal and flight resumed.

Schick (1953) claims that the dances described by Schneider (1949) are the "Zittertanze" observed by von Frisch (1923). These dances can be aroused by different substances and are the result of various disturbances to bees. Schick's conclusions were supported by Istomina-Tsvetkova (1957) who observed the "alarm" dances of Schneider (1949) performed by ordinary nectar gatherers.

Comb Building

Combs are constructed from beeswax that is secreted by wax glands possessed by workers only. Wax glands are best developed and most productive in bees 12 to 18 days old. Wax is secreted in the form of small, irregularly shaped oval flakes, or scales (Fig. 7), which project from between the over-lapped portions of the last four visible abdominal segments on the underside of the bee. Two scales are produced on each of these segments, one on either side of the midventral line, making eight in all. Wax is secreted primarily during warm weather when foraging is active. During wax secretion bees consume large amounts of honey or nectar. Studies by Whitcomb (1946) indicated that, in building comb from foundation, bees consumed approximately 3.8 kilograms of honey during the production of 453 grams of wax.

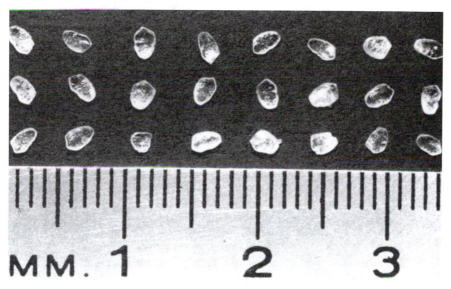

FIGURE 7. Tiny scales of beeswax taken directly from the wax glands. The smaller ruler divisions are millimeters. (Photo by Kenneth Lorenzen).

Experiments described by Taranov (1959) proved that protein food (pollen or "bee bread") is also of great importance during wax production. When young bees were fed sugar solution only, they lost up to 20% of their body protein in 15 days of intensive wax production. In his experiments a direct relationship was found between the quantity of wax obtained from a colony and the amount of pollen brought into the hive. Experiments by Freudenstein (1961) support these findings.

Workers actively engaged in secreting wax engorge themselves with honey and hang in festoons at or near the site of building operations. After about 24 hours they begin to build comb.

The following description of wax scale manipulation and comb building is based largely upon the excellent studies of Casteel (1912a). Wax scales are removed from their "pockets" (Fig. 8) by means of the spines on one or the

FIGURE 8. Ventral view of a worker bee removing a wax scale from the "pocket" at the wax gland. The bee is supported by the two middle legs and the right hind leg while the left hind leg removes the scale. (From Casteel, 1912a).

FIGURE 9. Ventral view showing the position of the wax scale just before it is grasped by the forelegs and the mandibles. The scale is still adhering to the spines of the pollen combs. (From Casteel, 1912a).

FIGURE 10. Side view of worker bee in the same position as that shown in Figure 9. (From Casteel, 1912a).

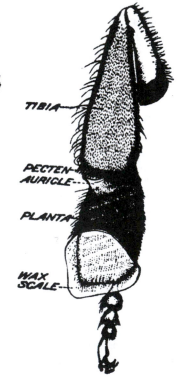

FIGURE 11. Inner surface of the left hind leg of a worker bee, showing the position of a wax scale immediately after it has been removed from the wax pocket. The scale has been pierced by seven of the spines of the pollen combs of the first tarsal segment (basitarsus or planta). (From Casteel, 1912a).

other of the hind tarsi (Fig. 11). While the scale is thus held, the hind leg bearing it is flexed toward the mouth, where the scale may be grasped conveniently by the forelegs or by the mandibles. (Figs. 9, 10). Bees do not follow any definite sequence in the order in which they remove scales from the wax pockets. Usually the forelegs are used in transferring the scale to the mandibles and in manipulating the scale during mastication by mandibles, just before the wax is affixed to the comb. Secretion of mandibular glands is used during mastication of the scale and building combs (Örösi-Pál, 1957). The freshly masticated wax is spongy and flaky when first deposited, but later it becomes smoother and more compact as it is manipulated into comb. The whole process of removing, masticating, and affixing one scale to the comb requires about four minutes.

In building combs, bees often appear to "work" in opposition to one another (Lindauer, 1953). One bee sticks a little ball of wax on the comb surface, molds it carefully, then a moment later another bee chews it off and reattaches it only a millimeter or so farther away. This is particularly noticeable when cells are being capped. If the amount of building wax is not quite sufficient, bees will collect some wax from neighboring cells. Hundreds of bees participate in the construction of a single cell; an individual worker may be active as little as half a minute (Meyer & Ulrich, 1952).

Gontarski (1949) ascertained that bees building combs can adjust themselves in the direction of gravity. Queen cells are also built under the influence of gravity. He stated that bees can perceive a slight deflection of their bodies from the vertical plane and can compensate for it. This has a great biological significance. Discovery of the organs of gravity by Lindauer & Nedel (1959) supports the assumption of Gontarski. Darchen (1959) demonstrated that the chains which bees form in their building cluster play an important part in regulating parallelism of the combs.

Brood Feeding

Young bees begin feeding brood when they are about three to 13 days old. Although workers beyond this age are still capable of feeding larvae to a limited extent, most of them begin other activities at about this age. The following description of the activities of nurse bees is based upon the observations of Nelson *et al.* (1924) and Lindauer (1953):

Nurse bees begin to visit cells as soon as eggs are laid and continue at frequent intervals throughout the duration of the egg and larval stages. Some of these visits are very short (at most about two to three seconds), but in others the larvae and the brood are examined longer with the antennae, the bee leaving the cell only after 10 or even 20 seconds. Each actual feeding is preceded by such an "examination." The time taken for one feeding (including inspection) is variable. Usually it is from one-half to two minutes; in exceptional cases even three minutes.

During the first two days after hatching, nurse bees continuously supply the tiny larvae with far more food than can be consumed, so that the larvae appear to float in the milky-white food. During the third day a larva in a worker cell continues to receive food, but consumes it as fast as it is deposited by the nurse bees.

During the eight-day period from the laying of the egg until the full-grown larva is enclosed in the capped cell, Lineburg (Nelson *et al.*, 1924) found that nurse bees visit individual larvae approximately 1,300 times daily, or more than 10,000 times throughout larval development. On the last day before the cells is capped, they visit it nearly 3,000 times, spending a total of approximately four-and-three-quarter hours within the cell. Lindauer (1953) found that the time that 2,785 nurse bees spent in rearing one larva from the time the egg was laid to capping of the cell was 10 hours, 16 minutes, and eight seconds!

Food Transmission

Food transmission, sometimes termed trophallaxis, is common between workers (Fig. 12) and also from worker to queen and drones (Fig. 13).

FIGURE 12. Transfer of food from worker (left) to worker (right, with the extended proboscis). Antennal tapping is active during the transfer. (Photo by Kenneth Lorenzen). (Shown as drawing in Figure 27,A).

Figure 13. Transfer of food from worker (left) to drone (right, with extended proboscis). Drones also can feed themselves from open cells of honey. (Photo by Kenneth Lorenzen).

Observations of individual bees showed that reciprocal feeding continues throughout the life of bees and that the frequency of food transmission appears to be related to age. This kind of behavior is difficult to study because exchange of food involves the elements of frequency as well as the actual amount of food exchanged at each feeding. Workers may appear to be actively exchanging food but the quantity may be low, and difficult to measure in

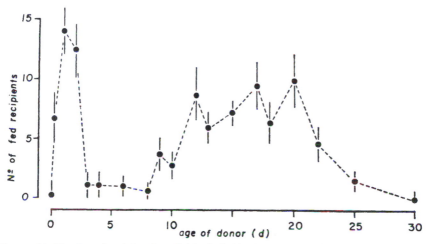

Figure 14. Number of recipient bees fed by individual donor bees of different ages. Vertical bars indicate standard error. (From Moritz & Hallmen, 1986).

context of a normal colony environment. Probably the majority of apparent trophallactic contacts have a communicative function (Korst & Velthuis, 1982). In a laboratory study by Moritz & Hallmen (1986) it was found that the frequency of food transmission by donor bees increases rapidly just after emergence, declines sharply between three to eight days of age, becomes very active again until bees are approximately 21 days old, and then declines to low levels for the remainder of their lives (Fig. 14).

Most feeding episodes last one to five seconds, some six to 10 seconds, and only a few are 20 seconds or more (Istomina-Tsvetkova, 1953). Transfer of food between two bees is preceded by either "begging" or "offering" behavior. Both types of behavior are directed more toward the head than to any other part of the bee's body (Free, 1959b). Antennal contact is important. During feeding, the antennae of both bees are in constant motion, continuously striking each other. This apparently facilitates orientation and communication. The odor of the bee's head is also important (Free, 1956). Individual bees vary greatly in their behavior.

Food transmission is a form of communication concerning the availability of food and water. In addition to functioning as a distribution system for nutrients in the colony, the behavior also serves as a medium for transmitting pheromones and possibly other substances important for the life and cohesion of the colony.

Colony Defense Activities

Honey bees have many "enemies," most of which are attracted to the hive entrance by the odors and activities of bees. Butler & Free (1952) observed that very few guard bees are present at the entrance during the honeyflow, if the colony is not disturbed. Any forager from another colony that is loaded with nectar or pollen may enter without being "examined" or "attacked." But when the colony is disturbed and thereby alerted, strange foragers entering the hive are likely to be intercepted and "examined," although they can still enter the hive. When very little nectar is available, guard bees are present continuously at the entrance, and they "examine" all bees coming to the hive. Robber bees are intercepted and quite often stung to death. Interestingly, when bees sting other bees or insects, they rarely lose their stings, and therefore rarely die. This is true because penetration of the sting is into the delicate intersegmental membranes which do not retain the barbed sting. Consequently, the sting is easily retracted without harming the stinging bee.

Other intruding insect species and animals are quickly "attacked" and usually stung to death. Some animals, such as bears and skunks, seem to tolerate bee stings. Others learn how to avoid the sting to some extent. In urban environments it is not uncommon to observe dogs routinely catching and eating large numbers of bees as they forage on flowers.

Guard bees "patrol" the alighting board and entrance area. When they are disturbed by an intruder, they assume a very typical posture (Fig. 15). They stand on their middle and hind legs with their forelegs lifted, their antennae held forward, and their mandibles closed. When there is greater excitement, they open their mandibles and spread their wings as if ready to attack. Buzzing "alarm sounds" are made, apparently as a means of communicating alarm. Guard bees actively examine bees entering the hive. Such "examinations" typically last from one to three seconds. The guard bee approaches and touches the incoming bee's body with her antennae and apparently recognizes the examined bee primarily by the sense of smell (Kalmus & Ribbands, 1952).

When a bee enters a strange colony and is approached by a guard, she either continues on her way, not responding to the guard "examination," or she stops and submits to a detailed "examination" or mauling. Young bees without pollen loads submit more readily to examination than older foragers carrying loads of nectar or pollen. Older bees usually assume a dominant

FIGURE 15. Two worker guard bees are shown. The one in the foreground is in a typical defensive stance with the forelegs uplifted and antennae outstretched. (Photo by Kenneth Lorenzen).

behavior toward the guards and quickly enter the hive without stopping, even if the guards run after them. If the bee submits to "examination" she behaves quite similarly to a bee being "massaged" (see "Massage" Dance in this chapter). However, in this case she "offers" a drop of fluid, presumably from the honey stomach, to the "massaging" guards.

Stinging is often considered to be an aggressive behavior. In reality it simply is a defensive behavior. Bees do not get "angry" or "seek revenge" because of human disturbance to colonies. They merely react instinctively in a predictable behavioral pattern to specific stimuli associated with the intruder. The stinging behavior of Africanized bees is quite different from European bees that have been traditionally kept in the United States, and is discussed elsewhere in this chapter (see "Africanized Bees"). For European bees, virtually all defensive behavior occurs in the immediate vicinity of the hive. Stinging away from the nest is an unusual event, an accident, in which man somehow physically contacts the bee, *e.g.*, colliding with a flying bee while riding a bicycle, or stepping on a bee in the lawn while barefoot. Bees that are actively foraging on flowers are not a sting threat under normal circumstances.

Less than a half percent of the bees in the hive are likely to sting, even when the colony is greatly stimulated by careless manipulation. The more defensively inclined bees seem to be located primarily near the entrance and can be identified by the typical guard bee position. Guard bees are not sensitive to man-made sounds, but they are extremely sensitive to vibrations, odors, and visual stimuli, such as movement and colors. A very powerful stimulus is exhaled breath, which contains chemicals, especially carbon dioxide, that elicit stinging. Hair and leather odors also elicit the stinging response. Combinations of these stimuli are especially effective in triggering defensive behavior. Once stimulated, the defensive bees fly near the entrance where they are further stimulated by movements and odors of the intruder. Darker colors are more likely to stimulate stinging behavior. After bees fly toward a moving object, such as one's arm, they grasp the object with their tarsi, and deposit the sting.

When a sting is deposited (Fig. 16), alarm pheromones are suddenly released from the structure. These odors linger at the sting site after the bee has departed, thus tagging the victim and exciting further stinging responses by other bees in the immediate area. Eventually the pheromones dissipate or can be removed by washing with water.

The most effective defense against stinging, once a colony is disturbed, is hasty departure from the defensive zone near the hive, providing one leaves at a speed that exceeds the flight speed of bees! Fortunately, such evasive action is rarely necessary if stinging behavior is controlled by the beekeeper before the hive is examined. Control is attained by puffing sufficient smoke into the hive

Figure 16. This worker bee has just deposited its sting. The barbs of the sting catch in the flesh and the venom sac separates from the bee's body as she pulls away. Although painful to the recipient, loss of the sting mechanism is always fatal to the bee. (Photo by Kenneth Lorenzen).

entrance at least one minute **before** the hive is opened and permitting it to circulate inside the hive. This time delay, prior to opening the colony, permits the bees to engorge with nectar or honey. The result is temporary disruption of defensive behavior in the colony, during which time the beekeeper may safely conduct the hive examination.

Very young bees less than a day old cannot sting. Their stings apparently are too soft and flexible to penetrate the flesh. Drones certainly cannot sting because they do not have a sting structure. Queens very rarely sting humans. I have been stung by two virgin queens, but only after holding other virgin queens first, thus transferring their pheromones to my fingers and making it likely that the stinging queens perceived my finger as a virgin queen, rather than as a finger!

Stinging behavior is strongly affected by genetic components and this has been studied extensively in recent years, *e.g.*, Breed *et al.* (1990), Moritz *et al.* (1987). It is common practice for beekeepers to replace the queens of colonies that become too defensive. After several weeks the old bees gradually are replaced by a new population that is less defensive and easier to manipulate. Stinging behavior is also affected by meteorological (Southwick & Moritz,

1987) and other environmental factors. Some of the circumstances that increase the intensity of stinging behavior are:

1) An increasing colony population.
2) The colony is manipulated too early or too late in the day, or at night.
3) Flight activity at the entrance is reduced owing to poor or changing foraging circumstances and/or inclement weather.
4) Smoke used for calming the bees is of poor quality, *i.e.*, not dense, white, and cool.
5) Bees and combs are crushed by careless handling of the equipment.
6) Colonies are subjected to frequent disturbance by animals or other pests, such as skunks or ants.
7) Visual elements that are enhanced by quick movements and wearing dark colors (these elements are not significant when they relate to foraging bees away from the hive vicinity).
8) Alarm pheromone is released from a prior sting.
9) An increase in the amount of comb with empty cells, thereby releasing more comb volatiles (Collins & Rinderer, 1985).

Colony defense is one of the most significant kinds of activity, not only because bees are able to protect themselves very effectively from many other animals, but because stinging behavior is one of the greatest deterrents to keeping bees. A thorough understanding of defensive behavior can greatly reduce the sting hazard and make beekeeping more pleasurable.

Robbing Behavior

Robbing is a form of foraging behavior in which bees from one hive collect the nectar and honey stores from another hive. Robbing is much more likely during dearth conditions than during a honeyflow.

Robbing activity is favored by keeping many hives close together in apiaries. When hives are opened, honey is exposed, and colony defenses are temporarily disorganized by the disturbance. Exposed honey is even more stimulating to scout bees than rich nectar sources. Thus, a large number of foragers are trained inadvertently by the beekeeper to forage for honey, instead of nectar in flowers. Since honey is a much richer source than nectar, additional foragers are rapidly recruited in a short time. Robbing can be a severe problem, especially when colonies are examined on successive days at about the same time of day, and hives are left open too long.

Robber bees are highly attracted to the odor of honey as it emanates from entrances or other hive openings. As these recruits approach the odor source, they behave similarly to inexperienced foragers approaching flowers for the first time. They tend to follow a weaving, serpentine flight pattern near the ground until they get very close to the food odor source. Then they hover and fly to and fro just prior to alighting. After alighting they are very sensitive to movements in the area and take flight readily. After a few trips they are much less sensitive to disturbances.

Similar behavior may be observed at hive entrances. However, in this context the guard bees "attack" them, thereby stimulating evasive behavior. Guard bees "recognize" them because their flight behavior and body odor are different from returning foragers. Guard bees may seize the robber bee and, in the fighting behavior that ensues, either of the bees may be stung to death.

Some robber bees gain entry into the hive, collect honey, and even establish a somewhat normal pattern of foraging behavior. This is probably facilitated by acquiring some of the hive odor. After they learn the entrance arrangement and appearance, they may enter the hive quickly in much the same movements as normal returning foragers. The extent to which chronic robbing behavior exists in an apiary is very difficult to assess. Such behavior has been reported frequently by beekeepers who describe the activity as "progressive robbing."

Robbing activity can become so very intense that the defenses collapse, all the honey and nectar are taken, and the colony perishes. For this reason, combs that are wet with honey, as a result of extracting honey during the

FIGURE 17. Bees fanning at the hive entrance, circulating air through the hive. Such ventilation cools the hive, facilitates the loss of water from nectar during processing into honey, and exhausts odors from the hive. The fanning bees are oriented in the same direction, toward the hive entrance. (Photo by Kenneth Lorenzen).

harvesting process, should not be placed outdoors for the bees to rob as a means of drying the combs of honey. This practice incites massive robbing, upsets the colonies, and greatly increases the sting risk for people and animals in the area.

The name "robber bee" is an unfortunate term. The connotation of the word suggests that such bees have criminal intent! Even some authors have written that these bees act "nervous" and show "guilt," as if an insect could know right from wrong!

In nature, where colonies are sparsely distributed and not disturbed or opened frequently, robbing behavior may be rare or perhaps non-existent, except when colonies are damaged by predation or die for some reason. Many aspects of robbing behavior are not yet understood, even though several studies have been made (Butler & Free, 1952; Free, 1954; Gary, 1966; Ribbands, 1954, 1955b).

Fanning

The nest environment is actively controlled by fanning behavior. Temperature and humidity control, pheromone distribution, evaporation of water from nectar, and regulating the concentration of respiratory gases are facilitated by fanning. Many studies of fanning have been conducted, *e.g.*, Southwick & Moritz (1987).

During hot weather, bees reduce the temperature within the hive by circulating air inside by fanning (Fig. 17). Tiny droplets of water are distributed throughout the hive interior and moving air causes evaporative cooling. The number of fanning bees varies with the hive condition, ranging from just a few individuals to several hundred bees. They occupy the hive entrance area with their heads oriented toward the rear of the hive and stand just far enough apart that they do not interfere with each other's movements. By fanning vigorously with their wings, they set up outgoing air currents on one side of the entrance. Two groups of fanning bees may be active simultaneously, the second group positioned on the other side of the entrance, and facing in the opposite direction. Thus, their fanning increases the circulation of air, which enters at one side of the entrance, circulates inside the hive, and goes out at the other side.

During the honeyflow season, air currents within the hive hasten the evaporation of excess moisture from nectar and unripe honey in the open cells, an important element in the processing of nectar into honey. Fanning activity is common throughout the summer, especially during late afternoons and early evenings after large quantities of nectar have been collected.

Respiratory gases and other odors produced inside the hive are exhausted by fanning and fresh air is brought into the central nest area. Honey bees have a high metabolic rate that requires sufficient exchange of respiratory gases.

FIGURE 18. A worker bee is scent fanning. Air is fanned over the exposed scent gland (a) and behind the bee. Other bees may align in the same direction behind the first bee, providing a directional air stream that contains stimulating, attractive odors (pheromones) used during orientation to the hive entrance or to a queen during swarming. (Photo by Kenneth Lorenzen).

Another kind of fanning activity, known as "orientation" or "scent" fanning, is shown in Fig. 18. The abdomen is held upward, the last tergite of the abdomen is bent down, and the scent gland (Nassanoff gland) on segment VI is opened. This exposes a membrane that is moist with a secretion that volatilizes rapidly. Chemicals in the pheromone secretion are geraniol, citral, nerol, farnesol, nerolic and geranic acids, and perhaps other unidentified components. The odor of these combined substances is highly attractive to bees in circumstances in which orientation behavior is involved. Scent fanning activity is especially evident in a swarm when bees are orienting to the odor of the queen, particularly when the swarm is entering a new nest for the first time. Scent fanning behavior can be induced at the hive entrance by placing an obstacle in front of the hive entrance for several minutes, then removing it. On alightment some of the bees pause a few seconds at the entrance and fan actively.

Fanning also distributes queen, brood, and alarm pheromones throughout the hive interior. This appears to be an important mechanism for communicating vital information quickly to all members of the colony.

"Rocking" or "Washboard" Behavior

Near the hive entrance, especially on warm evenings, bees "perform" an interesting activity on the front wall and alighting area of the hive. They stand on the second and third pairs of legs and face the entrance. Their heads are bent down and the front legs are also bent. They make "rocking" or "washboard" movements, thrusting their bodies forward and backward. At the same time they scrape the surface of the hive with their mandibles with a rapid shearing movement, sliding over the surface as if cleaning it. The tips of their antennae touch the surface, moving constantly. After a while some material accumulates on the lower edge of the mandibles. Periodically the bees clean their mandibles as well as the tarsi of the front legs. Occasionally they move and continue this activity at another place. These rocking movements probably serve as a cleaning process by which the bees scrape and polish the surfaces of the hive.

Nest Cleaning (House Cleaning) Behavior

Bees that die inside the hive are quickly removed and deposited outside. Approximately 90% of the old bees die outside the hive (Gary, 1960). Those that die inside the hive are dragged outside and dropped on the ground near the entrance. After the bodies begin to dry, and become lighter in weight, undertaker bees airlift them up to several hundred meters from the hive where they are dropped in flight. Such behavior insures that the bodies do not accumulate inside the hive where they could transmit diseases or attract scavengers. Brood that dies in the comb for any reason is also cleaned out (Fig. 19). Sometimes stored combs, or combs in colonies that die during the winter, become covered with mold growth; these combs are cleaned thoroughly when placed in an active colony. Any debris that enters the hive environment is either removed from the colony or, if too large, is encased in propolis.

Several studies of undertaking behavior (*e.g.*, Robinson & Page, 1988; Visscher, 1983) indicate that there is a strong genetic component in the tendency of bees of various genotypes to engage in housecleaning behavior. This is particularly true with respect to the tendency to remove dead larvae and pupae that are killed by diseases, such as American foulbrood (Rothenbuhler, 1964). Efficient housecleaning behavior contributes greatly to the resistance of bees to various diseases.

Working Habits of Field Bees

Bees may begin flying as early as the third or fourth day of life, but normally few foraging trips are made before they are about two weeks old. The choice of collecting nectar, pollen, propolis, or water by young foragers depends on the circumstances existing in the hive at any given time, the genetic makeup of each forager, the relative abundance of floral resources, and perhaps other environmental variables. Observations of Bonnier (1906) and

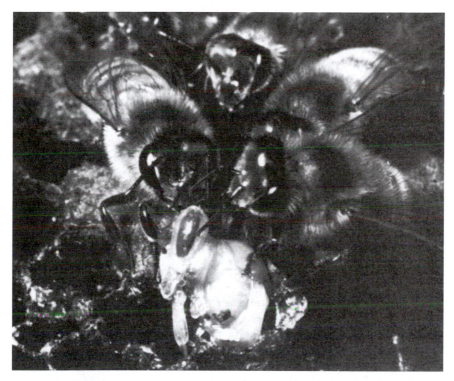

FIGURE 19. These worker bees are removing a dead pupa from its cell. (Photo by Kenneth Lorenzen).

Park (1949) showed that water, nectar, or pollen foragers continued their respective kinds of activity for days together. However, Ribbands (1949) observed that bees frequently changed from pollen to nectar collection, but never vice versa. The period of attachment to any particular crop varied considerably from a few days to 20 to 21 days. Bees that foraged on flowers which yielded nectar and pollen for a limited time each day, remained idle in the hive during the remaining portion of the day when the food was not present. Bees have a keen sense of time. This is of great biological significance because this enables them to remember the time of day when flowers of different species yield pollen, nectar, or both.

FORAGING ON SINGLE PLANT SPECIES

Even though bees in a given colony typically forage on many different species of flowers at any particular time, individual bees show a strong tendency to forage repeatedly on flowers of the same species for as long as they are available. However, mixed loads of pollen from more than one species are not uncommon. Maurizio (1953), during a three year study, found an average of 0.9% to 3% mixed loads. Mixed loads were evident throughout the season and were no more frequent at times of scanty forage than at times when pollen was abundant.

FORAGING AREA

Individual bees typically forage in a relatively small area, *e.g.*, a particular group of trees or even a single tree or bush (Eckert, 1933; Park, 1949; Perepelova, 1959; Singh, 1950). This phenomenon is sometimes referred to as "area fidelity." The size of the foraging area is quite variable, depending upon the number and density of available flowers, their nectar and pollen content, and the amount of competition provided by other foragers or other insects. Ribbands (1949) observed that, when a foraging source becomes unattractive, and is deserted in favor of another source, bees return with decreasing frequency to the original source as a mechanism of monitoring the foraging conditions. If conditions improve sufficiently, bees will resume foraging at the original source. When nectar and pollen production decline in a species, bees are more inclined to move around more within the area, rather than collecting from flowers in one small area.

One of the problems in studying the foraging area of individual bees is that the observer cannot follow the bee throughout the entire foraging trip. Thus, a bee that constantly returns to a small area conceivably could also be foraging at other distant areas. Large homogeneous areas, such as mono-cultured agricultural crops, may have insufficient landmarks for bees to orient to small areas within the field. In one such study, Gary *et al.* (1977a) found that most bees showed a tendency to forage in limited areas of a 3.2 hectare field of onion flowers. However, some bees foraged throughout the field.

Ribbands (1949) concluded that foraging behavior is not fixed but is adaptable. He considers "that the basic and cardinal feature of foraging behavior is the continuous exercise of choice and the comparison of the present with the memory of the past, and that it is the resultant of these factors whether the bee will remain attached to any particular crop and area."

RATE OF FORAGING ACTIVITY

The rate at which bees visit flowers, and the time they spend on each flower, depends on the amount of nectar and pollen present. This varies with the type of flower, stage of development, climatic conditions, and the degree of competition from other foraging insects. Park (1949) found that trips for pollen collection were considerably shorter in time than those for nectar collection. Actual times will vary according to many factors (*e.g.*, weather conditions, distance of the food source from the hive, and the differences in individual foragers).

Temperature greatly influences foraging activity. The minimum temperature for active foraging is approximately 13°C. At extremely high tempera-tures, *e.g.*, above 43°C, nectar and pollen foraging may cease, but water foraging continues.

LEARNING ABILITY

When bees first forage on a source a short time is required for them to learn how to locate and collect nectar or pollen. Each flower is different in structure. Reinhardt (1952), while conducting studies on alfalfa, found that young bees usually began by inserting the tongue into the open corolla. The sexual column is thus released and traps the proboscis of the bee. Bees sometimes had difficulty freeing themselves from such a situation. Soon the bees learned to insert their tongues between the petals of the standard and the keel at the base of the corolla tube, thus becoming "side or base workers." Their efficiency markedly improved. The time required for a bee to become a "side worker" depends on the individual adaptability of the bee in question.

Alfalfa pollen is more difficult to collect because bees normally have to trip a structure in the blossom in order to get pollen. In the process they are hit quite hard by this structure. After foraging experience, such bees become more efficient and quite successfully avoid being hit by this mechanism. Reinhardt (1952) concluded: "The change in habitual approach of bees to flowers, success in avoiding the trap, followed by pattern improvement and changes in working speed, lead to the conclusion that honey bees learn to work alfalfa flowers." The same learning pattern is apparent for bees foraging on hairy vetch. According to Weaver (1957), "the bees differ so widely from one another in their methods and approach to foraging that the individuality of the bees is the most obvious and striking phenomenon." There are efficient and inefficient workers.

Additional evidence of learning is routinely observed at experimental feeders where sugar syrup is offered as an artificial nectar to study foraging behavior. Round trip times between the hive and feeder gradually diminish and reach a stable minimum time. This is evidence that bees are learning behavioral patterns that require less time and energy expenditure, for example, flying the most direct pathway between the hive and feeder, or reducing the hovering time after arriving in the vicinity of the food.

Many excellent studies confirm the learning abilities of bees, especially during foraging activities (*e.g.*, Aschoff, 1986; Gould, 1984, 1985; Gould & Marler, 1984; Wells & Wells, 1983, 1984, 1985; Wells *et al.*, 1981, 1983). Almost all types of classical learning (adaptation, habituation, conditioned reflexes) are found in honey bees. Notably absent is the ability to learn by the intuitive, cognitive, or creative thought process used by humans. In other words, bees apparently have no imagination or foresight. Learning occurs only within the context of innate behavioral patterns. A bee cannot learn to do something completely new, something that no other bee has ever done before, such as constructing honeycomb cells twice as large as normal. In general, honey bees have the same learning potential as other insects and, in some kinds of learning, are comparable to higher animals (Bermant & Gary, 1966).

DISTRIBUTION OF FORAGERS

Beekeepers have always been curious to know where bees from their hives are actually foraging. How far do they fly and in which directions? Partial answers to these questions have been generated slowly by circumstantial evidence for many years.

In one of the first studies, Eckert (1933) observed that bees from hives placed in a desert will fly as much as 13.7 kilometers to a food source, if no other sources are closer to the hive. He also reported that bees have a tendency to fly in only one or two major lines of flight, neglecting similar forage plants in other directions. Levin (1959), working with colonies deprived of old field bees, found that young bees, undertaking foraging activities for the first time, dispersed similarly to the more experienced foragers from undisturbed colonies. Indications of foraging distribution were obtained in many other studies (*e.g.*, Lee, 1961; Levchenko, 1959; Levin, 1961a,b).

Clear and definitive evidence of the distribution of foraging bees in a variety of foraging systems was made possible for the first time by the development of a magnetic recapture system (Gary, 1971). In this system foraging bees are captured on flowers, narcotized with carbon dioxide for a few seconds while a ferrous metal, numbered identification tag is glued to the thorax or abdomen. After release, tagged bees return to their hives where magnets automatically collect the tags at the hive entrances, without harming the bees. Thus, the direction, distance, preference of floral species, and other related information are established for samples of bees that are representative of the populations in each hive.

Large scale distribution studies of foragers from hundreds of hives in many apiaries have shown that factors influencing distribution are more complex than previously realized (*e.g.*, Gary & Witherell, 1977; Gary *et al.*, 1973, 1975, 1976, 1977a,b; 1978a,b). In general, bees were found to have a **strong tendency to forage at the nearest source for each floral species in the area,** particularly if competition for the same species is greater at the more distant areas by virtue of other hives being located closer to those distant areas. There are situations where the average foraging distance may be reduced to several hundred meters. This occurs when multiple, small apiaries are distributed uniformly within the field or orchard (Gary *et al.*, 1973, 1976, 1978b). However, when the only available patch of a species is a great distance away, some bees nevertheless will fly that great distance. This is true even when competition from other hives is great at the distant source, and when nectar and pollen sources of other species closer to the hive would seem to be much more profitable. Thus, a fundamental aspect of foraging behavior is a "species diversity preference," in which bees may fly over and pass profitable food sources in route to other plant species, just because those sources are different, and perhaps more stimulating (Gary *et al.*, 1972). Presumably, nutrients from diverse flora are required by the colony, and this has influenced the evolution of this aspect of foraging behavior. A good example of this "distant flight" behavior is found in agricultural areas where attractive crops are planted in widely dispersed fields. Under these circumstances significant bee populations will fly at least six-and-one-half kilometers from the apiary to a crop such as safflower which produces an abundance of pollen and nectar. These findings indicate that the potential area accessible to bees from a given hive is at least 13,000 hectares! (See Fig. 20). The distribution of foraging bees over such large areas helps to explain their great potential for pollinating crops. Also, exposure to pesticides is potentially great. Severe bee mortality may be evident even when apiaries are located several kilometers from a treated area.

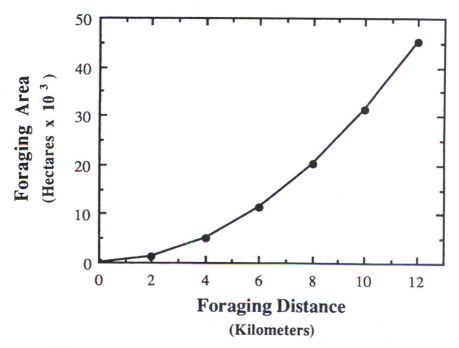

FIGURE 20. Relationship between foraging distance from the hive and foraging area.

In overview, the flight range in each area is quite variable, influenced primarily by the resource distribution, population levels of bees that are competing in the respective areas, and the productivity of the plant. Bees establish foraging territories within variable distances from their hives. For example, if there are multiple apiaries distributed within a very large alfalfa field, the area within several hundred meters from each apiary will be occupied primarily by foragers from the nearest apiary (Gary *et al.*, 1973, 1978a) (Figs. 21, 22). Establishment and dominance of such areas are probably determined primarily by competition, *i.e.*, depletion of nectar and pollen until the source is not sufficiently "profitable" (von Frisch, 1967a) to stimulate the recruitment of foragers to greater distances.

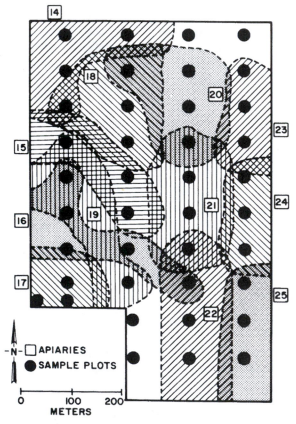

Figure 21. Map of an alfalfa field showing the "foraging territories" of bees from apiaries placed within the field. More than 50% of the bees in each shaded area originated from hives in the apiaries located in each of the shaded areas. (From Gary *et al.*, 1978).

Another potential mechanism in establishing dominance of an area could be active repulsion of "foreign" foragers that enter the area. When the available nectar decreased in a foraging area Weaver (1957) reported: "Often a bee was observed to collide with a near-by forager in a purposeful manner,

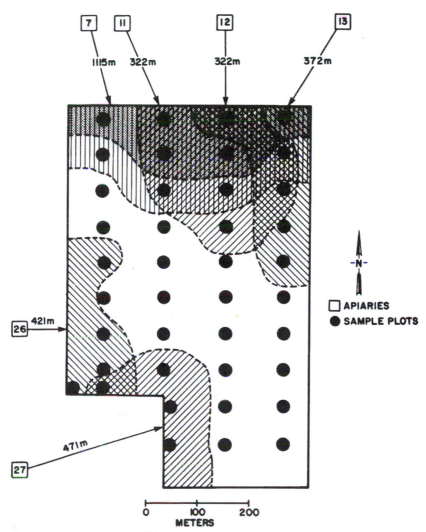

FIGURE 22. Map of the same alfalfa field as in Fig. 21 showing the "foraging territories" of bees from apiaries located various distances from the field edges. Approximately 50% of the bees originating from the distant apiaries were foraging in the respective shaded areas nearest the respective apiaries. (From Gary *et al.*, 1978).

or to fly threateningly toward another bee without actually touching her. Sometimes there is a prolonged struggle. This aggressive behavior of foraging bees is probably an aid in keeping a foraging area free of too much competition from other bees."

It is difficult to relate the absolute percentages of the total forager populations that are dispersing to various distances from their colonies. For example, what percent of the total population is foraging in the areas that are between 500 to 1,000 meters from the hive, compared to 1,000 to 1,500 meters? There may be fewer bees per unit area from a given hive at the greater distance even though a greater percentage of the population may be foraging at that range of distance.

SPEED OF FLIGHT

According to Park (1923a), the speed of flight for loaded bees varied from 20.9 to 25.7 kilometers per hour and averaged approximately 24 kilometers per hour. The speed of empty bees varied from 10.9 to 29.0 kilometers per hour and averaged two to four kilometers per hour less than that for the hive-bound bees. This suggests the possibility that a bee on her outward journey may not in all cases make a so-called "beeline" for the source of supply. The only case in which the outgoing bee made better time than the incoming one was when flying directly against the wind. Bees do not long continue to work in a wind blowing much over 24 kilometers per hour.

Gathering and Storing Pollen

The health of the honey bee colony is as dependent on pollen as it is upon honey. Pollen is virtually the sole source of proteins, fatty substances, minerals, and vitamins that are necessary during the production of larval food and for the development of newly emerged bees. A colony cannot rear brood if it does not have pollen (Haydak, 1937a). Older bees can rear brood without consuming pollen (Haydak, 1935, 1937b) but they do this at the expense of their own bodies and the amount of food produced is rather small. A populous colony typically collects about 35 kilograms of pollen during an entire season.

The manipulation of pollen from the time it is taken from the flower by a bee until it is stored within a cell in the comb was described by Casteel (1912b). From his observations of bees collecting from sweet corn, he stated that the bee alights on a tassel and crawls along the spike, clinging to the pendent anthers. The tongue and mandibles are used in licking and biting the anthers with the result that pollen grains stick to the mouthparts and become thoroughly moistened. Also, a considerable amount of pollen is dislodged from the anthers, and adheres to the hairy legs and body. The branched hairs of the bee are suited to retaining the pollen which is dry and powdery.

FIGURE 23. A flying bee (left) manipulating pollen with the forelegs and middle legs. The forelegs are removing the pollen from the mouthparts and face. The middle leg of the right side is transferring the pollen on its brush to the pollen combs of the left hind planta (basitarsus). A small amount of pollen already has been placed in the baskets. A flying bee (center) shows the position of the middle legs when they touch and pat down the pollen masses. A slight amount of pollen reaches the corbiculae through this movement. A bee in flight (right), showing the position of the hind legs during the basket-loading process. Pollen is being scraped by the pecten spines of the right hind leg from the pollen combs of the left hind planta (basitarsus). (From Casteel, 1912b).

After the bee has crawled over a few flowers, she begins to brush the pollen from her head, body and forward appendages and to transfer it to the posterior pair of legs (Fig. 23, left). This may be accomplished while she is resting on the flower, but more often while she hovers in the air before foraging for additional pollen. The wet pollen is removed from the mouthparts by the forelegs. The dry pollen clinging to the hairs of the head region also is removed by the forelegs, and added to the pollen moistened by the mouth.

The second pair of legs collects free pollen from the thorax, more particularly from the ventral region, and receives pollen collected by the first pair of legs. In taking pollen from the foreleg, the middle leg of the same side is extended forward and is either grasped by the flexed foreleg, or rubbed over it as the foreleg is bent downward and backward. Much sticky pollen is now assembled on the inner faces of the broad tarsal segments of the second pair of legs.

Pollen is transferred to the pollen baskets in at least two ways. A relatively small amount may reach the pollen baskets directly, as the middle legs sometimes are used to pat down the pollen accumulated there (Fig. 23, center). But by far the larger amount is first transferred onto the pollen combs on the inner surfaces of the hind legs. One of the middle legs and then the other alternately is grasped between the first tarsal segments of the hind legs and drawn forward and upward, thus combing the pollen from the middle legs. The pollen now held in the combs of the hind basitarsus is next transferred to the pollen baskets on the outer surfaces of the hind tibiae.

With the two hind legs drawn up beneath the abdomen, the pollen combs of one leg are scraped by the pecten spines of the opposite one as the legs are moved up and down in a sort of pumping action (Fig. 23, right). Thus the pollen removed from one basitarsus is caught on the outside of the pecten comb of the opposite leg, the two combs scraping alternately. The planta is gently bent backward bringing its auricular surface into contact with the outer side of the pecten comb. By this action, the pollen mass is pushed along the slightly sloping lower end of the tibia, and thence out onto the surface of the pollen basket at its lower end (Fig. 24). Each new addition of pollen is pushed against the last and, simultaneously, the masses of pollen on both legs grow upward, a very small amount being added at each stroke.

Finally, each leg is loaded with a mass of pollen, held in place by the long recurved hairs of the elevated margins of the tibiae. If the loads are very large, these hairs are pushed outward and become partly embedded in the pollen, allowing the mass to project beyond the margins of the tibiae. The bee accomplishes these brushing and combing actions so rapidly that the observer probably will fail to see some of the steps in the process without repeated observations.

When the bee is loaded, she returns to the hive. Some walk normally over the combs, while others appear to be greatly agitated, performing the charac-

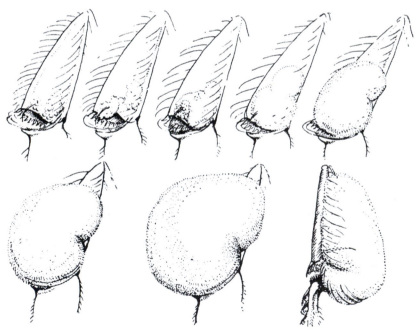

FIGURE 24. Progressive stages of pollen load packing. A single hair pointing downward in the corbicula is gradually pushed into a horizontal position and functions as a pin through the middle of the load. (From Hodges, 1974).

teristic wag-tail dance that communicates to the other field bees the existence of a source of pollen. Many pollen-bearing bees "solicit" nectar or honey from other workers, or take it directly from the cells.

Soon the pollen bearer puts her head into cell after cell as if looking for a suitable place to unload. For no apparent reason a certain cell will be selected, most often situated in the area immediately surrounding the brood, above and to the sides. The bee grasps one edge of the cell with her forelegs and arches her abdomen so that its posterior end rests on the opposite side of the cell. Her hind legs are thrust into the cell and hang freely within it. The middle leg of each side is raised and its basitarsus is brought into contact with the upper end of the tibia of the hind leg. The middle leg now is pushed between the pollen mass and the corbicular surface so that the mass is pried outward and downward and falls into the cell. The hind legs now execute cleansing movements that remove remaining bits of pollen.

After ridding herself of the two pellets, the bee usually leaves the cell. Parker (1926) gave a good description of what happens next. "Shortly afterward another bee, usually a house or younger bee, comes to the cell and examines it and its contents. On finding the loose pellets, it begins working them to the base of the cell by a pushing motion of the head and with the closed mandibles. When the pellets reach the base of the cell, where pollen may already be packed, they are broken up and incorporated into the mass, and the whole is smoothed off with the mandibles and tongue of the bee. During this process the bee often moistens the pellets with its tongue. The resultant mass takes on a more moist appearance and becomes darker." Casteel (1912b) found evidence that nectar and honey are added to the mass of stored pollen. Pollen stored in this way undergoes chemical changes and is called "bee bread."

Foraging trips of pollen gatherers are considerably shorter than those of nectar gatherers (Park, 1922; Ribbands, 1949). The number of flowers visited by pollen gatherers, the time spent in making a load, the number of trips per day, and the weight of pollen loads is variable, depending on the species and condition of the flowers, temperature, wind velocity, relative humidity, and possibly other factors. According to Vansell (1942), to collect a load of pollen, a bee visited 84 flowers of pear trees, and 100 flowers of dandelion. Ribbands (1949) gives only eight to 32 flowers for dandelion. To make a full load of pollen a bee may spend six to 10 minutes (Park, 1922) or as much as 187 minutes (Singh, 1950). The number of trips per day may be six to eight (Park, 1922), or up to 47 (Ribbands, 1949). The average probably is about 10 trips per day (Singh, 1950). Using pollen traps, Hirschfelder (1951) calculated that, during good gathering weather, between 50,000 to 54,000 pollen loads were brought into the hive daily. The weight of pollen loads ranged from 12 milligrams for elm to 29 milligrams for hard maple fresh weight (Park, 1922), or 8.4 to 21.4 milligrams dry weight (Maurizio, 1953). Marked differences were found in the amount and the character of pollen brought to the hive by various colonies of the same apiary (Free, 1959b; Louveaux, 1958, 1959; Maurizio, 1953). When foraging on flowers in which both pollen and nectar are available, a considerable number of foragers take both. Of the total of more than 13,000 bees observed, 25% were gathering pollen only, 58% nectar only, while 17% were collecting both on the same trip (Parker, 1926). Rashad (1957) observed that pollen was collected at temperatures as low as 8° to 11°C; above 35°C pollen collection was reduced. At wind velocities above 17.7 kilometers per hour the activity of pollen foragers slackened, and ceased at 33.8 kilometers per hour. High relative humidity decreased pollen collection.

Bees groom themselves after unloading pollen. In spite of intensive grooming, pollen grains in large quantities remain on the bodies of pollen collectors, *e.g.*, from 10,000 to 25,000 pollen grains per bee. The quantity of pollen carried on the body of the honey bee is larger than that of any other hairy insect (Lukoschus, 1957).

Gathering, Storing and Ripening Nectar

Nectar is a sweet liquid secreted by plant nectaries that typically are located inside flowers. Some species have extra-floral nectaries on the leaves, stems, or other parts. Nectar is the reward "offered" to bees and other insects in return for their indispensable services in cross-pollination. It is composed almost entirely of sugars and water. The proportion of these ingredients varies widely (Beutler, 1953; Free, 1960).

Sight and odor enable bees to locate sources of pollen and nectar. The behavior during flower visitation depends upon whether the forager collects pollen only, nectar only, or both, and differs also with the type and size of flower. If the size of the flower permits, the bee alights within the flower (Fig. 25). But if it is quite small, as in hard maple or sweet clover, she alights upon any convenient part of the plant that will support her weight. Upon alighting, the proboscis is brought forward from its inactive position beneath the "chin" and is inserted into that part of the flower where nectar accumulates. Typically this is at the bottom of the corolla, as in the florets of clover.

Observations on field bees at work suggest that a forager cannot sense the presence of nectar in a given blossom without inserting her proboscis. When flowers are open in structure, and the nectar is exposed, there is a possibility that bees can detect nectar by reflected ultraviolet light (Thorp *et al.*, 1975). Ribbands (1955a) suggested that foragers may avoid flowers visited recently by other bees, perhaps by recognizing the lingering odor of the previous forager. When nectar is found, the bee sucks until all nectar within reach of her

FIGURE 25. A worker bee foraging on a *Gazania* blossom. The pellet of pollen on the left hind leg is visible. Bees frequently collect both pollen and nectar on the same foraging trip. (Photo by Kenneth Lorenzen).

proboscis is taken up. If nectar is absent, the proboscis is withdrawn immediately and she moves quickly to another flower or floret.

Owing to the difficulty of following a bee throughout the entire course of a trip, exact data on the number of flower visits to obtain a load are lacking. Several hundred visits may be necessary to obtain a load of nectar from small flowers such as those of sweet clover (Ribbands, 1949). In *Limnanthes* a bee may have to visit 1,110 to 1,446 flowers to obtain a load. When nectar secretion is very active, large loads can be secured in a relatively short time, thus permitting numerous trips per day for each forager.

Park (1929) determined that foragers collecting nectar from sweet clover spend an average of from 27 to 45 minutes per trip. Approximately four minutes were spent in the hive between field trips. Ribbands (1949), however, observed individual bees working for 106 to 150 minutes to get a load of nectar from *Limnanthes* blossoms. The duration of a trip depends on the species of flowers visited and nectar abundance. The highest number of trips per day he observed for nectar carriers was 24, the average being between seven and 13. An average of 10 trips per day was observed by Heberle (1914) and Lundie (1925).

Very large loads of nectar may weigh as much as 70 milligrams, or 85% of the weight of the bee which, in the case of European bees, was found to be approximately 82 milligrams (Park, 1922, 1925a). Average loads of nectar during a honeyflow weighed about 40 milligrams. Lundie (1925) recorded 25.3 milligrams as an average load of nectar. Obviously there are great variations, depending upon the circumstances.

A forager loaded with nectar enters the hive and mingles with other workers on the comb. If the nectar availability is marginal she may not dance, in which case she walks about until she meets a house bee to which she gives part of her load. Occasionally she gives her entire load to a single house bee, but more often it is distributed among three or more. If the forager has been stimulated by a bountiful nectar source, the loaded nectar gatherer usually dances, as described earlier. At irregular intervals, the dancer pauses and "offers" a sample of the nectar to nearby workers. But soon she meets a house bee to which she gives a considerable portion of her load. As they approach each other, the field bee opens her mandibles wide apart and forces a droplet of nectar out over the upper surface of the proximal portion of her proboscis, the distal portion being folded back under the "chin." Assuming that the house bee approached is not already loaded to capacity, she stretches out her proboscis to full length and sips the nectar from between the mandibles of the forager (Fig. 27, A). While the nectar is being transferred in this manner, the antennae of both bees are in continual motion, and those of one bee are constantly striking those of the other. At the same time, the house bee may be seen to stroke the "cheeks" of the field bee with her forefeet. This may further stimulate unloading behavior.

Upon disposing of her load, a nectar forager sometimes leaves for the field immediately, but usually she pauses long enough to secure a small amount of honey. In any case her departure is immediately preceded by certain characteristic maneuvers. She first gives her proboscis a swipe between her forefeet, then rubs her eyes, and often cleans her antennae, then starts for the field. The entire process of disposing of her load often is accomplished in less time than it takes to describe it (Park, 1949).

STORING AND RIPENING

During the conversion of nectar into honey, two distinct processes are involved. One brings about a chemical change in the sugar and the other causes physical changes whereby surplus water is eliminated. Once the honey is "ripe," *i.e.*, reduced in water content to approximately 19% or less, it is sealed in cells with beeswax caps (Fig. 26).

The sugar content of nectars is highly variable. Sucrose (ordinary table sugar) is present in almost all nectars. Owing to the action of the enzyme invertase, sucrose is changed into two simple sugars, glucose (dextrose) and fructose (levulose). According to Park (1925b), when the house bee has

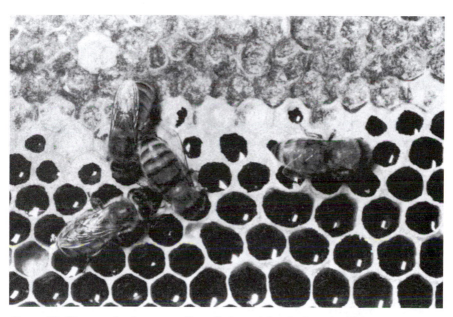

FIGURE 26. These worker bees are sealing cells (containing ripe honey) with beeswax caps. Air trapped under the caps of freshly capped cells gives the caps a whiter appearance that gradually diminishes in time, as shown in the upper, darker capped cells. The open cells below contain honey that is nearly ripe, except for a cell (lower left) which contains pollen. (Photo by Kenneth Lorenzen).

received the field bee's load, she moves about the hive to an uncrowded area. Here she usually takes up the characteristic position shown in Fig. 27,B, having the long axis of her body in a perpendicular position with head uppermost. She at once begins to go through a series of operations which are illustrated diagrammatically at right in Fig. 27. Starting with the mouthparts at rest, as shown in the step 1 diagram, the mandibles are opened wide and the whole proboscis is moved somewhat forward and downward. At the same time the distal portion of the proboscis is swung outward a little and a small droplet of nectar appears in the preoral cavity, as shown in step 2. The whole proboscis is then raised and retracted almost to the position of rest, but is depressed again, and is again raised as before, and so on. With each succeeding depression, the distal portion of the proboscis swings outward a little farther than before, but it makes only the beginning of a return to its position of rest.

Accompanying the second depression of the proboscis an increased amount of nectar appears in the preoral cavity, some of which begins to flow out over the upper surface of the proboscis. As the proboscis is raised and retracted the second time, the beginning of a droplet of nectar usually may be seen in the angle formed by its two major portions, as shown in step 3. This droplet increases in size each time the proboscis is alternately depressed and raised until a maximum droplet is produced, as illustrated in step 5. The bee then draws the entire droplet inside her body. As the nectar begins to be drawn in, the drop assumes a concave surface at its lower end, as shown at "a" in step 6. This distal portion of the proboscis is extended as at "b" until the droplet has

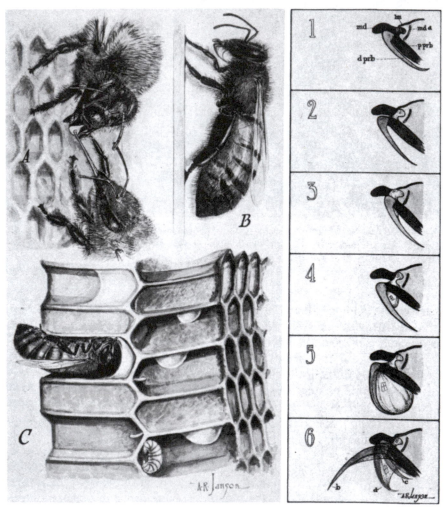

FIGURE 27. At left, drawings depicting three steps in the transfer of nectar from a field bee to a house bee, until the honey is stored in a cell. (A) Nectar being transferred from a loaded nectar carrier (upper) to a house bee (lower). (B) House bee ripening nectar. (C) House bee depositing nectar or unripened honey. At right, diagrammatic sketches of the mouthparts of a bee during the transformation of nectar into honey. (Drawings by A.R. Janson under the direction of O.W. Park).

disappeared, when it is again folded back to the position of rest indicated at "c."

A house bee commonly spends from five to 10 seconds in carrying out the series of activities illustrated at right in Fig. 27. This procedure is repeated with only brief pauses for about 20 minutes, although both of these intervals are subject to considerable variation. Upon the completion of this part of the ripening process, the bee locates a cell and deposits the droplet. Into this cell she crawls, ventral side uppermost, as shown in Fig. 27,C. This position is characteristic of a bee depositing unripe honey. If the cell is empty, she enters until her mandibles touch the upper rear angle of the cell. The nectar is forced out over the dorsal surface of the folded proboscis, between the mandibles, which are held well apart. Then, using the mouthparts as a brush, and turning her head from side to side, she "paints" the unripe honey across the upper wall of the cell so that it runs down and occupies the rear portion of the cell. But if the cell already contains honey, she dips her mandibles into it and adds her drop directly without the "painting" process.

When nectar is being collected in great quantities, and particularly if it is very high in moisture, house bees do not always immediately put it through the ripening process, but quickly deposit it in cells. Instead of depositing the entire load in a single cell, the house bee often distributes it by attaching a small drop to the roof of each of several cells as shown in three cells in Fig. 27,C. The hanging drop exposes a maximum surface which facilitates evaporation. Nectar, in the initial phase of the drying process, is found in greater abundance in cells within the brood nest, where the air is especially warm and dry. And droplets are as apt to be found in cells that contain eggs or young larvae (Fig. 27,C) as in empty ones. Later these droplets are collected and it is assumed that they are then put through the process of ripening by manipulation. The procedures described above result in the rapid evaporation of water from the freshly gathered nectar.

The other phase of the nectar-ripening process, the inversion of sugar, may be expedited by this process also. Although it was found that the inversion process begins while the nectar is being gathered and carried to the hive, additional invertase may be added by house bees while manipulating the nectar prior to depositing it in the comb.

Since the change from nectar to honey takes place gradually over a period of several days, Park (1949) suggested that the term nectar be restricted to the sugary liquid secreted by nectaries up to the time this product is deposited in the comb. Thereafter it may be referred to as unripe honey until its concentration approximates that of ripe honey.

It has been determined that the rate of ripening is greatly accelerated if there are sufficient empty combs for bees to spread the droplets about in many cells. It therefore is important for the beekeeper to place an abundance of empty combs on hives during periods of honey production. An added advantage is that the presence of empty comb also stimulates greater foraging activity for nectar (Rinderer, 1982; Rinderer & Hagstad, 1984).

When adequate comb space is available, few cells are more than half-filled with unripe honey at the close of a day of intense nectar collection, and many contain less. If such combs are held horizontally and shaken, unripe honey drips out freely. Examination early the next morning reveals important changes. The widely scattered cells that contained small amounts of unripe honey the preceding evening now are empty, while comb areas that were not quite full are now completely filled, and cells in adjacent areas are more full than they were. Scarcely a drop can be shaken from any comb.

For a time there was speculation that foraging bees eliminated some of the moisture from nectar during the foraging process and while transporting nectar to the hive. However, Park (1932) determined that this was not the case.

It is obvious that the rate at which water can be eliminated from fresh nectar and unripe honey is greatly influenced by a number of factors such as honeyflow conditions, colony population, amount and concentration of nectar brought in per unit of time, extent of available storage cells, temperature, humidity, and ventilation. In areas of high humidity the ripening process is slowed considerably.

Gathering and Storing Water

Water is actively collected (Fig. 28). It is required by nurse bees whenever it becomes necessary to thin honey in the processing of larval food. When fresh nectar is available this need is diminished. Caged workers or queens readily take water when it is offered to them and live longer than bees that lack water. This would indicate that adult bees need water (Woodrow, 1941), although the quantity of water required by adult bees has not been determined.

Water is also used for cooling and humidifying the interior hive environment. In hot, dry weather Parks (1929) noted that water is deposited on the tops of combs in small cell-like enclosures generally made of old wax and propolis. It is also deposited in indentations in the cappings of brood so that the comb looks as if it had been sprinkled with water. Lindauer (1954b) observed that tiny droplets of water are also placed inside the cells, especially those containing eggs and larvae, where it prevents drying of the larvae (Lindauer, 1951). Bees actively ventilate the hive interior by fanning. Also, bees unfold their proboscises and, while doing so, a droplet of water is drawn into a film by the movement. This process speeds up the evaporation and produces a cooling effect. Lindauer (1954b) showed that such movements, when used during nectar concentration, are also functional in regulating the temperature of the colony which is maintained at 34°C.

According to Lindauer (1955a), water foraging is initiated in the following way: The honey sac content of house bees and foragers is about the same due to food transmission. When a shortage of water appears in the hive and there is no new incoming nectar, the content of the honey sac becomes well concentrated. Soliciting for water may start in the hive and this may induce the water carriers to initiate collection of water.

Park (1923b) observed that returning water foragers dance vigorously and occasionally pause to transfer water to bees that are following the dancer. The entire load is distributed among as few as two or three individuals or to as many as 18 workers. After unloading, the water forager feeds on a small amount of honey and immediately departs on the next collecting trip.

According to Lindauer (1954b), water collection is regulated by the time required by the house bees to receive the load. When the delivery is completed within two minutes, water carrying is continued without interruption. If

FIGURE 28. Worker bees collecting water from a leaky faucet. Bees collect water from many different sources. (Photo by Kenneth Lorenzen).

unloading requires two to three minutes the bee continues to collect water, but for a short time remains inactive in the hive. If unloading the water takes longer, intervals between the foraging trips increase. Water collection stops entirely when the bee cannot unload within about 10 minutes.

Lindauer (1954b) also observed that the dancing of water foragers is closely related to the delivery time. When unloading takes less than 40 seconds, there is nearly always a dance. With the increase in delivery time, dancing decreases and completely stops when the delivery time takes more than two minutes. Lindauer (1955a) considers that "the delivery time itself gives only a rough indication of whether or not the bees in the hive are 'anxious' to have water. If there is a great scarcity, one can see the water carrier being met at the entrance by two or three bees; these eagerly take the water away from her, while carrier and receiver caress each other in a lively way with their antennae. If the need for water is only moderate, there is at best a single bee at the entrance, and the water carrier may have to walk around the comb, offering her load of water, until eventually it is taken from her, accompanied by only slight caressing with the antennae. If the need for water has been completely met, the carrier will offer her load in vain; now and again a house bee will taste from her glossal groove, but she immediately draws in her tongue, cleans it, and runs away; thus, it may be a good many minutes before the honey stomach is emptied. It may be that the water carrier adjusts her behavior not so much according to delivery time as according to the indifferent behavior of the house bees."

Water foragers tend to collect water from the nearest available supply, especially if the supply is continuous (Gary *et al.*, 1973, 1979). The mean time per water collection trip varies considerably. A bee commonly spends a minute or more in taking up a load of water, and it takes one minute for the bee to fly 0.4 kilometers. Of all field trips recorded by Park (1929), 67% were completed in three minutes or less, and 92% in 10 minutes or less. The time spent in the hive was two to three minutes as a rule, and seldom did one remain as long as five minutes. One hundred or more trips are made in a day by a single water carrier, but the average is probably about 50 trips. According to Park (1925c), a maximum load of water would weigh approximately 50 milligrams and an ordinary load about 25 milligrams. Thus, if the average amount of water brought into the hive in a day by one bee making 50 trips and carrying 25 milligrams per load is 1,250 milligrams, then approximately 800 water collectors would be required to bring in one kilogram (one liter) of water daily. The daily water requirement for an average colony during spring broodrearing has been estimated as approximately 150 grams under average conditions. For strong colonies, and especially under hot, dry conditions, bees are thought to collect up to one kilogram per day.

Bees do not store large amounts of water, but they can store a supply of water sufficient to last the colony from one flight day to the next. Park (1923b), using a one-frame observation hive, found that on such occasions the water was stored in the honey sacs of numerous bees of the colony. As the water was transferred, the abdomen of the water carrier decreased in size while that of the "reservoir-bee" became distended. The reservoir-bees were quite inactive and occupied places surrounding the brood area, rather than within it. When a good flight day was followed by a period of several days without access to more water, the abdomens of these reservoir-bees became greatly reduced in size. Then, on the first subsequent flight day, it became evident that they were being refilled.

There is no doubt that water is very important to the survival of the colony. Water deficiency can be most acute when hives of bees are confined for long periods of time during transportation, especially during hot weather. This is also true after hives are relocated. Time is required for foragers to find new water sources and to establish a normal foraging pattern again. In hot areas where water supplies are sparse, the consequences can be very serious unless water is supplied to the hives in individual or community feeders.

Propolis Gathering

Propolis is known as "bee glue." It is usually a brown, gummy, sticky, resinous, thermoplastic material that is obtained from the buds of trees and is used by the bees to coat the interior hive walls, seal cracks, reduce openings, and sometimes to encase objects that are too large for housecleaner bees to remove from the colony. At the time of Aristotle, the ancients were already aware of the presence of propolis in the hives of honey bees. The famous Roman naturalist, Pliny, in his *Historia Naturalis*, wrote extensively about propolis. The renowned French naturalist, Huber (1814), described in detail his investigations on the collection and use of propolis in the hive. His interesting series of observations is worth reading by anyone interested in the life of bees.

The description of the process of propolis collection differs somewhat in reports of various observers (Alfonsus, 1933; Meyer, 1954, 1956). Propolis foraging takes place only in warm weather when it is pliable. The bee bites into it with her mandibles and, using the first pair of legs, tears a little piece off. She may knead it between her mandibles and then, with one of the second legs, transfers it to the pollen basket on the same side. She may do this in a stationary position or in flight. Next, she places a piece of propolis in the pollen basket on the other side. The clumps of propolis are patted frequently with the second leg, putting them in proper shape. Such alternate loading continues until both baskets are fully loaded (Fig. 29). Approximately 15-60 minutes are required to accumulate a full load.

Figure 29. Worker bee returning to the hive with a load of propolis in her pollen basket. (Photo by Kenneth Lorenzen).

When a bee enters the hive with a load of propolis, she is unloaded by another worker bee that bites at the sticky, gummy material and pulls or tears off a little piece, then presses it firmly into place. While applying propolis, the "cementing bee" may mix some wax into it. The propolis carrier immediately makes the load smooth again by patting. She may be freed of her load in the course of an hour, or several hours, depending on the use of propolis in the hive. When bees have been freed of their loads, they again forage for more. Propolis gatherers remain "faithful" to their work, but there are very few of them in each colony.

When there is a dearth of nectar, propolis gatherers may become nectar foragers and later revert to the status of propolis gatherers. Meyer (1954) observed a few propolis foragers that "performed" the recruiting dance after bringing a load into the hive. Some of the bees followed the dancer, but none of them were recruited to propolis collection.

Swarming

Swarming is a form of reproduction at the colony level, as opposed to reproduction of individual bees within. By swarming, new colonies are established, thereby replacing those that perish from adverse living conditions. Swarming typically occurs during a particular season that varies considerably according to climate, but varies little within each geographic area. The "swarm season" is often a four- to six-week period, usually in late spring or early summer. It frequently occurs within a few weeks after nectar and pollen sources were abundant, and following an overwintering or prolonged dearth period.

Swarming is preceded by a great increase in egg laying. In one case the queen deposited 62 eggs in 45 minutes, *i.e.*, 1,984 eggs per day. The retinue of nurse bees around the queen becomes excited and they constantly and persistently offer food to her. After several weeks of intensified brood rearing there is a "population explosion" within the colony when massive quantities of brood emerges as adults at about the same time. At the time of swarming the colony population is made up of a large percentage of very young bees.

One of the events preceding swarming is that the queen lays eggs in the pre-constructed, thimble-shaped, queen cell cups that are common in the brood nest, but unused except at this time. The resulting larvae are fed lavishly with royal jelly throughout their larval development and are capped eight days after the egg was laid. Usually more than six cells are started. During this time the number of bees feeding the queen diminishes. Her ovaries shrink dramatically, reducing her abdomen size and weight to the extent that she can fly by the time the swarm departs. The large populations of young bees that are emerging are displaced from normal brood feeding behavior because of the sharp decline in egg laying. They fill all the available space in the hive, sometimes even hanging outside near the entrance. Taranov (1947) calls them the "active swarm bees" because they comprise the majority of the bees that leave with the swarm.

According to Allen (1956), about a week before the swarm issues from the hive, the queen may be pushed about and treated roughly, which tends to keep her moving. Sometimes workers even bite at her legs if she stops moving.

The queen may start "piping" (more recently called "tooting" by Wenner, 1962). Apparently contact with queen cells has some connection with "tooting" because one observer reported that, about an hour before the swarm left, the old queen "piped" 25 times during 25 minutes. On 14 occasions she was on a queen cell, six times she was near one, and the remaining five times she was elsewhere on the comb.

Several days before swarming, an abnormal number of bees may be seen "resting" quietly on the bottoms of combs. According to Lindauer (1955b), this is the time that scout bees may begin to look for a new nesting place. They perform wag-tail dances in the hive indicating the direction and the distance to the prospective new sites. The nest searchers, contrary to the food gatherers, do not interrupt the dance but continue dancing for hours, or even days, changing the direction of their dances according to the change in the position of the sun.

Shortly before the swarm departs, the bees engorge with honey. On a warm, calm, and sunny day searchers "perform" a special, characteristic "whir" dance (Lindauer, 1955b) that seems to incite 20,000 to 30,000 bees to swarm. Highly excited, the searchers force their way among the bees in zigzag

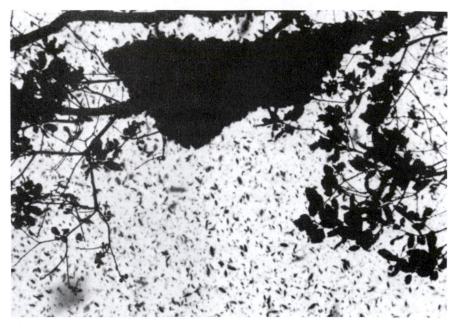

FIGURE 30. A swarm of bees in flight and beginning to cluster on a tree branch (above). (Photo by Kenneth Lorenzen).

running steps, vibrating their abdomens and producing a perceptible whir with their wings. One or two bees start this "whir" dance, but after a minute there are dozens. The number of whirring bees increases until the whole hive is in tumult. The swarm emerges in a frenzy of flight that can be terrifying to the uninitiated. Yet, during swarming behavior, bees are less defensive than at any other time. The number of bees going out with the swarm (Fig. 30) may represent half to 90% of the parent colony population. According to the majority of observers (Butler, 1940; Meyer, 1956), the age of bees in a prime swarm is mostly from four to 23 days, although bees of all ages, drones included, are commonly found in a swarm.

Within a few minutes the swarm, along with the queen, clusters in a dense mass on a nearby tree (Fig. 31), bush, or other somewhat protected site where there tends to be diminished light. Meyer (1956) found that a swarm consists of an outside, compact layer about three bees in thickness. The swarm interior is less compact and consists of hanging "chains" of bees that are connected with the outer layer in many places. The outer layer protects the swarm against outside influences, *e.g.*, rain, and provides the necessary mechanical strength to support the cluster.

According to Lindauer (1955b), after the swarm is settled the searchers perform "wag-tail" dances in various directions on the surface of the shell, thereby indicating different prospective nesting places in different directions. From all the available places, the searchers eventually "select" the best one. They prefer a wooden hive over a straw skep, a wind-protected location over one not protected, and a site far away rather than close, within certain limits.

FIGURE 31. A swarm of bees clustered on a branch. (Photo by Kenneth Lorenzen).

Also, nest volume, exposure to sun's rays, and infestation with ants play roles in the selection of a prospective nesting place. The most important factor is protection from wind.

The better the nesting place, the more lively is the dance of the searchers. Inferior nest sites are less stimulating and produce less vigorous dances, and such bees may be influenced by the more vigorous dances of the searchers from a better site. These bees may "inspect" the better nest sites and, upon returning, dance according to the new location.

The searchers also make repeated visits to the future nesting place. They may even stop "advertising" the place if the conditions become unfavorable. When there are two equally good places, two strong groups are dancing. If the bees do not come to an "understanding," the cluster may divide and start flying, but after a short time the parts come together again and the searchers may try to "agree" again. If this becomes impossible, or when nest sites are not available, then the swarm builds its nest where it has settled (Fig. 32).

When the searchers "agree" on the place, they start the "whir" dance, boring their way within the cluster. One can hear a very loud humming inside the cluster, the bees start to clean themselves, and start running to and fro, creating a tumult. When the excited running reaches its highest point, five to

FIGURE 32. These bees were not able to find a suitable nest site and constructed their combs on the tree limb where the swarm originally clustered. (Photo by Kenneth Lorenzen).

10 bees simultaneously fly out from the cluster, hundreds follow, and in a few seconds the whole cluster disbands. It appears that the swarm is led by about 100 bees which fly quickly toward the new nesting place, while the bulk of the swarm proceeds at a slower pace. The "leading bees" return, fly at the border of the swarm, and then "shoot" quickly to the fore. Searchers perform the "whir" dance also at the new location after the swarm starts to occupy it. The significance of this dance is not yet known. Within hours after the swarm enters the new nest cavity, new combs are under construction. Within a few days a new brood cycle is initiated and the new colony is successfully established.

Several theories are advanced for explanation of the swarming behavior of bees (Simpson, 1958). Probably both the internal and the external factors confronting the colony, coupled with the hereditary and physiological makeup of its members, play an important part in swarming preparations (Haydak, 1960).

From one of the queen cells in the original colony a new queen emerges. She usually kills all rival queens, then mates and initiates egg laying so that the original colony continues. Sometimes the first queen to emerge in the original colony leaves with a second swarm, mates after the swarm settles in the new nest, and two new colonies are created. Such swarms are known as "after-swarms." Sometimes there is even a third afterswarm if the original colony population was extremely large.

Excellent studies have been conducted on various aspects of swarming, *e.g.*, Lee & Winston (1987), Seeley *et al.* (1979), Visscher *et al.* (1985), and Winston (1980). Möbus (1987) and Severson (1984) also published reviews.

Activities of Queenless Bees

The queen has considerable influence on the orderly life processes in the colony. When a colony becomes queenless, bees become very agitated, start fanning, and most activities are disrupted. Within a few hours they initiate the construction of several queen cells. This kind of cell is termed an "emergency" queen cell because such cells are made from existing worker cells containing larvae, rather than from eggs laid in pre-constructed queen cell cups (Fig. 33). The few worker cells that are selected are enlarged, fed copious amounts of royal jelly, literally floating the larvae toward the cell opening. By this time the cell opening has been reconstructed and is oriented downward, the normal position for queen cells. When the new queen mates and initiates egg laying normal worker activities are resumed.

Colonies are termed "hopelessly queenless" when the workers fail to construct emergency queen cells and no fertilized eggs or larvae less than four days old remain as a potential source of constructing emergency queen cells. This condition also may develop if a queen is produced but is lost for any reason prior to laying eggs. The loss of queens during their mating flights is not uncommon.

Colonies without queens are in a stressed condition. Workers are more defensive and the number of guards is higher (Hoffman, 1961). In the absence of inhibitory queen pheromones, the ovaries of many workers enlarge and

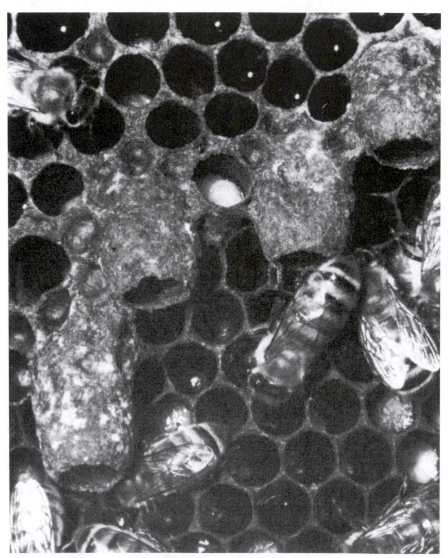

FIGURE 33. Emergency queen cells (four) which have been constructed over cells containing worker larvae less than three days of age. (Photo by Kenneth Lorenzen).

become functional. Such workers are known as laying workers because they lay eggs in the brood nest. This condition is easily identified. First, there is the absence of a normal pattern of eggs and brood produced from eggs laid by the queen before her loss. Also, the colony population is reduced, owing to the natural mortality of old bees and the interrupted supply of newly emerging worker bees. Eggs laid by the workers are somewhat scattered in the brood cells rather than being laid in the typical compact pattern of a queen. A primary symptom is that there are multiple eggs in many cells (Fig. 34). A normal queen rarely lays multiple eggs, except during situations, such as in early spring, when she is producing eggs very fast, but the cluster of workers is too small to incubate enough cells to accommodate the rate of egg production. Eggs produced by laying workers are rarely laid in the normal position, *i.e.*, standing on end and attached in the center of the cell bottom. Rather, they frequently are attached to the sidewalls of cells or at one side of the cell base. There is a high mortality of larvae that develop from these eggs. The surviving larvae are, of course, drones because workers cannot mate. Consequently, the worker cells are provided with protruding drone cell caps. Drones that emerge from these cells are dwarfed in size owing to their constrained development in the relatively small worker cells. Perhaps the high rate of mortality is related to poor nutrition. According to Haydak (1958), nurse bees in hopelessly queenless colonies do not recognize the sex of larvae, as they do in queenright colonies, and they feed the drone larvae as if they were female larvae.

FIGURE 34. Comb with multiple eggs per cell. Laying workers deposit more eggs in drone cells than in worker cells. (Photo by Kenneth Lorenzen).

According to Gontarski (1938), the first observations of laying workers were made in 1770 by von Riehm, and his observations were confirmed by Huber in 1788. Laying workers can also be found in normal queenright colonies, especially during the swarming season. Tuenin (1926) and Perepelova (1928) demonstrated that workers with developed ovaries (anatomical laying workers) were present in from 20% to 70% of colonies which showed signs of swarming. Koptev (1957) found that such workers are present during the entire season in normal colonies, especially in those colonies that have poor queens. Normal colonies at the end of the honeyflow had 7% to 45% anatomical laying workers.

Laying workers are sometimes called "false queens," particularly in cases where their queen-like characters are quite pronounced, and the behavior of other nearby workers greatly resembles the normal queen retinue (Sakagami 1958). A summary of the data presented by various investigators indicates the following profile of the activities of laying workers. A laying worker inspects the cell by dipping her head and thorax into it. Sometimes she even cleans the cell before laying. While laying, the worker inserts her abdomen, and sometimes part of her thorax, more or less deeply into the cell. Because she is shorter than the queen, she rarely can reach the cell bottom with the tip of her abdomen. During egg laying she orients her dorsal side toward the lower wall of the cell. While she is laying the egg, a circle of bees sometimes forms around her, but the retinue has fewer bees and is less stable than in the case of a normal queen. Bees in this entourage do not pay the same attention to the laying worker as they do to a queen. Other laying workers may be present in the circle. Such a worker lays only one egg to a cell. Other eggs in the same cell are laid by other laying workers. The time a laying worker spends in laying varies from 17 to 261 seconds, with an average of 50 to 70 seconds. This time greatly exceeds that of a normal queen. The behavior of the laying workers toward other bees appears normal. But with the onset of drone rearing, other workers show hostility toward them, mauling, pulling, and pursuing them through the hive (Sakagami, 1954; Hoffmann, 1961). Usually the laying worker makes no resistance, shows submissive behavior, strops her tongue characteristically, and moves around more when the mauling becomes too severe.

Sakagami (1959) and Hoffmann (1961) reported that in a colony with laying workers, age-related activities resemble those in normal colonies. Laying workers behaved like normal worker bees. Except for laying eggs they participated in all other activities of the hive, eating pollen and honey, and flying out. Older bees in a queenless hive participated in both brood rearing and foraging activities.

The queen is very attractive to bees (Fig. 41). Queenless bees are likely to desert their own colony, and join any nearby queenright colony (Butler,

1954). Comb building activity in a queenless hive is greatly impaired (Darchen, 1957; Genrikh, 1957). Drones are expelled from a laying worker colony in the same way as from normal colonies (Hoffman, 1961). Gontarski (1938) observed that in laying worker colonies bees stored honey and pollen below the brood and not above the brood as they do in normal colonies. When a queen was introduced, the method of storage became normal.

According to Genrikh (1957), removal of the queen greatly disturbed colonies. Flight activity diminished by 77%, the number of bees coming in with loads of pollen by 73%, the weight of pollen loads by 50%, the filling of the honey stomach by 62%, nectar income by 81%, and the rate of comb production from foundation by 73%.

Stray workers tend to join together to form a cluster, even in the absence of a queen (Lecomte, 1949). When 75 or more were present, they always formed a cluster. When only 50 bees were present clusters were formed in only half of the cases. When a smaller number was present, bees segregated in small groups of five to 10 individuals. Lecomte (1950) concluded that the sense of smell and the vibrations caused by the movements of bees are the stimuli for cluster formation. In a similar study Free & Butler (1955) concluded that cluster formation was also induced by heat produced by the bees in the cluster and the visual stimulus of the clustered bees.

Activities During the Winter

During the winter, brood production ceases except in warm climates. The broodless bees form a cluster when the temperature immediately surrounding the bees is approximately 14°C or lower (Phillips & Demuth, 1914). This cluster becomes more compact as the temperature goes down. When first formed, the cluster is usually located in the lower part of the hive, often near the front. The upper portion of the cluster is in contact with honey stores. As honey is consumed throughout the course of the winter, the cluster gradually moves upward and to the rear of the hive. Honey is converted into heat by metabolic processes of bees inside the cluster, and this heat is conserved by the insulating properties of the cluster as well as the comb.

Bees do not thermoregulate the space inside of the hive and surrounding the cluster. Heat regulation is within the cluster. Gates (1914) found that the temperature inside the broodless winter cluster varies between 20° and 30°C, even when the ambient outside temperatures are very low, *e.g.*, -31.7°C. The temperature of the thoraces of bees ranges between 20° and 36°C and is independent of the outside temperature. The temperature outside the cluster, but inside the hive, may be quite low. During severe cold the bees on the surface of the cluster bury their heads and thoraces inside the cluster so that only the abdomens are visible, and at the same time they move their abdomens with screwlike motions. From time to time some of the bees on the outside enter the core of the cluster and are replaced by other bees. Whenever the weather becomes warmer the bees spread out again. Bees in the winter cluster are not in a state of hibernation. They are capable of resuming foraging activities at any time when the temperature exceeds approximately 13°C and if foraging resources are available.

Thermoregulation requires that bees perceive the temperature accurately before they can make adjustments. In that sense every bee should be viewed as a tiny thermostat and also a heating unit. The fuel is honey. In the broodless winter cluster honey is consumed at varying rates, depending primarily upon fluctuations in the temperature (Ribbands, 1953).

Bees may be confined to the hive by cold weather for several months at a time. Normally they can defecate only while flying. During prolonged confinement fecal materials are stored in the rectum. At any time the temperature reaches approximately 7°C, bees take a brief defecation flight (Himmer, 1926), sometimes referred to as a "cleansing flight," in the vicinity of the hive and return quickly before they become too chilled to fly. Confinement of more than three to four months greatly stresses the colony and can cause colony mortality.

By late winter or early spring the cluster in a two-story hive is found most often in the upper story. Brood rearing is resumed in late winter, at which time the temperature inside the brood rearing cluster is maintained constantly at approximately 34°C. At that time the rate of honey consumption increases sharply, owing to the energy expended for incubation and the feeding of larvae.

For additional information, see Chapter 20, "Wintering Productive Colonies."

Activities of the Queen
Queen behavior begins with the emergence of the virgin from a queen cell. Newly emerged virgin queens are very active and, if not prevented by the workers, will destroy all other queen cells present in the hive. The queen's powerful mandibles are used to make an opening in the side wall of the cell. She inserts her abdomen into the opening and stings her "rival" to death. Worker bees dispose of the carcass and destroy the queen cell. During preparations for swarming, the virgin queen may be prevented from destroying the queen cells by the bees that cluster around them. From time to time the virgin queen pauses and, clinging to the comb, produces a sharp, high note resembling "tee-tee-tee-tee." The bases of her wings tremble at that time (Haydak, 1949). If one listens carefully, and usually the bees are then extremely quiet, one can hear another type of sound resembling "quahk-quahk-quahk" (the answers from virgins confined in the queen cells). These

sounds of communication, now termed "tooting" and "quacking," are very interesting and have been studied in great detail by Wenner (1962) and Michelsen *et al.* (1986). In the situation above, the first virgin to emerge may leave the hive with the second swarm or "afterswarm." If another virgin emerges while the first one is still in the hive, then a fight ensues and only one queen survives.

During approximately the first five days after emergence the queen matures as she interacts with workers in various ways, as described by Hammann (1957), Allen (1960), Ruttner (1956), and others. Sexual maturity is reached at five to six days of age. In the midday hours on the day of mating, the number of excited workers around the queen increases. Some of them run toward the entrance, and one can often see a continuous line of bees between the virgin and the entrance. Finally a group of fanning bees, with their scent glands exposed, gathers in front of the entrance, and normal flying and foraging are greatly reduced.

The virgin queen appears at the entrance accompanied by several workers. If she is hesitant to fly, workers seem to prevent her return into the hive and seemingly "try to force her" to flight. Thus, the mating of the queen not only involves the queen and drones, but the entire colony.

Ruttner (1957) considers that the mating flight is induced by the workers through their aggressive behavior and specific feeding. When the virgin was assaulted very vigorously by workers, she flew out for the first time much sooner. When she was not assaulted, she did not fly at all (Hammann, 1957).

Mating flights occur when the weather is warm (>18°C), the wind is fairly calm, and during the afternoon only. Oertel (1940) observed flights of 60 virgins and found that the duration of non-mating flights ranged from two to 30 minutes, and that of mating flights five to 30 minutes. Virgin queens frequently mated on the first flight. Thirty-two of 54 virgins mated within eight to nine days after emergence, 16 from six to seven days, and the remainder in 10 to 13 days.

Direct observations of mating under controlled conditions were made for the first time by Gary (1963) (see "Activities of Drones" in this chapter). The queen mates with six to 17 drones (Adams *et al.*, 1977; Peer, 1956; Taber & Wendel, 1958) at distances of up to several kilometers from the apiary (Peer, 1956, 1957; Ruttner & Ruttner, 1972). When the queen returns from a mating flight she is continuously followed by the excited workers which touch her and lick her "mating sign" (remains of part of the penis of the last drone to mate with her, and the coagulated mucous), or they may pull it out with their mandibles. Removal is also facilitated by convulsive contractions of her abdomen. The queen may fly out on another mating flight scarcely 10 minutes after returning from the prior one.

During multiple matings, the mating and separation of the queen and drone occurs within several seconds. Before the second and subsequent matings, the mating sign is removed from the sting chamber of the queen by being attached to the base of the penis of the next drone. At these successive matings the sting chamber of the queen remains open. Only at the last mating does the queen close the chamber, thus cutting off the bulb of the penis, and returns to the hive with a mating sign (Alber *et al.*, 1955). Gary & Marston (1971) succeeded in constructing model (or "dummy") queens that were sufficiently similar to living queens that drones mated with them.

After mating, the oviducts are filled with spermatozoa which migrate into the spermatheca, where approximately three to five million spermatozoa are stored for the life of the queen. Approximately three days after the last mating flight, the queen starts laying eggs. Before laying an egg, the queen walks over the comb and places her head in the cell, where she apparently senses that the cell has been prepared for oviposition (Fig. 35). Then she withdraws her head, curves her body, and quickly pushes her abdomen into the cell (Fig. 36). As the egg is laid she releases several sperm that fertilize the egg. Eggs laid in drone cells are not fertilized. In a few seconds she turns to the right or to the left and withdraws the abdomen out of the cell. The egg-laying process requires approximately nine to 12 seconds. After laying a number of eggs, the queen remains motionless, during which time she is groomed and fed by workers.

FIGURE 35. Queen, with head inserted into a cell, inspecting the cell prior to laying an egg. (Photo by Kenneth Lorenzen).

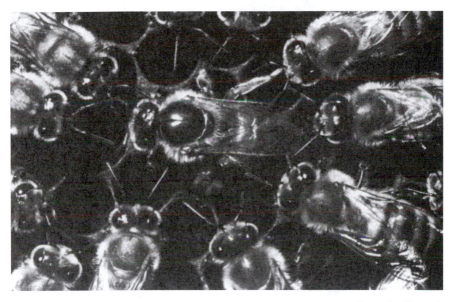

FIGURE 36. Queen, with abdomen inserted into a cell, laying an egg. (Photo by Kenneth Lorenzen).

There are various estimates as to the number of eggs a queen can lay daily. Nolan (1925) found that the highest daily average during any 12-day period was 1,587 eggs. Most queens probably lay between 1,000 and 1,500 eggs daily during the most active period of broodrearing. Egg production ultimately determines the maximum colony population that is attainable. For example, if workers lived an average of 30 days, then the maximum colony population, if the queen produces 1,000 eggs daily, is 30 X 1,000 = 30,000. This is why it is so very important to maintain a young, vigorous queen in managed colonies. A good queen in a populous colony may lay up to and over 200,000 eggs a year. As queens age, their egg production may decline. The average queen lives one to two years, but she is dependable as a vigorous egg layer for only one year. This is why requeening colonies at least once a year is very important.

Mated, laying queens do not leave the hive unless the colony swarms or absconds. The queen is not only an "egg-laying machine." She also produces and releases vital pheromones that exert a profound influence on the most important life activities of the colony.

Activities of Drones

Drones participate in almost none of the activities in the colony. Their primary contribution to the bee society is mating with the queen. When they are sexually mature, at approximately 10 days of age, they mate with virgin queens during flight away from the hive. Drones never mate inside the hive. They normally fly during the midafternoons. They fly in the morning only if they are departing the hive with a swarm. Drones are often considered by beekeepers as being somewhat parasitic to the colony. Yet, studies have shown that the productivity of colonies, in terms of honey at least, is not diminished by the presence of drones (Allen, 1963).

Drones develop from unfertilized eggs that are laid in drone cells. After emerging from the cell, approximately 24 days following oviposition, they remain stationary most of the time on the comb in the brood nest, frequently near other drones. When they are young they are fed by workers, apparently beginning with brood food and gradually changing to honey by the time they are approximately seven days old, at which time they feed themselves (Free, 1957; Örösi-Pál, 1959). Drones less than three days old are fed much more by nurse bees than are older drones (Jaycox, 1961).

The number of drones in a colony ranges from none in the winter to as many as 1,500 in the spring and summer. Populous colonies with older queens tend to produce more drones than small colonies with young queens. Drone populations are also limited to some extent by the amount of drone comb available to the queen in the brood nest. Older drones tend to congregate at the edges of the brood nest. At the end of the summer or early fall, or when food resources are critically short, drones can be observed at hive entrances and on the ground near the hive as they are actively expelled from the colony by workers (Levenets, 1956; Dathe, 1975) and die of starvation.

Mating flights have been studied extensively. Until the discovery that drones could be attracted to queen pheromones aerially in congregation areas (Gary, 1962) (Fig. 37), most of the information about drone flights was limited to observations at hive entrances. Another significant advance was the application of radar to follow the movements of flying drones (Loper *et al.*, 1987). Miniature transmitters are also under development for this purpose (McManus, 1988).

Mating flights have been studied by many researchers (*e.g.*, Gary, 1963; Howell & Usinger, 1933; Koeniger, 1986; Koeniger *et al.*, 1979; Kurennoi, 1954; Mikhailov, 1928; Oertel, 1956; Ruttner & Ruttner, 1972; Taber, 1964; Witherell, 1970). Most drones first fly when they are six to eight days of age. Before flying out, drones engorge on honey (Burgett, 1973) and groom their body, especially the antennae and eyes. The primary flight time is between 2:00 and 4:00 p.m. Orientation flights are six to 15 minutes in duration. Mating flights, taken by drones over 12 days of age when they are sexually mature (Kurennoi, 1953), are usually 25 to 57 minutes. Drones usually fly within about three kilometers from the apiary (Levenets, 1954). The relative importance of landmarks and sun position as orientation cues has not been

FIGURE 37. A group of drones (22), hovering around a caged queen that is suspended approximately 10 meters high. Drones locate the queen by perceiving her attractive pheromones. (Photo by Norman E. Gary).

determined. Drones tend to congregate at specific locations where they remain in constant flight, awaiting the arrival of a virgin queen. There appear to be "aerial pathways" that are consistently used en route to these congregation areas (Loper *et al.*, 1987). A drone's flying speed is nine to 16 kilometers per hour. On sunny days they make several mating flights when the temperature exceeds approximately 18°C and wind speed is less than 18 kilometers per hour.

After making direct observations of matings, Gary (1963) was able to accurately describe some of the details of mating behavior at drone congregation areas. During flight the drones, strongly stimulated by pheromones from the virgin queen, fly upwind until arriving sufficiently close to respond to the visual stimulus of the queen. The drone overtakes the queen and grasps the top of her abdomen with all six legs, his head extending over her thorax. The abdomen of the drone curls downward until it contacts the tip of the queen abdomen. If the queen opens her sting chamber, the penis everts and ejaculation occurs very quickly (Fig. 38,a). If the queen fails to open, the drone may remain in this position for several seconds or until another drone knocks him off. As soon as ejaculation occurs, the drone is instantly paralyzed. He releases the queen and topples over backwards, still attached by the genitals (Fig. 38,b). Approximately two to three seconds later a distinct "pop" sound can be heard as the two separate (Fig. 38,c). There are reports that the pair occasionally fall to the ground where they separate within a few minutes. This is probably a rare occurrence. Witherell (1965) determined that drones die quickly after mating, usually within about an hour.

Drones are strong fliers and literally carry the queen along with them in flight for a second or two before mating. In their mating efforts the drones are very aggressive. Gary (1963) observed one tethered queen suspended high in the air that mated with 11 drones in quick succession. This was the first time that multiple matings were actually observed, even though multiple matings were determined on the basis of other evidence (*e.g.*, Adams *et al.*, 1977; Peer, 1956; Taber, 1954; Triasko, 1956; Woyke, 1956). Multiple matings in quick succession are facilitated by the very short time (2.4 seconds) after mounting until eversion takes place (Gary & Marston, 1971).

Gary (1963) observed that eversion of the drone genitalia did not occur unless there was an opening at the tip of the abdomen. He also observed that the mating sign physically held the sting chamber open and stimulated such eversion. Drones even "mated" with dead queens if the tip of the abdomen were removed, leaving a cavity, or if a wooden queen model with a suitable opening was presented to flying drones (Gary & Marston, 1971). This aspect of mating behavior intrigued early investigators, who asked how successive drones could mate when each drone left his mating sign, consisting of mucous

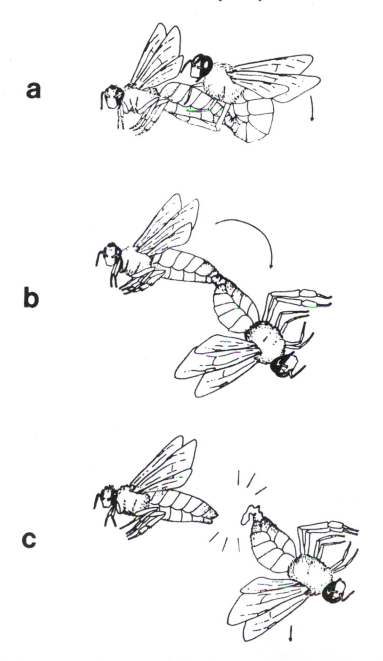

FIGURE 38. Diagrams showing three stages in the mating process. (a) Initial mounting position. (b) Intermediate position. (c) Drone separation by genital "explosion." (From Gary, 1963).

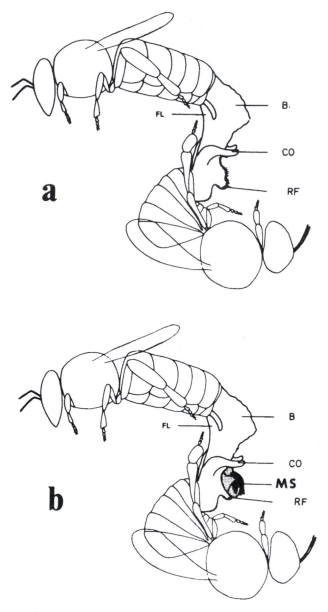

FIGURE 39. Diagrams depicting the removal by a subsequent drone of the mating sign left by the previous drone. (a) A drone has inserted his mating apparatus. (b) The next drone inserts his mating apparatus and simultaneously removes the mating sign of his predecessor. (B) bulb; (CO) cornua; (FL) fimbriate lobe; (MS) mating sign; (RF) rhomboid ventral hairy field. (From Koeniger, 1986).

from the male accessory glands and the expelled chitin plates of the drone's endophallus, in the queen's sting chamber. Koeniger (1986) excellently documented and described the process whereby each drone removes the mating sign of the previous drone during the process of eversion and mating (Fig. 39). Only the mating sign of the last drone remains when the queen returns to the hive.

Many researchers have investigated the "drifting" behavior of drones, *i.e.*, drones returning to the wrong hive after flights. The results of their experiments are highly variable. Under some circumstances very few drones drift, whereas in others the drift ranges up to 80%. The primary significance of drone drifting is the risk of vectoring various diseases and parasites from hive to hive.

Excellent reviews of the activities of drones were published by Currie (1987), Ruttner (1966), Washington (1967), and Witherell (1970).

Pheromones and Bee Activities

Virtually all bee activities are directly stimulated and coordinated to a large degree by pheromones. A pheromone is defined as a "chemical messenger" that is secreted externally by exocrine glands and elicits intra-specific behavioral and physiological responses. After being secreted as a liquid, these compounds are transmitted as a gas or liquid. Pheromones are detectable in extremely small quantities by honey bees, and the effects on behavior can be very dramatic. Unfortunately, humans cannot perceive most of these compounds, except with elaborate laboratory instrumentation. Consequently, the vital role of pheromones is not always fully understood or appreciated. Sources of pheromones in bees are shown in Fig. 40.

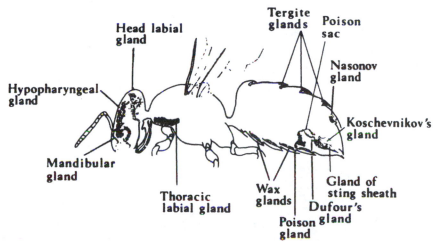

FIGURE 40. Diagram of a worker bee showing the location of pheromone-producing glands. (From Free, 1987).

Queen-produced pheromones, secreted primarily in the mandibular glands, consist of many compounds. One of the first pheromones to be discovered was called queen substance (9-oxo-2-decenoic acid), sometimes referred to as 9-ODA. This powerful compound, when combined with other pheromones, has multiple functions, *e.g.*, inhibiting the construction of queen cells, inhibiting the growth of worker ovaries, functioning as a chemical signal that attracts drones to virgin queens on the mating flight, and stimulating normal foraging behavior. Other queen pheromones attract workers to the queen, forming the familiar queen retinue, and stimulate feeding of the queen by nurse bees (Fig. 41). Our understanding of queen mandibular gland pheromones has been advanced significantly in recent research (Kaminski *et al.*, 1990; Naumann *et al.*, 1990; Slessor *et al.*, 1988, 1990; Winston, *et al.*, 1982, 1989, 1990).

Pheromones produced by worker bees are profoundly involved in nest defense. An alarm pheromone, isopentyl acetate, is associated with the sting. This pheromone alerts and alarms bees, and stimulates stinging behavior. When a bee stings, the sting structure, along with the pheromone, is left in the victim (Fig. 16). This causes a chain reaction of subsequent stings and releases of additional pheromone. To the human observer this alarm odor is detectable by holding a sting near the nose. It has an odor similar to banana oil. After

Figure 41. Worker bees encircle the queen, attracted to her by her pheromones. (Photo by Kenneth Lorenzen).

being stung it is prudent to wash away the odor with water before approaching the bee colony again. Otherwise, additional stings are likely.

Worker bees also produce pheromones, primarily in the Nassanoff gland, that function as attractants in various behavioral contexts, such as orientation to the queen, marking the entrance to the nest, or marking water or certain food sources. When releasing the pheromone complex, workers expose the Nassanoff gland, which is a "folded" membrane located between the sixth and seventh terga. While standing, the bee elevates her abdomen and fans, passing air over the exposed gland to facilitate pheromone dispersal (Fig. 18). The Nassanoff pheromone is known to contain geraniol, nerol, (E,E)-farnesol, (E)-citral, (Z)-citral, geranic acid, nerolic acid, and perhaps other compounds as well.

Drones apparently produce pheromones in their mandibular glands that are attractive to other drones on mating flights (Lensky *et al.*, 1985).

Brood pheromones are poorly defined at this time. Recognition of the various castes by nurse bees is presumably by pheromonal signals. Before nurse bees can feed the appropriate materials to developing larvae, they must first identify the caste and age. Furthermore, the presence of brood stimulates pollen collection (Free & Williams, 1975) and affects workers in other ways, *e.g.*, inhibiting ovary development.

Pheromones are evident in other sources within the colony. For example, pheromones contained in combs stimulate nectar gathering (Rinderer & Baxter, 1979) and increase defensive behavior (Collins & Rinderer, 1985). There are even footprint pheromones from queens that are reported to prevent the construction of queen cups for swarming (Lensky & Slabezki, 1981).

In summary, virtually every activity involves pheromones to some extent. Not only are individual compounds effective, but various mixtures, in which the respective components are variable quantitatively, have the potential for multitudes of possible "messages." The context within which pheromones are released is also vital to the nature of the message. For example, queen pheromones released inside the hive appear to be ignored by drones, but queen pheromones released during the mating flight are very stimulating to drones.

The successful manipulation of bee activities by using synthetic pheromones is an exciting possibility that awaits future research and development. As each pheromone is identified chemically and synthesized, there is the opportunity to seek practical applications. Already there are indications that worker bees can be directed to certain crops that require pollination (Mark Winston, personal communication). Queen substance has been synthesized successfully. Its first application was as a drone attractant in the first aerial drone traps, which I developed in 1964. Many other applications probably will be developed in the future to manipulate bee behavior.

Honey bee pheromones were reviewed by Free (1987) and Gary (1974).

AFRICANIZED BEES

The activities and behavior of Africanized bees are considerably different than for European bees, which have been the only bees we have known in the United States since the beginning of beekeeping, and which have been the subject of this chapter up to this point. A consideration of bee behavior would be incomplete without giving some attention to Africanized bees.

Historically, African bees (*Apis mellifera scutellata*) were brought to Brazil from South Africa in 1956 for the purpose of interbreeding them with European bees as a means of developing a race of bees better suited to tropical beekeeping conditions. European bees are adapted to a temperate climate, so they did not perform as well as desired in Brazil. Before the selection program was completed, African bees were released, either by accident or design, into the environment, where they immediately became established. They interbred freely with the meager populations of European bees, and the resulting stock retained almost all of its original behavioral characteristics. At first they were called "Brazilian bees." Eventually they were given the name "Africanized bees." Being quite dominant in virtually every respect, they reproduced and spread rapidly in all directions. Wherever they became established, they completely replaced European bees, usually within three years. Their rate of spread was approximately 320 kilometers per year. During this spread many people and animals were killed during fierce stinging episodes. The popular press dubbed them "Killer Bees," a highly inappropriate and prejudicial name that is greatly disliked by beekeepers, but nevertheless is commonly used by the press and understood by the general public.

Africanized bees reached the United States in 1990. They are expected to spread over a significant portion of the United States, particularly the southern states. Beekeeping practices and management will have to be changed dramatically to accommodate the very different behavior of these bees. One disturbing element is that Africanized bees are identical in appearance to European bees. Microscopic examination by experts is required for positive identification. Except for differences in behavior, which are extremely variable, beekeepers have no way to determine whether they are keeping European or Africanized bees. The following differences tend to characterize Africanized bees:

Reproduction

Colonies become established more quickly. Queens are very prolific and the brood develops more rapidly, *e.g.*, workers require only 18 to 20 days to reach maturity after the egg is laid (Winston, 1979). The "swarm season" is much longer and the frequency of swarming is much higher. New colonies increase rapidly in population and are routinely ready to swarm again within approximately two months after they were established. A 16-fold annual colony growth rate has been estimated (Otis, 1982). Great numbers of swarms establish feral colonies that intensely compete with European colonies for the limited food resources in the environment. Although beekeepers, with great and vigilant effort, may be able to keep European queens in their colonies, indications are that European colonies cannot compete effectively with Africanized colonies where the density of feral Africanized colonies is high.

Defensive Behavior

The intensive stinging behavior of Africanized bees is legend. Extreme variation between colonies is typical. Stinging episodes are highly probable, yet not always predictable. Colonies frequently are extremely sensitive to the slightest disturbance, such as an odor or a vibration. Thousands of defensive bees immediately leave the colony and, within seconds, somehow alert the other colonies in the apiary. In a short time they will deposit thousands of stings in any person or living animal **within approximately 400 meters in all directions** from the apiary. This has earned them the reputation as "killer bees." Without protective clothing, the only defense is to flee the area or to find protection inside a vehicle or building. Once initiated, the stinging "attack" continues for hours. After witnessing a stinging episode, it is very difficult to accept the fact that this is truly defensive behavior, and not aggressive behavior.

Stinging behavior of Africanized bees is likely to be a great threat to hobby and commercial beekeeping in the United States because of the difficulty in finding remote apiary locations away from people and animals, the litigious nature of our society, and the great expense of liability insurance.

Foraging

With minor differences, foraging behavior, and the resulting pollinating activity and honey production, are quite similar to European bees. Just like European bees, Africanized foraging bees away from the colony are not inclined to sting unless they are physically molested. Extensive foraging studies of European and Africanized bees in the same environments are required before accurate comparisons are possible. For example, early spring pollination during cool weather possibly could be a problem when using Africanized bee colonies.

Absconding

Africanized bee colonies tend to abscond, *i.e.*, the entire population permanently leaves the nest with the mated queen. Absconding can be triggered by disturbance to the colony and also by scarcity of nectar, pollen, or water. European bees very rarely abscond.

Mating Behavior

Africanized bees appear to have a great advantage over European bees. European queens in managed bee hives mate most often with Africanized drones. The mechanisms are not clear, but one obvious possibility is that Africanized drones simply outnumber European drones.

Disease Resistance

Africanized bees seem to be remarkably tolerant to pathogens and parasites that are great problems for European bees. Behavioral elements in resistance are not well defined at this time. Africanized bees seem to tolerate tracheal and *Varroa* mites without serious injury. Brood diseases, such as American foulbrood, are infrequent or absent in Africanized bees. Hopefully, these traits eventually can be selected and incorporated into future bee stock in the United States.

The eventual behavior and activities of bees in the United States will probably be altered significantly by Africanized bees. Beekeeping could become better or worse, depending upon the impact of Africanized bees and the degree to which research is supported for solving the problems that are sure to arise.

The subject of Africanized bees has been reviewed by Locke & Gary (1988) and Michener (1975).

REFERENCES CITED

Adams, J., E.D. Rothman, W.E. Kerr & Z.L. Paulino. (1977). Estimation of the number of sex alleles and queen matings from diploid male frequencies in a population of *Apis mellifera.* *Genetics 86*:583-596.

Alber, M., R. Jordan, F. Ruttner & H. Ruttner. (1955). Von der Paarung der Honigbiene. *Z. Bienenforsch. 3*:1-28.

Alfonsus, A. (1933). Zum Pollenverbrauch des Bienenvolkes. *Arch. Bienenk. 14*:220-223.

Allen, M.D. (1956). The behaviour of honeybees preparing to swarm. *Anim. Behav. 4*:14-22.

Allen, M.D. (1959a). The occurrence and possible significance of the "shaking" of honeybee queens by the workers. *Anim. Behav. 7*:66-69.

Allen, M.D. (1959b). The "shaking" of worker honeybees by other workers. *Anim. Behav. 7*:233-240.

Allen, M.D. (1960). The honeybee queen and her attendants. *Anim. Behav. 8*:201-208.

Allen, M.D. (1963). Drone production in honeybee colonies (*Apis mellifera* L.). *Nature 199*:109-119.

Aschoff, J. (1986). Anticipation of a daily meal: A process of 'learning' due to entrainment. *Monitore Zool. Ital. (N.S.) 20*:195-219.

Beecken, W. (1934). Über die Putz-und Säuberungshandlungen der Honigbiene. *Arch. Bienenk. 15*:213-275.

Bermant, G. & N.E. Gary. (1966). Discrimination training and reversal in groups of honey bees. *Psychon. Sci.* 5:179-180.

Beutler, R. (1953). Nectar. *Bee World* 34:106-116, 128-136, 156-162.

Boch, R. (1956). Die Tänze der Bienen bei nahen und fernen Trachtquellen. *Z. vergl. Physiol.* 38:136-167.

Bonnier, G. (1906). Sur la division du travail chez les abeilles. *C.R. Acad. Sci., Paris* 143:941-946.

Breed, M.D., G.E. Robinson & R.E. Page, Jr. (1990). Division of labor during honey bee colony defense. *Behav. Ecol. Sociobiol.* (in press).

Burgett, D.M. (1973). Drone engorgement in honey bee swarms. *Ann. Entomol. Soc. Am.* 66:1005-1006.

Butler, C.G. (1940). The ages of bees in a swarm. *Bee World* 21:9-10.

Butler, C.G. (1954). The method and importance of the recognition by a colony of honeybees (*A. mellifera*) of the presence of its queen. *Trans. Roy. Entomol. Soc. Lond.* 105:11-19.

Butler, C.G. & J.B. Free. (1952). The behaviour of worker honeybees at the hive entrance. *Behaviour* 4:262-292.

Calderone, N.W. & R.E. Page, Jr. (1988).Genotypic variability in age polyethism and task specialization in the honey bee, *Apis mellifera* (Hymenoptera: Apidae). *Behav. Ecol. Sociobiol.* 22:17-25.

Casteel, D.B. (1912a). The manipulation of the wax scales of the honey bee. *U.S.D.A. Bur. Entomol. Circ.* 161.

Casteel, D.B. (1912b). The behavior of honey bees in collecting. *U.S.D.A. Bur. Entomol. Bull.* 121.

Collins, A.M. & T.E. Rinderer. (1985). Effect of empty comb on defensive behavior of honeybees. *J. Chem. Ecol.* 11:333-338.

Collins, A.M., T.E. Rinderer, J.R. Harbo & A.B. Bolten. (1982). Colony defense by Africanized and European honey bees. *Science* 218:72-74.

Currie, R.W. (1987). The biology and behaviour of drones. *Bee World* 68:129-143.

Darchen, R. (1957). La reine d'*Apis mellifica*, les ouvrières pondueses et les constructions cirières. *Insectes Sociaux* 4:321-325.

Darchen, R. (1959). Un des rôles des chaimes d'abeilles: La torsion des rayons pour les rendre parallès entre eux. *Ann. Abeille* 2:193-209.

Dathe, R. (1975). [Studies on the behaviour of worker honeybees towards drones]. *Zool. Anz.* 195:51-63. (in German).

Dyer, F.C. & J.L. Gould. (1981). Honey bee orientation: A backup system for cloudy days. *Science* 214:1041-1042.

Dyer, F.C. & J.L. Gould. (1983). Honey bee navigation. *Am. Scient.* 71:587-597.

Eckert, J.E. (1933). The flight range of the honeybee. *J. Apic. Res.* 47:257-285.

Erickson, E.H. (1975). Surface electric potentials on worker honeybees leaving and entering the hive. *J. Apic. Res.* 14:141-147.

Esch, H. (1961). Über die Schallerzeugung beim Werbetanz der Honigbiene. *Z. vergl. Physiol.* 45:1-11.

Esch, H. (1963). Über die Auswirkung der Futterplatzqualität auf die Schallerzeugung im Werbetanz der Honigbiene. *Verh. dt. zool. Ges. Wien.* pp. 302-309.

Esch, H. (1964). Beiträge zum Problem der Entfernungsweisung in den Schwänzeltänzen der Honigbienen. *Z. vergl. Physiol.* 48:534-546.

Fletcher, D. (1975). Significance of dorsoventral abdominal vibration among honey-bees (*Apis Mellifera* L.). *Nature 256:*721-723.

Free, J.B. (1954). The behaviour of robber honeybees. *Behaviour 7:* 233-240.

Free, J.B. (1956). A study of the stimuli which release the food begging and offering responses of worker honeybees. *Anim. Behav. 4:*94-101.

Free, J.B. (1957). The food of adult drone honeybees (*Apis mellifera*). *Anim. Behav. 5:*7-11.

Free, J.B. (1959a). The effect of moving colonies of honeybees to new sites on their subsequent foraging behaviour. *J. Agric. Sci. 53:*1-9.

Free, J.B. (1959b). The transfer of food between the adult members of a honeybee community. *Bee World 40:*193-201.

Free, J.B. (1960). The pollination of fruit trees. *Bee World 41:*141-151, 169-186.

Free, J.B. (1965). The allocation of duties among worker honeybees. *Symp. Zool. Soc. Lond. 14:*39-59.

Free, J.B. (1987). Pheromones of Social Bees. Comstock Publ. Assoc., New York.

Free, J.B. & C.G. Butler. (1955). An analysis of the factors involved in the formation of a cluster of honeybees (*Apis mellifera*). *Behaviour 7:*304-316.

Free, J.B. & Y. Spencer-Booth. (1961). Further experiments on the drifting of honeybees. *J. Agric. Sci. 57:*153-158.

Free, J.B. & I.H. Williams. (1975). Factors determining the rearing and rejection of drones by the honeybee colony. *Anim. Behav. 23:*650-675.

Freudenstein, H. (1961). Über die Bedeutung der Pollennahrung für die Bautätigkeit der Honigbiene. *Arch. Bienenk. 38:*33-36.

Frummhoff, P.C. & J. Baker. (1988). A genetic component to division of labour within honey bee colonies. *Nature 333:*358-361.

Gahl, R.A. (1975). The shaking dance of honey bee workers: Evidence for age discrimination. *Anim. Behav. 23:*230-232.

Gary, N.E. (1960). A trap to quantitatively recover dead and abnormal honeybees from the hive. *J. Econ. Entomol. 53:*782-785.

Gary, N.E. (1962). Chemical mating attractants in the queen honey bee. *Science 136:*773-774.

Gary, N.E. (1963). Observations of mating behaviour in the honeybee. *J. Apic. Res. 2:*3-13.

Gary, N.E. (1966). Robbing behavior in the honeybee. *Am. Bee J. 106:*446-448.

Gary, N.E. (1971). Magnetic retrieval of ferrous labels in a capture-recapture system for honey bees and other insects. *J. Econ. Entomol. 64:*961-965.

Gary, N.E. (1974). Pheromones that affect the behavior and physiology of honey bees. *In:* Pheromones, M.C. Birch, *ed.* North-Holland Publ. Co., Amsterdam. pp. 200-221.

Gary, N.E. & J. Marston. (1971). Mating behaviour of drone honey bees with queen models (*Apis mellifera* L.). *Anim. Behav. 19:*299-304.

Gary, N.E. & P.C. Witherell. (1971). A method for training honey bees to forage at feeding stations. *Ann. Entomol. Soc. Am. 64:*448-449.

Gary, N.E. & P.C. Witherell. (1977). Distribution of foraging bees of three honey bee stocks located near onion and safflower fields. *Environ. Entomol. 6:*785-788.

Gary, N.E., P.C. Witherell & K. Lorenzen. (1978a). The distribution and foraging activities of common Italian and "Hy-Queen" honey bees during alfalfa pollination. *Environ. Entomol. 7:*233-240.

Gary, N.E., P.C. Witherell & K. Lorenzen. (1979). Distribution of honeybees during water collection. *J. Apic. Res. 18:*26-29.

Gary, N.E., P.C. Witherell, K. Lorenzen & J.M. Marston. (1977a). Area fidelity and intra-field distribution of honey bees during the pollination of onions. *Environ. Entomol.* 6:303-310.

Gary, N.E., P.C. Witherell, K. Lorenzen & J.M. Marston. (1977b). The interfield distribution of honey bees foraging on carrots, onions, and safflower. *Environ. Entomol.* 6:637-640.

Gary, N.E., P.C. Witherell & J.M. Marston. (1972). Foraging range and distribution of honey bees used for carrot and onion pollination. *Environ. Entomol.* 1:71-78.

Gary, N.E., P.C. Witherell & J.M. Marston. (1973). Distribution of foraging bees used to pollinate alfalfa. *Environ. Entomol.* 2:573-578.

Gary, N.E., P.C. Witherell & J.M. Marston. (1975). The distribution of foraging honey bees from colonies used for honeydew melon pollination. *Environ. Entomol.* 4:277-281.

Gary, N.E., P.C. Witherell & J.M. Marston. (1976). The inter- and intra-orchard distribution of honeybees during almond pollination. *J. Apic. Res.* 15:43-50.

Gary, N.E., P.C. Witherell & J.M. Marston. (1978b). Distribution and foraging activities of honeybees during almond pollination. *J. Apic. Res.* 17:188-194.

Gates, B.N. (1914). The temperature of the bee colony. *U.S.D.A. Bull. 96.*

Genrikh, V.G. (1957). [Influence of the queen on flight and hive activity of bees]. *Pchelovodstvo 34:*8-12. (in Russian).

Gontarski, H. (1938). Beobachtungen an Eierlegenden Arbeiterinnen. *Dt. Imkerführer 12:*107-113.

Gontarski, H. (1949). Über die Vertikalorientierung der Bienen beim Bau der Waben und bei der Anlage des Brutnestes. *Z. vergl. Physiol. 31:*652-670.

Goodwin, R.M. & R.D. Lewis. (1987). Honeybees use a biological clock to incorporate sun positions in their waggle dances after foraging under heavy overcast skies. *N. Z. Ent. 10:*138-140.

Gould, J.L. (1975a). Honey bee recruitment: The dance language controversy. *Science 189:*685-693.

Gould, J.L. (1975b). Communication of distance information by honey bees. *J. Comp. Physiol. 104:*161-173.

Gould, J.L. (1976). The dance-language controversy. *Quart. Rev. Biol. 51:*211-244.

Gould, J.L. (1980). Sun compensation by bees. *Science 207:*545-547.

Gould, J.L. (1984). Natural history of honey bee learning. *In*: The Biology of Learning, P. Marler & H.S. Terrace, *eds.* Springer-Verlag, New York. pp. 149-180.

Gould, J.L. (1985). How honey bees remember flower shapes. *Science 227:*1492-1494.

Gould, J.L. (1986). The locale map of honey bees: Do insects have cognitive maps? *Science 232:*861-863.

Gould, J.L., M. Henerey & M.C. MacLeod. (1970). Communication of direction by the honey bee. *Science 169:*544-554.

Gould, J.L. & P. Marler. (1984). Ethology and the natural history of learning. *In*: The Biology of Learning, P. Marler & H.S. Terrace, *eds.* Springer-Verlag, New York. pp. 47-74.

Hammann, E. (1957). Which takes the initiative in the virgin queen's flight, the queen or the workers? *Insectes Sociaux 4:*91-106.

Haydak, M.H. (1929). Několká Nová Pozorování ze Života Včel. *Cesky Včelar 63:*229-231.

Haydak, M.H. (1935). Brood rearing by honeybees confined to a pure carbohydrate diet. *J. Econ. Entomol. 28:*657-660.

Haydak, M.H. (1937a). Further contribution to the study of pollen substitutes. *Ann. Entomol. Soc. Am. 30:*258-262.

Haydak, M.H. (1937b). Changes in weight and nitrogen content of adult worker bees on a protein-free diet. *J. Agric. Res. 54*:791-796.

Haydak, M.H. (1945). The language of the honeybees. *Am. Bee J. 85*:316-317.

Haydak, M.H. (1949). The queen honeybee. *Iowa St. Apiarist Rep.* pp. 68-94.

Haydak, M.H. (1958). Do the nurse bees recognize the sex of larvae? *Science 127*:1113.

Haydak, M.H. (1960). Supersedure and swarming. *Glean. Bee Cult. 88*:265-269, 307.

Heberle, J.A. (1914). Notes from Germany: How many trips to the field does a bee make in a day? How long does it take to fetch one load? How long does a bee remain in the hive between trips? *Glean. Bee Cult. 42*:904-905.

Hellmich, R.L., J.M. Kulincevic & W.C. Rothenbuhler. (1985). Selection for high and low pollen hoarding honey bees. *J. Hered. 76*:155-158.

Himmer, A. (1926). Der soziale Wärmehaushalt der Honigbiene, I. Die Wärme im nichtbrütenden Wintervolk. *Erlanger Jb. Bienenk. 4*:1-51.

Hirschfelder, I. (1951). Quantitative Untersuchungen zum Polleneintragen der Bienenvolker. *Z. Bienenforsch. 1*:67-77.

Hodges, D. (1974). The Pollen Loads of the Honey Bee. Bee Research Association, London.

Hoffman, I. (1961). Über die Arbeitsteilung in weiselrichtigen und weisellosen Kleinvölkern der Honigbiene. *Z. Bienenforsch. 5*:267-279.

Howell, D.E. & R.L. Usinger. (1933). Observations on the flight and length of life of drone bees. *Ann. Entomol. Soc. Am. 26*:239-246.

Huber, F. (1814). New observations upon bees. Published in book form in 1926 by Dadant & Sons, Hamilton, Ill.

Istomina-Tsvetkova, K.P. (1953). [Reciprocal feeding of bees]. *Pchelovodstvo 30*:25-29. (in Russian).

Istomina-Tsvetkova, K.P. (1957). [Study of the activity of worker bees]. *Zool. Zuhr. 36*:1359-1370. (in Russian).

Jay, S.C. (1969). The problem of drifting in commercial apiaries. *Am. Bee J. 109*:178-179.

Jaycox, E.R. (1961). The effects of various foods and temperatures on sexual maturity of the drone honey bee (*Apis mellifera*). *Ann. Entomol. Soc. Am. 54*:519-523.

Johnson, D.L. (1967). Honey bees: Do they use the direction information contained in their dance maneuver? *Science 155*:844-847.

Kaiser, W. (1979). Circadian variations in the sensitivity of single visual interneurones of the bee *Apis mellifera carnica. Verh. dt. zool. Ges., 1979*:211.

Kalmus, H. (1957). Sun navigation of *Apis mellifera* L. in the southern hemisphere. *Bee World 38*:29-33.

Kalmus, H. & C.R. Ribbands. (1952). The origin of the odours by which honeybees distinguish their companions. *Proc. Roc. Soc. B 140*:50-59.

Kaminski, L.A., K.N. Slessor, M.L. Winston, N.W. Hay & J.H. Borden. (1990). Honey bee response to queen mandibular pheromone in laboratory bioassays. *J. Chem. Ecol. 16*:841-850.

Kappel, I. (1952). Die Form des Safthalters als Anreiz für die Sammeltätigkeit der Bienen. *Z. vergl. Physiol. 34*:539-546.

Kirchner, W.H. (1989). Freely flying honeybees use image motion to estimate object distance. *Naturwissenschaften 76*:281-282.

Kirchner, W.H., M. Lindauer & A. Michelsen. (1988). Honeybee dance communication: Acoustical indication of direction in round dances. *Naturwissenschaften 75*:629-630.

Koeniger, G. (1986). Mating sign and multiple mating in the honeybee. *Bee World 67*:141-150.

Koeniger, G., H. Koeniger & M. Fabritius. (1979). Some detailed observations of mating in the honeybee. *Bee World 60*:53-57.

Koptev, V.S. (1957). [Laying workers and swarming]. *Pchelovodstvo 34*:31-32. (in Russian).

Korst, P.J.A.M. & H.H.W. Velthuis. (1982). The nature of trophallaxis in honeybees. *Insectes Sociaux 29*:209-221.

Kurrenoi, N.M. (1953). [When are drones sexually mature?]. *Pchelovodstvo 30*:28-32. (in Russian).

Kurrenoi, N.M. (1954). [Flight activity and sexual maturity of drones]. *Pchelovodstvo 31*:24-28. (in Russian).

Lecomte, J. (1949). L'interattraction chez l'abeille. *C.R. Acad. Sci. 229*:857-858.

Lecomte, J. (1950). Sur le déterminisme de la formation de la grappe chez les abeilles. *Z. vergl. Physiol. 32*:499-506.

Lee, W.R. (1961). The nonrandom distribution of foraging honey bees between apiaries. *J. Econ. Entomol. 54*:928-933.

Lee, P.C. & M.L. Winston. (1987). Effects of reproductive timing and colony size on the survival, offspring colony size and drone production in the honey bee (*Apis mellifera*). *Ecol. Entomol. 12*:187-195.

Lensky, Y. & Y. Slabezki. (1981). The inhibiting effect of the queen bee (*Apis mellifera* L.) foot-print pheromone on the construction of swarming queen cups. *J. Insect Physiol. 27*:313-323.

Lensky, Y., P. Cassier, M. Nutkin, M. Delorme-Joulie & M. Levinson. (1985). Pheromonal activity and fine structure of the mandibular glands of the honeybee drones. (*Apis mellifera* L.)(Insecta, Hymenoptera, Apidae). *J. Insect Physiol. 31*:265-276.

Levchenko, I.A. (1959). [The distance bees fly for nectar]. *Pchelovodstvo 36*:37-38. (in Russian).

Levenets, I.P. (1954). [The flight range of drones]. *Pchelovodstvo 31*:36-38. (in Russian).

Levenets, I.P. (1956). [Observations on the expulsion of drones]. *Pchelovodstvo 33*:28-29. (in Russian).

Levin, M.D. (1959). Distribution patterns of young and experienced honey bees foraging on alfalfa. *J. Econ. Entomol. 52*:969-971.

Levin, M.D. (1961a). Distribution of foragers from honey bee colonies placed in the middle of a large field of alfalfa. *J. Econ. Entomol. 54*:431-434.

Levin, M.D. (1961b). The dispersion of field bees on alfalfa in relation to a neighboring apiary. *J. Econ. Entomol. 54*:482-484.

Lindauer, M. (1951). Die Temperaturregulierung der Bienen bei Stocküberhitzung. *Naturwissenschaften 38*:308-309.

Lindauer, M. (1953). Division of labor in the honeybee colony. *Bee World 34*:63-73, 85-90.

Lindauer, M. (1954a). Dauertänze im Bienenstock und ihre Beziehung zur Sonnenbahn. *Naturwissenschaften 41*:506-507.

Lindauer, M. (1954b). [The water economy and temperature regulation of the honeybee colony]. *Z. vergl. Physiol. 36*:391-432. (in German).

Lindauer, M. (1955a). The water economy and temperature regulation of the honeybee colony. *Bee World 36*:62-72, 81-92.

Lindauer, M. (1955b). Schwarmbienen auf Wohnungssuche. *Z. vergl. Physiol. 36*:263-324.

Lindauer, M. (1957). Sonnenorientierung der Bienen unter der Aequatorsonne und zur Nachtzeit. *Naturwissenschaften 44*:1-6.

Lindauer, M. (1959). Angeborene und erlernte Komponenten in der Sonnenorientierung der Bienen. *Z. vergl. Physiol. 42*:43-62.

Lindauer, M. (1971). Communication Among Social Bees. Harvard Univ. Press, Cambridge, Mass.

Lindauer, M. & J.O. Nedel. (1959). Ein Schweresinnesorgan der Honigbiene. *Z. vergl. Physiol. 42*:334-364.

Locke, S.J. & N.E. Gary. (1988). Africanized honey bee: Hybrids of *Apis mellifera scutellata* Lepeletier. *Calif. Dept. Food Agric. Exotic Pest Profile.*

Lopatina, N.G. (1956). [Bee dances]. *Pchelovodstvo 33*:19-24. (in Russian).

Loper, G.M., W.W. Wolf & O.R. Taylor, Jr. (1987). Detection and monitoring of honeybee drone congregation areas by radar. *Apidologie 18*:163-172.

Louveaux, J. (1958). Recherches sur la récolte du pollen par les abeilles (*Apis mellifica* L.). *Ann. Abeille 1*:113-188, 197-221.

Louveaux, J. (1959). Recherches sur la récolte du pollen par les abeilles (*Apis mellifica* L.). *Ann. Abeille 2*:13-111.

Lukoschus, F. (1957). Quantitative Untersuchungen über den Pollentransport im Haarkleid der Honigbiene. *Z. Bienenforsch. 4*:3-21.

Lundie, A.E. (1925). The flight activities of the honeybee. *U.S.D.A. Bull. 1328.*

MacDonald, J.L. (1968). The behavior of pollen foragers which lose their load. *J. Apic. Res. 7*:45-46.

Maurizio, A. (1953). Wertere Untersuchungen an Pollenhöschen. *Beih. Schweiz. Bienenztg. 2*:486-556.

McManus, J. (ed.). (1988). A way to bug the killer bees. *Time 132*:45.

Meyer, W. (1954). Die 'Kittharzbienen' un ihre Tätigkeiten. *Z. Bienenforsch. 2*:185-200.

Meyer, W. (1956). Arbeitsteilung im Bienenschwarm. *Insectes Sociaux 3*:303-324.

Meyer, W. & W. Ulrich. (1952). Zur Analyse der Bauinstinkte unserer Honigbiene. Untersuchungen über die "Kleinbauarbeiten." *Naturwissenschaften 39*:264.

Michelsen, A. & B.B. Anderson. (1989). Honeybees can be recruited by a mechanical model of a dancing bee. *Naturwissenschaften 76*:277-280.

Michelsen, A., W.H. Kirchner & M. Lindauer. (1986). Sound and vibrational signals in the dance language of the honeybee, *Apis mellifera. Behav. Ecol. Sociobiol. 18*:207-212.

Michelsen, A., W.F. Towne, W.H. Kirchner & P. Kryger. (1987). The acoustic near field of a dancing honeybee. *J. Comp. Physiol. A 161*:633-643.

Michener, C.D. (1975). The Brazilian bee problem. *Ann. Rev. Entomol. 20*:399-416.

Mikhailov, A.C. (1928). [On drone flight activity]. *Opyt. Paseka 3*:209-214. (in Russian).

Milum, V.G. (1947). Grooming dance and associated activities of the honeybee colony. *Trans. Ill. St. Acad. Sci. 40*:194-196.

Milum, V.G. (1955). Honeybee communication. *Am. Bee J. 95*:97-104.

Milum, V.G. (1956). The significance of some honeybee dances. *Proc. X Int. Congr. Entomol.* pp. 1085-1088.

Möbus, B. (1987). The swarm dance and other swarm phenomena. *Am. Bee J. 127*:249-251, 253-255, 356-362.

Moore, D. & M.A. Rankin. (1985). Circadian locomotor rhythms in individual bees. *Physiol. Entomol. 10*:191-197.

Moritz, R.F.A. & M. Hallmen. (1986). Trophallaxis of worker honeybees (*Apis mellifera* L.) of different ages. *Insectes Sociaux 33*:26-31.

Moritz, R.F.A. & E. Hillesheim. (1985). Inheritance of dominance in honeybees (*Apis mellifera capensis* Esch.). *Behav. Ecol. Sociobiol. 17*:87-89.

Moritz, R.F.A., E.E. Southwick & J.B. Harbo. (1987). Genetic analysis of defensive behaviour of honeybee colonies (*Apis mellifera* L.) in a field test. *Apidologie 18*:27-52.

Naumann, K., M.L. Winston, M.H. Wyborn & K.N. Slessor. (1990). Effects of synthetic honey bee (Hymenoptera: Apidae) queen mandibular gland pheromone on workers in packages. *J. Econ. Entomol. 83*:1271-1275.

Nelson, J., A.P. Sturtevant & B. Lineburg. (1924). Growth and feeding of honeybee larvae. *U.S.D.A. Bull. 1222.*

New, D.A.T., F.R. Burrows & A.F. Edgar. (1961). Honeybee communication when the sun is close to the zenith. *Nature 189*:155-156.

Nolan, J.W. (1925). The broodrearing cycle of the honeybee. *U.S.D.A. Bull. 1349.*

Nye, W.P. & O. Mackensen. (1968). Selective breeding of honeybees for alfalfa pollen collection: Fifth generation and backcrosses. *J. Apic. Res. 7*:21-27.

Örösi-Pál, Z. (1957). The role of the mandibular glands of the honeybee. *Bee World 38*:70-73.

Örösi-Pál, Z. (1959). The behavior and nutrition of drones. *Bee world 40*:141-146.

Oertel, E. (1940). Mating flights of queen bees. *Glean. Bee Cult. 68*:292-293, 333.

Oertel, E. (1956). Observations on the flight of drone honeybees. *Ann. Entomol. Soc. Am. 49*:497-500.

Otis, G.W. (1982). Population biology of the Africanized honey bee. *In*: Social Insects in the Tropics, P. Jaisson, *ed.* Proc. Symp. I.U.S.S.I. Cocoyoc, Mexico. pp. 209-219.

Park, O.W. (1922). Time and labor factors in honey and pollen gathering. *Am. Bee J. 62*:254-255.

Park, O.W. (1923a). Flight study of the honeybee. *Am. Bee J. 63*:71.

Park, O.W. (1923b). Behavior of water-carriers. *Am. Bee J. 63*:553.

Park, O.W. (1925a). The minimum flying weight of the honeybee. *Iowa St. Apiarist Rep.* pp. 83-90.

Park, O.W. (1925b). The storing and ripening of honey by honeybees. *J. Econ. Entomol. 18*:405-410.

Park, O.W. (1925c). Field bees and their work. *VII Int. Beekeep. Congr., Quebec.* pp. 472-478.

Park, O.W. (1929). Time factors in relation to the acquisition of food by the honeybee. *Iowa Agric. Exp. Sta. Res. Bull. 108.*

Park, O.W. (1932). Studies on the changes in nectar concentration produced by the honey bee, *Apis mellifera. Iowa Agric. Exp. Sta. Res. Bull. 151.* pp. 211-244.

Park, O.W. (1949). Activities of honey bees. *In*: The Hive and the Honey Bee, R.A. Grout, *ed.* Dadant & Sons, Hamilton, Ill. pp. 79-152.

Parker, R.L. (1926). The collection and utilization of pollen by the honey bee. *Cornell Univ. Agric. Exp. Sta. Mem. 98.*

Parks, H.B. (1929). Water storage by bees. *Iowa St. Apiarist Rep.* pp. 53-56.

Peer, D.F. (1956). Multiple mating of queen honey bees. *J. Econ. Entomol. 49*:741-743.

Peer, D.F. (1957). Further studies on the mating range of the honey bee, *Apis mellifera* L. *Can. Entomol. 89*:108-110.

Perepelova, L.I. (1928). [Laying workers, the egg-laying activity of the queen, and swarming]. *Opyt. Paseka 3*:214-217. (in Russian).

Perepelova, L.I. (1959). Priverzhennost pchel k mestu Vziatka. *Pchelovodstvo 36*:30-32.

Phillips, E.F. & G. Demuth. (1914). The temperature of the honeybee cluster in winter. *U.S.D.A. Bull. 93.* pp. 1-16.

Rashad, S.E.D. (1957). Some factors affecting pollen collection by honeybees and pollen as a limiting factor in brood rearing and honey production. Ph.D. Thesis, Kansas St. Univ.

Rauschmeyer, F. (1928). Das Verfliegen der Bienen und die optische Orientierung am Bienenstand. *Arch. Bienenk. 9*:249-322.

Reinhardt, J.F. (1952). Some responses of honey bees to alfalfa flowers. *Am. Naturalist 86*:257-275.

Renner, M. (1960). The contribution of the honey bee to the study of time-sense and astrological orientation. *Cold Spring Harb. Symp. Quant. Biol. 25*:361-367.

Ribbands, C.R. (1949). The foraging method of individual honeybees. *J. Anim. Ecol. 18*:47-66.

Ribbands, C.R. (1952). Division of labour in the honeybee community. *Proc. R. Soc. B 140*:32-43.

Ribbands, C.R. (1953). The Behaviour and Social Life of Honeybees. Jarrold & Sons, Norwich.

Ribbands, C.R. (1954). The defence of the honeybee community. *Proc. Roy. Soc. B 142*:514-524.

Ribbands, C.R. (1955a). The scent perception of the honeybee. *Proc. Roy. Soc. B 143*:367-379.

Ribbands, C.R. (1955b). Community defence against robber bees. *Am. Bee J. 95*:313, 320.

Rinderer, T.E. (1982). Regulated nectar harvesting by the honeybee. *J. Apic. Res. 21*:74-87.

Rinderer, T.E. & J.R. Baxter. (1979). Honey bee hoarding behaviour. Effects of previous stimulation by empty comb. *Anim. Behav. 27*:426-428.

Rinderer, T.E. & W.A. Hagstad. (1984). The effect of empty comb on the proportion of foraging honeybees collecting nectar. *J. Apic. Res. 23*:80-81.

Robinson, G.E. & R.E. Page, Jr. (1988). Genetic determination of guarding and undertaking in honey-bee colonies. *Nature 333*:356-358.

Robinson, G.E. & R.E. Page, Jr. (1989). Genetic determination of nectar foraging, pollen foraging, and nest-site scouting in honey bee colonies. *Behav. Ecol. Sociobiol. 24*:317-323.

Rösch, G.A. (1925). Untersuchungen über die Arbeitsteilung im Bienenstaat. *Z. vergl. Physiol. 2*:571-631.

Rösch, G.A. (1930). Untersuchungen über die Arbeitsteilung im Bienenstaat. *Z. vergl. Physiol. 12*:1-71.

Rosin, R. (1978). The honey bee "language" controversy. *J. Theoret. Biol. 72*:589-602.

Rosin, R. (1980a). Paradoxes of the honey-bee "dance language" hypothesis. *J. Theoret. Biol. 84*:775-800.

Rosin, R. (1980b). The honey-bee "dance language" hypothesis and the foundations of biology and behavior. *J. Theoret. Bio. 87*:457-481.

Rosin, R. (1988). Do honey bees still have a "dance language?" *Am. Bee J. 128*:267-268.

Rothenbuhler, W.C. (1964). Behaviour genetics of nest cleaning in honey bees. IV. Responses of F_1 and backcross generations to disease-killed brood. *Am. Zoologist 4*:111-123.

Rothenbuhler, W.C. (1968). Bee genetics. *Ann. Rev. Genet. 2*:413-438

Ruttner, F. (1956). The mating of the honeybee. *Bee World 37*:3-15, 23-24.

Ruttner, F. (1957). Die Sexualfunktionen der Honigbienen im Dienste ihrer sozialen Gemeinschaft. *Z. vergl. Physiol. 39*:577-600.

Ruttner, F. (1966). The life and flight activity of drones. *Bee World 47*:93-100.

Ruttner, H. & F. Ruttner. (1972). Untersuchungen über die Flugaktivität und das Paarungsverhalten der Drohnen. V. Drohnensammelplätze und Paarungsdistanz. *Apidologie 3*:203-232.

Sakagami, S.F. (1954). Occurrence of an aggressive behaviour in queenless hives with considerations on the social organization of honeybees. *Insectes Sociaux 1*:331-343.

Sakagami, S.F. (1958). The false-queen: Fourth adjustive response in dequeened honeybee colonies. *Behaviour 13*:280-296.

Sakagami, S.F. (1959). Arbeitsteilung in einem weisellosen Bienenvölkchen. *Z. Bienenforsch. 4*:186-193.

Schick, W. (1953). Über die Wirkung von Giftstoffen auf die Tänze der Bienen. *Z. vergl. Physiol. 35*:105-128.

Schneider, F. (1949). Über die Vergiftung der Bienen mit Dinitrokresol und das Auftreten von Tänzen als Reaktion auf die Verteilung des Giftes im Stock. *Mitt. Schweiz. Ent. Ges. 22*:293-308.

Schneider, S.S. (1987). The modulation of worker activity by the vibration dance of the honeybee, *Apis mellifera. Ethology 74*:211-218.

Schneider, S.S., J.A. Stamps & N.E. Gary. (1986a). The vibration dance of the honey bee. I Communication regulating foraging on two time scales. *Anim. Behav. 34*:377-385.

Schneider, S.S., J.A. Stamps & N.E. Gary. (1986b). The vibration dance of the honey bee. II. The effects of foraging success on daily patterns of vibration activity. *Anim. Behav. 34*:386-391.

Seeley, T.D. (1982). Adaptive significance of the age polyethism schedule in honeybee colonies. *Behav. Ecol. Sociobiol. 11*:287-293.

Seeley, T.D. (1985). Honeybee Ecology. Princeton Univ. Press, Princeton, New Jersey.

Seeley, T.D., R.A. Morse & P.K. Visscher. (1979). The natural history of the flight of honey bee swarms. *Psyche 86*:103-113.

Sekiguchi, K. & S.F. Sakagami. (1966). Structure of foraging population and related problems in the honeybee, with considerations on the division of labor in bee colonies. *Hokkaido Natn. Agric. Exp. Stn. Rep. 69.*

Severson, D.W. (1984). Swarming behavior of the honey bee. *Am. Bee J. 124*:204-210, 230-232.

Simpson, J. (1958). The factors which cause colonies of *Apis mellifera* to swarm. *Insectes Sociaux 5*:77-95.

Singh, S. (1950). Behavior studies of honeybees in gathering nectar and pollen. *Cornell Univ. Agric. Exp. Sta. Mem. 288.*

Slessor, K.N., L.A. Kaminski, G.G.S. King, J.H. Borden & M.L. Winston. (1988). Semiochemical basis of the retinue response to queen honey bees. *Nature 332*:354-356.

Slessor, K.N., L.A. Kaminski, G.G.S. King & M.L. Winston. (1990). Semiochemicals of the honey bee queen mandibular glands. *J. Chem. Ecol. 16*:851-860.

Smith, M.V. (1959). A note on the capping activities of an individual honeybee. *Bee World 40*:153-154.

Southwick, E.E. & R.F.A. Moritz. (1987). Effects of meteorological factors on defensive behaviour of honey bees. *Int. J. Biometerol. 31*:259-265.

Spangler, H.G. (1972). Daily activity rhythms of individual worker and drone honey bees. *Ann. Entomol. Soc. Am. 65*:1073-1076.

Spitzner, M.J.E. (1788). Ausführliche Beschreibung der Korbbienenzucht im sächsischen Churkreis, ihrer Dauer und ihres Nutzens, ohne künstliche Vermehrung nach den Gründen der Naturgeschichte und nach eigener langer Erfahrung. Leipzig.

Taber, S. (1954). The frequency of multiple mating of queen honey bees. *J. Econ. Entomol. 47*:995-998.

Taber, S. (1964). Factors influencing the circadian flight rhythm of drone honeybees. *Ann. Entomol. Soc. Am. 57*:769-775.

Taber, S. & J. Wendel. (1958). Concerning the number of times queen bees mate. *J. Econ. Entomol. 51*:786-789.

Taranov, G.F. (1947). [Occurrence and development of swarming instinct in bee colonies]. *Pchelovodstvo 24*:44-54. (in Russian).

Taranov, G.F. (1959). The production of wax in the honeybee colony. *Bee World 40*:113-121.

Thorp, R.W., D.L. Briggs, J.R. Estes & E.H. Erickson. (1975). Nectar fluorescence under ultraviolet irradiation. *Science 189*:476-478.

Towne, W.F. & W.H. Kirchner. (1989). Hearing in honey bees: Detection of air-particle oscillations. *Science 244*:686-688.

Triasko, V.V. (1956). [The mating sign of the queen and its characteristics]. *Pchelovodstvo 33*:43-50. (in Russian).

Tuenin, T.A. (1926). Concerning laying workers. *Bee World 8*:90-91.

Vansell, G.H. (1942). Factors affecting the usefulness of honeybees in pollination. *U.S.D.A. Circ. 650*.

Veldink, C. (1989). The honey-bee language controversy. *Interdisciplinary Science Rev. 14*:166-175.

Visscher, P.K. (1983). The honey bee way of death: Necrophoric behaviour in *Apis mellifera* colonies. *Anim. Behav. 31*:1070-1076.

Visscher, P.K., R.A. Morse & T.D. Seeley. (1985). Honey bees choosing a home prefer previously occupied cavities. *Insectes Sociaux 32*:217-220.

von Frisch, K. (1920). Über die "Sprache" der Bienen. *Münch. Med. Wschr.* pp. 566-569.

von Frisch, K. (1923). Über die "Sprache" der Bienen, eine tierpsychologische Untersuchung. *Zool. Jb., Abt. allg. Zool. Physiol. Tiere. 40*:1-186.

von Frisch, K. (1946). Die "Sprache" der Bienen und ihre Nutzanwendung in der Landwirtschaft. *Experientia 2*:397-404.

von Frisch, K. (1952). Orienting ability and communication among bees. *Bee World 33*:35-40.

von Frisch, K. (1958). The solar compass as the basis of communication in the colony. *Am. Bee J. 98*:100-101.

von Frisch, K. (1967a). The Dance Language and Orientation of Bees. The Belknap Press of Harvard Univ. Press, Cambridge, Mass.

von Frisch, K. (1967b). Honeybees: Do they use direction and distance information provided by their dancers? *Science 158*:1072-1075.

von Frisch, K. & R. Jander. (1957). Über den Schwänzeltanz der Bienen. *Z. vergl. Physiol. 40*:239-263.

von Frisch, K. & M. Lindauer. (1954). Himmel und Erde in Konkurrenz bei der Orientierung der Bienen. *Naturwissenschaften 41*:245-253.

Washington, R.J. (1967). Drones. *Am. Bee J. 107*:7-9.

Weaver, N. (1957). The foraging behavior of honeybees on hairy vetch. *Insectes Sociaux 4*:43-57.

Wellington, W.G. & D. Cmiralova. (1979). Communication of height by foraging honey bees. *Apis mellifera ligustica* (Hymenoptera: Apidae). *Ann. Entomol. Soc. Am.* 72:167-170.

Wells, H. & P.H. Wells. (1983). Honey bee foraging ecology: Optimal diet, minimal uncertainty or individual constancy? *J. Anim. Ecol.* 52:829-836.

Wells, P.H. & H. Wells. (1984). Can honey bees change foraging patterns? *Ecol. Entomol.* 9:467-473.

Wells, P.H. & H. Wells. (1985). Ethological isolation of plants. 2. Odour selection by honeybees. *J. Apic. Res.* 24:86-92.

Wells H., P.H. Wells & D.M. Smith. (1981). Honeybee responses to reward size and colour in an artificial flower patch. *J. Apic. Res.* 20:172-179.

Wells, H., P.H. Wells & D.M. Smith. (1983). Ethological isolation of plants. 1. Colour selection by honeybees. *J. Apic. Res.* 22:33-44.

Wenner, A.M. (1961a). Division of labor in a honey bee colony — A Markov process? *J. Theoret. Biol.* 1:324-327.

Wenner, A.M. (1961b). A method of training bees to visit a feeding station. *Bee World* 42:8-11.

Wenner, A.M. (1962). Sound production during the waggle dance of honey bees. *Anim. Behav.* 10:79-95.

Wenner, A.M. (1967). Honeybees: Do they use the distance information contained in their dance maneuver? *Science* 155:847-849.

Wenner, A.M. (1971). The Bee Language Controversy. An Experience in Science. Educ. Prog. Improv. Corp., Boulder, Colorado.

Wenner, A.M. & D.L. Johnson. (1967). Honey bees: Do they use the direction and distance information provided by their dancers? *Science* 158:1076-1077.

Wenner, A.M. & P.H. Wells. (1987). The honey bee dance language controversy: The search for "truth" vs. the search for useful information. *Am. Bee J.* 127:130-131.

Wenner, A.M. & P.H. Wells, (1990). Anatomy of a Controversy. Columbia Univ. Press, New York.

Whitcomb, W., Jr. (1946). Feeding bees for comb production. *Glean. Bee Cult* 74:198-292, 247.

Winston, M.L. (1979). Intra-colony demography and reproductive rate of the Africanized honeybee in South America. *Behav. Ecol. Sociobiol.* 4:279-292.

Winston, M.L. (1980). Swarming, afterswarming, and reproductive rate of unmanaged honeybee colonies (*Apis mellifera*). *Insectes Sociaux* 27:391-398.

Winston, M.L. & L.A. Fergusson. (1986). Influence of the amount of eggs and larvae in honeybee colonies on temporal division of labour. *J. Apic. Res.* 25:238-241.

Winston, M.L. & S.J. Katz. (1982). Foraging differences between cross-fostered honeybee workers (*Apis mellifera*) of European and African races. *Behav. Ecol. Sociobiol.* 10:125-129.

Winston, M.L. & E.N. Punnett. (1982). Factors determining temporal division of labor in honeybees. *Can. J. Zool.* 60:2947-2952.

Winston, M.L., H.A. Higo & K.N. Slessor. (1990). Effect of various dosages of queen mandibular gland pheromone on the inhibition of queen rearing in the honey bee (Hymenoptera: Apidae). *Ann. Entomol. Soc. Am.* 83:234-238.

Winston, M.L., K.N. Slessor, M.J. Smirle & A.A. Kandil. (1982). The influence of a queen-produced substance, 9HDA, on swarm clustering behavior in the honeybee *Apis mellifera* L. *J. Chem. Ecol.* 8:1283-1288.

Winston, M.L., K.N. Slessor, L.G. Willis, K. Naumann, H.A. Higo, M.H. Wyborn & L.A. Kaminski. (1989). The influence of queen mandibular pheromones on worker attraction to swarm clusters and inhibition of queen rearing in the honey bee (*Apis mellifera* L.). *Insectes Sociaux 36*:15-27.

Witherell, P.C. (1965). Survival of drones following eversion. *Ann. Abeille 8*:317-320.

Witherell, P.C. (1970). Analysis of some aspects of flight behavior of the drone honey bee (*Apis mellifera* L.), with particular reference to the effects of age and genetic stock. M.S. Thesis, Univ. Calif., Davis.

Wittekindt, W. (1961). An understanding of dancing behavior. *Am. Bee J. 101*:434-436.

Woodrow, A.W. (1941). Drying and storing pollen trapped from honey bee colonies. *U.S.D.A. Publ. E-529.*

Woyke, J. (1956). Mnogokratnoe Sparivanie Pchelinoi matki vo vremia odnogo brachnogo vyleta. *Pchelovodstvo 33*:32-36.

CHAPTER 9

HONEY BEE PHEROMONES

by MURRAY S. BLUM*

W hile it is obvious that animals must communicate in order to integrate their social activities, it has only been recently that we have realized that, for the most part, this communication is accomplished with chemical messengers. Indeed, in terms of the whole animal kingdom, visual and acoustical signals are utilized by a relatively small number of species, whereas chemical messages appear to be characteristic of most organisms. This is especially true of insects for which most of our understanding of information transfer with chemical agents has been derived. A multitude of intraspecific insect behaviors such as sexual attraction, dispersion, aggregation, trail following, and stimulation of aggression are now known to be regulated by chemical stimuli. It has become clear that many of the characteristic behaviors of insects and other animals, previously described in purely anecdotal terms, are triggered by volatile compounds secreted by one individual—the emitter—and detected by other individuals (receivers) of the same species. And among highly social species such as bees, ants, and termites, chemical signals have played a major role in making it possible for their complex societies to have achieved such spectacular successes. Furthermore, among invertebrates, it is not unlikely that the most highly evolved system of chemical communication is possessed by the honey bee.

Although honey bees transmit some information with acoustical and visual signals, most of their known messages are generated with chemical cues (Free, 1987). It is now possible to associate many of the characteristic behaviors of queens, workers, and drones with specific compounds that have been identified as key elements in the honey bee's chemical "language." The availability of these signaling agents explains how it is possible to rapidly galvanize the great resources of the colonial members in order to exhibit the group response, a hallmark of highly specialized insect societies. In essence, insects like the honey bee have evolved a chemisocial lifestyle that guarantees that the collective resources of the colony are available to be utilized whenever required.

Since chemical communication differs in so many particulars from other methods of information transfer (*e.g.*, acoustical), some of the characteristics of chemical signal systems will be examined in order to highlight their unique communication qualities.

---

*Murray S. Blum, Ph.D., Research Professor of Entomology, University of Georgia, Athens, GA 30602. Studies of chemical communication, physiology, and behavior of social insects. Author of over 300 scientific papers and books on bees and other insects.

PHEROMONES AS CHEMICAL MESSENGERS

In contrast to hormones, which are internal chemicals only affecting the individual that produces and secretes them, chemical compounds functioning as messengers between individuals of the same species are secreted to the exterior. Thus, these external signaling agents carry information to receiver individuals and their response to the chemical message is considered to be favorable to the emitter, the animal that produces the signal. These chemical messengers are called *pheromones*, a word derived from the Greek "pherein" meaning to transfer and "hormon" meaning to excite. A pheromone may consist of a single chemical compound or a mixture of compounds that is produced and secreted from a specific organ called an *exocrine gland* (see below). However, not all of the compounds produced in these glands function as chemical messengers and in many cases no function can be assigned to specific compounds.

All compounds produced in glands that secrete to the exterior are called *semiochemicals*, a term that includes the pheromones which are the chemicals utilized for communication between individuals of the same species (=intraspecific communication). This transfer of information between animals belonging to the same species is considered to be true communication in the sense that there is "intent" on the part of the emitter to transmit a specific message to receiver individuals. Furthermore, the chemical messages (pheromones) transmitted between individuals of the same species constitute a "language" for which only the receiver individuals can decipher the information encoded in the chemical signal. In a real sense, as will be described subsequently, specific pheromones are equivalent to words or phrases that can state, for example, "I am the queen," "Come to me," or "I am lost." Large societies such as that of the honey bee require an effective—and rapid—means of integrating societal activities, and pheromones have been demonstrated to function admirably in this context.

While pheromones constitute intraspecific semiochemicals, other compounds produced in exocrine glands are utilized in interactions between individuals belonging to different species. These compounds, the *allomones*, are utilized in interspecific situations and as such, are not regarded as true communication agents in the same sense as are pheromones. The toxins in the venom of worker bees are classified as allomones, as are the odorants from flowers that attract pollinators. In neither case can these products—venom or flower fragrance—be considered as constituting elements in a "language," as have been described for pheromones. Although worker bees stinging a bear clearly demonstrate their displeasure with the mammal, it can hardly be concluded that the injection of venom constitutes communication in the same sense as the pheromonal attraction of workers to the disturbance (=bear).

Honey bees "speak" to honey bees, not bears, but if they did, these insects would undoubtedly condemn these mammals for being such cruel honey plunderers.

Although allomones differ from pheromones in being interspecific agents whereas the latter are intraspecific, both classes of compounds share an important characteristic. For the emitter individual the secretion of either type of semiochemical results in reactions on the part of the target individuals that are favorable to the emitter. For a guard bee at the colonial entrance, the use of pheromones to attract other workers to intruders is a highly favorable act. Similarly, painting a defensive (repellent) allomone on a predatory wasp can be considered a highly beneficial act for a guard bee, especially if the intruder is repelled.

In discussing communication with pheromones, it should be emphasized that this type of information transfer occurs, for the most part, in the milieu of the colony where social interactions are maximal. Although some pheromonal interactions occur relatively far from the nest, maximal use of these chemical signals is realized where the population density is greatest, in the vicinity of the nest. This is highly adaptive since it ensures, for example, that the great defensive resources of the colony can be rapidly mobilized to defend the nest, and its inhabitants, against aggressive predators. Colonial defense, which is a hallmark of social insects, is identified with protection of the nest and ultimately the queen, whose survival is equivalent to that of the colony. It is therefore no surprise that stinging incidents with bees, ants, and wasps are most frequently encountered in the vicinity of their nests.

CATEGORIES OF PHEROMONES

Pheromones perceived by specialized sensory cells called chemoreceptors may be detected as an odor (olfaction) or a taste (gustation). In insects the primary olfactory organs are the antennae which usually contain chemoreceptor cells that respond specifically to pheromones that are often fairly volatile and diffuse rapidly. Some pheromones are much less volatile and are primarily detected through oral contact with chemoreceptors distributed around the mouth. These pheromones are sometimes referred to as contact pheromones.

In general, two broad classes of pheromones are recognized by behaviorists. *Releaser pheromones* trigger an almost immediate behavioral response in receiving individuals that is mediated wholly by the nervous system. These pheromones are referred to as chemical releasers of behavior whose actions can be rapidly monitored in terms of a stimulus (pheromone) - response (behavior). Many compounds that function as attractants (*e.g.*, sex attractants) are classified as releaser pheromones. Most of the known honey bee pheromones are placed in this category.

In addition to these chemical releasers, a second class of compounds is referred to as *primer pheromones*. These compounds are also detected by chemosensory organs which probably transmit their messages directly to the central nervous system. In a sense, these pheromones "prime" the receiver individual to exhibit an altered behavioral activity at some future time, but significantly, no immediate behavioral responses occur. Primer pheromones physiologically alter reproductive and endocrine (hormonal) systems so that eventually (*e.g.*, 2-3 days) the receiver animal exhibits new behaviors when exposed to certain stimuli. This class of pheromones, characteristically produced by mammalian species, such as mice, is exemplified in insects by the queen substance pheromone of the honey bee which inhibits ovarian development in workers (see below.)

While the division of pheromones into two major categories, releasers and primers, reflects our current understanding of the specific actions of these compounds, it is important to realize that a single pheromone can function as both a releaser and primer (see below). Animals appear to have increased their communicative potential by adapting pheromones to subserve an incredible variety of roles and this is particularly true of social insects. As will be described subsequently, the honey bee has proven to be an outstanding example of how a highly complex societal animal has considerably expanded its social horizons by utilizing pheromones with remarkable versatility (Free, 1987).

EXOCRINE GLANDS

Any gland whose products can be secreted to the exterior can be regarded as a potential exocrine gland. Pheromones and/or allomones may be in admixture with nonsemiochemicals such as waste products, as is the case for pheromones that are present in the urine of many animals. Indeed, glands with well-established functions, completely unrelated to chemical communication, have been subsequently demonstrated to produce pheromones as well. This is the case, for example, for the venom glands of some species of ants and the salivary glands of male bumble bees. Significantly, the finding that supposed nonexocrine glands are also utilized to generate external chemical signals emphasizes the great adaptability of animals such as social insects in exploiting these glands to function as social structures. This is a theme that we shall find well developed in the honey bee.

The known exocrine glands and related structures of a composite queen-worker honey bee, are presented in Figure 1, but it will not prove surprising if unrecognized exocrine glands are present as well. This Figure does not represent the drone bee which lacks most of these glands or possesses them as small or vestigial structures (*e.g.*, labial [=salivary] glands). Significantly, some of these glands may also be highly developed, vestigial, or absent in queens or workers. The consequences of these structural differences can be of great

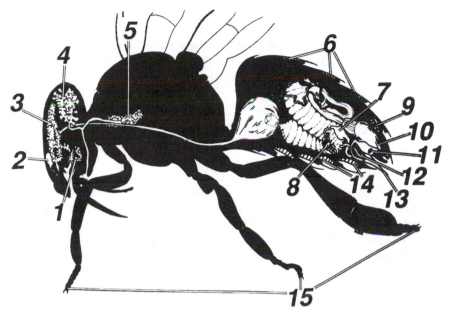

FIGURE 1. Exocrine glands and their reservoirs of a composite queen-worker honey bee (modified from Snodgrass, 1956, Wilson, 1971, and Michener, 1974). (1) Postgenal (=hypostomal) gland; (2) Mandibular glands; (3) Hypopharyngeal (=maxillary) gland; (4) Cephalic labial gland; (5) Thoracic labial gland; (6) Tergal glands; (7) Reservoir of poison gland; (8) Poison gland; (9) Nasonov gland; (10) Rectum; (11) Koschevnikov gland; (12) Dufour's gland; (13) Sting shaft with setose membrane; (14) Wax glands; (15) Tarsal (Arnhart) glands.

behavioral significance, especially since some of these glands are known to be social organs that generate important chemical signals (Free, 1987). On the other hand, several of these exocrine glands are not known to produce pheromones, and both their known functions and chemistry will only be briefly treated here before discussing the glands that are utilized for chemical communication.

Nonpheromone-Producing Exocrine Glands

The hypostomal (postgenal) (Fig. 1, No. 1), hypopharyngeal (maxillary) (Fig. 1, No. 3), and cephalic labial (salivary) (Fig. 4, No. 4), and thoracic labial glands (Fig. 1, No. 5) appear to be involved in the processing of either adult or larval food. The hypopharyngeal glands are relatively well developed in workers but not queens, a fact that undoubtedly reflects the utilization of these glands to provide food by workers but not queens. On the other hand, the hypostomal and labial glands are large in both female castes.

The hypopharyngeal glands, which are most highly developed in young workers, produce a variety of compounds that fortify food fed to larvae. In addition, these glands produce the enzyme sucrase (invertase), a key compound in the conversion of nectars into honey (Winston, 1987). The salivary glands appear to be involved in metabolizing sugars, probably enzymatically, cleaning the queen, and in the case of the workers, processing wax scales.

The two sting-associated glands have not been reported to produce pheromones in the case of either the queen or the workers. Both the poison gland (Fig. 1, No. 7) and Dufour's gland (Fig. 1, No. 12) are more highly developed in queens than workers, and this is especially true of the Dufour's gland. The poison gland, which was previously and incorrectly, called the acid gland, produces the proteins and peptides that fortify the venom secreted through the sting. The vesicle (Fig. 1, No. 8) attached to the poison gland appears to play a role in the acquisition of precursors from blood for the synthesis of venom compounds. Queens, but not workers, produce functional venom at the time of adult emergence, and the venom sac of the queen contains about 3 times more venom than that of the worker. Since virgin queens may frequently kill their sisters by stinging shortly after they emerge in the nest, the immediate availability of toxic venom is obviously essential.

Neither the function nor the chemistry of Dufour's gland (erroneously called the alkaline gland) are known.

Pheromone-Producing Exocrine Glands

A variety of glands, mainly of abdominal origin, are known to be utilized, for the production of chemical signals (Wilson, 1971; Michener, 1974; Free, 1987; Winston, 1987). With few known exceptions, these are extensively developed in either the queen or the worker, but seldom in both castes. The behavioral consequences of this glandular specialization for either queens or workers are considerable, and will be described as the chemistry and functions of the glands are individually treated.

The mandibular glands (Fig. 1, No. 2) of both queens and workers are well developed, and that of the queen is especially capacious. In the case of both of these female castes, this gland is a major social organ that is used for a variety of key functions. Indeed, the secretions of the paired mandibular glands appear to possess more roles as releasers of social behaviors than any other exocrine structures possessed by honey bees.

The tergal glands (Fig. 1, No. 6), located on abdominal tergites 4-6, are very well developed in young queens but not in workers. The wax glands (Fig. 1, No. 14), on the other hand, are limited to workers, occurring on ventral abdominal segments 4-7, as paired structures. Like the tergal glands, the wax glands actually constitute modified epidermal cells, other types of which may be eventually demonstrated to also produce pheromones.

The Nasonov gland (Fig. 1, No. 9), sometimes referred to as the scent gland because it produces a powerful and characteristic odor associated with

FIGURE 2. Lost honey bees exposing their Nasonov glands at the hive entrance

worker bees, is named after the Russian who described it in 1883. This gland, which is undeveloped in the queen, is present on the dorsal surface of the 7th or last abdominal segment, and its secretion is stored in a scent canal that is normally covered. The odorous secretion of the Nasonov gland constitutes another of the multifunctional pheromones produced by worker bees.

The rectum (Fig. 1, No. 10), the distal portion of the intestine, has recently been discovered to secrete an important pheromone that is produced by queens but not workers. Although the glandular source of this pheromone has not been determined, the rectum is classified as an exocrine structure because the pheromone is externalized in a rectal exudate. As indicated previously, this discovery further emphasizes that any exocrine structure can be a source of pheromones.

The Koschevnikov gland (Fig. 1, No. 11), consisting of a tiny cluster of cells in the sting chamber, was first described in 1899 by the Russian for whom it is named. This exocrine organ, which is not as well developed in workers as in queens, is reported to have different functions for each caste, although its precise roles still remain to be determined.

The setose membrane (Fig. 1, No. 13) at the base of the sting is the source of an important blend of pheromones that is released when the sting is everted. Although pheromones are liberated from the setose membrane, it has not been definitely established that this tissue constitutes the glandular source of this chemical signal that is so characteristic of disturbed bees.

Pheromones and Their Functions

Recent advances in analytical chemistry, in combination with the development of sensitive bioassays, have resulted in great progress being made in our understanding of chemical communication in honey bees. It is now possible to analyze the odorous world of this insect in terms of many identified chemical signaling agents and the behaviors that they regulate (Free, 1987).

MANDIBULAR GLAND PHEROMONES

The Worker

Nurse bees produce 10-hydroxy-(E)-2-decenoic acid in their paired mandibular glands (Butenandt and Rembold, 1957), and this compound is the main component of the brood food fed to larvae. In addition, simple fatty acids (*e.g.*, hexanoic, octanoic) are produced in this secretion, and it is suggested that these compounds may contribute to the antibiotic activity of royal jelly (Boch *et al.*, 1979), which was previously identified with 10-hydroxy-(E)-2-decenoic acid (Blum *et al.*, 1959). The latter acid is present in trace amounts in newly emerged bees, but the amount increases up to 60 μg/bee and is detectable throughout the life of the worker (Boch and Shearer, 1967). Isolated workers were also shown to produce less of this acid than bees maintained in groups (Arnold and Roger, 1979). It was also demonstrated that the concentration of this acid varied seasonally, being highest at the time of maximal brood rearing (Pain and Roger, 1970).

When workers become guard bees or begin foraging, they produce a very odorous compound, 2-heptanone (2-HP), in their mandibular glands (Shearer and Boch, 1965; Boch and Shearer, 1967). The odor of blue cheese, which also contains 2-HP, is quite similar to the mandibular gland secretion of the worker. The glandular content of this compound, which can reach 40 μg/bee, is dependent on the physiological rather than the chronological age of the bee, so that workers that are never allowed to forage produce very little of it even after three weeks of age. Thus, it appears that the synthesis of this compound reflects behavioral-biochemical changes that occur when a hive bee (*e.g.*, nurse bee) develops into a guard or forager, an event that normally occurs at about two weeks of age.

2-HP appears to play a role as a weak releaser of alarm behavior when presented on a cork at the hive entrance (Shearer and Boch, 1965; Free and Simpson, 1968). Alarm pheromones are compounds that usually attract excited workers to the emission source which is readily attacked. In a sense, these compounds signal danger, and after an alerted worker has perceived the pheromone, she will frequently secrete the alarm releaser herself, thus rapidly increasing its concentration which also serves to recruit more aggressive workers. While 2-HP possesses demonstrable activity as an alarm pheromone, it is 20-70 times less potent than an alarm pheromone derived from the sting apparatus (Boch *et al.*, 1970). The concentration of 2-HP was subsequently reported to be positively correlated with several manifestations of aggressive behavior at the hive entrance (Kerr *et al.*, 1974).

2-HP is also reported to possess a variety of other functions of considerable social significance. Crushed worker heads, as well as 2-HP, repel foraging workers (Simpson, 1966; Butler, 1966), and it is possible that this is the compound that workers utilize to mark flowers that do not provide nectar and pollen (Nuñez, 1967).

The repellent effects of 2-HP for bee workers can be dramatically demonstrated by applying this compound to the hands before opening a hive. At a concentration of ½ - 2%, applied as an aerosol, bees are effectively repelled, and they do not exhibit aggressive behavior (unpublished data). In our experience, it was possible to work bees without any major problems after treating our hands with a 1% spray of 2-HP. Obviously, this represents a much higher concentration of 2-HP than bees ever encounter in nature.

At low concentrations, 2-HP can act as an attractant at the hive entrance whereas a very high concentration acts as a powerful repellent (Boch *et al.*, 1970). This observation emphasizes the fact that the reactions of insects to pheromones are very dependent on concentration so that a compound that normally releases specific behaviors at low levels becomes a potent deterrent if the concentration is too great. In a sense an abnormally high concentration is overstimulatory for the receiver individuals; perfumers have long been aware of this phenomenon. When butyl mercaptan, an offensive component in the spray of the skunk, is present as a trace constituent in perfumes, it functions as a pleasant-smelling fixative.

Curiously, 2-HP stimulates food-hoarding behavior in bees (Rinderer, 1982) in much the same way as do volatile compounds in used comb (Free and Williams, 1972; Rinderer, 1981), but the real significance of this activity for 2-HP is not known. In addition, this compound is reported to inhibit workers near the queen in a swarm from secreting a pheromone that normally guides workers to the swarm (Morse, 1972).

Since 2-HP is produced in the mandibular glands of workers, it can be readily released during the act of biting with the mandibles. Thus, this compound could be used to "mark" foreign bees such as robbers (Simpson, 1966) and to attract workers to the intruder. Similarly, a foreign queen could be bitten and marked with 2-HP and thus labeled for attack by additional workers (Yadava and Smith, 1971). In addition, 2-HP probably functions as a defensive allomone because it acts as a topical irritant when applied to a bee or other insect. This compound is part of a defensive spray produced by cock-

FIGURE 3. Drone bees attracted to and trapped in a net containing "queen substance"

roaches, and there is no reason to believe that it does not have a similar function when applied to the body of an insect intruder by a bee worker.

The use of a pheromone for multiple functions has been called *pheromonal parsimony* (Blum, 1974), a term that is clearly identified with the roles that 2-HP plays in honey bee biology. This compound is used as a pheromone in different contexts that involve alarm behavior, attraction, and repellency but in addition, it is employed as a defensive allomone as well. It would be no exaggeration to describe 2-HP as a key compound in honey bee biology. The demonstration that 2-HP possesses considerable fungicidal activity (Cole *et al.*, 1973) raises the question of whether this compound could have an antibiotic role in the hive. At any rate, it will not prove surprising if 2-HP is subsequently demonstrated to possess additional functions as a honey bee semiochemical.

The Queen

Compounds in the mandibular gland secretion of the queen have been demonstrated to possess an incredible variety of functions, both as primer and

FIGURE 4. Honey bee workers attacking 2-heptanone-treated cork at the hive entrance

releaser pheromones. Although many of these functions have been studied in terms of single compounds, it is very likely that blends of compounds are utilized to regulate these activities. This is certainly the case for queen substance, (E)-9-oxo-2-decenoic acid (9-ODA) (Callow and Johnston, 1960; Barbier and Lederer, 1960), one of the major compounds found in the acid-rich secretion of the queen's mandibular glands (Pain *et al.*, 1962; Callow *et al.*, 1964).

Primer Activities—9-ODA is reported to inhibit both ovarian development of workers and queen rearing by workers (Butler *et al.*, 1961; Pain *et al.*, 1962). For both of these primer functions, it was demonstrated that other acids in the mandibular gland secretion are required for maximal expression of pheromonal activity. The mandibular glands of mated queens are more active than those of virgin females or superseded queens. It has also been reported that immature queens in sealed cells inhibit the initiation of queen rearing (Boch, 1979).

(E)-9-Hydroxy-2-decenoic acid (9-HDA), another compound produced in queen's mandibular glands, is reported to work in concert with 9-ODA in suppressing queen rearing (Butler and Callow, 1968). However, since a mated queen is more effective than a combination of 9-ODA and 9-HDA, it appears that additional pheromones are involved in this inhibition. As will be discussed later, abdominal, as well as mandibular gland pheromones, have been reported to suppress queen rearing (Velthuis, 1970a).

Releaser Activities—Many of the behaviors exhibited by workers in the presence of the queen are regulated by the latter's mandibular gland pheromones. When a worker encounters the queen, the former initially displays aggression and/or avoidance, followed by food offering, feeding, and finally retinue behavior (Velthuis, 1972). Although 9-ODA plays a major role in attracting workers to the queen and forming a retinue, maximal activity is realized when four other mandibular gland compounds are present (Slessor *et al.*, 1988). In addition to 9-ODA which constitutes more than two-thirds of the active mixture, full retinue behavior requires two forms of 9-HDA, methyl p-hydroxybenzoate, and 4-hydroxy-3-methoxyphenylethanol. It is very likely that this blend of compounds may also be utilized for regulating many of the worker behaviors observed in response to the mandibular gland secretion of the queen.

Queen mandibular gland secretions play important roles in regulating the movement, cohesion, and stability of swarms. 9-ODA is critical to the workers as a recognition cue for the presence of the queen, thus playing a role in guiding the workers in a queenright swarm and providing swarm cohesion (Butler and Simpson, 1967). On the other hand, 9-HDA is reported to promote the stability of the swarm (Butler *et al.*, 1964). However, since a mixture of these two compounds is not as active as a queen, it is evident that swarming is regulated by additional pheromonal stimuli (Free, 1987).

9-ODA is a powerful sex pheromone that readily attracts drones when presented on an elevated lure (Gary, 1962). Other compounds in the mandibular gland secretion may synergize the activity of 9-ODA, act to keep the drones in the area of the attractant source, or function as keeper substances (Boch *et al.*, 1975). 9-ODA is a very specific sex pheromone, related compounds being completely inactive as attractants for drone bees (Blum *et al.*, 1971).

Laying workers of *A. mellifera capensis* can function as false queens that produce 9-ODA and related compounds in their mandibular glands (Crewe and Velthuis, 1980). Queenless workers of *A. mellifera scutellata* produce 9-ODA and 9-HDA; laying workers of *A. mellifera mellifera* also produce these two pheromones (Crewe, 1987).

The ability of workers to discriminate their own queen from foreign queens is not based on the mandibular gland secretion of queens (Moritz and Crewe, 1988). Thus, kin recognition in honey bee workers does not appear to be associated with cephalic pheromones, but rather is related to volatile compounds produced in abdominal glands.

9-ODA moves from the head to the queen's abdomen either by surface translocation or internally (Butler *et al.*, 1974). Movement of 9-ODA on the surface provides bee workers with an opportunity to detect this compound (= the queen) on a much larger surface area than would be possible if this pheromone remained on the head after secretion. Workers injected with 9-ODA rapidly convert this compound to inactive products, thus ensuring that workers, separated from their queen, will rapidly experience queenlessness as a consequence of the absence of queen substance and related compounds (Johnston *et al.*, 1965).

Virgin queens produce a maximum amount of 9-ODA in the spring at the time of mating flights and swarming (Pain *et al.*, 1972). Significantly, virgins exhibit a 24 hour cycle in the production of 9-ODA, synthesizing most of this compound during the late morning and afternoon, the period during which nuptial flights occur (Pain and Roger, 1978). Therefore, the production of this pheromone is maximal during the period when the virgin female is utilizing it as a sex pheromone on mating flights.

NASONOV GLAND PHEROMONE

The pheromone produced by the Nasonov (=Nassanoff) gland constitutes a key attractant utilized by workers in a variety of situations. "Scenting bees" is a term that refers to the secretion of Nasonov pheromone and its dispersion by wing fanning, a behavior frequently seen at the hive entrance or in swarms. For the most part, it is the pleasant fragrance of the Nasonov bouquet, dominated by compounds identical to floral odorants, that beekeepers identify as the characteristic odor of honey bees. In a sense workers produce an attractant pheromone that chemically—and odorously—mimics the fragrances of many of the flowers to which they are attracted.

Chemistry

The chemistry of the Nasonov secretion was primarily elucidated by R. Boch and D.A. Shearer of Agriculture Canada between 1962-1964. These investigators showed that the major odorous compounds produced by this abdominal gland are oxygen-containing monoterpenes that function as powerful attractants. Geraniol, the major compound present (Boch and Shearer, 1962), which derives its name from gerani(um) (alcoh)ol, is also the chief part of oil of rose, and is thus described as having a sweet rose odor. This fragrant alcohol, which is an important perfume ingredient, is not produced by very young bees and reaches its maximum after workers begin to forage (Boch and Shearer, 1963). Geraniol, which is moderately attractive to foraging workers (Butler and Calam, 1969), enhances the activity of a synthetic Nasonov blend, but is not as attractive as other compounds produced by the gland (Williams *et al.*, 1981).

Two other alcohols are present in the Nasonov secretion. One of these, nerol, is very closely related to geraniol with which it shares an odor of rose. It is a very minor constituent (Pickett *et al.*, 1980) which, while it enhances the attractancy of a synthetic Nasonov pheromone, is not very attractive by itself (Williams *et al.*, 1981).

A third alcohol, (E,E)-farnesol, the least volatile of the Nasonov alcohols, is present at about half the concentration of geraniol (Pickett *et al.*, 1980). This compound, which is used in the perfume industry to emphasize the odor of sweet floral perfumes, is not especially attractive to workers, but, in combination with other Nasonov pheromones, contributes to the attractiveness of the mixture (Williams *et al.*, 1981).

Two forms (isomers) of citral, oxidized forms of geraniol and nerol, are also present in the Nasonov secretion. Oxidation of geraniol, presumably an enzymatic process (Pickett *et al.*, 1981), yields (E)-citral (geranial) whereas the oxidation of nerol results in (Z)-citral (neral). Citral was first identified as a Nasonov product by Shearer and Boch (1966) who showed that it was highly attractive to workers. Although the two isomers of citral are relatively minor constituents in the Nasonov secretion, they are powerful attractants for workers (Shearer and Boch, 1966; Butler and Calam, 1969). Williams *et al.* (1981) reported that (E)-citral is as attractive as any of the Nasonov constituents. The odor of citral, which is the dominant compound in oil of lemon grass and an important odorant in oil of lemon and orange, becomes quite conspicuous when workers begin to forage at which time its secretion is maximum (Williams *et al.*, 1981).

Oxidation of geranial ([E]-citral) produces geranic acid whereas the oxidation of neral ([Z]-citral) yields nerolic acid. These two monoterpenes are produced in the Nasonov secretion (Boch and Shearer, 1964; Picket *et al.*, 1980) and they add considerably to the attractancy of this exudate. Boch and Shearer (1964) reported that a mixture of the two acids, plus geraniol, was almost as attractive to workers as the natural Nasonov secretion itself.

Functions

The Nasonov secretion is clearly utilized as a powerful orientation signal when workers cannot readily locate their nest entrance (Ribbands and Speirs, 1953). Workers which perceive this pheromone expose their own Nasonov glands and thus augment the signal, as observed with an increase in the number of scenting bees at new nest sites or after finally entering a nest to which access had been denied (Ferguson and Free, 1981). Exposure of the Nasonov gland is induced by many stimuli, including a live queen, pollen, propolis, and a mandibular gland pheromone of the queen, 9-HDA (Ferguson and Free, 1981).

Foraging honey bees may expose their Nasonov gland when flying over artificial food (*e.g.*, sugar water) or when they begin to feed (von Frisch, 1923). The odor from the gland can rapidly attract other foragers to the food find. Free (1968) reported that foragers did not expose their Nasonov glands until they had visited an artificial food source several times, even when the bait had been pre-treated with Nasonov secretion. Foragers have also been observed to expose their Nasonov glands after collecting nectar from flowers (von Frisch and Rösch, 1926; Free, 1968), but this behavior appears to be clearly exceptional and does not reflect the normal activity of workers visiting flowers under field conditions.

The Nasonov gland secretion is of critical importance in regulating the movement and formation of swarms. This pheromone works in conjunction with queen substance, 9-ODA, to stabilize swarms (Morse and Boch, 1971). Breaking activity of a clustered swarm follows the temporary loss of the queen, who attracts the airborne workers with queen pheromones. These workers scent and fan, attracting more workers which also begin to scent. Some scenting workers return to the queenless cluster and activate the bees there for further searching and movement (Morse and Boch, 1971). Nasonov scent thus plays a vital role in a) causing bees in queenless clusters to become airborne in search of the queen, and b) to orient these bees to the queen's location (Mautz *et al.*, 1972). The critical importance of the interplay of queen pheromones (*e.g*, 9-ODA) and Nasonov scent is illustrated by the fact that swarms can be "lead" with synthetic Nasonov pheromone only if the queen is free and can fly with the swarm (Avitabile *et al.*, 1975).

The formation of stable queenless clusters can result from a mixture of synthetic queen substance (9-ODA) and synthetic Nasonov pheromone (Ferguson *et al.*, 1979). On the other hand, the additional presence of synthetic 9-HDA, another queen mandibular gland pheromone, actually reduces cluster formation. It thus appears that other unknown pheromones in the queen's mandibular glands are required for optimal cluster formation since crushed queen mandibular glands are more effective in initiating clusters than synthetic queen pheromones.

KOSCHEVNIKOV GLAND

The Queen

Butler and Simpson (1965) reported that the Koschevnikov gland (Fig. 1, No. 11) of the mated queen produced pheromones that are highly attractive to workers. Although nothing is known about the chemistry of this gland, it has been established that the secretion is transferred to the setose membrane of the sting shaft, to be externalized when the sting is everted (Grandperrin and Cassier, 1983). The gland degenerates in one-year-old mated queens.

The Worker

Mauchamp and Cassier (1982) present evidence that the Koschevnikov gland is the source of the powerful alarm pheromones released by alerted workers when the sting is everted. This alarm pheromone accumulates on the setose membrane of the sting shaft (Ghent and Gary, 1962) and is thus still functional when the sting is torn from the worker and impaled in the victim after stinging. The tendency of flying workers to be attracted to the alarm pheromones diffusing from the severed sting results in the well-known phenomenon of aggressive bees stinging close to the site of the impaled sting apparatus. Thus, the severed sting effectively marks the intruder for attack by the aroused workers. Other factors such as the intruder's odor, color, movement, and temperature also affect the defensive behavior of pheromonally alarmed bees (Free, 1961). Higher temperatures increase the probability, speed, intensity, and duration of response to alarm pheromones; higher humidities only increase the intensity of the response (Collins, 1981).

Isopentyl (=isoamyl) acetate (IPA) was the first compound identified as part of the sting pheromone (Boch et al., 1962). IPA is 20-70 times as active as an alarm pheromone as 2-heptanone, the mandibular gland product of workers (Boch et al., 1970). The banana-like odor of this compound is very familiar to beekeepers when encountering aroused bees which have everted their stings and are dispersing the alarm pheromones by fanning. IPA has also been identified as an alarm pheromone of three tropical species of Apis (Morse et al., 1967). The content of IPA is maximum when workers become guards or begin to forage (Boch and Shearer, 1966). Queens do not produce IPA but rather, a series of long-chain esters have been identified as characteristic sting-shaft compounds (Blum et al., 1983).

In addition to IPA, 13 other esters have been identified as sting shaft volatiles (Blum et al., 1978; Blum and Fales, 1988). These esters are accompanied by a large series of alcohols (Blum et al., 1978; Blum, 1982; Pickett et al., 1982), as well as several acids (Collins and Blum, 1983; Blum and Fales, 1988). A list of the major pheromones identified among the more than 40 compounds in extracts of the sting apparatus of foragers are presented in Table 1. Some of these compounds possess strong floral odors such as lavender

Table 1. Major Compounds Identified as Part of the Sting (Alarm) Pheromone of Mature Worker Bees.

Compound	Relative Proportion
Isopentyl acetate	+++
2-Octen-l-yl acetate	+++
2-Nonyl acetate	+++
2-Nonanol	++++
9-Octadecen-1-ol	+++
(Z)-11-Eicosen-l-ol	+++++

FIGURE 5. Honey bee workers attacking a cork treated with the sting alarm pheromone, isopentyl acetate

(2-nonyl acetate) and jasmin (2-nonanol) that add a distinctive note to the odor signal generated by alarmed workers.

Collins and Blum (1982, 1983) bioassayed a large number of the identified sting volatiles as alarm pheromones for caged bees. In terms of frequency, speed, duration, and intensity of reaction, many of the compounds functioned as effective alarm pheromones. Similarly, (Z)-11-eicosen-1-ol has been reported to possess alarm-releasing activity in addition to prolonging the effectiveness of more volatile compounds such as IPA (Pickett *et al.*, 1982). Another alcohol, 2-nonanol, appears to be of equivalent activity to IPA as a chemical releaser of alarm behavior (Collins and Blum, 1982). It seems likely that the diversity of compounds utilized to produce an alarm message is more active than a single compound and in addition, may serve to sustain the signal. Grandperrin (1983) found that a synthetic blend of sting volatiles was highly active as an alarm pheromone and had a stronger recruiting effect than IPA. Free *et al.* (1983) reported that several sting volatiles elicited attack and stinging by workers at the hive entrance.

IPA inhibits bees from scenting with the Nasonov pheromone. Since workers with a queenless swarm scent with the Nasonov pheromone when they locate their queen but liberate alarm pheromone when they locate a foreign queen, IPA can be utilized to release aggression against an alien queen (Morse, 1972). Similarly, Free *et al.* (1983) found that several of the sting volatiles repelled workers and inhibited scenting at the hive entrance.

TERGITE GLAND PHEROMONES

The abdominal tergite glands of the queen (Fig. 1, No. 6) produce pheromones that function as queen recognition signals for workers, inhibit queen cell construction, and inhibit worker ovarian development (Velthuis, 1970b). If the mandibular glands are removed from the queen, she is still accepted by her colony, and workers display typical retinue behavior. Thus, tergite gland pheromones possess many of the functional properties of queen mandibular gland pheromones and it is significant that secretions of both glands appear to interact to inhibit ovarian development of workers (Velthuis, 1970b). Vierling and Renner (1977) showed that young workers are strongly attracted to the tergite gland secretion which is only perceived by contact. On the other hand, the mandibular gland secretion of the queen contains volatile worker attractants (*e.g.*, 9-ODA), demonstrating that the secretions of both glands are required for maximal attraction. The tergite gland pheromone appears to be critical in stabilizing the queen's "court" (Vierling and Renner, 1977).

Similarly, the tergite gland secretion interacts with the mandibular gland secretion in attracting drones and inducing copulation (Renner and Vierling, 1977). Whereas, mandibular gland pheromones attract drones from a distance of 50 meters (160 feet) or more, the activity of the tergite gland pheromones predominates when the drones are within 30 cm (1 foot) of the queen. Furthermore, the tergite gland pheromone releases copulatory activity in drones (Renner and Vierling, 1977).

The chemical identity of the tergite gland secretion is unknown.

TARSAL (ARNHART) GLAND PHEROMONE

Secretions with very different functions are deposited by the tarsi of both queens and workers. These exudates, which are sometimes termed footprint pheromones, have not been chemically identified, but it has been clearly established that they play key social roles for both female castes.

The Queen

The oily secretion from the queen's tarsal glands is deposited by the foot pads upon the surface of the comb. This pheromone, in conjunction with mandibular gland secretion, when applied to the bottom edges of comb, inhibits queen cell construction in overcrowded colonies (Lensky and Slabeszki, 1981). Inhibition requires the presence of both glandular secretions, neither being active alone. In crowded colonies, the queen may not be able to move along the bottom of combs and deposit the secretions of the tarsal and mandibular glands. This could result in queen cell construction leading to the rearing of new queens and swarming.

The secretion of six-month-old queens is greater than two-year-old queens; the rate of secretion of the workers' glands is 10-15 times less than that of the queens' (Lensky *et al.*, 1984).

The Worker

Workers deposit a persistent trail pheromone at their hive entrance and the attractiveness of this secretion increases with the number of workers depositing it (Butler *et al.*, 1970). It appears that bees also mark forage sites with the trail pheromone, thus increasing their attractiveness to other foragers (Ferguson and Free, 1979). Thus, flowers and sites containing artificial lures are more attractive to other workers than similar resources that have not been marked with footprint pheromone.

It has been suggested that the trail pheromone of the worker, while it may be deposited by the tarsi, may not have originated in the tarsal glands. Ferguson and Free (1979) reported that the odors of the head, thorax, and abdomen are very active in inducing landing by workers searching for food. Thus, it is possible that this pheromone, while it is deposited by the feet, originates elsewhere on the body.

The trail pheromone is capable of inducing disoriented workers to expose their Nasonov glands (Ferguson and Free, 1981). Thus, this pheromone can work in concert with the Nasonov scent to aid workers which have become temporarily disoriented in the vicinity of the hive entrance.

WORKER REPELLENT PHEROMONE

When virgin queens are about 24 hr old, they produce a pheromone that repels workers and other queens. This pheromone, which is produced for about two weeks, during the period when the queen may have hostile encounters with workers or sister queens in the hive, is discharged as a fecal exudate from the rectum (Fig. 1, No. 10) (Post *et al.*, 1987). Workers, when exposed to this fecal material, are repelled and exhibit autogrooming. Significantly, after perceiving the rectal pheromone, workers quickly retreat from its source, no longer exhibiting any signs of aggressive behavior. In a sense, the rectal pheromone functions as a tranquilizer.

The repellent pheromone has been identified as o-aminoacetophenone, a minor component in the rectal exudate (Page *et al.*, 1988). Although this compound is readily detected in the fluid in the rectal region of the queen's hind gut, the glandular source of the pheromone is unknown. This compound, which has the grape-like odor that characterizes the queen's feces, has not been detected in the feces of freshly emerged queens or those that were more than 14 days old. No evidence was found for the presence of this pheromone in the feces of workers or drones.

o-Aminoacetophenone does not increase autogrooming in workers, indicating that other pheromones in the rectal exudate cause the grooming observed in worker bees after they are exposed to the fecal secretion (Page *et al.*, 1988). The rectal fluid is dominated by a series of characteristic esters, hydrocarbons, as well as distinctive alcohols and acids. While it seems likely that the hydrocarbons are responsible for the observed autogrooming, the presence of a large number of novel compounds in the fecal exudate—of unknown significance—raises the possibility that this secretion possesses additional pheromonal functions. This conclusion is supported by the fact that o-aminoacetophenone constitutes only about 0.5% of the volatiles detected in the secretion, which is fortified with a variety of compounds that are not characteristic waste products.

BEESWAX PHEROMONES

Workers synthesize several oxygenated compounds that can be readily detected in the comb wax produced by nonforaging bees. These compounds—octanal, nonanal, decanal, furfural, benzaldehyde, and 1-decanol—appear to be responsible for the characteristic odor of freshly prepared comb wax (Blum *et al.*, 1988). Since empty comb, and volatiles from empty comb, can stimulate the hoarding behavior of foraging bees observed by Rinderer and Baxter (1978) and Rinderer (1981), these volatile aldehydes and 1-decanol were evaluated as stimulators of this characteristic behavior. Although these beeswax volatiles affected the intensity of hoarding behavior by either increasing or decreasing it, the exact roles of these pheromones have not been determined yet (Blum *et al.*, 1988).

DRONE PHEROMONES

The heads of drones are reported to contain an extractable pheromone that attracts flying drones in their congregation areas (Gerig, 1972). Such a pheromone, produced in the mandibular glands, could promote the formation of drone aggregations at sites that are very suitable for mating (Lensky *et al.*, 1985).

BROOD PHEROMONES

Larval and pupal bees produce several pheromones that significantly affect the behavior of worker bees. The important roles of these brood pheromones emphasize the fact that many behaviors observed in the honey bee society are regulated by chemical signals produced by all developmental stages.

Inhibitory Pheromone

Worker ovarian development is inhibited in small queenless colonies by worker larvae and pupae (Milojević and Filipović-Moskovljević, 1959). On the other hand, neither queen larvae nor pupae inhibited worker ovarian development in queenless and broodless colonies (Jay, 1970). The development of worker ovaries is minimal in colonies containing both brood and a queen; brood removal results in greatly increased ovarian development (Jay, 1972). In contrast, removal of the queen produces only a slight increase in worker ovaries, demonstrating that ovarian inhibition is more strongly influenced by the presence of brood than the queen. Indeed, worker larvae and pupae suppress the development of the ovaries of adult workers as effectively as a mated queen (Jay, 1970).

This brood pheromone, which has not been chemically identified, does not appear to be highly volatile, and maximal inhibition occurs only if workers have direct contact with the larvae as opposed to being separated from them by a wire screen (Jay, 1972). In general, larval pheromones of social insects have been demonstrated to be primarily perceived as contact pheromones that regulate many of the known adult-larval interactions.

Brood-Recognition Pheromone

The ability of workers to readily distinguish worker and drone larvae and pupae is correlated with the presence of brood-recognition pheromones (Free and Winder, 1983). The recognition pheromones must be perceived by contact, demonstrating that they are of low volatility. There is also evidence that workers are able to differentiate pupae of different ages, enabling the adults to selectively respond to pupae at different stages of development (Free and Winder, 1983).

Koeniger and Veith (1983) identified glyceryl-1, 2-dioleate-3-palmitate as a recognition pheromone of drone pupae. This compound, which causes clustering of workers, is also reported to be present in olive oil.

Foraging Stimulating Pheromone

Foraging is stimulated by the presence of brood, the amount of which determines the amount of pollen that workers collect (Free, 1967). A contact pheromone appears to stimulate foraging maximally, as indicated by the reduction in foraging that occurs when workers are excluded from the brood by wire screens. Brood odor is less effective in stimulating foraging than direct contact with the brood, but is considerably more stimulatory than controls lacking brood and its odor (Free, 1967).

It is not known if the foraging stimulator, ovarian inhibitor, and brood recognition pheromones actually are different compounds or if these activities actually represent different roles for the same pheromone.

UTILIZATION OF PHEROMONES IN APICULTURE

The pheromones evolved by honey bees as behavioral regulators offer promise as highly specific agents with which to manipulate honey bees. The identification of a large number of pheromones provides beekeepers with a variety of compounds that are candidates for manipulating all castes of honey bees in order to achieve specific apicultural objectives. At this juncture, it is obvious that the behavioral regulators are available, but their practical utilization has, for the most part, still to be established.

Free (1987) has provided an elegant program for evaluating honey bee pheromones in beekeeping applications. Pheromones that regulate queen rearing, worker ovarian development, worker foraging, drone mating, attraction of workers to crops, swarm trapping, and aggressive behavior, are analyzed in terms of their economic potential for apiculture. Although it is obvious that, in general, practical economic utilization of honey bee pheromones has yet to be realized, Free (1987) presents a persuasive case for regarding these chemical messengers as a real boon to apiculture.

Synthetic Nasonov pheromone is a highly effective lure for swarms (Free *et al.*, 1984) and promises to provide an effective means of easily trapping feral swarms. Schmidt and Thoenes (1987) have developed a pheromone-containing trap box which is very effective in trapping swarms. This system is being utilized to monitor Africanized bee swarms in Mexico and promises to provide a very efficient means of determining the rate of northward movement of these aggressive bees. In addition to attracting swarms, synthetic Nasonov scent may be utilized to attract foragers to crops so as to increase pollination of crops with high economic potential. Williams *et al.* (1981) have demonstrated that synthetic Nasonov pheromone can be utilized to effectively attract bees under field conditions. The potential of the Nasonov pheromone as a honey bee attractant has barely been explored, and deserves detailed study.

The utilization of 9-ODA (queen substance) for establishing controlled mating areas could provide a means of regulating reproduction of honey bees (Free, 1987). Furthermore, use of this pheromone would seem to be especially desirable for inhibiting the construction of queen cells and preventing swarm-

Figure 6. Honey bee swarm attracted to swarm trap baited with synthetic Nasonov gland pheromones

ing. Since swarming seems to occur when the concentration of queen substance is diminished (Butler, 1960), artificial elevation of the level of this pheromone could effectively inhibit swarm formation.

Honey bee pheromones would also seem to be ideal candidates for discovering new and safe repellents. High concentrations of 2-heptanone, the mandibular gland pheromone of workers, can repel workers quite effectively without inducing aggressive behavior. This commercially available—and inexpensive—compound should be a natural candidate as a repellent. The same can be said for o-aminoacetophenone, recently identified as a queen-produced tranquilizing repellent for workers (Page *et al.*, 1988).

At this juncture it is obvious that a host of queen and worker pheromones are available for evaluation as honey bee behavioral regulators. What is required are testing programs that are both imaginative and realistic in terms of economics and practicality. There is no reason to believe that the technology required to utilize these pheromones is not now readily available. Pheromonally speaking, for beekeepers the best is yet to come.

ACKNOWLEDGEMENTS

I thank Dr. H.R. Hermann for preparing the drawing of the honey bee exocrine glands.

The Hive and the Honey Bee

REFERENCES

Arnold, G. and B. Roger. (1979). Group effect on the content of 10-hydroxydec-2-enoic acid in the head of worker bees. *Apidologie 10*:35-42.

Avitabile, A., R.A. Morse, and R. Boch. (1975). Swarming honey bees guided by pheromones. *Ann. Ent. Soc. Amer. 69*:1079-1082.

Barbier, J. and E. Lederer. (1960). Structure chimique de la substance royale de la reine d'abeille (*Apis mellifica* L.) *Compt. Rend. Acad. Sci. (Paris) 251*:1131-1135.

Blum, M.S. (1974). Deciphering the communicative Rosetta Stone. *Bull. Ent. Soc. Amer. 20*:30-35.

Blum, M.S. (1982). Pheromonal bases of insect sociality: communications, conundrums and caveats. *In*: Les Médiateurs Chimiques. Les Colloques de l'INRA, Versailles, pp. 149-162.

Blum, M.S. and H.M. Fales. (1988). Chemical releasers of alarm behavior in the honey bee: informational plethora of the sting apparatus signal. *In*: Africanized Honey Bees and Bee Mites, G.R. Needham, R.E. Page, M.Delfinado-Baker, and C.E. Bowman, eds. pp. 141-148. Halsted Press, New York.

Blum, M.S., A.F. Novak, and S. Taber, III. (1959). 10-Hydroxy-delta^2-decenoic acid, an antibiotic found in royal jelly. *Science 130*: 452-453.

Blum, M.S., H.M. Fales, K.W. Tucker, and A.M. Collins. (1978). Chemistry of the sting apparatus of the worker honeybee. *J. Apic. Res. 17*:218-221.

Blum, M.S., R. Boch, R.E. Doolittle, M.T. Tribble, and J.G. Traynham. (1971). Honey bee sex attractant; conformational analysis, structural specificity and lack of masking activity of congeners. *J. Insect Physiol. 17*:349-364.

Blum, M.S., H.M. Fales, T.H. Jones, T.E. Rinderer, and K.W. Tucker. (1983). Caste-specific esters derived from the queen honey bee sting apparatus. *Comp. Biochem. Physiol. 75B*:237-238.

Blum, M.S., T.H. Jones, T.E. Rinderer, and H.A. Sylvester. (1988). Oxygenated compounds in beeswax: identification and possible significance. *Comp. Biochem. Physiol. 91B*:581-583.

Boch, R. (1979). Queen substance pheromone produced by immature queen honeybees. *J. Apic. Res. 18*:12-15.

Boch, R. and D.A. Shearer. (1962). Identification of geraniol as the active component in the Nasanoff pheromone of the honeybee. *Nature 194*: 704-706.

Boch, R. and D.A. Shearer. (1963). Production of geraniol by honey bees of various ages. *J. Insect Physiol. 9*: 431-434.

Boch, R. and D.A. Shearer. (1964). Identification of nerolic and geranic acids in the Nasanoff pheromone of the honey bee. *Nature 202*:320-321.

Boch, R. and D.A. Shearer. (1966). Iso-pentyl acetate in stings of honeybees of different ages. *J. Apic. Res. 5*:65-70.

Boch, R. and D.A. Shearer. (1967). 2-Heptanone and 10-hydroxy-trans-dec-2-enoic acid in the mandibular glands of worker honey bees of different ages. *Z. Vergl. Physiol. 54*:1-11.

Boch, R., D.A. Shearer, and A. Petrasovits. (1970). Efficacies of two alarm substances of the honey bee. *J. Insect Physiol. 16*:17-24.

Boch, R., D.A. Shearer, and R.W. Shuel. (1979). Octanoic and other volatile acids in the mandibular glands of the honeybee and in royal jelly. *J. Apic. Res. 18*:250-253.

Boch, R., D.A. Shearer, and B.C. Stone. (1962). Identification of iso-amyl acetate as an active component in the sting pheromone of the honeybee. *Nature 195*:1018-1020.

Boch, R., D.A. Shearer, and J.C. Young. (1975). Honey bee pheromones: field tests of natural and artificial queen substance. *J. Chem. Ecol. 1*:133-148.

Butenandt, A. and H. Rembold. (1957). Über den Weiselzellenfutter-saft der Honigbiene, I. Isolierung, Konstitutionsermittlung und Vorkommen der 10-Hydroxy-2-decensäure. *Hoppe-Seyl Z. 308*:284-289.

Butler, C.G. (1960). The Significance of queen substance in swarming and supersedure in honeybee (*Apis mellifera* L.) colonies. *Proc. Roy. Ent. Soc. London, Ser. A 35*:129-132.

Butler, C.G. (1966). Mandibular gland pheromone of worker honey bee. *Nature 212*:530.

Butler, C.G. and D.H. Calam. (1969). Pheromones of the honey bee—the secretion of the Nasanoff gland of the worker. *J. Insect Physiol. 15*:237-244.

Butler, C.G. and R.K. Callow. (1968). Pheromones of the honeybee (*Apis mellifera* L.): the "inhibitory scent" of the queen. *Proc. Roy. Ent. Soc. London (A) 43*:62-65.

Butler, C.G. and J. Simpson. (1965). Pheromones of the honeybee (*Apis mellifera* L.) an olfactory pheromone from the Koschewnikow gland of the queen. *Vedecké Práce,* Sci. Studies, Univ. Libcice, Czech. *4*:33-36.

Butler, C.G. and J. Simpson. (1967). Pheromones of the queen honey bee (*Apis mellifera* L.) which enable her workers to follow her when swarming. *Proc. Roy. Ent. Soc. London (A) 42*:149-154.

Butler, C.G., R.K. Callow, and J.R. Chapman. (1964). 9-Hydroxydec-trans-2-enoic acid, a pheromone stabilizing honey bee swarms. *Nature 201*:733.

Butler, C.G., R.K. Callow, and N.C. Johnston. (1961). The isolation and synthesis of queen substance, 9-oxodec-trans-2-enoic acid, a honeybee pheromone. *Proc. Roy. Ent. Soc. 155*:417-432.

Butler, C.G., D.J.C. Fletcher, and D. Watler. (1970). Hive entrance finding by honeybee (*Apis mellifera*) foragers. *Anim. Behav. 18*:78-91.

Butler, C.G., R.K. Callow, A.R. Greenway, and J. Simpson. (1974). Movement of the pheromone, 9-oxodec-2-enoic acid, applied to the body surfaces of honey bees (*Apis mellifera*). *Ent. Exper. Appl. 17*:112-116.

Callow, R.K. and N.C. Johnston. (1960). The chemical constitution and synthesis of queen substances of honeybees (*Apis mellifera* L.). *Bee World 41*:152-153.

Callow, R.K., J.R. Chapman, and P.N. Paton. (1964). Pheromones of the honeybee: Chemical studies of the mandibular gland secretion of the queen. *J. Apic. Res. 3*:77-89.

Cole, L.K., M.S. Blum, and R.W. Roncadori. (1975). Antifungal properties of the insect alarm pheromones citral, 2-heptanone, and 4-methyl-3-heptanone. *Mycologia 67*:701-708.

Collins, A.M. (1981). Effects of temperature and humidity on honeybee response to alarm pheromones. *J. Apic. Res. 20*:13-18.

Collins, A.M. and M.S. Blum. (1982). Bioassay of compounds derived from the honeybee sting. *J. Chem. Ecol. 8*:463-470.

Collins, A.M. and M.S. Blum. (1983). Alarm responses caused by newly identified compounds derived from the honeybee sting. *J. Chem. Ecol. 9*:57-65.

Crewe, R.M. (1987). Lability of the mandibular gland signal of three races of African honey bees. *In*: Chemistry and Biology of Social Insects, J. Eder and H. Rembold, *eds.* pp. 433-434. Verlag J. Peperny, Munich.

Crewe, R.M. and H.H.W. Velthuis. (1980). False queens: a consequence of mandibular gland signals in worker honeybees. *Naturwissenschaften 67*:467-469.

Ferguson, A.W. and J.B. Free. (1979). Production of a forage-marking pheromone by the honeybee. *J. Apic. Res. 18*:128-135.

Ferguson, A.W. and J.B. Free. (1981). Factors determining the release of Nasonov pheromone by honeybees at the nest entrance. *Physiol. Ent.* 6:15-19.

Ferguson, A.W., J.B. Free, J.A. Pickett, and M. Winder. (1979). Techniques for studying honeybee pheromones involved in clustering and experiments on the effect of Nasonov and queen pheromones. *Physiol. Ent.* 4:339-344.

Free, J.B. (1961). The stimuli releasing the stinging response of honeybees. *Anim. Behav.* 9:193-196.

Free, J.B. (1967). Factors determining the collection of pollen by honeybee foragers. *Anim. Behav.* 15:134-144.

Free, J.B. (1968). The conditions under which foraging honeybees expose their Nasonov gland. *J. Apic. Res.* 7:139-145.

Free, J.B. (1987). Pheromones of Social Bees. Cornell Univ. Press, Ithaca, N.Y.

Free, J.B. and I.H. Williams. (1972). Hoarding by honeybees (*Apis mellifera* L.) *Anim. Behav.* 20:327-334.

Free, J.B. and M.E. Winder. (1983). Brood recognition by honeybee (*Apis mellifera*) workers. *Anim. Behav.* 31:539-545.

Free, J.B. and J. Simpson. (1968). The alerting pheromones of the honeybee. *Z. Vergl. Physiol.* 61:361-365.

Free, J.B., A.W. Ferguson, J.R. Simpkins, and B.N. Al-Saad. (1983). Effect of honeybee Nasonov and alarm pheromone components on behaviour at the nest entrance. *J. Apic. Res.* 22:214-223.

Free, J.B., J.A. Pickett, A.W. Ferguson, J.R. Simpkins, and C. Williams. (1984). Honeybee Nasonov pheromone lure. *Bee World* 65:175-181.

von Frisch, K. (1923). Über die 'Sprache' die Bienen, eine Tierpsychologische Untersuchung. *Zool. Jahrbuch* (Physiol.) 40:1-186.

von Frisch, K. and G.A. Rösch. (1926). Neue Versuch über die Bedeutung von Duftorgan und Pollenduft fur die Verständigung im Bienenvolk. *Z. Vergl. Physiol.* 4:1-21.

Gary, N.E. (1962). Chemical mating attractants in the queen honey bee. *Science* 136:773-774.

Gerig, L. (1972). Ein weiterer Duftstoff zur Anlockung der Drohnen von *Apis mellifica* (L.). *Z. Ang. Ent.* 70:286-289.

Ghent, R.L. and N.E. Gary. (1962). A chemical alarm releaser in honey bee stings (*Apis mellifera* L.). *Psyche* 69:1-6.

Grandperrin, D. (1983). Sting alarm pheromone of the honeybee: the recruiting effect of an artificial blend of volatile compounds of the worker sting (*Apis mellifica* L., Hymenoptera, Apidae). *Experientia* 39:219-221.

Grandperrin, D. and P. Cassier. (1983). Anatomy and ultrastructure of the Koschewnikow's gland of the honey bee, *Apis mellifera* L. (Hymenoptera: Apidae). *Int. J. Morphol. Embryol.* 12:25-42.

Jay, S.C. (1970). The effect of various combinations of immature queen and worker bees on the ovary development of worker honeybees in colonies with and without queens. *Can. J. Zool.* 48:169-173.

Jay, S.C. (1972). Ovary development of worker honeybees when separated from worker brood by various methods. *Can. J. Zool.* 50:661-664.

Johnston, N.C., J.H. Law, and N. Weaver. (1965). Metabolism of 9-ketodec-2-enoic acid by worker honeybees (*Apis mellifera* L.) *Biochemistry* 4:1615-1621.

Kerr, W.E., M.S. Blum, J.F. Pisani, and A.C. Stort. (1974). Correlation between amounts of 2-heptanone and iso-amyl acetate in honeybees and their aggressive behaviour. *J. Apic. Res.* 13:173-176.

Koeniger, N. and H.J. Veith. (1983). Glyceryl-1, 2-dioleate-3-palmitate, a brood pheromone of the honeybee (*Apis mellifera* L.). *Experientia 39*:1051-1052.

Lensky, Y. and Y. Slabezki. (1981). The inhibiting effect of the queen bee (*Apis mellifera* L.) foot-print pheromone on the construction of swarming queen cups. *J. Insect Physiol. 27*:313-323.

Lensky, Y., P. Cassier, M. Notkin, C. Delorme-Joulie, and M. Levinsohn. (1985). Pheromonal activity and fine structure of the mandibular glands of honeybee drones (*Apis mellifera* L.) (Insecta, Hymenoptera, Apidae). *J. Insect Physiol. 31*:265-276.

Lensky, Y., P. Cassier, A. Finkel, A. Teeshbee, R. Schlesinger, C. Delorme-Joulie, and M. Levinsohn. (1984). Les glandes tarsales de l'abeille mellifique (*Apis mellifera* L.) reines, ouvrières et faux-bourdons (Hymenoptera, Apidae). *Ann. Sci. Naturelle, Zool. Paris 6*:167-175.

Mauchamp, B. and D. Grandperrin. (1982). Chromatographie en phase gazeuse des composés volatils des glandes à phéromones des abeilles: méthodes d'analyse directe. *Apidologie 13*:29-37.

Mautz, D., R. Boch, and R.A. Morse. (1972). Queen finding by swarming honey bees. *Ann. Ent. Soc. Amer. 65*:440-443.

Michener, C.D. (1974). The Social Behavior of the Bees. Harvard Univ. Press, Cambridge, Mass.

Milojević, B.D. and V. Filipović-Moskovljević. (1959). Gruppeneffekt bei Honigbienen. II. Eierstock-entwicklung bei Arbeitsbienen im Kleinvolk. *Bull. Acad. Serbe Sci. Classe Sci. Math. Naturelles 25*:131-138.

Moritz, R.F.A. and R.M. Crewe. (1988). Chemical signals of queens in kin recognition of honeybees, *Apis mellifera* L. J. Comp. *Physiol. A. 164*:83-89.

Morse, R.A. (1972). Honey bee alarm pheromone: another function. *Ann. Ent. Soc. Amer. 65*:1430.

Morse, R.A. and R. Boch. (1971). Pheromone concert in swarming honeybees. *Ann. Ent. Soc. Amer. 64*:1414-1417.

Morse, R.A., D.A. Shearer, R. Boch, and A.W. Benton. (1967). Observations on alarm substances in the genus *Apis. J. Apic. Res. 6*:113-118.

Nuñez, J.A. (1967). Sammelbienen markieren versiegte Futter-guellen durch Duft. *Naturwissenschaften 54*:322-323.

Page, R.E., M.S. Blum, and H.M. Fales. (1988). o-Aminoacetophenone, a pheromone that repels honeybees, (*Apis mellifera* L.). *Experientia 44*:270-271.

Pain, J. and B. Roger. (1970). Variation annuelle de l'acide hydroxy-10 décène-2 oïque dans les têtes d'abeilles. *Apidologie 1*:29-54.

Pain, J. and B. Roger. (1978). Rhythme circadien des acides céto-9-décène-2-oïque, phéromone de la reine, et hydroxy-10 décène-2-oïque des ouvrieres d'abeilles, *Apis mellifica ligustica* A. *Apidolgie 9*:263-272.

Pain, J., B. Roger, and J. Theurkauff. (1972). Sur l'existence d'un cycle annuel de la production de phéromone (acide céto-9-décène-2-oïque) chez les reines d'abeilles (*Apis mellifica ligustica Spinola*). *Compt. Rend. Acad. Sci.* (Paris) *275*:2399-2402.

Pain, J., M. Barbier, D. Bogdanovsky and E. Lederer. (1962) Chemistry and biological activity of the secretions of queen and worker honeybees (*Apis mellifica* L.). *Comp. Biochem. Physiol. 6*:233-241.

Pickett, J.A., I.H. Williams, and A.P. Martin. (1982). (Z)-11-eicosen-1-ol, an important new pheromonal component from the sting of the honeybee, *Apis mellifera* L. (Hymenoptera: Apidae). *J. Chem. Ecol. 8*:163-175.

Pickett, J.A., I.H. Williams, A.P. Martin, and M.P. Smith (1980). Nasonov pheromone of the honeybee, *Apis mellifera* L. (Hymenoptera: Apidae) Part I. Chemical characterization. *J. Chem. Ecol. 6:*425-434.

Pickett, J.A., I.H. Williams, M.C. Smith, and A.P. Martin. (1981). The Nasonov pheromone of the honey bee, *Apis mellifera* L. (Hymenoptera: Apidae) part III. Regulation of pheromone composition and production. *J. Chem. Ecol. 7:*543-554.

Post, D.C., R.E. Page, and E.H. Erickson. (1985). Honeybee (*Apis mellifera* L.) queen feces: source of a pheromone that repels worker bees. *J. Chem. Ecol. 13:*583-591.

Renner, M. and G. Vierling. (1977). Die Rolle des Taschendrüsenpheromons beim Hochzeits flug der Bieninkönigin. *Behav. Ecol. Sociobiol. 2:*329-338.

Ribbands, C.R. and N. Speirs. (1953). The adaptability of the homecoming honeybee. *Brit. J. Anim. Behav. 1:*59-66.

Rinderer, T.E. (1981). Volatiles from empty comb increase hoarding by the honey bee. *Anim. Behav. 29:*1275-1276.

Rinderer, T.E. (1982). Maximal stimulation by comb of honey bee (*Apis mellifera*) hoarding behaviour. *Ann. Ent. Soc. Amer. 75:*311-312.

Rinderer, T.E. and J.R. Baxter. (1978). Effect of empty comb on hoarding behavior and honey production of the honey bee. *J. Econ. Ent. 71:*757-759.

Schmidt, J.O. and S.C. Thoenes. (1987). Honey bee swarm capture with pheromone-containing trap boxes. *Amer. Bee J. 127:*435-438.

Shearer, D.A. and R. Boch. (1965). 2-Heptanone in the mandibular gland secretion of the honey-bee. *Nature 206:*530.

Shearer, D.A. and R. Boch. (1966). Citral in the Nasanoff pheromone of the honey bee. *J. Insect Physiol. 12:*1513-1521.

Simpson, J. (1966). Repellency of the mandibular gland scent of worker honey bees. *Nature 209:*531-532.

Slessor, K.N., L.-A. Kaminski, G.G.S. King, J.H. Borden, and M.L. Winston. (1988). Semiochemical basis of the retinue response to queen honey bees. *Nature 332:*354-356.

Snodgrass, R.E. (1956). The Anatomy of the Honey Bee. Cornell Univ. Press, Ithaca, N.Y.

Velthuis, H.H.W. (1970a). Queen substance from the abdomen of the honey bee queen. *Z. Vergl. Physiol. 70:*210-222.

Velthuis, H.H.W. (1970b). Ovarian development in *Apis mellifera* worker bees. *Ent. Exp. Appl. 13:*377-394.

Velthuis, H.H.W. (1972). Observations on the transmission of queen substances in the honey bee colony by the attendants of the queen. *Behaviour 41:*105-129.

Vierling, G. and M. Renner. (1977). Die Bedeutung des Sekretes der Tergittaschen-drüsen für die Attraktivität der Bienenkönigin gegenüber junger Arbeiterinnin. *Behav. Ecol. Sociobiol. 2:*185-200.

Williams, I.H., J.A. Pickett, and A.P. Martin. (1981). The Nasonov pheromone of the honeybee *Apis mellifera* L. (Hymenoptera: Apidae) part II. Bioassay of the components using foragers. *J. Chem. Ecol. 7:*225-237.

Wilson, E.O. (1971). The Insect Societies. Harvard Univ. Press, Cambridge, Mass.

Winston, M.L. (1987). The Biology of the Honey Bee. Harvard Univ. Press, Cambridge, Mass.

Yadava, R.P.S. and M.V. Smith. (1971). Aggressive behaviour of *Apis mellifera* L. workers towards introduced queens. I. Behavioural mechanisms involved in the release of worker aggression. *Behaviour 39:*212-226.

CHAPTER 10

THE PRODUCTION OF NECTAR AND POLLEN

by R. W. SHUEL*

Honey has been a food for humans for thousands of years and until recent centuries it was the chief source of sugar available to most people in the world's temperate regions. The importance of honey-bee activity in collecting nectar and transforming it into honey is well-known. Less appreciated is the role that bees and other biological agents play in the pollination of flowering plants (Meeuse, 1961; Free, 1970; McGregor, 1976). The special relationship that exists between plants and their pollinators is a very ancient one, extending back for more than 100 million years (Kevan and Baker, 1983). It has benefits for both sides. The pollinator gains some form of sustenance from the flowers and, while the individual plant receives no benefit, the reproduction of the species to which it belongs is assured. The most common reward to the pollinator is nectar; in seeking it the visitor incidentally picks up pollen on its body and later deposits it, also incidentally, on the stigma of another flower of the same species. There the pollen germinates and sets in motion a train of events that culminates in the development of a fruit. In other plants where nectar is not produced, the pollinator's reward is the pollen itself, some of which is transferred directly to the stigma of another flower without the mediation of nectar. The tendency of individual foragers to limit their visits to one species at a particular time greatly increases the chances of pollen being carried to a flower which it can fertilize (Kevan and Baker, 1983). Some species are self-sterile and wholly dependent on pollinators for reproduction; others, though self-fruitful, produce more vigorous offspring if outcrossed. Cross-pollination has led to the development of the rich diversity of plants existing today. Many agricultural plants cross-pollinated by honey bees or other pollinators provide food, fiber or oil for us; on a world scale it is estimated that about 15 percent of the human diet is supplied by cultivated plants that are dependent for full yield and quality on the activities of biological agents (McGregor, 1976).

NECTAR

Nectaries

Occurrence and Structure

Secretory glands are widely distributed in the plant kingdom. Nectaries are glands that are specialized for the export of sugars. Other kinds include salt-secreting glands in plants adapted to high-salt habitats, glands of carnivorous plants that secrete digestive enzymes and glands that secrete resins or

*R. W. Shuel, Ph.D., Professor Emeritus, Department of Environmental Biology, University of Guelph, Guelph, Ontario, Canada.

waxes (Lüttge, 1971). Nectaries occur in a wide range of species of both monocyledons and dicotyledons, in fungal hyphae (Beutler, 1953) and in the bracken *Pteridium aquilinum* (Percival, 1965). Nectaries were originally classified as *floral* if they occurred in any part of the flower and *extrafloral* if they were situated on vegetative aerial parts (Fahn, 1979). However, nectaries may logically be regarded as functional rather than structural entities. A more recent classification limits the term floral to those which are located in the inner part of the flower and are directly associated with pollination, and designates others as extrafloral. In the amended classification nectaries on the sepals, for instance, are listed as extrafloral. The names *nuptial* and *extranuptial*, in use in Europe, are synonymous with the amended floral and extrafloral designations, respectively.

A common feature of nectaries is that they occur near plant surfaces—at surface level, embedded in the surface, or raised above it. Within this position they vary widely with respect to location, appearance and complexity. Some consist of simple groups of cells with little to distinguish them to the unaided eye, from adjacent tissue. At the other extreme they may be conspicuous structures as in poinsettia (*Euphorbia pulcherrima*). Often highly developed nectaries are more richly pigmented than adjacent tissues, being a brighter green or a different color; orange-gold, yellow-gold, yellow-white or white nectaries occur in different species of the genus *Lamium* (Gulyas, 1967).

Both floral and extrafloral nectaries may occur on the same plant, *e.g.*, as in cotton, *Gossypium* spp. (Butler *et al.*, 1975), broad bean, *Vicia faba* (Davis *et al.*, 1988a), and *Turnera ulmifolia* (Elias *et al.*, 1975). More than one form of extrafloral nectary may be present; cotton has extrafloral nectaries in several locations, and trumpet creeper (*Campsis radicans*) has four extrafloral nectary systems in addition to the floral system (Elias and Gelband, 1975). Earlier Frey-Wyssling and Haüsermann (1960) considered extra-floral nectaries to be simpler in structure and probably more primitive than floral types. Although some extrafloral ones are simple and inconspicuous, it is now evident that many possess a highly complex structure and physiology (Elias, 1972; Elias *et al.*, 1975; Durkee, 1983; Nichol and Hall, 1988).

Where both floral and extrafloral nectaries are present on the same plant, the latter may attract ants which serve to protect the flowers from grazing animals or from seed predators (Bentley, 1977; Keeler, 1977, 1981; Inouye and Taylor, 1979). They may also be visited by nectar foragers including honey bees. Extrafloral nectaries on the rubber tree (*Hevea brasiliensis*) are an important commercial source of nectar for honey in some countries (Crane *et al.*, 1984). Table 1 lists some locations of both floral and extrafloral nectaries. The significance, in relation to pollination, of nectary location and other special features of flowers is discussed by Meeuse (1961), Percival (1965) and

Table 1.   **Some Locations of Nectaries**

Location of Nectary	Representative Species	References
On receptacle	Cherry **(Prunus serotina)**	Brown, 1938
Ring at base of ovary	**Phlox drummondi** **Citrus** spp. Snapdragon **(Antirrhinum majus)**	Brown, 1938 Shuel, 1956
Disc above ovary	**Viburnum** spp.	Brown, 1938
Apex of ovary	**Hyacinthus orientalis**	Howes, 1945
Ring below stamens	**Portulaca** spp.	Brown, 1938
On petals	Mullein **(Verbascum** spp.) Tulip tree **(Liriodendron tulipfera)**	Howes, 1945
On petals modified as spurs	Garden nasturtium **(Tropaeolum majus)**	Rachmilevitz and Fahn, 1975
On sepals		Howes, 1945 Percival, 1965
	Linden **(Tilia** spp.) **Abutilon striatum**	
Between or below bracts	Cotton **(Gossypium** spp.)	Butler et al., 1972
Petiole		Elias, 1972 Elias et al., 1975
	Pithecellobium macridenium **Turnera ulmifolia**	
Leaf undersurface, on midrib	Cotton **(Gossypium** spp.)	Butler et al., 1972
Leaf undersurface, between midrib and main veins	Cherry laurel **(Prunus laurocerasus)**	Percival, 1965

Kevan and Baker (1983). Detailed descriptions of nectary morphology may be found in papers by Brown (1938), Ziegler (1968) and Fahn (1952).

Nectary Anatomy and Vascular Supply

The microscopic anatomy of the nectary tissue appears to be essentially the same for all species that have been studied (Fahn, 1952) and similar to that of other plant glands (Lüttge, 1971). At the magnification of the light microscope (Fig. 1) the cells are relatively small, isodiametric, closely packed, densely staining and either lacking vacuoles or with very small ones (Lüttge, 1971). The glandular tissue may be richer in protein than adjacent nonsecretory tissue (Shuel and Tsao, 1978). The epidermal cells usually are covered by a cuticle, often with stomates (Fig. 2). The conductive system that supplies the nectary varies with the species. It may contain mostly phloem (food-conducting vessels), xylem (water-conducting vessels) or similar

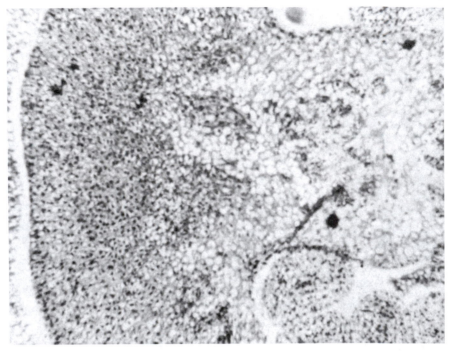

Figure 1. Cross-section through a nectary of *Streptosolen jamesonii,* Miers.

amounts of both (Agthe, 1951; Frei, 1955). The ratio of phloem to xylem appears to determine the concentration of the nectar as it is expelled, a high ratio being associated with relatively concentrated nectar (Agthe, 1951). In some species the conductive elements innervate the nectary tissue (Frei, 1955), but more commonly they appear to terminate short of it (Durkee, 1983). In most nectaries there is present below the secretory tissue subglandular tissue composed of cells larger than secretory cells and less densely packed; these lie between the vascular supply and secretory cells (Durkee, 1983).

High rates of oxygen uptake have been found for nectaries (Ziegler, 1956) and intense acid phosphatase activity, suggestive of high rates of energy utilization, seems to be typical (Ziegler, 1955; Cotti, 1962). Such observations give the impression that nectaries are centers of intense metabolic activity, an impression supported by numerous electron microscope studies of the ultrastructure of nectary cells carried out in recent years (*e.g.,* Schnepf, 1969; Findlay and Mercer, 1971a,b; Eriksson, 1977; Rachmilevitz and Fahn, 1975; Fahn, 1979; Durkee, 1983). The typical dense appearance under the light microscope has been resolved at the higher magnification of transmission electron microscopy (TEM) into a rich assembly of membranes and orga-

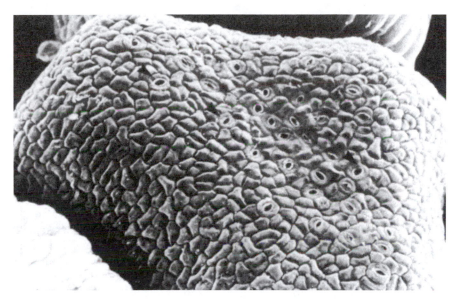

FIGURE 2. Scanning electron microscope of a lateral nectary of rapeseed, *Brassica napus*, showing stomates through which nectar leaves (Photo courtesy of A.R. Davis).

nelles including smooth and rough endoplasmic reticulum, ribosomes and mitochondria. Endoplasmic reticulum is associated with transportation and storage in the cell and ribosomes with protein synthesis, and mitochondria are the site of oxidative phosphorylation in which high-energy phosphate bonds are produced as adenosine triphosphate (ATP). Golgi bodies, associated with secretion in both plant and animal cells, have been observed in some nectaries, usually prior to secretion. Secretory cells are connected with one another and with cells of the adjacent subglandular tissue by plasmodesmata, cytoplasmic strands through which sugar solution can move from cell to cell.

Composition of Nectar

Chemically, nectar is a solution of sugars in water with minor amounts of numerous other constituents which include amino acids, organic acids, proteins, lipids, anti-oxidants, dextrins and minerals (Baker and Baker, 1983), and enzymes (Frey-Wyssling *et al.*, 1954).

Depending on species and on environmental conditions, sugar concentration in nectar at the time of collection may vary from 4 or 5 percent to more than 60 percent. Concentrations of total solids as high as 92% were found in nectar of dwarf mistletoe, *Arceuthobium abietinum* (Brewer *et al.*, 1974). Earlier semi-quantitative studies of nectar sugars in a large number of species, using filter-paper chromatography, indicated that the main sugars present were sucrose, glucose and fructose (Wykes, 1952; Percival, 1961). In a survey of 889 species of angiosperms Percival identified three general patterns of sugar composition: a) dominant sucrose nectar, b) "balanced" nectar containing similar amounts of sucrose, glucose and fructose, and c) nectar with hexoses dominant. Sucrose dominance was associated with long-tubed flowers in which nectar was protected and hexose-dominant nectar with open flowers such as those of the mustard family. Maltose was tentatively identified in a few nectars. Van Handel *et al.* (1972), however, detected no maltose by microchemical analysis of nectars of 35 species distributed among 19 plant families.

Much more sensitive methods are now available for quantitative determination of sugars in nectar samples in the microliter range. Gas-liquid chromatography (GLC) was used by Kleinschmidt *et al.* (1968), Bosi (1973), Battaglini *et al.* (1973), Bieleski and Redgewell (1980) and Tanowitz and Smith (1984) and high-performance liquid chromatography (HPLC) by Erickson and associates (1979), Southwick *et al.* (1981) and Severson and Erickson (1983). The general patterns of sucrose, glucose and fructose identified by Percival (1965) have been corroborated by Baker and Baker (1983) for 765 species; they amended Percival's classification to "sucrose-dominant", "sucrose-rich", "hexose-rich" and "hexose-dominant", providing quantitative limits for each. The rarity of maltose has been confirmed by GLC (Battaglini *et al.*, 1973). A number of new sugars have been found, including raffinose—a major sugar in the nectar of sunflower, *Helianthus annuus*, and present as well in the nectar of an orchid, *Eupactis atropurpurea* (Pais and Chaves des Neves, 1980). The latter also contained cellobiose, mellibiose and the sugar alcohol sorbitol. Galactose has been tentatively identified in some nectars (Battaglini *et al.*, 1973). Nectar-sugar patterns of floral and extrafloral nectaries present on the same plant may differ. Floral nectaries of the neotropical shrub *Turnera ulmifolia* secreted a sucrose-dominant nectar and foliar nectaries a balanced nectar (Elias *et al.*, 1975). Floral nectar of faba bean (*Vicia faba*) contained only sucrose, whereas extrafloral nectar had mostly glucose and fructose (Davis *et al.*, 1988a).

The sucrose-hexose ratio evidently has ecological significance. Baker and Baker (1983) have demonstrated a relationship between sugar-ratio category and principal type of pollinator visiting the flower. In feeding experiments with sugar solutions honey bees indicated a preference for sucrose-rich solutions (Waller, 1972). Other observations, however, suggested that honey bees tend to be opportunists, gathering nectar from flowers in which it is richest and most accessible, rather than selecting on the basis of sugar composition (Southwick *et al.*, 1981).

The amino acid content of nectar is too low to contribute significantly to honey bee nutrition; in a study of 13 species Mostowska (1965) found that it amounted only to 0.002 to 4.8 mg per 100 mg total solids. However, as amino acids have now been detected in almost all nectars tested for them and the amino acid complement appears to be characteristic of the plant species, it is considered likely that they have been significant in the co-evolution of plants with butterflies and moths (Baker and Baker, 1975, 1977). As with sugars, floral and extrafloral nectars produced on the same plant may differ in their amino acids (Baker *et al.*, 1978).

Lipids appear to be of frequent, though by no means universal, occurrence in nectar (Baker and Baker, 1983), and may influence post-secretion changes in concentration of nectar solids (Corbet *et al.*, 1979b). Organic acids and volatile chemicals may be significant in contributing to the flavor of honey. Some 52 volatile components of Piedmontese honeys in Italy were identified by capillary GLC (Bicchi *et al.*, 1983); presumably the majority of these, if not all, came from nectar.

Nectar and flower aroma are produced concurrently, though the physiological connection between them is not clear. Part of the aroma may come from essential oils in the nectar. As honey bees are easily trained to associate a sugar source with an aroma (Frisch, 1967), aroma presumably plays a part in recruitment of bees to flowers. Bees were found to select among alfalfa clones on the basis of their aromas (Loper and Waller, 1970); some of the aroma components were later identified (Loper *et al.*, 1972).

A number of substances that repel bees or are toxic to them have been identified. Non-protein amino acids, phenolic compounds and alkaloids are common nectar toxins (Rhoades and Bergdahl, 1981). Nectar of timber milkvetch (*Astragalus miser* var. *serotinus*) growing in British Columbia, Canada was found to contain a glycoside, miserotoxin, of which part of the molecule, 3-mitro-1-propanol, was poisonous to bees (Majak *et al.*, 1980). Nectar of the related species spotted loco (*Astragalus lentiginosus*) is also poisonous (Vansell and Watkins, 1934). Minute amounts of the nectar of a New Zealand plant, the yellow Kowhai (*Sophora microphylla*) caused narcosis in bees, owing evidently to the presence of an alkaloid (Clinch *et al.*, 1972). Not all toxic substances in nectar are of natural occurrence. Systemic pesticides applied to the plant before flowering may be secreted in nectar (Glynne-Jones and Thomas, 1953; Jaycox, 1964; Lord *et al.*, 1968; Barker *et al.*, 1980; Fiedler and Drescher, 1984; Loper *et al.*, 1987; Davis *et al.*, 1988a). Contaminated nectar may be carried in the foragers' honey sac to the hive and enter the brood food (Jaycox, 1984). The systemic insecticides dimethoate and carbofuran interfered with brood development at concentrations that were sublethal for adult workers (Davis *et al.*, 1988c). A reduction in colony population from such a cause would not be readily detected.

Quantitative Nectar Measurement

In studies of nectar secretion where quantity rather than detailed composition is of interest, nectar may be extracted from flowers by centrifugation (Swanson and Shuel, 1950) or withdrawn in capillaries (*e.g.*, Drummond Microcaps[R]) and measured volumetrically or by weighing. The percentage concentration of nectar solids is read with a refractometer (the refractometer scale is actually for pure *sucrose*, so there is a small error). Where amounts of nectar are very minute, they may be extracted by leaching in water, followed by chemical determination of sugars (*e.g.*, Käpylä, 1978; Roberts, 1979). Sampling should be standardized with respect to time of day and age of flowers.

The Mechanism of Secretion

Nectar is basically phloem sap which has undergone some alteration during the secretory process. Depending on the plant species, xylem sap may be added. The overall sequence includes the unloading of sap from the phloem sieve tubes, passage of the sap (sometimes called *pre-nectar*) from cell to cell through the sub-glandular tissue and into and through the secretory cells, and finally the expulsion of the finished nectar to the outside. The final step has been well studied. Nectar may pass out in one of a number of ways: through epidermal cells where a cuticle is absent, through trichomes, through thinly-stretched cuticle or ruptured cuticle (Daumann, 1935; Fahn, 1952), or through stomates (Fig. 2). The latter are a particularly common form of exit; familiar examples of nectaries with stomates include cucumber (Collison, 1973), red clover (Eriksson, 1977), alfalfa (Teuber *et al.*, 1980), birdsfoot trefoil (Murrell *et al.*, 1982), sunflower (Sammataro *et al.*, 1985), rapeseed (Davis *et al.*, 1986), fababean (Davis *et al.*, 1988a) and many others. Cell-to-cell movement of pre-nectar is probably through plasmodesmata. Of the remaining steps in secretion much still remains to be learned. Like other secretory processes, nectar secretion appears from the evidence of many studies to be an "active" process, requiring energy built up in respiration and released by the hydrolysis of ATP. The way in which energy is linked to secretion, and the stage at which it is used, are not known. It appears unlikely that energy is supplied directly to the sugar molecules by way of phosphorylation and dephosphorylation since the ATP available is not sufficient to support observed rates of secretion (Gunning and Hughes, 1976). More likely it is supplied to membranes. One approach to the problem has been to examine the ultrastructure of developing nectaries by TEM and to correlate changes in structure with secretion (*e.g.*, Schnepf, 1964; Findlay and Mercer, 1971b; Rachmilevitz and Fahn, 1975; Eriksson, 1977). On the basis of such studies two contrasting mechanisms have been proposed. According to the theory of *granulocrine* secretion nectar is transported as a solution in vesicles to the cell membrane (plasmalemma) adjacent to the cell wall of the secretory cell; there the vesicle fuses with the plasmalemma, releasing its contents into the cell wall area, from where it passes to the outside (see Rachmilevitz and Fahn, 1975; Fahn, 1979; Davis *et al.*, 1988a). In the *eccrine* secretion model, individual sugar molecules are thought to be transported across the membrane of the gland cell, possibly by a carrier molecule (see Schnepf, 1964; Eriksson, 1977; Davis *et al.*, 1986). In light of the evidence it is likely that both mechanisms exist, depending on the plant species.

Several other features of nectar secretion remain to be explained: 1) the movement of pre-nectar to the secretory cells appears to be directed, 2) in the great majority of species examined (Ziegler, 1956), sucrose is the only sugar found in phloem sap, yet nectar contains mixtures of sugars characteristic of the species; in the woody Rosaceae a sugar alcohol, sorbitol, is the major transport form of sugar, but the nectar contains mixtures of sucrose, glucose and fructose (Bieleski and Redgewell, 1980), and 3) secreted nectar components may be resorbed by the nectary (Pedersen *et al.*, 1959; Shuel, 1961b; Lüttge, 1962; Collison, 1973).

Excellent discussions of the physiology of nectar secretion and its relation to nectary ultrastructure are presented by Lüttge (1971) Fahn (1979) and Durkee (1983).

Factors Influencing Nectar Production

In discussing the influence of various factors on nectar production the main emphasis will be placed on the most important aspect for honey production: quantity of nectar sugar, which is a function of the average amount of sugar secreted per flower and the number of flowers. Effects on other aspects, such as nectar composition and timing and patterns of secretion, will be treated in less detail.

Potential nectar yield is set by heredity; the extent to which the potential is realized depends on environmental conditions of weather and soil. In the context of the physiology of secretion, yield depends on two basic factors: the secretory capacity of the nectary and the quantity of sugar delivered to it. The quantity delivered is in turn influenced by synthesis and transport, and also on competition for sugar from other processes. Some of the effects of heredity and environment on nectar yield can be related to these basic internal factors and conditions—photosynthesis, movement of sugar (and water) and sugar utilization.

In herbaceous plants nectar sugar is likely to be of recent origin, while in trees and shrubs it may be derived from stored carbohydrates as well. Girnik (1958), using radioactive isotopes, obtained evidence that some nectar secreted by linden trees (*Tilia* spp.) came from stored carbohydrate. Sugar is translocated from the site of photosynthesis to the site of utilization through the phloem, a living system of conductive cells (sieve tubes) that ramifies throughout the plant. It moves as a solution whose concentration varies with species and environmental conditions but might commonly be of the order of 5 to 20 percent (Crafts and Crisp, 1971). Transport through a living system is referred to as *symplastic*. There is still no universal agreement on the mechanism of translocation, but the theory most widely held is that a pressure difference established between regions of high sugar concentration (regions of synthesis) and regions where the concentration is kept lower by sugar utilization provides the driving force (the "mass-flow" hypothesis). Regions of synthesis are called *sources*, regions of utilization, *sinks*. Growth of a fruit creates a sink to which sugar moves. Likewise an actively secreting nectary, by expelling nectar to the exterior, provides a sink. According to Wardlaw (1968) the distribution pattern of sugars within the plant depends on the relative abilities of different plant organs to utilize the available supply.

Water enters the roots from the soil and moves upward through the plant as a dilute solution of not more than 1 percent through the xylem, a conductive system made up of non-living vessels, under the pull of transpiration from the leaves. Transport in the xylem is called *apoplastic*.

Hereditary and Internal Factors

Substantial differences among varieties or clones have been reported for red clover (Ryle, 1954), alfalfa (Pedersen, 1951; Kauffeld, 1967; Barnes and Furgala, 1978; Teuber, 1978; Teuber *et al.*, 1980, 1983), white clover (Oertel, 1956), linden (Demianowicz and Hlyn, 1960), cucumber (Collison, 1973) and rapeseed (Szabo, 1982) among many other crops. Hereditary variation in nectar yield has been linked to such anatomical features as volume of nectariferous tissue (Fahn, 1949), nectary size (Collison, 1973), receptacle diameter (Teuber *et al.*, 1980; Teuber *et al.*, 1983) and numbers of nectary stomates (Teuber *et al.*, 1980; Sammataro *et al.*, 1985). Nectar yield in birdsfoot trefoil was positively correlated with volume of functional phloem (phloem directly involved in transport) in the vascular supply to the nectary (Murrell *et al.*, 1982a). Lateral nectaries of rapeseed, which secrete much more nectar than median nectaries, are also much more richly supplied with phloem (Davis *et al.*, 1986). Heredity may also affect nectar yields through regulation of photosynthetic activity or the enzyme complement of the nectary, though such effects have not been documented. One would expect the inheritance of nectar yield potential to be complex and this appears to be the case. Multifactorial inheritance with additive gene effects has been postulated for alfalfa (Pedersen, 1953b; Teuber, 1978), red clover (Bond and Fyfe, 1968) and birdsfoot trefoil (Murrell *et al.*, 1982).

Where male and female flowers are present on the same plant (monoecious plants) nectar yields may be unequal for the two sexes. Female cucumber flowers secreted much more nectar than male flowers (Collison, 1973). Male flowers of banana (*Musa paradisica*), however, yielded several times as much nectar as female flowers (Fahn, 1949) and male flowers of several willows (*Salix* spp.) produced more than female flowers (Kropácová and Nedbalová, 1970). The cause of such differences is not known, but may be related to nectary size (Collison, 1973).

Secretory rates may differ with flower age, or periodic variation may occur. Cucumber flowers produced most of their nectar on the day of anthesis, with the majority of flowers producing little or none on the second day (Collison, 1973). Dandelion (*Taraxacum* spp.) secreted on two successive mornings, with a larger yield on the second day (Szabo, 1984). Flowers of the common milkweed (*Asclepias syriaca*), protected from insects, reached a peak of secretion 50 hours after anthesis, and ceased to secrete after 120 hours (Southwick and Southwick, 1983). A diurnal pattern was also observed, with peak rates between 4 A.M. and 8 A.M. (Southwick, 1983). By contrast, flowers of horse chestnut (*Aesculus hippocastanum*) protected from insects secreted fairly uniformly over a period of six days, then declined sharply (Beutler, 1953). Percival (1946) observed a daily rhythm of secretion in raspberry (*Rubus fruticosus*), with terminal flowers secreting about three times as much nectar during the day as at night, over a 90-hour period. Temporal patterns of secretion may be co-ordinated with foraging habits of pollinators. Flowers of plants that are pollinated by bees, flies, butterflies or birds, which are active in the daytime, are diurnal in their flowering, whereas species pollinated by night-flying pollinators, such as moths or bats, tend to be nocturnal in their secretory pattern. Amongst species that flower during the day, variation has been observed in the time of day at which bee visits occur, suggesting a variation in the rate of nectar production (Meeuse, 1961). This is exemplified by four species of the genus *Agave* in Arizona which differ with respect to both secretory pattern and pollinators (Schaffer and Schaffer, 1977).

Sugar composition also may be affected by flower age. The percentage of sucrose relative to hexoses in the nectar of three citrus species declined as flowers matured (Loper *et al.*, 1976).

The Effect of Pollination and Fertilization
Various events that occur during flowering appear to be co-ordinated, probably by hormones (Shuel, 1961b). Nectar secretion begins about the time the flower opens, pollen ripens, and the stigma becomes receptive. As a result of this timing pollination is successful in that it leads to fertilization. Fertilization in turn seems to activate a feedback mechanism in the flower which switches off nectar secretion. This phenomenon was noted by Bonnier in 1879 and has been observed in many species since. Barbier (1962) reported

that flowers of lavender quickly wilted and ceased to produce nectar shortly after they had been visited by bees. Pankiw and Bolton (1965) noted that in florets of alfalfa nectar ceased to accumulate after tripping, regardless of floret age. In cucumber fertilization, which followed within 6 to 16 hours of pollination, terminated secretion (Collison, 1973). Nectar secretion in cotton, which lasts for a day, was accompanied by a progressive change in petal color from white to red. During this time pollen viability dropped rapidly. The red color was thought to be a signal to foragers that the nectar source was no longer worth exploiting. The color change was observed to begin earlier and develop faster with repeated visits from foragers (Eisikowitch and Loper, 1984). Secretion of radioactive sugars supplied to snapdragon flowers dropped by 60 percent within 24 hours after either pollination or treatment with the plant hormone 3-indoleacetic acid; at the same time there was a breakdown of nectary tissue protein and its redirection into fruit growth (Shuel and Tsao, 1978). In the context of plant reproduction, nectar ceases to be of any consequence once fertilization has been achieved, and the energy allotted to it earlier is now directed to other uses. There is evidence, however, that while secretion lasts nectaries are an important "sink" for sugars. Southwick (1984) calculated the amount of energy allocated to nectar production in the common milkweed, *Asclepias syriaca*, and found that during flowering 4 to 37 percent of the daily photosynthate was secreted in nectar. Similar calculations for alfalfa indicated that the total energy content of nectar was almost double that of the seed produced. Pleasants and Chaplin (1983) estimated that about 30 percent of the total energy invested in flower production in the milkweed *Asclepias quadrifolia* was utilized in nectar secretion. In tests with radioactive sugars, snapdragon nectaries competed strongly with other flower parts for sugar even when the latter were undersupplied (Shuel, 1978).

The pollination effect on secretion may introduce a bias into the estimation of potential nectar yields, particularly for species that secrete for long periods in the absence of pollinators. Potential yields are commonly estimated by sampling from flowers protected from insect visits by cheesecloth or glassine bags. As protected flowers secrete longer and produce more nectar than those left open to visitors, their nectar yields may be overestimated.

The Influence of External Factors

External factors influencing nectar production are those of weather and soil. Weather is a complex of interrelated simple factors, and under field conditions it is usually difficult to separate the individual components. Controlled experiments in which one factor can be varied independently of the rest provide more specific data; most valuable of all are studies carried out in a controlled environment chamber, or *phytotron* which permits simultaneous variation in more than one factor (*e.g.*, Robacker *et al.*, 1983). The same considerations apply to soil factors.

Sunlight

There can be little doubt of the primary importance in herbaceous plants, where nectar is likely to be of recent origin, of enough sunlight during flowering to support a high level of photosynthesis. Multiple regression analysis of honey production in a 10-year period in Saskatchewan, Canada showed hours of sunshine to be the most influential weather factor (Hicks, 1977). Scale hives situated in white clover fields in New Zealand lost weight when solar radiation fell below about 50 percent of the average level (Walton, 1977). Long-term honey production records in the United States indicate the importance of clear weather (Kenoyer, 1916; Jorgensen and Markham, 1953; Moffett and Parker, 1953). Sunlight was the most important factor determining nectar yield in alfalfa (Pedersen, 1953a). A close association was found between the amount of solar energy reaching plants in the 24-hour period immediately preceding nectar collection and nectar yield in red clover (Shuel, 1952) and alsike (Shuel, 1957; see Fig. 3) and between hours of sunshine for a similar period and nectar yields of sainfoin and white clover (Kropáčová and Haslbachová, 1970). Daily nectar yields of red clover in the greenhouse, where temperature and soil conditions remained fairly uniform and favorable to secretion, varied as much as 300 percent with fluctuations in incident solar radiation (Shuel, 1952).

The influence of sunlight on nectar production in trees and shrubs may be much less immediate than for herbaceous plants. Girnik's work (1958) has already been mentioned. Percival (1946) found no obvious relationship between daily nectar yields in raspberry and current weather conditions. Here nectar yields may have been partly dependent on stored carbohydrate and hence on the amount of solar energy received in the previous season.

Air Temperature

Nectar production involves many individual processes within the plant, each one of which is probably affected by temperature, and few unconditional statements can be made about overall temperature effects. Records of honey production in the United States (Jorgensen and Markham, 1953) and Norway (Kierulf, 1957 and Ukkelberg, 1960) indicated a direct correlation between honey crops and air temperatures. However, in records of this kind, temperature effects can easily be confused with those of light intensity; low nectar yields in cool weather may result from low photosynthetic activity accompanying overcast skies. In temperate regions photosynthesis is more commonly restricted by low light intensity than by low temperature (Thomas and Hill, 1949). Alternation of high day and low night temperatures has been thought to promote good nectar flows (Bonnier, 1879; Kenoyer, 1916). Some evidence for this effect was found in field studies with white clover, though not with sainfoin (Kropácová and Haslbachová, 1970). In controlled-temperature experiments with red clover no effect of a difference between day and night temperatures was observed (Shuel, 1952), and in similar tests with alfalfa in which a constant 25°C day-night temperature regime was compared with a 32°-18° alternation, with total degree hours held the same for both regimes, nectar yield was significantly higher under the constant temperature (Walker *et al.*, 1974). What was thought earlier to be a favorable effect of wide diurnal variation in nature may have simply been an effect of fine sunny weather. Clear summer daytime skies are often accompanied by cool nights.

Minimum or threshold temperatures for secretion have been reported for various species. Basswood (*Tilia americana*) flowers began to secrete at about

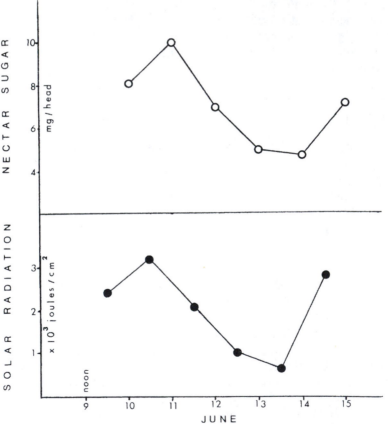

Figure 3. Relationship of nectar yield in alsike clover to solar radiation received in the 24-hour period preceding nectar measurement.

18°C (Demuth, 1923), bird cherry (*Prunus avium*) at 8° and cherry laurel (*Prunus laurocerasus*) at 18° - 20° (Beutler, 1953), cucumber (*Cucumis sativa*) at 17° - 21° (Collison, 1973) and several cultivars of soybean (*Glycine max*) at 21° (Erickson, 1975). The mechanism of the threshold effect is not understood; it may be related to the activation of an enzyme system involved in secretion. Studies by Demianowicz and Hlyn (1960) of linden species native to regions with different mean temperatures suggested that the temperature at which secretion begins may depend on the climate to which the species is adapted.

One would also expect an optimum temperature range and an upper temperature limit for nectar secretion for each species, but there is little information on the subject. Secretion in soybean plants grown under controlled conditions in a phytotron increased progressively with daytime temperature in the range 20° to 32°C (Robacker *et al.*, 1983). Prolonged high temperatures in the absence of adequate soil moisture can affect nectar production adversely by causing a moisture stress in the plant; under such conditions photosynthesis is reduced and sugar transport as well (Crafts and Crisp, 1971) and secretion may cease completely. Where moisture is plentiful, continuous hot weather has the effect of accelerating flower development so that the nectar flow may be of shorter duration, but more intense, than it would otherwise be.

Temperatures to which plants are exposed prior to flowering may affect the number of flowers produced and hence the total nectar yield per plant. In the soybean experiments referred to above, a daytime temperature increase from 28° to 32° reduced flower production by more than 50 percent (Robacker *et al.*, 1983).

Where honey crops are obtained from shrub or tree species temperature effects, like those of sunshine, are extremely complex and difficult to predict. Nectar flows in red ironbark, *Eucalyptus sideroxylon*, for example, were reported to be influenced by long-term weather conditions over several years preceding the honey harvest. Cool temperatures at the time of flowering favored the nectar flow (Porter, 1978).

Humidity

Although there is no evidence that atmospheric humidity affects secretion directly, it may have an indirect effect similar to that of temperature through its influence on rate of water loss from the plant by transpiration. Nectar-sugar yields in cotton in Arizona decreased with decreasing relative humidity (R.H.) as the day advanced (Butler *et al.*, 1972) probably owing to water stress. By contrast, nectar-sugar yields in detached flowers cultured on sugar solutions increased as ambient R.H. was reduced (Shuel, 1956). In the latter situation lowering the humidity enhanced the flow of solution to the nectaries in the absence of a moisture stress.

The most important effect of humidity on nectar production is manifested as an inverse correlation with concentration of solids (Park, 1929); Scullen, 1942; Shuel, 1952; Corbet *et al.*, 1979a,b). The effect is chiefly physical, operating in the following manner: as nectar is secreted it begins to undergo an exchange of water molecules with the surrounding atmosphere, tending to approach an equilibrium with it, but not attaining it (Corbet *et al.*, 1979b). Unless atmospheric humidity is very high, the result will be a net loss of water molecules from the nectar and an increase in sugar concentration. The rate of increase depends on humidity, air movement, temperature and the degree to which the nectar is protected by the flower parts. In some nectars evaporation may be reduced by lipids which form a thin film over the surface (Corbet *et al.*, 1979b). Nectar concentration and volume in fireweed (*Epilobium angusti-folium*) were observed to change in response to changes in ambient humidity in accordance with rates predicted from the nectar sugar concentration (Bertsch, 1983).

At night, nectar standing in the flower may become less concentrated owing to re-absorption of sugar into the nectary (Collison, 1973).

Soil Physical Conditions and Fertility

The physical state of the soil, with respect to water supply, temperature and aeration is an important element in establishing the base for a honey crop. The best level of precipitation varies with soil type. Light soils tend to be well drained and aerated and warm, and in seasons of regular rainfall they provide an ideal physical environment for root growth. Their low water-retaining capacity, however, is a disadvantage in dry seasons. Heavier soils retain water better in dry years, but are likely to have poorer drainage, which in wet years may reduce temperatures and aeration below optimal levels. Thus sandy soils were reported to support good nectar yields in white clover in New Zealand except in dry years, when heavier soils were superior (Johnson, 1946). A number of long-term honey production records indicate that the best crops occur in seasons that are slightly wetter than average and follow a season of more than average precipitation (Kenoyer, 1916; Bulanov, 1956; Heidemann, 1958). Both nectar and flower production were reduced in borage plants growing in dry soil (Czarnowski, 1953) and a seasonal difference in nectar volume yield in soybean cultivars was attributed to moisture stress in a dry season (Severson *et al.*, 1987). Water deficiency also promoted a decrease in nectar sucrose and an increase in glucose and fructose. Nectar production in soybeans is favored by heavy soil types with good water-holding capacity (Erickson and Robins, 1979).

Soil temperature influenced the number of flowers produced by soybean plants grown under controlled conditions (Robacker *et al.*, 1983). Raising the temperature from 20° to 24°C resulted in an increase of nearly 60 percent. Soil conditions during flowering affected nectar secretion in snapdragon. Less nectar was produced at a soil temperature of 16° than at 20°, and yield was also reduced in soil of water content either above or below the optimal, the field capacity (Shuel and Shivas, 1953). Nectar yields of black locust (*Robinia pseudoacacia*) in Europe are said to be better in sandy soils than in "good" soils (Sanduleac, 1961). This effect may be related to soil temperature and aeration.

The relationship of nectar production to soil fertility is extremely complex and data obtained by different researchers often appear to be conflicting. Both nectar secretion and flower production are affected. Benefits of fertilizer additions, particularly of phosphorus and potassium, have been reported for a number of crops (Ewert, 1935; Plass, 1952; Monokova and Chebotnikov, 1955; Kaziev, 1959). Other reports have shown negligible benefits for nectar, but enhanced flower production (Schöntag, 1952; Hasler and Maurizio, 1961). Ryle (1953) studied the interactions of nitrogen, phosphorus and potassium and concluded that (1) when potassium is limiting to growth, nectar yields are poor and (2) when nitrogen or phosphorus is limiting, a relative surplus of potassium is beneficial. It is important to be able to relate the effects of an element under test to its initial concentration in the soil before the test, or to some other measure of sufficiency or deficiency. Thus, Hasler and Maurizio (1961) noted poor nectar yields in soils deficient in potassium. Shuel (1957) observed interactions of potassium and phosphorus on nectar secretion, growth and flowering. The ratio of phosphorus to potassium regulated the expression of vegetative and reproductive growth, high potassium favoring vegetative growth and high phosphorus promoting flower production. As in Ryle's experiments, a drastic limitation on potassium supply reduced the amount of nectar secreted per flower, as did a high level of phosphorus. The best nectar plants were produced by a reasonable balance between the two elements. A balanced supply of nitrogen, phosphorus and potassium was most favorable for flower production in dragon's head (*Dracocephalum moldavicum*) grown in silica sand (Sklanowska, 1967). Excessive potassium shortened the flowering period. In tests with soybeans grown in a controlled environment at two levels each of nitrogen, phosphorus and potassium the best flower production was obtained with the higher level of nitrogen and the lower level of phosphorus. Potassium showed little effect (Robacker *et al.*, 1983). In experiments with red clover (Ryle, 1954; Shuel, 1955) and alfalfa (Pedersen, 1957) excessive vegetative growth accompanying an oversupply of nitrogen was detrimental to nectar production, especially in cloudy weather.

A common effect of high potassium, still unexplained, was a lower than normal nectar sugar concentration (Ewert, 1935; Shaw *et al.*, 1957; Shuel, 1957).

Good nectar yields have been reported for leguminous crops growing in calcareous soils, suggesting that calcium or magnesium, or high soil pH, may favor secretion. Carlisle and Ryle (1958) noted a beneficial effect of liming, apparently not related to soil pH, on nectar secretion in red clover. Oertel (1931) found no correlation between soil pH and nectar production in white clover. Extremes of nectar yield per flower in red clover in sand culture differed by about 100 percent with a 16-fold concentration range of calcium and magnesium in the nutrient-element supply (Shuel, 1961a). An intermediate level of calcium and a high level of magnesium were best for both nectar secretion and flower production. The calcium-magnesium effect was not related to pH, which was controlled at the same level in all treatments.

Application of boron to a boron-deficient soil enhanced the attractiveness of alsike and raspberry flowers to honey bees (Holmes, 1960). Smith and Johnson (1969) obtained a slight increase in nectar sugar concentration with the addition of boron to low-boron soil or sand cultures of white clover. The percentage of sucrose was increased and that of fructose reduced.

The Relative Importance of Various Factors

It is evident from the foregoing that nectar production is affected by numerous factors. Complex interrelationships with respect to both secretion per flower and number of flowers have been observed in controlled experiments. Exact statements can seldom be made about the importance of one factor without referring to others. Nonetheless, in particular situations some are more likely than others to restrict nectar yields. On a world basis it is likely that soil water supply is most often limiting to nectar production, as it is to plant growth (Smith, 1960). In temperate regions where cloudy weather is common, as in northern Europe, the amount of solar radiation received during flowering may often be limiting (Fig. 3). The importance of soil fertility is especially difficult to assess, but is probably much less in nature than under experimental conditions.

The difference in nectar productivity of dandelion in consecutive seasons in northern Alberta was estimated to be about 15-fold (Szabo, 1984). As the same plots were studied in both years, the genetic contribution was constant, and the seasonal differences reflected the combined effects of all environmental factors. Estimates of weather and soil contributions to total variation in scale-hive gains in white clover in New Zealand were 58 percent and 86 percent in consecutive years (Walton, 1977). Similar estimates for hives situated in fields of forage legumes and rapeseed were 58 and 89 percent (Hicks, 1977).

Heritable variation in nectar yield among cultivars of some common agricultural crops appears to be of the order of 100 percent (Shuel, 1989).

Some Important Honey Plants and Their Nectar Yields

Over 400 significant nectar-producing species belonging to more than 60 families have been listed for Great Britain (Howes, 1945). Six families are especially prominent for their contribution to honey crops: ROSACEAE (the rose family, which includes most temperate zone bush and tree fruits), LEGUMINOSAE (legumes, which include clovers, vetches and such tree species as black locust), LABIATAE (the mint family, in which are found many garden herbs), SCROPHULARIACEAE (the figworts, including figwort and toad flax), CRUCIFERAE (the mustard family) and COMPOSITAE (the composites, among which are thistles and dandelion). The same families, together with ERICACEAE (the heath family), MYRTACEAE (represented by Eucalyptus, a major nectar source in Australia and South Africa) and RUTACEAE (citrus species) are also prominent in world lists (*e.g.*, Crane, 1975; Crane *et al.*, 1984). More than 1000 species of North

Table 2.   **Some Estimated Nectar Sugar Yields**

Species	Common Name	Country	Estimated Yield kg/ha*	Reference
Phacelia tanacetifolia	Phacelia	Poland	183-1130	Demianowicz et al. (1963)
Medicago sativa	Lucerne, Alfalfa	United States	250**	McGregor and Todd (1952)
Tilia cordata	Linden	Poland Europe	125 500	Maksymiuk (1960) Crane (1975)
Melilotus alba	Sweet clover	Russia	up to 400	Girnik (1969)
Solidago serotina	Goldenrod	Poland	56-294	Demianowicz and Jablonski (1966)
Asclepias syriaca	Milkweed	Poland	187-576 (honey)	Demianowicz (1962)
Echium vulgare	Blueweed	Poland	182-429 (honey)	Demianowicz et al. (1963)
Trifolium pratense	Red clover	Alberta, Canada	880	Szabo and Nadja (1985)
Taraxacum spp.	Dandelion	Alberta, Canada	5-72	Szabo (1984)

* 1 kg/ha = 0.89 lb/acre
** One crop in a 5-day period

American nectar plants are included in the standard reference works of Lovell (1926) and Pellett (1947).

Records of actual honey yields from pure stands of single species are rather uncommon. Potential yields, on a hectare or acre basis, have been estimated for many species from nectar sampling and flower counts. Some examples are given in Table 2. The following comments on the data are appropriate. (1) In most instances the figures are estimates of potential nectar sugar production, not of potential stored honey. (2) There is a wide range within each species, reflecting the importance of environmental influences. (3) Variation between species is a good measure of relative nectar productivity; absolute values, however, should be accepted with caution because they are subject to errors from two sources, sampling errors owing to the small size of samples relative to total flower populations, and the effect, already mentioned, of pollination and fertilization in terminating secretion.

The Improvement of Nectar Yield

Sustained large increases in nectar yields, year after year, are not feasible because production is always subject to the vagaries of the weather, the major source of uncertainty. However, a substantial increase in potential yield which could be exploited under favorable weather conditions is possible with the application of present knowledge. There are two promising approaches. The first is to take advantage of hereditary variation in a program of selection and breeding. This procedure could be used with agricultural crop plants which have an important use other than honey production, *e.g.*, forage legumes grown for seed, or oilseed crops such as soybean, rapeseed and sunflower. The second approach is to increase nectar sources with permanent plantations of selected species along roadsides, railroad rights-of-way, power transmission lines, and in wastelands and land of low agricultural value. This method has been called "fixed-land" honey production (Ayers *et al.*, 1987).

POLLEN

Pollen production takes in a much broader range of plant species than nectar, including angiosperms that produce no nectar but are insect-pollinated, angiosperms that are wind-pollinated and conifers, which belong to the gymnosperms. Although bees have no ecological relationship with the last two groups, they obtain pollen from them. Pollen, like nectar, is a key element in pollinator-flower relationships. Like nectar, it is essential to honey bee nutrition (providing protein, minerals, sterols and vitamins). In the biology of the plant, however, nectar and pollen are in different categories; nectar is a product of secretory activity but a pollen grain is an integral part of the plant, a spore called the *microspore*, or *male gametophyte*. In flowering plants pollen is produced in the upper portion of the stamen, the anther, in four pollen sacs. (The corresponding female gametophyte, called the *megaspore* or *embryo sac* is formed in the ovary at the base of the pistil). Each pollen sac contains numerous pollen mother cells which undergo meiotic cell division to produce haploid microspores. The spore walls thicken to develop the sculptured appearance characteristic of the species. The nucleus divides mitotically to produce the *tube* nucleus (which controls pollen-tube growth) and the *generative* nucleus. The final result is a thickwalled, binucleate cell with a haploid chromosome number (Burns, 1974). Male gametes, or sperm cells, are formed either in the pollen grain or in the pollen tube produced by the germinating pollen grain. The opening of the anthers to release mature pollen is known as *dehiscence.*

In conifers, pollen is produced on cones, or *stroboli.*

Physical and Chemical Characteristics of Pollen

The pollen grain is remarkable in many respects. The wall, the most complex of any plant-cell wall, is stratified. The inner layer, the *intine*, is composed largely of cellulose and the outer layer, the *exine*, of sporopollenin which is believed to be formed by the oxidative polymerization of carotenoids (Stanley and Linskens, 1974). The exine confers on the wall a high resistance to mechanical or chemical damage. Under dry conditions the pollen grain is in a state of temporary dormancy with a vanishingly low respiratory rate owing to its low water content (10 to 15 percent). When deposited on the surface of a compatible stigma, the grain quickly becomes hydrated, the pollen tube is formed under the control of the tube nucleus and the generative nucleus migrates through the tube, eventually to combine with the female gamete and initiate the development of a diploid plant. At least some pollen grains also have the potential to develop into haploid plants. The production of haploid plants from pollen grains is a recent application of biotechnology to plant breeding.

Pollen grains vary in size with species (and to some extent with weather conditions) from less than 5 μm to more than 200 μm (Stanley and Linskens, 1974). The two most common modes of pollen transfer are by wind and by biological agents of which insects are the most numerous. Wind-borne (anemophilous) and insect-transferred (entomophilous) pollens differ in a number of characteristics. Wind-borne grains tend to be comparatively light, dry and drab in color. Pollen dispersed by insects is commonly heavier, somewhat moist and sticky and colored in various shades of yellow, brown, orange or red (Maheshwari, 1950). Colors are produced by two groups of pigments in the exine, the water-soluble flavonoids and the fat-soluble derivatives of carotene, the carotenoids and xanthophylls (Stanley and Linskens, 1974). Pollen loads of honey bees come in various colors which may vary with weather conditions (Hodges, 1952). Willow pollen, for instance, may change from golden to pale-yellow, then to green, after exposure to sunlight. After a rain or frost, or in the early morning, the color is darker.

Pollen contains a rich store of nutrients, as would be expected in light of its role in plant reproduction. Considerable quantitative variation exists between species. A range of 11.4 to 29.9 percent was found for protein content in 32 species (Todd and Bretherick, 1942) and a comparable range (6.2 to 20.7 percent) for six citrus cultivars (Gilliam *et al.*, 1980). The amino acid complement—in the free and combined forms—is generally well-suited to the nutritional needs of newly emerged honey bees, each of the 10 essential amino acids being present in excess of the minimum for optimal growth (Groot, 1953). Nineteen free amino acids were found by Kauffeld (1980) in bee-collected pollen; the individual levels of some varied during the season but the quantitative composition remained the same. Characteristic of most pollens is a high concentration of proline (often in excess of 1 percent), the significance of which is not clear. Citrus pollens were an exception to this rule, and a number of desert species contained high levels of glutamic and aspartic acids as well as proline (McCaughey *et al.*, 1980). Some pollens have a substantial lipid content (Todd and Bretherick, 1942). Except for sterols, lipids are not known to be essential for insects. Two sterols that are frequently dominant in pollen are 24-methylene cholesterol (Hügel, 1962) and B-sitosterol (Farag *et al.*, 1980). A number of long-chain fatty acids have been identified (Farag *et al.*, 1978). An unsaturated 18-carbon acid, similar in structure to linolenic acid, was isolated from pollen and found to be attractive to honey bees (Hopkins *et al.*, 1969). Pollen contains a large variety of sugars, including some that are very uncommon (Stanley and Linskens, 1974). The starch content varies widely with species. From an examination of 990 angiosperm species it was concluded that pollen of species visited by bees and flies is often starchless and has a high proportion of the energy content stored as oils, whereas wind-borne pollen usually contains starch (Baker and Baker, 1979). Apparently starch is broken down when pollen is stored in the hive as bee bread (Herbert and Shimanuki, 1978). Pollen is a rich source of B-complex vitamins (Vivino and Palmer, 1944) and is quite high in ascorbic acid (Dietz, 1975). The ash content is generally between 2.5 and 6.5 percent of the dry weight (Stanley and Linskens, 1974).

Pollen may contain natural substances such as alkaloids or saponins that are toxic to honey bees, though these are rare (Stanley and Linskens, 1974). A small amount of the systemic insecticide carbaryl has been reported in pollen of grape (*Vitis vinifera*) sprayed shortly before flowering (Muller, *et al.*, 1975). The systemic insecticide dimethoate was found in alfalfa pollen (Barker *et al.*, 1980). Microencapsulated insecticides may be collected by foraging bees and become mixed with pollen (Atkins and Kellum, 1984). A microbial flora may also be present. Forty-one bacterial species of the genus *Bacillus* were isolated from samples of almond pollen (*Prunus* spp.) taken from flowers, pellets or storage in beehives placed in an almond orchard (Gilliam, 1979). It was not certain whether all bacterial species had originated in the flowers.

Detailed analytical data for pollen are presented in the comprehensive treatise by Stanley and Linskens (1974).

Nutritive Value

Pollens differ greatly in both nutritive value and attractiveness to bees. Maurizio (1953) evaluated pollen sources for their efficiency in supporting development of hypopharyngeal glands and fat bodies and classified them as highly nutritious, nutritious, fair or poor. Pollen from fruit trees, willows, maize and clovers was among the best, dandelion pollen was in the second rank and pollen of pine and other conifers was the poorest. In a test by Standifer (1967) in which comparisons were made after adjusting the data for protein content, relative nutritional values were changed somewhat, with dandelion scoring higher than maize. In other tests based on brood produced per unit of diet consumed, differences were found even among closely related pollen species (Loper and Berdel, 1980). Pollen of desert species was found to have a low protein content but a good amino acid balance, which resulted in a significant correlation between hypopharyngeal gland development and quantity of protein ingested (McCaughey *et al.*, 1980). A similar correlation was obtained for hypopharyngeal gland development and amino acid level in another study (Kauffeld, 1980). Mixed pollen usually seems to be preferred to single sources (Doull, 1966; Schmidt, 1984).

It appears likely from the results of various feeding tests that amino acid balance is a major determinant of the nutritional efficiency of pollen.

Timing of Pollen Development and Availability

The general chronological order for pollen availability in plants of north temperate regions is: trees in early spring, shrubs in late spring and herbaceous species in summer (Percival, 1955; Severson and Parry, 1981). Within the constraints imposed by heredity, environmental factors, especially air temperature, cause some seasonal variation in time of pollen maturation. In a 10-year study of long leaf pine (*Pinus palustris*), in southern Alabama the dates of maximum pollen availability were observed to vary over a period of nearly 6 weeks; the timing was closely related to the accumulation of heat units above a base level of 10°C received after January 1 (Boyer, 1973). The concept of developmental time as a function of quantity of energy received applies to flowering plants as well, and seasonal variation in flowering dates of a week or longer is common for many temperate-zone species.

The time of day at which pollen is released is a species characteristic. Percival (1955) divided 81 species of angiosperms into six categories according to time of pollen presentation; times ranged from early morning (*e.g.*, sunflower, *Helianthus annuus* and rapeseed, *Brassica oleracea*; see also Murrell and Nash, 1981) to late afternoon (*e.g.*, broad bean, *Vicia faba*). In some plants release was continuous throughout the day. The degree of synchrony of pollen release among flowers within a species varied widely. In black currant (*Ribes nigrum*), plantain (*Plantago lanceolata*) and maize (*Zea mays*) release was almost simultaneous for stamens of all flowers while in others it extended over a period of hours and in a few (including raspberry *Rubus fruticosus*) over several days. Where individual flowers last for several days pollen release may be periodic; Synge (1947) noted that flowers of *Vicia faba* opened on each of three successive days, closing on the first two nights. The hour of opening was progressively earlier each day. A diel periodicity on a crop basis was found in sampling for concentration of airborne oil-seed rape (*Brassica napus*) pollen, with peaks in late morning and early afternoon and very low concentrations at night and midday (Williams, 1984). Timing of anther dehiscence in relation to flower opening is also variable. In Synge's study (1947) dehiscence of legume anthers—red clover, white clover and broad bean—occurred in the closed bud, in fireweed (*Epilobium angusti-folium*) it occurred after the flowers opened, and in a third group which included white mustard (*Brassica alba*) dehiscence was almost simultaneous with flower opening. Collection of pollen by insects was limited by whichever event, anthesis or dehiscence, occurred last.

Low temperatures, by retarding pollen ripening, delay dehiscence (Percival, 1955). The minimum temperature for dehiscence varies with species and may be quite low for those that flower in spring; anthers of apple dehisced at temperatures as low as 5°C (Percival, 1955). Considerable pollen is aborted at temperatures of 10° or lower and in cool seasons the incidence of pollen sterility is high (Younger, 1961). Anthers will not dehisce in the presence of free water, but in some species will do so at relative humidities in excess of 95 percent (Percival, 1955).

Factors Influencing Pollen Collection by Honey Bees

Pollen collection appears to be influenced by a number of factors both internal to the colony and external. The colony's need, or perceived need, for protein is important. The presence of brood stimulates pollen foraging (Free, 1967; Barker, 1971) and foraging is intense during population buildup in the spring (Krell *et al.*, 1981). Absence of the queen inhibits it (Louveaux, 1958/1959). Temperature appears to be the most important meteorological factor influencing the amount collected. Synge (1947) found a positive correlation between daily maximum air temperatures and number of loads of red clover pollen collected and attributed it mainly to the acceleration of anther dehiscence at higher temperatures. Louveaux (1958/1959) observed no collecting below 10°C. Above 10° there was no well-defined proportionality between temperature and quantity of pollen gathered, but in March twice as much pollen was gathered at 17° as at 15°. In Argentina peak activity in pollen foraging was recorded at 25° and 61 to 62 percent relative humidity (Camareno and Pecho, 1973).

Choice of pollen species for collection is also subject to many influences. Individual colonies have a characteristic selection which may differ even between adjacent colonies (Maurizio, 1953; Louveaux, 1958/1959; Adams *et al.*, 1979). No close relationship between pollen collection and pollen phenology was found by Louveaux (1958). In Percival's (1955) study pollen was collected almost as soon as it became available in the months of February, March and April; later in the season there was no close correspondence between time of presentation and time of collection. The quantity of pollen available in the flowers of a particular species was much less important in determining the choice than the presence or absence of nectar. It has been suggested, though not well substantiated, that bees may choose pollen that is relatively nutritious. A genetic influence is involved in preference for some pollens (Nye and Mackensen, 1970). Sometimes there may be a positive aversion to a pollen; pollen of cotton is not only avoided, but cleaned from the bees' body if picked up by accident (Loper, 1986). Botanical analyses of pollen in beehives have indicated that the bulk of a colony's pollen usually comes from a relatively small number of plant species (Maurizio, 1953; Percival, 1955; Louveaux, 1958/1959; Murrell and Szabo, 1981). It was concluded by Percival (1955) and Severson and Parry (1981) that bees forage chiefly on the most abundant species within their flight range. Evidently a common principle of economy of effort—obtaining the maximum benefit per unit of energy expended—underlies both nectar and pollen foraging (Kevan and Baker, 1983).

Colony Needs, Pollen Yields and Possible Increase

Quantitative data for pollen production are much fewer than for nectar. Wind-pollinated flowers in general produce a greater abundance than those pollinated by insects. Percival (1955) recorded pollen yields for individual flowers ranging from about 1/100 mg in a thistle, *Sonchus oleraceus* to 404 mg in maize. Cultivars of sour cherry in Poland produced from about 3 to 6.5 mg pollen/10 flowers; an orchard containing a mixture of cultivars was estimated to be capable of producing 3 to 4 kg/ha (Sklanowska and Pluta, 1984). Annual pollen additions to biomass of as much as 10kg/ha have been calculated for stands of alder, hazel and rye (Stanley and Linskens, 1974). Based on Percival's (1955) value of 404 mg for a single plant, a yield of 11 or 12 kg/ha should be possible for maize. Szabo (1985) estimated pollen yields of 9.3 and 20.2 kg/ha, respectively, for two species of rapeseed, *Brassica napus* and *B. campestris*, growing in the Peace River region of Alberta. Relative productivity has also been measured as numbers of grains per anther. Traynor (1981) used a haemacytometer to compare cultivars of almond and alfalfa in California and obtained a range of about 33,000 to 68,000 grains per flower for almond flowers and 2100 to 4600 grains per anther for alfalfa. The same method was used to compare species and cultivars of forage legumes in northern Alberta (Szabo and Nadja, 1985). Anthers of birdsfoot trefoil had considerably more pollen than anthers of red clover, alsike, white clover or alfalfa. One trefoil cultivar, Leo, averaged 13,570 grains per anther.

Pollen harvests of 25 to 30 kg in a season were recorded for honey bee colonies in France, with occasional daily collections by populous colonies of 2 kg (Louveaux, 1959). Harvests of 51 and 55 kg in successive years were recorded for a colony in California (Eckert, 1942). If 8 - 10 kg/ha is a realistic estimate of annual pollen yield, a single honey bee colony should be able to consume the pollen produced on a crop area of 3 or 4 hectares. The corresponding value for honey consumption is difficult to estimate accurately, but figures of 80 kg (176 lbs) and 60 kg (152 lbs) have been proposed (Dietz, 1975). Many crops are capable of producing this much on less than a hectare. Based on these estimates, the area needed to satisfy pollen requirements appears considerably greater than for nectar.

There is very little information on which to base recommendations for improving pollen yield. Research on pollen production in relation to beekeeping has been limited to a small number of field studies comparing yields of cultivars; there have been few if any controlled experiments on the influence of weather and soil factors. Results of work with forest species have indicated that pollen yield is largely under genetic and physiological control (Stanley and Linskens, 1974). A modest improvement in pollen resources might be obtained with plantations of spring-flowering trees and shrubs, such as willow and alder, at permanent apiary sites.

REFERENCES

Adams, R.J., Smith, M.V. and Townsend, G.F. (1979). Identification of honey sources by pollen analysis of nectar from the hive. *J. apic Res. 18*:292-297.

Agthe, C. (1951). Ueber die physiologische Herkunft des Pflanzennektars. *Promotionsarbeit, E.T.H.,* Zurich. *2017*:240-274.

Atkins, E.L. and Kellum, D. (1984). Microencapsulated pesticides: visual microscopical detection of capsules; quantification of residue in honey and pollen. *Am. Bee J. 124*:800-804.

Ayers, G.S., Hoopingarner, R.A. and Howitt, A.I. (1987). Testing potential bee forage for attractiveness to honey bees. *Am. Bee J. 127*:91-98.

Baker, H.G. and Baker, I. (1975). Nectar constitution and pollinator-plant coevolution. *In*: Gilbert, L.E., and Raven, P.A. Animal and Plant Coevolution. pp. 100-140. University of Texas Press, Austin, Texas.

Baker, H.G. and Baker, I. (1977). Intraspecific constancy of floral nectar amino acid complements. *Bot. Gaz. 138*:183-191.

Baker, H.G. and Baker, I. (1979). Starch in angiosperm pollen grains and its evolutionary significance. *Am. J. Bot. 66*:591-600.

Baker, H.G. and Baker, I. (1983). A brief historical review of the chemistry of floral nectar. *In*: Bentley, B. and Elias, T. The Biology of Nectaries. pp. 129-152. Columbia University Press, New York.

Baker, H.G., Opler, P.A. and Baker, I. (1978). A comparison of the amino acid complements of floral and extrafloral nectars. *Bot. Gaz. 139*:322-332.

Barbier, E.C. (1963). Quelques facteurs de la productivité quantitative et qualitatives des essences chez les lavandes (*Lavandula* Tourn). *Annls Abeille 5*:265-379.

Barker, R.J. (1971). The influence of food inside the hive on pollen collection by a honeybee colony. *J. apic. Res. 10*:23-26.

Barker, R.J., Lehner, Y. and Kunzmann, M.R. (1980). Pesticides and honey bees: nectar and pollen contamination in alfalfa treated with dimethoate. *Arch envir. Contam. Toxic. 9*:125-133.

Barnes, D.K. and Furgala, B. (1978). Nectar characteristics associated with sources of alfalfa germplasm. *Crop Sci. 18*:1087-1089.

Battaglini, M., Bosi, G., and Grandi, A. (1973). Considerations of the glucidic fractions of the nectars of 57 honey plants of central Italy. *Proc. XXIV Int. Beekeep. Congr.*, Buenos Aires :493-500.

Bentley, B. (1977). The protective function of ants visiting the extrafloral nectaries of *Bixa orellana* (Bixaceae). *J. Ecol. 65*:27-38.

Bertsch, A. (1983). Nectar production of *Epilobium angustifolium* at different air humidities: nectar sugar in individual flowers and the optimal foraging theory. *Oecologia 59*:40-48.

Beutler, R. (1953). Nectar. *Bee Wld 34*:106-116, 123-136, 156-162.

Bicchi, C., Belliardo, F. and Frattini, C. (1983). Identification of the volatile components of some Piedmontese honeys. *J. apic. Res. 22*:130-136.

Bieleski, R.L. and Redgewell, R.J. (1980). Sorbitol metabolism in nectaries from flowers of Rosaceae. *Aust. J. Pl. Physiol. 7*:15-25.

Bond, D.A. and Fyfe, J.L. (1968). Corolla tube length and nectar height of F_1, red clover plants (*Trifolium pratense*) and their seed yield following honey bee pollination. *J. agric. Sci. Camb. 70*:5-10.

Bonnier, G. (1879). Les Nectaires. *Ann. Sci. Nat.* (Bot. Series V1) *8*:5-212.

Bosi, G. (1973). Méthode rapide pour la détermination par chromatographie en phase gazeuse les glucides du nectar: technique de prélèvement du nectar et de préparation du nectar et de préparation des éthers trimethylsilyles en présence d'eau. *Apidologie 4*:57-64.

Boyer, W.D. (1973). Air temperature, heat sums and pollen shedding phenology of long-leaf pine. *Ecology 54*:420-426.

Brewer, J.W., Collyard, K.J. and Lott, C.E.,Jr. (1974). Analysis of sugars in dwarf mistletoe nectar. *Can. J. Bot. 52*:2533-2538.

Brown, W.H. (1938). The bearing of nectaries on the phylogeny of flowering plants. *Proc. Am phil. Soc. 79*:549-595.

Bulanov, N.A. (1956). [The influence of precipitation on flowering and nectar production in linden. In Russian.] *Pchelovodstvo 33*:33-34.

Burns, G. (1974). The Plant Kingdom. Macmillan Publishing Co., Inc., New York.

Butler, G.D., Loper, G.M., McGregor, S.E., Webster, J.L. and Margolis, H. (1972). Amounts and kinds of sugar in the nectars of cotton (*Gossypium* spp.), and the time of their secretion. *Agron. J. 64*:364-368.

Camareno, J.E. and Pecho, I. Meza. (1973). *"Arginea maritima"*(L) Bak (*Scilla maritima*) evaluated by statistical methods and by bee behavior to it. Proc. XXIV Int. Beekeep. Congr., Buenos Aires :484-490.

Carlisle, E. and Ryle, M. (1958). The influence of nitrogen, phosphate, potash and lime on the secretion of nectar by red clover in the field. *Emp. J. exp. Agric. 23*:126-130.

Clinch, P.G., Palmer-Jones, T. and Foster, I.W. (1972). Effect on honey bees of nectar from the yellow Kowhai *(Sophora microphylla* Ait.) *N.Z. Jl agric. Res. 15*:194-201.

Collison, C.H. (1973). Nectar secretion and how it affects the activity of honey bees in the pollination of hybrid pickling cucumbers, *Cucumis sativus* L. M.Sc. Thesis, Michigan State University.

Corbet, S.A., Unwin, D.M. and Prŷs-Jones, O.E. (1979a) Humidity, nectar and insect visits to flowers, with special reference to *Crataegus, Tilia* and *Echium. Ecol. Ent. 4*:9-22.

Corbet, S.A., Wilman, P.G., Beament, J.W.L., Unwin, D.M. and Prŷs-Jones, O.E. (1979b) Post secretory determinants of sugar concentration in nectar. *Plant Cell Environ. 2*:293-308.

Cotti, T. (1962). Ueber die quantitative Messung der Phosphataseaktivität in Nektarien. Ber. schweiz. *Bot. Ges. 72*:306-332.

Crafts, A.S. and Crisp, C.E. (1971). Phloem Transport in Plants. W. H. Freeman and Company, San Francisco.

Crane, Eva. (1975). The flowers honey comes from. *In*: Crane, E., Honey: a Comprehensive Treatise. pp. 3-76. Heinemann in co-operation with International Bee Research Association, London.

Crane, E., Walker, W. and Day, R. (1984). Directory of Important World Honey Sources. International Bee Research Association, London.

Czarnowski, C. von. (1953). Ueber die Eineverkung unterschiedlicher Wasserorgung auf die Nektarsekretion beim Boretsch (*Borago*). *Z. Bienenforsch 2*:89-91.

Daumann, E. (1935). Die systematische Bedeutung des Blutennektariums der Gattung Iris. *Bot. Zlb. Beih. B53*:525-625.

Davis, A.R., Peterson, R.L. and Shuel, R.W. (1986). Anatomy and vasculature of the floral nectaries of *Brassica napus* (Brassicaceae). *Can. J. Bot. 64*:2508-2516.

Davis, A.R., Peterson, R.L. and Shuel, R.W. (1988a). Vasculature and ultrastructure of the floral and stipular nectaries of *Vicia faba* (Leguminosae). *Can. J. Bot. 66*:1435-1448.

Davis, A.R., Peterson, R.L. and Shuel, R.W. (1988b). Distribution of carbofuran and dimethoate in flowers and their secretion in nectar as related to vascular supply. *Can. J. Bot. 66*:1248-1255.

Davis, A.R., Solomon, K.R. and Shuel, R.W. (1988c). Laborbory studies of honeybee larval growth and development as affected by systemic insecticides at adult-sublethal levels. *J. apic. Res. 27*:146-161.

Demianowicz, Z. (1962). [Factors affecting the nectar production of plants. In Polish]. *Pchelarstvo 13*:172-174.

Demianowicz, Z. and Hlyn, M. (1960). [Comparative studies of nectar secretion in seventeen varieties of lime. In Polish]. *Pszczel. Zesz. Nauk. 4*:113-152.

Demianowicz, Z. and Jablonski, B. (1966). [Nectar secretion and honey yields of plants with small flowers. In Polish]. *Pszczel. Zesz. Nauk. 10*:87-94.

Demianowicz, Z., Jablonski, B., Ostrowska, W., and Szybowski, S. (1963). [The honey yield of the main honey plants in Polish conditions. II. In Polish]. *Pszczel. Zesz. Nauk.* 7:95-111.

Demuth, G. (1923). Temperature and nectar secretion. *Glean. Bee Cult.* 51:582.

Dietz, A. (1975). Nutrition of the adult honey bee. *In*: Dadant & Sons. The Hive and the Honey Bee. Dadant and Sons, Inc., Hamilton, Ill. pp. 125-156.

Doull, K.M. (1966). The relative attractiveness to pollen collecting honey bees of some different pollens. *J. apic Res.* 5:9-14.

Durkee, L.T. (1983). The ultrastructure of floral and extrafloral nectaries. *In*: Bentley, B. and Elias, T. The Biology of Nectaries. pp. 1-29. Columbia University Press, New York.

Eckert, J. (1942). The pollen required by a colony of honey bees. *J. econ. Ent.* 35:309-311.

Eisikowitch, D. and Loper, G.M. (1984). Some aspects of flower biology and bee activity on hybrid cotton in Arizona, USA. *J. apic. Res.* 23:243-248.

Elias, T.S. (1972). Morphology and anatomy of foliar nectaries of *Pithecellobium macradenium* (Leguminosae). *Bot. Gaz.* 133:38-42.

Elias, T.S. and Gelband, H. (1975). Nectar: its production and functions in trumpet creeper. *Science* 189:289-291.

Elias, T.S., Rozich, W.R. and Newcombe, L. (1975). The foliar and floral nectaries of *Turnera ulmifolia* L. *Am. J. Bot.* 62:570-576.

Erickson, E.H. (1975). Variability of floral characteristics influences honey bee visitation to soybean blossoms. *Crop Sci.* 15:767-771.

Erickson, E.H. and Robins, J.M. (1979). Honey from soybeans: The influence of soil conditions. *Am. Bee J.* 119:444-445, 448-450.

Erickson, E.H., Jr., Thorp, R.W., Briggs, D.L., Estes, J.R., Daun, R.J., Marks, M. and Schroeder, C.H. (1979). Characterization of floral nectars by high-performance liquid chromatography. *J. apic. Res.* 18:148-152.

Eriksson, M. (1977). The ultrastructure of the nectary of red clover (*Trifolium pratense*). *J. apic. Res.* 16:184-193.

Ewert, R. (1935). Die Forderung der Nektar abscheidung bei Raps, Bucheweisen, und Rotklee durch Kalidungung. *Dtsch. Imkerführer* 9:63-66.

Fahn, A. (1949). Studies in the ecology of nectar secretion. *Palest. J. Bot. Jerusalem Ser.* 4:207-224.

Fahn, A. (1952). On the structure of floral nectaries. *Bot. Gaz.* 113:464-470.

Fahn, A. (1979). Ultrastructure of nectaries in relation to nectar secretion. *Am. J. Bot.* 66:977-985.

Farag, R.S., Youssef, A.M., Ewies, M.A. and Hallabo, S.A.S. (1978). Long-chain fatty acids of six pollens collected by honeybees in Egypt. *J. apic. Res.* 17:100-104.

Farag, R.S., Ahmed, A.I., Rashad, S.E. and Ewies, M.A. (1980). Unsaponifiable matter of six pollens collected by honeybees in Egypt. *J. apic. Res.* 19:248-254.

Fiedler, L. and Drescher, W. (1984). Residue analysis of insecticides in nectar: possible contamination after pre-blossom treatment. *Ed.* J.N. Tasai, Proc. Fifth Int. Symp. Pollin. Versailles. pp. 209-213 International Bee Research Association, London.

Findlay, N. and F.V. Mercer. (1971a). Nectar production in *Abutilon* I. Movement of nectar through the cuticle. *Aust. J. biol. Sci.* 24:647-656.

Findlay, N. and F.V. Mercer (1971b). Nectar production in *Abutilon* II. Submicroscopic structure of the nectary. *Aust. J. biol. Sci.* 24:657-664.

Findlay, N., Reed, M.L. and Mercer, E.V. (1971). Nectar production in *Abutilon* III. Sugar secretion. *Aust. J. biol. Sci. 24*:665-675.

Free, J.B. (1967). Factors determining the collection of pollen by honeybee foragers. *Anim. Behav. 15*:134-144.

Free, J.B. (1970). Insect pollination of crop plants. Academic Press, Inc., New York and London.

Frei, E. (1955). Die Innervierung der Floralen Nektarien dikotyler Pflanzenfamilien. *Ber. Schweiz. bot. Ges. 65*:60-115.

Frey-Wyssling, A. and Hausermann, E. (1960). Deutung gestaltlosen Nektarien. *Ber. Schweiz. bot. Ges. 70*:150-160.

Frey-Wyssling, A., Zimmermann, M. and Maurizio, A. (1954). Ueber den Enzymatischen Zuckerumbau in Nektarien. *Experientia 10*:490-492.

Frisch, K. von. (1967). The Dance Language and Orientation of Bees. Harvard University Press, Cambridge, Mass.

Gilliam, M. (1979). Microbiology of pollen and bee bread: the genus *Bacillus. Apidologie 10*:269-274.

Gilliam, M., McCaughey, W.F. and Wintermute, B. (1980). Amino acids in pollens and nectars of citrus cultivars from honeybee colonies in citrus groves. *J. apic. Res. 19*:64-72.

Girnik, D.U. (1958). [Use of isotopes to study the flowering and nectar secretion of linden (*Tilia*). In Russian]. *Fiziologiya Rast* :143-147.

Glynne-Jones, G.D. and Thomas, W.D.E. (1953). Experiments on the possible contamination of honey with schradan. *Ann. appl. Biol. 40*:546-555.

Groot, A.P.de. (1953). Protein and amino acid requirements of the honeybee (*Apis mellifera* L.). *Physiologia comp. Oecel. 3*(2,3) 90 pp.

Gulyas, S. (1967). Zuzammenhang zwischen struktur und Produktion in den Nektarien einiger Lamium-arten. *Acta biol., Szeged 13*:4-10.

Gunning, B.E.S. and Hughes, J.E. (1976). Quantitative assessment of transport of pre-nectar in the trichomes of *Abutilon* nectaries. *Aust. J. Pl. Physiol. 3*:619-637.

Hasler, A. and Maurizio, A. (1951). Ueber die Einfluss verschiedener Nahrstoffe auf Blutensatz, Nektarsekretion, und Samenertrag von honigenden Pflanzen, speziell von Sommerrraps (*Brassica napus* L.). *Schweiz. Bienenztg. 74*:208-219.

Heidemann, K. (1958). Heidetracht, ein undefinierbar Faktor in der Imkerei. *Bienen Bundesg. 11*:251-254.

Herbert, E.W. and Shimanuki, H. (1978). Chemical composition and nutritive value of bee-collected and bee-stored pollen. *Apidologie 9*:33-40.

Hicks, G.R. (1977). An investigation into the role of climate in explaining the variation in honey production between various localities in Saskatchewan. B.A. Thesis, Wilfrid Laurier University, Waterloo, Canada.

Hodges, Dorothy. (1952). The Pollen Loads of the Honeybee. Bee Research Association, London.

Holmes, F.O. (1960). Boron deficiency as a probable cause of the failure of bees to visit certain flowers. *Am. Bee J. 100*:102-103.

Hopkins, C.Y., Jevans, A.W. and Boch, R. (1969). Occurrence of octadecatrans-2, cis-9, cis-12-trienoic acid in pollen attractive to the honey bee. *Can. J. Biochem. 47*:433-436.

Howes, F.N. (1945). Plants and Beekeeping. Faber and Faber, London.

Hügel, M.F. (1962). Etudes de quelques constituants du pollen. *Annls Abeille 5*:97-133.

Inouye, D.W. and Taylor, O.R. (1979). A temperate region plant-ant-seed predator system: consequences of extrafloral nectar secretion by *Helianthus quinquenervis*. *Ecology* 60:1-7.

Jaycox, E.R. (1964). Effect on honeybees of nectar from systemic insecticide treated plants. *J. econ. Ent.* 57:31-35.

Johnson, L.H. (1946). Nectar secretion in clover. Effect of soil and climate on nectar production. *N.Z. Jl Agric.* 73:111-112.

Jorgensen, C. and Markham, F. (1953). Weather factors influencing honey production. Mich. agic. Exp. Stn Spec. Bull 340.

Käpylä, M. (1978). Amount and type of nectar sugar in some wild flowers in Finland. *Ann. Bot. Fenn.* 15:85-88.

Kauffeld, N.M. (1967). Clonal preference factors related to honeybee foraging on alfalfa. Ph.D. Dissertation, Kansas State University.

Kauffeld, N.M. (1980). Chemical analysis of Louisiana pollen and colony conditions during a year. *Apidologie* 11:47-55.

Kaziev, T. (1959). [Influence of fertilizer application and watering on nectar secretion in cotton. In Russian]. *Pchelovodstvo* 36:25-38.

Keeler, K.H. (1977). The extrafloral nectaries of *Ipomea carnea* (Convolvulaceae). *Am. J. Bot.* 64:1182-1188.

Keeler, K.H. (1981). Function of *Mentzelia nuda* (Loasaceae) postfloral nectaries in seed defense. *Am. J. Bot.* 68:295-299.

Kenoyer, L.A. (1916). Environmental influences on nectar secretion. *Bot. Gaz.* 63:249-265.

Kevan, P.G. and Baker, H.G. (1983). Insects as flower visitors and pollinators. *A. Rev. Ent.* 28:407-453.

Kierulf, B. (1957). [Is it possible to predict the heather flow? In Norwegian]. *Nord. Bitidskr.* 9:42-44.

Kleinschmidt, M.G., Dobrenz, A.K. and McMahon, V.A. (1968). Gas chromatography of carbohydrates in alfalfa nectar. *Pl. Physiol.* 43:665-667.

Krell, R., Dietz, A. and Couvillon, G.A. (1981). Seasonal pollen production by colonies of honey bees in the coastal zone of Georgia. *Proc. XXVIII Int. Beekeep. Congr. Acapulco:* 391-393.

Kropácová, S. and Haslbachová. (1970). A study of some climatic factors on nectar secretion in sainfoin (*Onobrychis viciaefolia* v. *sativa* Thell). *Sb. vys. Sk. Zemed. v Brne A* 18:613-620.

Kropácová, S. and Nedbalová, V. (1970). [The nectar-bearing capacity of some willow species (*Salix* sp.). In Czech]. *Lesnictvi Ustav. vedekotechnick Infor.* Prague 16(43):1095-1100.

Loper, G.M. (1986). Cotton pollen: honeybee avoidance and absence of gossypol. *J. econ. Ent.* 79:103-106.

Loper, G.M. and Berdel, R.L. (1980). A nutritional bioassay of honeybee brood-rearing potential. *Apidologie* 11:181-189.

Loper, G.M. and Waller, G.D. (1970). Alfalfa flower aroma and flower selection by honey bees. *Crop Sci.* 10:66-68.

Loper, G.M., Flath, R.A. and Webster, J.L. (1972). Identification of ocimene in alfalfa by combined GC-mass spectometry. *Crop Sci.* 11:61-63.

Loper, G.M., Olvey, J. and Berdel, R.L. (1987). Concentration of the systemic gametocide TD1123 in cotton nectar, and honeybee response. *Crop Sci.* 27:558-561.

Loper, G.M., Waller, G.D. and Berdel, R.L. (1976). Effect of flower age on sucrose content in nectar of citrus. *HortSci.* 11:416-417.

Lord, K.A., May, M.A. and Stevenson, J.H. (1968). The secretion of the systemic insecticides dimethoate and phorate into nectar. *Ann. appl. Biol. 61*:19-27.

Louveaux, J. (1958/1959). Recherches sur la récolte du pollen par les abeilles (*Apis mellifica* L.). *Annls Abeille 1*:113-188, 197-221; *2*:13-111.

Lovell, J.H. (1926). Honey plants of North America. The A.I. Root Co., Medina, Ohio.

Lüttge, U. (1962). Ueber die Zusammensetzung des Nektars und den Mechanismus seiner Sekretion II. Der Kationgehalt des Nektars und die Bedeutung des Verhaltnisses Mg^{++} und Ca^{++} im Drusengewebe für die Sekretion. *Planta 59*:108-114.

Lüttge, U. (1971). Structure and function of plant glands. *A. Rev. Pl. Physiol. 22*:23-44.

Lüttge, U. (1977). Nectar composition and membrane transport of sugars and amino acids: a review of the present state of nectar research. *Apidologie 8*:305-319.

Maheshwari, P. (1950). An Introduction to the Physiology of Angiosperms. McGraw-Hill, New York.

Majak, W., Neufeld R. and Corner, J. (1980). Toxicity of *Astragalus miser* V. *serotinus* to the honeybee. *J. apic Res. 19*:196-199.

Maksymiuk, I. (1960). [The nectar secretion of linden *Tilia cordata* at Reserve Obrozyska near Musyna (Carpathians). In Polish]. *Pczczel. Zeszel. Nauk. 4*:105-125.

Maurizio, A. (1953). Weitere Untersuchungen an Pollenhöschen. *Beih. schweiz. Bienenztg. 2*:486-556.

McCaughey, W.F., Gilliam, M. and Standifer, L.N. (1980). Amino acids and protein adequacy for honeybees of pollens from desert plants and other floral sources. *Apidologie 11*:78-86.

McGregor, S.E. (1976). Insect pollination of cultivated crop plants. U.S. Dep. Agric. Handb. 496. ARS-USDA, Washington, D.C.

McGregor, S.E. and Todd, F.E. (1952). Canteloup production with honeybees. *J. econ. Ent. 45*:43-47.

Meeuse, B.J.D. (1961). The Story of Pollination. The Ronald Press Company, New York.

Moffett, J.O. and Parker, R.L. (1953). Relation of weather factors to nectar flow in honey production. Kans. agric. Exp. Stn Bull. 74.

Monokova, N. and Chebotnikov, K.M. (1955). [Increasing the nectar production of honey plants by means of various fertilizers. In Russian]. *Pchelovodstvo 8*:44-46.

Mostowska, I. 1965. [Amino acids in nectars and honeys. In Polish]. *Zesz. nauk. Wyzsz. Szk. roln. Olsztyn 20*:413-432.

Muller, P., Stecke, W. and Gebbing H. (1975). Spreading of carbaryl in vine (*Vitis vinifera* L.) and its accumulation in pollen. *Proc. XXV Int. Beekeep. Congr. Grenoble* :166-167.

Murrell, D.M. and Nash, W. (1981). Nectar secretion by Toria (*Brassica campestris* L. var. *Toria*) and foraging behavior of *Apis* species in Bangladesh. *J. apic. Res. 20*:34-38.

Murrell, D.M. and Szabo, T.I. (1981). Pollen collection by honey bees at Beaverlodge, Alberta. *Am. Bee J. 121*:885-888.

Murrell, D.M., Shuel, R.W. and Tomes, D.T. (1982a). Nectar production and floral characteristics in birdsfoot trefoil (*Lotus corniculatus* L.). *Can. J. Plant Sci. 62*:361-371.

Murrell, D.M., Tomes, D.T. and Shuel, R.W. (1982b). Inheritance of nectar production in birdsfoot trefoil. *Can. J. Plant Sci. 62*:101-105.

Nichol, P. and Hall, J.L. (1988). Characteristics of nectar secretion by the extrafloral nectaries of *Ricinus communis. J. exp. Bot. 39*:573-586.

Nye, W.P. and Mackensen, O. (1970). Selective breeding of honeybees for alfalfa pollen collection—with tests in high and low alfalfa pollen collection regions. *J. apic. Res. 9*:61-64.

Oertel, E. (1931). Hydrogen-ion concentration of soils and its relation to the importance of white clover as a honey plant. *J. econ. Ent. 24*:627-632.

Oertel, E. (1956). Nectar production of white clover. *Glean. Bee Cult. 84*:461-463.

Pais, M.S.S. and Chaves des Neves, H.J. (1980). Sugar content of the nectary exudate of *Epipactis atropurpurea* Refin. *Apidologie 11*:39-45.

Pankiw, P. and Bolton, J.L. (1965). Characteristics of alfalfa flowers and their effects on seed production. *Can. J. Pl. Sci. 45*:333-342.

Park, O.W. (1929). The influence of humidity upon sugar concentration in the nectar of various plants. *J. econ. Ent. 22*:534-544.

Pedersen, M.W. (1951). Nectar production in alfalfa clones as related to bee visitation and seed production including a study of techniques for measuring nectar. Ph.D. Dissertation, University of Minnesota.

Pedersen, M.W. (1953a). Environmental factors affecting nectar secretion and seed production in alfalfa. *Agron. J. 45*:359-361.

Pedersen, M.W. (1953b). Seed production in alfalfa as related to nectar production and honeybee visitation. *Bot Gaz. 115*:129-138.

Pedersen, M.W. (1957). Effects of thinning and established stand, of nitrogen supply, and of temperature on nectar secretion in alfalfa. *Bot. Gaz. 119*:119-122.

Pedersen, M.W., LeFevre, C.W. and Wiebe, H.H. (1958). Absorption of C^{14}-labeled sucrose by alfalfa nectaries. *Science 127*:758-759.

Pellett, F.C. (1947). American Honey Plants. 4th Ed. Orange Judd Publishing Co. Inc., New York.

Percival, M.S. (1946). Observations on the flowering and nectar secretion of *Rubus fruticosus* (AGC). *New Phytol. 45*:111-123.

Percival, M.S. (1955). The presentation of pollen in certain angiosperms and its collection by *Apis mellifera. New Phytol. 54*:353-368.

Percival, M.S. (1961). Types of nectar in angiosperms. *New Phytol. 60*:235-281.

Percival, M.S. (1965). Floral Biology. Pergamon Press, Oxford.

Plass, F. (1952). Versuche zur Festellung des Einflusses der Mineral Dungung auf die Nektarabsonderung der Obstgenochse. *Bienenzucht 5*:270-271.

Pleasants, J.M. and Chaplin, S.J. (1983). Nectar production rates of *Asclepias quadrifolia*; causes and consequences of individual variation. *Oecologia 59*:232-238.

Porter, J.W. (1978). Relationship between flowering and honey production of red ironbark (*Eucalyptus sideroxylon* (A. Cunn) Benth. and climate in the Bendigo district of Victoria. *Austr. J. agric. Res. 29*:815-829.

Rachmilevitz, T. and Fahn, A. (1975). The floral nectary of *Tropaeolum majus* L. The nature of the secretory cells and the manner of nectar secretion. *Ann. Bot. 39*:721-728.

Rhoades, D.F. and Bergdahl, J.C. (1981). Adaptive significance of toxic nectar. *Am. Nat. 117*:798-803.

Robacker, D.C., Flottum, P.K., Sammataro, D. and Erickson, E.H. (1983). Effects of climatic and edaphic factors on soybean flowers and on the subsequent attractiveness of the plants to honey bees. *Field Crops Res. 6*:267-278.

Roberts, R.S. (1979). Spectrophotometric analysis of sugar produced by plants and harvested by insects. *J. apic. Res. 18*:191-195.

Ryle, M. (1954). The influence of nitrogen, phosphate and potash on the secretion of nectar. *J. agric. Sci. Camb. 44*:400-407, 408-419.

Sammataro, D., Erickson, E.H. Jr. and Garment, M.B. (1985). Ultrastructure of the sunflower nectary. *J. apic Res. 24*:150-160.

Sanduleac, E.V. (1961). Die Robinie, der wichstige nektarspendende Baum der Rumanischen Volksrepublik. *Arch. Geflügelz. Kleintierke 10*:220-233.

Schaffer, W.M. and Schaffer, M.V. (1977). The reproductive biology of Agavaceae I. Pollen and nectar production in four Arizona agaves. SWest Nat. 22:157-167.

Schmidt, J.O. (1984). Feeding preferences of *Apis mellifera* L. (Hymenoptera:Apidae): individual versus mixed pollen species. *J. Kans. ent. Soc. 57*:323-327.

Schnepf, E. (1969). Sekretion und Exkretion bei Pflanzen. Springer-Verlag, Vienna.

Schöntag, A. (1952). Der Einfluss von Mineralstoffen auf die Nektarabscheidung durch die Planzen. *Naturwissenschaften 39*:304-305.

Scullen, H.A. (1942). Observations on the relationship of alsike clover nectar to relative humidity. *J. econ. Ent. 35*:453-454.

Severson, D.W. and Erickson, E.H. Jr. (1983). High-performance liquid chromatography of carbohydrates in cucumber nectar. *J. apic. Res. 22*:158-162.

Severson, D.W. and Parry, J.E. (1981). A chronology of pollen collection by honeybees. *J. apic. Res. 20*:97-103.

Severson, D.W., Nordheim, E.V. and Erickson, E.H. Jr. (1987). Variation in nectar characteristics within soyabean cultivars. *J. apic. Res. 26*:156-164.

Shaw, F., Bourne, A. and Migliorani, R. (1957). The effect of various levels of fertilizers on the growth and nectar secretion of snapdragons (*Antirrhinum majus* L.). *Glean. Bee Cult. 85*:598-599.

Shuel, R.W. (1952). Some factors affecting nectar yield in red clover. *Pl. Physiol. 27*:95-110.

Shuel, R.W. (1955). Nectar secretion in relation to nitrogen supply, nutritional status, and growth of the plant. *Can. J. agric. Sci. 35*:124-138.

Shuel, R.W. (1956). Studies of nectar secretion in excised flowers. I. The influence of cultural conditions on quantity and composition of nectar. *Can. J. Bot. 34*:142-153.

Shuel, R.W. (1957). Some aspects of the relation between nectar secretion and nitrogen, phosphorus and potassium nutrition. *Can. J. Pl. Sci. 37*:220-236.

Shuel, R.W. (1961a). The influence of calcium and magnesium supply on nectar production in red clover and snapdragon. *Can. J. Pl. Sci. 41*:50-58.

Shuel, R.W. (1961b). Influence of reproductive organs on secretion of sugars in flowers of *Streptosolen jamesonii* Miers. *Pl. Physiol. 36*:265-271.

Shuel, R.W. (1978). Nectar secretion in excised flowers. V. Effects of indoleacetic acid and sugar supply on distribution of [^{14}C] sucrose in flower tissues and nectar. *Can. J. Bot. 56*:566-571.

Shuel, R.W. (1989). Improving honey production through plant breeding. *Bee Wld. 70*:36-45.

Shuel, R.W. and Shivas, J.A. (1953). The influence of soil physical condition during the flowering period on nectar production in snapdragon. *Pl. Physiol. 28*:645-651.

Shuel, R.W. and Tsao, W. (1978). Nectar secretion in excised flowers. VI. Relationship of secretion to protein metabolism. *Can. J. Bot. 566*:833-842.

Sklanowska, K. (1967). [Influence of potassium fertilization on the nectar yield of *Dracocephalum moldavicum* L. In Polish]. *Annls Univ. Mariae Curie—Sklodowska 20*:107-121.

Sklanowska, K. and Pluta, S. (1984). [Pollen productivity of the sour cherry cultivars Kerezer, Nefus and Lutowke. In Polish]. *Pszczel. Zeszel. Nauk 28*:163-174.

Smith, F.G. (1960). Beekeeping in the Tropics. Longmans, London.

Smith, R.H. and Johnson, W.C. (1969). Effect of boron on white clover nectar production. *Crop Sci. 9*:75-76.

Southwick, E.E. (1983). Nectar biology and nectar feeders of common milkweed *Asclepias syriaca* L. *Bull. Torrey bot. Club 110*:324-334.

Southwick, E.E. (1984). Photosynthate allocation to floral nectar: a neglected energy investment. *Ecology 65*:1775-1779.

Southwick, A.K. and Southwick, E.E. (1983). Aging effects on nectar production in two clones of *Asclepias syriaca. Oecologia 56*:121-125.

Southwick, E.E., Loper, G.M. and Sadwick, S.E. (1981). Nectar production, composition, energetics and pollinator attractiveness in spring flowers of western New York. *Am. J. Bot. 68*:994-1002.

Standifer, L.N. (1967). A comparison of the protein quality of pollens on development of the hypopharyngeal glands and longevity of honey bees *Apis mellifera* L. *Insectes soc. 14*:415-425.

Stanley, R.G. and Linskens, H.F. (1974). Pollen: Biochemistry Biology and Management. Springer-Verlag, Berlin, Heidelberg and New York.

Swanson, C.A. and Shuel, R.W. (1950). The centrifuge method for measuring nectar yield. *Pl. Physiol. 25*:513-520.

Synge, A.D. (1947). Pollen collection by honeybees. *J. Anim. Ecol. 16*:122-138.

Szabo, T.I. (1982). Nectar secretion by 28 varieties and breeders' lines of two species of rapeseed. *Am. Bee J. 122*:645-647.

Szabo, T.I. (1984). Nectar secretion in dandelion. *J. apic. Res. 23*:204-208.

Szabo, T.I. (1985). Variability of flower, nectar, pollen and seed production in some Canadian canola (rapeseed) varieties. *Am. Bee J. 125*:351-354.

Szabo, T.I. and Najda, H.G. (1985). Flowering, nectar secretion and pollen production of some legumes in the Peace River region of Alberta, Canada. *J. apic. Res. 24*:102-106.

Tanowitz, B.D. and Smith, D.M. (1984). A rapid method for qualitative and quantitative analyses of simple carbohydrates in nectar. *Ann. Bot. 53*:453-456.

Teuber, L.R. (1978). Characteristics and inheritance of nectar volume in alfalfa (*Medicago sativa* L.). Ph.D. Thesis, University of Minnesota.

Teuber, L.R., Barnes, D.K. and Rincker, C.M. (1983). Effectiveness of selection for nectar volume, receptacle diameter and seed yield characteristics in alfalfa. *Crop Sci. 23*:283-289.

Teuber, L.R., Albertsen, M.C., Barnes, D.K. and Heichel, G.H. (1980). Structure of floral nectaries (*Medicago sativa* L.) in relation to nectar production. *Am. J. Bot. 67*:433-439.

Thomas, M.D. and Hill, G.R. (1949). Photosynthesis under field conditions. *In*: Franck, J. and Loomis, W.E. Photosynthesis in Plants. pp. 19-52. The Iowa State College Press, Ames.

Todd, F.E. and Bretherick, O. (1942). The composition of pollens. *J. econ. Ent. 35*:312-317.

Traynor, J. (1981). Use of a fast and accurate method for evaluating pollen production of alfalfa and almond flowers. *Am. Bee J. 121*:23-25.

Ukkelberg, B. (1960). [Temperature and the heather flow. In Norwegian]. *Birøkteren 76*:42-43.

Van Handel, E., Haeger, J.S. and Hansen, C.W. (1972). The sugars of some Florida nectars. *Am. J. Bot. 59*:1030-1032.

Vansell, G.H. and Watkins, W.G. (1934). Adult bees found dying on spotted loco. *J. econ. Ent. 27*:635-637.

Vivino, A.E. and Palmer, L.S. (1944). The chemical composition and nutritional value of pollens collected by bees. *Archs Biochem* 4:129-136.

Walker, A.K., Barnes, D.K. and Furgala, B. (1974). Genetic and environmental effects on quantity and quality of alfalfa nectar. *Crop Sci* 14:235-238.

Waller, G.D. (1972). Evaluating responses of honey bees to sugar solutions using an artificial flower feeder. *Ann. ent. Soc. Am.* 65:857-862.

Walton, G.M. (1977). Weather factors influencing the production of white clover (*Trifolium repens* L.) honey in southern New Zealand. *Proc. XXVI Int. Beekeep. Congr. Adelaide* :429-432.

Wardlaw, I. (1968). The control and pattern of movement of carbohydrates in plants. *Bot. Rev.* 34:79-105.

Williams, I.H. (1984). The concentration of air-borne rape pollen over a crop of oil-seed rape (*Brassica napus* L.). *J. agric. Sci.* UK 103:353-357.

Wykes, G.R. (1952). An investigation of the sugars present in the nectar of flowers of various species. *New Phytol.* 51:210-215.

Younger, V.B. (1961). Low temperature induced male sterility in male-fertile *Pennisetum clandestinum. Science* 133:577-578.

Ziegler, H. (1955). Phosphataseaktivität und Sauerstoffverbrauch des Nektariums von *Abutilon striatum* Dicks. *Naturwissenschaften* 42:259-260.

Ziegler, H. (1968). Untersuchungen über die Leitung und Sekretion der Assimilate. *Planta* 47:447-500.

Ziegler, H. (1968). La sécrétion du nectar. *In*: Chauvin, R. Traité de Biologie de l'Abeille. pp. 218-248. Masson et Cie, Paris.

CHAPTER 11

BEE FORAGE OF NORTH AMERICA AND THE POTENTIAL FOR PLANTING FOR BEES

by GEORGE S. AYERS[1]
and JAY R. HARMAN[2]

INVENTORY AND RELATIVE IMPORTANCE
OF NECTAR AND POLLEN PLANTS OF NORTH AMERICA
Introduction

No recent comprehensive survey of the important bee forages of North America has been done. Based on his correspondence and numerous travels, Pellett (1920) included in his *American Honey Plants* a state and provincial accounting of the important bee forages of the United States and Canada. Minor changes were made in the 1923 edition of this work. In the 1930 edition, numerous comments on the climate and geography as well as modest additions to the plant lists were made. Few changes were made in the 1947 edition and none were made in the 1976 edition. Lovell (1926) produced a similar description, by state, of the important bee forages of the United States based upon both his correspondence with numerous beekeepers and personal travels. Oertel (1939) summarized the results of a survey based on questionnaires returned from 710 beekeepers from around the United States. This report has appeared with minor changes several times (Robinson and Oertel, 1975 and Oertel, 1980) but how the updating was accomplished is unclear. Ramsay (1987), largely by studying the available literature, reviewed the available and potential bee forages of Canada.

Preparation of this Report

Two decisions were made at the beginning of this study that significantly affected the data gathered for its preparation. First, the United States and Canada were divided into 14 sections based on natural floristic and land use patterns. In this division, established political boundries were largely ignored. Although a considerable departure from schemes employed in most earlier accounts of North American bee forage, our approach has several precedents in ecological literature (Hunt, 1967, and Gleason and Cronquist, 1964). Second, rather than collecting data from a large number of contributors about whom very little was known, we contacted a much smaller group of individu-

---

[1] George S. Ayers, Ph.D., Professor, Department of Entomology, Michigan State University.

[2] Jay R. Harman, Ph.D., Professor, Department of Geography, Michigan State University.

als who either had a reputation as being knowledgeable about the bee forage of an area or were in a position to gather reliable information from others. These individuals were identified by querying the members of various institutions including the staff of the *American Bee Journal*, the staff of state Departments of Agriculture, Cooperative Extension and research personnel at various universities and research stations, and the presidents and secretaries of the various state and provincial beekeeping organizations.

Different questionnaires were prepared for the various geographical regions represented. These questionnaires were essentially checklists developed from the earlier reviews described above. Space was provided for the participants to add additional species to the questionnaire. The participants were instructed to separate species into three categories according to the following definitions:

1. Important plants—those plants whose removal from the area under consideration would cause at least 10% of the beekeepers to suffer a noticeable loss in production. This loss could come about as a direct loss of nectar during a major nectar flow or could result from a poor population buildup prior to a major honey flow.

2. Very important plants—those relatively few species that reliably produce a large percent of the harvested honey in the region under consideration. The word "harvested" is important, for population build-up species were not included in this group.

3. Minor or unimportant plants—plants that did not meet either of the above criteria were considered unimportant.

For all plants considered important or very important, participants were asked to indicate the initial and terminating dates of the average blooming period. Participants were also asked to identify crops from a check list, again with space for additions, for which at least 4 or 5 beekeepers of the area under consideration received payment for pollination services. All participants were contacted by phone for a verbal commitment to complete the questionnaire. The return rate was slightly better than 98%. In all cases, after the initial questionnaire was returned, follow-up inquiries to address areas of ambiguity that arose in the initial questionnaire were done either by phone or through one or more additional questionnaires. Where the data returned from an area appeared qualitatively different from that received from surrounding areas, additional participants in the area were identified and sent questionnaires.

Use of Table 1

Table 1 and Fig. 1 are meant to be used together. Fig. 1 depicts the division of the United States and Canada and allows the reader to see which states and provinces or parts thereof lie within each of the 14 regions. Table 1 summarizes the results that were returned from the respondents of each of these parts. Where the boundaries for the different regions divided states or provinces into two or more parts, data were collected from each of the individual parts rather than from the state or province as a whole. In some instances these parts are quite small, as, for example, the southern tip of Illinois that falls into region XII.

Table 1 is a matrix of rows and columns representing plant species or genera (listed in the left-hand column) as rows and regions (listed along the top) as columns. The intersection of a particular column and row forms a box into which the data for that particular species or genus in that particular region are placed. If the returned data indicated that the particular taxonomic entity was important (see definition of importance in the "Preparation of this report") for a particular state or province or part thereof, the state or province is listed in the box by a two-letter abbreviation. If the data indicated that the taxonomic entity was very important (see definition of very important above) in that location, a "*" precedes the two letter abbreviation. If the species is commercially pollinated in that location, the two letter abbreviation is followed by a "p." Within the table both the major regions as well as their divisions are listed principally in the west to east and secondarily in the north to south arrangement. In both cases the extreme northwest region is listed first followed by the tier of regions that fall under it in the north to south direction. This listing is followed by a similar listing of the tier immediately to the east of the first, until all the divisions of a region are listed. The only exception to this format is within region XII where major ordering runs from west to east as usual, but secondarily runs south to north along the eastern sea coast.

The dates represent the range of blooming period within a region and are a composite of data from the questionnaire returns as well as from major published flora and apicultural botanical writings of the different regions. The references used are listed in the literature citations at the end of the chapter. For several reasons, these ranges are generally broader than what a beekeeper would expect to observe in a single location. Because the blooming date ranges are given for an entire area, they extend over longer periods of time than the corresponding range within any subunit of that area. Where the range given is for a genus composed of several to many species, the range will be broader than the range of any one species. Generally, within a species flowering times will be later northward for spring-blooming plants but earlier northward for those that flower in late summer or autumn. Finally, because the data are often drawn from records that extend over multiple years, the ranges are broader than would be found for any single year.

The data in the native and introduced column are derived from *Hortus Third* (Liberty Hyde Bailey Hortorium, 1976) and major published flora of the different areas. The data found in the nectar and pollen column are derived largely from bee forage publications referenced within the table.

Discussion of Survey Results

A high percentage of the harvested honey of both countries comes from a taxonomically small group of plants. While the taxa (species or genera) identified as being important bee forages came from 70 families, approximately 45% of this list came from only 5 families (Sunflower, Pea, Rose, Mint, and Mustard families, listed in descending order of importance). These 5 families also contain approximately 48% of those taxa identified by one or more respondents as being "very important." Of the 77 taxa identified by one or more respondents as being in this category, the clovers, sweet clover, alfalfa, and soybean (all Pea Family) accounted for about 40% of total number of "very important" responses obtained in the questionnaire returns.

A high percentage of the honey produced in the United States and Canada comes from introduced species. Approximately 29% of those taxa identified as "important" were from genera largely or totally composed of introduced species, whereas approximately 30% of those identified as "very important" were from this group. Almost half (49%) of all the "very important" questionnaire responses referred to plants in this group.

This heavy reliance on introduced species is not typical of all the regions, however, and seems related to both the percentage of land under cultivation and the forage available in the uncultivated areas. In New England and the Atlantic and Gulf Coastal Plain, for example, both regions still heavily forested, many old-field and wetland species provide local surpluses. The Intermountain Region, on the other hand, has a poorer native flora and surpluses from the large tracts of nonagricultural land are less reliable and more dependent on a relatively few key species.

A group of generally spring flowering plants, while not listed in table 1 as being "very important," are, in fact, very important to the beekeeper. They do not appear in the "very important" category because usually no honey is harvested from them and some produce only pollen. The products of these species, however, promote buildup of the large foraging force responsible for effectively harvesting the major summer honey flows that follow. The significance of dandelion, maple, elm, willow, and poplar can be appreciated by observing the total number of times they are listed in Table 1, and their importance can scarcely be overestimated.

Trends in Bee Forage

The most inescapable conclusion derived from this survey is that, despite its central importance to the beekeeping industry, the subject of bee forage has become the neglected half of beekeeping, particularly in the United States. Many beekeepers, as well as the state and university personnel upon whom

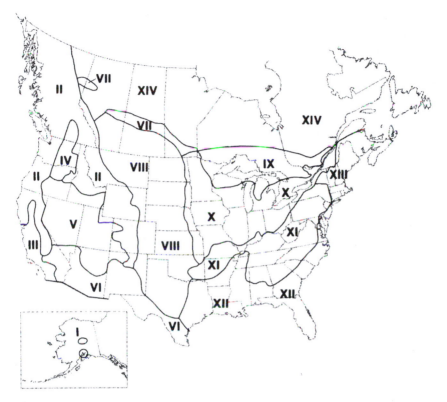

FIGURE 1. Bee forage regions of North America, based on native floral distribution and land use patterns. The regions depicted in the map correspond to those of table 1 and are referred to in the text as follows:

Region	Name
I	Alaska
II	Western Mountains
III	South and Central California
IV	Interior Northwest
V	Intermountain Region
VI	Southwest Deserts
VII	Aspen Parkland and Peace River Valley
VIII	Great Plains
IX	Northern Great Lakes and Southern Canada
X	Agricultural Interior
XI	Appalachian-Ozark Upland
XII	Atlantic and Gulf Coastal Plain
XIII	New England and Atlantic Canada
XIV	Boreal Forest

Region XIV has been omitted from Table 1 because little beekeeping occurs in the Boreal Forest.

they rely for information, do not know, much less understand, the bee forage of their area. This situation, particularly in the United States, will probably worsen with the continued retirement of the state, federal, and university personnel who are still knowledgeable in this area. Many of the respondents upon whom we relied heavily for data are either close to retirement or have already retired. In many cases, these individuals are either not being replaced or are being replaced by personnel with only modest training in apiculture and no training in apicultural botany. Simultaneously, the appearance of quality, up-to-date bee forage literature has slowed. Why this situation appears somewhat better in Canada is unclear. The bee flora in Canada is less complex than in many sections of the United States, and the Canadian nectar flows, particularly in western Canada, come from more obvious and very concentrated sources (rape and agricultural legumes) rather than from the more varied wild sources typical in the United States. Beyond this, basic interest in the subject appears to be higher than in the United States.

The amount of quality bee forage in the United States has been declining for many years. Much of this decline has resulted from the adoption of modern agricultural practices that are not likely to disappear. Like farm acreage in general, the land devoted to crops that provide quality bee forage also has diminished. The production of buckwheat, for example, has almost ceased. Because of specialization, farmers are less likely to rotate legumes such as alfalfa, clover, and sweet clover into their cropping systems. The reduction of pasturing in the dairy industry has also had a detrimental effect on honey bee forage.

Modern herbicides have greatly improved weed control in rural areas. Farmland that once supported a primary crop as well as bee forage in the form of weeds now supports only the crop. In a similar way, weeds in fence rows are less tolerated than they once were.

Scheduling alfalfa harvest to optimize its value as large animal forage also has had a major negative impact on bee forage. The protein content of alfalfa peaks just as the plant begins to bloom. In addition, early cutting during the first bloom is used in some parts of North America for control of the alfalfa weevil. Unfortunately, when alfalfa is cut at this time, it is of little benefit to bees.

The use of modern insecticides often adversely affects the quality of bee forage remaining in the modern rural landscape. In the absence of other suitable forage, this source of nectar and pollen, when it becomes contaminated with these agents, becomes an irresistible death-trap for bees.

Reforestation is also making large tracts of land unprofitable for honey production. Because the process is so slow and natural, many beekeepers do not recognize the negative impact it is having on their honey production.

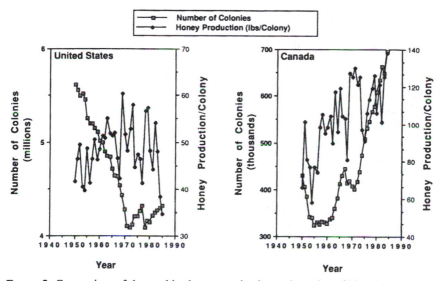

FIGURE 2. Comparison of the per hive honey production and number of hives in production between the United States and Canada. The trends in the United States are viewed by the authors as largely the result of declining bee forage, whereas those in Canada may result from a more complex situation where adequate amounts of high quality bee forage have allowed per hive production to increase despite an increase in the number of hives in production.

Abandoned farmland and harvested forests go through a natural sequential process known as succession. Initially, the barren land supports a flush of herbaceous plant species. During this stage the amount of bee forage in the area often increases. As the process continues, however, the herbaceous species are replaced by woody species and the amount of available bee forage declines. Large tracts of land in various parts of North America have entered this stage of succession resulting in a slow diminution of available bee forage and concomitant decline in potential honey production.

Finally, numerous species of bee forage may be considered noxious weeds, and the beekeeper often finds that an important species has become the target of an eradication program. Purple loosestrife, knapweed, mesquite, and yellow star thistle are examples.

The foregoing changes in the rural landscape are relatively slow and their impact on honey production is difficult to gauge on a yearly basis. They have, however, been occurring over a sufficiently long period that their effect should be evident in the historical data. The graph in Fig. 2a depicts both the average per hive production as well as the number of hives in production in the United States from 1950 to 1985. If only the per hive production data are considered, the changing rural landscape seems to have affected honey production very little, since production essentially shows no downward trend. The number of hives in production over that period, however, has diminished by nearly 30%. Apparently, despite the decreasing foraging competition, the beekeepers of the United States have only been able to maintain their 1950 per hive production levels. While other factors are undoubtedly involved, some of this failure to increase production must be attributable to a decrease in available bee forage.

This rather bleak overall picture of U.S. honey production is to some extent offset by the trends for urban and suburban apiculture. In many instances these beekeepers report substantially greater per hive yields than do their rural counterparts, apparently because the communities in which these beekeepers reside plant ornamentals that provide good bee forage. Secondarily, the number of hives per unit area is relatively low and thus the foraging competition is also low. While the production from these areas contributes very little to the national honey production, it does lend credence to the idea that planting for bees can be profitable if land is available.

The national honey production data of Canada are considerably brighter. Fig. 2b shows an upward trend in the per hive production even though the number of hives in production during this same period increased by approximately 70%. While the movement toward overwintering colonies rather than starting anew each spring has done much to foster this trend, increased production would have been impossible without adequate bee forage. The culture of canola (rape), for example, has undoubtedly had a positive impact on the Canadian honey industry. In addition, in parts of the Prairie Provinces where much of the increased production has occurred, extensive rotation with legumes is practiced.

Important Nectar
and
Pollen Plants
of the
Continental United States
and
Canada[1]

Table 1.
Important Nectar and Pollen Plants of the Continental United States and Canada

Family—Common family name Scientific name Common names [Nectar and or pollen] [1]	NAT OR INT [2]	Geographic Regions [3, 4, 5]						
		I	II	III	IV	V	VI	VII
Family Aceraceae—Maple Family *Acer negundo* L. box elder, ash-leaved maple [NP:2, 10, 11,; N?P:B8; P:7]	N				OR — AP-JN	CO — AP-MY		MB – MY
Acer spp. maple [NP:2, 3, 4, 7, 8, 10, 11]	LgN	WA OR ID UT ——— MR-MY		BC WA — MR-JN				MB – MY-JN
Family Agavaceae—Century Plant Family *Agave* spp. century plant, maguey, American aloe, stitchwort [NP:3]	LgN						TX 1AP-JL	
Family Anacardiaceae—Cashew Family *Rhus* spp. sumac, sugar bush, lemonade berry [NP:4, 7, 8, 10, 11; N:10]	LgN			F-MY			AZ NM ——— MR-S	
Schinus terebinthifolius Raddi. Brazilian pepper tree, Florida holly, Christmasberry tree	I							
Toxicodendron spp. poison ivy and poison oak [NP:2, 8, 11]	N	CA ——— AP-MY		AP-MY				
Family Apiaceae—Carrot Family *Daucus carota* L. carrot [NP:2, 3, 7, 8, 11]	I	WAp ——— JN-AU			WAp ORp — JN-AU	ORp IDp ——— JN-AU		
Eryngium leavenworthii Torr. & A. Gray Leavenworth's eryngo	N							
Petroselinum crispum (Mill.) Nyman ex A. W. Hill parsley [NP:2]	I				ORp — JN-JL	ORp ——— JN-JL		

Geographic Regions [3, 4, 5]

VIII	IX	X	XI	XII	XIII
CO MB ND NE ——— AP-eJN		MO NE KS WI MI ——— AP-MY	GA ——— MR-AP		MA ——— AP-MY
MB NE ——— AP-eJN	ON WI MI QU ——— AP-eJN	MB SD NE KS IA MO WI IL MI IN OH ON PA NY QU ——— MR-MY	OK MO AR IL IN KY TN MS NY PA NJ MD WV NC SC ——— F-JN	AR LA MO KY TN MS AL FL GA SC NC VA MD DE NJ ——— JA-MY	QU NY NH VT MA CT RI NB ME PE NS ——— MR-MY
WY NE KS ——— AP-JN	MI QU ——— JN-AU	NE KS OK WI MI OH ON *PA NY QU ——— MY-AU	OK MO AR IL IN OH KY TN MS *AL GA NY PA NJ *MD *WV VA NC SC ——— MY-AU	LA MO KY TN *MS AL GA SC NC VA *MD DE NJ ——— MY-S	*QU NY *NH *VT *MA *CT *RI NB ME ——— JN-AU
				*FL ——— AU-O	
			OK IL IN KY TN MS AL MD VA ——— AP-JL	AR *LA TN MS AL MD ——— MR-JN	
			OK ——— JL-O		

Family—Common family name Scientific name Common names [Nectar and or pollen] [1]	NAT OR INT [2]	Geographic Regions [3, 4, 5]						
		I	II	III	IV	V	VI	VII
Family Apocynaceae—Dogbane Family *Apocynum* spp. dogbane [NP:2; —:11]	LgN							
Family Aquifoliaceae—Holly Family *Ilex glabra* (L.) A. Gray gallberry, inkberry, winterberry, Appalachian tea [NP: 4, 5]	N							
Ilex spp. holly, yaupon [NP: 2, 3, 4, 10, 11]	LgN		BC WAp ORp ———— AP-MY					
Family Aracaea—Arum Family *Lysichiton americanum* Hulten & St. John skunk cabbage, yellow skunk cabbage [P: 2, 11]	N	MY						
Symplocarpus foetidus (L.) Nutt. skunk cabbage, polecat weed [N? P: 7]	N							
Family Araliaceae—Aralia or Ginseng Family *Aralia spinosa* L. Devil's-walking stick, Hercules'-club, prickly ash, angelica tree [NP?: 10]	N							
Family Arecaceae—Palm Family *Sabal* spp. palmetto, cabbage palm [NP:4]	LgN							
Serenoa repens (Bartr.) Small saw palmetto, scrub palmetto [NP:3]	N							
Family Asclepiadaceae—Milkweed Family *Asclepias* spp. milkweed, silkweed, butterfly flower [NP?:3, 10; N:2, 8, 11]	LgN		CA ———— MY-JL	MY-AU		CA ———— MY-JL		
Cynanchum laeve (Michx.) Pers. blue vine, honey vine, sand vine [NP: 7]	N							

Geographic Regions [3, 4, 5]

VIII	IX	X	XI	XII	XIII
	WI --- JN-AU	OH PA --- JN-S			ME --- JN-JL
				*MS *AL *FL *GA SC NC --- MR-JN	
			OK TN AL MD *NC *SC --- MR-JN	*TX *LA MO KY TN MS AL *SC *NC VA *MD --- AP-JN	
			MD --- F-MY	MD --- F-AP	CT --- MR-AP
				LA --- JN-AU	
				LA AL *FL *GA SC --- MY-JL	
				AL *FL --- MY-JL	
WY KS --- MY-AU	ON MI QU --- JN-eS	NE KS OK WI MI OH ON *PA NY *QU --- MY-AU	*OK IL IN KY TN MS NY PA MD NC SC --- MY-S	SC NC VA MD DE NJ --- JN-eAU	QU NY NH *VT MA CT RI NB ME --- JN-AU
		*KS *MO *IL *IN --- JL-AU		AR SC NC --- lJN-eS	

Family—Common family name Scientific name Common names [Nectar and or pollen] [1]	NAT OR INT [2]	Geographic Regions [3, 4, 5]						
		I	II	III	IV	V	VI	VII
Family Asteraceae—Sunflower Family *Ambrosia* spp. ragweed [P: 7]	N							
Anaphalis margaritacea (L.) Benth. & Hook pearly-everlasting [NP: 2, 11]	N		BC ——— 1JL-S					
Aster spp. aster (many species) [NP: 2, 4, 7, 8, 10, 11]	LgN		CO NM ——— JL-O			*AZ ——— JL-S	NV *NM TX ——— MR-D	MB - JL- S
Baccharis spp. baccharis, mule fat, groundsel, sea myrtle, yerba dulce, coyote brush [NP: 2, 3, 11; NP?: 8]	N						AZ NM ——— AP-O	
Balsamorhiza spp. balsamroot, sunflower [NP: 2, 8, 11]	N		WA ID UT CO ——— AP-JL		WA — AP- JL	ID UT CO ——— AP-JN		
Bidens spp. Spanish-needles, beggar-ticks, bur marigold, stick-tights, pitchforks, tickseed [NP: 2, 3, 7, 8, 11]	LgN							
Carthamus tinctorius L. safflower, false or bastard saffron [NP: 7, 10]	I			JL- AU			AZ ——— JN-AU	
Centaurea spp. star thistle, knapweed [NP:2, 3, 7, 10, 11; NP?: 2]	LgI		*BC *WA *OR *CA ID *MT WY CO ——— JN-O	* - MY- O	*BC *WA *OR — JL-O	OR CA *AZ CO ——— JL-S	NV ——— MY-S	
Chrysothamnus spp. rabbitbrush [NP: 2, 8, 11]	N		OR CA ID UT CO NM ——— JL-O		OR — AU-S	OR *CA NV ID UT *AZ *CO ——— JL-O	NV AZ NM ——— JL-D	
Cichorium intybus L. chicory, succory, witloof, blue-sailors [NP: 2, 7, 8, 11; N:10]	I						NM ——— JN-AU	

Geographic Regions [3, 4, 5]

VIII	IX	X	XI	XII	XIII
WY ND NE KS		NE KS WI	OK IL IN TN	AR LA	NH CT
JL-O		JL-O	JL-O	JL-O	JL-S
MB	ON WI *QU	MB *OK *IA *MO WI *IL *MI *IN OH ON *PA *NY *QU	*OK MO AR IL IN *OH *KY *TN MS AL GA NY *PA *NJ MD WV *VA *NC *SC	TX *LA MO KY TN MS AL GA SC NC *VA *MD *DE *NJ	QU NY *NH VT MA CT RI *NB ME *PE *NS
JL-S	JL-S	JL-O	JN-N	MY-N	AU-O
			MS	MS	
			O	O-N	
NE	MI	NE KS MO WI *IL MI *IN	OK MO AR IL IN *KY TN MS NJ MD VA NC SC	LA MO KY TN AL FL SC NC VA MD DE NJ	
lJL-eO	JL-S	lJL-O	JN-O	MY-N	
	*MI	WI MI	PA NJ	GA NJ	*VT NB *NS
	JL-S	JL-S	JN-S	JN-O	JN-S
NE		NE MI OH	MO AR OH VA		
JL-S		JL-S	MY-O		

Family—Common family name Scientific name Common names [Nectar and or pollen]₁	NAT OR INT₂	Geographic Regions ₃, ₄, ₅						
		I	II	III	IV	V	VI	VII
Cirsium spp.; to lesser extent *Carduus spp.* various thistles, plume thistle; plumeless thistle, musk thistle [NP:2, 7, 8, 10, 11 (*Cirsium spp.*)]	N & I; I		BC CO ——— JN-S		BC — MY-S	CO ——— JN-O		BC SA MB – JL- AU
Coreopsis lanceolata L. tickseed coreopsis [NP:10]	N							
Crepis tectorum L. narrow-leaved hawk's-beard;	I							SA – JN- JL
Eupatorium spp. boneset, joe-pye weed [NP:4, 7, 10]	LgN							
Euthamia minor (Michx.) Greene flat-topped goldenrod	N							
Gaillardia pulchella Foug. marigold, Indian blanket, blanket flower [NP:5, 6]	N							
Grindelia spp. resinweed, rosinweed, tarweed gumweed, gum plant, sticky-heads [NP:2, 8, 11]	LgN		ID UT CO ——— JL-S			NV ID UT CO ——— JN-O	NV TX ——— JN-S	
Gutierrezia spp. *G. sarothrae* (Pursh) Britt. & Rusby matchweed, snakeweed [NP:8] and *G. texana* (DC.) T. & G. broomweed most often listed	N		CO ——— JL-S			AZ ——— JL-O	NV TX MY-O AU-N	
Haplopappus tenuisectus (Greene) Blake burroweed, goldenweed	N						AZ •NM ——— AU-O	
Helenium spp. *H. amarum* (Raf.) Rock often reported bitterweed, sneezeweed [NP:4]	N							

Geographic Regions [3, 4, 5]

VIII	IX	X	XI	XII	XIII
SA WY CO MB SD NE	ON WI MI	MB ND SD NE OK WI MI OH ON	TN NY PA *MD *VA NC SC	AL GA NC VA MD DE NJ	NY NB ME
JN-S; JN-S	JN-S	JN-O; JN-O	JN-FROST	AP-O	JL-S
				LA	
				AP-JN	
NE	ON	NE ON PA	OK MO AR KY MS AL VA NC SC	LA MS AL SC NC VA MD DE NJ	CT ME
AU-S	JL-S	JL-S	JN-O	JN-N	JL-S
				FL	
				S-N	
*TX				TX LA	
MY-JN				AP-O	
ND NE		NE			
JN-S		JL-S			
ND OK TX			*OK	TX AR	
AU-O			JL-O Probably G. dracunculoides date is for this species	AU-O	
JL-O				JL-O	
		OK	OK MO AR MS AL NC SC	LA MS AL SC NC	
		JN-O	MY-FROST	MY-N	

Family—Common family name Scientific name Common names [Nectar and or pollen] [1]	NAT OR INT [2]	Geographic Regions [3, 4, 5]						
		I	II	III	IV	V	VI	VII
Helianthus spp. sunflower [NP:2, 3, 4, 7, 8, 10, 11]	LgN		CO ——— JN-S	p – JN-O	BC — JN-S	IDp *AZ CO ——— JL-O	NV NM TX ——— MR-O	MB – AU-S
Heterotheca subaxillaris (Lam.) Britt, & Rusby camphorweed, yellow flower	N						AZ ——— MR-N	
Hieracium spp. hawkweed [NP:5]	N&I							
Layia platyglossa (F. & M.) Gray tidy tips	N			MY- JN				
Machaeranthera bigelovii (A. Gray) Greene purple aster	N						*NM ——— MY-O	
Mikania scandens (L.) Willd. snowvine, climbing hempweed duckblind [NP:3]	N							
Pluchea sericea (Nutt.) Cov. arrowweed [NP:8]	N						NV AZ ——— MR-JN	
Rudbeckia spp. cone flower [NP:2, 11; N:10]	N					*CO ——— JL-S		
Senecio spp. groundsel, butterweed, golden ragwort [NP:2, 11]	LgN						NM ——— AP-O	
Solidago spp. goldenrod [NP:2, 4, 7, 8, 10, 11]	LgN		*WA ID UT CO NM ———— JL-O		WA — JL-O	AZ ——— JL-S	*NM TX ——— JL-N	MB – JL-S
Sonchus arvensis L. sow thistle	I							SA MB - JN-S

Geographic Regions [3, 4, 5]

VIII	IX	X	XI	XII	XIII
SA WY CO *MB *ND *SD NE *KS TXp ——— JN-O	ON ——— JL-S	*MB *NDp *SD NE OK MNp *IAp *MOp WI *ILp *INp OHp ON PA ——— JL-S	OK TN AL PA NCp SC ——— JN-O	LA AL FL SCp NCp VA MD ——— JN-N	
		NY ——— lMY-S			
			KY ——— JL-O	LA MS ——— JL-FROST	
				LA ——— AP-S	
				LA ——— F-MY	
CO MB ND SD NE KS ——— JL-O	ON *WI MI *QU ——— JL-O	MB SD NE KS *OK *IA *MO *WI *IL *MI *IN OH *ON *PA *NY *QU ——— JN-O	OK MO AR IL IN *OH KY *TN MS AL GA NY *PA *NJ MD *WV VA *NC *SC ——— JL-N	TX *LA MO KY TN MS AL GA *SC *NC *VA *MD DE *NJ ——— JL-N	QU NY *NH *VT *MA *CT *RI *NB ME PE *NS ——— JL-O
MB ——— JN-S		MB ——— JN-S			

Family—Common family name Scientific name Common names [Nectar and or pollen] [1]	NAT OR INT [2]	Geographic Regions [3, 4, 5]						
		I	II	III	IV	V	VI	VII
Taraxacum spp. dandelion, blow-balls [NP:2, 3, 4, 7, 8, 10, 11]	N&I	MY- SUM	BC WA OR ID MT WY UT CO NM ———— MR-ISUM		BC WA OR — AP-S	OR CA NV ID UT AZ CO NM ———— MR-S		BC AB SA MB — AP- S
Verbesina alternifolia (L.) Britt. ex C. Mohr wingstem, golden honey plant yellow ironweed [NP:7, 10]	N							
Verbesina spp. crown-beard [NP:10]	LgN						NM ———— JN-S	
Vernonia spp. ironweed [NP:4, 7]	LgN							
Wyethia spp. mule-ears [NP:2, 8, 11]	N		CA CO ———— MR-AU			NV ID UT CO ———— MY-JL		
Family Avicenniaceae—Black Mangrove Family *Avicennia germinans* (L.) L. black mangrove [NP:3]	N							
Family Balsaminaceae—Balsam or Touch-Me-Not Family *Impatiens pallida* Nutt. jewelweed, snapweed, touch-me-not (other species likely; blooming data for *I. pallida*)	N							
Family Berberidaceae—Barberry Family *Mahonia* spp. Oregon grape, holly grape [NP:2, 10, 11]	LgN		WA ———— AP-MY		WA — AP- MY			
Family Betulaceae—Birch Family *Alnus* spp. alder [N?P:11; P:2, 4, 7, 8, 10]	LgN	AP- JN	WA OR CA ———— JA-JN	JA- AP				

Geographic Regions [3, 4, 5]

VIII	IX	X	XI	XII	XIII
AB SA MT WY CO NM MB ND NE KS	ON MI QU	MB ND NE KS OK MN IA MO WI IL MI IN OH ON PA NY QU	OK MO AR IL IN OH KY TN MS AL GA NY PA NJ MD VA NC SC	AR LA MO KY TN MS AL SC NC VA MD DE NJ	QU NY NH VT MA CT RI NB ME PE NS
AP-O	AP-O	MR-S	F-FROST	F-S	AP-S
		OK	KY WV		
		AU-O	AU-O		
			KY TN	LA	
			AU-O	AU-O	
		PA	AL	AL	
		JL-S	IJN-O	JN-O	
				FL	
				JN-AU	
	ON				ME
	JN-S				JL-S
	ON QU	ON QU	OK AL MD VA NC SC	MS AL SC NC VA MD DE NJ	QU CT NB PE NS
	AP-JN	MR-AP	F-MY	JA-JN	MR-MY

Family—Common family name Scientific name Common names [Nectar and or pollen] [1]	NAT OR INT [2]	Geographic Regions [3, 4, 5]						
		I	II	III	IV	V	VI	VII
Corylus spp. filbert, hazel, hazelnut [P:2, 7, 10]	N&I		WA OR ——— JA-MR					
Family Bignoniaceae—Bignonia Family *Catalpa* spp. catalpa, catawaba, indian bean [NP:2, 7, 10, 11]	LgN							
Family Boraginaceae—Borage Family *Amsinckia* spp. amsinckia, leather breeches, fire weed, fiddle-neck, tarweed [NP:2, 11]	N			MR-JN				
Echium vulgare L. viper's bugloss, blueweed, blue-devil [NP:3, 10]	I							
Family Brassicaceae—Mustard Family *Barbarea vulgaris* R. Br. yellow rocket, winter cress [NP:3]	I							
Brassica oleracea L. broccoli (br) cabbage (cb) cauliflower (cf) [NP:2, 11]	I		ORp (br cb) WAp (cb) ——— AP-JL		ORp (cb) — JN-AU	ORp (cb) ——— JN-AU	AZp (br cf) ——— F-MR	
Brassica rapa L. & *B. napus* L. rape, canola [NP:2, 3, 10, 11]	I		ORp WA ——— MY		ORp — MY			*BC *AB *SA *MB — MY-AU
Brassica spp. mustard [NP:2, 3, 7, 8, 10, 11]	I		BC WA CA ——— F-AU	JA-S		NV ID UT CO ——— MY-S	AZ TX ——— JA-S	*BC MB - MY-AU
Descurainia spp. tansy mustard [NP:2]	LgN						NM ——— AP-AU	
Lesquerella gordonii (A. Gray) S. Wats. bladderpod	N						AZ NM ——— F-MY	

Geographic Regions [3, 4, 5]

VIII	IX	X	XI	XII	XIII
			TN ––– MY-JN	TN ––– MY-JN	
		•ON ––– JN-S	WV VA ––– JN-S		
	WI MI ––– MY-JL	MO WI IL MI IN OH PA NY ––– AP-JN	MO AR KY NY PA NJ VA NC SC ––– MY-JN	NC VA MD DE NJ ––– AP-JN	NY VT ME ––– MY-JN
•AB •MB ND KS ––– JL-AU	•ON ––– MY-JN	•MB ND ILp•ONp ––– MY-AU		MOp KYp ––– MR-MY	
SA MB ––– JN-AU	ON WI QU ––– MY-O	MB IA MO WI IL MI IN •OH ON QU ––– AP-O	OK MO AR IN KY MS NJ MD WV SC ––– MR-N	MO KY AL SC MD NJ ––– MR-S	QU VT RI NB NS ––– MY-S

Family—Common family name Scientific name Common names [Nectar and or pollen] [1]	NAT OR INT[2]	Geographic Regions [3, 4, 5]						
		I	II	III	IV	V	VI	VII
Raphanus raphanistrum L. wild radish, jointed charlock [NP:6, 9]	I		BC ——— MY-JL					
Raphanus sativus L. radish [NP:2, 11]	I			F- JN	WAp ORp — MY- JL		AZp ——— F-MR	
Sisymbrium irio L. London rocket, yellow rocket	I						NV AZ ——— JA-AP	
Stanleya pinnata (Pursh) Britt. prince's plume, desert plume	N					CO ——— MY-AU	NV ——— AP-JN	
Family Cactaceae—Cactus Family Cactaceae cactus [NP:8, 11]	LgN						AZ NM TX ——— MR-AU	
Carnegiea gigantea (Engelm.) Britt. & Rose saguaro, sahuaro, giant cactus [NP:3]	N						AZ ——— MY-JN	
Opuntia spp. prickly pear, tuna, cholla [NP:2, 3, 8, 10]	LgN					CO ——— MY-JL	TX ——— AP-JN	
Family Capparaceae—Caper Family *Cleome lutea* Hook. yellow bee plant, yellow cleome [NP:2, 10, 11]	N					UT ——— MY-AU		
Cleome serrulata Pursh Rocky Mountain bee plant, stinking clover [NP:2, 8, 10, 11]	N		CO NM ——— JN-AU			UT *AZ CO ——— JN-S		
Family Caprifoliceae—Honeysuckle Family *Lonicera* spp. honeysuckle [NP:7, 8, 10, 11; N:2]	N&I		WA ——— AP-JN		WA — MY- AU			
Sambucus spp. elderberry, elder [NP:2, 11; P:7]	LgN		WA ——— MR-JL					

Geographic Regions 3, 4, 5

VIII	IX	X	XI	XII	XIII
			TN ——— JN-AU	GA ——— MY-JN	
WY ——— MY-JL					
WY CO SD NE ——— JN-S					
		WI PA ——— MY-JN	OK IL IN KY TN MS VA NC SC ——— AP-S	MO KY TN MS SC NC VA MD DE NJ ——— AP-AU	NH MA CT RI ——— MY-JN
		OK ——— MY-JL	OK AL NC SC ——— AP-JL	LA AL NC VA MD ——— AP-JL	NH RI PE ——— MY-AU

Family—Common family name Scientific name Common names [Nectar and or pollen] [1]	NAT OR INT [2]	Geographic Regions [3, 4, 5]						
		I	II	III	IV	V	VI	VII
Symphoricarpos spp. snowberry, wolfberry, waxberry, indian currant, coralberry [NP:3, 7, 8, 10; N:2, 11]	N		*BC *WA OR CA ID MT UT ——— MY-AU		*WA OR — MY-AU	NV ID UT CO ——— MY-AU		
Family Caryolphyllaceae—Pink Family *Stellaria* spp. chickweed, stitchwort [NP:2, 11]	N&I							
Family Celastraceae—Staff Tree Family *Celastrus scandens* L. bittersweet, climbing bittersweet waxwork [NP:7; N:10]	N							
Mortonia scabrella Gray sandpaper bush	N						AZ ——— MR-S	
Family Chenopodiaceae—Goosefoot Family *Salsola* spp. Russian thistle, saltwort, tumbleweed [NP:2; P:11]	I		CO ——— IJL-S		OR — JN-AU	OR AZ CO ——— JN-O		
Sarcobatus vermiculatus (Hook.) Torr. greasewood [P:2, 8, 11]	N				NV ID UT ——— MY-AU			
Family Clethraceae—White Alder Family *Clethra alnifolia* L. clethra, sweet pepperbush, summer-sweet [NP:3, 10]	N							
Family Convolvulaceae—Morning-Glory Family *Convolvulus* spp. bindweed [NP:2, 8, 11]	N&I		WA CO NM ——— MY-AU		WA — MY-S	CO ——— JN-S	NM TX ——— MR-S	
Family Cornaceae—Dogwood Family *Cornus alternifolia* L.f. pagoda dogwood, green osier (other species likely; blooming data for *C. alternifolia*)	N							
Cornus florida L. flowering dogwood [NP:7]	N							

Geographic Regions [3, 4, 5]

VIII	IX	X	XI	XII	XIII
CO NE ——— JN-AU		KS ——— JN-AU	KY ——— JN-AU		
		OK ——— MUCH OF YEAR	KY MD VA NC SC ——— MUCH OF YEAR	LA SC MD ——— MUCH OF YEAR	
					CT ——— JN-JL
WY ND KS ——— AU-O					
			VA ——— JL-AU	MS SC NC *NJ ——— JN-S	*MA *CT *RI ME ——— JL-S
				LA MO KY ——— AP-S	
					CT ——— MY-JN
				LA ——— MR-AP	

Family—Common family name Scientific name Common names [Nectar and or pollen] [1]	NAT OR INT [2]	Geographic Regions [3, 4, 5]						
		I	II	III	IV	V	VI	VII
Family Cucurbitaceae—Gourd Family *Citrullus lanatus* (Thumb.) Matsum. & Nakai watermelon [NP:2, 7]	I			p — MY-AU	WAp ORp — JN-JL	UTp ——— JN-JL	CAp AZp TXp ——— MR-JN S-N	
Cucumis melo L. cantaloupe, muskmelon, casaba honeydew [NP:2, 3, 7, 11]	I			p — MY-AU	WAp — JN-eS		CAp TXp ——— MR-JN S-N	
Cucumis sativa L. cucumber [NP:2, 7, 8, 11]	I		WAp ORp ———— IJN-AU		WAp — JN-eS		TXp ———— MR-AP O-N	
Cucurbita spp. includes pumpkin, squash (summer and winter) and gourd [NP:2, 3, 7, 11]	N&I ?		ORp ——— JN-S	p — MY-AU	ORp — JN-S			
Family Cyrillaceae—Cyrilla Family *Cliftonia monophylla* (Lam.) Britt. ex Sarg. titi, buckwheat tree, ironwood, buckwheat bush, black titi, spring titi [NP:4]	N							
Cyrilla racemiflora L. titi, leatherwood, ironwood, summer titi [NP:1]	N							
Family Ebenaceae—Ebony Family *Diospyros virginiana* L. persimmon, possumwood, possum apple, date plum [NP:7]	N							
Family Elaeagnaceae—Oleaster Family *Elaeagnus angustifolia* L., and probably *E.* *commutata* Bernh. in parts of north & west Russian olive, oleaster, wild olive, silverberry [NP:2, 10, 11 *(E. angustifolia)]*	I, N	JN	CO ——— MY-JN			CO ——— MY-JL	NV NM ——— AP-JN	

Geographic Regions [3, 4, 5]

VIII	IX	X	XI	XII	XIII
COp NEp OKp TXp ——— MY-JL		NEp OK MOp ILp MIp INp OHp PAp ——— MY-AU	OKp ILp INp TN GA NJp MDp NCp SCp ——— MY-AU	TXp MOp KYp MSp ALp FLp GAp SCp NCp VAp MDp DE NJp ——— MY-eAU	
COp NEp OKp TXp ——— MY-AU	QUp ——— JL-AU	NEp KS OK MOp ILp MIp INp OHp NYp QUp ——— MY-AU	ILp INp TN NYp PAp MDp VA NCp SCp ——— MY-AU	TXp MOp KYp FLp GAp SCp NCp VAp MDp DEp NJp ——— F-eAU	NYp MAp RI ——— JL-AU
COp NE OKp TXp ——— MY-AU	WIp MIp QUp ——— JN-AU	NE KS OK MNp MOp WIp ILp MIp INp OHp ONp NYp QUp ——— MY-S	OK ILp INp KY TN NYp PAp MDp NCp SCp ——— JN-S	TXp MOp KYp MSp ALp FLp GAp SCp NCp VAp MDp DEp NJp ——— ALL YEAR	NYp NHp MAp RI NBp ——— JL-AU
COp NEp ——— JN-AU		NEp KS OK WI ILp MIp INp OHp PAp NYp ——— JN-AU	ILp INp KY TN NYp PAp NJp MDp VAp NCp SCp ——— MY-S	MOp KYp FLp GAp SCp NCp VAp MDp DEp NJp ——— ALL YEAR	NYp NH MAp CTp RI NBp MEp ——— IJN-AU
			AL ——— MR-AP	*MS AL FL *GA ——— F-AP	
				LA *MS SC NC ——— AP-JL	
		KS OK ——— MY-JN	OK MO AR KY TN VA NC SC ——— MY-JN	AR LA TN MS SC NC VA MD DE NJ ——— MR-JN	
*NE KS ——— MY-JN		NE ——— MY-JN			

Family—Common family name Scientific name Common names [Nectar and or pollen] [1]	NAT OR INT [2]	Geographic Regions [3, 4, 5]						
		I	II	III	IV	V	VI	VII
Elaeagnus umbellata Thumb. autumn olive	I							
Family Ephedraceae—Joint Fir Family *Ephedra* spp. joint-fir, Mormon tea [NP:6]	N						AZ ——— AP-JN	
Family Ericaceae—Heath Family *Arbutus menziesii* Pursh. arbutus, madrona, madrone, madrono [NP:2, 9]	N		BC CA ——— MR-MY	MR-MY				
Arctostaphylos spp. manzanita, bearberry [NP:2, 8, 10, 11]	N		WA OR CA ——— F-MY	• — JA-MY		AZ ——— AP-JN	AZ ——— F-JN	
Gaultheria shallon Pursh. salal, shallon [N:2, 10, 11]	N		BC ——— JN-JL	AP-JL				
Gaylussacia spp. huckleberry [NP:10]	N							
Ledum groenlandicum Ded. Labrador tea [N:10]	N	JN-JL						BC - MY-JL
Oxydendrum arboreum (L.) DC. sourwood, sorrel tree, titi [NP:4; N:3, 10]	N							
Vaccinium macrocarpon Ait. cranberry [NP:10]	N		BCp WAp ORp ——— MY-JN					
Vaccinium spp. blueberry, huckleberry [NP:10; N:2, 4, 11]	LgN	MY-JN	BCp •WAp ORp ID UT ——— AP-JN		WA — MY-JL	ID ——— JN-JL		
Family Euphorbiaceae—Spurge Family *Croton texensis* (Klotzsch) Muell. Arg. Texas croton	N							

Geographic Regions [3, 4, 5]

VIII	IX	X	XI	XII	XIII
					CT
					MY-JN
			OK GA VA	LA SC NC VA MD DE NJ	
			AP-JN	AP-JN	
			*KY TN MS AL *WV *VA *NC *SC	*GA *SC *NC *VA MD	
			JN-JL	MY-JL	
	WIp	WIp		NJp	MAp RIp NBp
	JN-JL	JL		JN-JL	IJN-JL
	ON	OK ILp MIp INp ONp PAp NYp	TN AL GA NYp VA NCp SCp	TXp LA AL FLp GAp SCp NCp VAp MD DEp *NJp	QUp NYp VTp MAp CTp RIp NBp MEp PEp NSp
	MY-JL	AP-JN	F-JN	JA-JN	MY-JN
			OK		
			JN-O		

Family—Common family name Scientific name Common names [Nectar and or pollen] [1]	NAT OR INT [2]	Geographic Regions [3, 4, 5]						
		I	II	III	IV	V	VI	VII
Euphorbia esula L. leafy spurge, wolf's milk	I		CO ——— MY-AU					
Euphorbia marginata Pursh snow-on-the-mountain, ghostweed	N							
Sapium sebiferum (L.) Roxb. Chinese tallow tree vegetable-tallow [N:4]	I						TX ——— MR-MY	
Family Fabaceae—Pea or Pulse Family *Acacia berlandieri* Benth. huajillo, guajillo	N						*TX ——— MR-JN	
Acacia greggii A. Gray catclaw, devil's claw, Texas mimosa [NP:3, 8]	N					UT ——— SPR	*NV AZ NM *TX ——— AP-S	
Astragalus spp. milk-vetch, loco-weed [NP:2, 3, 8, 10, 11]	LgN		CO ——— MY-AU			AZ CO ——— AP-S	TX ——— F-S	
Calliandra spp. fairy duster, false or mock mesquite [NP:3]	N						AZ ——— F-S	
Caragana arborescens Lam. caragana, Siberian pea-shrub, Siberian pea-tree [NP:10]	I							SA – MY- JN
Cassia spp. partridge-pea, senna [NP:7]	LgN							
Cercidium spp. palo-verde [NP:2]	N						AZ ——— MR-MY	
Cercis spp. largely *C. canadensis* L. but *C. occidentalis* Torr. in CA redbud, Judas tree [NP:4, 7; NP?:10]	LgN		CA ——— MR-AP	MR- AP				
Coronilla varia L. crown vetch	I							

Geographic Regions 3, 4, 5

VIII	IX	X	XI	XII	XIII
ND NE ——— MY-S		ND NE OK ——— MY-AU			
			OK ——— MY-O		
			AL ——— MY-JN	*TX LA MS AL SC ——— MY-JN	
				TX ——— AP-JL	
WY SPR & SUM					
SA ND ——— MY-JN		ND ——— MY-JN			
			NC SC ——— JN-O	LA FL SC NC VA ——— JN-O	
		NE KS OK MO IL IN ——— AP-MY	OK MO AR IL IN KY TN MS AL WV VA NC SC ——— MR-MY	AR LA MO KY TN AL SC NC VA ——— MR-MY	
			PAp ——— JL-AU		

Family—Common family name Scientific name Common names [Nectar and or pollen] ₁	NAT OR INT₂	Geographic Regions ₃, ₄, ₅						
		I	II	III	IV	V	VI	VII
Dalea scoparia Gray indigo bush, purple sage	N						•NM ——— JN-S	
Dalea spp. prairie clover, summer farewell [NP:9]	N							
Gleditsia triacanthos L. honey locust, sweet locust, honeyshuck [NP:2, 3, 7, 8, 11; N:10]	N							
Glycine max (L.) Merrill soybean, soja bean, soya bean [NP:4, 7; NP?:3; N:11]	I							
Lespedeza spp. lespedeza [NP:4, 7; N:10]	N&I							
Lotus corniculatus L. bird's foot trefoil [NP:2, 3, 10]	I		•WAp CA ——— S		WA — MY-S			
Lotus spp. deer-vetch, wild alfalfa [NP:2]	LgN			MR -S				
Lupinus spp. lupine [NP:2, 11; N?P:8; P:7, 10]	LgN		CO MY-AU			AZ CO MY-S		
Medicago sativa L. alfalfa, lucerne [NP:3, 4, 7, 8, 10; P:2]	I		•BC •OR •CA •IDp •MT •WY •UTp •CO •MN ——— MY-O	• _ MY- S	•BC •WA •OR — JN-S	•OR •CA •NVp •IDp •UTp •AZ •CO •NM ——— MY-S	•CAp •NV AZ •NM TX ——— AP-O	•BC AB •SA •MB – JN- S
Melilotus spp. sweet clover, melilot largely *M. alba* Desr. and *M. officinalis* (L.) Pall. (white and yellow sweet clover respectively) [NP:2, 3, 4, 7, 8, 10]	I	• _ JL- AU	•WA OR CA AB •ID •MT •WY •UT •CO •NM ——— MY-O	MY- S	•WA OR — JN- AU	•OR CA •NV •ID •UT •AZ •CO •NM ——— MY-S	NV •NM ——— AP-O	•BC •AB •SA •MB – JN- S

Geographic Regions 3, 4, 5

VIII	IX	X	XI	XII	XIII
WY NE —— MY-S		NE —— MY-AU		FL —— S-O	
		OK —— MY-JN	OK TN MS NC SC —— AP-JN	MO KY TN SC NC VA —— AP-JN	
ND NE *KS —— JL-AU		NE *KS OK *IA *MOp *IL *IN OH —— JN-AU	IN *KY TN *MS AL MD —— JL-O	*AR *MOp *KYp *TN *MS AL SCp NCp VAp MD —— JL-O	
			AL —— JN-O	AL —— JN-O	
	*ON *WIp *MIp —— JN-S	ND MN *ON PA NY *QU —— JN-AU	NY PA —— JN-AU		NYp VT CT NB —— JN-S
					PE —— MY-JL
*AB *SA *MT *WYp *CO *NM *MB *ND *SD *NEp *KSp *OK *TX —— MY-S	*ON *WI *MI QU —— JN-S	*MB *ND *SD *NE KS *OK *MN *IA *MO *WI *IL *MI *IN OH *ON NY *QU —— MY-S	OK TN AL NY PA NJ —— AP-S	AR NJ —— MY-O	QU NY *VT ME —— JN-AU
AB *SA *MT *WY CO *NM *MB *ND *SD *NE *KS *OK TX —— MY-O	ON MN WI *MI QU —— JN-S	*MB *ND *SD *NE *KS *OK *MN *IA *MO *WI *IL *MI *IN *OH ON *PA *NY *QU —— MY-O	*OK *MO *AR *IL *IN *OH *KY *TN AL NY *PA NJ MD VA NC SC —— AP-O	*TX LA SC NC VA *MD DE NJ —— AP-O	*QU NY NH VT MA CT RI NB ME —— JN-S

Family—Common family name Scientific name Common names [Nectar and or pollen] [1]	NAT OR INT [2]	Geographic Regions [3, 4, 5]						
		I	II	III	IV	V	VI	VII
Onobrychis spp. largely *O. viciifolia* Scop. sainfoin, esparcet, holy clover [NP:3, 7, 10; NP?:2, 11]	I		CO — MY-AU			CO — MY-AU		
Parkinsonia aculeata L. retama, Jerusalem-thorn, Mexican palo verde, horse bean [NP:3]	I?						TX MY-S	
Phaseolus limensis Macfady. lima bean [NP:7; NP?:2]	I				WA — JL- AU			
Prosopis glandulosa Torr. mesquite, honey mesquite [NP:3]	N					UT —— MY-JN	*NV AZ *NM *TX —— AP-JN	
Robinia pseudoacacia L. black locust, false acacia, yellow locust [NP:2, 3, 4, 7, 10; NP?:11; N:8]	N		BC WA OR ID UT CO —— MY-JL		BC *WA — MY- JN		NV *NM AP-JN	
Trifolium campestre Schreb. low hop clover, large hop clover and/or *Trifolium agrarium* L. hop clover, yellow clover	I I							
Trifolium hybridum L. alsike clover [NP:3, 7, 8, 10]	I	JL- AU	*WA ID MT WY UT CO —— MY-S		WA — JN- JL	IDp CO —— MY-S		*BC *AB MB – JN- S
Trifolium incarnatum L. crimson clover, Italian clover [NP:2, 3, 4, 10, 11]	I		*ORp —— MY-JN					
Trifolium nigrescens Viv. ball clover [NP:4]	I							

Geographic Regions [3, 4, 5]

VIII	IX	X	XI	XII	XIII
AB NM ——— MY-AU					
				SC NC •MD DEp ——— JN-JL	
NM •TX ——— MY-JL				TX ——— AP-JN	
CO OK ——— AP-eJN		NE KS OK •WI •MI OH ON •PA NY ——— MY-JN	•OK IL IN •OH •KY •TN AL •PA NJ •MD •WV •VA •NC SC ——— AP-JN	LA TN MS AL SC NC VA •MD DE NJ ——— MR-JN	•NH VT •MA CT •RI ME ——— MY-JL
			MO AR ——— MY-S		
WY MB NE ——— MY-O	•ON MN •WI •MI •QU ——— JN-O	MB NE KS OK •IAp •MOp •WI •ILp •MI •INp OH •ON •NY •QU ——— MY-S	•MO •AR •IL •IN NY NJ NC SC ——— AP-S	SC NC VA NJ ——— AP-S	•QU NY NH VT •MA CT RI •NB •ME •NS ——— MY-S
			OK TN MS •AL GA VA NC SC ——— AP-JL	•TX LA TN MS •AL GA SC NC ——— AP-JN	
			AL ——— MR-MY	AL ——— MR-MY	

Family—Common family name Scientific name Common names [Nectar and or pollen] [1]	NAT OR INT [2]	I	II	III	IV	V	VI	VII
Trifolium pratense L. red clover [NP:2, 3, 4, 7, 8, 10, 11]	I	JL-AU	*WAp ORp ID UT ———— MY-AU		*WAp ORp — MY-AU	IDp CO ———— MY-S		*BC *AB *SA – JN-S
Trifolium repens L. white clover, white Dutch clover ladino clover [NP:2, 3, 4, 7, 8, 10, 11]	I	* – JN-S	*BCp WA ORp ID UT *CO ———— AP-S		WAp OR — AP-S	OR *NV *IDp *UT CO ———— AP-S		MB – JN-S
Trifolium resupinatum L. Persian clover, reversed clover [NP:3]	I							
Trifolium vesiculosum Savi arrowleaf clover, yuchee clover [NP:2, 4]	I							
Vicia spp. vetch, tare [NP:2, 4, 7; - -:2]	N&I	JL-AU	BC *WA *OR CA CO ———— MY-AU	AP-JL	BC — MY-AU	CA *CO ———— MY-AU		BC SA – JN-AU
Vigna unguiculata (L.) Walp. cowpea, black-eyed pea [NP:7]	I							
Family Fagaceae—Beech Family *Fagus* spp. Beech [P:7, 10]	LgN							
Quercus spp. oak [N?P:10; P:2, 7, 8, 11]	LgN		CA CO ———— MR-JL		OR — AP-MY	CO ———— MR-JN		
Family Geraniaceae—Geranium Family *Erodium* spp. filaree, heron's-bill, storksbill, pin-clover, alfileria, clocks [NP:2, 8, 11]	N&I		CA CO ———— F-AU	F-JN		AZ CO ———— MY-AU	NV AZ ———— F-MY	
Geranium spp. wild geranium [NP:3, 7, 8, 10]	LgN	JN-JL				CO ———— JN-AU		

Geographic Regions [3, 4, 5]

VIII	IX	X	XI	XII	XIII
AB *WY ——— JN-S	ON *WI *MI *QU ——— 1MY-S	NE KS MN *IAp *MOp WI *ILp *MI *INp *OHp ON *QU ——— MY-S	OK IL IN TN MS AL VA NC SC ——— AP-S	LA TN MS AL SC NC VA ——— AP-S	*QU *NH MA RI NB ME PE NS ——— MY-S
WY CO MB SD *KS ——— MY-S	ON *WI *MI *QU ——— MY-O	MB ND SD *NE *KS *OK *MN *IAp *MOp *WI *ILp *MI *INp *OH *ON *PA *NY *QU ——— MY-O	*OK *MO *AR *IL *IN *KY *TN *MS *AL GA NY PA NJ *MD *VA NC SC ——— AP-O	*TX AR *LA MO KY TN *MS *AL GA SC NC VA *MD DE NJ ——— AP-O	*QU NY NH VT *MA *CT RI *NB *ME *PE *NS ——— MY-S
			TN MS ——— AP-AU	*AR *LA TN MS ——— AP-S	
			*OKp *AL ——— AP-JL	TX *AL ——— AP-JL	
SA WY CO ND NE OK ——— AP-AU	ON MI QU ——— JN-AU	ND *NE *OK WI MI ON *QU ——— MY-AU	*OK TN MS AL NY NJ NC SC ——— MR-O	*TX *AR *LA MO KY TN MS AL GA SC NC VA MD NJ ——— AP-S	*QU NY NH VT RI NB ME NS ——— JN-AU
			NC SC ——— JL-AU	LA MOp KYp SC NC VA ——— JN-S	
					ME ——— MY-eJN
		WI ——— AP-MY	OK IL IN TN MS NJ NC SC ——— AP-MY	TX AR LA TN MS SC NC VA MD DE NJ ——— MR-MY	

Family—Common family name Scientific name Common names [Nectar and or pollen] [1]	NAT OR INT [2]	Geographic Regions [3, 4, 5]						
		I	II	III	IV	V	VI	VII
Family Hippocastanaceae—Horse Chestnut Family *Aesculus californica* (Spach) Nutt. California buckeye, California horse chestnut [NP:2, 11]	N		CA ——— JN-JL	MY-JN				
Aesculus spp. (other than *A. californica*) buckeyes and horse chestnuts [NP:7, 10]	LgN							
Family Hydrophyllaceae—Waterleaf Family *Eriodictyon angustifolium* Nutt. yerba santa, mountain-balm [NP:8]	N		CA ——— MY-AU	MY-JL				
Hydrophyllum spp. and *Phacelia* spp. waterleaf and phacelia	N.					CO ——— MY-JL		
Family Lamiaceae—Mint Family *Agastache* spp. anise hyssop, blue hyssop, giant hyssop [NP:7, 8; NP?:10; N:2, 11]	LgN							
Lamium spp. dead nettle, henbit, dumb nettle [NP:2; N?P:5, 9]	I							
Marrubium vulgare L. horehound, white horehound [NP:3; N:2, 8, 10, 11]	I					CO ——— JN-S		
Mentha spp. mint [NP:4, 7; NP?:2, 8, 11; N:2, 10]	I		OR CA ——— JN-O	JN-O	*WA OR ——— JN-S			
Monarda spp. Horsemint, wild bergamot, bee-balm [NP:7]	N					CO ——— JN-AU	AZ TX ——— MY-S	
Nepeta cataria L. catnip, catmint [N:2, 7, 10, 11]	I							
Pycnanthemum spp. basil, mountain mint [NP:7, 10]	N							

Geographic Regions [3, 4, 5]

VIII	IX	X	XI	XII	XIII
		KS	NC SC	SC NC	
		AP-MY	AP-JN	AP-MY	
			MD NC SC	MO NC	
			JL-S	JN-S	
			OK KY	AR	
			F-O	F-O	
		MI		LA	RI
		JL-O		JN-S	JL-S
*NE OK *TX		OK	*OK AL MD VA NC SC	*TX AR LA AL SC NC VA MD DE NJ	
MY-AU		MY-S	MY-S	AP-O	
NE		NE WI			
JN-O		JN-S			
		NE	KY		
		JL-S	JN-S		

Family—Common family name Scientific name Common names [Nectar and or pollen] [1]	NAT OR INT [2]	Geographic Regions [3, 4, 5]						
		I	II	III	IV	V	VI	VII
Salvia spp. sage, ramona [NP:3, 7; N:2, 10, 11]	N&I		OR ———— JN-JL	• — MR-JL		ID ———— MY-AU		
Thymus spp. thyme [NP:3; N:10]	I							
Trichostema spp. blue-curls [NP:2, 3, 11]	N			JL-O				
Family Liliaceae—Lily Family *Allium cepa* L. onion [NP:2, 7, 8, 10, 11]	I			p – AP-MY	WAp ORp ORp — MY-JL	ORp IDp ———— JN-AU	CAp AZp ———— MR-MY	
Odontostomum hartwegii Torr. Hartweg's odonstoman	N		CA ———— AP-MY	AP-MY				
Family Limnanthaceae—Meadow-Foam Family *Limnanthes douglasii* R. Br. meadow-foam	N			MR-MY				
Family Lythraceae—Loosestrife Family *Lythrum salicaria* L. loosestrife, purple loosestrife [NP:3, 7, 10]	I							
Family Magnoliaceae—Magnolia Family *Liriodendron tulipifera* L. tulip tree, tulip poplar, whitewood [NP:2, 3, 4, 7, 10, 11]	N							
Family Malvaceae—Mallow Family *Gossypium* spp. cotton [NP:3, 4]	N&I			JL-S			CA AZ NM TX ———— JN-1S	
Sphaeralcea spp. globe mallow [NP:8, 12]	N					CO ———— MY-AU	NV AZ ———— MR-O	

Geographic Regions [3, 4, 5]

VIII	IX	X	XI	XII	XIII
SD<hr>JN-S			NC SC<hr>AP-JN	NCp<hr>AP-MY	RI<hr>SUM
					MA<hr>JL-S
	WI<hr>JL-S	WI *MI OH *ON<hr>JL-S	NY NJ WV NC SC<hr>JN-S	SC NC NJ<hr>JN-S	NY NH *MA CT ME PE<hr>JN-S
		MO IL MI IN *PA<hr>AP-JN	*IL *IN *OH *KY *TN *MS *AL *GA *PA *NJ *MD *WV *VA *NC *SC<hr>AP-JN	*TN *MS *AL *GA *SC *NC *VA *MD DE NJ<hr>AP-JN	*CT<hr>IMY-eJL
KS *OK *TX<hr>JN-eO			MS AL SC<hr>JN-O	TX *AR LA *MOp *KYp *TN *MS AL SCp NCp<hr>JN-S	

Family—Common family name Scientific name Common names [Nectar and or pollen] ₁	NAT OR INT₂	Geographic Regions ₃, ₄, ₅						
		I	II	III	IV	V	VI	VII
Family Myrtaceae—Myrtle Family *Eucalyptus* spp. eucalyptus, eucalypt, gum tree, Australian gum tree, ironbark, stringybark [NP:3]	I			D-JL				
Melaleuca quinquenervia (Cav.) S. T. Blake cajeput, punk tree, paperbark tree, tea tree, swamp tea tree [NP:3]	I							
Family Nyssaceae—Tupelo or Sour-gum Family *Nyssa* spp. tupelo, blackgum, ogeechee-lime [NP:3, 4, 7; N:3]	N							
Family Oleaceae—Olive Family *Forestiera neomexicana* Gray New Mexico olive, desert olive	N						NM ⎯ AP-MY	
Fraxinus spp. ash [P:2, 7, 11]	LgN		CO ⎯ AP-MY			CO ⎯ AP-MY		
Ligustrum spp. privet, hedge plant [NP:2, 3, 4, 7, 10, 11]	N&I							
Family Onagraceae—Evening Primrose Family *Epilobium angustifolium* L. fireweed, willow herb, wickup [NP:2, 3, 8, 10]	N	* _ JN-AU	*BC *WA *OR AB ID ⎯ JN-AU		BC ⎯ JN-S			
Family Papaveraceae—Poppy Family *Eschscholzia californica* Cham. California poppy [P:2, 11]	N			F-S				
Family Pinaceae—Pine Family *Picea* spp. spruce	LgN	$PR						
Pinus spp. pine [P:2, 7, 8, 11]	LgN		WA ⎯ MY-JL		WA ⎯ AP-JN			

Geographic Regions [3, 4, 5]

VIII	IX	X	XI	XII	XIII
				*FL MOST OF YEAR	
			OK MS GA ——— AP-JN	LA MS *AL *FL GA *SC NC ——— MR-JN	CT ——— MY-JN
			TN MS VA ——— MR-MY	LA TN MS ——— MR-MY	
		OK ——— JN-JL	OK TN MS AL *GA *NJ NC SC ——— AP-JL	TX LA TN MS AL GA SC NC VA NJ ——— AP-JL	MA RI ——— JN-eJL
	ON MN WI ——— JL-S				RI PE ——— JN-AU
			IN AL ——— F-AP	AL ——— F-AP	NH ——— eMY-eJN

Family—Common family name Scientific name Common names [Nectar and or pollen]₁	NAT OR INT₂	I	II	III	IV	V	VI	VII

Family Plantaginaceae—Plantain Family
Plantago lanceolata L.
  narrow-leaved plantain, ribgrass,
  buckhorn, ripplegrass
  other species likely; blooming and
  N&P data for *P. lanceolata*
  [P:2, 8]
	I					CO
						——
						MY-S		

Family Poaceae—Grass Family *Cynodon dactylon* (L.) Pers. Bermuda grass, scutch-grass	I						NV —— IAP-S	

Sorghum bicolor (L.) Moench
  sorghum, broom-corn
  [P:9]
	I						AZ
							——
							JL-AU	

Sorghum halepense (L.) Pers.
  Johnson-grass, aleppo grass
  means-grass, Egyptian millet
	I						CA AZ
							——
							AP-N	

Zea mays L.
  corn, maize
  [P:2, 4, 7, 8, 10, 11]
	I?		WA OR CO
			——
			JL-AU
					WA
					—
					JL-AU
						CO
						——
						JL-AU
							NM
							——
							JL-S	

Family Polygonaceae—Buckwheat Family
Brunnichia ovata (Walt.) Shinners
  eardrop vine, ladies'-eardrops,
  buckwheat vine
	N							

Eriogonum spp.
  wild buckwheat, umbrella plant
  [NP:2, 11]
	N		WA CA CO
			————
			MY-O
				AP-N	WA
					—
					MY-AU
						*CA AZ
						——
						MR-O
							CA AZ
							——
							MR-O	

Fagopyrum esculentum Moench
  buckwheat, brank
  [NP:2, 7, 10, 11; N:4]
	I		ORp
			——
			SUM
								SA
								MB
								–
								JL-S

Polygonum cuspidatum Siebold & Zucc.
  Japanese knotweed, Mexican bamboo
  [N:10]
	I		BC
			——
			AU-S					

Polygonum spp.
  smartweed, knotweed, fleece flower
  [NP:3, 4, 7; NP?:8; N:2, 11]
	N&I		WA
			——
			MY-O
					WA
					—
					JN-S
							NV
							——
							JN-S	

Geographic Regions [3, 4, 5]

VIII	IX	X	XI	XII	XIII
			OK MS ——— JN-N	AR LA MO KY MS GA ——— MY-N	
NE KS TX ——— JN-AU		NE KS ——— JL-AU	OK ——— JL-AU	TX MO KY MS ——— MY-O	
			MS ——— MY-N	LA MO KY ——— AP-N	
WY CO ND NE KS ——— JL-eS		ND NE KS MN WI MI PA NY ——— JL-S	OK MO AR KY TN MS AL PA VA NC SC ——— JN-S	AR LA MO KY TN MS AL SC NC VA ——— AP-S	VT MA CT RI ME ——— JL-S
				MO KY ——— MY-AU	
MB ——— AU-eS	QUp ——— AU	MB ND SD OK MN WI ON PA NY *QUp ——— JN-S	KY AL MD WV NC ——— MY-FROST	AL NC MD DE NJ ——— JN-FROST	NH *ME ——— JL-S
					VT MA ——— AU-S
CO *NE *KS TX ——— AU-S		ND *NE *KS OK MN *IA *MO WI *IL MI *IN *OH NY ——— MY-O	OK IL IN KY TN AL WV VA NC SC ——— MY-N	*AR LA MO KY TN MS AL SC NC VA MD DE NJ ——— MY-N	MA ME ——— JL-S

Family—Common family name Scientific name Common names [Nectar and or pollen]₁	NAT OR INT₂	Geographic Regions ₃, ₄, ₅						
		I	II	III	IV	V	VI	VII
Rumex acetosella L. sheep sorrel, red sorrel, common sorrel [P:2, 11]	LgN							
Family Ranunculaceae—Crowfoot or Buttercup Family *Clematis virginiana* L. virgin's bower, woodbine, devil's darning-needle [NP:7, 10]	N							
Family Rhamnaceae—Buckthorn Family *Berchemia scandens* (J. Hill) C. Koch rattan vine, supplejack	N							
Ceanothus spp. New Jersey tea, redroot, wild-lilac buck-brush [NP:2, 8, 10, 11]	N		OR ID UT ———— AP-AU	MR-NV ID JN		AZ UT AZ ———— MY-O	———— MR-JN	
Condalia hookeri M. C. Johnst. Brazil, logwood, capul negro	N						TX ———— JL-AU	
Rhamnus purshiana DC. sagrada, bearberry, cascara [NP:2, 10, 11]	N		WA OR ID UT ———— AP-JN	AP- JN				
Rhamnus spp. buckthorn [NP:2, 10]	LgN		CA ———— MY-JL	MR- JL				
Family Rosaceae—Rose Family *Adenostoma fasciculatum* H. & A. chamise, greasewood	N			AP- JN				
Amelanchier spp. service-berry, juneberry, shadbush [NP:8, 10, 11; N?P:2]	LgN		BC ———— MY-JN	AP- JN				BC AB SA – MY-JN
Crataegus spp. hawthorn [NP:2, 7, 8, 10, 11]	LgN		WA ———— MY-JN		WA — AP-JN			

Geographic Regions [3, 4, 5]

VIII	IX	X	XI	XII	XIII
					CT — MY-AU
				LA — JL-S	
			*OK *MS — AP-JN	*TX AR *LA *MS — MR-JN	
ND — 1MY-JN			OK — MY-JN	LA — AP-JN	
					ME — MY-JN
	ON QU — MY-JN	WI MI ON PA QU — MY-JN	OK PA VA — AP-JN	LA MS NC VA — F-JN	QU NH RI ME NS — MY-JN

Family—Common family name Scientific name Common names [Nectar and or pollen] [1]	NAT OR INT [2]	Geographic Regions [3,4,5]						
		I	II	III	IV	V	VI	VII
Fragaria x *ananassa* Duchesne strawberry [NP:2, 7, 8, 11]	LgN							
Heteromeles arbutifolia (Ait.) M.J. Roem. toyon, Christmas berry	N		CA ——— JN-JL	JN-JL				
Malus baccata (L.) Borkh. Siberian almond, Siberian crabapple [NP:3, 10]	I							
Malus spp. apple [NP:2, 3, 4, 7, 8, 10, 11]	LgI		WAp ORp IDp MT UTp COp NMp ——— AP-eJN		BCp WAp ORp ——— AP-MY	NVp IDp UTp COp NM ——— AP-MY	AZp ——— IMR-lAP	
Prunus armeniaca L. apricot [NP:2, 7, 8, 10, 11]	I		CO NM ——— MR-MY		WAp — MR-AP	ID UT COp ——— IMR-MY	NV NM ——— F-MR	
Prunus dulcis (Mill.) D.A. Webb almond [NP:10, 11]	I			p – F-MR			NV ——— F-MR	
Prunus persica (L.) Batsch. peach [NP:2, 4, 7, 8, 10, 11]	I		CO NM ——— AP-MY		WAp — IMR-AP	ID UT CO ——— AP-MY	NV NM ——— F-AP	
Prunus spp. cherry (cultivated) [NP:2, 7, 8, 10, 11]	I		WAp ORp MTp UTp COp ——— AP-MY	p – MR	BCp WAp ORp — IMR eMY	COp ——— AP-MY		
Prunus spp. cherry (various uncultivated species) [NP:7, 8, 10; NP?:2, 11]	N&I		WA ID MT WY UT ——— AP-JN		WA — AP-JN	ID UT MY-JN		SA – MY-JN
Prunus spp. plum (cultivated) [NP:2, 4, 7, 10, 11]	N&I		COp ——— MY-JN	p – F	BCp WAp — AP-MY	COp ——— AP-MY	NV ——— F-MR	

Geographic Regions 3, 4, 5

VIII	IX	X	XI	XII	XIII
	WIp ———— AP-MY	MNp MOp WIp ILp INp ONp PAp ———— AP-JN	ILp INp PAp NJp MDp VAp NCp SCp ———— AP-MY	FLp SCp NCp VAp MDp NJp ———— D-MY	NBp MEp NSp ———— MY-JN
					ME ———— MY-JN
CO ND KS ———— AP-eJN	WIp MI QUp ———— MY-JN	NEp KSp OK MNp IAp MOp WIp ILp MIp INp OHp ONp PAp NYp *QUp ———— AP-eJN	MOp ARp ILp INp OHp KY TN MS ALp GAp NYp PAp NJp MDp WV VAp NCp SCp ———— MR-MY	TN MS NCp VA MD NJp ———— MR-MY	QU NYp NH VTp MAp CTp RIp NBp MEp NSp ———— MY-JN
		KS MOp ILp MI INp ———— AP			VT ———— MY
		KS OKp MOp ILp MI INp ———— MR-MY	OK ILp INp KY TN MSp ALp MD WVp VA ———— F-MY	LA MOp TN MSp AL ———— F-AP	RI ———— AP-MY
	MIp QU ———— AP-MY	NE KS OK IAp MOp WIp ILp MIp INp OHp ONp PAp NYp QU ———— eAP-eJN	TN MS AL NYp PAp MD VAp NCp ———— MR-MY	AL NC ———— MR-MY	NYp RI NS ———— MY
MT CO ND ———— MY-JN	WI MI QU ———— MY-JN	ND IA MO WI IL IN PA QU ———— AP-eJN	OK MO AR IN KY TN MS GA PA MD VA NC SC ———— AP-MY	GA SC NC VA MD ———— MR-MY	QU NH MA CT RI NB ME ———— MY-JN
	QU ———— MY-JN	OK IAp MOp WI ILp MI INp OHp ONp NYp QU ———— MR-eJN	MS AL PAp VA NC SC ———— MR-MY	AR LA MS AL SC NC VA MD DE NJ ———— MR-AP	CTp RI PEp NSp ———— lAP-eJN

Family—Common family name Scientific name Common names [Nectar and or pollen] [1]	NAT OR INT [2]	Geographic Regions [3, 4, 5]						
		I	II	III	IV	V	VI	VII
Prunus spp. plum (uncultivated) [NP:8, 10]	N&I							
Pyrus spp. pear [NP:2, 4, 7, 10, 11]	I		WAp ORp ——— AP-MY		BCp WAp OR — AP-eMY	COp AP-MY		
Rosa spp. rose (uncultivated) [NP:2, 8, 11; P:7, 10]	LgN	JN-JL						
Rubus spp. blackberry [NP:2, 3, 4, 7, 8, 10, 11]	LgN		*BCp *WAp *ORp *CA ID UT ——— AP-JL	MR-JL		IDp ——— MY-JN		
Rubus spp. raspberry [NP:2, 3, 7, 8, 10, 11]	N&I	JN-JL	*BCp *WAp ORp CA IDp UTp ——— AP-JL	AP-JL		ID UTp CO ——— MY-JL		
Family Rubiaceae—Madder Family *Cephalanthus occidentalis* L. buttonbush [NP:7; NP?:10]	N			JN-AU				
Richardia scabra L. Mexican clover, Florida pusley [N:4]	I							
Family Rutaceae—Rue Family *Citrus* spp. citrus [NP:3]	I				* – MR-MY		CA AZ *TX ——— F-JN	
Family Salicaceae—Willow Family *Populus* spp. cottonwood, poplar, aspen [N?P:8; P:2, 4, 7, 10, 11]	LgN	MY-JN	WA AB ID MT WY CO NM ——— MR-JN		WA — eSPR	AZ CO NM ——— MR-JN	AZ NM TX ——— F-JN	AB – AP-MY

Geographic Regions [3, 4, 5]

VIII	IX	X	XI	XII	XIII
SD KS <hr> AP-MY		SD NE KS <hr> AP-MY	OK <hr> AP-MY		NB <hr> MY
		NE KS OK IAp MOp ILp MIp INp OHp ONp NYp <hr> AP-eJN	TN AL NYp PAp MD <hr> AP-MY	AL NC VA <hr> AP-MY	NYp CTp RI NSp <hr> MY-eJN
	*WI <hr> JN-AU	NE KS OK MO WI IL MI IN *PA <hr> AP-JL	OK MO AR IL IN *OH KY TN MS AL GA PA *MD *VA *NCp *SCp <hr> MR-JN	LA MO KY TN MS AL SC NC VA *MDp DE NJp <hr> MR-JN	NH VT MA CT RI ME *NS <hr> MY-JL
	ON MN WI MI QUp <hr> MY-JL	OK MNp IAp MOp WIp ILp MI INp ONp *PA QUp <hr> MY-JL	IN *OH NY PA NJp <hr> MY-JL	NJp <hr> MY-JN	*QUp NY NH VT MA CT RI *NBp *MEp *NS <hr> MY-JL
		NE MI <hr> JN-AU	OK KY <hr> JN-AU	LA MS <hr> JN-S	*MA RI ME <hr> JL-AU
			AL <hr> JN-FROST	AL <hr> JN-FROST	
				LA *FLp <hr> F-eMY	
AB WY CO ND NE OK <hr> MR-JN	WI MI <hr> AP-JN	ND NE KS WI MI <hr> MR-MY	IL IN AL SC <hr> MR-MY	AR LA MO KY TN AL SC <hr> MR-MY	NH VT MA CT NB ME PE NS <hr> AP-MY

Family—Common family name Scientific name Common names [Nectar and or pollen] [1]	NAT OR INT [2]	I	II	III	IV	V	VI	VII
			Geographic Regions [3, 4, 5]					
Salix spp. willow, osier [NP:2, 3, 4, 7, 8, 10, 11]	LgN	* – AP-JN	BC WA OR CA AB ID MT WY UT CO NM ———— F-AU	JA-JN	BC OR — F-MY	OR NV ID UT CO ———— F-AU	NM ———— AP-MY	BC AB SA MB – AP-JN
Family Sapindaceae—Soapberry Family *Cardiospermum halicacabum* L. balloon vine, heart pea, winter cherry	I							
Family Saxifragaceae—Saxifrage Family *Ribes* spp. currant [NP:2, 7, 8, 10, 11]	N&I	MY-JN	ID UT CO ———— MY-JL			CO ———— MY-JL		
Family Scrophulariaceae—Figwort Family *Orthocarpus erianthus* Benth. Johnny-tuck, butter and eggs	N			MR-MY				
Scrophularia californica Cham. & Schlect. figwort, California honey plant [N:2]	N			F-JL				
Family Styracaceae—Storax Family *Styrax officinalis* L. var. californicus (Torr.) Rehd. snowdrop bush, styrax, storax	N			AP-MY				
Family Tamaricaceae—Tamarisk Family *Tamarix* spp. tamarix, tamarisk, salt cedar, athel several difficult to distinguish species [NP:2, 8, 10, 11]	I		CO NM ———— AP-S			UT *AZ CO NM ———— AP-S	*NV AZ *NM *TX ———— MR-O	
Family Tiliaceae—Basswood or Linden Family *Tilia* spp. basswood, linden, lime tree, whitewood [NP:2, 3, 4, 7, 8, 10, 11]	LgN							
Family Ulmaceae—Elm Family *Ulmus serotina* Sarg. red elm, September elm	N							

Geographic Regions [3, 4, 5]

VIII	IX	X	XI	XII	XIII
AB SA MT WY CO MB SD NE OK ———— AP-JN	ON WI QU ———— AP-eJL	MB ND SD NE KS MN IA MO WI IL MI IN OH ON PA NY QU ———— AP-JN	OK KY TN MS AL NY PA MD VA NC SC ———— MR-JN	AR LA MO KY TN MS AL FL GA SC NC VA MD DE NJ ———— F-JN	QU NY NH VT MA CT RI NB ME PE NS ———— AP-eJN
				*AR ———— JN-AU	
			NC SC ———— AP-JN		
NM OK *TX ———— MY-S					
	*WI *MI *QU ———— JL	*ND *NE KS *MN *WI MI *OH *ON *PA NY *QU ———— MY-JL	OK KY *TN AL *GA NY *PA *NJ *MD *WV *VA *NC SC ———— MY-JL	LA AL NC VA MD DE NJ ———— JN-JL	QU NY *NH VT MA RI *NB ME ———— JN-JL
			OK ———— FALL		

Family—Common family name Scientific name Common names [Nectar and or pollen] [1]	NAT OR INT [2]	Geographic Regions [3, 4, 5]						
		I	II	III	IV	V	VI	VII
Ulmus spp. elm [NP:10; N?P:8; P:2, 4, 7, 11]	LgN		CO NM ——— MR-AP			NV ID UT CO NM ——— MR-MY	NM ——— F-MR	MB – AP-MY
Family Verbenaceae—Vervain or Verbena Family *Aloysia* spp. white brush, bee-brush, white sage [NP:3]	N						AZ *TX ——— MY-S	
Callicarpa americana L. American beautyberry, French mulberry, beautyberry	N							
Phyla nodiflora (L.) Greene carpet grass, matgrass, fogfruit, capeweed, turkey tangle	N							
Verbena spp. vervain [NP:4, 7; N:2, 11]	LgN							
Vitex spp. [NP:4, 7; NP?:10, N:2]	I							
Family Vitaceae—Grape or Vine Family *Ampelopsis* spp. pepper vine, snowvine [NP:7]	LgN							
Parthenocissus quinquefolia (L.) Planch. Virginia creeper, woodbine American ivy, five-leaved ivy [NP:2, 3, 10, 11]	N							
Vitus spp. grape [NP:2, 7, 11; N:10]	LgN							
Family Zygophyllaceae—Caltrop Family *Larrea tridenta* (DC.) Cov. creosote bush [NP:9]	N					CA ——— AP-MY	CA *NV AZ TX ——— F-O	
Mixed Flowers and vegetables for seed	N&I						CAp ——— MR-AP	

Geographic Regions [3, 4, 5]

VIII	IX	X	XI	XII	XIII
CO NM MB ND SD NE OK ——— F-MY		MB ND SD NE KS WI ON ——— 1F-MY	OK MO AR IL IN KY TN MS AL GA MD VA ——— F-AP	TX AR LA TN MS AL ——— F-AP	MA ——— AP-MY
				LA ——— MY-JN	
				GA ——— MY-FROST	
		ON ——— JN-S	MS AL ——— MR-O	*LA MS AL SC NC ——— MR-O	
			KY AL ——— JN-AU	AL ——— JN-JL	
			MS ——— MY-AU	*AR LA MS ——— AP-AU	
				AR ——— MY-AU	
		OK ——— MY-JL	OK TN MS VA ——— MY-JN	TN MS ——— MY-JL	CT ——— JN

Footnotes for Table 1

1. Information enclosed in [ ] indicates nectar and pollen production.

   Nectar and pollen production categories.

   N= Nectar production.
   P= Pollen production.

   ? following a letter indicates production represented by the letter preceding the ? is in question.

   – indicates no production for the missing letter.

   No data for a taxon indicates none was found in the references consulted.

   References

   Numbers following a N/P category indicate references from which data were obtained.

   Example: [NP:2, 4; NP?:10; N–:6] indicates references 2 and 4 indicated nectar and pollen production; reference 10 indicated the production of nectar but production of pollen was questionable and reference 6 indicated only nectar was produced.

   References

   1. Arnold, 1954.
   2. Burgett, Stringer and Johnson, 1989.
   3. Crane, Walker and Day, 1984.
   4. Dennis, 1983.
   5. Howes, 1979.
   6. Lovell, 1926.
   7. Milum, 1957.
   8. Nye, 1971.
   9. Pellett, 1976.
   10. Ramsay, 1987.
   11. Scullen and Vansell, 1942.

   Complete citations for these references are included in the References section at the end of the chapter.

2. The Nat/Int column indicates plant origin.

   N= native.
   I= introduced.

   Lg preceding either N or I means largely native and largely introduced, respectively.

   ? indicates uncertainty of origin.

3. Reported geographical importance and date of bloom.

   A. Where more than one state or province is involved.

      Each cell is divided into two parts. The upper part indicates states and/or provinces within the region where a given taxon was reported to be important. The bottom part of the cell indicates blooming date range for the taxon within the geographical region. Blooming date data generally reflect only the range indicated in the top part of the cell rather than from the entire region.

Top part of cell.

State and Provincial abbreviations.

States—States abbreviations are the two letter codes used by the U.S. Postal Service.

Provinces—Provincial abbreviations are:

BC = British Columbia
AB = Alberta
SA = Saskatchewan
MB = Manitoba
ON = Ontario
QU = Quebec
NB = New Brunswick
PE = Prince Edward Island
NS = Nova Scotia

* preceding a state or province indicates that the particular taxon was considered "very important" in the part of the state or province found within the region under consideration.

p following a state or province indicates that one or more respondents indicated that the particular taxon is commercially pollinated in that part of the state or province within the region under consideration.

Bottom part of cell (blooming date ranges).

Abbreviations of Months.

JA = January
F = February
MR = March
AP = April
MY = May
JN = June
JL = July
AU = August
S = September
O = October
N = November
D = December
SPR = Spring
SUM = Summer

Abbreviations preceding months or seasons.

e = early
m = mid
l = late

B. Where a region is made up of only one state (Regions 1 and 3).

Because only one state is involved, no state designations are used. The individual cells are divided only if the taxa involved are either "very important" or are commercially pollinated. In these cases * and p indicate "very important" and commercial pollination. The blooming date range designations are the same as elsewhere in the table.

4. Comment about pollination data:

There are instances where commercial pollination occurs within a state or province and is not listed. In these instances, data indicated that fewer than five participating beekeepers were involved. These data could be misleading for two reasons: (1) they are somewhat subjective being based on the opinions of one or more respondents, (2) a single beekeeper can participate in the commercial pollination of many farms and significant amounts of pollination could be accomplished by fewer than five beekeepers.

5. Comment about blooming dates:

The blooming dates reported in this table are broader than will be experienced by a single beekeeper since they often represent ranges that have occurred over considerable time periods for an entire geographical region. This is most evident where the geographical region extends over a large latitudinal range. In Region XII, for example, the blooming dates in Florida are quite different than those in New Jersey and to the New Jersey beekeeper the reporting of pollination of cucumbers over the entire year may at first seem strange.

FIGURE 3. City street planted with *Tilia cordata* (small leaf linden). These particular trees are exceptionally floriferous year after year (see inset photo) and have been planted widely throughout the central city. Such plantings are surely beneficial to urban beekeepers.

BEE FORAGE REGIONS OF NORTH AMERICA

Introduction

The present distribution of bee plants in North America reflects both the regional ecology and land use pattern, since bees obtain nectar and pollen from a wide range of native, naturalized, and cultivated plants. The regional ecology, in turn, has adjusted to climate, soils, geology, and topography, while land use patterns are influenced by cultural preferences, since what farmers raise is partly governed by what they know how to raise. In addition, the contribution of both native and cultivated bee plants continues to change— some native plants are encouraged by land use practices that increase their available habitat, whereas the habitat of others has been reduced. And, of course, the constantly changing economics of farming produce land use changes that further alter the bee forage picture.

The boundaries on the accompanying map respect both natural and land use patterns and are an attempt to depict existing forage regions more realistically. Two principles guided us as we drew and redrew them:

1. Climate and geology are primary controls. Climate determines present-day floristic patterns as well as the overall location of agricultural regions by allowing certain crops to be grown most economically in certain areas. However, within these areas, production of specific crops concentrates on lands best suited for them, particularly when farming costs are high, and the location of these lands is influenced greatly by geology, landforms, and soils.

2. Irrigation may improve bee forage opportunities otherwise limited by a poor native flora or a dry summer climate. In many areas of the West, widespread irrigation has created bee forage landscapes that exist independently of the local flora, and the success of a particular beekeeping operation may be determined by access to this irrigated cropland.

We are lumpers, not splitters. Many details of local interest have been omitted or lost through generalization because we wished to restrict the number of floral regions to something manageable. We have attempted to address the most important of these map omissions in the text, but apologize to any users who do not find their local regions adequately treated.

The discussion of each region includes, first, a description of the important criteria we used to define it and, second, the important bee forage plants on both agricultural and non-agricultural land.

Description of the Regions

I. Alaska

Beekeeping in this most northerly apicultural region is limited more by unfavorable foraging weather than by the availablility of forage itself. In fact, vast expanses of largely contamination-free forage plants go unutilized because of the difficulty of over-wintering bees in this harsh, long-winter climate. Most beekeepers start each spring with fresh packages shipped from the continental United States rather than try to over-winter their colonies.

Nonagricultural Areas

Dandelion, crabapple, poplar, alder, and willow provide abundant basis for spring build-up in May. These are soon followed by mustard, blueberry, Labrador tea, vetch, several species of *Rubus*, fireweed, and escaped sweet clover in combination with other minor plants that represent a variety of potential nectar sources. Because of the abundance of available forage, honey quality is quite variable from season to season, depending on weather and the specific plants available.

Agricultural Areas

Dairying is the most widespread agricultural activity and, of course, is centered near the principal urban markets of Fairbanks, Anchorage, and Juneau. Leguminous hay crops that include red, alsike, and white clovers, sweet clover, and alfalfa constitute the major sources of nectar, although native plants on idle or unused land within the agricultural areas may still contribute significantly. Thus, Alaskan honeys often consist of a complex mixture of nectar sources.

II. The Western Mountains

This region embraces a great variety of landscapes and climates, as altitude, geology, wind exposure, and native vegetation patterns all complicate the picture. Furthermore, the vegetation in this large, complicated area has a complex history and consists of many different species whose individual ranges do not necessarily overlap. In most areas, however, beekeeping is limited in upper elevations by cool summer nights, which limit nectar secretion or shorten bee foraging periods, and by cold winters that affect overwintering. So the best beekeeping is found at altitudes below closed coniferous forest.

Nonagricultural Areas

These are the rule in this area, with extensive mainly needleleaf forests in the cooler, moister high elevations, and semi-desert in the low valleys (4,000 ft. to 7,000 ft., depending upon latitude and local climate). Spring build-up is based mainly upon flowering willow, box elder, and poplar growing near water courses, and dandelion. Summer forage is spotty in the valleys and often limited by dry weather. At intermediate elevations (ponderosa or lodgepole pine zones), forage is still limited, but old fields supporting members of the raspberry/blackberry genus, *Cirsium* (thistle), blueberry, and vetch provide some summer nectar. Scattered stands of fireweed, more prevalent at higher elevations, may also yield, as might escaped sweet clover growing along fencerows or roadsides. Some *Centaurea* species (knapweed or thistle) may be locally important, but are more widespread in the Intermountain area. Local stands of *Ceanothus, Rhamnus, Arctostaphylos*, and *Symphoricarpos* species along with balsamroot, resinweed, rabbit-brush, and goldenrod can provide surpluses, as well.

Agricultural Areas

Primary beekeeping areas in this region are mostly restricted to irrigated alfalfa, alsike, red, and white Dutch clover, or orchards located in the lower elevations that provide springtime nectar and pollen. Availability of level land and water largely govern the location of these activities, and where practiced such farming may be intensively managed. If so, herbicidal use on weedy areas near fields may reduce forage opportunities on escaped alfalfa and sweet clover, and early cutting of the alfalfa fields before flower further restricts what is left to the honey bee. Legumes raised for seed, on the other hand, represent excellent nectar sources. Thus, not only are the locations of optimum forage spotty, but varying degrees of farm mangement complicate the picture. In other words, similar agricultural areas may provide very different bee forage potential.

III. South and Central California

Bordered on the east by the Southwest Deserts and Western Mountains, this region embraces two very distinct subregions—the coastal mountains and the Central Valley. The climate of both features winter rains and summer drought, although the Central Valley is drier. This regime in combination with mild winter temperatures produces a Mediterranean-type climate found nowhere else in North America that permits the survival, particularly in the coastal mountains, of a unique assemblage of plants. Consequently, the bee flora of southern California contains native plants not found in other regions. Much of the topography is mountainous, creating a great variety of climate, vegetation, and agricultural environments. In addition, the warm climate and large market for the various crops that can be raised here have favored the development of specialized agriculture; beekeepers may, therefore, practice migratory beekeeping in order to keep up with the flowering of the varied crops.

Nonagricultural Areas

Coastal Mountains.

The vegetation here is predominatly shrubby chaparral with patches of intensive agriculture in some of the valleys and mixed broadleaf/needleleaf forests at higher, moister elevations. Attractive shrubs of the chaparral include manzanita (*Arctostaphylos* spp.), sage (*Salvia* spp.), buckwheat (*Eriogonum*), toyon (*Photinia*), and blue curls (*Trichostema*). Because of cool, foggy summer weather, some choice forage near the coast may not be utilized by the bees, but inland where summers are warmer some high quality honey,

FIGURE 4. Tamarisk or salt cedar—several species of this group have been introduced from Europe, Asia, and Africa into the southwestern United States where they have become naturalized and "very important" honey producers in several states. The blooming period is lengthy, but the quality of the honey is poor and not suited for table use. (*Photo by Gordon Waller*)

particularly from *Salvia,* is obtained. *Eucalyptus* species have been widely planted for soil stabilization in the areas least prone to frosts, and these plants provide forage, as well.

Central Valley.

    Originally a semi-arid to arid grassland, much of the Central Valley is now agricultural, and the remaining native vegetation is not rich in important bee plants. However, in the moister foothills of the Sierra Nevada, where chaparral or savanna woodland occurs, some of the same species seen in the coastal mountains are again important, including some *Arctostaphylos* species and toyon. Yellow star thistle (*Centaurea*) is locally important, especially in the north.

Agricultural Areas

Coastal Mountains

    Lowland pockets of diverse, intensive agriculture provide bee forage very different from that available from nonagricultural areas. Citrus, avocado, beans, and almonds (especially inland) are all available locally for nectar and/or pollen.

Central Valley

Extensive, often irrigated fruits, nuts, cotton, alfalfa, sweet clover, and safflower are all locally important. Citrus is prominent in the San Joaquin Valley, with the cold-hardy temperate fruits (peaches, cherries, apricots, and apples) in the Sacramento Valley toward the north.

IV. Interior Northwest

Lying behind the Cascade Mountains in eastern Washington and north-central Oregon and including some interior semi-arid valleys such as the Okanogan in British Columbia, this portion of the Columbia River basin has warm, dry summers but cold winters that contrast with the moister mountains to the west, north, and east, and the drier intermountain desert to the south. Much of the original prairie vegetation has been replaced by wheat fields or ranching. Temperate fruits and vegetables are widely grown in the valleys where ample irrigation water is available from the melting of the heavy winter snowpacks in the nearby mountains.

FIGURE 5. Catclaw—several of the native acacias have curved thorns resembling the claws of a cat and are often referred to collectively as catclaw. These species are "very important" honey producers in parts of the Southwest. Because of their brushy nature, catclaw, mesquite, and a few other species produce what is sometimes referred to collectively as the "brush honeys." *(Photo by Gordon Waller)*

Nonagricultural Areas

Much of this area has limited bee forage due to the dry summer climate. Following buildup from dandelion, *Brassica* species, *Symphoricarpos,* and spring-flowering broadleaf trees such as willow and poplar, the summer flow is slow and irregular. Sources of possible surplus include escaped sweet clover, yellow star thistle (*Centaurea solstitatis*), black locust (introduced from the eastern United States), Russian knapweed (*Centaurea repens*), vetch (*Vicia* species), and species of *Rubus* (raspberry/blackberry).

FIGURE 6. Mesquite—this native group of plants constitutes one of the "very important" honey sources of the Southwest. Mesquite honey is one of the "very important" regional honeys that, because of local demand, is rarely sold outside of the area of production in pure form. *(Photo by Gordon Waller)*

Agricultural Areas

Primary agricultural areas within this region are the Yakima, Snake, and Wenatchee valleys in Washington and the Okanogan Valley in British Columbia, the premier fruit-producing area in western Canada, and together these offer the best beekeeping potential. Beginning with spring fruitbloom, which rarely provides a surplus because of cool weather, and extending through the summer when alfalfa, clover (red, white Dutch, sweet), mint, beans, and cucurbits are in flower, the availability of nectar is determined by

FIGURE 7. Canola—this small group of plants is sometimes referred to by its older but more stigmatized name, rape. Canola plays a very important part in Canadian honey production and, because of the high unsaturated fat content of its oil, is currently generating much interest in the United States. It produces a high quality honey that, because of its great tendency to crystallize, requires special attention during harvesting and processing. *(Photo by Paul Love)*

both the weather and farming practices. Early cutting of alfalfa, heavy pesticide use, and chemical control of weeds along field edges all reduce the available forage. Where farmers are less diligent or the clover and alfalfa are raised for seed, high quality, low moisture honey is obtained.

V. The Intermountain Region

This region encompasses much of the cold desert and semi-desert lying between the Rocky Mountains to the east and the Sierra Nevada-Cascade Mountains to the west. Summer temperatures frequently rise to the 80's and 90's (F) or higher, but winters are subfreezing, and snow is not uncommon. Habitat variety is rich and depends upon elevation, exposure, and latitude. Overall, however, precipitation is light and falls mainly in the winter, restricting the summer nectar flow from the few important forage plants.

Nonagricultural Areas

Bee forage is sparse; most areas have long summer periods when no nectar or pollen are available. Moister valley areas provide the best forage, particularly in spring, where local dandelion, poplar, willow, box elder, elm, hackberry, balsam-root, mule-ears, and *Ceanothus* are possible sources. Summer sources are scant except during unusual wet weather; vetch, star thistle, rabbit-brush, resinweed, coneflower, *Symphoricarpos*, Rocky Mountain bee plant, and aster can be locally important.

Agricultural Areas

Best beekeeping opportunities again are afforded by irrigated agriculture. After spring buildup from fruitbloom, alfalfa, and white Dutch clover, mixed with roadside sweet clover, have historically contributed the bulk of the summer nectar flow, particularly if the crop is grown for seed, although sunflower acreage is increasing. Agricultural operations here are often closely managed, however, and spraying losses of honey bees may be high. Dry summer weather results in low moisture honey most years. Important agricultural concentrations are found in the Snake River Valley of Idaho and the Wasatch, Uinta, and Delta regions of Utah.

VI. The Southwest Deserts

Lying at a generally lower elevation than the intermountain (cold) deserts, this region includes a mixture of desert types defined according to amount and timing of annual rainfall and each type has its distinctive plants. As a whole, these deserts are characterized by a milder, shorter winter and more summer rainfall than that area, although the region is complex. It is bordered on the east by the Great Plains, the mountains or intermountain desert to the north, and the Sierra Nevada or California coastal mountains to the west. Taken together, these deserts offer a wider variety of native plants and agricultural landscapes than does the previous region.

Nonagricultural Areas

Among the most important honey and pollen plants, mesquite and species of *Acacia* such as catclaw are probably the most widespread. Mesquite, particularly, produces a light, high quality honey, but like most native desert plants, will yield variably depending on the weather. Early build-up is provided by poplar, willow, paloverde, and a variety of herbaceous sources. Locally important nectar and pollen plants include greasewood, tamarix, creosote bush, paloverde, some cacti, black locust (planted in parts of northern Arizona and more widely in New Mexico), sumac, burrowseed, goldenrod, and aster. Manzanita (*Arctostaphylos*) may yield where it occurs in the lower mountains.

FIGURE 8. Sweet clover—new beekeepers are commonly surprised to learn that the sweet clovers resemble very little the *Trifolium* clovers with which they are more familiar. Collectively, the sweet clovers, alfalfa, and the *Trifolium* clovers constitute an exceptionally important honey production group. Members or combinations of this group are frequently recommended to beekeepers who wish to make plantings for honey production.

Agricultural Areas

Here, too, the pattern is complex, with irrigated clover, sweet clover, alfalfa, cotton, or citrus areas offering some of the best beekeeping opportunities. In Arizona, the principal irrigated regions are in the Tucson-Phoenix-Yuma areas along the Santa Cruz, Salt, and Gila rivers, and near the Colorado River and Salton Sea in southwest Arizona and adjacent southeast California. Major irrigated agricultural areas in New Mexico are found along the Pecos and San Juan and Rio Grande rivers, where fruit and some alfalfa and clovers are grown. In Texas, irrigation in the lower Rio Grande Valley supports some cotton and citrus.

VII. The Aspen Parkland and Peace River Valley

Sandwiched between the drier Great Plains to the south and the cooler boreal forest to the north, this region collectively represents one of the leading beekeeping areas in Canada. Originally a mixture of forest and prairie, this region has undergone heavy agricultural development because of favorable soils and a warm summer climate in spite of the long, cold winters. Long hours of daylight in summer permit prolonged bee foraging, and the nectar flow can be very heavy when the weather cooperates.

Nonagricultural Areas

Willow, dandelion, box elder, elm, serviceberry, cherry, and poplar provide build-up and perhaps some surplus in spring; local summer flows come from fireweed (especially in the west) and *Agastache* (in the east), *Cirsium* species, sow thistle, and vetch. Where sufficient old fields are present, aster and goldenrod may support a late flow in the east.

Agricultural Areas

By far the best beekeeping is found in the agricultural areas where canola (rape), alsike, white Dutch and red clovers, alfalfa, buckwheat, and sweet clover are grown, often commercially for seed. Locally, extensive fields of these crops may provide almost unlimited forage. Sunflowers are an important crop in southern Manitoba, where they may provide both nectar and pollen.

Detached pockets of somewhat similar land use are found in north-central and northern Alberta where commercial agricultural areas continue to be developed, of which the Peace River Valley is the largest example. Here, crops similar to those raised in the Parkland are produced; these, in combination with the very long day length, favorable soils, and a brief but warm summer, permit very heavy nectar flows and some of the highest total yields per hive in North America. Important honey plants include sweet clover, red, white, and alsike clovers (often for seed), and canola.

VIII. The Great Plains

Extending southward from south-central Canada almost to the United States border with Mexico, this region includes a variety of landscapes that all share three similar characteristics—generally level to somewhat hilly topography, semi-arid climate, and non-forest native vegetation. Islands of forest occur on enclaves of higher, wetter terrain, but generally grasses and forbs (broad-leaved herbaceous plants) dominated this area at the time European settlers arrived. This region is bordered on the west by the Rocky Mountains and Southwest Deserts, on the north by the Aspen Parkland, and on the east by the Agricultural Interior, Ozark Upland, and Gulf Coastal Plain regions. Throughout, it is characterized by relatively hot, dry summers, cold, changeable winters, and an open, sunny, windy aspect that grows drier as you move west toward the mountains.

Nonagricultural Areas

Spring nectar and pollen are available from dandelion and to a lesser degree from hardwood forest strips lining the river valleys, and, in remnants of the original prairies, species belonging chiefly to the sunflower family provide some nectar in summer. In these landscapes, willow, poplar, elm, *Polygonum* species, vetch, Russian olive, milkweed, *Cirsium* species, resinweed, *Monarda* species, Rocky Mountain bee plant (in only the western areas), and goldenrod may be important. Stands of naturalized sweet clover yield heavily during some years, but, overall, wildflower nectar sources are limited by the dry, hot weather typical of summer, especially westward.

Agricultural Areas

These areas again offer the best beekeeping opportunities. Few agricultural crops except locally important fruit orchards provide nectar and pollen for spring build-up. During summer, however, leguminous crops such as alfalfa, sweet clover, and, locally, red and white Dutch clovers yield very well under the sunny, hot, dry climate if soil moisture is not limiting, which it often is. Irrigation, particularly for cotton in the southern areas, may promote nectar flow. Other summer sources are sunflowers, rape, some commercial cucurbit acreage (mostly pollen), and soybeans. Because of the dry summer climate, water content of honey from this region is often lower than the national average.

IX. The Northern Great Lakes and Southern Canada

Lying between the boreal evergreen forest in south-central Canada and the agricultural interior to the south, this region represents the most northerly area in east-central North America where commercial beekeeping is generally practiced. Much of the present landscape is forested, but has a history of deforestation and exploitation. Unsuccessful attempts at commercial farming followed logging and were later abandoned; today, the percentage of forested land in this region is increasing as old fields are reverting to woodland. Pines are characteristic old field invaders and thus dominate large tracts; mature forest consists of a mixture of broadleaf and needleleaf species.

Nonagricultural Areas

Spring build-up is fueled by an abundance of dandelion, fruit bloom, hawthorn, red and sugar maples, poplar, yellow rocket, mustard, and willow. Old fields of milkweed, sumac, star thistle, pearly everlasting, Canada thistle, raspberry/blackberry, blueberry vetch, and, locally, fireweed may yield a surplus in summer, and basswood flows are important locally, especially in the western portion of the region. Where basswood is absent and mature forests dominate the landscape, summer flows are scant and a surplus is less likely. Late summer stands of *Bidens,* golden ragwort, goldenrod, loosestrife, and aster may yield a surplus.

Agricultural Areas

Dairying is the most common agricultural activity, mostly near the scattered urban centers. Clovers and trefoil provide hay and a summer flow if they are allowed to flower. Spring build-up is seldom limited by available plants because even in the more heavily agricultural areas enough woodland and old fields remain to supply nectar and pollen; weather is more often a limiting factor. These same remnant woodlots may also provide a basswood flow if the

FIGURE 9. Sourwood—this native species is one of the locally "very important" bee forages. Its reputation for producing exceptional quality honey is both widespread and well deserved. Unfortunately, it rarely reaches markets in pure state outside its region of production, the southern Appalachians. *(Photo by Paul Love).*

species is abundant. Otherwise, any summer surplus is based mainly on flowering leguminous hay crops such as alsike and Dutch white clovers, red clover, sweet clover, and trefoil.

X. The Agricultural Interior

Stretching from the St. Lawrence valley on the east through the southern Great Lakes to the drier Great Plains wheat and range lands in the west, this region is the most important expanse of largely nonirrigated upland agriculture in North America. Although the heart of this region in Iowa and Illinois occupies former native prairie, its eastern portion, from Indiana eastward, has developed on former broad-leaved forest landscapes. It is limited to the south by the Ozarks and Appalachians, where mechanized agriculture is restricted by topography and poor soils, and to the north by poorer soils and cooler climate. Principal crops include corn, soybeans, winter wheat, and vegetables near urban markets; dairying and livestock raising are common throughout.

Nonagricultural Areas

Where this region occupies former prairie, much of the original vegetation has been destroyed and bee forage is limited to roadsides and fencerows, where escaped sweet clover mixed with assorted native and nonnative plants provide some pollen and nectar. Woodlands remaining along streams provide springtime nectar and pollen from oaks, maples, elm, willows, pearly everlast-

FIGURE 10. Citrus—Citrus is one of the "very important" bee forages in the most southern parts of the United States. Although several species are involved, most citrus honey is marketed as orange honey. Despite the fact that, under proper conditions, few plants yield more copiously than citrus, few varieties require insect pollination.

ing, and dandelion, and a local basswood flow may occur in early or mid summer. Old fields of *Rubus* species and blueberry (on acid sites only) are heavily worked by the bees, as well. Summer sources include sumac, milkweed, Canada thistle, goldenrod, aster, vetch, loosestrife (wet sites), and *Polygonum* species.

In the eastern, formerly forested parts of this region, more remnant woodlots remain and bee forage is better. In addition to the springtime sources of maples, oaks, and willows, black locust and tulip-tree contribute surpluses in more southern areas.

Agricultural Areas

The premier beekeeping areas within this region are agricultural, but the success of individual bee yards is largely determined by the local availability of particular crops. After spring buildup on fruit bloom and *Brassica* species, fine summer flows are obtained from alfalfa (where grown and left to flower), soybeans, possibly red clover, and, locally, sweet clover where surviving stands remain. Alsike and white Dutch clovers and trefoil become important locally, especially eastward in Ohio and the St. Lawrence Valley. Forage is less available where corn and soybeans constitute the main crops than where dairying is a major land use. Buckwheat and members of the *Cucurbita* may be local nectar and pollen sources, respectively. In the western fringes of the region, sunflowers have become an important forage plant.

XI. The Appalachian-Ozark Upland

This large and complex region is bounded on the southeast and south by the Atlantic and Gulf coastal plain, and on the west and north by more level agricultural landscapes. The topography of much of this region is similar to that of New England, with steeply rolling to mountainous terrain separating more level agricultural valleys, but it differs from New England in having hotter summers, milder winters, and a much longer growing season. The more rugged landscapes are usually covered with second or third growth hardwood forests that in the southern Appalachians attain extraordinary richness. The somewhat drier Ozark landscape supports a less diverse native flora, with correspondingly fewer bee plants.

Nonagricultural Areas

Throughout the region, poplar, redbud, mustard, red and sugar maple (mostly in the Appalachians), oaks, elm, dandelion, and yellow rocket are isolated to generally important springtime nectar and pollen sources. Later spring nectar flows from black locust, persimmon, and, in the Appalachians, honeysuckle, tuliptree, basswood, and sourwood may provide heavy surpluses. Sourwood flowers are later than most other trees and where locally abundant, yields high quality honey much in demand. Ozark forests are drier, dominated by oaks and hickories, and generally yield poorly in the summer. However, old fields throughout this region not yet reverted to woodland support a mixture of herbs that provide some darker summer honey. Occasional and locally important sources include the mountain mints, escaped sweet clover, several members of the raspberry/blackberry genus, species of *Bidens, Cirsium,* and *Polygonum,* sumac, blueberry, huckleberry, and in some years asters and goldenrods during autumn.

Agricultural Areas

Farming is widespread but generally patchy and limited to more level areas of better soil. Concentrated and sometimes specialized agriculture has developed in the bluegrass area centered on Lexington, Kentucky, the Nashville, Tennessee area, and in the wider Appalachian Valleys along the east edge of the region stretching southward from Maryland. Dairying is the most common farming activity, with pockets of tobacco, cotton, fruits (apples, peaches), and vegetables. Dairy farming creates a demand for hay crops often filled by legumes, so alfalfa, sweet clover (often escaped), and other clovers (alsike, crimson, white Dutch, and arrowleaf) are local light honey sources. Corn is a summer pollen source where grown.

XII. The Atlantic and Gulf Coastal Plain

A rolling landscape with highly variable soils and a complex native flora, this region as a whole has seen a decline in agricultural acreage over the last century, with much of the old farmland reverting to forest. Former corn, cotton, and tobacco lands now often support pine stands; thus, overall bee forage acreage has decreased. Today, agriculture is practiced primarily on the better soils, with woodland occupying the greatest amount of landscape. However, the Mississippi Valley, formerly occupied by rich bottom land forests, is being increasingly farmed as the original woodlands are cleared, drained, and replaced by tilled fields. Long, hot summers and mild, brief winters permit an extended foraging season, but little honey is produced through much of the summer when native sources are sparse, unless agricultural land is located nearby.

Nonagricultural Areas

Summertime bee forage is particularly limited in the pinelands. However, springtime nectar and pollen are readily available from a variety of tree and herbaceous sources in both the pinelands and hardwood forests. Arboreal sources include maple, alder, persimmon, sparkleberry, redbud, black locust, the oaks, Ohio buckeye, escaped privet (*Ligustrum*), willow, and poplar. Where present, sourwood, a summer-flowering tree, and basswood may be very important in the areas adjacent to the Appalachians. Shrub or herbaceous sources are equally diverse and include dandelion, mustard, thistle, honeysuckle, huckleberry (*Gaylussacia*), blueberry, sumac, *Berchemia*, vetch, and species of *Rubus.* Some of the best nectar sources are found in moist or river bottom areas, where surpluses are obtained from *Ilex* species (gallberry), black titi (*Cliftonia*), tulip-tree, and white tupelo, and in Florida from palmetto, cabbage palm, Brazilian pepper, and black mangrove where they occur. Important late summer and autumn sources include aster, spanish needle (*Bidens* spp.), boneset, sneezeweed, sunflower, goldenrod, horsemint, and vervain.

Agricultural Areas

These provide the best beekeeping opportunities. Dairying is the most widespread farming activity, and high quality surpluses are obtained from

FIGURE 11. Gallberry in its natural habitat—this dioecious, native species has long been a "very important" producer of high quality honey on the acid soils of Region XII (the Atlantic and Gulf Coastal Plain) where alfalfa and clovers do poorly. The honey is very slow to crystallize. *(Photos by Jim Castner)*

cultivated legumes, particularly crimson, Dutch, white, and red clovers. Local stands of naturalized sweet clover may be important. Corn is a pollen source, and cotton and soybeans may be very important in parts of the Mississippi Valley. Some concentrations of vegetable crops, particularly in the mid-Atlantic region, provide pollination opportunities if the cucurbits (squash, melons, gourds) are raised. The central Florida citrus belt is a renown source of specialty honey.

XIII. New England and Atlantic Canada

A varied area consisting of forested interior mountains mixed with sometimes broad agricultural valleys and more level coastal lowlands, this region is bordered on the northwest by the St. Lawrence lowland and on the southwest by the Hudson-Mohawk valley system. In addition to general farming in the valleys, isolated pockets of specialty crops such as potatoes, tobacco, or vegetables are found on better soils or near some larger urban areas. Long, cold winters and relatively brief, mild summers mean a short foraging season, so that permissible foraging weather is as much a limitation on honey production as is available forage.

Nonagricultural Areas

Mixed needleleaf and broadleaf forests provide limited forage. Springtime sources include dandelion, red, sugar, and silver maples, willow and poplar species, blueberry, alder, mustard, and black locust. Tulip-tree, basswood, and vetch may be important summer sources where they occur. Some of the better sites are old abandoned farmland, where milkweed, raspberry, blackberry, sumac, goldenrod, aster, and star thistle are locally important nectar sources. Wet sites contain loosestrife, buttonbush, and, to the south, pepperbush.

Agricultural Areas

Agricultural concentrations in the larger valleys such as those of the Hudson-Mohawk, Connecticut, Champlain, and Aroostook-St. Johns rivers, as well as enclaves of better upland soil such as Prince Edward Island, provide some of the best beekeeping opportunities. Important nectar sources include fruit bloom, Dutch, red, and alsike clovers, alfalfa, trefoil, rape, and escaped sweet clover. Buckwheat is declining in importance but remains a local source.

XIV. Boreal Forest

Except for scattered hobbyists, beekeeping is generally unimportant in the boreal forest. Available bee forage is similar to that found in Alaska, but long, cold winters and a short foraging season often characterized by uncooperative weather are serious hindrances to beekeeping. The greatest concentration of hobbyists is located in the the dairying region of the Ontario claybelt around the site of Cochrane.

PLANTING FOR BEES

Introduction

Interest in planting for bees may be nearly as old as beekeeping, itself, and periodically the subject receives much attention in the beekeeping literature. In the United States, several species have been named after beekeepers who championed their culture. Chapman honey plant (*Echinops sphaerocephalus*), Simpson honey plant (*Scrophularia marilandica*), and Pellett clover (*Trifolium ambiguum*) are examples. While these attempts at planting for bees met with varying degrees of success, none are being actively pursued today by large segments of the beekeeping community. Indeed, many apicultural experts believe that planting only for bees is not practical. In their opinion, only when plants are to be established for some other purpose and can also serve as bee forage should the possibility of a dual purpose planting be considered. This idea originated during a time of plentiful, pesticide-free bee forage, a condition quite different from that of today. We believe it is now time to re-examine this idea and that at least three situations exist where bee forage could be planted profitably.

The first of these occurs when the beekeeper owns land and wishes to plant something on it, but does not intend to manage the land for conventional agricultural production. The second occurs when land adjacent to an apiary is not owned by the beekeeper, but the owner is willing to have it planted to bee forage. This situation is probably quite common, being represented by road, power line, and pipe line rights-of-way, ditch banks, wet areas, and abandoned fields. Such plantings, in addition to being aesthetically pleasing, might prevent erosion and provide food and cover for wildlife, and thus actually increase the value of the land. The third situation occurs where the bees of a particular apiary are routinely poisoned by agricultural pesticides, but the beekeeper has a compelling reason for maintaining an apiary in that particular location. Often the beekeeper owns the land on which the apiary is located. Here, a bee forage planting could help divert bees from pesticide-laden agricultural areas during high risk periods. This procedure could reduce the losses of the beekeeper and at the same time provide a source of high quality bee forage. This solution places the resolution of the bee kill problem directly in the hands of the beekeeper and not in the hands of some disinterested party.

While it is probably not practical to purchase or rent land expressly for planting bee forage, in two of the above cases the land is already essentially a fixed cost of the beekeeping operations and is, in a sense, essentially free. In the third, it is free.

Desirable Properties of a Bee Forage Planting

The authors believe that it is time to re-examine the proposition that planting exclusively for bees is not profitable. This part of beekeeping has essentially been ignored, and the beekeeper who sets out to attempt it is venturing into largely uncharted territory. For this reason, much of what follows is not tried and tested in the sense that the regimen for producing corn or soybeans is tried and tested. We, nevertheless, have had some experience with all of the topics discussed and, while some of it may be somewhat theoretical, we believe, based on our experience, that all we propose is possible. In most cases, we present guidelines that are based on this experience. Because fixed-land honey production is a relatively undeveloped discipline, the beekeeper is advised to proceed slowly and cautiously. The motto should be, start small (perhaps with only a flower bed) and expand cautiously. Above all, the planting should produce high quality honey, be as productive as possible, and require a minimum of maintenance.

High Quality Honey

Where the beekeeper goes to the trouble and expense of planting for bees, the planting should produce honey of high quality so that a good price may be demanded for it. For this reason, especially the small beekeeper might consider producing a specialty honey. Fortunately, many plants are available for this end.

Considerations for Achieving High Productivity

The design characteristics of a foraging system intended to augment existing forage are not well understood. Based on our current understanding of how the hive allocates foragers to a particular floral patch (Seeley, 1986, and Seeley and Levien, 1987), however, we believe that selection of the most efficient strategy for increasing productivity may depend upon the apiary setting. At one extreme, the existing forage supplies adequate nectar and pollen during much of the season except for one or more periods when the nectar and/or pollen flow dwindles. At the other end of the continuum, the area surrounding the apiary is relatively unproductive during much of the season.

In the first case, where flows are satisfactory during large parts of the year, planting highly productive species that bloom during these flows may be inefficient. Since bees preferentially forage highly productive areas, these plantings might do little more than divert bees from the original surrounding forage. In this case, the production of the new planting and that of the pre-existing forage would not be additive, and, unless the new planting were quite large, the beekeeper might receive very little for the effort expended in establishing the new planting. Plantings that bloomed when nectar and/or pollen flows were very low, however, might be more productive since the productivity of the surrounding forage would not be lost.

In the second situation, where large parts of the summer produce little nectar or pollen, the planting should be productive for as much of the season as possible. This goal can be achieved in two ways. In the most obvious solution, the land could be planted to long-blooming and highly productive species. In the second and more interesting solution, the system could be planted in some multilayered design, as, for example, a herbaceous layer with a partial tree canopy layer (Fig. 12). In this system, even though the different layers might not be made up of extremely long-blooming species, the unit of land they occupy could remain productive over prolonged periods if the species were chosen correctly. A second desirable feature of a multilayered system is that the area occupied by the forage can theoretically be increased several fold over that of a simple monolayered system. For this reason multilayered systems should also be considered for those situations where the supply of nectar and pollen needs to be augmented for only part of the season. Careful attention

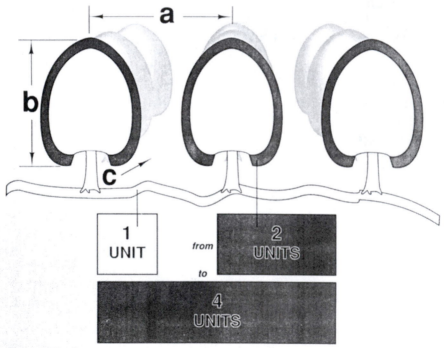

FIGURE 12. A theoretical, three dimensional honey production system that would allow increasing the area of bee forage over that of a simple herbaceous ground cover. The top of the diagram represents a cross-sectional view of a planting that is comprised of rows of trees and an herbaceous understory, both of which produce nectar. The ground cover represents one unit area of production while the trees represent additional surface area that could easily increase the total production area by an additional two to four units, depending on (a) row spacing, (b) tree height, and (c) tree spacing. The bottom part of the diagram depicts the area contributed by each forage type. By judicious selection of the dimensional components, a, b, and c, the beekeeper could significantly increase the total honey production from a plot of land.

should be given to the blooming dates of the various components of the planting. If plants such as shrubs and vines that occupy ecological niches different from those of the herbaceous and trees layers are added to the planting, multilayered systems could become quite complex and presumably very productive.

The second consideration of paramount importance when planning a bee forage system is that the different constituents of the system should be productive in the environment in which they will be planted. Unfortunately, with the exception of a few agricultural crops, not much is known about the effect of niche characteristics on nectar production. Soybeans are known to be productive on some soils and not on others (Erickson, 1984). The same is true of alfalfa. In many cases the best indication available of where a nonagricultural plant will be productive will come from observations of that plant in its native habitat.

For the system to be as efficient as possible, the individual constituents of the system should be as productive as possible under local climatic conditions. If the area commonly experiences drought during mid summer, the plants that bloom then should be drought tolerant. Conversely, if much rain is experienced during a particular season, the appropriate response by the beekeeper is more complex. If few flight days are expected, the beekeeper may choose to not plant anything for that season since the productivity of the planting would be largely wasted. If modest amounts of flight time are expected and it is particularly important that something come into the hive during that period, species such as *Scrophularia* that protect their nectaries from "wash-out" should be chosen. Where the relationship between temperature and nectar production is known, plants should be chosen for the different temperature regimes experienced during the various parts of the season. Unfortunately, our knowledge of this relationship is very limited. For more information on the effect of environment on nectar production, the reader should refer to Chapter 10.

Finally, the possibility of competition from feral colonies of those of another beekeeper should be considered, also. Feeding bees that are not your own does not make good economic sense. While some states have laws that protect the beekeeper from encroachments by other beekeepers, many do not.

Considerations for Development of Low Maintenance Systems

The beekeeper should view the planting of bee forage much like depositing money in a bank. The initial planting is analogous to the original bank deposit. Just as interest accrues from the initial bank deposit, returns from a planting of bee forage should also continue for many years with as little effort as possible on the part of the beekeeper.

The first rule of producing a low maintenance system is that the plants should be adapted to the niche into which they are planted. While nonagricultural plants can often be grown in inappropriate habitats, the extra maintenance needed to keep them there almost always outweight the benefits. The first estimate of where a plant will thrive often comes from a knowledge of where the plant grows in nature, and both biotic and abiotic factors are involved. Soil type and fertility, available moisture, and temperature are the major features of the abiotic environment. Plant competition is generally the primary biotic factor concern, but herbivores and plant diseases can also be important limiting factors. The two sets of factors usually work together. Often, for example, the abiotic conditions will be satisfactory for the growth of a planting and the species will initially appear to thrive, only to be choked out by weeds later. As an example, the excellent bee forage, butterfly weed (*Asclepias tuberosa*), grows natively in sandy, relatively dry conditions where it has little competition. It will do well in more hospitable conditions if weeded. If it is not, it will eventually be overrun by weeds and disappear. In addition to observing the plant in nature, the beekeeper can also determine whether a plant will thrive in a particular location by establishing small test plantings at the site. The value of this step cannot be overemphasized, since it will alert the beekeeper to many of the problems encountered in larger plantings of a particular species before extensive outlays of time and money have been made.

Fitting agricultural species to a particular niche is less of a problem than with nonagricultural species. Many crops have been bred to grow well in a variety of soil types. In addition, herbicides are also available to protect them from competition from weeds, a particularly important precaution during their establishment, but useful in their maintenance, as well.

In general, if the planting is not to be harvested for some other purpose, the species chosen for the planting should be long lived perennials so that they will not need to be planted often. While annuals and biennials will often reseed themselves, they usually reseed to a lesser stand than the original. The species should be aggressive enough to quickly occupy the initial interplanting spacing. This characteristic will do much to eliminate weed problems.

Herbicides are available for most agricultural crops and the cooperative extension literature of the beekeeper's state should be consulted for recommendations of herbicide selection and application when agricultural crops are chosen for the bee forage planting. In general, if the species chosen are not agricultural crops, weeds will be a problem because no herbicides are available for selective use on most of the wild species that make excellent bee forage. For this reason, the area in which nonagricultural species are to be planted should be made as weed free as possible before planting. Herbicides such as glyphosate or metam-sodium might be considered for this purpose. The first of these kills most green plants, but does not affect the dormant seed that remains in the soil and some provisions should be made to kill these as they germinate, as through later cultivation. If the planting were done in the fall, emerging annual weeds might be killed by frost. In any case some cultivation will probably be needed the following year. Metam-sodium, on the other hand, is a soil sterilant that kills most living organisms including dormant seed. On the negative side, the compound is an unpleasant one with which to work and must be applied at high rates. For best results the treated area should then be covered either with a tarp or sealed with a water cap. If these herbicides are objectionable, the beekeeper should consider the possibility of summer fallowing followed by cultivation after planting. Currently, the only postemergence herbicides that could be recommended for most nonagricultural bee plants that will not damage the planting are fluazifop-P-butyl and similar materials. This group of compounds is good, however, for controlling only grasses.[1]

Establishment should be made as a dense stand to provide maximum competition for weed species. The beekeeper should keep in mind that, unless the seedbed has been sterilized, some form of weed control will probably be necessary during at least the year of establishment. Since, for the foreseeable future, cultivation is likely to be the only choice for postemergence broadleaf weed control for nonagricultural crops, broadcasting as a planting procedure should be ruled out in favor of some form of row planting.

Finally, plant species should be chosen that will compete well with weeds. Again, establishing a test planting in the area under consideration is the best way of choosing such plants.

Woody plants possess many of the low maintenance characteristics desired in a bee forage system since, once established, they often remain in the system for many years. They suffer, however, from two main drawbacks. The time interval between establishment and first bloom can be quite long, sometimes many years. In addition, in temperate zones they often remain in bloom each year for only short periods of time. Because of their low maintenance characteristics, and because they can provide additional layers of bee forage to make the planting more productive, they should, nevertheless, be considered among the prospective bee forage candidates.

If the planting is an agricultural crop intended to be harvested, harvesting considerations will usually dictate that the planting be made as solid stands. Otherwise, for a variety of reasons, the planting should exhibit considerable

---

[1] The use of herbicides on plants for which they are not actually registered is strictly illegal and state authorities should be consulted before using them.

diversity. This property will do much to ameliorate insect and plant pathogen problems. Pest management consideration alone should dictate that such plantings are not monocultures. Species diversity will also serve as a hedge against meteorological vagaries. Where one species might be adversely

FIGURE 13. Anise Hyssop—a relatively easy native plant to grow that has a good reputation for being a highly productive bee forage. Frank Pellett, author of *American Honey Plants*, referred to *Agastache foeniculum* as his "wonder honey plant."

affected by drought, another may not and thus tide the bees over till more favorable conditions arrive. Diversity also will provide the variety that bees seem to want and therefore help keep the foragers close to the home apiary. This aspect may be extremely important in diversionary planting. Diverse plantings will also allow for multilayered systems with continuously overlapping blooming periods.

The final characteristic of a good bee forage system involves being a good neighbor. The species chosen should not become obnoxious weeds for someone else. Careful consideration should therefore be given to native plant species or at least to species that are already established in the area. Even these species might cause considerable consternation among neighbors if they were already considered obnoxious weeds. Purple loosestrife is a good example of such a plant.

SOME PLANTS TO CONSIDER

Bee Forages with Other Commercial Value

Planting bee forages with a second commercial value has obvious benefits. If the local conditions are satisfactory and the crop can be harvested and utilized, the beekeeper should certainly consider planting one of more of these alternatives. Often, however, the beekeeper is ill equipped to plant, care for, and harvest the crop. Sometimes local conditions also rule out these endeavors. Alfalfa seed, for example, is produced profitably in only the dryer regions of the country where the potential for diseases and seed germination in the pod is low. Sometimes, even where the other conditions are met, no means are available for utilizing the crop. For example, oil seed rape could be grown in many parts of the United States, but the unavailability of processing facilities has slowed its spread.

Where all the conditions can be met, seed crops such as canola, large animal forage such as alfalfa and various clovers, sunflower, and some of the beans should be considered. Multi-cut leguminous large animal forages are also good candidates. These crops have the advantage of producing nectar and pollen over large parts of the summer if they are managed correctly. Often, local farmers are willing to purchase fields of alfalfa and clover as standing crops for their own animal production. They may be reluctant, however, to wait until late in the blooming period before harvesting. The beekeeper should assess local demand before making such plantings with the intention of selling the crop. The planting of a woodlot that could be sold for firewood, fence posts, or timber is often overlooked.

Planting Exclusively for Bees

Multi-Cut and Multi-Cropped Legumes

Where they produce honey, the multi-cut legumes (alfalfa, sainfoin, trefoil, and some clovers) should be considered. These plants can be made to bloom more than once if cutting is timed correctly and thus can provide resources for the bees over large portions of the summer. If the cuttings of different parts of the planting are staggered, the nectar and pollen supply can be continuous.

Where multicutting is impractical, the beekeeper should consider planting mixtures of leguminous large animal forages in order to extend the period of bloom. Mixtures of *Trifolium* clovers, alfalfa, and sweet clovers are often recommended.

A great deal is known about cultivating these species and herbicides are available for their establishment. They, therefore, present less of a risk than many of the suggested plants that follow, even though they may not be as productive. While not permanent, these leguminous species will produce for many years. The beekeeper with land to plant might consider them initially along with test plots of more productive species. As the leguminous plots deteriorate, and as the beekeeper gains more experience with the more productive species, the plots of the latter could be expanded.

Other Recommendations
Anise Hyssop
    Several species of the perennial genus *Agastache* are called hyssops. They go by several names such as anise hyssop, giant hyssop, blue hyssop, etc. With the exception of *Agastache rugosa*, the members of this genus are native to North America. Many are very productive. Pellett, for example, called *A. foeniculum* his wonder honey plant. Many accessions of this genus bloom over an extended period of the summer and can be made to bloom again if cut. As small seeded species go, they are relatively easy to establish. Small quantities of seed can often be purchased from large flower seed companies. Since they often bloom the first year and since they produce prodigious amounts of seed, the beekeeper can make rapid seed increases from a single small pack of seed. The honey of at least *A. foeniculum* is reported to be of good quality, sometimes with a slight hint of mint.

Mountain Mint
    The genus *Pycnanthemum* is generally known as mountain mint, but also sometimes as basil, though it should not be confused with the culinary herb of that name. This is a North American genus with approximately 20 species. The ten species with which we have worked are all attractive to bees. Depending on the species, they bloom from mid July to frost (in Michigan). Pellett thought highly of *P. pilosum* and claimed that it produced high quality honey. Seeds in this genus can be purchased from large flower seed companies. While the plants produce fair amounts of seed, they do not usually bloom during the first year, and seed increases cannot be made as rapidly as for the

genus *Agastache*. These plants are also a little more difficult to establish as dense stands than is *Agastache*.

Globe Thistles

The genus *Echinops*, usually known as globe thistles, contains many species that are highly attractive to bees. *Echinops sphaerocephalus*, championed by Hiram Chapman for its attractiveness to bees, is sometimes known as the Chapman honey plant. None of the species are native to North America

FIGURE 14. Globe or *Echinops* thistle—these relatively large seeded species are very easy to grow and have a long and interesting history in the American beekeeping literature. They are, however, not native North American species and some might become weeds.

and in our experience some, though not all, have the potential for becoming serious weeds. Beekeepers who plant this genus should be careful that an accession has been chosen that will not become a serious weed. The seeds of these plants are large, some as large as barley, and are easy to establish with conventional planting equipment. Seed of this genus usually advertised as *E. ritro*, or blue globe thistle, can be purchased from large flower seed companies. For some reason, some accessions produce very little or no seed. While these accessions are not likely to become weeds, increasing the seed is difficult or even impossible. We have found many of our accessions to be highly productive honey plants and have never noticed a stupefying effect on the bees as reported by C.C. Miller (1918). Honey quality is unknown to us.

Oregano

The perennial *Origanum vulgare* (known as oregano or marjoram) is much liked by bees. It blooms during mid and late summer and is said to produce honey of fine quality. The floral color ranges from white to dark pink. While it spreads slowly, it is quite resistant to invasion by other plants. This feature, plus its small stature (usually under 2½ ft tall), and preference for limed soils make it a good candidate for road-side plantings. The seed is available commercially from many seed companies.

Buttonbush

Buttonbush, *Cephalanthus occidentalis*, is a native shrub of general distribution in the wet areas of Eastern North America west to Texas. Smaller pockets of this species also occur naturally in several western states, as well. The honey is said to be light in color and of good flavor. Those beekeepers with large, accessible wet areas that they would like to make more productive should give this species careful consideration. Those who dislike purple loosestrife (*Lythrum salicaria*) because it is a foreign species now invading North American wetlands should not object to the propagation and spread of this species by beekeepers. The species is easy to propagate from seed or from cuttings. Seed can be collected from the wild or obtained from botanic gardens and arboreta.

Chinese Tallow or Vegetable-Tallow Tree

Chinese tallow tree or vegetable-tallow, *Sapium sebiferum*, is a rapid growing tropical tree highly revered as bee forage in some of the more southern states of the U.S. (Hayes, 1979). The honey is said to be amber with a mild but excellent flavor. Seed is commercially available. Trees begin blooming in 3 to 4 years from seed.

Black Locust

Black locust, *Robinia pseudoacacia*, is native to the Appalachians, but has been naturalized over extensive parts of North America. Though it usually blooms over a very short period of time, the species can be very productive. In northern climates, because of the relatively early blooming period, the honey is often not harvested, but is used for brood rearing. The honey flow in northern climates is also frequently adversely affected by late freezes and frosts. These may be reasons that this species is often considered an unreliable source of nectar. A great deal of genetic variability exists in this species, however, and accessions are available that bloom one or two months later than the average. Other varieties exist that bloom over extended periods of time. Bees often seem to prefer the softer flowered accessions that allow easy access to the nectaries. The species will grow in nearly any reasonably drained soil. The wood is dimensionally very stable, quite resistant to decay, and makes excellent fence posts. As a source of firewood, it is one of the more productive species that can be grown in northern climates. The wood is dense, has a heat content rivaling hickory and apple, and it burns well. One of the plant's major drawbacks is its susceptibility to the locust borer (*Megacyllene robiniae*). Seed and numerous cultivars of the species are available commercially. Accessions exist that bloom profusely by their third year from seed.

Vitex

Plants of the genus *Vitex*, especially *V. negundo* var. heterophylla (sometimes called incisa), have a very good reputation in the southern U.S. as a bee plant (Croley, 1966; Pellett, 1967; Pellett, 1979). *V. negundo*, growing to a height of 20 ft., is an unusual woody plant, in that it blooms both early in life, sometimes the first year from seed, and over a long period of the summer. The honey seems to be of good flavor, though there is some disagreement in the literature about its color. While the tree can be grown in the Northern States, it generally dies back to the roots during the winter and flowers relatively late in the season. The plant is said to be quite drought resistant once established. It can be propagated from seed or by cuttings. Seed is available commercially.

Basswood or Linden

Where sufficient quantities of American basswood (*Tilia americana*) still exist, the species provides beekeepers with large quantities of white honey with a distinctive flavor that is considered by some to be strong, but by others to be mild. In general, the American basswood is not a dependable honey source each year, but accessions exist that are much more dependable than the average. Unfortunately, it blooms for only a short period of time. It is possible, however, to develop a collection of basswoods, drawn from the 30 or so species around the world, that would bloom during much of the summer. Just as the American basswoods have a good reputation in North America of being highly productive bee forage, many of the species that would be used to make up this collection also have good reputations in their areas of origin. In our experience, some of these species bloom profusely every year. The little leafed linden (*T. cordata*) is being planted sufficiently in North American cities so that urban beekeepers sometimes produce basswood honey. Some of the species, particularly those with thick seed coats, are notoriously difficult to germinate and may require several sequences of warm followed by cold stratification to break dormancy. Seed of several species is available commercially as are several cultivars of *T. americana* and *T. cordata*. Woodlots planted to species of *Tilia* would accrue in value since there is some demand for the lumber.

FIGURE 15. *Vitex*—this shrub of warmer climates has a good reputation in the southern United States as a honey producer. For a woody plant suited for the temperate regions of the world, it blooms over an unusually long period of time. In the northern parts of the United States, it dies back to the ground each winter. Although flowering each year from new regrowth, it blooms too late in northern states to be of much use in honey production.

CONTRIBUTORS

The following individuals took the time to freely contribute information about bee forage in geographical areas with which they were familiar. Without their contributions, this writing would not have been possible. (Arranged alphabetically by last name.)

John T. Ambrose, Professor of Entomology and Extension Apiculturist, North Carolina State University.

Dennis Arp, Mountain Top Honey Co.; Flagstaff, AZ.

Delvin Ashurst, American Honey Co.; Westmoreland, CA.

Jim Bigelow, Director, Technical Services, Wyoming State Dept. of Agriculture.

M.J. and Dorothy Brister Bragg, Brister Honey Co.; Caddo, OK.

Michael D. Burgett, Professor, Dept. of Entomology, Oregon State University.

Alfred R. Carl, Chief Apiary Inspector, Mass. Dept. of Food & Agriculture.

Matthew E. Cochran, State Bee Specialist, West Virginia Dept. of Agriculture.

Jerry Cole, New Mexico Beekeepers' Association; Bosque Farms, NM.

Clarence Collison, Extension Entomologist, Dept. of Entomology, Pennsylvania State University.

Douglas A. Colter, Chief Apiculture Inspector, Alberta Agriculture.

Lawrence J. Connor, Beekeeping Education Service and Wicwas Press; Cheshire, CT.

John W. Cramer, Plant Industry Supervisor, State of Alaska Dept. of Natural Resources, Division of Agriculture.

Beth Schmidt Crowder, Sparrow Hawk Farm; Bosque, NM.

Lorne Crozier, Extension Entomologist, Nova Scotia Dept. of Agriculture and Marketing.

Carl Dennis, Apiculturist (retired), Alabama Cooperative Extension Service, Auburn University.

Alfred Dietz, Professor, Dept. of Entomology, University of Georgia.

Don Dixon, Provincial Apiarist, Manitoba Agriculture.

Vern D. Dover, Prairie Gold Honey Co.; Hebron, ND.

Richard Drutchas, State Apiarist, Vermont Dept. of Agriculture.

William G. Eaton, Bees and Honey Specialist (retired), Kentucky Dept. of Agriculture.

Marion D. Ellis, State Apiarist, Nebraska Dept. of Agriculture.

Norman E. Farmer, Connecticut Beekeepers' Association; Bristol, CT.

C. E. Feeney, Ozark Beekeepers' Association; Springfield, MO.

Jerry Fischer, Apiary Inspector, Maryland Dept. of Agriculture.

Adam C. Fuller, Eastern Conn. Beekeepers' Association; North Windham, CT.

Harry R. Fulton, State Apiarist, Mississippi Dept. of Agriculture and Commerce, Division of Plant Industry.

Basil Furgala, Professor of Apiculture, University of Minnesota.

L. R. Goltz, Editor (retired), Gleanings in Bee Culture.

James E. Grayson, Grayson's Three Bees; Owasso, OK.

Pete Gregoire, North Dakota Beekeepers' Association; Horace, ND.

John Gruszka, Provincial Apiarist, Saskatchewan Agriculture.

John Haefeli, Haefeli's Honey Farms; Monte Vista, CO and Presidio, TX.

Truman C. Hardin, Ozark Beekeepers' Association; Springfield, MO.

Thomas C. Hart, State Apiarist, Tennessee Dept. of Agriculture/Apiary Section.

Milton R. Henderson, Bee Inspector, Mississippi Dept. of Agriculture and Commerce, Division of Plant Industry.

Richard C. Hicks, Chief of Vector Control, Clark County Public Works, Las Vegas, NV.

Michael Holmes, Michigan State Beekeepers' Association; Iron Mountain, MI.

Roger A. Hoopingarner, Professor, Dept. of Entomology, Michigan State University.

Frank J. Howard, Beekeeping Specialist (retired), Clemson University

Charles Howe, Jr., Connecticut Beekeepers' Association; Goshen, CT.

K. E. Hyland, Professor of Zoology and Rhode Island State Apiary Inspector, University of Rhode Island.

Anthony M. Jadczak, Apiculturist, Maine Dept. of Agriculture.

Carl A. Johansen, Professor Emeritus, Washington State University.

Guy W. Karr, Plant Pest Administrator, Alabama Dept. of Agriculture and Industries.

Art W. Kehl, Wisconsin Branch Manager (retired), Dadant and Sons.

Walter R. Kenney, Landscape Designer, Connecticut Dept. of Transportation.

Eugene Killion, Apicultural Extension Specialist, University of Illinois.

Willard Kissinger, State Apiarist, Montana Dept. of Agriculture.

M. L. Lashbrook, Union Valley Apiaries; Stonewall, OK.

Paul and Nanci Limbach, Western Colorado Honey; Silt, CO.

John A. Mathewson, Associate Professor of Zoology Emeritus, University of Rhode Island.

J. C. Matthenius, State Apiarist, New Jersey Dept. of Agriculture, Division of Plant Industry.

Daniel F. Mayer, Entomologist, Washington State University.

Byrd McCalman, American Honey Producers' Association; Bradley, AR.

Douglas M. McCutcheon, Supervisor Apiculture Program, British Columbia Ministry of Agriculture and Fisheries.

Stephen McDaniel, Central Maryland Beekeepers' Association; Manchester, MD.

Curtis Meier, Texas Branch Manager, Dadant & Sons.

Roger A. Morse, Professor of Apiculture, Cornell University.

Raymond Nabors, Delta Center Entomologist, University of Missouri/Delta Center.

Henry C. Neuhauser, Connecticut Beekeepers' Association; Farmington, CT.

William P. Nye, Research Apiculturist (retired), Bee Biology/Systematics Laboratory, U.S.D.A./Utah State University.

Bruce D. Palmer, Chief Apiary Inspector, New Brunswick Dept. of Agriculture.

Dale K. Pollet, Apiculture Specialist, Louisiana State University.

Homer P. Powers Jr., Virginia State Apiarist (retired), Virginia State Dept. of Agriculture.

Chris Prouse, Horticulture Specialist, Prince Edward Island Dept. of Agriculture.

Mike Roling, Ozark Beekeepers' Association; Springfield, MO.

Gary R. Ross, State Apiarist, Kansas State Board of Agriculture, Plant Protection Section.

Jim M. Ross, Ross Honey Co.; Minco, OK.

Joseph F. Rossman, Rossman Apiaries, Inc.; Moultrie, GA.

Malcom T. Sanford, Extension Apiculturist, University of Florida.

Rudolph A. Scheibner, Professor, Dept. of Entomology, University of Kentucky.

Charles Shanks, North East Arkansas Beekeepers' Association; Tyronza, AR.

Rev. Mother Agnes Shaw O.S.B., Abbey of Regina Laudis; Bethlehem, CT.

R. W. Shuel, Professor Emeritus, Dept. of Environmental Biology, University of Guelph (Ontario).

Robert G. Simpson, Professor Emeritus, Dept. of Entomology, Colorado State University.

Thomas E. Smigel, Deputy Director—Plant Industry, Nevada State Dept. of Agriculture.

I. Barton Smith, Maryland State Apiarist, Maryland Dept. of Agriculture.

Edward E. Southwick, Professor, Dept. of Biological Sciences, State University of NY/Brockport.

B. A. Stringer, Research Assistant, Dept. of Entomology, Oregon State University.

Sonny Swords, Swords Apiaries; Moultrie, GA.

Siegfried E. Thewke, State Entomologist, New Hampshire Dept. of Agriculture.

L. Dooley Toyne, Colorado Beekeepers' Association; Sedgwick, CO.

Ralph T. Vale, Master Gardner, University of Rhode Island.

Jean-Louis Villeneuve, Chief of Apiculture (retired), Quebec Ministry of Agriculture and Fisheries.

E. L. Walker, Walker Honey Co.; Mission, TX.

Gorden D. Waller, Research Entomologist, Carl Hayden Bee Research Center, U.S.D.A.

Robert J. Walstrom, Professor Emeritus, Dept. of Entomology, South Dakota State University.

Douglas R. Warner, Market Development Specialist, State of Alaska Dept. of Natural Resources, Division of Agriculture.

Edward A. Weiss, Western Connecticut Beekeepers' Association; Wilton, CT.

P. N. Williams, Georgia Beekeepers' Association, Forest Park, GA.

Charles Wills, Ozark Beekeepers' Association; Springfield, MO.

The authors also wish to express thanks to Paul Love and John Beaman. Paul spent many hours pouring over Table 1, greatly reducing the number of mistakes that found their way into it. John was instrumental in providing solutions to taxonomic quandaries that occurred from time to time in the preparation of Table 1. Their help has greatly improved that part of this work.

REFERENCES

Abrams, G.J. (1957). Beekeeping in Maryland. *Glean. Bee Cult.* 85:34-36.

**Alex, J.F. and C.M. Switzer. (1970). Ontario weeds: descriptions, illustrations and keys to their identification. Publication 505, Ontario Ministry of Agriculture and Food. Toronto, Ontario.

Anonymous. (1975). Beekeeping in North Carolina. *Glean. Bee Cult.* 103:236.

Arnold, L.E. (1954). Some honey plants of Florida. University of Florida Agricultural Experiment Stations Bulletin No. 548. Gainesville, FL.

**Arnow, L., B. Albee and A. Wyckoff. (1980). Flora of the central Wasatch Front, Utah. Second ed., revised. University of Utah Printing Service, Salt Lake City, UT.

Bacon, M.R. (1955). Beekeeping in Massachusetts. *Glean. Bee Cult.* 83:838-529, 573.

Barrett, D.P. (1955). Beekeeping in Michigan. *Glean. Bee Cult.* 83:20-21.

**Beatley, J.C. (1976). Vascular plants of the Nevada test site and central-southern Nevada: ecologic and geographic distributions. Technical Information Center, Office of Technical Information, Energy Research and Development Administration, Oak Ridge, TN.

Black, A.B. (1951). Beekeeping in Oregon. *Glean. Bee Cult.* 79:146-150.

Briggs, D. (1975). Beekeeping in California. *Glean. Bee Cult.* 103:259, 273.

**Brown, C.A. (1972). Wildflowers of Louisiana and adjoining states. Louisiana State University Press, Baton Rouge, LA.

**Brown, M.L. and R.G. Brown. (1984). Herbaceous plants of Maryland. Book Center, University of Maryland, College Park, MD.

**Burbridge, J. (1989). Wild flowers of the southern interior of British Columbia and adjacent parts of Washington, Idaho and Montana. University of British Columbia Press, Vancouver, BC.

Burgett, D.M., B.A. Stringer and L.D. Johnston. (1989). Nectar and pollen plants of Oregon and the Pacific Northwest. Honeystone Press, Blodgett, OR.

Burgin, C.J. (1955). Beekeeping in Texas. *Glean. Bee Cult.* 83:596-597.

Burgin, C.J. (1974). Beekeeping in Texas. *Glean. Bee Cult.* 102:305, 323.

Cale, G.H. (1956). Beekeeping in Illinois. *Glean. Bee Cult.* 84:396-397, 445.

Caron, D.M. (1975). Beekeeping in Maryland and Delaware. *Glean. Bee Cult.* 103:300, 307.

Caron, D.M. (1985). Beekeeping in Delaware. *Glean. Bee Cult.* 113:634-635.

**Carrell, D.S. and M.C. Johnston. (1970). Manual of the vascular plants of Texas. Texas Research Foundation, Renner, TX.

Clarke, W.W., Jr. (1955). Beekeeping in Pennsylvania. *Glean. Bee Cult.* 83:658-659.

Clarke, W.W. (1974). Beekeeping in Pennsylvania. *Gleanings in Bee Culture* 102:380.

**Clewell, A.F. (1985). Guide to the vascular plants of the Florida panhandle. University Presses of Florida, Gainesville, FL.

**Cormack, R.G.H. (1977). Wild flowers of Alberta. Hurtig Publishers, Edmonton, Alberta.

Couture, J.M. (1959). Beekeeping in Quebec. *Glean. Bee Cult.* 87:463-467.

Cox, O.S. (1953). Beekeeping in Idaho. *Glean. Bee Cult.* 81:340-341.

Crane, E., P. Walker and R. Day. (1984). Directory of important world honey sources. International Bee Research Association, London.

Croley, V.A. (1966). Ton of Honey Per Acre with Vitex. *Am. Bee J.* 106:455.

**Cronquist. A., A.H. Holmgren, N.H. Holmgren, J.L. Reveal and P.K. Holmgren. (1972-1984). Intermountain flora. Columbia University Press, New York, NY. 3 vol to date.

Davis, J.H. (1951). Beekeeping in Arkansas. *Glean. Bee Cult.* 79:534-536, 572.

**Deam, C.C. (1970). Flora of Indiana. Lubrecht & Cramer, Forestburgh, NY.

Dennis, C. (1983). Nectar and pollen producing plants of Alabama. A Guide for beekeepers. Auburn University Circular No. ANR-351. Alabama Cooperative Extension Service, Auburn University, Auburn, AL.

Dietz, A. and R.M. Burrill. (1976). Beekeeping Regions of Georgia. *Glean. Bee Cult.* 104:13, 32.

---

* & ** Indicate purely botanical works that were used in determining plant distribution, blooming period and sometimes taxonomic status.

** Indicates works that were useful for determining both distribution and blooming periods.

* Indicates works that gave no blooming date data.

Dirks, C.O. (1956). Beekeeping in Maine. *Glean. Bee Cult.* 84:78-79.

Edmunds, J.W. (1957). Beekeeping in the Peace River Country. *Glean. Bee Cult.* 85:274-276.

**Elliott, S. (1971). A sketch of the botany of South Carolina and Georgia. Hafner Publishing Co. Inc., New York, NY. 2 vol.

Erickson, E.H. (1984). Soybean pollination and honey production—a research progress report. *Glean. Bee Cult.* 112:575-579.

Feller-Demalsy, M.-J. and Y. Lamontangne. (1979). Analyse pollinique des miels du Quebec. *Apidologie* 10:313-340.

Feller-Demalsy, M.-J. (1983). Le spectre pollinique des miels du Quebec. *Apidologie* 14:147-174.

Ferguson, S.B. (1975). Beekeeping in Wisconsin. *Glean. Bee Cult.* 103:40.

**Fernald, M.L. (1970). Gray's Manual of botany. Eight ed. D. Van Nostrand Co., New York, NY.

Fowell's, H.A. (1965). Silvics of Forest Trees of the United States. Agriculture Handbook No. 271. Forest Service, United States Department of Agriculture. Washington, DC.

**Gleason, H.A. (1952). The new Britton and Brown illustrated flora of the northeastern United States and adjacent Canada. New York Botanical Garden, New York, NY. 3 vol.

Gleason, H.A. and A. Cronquist. (1964). The natural geography of plants. Columbia University Press, New York, NY.

Goltz, L. (1975). Beekeeping in Ohio. *Glean. Bee Cult.* 103:22, 25.

Graves, J. (1976). Amateur Beekeeping in North Central Texas. *Glean. Bee Cult.* 104:410-411.

**Great Plains Flora Association. (1986). Flora of the Great Plains. University Press of Kansas, Lawrence, KS.

Griffith, J. (1961). Beekeeping in New Mexico. *Glean. Bee Cult.* 89:336-339.

Haydak, M.H. and C.D. Floyd. (1955). Beekeeping in Minnesota. *Glean. Bee Cult.* 832:464-465, 508.

Hayes, B. (1979). The Chinese tallow tree—artificial bee pasturage success story. *Am. Bee J.* 119:848-849.

Hepler, J.R. (1956). Beekeeping in New Hampshire. *Glean. Bee Cult.* 84:280-281.

Hewitt, P.J., Jr. (1955). Beekeeping in Connecticut. *Glean. Bee Cult.* 83:75, 122.

Hilbig, F. (1958). Beekeeping in Nevada. *Glean. Bee Cult.* 86:24, 60.

Hines, T. (1962). Beekeeping in Kentucky. *Glean. Bee Cult.* 90:338-340, 380.

**Hitchcock, A.S. and P.C. Standley. (1919). Flora of the District of Columbia and vicinity. Contributions from the United States National Herbarium, Vol. 21. Washington, DC.

**Hitchcock, C.L., A. Cronquist, M. Ownbey and J.W. Thompson. (1955-1969). Vascular Plants of the Pacific Northwest. University of Washington Press, Seattle, WA. 5 vol.

Holcombe, P. (1955). Beekeeping in New Jersey. *Glean. Bee Cult.* 83:341, 381.

Holmes, F.O. (1976). Beekeeping in New Hampshire. *Glean. Bee Cult.* 104:257, 272.

Holzberlein, J.W. (1957). Beekeeping in Colorado. *Glean. Bee Cult.* 85:408-409.

Howes, F.N. (1979). Plants and Beekeeping. Revised ed. Farber and Farber Ltd., London.

Hunt, C.B. (1974). Natural Regions of the United States and Canada. W.H. Freeman and Co., San Francisco, CA.

Jaycox, E.R. (1954). Beekeeping Regions of California. *Glean. Bee Cult.* 82:18-19.

Johansen, C. (1974). Beekeeping in Washington. *Glean. Bee Cult.* 102:245, 256.

**Jones, G.N., and G.D. Fuller. (1955). Vascular Plants of Illinois. University of Illinois Press, Urbana, IL.

**Kearney, T.H. and R.H. Peebles. (1960). Arizona Flora. University of California Press, Berkeley and Los Angeles, CA.

Killion, E.E. (1974). Beekeeping in Illinois. *Glean. Bee Cult. 102*:288, 290.

Kissinger, W.A. (1975). Beekeeping in Montana. *Glean. Bee Cult. 103*:333, 339.

**Kuijt, J. (1982). A flora of Waterton Lakes National Park. University of Alberta Press, Edmonton, Alberta.

**Lakela, O. (1965). A flora of northeastern Minnesota. University of Minnesota Press, Minneapolis, MN.

Latham, G.H. (1956). Beekeeping in Delaware. *Glean. Bee Cult. 84*:532-533.

LeMaistre, W.G. (1952). Beekeeping in Alberta, Canada. *Glean. Bee Cult. 80*:336-341.

Lesher, C. (1954). Beekeeping Regions of North Dakota. *Glean. Bee Cult. 82*:408-409.

Levin, M.D. (1958). Beekeeping in the Beehive State. *Glean. Bee Cult. 86*:528-531.

Liberty Hyde Bailey Hortorium. (1976). Hortus Third. McMillan Publishing Co. Inc., New York, NY.

Little, E.L., Jr. (1971). Atlas of United States Trees. Volume 1. Conifers and Important Hardwoods. Miscellaneous Publication No. 1146, U.S. Department of Agriculture. Washington, DC.

Little, L.H. (1954). Tennessee honey producing areas. *Glean. Bee Cult. 82*:535.

Livingston, H. (1968). Beekeeping—Fairbanks 1967-68. *Glean. Bee Cult. 96*:397, 439.

Livingston, H. (1981). Beekeeping in Alaska. *Western Apicultural Society Journal 4*:108-111.

**Lommasson, R.C. (1973). Nebraska wild flowers. University of Nebraska Press, Lincoln, NE.

**Long, R.W. and O. Lakela. (1971). A flora of tropical Florida. University of Miami Press, Coral Gables, FL.

Lord, W.G. (1979). Beekeeping in the North Carolina Coastal Plain. *Am. Bee J. 119*:112-113.

Lovell, H.B. (1955). Beekeeping in Kentucky. *Glean. Bee Cult. 83*:281-282.

Lovell, J.H. (1926). Honey plants of North America. A.I. Root Co., Medina, OH.

MacLean, D. (1961). Beekeeping on Prince Edward Island. *Glean. Bee Cult. 89*:662-663, 701.

Maher, J. (1975). Beekeeping in Missouri. *Glean. Bee Cult. 103*:393, 407.

**Martin, W.C. and C.R. Hutchins. (1980 and 1981). A flora of New Mexico. Lubrecht & Cramer, Forestburgh, NY. 2 vol.

McCutcheon, D.M. (1958). Beekeeping in Saskatchewan. *Glean. Bee Cult. 86*:329-331, 380.

McGinnies, William G. Flowering Periods for Common Desert Plants. University of Arizona Press. (a chart).

**McKay, S.M. and P.M. Catling. (1979). Trees, shrubs and flowers to know in Ontario. J.M. Dent and Sons, (Canada) Ltd., Don Mills, Ontario.

Mead, R.M. (1954). Beekeeping areas of Vermont. *Glean. Bee Cult. 82*:470-471.

Miller, C.C. (1918). Stray straws. *Glean. Bee Cult. 46*:727-728.

Miller, W. (1955). Beekeeping in South Dakota. *Glean. Bee Cult. 83*:144-145.

Milum, V.G. (1957). Illinois Honey and Pollen Plants. Contribution from the Department of Horticulture, University of Illinois, Urbana, IL.

Mitchener, A.V. (1948). Nectar and Pollen Producing Plants in Manitoba. *Scientific Agriculture (Ottawa)* 28:475-480.

**Mohlenbrock, R.H. (1986). Guide to the vascular flora of Illinois. Revised ed. Southern Illinois University Press, Carbondale, IL.

Morse, R.A. (1974). Beekeeping in the Empire State. *Glean. Bee Cult.* 102:211-212, 227.

Morton, J.F. (1964). Honeybee plants of South Florida. *Florida State Horticultural Society Proceedings* 77:415-436.

**Munz, P.A. (1965). A California Flora. University of California Press, Berkeley and Los Angeles, CA.

Noetzel, D.M. (1974). Beekeeping in Minnesota. *Glean. Bee Cult.* 102:350, 355.

Nye, W.P. (1971). Nectar and Pollen Plants of Utah. *Monograph Series, Utah State University Press* 18:(3). Logan, UT.

Nye, W.P. (1976). Beekeeping in Utah. *Glean. Bee Cult.* 104:133-134.

Oertel, E. (1939). Honey and pollen plants of the United States. United States Department of Agriculture Circular No. 554. Washington, D.C.

Oertel, E. (1957). Beekeeping in Louisiana. *Glean. Bee Cult.* 85:92-95.

Oertel, E. (1975). Beekeeping in Louisiana. *Glean. Bee Cult.* 103:356-357.

Oertel, E. (1980). Nectar and pollen plants. pp. 16-23. In Beekeeping in the United States. United States Department of Agriculture Handbook No. 335. Washington, D.C.

Paddock, F.B. (1956). Beekeeping in Iowa. *Glean. Bee Cult.* 84:474-477.

Parker, R.L. (1957). Beekeeping in Kansas. *Glean. Bee Cult.* 85:152-154.

Pellett, F.C. (1920). American honey plants, together with those which are of special value to the beekeeper as sources of pollen. American Bee Journal, Hamilton, IL.

Pellett, F.C. (1923). American honey plants, together with those which are of special value to the beekeeper as sources of pollen. Second ed. American Bee Journal, Hamilton, IL.

Pellett, F.C. (1930). American honey plants, together with those which are of special value to the beekeeper as sources of pollen. Third ed. American Bee Journal, Hamilton, IL.

Pellett, F.C. (1947). American honey plants, together with those which are of special value to the beekeeper as sources of pollen. Fourth ed. Orange Judd Publishing Co., New York, NY.

Pellett, F.C. (1947). Mountain Mint. *Am. Bee J.* 87:172-173.

Pellett, F.C. (1976). American Honey Plants, together with those which are of special value to the beekeeper as sources of pollen. Fifth ed. Dadant & Sons. Hamilton, IL.

Pellett, M. (1967). Vitex negando incisa. *Am. Bee J.* 107:177.

Pellett, M. (1979). Vitex blooms for months. *Am. Bee J.* 119:754-755.

Perigo, G. (1956). Beekeeping in Indiana. *Glean. Bee Cult.* 84:722-723.

Pike, H.A. (1956). Beekeeping in Rhode Island. *Glean. Bee Cult.* 84:360.

Post, R.L. and H.R. Cree (1976). Beekeeping in North Dakota. *Glean. Bee Cult.* 104:214, 235.

Purser, W.H. (1954). Beekeeping in South Carolina. *Glean. Bee Cult.* 82:585-587.

**Radford, A.E., H.E. Ahles and C.R. Bell. (1968). Manual of the vascular flora of the Carolinas. University of North Carolina Press, Chapel Hill, NC.

Rahmlow, H.J. (1962). Beekeeping in the Sunshine State. *Glean. Bee Cult.* 90:486-489, 500.

Ramsay, J. (1987). Plants for Beekeeping in Canada and the northern USA: a directory of nectar and pollen sources found in Canada and the northern USA. International Bee Research Assoc., London.

**Rickett, H.W. (1966-1973). Wild flowers of the United States. McGraw-Hill Book Co., New York, NY. 6 Vol. in 14 parts. Plus index.

Robinson, F.A. and E. Oertel. (1975). Sources of nectar and pollen, pp. 283-302. *In* Dadant and Sons (*ed.*), The Hive and the Honey Bee. Dadant and Sons. Hamilton, IL.

**Roland, A.E. Flora of Nova Scotia. Truto Printing and Publishing Co. Ltd., Truro, Nova Scotia.

**Rydbergh, P.A. (1922). Flora of the Rocky Mountains and adjacent Plains. The author, New York, NY.

**Rydbergh, P.A. (1932). Flora of the Prairies and Plains of Central North America. New York Botanical Garden, New York, NY.

Sagunsky, W.G. (1957). Beekeeping in Montana. *Glean. Bee Cult. 85*:280-281.

Sanborn, C.E. and E.E. Scholl. (1908). Texas honey plants. Texas Agriculture Experiment Station Bul. No. 102. College Station, TX.

Sanford, M.T. Florida Bee Botany. Florida Cooperative Extension Service, University of Florida, Institute of Food and Agricultural Sciences Circular 686. Gainesville, FL.

Scullen, H.A. and G.A. Vansell. (1942). Nectar and Pollen Plants of Oregon. Oregon State Agricultural Experiment Station Bulletin No. 412. Corvallis, OR.

Seeley, T.D. (1986). Social foraging by honey bees: how colonies allocate foragers among patches of flowers. *Behav. Ecol. Sociobiol. 19*:343-354.

Seeley, T.D. and R.A. Levien. (1987). Social foraging by honey bees: how a colony tracks rich sources of nectar, pp. 38-53. *In* Menzel, R. and A. Mercer (*eds.*), Neurobiology and behavior of honeybees. Springer-Verlag, New York, NY.

**Seymour, F.C. (1982). The flora of New England. Second ed. Harold N. Moldenke and Alma L. Moldenke, Plainfield, NJ.

**Shaw, R.J. (1989). Vascular plants of northern Utah. Utah State University Press, Logan, UT.

**Small, J.K. (1913). Flora of the Southeastern United States. Second ed. The author, New York, NY.

Stanger, W. and R.W. Thorp. (1967). Honeybee Pollination in California. *Glean. Bee Cult. 95*:457-461, 507.

Stanley, G.L. (1975). Beekeeping in Iowa. *Glean. Bee Cult. 103*:117.

**Stemen, T.R. and W.S. Myers. (1937). Oklahoma flora. Harlow Publishing Corp., Oklahoma City, OK.

Stephen, W.A. (1957). Beekeeping in North Carolina. *Glean. Bee Cult. 85*:338-340.

Stephen, W.A. (1956). Beekeeping in North Carolina. *Glean. Bee Cult. 84*:668-670.

**Stevens, O.A. (1963). Handbook of North Dakota plants. North Dakota Institute for Regional Studies, Fargo, ND.

**Stevens, W.C. (1948). Kansas wild flowers. University of Kansas Press, Lawrence, KS.

**Steyermark, J.A. (1963). Flora of Missouri. Iowa State University Press, Ames, Iowa.

**Strausbaugh, P.D. and E.L. Cove. (1978). Flora of West Virginia. Second ed. Seneca Books, Inc., Grantsville, WV.

Stricker, M. (1956). Beekeeping in Rhode Island. *Glean. Bee Cult. 84*:360.

Stricker, M.H. (1976). Beekeeping in New Jersey. *Glean. Bee Cult. 104*:292-293.

Tate, H. (1956). Beekeeping in Mississippi. *Glean. Bee Cult. 84*:21, 60.

**Tatnall, R.R. (1946). Flora of Delaware and the Eastern Shore. Society of Natural History of Delaware, Wilmington, DE.

**Taylor, R.L. and B. MacBryde. (1977). The vascular plants of British Columbia: A descriptive resource inventory. University of British Columbia Press, Vancouver, British Columbia.

Telford, H.S. (1955). Beekeeping in Washington. *Glean. Bee Cult.* 83:395-396.

Turnbull, W.H. and B.C. Venon. (1951). Beekeeping in British Columbia. *Glean. Bee Cult.* 79:329-334.

**Van Bruggen, T. (1985). The vascular plants of South Dakota. Second ed. Iowa State University Press, Ames, IA.

Vance, F.R., J.R. Jowsey and J.S. McLean. (1984). Wildflowers of the northern Great Plains. University of Minnesota Press, Minneapolis, MN.

Vansell, G.H. (1931). Nectar and Pollen Plants of California. University of California Agricultural Experiment Station Bulletin 517. Berkeley, CA.

Vansell, G.H. (1949). Pollen and Nectar Plants of Utah. Utah Agricultural Experiment Station Circular 124. Logan, UT.

Viereck, L.A. and E.L. Little. (1972). Alaska trees and shrubs. Agriculture handbook No. 410. Forest Service, United States Department of Agriculture, Washington, D.C.

Wade, C. (1975). Beekeeping in Indiana. *Glean. Bee Cult.* 103:92.

Walstrom, C.J. (1954). Beekeeping Regions of Nebraska. *Glean. Bee Cult.* 82:88-89.

Walstrom, C.J. (1975). Beekeeping in Nebraska (Part II). *Am. Bee J.* 115:305, 313.

Washburn, R.H. (1961). Beekeeping in the Land of the Midnight Sun. *Glean. Bee Cult.* 89:720-723, 756.

Washburn, R.H. (1974). Beekeeping in Alaska. *Agroborealis* 6:23-24.

Weatherford, H.W. (1957). Beekeeping in Virginia. *Glean. Bee Cult.* 85:220-221.

Weaver, W.F. (1962). Beekeeping in West Virginia. *Glean. Bee Cult.* 90:592-593, 635.

*Welsh, S.L. (1974). Anderson's flora of Alaska and adjacent parts of Canada. Brigham Young University Press, Provo, UT.

**Wiegand, K.M. and A.J. Eames. (1925). The flora of the Cayuga Lake Basin, New York vascular plants. Cornell University Agriculture Experiment Station memoir No. 92. Ithaca, NY.

Wilkins, H.A. (1974). Bay State Beekeeping Thoughts. *Glean. Bee Cult.* 102:59.

Wilson, A. (1959). Beekeeping in New Brunswick. *Glean. Bee Cult.* 87:24-26.

Wilson, W.T., J.O. Moffett and H.D. Harrington. (1958). Nectar and Pollen Plants of Colorado. Colorado State University Experiment Station Bulletin 503-S. Fort Collins, CO.

**Wofford, B.E. (1989). Guide to the vascular plants of the Blue Ridge. University of Georgia Press, Athens, GA.

**Wunderlin, R.P. (1982). Guide to the vascular plants of central Florida. University Presses of Florida, Gainesville, FL.

CHAPTER 12

BEEKEEPING EQUIPMENT

by CHARLES C. DADANT*

It is significant that beekeeping equipment and appliances remained basically the same from early recorded history until the mid 19th century. Equally surprising is the fact many portions of the world still keep bees and produce honey with the same primitive hives. Straw skeps, once the vogue throughout central Europe, are now primarily relegated to roadside stands and ornamental displays, while log gums and clay pots sealed on one end and with only a narrow opening on the other, remain in use throughout large portions of the world, Africa in particular. Most of these are suspended from the limbs of trees, away from predators, and although primitive and inefficient, substantial quantities of honey and beeswax are produced from sheer numbers. After rudimentary processing, the beeswax moves to collecting points where large cakes are cast after melting and cleaning, and ultimately sizable shipments move abroad as an international raw commodity. The vast proportion of the honey, as it has since time immemorial, serves as the major ingredient of the particular alcoholic beverage of the area. Modern equipment and methods are being introduced and as conditions permit, these older practices and tools will be modernized.

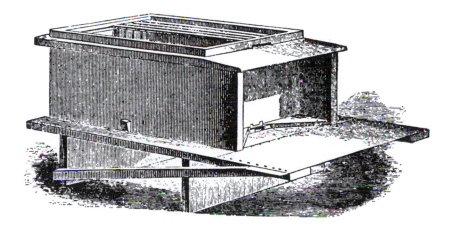

FIGURE 1. The original Langstroth hive, the first top-opening, movable frame hive that provided a bee space between frames and other hive parts, making possible the present extent of beekeeping.

---

*Charles C. Dadant, Chairman of Dadant & Sons, Inc., Hamilton, Illinois. Manufacturers of beekeeping supplies, Publisher of the *American Bee Journal* and this book, *The Hive and the Honey Bee*.

In 1851, a remarkably simple discovery permitted a revolution in hive construction and design, (Figure 1) and led to creation of an agricultural function in which the beekeeper guides the progress of the colony, not the bees. Formerly, the honey could be harvested only by excising portions of the comb, and the honey it contained, from within the hive, then heating the honey and the comb, or squeezing the honey out into a liquid condition and salvaging the wax for other uses. Only small colonies could be reared with these primitive hives and the honey they yielded was of equally modest means. The Rev. L. L. Langstroth, Philadelphia, first determined that if all interior portions of the hive were spaced 3/8 inch (9.525 mm) apart, bees would construct their comb in a uniform manner, without brace and bridge combs between, and permit the parts to be interchangeable and movable, and permit additional storage areas to be readily stacked on top and removed at will. The principle of his hive formed the basis for the original Langstroth book, *The Hive and the Honey Bee*, the predecessor of this revised edition, and remains the principle incorporated in all modern hives.

Two other major developments contributed substantially toward the modernization of honey production in the mid 1850's. About 1843, Gottlieb Kretchmer produced a comb base and Johannes Mehring made comb foundation on a flat press in 1857. In 1865, Franz von Hruschka invented the honey extractor. Both of these changes permitted the beekeeper to produce and sell liquid honey, while at the same time offering the bees a base upon which they could establish uniform combs, and to return these combs intact to the bees for refilling with liquid honey that same year and for many years in the future.

Beekeeping periodicals immediately began lengthy discussions of the pros and cons of the Langstroth principle, and soon gravitated into a full-blown controversy with respect to hive size.

Initially the public favored smaller units and such proponents as Heddon, Bingham, and Danzenbaker promoted the advantages of contracting the brood nest and restricting hive size. Notable among the defenders of the Langstroth concept of the large hive were Charles Dadant and his son, C. P. Dadant. They recorded lengthy experiments with hives containing from 8 to 20 frames and of varying frame size. Charles Dadant wrote prolifically for beekeeping periodicals published in his native France, and it was largely through his efforts that the Langstroth concept of bee space and the large hive became standard throughout most of Europe, where the Dadant hive is still known and widely used. Eventually this controversy subsided and the world has come to accept the principle of the large hive; there remains only some variation on the exact units, and more importantly, on the number of units that comprise a big hive to produce big crops.

THE MODERN HIVE

The Original Langstroth dimensions have survived the test of time. The accompanying Figure 2 illustrates the basic components required. One, two, or three hive bodies are generally devoted to brood rearing space and considered part of the colony the year around. More often than not, two hive bodies with interior dimensions of 9-19/32 inches (24.368 cm) in depth, 14-11/16 inches (37.306 cm) in width, and 18-5/16 inches (46.514 cm) in length are considered ideal for brood space. Exterior dimensions have undergone periodic changes as Western Pine Association regulations have permitted reductions in the thickness of an inch board until now it measures only 3/4 inch (19.05 mm). Official changes in the width of the board have also been sanctioned from time to time and it is anticipated that at some point the depth of the hive body and the frames inside may have to be adjusted to compensate. A recessed portion of 5/8 inch (15.875 mm), commonly referred to as a rabbet, is provided at the top of the interior portion of each end of the hive body as a resting place for the free-hanging frames inside.

The joints are normally dovetailed at each of the four corners for a sturdy moisture-proof fit, and nailed in both directions on all corners. A recessed portion about one-third of the distance from the top is made available on the outside of all exposed parts to provide a handle for lifting and transporting each unit. Ten frames spaced 1-3/8 inches (34.925 mm) center to center, with the top measuring 19 inches (48.26 cm) overall hang inside. Interior dimensions of the rectangle comprising the frame for the hive body measure 8 inches (20.32 cm) in depth by 17 inches (43.18 cm) in length. Normally the lower two-thirds of the end bars of the rectangle are scalloped to only 1-1/8 inch (28.575 mm) in width instead of the 1-3/8 inch (34.925 mm) on the upper one-third, to provide for more ready access by the bees from frame to frame.

Many beekeepers, especially those operating commercially, are sufficiently concerned with standardization and interchangeability, that both brood and supering areas, where the honey is stored, will consist of nothing but hive body depth. Popular variation on the supering dimensions include, in order of importance, the Dadant super of an overall depth of 6-5/8 inches (16.828 cm), 5-11/16 (14.446 cm), and the 7-5/8 (19.368 cm) supers. Frame depth, of course, is proportionately less for each. The shallow sizes have some popularity because of their lesser weight, in areas where honeyflows may be moderate to poor, and for the production of chunk comb honey. Brood areas consistently utilize 10 frames, although frequently supering frames are reduced to nine and sometimes eight per unit. Beekeepers theorize a thicker comb, more honey, and substantially more beeswax will be harvested with this approach.

FIGURE 2. Cut away view of the basic components of the modern hive.

COMPONENTS OF A MODERN HIVE

Think of a modern honey bee hive as a multi-storied factory with each section of the hive having a specific function. These sections work together in an orderly fashion to provide a highly efficient house for bees and "surplus" honey for the beekeeper.

A) **Hive Cover** - Shown here is a telescoping cover which "telescopes" well down over the sides of the top super and protects the hive from rain and wind. Galvanized covering protects the wood. Flat hive covers are also available.

B) **Inner Cover** - Creates a dead air space for insulation from heat and cold. Center hole may be fitted with a bee escape to aid in removing bees from honey supers.

C) **Shallow Supers** - For "surplus"honey production. These are the supers that the bees work hard to fill with honey for the beekeeper. 6-5/8 inch, 5-11/16 inch supers, or even hive bodies may be used for surplus honey production.

D) **Queen Excluder** - Openings in the unit allow the smaller worker bees to pass through easily, but prevent the larger queen from climbing higher in the colony and makes her lay her eggs below.

E) **Standard Hive Bodies** - These provide the living quarters for the bees. The queen lays eggs in these chambers and brood is raised here. Honey is also stored for the hive's consumption.

F) **Standard Bottom Board** - It forms the floor of the hive and is shown here with a wooden entrance reducer in place to keep mice and some cold winds out during winter. The board is reversible.

G) **The Hive Stand** - Supports the hive off the ground to keep hive bottom dry and provide an insulating factor. Built-in angled board provides an efficient landing area for bees.

Successful beekeeping means easy manipulation of the frames of brood and honey to provide a "surplus" of honey beyond that needed by the bees to live on and rear their replacements. It is this "surplus" that the beekeeper removes and markets.

The construction described offers complete control of the combs which are the heart of beekeeping. It is also sufficiently large to reduce crowding, minimize swarming, accommodate a prolific queen and her brood, and store two foods - pollen and honey.

Comb honey supers offer the opportunity to produce honey in the comb. The overall depth of the unit is 4-13/16 inches (12.224 cm) with the same 5/8 inch (15.875 mm) rabbet on the upper interior of each end. A flat tin 3/4 inch (19.05 mm) wide by approximately 16-1/8 inches (40.958 cm) long is nailed on the bottom of either end so that 1/2 inch (12.7 mm) extends across the inside portion of the end. A three-sided section holder 4-1/2 inches (11.43 cm) in depth by 18-1/8 inches (46.038 cm) overall length holds four square sections 4-1/4 x 4-1/4 x 1-7/8 inches (10.795 cm x 10.795 cm x 47.625 mm), each with delicate sheets of thin surplus foundation fitted completely in the interior. Seven of these, or a total of 28 sections, rest on the tins with 8 separators varying from 1/10 inch (2.54 mm) to 1/16 inch (1.588 mm) in thickness on the outside and between each four-section unit. Two cleats 25/32 inch (19.844 mm) by 4-1/2 inches (11.43 cm) and approximately 7/16 inch (11.113 mm) thick are inserted between the separator and the sidewall of one interior of the hive. Three super springs are forced into position vertically on the opposite side between the wall and that separator and two cleats 25/32 inch (19.844 mm) by approximately 14-5/8 inches (37.148 cm) by 9/32 inch (7.144 mm) thick are wedged into position in the rabbet on either end. The total combines to rigidly lock all components in place, and restrict the bees to only the section where honey is desired. A natural reluctance to work in such confined areas requires crowding the bees carefully, and, of course, can be successfully completed only when a strong honeyflow exists. It is not a procedure recommended for the beginner.

Section comb honey may also be produced through the use of some rather ingenius round plastic rings designed to fit in half frames, which in turn go into conventional wood comb honey supers. Upon completion by the bees, special covers are used to complete the package and prepare it for market. Acceptance by the bees is generally quite good, though it should be pointed out that as with the wood sections, optimum performance will result when colonies are strong, reasonably confined, and most importantly honey flows are good. This development could have only been created through the design characteristics permitted through the use of plastics, and is perhaps the most notable use of plastics to replace wood parts. A recent addition to comb honey equipment, developed by Dr. John Hogg, Ph.D., a beekeeper in Michigan, is a half comb concept marketed under the terminology of a cassette (Figure 3). This clever approach results in a deep half comb of honey in a marketable tray complete with lid. This approach may further revolutionize comb honey production and further enhance the popularity of this type of honey.

Conventional supers of 6-5/8 inches (16.828 cm) in depth or 5-11/16 inches (14.446 cm) in depth are generally preferred when chunk comb honey is desired. Normal frames are used and a cut-comb type foundation is inserted

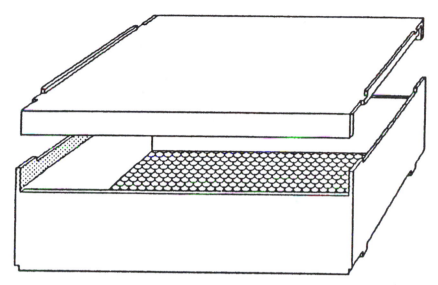

FIGURE 3. Line drawing of half-comb cassette, complete with form-fitting lid.

with the net result being a chunk of honey some 5 inches (12.7 cm) in depth and 17 inches (43.18 cm) in length which can be cut into squares and served on the table.

A hive stand or alighting board, frequently rests on the ground, elevating other parts above the grass and weeds, and making them easier to mow and keep clean. In addition, it reduces the dampness within the hive. These are cumbersome and difficult to move and generally are not in use when bees are operated on a commercial scale. Sometimes permanent locations are elevated on concrete blocks or cement stands; corrugated metal flat on the ground often suffices.

The bottom board, itself, on which the hive bodies rest, consists of a flat floor 15-3/16 inches (38.576 cm) in width by 22-1/16 inches (56.039 cm) in length mounted in side rails 1-5/16 inches (33.338 mm) in depth. These rails have a 1/4 inch (6.35 mm) groove recessed in either side 3/8 inch (9.525 mm) from one end of the rail, leaving approximately 7/8 inch (22.225 mm) on the other. The resultant platform measures 16-1/4 x 22-1/16 inches (41.275 cm x 56.039 cm) overall. Turning the unit over will thus provide an entrance for the bees of either 3/8 inch (9.525 mm) or 7/8 inch (22.225 mm) and an entrance reducer of 7/8 inch is normally used to permit modification of the size of the entrance at the front of the hive from season to season as more or less is required. Two cleats, 3/8 inch and 7/8 inch both by 14-11/16 inches (37.306 cm) enclose the rear of the bottom. Generally speaking, when the colony is small, winter and spring, the opening should be restricted in the front and enlarged as the population grows or during warm weather.

The top of the hive is normally fitted with an inner cover, most frequently a 1/8 inch (3.175 mm) sheet of tempered hardboard housed in a rim approximately 7/8 inch (22.225 mm) thick slightly off-centered to provide more space on one side than the other. An oblong hole is notched clear through the center of the hardboard to accept a bee escape and the unit is sometimes referred to as an escape board since that's the function it performs. In addition to offering slightly more access for the bees across the top of the frames at this extreme top portion of the hive, it may also be inserted with the bee escape in place between supers of capped honey that are to be removed. The escape permits the bees to move down the hive but not back up, and after a period of time the super will be free of bees. The procedure is slow and oftentimes other more rapid methods are used for removing the honey.

An outer cover fits on top of all these components that are stacked below. Sometimes they are of a flat wood type cleated front and back, and flush on either side, more often a 2-1/2 inch (63.5 mm) rim is built on all four sides so that the unit telescopes over the top of the hive about 1-1/2 inches (38.1 mm). Generally, a fitted galvanized or aluminum pan then fits over all of the top to provide strength and better weathering capabilities.

Painting the outside parts of the hive is recommended. A good grade of primer plus two additional coats of white paint is recommended on new wood prior to use. Some prefer a good grade of aluminum paint instead. Lighter colors of paint are highly recommended since the darker shades absorb the rays of the sun and further contribute to overheating and the demoralization of the colony when the weather warms. Interior portions of the hive and interior fixtures should not be painted. Ponderosa pine is generally accepted as the preferred species of lumber for hive and hive part construction. Spruce and Idaho white pine may be used, but are less desirable for interior parts. Original growth cypress is an excellent lumber to use but is now scarce. While second growth cypress is fairly common, it is generally considered no more durable than the pine parts that are easier to obtain and easier to machine and use. Southern pines generally shatter too much on cutting. Basswood is conceded to be far superior to any other material for the production of the small wood comb honey sections.

Testing continues on the possible replacement of the various hive components formed from one or more of the wide range of plastic materials currently available. Generally speaking, any exterior application must still be viewed in the experimental stage since the longevity of the materials acceptable price-wise have not yet been accurately determined. Sophisticated substitutes for wood, of course exist, but the price precludes consideration. Wood, except for its limitation in design, figures nearly perfect for the desired requirements of hive construction. Much depends, with plastics in particular, on the desired life span. It is generally assumed price should equate to the life expectancy of wood and ultimately these answers should be available.

The Modified Dadant size hive still remains in moderate use and the principle is still considered sound for apiaries in permanent locations and seldom if ever moved. This 11-frame principle has frames of the same interior length as the Langstroth, but 10-1/8 inches (25.718 cm) interior depth and are spaced 1-1/2 inches (38.1 mm) center to center. The hive body, itself, has an overall depth of 11-5/8 inches (29.528 cm), 18-5/16 inches in length (46.514 cm), 16-15/16 inches (43.02 cm) in width, all interior spaces.

The 8-frame Langstroth hive remains in limited use and generally for comb honey production only. This style parallels Langstroth except it accommodates two fewer frames and has an interior width of 12-3/16 inches (30.956 cm).

The jumbo hive has been nearly phased out, but a few remain in service. This unit has the 10-frame Langstroth width, the Modified Dadant depth, and only 10 frames per unit so that exterior dimensions fit the 10-frame hive.

Standardization yields many advantages, and very little remains untried with respect to hive sizes per se.

Generally speaking, exterior hive parts are assembled with 7d cement-coated nails. Most of the frame parts hold well when 1-1/4 inch by 17 gauge cement-coated nails are used. Many commercial operators now prefer to equip themselves with a compressor and air gun and assemble all wood parts with a nail type staple of those approximate dimensions.

Frame construction performs the basic function of holding a sheet of comb foundation in place while the bees draw out a finished comb, and yet it is a sufficiently sturdy rectangle to withstand the rigors of transporting and extracting. The top bar is usually constructed with a removable wedge or a 9/64 (3.572 mm) groove in the center, both recessed to a depth of approximately 5/16 of an inch (7.938 mm). Bottom bars are usually 3/8 inch (9.525 mm) thick, 25/32 inch (19.844 mm) wide, and 17-11/16 inches (44.926 cm) in length. They may be of solid construction, two pieces of only 5/16 inch (7.938 mm) width, a single piece with a slot clear through the center and within 7/16 of an inch (11.113 mm) at either end, or grooved to a depth of 9/32 inch (7.144 mm). The end bars usually have 4 holes drilled to accommodate horizontal wires for additional strength and rigidity and to hold the foundation centered in the frame.

COMB FOUNDATION

The second major development of the mid 19th century was the advent of comb foundation. This is comprised of a sheet of pure beeswax embossed on both sides with the bases and the beginnings of the cell walls of the comb of the honey bee. It is inserted in the frame and placed in the hive where it becomes a midrib or base of the comb which the bees complete. It is astounding as well as pleasing to see how quickly a colony of bees will build its combs from foundation.

The cells of comb foundation are ordinarily made of worker bee size, as a sufficient population of drones is provided by a few small areas of drone cells which the bees usually construct along the bottoms and corners of the combs (Fig.4). Inasmuch as the bees draw out the foundation according to the size of cell embossed on it, a large force of worker bees is obtained. In fact, when frames are furnished with full sheets of worker size comb foundation, colonies can be induced to abandon their natural trait of building 1/4 or more drone size cells.

The proper use of comb foundation has many advantages. Straight combs are obtained which permit easy and rapid manipulation of the colonies; the

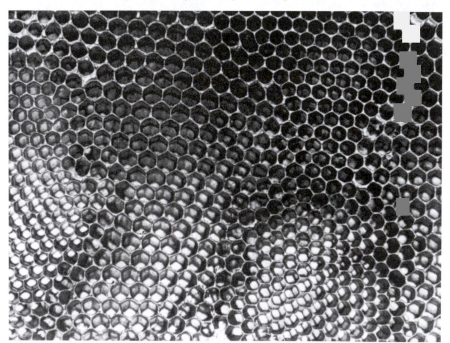

FIGURE 4. Worker and drone cells showing many odd shaped transition cells in between. About actual size.

removal of honey from the supers is greatly facilitated. At least half of the honey and much of the labor required by the bees in the construction of combs is saved. These advantages, plus the control of the desirable population of worker bees, in great measure makes commercial honey production possible.

A few sheets of drone cell foundation may be used to advantage in queen rearing yards where a large force of drones of selected stock is desired. There has been some contention that, due to the larger cells, drone comb facilitates honey storage, rapid evaporation of nectar, and greater ease of removal of honey from the cells when extracting. The difficulty of excluding the laying queen from large areas of drone comb, however, discourages its use. In any case, a large number of drones is a detriment to the colony because they do not work and consume large amounts of pollen and honey. Intermediate sized foundation larger than worker cells, but smaller than drone, are offered for commercial sale in a limited number of sizes and do, to a degree, contribute toward the elimination of brood rearing in the area and enhance the production of honey. Since they vary from standard and preclude any future use in brood areas, they are not popular and are more of an experimental tool.

Attempts have been made to incorporate higher melting point waxes in comb foundation, but the practice has been almost totally abandoned by most reputable manufacturers of the product. Theoretically, microcrystalline wax, ceresin, carnauba, or hydrogenated castor oil, elevate the melting point and stiffen the final sheet, but two major problems ensue. If the addition is considerable, the bees notice the difference and show a decided preference for pure beeswax foundation. When nectar is not available or in limited supply, the problem becomes more acute. Even more important, the major uses for beeswax are other than in comb foundation, and the manufacturers of candles, cosmetics, pharmaceutical and industrial products vigorously resist any beeswax offered that does not meet their exacting tests for purity. Since the vast majority of comb foundation is a recycling process in which the beeswax eventually is reclaimed and reprocessed, protecting the integrity of the pure beeswax in the foundation sheet has become vital. Beekeepers and manufacturers alike share a joint responsibility in rejecting the addition of other waxes or waxlike compounds which become inseparable. Chemical and physical tests can detect the impurity of these unknown mixtures.

In the manufacture of comb foundation, the beeswax is sorted into two grades. The deep yellow and brown shades are used for making comb foundation for brood and extracting frames. These darker shades of beeswax mainly come from the rendering of old combs and frame scrapings; the lighter shades are used for making foundation for comb honey sections and bulk honey frames. The light colored beeswax invariably is obtained from cappings and new combs.

The beeswax is carefully refined until free of all separable materials. It is then made into sheets which are run through milling rolls which emboss the sheets of wax with cell shapes. Brood foundation is made on a mill which makes foundation having a thick base and a deep cell wall which the bees draw into combs more readily. Comb honey foundation is made on a mill which make the bases and wall so fragile that the foundation will not stand the weight of the bees in a full size frame.

Comb foundation for section comb honey must be as light as the finest machine can make it to avoid what is called the "fishbone", a heavy central rib of wax found in honeycomb built on foundation that is too heavy. This type sheet, known as thin surplus, averages 25 sheets per pound in size 4-1/8 x 17 inches (10.478 cm x 43.18 cm), over 12 square feet (1.115 sq. meter) from only one pound of beeswax. As with other foundations, the bees utilize this base provided and extend it partially into cell walls, then add new beeswax they have secreted to finish the comb. This surplus is not built up rapidly because the bees have to add more of their own beeswax and this, of course, slows down comb production.

Foundation for bulk comb honey is made a little heavier, slightly heavier walls and cell bases than for section comb honey. A sheet size 4-1/2 x 16-1/2 inches (11.43 cm x 41.91 cm) runs 20 sheets per pound. A slightly heavier weight becomes necessary since the span of the frame in the hive is nearly four times that of the section comb honey and the extra weight and stress necessitates the heavier construction.

Medium brood plain foundation, made from the darker grades of beeswax, is the lightest weight used in the brood and extracting areas of the hive. A sheet 8 x 16-3/4 inches (20.32 cm x 42.545 cm), normal for the Langstroth brood frame, is made 8 sheets per pound. Reasonably good performance can be expected under minimum conditions of stress, but high speed extracting, consistent moving, or heavy bodied honey, all contribute to a decrease in the longevity expected from combs based on this construction. As a general rule medium brood plain foundation should be reinforced in brood frames with four longitudinal, malleable tin wires of 28 or 30 gauge. The wires are threaded through the holes in the end bars of the frame, pulled tight and fastened. The foundation then is inserted in the frame and secured by nailing the wedge in place in the top bar. Then, with a sheet of foundation underneath, the wires are embedded by an electrical device or by a spur embedder.

Reinforced comb foundation for brood or extracting purposes first became practical in 1921 when Dadant & Sons perfected a method of wiring which consisted of vertical crimped wires woven into the foundation by machines (Figure 5). Crimped wires proved better than straight ones, for the shoulders of the support radiate reinforcement between the wires and prevent

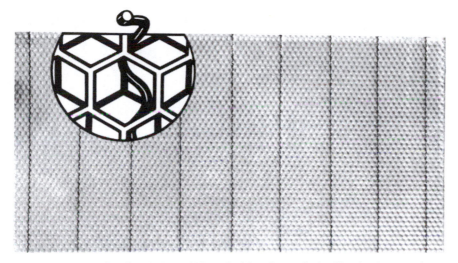

FIGURE 5. Crimp-wired foundation, reinforced with spring steel wire. Note hook at top edge to fit under wedge and hold sheet securely in place.

the beeswax from slipping downward when soft from heat. Nine or 10 vertical crimped wires are ample to prevent sagging. When proper care is taken in handling and assembling the foundation, and when the hives are level, the vertical wires alone may be sufficient to hold the foundation in the central plane of the frame and also to prevent sagging. Generally, two to four longitudinal wires are used in addition to the vertical as a support to prevent the foundation from swinging in the frame. This double wiring results in a rigid straight comb, and permits rapid handling and long distance hauling with little or no damage to combs, even though newly built and heavy with honey. Crimp-wired foundation, and subsequent variations upon it, have since been the most popular reinforced foundation used.

In 1963 Dadant & Sons introduced a plastic base comb foundation (Figure 6) that has subsequently become extremely popular. A thin film of plastic material between layers of beeswax, then embossed with the cell of the honeycomb, is readily accepted by the bees. This item, with the trade name Duracomb, or Duragilt with a metal strip along either short end, conveniently lends itself to ease of assembly and maximum longevity. Oftentimes, the Duragilt requires nothing but assembly of the frame around the foundation, although sometimes the top and bottom are further wedged in the grooved top and grooved bottom bars with a v-shaped metal wedge, using a special inserting device that's quick and economical to operate. Duracomb is frequently stapled into position since the plastic sheet forms a base for a staple and almost positive attachment to the frame. Although the plastic and wax are

FIGURE 6. Duragilt plastic base foundation with liberal beeswax coating, notched for bee access.

chemically comparable, there is no combining of the two and when immersed in boiling water or subjected to live steam the wax separates in pure form.

Other attempts have been made, and continue to be made, with other methods of reinforcement. The addition of high melting point adulterating materials has been largely discontinued. Aluminum base, wood base and innumerable other materials have been tried from time to time. The selective nature of the bees, themselves, restricts many efforts and instinctively they attempt to remove any foreign material in comb building. Metals prove to be too good a conductor of heat and cold.

A fairly recent development that has found wide favor is the so-called "Hard Combs" (Figure 7). Most often used in supers and for the surplus storage of honey, they are frames molded of plastic complete with cell impressions varying in depth, and either used as is, or with a thin coating of pure beeswax added to enhance bee acceptance. Several manufacturers of this type product now offer their combs for sale, but the cell configuration differs widely and bee acceptance, in turn, is not the same. Some are offered with the

FIGURE 7. Semi-horizontal view of plasticell hard comb foundation, showing cell depth.

comb, itself, molded as part of a plastic frame, thus totally replacing wood and wax for the frame and foundation. One disadvantage is price, for hard combs generally command a premium investment cost. Varying reports comment on acceptability, but there seems to be a consensus that these materials also need a good flow for initial drawing. Longevity is, as yet, an unanswered question. When one considers that supering equipment using wood frames and either all wax or mostly wax foundation have provided a score or more years of continual use, it follows that only additional time will determine if the long-range effectiveness of hard combs is commensurate with the substantially higher initial cost.

If properly stored, comb foundation will last for years. When cold, it becomes brittle and the least handling will crack it. If handled roughly in shipping containers when cold, the jarring of the boxes will crack the sheets of foundation imperceptibly, and, although they may appear perfect, when handled they will fall into numerous pieces. Too much heat causes the foundation to become too soft and ductile. The bees can manipulate the beeswax best at about 90 degrees F (32.2° C). This is also a good temperature for embedding wires with a spur embedder, although a few degrees lower will not be injurious.

In nature honey bees build their comb from nothing but pure beeswax. Usually this is secreted fresh for the purpose by the worker bees. Both in nature and on comb foundation, construction will start at or near the top and proceed downward. The attachment of the comb is extensive at the top, the point of attachment, and the average thickness of the comb will measure approximately 1 inch. In nature the finished comb is a broad U shape (Figure 8) and highly irregular. In nature both bridge and brace combs are constructed between these U-shaped configurations, serving to do as the name implies, brace the separate segments, and offer a bridge for the bees to pass from one portion to another. Both brace and bridge combs may serve as storage areas for nectar during the flow, but can become too plentiful and interfere with ventilation. Burr combs are usually built on flat surfaces of the interior portions and are, as the name implies, burr-like in appearance. Individual colonies, as well as strains of bees, vary in their tendency to build supporting combs of this type. With the use of comb foundation, the U-shaped configuration, to a degree, is utilized effectively, and brace, bridge and burr combs are eliminated almost entirely.

The first requisite for the secretion of beeswax is a stomach well filled with nectar or honey. It is an interesting fact that comb building and honey gathering proceed simultaneously and that when one stops, the other stops also. As soon as the nectar flow slackens to a point where the consumption of honey or nectar exceeds the surplus, bees cease to build new combs, even

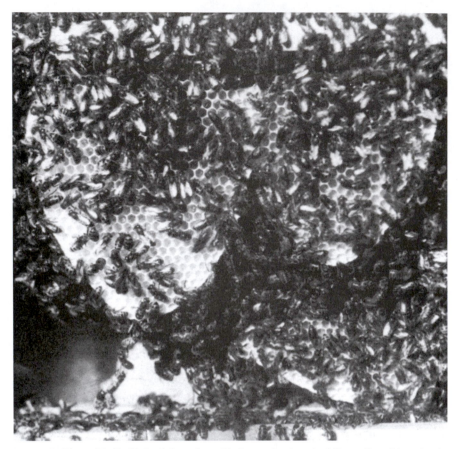

FIGURE 8. Natural built, U-shaped combs, with the worker bees building cells and hanging in festoons while secreting beeswax.

though large portions of their hive are unfilled. Langstroth said, "When honey no longer abounds in the field, it is wisely ordered that they should not consume in comb building the treasure which may be needed for winter use."

There is evidence that the secretion of beeswax is involuntary during the honeyflow. Although the amount of beeswax secreted may exceed that required for extending the cell walls and capping the cells of the combs, wax scales seldom are wasted in noticeable quantities about the hive. Usually bees secrete beeswax in proportion to their needs. A swarm expanding to the necessity of building new combs has a supply of wax scales ready for the purpose. If hived on fully drawn combs, some of the wax scales may be wasted and wax secretion diminishes rapidly. If hived in an empty box or on full sheets of comb foundation, the bees immediately make use of the beeswax they are secreting and continue to secrete more until a sufficient number of combs have been constructed for use by the colony.

Simply supplying full sheets of comb foundation should be combined with favorable conditions to insure the best combs. On established colonies, new foundation should be offered to the bees directly over the brood nest, and preferably during a strong honeyflow for maximum results. Bees hived on nothing but foundation should be stimulated with ample feeding of syrup for optimum wax secretion.

At the same time, regular inspection of combs in use, and elimination of those whose effectiveness has deteriorated with age, called comb culling, should be practiced. An adequate amount of drone cells will be provided by the bees, as well as accommodation cells, usually odd-shaped, found at the borders of combs. Transition cells are built where a change from the worker size to the drone cell size occurs. Maximum worker cells is the desirable result and melting up of distorted or defective combs and replacing them with new sheets of reinforced foundation, will prove a profitable management practice.

CELL SIZE AND SHAPE

The cell of the honey bee is a hexagonal-shaped tube, consisting of six walls each of which forms a proportionate part of another cell, closed at its base by three rhomboids which form an inverted pyramid. Each of the three rhomboids forms a third of the base of a cell on the opposite side of the comb. While combs are built vertically, or practically so, the cells are not built at a right angle to the vertical, but slope upward from the central plane of the base. The angle of the upward slope varies from 9° to 14°. This sloping apparently tends to prevent the larvae from sliding out the mouth of the cells before they are sealed, and aids in containing the food placed there by the worker bees. Cell base thickness in new brood foundation averages 0.025 inch (0.635 mm) before it is offered to the bees and 0.008 inch (0.2032 mm) after the comb is drawn out. The cell walls, themselves, are manufactured and generally remain 0.0025 inch (0.0635 mm). The thin surplus foundation has a cell base as manufactured 0.011 inch (0.2794 mm) and after drawing out 0.005 inch (0.127 mm). Cell walls remain 0.0025 inch (0.0635 mm). In naturally drawn combs without the foundation guide, bases approximate 0.0035 inch (0.0889 mm) and the cell walls 0.0025 inch (0.0635 mm).

Worker bee cell size in brood comb foundation is usually made on dies providing 857 cells per decimeter. Cells from the resultant product may vary from this and usually are larger due to stretching of the wax in manufacture. The following table records the approximate number of cells on both sides of the comb, worker size, for various races of *Apis mellifera*, as well as *Apis cerana*, the honey bees of the Far East, *Apis dorsata*, the large bee of the Far East, and *Apis mellifera scutellata* as this race is found in Africa.

Table I. Approximate Number of Cells on Both Sides of the Comb, Worker Size

Race of Bees	Cells per Square Decimeter	Cells per Square Inch
Italian	857	55.3
Caucasian	857	55.3
Carniolan	857	55.3
Italian (Drone)	520	33.5
Native German	897	57.9
Apis cerana	1243	80.0
Apis dorsata	787	50.8
Apis mellifera scutellata	1000	64.4

There is considerable conjecture that brood cells reduce in size with age and use. The accumulation of cocoons, cast-off larval and pupal skins, and the treatment given the cell in preparation for the next cycle of brood may result in a thickening of the cells over a period of years, that may contribute to a reduction in cell size and the subsequent bee emerging from this restricted environment. At the same time there is literature, Taber and Owens in 1970*, suggesting bees measured in the late 1960's produced worker cells 814 to the square decimeter and drone cells 540, the former larger than the table above and the drone cells smaller. There is speculation that scientific queen rearing has developed a genetic characteristic enlarging upon the size of the *Apis mellifera* in its native environment, and contributing to both a larger version of these same species and a larger cell size as a result. There is evidence to suggest that bees reared from the larger cells will carry larger loads of nectar and produce more honey as a result.

Recent attention to the necessity of finding ways and means of controlling, hopefully eliminating, the scourge of honey-bee mites, has resulted in more attention to cell construction. European proponents of a special cell configuration suggest that adequate mite control may result. At the present time there is no consensus that this is so, and in particular no explanation as to why it might be true. Suffice it to say that as the mite continues to involve beekeeping throughout the world, and as environmental considerations discourage a chemical means of control, much effort and observation will obviously take place in this and in other means and methods of mite control.

---

*Taber, Stephen III, and Charles D. Owen. (1970). Colony founding and initial nest design of honey bees, *(Apis mellifera L) Anim. Behav., 18* (4):631.

EXTRACTORS AND APIARY EQUIPMENT

The von Hruschka honey extractor of 1865, employing the centrifuge principle, continues to be used today. After the frames have been filled with honey and ultimately capped with wax, these are removed from the hive and taken to an area screened from bees. A sharp and/or hot knife, or machine embodying that same principle, is used to slice a thin layer of wax protecting the combs, and exposing the honey beneath. These are placed in an extractor of varying size and whirled. The honey is thrown free of the cells to the side of the tank, drips down and accumulates in the bottom, hence is removed for storage or processing. The combs, intact, go back into the super or hive body for reuse or storage until the next year.

Modern extractors (Figure 9) vary in size from small 2-frame units with a tank 14 inches (35.56 cm) in diameter and 21 inches (53.34 cm) in height,

FIGURE 9. Small hand-operated honey extractor, normally several sizes available. Shows baskets and also a small power extractor.

equipped with either a hand crank or a small motor drive, up to radial units (Figure 10) always motor driven, with a tank height of 36 inches (91.44 cm) and a diameter of 52 inches (132.08 cm) and holding 50 to 85 frames. Models of varying capacity exist in between, the selection usually dictated by cost and the volume of honey to process each year. Operation of these machines, particularly the larger units where the circumference is wider and the force is greater, necessitates operation at slow speed initially with gradual acceleration to a top speed of some 300 RPM over a period of 10 to 15 minutes, depending on the viscosity of the honey. This will find the bulk of the honey being thrown out at low speed and the high speed cleansing the combs. The extractor accelerates slow enough that combs are not destroyed from the force of the centrifuge in the early state when the honey is in the comb. Mechanical devices

FIGURE 10. Large radial honey extractor using centrifugal force to expel honey from combs.

are available to permit the larger units to gradually bring up the speed, and new electronic devices are now available to simplify that function even further. Machines are now available in which the axis of the unit is horizontal rather than perpendicular, and the movement of the centrifuge is vertical rather than horizontal. These machines incorporate the uncapping and loading feature and reduce the labor involved, but cost precludes their use except on operations of substantial size.

UNCAPPING AND WAX HANDLING

The uncapping function, removing the thin wax coat in the completed comb, can be done in small amounts with a plain two-bladed knife occasion-

FIGURE 11. Honey-wax separator. Operates continuously, capacity about 1-1/2 tons per hour. Centrifugal force throws honey out, and clear wax is dropped down at the center.

ally dipped in hot water, to speed the cut and keep the unit clean. Hand knives heated with either steam or electric current remain popular, and small, comparatively economical power driven 2-edged heated knives, increase the productivity of this function. Different models of large machines are available, all of them comparatively expensive, whereby the operator loosens and cleans the frames, then feeds them into a hopper where they pass between heated vibrating knives or oscillating flails, then are ejected on to a belt with the honey exposed and ready to be loaded into the extractor.

The wax cappings, with honey adhering, are further handled in one of a number of means. The small operator can place them into a solar wax melter, a frame with a black background and a glass top, where wax and honey separate and drain off into different containers, all activated by the rays of the sun. Other methods of separating the two with means of a small metal container are also available. Again, heat is the activating medium melting the wax, separating the honey and wax, and draining the two into separate receptacles. Larger Brand melters or cappings separators, heated by steam and equipped with copper coils, permit the injection of cappings via a hopper below the coils and a gradual warming process that drains the honey off before the heat darkens it and frees the wax to separate into another container. The separator (Figure 11) for the large operator is a centrifuge type spinner into which homogenized honey and cappings flow. The force spins the honey out and the wax, in semi-dry form, falls free.

Ultimately these cappings are heated to separate the remaining honey and liquefy the wax or put into warming chambers where the remaining honey drips free, then the wax is melted and poured into blocks.

HONEY FILLING DEVICES

Honey filling devices (Figure 12) take many different shapes and sizes. A small tank with a bottom gate may work adequately with small amounts. As the capacity of honey to be filled increases, brass bottle fillers may be preferred. Stainless steel valves are the next step up, often used in conjunction with water jacketed tanks. Immersion heaters control the temperature required, while stirring and blending units can be installed and stainless steel honey filters with micron-rated filter bags also enhance the quality of the final product. Flash heaters designed for modest-sized units are now on the market and accurate no-drip honey measuring dispensers offer quick and accurate filling. These reduce bottling time, eliminate overfills, can be used with hot or cold honey, and, in turn, may be adjusted for various filling volumes. High speed equipment is, of course, available from manufacturers of food processing machinery, but these generally have to be tailored to fit individual need and are extremely expensive to install.

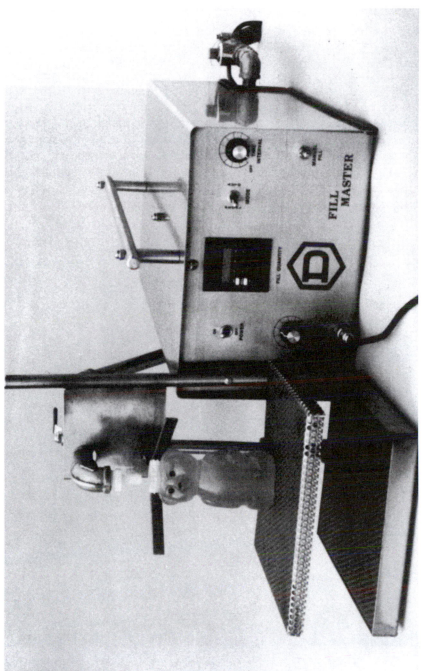

FIGURE 12. Modern no-drip filling machine, accurate with either hot or cold honey.

Dehydration of honey has become an important consideration. Federal programs requiring standards of moisture content often necessitate drying honey by the producer. These units (Figure 13) are complete except for storage tanks and, of course, the honey itself, and include pumps, heaters, thermometers and controls, covers and stands, flash heater, energy reclaimer, vacuum source, vacuum processing chamber, and honey level sensor. One hundred amp electrical service is required, but the unit needs only 6½ x 6 feet of floor space. Low-temperature processing results in a high quality product and permits removal of 1½ to 2 points of moisture from up to 2 barrels of honey per hour. The units are fully controllable as to amount of moisture removed and speed of removal. They are not affected by outside temperature and humidity and are inexpensive to operate. The resulting product has been clarified, fermentation removed, and crystallization has been slowed.

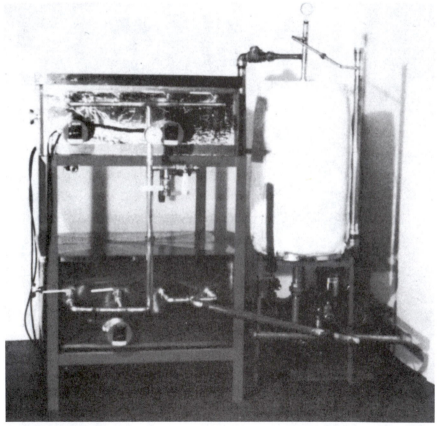

FIGURE 13. Comprehensive view of a modern honey-drying machine to remove excess moisture.

MAJOR ACCESSORIES AND TOOLS

Smokers

The bee smoker (Figure 14) is considered the one real necessity in handling bees. The most commonly accepted theory is that smoke both masks the alarm pheromone and alerts the bees, causing them to gorge themselves with nectar or honey, preparatory to abandoning the hive and avoiding the threatening source of the smoke. The result is that they are less apt to sting. The smoker, invented by T.F. Bingham, consists of a metal fire pot with a directional funnel at the top, plus a bellows for injecting oxygen into the base of the unit and blowing smoke out at the top. The objective is to produce a maximum amount of cool smoke, and a wide range of fuels continue to be used, largely dependent on whatever is most readily available. Probably the most popular is a common burlap bag cut into strips of 4 to 6 inches (10.16 cm

FIGURE 14. Illustrates both stainless steel and tinplate models of bee smokers, complete with bellows.

Figure 15. View of one type (folding) veil, helmet—plus coveralls.

to 15.24 cm), loosely rolled and then lighted at the frayed end. Dried sumac heads, decayed wood, wood shavings, corn cobs, tightly rolled corrugated paper, or dry pine needles are often used. Extreme care should be exercised so that the smoker, itself, does not become so hot to be troublesome to the operator. Care should be exercised also to be sure sparks don't fly out the top from the blast of air, and to be certain that every vestige of combustion is removed or extinguished after the unit has been used. Easier lighting can be accomplished by dipping the materials used in a solution of saltpeter and allowing it to dry.

Stewart Taylor of eastern Illinois, in the 1950's developed a hardwood smoke concentrate in an aerosol container to simulate smoke. Expense precludes constant use on a commercial scale, but many beekeepers, including commercial men, sometimes keep a can or two for emergencies for intermittent use. Just a few sprays with the solution and the bees are reasonably controlled.

Veils

Bee Veils (Figure 15) are a popular protective device and considered essential in handling bees. They vary widely in materials and design, but essentially fit snugly around a hat or cover the top of the head, and should be made to draw up tight around the neck and shoulders. Black screen wire of 12 mesh seems to provide the best visibility and whether they are square or round is a matter of personal preference. A tulle veil, a sheer fabric of different types, is lightweight and easy to carry and use, but oftentimes wire panels are preferred because they both last longer and do not blow against the face. Wide brim hats are desired and modifications of the pith helmet are most widely used. Lighter colors are preferred since they normally create less antagonism from the bees and, of course, are cooler for the operator to wear.

Helmets are available in two styles. A molded plastic unit with side louvers allows maximum circulation of air around the head. An adjustable head band fits all sizes and holds the helmet away from the head for more circulation. A woven mesh material design, is comparatively strong and light weight. Less durable than the plastic model, both units work quite well with all style veils.

Hive Tools

The hive tool is so handy that it almost becomes an essential device (Figure 16). Normally 10 inches (25.4 cm) long, with a sharpened surface on one end, with the other end bent approximately one-half inch at a right angle and sharpened on the end, plus a punched hole to serve as a nail pulling device, it finds countless uses working with bees. High quality spring steel provides long years of service and the prying and moving to manipulate supers and handle frames becomes easy with this little device. It serves countless other purposes in and around the home as well as serving as a necessity in handling bees.

FIGURE 16. 10 inch (25.4 cm) tempered hive tool, note sharpened edges for prying and notch for nail pulling.

Gloves and Clothing

Bee gloves are a desirable protective device for the inexperienced bee-keeper who wishes to avoid stings. A cloth gauntlet from wrist to elbow with either cloth, plastic or leather band, is ideal for handling frames. Experienced beekeepers keep a pair handy for the occasional cross colony; the hobbyist finds them a matter of routine.

White coveralls complete the ensemble. It is recommended that the bottom of the trousers be fastened tight at the ankles, or tucked into boots or loose-fitting socks of light color. Any loose-fitting clothing, cool to wear, is acceptable, but the light colors are always preferred.

Bee Blower

The bee blower (Figure 17) is a comparatively new device, but serves the function the name adequately describes. A motor driven impeller in a housing generates air that is blown out a metal tube and through a neoprene hose with a directional plastic nozzle. Although there are many refined functions it can perform, basically it is used to remove the bees from supers prior to bringing them into the honey house to extract. These blowers operate economically

FIGURE 17. Bee blower, complete with gas powered engine, plastic hose and spout.

and quickly and are lightweight for transporting to and from, as well as within the apiary site.

Queen Excluders

The queen excluder is either a sheet of perforated metal, plastic or a wire grid housed in a metal or wood binding that is placed between the brood and super spaces to prevent the queen, but not the worker bees, from moving up in the hive. The openings should be 0.163 inch (4.14 mm) to allow free access to the worker bees, but to restrict both queen and drones. Their use is a matter of debate among commercial operators, but the trend seems to be toward using them extensively again and restricting the queen and subsequently brood from areas where the surplus honey crop will be stored.

Feeders

Feeding devices are available when sugar syrup or fructose must be offered to bees. A small entrance block with a perforated metal cap, built to fit on a glass jar, can be inverted at the entrance of the hive. Plastic containers resembling frames can be filled with syrup and placed in the hive replacing a frame, and oftentimes syrup is placed directly over the colony in a 5 (2.268 kilo) or 10 pound (4.536 kilo) friction-top metal pail with several holes punched in the lid. The latter method requires a housing of an empty super or hive body and then the cover on top. Oftentimes, the pail will be inverted over the open hole of the inner cover or placed directly on top of the frames.

Drugs and Chemicals

Sulfathiazole was first developed to control American foulbrood in the late 1940's. Regulatory control no longer permits the use of this drug. Presently, Terramycin* is the recommended drug for control of both European and American foulbrood. The drug is available for mixing with either powdered sugar or sugar syrup and may be fed in combination with other drugs. Early spring feeding is recommended with preferably three applications at four or five day intervals, but never fed during or 30 days prior to a honey flow. A mixture of a ready-to-feed material including Terramycin is available. As with all drugs, directions should be carefully followed. Fumidil-B* is the only known drug effective for the control of nosema, a disease that causes severe loss of honey crop if not controlled. It is particularly useful for package bee colonies as well as queen-producing units. Quik-Gro is one example of bee foods available to supplement the nutrition requirements of the developing colony. Containing both carbohydrates and proteins in carefully formulated amounts, this is most often required in early spring when the colonies are rapidly building ahead of anticipated flow conditions.

---

*Terramycin is a registered trademark of the Pfizer Co.

*Fumidil-B is a registered trademark of Abbott Laboratories

The discovery of tracheal mites and subsequently *Varroa* mites in the Western Hemisphere has led to a search for chemical control. Both federal and state regulatory agencies exercise control over the authorization of use of these materials with widely varying and seemingly constantly changing regulations. Individual state departments of agriculture and often bee supply houses can offer updated information. Menthol crystals are most widely used for tracheal mites; Apistan* for *Varroa*. Formic acid, currently used to some degree in Europe and Canada, is also being considered as an acaricide for use in the U.S.

Moth control can be achieved through the use of paradichlorobenzene crystals which should be used only on empty stored combs and never on combs containing honey. A biological insecticide for moth control called Certan was once widely advertised, but slow sales caused manufacturers to withdraw the product from the marketplace. Chemicals for removing bees from supers of honey include benzaldehyde (not currently registered in the U.S.) and a material marketed under the name Bee-Go (butyric anhydride). Both are used with a fume pad constructed of 2¼ inch wood frame with a metal cover and heavy flannel-coated canvas underneath. Carefully controlled amounts of the chemical are placed on the canvas. Then, it is inverted over the top of the super to be removed. For the hobbyist seeking quick relief from stings, swabs are offered under the trade name Sting-Kill. They also prove effective for the relief of mosquito and chigger bites.

Pollen Traps

The availability of pollen is a critical necessity in colony development. Rich in amino acids, the building blocks of protein, it is an absolute necessity for the elaboration of brood food. Traps of widely varying sizes and designs, one of which is illustrated (Figure 18), are in use, each constructed to suit the particular necessity of the beekeeper. All of them are so constructed that as the bees come into the hive loaded with pollen on their legs, the pollen will be dislodged and dropped into a collecting box. Intermittent use is recommended so that the colony is not permanently deprived of pollen. The pollen should be removed regularly, dried, cleaned, then frozen for storage and future use. The role of pollen in queen rearing is extremely important. The developing queen and larvae require tremendous amounts of royal jelly and this in turn comes from the pharyngeal glands that will not continue to function without large amounts of protein food. Pollen, a dietary supplement for human beings, is in widespread use in Europe and to some degree in the Western Hemisphere.

---

*Apistan is a registered trade name of Zoecon, Inc.

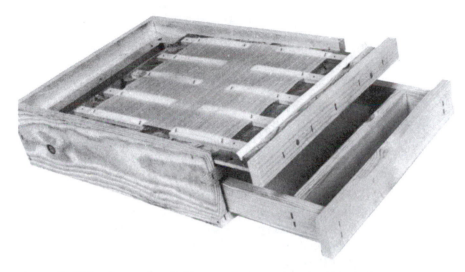

Figure 18. This pollen trap installed between the bottom board and hive body has an easy slide drawer for collecting pollen.

Honey Grading Equipment

Previously, honey grading equipment was sufficiently expensive that the honey producer relied almost entirely on the honey packer for moisture and color grading. The honey refractometer (Figure 19), an optical instrument, quickly and accurately registers moisture content. One or two drops of honey are placed on the prism and spread over the surface. It is simple and easy to use and with a clean tissue paper and water, one sample can be removed quickly and the instrument readied for another. The Lovibond color grader, slightly over one-half pound in weight, operates with discs that are simple and easy to use. Separate discs may be obtained for the grading of colors of a wide range of materials including butter, coconut oil, rubber latex and many more. Similarly

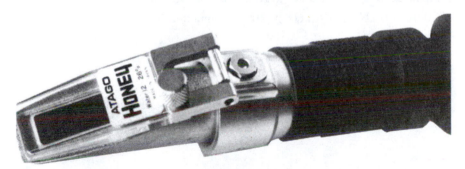

Figure 19. Refractometer, about size of hand, reads moisture range from 12.5% to 26%.

the Pfund color grader not only works well with honey, but with other liquids of similar amber color characteristics. Complete with an illuminated lamp arm and a trough to hold the honey by means of a rack and gear arrangement, the slide is adjusted until the color match is obtained and an accurate reading registered on the millimeter scale.

MISCELLANEOUS ACCESSORIES AND TOOLS

BEE BRUSH—This is a long-handled wood brush with soft flexible bristles covering about one-half of the length. It is ideal for brushing bees off queen cells, yet firm enough to remove bees from frames, supers or clothes. It will not crush or injure bees.

BEE ESCAPES—A metal or plastic device designed to place in the hole of inner covers which in turn go under supers of honey to be removed. Bees can move down through the escape and not back up. Usually supers will be free of bees the following day.

BEES AND QUEENS—Commercially available in the South and western parts of the country and from selective producers elsewhere, queens may be purchased individually or in battery cages of dozens or more. Normally, queens are delivered in mailing cages with a few attendant worker bees. Packages of bees are shipped in wire screen cages and sold usually in 2 pounds, 2½ pounds or 3 pounds at a time, most generally with an attendant queen.

BOOKS—A wide range of books on bees are available including this comprehensive textbook *The Hive and the Honey Bee*. The claim has been made, without verification, that more has been written about honey bees than on any subject with the exception of religion. These include books on first lessons, cook books, honey plants, questions and answers, comb honey production, laws relating to beekeeping, bee and brood diseases, queen rearing, and many, many more. Monthly magazines on beekeeping are available in most countries of the world. The *American Bee Journal*, a companion of *The Hive and the Honey Bee*, is but one example.

BOOT BANDS—An elastic strap designed to fit around the ankles and secure the trouser legs.

FIGURE 20. Branding iron to mark wood hive parts for identification.

BRANDING IRONS—Many inspection services highly recommend the use of hive branding irons (Figure 20) to burn some means of identification into the wood parts of the equipment. Available in gas, propane, or electric models, and normally including up to five letters, not over one inch in height, the brands themselves can be individually styled.

CAPPINGS SCRATCHERS—A small wood or plastic handle with about 20 sharp needles, for use in scratching low spots on combs while extracting to make sure the honey flows freely.

CONTAINERS AND LABELS—A wide range of glass, plastic, and tin containers are available for marketing, as well as transporting and storage of honey. These include small square cartons to hold wood sections filled with honey, attractive plastic squeeze dispensers (Figure 21) in the form of a honey bear, plastic and glass containers holding one or two pounds of honey, plus assorted sizes of cansisters and pails are readily available. Attractive honey labels may be purchased to complete the package and for appeal and marketability.

SPUR EMBEDDERS—A wood handle with wide tooth metal spur points that are shaped like a roller, slightly warmed and with the use of slight hand pressure, they press horizontal wires into bee comb foundation to center the sheet in the wood frame.

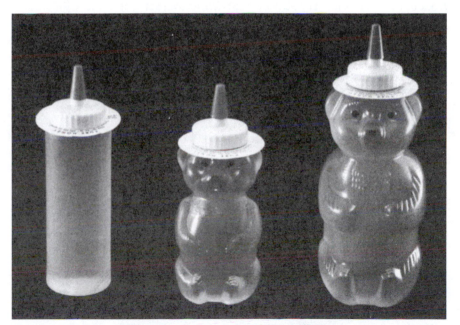

FIGURE 21. Squeeze dispensers shown above, attractive and efficient.

EYELETS AND TOOLS—Small brass eyelets with the use of a metal punch, are normally inserted into the holes in the ends of the end bars of brood and super frames. These permit wiring of the frames horizontally without cutting into the wood and losing the tension necessary to properly perform the function.

FRAME CLEANER—A small metal tool with a bent handle and offset at one end to remove wax and propolis from grooves and frames, particularly useful in comb honey production.

FRAME GRIP—A 2-piece metal device hinged in the center with a spring in the handle to permit one-hand operation of gripping the top bar of frames for loosening and removing combs.

FRAME WIRING BOARD—28-oz tinned wire made especially for the beekeeping industry is nonrusting and has great tensile strength. In conjunction with a board designed to hold the frames securely in place while the wiring is performed, horizontal wiring using a minimum of two and sometimes four horizontal wires, is a common practice.

HONEY GATES—Several sizes of brass seated gates along with newer plastic units are available for attaching to tanks for syphoning honey.

DRUM HEATER—Electrically operated and thermostatically controlled, these belts fit over 55-gal. steel drums and serve to liquefy the honey they contain.

INSEMINATION DEVICE—A scientific instrument designed for the artificial insemination of queen bees. It comes complete with instruction manual, stainless steel syringe, sting-hook, ventral plate hook, sting depressor, valve fold lifter, queen holder and backup tube, in an operating stage. It must be used with a binocular microscope of about 12X magnification and a source of CO_2 gas.

OBSERVATION HIVE—A simple device that can provide countless hours of pleasure to the beginnner, the student, and even the experienced beekeeper. Just one frame in width and 1½ or 2 frames in height, the unit is designed to be fitted with glass sides so that the queen and all activities of the bees may be observed at all times. These units are normally placed in or near a window with a direct opening to the outside, or with a piece of clear plastic tubing from the entrance of the observation hive to the out-of-doors. In the spring the unit should be established with a comb of brood plus adhering bees and a queen. Provisions for stimulative feeding can be made available through the top.

QUEEN AND DRONE TRAPS—Still in limited use, these traps are designed to fit in the entrance of a hive, permitting free access to the worker bees while drones and queens that may emerge are trapped in the upper section. Theoretically, undesired drones can be trapped and removed and if the swarm emerges, the queen will be retained and when the bees return, the queen and bees then hived. Many claim the device hinders access, reduces the honey crop, and severely impairs ventilation within the hive. Comb culling to eliminate drones and conventional swarm procedures are far preferred. Young clipped queens, adequate space, and proper ventilation achieve better results.

QUEEN SUPPLIES—The jelly spoon, as the name implies, is shaped to move royal jelly from cell to cell. One or more types of transferring needles are available to move larvae. Wax and wood cell cups are designed to provide the base for rearing queens and clipping scissors are, as the name implies, to clip portions of the wing to prevent free flight by the queen. Marking fluids are available to mark queens for identification purposes within the hive.

ROPE—Normally, a premium grade nylon rope, firm solid braid, of 2000 lb. test and roughly 5/16 in diameter. This quality will not be damaged by weather, oil or gasoline.

SKEPS—Infrequently available and normally relegated to ornamental displays and roadside stands, these conical straw skeps are attractive and invariably gain attention. Straw shaped in ropes and made secure with blackberry bindings are the traditional materials and usually vary from approximately 15 inches in width and 20 inches in depth. Some traditionalists shape exterior coverings in various forms to enhance the appeal. A popular cover is the form of a clergyman with a scepter to resemble St. Ambrose, the Patron Saint of Beekeeping.

SPACERS—These are usually metal devices notched to receive frames and space them evenly. They fit over the frame support area of the hive bodies and supers and are most often of either eight comb or nine comb design.

SPLIT SECTION DEVICE—This is a metal device to hold 4 split sections with a slot in the center so that the sections will be spread apart to receive a sheet of thin bee comb foundation.

STAPLES AND STAPLERS—Coppered wire staples of 12 gauge, 3/4 inch points about 2 inches apart, are a quick and easy way to fasten a bottom board to the hive body or various hive parts together. Conventional nails and hammer are now often replaced by strip loading pneumatic staplers operated by compressed air. Different size brads may be used to staple a hive body and super parts with one size and frame units with another.

SWEAT BANDS—A sponge strip with elastic band to fit around the head works well in hot weather to prevent perspiration from impairing vision while working with the bees.

TANKS—Heavy duty all-welded food-grade stainless steel tanks are available in a wide assortment of sizes of varying capacity from a small 14-gallon unit to 285-gallon unit or more. Tank covers and straining screens are available in sizes to fit (Figure 22).

FIGURE 22. Seven sizes of stainless steel tanks shown above, all welded type 304 stainless steel.

TRUCKS—Available in an assortment of designs for various purposes, these range from a utility unit with 10 inch ball bearing wheels and a sturdy 9 inch shovel nose up to pallet trucks available in either 10 inch solid tires or 12 inch air tires. Heavy duty drum-handling trucks simplify movement within and around the honey house.

WAX TUBE FASTENER—These are ideal for securing foundation into grooved top bar frames. The metal cylinder, fitted to a wood handle, is filled by lowering it into a container of hot liquid beeswax. Wax is retained in the cylinder by placing a finger over the air hole. A small amount of wax may be released by removing the finger from the hole and running the fastener along the top bar groove.

An interesting side note deals with hive sizes. Most sections of the world, and in particular the countries that may be considered major producers of honey and beeswax, have standardized their hive sizes to conform to Langstroth dimensions. While there may be some movement in that direction in Europe, there is still a wide range of hive sizes of varying configuration still in use. Aesthetics may play a small part, as may dictates of floral sources and types of flow, but probably the most important contribution to the maze of shapes and designs is the fact that American producers, both large and small, follow the success of the larger commercial operations that produce most of the honey. European beekeepers thoroughly enjoy an entrepreneural spirit of experimentation. There is, indeed, much good to be found in this approach for there remains a wide difference between a beekeeper and a honey producer. Commercial operations must of a necessity be conducted by the numbers, while the student of bees finds his greatest pleasure in observing the actions and behavior of the world's most famous insect. This is very likely the reason why bees and beekeeping have such a universal appeal to old and young, large and small, men and women, and individuals from all walks of life.

CHAPTER 13

FOR THE BEGINNER

by ALPHONSE AVITABILE*

The beginning beekeeper needs to concentrate on the fundamentals of beekeeping in order to achieve early success in the husbandry of keeping bees. A knowledge of the biology of the honey bee, including its behavior, will permit the beginner to make rapid advances while reducing the setbacks, which are common in all new undertakings. Most, if not all, authors of books on beekeeping repeatedly stress certain needs, requirements and conditions which permit the bee colony to survive and prosper. The wisdom of these authors should be adopted and practiced by all beginners.

How does the beginner get started, how many colonies should he begin with, and where does he obtain his bees and equipment?

Although there is no standard prescription, some general guidelines are available. Many books have been written on bees. Over thirty have been written in the last two decades by North American authors alone. These books are available in both public and college libraries. Many countries have monthly journals and magazines which devote themselves to articles, stories, and up-to-date research work on honey bees. Throughout the world you can find local, county, parish, state, provincial, regional, national, and international organizations that are devoted to the promotion and understanding of the honey bee. These organizations are anxious to welcome new members to their fold, and by joining, beginners can quickly advance themselves by attending lectures and discussing matters with other members.

There are colleges that offer formal and correspondence courses on beekeeping. In addition, there are opportunities to enroll in short courses that run for several consecutive days or meet periodically for a short period. Beginners should look for information on local beekeepers and may receive permission to visit an established beekeeper's yard to see firsthand what is really involved in keeping bees. Firsthand experience is the most valuable; thus, beginners should make every effort to learn from other beekeepers in their area and to make their acquaintance.

Once the beginner is ready, equipment and bees have to be obtained. We recommend at the onset that the beginner purchase new hive bodies, frames, and other woodenware. Although we cannot absolutely say this advice carries a guarantee for success, we can state that, in general, it seems to be the best way to proceed for a host of reasons. New hive equipment, when obtained,

---

* Alphonse Avitabile, Professor Emeritus, Department of Ecology and Evolutionary Biology, The University of Connecticut at Waterbury, Waterbury, Connecticut.

comes unassembled, giving the beginner the opportunity to become familiar with the equipment, to learn how to assemble the various parts, and in the process, to get to know what is involved in assembling both the shell and inner parts of the hive. An appreciation of how hive bodies, outer covers, inner covers, and bottom boards are assembled, and of how the frames and wax foundation are put together, is an important part of becoming an educated beekeeper. After beginning this way and after becoming more sophisticated by understanding the problems that may be encountered with the purchase of used equipment, primarily the fact that such equipment may harbor bee diseases, you can purchase, if prepared to face the consequences, used equipment that has been certified to be disease free.

FIGURE 1. Common types of protective clothing worn by beekeepers. Proper head and neck protection is vitally important since these areas of the body usually are more sensitive to bee stings.

In addition to hive equipment, every beekeeper needs a bee veil, a smoker, a hive tool and a hive scale. An assortment of veils and head gear is available, but in all cases, the see-through portion is black in color, because black veils permit the clearest vision when looking through a veil. Although there are times when bees may be worked without a veil, for instance, during a strong honey flow, all prudent beekeepers always enter a bee yard wearing one. The veil, when properly worn, protects the beekeeper's head and neck. In addition to a veil, light color clothing should be worn to protect the rest of the body. Garments terminating at the wrist and ankles should be sealed off to prevent bees from getting underneath. Many beekeepers wear complete suits covering themselves from head to foot, and cover their hands with gloves. Some find such apparel excessive and uncomfortable. Gloves make the removal and replacement of frames difficult and, thus, many beekeepers would rather trade some stings on the hands for the restrictive quality of gloves.

The smoking torch used by the early honey gatherers has been replaced in most parts of the world by the modern-day smoker, which consists of two parts. The first is a cylindrical part called the firepot, where the fuel for making smoke is placed and ignited. Its top consists of a hinged lid, somewhat funnel shaped, with a hole that permits smoke to escape. Behind the firepot is a bellows which blows air over the fuel to keep it burning, while at the same time expelling smoke from the firepot.

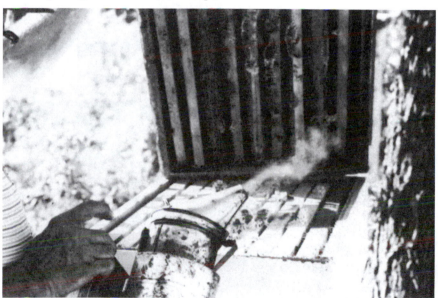

FIGURE 2. Learn how to use your hive tool and just enough smoke to examine your colonies without unduly disturbing them.

A hive tool is an instrument made of steel, about 5 inches long, and about 1-1/2 inches wide. Its length permits sufficient leverage for prying supers apart, and its width allows separation of hive bodies without gouging the wood. The need for such a tool is due to the fact that bees collect a sticky, orangey-brown to red in color, resinous substance from woody plants called propolis. The returning foragers carry this material back to the hive, loaded on their pollen baskets, where it appears to be in a liquid-like form. This substance is used by bees to waterproof their nest, coating brood combs and sealing cracks, including the spaces which are naturally created when hive bodies sit one on top of one another. This material also inhibits the growth of fungi and bacteria. All of these uses of propolis by bees would be of little matter to the beekeeper, except for the fact that this substance becomes stiff and sticky once applied in the hive by bees. Thus the inner and outer hive cover, the supers adjacent to one another, and the one sitting on the bottom board become fastened to each other by the propolis. Further, bees apply propolis to the frames, thus gluing them to one another and their hive bodies. The hive tool is the proper instrument needed to break the propolis seal between supers, between frames and supers, and between other hive parts. As the propolis seal is broken, individual supers can be removed and frames in each super can be inspected one at a time, if necessary.

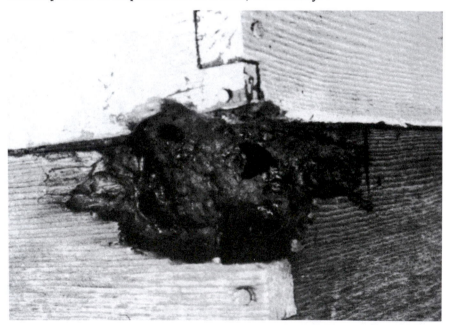

FIGURE 3. Honey bees fill empty cracks or holes in their home using their own special cement (propolis).

Hive scales are used to measure the weight of a hive, and its contents including the bees. They permit the beekeeper to check the daily weight fluctuation of a colony. By charting the hive's weight gains and losses, the beekeeper will obtain invaluable knowledge about each colony and also learn when the major honey flow periods (time of maximum nectar gathering opportunities) occur in the area. This knowledge will permit the beekeeper to

FIGURE 4. Hives on Scales are helpful since they record daily gains or losses in honey stores.

make vital and critical colony manipulations at the most opportune times. Such activities, and the resulting data, will go a long way in guaranteeing a successful enterprise. If, by charting the major honey flows over several seasons, you learn that the major honey flow begins in the middle of May, you will have the appropriate number of shallow supers in place for the bees to use to store the incoming nectar. During such flows, bee hoarding instinct is stimulated by the presence of ample storage space. Bees appear to produce more wax during flows and thus it is an optimum time to place frames with foundation on hives for conversion into superior comb. In many ways, the data collected from hive scales will permit beekeepers to map out a precise calendar of activities for each season of the year. The seasons follow an orderly progression; successful bee management requires that the beekeeper know exactly the time of the year that each critical manipulation should be done in order to maximize honey yields and insure the colony's health and well-being.

Once the hives have been assembled and painted, the beginner needs to obtain the bees. We suggest that the beginner purchase package bees or a nucleus colony. Package bees are bees that are removed from a colony by shaking them off of their combs and into a screened box, which usually includes a queen in a separate cage called a queen cage. These packages vary in the number of bees they contain, but the average package is three pounds; and, since there are approximately 3,500 bees to the pound, a three pound package contains approximately 10,500 bees as well as a caged queen. Before continuing our discussion of package bees, let us digress and discuss the members of a honey bee colony.

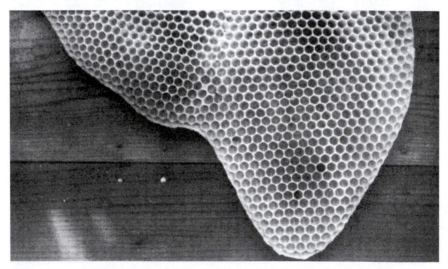

FIGURE 5. Naturally constructed comb.

A normal colony of honey bees consists of two female castes, a single queen bee, numerous female bees called workers, and some male bees called drones. However, unlike the female castes, upon the approach of the winter months or whenever there are long pauses in the nectar flow, the adult drones and pupae are expelled from the colony by the workers, leaving the colony with only its female inhabitants.

Coinciding with the period when the drones are expelled, the queen's egg laying activity is drastically reduced or ceases. Shortly after the winter solstice, the queen begins laying fertilized eggs again and the production of new workers begins. As the spring months approach, the queen also resumes laying unfertilized eggs in drone cells; these eggs, upon hatching, will eventually produce the next generation of drones. Since these eggs are unfertilized, drone honey bees are haploid, having only half the chromosomes of their mother and sisters. The primary function, and the only well defined one, for the drones, is for them to seek young nubile queens in flight, and to copulate with them in order to provide them with sufficient semen for the fertilization of eggs.

All fertilized eggs will produce female members of the colony. However, whether these females will be workers or queens depends upon the diet fed to them. Female larvae that receive a constant diet of royal jelly during larval life will develop into queens, and those whose diet, after the third day of larval life, is switched from royal jelly to honey and pollen will become workers. Thus, all fertilized eggs upon hatching have the potential to develop into workers or queens, and diet, among other factors, gives direction to the kind of adult which will finally develop from what are genetically similar female larvae.

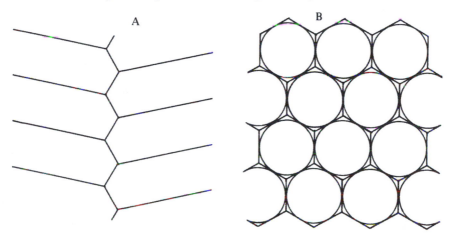

FIGURE 6. a. Diagram showing slope of cells from front to middle of comb. b. Diagram showing the economy of the hexagonal shape for making honey comb cells, no wasted space and shared cell walls economize wax.

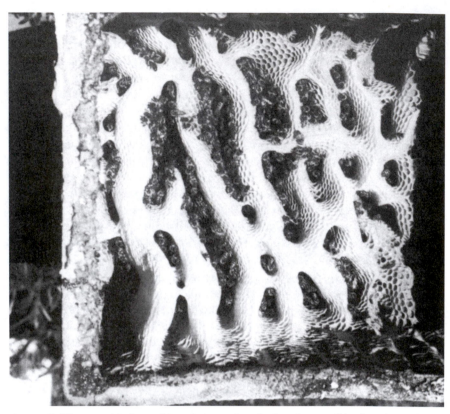

FIGURE 7. The movable-frame hive made modern beekeeping possible. Imagine trying to examine this colony which has completely burrcombed a frameless hive body!

One example which proves this point is that female larvae, under three days of age, removed from worker cells and transferred to queen cells, which are then placed in queenless colonies, will become queens, although their original destiny, as they lay in worker cells, was to become worker bees. In some ways still not fully understood, the royal jelly diet silences certain genes that would, if active, produce a worker, and causes other genes that are silent in workers to be active in female larvae, thus producing physiological, morphological and behavioral characteristics that belong to queen honey bees.

The queen is responsible, when present and healthy, for laying all of the eggs in the colony—up to 2000 per day during the height of the egg-laying period. Normally there is only one queen per colony. In addition to her egg production, she produces pheromones that help to maintain colony unity and order. These pheromones, discovered and identified by Dr. Colen Butler and others, are fascinating private odors that enable workers to recognize their own queen. In addition, these odors control the ovaries of workers by supressing their maturation, and inhibit worker bees from constructing queen cells which could ultimately lead to the production of new queens. The construction of queen cells and the rearing of queens in these cells are signs that the colony is either preparing to cast a swarm or to replace the existing queen. Young queens embarking on mating flights in the open air are detected by drones when her pheromones are carried downwind to them as they hover in drone congregating areas. Once a queen's pheromones are detected, drones fly upwind towards her, where up to 10 or more drones may mount her in succession. These same queen pheromones also help settle a swarm into a stable, quiet, swarm cluster. You can easily see why many have referred to the queen as "the spirit of the hive."

Let us now return to our account of package bees and reasons why beginners should start with package bees. The first reason is that the bees are certified to be disease free, which is very important now that bees may harbor two parasitic mites as well as a bacterial disease, all of which can adversely affect their health. Another reason is that a package of 10,500 bees is a number that is manageable for the beginner, and in the six weeks following the installation of the package, the colony's numbers, due to attrition, will decline significantly. This negative effect has a positive side since it will permit the beginner to inspect a colony that will not be as populous as established colonies, thus giving the beginner the time needed to become familiar with hive inspection procedures, to observe and recognize cells containing eggs, larvae, and sealed brood, to distinguish worker, drone, and queen cells, to see adult workers and drones, and to find and identify the queen, as well as recognize cells containing pollen and honey. A single brood chamber contains

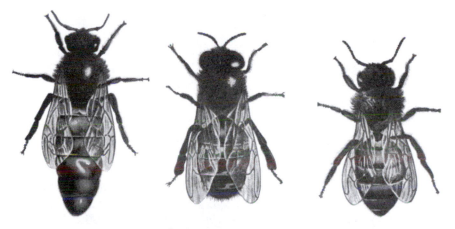

FIGURE 8. Queen, drone and worker honey bees.

FIGURE 9. Drones are removed from hives by the worker bees when day length decreases in the fall.

more than ample room for the installation of a package of bees and consequently, the beginnner is not overwhelmed with having to remove and/or inspect several shallow supers and at least two brood chambers, each containing 10 frames, as would be the case with an established colony.

Firms that provide package bees exist primarily in the southern parts of the United States, because it is there that bee populations increase rapidly in the late winter and spring months due to the climate. These firms shake bees off combs into packages for shipping to the northern states and Canada when the weather is suitable for their shipping and installation.

Although bees shaken into a package are usually all from the same colony, the queen that is added to the package comes from a different colony. She is usually a young mated queen, raised in a queen rearing yard specifically for inclusion with the package of bees.

Most beginners start with a three-pound package of bees, although two-, four-, and five-pound packages are also available. The package is a wooden box, usually with wire screening on two sides, and contains a small can of sugar water with two to four small holes in it. Bees are able to obtain the sugar syrup through these holes with their tongues while the package is in transit.

When the package arrives, the beginner should check to see that most of the bees are alive; some dead bees on the bottom of the package should not alarm the novice. The package should be placed in a dark, cool not cold, area and, if the bees are not formed in a calm cluster around the can of sugar water, some sugar water, mixed in advance, can be painted or sprayed on the wire screens to settle the bees.

The hive for these bees should be ready at its location. Towards dusk, but while temperatures are still reasonably warm, at least in the fifties, the bees can be installed in their new home. Before removing the lid, jar the package gently so that the bees will fall to the bottom, remove the lid, then remove the queen cage and replace the lid. Check to see if the the queen is alive; besides the queen there will usually be three to eight worker bees in her cage. The queen cage usually has its two openings corked, and behind one of the corks is a plug of candied sugar; remove this cork and then with a nail, make a hole equal to the nail's diameter through the candied sugar, being careful not to injure the queen. If the plug is rock hard, make the hole slightly larger. This hole will facilitate the bees in removing the candy and eventually free the queen from her cage. The candied plug permits a delayed release of the queen. This slow release provides more time for the bees to come to know and accept this queen as their own.

Now, counting in from either side of the hive to the fourth and fifth frame, suspend the queen cage between these frames, locating the cage several inches

FIGURE 10. A nearly solid frame of sealed worker cells usually tells the beekeeper that the queen in this hive is productive and that the colony appears free of brood diseases.

behind the oval hole of a sitting inner cover. With the cage hanging candy-side-up between the frames, jar the package again, remove the lid and shake the equivalent of two cups of bees onto the queen cage, replacing the lid on the package.

After the bees have moved around the queen cage and between the frames, put the inner cover over the frames and invert a gallon of sugar water (medicated) over the inner cover hole. Place an empty deep super above the inner cover, and cover it with a hive cover, then place a heavy object on the top of the cover so it will not be dislodged during heavy winds.

Now to the package again; jar it, remove the lid, and shake 1/3 of the bees at the front entrance, replacing the lid. Bees will be confused momentarily, but soon will begin to enter the hive; some bees will begin scenting, which means they lift their abdomen exposing at their dorsal tip a gland called the Nasanoff gland, which elaborates a pheromone that smells like lemon oil; when bees smell this odor they move toward it. Soon, if all goes well, bees will begin to flow towards the entrance. To witness this fairly uniform movement of thousands of individual bees is truly an exciting moment for beginners and

FIGURE 11. A marked queen prepares to lay an egg. (Photo by O.W. Park).

FIGURE 12. A queen cell with capping still hanging, from which a virgin queen has emerged.

FIGURE 13. Cutaway view of freshly laid eggs in brood cells (Photo by O.W. Park).

FIGURE 14. A typical sign of a laying worker, queenless hive situation—scattered drone brood and no worker brood.

professionals alike. When this activity is under way, jar the package again and then shake the remaining bees near the entrance.

While the bees are being liberated, one will notice bees expelling a yellow liquid-like substance. This material is bee feces which the bees retained in their gut during their confinement. Fecal spotting may also be seen on the hive and other objects for the next several days.

By installing the bees near dusk, the bees will be less likely to drift or wander away and be lost. Hopefully, the next morning will see a slow warming so that the bees will gradually begin to fly out and become oriented to their new surroundings, and losses will be minimized as they begin to forage. For the next 10 days, the inner cover should remain in place and the only disturbance to the hive should be the adding of additional medicated sugar syrup whenever it is required. After 10 days, the hive should be opened and checked to learn whether the queen was released and that she is laying, which can be determined by finding eggs and larvae in the honeycomb's cells. If she is still confined and alive, the cork opposite the candied plug should be removed, the cage returned quickly to its position, and the hive closed and then checked again in five days.

The method of opening and inspecting a colony should remain relatively the same each time that the colony requires attention. You should enter the bee yard, properly attired with your hive tool and lit smoker. Standing on one side of the hive, compress the bellows of the smoker, directing three or four puffs of smoke into the colony's entrance. Pause for two minutes to give the smoke a chance to pervade the hive, then begin to pry off the outer cover, puffing smoke underneath it as it is being removed. Place it underside up on the ground or on a stand. Now, puff some smoke into the oval hole by first removing the inverted gallon of sugar syrup. Next, remove the inner cover, where bees are usally clinging to the underside, which makes it prudent to lean it against the hive or some other object. With the removal of the inner cover, the tops of the ten frames are exposed and usually there are bees covering most of the frames' top bars. Again, gently puff some smoke over the frames, which will drive the bees down among the frames and clear the top bars of bees. With the hive tool inserted between the side bar of the frame closest to you and the hive body, slowly leverage the frames away from you and towards the opposite side of the hive. Repeat this action with the frame's other side bar, creating some space between the frames and the wall of the hive body.

Now move the frame closest to you towards the wall of the hive, breaking the propolis if necessary, and then gently withdraw the frame from the hive. Often this frame has fewer bees than those closer to the center of the hive, and in a hive with a newly installed package, the outer frames are usually not covered with bees. If the frames were started with foundation (flat sheets of

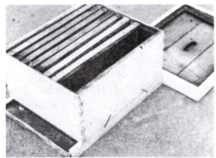

FIGURE 15. This hive with reduced entrance is ready to receive its package of bees. It has frames of comb rather than foundation.

FIGURE 16. Removing the package lid. The can of syrup and the small cage holding the queen are directly under it.

FIGURE 17. Next the queen cage is removed, shaking the bees clinging to it off, to view the queen inside it.

FIGURE 18. As soon as all the bees are shaken between the combs where the queen is located, the combs are carefully pushed together.

FIGURE 19. The hive, filled with combs, is now ready to have the inner cover and the outer cover replaced.

FIGURE 20. The entrance feeder is in place with a full jar of syrup. A little grass that can quickly be removed by the bees is placed in the entrance to keep the bees inside until they have time to organize themselves.

wax without bee cells), the bees may not have transformed by adding wax, the foundation into comb with wax cells. Nevertheless, after inspecting the frame, you can lean it against the hive or place it in an empty hive body. Continue the inspection by breaking the propolis seal between the next two frames and moving the frame closest to you into the empty space resulting from the removal of the first frame; then withdraw the second frame. Continue your inspection until you have accomplished the purpose of your inspection. Return the frames to their former positions in the hive, then close the hive, and add more sugar syrup as needed.

There is no safe rule of thumb as to how many gallons one should feed a package before they should be left on their own. Weather, the availability of nectar, the hive's population, and the hive's store of honey will all be factors that will require assessment before terminating feeding. Bees can always use food and will store the excess. Therefore, feeding is the safer course to take when in doubt.

This first hive inspection is necessary in order to learn that the queen was freed from her cage, was accepted by the bees, and is laying eggs, and as a consequence, that larvae may be present as well. During the first inspection, while it is not necessary, great satisfaction is derived if the beginner spots the queen. However, the presence of eggs and larvae in hundreds of cells adjacent to one another is sufficient evidence that the queen is present and performing her tasks.

Learning this, it is best to cut the inspection time short, leaving the spotting of the queen for another day, and placing each frame in its former position in

FIGURE 21. Bees fanning at the entrance to pull air currents through the hive. This cools the hive and helps to dry the nectar and transform it into honey. (Photo by Wolfgang Wittekindt)

Figure 22. Swarms are not always found on tree limbs.

the hive, maintaining the original nest structure. By minimizing your inspection, you are permitting colony activities to return to normal earlier, and in the early part of the season, you are also avoiding the possibility of having the brood chilled. In fact, all inspections for whatever reason, need not require that the hive be torn apart from top to bottom. It is far better to learn what needs to be known with minimum disturbance to the colony, and upon being satisfied with your observations, prepare to reassemble and close the hive up as soon as it is feasible.

Once the beginner becomes sophisticated and at ease while working with bees, then other methods of obtaining bees can be considered. Besides package bees, bees may be obtained by removing colonies that have taken up residence in buildings and trees. Swarms can be obtained by listing your name with local fire and police departments. You can also purchase established colonies from other beekeepers. An alternative to beginning with package bees is to purchase nucleus colonies, which usually consist of four to five frames of bees and a queen. These small colonies can be transferred to a deep brood chamber. Their development and growth will be faster than with package bees, because nucleus colonies already have comb with eggs, larvae, and sealed brood.

Although cost and other factors may determine the initial number of colonies you should have, most beekeepers agree that an inexperienced beginner should not plunge into the enterprise by purchasing either a large

FIGURE 23. Hiving your first swarm of bees is one of the pleasures of beginning beekeeping.

number of colonies, or on the other hand, a single colony. Even though a magic number may not exist, three colonies is a good number to choose. Only by working with bees and learning of the kind of profit that you may derive from them, will you truly be able to determine whether you desire to augment your activity, continue it at its present level, or even at some point in time, phase it out.

Three colonies should also help assure the beginner of some minimum success in the early years, and serve as an index to the potential honey production available in the area. Should misfortune fall upon one of the colonies, the remaining ones will serve to encourage the beginner to continue, as well as provide the opportunity for the novice to use a portion of the bees from one of the surviving colonies to restock the dead colony, provided that the colony's demise was not the result of a disease that may infect other bees.

The selection of a site and the location of the colonies at the site, once it is chosen, require your attention and thought. Although it has been practical for man to construct roads and build houses in row upon row, we should not consider this neat and economically profitable way to parcel land as the mode for positioning beehives. Beehives located in rows may look neat and tidy, minimize the space needed for the colonies, and make them more convenient to manage, but other factors, including the behavior of bees, preclude you from considering locating hives in this manner.

FIGURE 24. Feeding colonies is often a necessity in the fall, late winter and early spring. Five-pound friction-top metal cans or plastic pails make convenient feeders. Just drill two small holes in the lid and invert the pail over the colony surrounded with an empty super and the hive cover on top.

FIGURE 25. Feeding dry sugar to prevent starvation in an emergency situation.

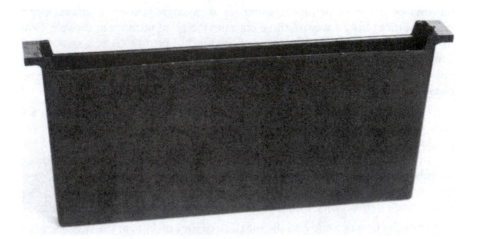

FIGURE 26. Plastic division board feeders replace a frame in the hive body. They can be filled with syrup or dry sugar.

Hives should be randomly located in the bee yard. The reason for this relates to bee behavior as well as environmental factors. When hives are placed in rows, breezes will blow bees away from their hives, and, under certain conditions, these bees will enter and become members of other hives. Should winds prevail for long periods during the flying season, some hive populations will expand, artificially in a sense, while others will be reduced. This imbalance will weaken some hives, strengthen others, and make it difficult for the beekeeper to assess the value of each hive's queen or the true honey production capabilities of each colony.

The movement of bees from one colony to another is referred to as drifting. It may result from bees being pushed from their entrance to another colony by air flow, or other factors, including failure to differentiate their own colony from another one, especially if the hives are in rows and are all painted the same color. Should bees carrying diseases or parasites drift from one colony to another, the predators may be transmitted to what was a disease-free colony. Further, with colonies in row, it may be more likely for a colony weakened by disease to have a bee drift into it, rob the honey, return to its own colony, and report its finding, resulting in many bees being recruited to rob the weak colony. As the bees return to their own colony with the booty, they also may be carrying back diseases.

In addition to all these disadvantages to locating bees in a row, still another one needs to be emphasized. When a colony of bees is disturbed, particularly when a beekeeper begins to inspect the colony, guard bees release an alarm odor which smells like banana oil. As the beekeeper works his way down a row, this alarm odor is likely to drift to successive colonies as well as to slowly pervade the apiary. The result often is that each succeeding colony will become more difficult to work, because the aggressive behavior stimulated by the alarm odor is intensified as the alarm odor level heightens in the bee yard. This odor is one of several odors collectively called pheromones. Pheromones are specific chemicals that convey messages to other members of the same species of organism. This alarm odor will incite bees to fly around the intruder and often to sting it as well.

Someone, many centuries ago, discovered that when smoke was administered to bees, the likelihood of being stung was reduced. Although the individual who made this observation was unable to learn why smoke reduced stinging incidents, the fact was passed on from that time forward. We wonder if some early bee gatherer using a torch to either light his way to a colony or to burn the bees, arrived at the colony with a torch that was smoking excessively and, as the gatherer began to remove the colony's comb, it was observed that the bees were relatively calm and stinging was far below normal. This early honey gatherer obviously connected the fact that smoke somehow changed the bees' normal response to having their nest robbed.

We now know why smoke reduces the incidents of aggression and stinging by bees. When bee colonies are being manipulated, guard bees, as a result of the colony being disturbed, release alarm pheromones, but the smoke's odor masks the alarm pheromone, thus other bees fail to respond, since the message to attack the intruder is short-circuited by the smoke. Smoke not only masks the alarm odor, it causes bees to move towards honey cells and begin to engorge the honey. The failure to smell the alarm odor because of the smoke and the fact that smoke stimulates bees to devour honey permits colony manipulation to take place while minimizing the chances of being stung. Thus all successful and intelligent beekeepers armed with this knowledge always use smoke as a standard operating procedure when planning to manipulate their bees. The smoker, the instrument used to make the smoke, has been likened to the lion tamer's whip. The tamer doesn't go into the lion cage without it; the beekeeper should never go to his bee yard without a working smoker. Just as man has not domesticated the lion, neither has man domesticated the bee.

Randomly located hives minimize, to a great extent, the problems that prevail when colonies are located in rows. Therefore, it should be clear that it is not to the beekeeper's advantage to locate hives in rows. It is also clearly demonstrated that, when bees begin their foraging activities, nearly one-third of the inexperienced foragers are lost. Hives located in rows may contribute to or add to this loss because of the difficulty young foragers may have in distinguishing their hives from one another. One way to assist bees in locating their hives and preventing drifting is by painting each hive a different color, or, if this does not seem practical, one may paint or hang different colored designs, such as triangles, squares, rectangles, and circles, near the hive entrance.

Returning to site selection, other relevant environmental factors to consider include the need to have a location that: is not wet; is protected from winds (particularly during winter months); has good air drainage, *i.e.* a site where air flows rather than stagnates; and is sunny throughout the year. In areas of extreme heat, where temperatures exceed 95°F for long periods, shade will be required in the summer months.

Another bonus at a site location would be the availability of water. However, having the good fortune to locate colonies close to water will be an uncommon opportunity. Nevertheless, since bees require water to cool their hives and dilute honey, they will search for and find water. Beekeepers should make an effort to avoid problems with their neighbors, especially in urban areas where swimming pools, bird baths, and other water containers will attract bees searching for water. The best way to keep bees from obtaining water from your neighbor's water containers is by supplying water to your bees.

FIGURE 27. Fences, trees, bushes or buildings provide excellent protection against cold winter winds and drifting snow.

The most challenging season for beekeepers is, without question, the winter. When the air temperature reaches 45°F - 57°F, bees begin to form, inside their hives, the winter cluster which is usually ellipsoidal in shape. The cluster becomes more compacted as temperatures drop. Depending upon latitude, such clusters may last for brief periods or for months at higher ones. The true measure of a beekeepers's ability should be how successful were his efforts in carrying full colonies through the winter into the spring alive, in good health, and capable of producing a sizable honey crop. Many papers and chapters have been devoted to the steps that need to be taken to insure the colony's health and survival through the winter months, but, alas, few bee-keepers have adopted these sound and tested principles.

Colonies kept outside must be readied for the winter in the late summer and early months of the fall season. Colonies must enter the winter with adequate bee populations, must be provided with medication and methods to control mites in order to keep them disease free, must be protected from wind, must have access to adequate food stores, must be queenright, must be kept dry, must have their main entrances reduced to keep out predators and chilling winds, and must have one opening in the second super or an escape in the inner cover, which permits flight and provides ventilation. Without diligent attention to these points, winter losses can be guaranteed.

Let us elaborate on the points made in the previous paragraph. The winter bee cluster is a living, eating, moving, brood rearing, and heat generating mass. In order for this mass to maintain itself, it must have ample food stores and be in continuous contact with these stores. In order to achieve this goal it must have a population of 20,000 to 30,000 which ensures its ability to cover a sufficient comb area containing food. Food will sustain this mass, and certain molecules, manufactured from the food, release heat energy when broken down during muscle contraction. This heat, a by-product of muscle contraction, is essential for the bees if they are to stay alive, sustain their brood, and remain mobile. A loss of contact with the food will, therefore, bring about the demise of the colony. The consumption of food, the building up and breaking down of food molelcules by bees, produces CO_2 and water vapor. The water vapor and excessive amounts of CO_2 must escape from the hive in order to keep the bees dry and the hive air clean. A hole in the second hive body will help, however, some moisture absorbing material above the second hive body, such as straw or an insulate board, is recommended. In addition, reversing the inner cover, so that the half moon cut in the inner cover is open, will also aid in ridding the colony of excessive moisture and CO_2 as well as permitting bees a flight hole.

Given an adequate population with sufficient stores, the need to maintain that population in good health is a necessity. Otherwise, diseases or mites will decrease the population and could reduce numbers so that the remaining bees can neither cover combs containing food nor generate sufficient heat to sustain the colony. Therefore, the colonies being prepared for the winter season should be given medication to prevent American foulbrood and Nosema diseases. Medications can be administered while feeding bees sugar syrup, which serves a two-fold purpose: The syrup will add to the colony's stores, since each colony should have 90 pounds of food for successful wintering, and as the bees draw and feed on the syrup, they also are partaking in the medication. In addition to American foulbrood and Nosema diseases,

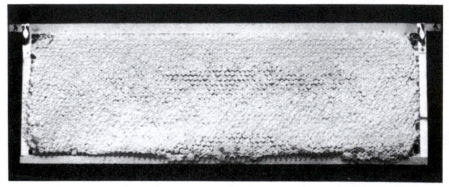

FIGURE 28. A completed shallow frame of comb honey may be consumed, sold as is or cut into sections as shown below.

two recently, accidentally introduced mites (*Varroa jacobsoni* and *Acarapis woodi*) have placed an additional stress on the honey bee. These mites shorten the life span of the honey bee and high populations of *Varroa* ultimately bring about the death of the colony. Acaricides are being used to drastically reduce populations of these mites in many parts of the world, and tests on their effectiveness and whether or not they adulterate honey and beeswax are now being conducted in the States. There is little doubt that colonies with sizable mite populations will not be able to survive on their own. The open question at this time is whether to use acaricides or follow some form of biological control in order to keep mite populations low or nonexistent. In any case, bee mites present in wintering bee colonies will undoubtedly increase winter losses which only further underlines the need for taking seriously the importance of preparing colonies for their passage into winter.

The main entrance of the hive needs to be reduced with an entrance cleat, which will block out winds, snow, and rain and also keep small mammals, especially mice, from entering the hive. Mice are more than just a small nuisance in a winter hive. In order to build their nests, they will remove a portion of comb from several frames and drag into the hive an assortment of debris, creating a nest of grass, paper, and other materials in the space formed by removing comb. Often in the spring, they are still found in the hive with their young. Some beekeepers place quarter inch hardware cloth over the entrance to keep mice from entering the hive. With a reduced main entrance and with the possibility of it being blocked by snow, ice, or dead bees during the winter, an entrance hole in the second hive body is essential. Some beekeepers are extremely reluctant to drill a hole in the second brood chamber, yet such an entrance will provide ventilation and an exit hole for bees when winter temperatures rise, to permit bee flights.

FIGURE 29. This is what it is all about! You'll be proud of your first comb honey or attractively bottled extracted honey.

Although the colony population will be reduced during the winter due to attrition, the queen will resume egg laying in late December or early January and, thus, replenish some of the lost bees. Colonies, therefore, need to be queenright in the fall; otherwise, as the population decreases in the absence of a laying queen, the colony will either die or be so diminished by the spring that its recovery, even if successfully requeened, will be unlikely to produce a honey crop.

After the winter passes, the great renewal begins with most of nature's animals and plants. We, as mankind, are not silent partners to this renewal; in fact, we delight in its arrival. As beekeepers, we await our first inspection of the colonies and begin to think seriously about the season ahead. We come to realize our work and our bees are not separate from the rest of the fabric of nature, but directly dependent upon it. Rain, sunlight, temperature, humidity, and flower abundance, to name a few, will all play a role in how the colonies will prosper. None of the meteorological phenomenon will be the same from year to year, and thus, as we strive to master the management of the honey bee, we must always be aware that no two seasons are alike. Each new season will bring to us new problems as well as new opportunities. Only through constant attention to changing conditions can the well-being of bees be maintained, and, in turn, our minds will remain active. This is one of the great rewards given to those who, through their work with bees, continue to learn and grow, as they strive to match nature's harmony with bee management.

CHAPTER 14

MANAGEMENT FOR HONEY PRODUCTION

by JOHN T. AMBROSE[*]

Beekeeping and bee management for honey production is an art not a science. It may be a science one day when bee researchers and beekeepers have a more complete understanding of bee biology and behavior and the environment in which bees live, but that is not yet the case.

The honey bee is not a domesticated animal and most of the "bee management" practiced by beekeepers is really just a matter of accommodating the natural biology and behavior of the bee. F.E. Moeller, in the 1980 issue of the USDA Beekeeping Handbook #335, stated: "Beekeepers in managing or manipulating colonies, are merely facilitating normal biological colony changes to suit their purposes." Management for honey production definitely fits into his description.

Honey bees naturally collect nectar and store surplus honey. Management for honey production by the beekeeper is in reality a system that assists the bees in the process. By providing adequate storage space for the honey surplus, by insuring that the colony has young viable queens, by minimizing bee losses due to diseases, pests and pesticides, and by other similar activities, the beekeeper is enhancing the honey bee colony's ability to make surplus honey. This necessitates the beekeeper having some knowledge of bee biology and behavior as well as some understanding of local plant (floral) sources in order to be a really successful honey producer.

The more the beekeeper knows about his bees, the better he will be in "managing" his bees. For example, swarm prevention is an essential part of good beekeeping management and the successful beekeeper will develop an understanding of what conditions lead to swarming. Time of year, nectar and pollen availability, and congestion in the hive are just some of the factors contributing to swarming. Unsuccessful swarm prevention management results in decreased honey production because of the decrease in colony size. This can be minimized with a good management system and an understanding of the bees.

There is no set formula or recipe for good bee management, but there are general procedures that should be considered and understood. This chapter is designed to assist the beekeeper in developing a management system that works for him in his area. One consideration is that nectar- and pollen-producing flora vary considerably from area to area and even within areas due

---

[*]John T. Ambrose, Ph.D., Professor of Entomology and Extension Apiculturist, North Carolina State University, Raleigh, N. C.

to elevation and other factors. However, in order for the bees to make surplus honey they must be able to take advantage of the nectar and pollen flows that do occur. Each beekeeper must take those flows into consideration in developing a management system.

For ease of development and to meet the varied needs of beekeepers across the country, this chapter is divided into sections which can be read separately or in conjunction with each other. The sections are arranged into a general seasonal format starting in the spring of the year with the establishment of an apiary and of individual colonies of bees from either packages, nucs or swarms. This is followed by a general description of seasonal management and related topics *i.e.,* swarm prevention and control and the value of superior stock and good queens. Then a section is devoted to the specialized two-queen management system. This is followed by a variety of topics that should be considered throughout the year by the beekeeper in developing a management system for honey production: feeding bees; providing water for the bees; ventilation and shade; moving bees; drifting; and robbing. The last brief section deals with the development of a record-keeping system for the beekeeper.

SITE SELECTION

A statement frequently made in beekeeping texts is "Bees may be kept almost anywhere" and that statement is basically true, but not every location will provide the beekeeper with a surplus of honey. The location that may be appropriate for maintaining a few colonies of bees on a permanent basis may not be suitable for producing a surplus honey crop with a large apiary.

Meeting all of the criteria for an ideal apiary location may not be possible in most situations, but all of the following factors should be considered:

Nectar and Pollen Availability

Almost any location from a roof garden in New York City to the desert areas of the southwest United States have some nectar and pollen sources, but the ideal location is one that has enough bee forage for a large number of hives. In general this would mean the availability of a good mix of plants and trees that produce adequate pollen and nectar for the colonies to develop strong bee populations in the spring and one or more major nectar flows during the remainder of the year that provides the bees with a large surplus of honey. There is no set formula on the number and duration of nectar flows, but it is best not to have long periods of little or no flow during the active season.

Quantity of the nectar flow is important to the beekeeper, but so is quality. Some honeys are considered to be superior to other types and regardless of personal opinion on this subject, the beekeeper is probably in business to sell his honey. Thus, if the public is willing to pay a premium price for certain honeys such as basswood or sage honey, then it might be worthwhile for the beekeeper to accept small honey surpluses if he can sell the crop for a premium price. On the other hand, some honeys such as aster tend to present problems in that they will granulate very rapidly. This can present a problem if the beekeeper is using the honey for a winter feed for the bees or if he is bottling it for sale. Aster may produce large quantities of honey in some areas, but the quality must be considered.

The availability of nectar is obviously of primary concern in selecting an apiary site, but it is also important to consider the availability of pollen. Bees must satisfy their protein needs and brood production will suffer if pollen is not available in adequate amounts. Early spring is a crucial time in the development of the colony and pollen is essential. Once the bees have exhausted their overwintering pollen stores, they must have access to pollen to continue brood rearing. The beekeeper can feed pollen substitutes or supplements to the bees, but this is an extra expense, not to mention the time involved. When selecting a site, the beekeeper should keep in mind that pollen may be available to the bees from both nectar and non-nectar producing plants. See Chapter 11, for information on pollen and nectar sources in the various regions of the United States.

Water Availability

An adequate water supply is essential to the survival and success of an apiary. Bees require water to cool the hive and to dilute food for feeding of the brood. See the "Providing Water for the Bees" section of this chapter for more information.

Exposure to Sun and Weather

Bees are cold blooded insects and they will only fly when it is warm enough for them to leave the hive. It is generally assumed that 70°F or better temperatures are necessary for good bee flight from the hive. The location of the apiary site can have a substantial impact on bee activity. Locate the hives so they receive morning sun and face the hives toward the south or east if possible. In some areas with high daytime temperatures, it is also advisable to combine afternoon shade with the morning sun exposure.

The availablility of natural windbreaks also should be considered. Try to locate the apiary on dry ground with good drainage and so that the hives are not facing into the prevailing winds unless a substantial windbreak is present. As a rule of thumb avoid extremes: don't place the hive on a hilltop, but also avoid low lying areas or depressions. Cold air and moisture can settle in such areas and are detrimental to the bees. Bees in a cold, wet location tend to be more aggressive and more prone to stress such as nosema and European foulbrood.

Access

Some locations may be terrific locations for honey production, but the lack of access to the beeyard can be a serious problem. If the beekeeper cannot get to the yard for routine management or honey removal, then the yard may not be in a suitable location. Access includes getting to the apiary when the weather is not perfect.

Pesticides

Each year a substantial number of bees and bee colonies are destroyed by pesticides. When selecting an apiary site, the beekeeper should check on pesticide usage in the area. In many situations it is impossible to locate in an area where there is no pesticide usage, but the beekeeper should definitely avoid establishing an apiary next to a field that receives aerial spraying of insecticides that are toxic to bees. Check with the local agricultural extension service for information on pesticide usage.

Vandalism and Theft

A hive of bees is a valuable commodity and "bee rustling" does occur. If the apiary is an outyard where the beekeeper cannot make frequent checks, then it is important to make the apiary as inconspicuous as possible. Avoid locations that can be seen from nearby roads and consider painting the hives with natural colors so they will blend into the background.

Pests and Predators

Every area of the country has at least some minor occasional pests or predators that affect honey bee colonies. Skunks, yellowjackets, and mice are examples of beekeeping problems which normally can be handled without too much difficulty and are really just nuisances. However, some problems such as bears are a more serious consideration. Bears will eat honey and bees and can destroy entire apiaries in the process. Apiaries can be protected with electric fences, but the beekeeper must decide if the location is worth the expense and maintenance associated with an electric fence.

STARTING WITH PACKAGE BEES

The use of package bees is one of the most common and perhaps enjoyable methods for both beginning and experienced beekeepers to establish or increase the number of colonies in the spring. A package is a wooden box with wire screen on both sides that contains a mated queen, usually 2 or 3 pounds of bees, and a small amount of food for the bees to eat while enroute to their new homesite (see Fig. 1).

The package bee business is a major industry that has developed in several of the southern and western states to provide starter bees for the rest of the country and a number of foreign countries. Package bees are used to start new colonies, either replacement or increases. Most package bees are purchased in

FIGURE 1. Every year commercial beekeepers ship large numbers of package bees to beekeepers who are starting new colonies or replacing lost ones.

early spring, but there are some sales throughout most of the spring and summer.

The best time to install a package of bees is when the build-up of nectar is underway in your area. This should provide some time for the package to develop into a working hive so that it can take advantage of the primary or major nectar flow(s). It is recommended that orders for packages be placed by January of the year in which you want delivery. Because of the variation in nectar flows among regions and the variations from year to year, it would be arbitrary and unreliable to recommend specific dates for package deliveries. Based on an area's general honey flow history, the beekeeper can select the best time for package delivery. It is better to have the bees arrive on the late side than on the early side because cold weather or lack of nectar will work a hardship on the bees.

Packages may be ordered as 2, 3 or even 4 lb. containers. The best size depends on area conditions and whether the bees will be placed on drawn comb or on foundation. In areas where there is only a short buildup (nectar flow) before the major flow, the three pound package is recommended, whereas a longer build-up period lends itself to economical use of a two-pound package. The other consideration is comb availability. The package bees will become established and build up more rapidly on drawn comb than on foundation. In order to use foundation, the bees must first draw out some comb for nectar storage and egg laying by the queen. The delay time in producing wax and building storage space will delay colony development. To allow for such delay, the beekeeper should order a 3 pound package of bees.

Package bees are shipped through the U. S. mail system and contrary to some opinions, the U. S. Post Office does a good job with package bees. The alternative to receiving bees through the mail is to make a trip to the package bee supplier and pick up the bees. This may be economical and time efficient if a large number of packages are being purchased either by a beekeeper or a group of beekeepers. The actual package, regardless of how it is shipped, will contain the specified pounds of bees with an average of 3,500 bees to the pound so a 3 pound package has about 10,500 bees. Most of the bees will be worker bees with only a few drones, if any. The queen that is shipped with the package will be a mated queen (unless otherwise noted), but she will come from a different hive than did the package bees. The package bees were taken from a colony with surplus bees, but the queen was taken from a mating nuc. The queen will be placed in a small cage called a queen cage (see Fig. 2) and several attendant bees from her mating nuc will be placed in the small cage with her. That cage will be fastened inside the larger package and some sort of a feeding container will be installed so the bees have food during the shipping process.

Normally it takes only a few days for the bees to arrive after they have been mailed and most beekeepers make arrangements with the post office to

FIGURE 2. The beekeeper examines a "Benton" queen cage which contains a newly mated queen and several attendant bees which will care for the queen. *(Photo by S. Bambara)*

FIGURE 3. The U.S. Post Office will ship packages of bees and the beekeeper can either pick them up at the local post office or have them delivered to his home with the regular mail. *(Photo by S. Bambara)*

call them when the packages have arrived. The package can be picked up at the local post office or the route mailman will bring them with the regular mail (see Fig. 3). Upon receipt of the bees the beekeeper should check the package for bee loss during shipment. Normally some of the bees will have died during shipment due to normal longevity if nothing else and the beekeeper should expect to see some dead bees on the bottom of the package. If the dead bees are no more than 1/2 inch in depth, then there is probably no real problem. At this stage it is impossible to observe the queen's condition without opening the package and this is not advisable from the mailman's perspective. Regardless of the condition of the bees, it is best to accept the shipment and then to contact the supplier if there is a problem. The U.S. Post Office will not insure the package bees and you may not make a claim against the U.S. Post Office unless there was obvious damage to the package, itself, such as it being broken. A word of caution: the bees in the package will give off attractant chemicals call pheromones and these may attract other honey bees to cluster on the outside of the package. Do not assume that the package has a leak just because there are bees on the outside of the screen.

Upon accepting the package, the next step is to acclimate the bees for introduction into their new hive. The bees have been under stress during the shipping process and the best thing you can do is to feed them a light mixture

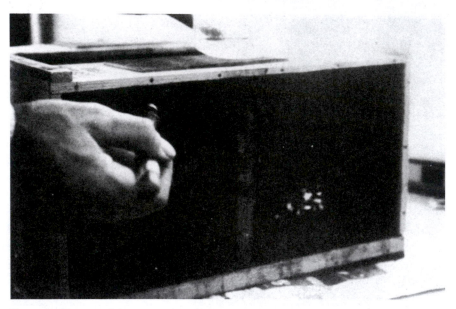

FIGURE 4. Upon receiving a package of bees, the beekeeper should feed them a light (1 part sugar to 2 parts water) sugar syrup. The best feeding method is to brush the syrup onto the side of the cage and continue to do so as long as the bees will feed on the syrup. *(Photo by S. Bambara)*

of sugar syrup (one part sugar to one part water) (see Fig. 4) and then place the package in a dark room at room temperature. See "Feeding Bees" in this chapter for more details on the feeding and care of package bees. It is assumed that the beekeeper has prepared his equipment for hiving the package bees before their arrival. The necessary equipment will vary, depending on what the beekeeper has available, but the preferred equipment would be a deep super (brood chamber) with drawn frames, a frame of honey, a frame of capped brood, and the necessary miscellaneous hive equipment.

The best time to install the package of bees is in the late afternoon. This will reduce the possibility that some of the bees will fly off in search of food and become lost. The bees will have time to acclimate to their new hive site if they are installed late in the day and have little flight activity until the next day. Reducing the stress on the bees is still a prime consideration and the bees should be fed just before installation in the hive. A thin sugar syrup of one part sugar to one part water is again recommended.

After the bees have been fed, carry the package out into the beeyard. The beekeeper should wear a veil and it is recommended that the beekeeper's pantlegs be tied so that bees will not inadvertently crawl up his legs during the installation process. The use of smoke is not advised. Smoke works on bees primarily because it causes them to engorge on stored food in the hive, and since the package bees do not have any stored food, the smoke might serve as a detriment in that it might delay their transfer into the hive.

After preparing the hive equipment and feeding the package bees, the next step is to open the package and check on the queen's condition. The most important consideration is whether or not she is alive. The installation process will continue whether she is alive or not, but the beekeeper should obtain a replacement queen as quickly as possible if she's dead. The actual installation process is described in Chapter 13.

As mentioned before, it is recommended that the beekeeper use drawn comb, a frame of stored honey, and a frame of capped brood if they are available. The advantage of drawn comb and stored food and brood is that it will shorten the time for colony development and reduce the possibility that the new colony might die in the transition period. The package of bees contains only adult bees and no brood. This means that it will be at least 21 days before any new adult bees emerge, assuming the queen starts to lay eggs immediately (thus the value of drawn comb). Fig. 5 shows the typical development of a bee colony made from a three-pound package of bees in a

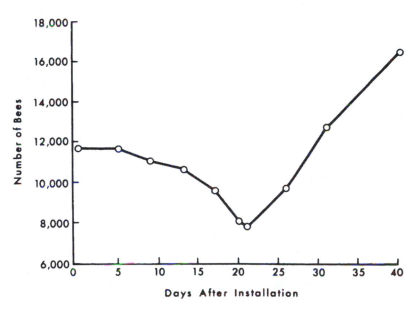

FIGURE 5. The typical growth of a three pound package of bees during the first 40 days after package installation. Note the decline in bee numbers for the first three weeks since the colony was started with only adult bees and no new bees will emerge until after the initial 21-day period. (*after Cooperative Extension Service, Cornell University*)

study by the Cooperative Extension Service at Cornell University. It is very important to note that the bee population actually declines for the first three weeks as adult bees die when there are no emerging adult bees to take their place. After three weeks, new adult bees begin to emerge, assuming the queen began to lay eggs immediately in the new hive. The installation of a frame of capped brood provides replacement bees during most of that initial three-week period. It is important that capped brood and not uncapped brood be used because the uncapped brood would place additional stress on the bees due to their food and heat needs.

Depending on where the package colony is established, the beekeeper may or may not expect to make any surplus the first year. In those areas with a long period of nectar flow, the colony may make a surplus for the beekeeper, but in other areas the goal is for the bees to make enough honey just to overwinter the colony the first year.

Two additional considerations for setting up colonies from package bees are bee diseases and pests and the danger of robbing and drifting. When buying bees the beekeeper should always insist on getting a certificate stating the bees are disease-free. Most states require such a certificate before the bees can be shipped into the state. The beekeeper should also consider whether or not the shipping state has a program to control such bee pests as *Varroa* and tracheal mites. It would be unfortunate to contaminate an apiary with one of these pests if it was not already established in the beekeeper's area.

The other consideration is drifting and robbing. Strong colonies of bees may attempt to rob the newly established package colony of its food stores, particularly if it was fed sugar water and it leaked onto the ground. Drifting is also a concern if the package colony is established in an existing apiary. The best remedy is to initially establish any package colonies in a distinct apiary with only other package colonies until they are well developed. At that time the hives can be moved to an existing apiary with other colonies of bees.

Management of the colony started from package bees is similar to that of other colonies once the colony is well established. The primary goal is to increase colony size so that it can produce a surplus of honey.

STARTING WITH NUC HIVES

Recently it has become more popular to start a hive of bees with a nuc hive. A nuc generally contains several frames of brood, a laying queen, and some stored food: both pollen and nectar. A nuc hive has all of the features of an overwintered colony of bees except on a reduced scale. There are obvious advantages to starting a colony with a nuc compared to a package, but there are also disadvantages.

The most obvious advantage to a nuc over a package of bees is that the nuc contains brood and food and the queen is already laying. If the beekeeper refers to Fig. 5, it is obvious that the lack of existing brood is a hindrance and perhaps even a threat to the development of a package colony of bees. That problem can be reduced by adding a frame of brood to the hive when the package is installed, but there is still at least a temporary delay before the queen begins to lay eggs. Perhaps just as important is the fact that the nuc hive already has food stores and those stores are an integral part of the colony unit. Adding brood and food to a package is not quite as good as having a unit (nuc) that has incorporated sources of stored food and brood.

The disadvantages to starting with nuc hives are primarily cost and shipping. The cost of a nuc is generally at least 25 - 50% more than the cost of a package of bees. In addition, the beekeeper will normally have to make arrangements to pick up the nuc(s) or have them delivered to his location. Some beekeepers also worry about the threat of diseases, particularly brood diseases. Since a package has no brood, there can be no brood disease, but a nuc does have brood. The solution is to buy nuc hives from reputable sources where the local state department of agriculture or its equivalent has certified that the bees are disease free.

If the decision is made to purchase a nuc hive(s) to start new colonies or replacement colonies, then the beekeeper still has the question of whether to use drawn frames or foundation to fill the hive body after installing the four or five frames from the nuc or to use foundation. As always, drawn comb is the preferred situation in that it permits a faster buildup by the bee colony. However, the nuc colony can more readily deal with foundation than can a package bee colony because the nuc colony is already on its original drawn comb and has a supply of brood. That difference is one more advantage of a nuc hive over a package of bees when starting a colony.

Given a choice between starting a colony of bees from a nuc hive or from a package of bees, the nuc is the preferred choice from a colony development and honey production standpoint, assuming both types of units are available at the same approximate time and the price difference is not too great. The one major advantage of starting with a package colony is for the beginning beekeeper since he can grow in beekeeping experience as the colony grows in complexity and in size.

STARTING WITH A SWARM OF BEES

Many beekeepers have obtained their start in beekeeping by capturing and hiving a swarm (see Fig. 6). There is no doubt that this is a very satisfying way to start because you have "captured" the bees and there is also the obvious advantage that the bees are free if the beekeeper doesn't count his time, gasoline, and other related expenses.

FIGURE 6. A swarm of bees. Collecting such swarms can be a very enjoyable and profitable springtime activity for the beekeepers. *(Photo by J. Ambrose)*

Basically, starting with a swarm is very similar to starting with a package of bees. The same equipment is needed to hive the bees and the beekeeper is dealing with only adult bees so the addition of brood to the hive is an important consideration. The one major advantage of a swarm over a package, other than cost, is that the swarm queen is really the swarm's own queen and not a queen that was added to a package of bees. This association reduces the small chance that the queen will be rejected after the bees are hived.

Management of the swarm colony is similar to the management of the package colony or the nuc colony once the unit is established with the primary goal of increasing colony size or strength and honey production. One important caution to the beekeeper is that not all swarms are equal. Swarming tends to occur just prior to or during the beginning of the major nectar flow(s), but that is not always the case. Some colonies will put out swarms as late as June or July and some colonies will even produce swarms in the late summer or early fall. Such late-season swarms are not a bargain for the beekeeper even if the bees are free. The bees must develop sufficient food and brood stores to overwinter and that may involve considerable expense to the beekeeper if he has to feed large amounts of sugar or other honey substitutes to the swarm colonies.

Location, climate and nectar flows determine whether or not a swarm is a bargain. A swarm of bees that issues from a hive in upper New York state in

June would probably have sufficient time and nectar flow to store enough surplus for successful overwintering, but a June swarm in eastern North Carolina would simply be an expense to the beekeeper. The old saying:

A swarm in May is worth a load of hay
A swarm in June is worth a silver spoon
A swarm in July isn't worth a fly

is true in certain areas and untrue in others. Southern beekeepers need to modify the refrain to describe the value of a swarm in April and the expense of a swarm in June.

SEASONAL MANAGEMENT

Because of the great diversity in beekeeping conditions throughout the United States, it would serve little purpose to propose an exact scheme of seasonal management based on the calendar that would apply to all beekeepers. Some areas produce their major honey crops in the late spring and early summer, while other areas have major flows spaced more regularly throughout the year. Nevertheless, there are similarities in all of the management schemes and some general management practices can be developed. For ease of discussion, seasonal management is divided into the following four periods: 1) Early Spring Management of Overwintered Colonies; 2) Spring Management; 3) Summer Honey Flow Management; and 4) Fall/Winter Preparation Management.

Early Spring Management of Overwintered Colonies

The primary purpose of the late winter/early spring inspection is to make sure that the overwintered colonies have enough food stores so that they will not starve before the nectar flow starts in earnest. This is also the time to remove those hives from the apiary where the colonies have died.

Most bee colonies will begin brood rearing in the late winter/early spring before the bees begin to forage and to collect pollen. The bees use pollen which was stored in the hive from the previous year. This resumption of brood rearing after a winter break is a crucial time for the colony. Large amounts of honey (perhaps 10 lbs. per week) and pollen are necessary to continue the process. If the bees run out of stored food before foraging conditions permit adequate food collection, then the colony may starve. The beekeeper's job at this time is to check on food stores in the colony.

This early inspection is not an intensive manipulation of the hive, but chilling of the brood must be avoided. The preferred minimum temperature for this inspection is at least 40°F. An experienced beekeeper can get a fairly good estimate of the colony's food stores by lifting one corner of the hive. The colony should have a minimum of 30 pounds of stored honey in most areas to insure survival. A good way to estimate the amount of food stores is to count frames of honey. A deep frame that is basically filled with capped honey will contain 5 to 6 pounds of food, while a shallow frame similarly filled will have about 3 pounds. The beekeeper should not pull out individual frames to determine food stores at this time, but rather lift the cover and look down between the frames or crack the supers and look between the supers for frames of honey.

In addition to determining if there are adequate food stores, it is also important to check the location and arrangement of the food stores. As the bee

FIGURE 7. Honey bees can consume dry sugar and survive in emergency winter starvation situations. However, colonies are not as stimulated to rear brood as when liquid feeds are fed to colonies by the beekeeper in the spring.

cluster goes through the winter, it will tend to move upwards in the hive. There is a definite reluctance by the bees to move downward in the hive in search of food during the winter. If the cluster is found to be located just below the hive cover, then there may be a problem, even though the hive has adequate food stores for the colony. It is important to move food close to this cluster, but every effort should be made to avoid breaking or disrupting the cluster of bees.

In the situation where the cluster is just below the hive cover, the best remedy is to place several frames of honey on both sides of the cluster without causing any undue disruption to the colony. These honey frames can be obtained from the same hive or from colonies that have surplus food stores. Whatever the source, the beekeeper should be sure that the honey frames are free of disease.

If stored honey frames are not available, then the beekeeper will have to provide an alternative sugar source to those colonies that have inadequate stores. For a description of feeding bees, see the section "Feeding Bees" in this chapter. The primary consideration in feeding bees at this time of year is that it is very difficult, if not impossible, for them to break their cluster. The food must be located near the cluster. Some beekeepers will avoid the use of sugar syrup at this time because it does place some stress on the bees, since they have to evaporate the excess water and convert the sucrose (table sugar) to glucose and fructose. An alternative is to feed dry sugar or bee candy (see Fig. 7). Whatever method is used, only feed inside the hive.

In addition to insuring that adequate food stores are available to the bees, the beekeeper should also use this inspection to remove any hives from the apiary where the colonies have died. If possible, try to determine why the bees died so that equipment from potentially disease-killed colonies is not put back into use where other bees may be infested. Dispose of the dead bees at a distance from the apiary and then transfer the equipment into your equipment storage area. Remember that any stored honey in the frames may ferment and wet equipment will grow mold.

Spring Management

The first spring management inspection should be performed when the bees think it is spring. This is one of those situations where the bees and their beekeeper may have different opinions on the subject. As a general rule the beekeeper can assume it is spring time for his bees when they have begun to forage for pollen and are seen bringing it into the hive. This collection of pollen indicates that brood rearing is definitely underway in the hive.

This inspection will be a fairly intensive inspection so it is important that weather conditions be favorable. Brood can be very quickly examined when the temperature is in the low 50 degree range with no wind, but more involved manipulations really require temperatures of at least 65 degrees. This inspection will be used to again evaluate food stores in the hive, to clean colonies if necessary, to remove any exterior hive wrapping that may have been used for overwintering, to check the queeen's condition, to check for diseases, to provide room for the future honey crop, and to begin seasonal swarm control preparations.

Food stores are still an important consideration for the survival of the colony. Since the bees are foraging, there is probably some nectar and pollen coming into the hive. However, weather conditions are often unpredictable in the spring and a period of cold or wet weather may stop food collection. The beekeeper should make sure that his bees have enough stores to "weather" such a crisis. A surplus of 30 pounds is still a good goal. Colonies that are borderline in stores should be fed. Feeding might also include providing pollen substitute or supplement, particularly in those locations where pollen availability is low or erratic in the spring.

Typical housekeeping chores should also be done during this inspection. Replace any frames that are damaged, such as comb destruction caused by mice that overwintered in the colony. If there has been any moisture accumulation in the hive, then dry or replace any wet equipment and consider the use of an auger hole in the upper hive body to increase ventilation. As with all inspections, the beekeeper should be alert to any disease problems. The presence of nosema may be indicated by the signs of dysentery on the outside of the hive or on the ground. In the past, it has been cautioned that the presence of crawling bees in front of the hive is also a sign of nosema, but in today's beekeeping environment, it may also indicate the presence of a heavy infestation of tracheal mites. In any case, the cause should be determined as quickly as possible and the state department of agriculture will normally provide a bee disease inspection for this purpose.

Some beekeepers use the spring inspection process to medicate their bees against potential disease problems. Fumidil-B® may be fed to the bees in sugar syrup for nosema control. Terramycin® may be fed to the bees in patties or dry sugar to prevent American or European foulbrood. In addition, control measures may be performed for the control of tracheal mites and *Varroa* mites. If any medication or acaricides are used, it is essential that all treatments stop at least four weeks before the beginning of the major nectar flow so that the honey crop is not contaminated.

The success of the colony is directly linked to the queen's condition. This inspection should involve a look at the queen. The queen does not have to be visually sighted to determine her condition, but it is helpful. A colony that has a good brood pattern with brood in all stages, particularly eggs, has a queen which has been there within the last three days. In order for the colony to make a good honey crop, it must produce a large number of bees and the queen is essential. If the queen is too old or has developed a problem, then she should be replaced. An excellent management strategy is to requeen the colony when the queen reaches two years of age. This requeening can be done in the spring or fall or even after a summer honey flow, but it should be done on a routine basis. This process insures that you have a viable queen that can lay a large number of fertilized eggs and it will also reduce the tendency of the bees to swarm. Color coding the queens will be a real asset in locating and determining the age of the queens.

The remaining two goals of this inspection are to control swarming and to provide adequate storage space for the coming year's honey crop. These two items are related in that storage space for brood and food is needed to control swarming, as is the regular replacement of queens. One of the steps which is usually beneficial in spring management is hive "reversal" or the rearrangement of the hive bodies.

As the bee cluster goes through the winter, the bees usually move into the upper unit of the overwintered hive. When brood rearing is initiated, it usually occurs in the top hive body which now becomes a brood chamber. If the bees were overwintered in two hive bodies, then the bottom unit may now be empty (of brood or food) and it will probably remain so. Bees do not tend to store food below the brood nest and this means that the bottom chamber may

FIGURE 8. Reversing the hive bodies of a beehive in the early spring is a simple and effective method of swarm control. The goal is to place the hive body with the brood and the queen beneath any hive bodies that are broodless. *(Photo by S. Bambara)*

not be used by the bees. This effectively reduces the "useable"size of the hive by one-half and may contribute to swarming. The solution is simple, reverse the hive units.

Reversing (see Fig. 8) is one of the easiest and perhaps one of the most effective swarm control processes that the beekeeper can perform. There is generally no need to switch or rearrange frames. The beekeeper simply moves the upper hive body to the bottom of the hive and places the empty hive body on top of the hive. Depending on the bees and the progression of the season, this step may have to be repeated in another two weeks or so. Simply check the colony to see if the brood cluster has moved to the upper chamber of the hive and the bottom super is not being used. Usually a second reversal is all that is needed, but sometimes a third reversal is helpful.

Once the hives have been reversed and it has been determined that the colony has a good queen and food stores, then is the time to begin providing storage space for the coming honey crop. Some beekeepers will place a queen excluder above the second hive body and this keeps the brood chamber in the two bottom units of the hive. This eliminates the possible mixing of brood with stored honey and makes the honey extraction process simpler. Other beekeepers believe that a queen excluder acts as a "honey excluder" in that it slows down honey storage above the excluder. Bees can be managed with or without an excluder and it varies from beekeeper to beekeeper.

This is now the time to place supers of drawn combs on the hive for the honey surplus. The number of supers to use is still a matter of discussion among some beekeepers. Some individuals believe it is best to add one super at a time while others will add multiple supers. Researchers for the U.S. Dept. of Agriculture conducted some very practical research in which they demonstrated that the "honey hoarding" instinct of the bees was actually increased if the amount of storage space (drawn comb) was increased. A colony with two or three empty supers of drawn comb will store more honey than a colony with one super of drawn comb during the same period of time, assuming that the colonies are of equal size. Some beekeepers worry that wax moths may damage the surplus comb if the honey flow stops, but this can be incorporated into the beekeeper's management system. An important consideration is that the hoarding instinct of the bees is increased only when drawn comb is used, and the use of foundation does not show any positive effect.

If the beekeeper follows the USDA recommendation to place all the available drawn comb on the hives, then the best approach may be to oversuper in the spring and to undersuper at the end of the honey flow. This will encourage rapid filling of the frames at the beginning of the major flows and will also encourage consolidation of the honey stores at the end of the flow.

Whether the beekeeper decides to "oversuper" or to add one super at a time, he has one other decision to make. Should he use supers with 10 frames or supers with nine or even eight frames? Even though the standard supers are made to accommodate 10 frames, they can be managed with eight or nine frames, and the resulting honey-filled frames will be easier to uncap and may even contain more honey than would the standard 10 frames. Honey bees will always honor bee space which is to say that they will maintain a space of 1/4 to 3/8 of an inch between all surfaces including the space between frames filled with honey or brood. Thus, a honey super that has nine frames will tend to have frames which are drawn out further and may contain more honey on a frame per frame basis than do the frames from a honey super with 10 frames. Because the bees tend to draw out the combs further in a nine frame super than a 10 frame super, the honey frames are also generally easier to extract because it is easier to cut off cell cappings which tend to be elevated above the wooden frame edges.

The advantages of a nine frame honey super are that it is probably easier to manipulate the frames in the hive, the frames are generally easier to extract, there may be more honey in the super because more space is available for honey storage, and the beekeeper has one less frame to buy for each super. These advantages can be increased even more by managing standard size supers with eight frames, but it does take more management on the beekeeper's part to insure that there is always equal spacing between the frames. Some beekeepers will also use a nine frame system in the brood chamber. This system reduces the number of cells for egg laying, but it does make frame manipulation easier. The increased ease of frame manipulation means that there is a reduced chance of accidentally killing (crushing) the queen when manipulating the brood nest. This author recommends nine frames in brood chambers and honey supers.

One last consideration is that some colonies may be underpopulated or weak at this time of year. The beekeeper who wishes to make a good honey crop wants as many strong colonies of bees as he can and now is the time to consider strengthening those weak colonies. Generally, it is not advisable to combine two obviously weak colonies, but the combination of a weak and a moderately sized colony is practical. The use of a newspaper sheet in the uniting process is advisable. Other possibilities include switching the position of a weak colony with a strong colony while the bees are foraging so that the foragers from the strong colony will join the weaker colony. Likewise, frames with capped brood can be transferred from a strong colony to a weak colony. Either of these last two procedures may serve to increase the population of the weak colony, but if the weakness is due to a failing queen or a disease problem, then the work may be of no real value to the beekeeper.

Summer Management

The primary purpose of summer management is to insure that the bees have adequate space for brood production and honey storage. Of course, the overall health and well being of the colony is also important.

If the beekeeper follows the advice of placing all of his drawn comb on the hives during the spring inspection, then most of the summer management may be already accomplished. If only one super of drawn comb was used during the spring inspection or the beekeeper has mostly foundation available for honey production, then more work is necessary.

Honey bees will draw out foundation during a honey flow and beekeepers can count on this behavior to obtain drawn combs. However, it is a mistake to give the bees too much foundation at one time. Never use more than one super of foundation on a hive during the nectar flow, and if drawn comb is available, then some drawn combs should be mixed with the foundation-filled frames.

Bees need space to store their ripened honey, but they need even more space to store the nectar when it is first collected. Capped honey will have a moisture content generally below 19%, but the moisture content of nectar may reach 80%. Considerable space is needed for storing and processing the honey. The beekeeper should attempt to have at least one empty super on the hive during the honey flow at all times.

One topic of discussion among beekeepers is how to add extra supers to the hive. The choices are between top supering and bottom supering. Top supering is simply placing empty supers above the filled honey supers or the brood nest if no honey supers are present. Bottom supering involves lifting the filled or partially filled honey supers and placing one or more empty supers beneath the filled supers. This does have the advantage of reducing bee traffic over filled supers at the top of the hive and makes for cleaner (whiter) comb. Some beekeepers also believe that the presence of the filled or partially filled supers at the top of the hive encourages the bees to enter the lower empty supers.

There are three drawbacks to bottom supering. The first is that it is more work than top supering. The second is that the queen may enter the empty super if there is no queen excluder and begin to lay eggs. The third consideration is that if the honey flow stops before the supers are filled, then there may be a large number of supers only partially filled with honey.

In addition to insuring that adequate space is available for the bees, the beekeeper should also check on the overall condition of the colony and conditions which might affect it. Inspections for disease, including mites, are always a consideration in any inspection process. Summer is also a time when pesticide problems may result. The beekeeper should be alert for these and any other potential problems such as attacks by skunks, bears, and other pests or predators.

Fall/Winter Preparation Management

Fall is often the time when many beekeepers remove the crop. In some areas this has been done earlier in the season, but the procedures are similar. The beekeeper has several alternatives available and the best one depends on local conditions and the number of colonies involved.

Brushing bees off of the frames, the use of bee escapes, fume boards, and bee blowers are all options in removing or robbing the crop. A beekeeper with a few hives of bees or a beekeeper in a hurry might well prefer the use of the bee brush. This can be done at any time of the day and either a few frames or entire supers can be removed by this process. Brushing is usually a combination of brushing some of the bees off of the frames and some shaking of the frames. The key is speed and keeping the beeless frames from attracting bees.

Bee escapes have been in use for some time and can be very effective in some situations. The bee escape is really an insert that allows bee movement in only one direction. The bee escape is placed in the center hole of an inner cover and that cover, now called an escape board, is placed beneath the supers of honey to be removed from the hive. This technique works best on cold or cool nights because the bees in the honey supers leave them to cluster with the bees in the brood nest and then cannot return to the honey supers. If the nighttime temperatures are warm then the devices work very slowly, if at all. Bee escapes are seldom used in the south.

Fume boards are another fairly popular way of removing or forcing the bees to leave the honey supers. This process involves the use of a board the width of the hive body which is covered on one side with a felt pad. A chemical is placed on the pad and the fumes from the process force the bees to move away from the board which is placed above the honey supers to be removed. Two chemicals are labelled for use in this process. The first is benzaldehyde and it works best when the temperature range is between 60 -70°F. The second product is butyric (proprionic) anhydride and it works best at temperatures between 80 - 95°F. This is a relatively inexpensive way to force bees out of the honey supers, but it is a temperature-dependent process.

The last and one of the most universally popular methods of removing bees from a large number of colonies is the use of a bee blower (see Fig. 9). There are no chemicals involved in this process and it uses only the force of blowing air to remove the bees from the supers. An effective way to use this device is to have one person tilt the honey super on end and the second person uses the blower hose to blow all of the bees off of the honey frames onto the

FIGURE 9. A bee blower can be a very efficient and rapid technique for removing bees from a large number of hives. *(Photo by S. Bambara)*

ground in front of the hive. The process is fast and does not cause excessive agitation among the bees if done in good weather and handled efficiently.

Once the honey crop is removed then the primary purpose of fall management is to prepare the bees for winter. Winter is the time when most bee colonies are lost and adequate fall preparation will help to reduce such losses.

Fall preparation involves several factors including colony strength, food stores, hive ventilation, and disease and pest control. Disease and pest control practices often include the use of medications or pesticides at this time of year. The use of Fumidil-B® for nosema control and menthol for tracheal mite control are fairly common. Local conditions and the history of the apiary will determine such usage.

Colony strength is very important. Strong colonies with a large number of young bees and a productive queen are essential. If the queen is more than two years old, then this may be the time to requeen the colony. A young actively laying queen will insure that the colony has a large number of young bees which will help insure survival over the winter months. Winter is often five to six months long and the importance of young bees becomes more essential in those areas with long winter periods.

Food stores are a major consideration for overwintering success. The required amount of food stores will vary from area to area and with the length of the winter. A colony in the north may require 100+ pounds. Check with your agricultural extension service or the apiculturist in your state for overwintering recommendations. If the food stores are inadequate, then supplemental feeding is necessary. See the section on "Feeding Bees" in this chapter. Fall feeding can consist of honey or sugar syrup. The sugar syrup used in the fall should be a concentrated mixture of 2 parts sugar to 1 part water. Do not feed any substances such as molasses or products that contain non-digestible (by the bees) material.

The quality of the food must also be considered. Some honeys such as aster are not good overwintering food stores unless they are mixed with other floral sources. Aster honey has a tendency to granulate in the comb so that the bees cannot use it during the cold winter months and it may cause dysentery. In some areas aster is a major nectar flow in the fall and the bees will collect large amounts. The honey can be a part of the winter stores, but it should not be the predominant portion. If nothing else, sugar syrup can be fed to the bees so that they can incorporate it into their food stores with the aster.

The arrangement of the food stores is also important, as is the bees' ability to reach it. Queen excluders should definitely be removed so that they are not between the bee cluster and the stored food. The food stores should be distributed among the hive bodies. If two hive bodies are used for overwintering, then approximately 1/2 of the food stores should be in the top super. The bees will tend to move upwards in the hive as winter progresses and this is an effective arrangement of food stores. It is also helpful to arrange the frames so that the two middlemost frames in the upper super are only partially filled with honey. This will make it easier for the winter cluster to make the transition from the bottom unit to the top unit during the winter.

One of the most important considerations for the overwintering colony is ventilation. Cold, dry weather is not a real problem to overwintering bees, but wet conditions can create serious problems (see the section on "Ventilation and Shade" in this chapter for additional information). Some beekeepers, particularly those in the northern states, wrap their hives for overwintering. This practice varies considerably even within the far northern states.

SWARMING: PREVENTION AND CONTROL

There are fewer things more rewarding or satisfying to a beekeeper than capturing and hiving a swarm. The satisfying part is probably almost always true, but the rewarding part depends upon whether the swarm issued from your hive or someone else's hive.

Until the mid 1800's beekeepers actually managed their colonies to encourage swarming. In those times, the parent colony was killed for its honey stores and the swarm was a replacement colony for the coming year. Until L.L. Langstroth revolutionized beekeeping with his discovery of bee space and the development of the moveable frame hive, swarming was encouraged. With the advent of the moveable frame hive, beekeepers could manipulate their hives and remove the honey crop without killing the bees, and swarm control became a way to increase honey production.

Table 1: Honey Production by Various Sizes of Colonies

Bees	Brood	Ratio of Brood Bee %	Lbs. Honey
15,000	11,850	79	25
30,000	18,300	60	68
60,000	15,000	12	154

After W.L. Gojmerac's Honey Guidelines for Efficient Production, Wisconsin Cooperative Extension Service.

If a beekeeper is in business to make honey, then he should exercise effective methods of reducing the bee's efforts to swarm (swarm prevention) and attempt to stop swarming once the bees have begun to make swarm preparations (swarm control). Swarm prevention does take some work and planning and the beekeeper should understand the value of this approach. Basically, it comes down to the fact that stronger colonies make more honey than do weaker colonies and two weak colonies do not make as much honey as does one strong colony.

Strong colonies are more effective honey producers than weak colonies because the stronger colony has a larger percentage of adult bees available for honey collection. As a colony increases in size, the number of bees required to take care of the household chores in the hive does not increase accordingly. Table 1 by W. Gojmerac shows the relationship between colony size, the percentage of bees that are brood bees (house bees) and the honey production of the various size colonies. A colony with 60,000 bees will make more than twice as much honey as will two colonies each of which has 30,000 bees. The implications for swarming are evident from this data. If a beekeeper has a colony of 60,000 bees and it swarms and the beekeeper catches the swarm and hives it, then he has two colonies each of which has about 30,000 bees. However, his honey production from those two units will be less than his expected production from the large parent unit if it hadn't swarmed.

Yes, it is advantageous to catch swarms even if they are your own swarms, otherwise you lose half of the bees. But the good beekeeper should practice swarm prevention and spend his free time collecting swarms from his neighbor's colonies.

Swarm prevention can be readily incorporated into a good bee management program and the benefits are usually spread over the entire management operation. Swarming is a natural process of bees where the colonies divide, and these divisions make up for colonies lost to disease, starvation, pests, and natural disasters. The beekeeper, however, is concerned with keeping his colonies strong and flourishing and not populating all of the surrounding areas with bee colonies. It is probably impossible to stop all swarming, but some progress can be made.

One of the most important contributions to swarming is the age and condition of the queen. Bees will replace a failing queen, usually by supersedure, but swarming also enters the picture. Research has shown that a colony with a three-year-old queen is about twice as likely to swarm as is a colony with a two-year-old queen. A similar comparison exists between a two-year-old and a one-year-old queen, but the difference is not as great. From an economic standpoint, the beekeeper should requeen the colony when the queen reaches two years of age. This can be done in the spring, the fall, or after a major honey flow, depending on local conditions. Requeening on a two-year cycle has two advantages to the beekeeper: it reduces swarming and it increases the probability that the queen will be a viable and reliable egg-laying queen.

Several factors affect the beekeeper's success in requeening. The new queen should be similar in condition to the old queen. In other words, it is better to replace a mated laying queen with a mated laying queen than with a virgin queen. The bees will more readily accept a new queen in the spring or in the fall when the colony population is not at its peak. The populous mid-season colony will more readily accept a queen cell or a queen that has already been laying in a nuc hive. The requeening process should be done when a natural nectar flow is in progress, if possible. If this is not possible, then the bees should be fed sugar syrup to approximate a flow.

The actual requeening process is fairly simple. Buy or obtain your new queen. Locate and remove the old queen from the colony, but not before you have obtained the new queen. Wait several hours before introducing the new queen. One of the most common introduction procedures is to use a queen mailing cage. When the queen was purchased, she probably came in such a cage and this may be used. Several attendant bees were probably shipped with the queen. These should be removed from the cage, preferably inside of a building or vehicle so that the queen does not accidentally escape. The cage will normally have a candy plug at one end. Remove the cork from that end and place the cage between two frames of uncapped brood in the colony with the candy end up (see Fig. 10). Within a few days the bees should have eaten the candy and released the new queen.

If the beekeeper follows the two-year requeening recommendation, then he should consider marking his queens. This serves several purposes. It makes

FIGURE 10. Using a queen cage is generally a very effective method of introducing a queen to a colony or of requeening the colony. Place the cage with the candy end up and with the screened portion of the cage perpendicular to the hive frames. *(Photo by S. Bambara)*

it easier to locate queens in the colony and by color coding the queens, the beekeeper can tell the age of each queen. The marking system will also alert the beekeeper to any supersedures that might occur in the hive because the marked queen will be replaced with an unmarked queen. The International Color Code for Queens is as follows:

> 0 & 5 - Blue
> 1 & 6 - White
> 2 & 7 - Yellow
> 3 & 8 - Red
> 4 & 9 - Green

Thus, if a beekeeper is following the system, his queens produced in 1991 should be marked with white paint and the queens for 1992 should be marked with yellow paint.

When ordering queens, the beekeeper can request that the queens be color coded. There is a small fee, but it is worth the cost. The alternative is for the beekeeper to color code his own queens, particularly those that he produces himself. It takes a little practice, but the coding is not really too difficult. Use any quick drying paint such as those sold for model airplane kits and a very fine brush. The inexperienced beekeeper can practice on drones. Hold the bee between the thumb and the forefinger and place a very small dab of paint on the bee's thorax (see Fig. 11). Do not blow on the paint to dry it. Wait a few

FIGURE 11. Marking the queen with a brightly colored paint will aid in locating the queen in the dark hive and will also help in management in that you can color code queens to keep track of their age. Use any quick drying paint for the process.

seconds and then put the bee back on one of the frames in the hive. If this process is done while the queen is in the queen cage, then it is not necessary to actually handle the queen. Simply stretch the wires on the cage apart and touch the brush to the queen's thorax.

Some beekeepers will clip one of the wings on a new queen in an attempt to prevent swarming and also to mark the queen. In odd numbered years the left wings are clipped and in even numbered years the right wings are clipped. It is assumed that because the queen cannot fly with the clipped wings, that the colony will not swarm. Unfortunately, this is not true. When a colony swarms it is usually because there is some kind of congestion in the colony and the bees actually force the queen to accompany them. A queen that is forced out of the hive entrance who cannot fly may be damaged. A swarm, upon leaving the hive, will settle at a temporary site to determine if the queen is with them. If the damaged (clipped wing) queen was lost, then the swarm will return to the hive and swarm a few days later with a virgin queen. The clipping process only slowed down the swarming process and may have resulted in the loss of a queen.

A second management technique that will help in reducing swarming is for the beekeeper to reverse his hive bodies in the springtime. See the "Spring Management" portion of the Seasonal Management section of this chapter for more details. The reversal process results in an empty hive body being placed above the brood chamber so that the queen will have ample space to lay eggs and the bees can store pollen and honey. Congestion or crowding is thought to play a prominent role in swarming and this process helps to eliminate congestion without adding or subtracting equipment to the hive. This process may have to be repeated several times in the spring.

The third management consideration is to always insure that the bees have ample room for food storage and egg laying by the queen. This is discussed more fully in the "Spring Management" portion of the Seasonal Management section of this chapter. Providing adequate space, requeening on a two-year cycle, and performing reversals on the hive bodies in the spring should keep swarming to a minimum.

There are steps that the beekeeper can take to control swarming, but the consequence is that swarming is usually reduced at the expense of the honey crop. Swarming can be stopped in a number of ways, but they are not appropriate if the goal is honey production. One recommendation is to cage queens when the colony becomes too crowded and is raising queen cells. The queen can be caged in a Benton mailing cage or a push-in wire cage (wire mesh measuring about 4 inches square with the edges pushed down [see Fig. 12]). Caging the queen for a week or so will normally relieve congestion in the hive

FIGURE 12. Two common kinds of queen cages are the "Benton" mailing cage and the larger push-in wire cage. The latter cage generally measures about 4 inches square and is made by folding wire screen. *(Photo by S. Bambara)*

and may stop the swarming process if it is accompanied by the destruction of queen cells. However, there is no brood production during that period and those potential bees will not be available to the hive.

Another approach to swarm control is to actually split the colonies. If the colony has queen cells, then the colony can be divided into two units and neither unit will swarm. The division is usually made by transferring the old queen and approximately 1/2 of the brood and food to a new hive. This unit is then established in a new location, preferably some distance from the parent hive. The old hive is either requeened or allowed to rear a new queen from one of the queen cells that have already been constructed. This unit is left in the hive's original location so that the forager bees in the field will return to it and keep its adult population at a high level. This approach will normally stop swarming, but honey production will suffer.

A short-term method of stopping swarming is to cut queen cells. This will work temporarily, but the long-term prospects are not good. Destroying queen cells is similar to treating a headache by taking a pain reliever. The underlying cause of the problem is still there. Assuming that congestion is the cause of the swarming, then queen cell destruction is a temporary measure. The bees will construct new queen cells and the process will continue. It is conceivable that the beekeeper can go back and cut cells every 8 - 10 days (see Fig. 13), but it is likely that the bees have more patience than does the beekeeper.

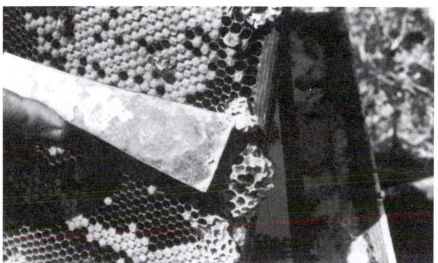

FIGURE 13. Using a hive tool to cut queen cells. If the beekeeper simply pokes a hole in the queen cell, then the bees in the hive will complete the process and destroy the queen larvae or pupae. *(Photo by S. Bambara)*

There is one method of swarm control that will work after the bees have constructed queen cells, but it is a labor intensive approach and is generally recommended only for hobby and side-line beekeepers with a small number of hives. This is the Demaree method named after George Demaree who first explained the process in an article in the *American Bee Journal* in 1884 with an updated version described in 1892.

The procedure is really not as complicated as it sounds. The process is initiated in the spring when the beekeeper finds queen cells and swarm preparations in a colony prior to a period when there should be a good honey flow. This procedure serves to maintain and produce a very strong colony that can take advantage of a nectar flow, but the beekeeper may wish to reconsider doing this process if a good flow is not anticipated.

For descriptive purposes let us assume that the colony is in two brood chambers and has one honey super. The beekeeper will need two additional brood chambers, preferably with drawn comb. Actually, the beekeeper will need an amount of equipment which is equal to the current brood capacity of the hive. He will also need a queen excluder.

The steps in the Demaree process are as follows:

1. Destroy all of the queen cells in the hive.
2. Transfer all of the frames with brood (capped or uncapped) from the hive's present brood chambers into the new brood chambers.
3. The queen will be kept in the hive's current brood chambers and it may be best to cage her during the transfer process so that she is not accidentally damaged or transferred with the brood.
4. Empty frames (preferably drawn combs) are placed in the original brood chambers from which all of the brood frames were removed.
5. The queen is placed in the bottommost brood chamber which is now empty of brood.
6. The queen excluder is placed above the now broodless brood chambers. This is not absolutely necessary and a super of honey above the brood chamber will serve the same purpose in that it will stop the upward movement and egg laying of the queen.
7. Place the honey super above the broodless chambers.
8. Place the two "new" brood chambers containing all of the brood on the very top of the hive.
9. Close the hive.
10. 7 - 8 days later, return to the hive and destroy any queen cells you find in the upper brood chambers. There shouldn't be any queen cells in the

bottom units. The bees in the upper chambers were so far removed from the queen, they thought themselves queenless and may have begun to construct queen cells. The bees in the upper hive body units will be unable to construct any more queen cells because the brood is now too old. (Queen rearing requires larvae no more than 2 or 3 days of age and all of the larvae is at least 4 to 5 days old now).

The bees will emerge as adults from the upper hive bodies and the bees will store honey in the empty cells. None of the bees were lost to the colony and it should make a surplus of honey if there is a good nectar flow. Figure 14 depicts a hive after it has gone through the Demaree process.

VALUE OF SUPERIOR STOCK AND GOOD QUEENS

All of the bees in a colony are the offspring of the queen and she is an essential element in a good beekeeping management system.

There are variations between and among various bee lines. Many claims are made that one race or hybrid of bees does better in one area than another, and there is some truth to those claims. However, the characteristics of the bees are obtained through the queen and she is only as good (reliable) as the selection and breeding program used to produce the queens. The best advice to

FIGURE 14. The end result of the Demaree process of swarm control. The process does not include the removal of any adult bees, brood or honey stores from the hive, but it does require some intensive manipulation of the hive. This can be a very effective method of swarm control and honey production for the beekeeper with a few colonies of bees. *(Photo by S. Bambara)*

the beekeeper is to learn about the various lines of available bees and then to test the most promising in his own colonies.

Maintenance of good queens involves both selection and replacement. One of the general recommendations in beekeeping that has real universal validity is to "requeen the colony when the queen is two years old." This management principle will help insure that the colony always has a viable queen with excellent egg-laying capability and it will also reduce swarming.

One point that requires some explanation is the term "superior stock." The term means different things to different beekeepers, but there is some general agreement. A starting point in defining the term would be a description given by F. E. Moeller in a 1961 USDA publication entitled "The Relationship Between Colony Populations and Honey Production as Affected by Honey Bee Stock Lines."

It included the following characteristics:

> reasonably gentle
> not prone to excessive swarming
> winters well
> ripens its honey rapidly
> caps the combs white
> uses a minimum of burr comb

Under today's beekeeping conditions we might add:

> demonstrates resistance or minimal susceptibility to tracheal and *Varroa* mites
>
> disease resistance including diseases presently undetected in U.S. bees
>
> competitive with Africanized bees

The above characteristics may be selected, but it is doubtful that there will ever be one line of bees which meets all of the "needs" of all beekeepers. The United States is a big country with considerable variation and the bee that does well in one area may not do well in another. Each beekeeper should decide what "bee" is best for his operation.

TWO-QUEEN MANAGEMENT SYSTEM

One of the approaches to management for honey production that should receive attention is the two-queen management system. This is a system that has been advocated in various forms since the late 1800's and it has a number of forceful advocates among commercial beekeepers and bee researchers.

There is a general agreement among beekeepers that strong or populous colonies of bees are more efficient in producing honey than are weaker colonies. Table 1 in this chapter, entitled "Honey Produced by Various Sizes of Colonies" shows a definite increase in the efficiency of honey production as the colony increases in size. Thus, is would seem to follow that colony efficiency in honey production could be increased even more by substantially increasing colony populations. One suggested method of obtaining these increases is to switch to a two-queen management system.

In a two-queen system it is supposed that two queens will lay more eggs than will one queen and this will lead to more rapid colony growth. It is recognized that colony growth is also regulated by available space for brood and food, but the two-queen systems take this need for extra space into account. These systems have been promoted extensively, but most of the larger commercial bee operations do not use a two-queen system because of the time involved in establishing and managing the colonies. In addition, it should be noted that most of the research on two-queen systems was conducted in the northern states and the findings may not apply to beekeeping conditions in the south where the honey flows are less intense and more erratic in start and stop dates.

The basic reference for two-queen management systems is a 1976 USDA report by F.E. Moeller entitled, "Two-Queen System of Honey Bee Colony Management." This system is also described in the USDA Handbook #335 entitled, "Beekeeping in the United States."

F.E. Moeller is quoted as saying, "The establishment of a two-queen colony is based on the harmonious existence of two queens in a colony unit. Any system that ensures egg production of two queens in a single colony for about two months before the honey flow will boost honey production." Moeller goes on to state that the population of such a two-queen unit may be twice that of a single queen unit and will produce honey more efficiently. He also claimed that the two-queen colony would enter the winter with increased pollen stores which encourages the production of a larger population of young bees in the spring and thus was an ideal unit for producing another two-queen colony the following year. Moeller's claims are the basis for most of the two-queen systems in use today.

Using Moeller's work as a standard the following general procedures can be followed in developing a two-queen system.

1. Start with strong overwintered colonies.
2. Build the colonies to maximum strength in early spring by feeding pollen patties and sugar, if appropriate.
3. Approximately two months before the start of the major flow the colony is divided and a new queen is introduced into the queenless portion. The old queen, most of the younger brood, and about 1/2 of the adult bees are placed in the bottom section of the hive. Either an

inner cover with the hole plugged or a thin board is placed over the bottom unit. The new queen, most of the capped brood, and the remaining bees are established above the division board.

4. The upper unit is provided with an auger hole which serves as a hive entrance for those bees.

5. Two weeks after the new queen's introduction the division board is replaced with a queen excluder.

6. During the initial division the bottom unit was provided with an empty brood chamber so that the old queen would have plenty of room to lay eggs. After the new queen begins laying, an empty brood chamber is also added to the upper unit.

7. Brood production may advance quite rapidly and it may be necessary to practice reversals (of the brood chambers) in both the upper and lower units. This may have to be accomplished every 8 - 10 days until about 4 weeks before the end of the major nectar flow.

8. Honey supers are provided to the upper and lower units as needed for honey storage. Extracted supers are returned to the units for additional filling. Moeller cautions that a two-queen unit requires considerable storage space and when a one-queen colony might require one super, then a two-queen colony might require two or three empty supers.

9. The advantage of a two-queen system is lost about four weeks before the end of the major flow. Eggs being laid at this time will take at least four to five weeks before they become foragers that can collect honey. The queen excluder is removed and the units are combined. The result, according to Moeller, is usually the death of the older queen. This insures that the colony will normally go into the winter with a young queen.

Recent research by the University of Minnesota by Duff and Furgala demonstrated that there is value to a two-queen management system under the conditions found in east central Minnesota. They also tested a variation of the two-queen systems which they called the horizontal two-queen system and that system demonstrated the greatest weight gain in honey stores.

The two-queen system does seem to demonstrate some advantages under beekeeping conditions found in the northern states where there is a fairly long (about two months), intense and predictable honey flow. The advantages are not as obvious to southern beekeepers or any beekeeper where the honey flows are shorter, less intense and more unpredictable. A summary of the advantages and disadvantages of the system are provided for those who might be interested in experimenting with this production system.

Advantages:

1. The beekeeper tends to equalize the colonies during the set-up process, insuring that all have adequate bees to take advantage of the coming nectar flow.
2. Swarming and supersedure tend to be reduced because of the regular introduction of new queens and the swarm prevention practiced in the management system.
3. The system produces colonies that should overwinter well because of the presence of young queens and good food stores.
4. Most beekeepers say they get good queen acceptance using this system.

Disadvantages:

1. The two-queen system may not be advantageous for southern beekeepers or beekeepers in areas that do not have long-lasting, intense nectar flows with predictable start and stop times.
2. Significant labor and timing are necessary for a successful system.
3. Since the units may be very productive, they may become quite tall as honey supers are added and the risk of hive tipping is of concern.

The best advice to the beekeeper considering a two-queen system is to experiment with a few hives of bees and see how the system works for him.

FEEDING BEES

It varies from region to region and beekeeper to beekeeper, but 10% to 20% of bee colonies die each year. The general average is probably closer to 10%, but that is still a large loss and much of it is due to starvation.

Normally a well managed colony will require little supplemental feeding, but such feeding, when it is necessary, may determine the survival of the colony. Insuring that the bees always have adequate food stores is a form of insurance and it is generally a profitable investment. A colony which has a surplus of food, either from foraging natural sources or from supplemental feeding, will not eat the surplus just because it is available; however, the colony that is just a few ounces short of necessary food stores will starve to death.

Feeding bees can be done in several different ways and there are a small variety of products that may be fed to bees. Bees may be fed stored honey, isomerized corn syrup or high fructose corn syrup, and of course, sucrose or table sugar. In addition, the bees may need a pollen substitute or a supplement particularly in the spring for those areas where pollen sources may be in short supply. See the section on "Seasonal Management" for recommendations on when to consider supplemental feeding of your bees.

Honey is the simplest product to feed the bees when their food stores are in short supply. Some beekeepers will save frames filled with honey so that they will be able to feed their bees when necessary. The feeding of honey already in frames requires very little preparation and the frames can be placed in the hive near the bee cluster. The honey frames used as feeding supplements should be collected from honey supers that had nine or ten frames. Honey supers with eight frames tend to produce frames that are very thick. The thinner frames from the nine or ten frame hive will be easier to insert around the brood nest of the colony that requires feeding.

Extracted honey may also be fed to bees, but it requires more work than using honey already in combs. If extracted honey is used, then the beekeeper should be certain that it came from a disease-free source. Honey can contain the spores for American foulbrood, and honey from infested colonies can spread the disease. If extracted honey is used, it can be fed by mixing it with water and feeding it in a container as you would do with sugar syrup.

Isomerized corn syrup, commonly called high fructose corn syrup, is also a substance that can be fed to bees. This is a relatively new product that was first produced in 1969 and it is made by converting starch to glucose and then some of the glucose to fructose. The fact that the product contains both fructose and glucose makes its composition similar to that found in real honey. The product is available as Isomerose 100 (42% of the sugar is fructose), Isomerose 550 (55% is fructose) and Isomerose 900 (90% is fructose). The importance of the numbers is that the higher the concentration of fructose, then the more slowly it will granulate, but higher fructose levels also cost more. A fructose content of 55% probably works best for the beekeeper as a feed for the bees. The actual methods of feeding the isomerized corn syrup to the bees are similar to those used in feeding sugar syrups.

Sucrose or table sugar is still the most commonly used bee feed. The sucrose may be fed in solid (dry), liquid, or in candy type forms. The form is influenced by the condition of the colony and the weather or time of year. One word of caution, bees can completely digest sucrose, but sugar products that have other components such as starches and sugars other than sucrose, may present problems to the bees. If the bees are fed a product that has indigestible components, then this will increase the chance that the bees (particularly overwintering bees) may develop dysentery. For this reason the beekeeper should not feed pancake syrups, molasses, candy residues and similar products to the bees, particularly as winter feed.

Sucrose may be fed to the bees as a dry sugar, sugar candy, or a sugar and water mixture. The dry sugar is very easy to use and some beekeepers will use it as an emergency winter feed as well as feeding in the spring and fall. The problem is that the bees need a water source to utilize the dry sugar and some bees will treat the dry sugar as debris and throw it out of the hive. In spite of those limitations, it may be the only readily available food source for an overwintering colony which is very low on food stores in the late winter months. Some beekeepers believe that liquid sugar or even honey at that time of year may stimulate brood rearing which is very costly to the hive. Dry sugar does not seem to initiate the brood rearing response nor does sugar candy which is described later in this section.

If dry sugar is fed to the bees, then the simplest ways are to either pour an amount around the top hole of the inner cover or place some sugar on a piece of wax paper on top of the brood combs. The bees do need water to process the dry sugar and that can be a problem if bee flight is limited or prohibited due to weather conditions. In some cases condensation caused by winter cluster respiration provides this needed water. Of course, the bees could mix nectar or unripened honey with the sugar, but if those products are available, they probably wouldn't need the sugar.

Sugar candy is made by mixing sucrose with a smaller amount of glucose or white corn syrup and water and a small amount of cream of tartar. The mixture forms a paste which can be placed near the hole of the inner cover or on a piece of wax paper above the brood combs. The advantage of this mixture over dry sugar is that the bees do not have to collect much, if any water. It has the same advantage as does dry sugar in that it doesn't seem to initiate brood rearing by the bees when it is used as an emergency feeding to colonies in the late winter. The following recipe (Table 2) is repeated from Clarence Collison's Pennsylvania Extension Service Publication, "Fundamentals of Beekeeping."

Table 2: Sugar Candy

15 pounds sugar
3 pounds glucose or white syrup
4 cups water
1/2 teaspoon cream of tartar

Dissolve the sugar in the water by stirring and boiling the mixture until the temperature of the syrup rises to 242°F. Let syrup cool to 180°F, then beat until thick. Pour the candy into molds lined with wax paper. Place a cake of sugar on two small, 1/2-inch square strips of wood in an empty super above the cluster of bees. Cover the candy and the space around it with cloth or newspaper to keep it warm. Remove any remaining candy and feed syrup when the weather gets warm.

Sugar syrup is probably the most popular way to fed bees that are low in food stores, though the use of high fructose corn syrup is becoming more common among commercial beekeepers. Sugar syrup is basically a mixture of sucrose in water with an optional amount of material like cider vinegar to retard granulation. The mixtures are generally one part sugar to one part water for a thin mixture or two parts sugar to one part water for a heavy mixture. The amounts of sugar and water may be measured by weight or by volume. The best way to prepare either mixture is to bring the water to a boil and then mix in the sugar after the water is removed from the heat source. The sugar and water can be heated together, but the mixture may burn or caramelize if it is not stirred continuously during the heating process.

A heavy sugar mixture (2 sugar to 1 water) is generally used for fall feeding when the beekeeper is trying to increase the colony's stores for overwintering, and 1 gallon of 2:1 sugar water adds about 7 lbs. of stored food. The thin solution is more commonly used for spring feeding and as a starter food for package bees. The thin solution tends to increase the humidity in the hive which is beneficial to brood rearing by the spring and package colonies. Fumidil-B® may be added to either the thin or the heavy mixture for nosema control, but the use of Terramycin® is not recommended because the antibiotic breaks down very quickly in a sugar water solution.

Beekeepers have devised numerous methods of feeding the sugar solutions to the bees and the individual should select the one or two methods that work best for him. These methods include the use of external hive feeders such as the Boardman feeder, the use of frame (division board) feeders, the use of pails or buckets as internal feeders, the use of Miller-type super feeders, and community feeders in the beeyard. All of these methods or devices have some advantage, but the Boardman external feeder and the use of communal feeders in the beeyard are strongly discouraged.

A Boardman feeder is a quart jar that attaches to the outside of the hive and allows the bees to collect sugar water at the hive entrance (see Fig. 15). The feeder is easy to use and to observe, but its disadvantages outweigh these advantages. The major problem is size, the jar only holds a quart of liquid, whereas a colony in need of supplemental feeding probably requires several gallons of syrup. The second problem is that the bees will probably not collect syrup from it on cold days or at night. The final problem is that it may leak and attract other insects such as yellow jackets or robber bees.

Communal or bulk outdoor feeders in an apiary would seem to be an easy way to supplement the needs of many colonies of bees at one time. Such feeders may be barrels or tanks that are filled with large quantities of sugar syrup. The problem is that the weaker colonies do not seem to get nearly as much advantage as the strong colonies in the yard from this feeding arrangement. In addition, communal feeding may encourage robbing and the weaker colonies will end up the losers.

Frame or division board feeders are a very attractive way to supplement the food supply of bees. These feeders look like a typical deep frame without a top (see Fig. 16). The feeder is made of either wood or plastic and it actually

FIGURE 15. A Boardman feeder being used to provide supplemental food (sugar syrup) to a colony of bees. The use of external hive feeders is generally not a recommended practice.

FIGURE 16. A frame or division board feeder is as very useful tool in bee management. It replaces a frame in the hive and can be left in place on a permanent basis to be used whenever the bees need supplemental feeding.

replaces one of the frames in the hive, frequently in the brood chamber. Such a feeder will hold about a gallon of sugar water. The feeders should be equipped with a float or have a wire screen attached to the sides to prevent bees from drowning while they are collecting the syrup. This device has two major advantages: it holds a good amount of syrup (about one gallon) and it can be left in the hive on a permanent basis. Many beekeepers leave the division board feeder in place throughout the year so they can feed whenever it is necessary. The feeder is filled by simply pouring syrup into the top of the feeder.

An alternative to division board feeders is the use of friction top containers or any suitable container that has a reasonable lid. Such containers should be large enough to hold a reasonable amount of syrup, from 5 to 10 pounds. The use of pails with friction type lids has been popular for a long time. The pails are filled with sugar syrup and then 1/2 dozen or more holes are made in the lid with a 4d nail. The pail is then inverted over the top of the frames of the upper hive body or placed above the hole in the inner cover (see Fig. 17). The container should be placed as close to the bee cluster as possible. Glass jars with removable plastic or metal lids are most commonly used. Glass jars have the advantage that you can view the amount of syrup still in the container without removing the container from the hive. This feeding method is very

FIGURE 17. Friction pail feeders are a commonly used technique for providing a colony of bees with sugar syrup. Glass or plastic jars do have the advantage that the beekeeper can simply look at the jar and determine the amount of sugar syrup that the bees have consumed instead of being forced to lift the pail during each inspection.

often used when starting a package of bees. Whether the container is placed directly on the top frames or above the hole in the inner cover, it should be covered by a suitable size hive body and a cover placed on top to prevent robbing of the food and to protect the hive.

The bees in the hive will collect syrup from the container by sticking their tongues into the holes and sucking out the syrup. A 4d hole is a good size for this process in that it is big enough for the bees to use, but not so big that the syrup will drip out of the container. Because the container can be placed near the bee cluster, the bees can work it day and night which increases the efficiency of the food storage process.

Miller-type feeders were developed by Dr. C. C. Miller of comb honey fame. These feeders are the approximate size of a half-depth honey super and fit on top of the hive. The box contains two pans or open-top trays that can be filled with sugar syrup. Crawl spaces are provided so that the bees can enter the box and reach the syrup. Such feeders can hold large amounts of syrup and are easily filled by just lifting the hive cover. The primary disadvantages are that these feeders tend to be some distance from the bee cluster and are not as efficient as jar feeders above the brood nest or division board feeders, and the Miller-type feeders may leak. In addition, some bees may propolize the entrances to the feeders so that the bees cannot reach the stored syrup.

In addition to a carbohydrate source (honey or sugar), the bees also need a supply of pollen to meet their protein and fat requirements. Bees will normally store enough pollen to take care of their needs, but they may run short because of unusual weather conditions which stop bee foraging or because the bees are located in an area of low pollen availability. Colonies that are short on pollen may be fed pollen supplements that contain natural pollen or pollen substitutes.

The most common time to feed pollen substitutes or supplements is in the early spring. This may be particularly important if the bees have started intensive brood rearing using stored pollen before natural pollen sources were available to foraging bees or in areas that are known to have low pollen availability. There are several pollen substitutes on the market and they are available from most bee supply dealers. Such substitutes contain proteins from non-pollen sources mixed with large amounts of sugars to make them attractive to the bees. An alternative to buying a substitute is to make a pollen supplement. There are a number of recipes, but most consist of mixing one part water with three parts soybean flour (expeller process) in a heavy sugar syrup (2 parts sugar to 1 part water). The best pollen to use for this process is stored pollen which the beekeeper has trapped from the previous year.

Table 3 lists the ingredients for making either 1 or 32 pollen cakes to feed the bees. This mixture is pressed between sheets of wax paper to form a patty about 3/4 inch thick and weighing about 3/4 pounds. One side of the patty is then left uncovered with the wax paper and the patty is placed (paper side down) on the top bars of the hive just above the brood nest. The bees will normally be very attracted to this mixture in the early spring or fall and will probably consume it in about a 10-day period. Replace the patties until a good natural pollen flow starts in the spring or the bees stop working it in the fall. Medications for both nosema and foulbroods (AFB & EFB) may be incorporated into the patties.

Table 3: Ingredients for Preparing Pollen Supplement Patties

One Cake	Thirty-two Cakes
2 ounces pollen	4 pounds pollen
6 ounces soybean flour	12 pounds soybean flour
5½ ounces water	11 pounds water
10½ ounces sugar	21 pounds sugar

After C. Collison in Pa. Extension Service Publication, Fundamentals of Beekeeping.

PROVIDING WATER FOR THE BEES

Bees are like most animals in that they require water, but their uses for it may seem a little unusual.

Bees require water for two basic purposes: to dilute the brood food and to cool or air condition the hive. Water is not stored in the hive as is honey and pollen, but a small number of worker bees will keep some water in their honey stomachs as a sort of water reservoir for the colony. This water is mostly used to keep the larvae from drying out or becoming desiccated. These bees will place a drop of water in the cell with the brood when the temperatures are very high. The water used for cooling the hive and diluting the food is collected as needed by forager bees.

One of the commonly asked questions about honey bees is how do they know when to perform certain tasks. In the case of water collection the process is probably known. After forager bees collect nectar, they transfer it to a house bee when they return to the hive. The forager does not place the nectar directly into a cell in the hive. Thus the house bee has some control over what the forager collects. If the temperature goes above 92°F in the hive, then it has become too high for the rearing of the brood. Under such conditions the house bees will more readily accept nectar that has a high water content over a nectar source that has a lower water content. This results in the foragers collecting more dilute nectar and some will begin to collect water.

The water that comes into the hives is distributed over the surfaces of the comb and the interior hive walls so that there is a thin layer of water. Other bees in the hive are fanning their wings to set up an air current and the current causes the water to evaporate which works like a water-cooled air conditioner. This process lowers the temperature in the hive and prevents heat damage to the developing larvae. The foragers will collect water as long as it is necessary for the cooling process to continue.

Bees learn the locations of water sources and will continue to collect water from those sources throughout the year. Forager bees will also recruit additional bees to those water sources using dances similar to those used for nectar and pollen sources.

The location of available water sources for the bees is of importance to the beekeeper. Usually the number of bees collecting water is not very large, but it should be remembered that water-collecting bees have switched from nectar collection and so nectar collection is reduced. In addition, the bees that fly to the water source(s) must expend energy or consume food to collect the water. Therefore, it is fairly obvious that a good management scheme would include the provision of a water source which is reasonably close to the hive. Reasonably close might be a 1/4 mile in a rural situation, but only a few feet in an urban or suburban situation.

Honey bees do not respect property rights and they will collect water from a neighbor's yard more readily than the beekeeper's yard if the former water source is closer. If the neighbor's water source is a birdbath or a swimming pool, then there may be problems. One of the most common complaints made about urban and suburban beekeepers is that their bees are chasing birds from someone's birdbath or chasing people from their swimming pool. These can be serious problems and the best solution is prevention.

If a beekeeper's bees are visiting the neighbor's birdbath, then the situation can usually be corrected. The solution is to drain the birdbath or cover it for a few days, while the beekeeper establishes a water source in his own yard that is closer to the beehive(s). After a few days the bees will have switched to the new source and the birdbath can be refilled. That was a minor problem to solve, but what can be done about the swimming pool? Once bees have learned a water source, they will continue to work it even though a new source is established closer to the hive. A new source might attract new foragers, but the older foragers will still be a problem. This is the kind of problem that leads to municipalities passing ordinances against keeping bees in the municipal limits. The solution would have been prevention of the problem by insuring an appropriate water source when the hive(s) was established in the beekeeper's yard.

The establishment of a water source for the bees in an urban environment can take many forms, but there is some evidence that a running source is better than a stagnant or a still source. The visits of large numbers of bees to a stagnant water source may result in the spread of bee diseases such as nosema.

The provision of a dripping water source and an inclined surface for the water to flow will eliminate this potential problem. However the water source is created, the beekeeper should keep in mind that bees are not good swimmers. The water source should either be very shallow or have scattered or floating objects on which the bees can land for collecting the water.

VENTILATION AND SHADE

One of the major differences between honey bees and most other insects is that the bees tend to control or at least to modify their environment. Honey bees regulate the temperature in the hive which permits the colony to survive on a perennial basis. The key to that success is that honey bees are highly social animals and there is a significant degree of cooperation among the colony members. The bees produce heat by clustering and giving off body heat, and they cool the hive by a combination of fanning their wings and evaporating collected water to air condition the hive (see section on "Providing Water for the Bees" in this chapter). The bees also exercise control over the humidity in the hive by their fanning and water-collecting activities. Humidity control is necessary to change the nectar into honey (reduce water content) and to keep the larvae from desiccating.

From the beekeeper's standpoint the major consideration is that the bees must expend energy to control the hive's temperature and humidity and that such energy expenditures may reduce the production of honey stores. Therefore, it is to the beekeeper's advantage to assist his bees in these endeavors when he develops his management strategy. Temperature control is one of the areas where the beekeeper can assist his bees with minimal effort on his part. The beekeeper should consider that the bees need relatively little assistance in keeping the hive warm enough for brood rearing and colony survival, but they can use some help in cooling the hive or keeping it from overheating. At temperatures over 92°F, the bee brood will begin to overheat and die. Preventing such losses and minimizing the energy expenditures by the bees in preventing high temperature extremes will benefit the beekeeper and increase his honey crop.

The beekeeper's contributions to temperature control are basically in apiary selection, painting the hives, providing an adequate water source, and providing shade when appropriate. Hive or apiary location is one of the most important considerations in reducing temperature extremes in the hive. A hive is obviously affected by its environment which includes exposure to the sun. Morning sun exposure is probably beneficial to all colonies of bees in that heat from the sun permits early morning flight by the bees which are cold-blooded insects. On the other hand, afternoon shade will not affect (reduce) afternoon foraging once the bees have started to fly. The beekeeper, in placing his apiary, should locate them to receive morning sun and afternoon shade. In addition, it is usually beneficial to face the hives towards the southeast so that the morning sun will fall on the hive entrance and yet the colonies will be facing away from cold, northern winds.

In some areas such as the great plains region, the desert areas, and other treeless locations, it may be difficult to provide afternoon shade for the bees from natural sources such as trees. In those instances it may be beneficial to construct a structure to provide artificial shade. Obviously, there is some cost involved in such construction, but it may pay for itself in increased honey production. Contact your state extension service at the land grant university, the bee section of the state department of agriculture, or an experienced beekeeper in your area if you think shade is a limiting factor in your bee operation and inquire about the construction of a shading structure.

Along with providing any necessary shade for his bees, the beekeeper should also insure that an adequate water source is available. Bees will collect large quantities of water to cool the hive and such activity is often at the expense of nectar collection. See the section on "Providing Water for the Bees" in this chapter.

Painting the hives is a good investment in that it prolongs the life of the equipment and the use of appropriate colors will affect temperature control by the bees. A light colored paint will absorb less leat than a dark color and may aid in temperature control. The selection of a color is influenced by several factors including the beekeeper's desire to have his hives blend into the surroundings to reduce theft and vandalism. Nevertheless, temperature control should be a consideration.

Earlier it was stated that the bees require little assistance in maintaining adequate temperatures for brood rearing and the survival of the winter cluster and that does seem to be the case. However, moisture can be a problem. Bees seem to have little trouble with cold temperatures, but cold temperatures, in combination with moisture or high humidity, do present a problem. Therefore, it is to the beekeeper's advantage to provide adequate ventilation in the hive to reduce moisture buildup. This is particularly true in the winter when the bees cannot break their cluster to dry out the hive interior by fanning their wings.

The bees are able to deal with some moisture in the winter even without being able to break cluster and fan their wings. The heat given off by the cluster tends to rise in the hive and will carry some excess moisture with it. This ventilation flow can be augmented if the beekeeper maintains an auger hole in one of the upper hive bodies. This hole can be anywhere from 3/4 of an inch to an inch in diameter and it should be on the front of the hive. This hole can do double duty as an upper hive entrance in the winter if the regular entrance is covered with snow. Air flow should follow a pattern of entering the hive through the normal entrance, then rising in the hive due to the small amount of heat lost from the bee cluster and then flow out through the upper auger hole. Air will also rise in the hive and exit through the inner and outer covers if they are not airtight.

The beekeeper can also reduce moisture problems in the hive by facing the winter colonies away from the prevailing winds that may carry rain or snow, or by providing a windbreak which may be natural vegetation. An extra but valuable precaution is to raise the back of the hive about an inch or two above the height of the front of the hive. Then, if moisture does accumulate in the hive, it will be able to flow out over the bottom board. This precaution is most important during the winter months.

A strong colony that stays dry will be able to survive cold weather conditions and the beekeeper's main concern is keeping the bees dry not warm.

MOVING BEES

If life were simple, then one of the advantages would be that an apiary could be established and the beehives would never have to be moved. But life isn't simple and beekeepers have to move their bees for various reasons. It may be to make a special honey crop such as sourwood, orange blossom, or sage honey; or to pollinate a crop such as apples or almonds; or to establish a new apiary; or any number of other reasons. Whatever the reason, there are a number of steps the beekeeper can take to make the move more efficient from his standpoint and less disruptive to the bees.

The first condition is whether the move is a short one or involves a long distance. The other two considerations are the number of hives involved and the time (season) of the year. The approach to moving one or two hives within an apiary is different than moving a number of hives a distance of several miles or more. The movement or relocation of a hive or two within the apiary or yard can be handled in one of two basic ways. Either move the hive in small increments over time to its new location or move the hive to an interim location which is several miles removed and then move the hive to its new permanent location after three or four days.

Honey bees learn the location of their hive and a movement of the hive that does not consider the bee's biology and behavior can result in the loss of some of the field bees. If a hive is moved in small increments over a period of time, then most of the foragers who return to the hive at the end of their foraging flight should be able to find the new location. The best way to determine what distances are acceptable is trial and error. Move the hive a few feet and then watch to see if a large number of the hive's foragers are returning to the former hive site and flying about that location instead of entering the hive in its new location. Generally, the bees will be less confused by moves that are forward or backward as opposed to sideways moves. The probable reason for the foragers' confusion is that bees tend to learn the location of the hive based on the landmarks they see around the hive. If the hive is moved only a short distance, then the bees do not realize they are in a new location and will not "learn" the new site before going off on a foraging flight. On the other hand, if the bees are moved a large distance, then they will generally learn the new location when they exit the hive on a flight. This is particularly true if the move occurs at night.

The idea behind moving the hive to a temporary location is that the bees will learn a new location and "forget" the old location and its landmarks. The forgetting-learning process will occur in several days of active bee flight. Then the bees can be moved to the new permanent location even though it is close to the original hive location. The forager bees will again learn the new location and there should be minimal confusion in flying back to the original location of the hive. This is not a foolproof strategy, but it does reduce forager loss.

Movement of hives to a new distant location is simpler than a short distance move in that there is less chance of forager loss if certain precautions are taken. It is best to move the bees at night (when all of the bees are in the hive) or during inclement or cold weather when the bees are not flying. If the bees are moved when they are active, then they should be confined until late afternoon or early evening before they are allowed to fly from the hives at the new site, otherwise some of the foragers may become confused and not return to the hives.

Whatever distance the bees are moved, they should be prepared for the move unless it is to be in small increments where the beekeeper is manually carrying the hive(s). All cracks and crevices in the hives should be sealed or closed. The hive bodies and bottom boards should be fastened together and stapling is probably the simplest procedure. Hive staples are available for this purpose and may be purchased from bee supply dealers. An excellent precaution is to use two staples on each side of each juncture (*i.e.* bottom board to hive body or hive body to hive body) and to slant the staples in opposite directions. This will prevent the equipment from shifting when it is being lifted or carried. An alternative to hive staples is to use metal or plastic bands around the hive.

In addition to securing the hive equipment together for the move and eliminating any cracks and crevices, the beekeeper should also provide for ventilation in the hive. A screen placed in the hive entrance will keep the bees in the hive, but it will also allow for air circulation. In addition, the beekeeper should consider using top-moving screens for additional ventilation during hot weather. These screens cover the entire top of the hive and provide several inches of clustering space for the bees. The consequence of not providing adequate ventilation if the bees are being transported during the heat of the day is that combs may melt and buckle and bees may be killed.

The best time of day to move bees is in the evening. The bees will be in the hive and the temperatures are usually cooler than daytime temperatures. The hives should be prepared for movement the day before which includes stapling the hive parts together or using plastic or metal bands for this purpose. Whatever type of vehicles are used to transport the bees, it is best to have the engine running during the loading process and to keep it running until after the bees have been unloaded at the new apiary site. The vibrations caused by the engine have an effect on the bees which causes them to move up and away from the hive entrance. This makes it less likely the bees will fly out of the hives during the moving or loading and unloading process.

An extra precaution that many beekeepers take is to cover the entire load of beehives with some type of wire or screen netting. This keeps the bees on the truck and reduces the possibility that stinging episodes will occur enroute to the new location. This can be particularly important if the beekeeper has to stop enroute to buy gas or stop for any purpose. If a stop is necessary, it is recommended that the engine be left running unless it is for a long period of time.

The actual loading and unloading process of hives can be done by hand or with the aid of a hive loader. A number of loaders are available and they do make life easier for the beekeeper (see Fig. 18). A well designed loader can enable one person to load and unload two- and three-story hives of bees. A loader can be a particularly appealing piece of equipment for the beekeeper who is moving large numbers of hives for pollination.

The unloading of the bees at the new site can be made less stressful on the bees and the beekeeper if a few precautions are taken. Do not turn off the truck engine upon arriving at the site. The vibrations will work to keep the bees from flying out of the hives as they are being unloaded. If possible, it is best to place the hives in the new location late in the day or at night. This will reduce the tendency of some of the foragers to fly away from the hive in search of food before they have learned the new location of the hive. Unload all of the hives in the new location and wait a few minutes to remove the entrance screens. Then, remove the screens with as little disturbance as possible so that the bees do not fly out of the hive and become lost or sting the beekeeper. Some beekeepers will reinforce the new site location by placing a board against the hive entrance. This will tend to confuse the foragers and encourage them to take orientation flights and learn the new hive location.

FIGURE 18. Hive loaders come in many designs and sizes. The primary advantage is that it allows one person to manage heavy hives without assistance.

If the bees can be moved during the winter months when they are not normally flying or during inclement weather, then few precautions are necessary to reduce bee loss enroute to the new site or upon arrival at the new site. However, a few precautions are probably always advisable such as covering the entire load of beehives with a screen netting to reduce the chances of stinging episodes enroute to the new site.

The beekeeper who is moving bees in his yard or moving to a new site should be especially sensitive in his precautions if he has neighbors near the old or the new site. It is just common sense to manage your bees so that they do not disrupt your neighbors. This means that bees should not be moved from one site to another during the day unless the movements are the small incremental moves discussed at the beginning of this section. In addition, the beekeeper should not locate the bees where they might become a nuisance to the neighbors. Locate the hives at least 50 feet inside your property line and avoid facing the hives so that bee flight will be over walkways or areas of human activity such as swimming pools, patios, etc. If the bees are moved to a field for pollination, then locate or face the hives away from roads and paths through the fields. For additional information see the section on locating an apiary in this chapter.

DRIFTING

It is generally acknowledged that honey bees drift considerably within an apiary. That is to say that a bee will inadvertently enter the wrong hive and become a member of that colony. This is usually not a serious problem, but it can reduce the population of some colonies which will affect their honey production capability, and it may result in the spread of bee diseases and pests such as tracheal or *Varroa* mites. As with most bee problems, the best control is prevention. Thus, a good management system should include provisions to minimize drifting.

Before the beekeeper can develop a management system which reduces the potential for drifting by his bees, he should understand what factors contribute to the phenomenon. Strong winds, the arrangement of the hives in the apiary, and the inexperience of young forager bees in locating their own hives all contribute to drifting. However, before dealing with these factors it may be interesting to examine the bee behavior which permits drifting to occur.

In order to help insure the survival of their colony, honey bees will attempt to stop bees from other colonies from entering their hive. Such a procedure is obviously beneficial in preventing robbing of stored food. The bees in a colony recognize bees from other colonies as "foreigners" by a combination of factors including hive odor, which is distinctive to each hive, and by the behavior of the bees. Robber bees generally exhibit characteristic behaviors when entering a hive that identify them as robbers. This is not the case with a bee that has drifted into the wrong colony. In most cases the bee will be carrying food and will be accepted by the new hive as an asset. A bee that comes into a hive carrying nectar or pollen is obviously not going to be a successful robber of that hive.

The arrival of the drifted bee into the new hive will benefit the new colony, but it will also have an effect on the original colony. The beekeeper wants strong colonies which will make honey surpluses and the weakening of some colonies will probably reduce their honey-making capability. Drifting would not be much of a problem regarding honey production if all of the hives in an apiary drifted equally, but that is generally not the case.

Beekeepers tend to arrange apiaries in a series of rows for the beekeepers benefit, but this can create problems for the bees. Honey bees tend to learn the location of their own hive by performing an orientation flight and they do learn the general area and some landmarks. However, if all of the hives are lined up in a straight row and all of the hives are painted alike, then the bees can become confused. The hives at the end of the rows are more distinctive than the hives closer to the middle, and the rows to the front or back of an apiary are more distinctive than the rows in the middle. The common result is for some of the bees to enter the wrong hive and the hives on either end of a row and the outer rows of hives will receive the greatest number of bees. The bees from the outer hives and outer rows will more readily recognize their own hives and the other bees will distribute in a more random manner if they are unable to locate their own hives. This drifting is particularly prevalent among new foragers on their orientation and first foraging flights and by drones. Even though drone bees do not bring food into the new hive, they are generally accepted if foraging conditions are good.

In addition to the bees' difficulty in recognizing their own hive because of the arrangement of the hives in the apiary must be added the effect of winds. A strong prevailing wind may force the foraging bees down to the end of a row. A heavily laden bee may not make the effort to fly against the wind and enter a hive other than its own. These bees are readily accepted by the new colonies at the end of the row.

There are several steps the beekeeper can take to reduce drifting. One of the most immediate steps is to correct the problem after it occurs. Hives that have lost some of their bees to other hives can be switched so that the weak hive and the strong hive have their positions interchanged. Thus, the foragers returning to the strong colony will enter the hive of the weaker colony if the switch is made while the bees are flying. Another solution is to transfer some of the frames of brood with attached bees from the stronger colony to the weaker colony. These are temporary measures and not long-term solutions to the problem. The recommended approach is to arrange the apiary to reduce drifting and to provide aids to the bees so that they can more readily locate their own hives.

Instead of placing the hives in a series of rows, the beekeeper can arrange them in a U-shape or a semicircular pattern. Such an arrangement will not normally reduce the beekeeper's ease of manipulation in managing the hives. Arranging the hives in groups of two with spaces at least 8 feet between the pairs of hives will also help in reducing drifting. The bees in the paired hives do not seem to have a problem in returning to their own hive instead of the paired hive in this arrangement. Another consideration is to face each pair of hives in a slightly different direction. If more then one row of hives is established, then a minimum distance of 10 feet should be established between the rows, and the beekeeper should consider spacing the hives so that the hives in one row are not directly behind the hives in the row to the front of it. Some beekeepers, (particularly European beekeepers) rely on symbols and colors to reduce drifting. This includes the use of more than one color paint for the hives in an apiary. Bees can learn to recognize colors and shades of gray (for colors they cannot distinguish) and the beekeeper should consider this approach. In addition, some beekeepers will place or mark the hive entrances with small symbols, such as triangles, circles and squares. Bees can learn to recognize such markings and this may reduce drifting. If a beekeeper is considering the use of symbols, then he should understand that a bee's vision does differ from man's vision and bees have difficulty in distinguishing between some shapes. Bees can distinguish between a circle, a triangle, and a square, but not between a square and a diamond. A diamond is really a square that has been rotated 45° and the two shapes would not be distinctive to a bee. The use of figures in conjunction with colors for painting the hives is worthy of consideration by the beekeeper in developing a management scheme to reduce drifting.

One special case to consider in regard to drifting is the establishment of package bees in a hive or the movement of a hive to a new location during the day. The best time of day to establish a package colony in a hive is in the late afternoon, otherwise some of the bees may fly off and become lost to the new hive site. By installing the package later in the day, this retards flight on the first day and provides more time for the bees to learn the location of their new homesite. A similar problem can occur if a foraging colony of bees is moved to a new location early in the day. Moving the hive at night will reduce the loss of foraging bees and generally will be less disruptive on the overall operation of the colony.

ROBBING

Honey bees do not respect the property rights of each other and will attempt to steal or rob honey and nectar from other colonies, particularly during periods of limited nectar supply or dearth. This robbing instinct can develop into a serious problem for the beekeeper resulting in the loss of honey stores in some hives and even the loss of some colonies.

Scout bees instinctively search for rich nectar sources and under the proper conditions they will initiate the robbing of nectar and honey stores in the beekeeper's hives. This problem can occur at anytime during the year, but it is most prevalent during periods of dearth such as in the fall when the nectar flow has stopped or during the active season between nectar flows. Often the resumption of a nectar flow or the start of a new one will stop the robbing problem, but some bees will continue to rob even though nectar is available. The most common scenario is for strong colonies of bees to rob weaker colonies of bees.

The robbing process can cause a number of problems. At the very least it disrupts some of the hives in the apiary as the weaker colonies are forced to defend themselves against marauding attacks of the stronger colonies. Bees in the colonies being robbed must devote some of their potential forager or older house bees to the job of defending the hive against the robbers. This situation obviously increases the aggressiveness of the bees and increases the beekeeper's chances of being stung when he works the hives. Far more serious is the possibility that the robbing will progress to a point where some of the weaker colonies are completely destroyed by the robbing bees. The term weak colonies includes colonies newly established from packages, nucs or divides.

The best control of robbing is prevention, but it is also important for the beekeeper to recognize when robbing occurs. Robbing will usually start with a small number of bees, but it will quickly grow as the robbers recruit additional bees from their own colony. The number of bees at the entrance of the hive being robbed will increase and some fighting will usually be apparent. An increase in bee activity at the entrance of a weak colony or a newly established colony, particularly during a period of dearth, should alert the beekeeper to the possibility of robbing. On the other hand, the beekeeper should not confuse all signs of congestion at the hive entrance with robbing. On warm days young adult bees will conduct play or orientation flights in front of the hive. These bees are learning the location of their hive by performing orientation flights, but there is no associated fighting at the entrance.

Robbing cannot be completely eliminated by the beekeeper, but he can take preventive measures to reduce it. The primary cause for most robbing in the apiary is the exposure of nectar, honey or some form of sugar. During periods of dearth be particularly careful not to spill honey or sugar on the ground or on the hives. The beekeeper should also keep his manipulations of the hives to a minimum. If a hive must be opened during a period of potential robbing, then every attempt should be made to minimize the exposure of any combs containing stored food. Some beekeepers will even take the step of covering the combs with a wet cloth while they are manipulating the beehive. Remember that honey bees have a highly developed sense of smell and potential robber bees can detect the smell of exposed honey, particularly if the cappings are broken or the honey is uncapped. As a general rule, never use communal (outside-open) feeders for the apiary during a period of dearth. If feeding is necessary, do it inside the hive.

Once robbing has started it is difficult to stop, but some actions are helpful. The best action is to improve the defensive posture of the bees that are being robbed just as you would assist them against attacks by yellow jackets, hornets, or other similar pests. Reduce the entrance and seal all cracks or openings in the hive. A reduced entrance should make it easier for the bees to defend their hive. Another recommendation is to place a wide board across the top of the hive's bottom board. This creates a wide tunnel-like arrangement through which the robber bees must pass on their way to the hive and increases the effectiveness of the guard bees at the hive entrance. This latter arrangement may be supplemented by placing handfuls of grass or straw in front of the hive entrance.

If the robbing is really serious, then it may be worthwhile to move any weak or newly established colonies to a new location. The new location should be at least a mile away. The beekeeper should give some thought to starting all new colonies in a separate apiary until they are strong enough to be put in his established honey-producing apiaries. A short-term emergency procedure for dealing with a robbing situation is to set up a water sprinkler so that water is spraying on the hive entrance. This may discourage the robbers temporarily, but it is only useable when the hives are located near a suitable water source.

Some beekeepers describe a robbing situation called progressive robbing. This involves only a small number of robber bees and there is little fighting, so it is difficult to detect. The robber bees attempt to sneak by the hive guard bees or enter the hive through cracks and crevices. Because it only involves a small number of robber bees, it is not thought to be a serious problem and little attempt is made at control. However, good management practices such as eliminating all cracks and crevices in the hive and eliminating the exposure of honey or spilling honey and sugar during periods of dearth will help to prevent all types of robbing.

RECORD KEEPING

Record keeping is a chore to most people and beekeepers are no different. The beekeeper who has only a few hives of bees and does not sell any of the honey or other hive products does not have to keep records, unless he wants to learn more about his bees. Record keeping over several years can reveal trends in nectar and pollen flows, swarming cycles and other topics of interest to the beekeeper. However, the beekeeper who sells any of his product and treats beekeeping as either a hobby or a business for tax purposes definitely needs to keep records.

Record keeping may be divided into two general types for the beekeeper: financial records and management records. There is obvious overlap between the two types of records and the actual record-keeping processes may be combined by the beekeeper. The general distinction is that financial records are the basis for the profitability of the bee operation and are used for tax purposes. Management records deal with such things as colony weight gains during the year, overwintering success, requeening schedules, and general historical background information that can be used in making management decisions.

Financial records would include such things as the cost of equipment, queens, feed supplies, transportation, equipment purchase, *i.e.* extractors and vehicles, and other related items and activities. The beekeeper may also wish to keep a log of the time he spends in each apiary to determine if a location is profitable based on the demands of his time. Tax rules change from year to year and it is important for the beekeeper to know what signifies a hobby vs. a business for tax purposes. Consult with your local IRS office, agricultural extension office, or an accountant to discuss those matters.

Management or operational records that are maintained over a period of years will assist the beekeeper in developing the seasonal management system that works best in his area. Conditions do change from year to year, but general trends remain fairly constant. These records include both colony and beeyard information.

Records on honey production in an apiary will assist the beekeeper in determining what is the optimal number of hives to have in an apiary for honey production. Many beekeepers use some sort of a scale to weigh individual hives and it is probably a good investment to purchase a hive scale and have a scale colony in at least the beekeeper's home yard.

The key to success in record keeping is to make the records serve a purpose and for the information to be useful with little manipulation of the material. A number of record systems, including computer-based systems, are available. One source of information is the local agricultural extension service.

REFERENCES

Cale, G.H., R. Banker and J. Powers. (1975). Management for Honey Production in Hive and Honey Bee, Dadant and Sons, Hamilton, IL 355-412.

Collison, C.H. (1984). Fundamentals of Beekeeping. Pennsylvania State Extension Service. 66p.

Duff, S.R. and B. Furgala. (1990). A comparison of three non-migratory systems for managing honey bees (*Apis mellifera* L.) in Minnesota. *Am. Bee J. 130(1)*:44-48 and *130(2)*:121-126.

Gojmerac, W.L. (1980). Bees, Beekeeping, Honey and Pollination. AVI Publishing Co. Westport, Ct. 192p.

Moeller, F.E. (1961). The Relationship Between Colony Populations and Honey Production as Affected by Honey Bee Stock Lines. U.S. Dept. of Agriculture Production Research Report 55:20p.

Moeller, F.E. (1976). Two-Queen System of Honey Bee Colony Management. U.S. Dept. of Agriculture Production Research Report 161:11p.

Moeller, F.E. (1980). Managing colonies for high honey yields. *In:* Beekeeping in the U.S., U.S. Dept. of Agriculture Handbook #335:64-72.

CHAPTER 15

HONEY AND WAX—A CONSIDERATION OF PRODUCTION, PROCESSING AND PACKAGING TECHNIQUES

by JAMES E. TEW*

INTRODUCTION

A honey crop is the primary reward of beekeeping. All beekeepers are thrilled upon finding supers that are filling quickly with thin, clear nectar that will, within a short time, become delicious honey. The sweet product, confined in beeswax cells behind brilliant, white wax cappings, is the essence of keeping bees.

Eager beginning beekeepers frequently find that their colony numbers have grown beyond the capacity of their small extracting operations, changing what had been a pleasant hobby into a task comprised of long hours of monotonous, hard labor. The extracting operation must develop as the colony numbers grow. The proper hive and extracting equipment usually requires significant financial investments and long-term commitments to beekeeping.

THE HONEY SUPERS AND THEIR CONDITION

Removing honey from colonies that is stored by the bees in supers that are clean and appropriately sized for the specific operation is the first component of an efficient and successful operation. It has frequently been said by many beekeepers that, "The bees don't care what kind of equipment they are kept in." Consequently, it is not uncommon to encounter beekeepers whose bee hives and supers are not well maintained. It does seem to be true that bees seem to take little notice of the condition of the equipment; however, shoddy supers and other poorly maintained field equipment can cause problems when the beekeeper tries to remove, transport, and process the honey crop stored in such equipment.

For record-keeping purposes, the useful life of outdoor, wooden equipment seems to be about seven years (assuming equipment is in constant use). To be functional for many years, wooden equipment will require frequent maintenance. Latex paint, applied only to outer surfaces, is the most common protective coating. Penetrating oils, advertised for beehive use, are commercially available and offer a high degree of woodenware protection.

Another technique, used by beekeepers to protect equipment from weathering and decay, is submerging wooden equipment into a mixture of hot rosin

*James E. Tew, Ph.D., The Agricultural Technical Institute and the Ohio Cooperative Extension Service, The Ohio State University, Wooster, Ohio 44691.

and beeswax. In future years, as equipment requires recoating, simply submerging the wooden equipment a second time into the hot mixture will recoat the equipment and remove propolis and burr comb accumulation.

THE APIARY

The characteristics of a potential apiary location ranges from scenic to practical. Hobby beekeepers normally want to keep their colonies nearby and may pay little attention to considerations that would be important to commercial honey producers. Since hobby colonies are for enjoyment, garden pollination, and modest honey production, the apiary site may not be the most significant factor in their operation. Commercial honey producers, however, will normally choose locations for specific nectar flows, yard accessibility, or other practical purposes. There are several factors to consider when selecting an apiary site for honey production.

Nectar and Pollen Flow

The main purpose of a good location is to produce a honey crop; therefore, the area should provide a dependable nectar flow. There are a few ways to determine the nectar potential of an area and all of them are slow: (a) ask local beekeepers for advice; (b) put a small yard on the location for a few years; and (c) canvass the surrounding area for several miles in all directions to determine the extensiveness of nectar and pollen plant populations. There really is no easy way to determine the quality of a location or the number of hives that can be productively supported without experimenting with colonies on the location. In fact, when commercial operations are sold, the advertisement will frequently have the notation "on locations" indicating that the seller knows the potential for production of the locations and can probably offer some assistance.

In general, 15-20 colonies can profitably be kept at one site in the New England states while 50-100 colonies can be kept in one location in the Western states (Eckert and Shaw, 1960). There should be enough pollen at the appropriate times of the year to insure colony population development. If not, the beekeeper will have to consider feeding a protein substitute to offset pollen shortages.

Yard Accessibility

The yard needs to be approachable throughout the year even though most beekeeping work is done at specific times of the year. The yard should be somewhat protected from prevailing winds, animals (both domestic and wild), and people. It should have a firm path wide enough for an automobile to drive directly into the bee yard. Additionally, the beekeeper should try to have keys to any gates or locks.

Landowner Commitments

Agreements always change, but try to get assurance that the yard will be available over a long period. It is tiring to clean yards and establish their locations with bee inspectors, helpers, and others, only to be asked to move the yard. Ask the landowner to sign legal papers indicating that the bee colonies are not part of the landowner's property in case ownership changes.

Water Resources

In most parts of the United States, an ample water supply is not a problem. Clean water should be readily available throughout the year.

Pesticide Exposure

Unless the beekeeper is prepared to move the colonies during the chemical application periods, locations should be away from harmful pesticide application areas.

Scenic Considerations

Most of the time, scenic considerations are not important, but working in pleasant surroundings makes a beekeeping task more palatable.

Many other factors will have to be considered when beekeepers search for sites in various parts of the United States. If the potential location has heavy underbrush growth, clearing may be required to lessen the threat of fire and, in some areas, to lessen the effects of ant and other insect predation. In other areas, locating on high ground away from potential flooding, may be a major consideration when selecting a location.

It is also helpful if the area is flat and free of stones and other obstacles that would affect the use of a hand truck or other carts needed to transport supers within the yard.

TECHNIQUES FOR REMOVING SUPERS FROM COLONIES

Protective Equipment

No matter how well the beekeeper may have managed the hives, honey bees are generally not enthusiastic about giving up their honey crop. The beekeeper should be well protected from the defensiveness of the bees. A smoker, veil, bee suit, and gloves are the usual garb of a prepared beekeeper (Figure 1). If the work to be done is light, the protective gear can be light. However, if a considerable number of colonies are to be manipulated, the beekeeper should be prepared.

Several companies are currently manufacturing high quality bee protective clothing. Bee suits may be either waist length or full length and in many different styles. Cotton fabric has been the suit standard for many years. The cotton suit is durable and washable. One specific type of cotton suit, manufactured in England, has an attached collapsible veil (Sherriff, 1990).

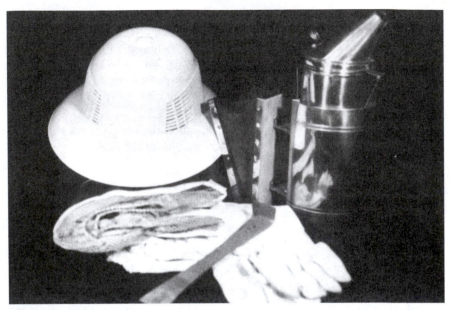

FIGURE 1. The basic protective equipment for a beekeeper.

Nylon is another material used to manufacture bee suits. Rip-stop nylon has a smooth surface which makes stinging difficult for bees. Such suits are durable and economically priced. Another style with an attached veil is made of 1/4 inch (6.4 mm) porous nylon. The major advertised advantage of this suit over other designs is complete ventilation. Also, due to the thickness of nylon, the suit is highly sting resistant (Lemoine, 1990).

Bee suits made of reinforced paper are inexpensive and are frequently used in bee classes or for other occasional uses. For limited use, paper suits have proven to be a viable alternative to more costly suits.

For use in hot climates, Tew (1989) has preliminarily tested a water cooled vest worn beneath a full length bee suit. The vest, originally designed to be used beneath suits required in areas of potential radioactive contamination, has a battery powered pump that circulates cool water on the front and back of the wearer. The vest works reasonably well at keeping the wearer cooler, but the usable life of the rechargeable batteries is limited and the bulky vest makes routine movements more cumbersome.

Removing Bees from Supers

Many beekeepers have developed specific techniques for "robbing" their bees—some more complicated than others. Techniques used to remove the crop vary from one operation to another, but the number of colonies from which honey is taken is a major consideration. The new honey crop should be removed as quickly as possible for several reasons. It prevents the bees from tracking across the new, white wax cappings and darkening them with pollen and propolis residues. Some honeys crystallize quickly. Getting such honey from the combs before crystallization will enable the beekeeper to reliquefy the crop as needed. If multiple flows are a characteristic of the apiary site, prompt removal of one honey crop will prevent mixing with the second crop.

If a beekeeper is removing honey from only a few colonies (or frames), the bees may simply be brushed from the combs with a stiff bristle brush. Appropriate brushes purchased from a bee supply company or snow brushes and other household brushes will work well. Brushing bees from frames is time consuming and causes the bees to become defensive. Smoking the colony while brushing bees will make the task of removing surplus honey proceed more smoothly.

Since most bees will be clustered below the surplus honey crop during cool weather, removal at such times is relatively simple. The few remaining bees in the supers can be brushed back into the colony with little effort. However, the temperature should not be frigid; too many bees will die. There are other problems associated with cool weather. If it is not allowed to warm back to 70-80°F (21-26.7°C) cold honey will be thick and difficult to extract. Additionally, crystallization will be a problem with some honeys that are left on until the weather cools.

The bee escape, a simple device that lets bees out of supers but will not allow them to re-enter the super, has been used by beekeepers for many years. The bee escape fits in the hand hold of an inner cover. The combination of these two pieces of equipment (the bee escape and the inner cover) forms the "escape board." The escape board is then placed beneath the supers, but above the brood nest. The board works best on cool nights when the bees in supers move through the bee escape into the brood nest. The winter entrance that some inner covers have, must be closed when the inner cover is being used as an escape board. The advantage of the escape board concept is that bees are not aggravated. Since the beekeeper is not exposed to the bee colony population, the entire super removal process moves rather quietly.

However, there are some problems that prevent the bee escape board from being the ideal solution to super removal. In order to allow the bees time to exit supers, the bee escape board requires another trip to the apiary a day or so later. If a beekeeper is managing many colonies or the yard is some distance away, this requirement can be inconvenient at best. Secondly, the supers must be in good condition or robbing may be encouraged in supers that cannot be defended by the bees. Another consideration is that bees will not readily leave brood that the supers may contain. Finally, the bee escape can become clogged with bees which plugs the device and prevents bees from exiting correctly.

For many years, beekeepers have known that various chemicals repel bees from supers. However, concerns about safe use and repellent residues in the honey crop have prevented some compounds from being approved for use. Currently, two chemicals are approved for beekeeping use. Benzaldehyde (oil of almonds) works well in the 65-80°F (18.3-26.7°C) temperature range. Benzaldehyde, when exposed to air, oxidizes quickly to form benzoic acid. The other chemical, butyric anhydride, is commercially available under several trade names. This material works at a higher temperature, 80-100°F (26.7-37.8°C), than benzaldehyde, but works in a similar way.

Repellents are sprinkled on absorbent pads or thick cloth stapled to inner covers. Inner covers modified in this way or boards built specifically for this purpose are called fume boards. Depending on weather conditions, the chemical is sprinkled on the pads, placed on top of the colonies, and left there for a few minutes. In some parts of the United States, the backs of fume boards are painted black to absorb heat that will aid the vaporization process. Pads should stay in place only long enough to start the bees moving downward. Puffing smoke into the colony first will help the repellent begin to work. Rarely, would more than 2 tablespoons (30 ml) of chemical be required per pad to start the process. Pads should be recharged when the bees are no longer responding to the fume boards.

The chemically impregnated pads should not be left on the colonies any longer than necessary and repellents should never be allowed to come in contact with honey. If bees begin to come out around the pad or out the front of the colony, too much repellent has been used or the pads have been left on too long. The beekeeper would be well advised to apply too little material rather than too much.

Low pressure, high volume forced air is a popular removal technique used by many beekeepers—especially commercial beekeepers. "Bee blower" units that are designed for beekeeping purposes are available commercially (Figure 2). Many home designed models exist that are quite functional. The blowers are efficient in that one trip to an apiary is all that is required. The procedure is fast and works well under ideal conditions. When combs are not capped, bees that are engorging on honey are difficult to blow from supers. Further, beekeepers removing honey during a dearth with a blower may find that robbing is a problem. During the process, the odor of honey is blown across the apiary enticing other bees into the area. The noise and fumes of the gasoline engine can be annoying, but conversely, the cooling breeze of the blower can be a blessing to the beekeeper on hot days.

In an attempt to remove remaining bees from supers, beekeepers occasionally transport full uncovered honey supers to the processing facility in the

FIGURE 2. A portable, gasoline powered bee blower.

back of open trucks. To prevent extraneous dust and other contaminants from entering the honey, this procedure is not recommended.

No procedure for removing bees from supers is perfect. In fact, a combination of procedures will probably offer the best results. The beekeeper should always expect a few bees to remain in the supers regardless of the removal system(s) used.

PROCESSING THE CROP

A custom designed honey processing facility is an individualistic achievement. No two processing plants are alike, yet all are comprised of the same basic equipment components. Selecting and connecting the most suitable processing equipment for a specific operation can be a challenge. The extracting capacity of the processing equipment should be comparable to the size of the beekeeping operation. Except for processing capacity and complexity, extracting equipment for hobby or sideline beekeepers is similar to that used by larger commercial operations.

The Hobby Beekeeper's Extracting Room

The room selected for processing honey is a basic component of the small extracting operation. All the components of the perfect room are rarely available which means the beekeeper must be resourceful. A screened porch, an enclosed garage, or a basement room are all common possibilities for the beekeeper with a few hives. Frequently, the home kitchen has been the center for the extracting process. The availability of hot and cold water for clean up and a well-lighted work area add greatly to the efficiency of the hobby operation. Since honey spills are almost inevitable, a tile or linoleum floor surface is desirable (Figure 3). Commercial extracting facilities are described elsewhere in this chapter.

After removing supers and transporting them to the extracting room, care should be taken that honey leaking from broken burr combs does not drain on the floor. Constructing simple drip boards is a practical solution for preventing the resulting mess (Figure 4).

Removing Moisture from Honey

The hobby beekeeper rarely considers the moisture content of honey at the time of super removal. High moisture honey (greater than 18% water) will ferment in storage after extracting. In general, it is safe to assume that capped honey has a low enough moisture content to prevent fermentation from readily occurring.

Though various moisture removal systems have been developed for hobby and sideline beekeepers, they have not been widely accepted. Many of the designs have centered around various heating devices that blow warm air through supers of high moisture honey. Controlling the humidity in the immediate vicinity of the drying process has always been one of the problems. If only a few supers are to be processed, the time and effort required to remove excess moisture may be too great for a hobby or sideline beekeeper. It would

FIGURE 3. Honey extracting set-up for the hobby beekeeper.

FIGURE 4. A drip board to be used beneath honey supers that are waiting for extracting. Note the slot underneath the drip board for convenient movement with a hand truck.

be simpler for the hobby beekeeper to leave combs that are not completely capped on the colonies and only extract combs that are fully capped. Procedures such as heating a room to 85-95°F (29.4-35°C) and holding relative humidity to 0-20% (Martin, 1939) while moving air across and through supers will remove moisture slowly.

The use of drying rooms is not uncommon in commercial operations. Commercial honey drying rooms usually have low ceilings with ventilation systems to keep air moving and dehumidified.

A dryer developed by Dadant & Sons, Inc. is reported to be able to remove 1.5 - 2 points of moisture within one hour from 2 barrels (110 gal., 417 liters) of honey (Figure 5). The process also clears the honey considerably, and slows crystallization. The system works at a fairly low temperature so the honey is not damaged by high heat applications. The enclosed device is not affected by environmental humidity, a problem more common with other systems. Additionally, the system eliminates fermentation problems from high-moisture honey even when fermentation has already started because the alcohol is boiled off along with excess moisture in a vacuum under low heat (Dadant & Sons, 1989).

Another device is available that can be installed above the honey storage tanks. Honey being extracted can flow through the honey moisture removing unit and into the storage tanks. Honey should be at about 90°F (32.2°C) when entering this unit. The unit has a capacity of approximately two drums

FIGURE 5. Dadant honey dryer.

per hour. During that period it will take honey at 19.5% moisture and reduce it to 17.8% moisture (Kehl, 1990).

Uncapping Devices

Combs can be uncapped using little more than kitchen knives that have been heated in warm water. Knives that have cooled are exchanged for newly heated knives as needed. Electric or steam-heated knives are designed to supply constant heat to the cutting edges of the knife. Most electric knives have thermostats to control knife temperature. The cutting temperature should be no more than required to easily cut cappings from combs. Heated knives can cause burns to the beekeeper either by contacting the knife directly or from hot honey or wax.

The commercial beekeeper may choose from several power uncappers. A common design has a pair of heated, serrated knives. The horizontally mounted knives saw through the honey cappings on each side of the frame as it

FIGURE 6. An uncapper that incorporates a pair of steam heated knives that cut cappings from honey comb.

passes between the reciprocating knives (Figure 6). Various uncappers of this type will uncap from 6 - 11 combs per minute (Cowan, 1989).

Other designs use a series of small, short chains attached on one end to a rotating drum. The opposite ends of the chains barely touch the wax cappings as the drum rotates rapidly perforating the cappings.

In recent years, the Dakota Gunness Uncapper® (Figure 7) was introduced to the beekeeping industry. The uncapper is designed to completely remove wax cappings by horizontally[1] placing combs on a stainless steel wire conveyer. While on the conveyer, frames are fed rapidly through two stainless steel flails, one above combs and one beneath combs (Figure 8). The uncapper is adjusted so only the very tip ends of the two flails lightly strike the comb's wax surface. Neither frames nor combs are damaged by the procedure. The conveyer will handle two full depth or three shallow depth frames at one pass. The model 200 stainless steel uncapper will uncap 24 deep frames per minute or 36 shallow frames per minute. Cappings drop into various containers or wax melters beneath the uncapper (Gunness, 1990). Since wax cappings are mixed with the extracted honey, uncappers of the chain flail design require a good straining system somewhere in the honey processing system. Flail type uncapping systems do not require heat to complete the uncapping operation.

---

[1]Most uncappers require that the frames be positioned vertically within the uncapper.

FIGURE 7. A chain flail-type uncapper.

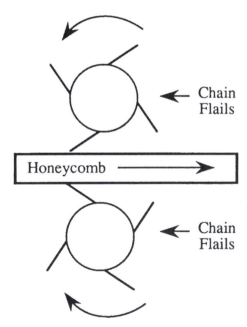

FIGURE 8. Flail mechanism diagram.

Another commercial system commonly used by commercial beekeepers is the rotary knife uncapper[2] that consists of rotating knives that are adjusted in a manner similar to the "chain flail" models and can process nine frames a minute (Kehl, 1990).

A few extracting sequences require the use of devices to temporarily hold uncapped frames while awaiting the next extractor cycle. One common container is a stainless steel, rectangular box having a slanted bottom and a drain spout on the lower end. Uncapped frames, having honey dripping from them, are temporarily allowed to drain in these boxes until they can be extracted.

A second device, called a merry-go-round, is circular and also made of stainless steel. It is a large, saucer-shaped stainless steel dish and has a drain spout in the center. The dish, which can be pivoted around to any selected position, is encircled with a stainless steel rack that is sized to hold uncapped frames that are sitting on their end bars. The uncapper operator can load the merry-go-round on the uncapper side, while the extractor operator can turn the dish around to the uncapped frames and more conveniently load the extractor. All such devices have their honey drains attached to the sump or some other part of the honey plumbing system.

Devices for Processing Honey Cappings

Honey cappings contain a large quantity of honey; therefore, comb uncapping should occur over a container having a drain plug. Every effort should be made to salvage cappings honey. Cappings containers of various sizes are available commercially or can be improvised. Only plastic or stainless steel are appropriate containers for processing honey. Honey in cappings may take several days to completely drain, depending on the air temperature and the quantity of cappings to be drained.

The spin-float honey/wax separator, sometimes called a *wax spinner*, is a centrifuge type cappings processor that is designed for slinging honey from cappings. The machine works on the same principle as a cream separator. The honey/wax cappings mixture is dropped into the center of a revolving drum positioned within a slightly larger stationary drum. For the most efficient operation of the wax spinner, the honey/wax mixture should be at least 95-105°F (35-40.6°C) when delivered to the spinner. The honey-wax mixture is thrown to the perforated wall of the revolving drum, where separation takes place. The clear honey is directed into the settling tank while dry cappings, after being cut out with a rotary cutter, drop from beneath the cappings spinner into containers for melting later (Kehl, 1990).

---

[2] This may be known as a Bogenschutz uncapper or as a Sioux Bee Automatic Uncapper.

FIGURE 9. An uncapper having an electrically powered cappings melter underneath.

Uncappers, having either melters or cappings spinners located directly underneath them are efficient because they eliminate extra cappings movement to other devices (Figure 9). Otherwise, an auger or conveyer is used to move cappings from the uncapper to the rendering equipment.

A unique cappings compressing system, designed for use with uncappers using reciprocating knives, has been developed that mechanically squeezes all but a small percentage of honey from cappings. The cappings compressor is essentially two specially designed stainless steel aprons running very slowly that form the sides of a rectangular tunnel. Cappings, having dropped into a chute beneath the uncapper, start at the lower, wide end and are moved, on a conveyer, to the narrower end. As the tunnel gradually narrows, the perforated, stainless steel walls of the tunnel gradually applies pressure and squeezes out most of the honey. Honey extracted from the cappings drains into a tray beneath the tunnel apparatus and drains back into the sump where it is pumped into the processing system. Moist, irregularly-shaped cakes of compressed wax cappings are then transferred to a wax melter for rendering (Fager, 1989).

Small extracting operations requiring cappings movement to a melter may use nothing more than buckets; however, commercial systems may use large augers or other transportation devices to move cappings from beneath uncappers to the melter. A "Goose Neck" conveyer has been developed by the Fager Corporation (1989) that relocates cappings collected from beneath uncappers and moves them up along an elevated conveyer where the cappings/honey mixture drops into a melter.

Wax Melting Devices

Various models of wax melters are commercially available, having a wide range of capacities and outputs. Small units are available for the hobby beekeeper, while much larger units are designed for the commercial producer. Such units may be positioned under an uncapper or may be a free-standing unit. Melting units use radiant heat, indirect steam, hot water, hot air, or some combination of these procedures. (Figure 10) Wax melts at 147.9° ± 1.0°F (62-65°C) (Bission, *et al.*, 1940). Honey coming from some wax melters may be darkened and caramelized and should not be mixed with the remainder of the crop (Townsend, 1969).

Figure 10. A hot air type cappings/wax melter. The thermostatically controlled unit has a plastic catch pan underneath for liquid wax.

Honey Extracting Devices

The extractor, always the focal point of any extracting operation, is normally in the center of the processing system. Small extractors, those having a capacity 2-6 frames, are popular with hobby beekeepers. Manual or electrically powered machines, manufactured in many different sizes, are available. The current range of extractor sizes in the United States are 2 frames, the smallest, up to 240 frames, the largest available at this time. Extractors are of

FIGURE 11. An extractor of the parallel design. The unit loads from the top.

two basic designs, either having a vertical shaft or a horizontal shaft [3] (Figure 11). Small extractors may process approximately 90 pounds (40.5 kg) per day, while a large commercial unit may process several tons per day (Townsend, 1976). A high quality extractor, made of stainless steel and correctly maintained will last for many years and is a valuable beekeeping investment (Figures 12 and 13).

The prominent problem with most extractors is vibration. To help control extractor sway, the operator should spin frames slowly initially, then increase extractor speed as the honey frames become more equal in weight. Addition-

[3]Extractors having a horizontal shaft may be called parallel extractors.

ally, turn-buckles or ropes may be used to attach the extractor to a heavy table, an extractor stand, or some other added weight appropriate for the task. Large commercial extractors are bolted directly to the extracting room floor.

FIGURE 12. A modern stainless steel 32 frame radial extractor.

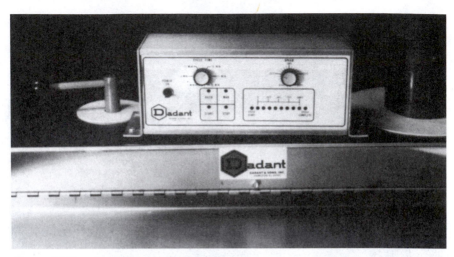

FIGURE 13. Control panel for 32 frame radial extractor.

A prototypic extraction unit is currently under development that uses solid plastic combs. These combs, able to withstand a great deal more stress than wooden frames filled with comb and honey, are currently available and work well in regular extractors. If successfully developed, these high-speed extractors, driven by a 2-10 HP electric motors, will rotate at speeds (1100-2000 rpm) great enough to force honey through natural beeswax cappings. The process of uncapping combs before extracting is eliminated. Four, 10, and 36 frame high-speed parallel extractors are being developed. A procedure currently under development will allow the extractor to be housed in a mobile extracting facility and moved to a hive location where the extraction process will occur. Prototypic machines are very fast and have been reported to be able to extract cool honey with little vibration. At such high extraction speeds, even crystallized honey can be removed from combs. Regular frames cannot be extracted in this machine; wooden frames full of honey will frequently break when spun at high speeds in regular extractors (Drapkin, 1990).

WARMING HONEY FOR PUMPING, STRAINING AND FILTERING

Cool honey, that resists flow, is more difficult to process than warm honey. Each year, however, considerable quantities of honey are processed without any heat whatsoever being added. Heat is only added for processing purposes. Townsend (1965) reported that 100-110°F (37.8-43.3°C) was necessary for honey to be conveniently processed. There are few reasons to heat honey higher than 150°F (66°C) (Dyce, 1931).

Due to the aeration of honey during extraction, most hot room advantages are lost. Consequently, honey should have modest heat added somewhere else within the processing cycle. Heat could be added within the extractor or in the sump near the extractor by using a *flash heating device*[4] or various other commercial honey heating devices. Extractors have not commonly had heating devices added to them. Occasionally beekeepers have added heating coils to the side and bottom of their extractors to cause the honey to flow from the extractor faster. Recently, Wood (1990) has suggested a procedure for heating honey within the extractor. A *flexible PVC heater*, thirty feet (9 m) long, is wrapped around the outside of the extractor, being careful not to overlap the heater coil on itself. Since honey flows faster from a warm extractor, beekeepers using this concept have reported a decrease in extracting time. Offered to the hobby beekeeper for use on the two and four frame metal extractors, the heating device will heat the interior of the cylinder to about 100°F (37.8°C) in about one hour (depending on conditions). Flexible heaters are also available for a 40 frame radial extractor. Since the heating cable is simply wrapped around the exterior of the extractor, the heating cable attachment is semi-permanent. The extractor should be electrically grounded to insure safe operation.

The heating cable can also be used to heat honey supply lines connecting sumps, filters, or settling tanks. The heat cable supplies 120°F (48.9°C) on the band itself and approximately 100°F (37.8°C) on the honey inside the pipe, depending on conditions, and is moisture resistant.

Heat exchanger units, designed to heat honey under controlled conditions, are available in several different styles for the commercial beekeeper. Several designs of heat exchanger units are currently in use within the United States' beekeeping industry. A popular type of exchanger is made from aluminum pipes. Heated water is pumped through a large pipe, approximately 6 inches (15.24 cm) in diameter. Honey flows through smaller tubes, also made of aluminum or stainless steel, inside this larger pipe. Heat is transferred from the heated water in the large tube to the honey flowing in the smaller tubes. If they should become clogged, tubes are removable for cleaning or unplugging. Some units are reported to heat about 3000 pounds (1350 kg) per hour to the preset heat setting.

As honey flows from the extractor, generally into a sump, it will have particles of wax, bees, or other hive detritus in it. Strainers, though widely used, are notorious for clogging easily, are messy to maintain, and are the frequent cause of honey spills. Coarse nylon cloth makes an excellent filter.

---

[4] A flash heater is a honey heater that quickly heats honey using either radiant heat or hot water heat. Honey is heated to approximately 120°F (48.9°C) before passing through strainers or filters. Afterwards, it should be cooled as quickly as possible.

Appropriate nylon netting and strainers can be purchased from bee supply operations or from equipment manufacturers. Nylon paint strainers also work quite well and can be purchased at most paint supply stores.

FIGURE 14. An "in line" honey filter with internal components removed for viewing.

Strainers will be functional longer if they are first submerged in previously filtered honey contained in a settling tank[5]. As unfiltered honey is poured into the submerged filter, a comparable amount of filtered honey is drained from the settling tank, thereby maintaining a constant tank level. If the filter is simply suspended above the settling tank, wax particles clog the strainer as the honey drains from the filter. By submerging the strainer in filtered honey, particles are held in suspension within the unfiltered honey allowing the filter to function longer. Simply allowing honey held in settling tanks to sit in a warm room for a few days will result in most of the foreign material floating to the surface of the honey where it can be skimmed off. Strained honey can be poured directly into glass honey jars or can be stored in bulk containers (Figure 14).

Commercial processing plants frequently use high pressure filters containing a series of filtering paper sheets. Filter sheets begin the series coarsely and successively become finer. The addition of diatomaceous earth[6] makes the filter operate more efficiently (Figure 15).

THE COMMERCIAL EXTRACTING AND PROCESSING FACILITY

The development of an efficient processing facility depends on the ability of the beekeeper to do required ancillary work. For example, one needs to be somewhat acquainted with principles of electricity and construction. Additionally, knowledge of welding, pipe fitting (both steam and water), and general woodworking are also invaluable. All honey processing facilities should be constructed or modified with current and future health standards in mind. Check with local and state authorities before beginning.

Naturally, smaller extracting operations are easier to design and require less investment capital (Figure 16). One of the common attributes of beekeeping is that each beekeeper operates the beekeeping business in a different manner. Many beekeepers simply modify an existing structure and make it work as well as possible. In the following sections, discussions are presented that describe proper construction of the facility. If most of the concepts are incorporated in the facility, an efficient and functional processing plant should result.

If the decision is made to construct a new building, the beekeeper must decide if the facility will be on one level or multi-levels. Extracting plants that

---

[5]The filtering process should be started with clean, filtered honey rather than unfiltered honey.

[6]Diatomaceous earth, a white or cream colored filtering agent, is composed of minute siliceous skeletons of unicellular or colonial algae (Bacillariophyceae).

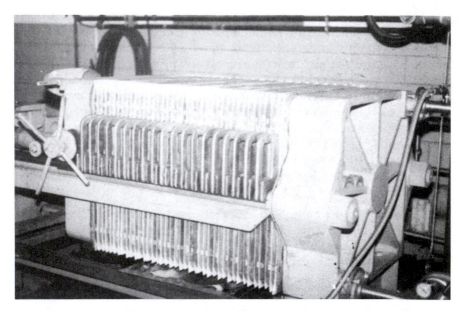

FIGURE 15. A commercial type pressure honey filter that requires the use of paper filters.

are single level require honey pumps to move the honey crop through the processing system. However, single level processing plants are cheaper to construct in most cases.

A multi-level extracting plant, having the extracting room higher than the honey storage room only requires the placement of holes in the extracting room floor for piping the extracted honey to the storage tanks on the lower level. A potential disadvantage that must be addressed when constructing a multi-level facility is the unloading of supers. If a suitable hillside is available on which the extracting plant can be constructed, the extracting room should be located above the honey storage room. In essence, trucks are unloading at ground level at the extracting room—but higher up the hill. If the facility is being constructed in areas not having appropriate hillsides, arrangements should be made for the installation of elevators or ramps that allow for unloading honey supers at the extracting room level.

A few beekeepers have designed mobile extracting facilities that are moved to the yard rather than bringing full honey supers to an extracting plant. After extracting, empty frames and supers can be returned immediately to the hives. These units reduce the amount of time spent handling and transporting supers. When honey is on board, however, these extracting units are quite heavy. They also require firm pathways and space for maneuvering.

Commercial operations vary greatly in design and efficiency. There are no

FIGURE 16. Partial view of the extracting room at the Agricultural Technical Institute of the Ohio State University, Wooster, Ohio.

universal plans available for assembling a "standard" honey processing plant. A generalized honey processing pathway has been included for consideration (Figure 17). And, four extracting plant floor plans are included for review. Floor plans are presented as suggestions and are not intended to be used as construction plans. The plans are for single- story construction. The positioning of processing equipment is estimated and subject to personal preferences. Symbols used are intended to depict general equipment placement and are not intended to show equipment of varying capacities or designs (Figures 18-21).

Extracting Facility Floors

The floors throughout the extracting facility should be seamless and smooth to allow scraping wax and cleaning honey spills from the floor during and after the extracting process. Honey is acidic, and over the years, will etch an uncoated cement floor. Installing a ground tile floor (seamless) or coating a cement floor with a clear acrylic finish are good honey house floor selections. Linoleum, if replaced as it wears, is also appropriate in some cases. Smooth floors, when covered with a film of honey, can be quite slippery. Removable rubber floor mats can be used in high traffic areas. They are easily cleaned and offer good traction. Floors should be correctly sloped and have drains for piping away wash water. Drain traps, required for removing wax accumulation, should be easily accessible.

Depending on future development of the beekeeping operation, the floor supports should be designed to handle the heavy weight of drums and honey supers. Extracting facilities that will use fork lifts and other heavy machinery should be constructed on cement floors.

A GENERALIZED HONEY PROCESSING PATHWAY

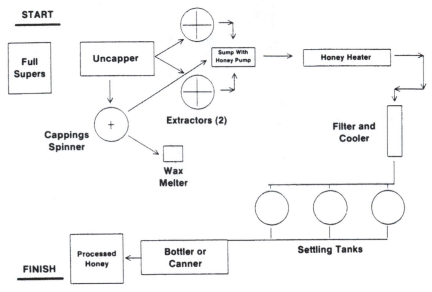

FIGURE 17. A generalized honey processing pathway.

Extracting Facility Walls, Ceilings, Doors and Windows

Generally, cement block walls function well in the extracting room. Wood walls may offer easier attachment of honey lines, wiring, and other supports, but probably won't tolerate frequent scouring as well as cement block walls. Ceiling height should be a minimum of 10 feet (3 m) to allow for installation of piping and honey heating devices.

The walls of an extracting operation should be painted with an epoxy or oil finish[7] from ceiling to floor. Such a finish can be frequently cleaned using pressure washers or other cleaning systems. Ceilings are best made of painted exterior plywood that will allow for future attachments of honey lines, wiring, supports, and lighting fixtures.

Single doors should be wide enough and tall enough to allow for tall hand trucks hauling high stacks of supers to pass through easily. The installation of metal plating along the lower halves of the doors protect them from the wear of hand trucks and pallets. In general, doors should be "bee tight" and swing freely even though they are frequently kept open. If mechanized fork lifts are used, roll-up doors are better. Since these doors avoid routine contact with

---

[7] All paint finishes should be approved for food facility use.

Floor Plan Legend

Equipment Symbols

Uncapper

Sump/Pump

Honey Heater
Honey Filter

Merry-go-round

Honey Cooler

Extractor

Conveyor

Wax Melter

Settling Tank

Blending Tank

Building Symbols

Overhead Door

Door

Sink

Window

1" = 10'

Figure 18. Floor plan legend.

fork lifts, they last longer; however, they are not usually bee tight. Additionally, heating and cooling considerations may dictate that double swinging doors be used.

Windows, while providing employees with natural light and an improved work environment, are not required in extracting plants having an air conditioning system. Window positioning is arbitrary. Since bees are attracted from both inside and outside to honey house windows, they must be screened. Windows require frequent cleaning to remove bee defecation and insect

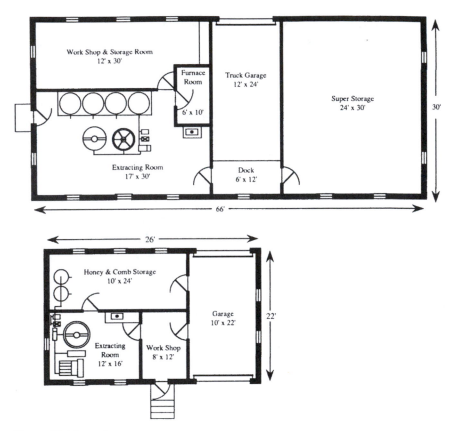

Figure 19. Floor plan.

carcasses which accumulate on window sills. Occasionally, beekeepers install bee escape devices on windows to allow bees trapped inside the honey house to escape. Escapes used in this fashion are reasonably effective, but many bees will still remain within the extracting facility.

Insect electrocution grids operating in conjunction with "black lights" (available commercially) are effective control devices for removing extraneous bees and insects from the honey house without the use of insecticides. Additionally, positioning a black light near an exhaust fan located in the wall or ceiling of the honey house works well for removing bees and insects from the facility. Both devices work best at night.

Electrical Wiring

Commercial honey processing plants require a heavy electrically wired building. In most cases, an individual electrical circuit should have only two or three electrical outlets on it. Additionally, several outlets providing 220 volts

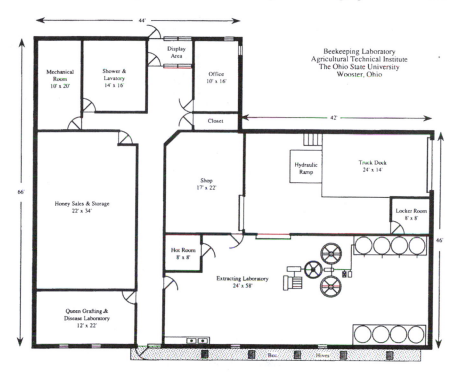

FIGURE 20. Floor plan.

should be available for the occasional use of welders or power washing devices. Since most extracting operations periodically add new machinery, a larger electrical panel than is initially required should be installed. Due to the fire hazard from wax combs, electrical services and lines should be installed correctly and should be encased in conduit. During cleaning procedures, it is helpful if provisions are made for turning off those circuits providing electrical power to processing machinery. In this manner, pressurized power washers, used for cleaning equipment and facilities, can be used more safely.

The extracting room should be well lighted. Fluorescent tubes are preferred but incandescent bulbs are appropriate, also. Lighting should be high enough that bees attracted to lights are far above employees' heads.

Plumbing

The facility will need a large, deep sink for washing various pieces of processing equipment. Stainless steel or plastic sinks are better than galvanized or ceramic sinks. Additionally, for cleaning and maintaining equipment, hot and cold water sources should be located in at least one other place within the extracting room.

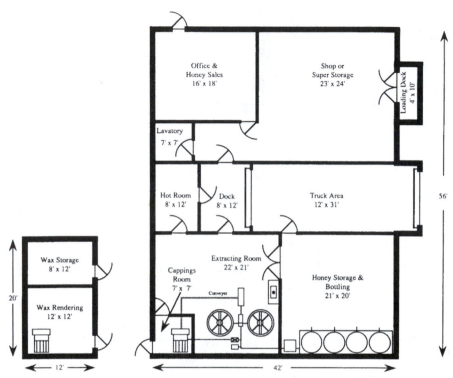

Figure 21. Floor plan.

Drainage lines need to have traps for cleaning wax from the system. Wax does not degrade in water and will frequently plug the drainage system.

Loading Docks and Truck Bays

Where possible, the addition of an enclosed truck bay adds greatly to the efficiency of a processing plant. Frequently, beekeepers construct truck bays so that loading/unloading can occur from the sides of the truck as well as the rear. Drive-through bays are shown in some of the honey house floor plans (Figure 19). Being able to enclose the loaded truck within the bay prevents robber bees from becoming pests around the honey house.

Office and Sales Room

An office and sales room are common components of a honey extracting plant. These rooms are commonly visited by the general public who are not usually familiar with honey processing and packing. Consequently, the sales-room should be immaculate and reflect a modern food sales facility. Shelves should be well stocked with a variety of honey and honey products.

Honey Extracting Plant Hygienic Considerations

Processing honey and wax is potentially a messy task. Even the most careful processor will have an occasional honey spill. Wax, when pressed underfoot, is extremely difficult to remove from floors and processing equipment with anything but a scraper. Additionally, it is resistant to most chemical solvents. When these conditions are combined with the usual accumulation of broken frames, dead bees and bits of propolis, conditions exist that could disturb some health inspectors and many potential honey customers. A honey processing plant can never be too clean. Even if extensive cleaning requires a great deal of time, it will always be worth it. The honey processing plant should be designed and constructed with hygienic concerns in mind. Since wood absorbs honey that will ferment later, the use of wood in the processing plant should be minimized. Currently, many states do not have honey house standards, but there is a general beekeeping industry concern for the good reputation of honey and a feeling among many beekeepers that honey house standards will be imposed within a few years. Honey processing facilities should be designed with that consideration in mind.

OTHER BUILDINGS ASSOCIATED WITH THE EXTRACTING FACILITY

Wood/Metal Shop

The inclusion of a wood and metal working shop within the extracting facility has obvious benefits. Many extracting plants physically attach the shop to the extracting plant. Wood equipment repairs can quickly and conveniently be made without having to relocate to other areas. However, there are reasons for having the shop space located in a separate building. Saw dust, metal filings, or other shop residues are not put into the air or tracked back into the honey processing area. Any dust particles that settle in honey are ideal crystallization nuclei. The shop should be equipped appropriately to meet the needs of the extracting facility. Obviously, the beekeeper must decide, for reasons pertinent to the specific operation, where to locate the repair facility and how well equipped it should be.

Super Storage Building

Due to fire hazards, combs and supers are best stored in another building away from the extracting facility. In general, an unheated building having a cement floor will suffice. Wide doors and good lighting are also common considerations. A few storage buildings have special rooms for fumigating equipment or otherwise destroying wax moths and other pests.

In storage areas, mice, roaches and other vermin are frequently problems. If poisons or other chemical controls are used, be particularly careful. Wax combs will absorb many chemicals easily. Be certain that chemical controls are approved for use around honey comb.

Personal Residence

Even if it is convenient for the beekeeper's home to be on the same property as the processing plant, many beekeepers choose to have their personal dwelling at another location. If the extracting plant is ever sold, it is generally accepted that the property sells better when the personal residence is detached from the processing complex.

Wax Extracting Facility

If wax rendering (old combs, slumgum, or previously rendered cappings that require more processing) is done on a large scale, it is usually done in a separate building due to the general clutter and fire hazard generated by the wax melting procedure.[8] Beeswax is extremely flammable and should never be heated with a direct flame. Molten wax, floating on water, can splatter out of the vessel if the water beneath it is boiled. The wax splashing from the boiler could come in contact with the flame and ignite. Due to fire hazard, the wax rendering room should not be attached to the main extracting plant or other buildings.

The floor should be smooth cement allowing for frequent scraping to remove wax residue accumulation. A stiff paint scraper attached to a long handle or an ice spade make good scrapers for removing wax from floors.

TECHNIQUES FOR RENDERING WAX RESIDUE
Hot Water

Beekeepers having only a small amount of cappings or beeswax to render may simply put the wax residue in water that has been heated to approximately 150°F (66°C). Melted wax will float on water (specific gravity = 0.95) (Vansell and Bisson, 1940). While floating on the water, it can be ladled off or allowed to solidify and be removed as a cake. This procedure, though simple, is inefficient, leaving most of the wax in the slumgum. The bottom of the wax cake will have accumulated all the impurities contained in the wax residue and will have to be scraped off. The sweetened water, a by-product of this procedure, can be fed back to bees during warm months if diseases, such as American Foulbrood, have not been a problem. Since the mixture will ferment quickly, feed only in quantities that the colony can use swiftly. If too little water is used to melt the mixture, the bottom of the wax cake may have large, flaky crystals or small balls of beeswax.[9] This form of beeswax is difficult to convert into the normal beeswax form. Dry heat may be required to remelt the wax (Eckert and Shaw, 1960).

---

[8]Wax cappings are normally processed in the extracting room at the time of comb uncapping.

[9]Referred to as granular wax in some texts.

FIGURE 22. Frontal view of a solar wax melter.

Solar Wax Melters

Solar wax extractors[10] are cheap, in fact absolutely free, to operate. Additionally, wax collected from a solar melter is partially bleached. High quality solar extracted wax is in demand for use by individuals to make candles or other wax objects for show (Coggshall and Morse, 1984).

In principle, the solar wax melter is a box covered with glass (Figure 22). Much like a common greenhouse, the glass cover captures some of the sun's rays and heats the inside of the box. Inside the melter is a pan that has a drain

---

[10]May be called Sun Wax Extractors in some of the older literature.

on one end for melted wax to escape. The best vessels for use with wax melters are made from stainless steel, nickel, or aluminum (Corner, 1976). Sizes and specific designs may vary, but all solar wax melters use the same principle.

Solar wax melter designs should be painted black *outside* and white *inside* (Anderson, 1960). The black exterior absorbs more heat and the white color inside reflects some of the sun's rays—rays which are unable to escape due to the glass covering. Double glass panes, with a one half inch air space, are more efficient than a single glass pane and should be positioned about five inches from the top of the tray. Anderson (1960) suggests sloping the wax melter in order to help melted wax drain away from the unmelted material and to allow for a better angle with the sun. The overall size of the melter is not important as long as the tray is within five inches (12.7 cm) of the glass and wax collection pans and wax residue pans are appropriately sized for the specific melter. Other design variations include mounting the melter on a steel shaft. The melter can then be turned, as required, to maintain the most efficient exposure with the sun. Solar wax melters with a large capacity may have an improvised procedure that allows liquefied wax to drain from inside the melter to molds outside the melter where it will solidify into blocks.

The solar extractor is extremely inexpensive to operate; however, it is unfortunately inefficient because it leaves as much as one half to two thirds of the wax trapped in the slumgum (Coggshall and Morse, 1984). All solar extractor slumgum should be saved for subsequent processing in other ways. Additionally, any honey extracted from the wax using a solar wax melter is ruined. Such honey is not good for feeding to bees for winter feed.

Steam Chests

Popular in years past and still seen occasionally, steam chests are more efficient than solar melters, but accumulated slumgum still requires further processing to remove as much wax as possible. Steam chests usually have a screened or perforated slanted bottom that allows melted wax to pass through and away from the slumgum. The steam box is usually insulated and steam is injected in one port and allowed to escape through another. Since partial saponification[11] can occur, it is frequently recommended that wax not be heated directly with steam.

Wax Presses

Wax presses are not commonly available from bee supply sources anymore. Presses are still the most efficient procedure available for beekeepers managing small to medium operations to extract wax from old combs, burr combs, and frame scrapings. Wax presses used in conjunction with hot water

---

[11]Saponification is a chemical reaction with water (hydrolysis) that produces a free alcohol and an acid salt to make soap.

SLUMGUM[12]

1. Slumgum is the residue remaining after the wax rendering process. It is composed of wax moth cocoons and detritus from the brood nest and is usually dark brown to black.

2. The quantity of wax contained in slumgum is high unless it is derived from a highly efficient wax rendering procedure using steam and pressure. If low in wax, the slumgum crumbles when dry whereas slumgum high in wax will form a solid cake when cool.

3. Slumgum low in wax is generally discarded. It will burn well. Larger quantities of slumgum having high amounts of wax may be purchased by commercial wax extraction operations (depending on the current selling price of beeswax).

4. Slumgum has little attraction to bees and will not cause robbing.

5. Slumgum from efficient wax extraction devices (devices using hot water, steam and pressure) will destroy American foulbrood spores; however, slumgum from solar wax melters may still contain viable spores. Though it is not attractive to bees, slumgum from solar melters should be destroyed.

6. Slumgum is not particularly attractive to wax moths, especially if it has a low wax concentration.

---

[12]From *Beekeeping* by Eckert and Shaw, 1960

or steam are very effective wax rendering devices. The wax press is frequently used to process slumgum accumulated from other extraction procedures. Presses are easier to operate if the wax or slumgum is initially heated in hot water or steam before placing it in burlap or nylon bags. Filled bags are then placed under steadily increased screw torque while under hot water. The cylindrical walls of the press container are either perforated or slit to allow liquid wax to escape into the surrounding water where it rises to the surface. The press is efficient, but considerable time (maybe as long as 10 hours) and re-torquing are necessary to extract most of the wax from slumgum. To allow for thorough extraction, bags should not be very full when starting the procedure. If the process is done slowly and correctly, Coggshall and Morse (1984) have reported that only 0.5-2.5% of the final slumgum content will be beeswax. Efficient wax presses have been devised by ingenious individuals using hydraulic automobile jacks or modified apple juicers.

Solvents

There have been occasional reports of industrial solvents being successfully used to dissolve wax from slumgum. Since propolis and other wax contaminants are also dissolved, there are many problems with this technique for beekeepers. Research has been conducted on various chemical treatments to dissolve wax. Even so, beekeeper use of solvents is still impractical.

GRANULATION AND FERMENTATION OF HONEY

In practically all honeys after some time has passed, granulation, a natural phenomenon, should be expected. Additionally, after granulation has occurred, a second problem may soon follow—honey fermentation. Fermentation is a common problem for beekeepers processing liquid honey, especially if the honey contains more than 18% moisture. In years past, beekeepers had to contend with an occasional bulging drum or container—an outward symptom of the deteriorating state of the honey contained within the container. Early beekeepers did not know that fermentation was caused by yeast organisms, nor would they have known how to effectively deal with the problem had they known the cause.

In 1928, Dr. E.J. Dyce, Cornell University, began to thoroughly research the natural processes of fermentation and crystallization in honey. Dr. Dyce's findings published in 1931 are still the recognized standards for honey granulation and fermentation (Dyce, 1931). By analyzing different honeys from different parts of the United States, Dr. Dyce was able to show that most honey samples crystallized soon after extracting, while a few honey samples were highly resistant to crystallization. Dr. Dyce observed that the glucose content of a specific honey was the major factor in the crystallization process of honey. Glucose is a common sugar found in varying quantities in honey.

Honey processed with the use of extractors and honey pumps granulates faster than comb honey. Also, honey that has been through a honey pump generally produces a finer granulation than honey that is allowed to crystallize in the comb. Apparently, the pump serves to break crystals into smaller "seeds" which result in a more finely granulated product. Pure glucose crystals are highly effective (as is a "starter" of crystallized honey) in causing honey to granulate; however, any number of things can serve as honey crystallization nuclei.[13] Dust, pollen, wax particles, small pieces of propolis, or air bubbles all work quite well.

Most honeys are supersaturated with glucose and other sugars. As honey is extracted and stored, excess glucose, temporarily suspended in the honey, precipitates out in the form of glucose crystals. At any time after complete granulation has occurred, only about 15% of the honey is solid. Glucose crystals form a lattice-work within the honey which immobilizes other honey components into a suspension. As glucose forms a crystal lattice (glucose hydrate) within the honey, excess moisture is made available to the remaining honey constituents. Therefore, the liquid phase of granulated honey is a good medium for yeast cultures to grow. Most ripened honeys are about 18% water. Only a slight increase in water content is required for fermentation to begin. Therefore, high moisture (greater than 18%), not granulation, allows fermentation to proceed. The easiest way for beekeepers to control fermentation is to heat honey to 160°F (77°C) for 4-5 minutes. Unfortunately, high temperatures will quickly damage honey. Proper agitation of the honey and ensuing rapid cooling are required to handle heated honey properly.

Proper honey heating will control fermentation, but the natural granulation process, which occurs after heating, results in large coarse crystals. Leaving one with the sensation of eating sand, such large crystals do not make a pleasant food product.

Additionally, there are other problems with the use of high heat to control fermentation. The production of 5-hydroxymethylfurfuraldehyde (HMF), a breakdown product of sugar solutions containing glucose and fructose, results from the use of high heat to control fermentation. The formation of HMF is not unique to honey. Heating honey above 167°F (75°C) even for a few minutes or storing honey at temperatures above 80°F (27°C) for several months can cause HMF levels to exceed 40 mg/kg, which indicates "overheated honey" in the international export market. Naturally occurring levels of HMF are about 10 mg/kg (Crane, 1976).

Honey can be frozen to prevent fermentation. However, this has not generally been cost effective on a commercial scale. Beekeepers having a small

---

[13]The term *crystallization nuclei* refers to the initial starting point of a glucose crystal.

amount of honey or a special honey may find freezing to be a viable way to store honey for long periods.

IMPORTANT HONEY FERMENTATION TEMPERATURES[14]

	Degrees Farenheit	Degrees Celsius	Temperature Effects
Below	52	11	Fermentation yeasts unable to grow
Below	55-70	13-21	Low range optimum temperature for honey fermentation
Above	80	27	High range optimum temperature for honey fermentation
Above	160	71	Highest temperature necessary to control honey fermentation yeast
Above	167	75	Quick damage to honey flavor and chemistry

[14]From Dyce, 1931.

IMPORTANT GRANULATION TEMPERATURES FOR CREAMED HONEY PRODUCTION[15]

	Degrees Farenheit	Degrees Celsius	Temperature Effects
At	57	14	Best temperature for honey granulation
At	58-59	15	Best temperature range for low moisture honey granulation
Above	60	16	Most honeys will not granulate
Below	50	10	Crystallization greatly retarded
At	30	-1	No crystal growth after two years (even with crystallization seed added)

[15]From Dyce, 1931.

STORAGE, RELIQUEFYING AND PACKAGING EXTRACTED HONEY

Storing Extracted Honey

Fermentation is the major concern for honey in storage. In order to prevent fermentation as much as possible, only honey having approximately 17% moisture should be stored for any length of time.

Granulation may occur in honey that is in storage, but is not considered a problem if the honey moisture level is around 17%. In fact, some commercial packers prefer it granulated. If containers begin to leak or a storage container is accidentally punctured, granulated honey contained within the container will not drain. Since granulation is anticipated, even desired, storage containers should have tops that are completely removable in order to remove granulated honey for melting.

Under perfect conditions, honey crops would be sold during the same year as that of production. Obviously, such conditions are not always possible and honey must be stored for some length of time. Honey should be stored in tightly sealed containers in a dry, cool room. The temperature should be as cool as possible. At 52°F (11°C) and below, fermentation yeasts are not able to grow, thereby protecting honey from the problem of fermentation. As the temperature in the storage room is allowed to increase, there is a proportional decline in the quality of the stored honey. The warmer the room, the poorer the storage conditions. Over time, stored honey will darken and will undergo slight chemical changes that commercial honey packers are able to detect.

Liquefying Granulated Honey

Small quantities of honey, with lids loosened, can be put into a water bath having a temperature range of 95-120°F (35-48.9°C). The honey is darkened slightly each time this cycle is repeated; however, using heat is the only way to remelt granulated honey. Thermostatically controlled stainless steel water baths are available from bee supply companies (Figure 23). Such units usually will handle 3-4 five gallon pails. For five gallon (19 liters) pails containing

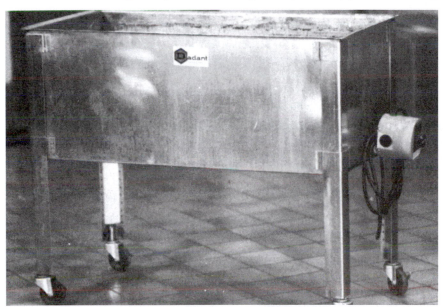

FIGURE 23. An electrically heated, water-jacketed honey liquefying tank. Such devices will liquefy 2 - 6 pails of honey during each cycle.

honey that is completely granulated, water baths will require operation over a 24 hour period to completely and safely melt honey.

Pail and drum heaters are available that heat honey held within the container using a heating band that wraps around the pail or drum (Figure 24). Bands used to heat drums are thermostatically controlled, but the beekeeper is expected to monitor the temperature until the correct thermostat setting has been established (Wood, 1990).

Economical 5 gallon (19 liter) pail heating bands are available. The 155 watt heating band may be placed around the bottom, middle or top of a round five gallon pail. The honey contained within will be reliquefied within six to eight hours. The unit is not thermostatically controlled. It is recommended that beekeepers, using a thermometer, monitor the reliquefaction of the first pail to establish how long it took for the honey to liquefy. A timer can then be used on subsequent containers in the same batch. Use the device carefully and follow heater instructions (Wood, 1990). Drum and pail heaters can also be used to heat other non-flammable liquids such as corn syrup, sugar syrup or plain water.

Commercial honey bottling operations have melting rooms that are kept at a high enough temperature to reliquefy drums or pails of honey. Water baths are more efficient, but they are normally impractical for use in melting rooms; therefore, melting rooms using hot, moving air are the most common. Drums of crystallized honey are up-ended on melting grids (pipes) having hot

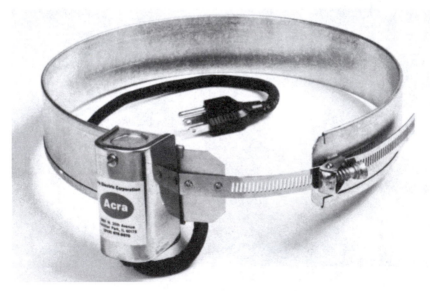

Figure 24. An electric band heater for liquefying honey in drums or cans.

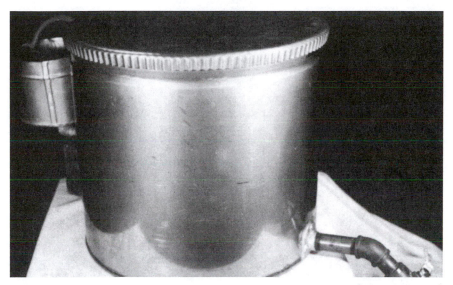

FIGURE 25. An electrically heated, thermostatically controlled bottling tank for the hobby beekeeper.

water or steam passing through them. Melting grids should not be more than 2.25 inches (5.7 cm) apart (Townsend, 1976). As honey is liquefied, it drains from the drums, through the melting grids, into trays which direct the partially melted honey from the melting room to other heating devices within the processing plant that complete the melting process. Occasionally, honey that is high in moisture may be restricted to the melting room in order to remove some of the excess moisture. Oppositely, small amounts of steam may be released in the melting room to put small amounts of moisture back into the honey. The temperature of the melting room is generally heated to 140-158°F (60-70°C) (Townsend, 1976).

Bottling Honey

Simple dispensers, usually made of food-grade plastic or a stainless steel tank, having a valve near the bottom of the tank, are all that is required to satisfy the bottling needs of most hobby beekeepers (Figure 25). Cool honey drains slowly. Thermostatically controlled, water-jacketed stainless steel tanks heat the honey to make it flow faster (Figure 26). Various measuring devices are available for dispensing pre-determined quantities of honey or the bottling tank can be manually operated.

For the larger operator, a no-drip, accurate bottling device (named Fill Master) (Figure 27) has been developed (Little, 1992). The filler reduces bottling labor, provides accurate fill volumes, and reduces fatigue associated with hand filling. Fill volume can be quickly changed to conform to the bottle

Figure 26. A double-walled electrically heated, thermostatically controlled bottling tank designed for the hobby beekeeper.

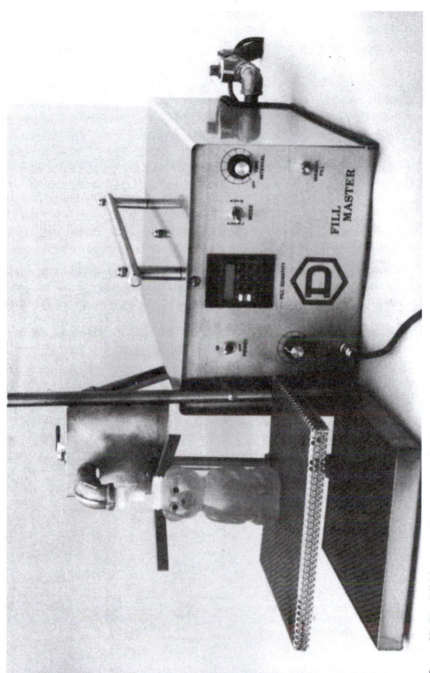

FIGURE 27. The *Fill Master* no-drip bottling device for beekeepers. Volume can be adjusted, depending on container size being filled.

currently being used. To produce exact volume filling, the rotation of an electrically driven positive displacement pump is precisely controlled. The fill volume is displayed on a digital numerical display for the convenience of the operator.

Due to the sophistication required to develop and construct a commercial honey bottling operation, most operations specialize in either honey production or honey processing. Industrial equipment, either specifically purchased or modified for honey bottling, makes the layout in almost all industrial honey bottling operations similar, but yet different in subtle ways. A diagram of a hypothetical operation is presented (Figure 28). In general, strained honey that has been heated, is pumped into tanks where the heat is maintained. In some instances, heat may be added at other locations in the honey flow line. If the honey was not filtered prior to placing in the bulk tanks, it is filtered through a pressure filter on the way to the bottling machine. Warmed, filtered honey is then pumped to the bottling machine on the conveyer line.

The bottle conveyer line begins with the placement of empty jars on an apron which arranges jars in a single line to pass the trip through the bottling cycle (Figure 29). The first device, using either a vacuum or forced air,

GENERALIZED COMMERCIAL HONEY BOTTLING PLANT LAYOUT

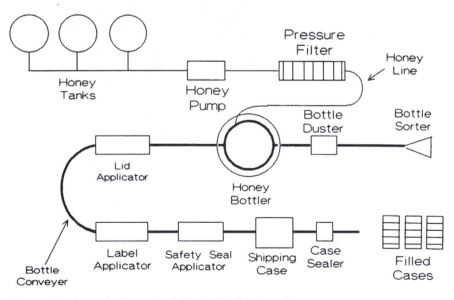

FIGURE 28. Generalized commercial honey bottling plant layout.

removes any debris from the jars before they are filled. The filler machine can either be a revolving ring having filler spouts mounted above the jars or the filler mechanism can be in the straight line style where several jars are filled at once. As the empty jar passes underneath, the predetermined volume of honey is placed in the jar. The filler apparatus, capable of filling four jars at once, may be in a straight line rather than a circle. The filled jars continue along the conveyer until they pass beneath a lid applicator (Figure 30) where a lid is screwed on. Never leaving the conveyer, the label is applied next. Some operations, to insure product safety, put a safety seal around the cap of the jar. Filled and labeled jars are then placed, usually by hand, into shipping cases. Next, full cases are automatically sealed and bottled and packed honey is either stored or is shipped to market. As would be expected, there is great variation from one operation to another. Most of the suggestions concerning the construction of a commercial extracting and processing facility are applicable to the development of a bottling operation.

CREAMED HONEY PRODUCTION

Creamed honey is finely granulated honey that has had the crystallization process carefully controlled. Creamed honey has a much smoother consistency than naturally granulated honey. Having the texture of butter, creamed honey is a pleasing food that has many devoted consumers (Figure 31). In fact, most of the world's people consume honey in the crystallized state, not in the liquid state.

FIGURE 29. A commercial honey bottling conveyor line.

FIGURE 30. A commercial lid applicator device.

FIGURE 31. Creamed honey on bread—just one of the many ways to eat honey. (Photo courtesy Stoller Honey Farms).

All honey can be creamed, but those higher in glucose granulate the fastest. The lightest, mildest flavored honey normally makes the best creamed honey, but darker, fuller flavored honeys will also work quite well.

The Procedure for Producing Creamed Honey

Mr. Darl Stoller of Stoller Honey Farms, Inc., Latty, Ohio, has commercially produced a superior creamed honey product for many years. Many of the following comments are from personal communication with Mr. Stoller concerning the production of creamed honey. The four important elements to consider when producing quality creamed honey are: heating, straining, mixing, and cooling. If any step is hurried or omitted, the final product will have less quality and appeal.

Selecting Honey for Creaming

Almost any honey can be finely granulated, including honey that has begun to ferment. Many consumers believe that creaming dark or strongly flavored honeys improves that particular honey's flavor. Honey for creaming should be in the 17.5% - 18% moisture range. Finely granulated honey made from honey in this moisture range will be neither too hard nor soft.

Heating the Honey for Creaming

The heating process destroys yeasts, organisms that cause fermentation, dissolves any coarse glucose crystals that may be present, and makes the honey thinner, thereby providing for easier straining. The honey should be heated to a maximum of 150°F (66°C). Dyce (1931) recommended heating the honey twice—once to 120°F (49°C) and then a second time to 150°F (66°C).

Straining the Honey That Is to Be Creamed

Honey should be thoroughly strained after each heating session to remove extraneous debris, pollen, or wax. Then it should be cooled as quickly as possible to room temperature (75°F, 24°C). Pressure filtering with special filters is not necessary, but honey does need to be carefully and thoroughly strained. Cold water-jacketed tanks and other coolant equipment have been developed by commercial producers for promptly cooling large quantities of heated honey. Discussion of such specialized and costly equipment is beyond the scope of this chapter.

Mixing the Crystallization Starter with the Honey

To the honey that has been heated and cooled, under strictly controlled conditions, add 5% to 10% (by weight) of the "starter" seed. The "starter" seed is honey that has been finely ground in a type of meat grinder or other specialized grinders, which breaks down the glucose crystals into an exceedingly fine state. Starter seed should be from finest, mildest honey available. Commercial producers always have a good supply of starter seed honey on hand and frequently regrind it to ensure fine crystals. Since warm honey may liquefy the delicate crystals in the seed honey, the honey to be creamed should be no warmer than 75°F (24°C), when the seed is added, but need not be as low as 57°F (14°C). However, if mixing is done by hand, cool honey may require more labor to mix with the starter honey. Completely blending the seed into the honey will help assure uniform crystallization of the final creamed product. Most large packers use large, modified, stainless steel dairy tanks, having a horizontal agitator, that efficiently mixes honey and crystallization seed. Care must be taken to incorporate as little air as possible during all aspects of the crystallization process. Air bubbles, rising during the early stages of the granulation process and leaving an objectionable "scum" on the surface of the honey, negatively affect the final product's appearance. After processing and granulation, even the best creamed honey producer's product will have a thin layer of air bubbles on the surface of the honey. This delicate "frosting," differing from "scum," is not considered objectionable.

Filling Creamed Honey Containers

After the honey is thoroughly mixed with the seeded honey, it is pumped into the bottling tank and allowed to set for an hour or so, thus giving time for air bubbles to rise to the surface, where they are skimmed off. Seeded honey is then transferred to various types of retail containers. The creamed honey should set up firmly within 4 to 6 days, and will then be ready for use.

The Cool Room

Honey in the newly filled containers is allowed to granulate in a cool, dry room (57°F, 24°C). Granulation room temperatures are generally held to be about 55°F (13°C). Dyce (1931) reported that the temperature should not go above 59°F (15°C) nor below 50°F (10°C). Most packers keep creamed honey in cool rooms until the honey is ready for shipment to customers or other outlets.

Creamed Honey Containers

Unless creamed honey is being prepared for a honey show, glass containers are rarely used for retail packaging. Quality creamed honey production is extremely demanding. Even though creamed honey may have been produced under exacting conditions, imperfections still exist. Glass containers exaggerate these imperfections. Consequently, opaque packages such as paper or plastic are normally used.

For aesthetic reasons, creamed honey, after undergoing the complete granulation process, is not transferred to other containers. Keeping the finished product in the original retail container keeps the product neat and pleasing.

Consumer Use of Creamed Honey

Creamed honey does not need to be refrigerated, but it may revert to a liquid state after being held at 90°F (32°C) for an extended period of time. It cannot be brought back to a creamed state, unless the complete granulation process is repeated.

If the consumer finds that a creamed honey is too thick or thin, he or she can simply warm or cool it slightly until the desired consistency is obtained. The ability to spread is the unique feature of creamed honey that makes it popular as a topping and on sandwiches.

Producing Creamed Honey in Small Quantities

If the general procedure as has been discussed is followed, creamed honey can be made in small amounts without elaborate equipment. Mix 1 cup approximately one pound (.45 kg), of quality creamed honey purchased from a local grocer with 10 lbs (4.5 kg) of quality liquid honey. After following the aforementioned steps of mixing and cooling, a quality creamed honey can be made in one's own kitchen. Different flavorings or small pieces of dried fruit can be added to the creamed honey to create an interesting product.

REFERENCES

Anderson, E.J. (1960). An improved solar wax extractor. Pennsylvania State University of Agriculture, Agricultural Experiment Station Progress Report 225. 5 pp.

Bission, C.S., G.H. Vansell, and W.B. Dye. (1940). Investigations on the physical and chemical properties of beeswax. U.S. Dept. Agr. Tech. Bull. 716.

Coggshall, W.L., and R.A. Morse. (1984). *Beeswax*. Wicwas Press, 425 Hanshaw Road, Ithaca, N.Y. 192 pp.

Cook and Beals, Inc. (1990). Promotional literature. Cook and Beals, 221 South 7th Street, Loup City, NB.

Corner, John. (1976). Handling Beeswax. Bee Master's Course, Simon Fraser University, Vancouver, B.C. 7 pp.

Cowan, J. (1989). Personal communication and promotional literature. Cowen Manufacturing Co., Inc., P.O. Box 399, Parowan, UT 84761-0399.

Crane, E. (1976). *Honey, A Comprehensive Survey.* Heinemann, London, England. 608 pp.

Dadant and Sons, Inc. (1990). Promotional literature and personal communication. Dadant and Sons, Inc., Hamilton, IL 62341.

Drapkin, H. (1990). Personal communication. Perma Comb Systems, 22543 Ventura Blvd., Suite 222A, Woodland Hills, CA 91364.

Dyce, E.J. (1931). Fermentation and crystallization of honey. Bull Cornell Agr. Exp. Sta. No. 528. 76 pp.

Eckert, J.E. and F.R. Shaw. (1960). *Beekeeping.* The MacMillan Company, N.Y. 536 pp.

Fager, O. (1989). Promotional literature from Fager Corporation, Rt. 3, Kewaunee, WI 54216.

Grout, R.A. (1963). Extracting the honey crop. *The Hive and the Honey Bee.* Dadant & Sons, Hamilton, IL pp. 303-322.

Gunness, D. (1990). Promotional literature from Dakota Gunness, Inc., P.O. Box 106, Abercrombie, ND 58001.

Lemoine, P. (1990). Promotional literature and personal communication. Golden Bee Products, 801 Little Farms Ave., Metairie, LA 70003.

Little, R. (1989). A no-drip measuring dispenser for the beekeeping industry. *Amer. B. Jour. 129* (8) p. 529.

Martin, E.C. (1939). The hygroscopic properties of honey. *J. Econ. Ent. 532*(5) 660-663.

Sherriff, B. (1990). Promotional literature and personal communication. B.J. Sherriff, P.O. Box 416, Nacoochee, GA 30571.

Stoller, D. (1990). Personal communication. Stoller Honey Farms. Latty, OH.

Townsend, G.F. (1976). Processing and storing liquid honey. From *Honey, A Comprehensive Survey, ed.* E. Crane (London: Heinemann). pp. 269-292.

Townsend, G.F. (1969). How the beekeeper can influence the quality of honey. XXII Int. Beekeep. Congr. Munich: 593-596.

Wood, T. (1990). Personal communication and promotional literature from Acra Electric Corporation, 3801 N. 25th Avenue, Schiller Park, IL 60176.

FLOOR PLANS

Floor Plan 1. (Undated). Modification of plans presented in *Honey Handling Catalog.* 16 pp. Dadant & Sons, Inc., Hamilton, IL 62341.

Floor Plan 2. (Undated). Modification of plans presented in *Honey Handling Catalog.* 16 pp. Dadant & Sons, Inc. Hamilton, IL 62341.

Floor Plan 3. 1990. Modification of extracting facility plan. Beekeeping Laboratory. The Agricultural Technical Institute, The Ohio State University, Wooster, OH 44691.

Floor Plan 4. Modification of extracting facility plan presented in *Selecting and Operating Beekeeping Equipment.* C.D. Owens and B.F. Detroy, Agricultural Research Service, USDA. 24 pp.

ACKNOWLEDGMENTS

Romig, R.L. (1990). The Agricultural Technical Institute, The Ohio State University, Wooster, Ohio. Production of computer generated graphics for extracting plant floor plans.

Stoller, D. 1990. Stoller Honey Farms, Latty, Ohio. Assistance with creamed honey production procedure.

CHAPTER 16

THE PRODUCTION OF COMB
AND BULK COMB HONEY

by Eugene E. Killion*

Comb honey is best described as the "natural" form of honey. Honey bees gather nectar from blossoms; store, ripen, and seal the crystal clear honey in fragile hexagonal-shaped cells of virgin beeswax. No other form of honey is as attractive as comb honey since it is not altered in any way by man. It remains just as the honey bees created it. The fresh honey with floral fragrance is sealed in each cell with beeswax giving comb honey an unimpaired flavor and aroma that cannot be matched by any other type of processed honey (Figure 1).

Over 160 years ago the Russian beekeeper, Peter Prokopovitch was one of the first to produce comb honey. The production of this marvelous honey continued until the invention of the honey extractor in 1865, leaving only a

---

*Eugene E. Killion, retired Illinois State Apiary Inspector and Illinois Extension Apiculturist. Author of the book *Honey in the Comb* and former commercial comb honey producer in Illinois, operating over 1,000 colonies with his father, Carl E. Killion, strictly for section comb honey production.

FIGURE 1. Fresh nectar stored in newly drawn honey comb

[705]

few prominent leaders in the field of comb honey production in the early 1900's such as Dr. C. C. Miller, G. F. Demuth, Charles Kruse, and Carl E. Killion. Today the majority of beekeepers extract the honey combs and market honey in liquid or crystallized form. Since the 1970's the production of comb honey has been increasing in popularity although production is still far below the consumer demand. This demand has created a better price and more return to the beekeeper than other forms of honey. Less equipment is needed in producing and preparing comb honey for market and with the development of new comb honey equipment and packaging materials, any beekeeper from the beginner to the veteran is now able to produce good quality comb honey. A beekeeper with an aptitude for orderly planning and exacting work can profit by producing honey in the comb. Certain areas are unsuitable for comb honey production such as locations near dense woodlands where excessive propolis may be gathered. Propolis creates more labor for the beekeeper in removing it from sections before the comb honey is ready for market. Locations with short overlapping honey flows of light-and-dark-colored honeys are inappropriate. Regions where honey flows are very slow and intermittent or have high humidity are less favorable than those ideal areas that have heavy honey flows with an abundance of nectar-secreting plants such as the various popular clovers.

COMB HONEY EQUIPMENT

Some of the early master comb honey producers such as Miller, Ill., Stevens, N.Y., and Kruse, Ill., used only 8-frame equipment, but the standard 10-frame size of today can be used for all types of comb honey production with success. Those beekeepers who want to increase their knowledge and skills of this beekeeping art can find detailed equipment and specifications described in *Honey in the Comb* published by Dadant & Sons, Inc., Hamilton, IL 62341. This book describes in detail the procedures of comb honey production and will assist those who seriously desire to produce the finest quality award-winning comb honey.

Standard 10-frame hive bodies are used for the broodnest in producing either comb or liquid honey. However, different supers with fixtures are used to produce the various forms of comb honey.

Two 10-frame width supers are favorites in producing bulk comb honey: the "Illinois" depth super of 6-5/8 inch and the shallow super of 5-11/16 inch. Three types of frame top bars are used in the bulk comb production: the wedge top bar, grooved top bar, and the slotted top bar. Foundation is secured in wedge top bars with either a wooden strip nailed or a slender V-shaped tin strip. For the slotted top bar, the foundation extends through the slot and is bent over while the foundation is warm to secure it in place. For the grooved top bar, melted beeswax is used.

Form boards are convenient for securing the foundation in the grooved top bars. The board is slightly smaller than the inside of a frame and thick enough to reach the back edge of the cut-out portion of the top bar groove when the frame is placed in position. Four of these boards can be fastened to a wooden reel to speed up the foundation fastening process. A sheet of warm foundation is placed on the board and into the frame groove with melted beeswax poured in the groove. A teaspoon with the tip pinched together to form a pouring spout with the handle bent up at a right angle to the bowl of the spoon makes an ideal instrument to direct the wax into the groove. The reel is then turned so that one end of the frame is slightly higher, allowing the melted wax to run the length of the groove. The foundation must be fastened solidly in the frame as the sheets will come loose with the weight of the bees clustering on them during hot weather.

Frames for bulk comb honey should not be filled too far in advance of the honey flow as the foundation tends to become wavy or crack and break loose from the top bars during cold weather. Only full sheets of foundation should be used. When just a small strip is used, bees tend to build drone size comb cells.

During the 1800's wide frames were used with two and three tiers of sections installed in each frame. Later G. M. Doolittle, Borodino, N.Y., introduced a wide frame using a single tier of sections. During this same era a honey rack was used to some degree. This consisted of only a flat board holding 27 square sections tied together with string placed above the broodnest. Later a wooden crate was designed with a slotted bottom to hold the sections which somewhat resembles modern-day comb supers. This so-called combined crate could be removed from the hive and the comb marketed right from the crate. As most beekeepers preferred to scrape propolis from the sections before taking the honey to market, the combined crate proved unpopular. Finally came the Heddon or Moore crate which was simply a shallow box, deeper by a bee space than a section with partitions for each row of sections. These partitions served to strengthen the crate and hold the sections square. Some beekeepers started using separators rather than fixed partitions between each row with the new idea of using bent tin strips upon which the separators and sections could rest. The "T" super received its name since the sections are supported by tins bent in the form of an inverted "T." Some changes have been made on the T-super since it was first shown to Dr. Miller while on display by D. A. Jones, Beetown, Ontario during the North American Beekeepers' Convention in Toronto, Canada in 1883. Today it is referred to as the Killion super since Carl Killion modified it extensively (Figure 2).

Wood sections are made from basswood lumber which has been selected for its texture, color, and smooth velvet surface. The white-colored basswood has tough fiber and the ability to take the high polish desired in top quality sections.

A 10-frame section comb super will hold twenty-eight 4-1/4″ x 4-1/4″ x 1-7/8″ (108 mm x 108 mm x 47.6 mm) square wooden sections. Eight-frame supers contain 24 of this size section. J.S. Harbison, a California beekeeper invented the square wooden section in 1857. Some supers are modified to hold straight sided 4″ x 5″ wooden sections. The 4-1/4 inch bee-way section, scalloped along the top and bottom for vertical passage of bees, can be purchased either plain or split on three sides. The split section cannot be used in the T-super.

In split sections, three of the sides have been cut lengthways down the center and placed in the section holder with the three split sides up. A full sheet of wax foundation can be inserted into four sections at a time.

Another type of super is called a half-comb cassette super. The half-comb

FIGURE 2. Adding the "Killion Super"

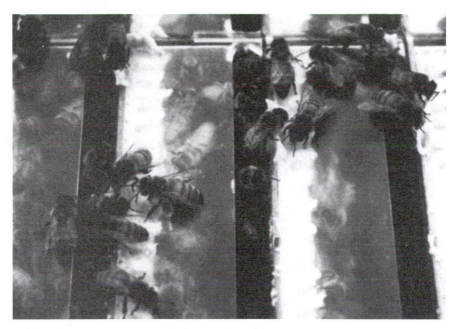

Figure 3. Bees working the Hogg Half-comb Cassette

is an inch deep clear polystyrene box with lid placed in units of 40 half-combs per hive (Figure 3). Each half-comb box has an embossed honeycomb pattern on the bottom coated with wax from which the bees construct their comb. Bees are induced to build comb on the image of combs seen through adjoining boxes.

The most popular comb honey super in the industry today is one containing round plastic sections. The first round sections were tried with success in 1888 when a beekeeper named Rambler used strips of wood shavings to form round sections. T. Boomer Chambers of England used glass rings cut from a round jar during that same era.

In the early 1950's a round plastic section was designed by the late Dr. W. S. Zbikowski for producing honey in circles approximately four inches in diameter. Dr. Zbikowski told of amazing results with these round sections using the Killion T-super chosen because of its excellent ventilation and size, capable of holding 36 round sections. The bees will complete round sections more quickly and completely than square ones since there are no corners to leave open (Figure 4). Round sections are held in plastic frames, four to each frame. Beekeepers use either nine or eight frames to a super. In using eight frames, a small board is inserted on each side and secured with two springs to take up the space of the ninth frame, creating better ventilation for the bees.

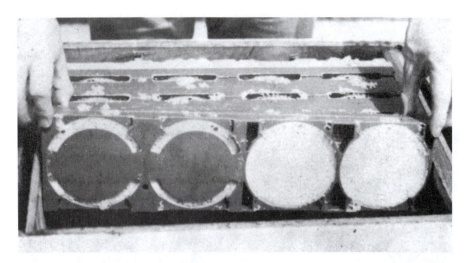

FIGURE 4. Frame of completed round sections

Foundation

Only the lightest colored grade of foundation should be used in sections and frames for producing comb honey since each portion of this foundation becomes a part of the honey comb sold for consumption. The very finest of virgin wax is used in the manufacture of high quality foundation. It is described in catalogs as thin surplus or thin super foundation.

Full sheets of foundation are used in frames for cut-comb, split and round sections. A metal split section device inserted at the top of four wooden split sections holds them open for easy and fast insertion of the full sheet. This allows the foundation to be installed properly with the rows of cells running horizontally with the hexagons pointing vertically as in brood or shallow frames. The unsplit sections used in T-supers require the foundation to be cut in individual pieces and fastened inside the sections. Some beekeepers make the mistake of fastening these pieces in the wrong direction, showing the cells running vertically.

A box for cutting these individual pieces to the desired size can be easily made by anyone with basic carpentry skills (Figure 5). The box can be nailed to a larger piece of plywood or masonite, allowing the unit to be fastened to a work table while cutting the foundation. There are four saw kerfs in the sides of the box spaced equally for the correct size cuts.

A thin knife, such as a scalloped slicing knife, is used with a sawing motion to cut several sheets of foundation at a time. The optimum temperature of 70°F is ideal for cutting foundation. Cooler temperatures tend to make foundation shatter along the cut edges.

FIGURE 5. A foundation-cutting box

Most beekeepers use a single square cut of foundation in each section. In early years of comb honey production different methods were tried such as using one small narrow strip or a single triangular piece fastened to the upper center of the section as a way to cut labor and costs.

One of the better methods and the one used by the Killion family of Illinois is the practice of using two pieces of foundation (a large one at the top and small piece at the bottom) in each unsplit section (Figure 6). The advantage is that the bees join the two pieces together as they construct the comb, creating more perfectly formed honey combs. This eliminates the large holes in the corners of the wood section which the bees create by chewing them out to be used as a communication hole or crawl space. The foundation is fastened in the sections using a multiple block board and aluminum hot plate. After the foundation is inserted or fastened in the various sections chosen by the beekeeper, the extra pieces of the supers are assembled. In the T-super both top and bottom of the sections are painted with paraffin. For the standard section super containing section holders, only the top of the sections can be painted. Coating these exposed surfaces of sections prevents the bees from propolizing and staining the soft white basswood surface and makes cleaning the sections much easier when preparing them for market.

Bait Sections

The tradition for many years was to insert a bait section in the first comb super placed upon the hive. These sections were obtained by allowing the bees

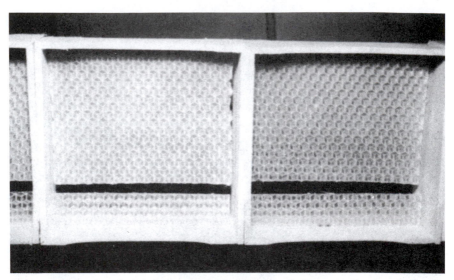

FIGURE 6. Two piece foundation method of section comb honey production. The main advantage is that the bees join the two pieces together as they construct the comb, creating better formed honey combs.

to rob out off-grade or partially filled sections the previous year. The drawn comb was believed to entice the bees to move up into the comb super more quickly. When preparing the supers, the bait section was marked with an "X" scratched into the wood on top of the section. Combs in bait sections were darker and tougher and not to be sold to the public, but melted up at the end of the season. Section-comb honey producers with large numbers of colonies consider this a wasteful practice and have discontinued using bait sections. Since only strong colonies should be used for comb honey, the bees move into the first super rapidly without the use of a bait section.

COLONY MANAGEMENT

In preparing colonies of bees for the summer honey flow, there is little difference in the preparation, whether one desires to produce liquid or comb honey. Since strong colonies are used, those that are mediocre in strength should be used for liquid production and weak ones united with others. In producing liquid honey, the supers are added just prior to the honey flow. At this same time beekeepers producing section or cut-comb honey change their management practice.

Many producers use a double brood chamber during spring buildup and continue to use them for comb honey production through the honey flow. During spring the two hive bodies should be reversed as often as possible, keeping the most brood down in the lower hive body. This continues to equalize the two brood chambers and somewhat delays the swarming impulse.

Articles and books have been written describing the best methods of producing comb honey such as "padgening" and "shook swarming." In "padgening" the beekeeper finds the hive whose colony has just swarmed and moves the hive to another location. On the original spot a full depth hive body or shallow extracting super containing foundation is placed. A queen excluder is then added with four comb honey supers placed above and the swarm introduced.

In the "shook swarming" method, the beekeeper chooses a colony that is preparing to swarm. He then shakes 1/2 to 2/3 of the bees from the colony on a cloth under which the queen of the colony has been confined in a cage over the cloth thus creating the "shook swarm." The hive is then moved to another location and a new hive put in its original place. The artificial swarm and queen are then introduced into the new hive and comb honey supers added. Sometimes the parent hive is just moved to the backside of the new hive with entrances in opposite directions.

FIGURE 7. Comb honey prepared for marketing

Of the various methods used in comb honey production, the Killion system has proven the best in producing the finest award-winning comb honey (Figure 7). Regardless of whether colonies are in two or three brood chambers at the beginning of the honey flow, each colony is reduced to one brood chamber and a single comb honey super is given whether producing section comb honey or comb honey in frames. When most of the bees including the queen are shaken into one hive body and a single comb honey super is added, bees are encouraged to swarm (Figure 8). This is a critical time for swarm control measures to be undertaken. Within a few days the colony will start building queen cells in preparation for swarming. At this time the queen is removed and all queen cells destroyed. The queenless colony will continue to build cells in an attempt to replace the lost queen. After four days all queen cells are once again destroyed. On the eighth day, after all cells are destroyed for the last time, a queen cell or mated queen is given. To find all the queen cells on the combs, it is necessary to shake most of the bees from each frame and look very carefully. Care should be used in shaking the frames at this time since fresh stored nectar will spatter out if the comb is shaken too vigorously. If one cell is missed and a new queen or cell is given, the colony will swarm. Since the colony now has a new queen, any crowding will not cause swarming. The beekeeper now has only to add and remove supers. The finest quality comb is produced at this time (Figure 9).

FIGURE 8. Preparing colony for comb honey production

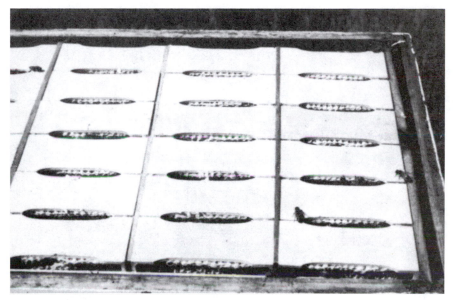

FIGURE 9. Well filled comb honey super

Supering

As the honey flow begins, beekeepers using the double brood chamber method for producing bulk or cut-comb honey usually add a queen excluder then a liquid super of drawn combs followed by a comb honey super on top. Bees store honey in the drawn combs before advancing up into the comb super to work. This method also reduces the bees' desire to swarm. A queen excluder must be used in producing bulk or cut-comb honey (Figure 10). None is needed when using square or round section supers.

One serious mistake made by beekeepers is adding comb supers too far ahead or too many at one time. Only one super should be added at the beginning of the honey flow. This is contrary to adding two, three or four supers of drawn combs as recommended for liquid honey production. The manipulation of supers for either sections or shallow frames is almost identical. As the first comb super becomes 1/2 or 2/3 full, a second one is added on top. Bees have a tendency to work in the back 2/3 of the super and reversing becomes necessary in order for the super to become uniform in fullness. As the first super appears almost full, it is placed on top of the second one and a third one added if necessary (Figure 11). Before giving the third super, the first should be nearly full and the second at least half full. Supers that are becoming full are raised to the top to prevent travel stain caused by the bees walking over the dark brood combs and then up onto the fresh new white cappings. It is wiser to crowd the colony with fewer supers than give extra supers too freely.

FIGURE 10. Colony prepared for cut-comb honey production

Honey flows may end abruptly leaving partially filled comb supers on the colonies and the beekeeper wondering what to do with them. Any partially filled sections or shallow frames should be melted to save the wax instead of being used the following year.

CARE OF COMB HONEY

When preparing to remove comb supers from the hives, it is important to make certain all cells of the comb are sealed (Figure 12). Since the bees seal the upper portion of frames and sections first, one only has to elevate the super and check the bottom rows of cells. If cells are not fully capped, leave the super on the colony a few more days. Unsealed cells may contain unripe honey which is high in moisture causing the honey to ferment (sour) if removed too early. In some areas of high humidity even though cells are capped, the honey will start fermenting since the bees are unable to ripen it fully. Comb honey in this condition begins to have wet-looking cappings that finally burst open causing the thin, soured honey to seep out. Removing these supers to a drying room, stacking them crosswise and using warm dry circulating air, plus a dehumidifying unit, will alleviate most of the fermentation problems and save lost supers for the beekeepers.

Some beekeepers use chemicals to remove comb honey supers, giving the

FIGURE 11. Three comb honey supers on hive. The top one is empty, the second is full and the lower one is partially full. This positioning scheme helps prevent travel stain on comb honey.

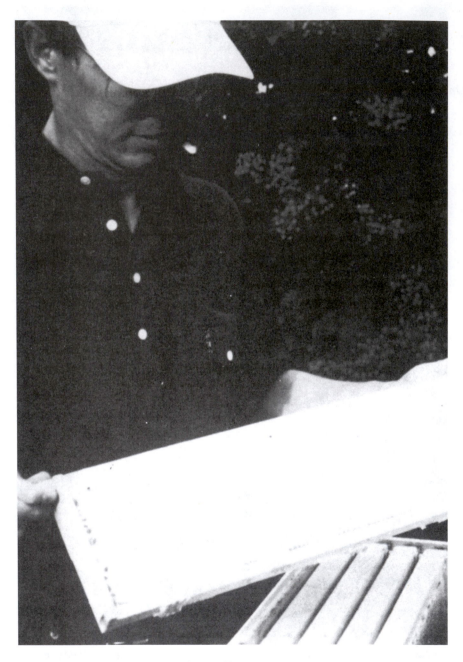

FIGURE 12. A well sealed cut-comb frame of honey

FIGURE 13. Cleaning section comb honey before packaging

FIGURE 14. Round sections of honey need only be trimmed of excess foundation and they are ready for marketing after lids and label have been added.

fresh new comb-honey an unpleasant odor. The best advice is to either use cool smoke and brush the bees off or place an inner cover with the bee escape under the completed supers. The quickest and most popular commercial removal method is the use of a bee blower.

As soon as supers are removed, comb honey must be treated for the prevention of wax moth damage. This can be done by putting the honey in plastic air-tight bags and storing it in a freezer overnight. This kills all stages of the wax moth. Comb honey may be stored over a length of time at 0°F (-18°C). It will retain its quality, does not granulate and the freezing even tends to whiten the cappings somewhat. When removed from the freezer, allow the honey to thaw at room temperature. The moisture will collect on the surface of the plastic bag rather than on the face of the comb.

At the present time the only fumigant recommended for wax moth control in comb honey for human consumption is carbon dioxide (CO_2). A few commercial producers have built large airtight fumigation rooms to treat large quantities of comb honey at a time. Carbon dioxide is placed with the supers in a space where the concentration can be kept at 98%, the temperature at 100°F, and the relative humidity at 50%. Four hours in this environment will destroy all stages of wax moth.

FIGURE 15. The popular round section comb honey ready for market

FIGURE 16. Cut-comb honey draining before packaging

FIGURE 17. Attractive chunk honey packs

PREPARING COMB HONEY FOR MARKETING

The preparation of section comb honey for market is a simple task compared to all the work and extra equipment needed in extracting honey from the combs. The most labor involved in section comb honey is removing the propolis and paraffin from the wooden sections (Figure 13). Paraffin is scraped from the tops and bottoms with a single-edge razor blade. The sections are then cleaned individually by scraping the edges and corners with a sharp, short-bladed knife. Section window front cartons are available for marketing square section.

Round plastic sections are usually free of excess propolis with only the small extra foundation protruding from the outer edge needing to be removed (Figure 14). A plastic bottom and cover is placed over the section and secured with a wrap-around label. This makes an attractive pack for marketing (Figure 15).

Bulk comb, sometimes called shallow frame honey, is produced in shallow supers containing 10 frames. Each frame weighs from three to five pounds, depending upon the thickness and depth of the comb. Square sections usually weigh slightly less than one pound with round sections being approximately eight or nine ounces. Whole shallow frames can be sold in individual cartons, but the favorite method of most beekeepers is to cut the comb in various sizes, let it drain (Figure 16) and place the honey in a clear plastic box or polyethylene bag, then sealed and inserted into a cardboard window front carton.

Various size pieces of comb are also cut from the frames and placed in glass jars with extracted honey surrounding the comb. This attractive combination pack is called "chunk honey."

After the comb has been inserted in the jar, liquid honey that has been heated to approximately 145°F then cooled to 128°F is added. Cooling the honey before filling protects the delicate comb from melting. Consumers who desire the taste of both comb and liquid honey will gladly pay the higher price for this fancy pack. Both liquid and comb for chunk honey must be of excellent quality, free of travel stain, pollen and foreign particles. Chunk honey is considered a seasonal pack as the honey tends to granulate more rapidly in cooler temperatures. It is a unique and sought after pack at fairs, roadside stands or festivals from late summer to early fall (Figure 17).

CHAPTER 17

BUSINESS PRACTICES AND PROFITABILITY

by MALCOLM T. SANFORD[1]
and ROGER A. HOOPINGARNER[2]

To men, time is money;
to bees, it is honey;
and all the accomplishments of
the hive should be such as
to economize it to the utmost.
            —L.L. Langstroth

INTRODUCTION

This is the first time a chapter on business practices and profitability in beekeeping has been published in this volume. Its presence suggests the time has come to explore in depth the economic feasibility of the craft. Agriculture has become much more competitive, and this demands that beekeepers begin to focus on all aspects of their business that contribute to profitability.

We now live in a smaller world. Some call it "spaceship Earth" to emphasize this fact. Prices paid for a product no longer reflect only local economic conditions, but are also influenced by international markets. Honey, the major product of the beekeeper, is a world commodity. There is no formalized futures market for honey, and prices do not always conform to what is expected in a free market. Many developing nations, in search of hard currency at any cost, have subsidized their labor-intensive beekeeping enterprises. Several have become major players in the world honey market.

Profitability in Beekeeping

According to several studies, it has not always been profitable to produce honey in the United States. Adams and Todd (1933) estimated it cost $.07 to produce one pound of extracted honey for which beekeepers received $.045. Coke (1966) analyzed 15 commercial honey-producing firms in Florida and found a net loss of $.0261 per pound of honey. Rodenberg (1967) declared that over a 10-year period the cost of the most important items used by beekeepers increased 18 percent, while the price of honey decreased 7 percent. A detailed study of southwestern beekeepers (Owens *et al.*, 1973) concluded:

[1]Malcolm T. Sanford, Ph.D., Extension Apiculturist, Department of Entomology & Nematology, University of Florida, Gainesville, Fla. 32611-0740.

[2]Roger Hoopingarner, Ph.D., Professor, Department of Entomology, Michigan State University, East Lansing, Mich. 48824-1115.

"Beekeeping for honey production in the United States is not profit-able. The unit price received by beekeepers for bulk, extracted honey has not changed in the last twenty-five years, while the cost of production has increased. Thus, beekeepers, who rely on honey production for income must supplement their income from other sources, such as crop pollination and outside employment."

Honey and the Sweetener Market

Profitability in beekeeping has traditionally been greatly influenced by the sweetener market. The major factors affecting this market are consumption, competition and price. The search for sweetness has influenced history in many ways. Honey was the first sweetener available to western civilization on any scale. This is a major reason for the long European beekeeping tradition, and the European market continues to have the highest per capita honey consumption in the world.

Development of the sugar trade in the West Indies drove much of the New World slave trade in the 18th and 19th centuries. Sugar effectively replaced honey as the world's sweetening agent and continued in that role until World War II. With a shortage of sugar, honey demand and prices rose for a period during the War, but new technology was also responsible for developing substitutes like glucose corn syrup. The rise of high fructose corn syrup (HFCS) in the 1970s provided even more competition for honey because it had similar properties and was cheaper. The sugar substitute called aspartame has also influenced the sweetener market in the 1980s. A glance at United States sweetener consumption over two decades is revealed in Fig. 1 (Modi-

FIGURE 1. U.S. Per Capita Sweetener Consumption

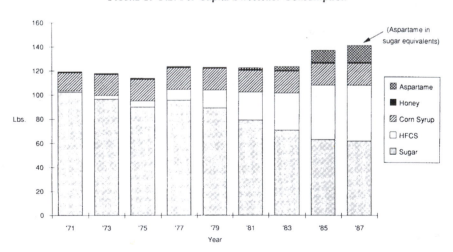

fied from Stoller, 1989). The trend is clear; per capita honey consumption has remained the same, while use of both aspartame (Nutra-sweet ®) and HFCS have increased. Nevertheless, honey consumption has not declined in the face of this competition, although use of sugar has. A principal reason for the popularity of glucose corn syrup and HFCS are changes in the industrial sweetener market where consistency of the product and low price are of utmost concern. The pricing disadvantage for honey when compared with other sweeteners is distinctly seen in Fig. 2 (Stoller, 1989). The highest relative prices per pound are shown by aspartame and honey. Although very expensive, aspartame's sweetness potential is about 200 times that of sugar, and thus cheaper than honey on a "relative sweetness" basis. Low honey prices coupled with low consumption are the major reasons the beekeeping industry voted for legislation creating the National Honey Board. This body uses funds generated by producer assessments to promote honey in the United States.

Honey Imports

With the exception of the 1940s, 1960 and 1961, the United States was a honey exporting nation until 1967. After that imports increased until the U.S. loan program allowed beekeepers to "buy back" the honey and market it in domestic channels. The trends are shown in Table 1.

Reasons behind the increase in honey imports in the early 1980s were a dramatic rise in world honey prices in the 1970s stimulated by short crop years and changes in U.S. sugar policy which increased that commodity's price. Developing countries quickly jumped on the honey production bandwagon; this was particularly true of mainland China. Imported honey continues to be

FIGURE 2. Cost Per Pound   "Relative Sweetness"

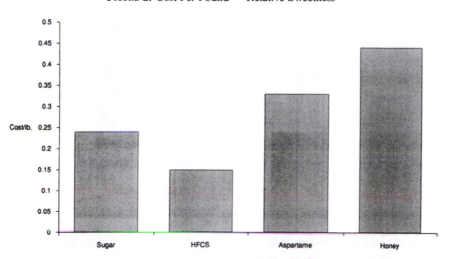

less expensive because the labor inputs in other honey-producing regions are less costly and U.S. tariffs are low. It should be remembered that average prices may cloud the real price of honey. Light colored honey produced in the United States will invariably command a higher price than dark colored honey from the tropics. However, China and Argentina are both major exporters of high quality, light colored honey. Finally, a high-valued dollar, with respect to other world currencies, limits U.S. export potential.

Honey Loan Program

During the 1980s, the Commodity Credit Corporation's (CCC) honey loan program became a price support program by default. The loan price on honey sometimes exceeded the world market price by as much as 40%. This created a great incentive for beekeepers to turn their honey over to the U.S. Government, while those packing the product purchased imported honey. By 1986, almost half of the U.S. production was making its way into government warehouses (Table 1). Subsequent innovative pricing by the U.S. Government, however, returned much of the honey to beekeepers at a loss. As a result, most of the U.S. government stored honey stock was eliminated.

Other Factors Affecting Honey

Honey is, by far, the major product of the U.S. beekeeping industry. Over the last decade, the national crop has averaged about 200 million pounds. As described above, changes in the sweetener market have markedly influenced the profitability of honey production. Beyond competing with honey in the bulk market because it has many of the same properties and is available at a lower cost, HFCS has also been used to purposefully adulterate honey. A self-policing policy by the industry and strict enforcement of food laws have so far prevented honey's reputation from being permanently damaged by this practice. A side benefit of HFCS to the industry is the availability of a less expensive alternative food to feed bee colonies.

Honey's reputation as a safe food has also been directly attacked. It has been implicated by the medical establishment in "Sudden Infant Death Syndrome." Botulism spores, resident in many fruits and vegetables, as well as honey, can germinate in the intestines of a very few infants causing botulism toxicity. This resulted in loss of the infant market. Although the honey industry supports warnings by pediatricians that honey should not be fed to infants under one year of age, the U.S. Food and Drug Administration does not require honey container labels to contain such a warning. A recent study has provided a review of research on the subject (Foster, 1988).

Although it continues to maintain its reputation as an organic, healthful product, honey is increasingly vulnerable to complaints of contamination. The public unrealistically demands a risk-free environment. Various chemicals are

Table 1. Honey: Yield, Supply, Disposition and CCC Program Activity (Million Pounds)

Calendar Crop Yr.	Production	Imports	Exports	Loans Made	CCC Takeover
1967	221.7	8.9	8.1	40.6	[1]
1970	215.8	16.8	11.7	31.	5.4
1971	197.8	11.4	7.6	22.9	0
1972	215.6	39.0	4.1	19.8	0
1973	239.1	10.7	17.6	12.1	0
1974	187.9	26.0	4.6	13.9	0
1975	199.2	46.4	4.0	[2]	0
1976	198.0	66.5	4.7	[2]	
1977	178.1	63.9	5.5	14.1	0
1978	231.5	56.0	8.0	40.5	0
1979	238.7	58.6	8.8	49.1	0
1980	199.8	49.0	8.5	41.1	6
1981	185.9	77.3	9.2	55.2	35.2
1982 [3]	230.0	92.0	8.5	88.4	74.5
1983 [3]	205.0	109.8	7.5	103.8	106.4
1984 [3]	165.1	128.7	7.5	109.9	105.8
1985 [3]	150.1	138.2	6.5	104.4	98.0
1986 [4]	200.0	118.4	9.2	154.5	41.9
1987	227.0	58.2	12.4	172.6	52.1 [5]
1988	214.1	55.9	13.9	198.4	32.7 [5]
1989	169.3	77.1	9.9	159.7	.3 [5]
1990	180.4	83.6	8.1	157.2 [5]	[5]

[1] 5,900

[2] Purchase agreements only, no loan program.

[3] Data other than imports, exports and price support activity are estimated by ASCS.

[4] Annual honey reporting reinstated.

[5] As of January, 1991.

Source: F. Hoff and J. Phillips, *Honey: Background for 1990 Farm Legislation,* USDA Economic Research Service, Staff Report No. AGES 89-43, September, 1989 and USDA *Honey Market News,* January, 1991.

used by beekeepers to remove and kill bees, control diseases and eradicate pests in stored comb. Using pesticides inside a living colony to control mites also inevitably increases pesticide levels in the bees' environment. Many pesticides are soluble in the wax on bees' bodies and may accumulate in the comb or honey. Product liability, thus, becomes an issue.

Future Profitability Outlook

It is risky to predict the future profitability potential in honey production, but it will continue to be closely tied to the world sweetener market. Over the last decade, the CCC's honey price support program has kept many honey producers solvent. How long this will continue is anyone's guess. It is not prudent, however, to fully rely on a subsidized price. Given this fact and the trends seen above in honey use and price, one should view the future conservatively. This does not mean there are no optimistic signs. The reduction of domestic product in U.S. Government warehouses, innovative promotions by the National Honey Board and the weakening of the U.S. dollar, which has awakened exports, are good signs that marketing potential exists.

Nevertheless, whether the beekeeper of the future will be able to rely as much on honey production as in the past is open to question. It is time that marketing strategies and the value of other beekeeping enterprises should be carefully examined. Profit margin, for example, may be much higher in commercial pollination at certain times and places. The same may be true for rearing queens, producing package bees or collecting and marketing other bee products.

One thing is sure; costs of production for all beekeeping enterprises will continue to rise. This is especially true for labor, and beekeeping continues to be very labor intensive. Apiculture has been bypassed by much of the mechanical age. Most technological innovations in the craft occurred in the nineteenth century. Only one, instrumental insemination, was developed in the twentieth century. There have been some developments in transporting and feeding bees, as well as in processing honey, but they have not been able to replace most of the sweat-producing, back-bending work beekeeping requires. Other costs which will inevitably rise in the future are associated with chemical treatments, legal liability and governmental regulation. These will be inevitable because of introduction of exotic parasitic bee mites and the arrival of the African honey bee. Finally, the general rise in interest rates over the last two decades continues to be a real barrier to entry into and/or expansion of the honey industry.

In the remainder of this chapter, we will discuss in depth many of the issues that affect profitability in beekeeping. Because of the diversity within the industry, what may be important for one segment may not be so for another, and what may work well for one operator will not necessarily be ideal for another. We can only try to give a balanced accounting and let you, the beekeeper, decide what is best.

The Beekeeping Industry

The U.S. beekeeping industry is an enigma. We do not know with any precision how many persons keep bees in the United States. Less is known about the underlying economics of the craft. Perhaps the most intensive investigation of the industry was published in the late 1960s.

> "...beekeeping is a very specialized type of farm enterprise. It is engaged in by only a few thousand professional and part-time bee-keepers, who have 3 million colonies and who provide practically all the honey bees used for pollinating commercial crops valued at $1 billion annually. They also produce an estimated 80 percent of the $45-million annual crop of honey and beeswax; the 185,000 bee hobbyists, who own 2 million colonies, produce the balance." (Anderson, 1969).

More recent estimates of colony numbers have shown an increase to 4.3 million in the early 1970s, dropping back to 3.2 million colonies* in the 1980s. (F.A. Robinson, Personal Communication). About 200,000 persons keep bees in the United States (Martin, 1980; Hoff, 1988).

Most large-scale honey producers move several times a year to different regions of the country in search of nectar flows. Many smaller beekeepers also move at least once during the season. The true extent of bee movement is not fully recognized by anyone not intimately associated with the industry. Much of it involves transporting bees from wintering grounds in the southern United States (California, Arizona, Texas, Louisiana, Mississippi, Alabama, Florida, Georgia) to nectar-producing areas in the Midwest (the Dakotas, Nebraska, Minnesota, Wisconsin).

Another enterprise which requires movement is commercial pollination. Although smaller than honey production, pollination as an enterprise is extremely important, and could well be the growth area in the future. Few figures exist on size, but large numbers of colonies (in the hundreds of thousands) are set in California almond groves each year (McDowell, 1984). In early spring, fruit producers in New Jersey, New York and Maine also rely on colonies from the southern United States. Some 2 million pollinating units are rented each year and a cumulative estimate of between 8 and 19 billion dollars has been placed on the total pollination value of honey bees in the United States (Levin, 1983; Robinson *et al.,* 1989).

---

*This figure represents only beekeepers with five or more colonies. The 4.3 million figure of the 1970s represented all colonies.

Queen and package bee production is a third major segment of the beekeeping industry. Traditionally, this activity exists in California, Texas and the southeastern states. The major markets are honey-producing areas in the Midwest and Canada. Queen and package production is a very different activity than honey production or pollination. It requires much less bee movement. This has brought it into contention with the other segments that rely on moving colonies. Detection of the honey bee tracheal mite, *Acarapis woodi* and introduction of the Asian bee mite, *Varroa jacobsoni*, has caused much of this friction. Movement quarantines proposed by most regulators to control spread of these pests have met with resistance from honey producers and pollinators. Queen and package bee producers, however, see spread of any disease or pest as completely unacceptable and advocate stiff movement regulations.

Another major player in the beekeeping industry is the processor-packer. Many processor-packers may also be producers. Finally, the equipment supplier exists to service the beekeeping industry. This enterprise has a set of problems all its own.

Unique to beekeeping as an agricultural enterprise is the presence of a large number of hobbyists and sideliners. The former are not in beekeeping for economic gain, but hobbyists often have more time to contribute to industry activities and many have made significant contributions. However, when it comes to major political decisions, large, commercial interests usually prevail. The consequence of this segmentation is an uneasy alliance of many groups, collectively called the beekeeping industry.

Enterprise Budgeting
As noted above, beekeeping has generally meant honey production. This single enterprise, however, can be broken down into many others. Therefore, the beekeeper has the option to produce extracted honey in the barrel or bottle it him/herself. Other products include creamed or spun, chunk, cut-comb and comb (section) honey.

There are several other ways a beekeeper can produce income. These include rearing queens, shaking package bees, trapping pollen, and collecting and selling propolis, royal jelly and venom and other bee products. Finally, there's the service of commercial pollination. To better identify all costs of each operation, economists recommend enterprise budgeting. This activity breaks down costs and returns by specific activities or jobs (enterprises) and can be used effectively to see where strengths and weaknesses are in an operation. It is important to remember that within a business that shows a profit as a whole can lie hidden individual enterprises that are not profitable. To be effective, enterprise budgeting requires adequate record keeping. (More on this later in the chapter).

Fixed and Variable Costs

Two kinds of costs are present in any operation, fixed and variable. Fixed costs are those funds expended whether or not any production or sale of products results. They include capitalization and depreciation of equipment, taxes, insurance and rent.

Variable costs fluctuate depending on the scale of the operation. Labor, for example, is one of the largest variable costs in a beekeeping operation. Labor costs increase as more honey is produced. Other variable costs are fuel, chemicals, supplies and bee feed.

A Typical Enterprise Budget

Both variable and fixed costs make up a typical enterprise budget. Table 2 budgets two enterprises, honey production and pollination. This table is an incomplete listing of variable and fixed costs. It is used to show how the items may be listed. It is important to know the cost/hive as this will allow you to know the break even point, *i.e.*, the amount of money a colony needs to produce before any profit can be seen.

The distribution of hives and costs in Table 2 reflect differences between two enterprises (honey production and pollination) in an operation of 3,000 total colonies. Because the number of colonies used in each enterprise is not equal, costs of many items differ. Although sugar cost per colony is the same, this budget shows that in aggregate it costs much more to feed honey-producing colonies simply because they are more numerous. Fairly large differences between enterprises also exist in projected transportation costs. Total variable costs here are shown to be $56,460 for honey production and less than half ($26,515) for pollination. This is in spite of the less than $2.00 differential in variable costs per colony. The third column representing percentage of total variable costs is also important. Notice that four items absorb the biggest part of variable costs, labor, sugar, honey packing and queens.

Fixed costs will also vary, but not on a per colony basis. Rather, they will depend more on the kind of enterprise. Apiary rental is projected to be less for pollination than for honey production at least for part of the year. Machinery repair is shown to be much higher in honey production, reflecting mostly repair and replacement in the honey house. Investment in equipment is much more for honey production (here a total of $80,000 divided by 10 years) considering all the items needed to efficiently extract and pack the crop. Finally, the investment in hives at $75 each, but spread over 10 years, is twice as much for honey production as for pollination (Reflecting the number of colonies/enterprise). Again notice the percentage of investment going to each category of fixed cost. The numbers presented here are estimates; they provide an illustration of differences that can exist between enterprises, and thus, the value of this kind of budgeting. Most apparent is that total costs are much

Table 2. Sample budget for Beekeeping Operation of 3000 Colonies

	Enterprises			
	Honey Prod.	*Pollination*	*Total Cost*	*% Total*
Number of Hives:	2000	1000		
Variable Costs:				
Sugar/HFCS	$ 8,000.00	$ 4,000.00	$12,000.00	14.46%
Labor $11/col	22,000.00	11,000.00	33,000.00	39.77%
Transport Pol. $6/col	0.00	6,000.00	6,000.00	7.23%
Transp. Hon. $2.50/col	5,000.00	0.00	5,000.00	6.03%
Queens $6/col/Trien.	4,000.00	2000.00	6000.00	10.85%
Honey Pack $6/col	12,000.00	0.00	12,000.00	14.46%
Advertising	200.00	600.00	800.00	0.96%
Office supplies	100.00	500.00	600.00	0.72%
Utilities				
Electricity	500.00	85.00	585.00	0.71%
Telephone	100.00	300.00	400.00	0.96%
Gas/Oil	800.00	0.00	800.00	0.96%
Association Meetings	300.00	300.00	600.00	0.72%
Hive Repair @10%	200.00	100.00	300.00	0.36%
Comb Replace @15%	1,260.00	630.00	1,890.00	2.28%
Total Variable Costs	$56,460.00	$26,515.00	$82,975.00	100.00%
Variable Costs/Hive	$28.23	$26.52	$27.66	
Fixed Costs:				
Rent	$ 1,000.00	$ 100.00	$ 1,100.00	2.57%
Machinery Repair	4,000.00	750.00	4,750.00	11.09%
Overhead	2,000.00	500.00	2,500.00	5.83%
Equipment/10 yr	8,000.00	4,000.00	12,000.00	28.00%
Hives @ $75/ea/10yr	15,000.00	7,500.00	22,500.00	52.51%
Total Fixed Costs	$30,000.00	$12,850.00	$42,850.00	100.00%
Fixed Cost/Hive	$15.00	$12.85	$14.28	
Total Costs/Hive	$43.23	$39.37	$41.94	

greater for honey production than for pollination, even though the costs per colony are similar. In essence, one could run half as many pollinating units with much less total cost than for honey production. This difference would not have been noticed if the whole beekeeping operation was treated as a simple enterprise.

PRODUCTION AND MARKETING STRATEGIES

Type of Colony

One of the first decisions in an operation is the kind of colony to use. Much of the information that follows is necessarily oriented toward honey production, because historically that has been the goal of most beekeepers and the research establishment. Collecting other bee products, producing queens and package bees and pollinating may require special kinds of hives and equipment. Consult other sections of this book for recommendations concerning these other enterprises.

First beekeeping decisions for honey production usually include whether to operate package or over-wintered colonies and/or use single-queen or two-queen colonies. An "economic" analysis of these choices was that of Moeller (1976), where he compared honey production and labor (minutes of time necessary) for a colony's manipulation over a six-year period. The results are summarized in Table 3.

The original analysis of this data did not include the last column. Large honey production was Dr. Moeller's concern. The pounds of honey per minute data indicate that package colonies produced more honey for each minute of labor and allow a beekeeper to operate nearly two and one half times more package colonies (for the same investment in time) than single-queen colonies. Even considering the much lower average yield for package colonies, this would give 50% more honey for the beekeeper. Many other factors, however, must be examined for a full analysis of the above situation. The major significance of this is to remember that one can be led astray if only information on yield is analyzed. The regions of the country where use of package bees has been most successful are places that have their major nectar

Table 3. Comparison of Honey Yield and Yearly Labor for Three Colony Types (Average 1967-1974, Wisconsin).

Colony Type	Lbs. Honey	Min./Col.	Lbs./Min.
Package Colony	117	12	9.75
Single-Queen	168	29	5.79
Two-Queen	280	43	6.51

flow late in the season. The later nectar flow allows the package colony time to develop the necessary population. Package colonies are not usually satisfactory for use in pollination rentals because they have much reduced populations about the time they are needed for early season pollination.

The Hive

The basic unit of a beekeeping enterprise is the hive, and for centuries it has been made of wood. However, recently other materials have been introduced. Chiefly, these have been plastics and cements. Nevertheless, wood continues to be preferred by most because it is inexpensive, easy to work and repairable. Wood is also permeable to moisture and acts as a good insulating agent. Disadvantages of wood are that it requires periodic maintenance and is subject to weathering, dry rot and termite damage.

Whether or not to manufacture one's own hives is often a major beekeeping decision. They must be made to accurate dimensions to maintain the bee space. Generally speaking, the beekeeper can save money by making supers, bottom boards and tops from scrap lumber. Because of the number of complex cuts, frames should usually be purchased. A possibility is to subcontract out equipment manufacture and/or frame assembly to firms which specialize in labor intensive activities, such as Goodwill Industries, Inc.

Another major decision is hive size. The vast majority of operations use the standard 10-frame Langstroth full-depth super as a brood chamber. Because it is standard, the equipment is compatible with most that's available for purchase. As a corollary it is easier to sell. Two brood chambers are used in the Midwest; only one is usually employed in the South. Some operators use 9 instead of 10 frames in the standard brood chamber. Fewer frames are easier to work, but this sacrifices storage space for the bees. Some beekeepers who have concerns about heavy lifting employ standard height eight-frame colonies. An increasing disadvantage of the standard, full depth super is that lumber in the width required is becoming more and more expensive.

The honey super is less standard in height than the brood chamber. Sizes range from 4-3/4 inches to 7-5/8 inches and all may be used in one operation. The full-depth (9-5/8 inches high) brood chamber can also be used as a super, but full of honey it can weigh up to 80 pounds. There is always a tradeoff between fewer combs which weigh more individually as opposed to numerically more frames that weigh less, but increase handling time.

In many areas, beekeepers have gone to a shallow super for both brood chamber and honey storage. A uniform size will standardize the outfit across the board. Where wintering is a concern, there is evidence to suggest that bees survive better in a brood chamber made up of shallow supers. The cluster can maintain its integrity through the more numerous spaces provided by shallower combs. Usually the size recommended is the 6-5/8 inch super (Farrar, 1968).

Nominal lumber size has been changed so that the standard 1 x 8 inch (actual size 3/4 inches by 7-5/8 inches) now is exactly the width required for the 7-5/8 inch super. This means very little wood waste in making the boxes.

Preserving Woodenware

Wooden boxes, tops and bottoms should be treated to maximize their life. Painting is traditionally the best preservation of the woodenware in beekeeping enterprises. Normally, supers are only painted with two coats on the outside. The inside is not painted to allow the wood to absorb moisture produced by a colony. Special attention should be given to the joints and exposed end grain.

Both water- and oil-base paint are effective. The former is used by most; water-base paint is more permeable to moisture than the oil-base variety. Hive body color appears to be of little importance to bees and most are painted white as a tradition. This color reflects sunlight and may keep colonies cooler in summer. In the North, darker colors may be more beneficial to help colonies survive cold winters and increase foraging on cool days.

Many beekeepers rely on wood preservatives and do not paint their colonies at all. Research indicates that copper napththenate, copper 8-quinolinolate and acid copper chromate are the best wood preservative options at present for beehives. Creosote, pentachlorophenol (PCP), tributyl tin oxide (TBTO) and chromated copper arsenate (CCA) have adverse effects on bees and can accumulate in wax (Kalnins & Detroy, 1984; Kalnins & Erickson, 1986). Most wood preservatives are classified as pesticides and must be used consistent with their label.

Constructing hives out of commercially treated lumber by the process known as "Wolmanizing ®" is not recommended. The process is variable and many different chemicals may be used. Evidence has accumulated that lumber can be made totally worthless for beehive construction when treated by Wolmanizing ®. Bees coming into contact with treated wood have died. It is also dangerous to humans; a sawdust mask is now recommended when working with treated lumber.

Frames (Combs)

Frames (Combs) are a major investment of any beekeeping outfit. They take time to assemble and require much bee energy to draw out the foundation. In the past, it was believed that well constructed frames (combs) should be expected to last forever. And in many operations there are combs over 50 years old. A recent survey showed that average comb replacement among 70 beekeepers was 4%, and the maximum replacement was 25% (Hoopingarner & Sanford, 1990).

Keeping combs for long periods is now questionable. There is evidence that periodic brood comb renovation improves the bees' environment. Brood comb cells become narrower with time, producing smaller bees. Old wax comb, loaded with impurities over the years, has been associated with increase in diseases like chalkbrood, nosema and foulbrood. Once the organisms responsible for these diseases becomes established, the comb becomes a constant source of potential re-infection. Finally, wax comb is a "sink," that over time can accumulate toxic levels of air pollutant particulates like lead and mercury, and pesticides.

Beeswax comb is a significant capital investment for any beekeeper. This is because of the cost of the foundation and labor to insert it into the frame, as well as the cost of the honey used by the bees in producing the comb from foundation. There is also the added expense of removing and rendering the comb at replacement time. Again, as suggested above, continual periodic renovation of brood comb should be seriously considered as cost effective. The results are strong, vigorous colonies that will be top producers.

Beeswax melted from old comb can and should be recycled. An easy way to obtain wax from old comb is using a solar wax melter. A survey of commercial beekeepers indicates that nearly two pounds of wax can be gleaned per colony over the course of a year (Hoopingarner & Sanford, 1990). There sometimes is local demand for beeswax from those making candles or doing batik. Alternatively, the beekeeper can produce and sell candles at premium prices. Beeswax rendering can also be accomplished in the off season when the beekeeper's work slows down.

Many beekeepers both nail and glue the wooden parts of the frame together and then wire them horizontally using special wiring boards. This provides the best framework to get strong, straight combs, especially in the brood nest. At the other extreme are those who simply provide a frame, believing that the resultant comb will provide all the support necessary.

Comb Foundation

Beeswax foundation is available in a variety of forms. Thick and regular foundation is acceptable for the brood nest and extracting supers. Thin super foundation is also used for cut-comb, chunk and comb honey. The most cost-effective foundation for commercial beekeepers has been the plastic base, beeswax coated type. One reason is that the foundation can be snapped into a slotted top bar and bottom bar by inexperienced help. The foundation also provides support without the added costs of wiring. On the other hand, once the wax is removed from the plastic base, the bees are reluctant to build on bare plastic.

The newest foundation technology involves a bewildering variety of plastic alternatives. Some plastic cells are partially drawn out and covered with thin beeswax. Others are fully drawn out with no wax covering. A great advantage of this kind of technology is the permanence of a frame that requires very little maintenance and cannot be destroyed by wax moths. On the other hand, there continues to be disagreement about the strength of these plastic frames and the degree to which the bees will build comb on, and/or rear brood and store honey in them. This is probably because conditions vary considerably among various regions. The maxim is that bees will accept and build on almost anything, as long as conditions are favorable. It is during marginal periods, not under control of the beekeeper, when the variability in acceptance and quality of the resultant comb must be judged.

Again, it is emphasized that most of the information presented above is oriented toward the honey-producing beekeeper. Other enterprises may require different comb technology. For example, at least one commercial queen rearing outfit uses little foundation (S. Taber, personal communication). The bees are guided by vertical dowels spaced in the frame. This technology is used only in the brood nest where drone comb is highly desirable.

Extracted Honey
Most beekeepers produce extracted honey. Its production requires minimal labor costs. A very important variable cost in extracted honey is the number of times the supers and frames are handled. Mechanical aids in reducing weight are also important, but it is critical to reduce the actual number of movements in the extracting process. Modern business practice often uses what are called "time and motion studies" to analyze every detail of the operation. Although this may go beyond the scope of a beekeeping operation in its infancy, this is not so after a certain amount of growth has occurred. It will pay big benefits to think about (and possibly write down) all of the times honey supers and frames are handled. Usually, one or more steps can be eliminated with substantial cost savings.

Beekeepers often use air blowers to remove bees from the supers even though this probably requires at least one additional handling. The compelling argument for using blowers has been that they can be used in all kinds of weather, whereas fume boards generally require sunshine to be effective. The perceived lack of success using fume boards may be because of: 1) inadequate scheduling, 2) impatience with the fume boards in the apiary, and/or 3) improper use (concentration and amount) of repellent. Proper use of fume boards is fast, requires less physical movement of supers, and is inexpensive. A potential problem with any kind of chemical use around beehives, however, is contamination of the crop. It is conceivable that in the future the use of chemicals to remove honey will be considered unacceptable by regulatory agencies.

Movement of honey supers onto and off of vehicles should be done with large-tire hand trucks, forklifts or similar labor-saving devices. The use of hand trucks can be enhanced if loading ramps are constructed in the apiary. A simple, but effective ramp can be made in each apiary by grading out a sloped depression that puts the truck bed at ground level. A load of gravel, or stones, should be placed in the bottom of the depression so that traction can be maintained during wet weather.

Having the proper size extractor is generally only a problem for the small, or growing, beekeeper. This is because extractors are manufactured with a discrete number of spaces or slots. Most common are 4, 12, 33 or 50-frame extractors. Large beekeepers have been remarkably consistent in the number of extractor slots that they have for their size operation, and the overall average was 0.1/colony (Hoopingarner & Sanford, 1990). That is, a beekeeper with 1000 colonies would have the equivalent of two 50-frame extractors. There are some beekeepers who pay too much in labor costs because they have too few extractor spaces, and, therefore, cannot get throughput of honey in a reasonable time.

Settling tanks like extractors continue to come in all sizes. A major problem to avoid is either extractors or tanks manufactured using lead solder. Most manufacturers no longer use this technology. Joints in older units may be covered up with food grade epoxy paint to provide some protection. If in doubt, consult food regulatory officials.

Numerous uncapping devices exist from the household butcher knife to electric uncappers available through major suppliers. Recently, a new kind of machine which uses flails instead of knives has come on the market.

Within the honey house the greatest problem usually revolves around the handling of beeswax cappings. There have been a few developments that have tried to address this bottleneck. One is the use of continuous flow cappings crushers. Another method that may offer some relief is removal, and extraction, of honey before it is capped. This eliminates most of the beeswax problem.

Processing "wet" honey requires precise water management using warming and drying rooms and a refractometer. There are additional benefits from this technology. It ensures that honey is at the required moisture level for government loans or grading, maintains good viscosity as a result of heating and moisture control, and increases honey yields by fine tuning moisture levels in the resultant honey crop (Murrell & Henley, 1987).

Use of heating devices, honey pumps, filters, cappings spinners and bottlers are all considerations of the larger-scale beekeeper. Specific details of all these are beyond the scope of this chapter. Standard references on these subjects are "Selecting and Operating Beekeeping Equipment," USDA Farmers Bulletin 2265, 1967 and "Honey Removal, Processing, and Packaging," USDA Agriculture Handbook 335, pp. 92-103, 1980.

Honey House

A first consideration in extraction is to find a space for processing the crop. Many small-scale beekeepers begin in their home's kitchen, but this is soon replaced by a honey house. A survey of commercial beekeepers shows that on the average honey houses had between 2.5 to 3.0 sq. ft./colony of storage and extracting space (Hoopingarner & Sanford, 1990). Some beekeepers need less room because they have storage houses at each apiary, or keep extra equipment above the inner cover of the colony when not in use.

Marketing Honey (Wholesale or Retail)

A significant decision the beekeeper must make is whether to market honey wholesale or retail. Generally speaking one can, and because of time constraints must, operate fewer colonies if one intends to market the honey oneself. Retail sales demand marketing and communication skills, a location close to markets, and additional equipment. The profit margin is increased several fold when honey is sold retail. In a national survey, the average retail figure was near 12 percent with only a few producers retailing the majority of their crop (Hoopingarner & Sanford, 1990).

Comb Honey

Increased costs and skill levels are needed to produce quality comb honey. Cut-comb honey production eliminates many of the expenses of producing quality comb sections, but it is still very labor intensive. More manipulation of colonies is also needed than that necessary to produce extracted honey.

The re-introduction of round sections, now made of plastic has boosted section comb honey production to some extent. Round plastic sections require less labor to install. They are also filled more uniformly by the bees, reducing the number of culled sections. Most comb honey and/or cut-comb producers appear to have no difficulty selling their entire crop at premium prices. Good comb honey sections, or comb that is to be cut, must be produced in an intense nectar flow. This ensures the sections will be uniform in color and capping profile. This kind of nectar flow is not always available in many beekeeping regions.

Commercial Pollination

A pollination service has often been looked at as a complement to honey production. Beekeepers may brag about the importance of honey bees in the necessary transfer of pollen, but many are not involved in the practical aspects of the service. A look at successful commercial beekeepers, however, shows that pollination contracts are an important part of their operations. Usually, these beekeepers are equipped to move colonies efficiently, which provides the necessary flexibility to service different crops during the year.

There are significant costs associated with a pollination service. Hive bodies must be reasonably secure (tight) and equipment either stapled, or banded together or palletized. Forklifts are necessary for pallets, and boom lifts for individual hives. There may be increased labor expenses associated with moving bees into tight locations near fields, and hives often need to be strategically placed throughout fields, rather than grouped in one convenient location. Pollination is called for early in the season when many other chores are facing the beekeeper. Colonies used for pollination must be fed very early to stimulate population buildup, often too far in advance to take advantage of any subsequent honeyflow. Bees cannot be expected to feed themselves, much less produce any surplus honey, while engaged in commercial pollination. On the other side, a commercial pollinating service gives the beekeeper a significant role in agricultural production, and more importantly a major boost in cash flow at a time of year (spring) critical to the business. A hive rented for only one crop needs to produce less honey per colony than one not used for pollination.

Once pollination becomes a major part of the business, beekeepers tend to specialize in the activity. A different set of skills is needed to be successful in the pollination business than in producing and selling bee products. The beekeeper must become oriented to marketing a service. In addition, one must be prepared to move bees on short notice and be knowledgeable about the crop to be pollinated. Finally, written contracts are a must. Many of these issues are addressed in the chapter on pollination in this volume.

Queen Rearing by the Beekeeper

The cost differential between the purchase of queen bees from a breeder and the raising by the beekeeper him/herself is probably in favor of purchase. There may be other beekeeping considerations that should weigh heavy in this decision, however. By careful selection the beekeeper may produce colonies that are better adapted to local nectar flows, and winter conditions. The beekeeper then increases production, as well as reduces winter losses and the cost of producing queens is well worth the effort. The point here is to make sure that careful selection is possible before undertaking queen production. It is easy to become enamored of queen rearing and lose sight of the real costs.

Many beekeepers keep laying queens in two- and three-frame nuclei within each apiary. Whenever they find a colony with a failing queen or a queenless colony, they simply insert two frames of bees with the queen in the center into the colony. The queen will almost always be accepted, hence an instant requeening. The nucleus is then stocked with new combs of eggs and larvae from a good colony and allowed to raise a new queen. With the number of nuclei equal to about 10 percent of the number of colonies in the apiary, there are usually enough extra laying queens. Having laying queens at crucial times is well worth the expense of providing the nuclei, even if stocked with queens purchased from a supplier.

LIMITING FACTORS AFFECTING BEEKEEPING

Colony Numbers

One should always keep in mind that a well managed, strong colony will produce more honey and pollinate more blooms than several weak ones. Many beekeepers are lulled into complacency by counting colonies and this concern about the physical number of colonies often takes precedence over other managerial decisions. This is not surprising, for often a beekeeper's reputation is based on the quantity of hives controlled. Pursuit of numbers for numbers' sake, however, should be assiduously avoided. The maxims that less is best, and to begin small, growing only after attaining experience generally hold.

The number of colonies that a beekeeper can feasibly operate can be estimated by multiplying the number of hives that can be manipulated per working day times the days between required visits. This will give the maximum number of colonies per beekeeper. For example, if a beekeeper can examine 100 colonies per day including travel time and the colonies are re-examined at approximately 10-day to two-week intervals, then 1000 colonies is the maximum per beekeeper. This figure doesn't include rain delays nor sickness. The number of colonies worked can be increased or decreased by using part-time help and/or labor saving devices. Fixed costs such as trucks, rent and machinery can also be spread over more hives.

Concerns About Business Growth

An often-asked question deals with how to measure growth in a beekeeping operation. There is no easy answer. Analyzing the rate of increase in colony numbers is certainly valid. However, the condition of each hive, time spent manipulating colonies and many other variables must also be considered. One of the advantages of beekeeping is that the number of colonies a person manages can be increased in small increments. Thus, a beekeeping business can grow more easily than almost any other agricultural enterprise. It is important to also consider another aspect of growth, that of profits. This should be analyzed first, and only after that the size of the operation or number of colonies being managed.

There is a noticeable lack of information on the economic growth of U.S. beekeeping operations. However, a Canadian study (Andruchow, 1982) confirms that efficiencies increase with size:

"As the size of the operation increases, cost savings can be expected in areas of operations and fixed investment components. For example, specialized bee equipment costs will be spread over a wider base as the number of hives increases. Operating cost savings can come about from the use of larger equipment, bulk buying, and more efficient use of labor and land. Lower production costs per pound of honey with increased size would result in increased returns to management."

Although increase in size usually results in more efficiency, a point is reached when growth can be counterproductive. One trend found in the above study was a general decrease in return per pound of honey as the size of the operation increased. The average prices received per pound of honey sold were $0.787, $0.616 and $0.581 for operations with fewer than 100 hives, between 100 and 699 hives, and over 700 hives, respectively. On the other hand investment costs per hive ranged from $546.58 for apiaries with fewer than 10 hives, to a low of $194.75 for operations with 700 to 999 hives. Operations with 1000 hives or more had significantly higher building and equipment costs (about 22%) when compared to those having 700 to 999 hives. This appears to be justified because of purchase of labor-saving devices such as forklifts and hive loaders.

Economies of size were most apparent for labor use in the Canadian study. Labor per pound ranged from $0.88 for operations with less than 10 hives to $0.14 for operations with 700 to 999 hives. Average vehicle expenses were $0.059, $0.055 and $0.037 for operations with fewer than 100, 100 to 699 and over 700 hives, respectively. Building and repair costs also declined with size, as did operating costs. Positive operating returns were only seen in the 700- to 999-hive class and a positive return on investment was only realized by operations having 400 or more hives. In summary, the 700- to 999-hive class operations had lowest investment, labor, operating and total production costs on a per pound of honey sold basis. In addition, the relatively high honey production of 151 pounds per hive yielded a profitable return to apiaries within this class.

Optimum Size of an Operation
    Although the Canadian study suggests that economies of scale favor operations of 700 to 999 hives, those data must be tempered by present and future conditions in the United States. Unlike the prairie provinces of Canada, many states are rapidly becoming urban with the concomitant problems of decrease in bee pasturage, increase in urban and agricultural pesticide usage and increase in potential for neighbor-to-neighbor conflicts resulting in bee-keeping ordinances. On the other hand, urban markets nearby may provide more retailing opportunity than in the past.

A survey of the beekeeping industry (Hoopingarner & Sanford, 1990) revealed an average number of colonies for 73 commercial beekeepers to be 2,442. The range was from 350 to 20,000. In the past the traditional line between part-time and full-time beekeepers, for statistical purposes, was usually 500 colonies. An informed guess, by us, concerning how many colonies might support one family on a full-time basis is four times that number, or 2,000 colonies.

Labor

Labor is the backbone of any beekeeping operation. Many beekeepers get away with inexpensive labor because family members are involved. Nevertheless, there still is what is called an opportunity cost involved. Family members may potentially make more income in some other occupation, but will not have the opportunity to do so, if employed in bee work.

As the business grows, short-term help can fill in the gaps. This labor is relatively cheap and accrues few benefits, but it is also not well trained. As growth occurs, experienced, trained labor becomes more necessary to a beekeeping operation. A look at enterprises that have kept their profit potential over the years indicates they possess enthusiastic, hard-working employees who are paid well. Skimping on rewarding employees is one of the best ways to go out of business. As pressures grow on profitability, it may be well to examine what many large concerns are doing to keep employees well remunerated. For example, giving employees part of the action through profit sharing or employee stock option plans shows trust and provides the best possible incentive to be productive.

Employing a trained, full-time work force, however, invariably converts the beekeeper into an employer with all the duties and responsibilities this entails. New management skills are required to get the most out of the labor force, paper work increases and governmental regulations tend to become complicated. Generally, benefits begin after 90 days full-time employment and costs can add up dramatically. These may include unemployment insurance, health insurance and social security (F.I.C.A.). It is best to consult the government section of your phone book for needed information. As a bare minimum, talk to the U.S. Internal Revenue Service's Taxpayer Service Office, your state's unemployment tax office and local health officials about guidelines to follow when hiring full-time employees.

Because of the potential amount of red tape and extra expense, hiring labor has been a hurdle many beekeepers have decided not to take. They have purposefully kept their outfits small enough to avoid taking on these added responsibilities. There is a tradeoff, however. It may mean that income from the operation must be supplemented by that from non-beekeeping sources.

Apiaries

Finding good apiary locations takes time. They must be monitored for a few years before a determination can be made as to their long-term suitability. Many factors are involved such as access by vehicles, availability of water, and predominant soil types and honey flora.

There is a general tendency for beekeepers to choose apiary sites based only on what is readily available or has the cheapest rent. In addition, the quality of an apiary site can change over time. The importance of keeping records of the number of colonies on the site, honey production, buildup in the spring (minor honey and pollen plants), as well as wintering success cannot be overemphasized. Keep these records on a chart so that any trends can be seen, and it may be fruitful to plot yields on an aerial map (see nectar resources).

Certainly, the apiary site needs good vehicle access throughout the entire season. Beekeepers are not usually required to maintain roads or access to all their apiaries. However, sometimes minor filling or grading will provide improved access to an apiary throughout the season.

Ideally, outyards should be close together and as near as possible to the extracting center. Situations where two beekeepers pass each other's apiaries to get to their own are not unknown. A simple trade of locations would have benefited both. Care should be taken when setting up apiaries and groups of apiaries. Groups of apiaries should be thought of in terms of working days, that is, how many colonies can be handled in a working day with a typical crew. Energy costs will go up in the future, relative to inflation, and a concentration of apiary locations will help reduce expenses. Grouping apiaries, however, has a potential cost because a crop failure in one area may mean the same for nearby areas. Beekeepers with large numbers of colonies often spread their apiary sites over a wide region to provide insurance against a crop failure in a smaller area. Again, there is a counter cost to this diversity in that there are greater transportation costs as the apiaries are located further from the honey extraction center.

Visits to an apiary are costly. There can also be a loss of honey if colonies swarm, have insufficient stores, become queenless or are diseased. Therefore, a balance must be struck between necessary visits and cost ineffective manipulations. In a survey of commercial beekeepers (Hoopingarner & Sanford, 1990), the average number of visits to an apiary during the year was 11. The range was from a low of 5 to a high of 50. Statistical analysis of the data showed no significant increase in honey with multiple visits. There are several possible reasons for the apparent contradiction that more visits do not necessarily mean higher honey production. The most obvious is that the better producing areas are occupied by beekeepers with the largest number of colonies. Beekeepers with a large number of colonies tend to not visit their apiaries as often. It seems to be fairly clear that many beekeepers examine their colonies more than is economically desirable. Evidence exists that examining a colony disrupts the bees' organization. It takes time to re-establish the delicate dynamics of a colony. One way to avoid wasteful visits is to always have securely in mind what must be accomplished before going to the bee yard. Most large-scale beekeepers seek to equalize colonies. Thus, they are able to do the same thing to each and every hive which conserves valuable time.

Nectar Resources

Nectar resources are changing in many areas of the nation. The majority of nectar plants are not cultivated, and forces are at work inexorably replacing or reducing nectar plant habitat. Urbanization has traded acres of wild bee forage for paved surfaces and/or wide rolling grass lawns. In some highly disturbed areas, introduced plants which are good, in some cases better, nectar producers than the native vegetation have gained a foothold.

Examples of the latter are the punk tree, *Melaleuca quinquenervia*, and Brazilian pepper, *Shinus terebinthifolius*, which produce great quantities of nectar in Florida. Unfortunately, these introduced species are considered "weeds" by many scientists and urban residents. They are blamed for disappearance of native flora, ecological changes and human health problems. Conflict is often caused when beekeepers locate their colonies in residential areas where many of these plants grow. Beekeepers also oppose eradication schemes to eliminate these "objectionable" plants from the environment.

Urbanization of agricultural areas means fewer beekeeping locations. Residents taking advantage of amenities in rural settings are not necessarily sympathetic to those who grow the nation's food supply. This is especially true for an enterprise that uses a stinging insect. Beekeepers have increased tensions, on occasion, by not properly informing urban neighbors of their activities.

Large-scale, modern agriculture practices have also unfavorably affected beekeeping. Farm managers, in an effort to produce as much as possible, exploit marginal lands. These were previously reserved by default for bee forage. Pesticides used on many crops, as well as railroad, highway and power line rights-of-way, have exacted an enormous price on bee colonies. This has happened through direct application of chemicals on field bees, and indirect damage to brood, by herbicides which also eliminate nectar resources. Finally, use of systemic insecticides that eventually show up in nectar is responsible for reducing foraging populations in certain areas.

Although the nectar resource is shifting, so are technologies to monitor the change. Gone are the days when the beekeeper had the opportunity to travel over large areas in search of nectar sources. Private land is fenced and posted as are many public landholdings. However, there are tools available that can be exploited to see the lay of the land. Most of the United States has been mapped by the U.S. Geological Survey in a number of scales. Any map library will have rectangles, sometimes called quadrangles, in 60, 30, 15 and 7-1/2 minute sizes. These can show potential nectar-producing sites. Other resources available to the modern beekeeper include aerial photography and satellite imagery. Most of these photos not only reveal details about contour, but also the kind of vegetation on the ground. The use of remote sensing using infrared photography, still in its infancy, may become a resource for the beekeeper of the future. This technology will not only find specific plants, but will also provide information on the physiological state of the vegetation. Actual data from colonies in imaged areas, over time, could be programmed into models predicting nectar flow potential.

Soil profile maps are another source of information regarding possible apiary locations. There are some areas that have soils more suitable to nectar-producing plants, and the same plants will produce variable quantities of nectar depending on soil type.

Migratory Beekeeping

The search for nectar resources invariably demands a hard look at migratory beekeeping. Large-scale beekeepers must move to survive. Full-time employees need year-round work. Anytime bees are on a location with no flow, they are consuming food. This is especially true in the Midwest in the winter. An alternative to wintering is to move south. The southern United States from Florida to Arizona continues to be a major area where beekeepers can move for the cold season. The bees may still need food, but not as much as it takes to get them through a northern winter.

Beyond simply providing an environment suitable for wintering bees, the southern locations allow for a good deal of manipulation. Many beekeepers requeen and make splits or nuclei which are then moved back to northern areas for the major honey flows.

Perhaps the heyday of the migratory beekeeper was the early 1970s, when honey prices were relatively high and fuel cheap. This is no longer the case. Fuel prices have risen dramatically while honey prices have dropped. Other costs have also skyrocketed. An additional cost in today's litigious environment is liability insurance. This insurance is not so important for permanent locations, but considered a must once colonies are trucked.

Equipment Costs

A major investment in beekeeping operations is equipment. This runs the gamut from the woodenware of a beehive to forklifts and semi-trailers for hauling bees. How equipment is purchased and maintained can mean the difference between success and failure of an operation. Many established beekeepers can afford to stay in business with a very small profit margin because their equipment is paid for. Those just entering the business will often have to take into consideration the costs of financing a considerable capital investment.

Regulations and Disease Treatment

Historically there have always been regulations to control bee diseases and pests. However, new and changing rules because of the introduction of the exotic bee mites, *Acarapis woodi* (honey bee tracheal mite) and *Varroa jacobsoni*, will continue to affect U.S. beekeepers. This is especially true for migratory honey producers and pollinators whose business is based on moving colonies.

Materials to treat maladies of bee colonies are costly to the beekeeper. In the past, only prophylactic feeding of the antibiotic Terramycin® was called for. Many beekeepers, especially in the southern United States, did not find it economical to feed fumagillin for nosema control. Nosema infestation, combined with new stresses, especially those posed by mites, however, may require rethinking.

Treatment for tracheal mites now must be considered, as well as detection and control of Varroa. The cost of the materials is significant, not to mention the time required to apply them. Some chemicals have shown repellency to bees. Others may have undesirable side effects on colonies which go unnoticed by the beekeeper. Controlling mites by putting insecticides inside a colony has added a new dimension to the risks employees now face. Many chemicals call for use of protective gloves or respirators.

Regulatory costs may also increase in the future. Most bee inspection services have found themselves deluged with new duties since introduction of tracheal and Varroa mites. Many states continue to provide basic inspection services free, and some pay an indemnity for colonies burned when infected with American foulbrood. How long such services will be supported by legislatures is not known. Look for the arrival of the African bee to increase inspection costs even more. The future will no doubt call for hard decisions by governments to collect fees for inspection and regulation of the beekeeping industry.

Now that it appears likely parasitic bee mites will be affecting U.S. beekeeping for a long time, the concept of a bee diagnostic service is gaining acceptance. Although beekeepers are currently doing much of the diagnosis themselves, a service using sophisticated equipment would be able to do more than look at mite populations in colonies. There is evidence that bee populations are affected by air and water pollution. Colonies in some areas are being used to monitor heavy metals and radiation present in the environment.

Among other things, a diagnostic service would be able to test colonies for protein levels and monitor other aspects of colony nutrition such as is currently done in Austrailia (Kleinschmidt, personal communication). Finally, there may be a need for certification of bee stock when the African honey bee becomes widespread in the United States.

Winter Loss

In a typical year, the U.S. beekeeping industry loses some 10 to 20 percent of its colonies during winter. Even a small reduction in this loss could be translated into a decrease in operating costs. There are several ways to reduce winter losses. The first of these, and the most beneficial, is just good beekeeping management. A strong, disease-free colony is capable of surviving a long and cold winter. Such a colony will in most cases also be a good producer the following year. Other techniques can be used to help colonies survive, but they can be costly. These include wrapping (several degrees of complexity), moving colonies indoors or migrating to warmer climates.

A convenient yardstick to lay against the cost of wintering is analysis of expenses associated with migrating to the South or using package bee colonies. The cost of migrating with colonies of bees is high. However, many beekeepers also use their bees for pollination or producing a crop of honey at the winter location. This certainly reduces the overall transportation cost. In this regard, it is also important to analyze all of the costs and benefits for each operation. For example, while a beekeeper may receive a fee for pollinating in California during the winter, there are some expenses associated with this service as well. The same could be said of any honey produced while on winter location. However, commercial pollinating or honey-producing colonies returning from winter locations in the South would be in better condition than the average colony that remained in the North.

The other option is to kill colonies in the fall, extract and sell most of the honey that would have been consumed during the winter months and start with package bees the following spring. This appears practical since the 60 or so pounds of honey that would have been consumed by an over-wintering colony more than offset the cost of the package of bees. The labor savings seen in Table 3 also support the conclusion that using package bees has advantages. When analyzed more completely, however, the cost savings from selling honey that would have been used in winter may be offset by the reduced success rate of colonies started from packages. Package bee colonies may also have reduced value in pollination and honey production as compared to an overwintered colony.

The ethics of killing bees is also a consideration. Deliberate destruction of a colony goes against many a beekeeper's philosophy. The knowledge that honey comes from colonies killed off by the beekeeper simply for economic gain may result in negative public relations, damaging efforts to promote and sell the crop.

Loans

Beekeeping enterprises often require infusions of cash to cover operating expenses and purchasing capital assets. Commercial loan sources such as banks, savings and loans and credit unions may make loans to agricultural enterprises. In addition, government sponsored loans are available as are funds from agricultural cooperatives. The following are possible sources of loans for beekeeping operations: Farmer's Home Administration (FmHA), Agricultural Stabilization and Conservation Service (ASCS), Small Business Administration (SBA), Production Credit Association (PCA), and Federal Land Bank Association (FLBA).

Each of the above organizations may have different collateral or security needs and/or disparate lending goals. In all cases, the loan officer will require documentation of net worth, production, sales, and potential growth. Because beekeepers are not required to be land owners, they often do not have this asset to use as collateral, limiting the amount agencies are willing to underwrite.

BUSINESS PRACTICES RELATING TO BEEKEEPING

Desktop Computers

It has already been mentioned that few innovations in beekeeping occurred in this century. The electronic age, however, has ushered in some significant changes which will aid beekeepers significantly in the future. Of special importance is the desktop computer. This flexible electronic device has several uses that can immediately be put into action by any agriculturalist.

A number of creative ways exist for word processing to help in the myriad office chores required by a modern business. No longer must letters be retyped after revision. Special features abound in most programs such as employing fancy fonts to design letterhead and merging addresses with form letters so that each looks like an original.

The beekeeper can use spreadsheets and database managers to record, revise, sort and analyze information about an operation. At least one spreadsheet program is currently available from the Cooperative Extension Service which helps the beekeeper in financial management (Sanford, 1986). Of special significance is the possibility to store detailed information about locations, individual colony performance, queen sources, nectar flows and the like. Even the smallest items that make up an extensive inventory of frames, hive bodies, tops and bottoms can now be cataloged quickly and efficiently.

Accounting and payroll computer programs provide a wealth of options to analyze costs. There is even the future possibility of looking at efficiencies through elaborate project management or expert system programs. Finally, one can analyze the dynamics of colonies themselves using sophisticated biological models.

The power is now there to do all of the above tasks and more. And the capability can be had for a very modest price. However, there is a considerable investment in time and energy required to become trained in the use of these machines. It will be up to each individual to determine how much effort he/she is willing to put forth in using this new technology.

Record Keeping

One of the most important aspects of any business is good record keeping. Unfortunately, because so much of beekeeping takes place outdoors, often in remote spots, record keeping is difficult. Yet it is required if adequate business expenses are to be documented. Many beekeepers have developed their business slowly via the sideline route where precise records were not so important.

Record keeping goes beyond the obvious receipts for gas, oil and equipment. Keeping track of apiaries or colonies for yields and bee genetic stock is just as important. Again, for many of these items a desktop computer serves very well. Spreadsheets, such as Lotus 1-2-3® or SuperCalc®, organize and sort data, and will graphically plot trends like yields, gas mileage or oil consumption. These data will often help make decisions such as whether to move an apiary or repair a truck. It is the acquisition and analysis of such timely knowledge that is often the difference between success and failure.

Inventory

Inventory records can be important in case of fire, theft or when collateral is needed for a loan. A complete inventory will also aid bee management. A knowledge of how many supers (of each type and condition) are located in specific apiaries or other locations will enable the beekeeper to better organize trips to outyards.

Business Expenses

Almost any expense associated with the business is legitimate to use as a deduction at income tax time. Often, however, significant items get overlooked. For example, smaller-scale operators may not deduct telephone expenses, even though they make legitimate long distance calls or have an extension phone specifically for the beekeeping operation. It is best to develop a list of recurring costs, or better yet keep a ledger so that each cost can be listed as it occurs.

Taxes

The subject of income taxes is beyond the scope of this chapter. The tax laws are continually becoming more complicated. Modern business practice insists that beekeepers consult with professional tax accountants. An indispensable guide for any agricultural enterprise is the *Farmer's Tax Guide*, published each year by the Internal Revenue Service (IRS). There are tax books sold by private publishers and short courses offered by the county extension offices and other organizations in each state.

State taxes are also payable by the beekeeper. Consult your local state taxing authority each year for up-to-date information. As an agricultural enterprise, beekeeping may be exempt from paying state sales or other taxes.

Depreciation

The role of depreciation must be considered in the capital investment of any beekeeping operation. Hives, frames, honey house equipment, trucks and other hardware are generally depreciable assets. The concept of investment credit may also apply. This has been eliminated during recent tax legislation, but could be put back in place at any time. Bees are usually not depreciable, but new purchases can be deducted as an expense. At least one IRS ruling to the authors' knowledge rules out the use of investment credit in purchasing queens.

A good rule is to depreciate equipment or buildings for tax purposes as fast as is allowed by law, even though you have to "correct" some of the depreciation when an item is later sold. In essence, you receive a no-interest loan from the federal government while using the money. It is important, however, not to confuse depreciation for tax purposes with actual life of the asset. The best way to do this is to record market value along with depreciated value in the balance sheet.

Managing Financial Resources

As noted elsewhere, because beekeeping enterprises are so variable, it is impossible to give detailed recommendations. This section is meant to provide a guide for the beekeeper in analyzing costs and returns from his/her operation. Much of the following information has been taken from the *Farm Finance Newsletter*, published by the Florida Cooperative Extension Service (van Blokland, 1985).

Conditions over the last decade have changed the face of agriculture and revolutionized its practice. We are now in a managerial revolution. Management begins with good record keeping and intensive financial analysis if profitability is to be maintained.

It's time to return to the basics for agricultural enterprises. That means gathering and recording information. A guide to financial information flow for any business is provided in Fig. 3. It is divided into two sides; one showing what happened in the firm (basic recording) and another looking at what might happen (managing). The right hand flow of information is recorded in the balance sheet and income statement.

Financial Statements

The balance sheet is a picture of the business on one day of the year, typically December 31. The picture will change the next day, but is fixed on the 31st. The balance sheet records assets, liabilities or debts, and net worth or equity at that time. Assets are what you have; liabilities are what you owe; and net worth is what you own or what's left after assets are used to pay off liabilities.

Balance sheets recording both cost and market value for assets and liabilities are strongly recommended to show actual depreciation versus that used for tax purposes. This provides another level of information in the same

Figure 3. GENERALIZED INFORMATION FLOW IN A BUSINESS
(READ DOWN)

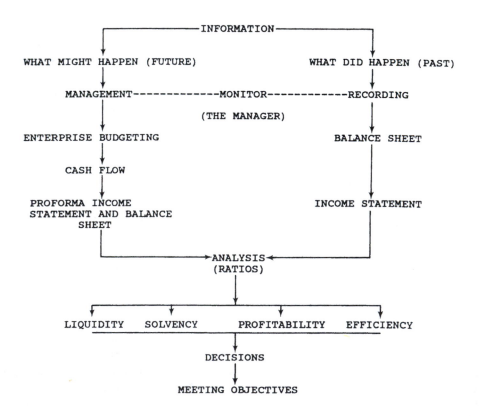

instrument. Analyzing the relationship between current and long-term assets and liabilities is also important. Current assets and liabilities are those that can readily be turned into cash in less than a year. If not, then these items become long-term (will take over a year to sell for cash).

The income statement reveals what a business did over a period of time, usually the calendar year. If the balance sheet can be looked at as a snapshot, then the income statement can be viewed as a movie. It shows three things—revenues, costs and income.

Many agricultural enterprises are weak in inventory control. This area must be watched closely on the income statement. It demands accrual accounting which provides more information to the manager. Beekeepers have generally used the cash method of accounting for tax reporting. Regardless of the tax reporting method, accrual accounting should be used to prepare income statements.

The manager now looks at what might happen in the operation by following information flow on the left side of Fig. 3. As previously mentioned, enterprise budgeting is an important tool in managing any operation. The costs for each enterprise are consolidated into a total operation budget.

The cash flow statement is perhaps the most important single financial instrument for an operation. It is a picture of what the business might do in the future in cash terms. The statement may show monthly and/or quarterly costs and reveals three things: (1) anticipated cash receipts; (2) anticipated cash costs including living expenses and (3) anticipated borrowing and payback. Because it only shows cash, the cash flow statement alone will not indicate whether a profit is made or not.

The cash flow is the manager's best tool to monitor and adjust predictions, depending on the real outcome. For best results, it should be done monthly. Information from the cash flow statement is used to make "proforma" or interim financial statements (income statements and balance sheets). These are predictions of what might occur and help the manager plan ahead.

Ratio Analysis

Information gained during recording and in developing cash flow and proforma statements is used to analyze the condition of a business. This usually is accomplished through ratio analysis. A great number of ratios can be calculated for any business. Fig. 3 lumps them into four categories—liquidity, solvency, profitability and efficiency.

It is important to remember that ratios are only as good as their interpretation by different interests in the business (owners, employees, investors, lenders). Average ratios have been developed for many businesses on a historical basis. They are routinely published by investment firms and other financial institutions. Unfortunately, representative ratios for specific types of agricultural businesses are unknown, as are those for beekeeping enterprises. This means agricultural managers must begin at the beginning and track the ratios in their own businesses over a period, before they can be used with any degree of confidence in analyzing and predicting performance.

REFERENCES

Adams, R.L. and F.E. Todd. (1933). Cost of producing extracted honey in California. Tech. Bull. No. 656. U.S.D.A.

Anderson, E.D. (1969). An appraisal of the beekeeping industry. U.S.D.A. ARS 42-150.

Andruchow, L. (1982). The Economics of Beekeeping in Alberta 1980. Agdex 821-16, Economic Services Division, Alberta Agriculture, September.

Coke, E.W. (1966). Apiary records of citrus and tupelo honey producers. 12th Annual Florida Beekeepers Inst. Florida Agr. Exten. Serv.

Farrar, C.L. (1968). Productive management of honey-bee colonies. *Amer. Bee Jour. 108*:3-10.

Foster, E.M. (1989). Honey in Relation to Infant Botulism. National Honey Board Food Technology Program. Mimeo. 5 pgs.

Hoff, F. (1988). Report on U.S. honey imports. U.S.D.A. Agr. Econ. Res. Serv. Mimeo.

Hoopingarner, R.A., and M.T. Sanford. (1990). The costs of beekeeping. I. Survey of commercial beekeepers. *Amer. Bee J. 130*:405-407.

Kalnins, M.A. and B. Detroy. (1984). Effect of wood preservative treatment on beehives on honey bees and hive products. *J. of Agric. and Food Chem. 32*:1176-1180.

Kalnins, M.A. and E. Erickson. (1986). Extending the life of beehives with and without preservatives. *Amer. Bee J. 126*:488-491.

Kleinschmidt, G. (Personal communication).

Levin, M.D. (1983). Value of bee pollination to U.S. agriculture. *Bull. Entomol. Soc. Amer. 29*:50-51.

Martin, E.C. (1980). Introduction to: Beekeeping in the United States, Agriculture Handbook 335, USDA pg. 1.

Moeller, F.E. (1976). Two-queen system of honey bee colony management. U.S.D.A. ARS Prod. Res. Rpt. No. 161.

McDowell, R. (1984). The Africanized honey bees in the United States. What will happen to the U.S. beekeeping industry. U.S.D.A. ERS Agr. Econ. Rpt. No. 519.

Murrell, D. and B. Henley. (1987). Drying honey in the hot room. Saskatchewan Agr. Extension Serv. & Soils and Crops Branch. 6 pgs.

Owens, C.D., T. Cleaver and R.E. Schneider. (1973). An analysis of beekeeping production costs and returns. U.S.D.A. Production Res. Rept. No. 151.

Robinson, F.A. (Personal communication).

Robinson, W.S., R. Nowogrodski and R. Morse. (1989). The value of honey bees as pollinators of U.S. crops. *Amer. Bee J. 129*:411-423, 477-486.

Rodenberg, H. (1967). Cost of production survey related to market price of honey. *Amer. Bee J. 107*:139.

Sanford, M.T. (1986). A study in profitability for a mid-sized beekeeping operation. Univ. Florida, Coop. Exten. Serv. Circular 692. 42 pgs. plus disk.

Stoller, D. (1989). Quality and competition in the sweetener's market. National Honey Board 2(3).

Taber, S. (Personal communication).

van Blokland, P.J. (1985). Florida Coop. Exten. Serv. Farm Finance Newsletter Vol. 2, June, mimeo.

ACKNOWLEDGEMENT

We wish to thank Drs. P.J. van Blokland and Stephen Ford of the Food and Resource Economics Department, IFAS, University of Florida for helpful criticisms and review of this chapter.

CHAPTER 18

SURPLUS HONEY AND THE SMALL BEEKEEPER

by GEORGE S. AYERS*

For many, beekeeping starts as a fascinating hobby perhaps with only a single hive. Hobbyists often find that the ensuing honey production is more than their immediate family can consume and are suddenly faced with the problem of what to do with the excess honey. There are five solutions to the dilemma.

1. The excess honey can be left with the bees;
2. It can be given to friends, relatives and acquaintances;
3. It can be packaged and sold retail;
4. It can be packaged and sold wholesale; or
5. It can be disposed of in bulk to a packer or through a government price support program. The continued existence of the latter program is, however, always uncertain.

Leaving the Excess with the Bees

While this solution is undoubtedly practiced by many backyard beekeepers, hives managed in this manner often become essentially abandoned and develop into reservoirs of disease. In addition, this management practice robs the beekeeper of one of the main joys of beekeeping, the management of a hive for top productivity.

Giving the Excess Away

Many hobbyists give their excess honey to friends and relatives. Although generosity is a desirable human trait, disposing of excess production in this manner should be done carefully. Subtle differences in the circumstances of the giving can profoundly affect the outcome of the unselfish act. When given as something special; as a present, or in repayment of a favor, the practice confers value to the honey in the mind of the recipient and the generous beekeeper has made a small but valuable contribution to the industry. On the other hand, if the beekeeper gives the excess away in a manner that connotes that the recipient is doing the beekeeper a favor by helping eliminate the excess, the practice tends to create the idea that honey is valueless. When honey is given in this way, the beekeeper unwittingly does a great disservice to the industry. While the actions of an individual may seem insignificant, if all

*George Ayers, Ph.D., Associate Professor, Department of Entomology, Michigan State University, East Lansing, Michigan.

the hobbyist beekeepers of North America would act as in the first instance, the result would be quite significant. Closely related to the hobbyist who gives the excess away, is the one who sells it for a pittance. This beekeeper also does a disservice to the industry by unwittingly helping drive the price of honey to unrealistically low levels. For the industry, it would be better if these individuals left their excess with the bees.

Deciding to Expand the Hobby into a Small Business

For the beekeeper contemplating selling the excess, the distinction needs to be drawn between selling the excess from one or two hives and expanding for the purposes of developing a small business. The first situation is a simple matter requiring only the simplest of marketing strategies. For the hobbyist bent on expansion, however, marketing often becomes the biggest and most perplexing problem to be solved. The expanding beekeeper soon discovers that developing and maintaining markets becomes as important and time consuming as the production itself. It is often not an easy transition and the hobbyist should not be lulled into thinking of the new business venture as an easy income based on the experience of selling the excess from the first one or two hives.

There is a second transition with which the small beekeeper often has trouble. While the packaging of honey for family and friends is not a critical matter, this attitude should not be carried into even the beginnings of a commercial venture. The products of the small beekeeper should always be packaged neatly and be of the highest quality to both ensure repeat sales and to perpetuate the high-quality image of the honey production industry as a whole. While the hobbyist beekeeper often is tempted to package honey in obviously recycled containers to reduce costs, this practice should be avoided. Not only does it look bad, but the small beekeeper is probably ill equipped to meet the regulations regarding recycling of food containers. Great care should be taken to guard against deterioration of appearance of the product on the shelf as a result of crystallization or worse, fermentation. The small beekeeper, in attempting to produce something special, should strive to set the highest quality of the industry and not its lowest acceptable limit.

Some Decisions That Should Be Made Early

Once the decision has been made to sell the excess honey, several important decisions need to be made that apply to both the retail and packaged wholesale trades.

Unless the decision is made to market the honey in bulk, it is often in the best interest of the small beekeeper to produce a specialty item. By producing the ordinary, the small beekeeper not only competes with the cheap honey substitutes (sugar, corn syrups, molasses, etc.), but also with the millions of pounds of extracted honey on the supermarket shelves. By producing a specialty product, much of this competition is eliminated. The second advantage to specialty production is that higher prices can be demanded for the special than for the mundane and ordinary.

The products of the small beekeeper can claim distinction in many ways. What is necessary is that it meets the needs and desires of a particular clientele. The honey produced could be from a special floral source for the honey connoisseur, from a particular location for the vacationer, or be produced organically without chemicals, heat or extensive filtration for the health food devotee. For those searching for something special or unusual, it could be produced in unusual forms such as creamed honey or as one of the types of comb honey. Also, it could be incorporated into some unusual honey-based product. Packaging is often as important as the product itself when creating an aura of distinction. Careful attention should be given to this matter, for subtle changes in packaging can make profound differences in sales volume. To be successful at specialty marketing, both the product and the packaging must be finely attuned to the needs and desires of the particular clientele being served. This point cannot be overemphasized for it is exactly this that makes something a specialty product.

Most hobbyists cannot resist the temptation to produce extracted honey. Before making this decision, careful consideration should be given to producing some form of comb honey. There are several advantages to this. There are no extractors, uncapping equipment, wax melters, holding and bottling tanks and honey filtering and liquefying equipment to be purchased and stored. In addition, comb honey production is often freer from state regulation than is extracted honey. The expense of complying with the additional state regulations associated with extracted honey may be significant. The beekeeper should contact the appropriate state authorities for this information. Finally, the market rarely seems saturated with comb honey.

Balanced against the advantages of producing specialty products are some disadvantages. Specialty items can't be sold profitably just anywhere. The beekeeper, while eliminating competition, must at the same time identify the clientele to whom a particular product can be sold. This often requires an element of creativity in the beekeeper's planning. In the case of comb honey, the public knows almost nothing about the product. Unless it is sold wholesale, the market will need to be created. Comb honey is, however, clearly special and distinctive and good markets can be created for it.

Some specialty markets are quite small and will accommodate only a small production. Sometimes as the business grows, production gets ahead of the demand of the original market, placing the beekeeper in the dilemma of deciding whether to abandon the original market and expand into a larger market, or add additional but different markets to the business. It is possible to expend inordinate amounts of energy maintaining several small but different specialty markets. As a hobby, this may be acceptable, but for efficiency, it would generally be better to expand into a larger market than to string along several small but very different specialty markets.

Specialty production may require additional skills, situations and operations that are not needed in the production of less distinctive products. The production of a high quality raspberry honey, for example, will require large relatively pure stands of raspberry and careful timing of the supering and harvesting. Section comb honey is more difficult to produce than extracted honey. Specialized equipment may also be needed. Production of section comb honey, for example, requires special supers. When done on a small scale, however, these specialized equipment needs are often quite modest.

Sometimes the per-hive yield can also be diminished, as in the case of comb honey production.

While the disadvantages to specialty production exist, they can often be offset by the higher price of the specialty product. For the small beekeeper who can efficiently fill a particular market niche, the benefits will often outweigh the disadvantages.

SELLING RETAIL

At the onset of considering a retailing venture, the beekeeper should be aware that this practice is extremely time consuming because it demands that the beekeeper be personally there. Selling from the home or at the job are exceptions since these practices require little extra time beyond normal commitments.

For many beekeepers, retail sales conjures up unpalatable visions of the stereotyped used-car salesman. This feeling is partly derived from the fact that real salesmanship in the modern commercial world is largely nonexistent. Sales personnel encountered at shopping malls and variety stores are usually little more than stock clerks. Except for the location of the products for which they are responsible, they know very little about the merchandise that they sell and they make no effort to understand the customers' requirements. A true salesperson knows his stock well and can quickly assess the needs of the customer. A good sale is making the best match between the merchandise available and the needs of the customer. While it may not be appreciated by the customer, the good sales person actually performs the valuable service of sending customers home with products with which they will remain happy. Sales, when viewed from this perspective, is among the most human of activities and can be very satisfying.

While the beekeeper generally knows his product well, this information often is not passed on to the consumer. Many beekeepers sit passively behind their booths at farmers' markets, bazaars, county fairs, etc., waiting for the customer to select a product. While they may offer a polite "May I help you?" when the customer is looking over the merchandise, and "Thank you," after the sale has been made, often no information other than pricing is transferred to the customer. Honey, especially specialty honey that the small beekeeper is in a good position to offer, has many attributes that will be useful to the customer—it isn't just sweet!

Examples:

"This is locally produced honey containing local pollens."

"I sell only raw unheated and unfiltered honey."

"My honey is produced without chemicals."

"This honey from (floral source supplied) I consider to be the Lafite-Rothschild of fine honeys."

"This comb honey is honey in its purest form; it's the way your grandparents enjoyed their honey."

"This (floral source supplied) honey is unique to the (area supplied) area. It would make a unique but inexpensive gift to take home."

"This (floral source supplied) honey is an exceptional baking honey that will improve any home baked bread."

"I have a honey glaze recipe that turns this (type of honey supplied) honey into a treat fit for the gods."

"My honey bear provides a neat way of dispensing honey on those biscuits you like so much."

"Creamed honey is very neat—did you know that most of the world uses honey primarily in this form?"

The above list can be expanded almost indefinitely and is limited largely by the beekeeper's imagination and creativity.

These statements would, of course, be inappropriate for all customers. The beekeeper, like any other good salesperson, should strive to learn about and understand the customer's needs. This takes considerable skill, for unless the beekeeper is dealing with a repeat customer, there are usually only a few minutes available for the process.

Getting To Know the Customer

A good deal can be known about the customer before the initial encounter. It helps, for example, to understand that customers react to a salesperson in several basic ways. Being able to assess the type of customer early in the transition can do much to cement a sale. The easiest customer to deal with is the one who already knows about honey, knows exactly how much is wanted and how it will be used. The salesperson has little to do in this situation except to accept the money, be courteous and perhaps introduce a new product or new use for the honey.

Then there is the friendly customer. This person likes to buy from a friend and appears to use lengthy, often wondering conversations to determine whether the particular beekeeper is a friend. This customer may not even be hunting for honey at the time of the encounter. In this situation it is the beekeeper's job to remain friendly and continually steer the conversation back to the honey and its properties in a way that will unveil the customer's honey needs.

There unfortunately is also the converse of the friendly customer—the hostile customer. This person demands proof that the honey is unadulterated, that it is 100% pure (floral source supplied), or complains that the price is too high. The questions are often ones for which the beekeeper has no way of definitively backing up his answers except to personally pledge that it is so. The hostile customer generally knows this and probably has no intention of purchasing honey in the first place. In this situation, the salesperson, in spite of first impulses, should remain courteous, be resolute in answering and if given the opportunity, probe to learn if there is reason for the hostility such as a past bad experience with honey. If there is, this will provide an excellent opportunity to meet the needs of this particular customer. If there is no apparent rational reason for the hostility, there may well be no immediate sale, and the beekeeper has to be content with knowing that a difficult situation was handled as well as possible.

Finally there is the customer who has the notion that it is honey that is wanted, but doesn't know what kind, how much, or how to use it. This situation represents the salesperson's finest opportunity to perform that special function of filling the needs of the customer. It is the salesperson's job through observation and friendly probing to assess the lifestyle and therefore potential uses for the product and finally to make suggestions that match the product with these findings.

The place of business can also tell a great deal about potential customers and their needs. Shoppers of farm markets are often looking for fresh and wholesome food. Those who make purchases in organic food stores are looking for freedom from pesticides and other chemicals. If the beekeeper is doing business in a tourist area, the clientele will be looking for items peculiar to the area, either to consume immediately or to take home as a present or as a reminder of a pleasant vacation. In some cases, customers will be looking for bargains while in others, price seems almost inconsequential. Often researching the situation by talking to local merchants prior to setting up shop will help provide a useful customer profile.

While developing customer profiles is very useful, for the salesperson to function truly effectively, some reasonably accurate assessment of individual customers must be made. This can only be done through careful observation and friendly conversation designed to inoffensively probe for the needs of the prospective customer. The beekeeper should constantly try to set up situations where this happens naturally. Techniques for doing this are discussed later in sections where they are appropriate.

The Needs of the Customer

The full needs of the customer that can be met by the beekeeper are often not apparent. First there are the physical needs. For what is the honey intended? Is it to be used for baking or to sweeten tea? Who uses the honey in the family, the parents or the children? Here not only the honey itself but also the packaging may be important as , for example, honey bears for the children or honey sticks (see fig. 1) for school lunches.

Then, there are the informational needs. How do you cook with honey? Can honey be used as a sweetener during canning? How is creamed honey

FIGURE 1. Sometimes packaging is paramount to the customer. Honey Stix™ provides an innovative package that fulfills the need of the parent who is looking for an alternative to candy for school lunches or snacks. This packaging also provides an excellent way to distribute honey samples as a form of advertising. *Photo by B.A. Stringer.*

used? What about infants, botulism and honey? For such questions the beekeeper should be ready with answers. Some of this information can be provided as information sheets given as a kind of lagniappe, something extra the customer receives when purchases are made from a particular beekeeper. Perhaps a few tried and true recipes might be made available as well. A small line of honey cookbooks that range in price should also be considered. All of this information should include the beekeeper's name, address and telephone number.

Then, there are the psychological needs. While these are harder to uncover, they are often the most important. Is pureness and wholesomeness exceptionally important to those responsible for the family diet? Is the bee-keeper dealing with a gourmet chef who takes great delight in the high quality of the products that go into a culinary masterpiece? Does the customer like to experiment with the unusual—perhaps desserts, or maybe salad dressings or barbecue sauces? Is the honey to be a special but inexpensive present? If it is, the beekeeper should be ready to supply reasons that the purchase is special in an attractive written form to accompany the gift. It is often important to the giver that the recipient appreciate the special qualities of the gift.

DIRECT RETAIL SALES SITUATIONS
Sales from the Home

For the beekeeper with only a few hives, surplus honey can often be sold directly from the home. This practice requires only small financial investments since all that is necessary is a sign stating that honey is for sale. It is also possible to dispense honey directly into the customer's container, thus keeping the stocking expense of the beekeeper and, therefore, the cost to the customer to a minimum. This practice pleases many customers. In addition to requiring only minimal financial outlays, selling from the home requires the smallest time investment of any retailing venture since sales occur only when a member of the beekeeper's family is home. Serious consideration, however, needs to be given to whether it is desirable to allow total strangers into the home for these transactions. This concern can be alleviated to some extent by selling from the honey house if one exists. Alternatively, it is sometimes possible to operate a self-serve honey stand (see fig. 2). Depending on the situation, the problem with theft may rule out this practice. Some operators of self-serve stands feel that a sign that conveys the message that "it is wrong to steal" helps prevent theft. Others use a sign that indicates that ringing the bell that is provided for the purpose or beeping the car horn will summon the beekeeper to assist with the transaction. This practice alerts a potential thief to the possibility that someone may be watching. The theft problem can be minimized by stocking the stand with minimal supplies and leaving only minimal change in the change box. While self-serve stands are not as common as they once were,

Figure 2. This relatively elaborate self-serve honey stand sells a considerable amount of honey for its owner. For this beekeeper theft has not been a problem.

there are still many beekeepers who do a profitable business through them. Provisions should be made to keep the honey out of direct sunlight since heat and sunlight deteriorate honey.

Sales at Work

Some beekeepers sell their entire surplus at work. This practice works well where the employer hires relatively large numbers of people as in a factory. Provisions obviously should be made so that the practice doesn't interfere with the job and thus irritate the employer. This often can be done by limiting the transactions to the lunch hour or to a period of time immediately after work, perhaps in the parking lot.

Farmers' Markets, County Fairs and Similar Situations

Farmers' markets, county fairs and similar settings provide good places for the beekeeper to sell honey. Many beekeepers who sell honey in these settings, however, do so passively, sitting behind their table waiting for the customer who wants to purchase honey. For every potential customer who comes to the booth, many, often hundreds, walk by without stopping. The beekeeper should try to find some acceptable way of engaging in meaningful dialogue about honey with a significant portion of this group. The word acceptable is very important! Hawking, for example, is usually neither acceptable nor particularly effective. The beekeeper should strive to find a way of getting some portion of this crowd to stop on its own. In some instances a sign stating something to the effect of "ask us about our honey", "ask us about our no-mess honey", "ask about our comb honey", "ask about our bulk honey prices", etc., may help. In some circumstances demonstrations are an appropriate way of catching the interest of the crowd. Honey extraction, hive demonstrations and observation hives always attract attention. They also generate a plethora of questions, many about bees and only a few about honey. For this technique to be effective the beekeeper must find ways of guiding the ensuing conversation toward the beekeeper's products and their desirable attributes. If the conversation is not successfully managed, all of the beekeeper's energies will be dissipated on extraneous questions and sales will not have been benefitted significantly. Careful prior consideration needs to be given to this matter for steering the conversations from the topics of stinging, Africanized bees, and swarming, will challenge the conversational skills of any beekeeper. Where demonstrations are employed, the beekeeper will probably need one or more trained sales personnel in attendance to help with the process. When employing demonstrations involving unpackaged honey or bees, extreme care should be taken to maintain clean conditions and avoid congregations of bees and yellow jackets that can affect the audience.

Often, selling honey alone at farmers' markets and similar situations is only a marginally profitable business, especially if the beekeeper places significant value on personal time. Selling something in addition to honey can increase the profitability of the operation in two ways. There is the obvious direct sales of the other item which sometimes generates greater revenues than the honey. Less obviously, the additional items will draw more customers to the booth and give the beekeeper an opportunity to open meaningful dialogue about the displayed honey and thus increase honey sales. The beekeeper might, for example, consider some other bee product such as beeswax candles. The customer's attention might then be drawn to the displayed honey by pointing out that "the delicate aroma of the candles is entirely natural, being derived from fine honey that was once stored in it." The beekeeper might also consider offering items that go well with honey, perhaps fresh fruit such as blueberries. Alternatively, the beekeeper might team up with a neighboring vendor of these products. Both might advertise and offer coupons for price reductions on the other's product. Selling flowers would provide not only a natural linkage to honey, but also a very attractive setting in which to feature the honey. Finally, items totally unrelated to honey might also be tried, but steering the conversation toward honey will be more difficult. Selling something made with honey deserves special consideration and is discussed more fully later in the chapter.

Selling Honey at Gala Events—Christmas Bazaars, Festivals and Tourist Events

These situations are similar to fairs and farmers' markets in that the crowd passes in front of the beekeeper's booth. The techniques discussed earlier for opening meaningful dialogue with a greater percentage of this crowd will be useful in these circumstances as well. The beekeeper's booth and persona should be attuned to the festivities at the time. Some beekeepers dress specifically for such events. At some festivals and tourist events, demonstrations such as honey harvesting and candle-making are appropriate. At Christmas bazaars fancy gift packs are in order. Depending on the event, producing a fancy packing might involve little more than a special label, a swatch of gingham or a fancy jar with a tie of ribbon (see fig. 3). Beeswax candles and Christmas tree ornaments will also often be appropriate.

At tourist events the honey should be presented as something special associated with the region and not simply as a jar of honey. In Arizona, honey is advertised as "Arizona Gold." The Maine beekeeper might feature blueberry honey from the wilds of Maine. As always the beekeeper should try to imagine the need or the customer. Here the need may be a fond reminder of a wonderful vacation or an inexpensive but personal gift to take home for someone special.

FIGURE 3. A simple but very attractive gift package that would be appropriate at festive events such as Christmas sales or harvest celebrations. If the honey was a specialty of the area, the package would also be very appropriate at a tourist gift shop.

Honey House Open Houses

A few beekeepers hold "open honey houses" during extracting season and turn the honey harvest season into a gala and educational event. During the Fall, many reminisce romantically about the bountiful harvest. Numerous small agricultural enterprises take advantage of this situation with cider pressing, pumpkin sales, etc. Honey production is both mysterious and interesting to large segments of society and provides the beekeeper with a ready-made situation to talk about honey. This situation also provides an excellent opportunity to generate goodwill for the industry in general by providing the public with a valuable educational experience about bees and beekeeping. Advertising for these events is usually done in the local papers, and advertising flyers. The event is unusual and timely enough so that it is often possible to get the news media involved as well. Honey harvesting demonstrations are also a natural to be done in conjunction with other local and festive harvest events such as cider pressing, old fashion apple butter making, etc.

By participating in such an event, the beekeeper symbolizes the standard of the industry to most of the audience. Every attempt should be made to ensure that this impression is favorable. The facilities should be immaculate and free of yellow jackets and bees (unless in an observation hive). Supers should be neat and clean and the frames attractive.

Direct Mail Order Sales for Gift Items

Some beekeepers pack gift packs for holidays. While many merchandise the packs through retail stores, others operate direct mail-order businesses. Generally this activity occurs at Christmas, but some produce packs appropriately decorated for other holidays such as Easter and Valentine's Day as well. The success of these operations hinges on selectively developing a list of customers who are likely to give such packages as gifts. Simply mailing flyers to the public at large would probably not be a very cost-effective way of developing this list. Lists of names, partly selected to interests and habits, can be purchased from other merchandisers or organizations who cater to selected clientele. These lists can be quite expensive, but might be generated least expensively from local organizations. Organizations should be selected where the members will naturally be interested in a particular product. The members of a naturalist organization, for example, might be interested in giving a gift of honeys from selected wild sources. Some beekeepers develop their lists, based on the tenet that those who purchase honey for their own use will also be likely to give it as a gift and find ways of developing lists of names and addresses of their regular customers. Having customers register for a free drawing is a useful way of developing such a list. Others simply make handbills available where they do business that advertise the availability of seasonal gifts. Since customers who go to the trouble of picking up these handbills are likely to be interested in giving the advertised package as a gift, this procedure is a very inexpensive way of developing a very productive list.

Mail order businesses usually upgrade their lists in some manner. At a minimum, names that have remained unproductive for a given period of time are removed. Some beekeepers upgrade their lists by adding to them the names and addresses of the recipients of the gifts.

If the advertising is to be mailed, the U.S. Postal Service should be consulted for the latest rules, regulations and rates for bulk mailing. Rates for packages and rules for the packaging of liquid items such as honey also should be investigated with the various carriers.

SELLING PACKAGED HONEY WHOLESALE

While retailing one's own honey can be a rewarding, human experience, it does consume large amounts of time. If the beekeeper places a high value on time and little on the human interactions, direct marketing often is only modestly profitable. Furthermore, as the beekeeper's operations continue to enlarge, the beekeeper, at some point, will not be able to market all of the excess honey directly and at least part of the production will need to be wholesaled either packaged or in bulk. For this reason many intermediate-sized beekeepers manage a mixed retail and wholesale operation.

Finding someone else to sell the excess honey offers great time advantages over direct retail sales. While the per unit gross income will not be as great as for direct retailing, the placing of even modest values on personal time often makes the practice quite competitive. In many cases, it will be desirable for the beekeeper to educate the third party about the product and the best ways to sell it. To be effective at this, personal experience in direct retailing will be exceptionally helpful. How this educational process can be best accomplished will depend on the particular circumstances and the personalities involved. In general, the larger the retail operation, the less time the retailer can afford to interact with customers, so that what is appropriate for a merchant at a farmers' market usually will not be appropriate for the manager of a grocery store. At a farmers' market, the beekeeper might volunteer to act as a salesperson to demonstrate how to draw customers to the stand as well as how to handle them once they arrive. Later, after the initial training process, the beekeeper might provide only the promotional materials that would help attract customers to the retailer's booth. The beekeeper might also provide written information about the product to be distributed to customers. This information should be tailored to the situation, stressing the properties of the honey that would interest the particular clientele of the retailer. At super-markets, the beekeeper might supply attractive displays.

SELECTING LOCATIONS

Success in wholesaling packaged honey is dependent on both choice of location and presentation of the product including the advertising. Success in choosing locations and making them profitable will depend on working creatively with what is at hand.

Local Supermarkets

While local supermarkets are generally the first place the small beekeeper considers for selling honey, attempts to market through this medium often border on disaster. Usually only a small space is allotted to the product. Often the product's appearance is not on the same par with the products of the large packers. The label may not be as nice, and the honey usually isn't as clear. The result is slow sales with the honey crystallizing, making it even less appealing to the customer. Finally it may begin to ferment right on the shelf. This experience usually convinces the store manager to stick with the large producer. Worse yet, potential customers have seen a honey product at something less than its best and the entire industry suffers.

If the small beekeeper decides to attempt marketing through grocery stores, it must be realized that these establishments are strictly profit-oriented organizations with limited shelving space. They stock only what sells and the first obstacle the small beekeeper encounters is getting the manager to take a chance on the product. If the market is part of a large chain where stocking is done from regional warehouses under the watchful eye of the computer, the manager may not even have the option of accepting a new item for trial. The manager generally will be more willing to take a chance if there is something about the product that is different than what is already being offered for sale. The difference may be in the product itself such as with creamed or comb honey, or may result from unusual and creative packaging. Because the small beekeeper sells through a limited number of stores, something special in the way of in-store advertising can often be offered that would be impossible for the large packer. Above all else, if there is anywhere that the small beekeeper should strive to set the standard of quality, it is in supermarket sales.

Farmers' Markets and Health Food Stores

Farm markets and health food stores are good locations for the small beekeeper to consider when planning a wholesaling strategy. It would be wise to consider ways of drawing attention to the apicultural products in ways similar to those described earlier under retailing through these outlets. It also would be prudent to provide the vendor with information relative to selling honey. In many instances these vendors will be quite busy and not be able to spend a lot of time in dialogue with customers about honey and wax products. Where this is the case, it would be wise to provide fact sheets for distribution to interested customers.

Gift, Tourist, Antique and Similar Shops

Gift and tourist shops are ready-made situations for small beekeepers for they can offer specialized products and services that wouldn't be profitable for large packers. Here, perhaps more than anywhere else, the key to success is catering to the special needs of the shops involved. Selling conventional extracted honey in conventional jars with conventional labeling won't be successful. Why should customers looking for something special in these shops purchase something they could buy in any supermarket? The beekeeper's products should proclaim the very essence of what the shopkeeper is trying to purvey. The package, the labeling and the in-store advertising should all harmonize toward that end. For the tourist shop, the label and in-store advertising might announce "A Gift from the Flowers of Nantucket Bay"; for the antique shop selling basswood sections, the message might be "Old-fashioned Comb Honey the Way Your Great Grandparents Enjoyed It!"; and for the gift shop it might be "Something Special for Your Favorite Sweet Tooth." It will take creativity to fulfill the needs of the particular circumstances, but the key will always be found in the feeling that "here is something special."

There is a common notion among beekeepers that honey sells itself. This is true only when honey is being sold in locations where the customers are looking for it. The gift market is not such a place and many beekeepers have become discouraged with their lack of success in it. In the supermarket, price is the first or second concern of the customer. In the gift market, presentation (the combination of packaging and in-store advertising), is everything and the beekeeper should expend the time and money that make the special difference. Success will depend on it. Sometimes the difference can be quite subtle and seemingly inconsequential. Some beekeepers have been denied entry into the gift market until they changed their label. The painting of the eyes of a honey bear candle, for example, can increase its sales many times.

Other Special Groups—Hunters, Campers and Nature Centers

After the more obvious opportunities for selling packaged honey have been considered, there will generally be a large group of less obvious opportunities. These situations will differ from location to location and will take creativity to both discover and develop. Here again, the key to success will be discovering, getting the attention of and then catering to the needs of a particular clientele. The following are two examples where this has been done.

Often in and around state and national parks and forests there are small country stores that cater to campers and hunters. As part of their communing with nature, these groups are often inclined to experiment with natural foods. One Michigan beekeeper takes advantage of this situation by making up "camp packs" containing honey and several camp staples that go with honey such as pancake and biscuit mix. The package comes complete with paper plates for easy disposal. This beekeeper, who has no apparent large clientele, has creatively taken advantage of a local situation. Quite probably jars of honey on a shelf in the same stores wouldn't sell well. In addition to the honey sales, the beekeeper can profit from a small markup on the remainder of the items in the pack. One final but not so obvious benefit from this practice is that honey converts are probably created by this practice. Food, including honey, never tastes so good as in the wild.

Nature centers are another local situation that is generally underexploited by the small beekeeper. These establishments are often strapped for funds and frequently operate gift shops to help with the expenses. They generally are eager to sell products on commission. One nature center maintains a bee tree with a plexiglass opening for viewing the bees. They sell section comb honey and effectively advertise it with a small sign attached to the bee tree that reads "ask about our comb honey." This situation might also create honey converts. Toward this end, each section comes with a short description of the history of comb honey and the uses for which it is best suited.

Many such local situations undoubtedly exist. They simply need to be ferreted out and then to be creatively developed. These situations are almost tailor-made for the small beekeeper for they would be impractical for the large packer to develop.

Selling Through Groups

Some beekeepers market honey through the money raising campaigns of groups such as Scouts and school organizations. A great deal more honey can undoubtedly be marketed this way. There is no fundamental reason, for example, that honey bears or section comb honey couldn't be as closely associated with Cub Scouts as cookies are with Girl Scouts. For those contemplating this type of marketing strategy, some form of training program should be worked into the planning. These groups generally will know very little about either honey or salesmanship and to be successful, the salesperson must understand both.

SELLING IN BULK WHOLESALE

Most large beekeepers dispose of large portions of their honey in bulk through either some USDA program and/or through a honey packer. For the hobbyist beekeeper with few hives neither alternative may be appropriate. Neither repays the beekeeper with the highest returns for the honey, especially if it could be sold retail as a specialty product. There may be special packaging and minimum quantity requirements with which the small beekeeper will have difficulty complying. With the loan program, there will be paperwork that most find unpalatable. The long-term structure and existence of governmental price support programs is always uncertain and the small beekeeper would be well advised not to become too dependent on them.

The main advantage of these two alternatives is that in one transaction an entire season's honey can be sold. As the beekeeper's business grows, this feature will become increasingly attractive. Marketing is a very time-consuming operation. While wholesaling is often a more efficient use of time than retailing, even here an individual or even a husband and wife team can service and maintain only so many accounts, especially when the needs and delivery schedules of the accounts are diverse. For this reason many large honey production operations make extensive use of governmental programs and/or packers. They simply leave the marketing to someone else. While the gross return is lower, the equipment, packing and marketing expense is also much lower. This is a critical issue for the beekeeper who is about to expand into a full-fledged business venture. See chapter 17 for details of these tradeoffs.

Several outlets for bulk honey sales that are appropriate for, but often overlooked by the small beekeeper, are businesses that use honey such as bakeries, restaurants and candy makers. The beekeeper should be ready to expand the perceptions of the owners of these businesses relative to the versatility of honey. Honey isn't just sweet, it can be the subtle difference that accentuates many a culinary delight.

Topics That Apply to Both Retail and Wholesale Marketing

SELLING SOMETHING MADE WITH HONEY

The beekeeping industry has not taken full advantage of adding value to honey by manufacturing products made from it. While there is unquestionably a lot of potential in this area, the independent beekeeper faces large obstacles in preparing for the processed food market. The products that come easily to mind (Syrups, toppings, spreads and sauces) already have extensive, well-financed competition from large industries that can afford massive and glitzy advertising campaigns. The small beekeeper selling a similar product on a grocery store shelf will find the competition stiff indeed.

For the beekeeper who wishes to make a product from honey, the key to success will be finding the right specialty market and developing it creatively. In one successful innovative approach, a husband and wife team operate side by side "his and her booths" at local public events. She sells baked goods advertised as being made from honey and he sells honey. Both booths give out recipes for the baked goods and have advertising suggesting that the customers could make the tasty products themselves if they would purchase honey as one of the ingredients. Similar schemes using fancy honey candies also seem potentially rewarding. Public barbecues might offer similar opportunities for honey barbecue sauces.

Before getting involved in making a honey product, there are several important aspects that the beekeeper should consider. First, there is the matter of time. Along with the already time-consuming activities of beekeeping and marketing, the beekeeper would need to find time for the manufacturing of the new items. Next the item, itself, should be carefully selected. The beekeeper should not base the new venture on the often heard premise that "honey sells itself" and develop a product that relies on the word honey for sales. Honey doesn't sell itself to the large portion of the public that hasn't developed a taste for it. It would, in most circumstances, be better to develop a product that sells itself to this segment of society. For this reason, fancy baked goods and candies seem a better bet than honey syrups and spreads. The beekeeper entering this market will be faced with the necessity of adjusting to not only new packaging procedures and regulations, but also to additional food processing legislation as well. Finally, there will be the problems of acquiring new equipment.

ADVERTISING

The public is barraged from all sides by fancy, expensive and sophisticated advertising campaigns. In this environment, not to advertise would appear tantamount to commercial suicide. Historically the honey industry has done very little advertising, and honey sales have remained relatively constant. With the initiation of the National Honey Board in 1986, hopefully, this will change allowing the entire industry to profit from the increased public awareness of honey brought about by this group's national advertising campaign. In addition to benefiting from this increased awareness, the small beekeeper should take full advantage of the professional advertising materials produced and made available by this organization.

Immersed in an ocean of fancy, expensive and sophisticated advertising, what type of advertising can the small beekeeper afford that will increase sales? Actually much of the previous discussion has been about advertising. Having the right product in the right place at the right time appropriately packaged is the first form of advertising on which the small beekeeper should concentrate. As the small honey producer ventures beyond this, the first problem encountered is being able to assess the benefits of the advertising. Unless the cost of a particular advertising venture is very small, careful consideration should be given to developing an assessment plan as part of the overall advertising design.

Advertising You Take With You

Some beekeepers wear clothing that advertises their honey. Some wear hats that state "I sell honey." These advertisements can be made inexpensively with permanent felt pens and textile paints or with a fancy-stitch sewing machine. It's surprising how often this form of advertising generates conversations about honey that lead to a sale. When this happens, assessing the effectiveness of the advertising is essentially automatic. For those who are reticent about taking their advertising wherever they go, remember that beekeeping is an occupation of which the beekeeper can be justifiably very proud. It hurts no one while at the same time providing a multitude of healthful and enjoyable benefits for society.

In a similar vein, the use of bumper stickers and signs for truck and car doors should be considered. Magnetic signs are available that can easily be removed in situations where this is desirable.

Store Advertising

The goal of store advertising is to draw the customers' attention to a particular product in such a way that they want to purchase it. Usually store advertising takes the form of attractive eye-catching signs and displays, neither of which need to be expensive. The signs can be done on poster board with felt pens or poster paints. The display needn't be much more than an attractive stack of honey packages or perhaps a shopping cart full of the beekeeper's product.

The beekeeper should always consider the potential for "tie-in" promotions. The tie-in promotions in figs. 4 & 5 suggest that the biscuits the shopper is about to purchase would be especially good with honey. The display in fig. 5 has the advantage of being mobile, enabling it to be moved to different parts of the store to get the attention of a wider cross-section of the store's clientele. With a small collection of signs, this display could become an effective tie-in promotion in several locations. One week, in front of the peaches, the sign might read "peaches and cream go great with honey." At another time, in front of blueberries, the sign might read "blueberries with milk and honey make a healthful delicious snack." Potential tie-ins exist with biscuits, teas, ice cream, peanut butter, barbecuing supplies, salad fixings, etc.

Where the store with which the beekeeper is dealing is located in a busy setting, store-front advertising might be useful not only in pulling passers-by into the store, but also to focus regular patrons' attention on the beekeeper's product as they enter.

All store advertising has one thing in common, it is aimed at a particular subset of customers, each with their own particular needs. The blueberry tie-in is designed for the parent who is looking for an alternative to candy for snacks. The tie-in with biscuits is directed toward the cook who derives great pleasure from serving something special. As with salesmanship, the person designing advertising must understand the customer who shops in a particular location. Where the beekeeper can't personally act as the salesperson and carefully probe for the needs of a particular customer, store advertising is the next best thing.

Some beekeepers give out samples in grocery stores. This can be of the honey, itself, or samples of another product in or on which honey has been used. While this practice is quite time consuming, those who do it report good results. In addition to increasing sales, it can provide valuable insights for

FIGURE 4. Most store owners would welcome this very simple and inexpensive tie-in promotional display. It will undoubtedly increase the sales of both the beekeeper's honey and the ready-to-bake biscuits.

future advertising and at the same time provide an enjoyable human experience closely related to that provided by retailing.

When approaching a store manager with a promotional idea the beekeeper should keep in mind that the manager's feelings are law. Some will be more cooperative than others. Those associated with big chains may be very

FIGURE 5. This simple and inexpensive display is easy to maintain. With a small collection of signs, the display could easily be moved to numerous tie-in products and thus catch the eye of a greater proportion of the store's patronage.

limited in what they may allow the beekeeper to try. Above all, it should be remembered that good store managers allocate space in proportion to the profit they believe will be realized from that allocation. If precious aisle-end space is used for a display, it has to be a good display, and it must sell the product!

One means of assessing the effectiveness of this type of advertising is to compare sales before and after the advertising. To be certain of the results of this type of assessment, it should be done more than once. This could be accomplished by using several stores, or alternatively, it could be done several times in the same store with appropriate intervening time periods that serve as the "without advertising parts" of the study. If the results of a high percentage of the trials indicate better sales, it can be presumed that the advertising is having a positive effect. If, however, a large majority of the trials don't provide positive results, the effect of the advertising is questionable. Here the apparent increases in sales are as likely to have come from chance as from the advertising. There simply is no objective way of telling.

Coupons

The advertising world relies heavily on coupons, so much so that many people won't make certain purchases without a coupon. Before implementing coupon advertising, careful attention should be given to the possibility of this development. The beekeeper gains little if regular customers begin to demand lower pricing as a result of the technique. A coupon campaign should be designed to entice patrons who would not normally purchase the product.

Much of the coupon world is outside the reach of the small beekeeper. Because large food chains use coupon sorting firms to process their coupons, these stores may not accept the coupons of the small beekeeper. Sometimes special arrangements can be made, especially with smaller stores, but it means special handling for the store.

Despite their disadvantages for the small honey producer, in conjunction with other forms of advertising, coupons can provide an effective method of evaluating the success of the associated advertising. As an example, a bee-keeper or a shop owner who stocks the beekeeper's products, may wish to provide pertinent literature to those who ask questions about the product. These questions might come about as the result of the signs discussed earlier that are designed to draw potential customers to the booth. If the literature contained a coupon, and the booth attendant were selective about the dispersal of the literature, giving it only to potential new customers, the returned coupons would give an indication of the cost effectiveness of the advertising literature.

Coupons can also be used in situations such as farmers' markets where reciprocal coupon accepting agreements have been struck with other vendors who sell products that go well with honey, fresh berries, for example.

The Label

The label is an important part of advertising. It is the last advertising that sways the choice of the customer. The small honey producer has two choices: use ready-made labels or design new ones. The printing of a new label can be a relatively expensive process, especially if it is designed by a professional, is an unusual shape, is a multi-colored product or is produced in small runs. For this reason, the small beekeeper should proceed cautiously when deciding to produce a new one since it may not be cost effective. There are many attractive ready-made labels that will be quite serviceable under many circumstances.

Whether a ready-made label or a new creation, the appropriateness of the label depends on the circumstances of the business location. A self-stick label with the weight penned in by hand is generally quite appropriate for a beekeeper's booth at a farmers' market. This label addresses the needs of the farmers' market shopper by proclaiming "I personally produced this honey, it's fresh and free of excessive processing." To the shopper of a large modern grocery store, its message is likely to be "I don't look as nice as my competition." The label meant to appear on the supermarket shelf should emanate the message "this is really good" or "this is a fun food." While the labels of many non-staples and newcomers to the supermarket shelf advertise themselves in this manner, few honey labels, unfortunately, attempt this message.

If there is anywhere that the ready-made label will not be appropriate, it is in the gift market. Often the beekeeper can't get his product into a gift shop with this type of labeling. To the shopper searching for a gift, such a label says "I'm not special enough." While the label for a gift package often need not be excessively fancy, it does need to emanate the feeling that "this is special enough to be a gift." The simple but tasteful label in Fig. 6 not only allowed its owner entrance into the gift market after being denied that right, but caused one gift shop owner to seek out the beekeeper to request the privilege of handling the product.

By using a ready made label, as for example, those available from the major beekeeping supply houses, meeting the requirements of labeling regulations is largely automatically accomplished. The only exception that would generally be encountered, would be with "blank" labels to which the beekeeper's name and address and the net weight of the package needs to be added. While not difficult, labeling regulations are quite lengthy and outside of the scope of this chapter. They may be obtained in the United States, depending on the state, from either the state Department of Agriculture or the state Department of Health and in Canada from Agriculture Canada. Printing companies that specialize in producing labels can also supply this information.

Pure

Honey

net wt.

Steelville, Mo.
65565
314 Quinnmoor
Ballwin, Mo.
63011
phone
394-5395

Gibbons Bee Farm

Figure 6. This simple but very tasteful label allowed its originator to enter the gift shop trade. It is printed in dark brown on a light brown stock and could be printed in small batches on a color photocopy machine. It seems to proclaim "Honey is fun; this is a nice gift." *Label submitted by Sharon Gibbons.*

If only a small number of special labels are to be produced, the small honey producer should consider producing them on a photocopy machine. Copying machines are now available that produce colored copies and quite tasteful results can be obtained by working with a colored stock. The label in Figure 6 could have been produced in this manner.

The labels of some packers now bear the Uniform Product Code. Storekeepers who utilize this pattern of bars and numbers that now appears on most food packaging may be reticent about taking packaging that does not display it. The code eliminates individual package pricing, speeds up the check-out procedure, and allows complete sales information on all products bearing the code. A set of exclusive product codes can be obtained by applying to the Uniform Code Council. The cost of obtaining these codes is dependent on the sales volume of the products involved. The cost for obtaining a set of codes will, however, generally preclude participation in the program by the small beekeeper. For those interested in obtaining more information about the code, the present address of the Uniform Code Council is included in the references.

Advertising in the Yellow Pages

Only a very small percentage of beekeepers advertise in the yellow pages. Most phone books list nothing under honey. Even large cities usually have only one or two listings. Yet this is the place where those interested in purchasing larger quantities of honey will first turn for information about suppliers.

The cost of advertising in the yellow pages varies considerably, depending on the circumstances. For a local, simple listing (name and phone number in regular type) under a honey heading, yellow pages advertising often requires only the expense of installing and maintaining a business phone. Many who advertise in the yellow pages under honey, however, indicate that if a business phone isn't practical for other reasons, maintaining one solely for yellow pages advertising would at best be a marginally profitable venture. Yellow pages advertising may be more profitable in some localities than in others. The method provides a simple way of assessing its effectiveness since phone inquiries about honey would have resulted primarily from the yellow pages advertising.

Radio, TV and Newspapers

Generally radio, television and newspapers are out of the financial reach of the small beekeeper as advertising media. Despite this fact, there are beekeepers who use these media in a powerful type of advertising at no cost. If some aspect of the beekeeping operation can be turned into a newsworthy event, these media often will send a reporter to interview the beekeeper for a local-interest story. Done correctly, harvesting honey can be just such a story. A word of caution: If the situation is a blatant attempt at advertising, it will neither be given air time nor be printed.

Beekeepers with a gift for writing sometimes write their own press releases for local papers. These releases usually report events such as honey contests, the results of baking contests where honey was used, or about presentations to public schools. Done well, these writings keep the beekeeper in the public eye and often translate into extra sales.

Sometimes radio stations as part of a fund-raising campaign will auction donated goods. The donor is usually given a short plug for the donation.

Unfortunately, the effectiveness of advertising in this media is difficult for the small beekeeper to evaluate. For this reason, it would be prudent to participate in such advertising only when the costs are minimal as in the situations just described.

Less Formal Types of Advertising

Some beekeepers go to great lengths to keep honey and honey products in front of their acquaintances. Their guests are served honey products or they pass out snacks made with honey at work. Some hand out honey treats at Halloween. Others generate interest in honey by sponsoring honey tastings, playing up the similarity with wine tasting by pointing out that "fine honeys are like wines—they vary from location to location and from season to season and that the current year is a vintage year."

Throughout this chapter the small beekeeper has been encouraged to be creative in both discovering and developing underexploited markets. While many suggestions have been made along these lines, small beekeepers should continue the process and develop lists of ideas that are tailored to their own particular situation. One of the references (Ayers, 1985), describes a powerful technique for continuing this process. Given a chance, the simple technique described there could pay the small beekeeper huge dividends in creatively finding and unlocking these markets. Two other references that deal with sales from slightly different perspectives are also provided. While they do not deal with honey per se, they offer much to small beekeepers if they are willing to creatively adapt the ideas found therein to the sale of honey.

REFERENCES

Ayers, G.S. (1985). How to Have a Hundred Dollar Idea. *American Bee Journal 125*:95.

Brisco, N.A., G. Griffith and O.P. Robinson. (1947). Store Salesmanship. 435pp. Prentice-Hall, Inc. New York.

Buzzotta, V.R., R.E. Lefton and M. Sherberg. (1972). Effective Selling Through Psychology. 387 pp. John Wiley & Sons, Inc. New York.

Address of Uniform Code Council
  Uniform Code Council, Inc.
  8163 Old Yankee Road
  Suite J
  Dayton, Ohio 45448
  (513) 435-3870

CHAPTER 19

MARKETING THE CROP OF THE COMMERCIAL BEEKEEPER

by HARRY RODENBERG*

Honey production was once an integral part of U.S. agriculture, and honey played an important role as a sweetener in the diet of rural America. Commercial beekeeping has evolved from a sideline, during the late 1800's, to a sophisticated and mobile industry, which nevertheless has only a relatively small number of operations. The United States Department of Agriculture reports that there are approximately 1,600 to 2,000 commercial beekeeping operations in the United States. According to industry sources, only 1 percent of all beekeepers in 1983 were classified as commercial. These commercial beekeepers maintained about 50 percent of the 3.3 million colonies (*based on beekeepers with five or more colonies*) and produced 60 percent of the honey.

Following the depression years of the 1930's and World War II, agriculture became mechanized, the application of chemicals for weed and insect control became commonplace, and the use of commercial fertilizers became the norm. Emphasis on breeding increased the production of livestock and poultry, while varieties of hybrid seeds were developed that increased production per acre. All of these advances helped to make the American farmer the most efficient in the world.

Similar advances have occurred within the beekeeping industry, but less rapidly than in many other agricultural areas. A considerable amount of physical effort, time and energy must still be devoted to management of individual bee colonies at critical periods during the season, but labor-saving devices have eliminated some of the physical effort of moving bees and handling the honey supers in the field and in the warehouse. Strides have been made toward improving the efficiency of the honey extracting process.

A pattern of specialization of the various elements associated with beekeeping has developed over the years as the beekeeper has become more sophisticated. These various elements associated with beekeeping can be divided into the following classifications (the lines between these elements are not rigid and may overlap):

1. honey production,
2. queens and package bees,
3. honey packing and marketing

*Harry Rodenberg, commercial beekeeper and director of the Sioux Honey Association for 33 years, Chairman of the National Honey Board from 1986-1990. Past Chairman of the Honey Industry Council, Vice President of the American Beekeeping Federation and President of the Montana Beekeepers Association.

4. bee supply manufacturers and distributors,
5. beeswax and beeswax products,
6. other bee products,
7. pollination services.

This portion of the chapter "Marketing the Crop" is devoted to honey packing and marketing, and to those options available to the commercial honey producer for marketing surplus honey. These options might include the following:

1. process, package and market the crop,
2. sell or consign to an independent honey packer or producer/packer,
3. join a cooperative honey marketing organization and deliver production under the guidelines of that organization,
4. sell directly to a honey exporter,
5. sell to an industrial or commercial honey user.
6. forfeiture of honey to the U.S. government.

Figure 1 shows a flow chart illustrating possible movements of honey from the producer through the various channels of commerce.

PROCESS, PACKAGE AND MARKET THE CROP

The management of bee colonies on a commercial scale leaves very little time for the majority of honey producers to devote to processing, packaging

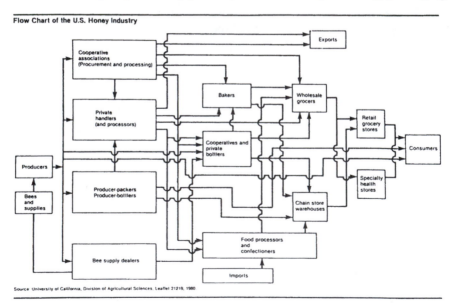

Flow Chart of the U.S. Honey Industry

Source: University of California, Division of Agricultural Sciences, Leaflet 21219, 1980.

FIGURE 1. *(Courtesy USDA)*

and marketing their crop to wholesale grocers or retailers. The investment required for facilities, processing equipment, container inventory, and the expertise required to successfully market the product discourages most honey producers from entering this field, except perhaps on a limited local basis. There are notable exceptions of honey producing operations that have successfully handled the production, packing and marketing of honey. In a few cases, honey producing operations have evolved into successful honey packing and marketing businesses and have discontinued production.

For the commercial honey producer who is determined to become involved in these operations, entry on a limited scale would be a reasonable approach. Before considering this avenue, the producer should decide these questions.

1. What is the potential of the marketplace that he could serve?
2. Can a consistent product (quality and quantity) be available?
3. What are the inbound freight costs on containers and supplies, and what will be the outbound delivery costs of the finished product?
4. What will be the cost of containers and supplies, and will the supply be readily available?
5. What investment in facilities and equipment will be necessary?
6. Will additional personnel be necessary? If so, at what cost?
7. What other costs, such as product insurance coverage, licenses, commissions, storage costs, interest, etc., will be incurred?
8. Decisions will have to be made whether contacts will be made to sell the honey through personal salesmanship, to wholesale houses, through food brokers supplying retail stores, through roadside stands, a mail order business, on consignment, a combination of the preceding, or in some other manner.
9. And the bottom line, will the profit be better than marketing bulk honey to a wholesale market?

FIGURE 2. Tractor-Trailer bulk tank load of honey *(Photo courtesy Sioux Honey Association)*

SELL OR CONSIGN TO AN INDEPENDENT
HONEY PACKER OR PRODUCER/PACKER

The 55-gallon steel drum with an open head and an approved Food and Drug Administration interior food liner has become the commonly accepted container for bulk raw honey being sold by the commercial honey producer. There are a few producers who deliver in tractor-trailer bulk tanks (see Figure 2).

The honey producer must initiate a dialogue with independent honey packers or producer/packers in order to sell honey (Figure 3). To locate a packer, the American Beekeeping Federation's Newletter has a honey packers and dealers directory. Neighboring beekeepers may also be a source of information for locating packers and producer/packers. In addition, beekeeping trade journals and newspapers usually carry want ads for honey. Also, an excellent opportunity exists to meet packers and producer/packers at local, state and national beekeeping meetings.

When a producer makes his contact, he should inform the packer of the amount and the floral source of the honey that he has for sale and the condition of the drums. If the packer is interested, he may ask the producer to send representative samples of his honey. If the packer needs the type of honey represented by the sample(s), a negotiation process should proceed regarding price, details of payment, a determination of the disposal of drums (whether they will be returned to the producer or consideration will be allowed for them), and an agreement on who will be responsible for transportation charges on the inbound honey as well as return of the drums. The packer will be interested in the integrity of the producer, and the producer should be interested in the integrity of the packer. A considerable amount of time and effort from both parties may be necessary to finalize a sale. Once a satisfactory transaction has been completed and a mutual respect and rapport has been

Figure 3. Modern independent honey packing facilities such as this one provide quality honey to both the retail and bakery trade. *(Photo courtesy Dutch Gold Honey, Inc.)*

developed between the producer and packer, future selling transactions may be simplified. In some cases, transactions between the same producer and packer may continue for years for a particular quality and floral source of honey.

Both the producer's and the packer's operations, to be enduring, must be founded on the premise of a profit motive. The producer's production may vary due to the uncertainty of the honey flow as affected by weather conditions and other variables over which he has little control. The producer must manage his operation so that he can live with these uncertainties. In turn, the packer is faced with a competitive marketplace and costs that may directly affect his limits of negotiating terms of sale. Each needs the other. Although their viewpoints differ, common sense and reason should prevail with each party.

COOPERATIVE MARKETING ORGANIZATIONS

The number of cooperative honey marketing organizations in the United States is limited. Most of them are small and some deal strictly in bulk honey, with one exception.

The Sioux Honey Association

This association was established in 1922 in Sioux City, Iowa, and it is the only cooperative marketing organization operating on a national basis, having plants in Anaheim, Calif., Sioux City, Ia., and Waycross, Ga. (Figure 4). Presently, the association has a membership of approximately 600 producing members located in 27 states.[1] Producing members pool their honey, resulting

[1] 1988 Sioux Honey Association Annual Report, Page 8.

FIGURE 4. The Sioux Honey Association packs and markets about 25 percent of the total yearly U.S. honey production. *(Photo courtesy Sioux Honey Association)*

in an annual average volume in excess of 45 million pounds or about one-fourth the production of the United States. The major brands marketed by the Association are "Sue Bee," "Aunt Sue," "Bradshaws" and "Clover Maid" labels.

Each year, the producing members receive earnings which are calculated after operating costs have been deducted from the selling price of the honey. A portion of each member's earnings is retained to supply the capital needs of the cooperative. Each member, upon their acceptance for membership by the board of directors, signs a marketing agreement with the cooperative, which provides the rules that each party is expected to honor.

The members of the board of directors of the cooperative are the elected representatives of the producer members, and since the board is comprised of producers, the marketing approach of the cooperative reflects a producer's perspective. A manager is hired by the board of directors and is responsible for implementing the board's policies. The manager is assisted by a staff and a sales force.

The marketing of a producing member's crop is handled by the cooperative, thereby relieving the producer of considerable time and effort. Additionally, each producing member owns a portion of the marketing cooperative in proportion to the amount of retained earnings. The cooperative marketing association does not fit the needs of everyone, but it does serve the needs of a segment of commercial producers.

The Mid-U.S. Honey Producers Marketing Association

In 1968, a honey producer cooperative was organized for producers from Iowa, Kansas, Minnesota, Montana, Nebraska, North Dakota, South Dakota and Wyoming. This producing area might be considered the "bread basket" of light colored, mild flavored, table grade honey. The organizers of the Mid-U.S. Honey Producers Marketing Association felt that, if the group worked together to provide production figures, input on production costs, and information on price offers, their members would be better informed and would have the leverage to make an impact on the U.S. honey market. The membership presently consists of 25 to 30 producers with a combined production of 5 to 7 million pounds of honey.[2] At their annual meetings, held in late August in Pierre, S.D., association members exchange information of crop estimates, cost of production, and evaluate the marketplace for their honey crop. The activities and recommendations are transmitted through a regular newsletter to the membership. At various marketing periods, the Mid-U.S. Honey Producers have had an impact upon the marketplace. During the late 1970's and early 1980's, however, as a greater volume of imported honey entered the

[2]Gary Reynolds, President, Mid-U.S. Honey Producers Marketing Assn., 1988.

country at a price far below the support price for honey, the leverage of this large block of quality honey diminished. During recent years, with the advent of the "buy back" implementation in the honey program, this association could again come to the forefront.

EXPORTING THE CROP

The export market provides another opportunity for the commercial honey producer to sell his crop. Because of its complexities, export marketing is probably the least known or understood by the average producer. In fact, some producers have had devastating experiences with this market, such as deterioration of quality specifications during transit, or inability to finalize a transaction after having invested considerable time and dollars toward securing and preparing the product for export. Except for 1973, the United States has been a net importer of honey since 1967. The dramatic rise in exports in 1973 resulted from a large domestic crop coupled with a decrease in world production, which resulted in an increase in world honey prices. Although the United States exports to many countries, West Germany, Saudi Arabia, the Netherlands, Japan, Kuwait, France, Hong Kong, and Belgium-Luxembourg have been the principal markets in recent years.

In the 1975 edition of *The Hive and the Honey Bee*, the chapter on marketing mentioned that exports peaked in 1953 at nearly 33 million pounds. At that time, an export subsidy program was being sponsored by the U.S. government.

Exports of honey are affected by government trade policies of the United States as well as by our trading partners. The fluctuation of the dollar versus foreign currencies is also a critical factor. Furthermore, technical criteria for honey by some countries has made it difficult for U.S. honey to penetrate their

Table 1

United States Honey Exports 1978 - 1989	
Year	**Millions of Pounds**
1978	8.0
1979	8.8
1980	8.5
1981	9.2
1982	8.5
1983	7.5
1984	7.5
1985	6.5
1986	9.2
1987	12.4
1988	14.1
1989	9.9

markets. Table 1, taken from USDA statistics, lists honey exports for the period 1978-1989.

The United States produces large marketable supplies of numerous specialty honeys of a floral source and quality which are in demand by the export market.

1. Citrus blossom honey is produced in Florida and California in large quantities of a comparable quality to any other production worldwide. Asian and Middle Eastern importers pay premium prices for such specialty honeys.
2. Light clover honey produced in the upper midwestern states, particularly in the Dakotas, has a reputation for excellent quality. Japanese buyers have sought supplies of this type of honey in the past.
3. Buckwheat and Tupelo honeys command a premium price in specific foreign markets.

Specialty markets for such U.S. honey tend to be niches with buyers interested in quality, brand names or specific floral sources. Long-term commitment to exporting is essential for successful penetration of foreign markets.[3] Direct sales require a significant commitment of internal resources, such as sales representatives or some other method of actively promoting the product.[4] The Foreign Agricultural Service's Agriculture Information and Marketing Services program is a cost effective way to make trade contracts and to look for trade leads.[5] It is also important to establish a relationship with a freight forwarder who can advise on freight rates, documentation, export packing and other technical aspects.[6] Alternatives for hiring intermediaries include management companies, export trading companies or piggyback arrangements with another company that has its own export department.[7] Export intermediaries are not generally interested in sporadic exporters or those who cannot provide continuity of supply, quality and dependability.[8]

Table 2 lists U.S. honey exports in metric tons for a five-year period 1983-1987 for various countries and regions of the world.

SELL TO AN INDUSTRIAL OR
COMMERCIAL HONEY USER

Bakeries, candy and confectionery manufacturers, and cereal manufacturers which incorporate honey in their formula would be considered industrial users of honey. Undoubtedly, there are industrial uses other than these mentioned.

The commercial market includes the foodservice sector which consists of

---

[3,4,5,6,7,8]Tracey L. Kennedy, Profiles of Potential Markets for U.S. Honey Exports, Preface.

Table 2

U.S. Agricultural Exports PERIOD: Jan. 1983-Dec.1987	United States Department Of Agriculture Foreign Agricultural Service				
Commodity Exported and Destination Areas/Countries	— CUMULATIVE TO DATE QUANTITY —				
Honey In Metric Tons*	1983	1984	1985	1986	1987
North America					
Canada	242	362	198	176	213
Mexico	1	6	79	0	0
Subtotal	243	368	277	176	213
Caribbean					
Bermuda	3	8	0	0	0
Bahamas	0	0	13	18	5
Jamacia	0	0	0	0	0
Cayman Islands	0	0	0	1	0
Dominican Republic	3	0	0	0	0
Leeward/Windward Is.	9	5	1	0	0
Barbados	27	16	13	9	9
Netherlands Antilles	14	25	6	4	4
French West Indies	4	6	14	25	20
Subtotal	60	60	47	57	38
Central America					
Panama	0	0	1	0	4
Subtotal	0	0	1	0	4
European Community					
United Kingdom	72	224	110	89	97
Netherlands	331	464	194	660	170
Belgium/Luxembourg	22	8	15	74	137
France	48	23	42	51	70
Germany, Fed. Rep. of	474	285	364	1442	2187
Spain	0	0	7	0	0
Subtotal	987	1043	777	2349	2680
Other Western Europe					
Iceland	0	2	0	0	0
Sweden	2	4	4	3	8
Norway	0	18	69	5	0
Switzerland	24	0	20	5	0
Cyprus	35	16	0	0	0
Subtotal	61	40	93	13	8
Middle East					
Lebanon	0	0	16	0	14
Oram	0	20	0	0	0
Israel	0	2	0	0	147
Jordan	3	1	0	1	0

* Multiply by 1.1 to obtain U.S. short tons.

Table 2 *Continued*

U.S. Agricultural Exports PERIOD: Jan. 1983-Dec.1987	United States Department Of Agriculture Foreign Agricultural Service			
Commodity Exported and **Destination Areas/Countries**	— CUMULATIVE TO DATE QUANTITY —			
Honey In Metric Tons*	**1983**	**1984**	**1985**	**1986**	**1987**
Kuwait	167	277	274	209	237
Saudi Arabia	1018	793	541	553	945
Qatar	2	16	7	3	27
United Arab Emirates	138	191	74	132	217
Yemen (Sana)	43	30	27	26	0
Yemen (Aden)	6	20	2	0	0
Oman	11	28	23	0	0
Bahrain	17	2	22	16	2
Subtotal	1405	1380	986	940	1589
North Africa					
Morocco	0	0	0	0	0
Subtotal	0	0	0	0	0
Other Africa					
Sudan	1	0	0	0	0
Nigeria	11	0	0	0	11
Angola	0	0	0	1	0
Liberia	6	2	7	1	4
Subtotal	18	2	7	2	15
South Asia					
Pakistan	0	0	1	0	0
Subtotal	0	0	1	0	0
Other Asia					
Thailand	4	3	1	2	49
Malaysia	29	40	15	14	2
Singapore	44	13	17	61	140
Indonesia	15	28	20	1	8
Brunei	0	0	0	2	0
Philippines	73	16	0	55	24
China (Mainland)	0	0	15	0	0
Korea, Republic of	10	7	6	24	71
Hong Kong	53	47	147	150	386
Japan	358	334	516	284	339
Subtotal	607	491	754	616	1049
Australia & Oceania					
Australia	0	8	3	29	15
Trust Terr Pac Is	0	16	1	0	0
Subtotal	0	24	4	29	15
Total World	3384	3411	2949	4181	5610

* Multiply by 1.1 to obtain U.S. short tons.

restaurant chains, independent restaurants, hotels, and institutional operations such as hospitals, retirement and nursing homes, and school foodservice institutions.

Foodservice operators usually purchase honey in sizes other than bulk containers for ease of handling and storing. For table servings they may use a drip-cut dispenser, a plastic squeeze container, or some type of portion pack. For refill or ingredient usage by the foodservice operator, the containers may range from a 5-pound jar to a 5-gallon plastic bucket, depending upon the volume of usage and the size of their storage cabinets.

The brokers that service the foodservice industry are not usually the same organizations that service the retail stores. They operate separately. The local area commercial producer or producer/packer may enjoy an advantage in penetrating the local market on a limited basis, whether it be restaurant or institution. The local or home-grown product, providing that it is a quality product, serves as a local attractive food feature. For those franchised or chain foodservice operators who have their own food distribution supply organization, the locally produced product's advantage diminishes.

The local producer or producer/packer may enjoy a similar preference with the local smaller bakeries or industrial user, but the larger bakeries or industrial user is likely to employ a bidding process to seek honey with a particular color, moisture content and flavor, and for a specified volume to be delivered over a defined period of time. The large industrial user may also contract for delivery in large tote tanks for his convenience.

FORFEITURE OF HONEY TO THE U.S. GOVERNMENT

Forfeiture is another, although rather undesirable, alternative for disposing of the crop, but it does not require a great deal of salesmanship or skill to finalize.

Nearly all commercial honey producers utilize the Honey Loan Program at the present time. The "buy back" feature, incorporated into the Honey Price Support Program in the 1985 Farm Bill, was designed to keep the U.S. Government from foreclosing on large volumes of honey held under the loan program and return that domestic honey production to the marketplace.

The commercial honey producer should exert every effort to sell his honey rather than forfeit his crop to cover the honey loan. To utilize the forfeiture feature of the program will hasten the inevitable termination of any type of government program for honey.

Later in this chapter, further details of the U.S. Honey Stabilization Program will be discussed.

PROFILE OF A HONEY USER[9]

In order to successfully market honey, a producer must understand his consumer and the consumer's use of his product. The following outlines the profile of a honey user based on research conducted in 1987:

Age of Householder. Homes with householders 40 years old or older were 13 percent above the U.S. average in honey use. Householders 34 years old or younger were 18 percent below average.

Annual Income. Households with incomes of $10,000 and over were between 11 and 31 percent above the average honey usage. (Middle income, in excess of $20,000 were greater users; however, there was a strong honey use reported in the $10,000 to $15,000 bracket probably due to the honey giveaway programs.)

Children. When children 6 to 17 years of age were in homes, the households were 10 to 13 percent above the U.S. average in honey purchases incidence. When children under 6 years of age were present, usage declines.

Household Size. Homes with two and four members purchased more honey than all other household size groups. Single member units were 25 percent below average.

Education. As education increases, so does honey consumption.

Market Size. Mid-size markets with populations of 50,000 to 1 million yield higher penetration levels than either the very large or very small areas.

Regions. The Mountains, the South Atlantic and the Northeast regions reflected the highest honey usage. The East North Central and East South Central areas have the lowest honey usage.

Other Survey Findings[10]

Approximately one-third of the U.S. households made one or more purchases of honey in the past year. The typical honey user purchased honey about every 4 to 5 months. Thirty-six percent of all honey buyers claimed only one honey purchase in the past year.

The winter months were identified as the peak usage period among 45.4 percent of the households studied. Nearly 49 percent of the homes, however, indicated a stable, year-round use of honey.

Honey is primarily used at breakfast, with 82 percent of the users claiming use at the morning meal. Honey at dinner was mentioned by 42.4 percent of the users.

Honey was frequently used when illness occurred (in 64 percent of the users' homes). Honey was also commonly served at non-holiday guest meals (42.4 percent of the households) and used as a gift (24.2 percent of the households).

[9,10]Mary Humann, *American Bee Journal*, Oct. 1987, Page 714.

Honey buyers said that they prefer glass containers in 55 percent of the cases and plastic containers in 42.4 percent. Their most recent purchases, however, were glass jars in 64 percent of the transactions and plastic containers in 33 percent.

Non-users of honey perceive honey to be used primarily as a spread/topping or as a sweetener. Ingredient usage accounted for 44 percent of non-user responses.

Seventy-two percent of the households use honey as an ingredient. Honey on meat/poultry products and in baking dominated the ingredient usage.

Honey purchasers use honey as a spread or topping in 90 percent of the households. Honey users said that they used honey most frequently to top biscuits, bread or waffles. Beverages are sweetened by honey in 41 percent of the user households, with hot tea being the most common beverage.

When asked why they did not buy honey more often, 64 percent of the honey users responded that "they didn't use that much" or "it goes a long way."

Honey users said that they like honey because of its taste and flavor and because it is natural. About one-third of non-users said that they did not like the taste of honey.

Over 90 percent of honey users said honey is convenient and easy to use, and honey users consider honey to be a good value.

A honey retail distribution study showed that honey was on display at virtually all large-volume, major-market supermarkets.

In 34.8 percent of all stores, honey was stocked next to syrups. Honey was also frequently stocked near jellies or on a shelf with or next to peanut butter.

Sticky honey containers were found in 21.8 percent of all stores, while packages showing crystallization of honey were on display in 28.4 percent of the stores.

MARKET CHANNELS TO THE HONEY CUSTOMER
The Supermarkets and Chain Stores

The majority of the table-grade production of the commercial producer finds its way to the consumer via the chain stores and supermarkets after first passing through either the producer/packer, the independent honey packer or a cooperative marketing organization for processing, blending, packaging and marketing.

Honey can be found on display at virtually all large-volume, major-market supermarkets, but that does not mean that honey receives a priority location or adequate display space. Honey is not a rapid turnover item, and therefore, it does not receive choice placement. Thousands of products are competing for shelf space in today's competitive marketplace.

Since 1983, it is estimated by honey industry leaders that honey sales through the retail market have decreased in excess of 30 percent, although this trend has begun to reverse itself in recent years with heavier promotion efforts. Many honey industry leaders believe that this reduction in retail sales resulted from the millions of pounds of honey forfeited to the government which were distributed to schools, institutions and needy families. This subject will be covered in more detail later in the chapter. Another possible contributing factor to the reduction of retail sales is the trend of many couples and families to eat an increased number of meals away from home. Today, consumers eat in excess of 20 percent of their meals away from home, spending nearly 40 percent of their food dollars for these meals.[11] The reduced sale of honey through retail sales outlets parallels a reduction of sales volume of food items.

Health Food Stores

Chemicals used for fertilizer, for preservation, for insect and weed control have been found as residues in our food production. A backlash from this situation has resulted in a segment of our society that supports "natural" foods, which are grown without the benefit of these chemicals. Honey is accepted as a natural food. Requirements of foods which are accepted by health food stores vary, but frequently "Raw Honey" is found on the label. Raw Honey could be defined in the most rigid terms as comb honey, but a more liberal definition would be honey that has had a minimum of heat applied in the extracting and processing procedure, and no filtering process to remove the pollen and minute wax particles. The wax particles are removed either through a settling method or minimal straining procedure.

Health food stores are found in larger shopping centers as individual stores or as separate sections of supermarkets. Depending on the volume of business, health food stores may purchase their supply directly from a packer or a producer/packer.

Gourmet and Specialty Shops

Honeys of distinctive floral sources such as citrus, tupelo, sourwood, fireweed, gallberry, raspberry, basswood, and undoubtedly many others, may be found in specialty and gourmet shops. Frequently, they are packaged in a unique or attractive container. The value of the packaging, in fact, may exceed the cost of the product. They are designed to attract the selective purchaser or appeal to that person looking for a special gift to bring home. These shops are found in high-traffic areas such as airport gift shops, hotel and motel lobbies, shopping centers, and along heavily traveled tourist routes. The suppliers of these specialty honey items may range from packers and producer/packers to multinational food processor/distributors.

---

[11]The Hale Group, Danvers, Mass., Survey, 1987.

FIGURE 5. Honey gift packs are essential for a successful mail-order business.

A number of producer/packers and packers have developed a successful mail-order business for honey gift packs. They are comprised of different flavors of honey, creme or spun honey, and perhaps comb honey. Honey candy and a cookbook emphasizing honey may be included (see Figure 5).

To stimulate mail order sales of honey gift packs, advertising is necessary. Often this advertising is accomplished through a catalog with other gift ideas, or through local or regional newspaper, radio or television advertising directed toward holiday seasons.

THE FOODSERVICE INDUSTRY

The foodservice industry is comprised of commercial operations (independent restaurants, fast-food and other chains) and the noncommercial segment (schools, nursing homes and other institutional operations). Approximately one-quarter to one-third of the honey produced in the United States is used by the foodservice industry. While honey is rarely mentioned on restaurant menus, it appears as an ingredient in meat glazes, salad dressings, desserts, marinades and beverages. The exploitation of this market has been approached in a less than concerted effort. Occasionally, sustained use of honey has occurred on a local or national basis. One problem in the foodservice industry is the lack of suitable drip-free dispensers. The squeeze container has been a step in the right direction, but does not completely eliminate the problem. The introduction of portion-pack honey opened another avenue to serving this market. There are some portion packs that preserve the quality of the honey in an excellent fashion, while others contribute to a deterioration in color and allow the moisture content to lower until the honey closely resembles taffy, which makes the product difficult to use.

The noncommercial foodservice market includes schools, business and industry foodservice operations, nursing homes, hospitals, rest homes, military personnel and penal institutions. In recent years, the supply of honey for schools came from the Commodity Credit Corporation as a result of forfeiture of honey placed under the Honey Loan Program. Hopefully, schools will continue to add honey to their menus when the government surplus is depleted. Some school foodservice personnel have expressed interest in continuing to use honey even though it may no longer be a free commodity.

A survey conducted by the Hale Group of Danvers, Mass., under the auspices of the National Honey Board, revealed that foodservice operators rank taste and use as a flavoring agent as the top characteristics of honey. They reported that price and quality were factors when they purchased honey.

The Hale Group study (as shown in Table 3) outlines the pack sizes used by foodservice commercial and non-commercial operators as well as the percentage of honey used at various meal times. Table 4 from the same survey

Table 3

Summary by pack size usage among honey users
Commercial operators (restaurants)

portion pack	40.1%	squeeze bottles	7.2%
bulk size	27.9%	don't know	2.6%
glass jars	29.6%		

Summary by pack size usage among honey users
Institutional operators (school, business & industry, hospital)

portion pack	37.0%	squeeze bottles	4.3%
bulk size	44.9%	don't know	7.6%
glass jars	21.7%		

Meal occasions honey is offered

	Commercial	Institutional
breakfast	59.4%	77.1%
lunch	55.0%	58.5%
dinner	55.5%	35.6%
non-specific	12.2%	10.2%

Table 4
Type of use of honey

ingredient	53%	topping	43%
dipping sauce	46%	sauce for entree	7%
sweetener	45%		

Use in food preparation

baking	53%	marinating	25%
meat glaze	36%	beverages	20%
salad dressing	25%	other	8%
desserts	25%	don't use in preparation	18%

Where honey is featured

not at all	78%	menu board	4%
menu	15%	both	3%

Totals add up to more than 100% because respondents could choose more than one category.

shows the percentage of usage of honey in various preparations (respondents could answer more than one response), and the percentage of times featured on the menu. These figures indicate that there is a potential for increased usage of honey in the menus of the foodservice industry.

THE INDUSTRIAL MARKET

The baking industry is the largest volume user of honey at the present time. Common honey bakery products include "Honey Grain Bread," "Light Honey Wheat Bread," "Milk and Honey Bread" and "Honey Graham Crackers." Honey enhances bakery products by preserving the moisture, thereby extending their freshness and shelf life.

Many breakfast cereals include honey in their name, such as "Honey Graham Chex," "Honey Raisin Bran," "Honey Nut Cheerios," "Honey Buck-Wheat Crisps" and "Nut and Honey Crunch." Despite the honey identification, many of these popular cereals contain less than 1 percent honey. Many other products with the name honey on the label contain honey as the least prominent sweetener.

Honey is used in some unusual products such as "Honey Tortilla Chips," "Dessert" sopaipillas, honey pita breads and a honey sweetened ice cream cone featured as an all-natural novelty. Honey candy makes an appearance frequently, some of which has honey as the major sweetener. Honey is also featured in a number of new sauces and salad dressings (Figure 6). One major

FIGURE 6. The magic of honey is found in a variety of foods, from traditional graham crackers and candy to the more adventurous sauces and honey roasted nuts. *(Photo courtesy National Honey Board)*

salad dressing uses 15 percent honey, and several ultra-premium barbecue sauces now use in excess of 25 percent honey. One limited-market barbecue sauce contains 48 percent honey.

GRADING, QUALITY CONTROL AND PACKAGING
Honey Grading
The U.S. standards for grades of extracted honey are established by the U.S. Department of Agriculture. The use of the U.S. standards or grades is confined to consumer packaged honey and bulk honey going into industrial channels. Virtually none of the crop sold by commercial producers at wholesale carries a grade. Technically, if the grades are stated on the label, the honey must meet those requirements, but food authorities who check honey are more concerned with possible contamination than factors of grade. The following descriptions of the grades, taken from a California publication [12] will acquaint the reader with the principal points in the standards.

---

[12]Frederick W. Bauer, 1960 California Agricultural Experiment Station Bulletin #776.

"U.S. grades are based on four factors: soluble solids (moisture content), flavor, absence of defects, and clarity. After determination of soluble solids, the specific grade is determined by means of a scoring system." The characteristics are weighed as follows: flavor is given 50 possible points, absence of defects is allowed a maximum of 40 points and clarity is permitted a maximum of 10 points, which makes a possible total score of 100 points. Each characteristic is judged as "good," "reasonably good" or "fairly good" in decreasing numbers of points.

U.S. GRADE or U.S. FANCY is a honey which contains not less than 81.4 percent soluble solids (18.6 percent moisture content); possesses a good flavor for the predominant floral source or, when blended, good flavor for the blend of floral sources; is free from defects; is clear and scores not less than 90 points.

U.S. GRADE B or U.S. CHOICE is a honey which contains not less than 81.4 percent soluble solids (18.6 percent moisture content); possesses a reasonably good flavor for the predominant floral source or, when blended, a reasonably good flavor for the blend of floral sources; is reasonably free from defects; is reasonably clear and scores not less than 80 points.

U.S. GRADE C or U.S. STANDARD is honey for reprocessing which contains not less than 80 percent soluble solids (20 percent moisture content); possesses a fairly good flavor for the predominant floral source or, when blended, a fairly good flavor for the blend of floral sources; is fairly free from defects; and is of such quality as to score not less than 70 points.

U.S. GRADE D or SUBSTANDARD is honey which fails to meet the requirements of U.S. Grade C or U.S. Standard.

Honey Grading Equipment

A frequently used instrument for grading the moisture content of honey is the honey refractometer. It is a small instrument, which can be held in the palm of the hand. A drop of honey is placed on the hinged portion, then pressed against a glass prism. Calibrated readings are obtained through the eyepiece and converted to actual moisture content by adding or subtracting the reading of the temperature correction thermometer. A reading can be obtained in seconds (see Figure 7, the refractometer).

The Pfund grader (Figure 8) is used for determining the color of honey. A wedge-shaped glass trough is filled with honey which is compared to a colored glass wedge. Either artificial light or natural light is good for the comparison. The color of honey is recorded in millimeters (see Table 5 and Table 6).

The United States Drug Administration Color Comparator (Figure 9) employs colored glass slides and also bottles filled with materials which provide different degrees of turbidity. Clear glass bottles contain the honey for

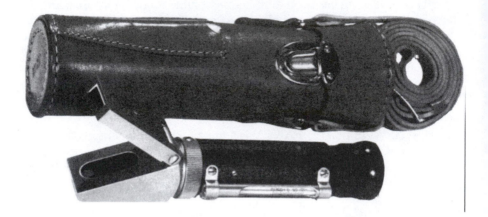

FIGURE **7.** The refractometer will read a moisture range from 12.5% to 26%.

Table 5

Color designations		
	Color range	**Optical density**
Water White	<8	0.0945
Extra White	9-17	0.189
White	17-34	0.378
Extra Light Amber	35-50	0.595
Light Amber	51-85	1.389
Amber	86-114	3.008
Dark Amber	>114	–

color comparison. The instrument provides for six classifications, as does the Pfund grader, and also allows for turbidity in the samples to be classified. The color comparator is not as precise as the Pfund grader in calibrating color.

The commercial honey producer must constantly be aware of the quality of his product by checking the moisture, the flavor and the color during the extracting process. It is to his advantage to separate various lots that differ in floral source, flavor or color. The utmost care should be given to the extracting process to maintain the quality of honey as nearly as possible to that removed from the hive.

Nearly all of the production of the commercial honey producer at the present time is channeled through the honey loan program. Honey is not eligible for a loan if the moisture content is 18.6 percent or higher. This factor alone illustrates the importance of checking the moisture content. In addition, the honey loan rates are based on differentials of color and/or class from White to Extra Light Amber, to Amber and to non-table honey. The commercial producer must be familiar with the grading of his production.

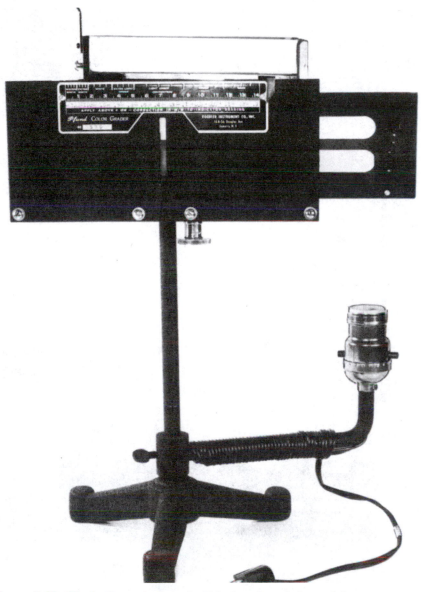

FIGURE 8. The Pfund color grader is used widely throughout the honey industry.

Most honey packers, including the cooperative marketing organization, pay premium prices for honeys that grade in the premium ranges.

Roy Grout in the 1946 edition of *The Hive and the Honey Bee* wrote, "Whoever sells honey must expect to service the trade constantly with an attractive high quality uniform product, skillfully advertised and merchandised at current price levels. It is only when departures are made that one fails to be an asset to the honey industry."

Table 6

FLORAL SOURCES — INFINITE, YET PREDICTABLE

			Prune	Buckwheat
	Mesquite		Raspberry	Heather
	Locust		Thyme	Tupelo
	Spearmint			Sunflower
Palmetto	Manzanita		Cranberry	Bergamot
Sage				
Mint	Basswood		Blackberry	
	Bamboo			Blueberry
Clover, hubam	Cantaloupe		Orange	
Acacia		Horsemint		
Clover, crimson				
Clover, sweet yellow				

Taste: Strong → Light (vertical axis)
Color: Light → Dark (horizontal axis)

Figure 9. The USDA color comparator enables the producer and the packer to classify honey as to color and turbidity.

Control of the quality of honey marketed has become a necessity, as consumers are more sophisticated and demand a superior and consistent product. Any changes in the blend, viscosity or flavor are not apt to pass unnoticed. The larger honey packers practice quality control, either constantly checking their finished product in their own laboratory or contracting with an independent laboratory or agency to monitor their end product.

The Food and Drug Administration personnel periodically inspect honey packing plants to check on the facilities and the conditions under which the honey is packed and processed. In addition, state public health departments make periodic inspections of extracting facilities and packing operations. The FDA may also pull packed containers of honey from the store shelves to check for adulteration or contamination.

PACKAGING AND LABELING

For years, tin and glass were the standard containers for packaging honey. As plastics have been improved, their light weight and improved strength have led to greater use. Plastic lends itself to creative molds of shapes and designs which are attractive to the consumer. The longevity of the little plastic squeeze container affirms its acceptance to the honey consumer. The trend in the marketplace has been toward smaller sized containers, which has created a movement away from the tin container to the glass or plastic.

Product tampering has made the honey industry aware that steps need to be taken to guard against such incidents, so an additional foil or paper seal under the lid has been used, or a shrink wrap plastic seal, which serves to alert the consumer if the seal has been broken. It would also be possible to package the honey under a partial vacuum process as the lid is applied; the lid carries an identifying portion which would indicate if that seal had been broken. Speed and versatility have been improved, and "back labeling" has allowed additional information or coupons to be placed on the label.

In 1970, the first feasibility studies for a product code system began. The new, computerized system for checkout and inventory control was initially designed for supermarkets, but has expanded to mass merchandise outlets.[13] The Uniform Product Code allows the use of automated checkout stands. As items are presented to the checker, they are passed over an optical scanner, which "reads" the UPC symbol, decodes it to the UPC number and transmits this code number to a small computer, which stores price and other information on all items carried in the store. The computer transmits back to the checkstand the item's price and description, which are instantly displayed and printed on the customer's receipt tape, as well as relevant information on taxability, food stamp acceptance, etc. At the same time, the computer

---

[13]About the Universal Code, Uniform Code Council, Inc., Dayton, Ohio.

captures and stores item movement information. This information can then be aggregated and summarized into a wide variety of control reports. Because it is possible to scan and bag an item faster than it is to enter department and price information on a keyboard, the checkout operation is speeded up as well as being made more accurate and fully recorded.

Most states have regulations requiring that the labels used on honey containers state the contents, the name and address of producer, the packer or distributor, the grade and the net weight.

Federal law requires "Country of Origin" labeling if imported honey is packed and/or commingled with domestic honey in the packing process.

Federal law makes a distinction in label requirements between bulk honey in drums, gallons or larger container sizes versus consumer size packages (under 4 pounds or under 1 gallon). For the bulk honey in drums or larger container sizes, there is an exemption.

"6. Exemption from labeling requirements.

  (a) Shipments of bulk honey in interstate commerce are exempt from above labeling requirements if (and only if) either of the following two conditions exist:

   (1) The shipper is the operator of the establishment where the honey is to be processed, labeled or repacked; or

   (2) In case the shipper is not such operator, such shipment or delivery is made to such establishment under a written agreement, signed by and containing the Post Office address of the shipper and such operator and containing such specifications for the processing, labeling or repacking, as the case may be, of such honey in such food establishment as will insure, if such specifications are followed, that such food will not be adulterated or misbranded upon completion of such processing, labeling or repacking. Both shipper and such operator must keep a copy of such agreement for 2 years after final shipment of honey from such establishment. They also must allow FDA inspectors to examine these copies."

The majority of the commercial honey producers are affected by the above exemptions for bulk honey shipments interstate. Producers who are interested in packing either bulk or consumer sizes of honey, since the labeling regulations are quite detailed and involved, should obtain a copy from the Food and Drug Administration (for interstate shipments), or from their state agency (for shipments within the confines of your state).

As this chapter is being edited, it is quite possible that either through the legislative process or through regulations additional information will be required on food labeling. Nutritional information could be required, as well as rules to standardize the "serving size" on labels.[14] It is not clear at this time how the product honey will be affected. It would be prudent advice to check to determine any changes that may have occurred in the regulations for labeling food products insofar as how they may affect your particular operation.

ADVERTISING AND PROMOTION

Commercial honey producers, as a general rule, promote honey to their circle of friends within the community. Word of mouth works well, but on a very limited basis. The time involved in honey production prohibits most commercial producers from becoming actively involved in promoting honey.

Most often, the commercial honey producer is outside the sales/promotion loop yet hears of "cents off" or "case free" deals offered to the chain store or supermarket by competitors within the industry. Of course, such offers are made to break into a competitor's market or to stimulate slow moving sales. The struggle for one honey packer to gain a market from another is a reality in the competitive marketplace. In the commercial honey producer's view, these types of promotions result in the producer becoming the loser because of the pressure that it puts on the wholesale market price of honey. The producer often would prefer promotional give-away dollars diverted to advertising dollars that would expand and broaden the honey market. The competition for markets through promotional "deals" creates a divisive force within the industry. The interests of the honey producer differ from that of the honey packer. Nevertheless, the economic health of each is essential for a viable industry as a whole.

Most of the honey packers of the United States supply a local or regional market, with one exception, Sioux Honey Cooperative Marketing Association. Advertising and promotion, therefore, is done on a local or regional basis. Advertising may consist of in-store promotional displays, newspaper advertising or couponing, radio advertising and, on a limited basis, television advertising. The consistency and dollar expenditure vary greatly, depending on the packer's needs. Local or regional packers have contributed money toward advertising, and markets have been developed by successfully using aggressive advertising tactics.

Over the years, the management and directors of Sioux Honey Cooperative Marketing Association have realized the value of budgeting funds for advertising and promotion of their brand names. By using the expertise of an advertising agency to channel dollars into magazines and newspaper advertising, radio advertising, limited television advertising, and the distribution of coupons, the cooperative can urge the purchaser to stock a brand name that is

---

[14]The New York Times, March 8, 1990.

nationally advertised. Even though the members, at times, have sacrificed several cents per pound from their earnings and received less than their non-cooperative counterparts, the members view the expenditure as an investment in the future.

THE AMERICAN HONEY QUEEN PROGRAM

For many years, The American Honey Queen Program was the only national, industry supported, promotional activity.

The suggested procedure for choosing an American Honey Queen was proposed in the late 1950's.[15] Since that time, the Queen Program has been sponsored by the American Beekeeping Federation. States affiliated with the Federation are encouraged to choose a state queen to enter the competition at the annual convention of the American Beekeeping Federation. The Queen program has grown and prospered under the guidance of a hard working chairperson and committee. This committee contributes its time and effort without compensation. The American Honey Queen attends fairs, conventions, food shows, and festivals expounding the virtues of honey. In addition, she appears on local radio and television programs. The appearances that each state queen makes during her reign enhance the honey image, and also, serves as a public relations link between the consumer and the beekeeper. The American Honey Queen program generates supportive backing from the industry as a whole, and there is little question that the promotional value that the program generates exceeds by many times the dollars devoted to the program.

THE NATIONAL HONEY BOARD

On Oct. 30, 1984, Public Law 98-590, "The Honey Research, Promotion, and Consumer Information Act," was enacted by the 98th Congress of the United States. The Act authorized the establishment of an orderly procedure for the development and financing, through an adequate assessment, of an effective and coordinated program of research, promotion, and consumer education designed to strengthen the position of the honey industry in the marketplace and to maintain, develop, and expand markets for honey and honey products.

From this enactment followed hearings, a favorable referendum by the producers, a meeting of the nominations committee (one person from each state and Puerto Rico) to select board members, and finally the organizational meeting held in Alexandria, Va., in September of 1986. Soon after, a manager was hired and an office was set up in Longmont, Colo. The first assessments were made on Feb. 2, 1987.

Annually, the National Honey Board develops a marketing plan with

---

[15]American Beekeeping Federation Newsletter, Vol. 15, No. 3-4, March-April, 1958, Pg. 3.

programs to promote honey to consumers, the foodservice industry and food manufacturers. The programs range from attendance at food shows to full-color advertisements in leading consumer, foodservice and technical publications. The Honey Board also works directly with newspaper and magazine food editors to develop special features about the many ways to use honey. To help encourage the use of honey in manufactured food products, the Board developed technical brochures about honey and hired a food scientist to monitor the honey hotline. The Honey Board is dedicated to increasing the demand for honey in the United State and for U.S. honey overseas.

The formation of the National Honey Board brought together various segments of the industry, such as honey producers, independent packers, a cooperative packer representative, importers, and a public member representing the consumer. Each of these segments have interests and objectives that do not always coincide with each other. Bringing together this diverse group as a board has enlightened the members to the problems of their counterparts. From this interaction has developed a group dedicated to working toward the common goal of increasing the awareness of an interest in honey and honey products.

The first few years of existence for the National Honey Board have been filled with activity. The mechanics of setting up a national office needed to be handled simultaneously with determining the most beneficial advertising and promotional dollar investments for the honey industry. The information gleaned from market research surveys (some of which was presented earlier in this chapter) helped to serve as a guide for expenditures.

At this writing, The National Honey Board is entering its fifth year of operation. Its impact on and reception by the honey industry has been as a whole positive. The demand for honey has increased, particularly with the commercial and industrial users of honey. The market price for wholesale or bulk honey has shown a favorable increase, which is an objective of the National Honey Board. The honey producers and importers overwhelmingly voted to continue the National Honey Board and eliminate the refund provision in 1991. The sunset provision of the law provides for a referendum vote each 5 years.

Targeted Export Assistance Program

The Food Security Act of 1985 authorized the Secretary of Agriculture the use of Commodity Credit Corporations commodities to conduct an intensive overseas market development program for specific U.S. agricultural commodities which have been adversely affected by foreign unfair trade practices. Commodities which can show damage caused by such unfair trade practices can apply for funds under the Targeted Export Assistance Program which is administered by the USDA's Foreign Agriculture Service.

The National Honey Board, in 1988, made application to participate in the TEA Program as a nonprofit agricultural organization. Funds were provided to the Honey Board to pursue honey market development in the Middle East, the Pacific Rim countries, and Western Europe. The role of the National Honey Board will be to promote U.S. honey, to educate and inform the consumer how to use honey, perhaps setting up trade offices in the foreign countries, and ultimately to assist in opening the door to U.S. honey in these countries.

Adverse Publicity

Publicity may originate outside of our industry that is less than favorable. The beekeeping/honey industry has faced many such incidents. If bad publicity is left unanswered, it can be damaging. The public sees the beekeeping/honey industry as a clean industry that does not damage the environment, but in fact, enhances the environment through honey bees as pollinators. It behooves the beekeeping/honey industry to be constantly alert and prepared to counteract or neutralize any developing issues which might tarnish the favorable image of beekeeping and honey.

UNITED STATES PRODUCTION AND CONSUMPTION

The number of bee colonies in the United States peaked in 1947 at 5,916,000. Bee colonies have decreased in number in recent years (see Table 7 and Figure 10).

In recent years, the per capita consumption of honey in the United States has hovered around 1.1 pounds per person. Average production in recent years has been approximately 190 million pounds (see Table 8 for production figures for 1978 through 1988).

Figure 11 vividly illustrates the range of fluctuation of U.S. honey production. Weather factors such as rainfall and temperatures during critical periods directly affect production. Furthermore, floral sources vary in their predominance and their ability to secrete nectar from year to year.

With decreased production in the United States and a decline in per capita consumption of honey since the 1950's, the country moved from a honey-exporting country to a honey-importing country to meet the population's demand. Table 8 illustrates the relationship between production, imports, exports, the disappearance, and ending stocks. This table also emphasizes the effect of an increased support price in accelerating imports. The reversal of imports is seen with the implementation of the "buy back" price in 1986, 1987 and 1988. Figure 12 shows the relationship of exports to imports for the period 1950-1983. Table 9 shows U.S. importation in metric tons on a July to June year from 1982 to 1986. Figure 13 illustrates graphically U.S. honey imports for a recent 10-year period.

HONEY LOAN PROGRAM

In the past, the U.S. Department of Agriculture has provided an immeasurable degree of assistance to the honey industry in an effort to stabilize the price of honey. The passage of the Agriculture Act of 1949 required the U.S. Secretary of Agriculture to support the price of honey. The program was established to ensure an adequate supply of honey bees because of their significant role in pollinating agricultural crops. Price support for honey is accomplished through loans to honey producers and honey marketing cooperatives. Under the loan program, producers may obtain a loan on honey they produce domestically, which will assist them in continuing their operations, by applying to the USDA's Agricultural Stabilization and Conservation Service. Applications must be made at one of more than 2,800 ASCS county offices. Participants can borrow an amount at the established loan rate per pound up to 90 percent of the honey pledged as collateral. The loan amount can be up to 95 percent if the honey collateral is stored in a warehouse approved by the USDA Commodity Credit Corporation.

The government did not acquire any honey through defaulted loans for a nine year period ending in 1979. Inflation, however, which affects the indices used for computing the support price, caused the support price to increase

Table 7. Colonies of Honey Bees, United States, 1946-1990 Crop Years[1]

Crop year	Colonies	Crop year	Colonies	Crop year	Colonies
	Thousands		Thousands		Thousands
1946	5,787	1961	4,992	1976	4,285
1947	5,916	1962	4,900	1977	4,346
1948	5,724	1963	4,849	1978	4,081
1949	5,578	1964	4,840	1979	4,155
1950	5,601	1965	4,718	1980	4,140
1951	5,546	1966	4,646	1981	4,213
1952	5,493	1967	4,635	1982[2]	4,250
1953	5,520	1968	4,539	1983[2]	4,275
1954	5,461	1969	4,433	1984[2]	4,300
1955	5,252	1970	4,285	1985[2]	4,325
1956	5,195	1971	4,107	1986[3]	3,205
1957	5,199	1972	4,085	1987[3]	3,190
1958	5,152	1973	4,124	1988[3]	3,219
1959	5,109	1974	4,210	1989[3]	3,311
1960	5,005	1975	4,206	1990[3]	3,322

[1] Data not reported from 1982-1985 by the Statistical Reporting Service, USDA.
[2] Estimated by the Agricultural Stabilization and Conservation Service, USDA.
[3] Large decrease due to different reporting method which does not count beekeepers with fewer than five colonies.
Source: Statistical Reporting Service, USDA.

Honeybee Colonies in the U.S., 1945-83

Million

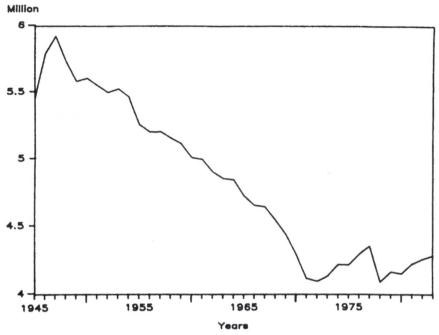

FIGURE 10. *(Courtesy USDA)*

rapidly since the mid-1970's and double from 32.7 cents per pound for the 1977 crop year to 65.8 cents per pound for the 1984 crop year. At the same time, world honey supplies increased more rapidly than demand, causing prices to drop. Support prices higher than world market prices, and the strength of the dollar, encouraged honey imports to the United States. From 1979 to 1983, annual honey imports nearly doubled to 109.8 million pounds (see Figure 14). Honey used as collateral for loans increased from 41.1 million pounds for the 1980 crop year to 113.6 million pounds for the 1983 crop year. Forfeiture of honey to the government increased (see Table 10). Table 11 lists further price support activity for the years 1950 through 1988. Figure 14 graphically illustrates the relationship of loan rates, market price and imports. Most of the foreited honey was distributed through government donation programs.

THE 1985 FARM BILL

The 1985 Farm Bill, enacted in the closing weeks of 1985, incorporated a provision for a honey loan pay back program. This provision has become known as the "buy back" provision. The U.S. Secretary of Agriculture has the discretionary authority to set a market clearing or redemption price at the world honey market level. Additionally, the legislation removed the tie of the support price from a percentage of parity to fixed price that would be decreased over the five-year period of the Farm Program. Since its implementation, there have been some technical changes, but it has greatly enhanced the return of domestic honey into the marketplace rather than into government channels of the give-away program. With the return of domestic honey to the marketplace, imports have been reduced considerably. Figure 13 and Table 9 indicate the decline of imports during 1986 and 1987. The increase in imports in 1989 reflects a shortage of honey in the United States resulting from poor honey crops. Table 12 lists the Loan Repayment Rates for 1986 and 1987.

The 1985 Farm Program expired in 1990. When the 1990 Farm Bill was written by Congress, it continued the honey loan program, but started a phase down on dollar loan limitations. In addition, a 1% assessment was levied on loans, thereby reducing the loan rate slightly. The 1% assessment was applied

Table 8.

Honey: Production, Imports, Exports, Disappearance, Ending Stocks, and Support Prices 1978-1988[1]

Fiscal Year	1978	1979	1980	1981	1982	1983
(Million Lbs.)						
Production	231.5	238.7	199.8	185.9	230.0	205.2
Imports	56.0	58.6	49.0	77.3	92.0	103.7
Exports	8.0	8.8	8.5	9.2	8.5	8.0
Disappearance	277.3	282.7	226.2	232.0	254.3	269.1
Ending Stocks	32.2	38.0	52.1	74.1	133.3	165.1
(Cents/Lbs.)						
Support Price	36.8	43.9	50.3	57.4	60.4	62.2
Buyback Price						

Fiscal Year	1984	1985	1986	1987	1988	
(Million Lbs.)						
Production	165.1	150.1	200.0	227.0	214.1	
Imports	124.0	141.0	130.4	65.0	60.0	
Exports	7.6	6.7	7.4	12.9	18.0	
Disappearance	251.8	256.1	291.9	344.5	296.0	
Ending Stocks	194.8	223.1	254.2	188.8	134.8	
(Cents/Lbs.)						
Support Price	65.8	65.3	64.0	63.0[2]	59.1	
Buyback Price			42.0	37.0	37.0	

[1] Figures for years 1982-1985 and 1988 are USDA estimates
[2] Support Price adjusted to 61.0 cents per pound on December 23, 1987.

Honey Production in the U.S., 1945-83

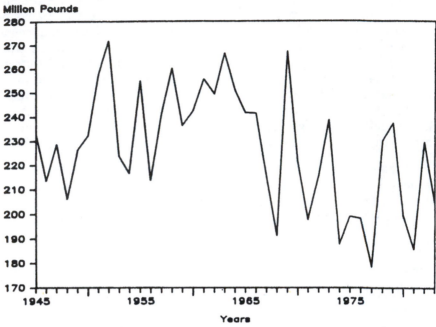

Figure 11. *(Courtesy USDA)*

Table 9.

U.S. IMPORTS OF HONEY, JULY/JUNE YEAR
(Metric Tons)

Country	July 82/ June 83	July 83/ June 84	July 84/ June 85	July 85/ June 86	July 86/ June 87	Percent Change, 85/86-86/87
China	9,803	6,278	6,807	15,479	11,047	-29%
Mexico	14,793	22,054	17,171	15,015	5,555	-63%
Canada	7,243	10,251	17,904	12,507	5,332	-57%
Argentina	7,913	10,480	12,650	14,944	4,826	-68%
Brazil	67	21	237	1,399	1,085	-22%
El Salvador	547	598	747	1,117	615	-45%
Australia	2,098	1,202	3,254	2,675	312	-88%
Rest-of-World	3,657	2,910	3,866	3,951	2,083	-47%
TOTAL	46,121	53,794	62,636	67,087	30,855	-54%

Source: U.S. Department of Commerce, Bureau of the Census.

to all commodity programs to reduce costs. As the United States has become more urbanized, support for farm programs has eroded in Congress. The story of the honey bee as a pollinator for fruit, nuts, and seed crops will have to be delivered effectively by the beekeeping/honey industry to those persons writing and enacting future farm legislation. The legislators will need to be convinced by their constituents that, without the beekeeping industry, the price they pay for food will increase because of the loss of the honey bee pollinators.

WORLD PRODUCTION AND MARKETS

As communications systems improve and as transportation modes expand, the world trade will economically affect virtually every commodity. The world supply and demand for honey determines the price of honey to a great extent, however, complicating the law of supply and demand are fluctuations in currency exchange rates, varying transportation rates, import and export tax levies, governmental assistance to honey production in various countries, nontariff trade barriers, and government trade agreements.

The World Honey Situation, published by the USDA's Foreign Agriculture Service in October, 1990 reports:

U.S. Honey Foreign Trade, 1950-83

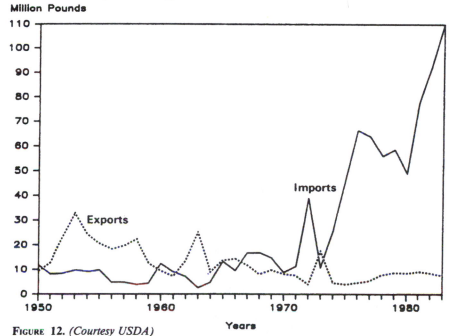

FIGURE 12. *(Courtesy USDA)*

U.S. Honey Imports, July/June Year
1977/78—1986/87

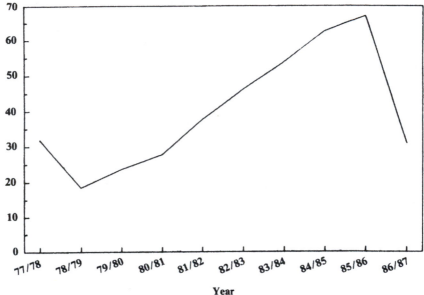

FIGURE 13. *(Courtesy USDA)*

Table 10.

Honey Used As Collateral for Government Loans and Forfeited
1980-83

Year	Estimated honey produced	Honey used as collateral	Honey forfeited	Acquisition costs
	(pounds)			
1980	199,756,000	41,135,000	5,327,000	$ 2,687,078
1981	185,927,000	55,168,000	35,154,000	19,968,612
1982	230,000,000	88,443,000	74,075,000	44,567,227
1983	205,000,000	113,629,000	105,987,000	65,741,578
Total	820,683,000	298,375,000	220,543,000	$132,964,495

  The aquisition cost is the value of the forfeited loan and does not include handling, transportation, or costs to reprocess the honey for the government's food donation programs. These costs also do not include any administrative or interest costs because, as mentioned earlier, USDA records do not identify these costs with the honey program.

"Honey production in 10 major producing countries for 1990 is forecast at 750,500 metric tons, up 3 percent from the revised 1989 estimate of 729,762 metric tons. These countries make up 66 percent of the Food and Agriculture Organization (FAO) 1989 estimate for world honey production. The projected rise is attributed to production increases in Brazil, Canada, Japan, Mexico, the Soviet Union, and the United States."

Table 13 shows 1985-1989 production, exports and imports of several major honey-producing or honey-consuming countries. Countries included are Argentina, Australia, Brazil, Canada, China, Federal Republic of Germany, Japan, Mexico, the United States and the USSR.

The USDA Foreign Agricultural Service's *World Honey Situation* October, 1990, states:

"For the countries covered in this report, aggregate exports in 1990 are expected to decline 1 percent while imports may increase 4 percent from 1989. West Germany ranks as the world's biggest importer of honey. German imports are expected to drop by 8 percent in 1990 due to carry-over of substantial stocks from 1989. Japan is the second largest importer of honey. In 1990, Japan's imports should show a dramatic 21 percent jump from 1989, paralleling increasing domestic consumption. China is the world's number one exporter of honey with 1990 exports forecast to rise slightly by 3 percent over 1989. This represents a decline of 8 percent from the peak of exports

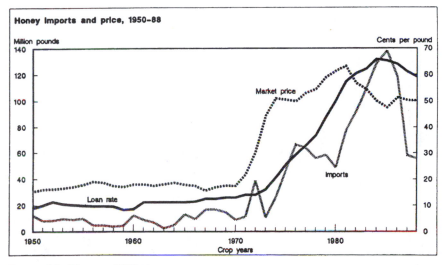

FIGURE 14. *(Courtesy USDA)*

Table 11: Honey price support rates and loan activity, 1950-88

Crop year	National average price support rate[1]	Parity price adjusted	Support rate as a percent of parity	Program activity		Net Government (return) or expenditure[2]
				Quantity placed under	CCC take over	
	— Cents/pound —		Percent	— Million pounds —		Million dollars
1950	9.0	15.0	60.0	[3]	7.4	NA
1951[4]	10.0	16.7	60.0	[3]	17.8	NA
1952	11.4	16.3	70.0	9.3	7.0	NA
1953	10.5	15.0	70.0	3.1	.5	NA
1954	10.2	17.0	60.0	1.5	0	NA
1955	9.9	13.2	75.0	1.8	0	NA
1956	9.7	13.9	70.0	1.6	0	NA
1957	9.7	13.9	70.0	2.9	.1	NA
1958	9.6	13.7	70.0	5.6	2.0	NA
1959	8.3	13.8	60.0	1.3	0	NA
1960	8.6	14.3	60.0	1.1	0	NA
1961	11.2	14.9	75.0	4.2	1.1	0
1962	11.2	15.1	74.0	3.4	0	.1
1963	11.2	16.7	67.0	3.2	0	(.1)
1964	11.2	17.2	65.0	9.5	2.2	0
1965	11.2	17.8	63.0	17.3	3.3	.7
1966	11.4	18.6	61.3	33.9	4.1	.1
1967	12.5	19.5	64.0	31.0	5.4	(.1)
1968	12.5	18.7	66.8	24.9	.1	.4
1969	13.0	19.5	66.7	45.7	3.5	(.9)

Continued on next Page

Table 11: Honey price support rates and loan activity, 1950-88 — Continued

Year						
					[5]	
1970	13.0	20.4	63.7	40.6		.8
1971	14.0	21.0	66.7	22.9	0	(.9)
1972	14.0	22.3	62.8	19.8	0	[6]
1973	16.1	26.7	60.2	12.1	0	0
1974	20.6	34.3	60.0	13.9	0	.3
1975	25.5	42.4	60.1	[7]	0	(.3)
1976	29.4	49.0	60.0	[7]	0	(.2)
1977	32.7	54.4	60.0	14.1	0	1.5
1978	36.8	61.3	60.0	40.5	0	3.5
1979	43.9	73.1	60.0	49.1	0	(1.7)
1980	50.3	83.9	60.0	41.1	6.0	8.7
1981	57.4	95.6	60.0	55.2	35.2	8.4
1982	60.4	100.7	60.0	88.4	74.5	27.4
1983	62.2	103.7	60.0	113.6	106.4	48.0
1984	65.8	109.7	60.0	107.5	105.8	90.2
1985	65.3	108.7	60.0	102.0	98.0	80.8
1986	64.0	108.7	[8]	179.3	41.9[10]	89.4
1987	61.0[9]	113.7	[8]	216.4	52.1[10]	72.6
1988	59.1	114.0	[8]	192.1	.3[10]	100.1

NA = Not available. [1] For extracted honey in 60-pound or larger container. [2] Fiscal year. [3] Direct packer purchase program. [4] On March 22, 1951, support for most flavors of honey was announced at 10 cents per pound with a dozen flavors of honey of limited domestic acceptability supported at 9 cents. On April 5, 1951, it was announced that the support price of honey of wide table acceptability would be increased from 10.0 to 10.1 cents per pound. [5] 5,900 pounds. [6] Less than $50,000. [7] Purchase agreements only, no loan program. [8] Purchase agreements only, no loan program. [9] Loan rate was reduced from 63 to 61 cents per pound on December 23, 1987, because of the Omnibus Budget Reconciliation Act of 1987. [10] As of May 3, 1989. (Source: USDA) in 1986. Mexico, the world's second largest exporter of honey in 1990, is expected to increase exports by 9 percent over 1989.

"Domestic consumption in the 10 countries is expected to rise 42,015 metric tons or 6 percent in 1990 over 1989. Japan and the Soviet Union are anticipated to account for 10 and 34 percent of total utilization, respectively.

"Ending stocks for the selected countries are expected to drop in 1990 by 27 percent from 1989 due to major reductions foreseen in China and West Germany."

Since the last edition of this chapter was written, China has become a major honey-producing country. China is the world's largest honey exporter and second largest producer after the Soviet Union. Forecasts indicate increased honey production capacity in China as colony numbers increase. China could conceivably sustain up to 30 million colonies, more than three times the current number.[16]

If we look at Table 13, *"Honey Production, Number of Colonies and Yield for Selected Producers, Calendar Year 1986-1990"*, we note that colony numbers in China have somewhat stabilized over the past 5 years, as well as exports; however, their domestic consumption has shown signs of increasing. Higher incomes have resulted in increased demand for pure honey and honey medicines/health products. In 1988 about 40 percent of the honey consumed

[16]FAS, USDA, AGR Number CH8515, 6-21-88.

Table 12

U.S. HONEY LOAN REPAYMENT RATES[1]
(cents per pound)

Date	1986 Crop	1987 Crop
January 1987	37.5	—
February 1987	37.5	—
March 1987	36.3	—
April 1987	36.3	42.5
May 1987	36.3	42.5
June 1987	36.3	42.5
July 16, 1987	36.3	36.8
July 30, 1987	33.3	37.3
August 1987	33.3	36.8
Average (July-December 1986)	40.5	—
Average, 1987	35.9	39.7
National average support price[2]	64.0	63.0

[1] Average of all grades of honey. [2] For extracted honey in 60 pounds or larger container.

Source: ASCS, USDA

in China was consumed as pure honey, 40 percent was used to produce Chinese medicines, and 20 percent went to food and beverage industries. China's consumption of honey is expected to continue to increase as its people's standard of living improves.

It is interesting to note that whereas colony numbers have increased in the Soviet Union, as well as their production, their domestic consumption has increased while their exports have decreased in the past five years. The Soviet Union remains the largest producer and consumer of honey in the world.

Although colony numbers in the United States have shown a slight increase, honey production has varied widely, with domestic consumption relatively stable.

Mexico, the fourth largest honey producer in the world has seen only a slight decrease in the number of colonies in the past five years, as well as a slight decline in production; whereas, their domestic consumption has shown an upturn.

Argentina colony numbers, honey production, and domestic consumption have all remained relatively stable during the past five years.

It is worthy to note that colony numbers, honey production, and domestic production have all increased rather dramatically in Brazil during the past five years.

Canada has seen rather a dramatic reduction in number of colonies during the past five years, which has somewhat affected their honey production, but any reduction in their domestic consumption has more than been offset by their increased exports.

Australia has a slightly reduced number of colonies during the past five years, but increased yields have helped to keep their total production somewhat better than level. Their domestic consumption has remained stable, and there has been a growth in their exports.

Japanese colony numbers and production have been relatively stable, but there has been a tremendous growth in their domestic consumption. That increase in domestic consumption has been offset with increased imports nearly every year in the past five years. Honey sweetened fruit or soft drinks are very popular in Japan, and their popularity has contributed to this increased usage of honey.

The colony numbers in West Germany have been quite stable, with an increase in colony yields resulting in increased honey production during the past five years. Their domestic consumption has shown solid growth, being filled by consistent imports, only falling off in 1990, apparently because of a sizeable carryover.

Table 13

HONEY PRODUCTION, COLONY NUMBERS AND YIELD FOR SELECTED PRODUCERS, CALENDAR YEAR 1985-1989[1]

Country	Year	Total[2] Colonies	Yield Per colony	Honey Production	Beginning[3] Stocks	Imports	Total Supply/ Distribution	Exports	Domestic Consumption	Ending Stocks
		(1,000)	(Kilograms)	(Metric Tons)						
Argentina	1985	1,500	30.0	45,000	4,006	0	49,006	42,173	5,000	1,833
	1986	1,500	24.0	36,000	1,833	0	37,833	30,918	5,000	1,915
	1987	1,500	29.3	44,000	1,915	0	45,915	36,273	6,500	3,142
	1988	1,500	30.7	46,000	3,142	0	49,142	41,458	6,500	1,184
	1989	1,500	25.3	38,000	1,184	0	39,184	30,000	6,500	2,684
Australia	1985	560	48.0	26,871	2,259	87	29,217	14,653	13,359	1,205
	1986	542	46.3	25,077	1,205	99	26,381	11,874	14,428	79
	1987	366	76.5	28,000	79	223	28,302	11,814	15,236	1,252
	1988	370	74.3	27,500	1,252	402	29,154	13,540	15,600	14
	1989	374	73.5	27,500	14	150	27,664	12,000	15,650	14
Brazil	1985	1,850	15.1	28,000	0	460	28,460	853	27,607	0
	1986	1,900	14.2	27,000	0	576	27,576	1,960	25,616	0
	1987	1,900	16.1	30,500	0	1,248	31,748	406	31,342	0
	1988	1,980	18.2	36,000	0	937	36,937	231	36,706	0
	1989	2,000	19.0	38,000	0	1,000	39,000	500	38,500	0
Canada	1985	694	52.0	36,120	6,513	246	42,879	17,278	20,601	5,000
	1986	707	48.1	34,041	5,000	265	39,306	11,843	21,463	6,000
	1987	699	56.9	39,776	6,000	391	46,167	10,903	23,264	12,000
	1988	608	60.5	36,805	12,000	465	49,270	14,208	24,062	11,000
	1989	605	52.1	31,500	11,000	500	43,000	15,000	23,000	5,000
China	1985	6,000	25.0	150,000	0	0	150,000	54,800	95,200	0
	1986	7,000	22.9	160,000	0	0	160,000	80,589	79,411	0
	1987	8,320	24.5	204,000	0	0	204,000	66,831	137,169	0
	1988	8,140	19.2	156,000	0	0	156,000	46,487	109,513	0
	1989	8,300	21.7	180,000	0	0	180,000	60,000	120,000	0
West Germany	1985	1,109	9.9	11,000	2,000	79,000	92,000	14,000	77,000	1,000
	1986	1,150	13.9	16,000	1,000	87,000	104,000	16,000	83,000	5,000
	1987	1,070	15.0	16,000	5,000	83,000	104,000	15,000	86,000	3,000

Table 13 *(Continued)*

Country	Year									
	1988	1,106	16.3	18,000	3,000	84,000	,105,000	15,000	87,000	3,000
	1989	1,100	16.4	18,000	3,000	85,000	106,000	15,000	88,000	3,000
Japan	1985	285	25.4	7,225	6,500	28,046	41,771	0	38,771	3,000
	1986	282	19.7	5,553	3,000	36,354	44,907	0	40,907	4,000
	1987	272	22.1	6,023	4,000	40,129	50,152	0	40,152	10,000
	1988	269	18.1	4,870	10,000	37,643	52,513	0	46,513	6,000
	1989	258	23.3	6,000	6,000	44,000	56,000	0	50,000	6,000
Mexico	1985	2,500	22.4	56,000	5,960	0	61,960	42,380	8,000	11,580
	1986	2,500	21.6	54,000	11,580	0	65,580	57,992	7,400	188
	1987	2,500	19.1	47,850	188	0	48,038	39,559	7,500	979
	1988	2,400	19.2	46,140	979	20	47,139	39,154	7,500	485
	1989	2,400	21.9	52,530	485	25	53,040	45,000	7,500	540
United States	1985	4,325	15.7	68,000	104,553	62,706	235,259	2,949	116,553	115,757
	1986	3,205	28.4	90,898	115,757	54,420	261,075	4,181	128,346	128,548
	1987	3,190	32.2	102,867	128,548	26,428	257,843	5,610	145,684	106,549
	1988	3,190	30.1	95,940	106,549	25,370	227,859	6,396	128,975	92,488
	1989	3,170	23.6	74,843	92,488	39,281	206,612	5,715	147,645	53,252
Soviet Union	1985	8,157	25.0	204,000	0	0	204,000	22,698	181,302	0
	1986	8,300	25.3	210,000	0	0	210,000	20,607	189,393	0
	1987	8,300	26.4	219,245	0	0	219,245	20,871	198,374	0
	1988	8,400	26.2	220,000	0	0	220,000	17,438	202,562	0
	1989	8,500	26.5	225,000	0	0	225,000	18,000	207,000	0
TOTAL	1985	26,980	23.4	632,216	131,791	170,545	934,552	211,784	583,393	139,375
	1986	27,086	24.3	658,569	139,375	178,714	976,658	235,964	594,964	145,730
	1987	28,117	26.3	738,261	145,730	151,419	1,035,410	207,267	691,221	136,922
	1988	27,963	24.6	687,255	136,922	148,837	973,014	193,912	664,931	114,171
	1989	28,207	24.5	691,373	114,171	169,956	975,500	201,215	703,795	70,490

¹ Except Australia where crop year is on a July/June basis, with the indicated year representing when the crop year begins. ² For the United States only colonies with 5 or more hives are included. ³ Stock data for China are not available and are included in consumption. Actual consumption is assumed to be rising steadily in China, despite the erratic movement in consumption statistics. Stock figures for the United States have been heavily revised and will not agree with those published earlier. (Source: USDA).

It is plausible that as the standard of living improves in countries around the world and as interest grows for the use of natural foods, that the interest in honey as a natural sweetener will result in increased usage of the product honey.

Future Honey Market

The economics of honey production in all producing countries and the ability of the consuming countries to absorb that production will determine the future market. Varroa and tracheal mite problems must be coped with in many of the honey-producing countries, including the United States. It is critical that the mite problems be dealt with in such a manner that the image of honey as a natural and wholesome sweetener will not be tarnished.

As developing countries attain a higher standard of living, honey may find a place in the diet of millions. The National Honey Board's efforts could open the door for thousands of new consumers and increased usage of current consumers. Other producing countries could be encouraged, by example, to follow and assist in carrying the honey story to their consumers.

Product Development

Product development within the honey industry has not progressed rapidly. There have been more new products incorporating the use of honey outside of the industry than within. While the amount of honey used as a sweetener in comparison to other sweeteners in many of these products is minimal, the mystical enhancement that the word "honey" lends to a product is the greatest reason why honey has been added to many of these products. To consumers, honey on a label implies "wholesome, pure and natural."

In recent years, the industry has developed new products which do use substantial amounts of honey.

1. A barbecue sauce, which contained 48 percent honey,
2. Honey Stix, "candy of the future,"
3. Honey candies, some of which have honey as the only sweetener, and
4. Honey in cosmetics, such as honey scented shampoos, soaps and hair sprays.[17]

The Honey Bear Logo[18]

The National Honey Board through its Food Technology programs has launched a logo campaign to encourage manufacturers to use honey as the primary sweetener. Its logo—a cuddly bear guzzling honey from a jug—can then be used on the label of products that meet the guidelines of the National Honey Board Food Technology criteria for honey usage in a product. Mary Humann, marketing director of the National Honey Board, said, "The cam-

---

[17]G. Toth, Cosmetic Use of Hive Products, Facts and Prospects, *American Bee Journal*, June 1988, Pg. 431.

[18]The Honeyboard Newsletter, Vol. 3, No. 3, Summer, 1990.

paign has a twofold purpose—to help consumers identify honey-sweetened products quickly by looking for the bear symbol and to target products that use the word 'honey' but only contain a minute amount or a honey substitute." By mid summer of 1990 more than 55 products which include wine, pet food, meat, bread, spread, salad dressing, and cereal had been approved for logo use.

The National Honey Board Food Technology Programs

The National Honey Board is assisted by its agent Thomas J. Payne & Associates in the food technology field. Technical and scientific questions regarding the usage of honey in manufactured products are answered by T.J. Payne & Associates through a *"Honey Hotline"* newsletter distributed to food manufacturers as well as an *"800"* telephone number for telephone inquiries.

T.J. Payne & Associates monitors new honey products introductions constantly. It is interesting to note in 1988 there were 156 new honey products. In 1989 212 new honey products were added. An additional 279 new honey products were added in 1990. These honey products introductions included bakery goods, beverages, gum/snacks, cereals, condiments, dairy products, spreads, prepared foods, prepared meats, miscellaneous, and non food products (See Figure 15).

By percentage of new honey product introductions in 1990 condiments represented 18%, followed by bakery 16% confectionaries/snacks 14%, miscellaneous 12%, spreads 10%, dairy 8%, processed meats 6%, beverages 5%, breakfast cereals 5%, prepared foods 3%, and nonfoods, 3% (See Figure 16).

Additional opportunities exist for using honey in beverages, snack foods, animal food, and nonfood usage. With imagination, research, financing, and proper promotion expanded usage of honey can become a reality.

THE CHANGING SOCIETY OF THE UNITED STATES

The dynamic demographics of the United States create a marketing challenge and opportunity for the honey industry.

Our society is growing older. The median age for whites is now 33 years, for blacks it is 27.2 years, the highest median age in U.S. history. Seniors are the fastest growing population. Older consumers are more concerned about nutrition than younger adults.

America has more households, but fewer people in each household, down from 2.75 persons per household in 1985 to 2.64 in 1987. Fifty-three percent of the population growth is occurring in Texas, California and Florida, and the largest gains are in the Hispanic population.[19]

In the modern family that has two working parents or a single parent, time and convenience are critical issues causing a decline in the number of family

---

[19]Evans Food Group, *Consumer Trends Update*, April 1988.

meals. Children, especially girls 12 - 19, are becoming the family food shopper, spending 6 hours per week shopping for and preparing food for themselves and family members. They use lists, but improvise and have brand discretion. Seventy percent of all U.S. households now have at least one microwave, with a projection of 90 percent by 1990. Kitchens are becoming "warming centers" during the week, and cooking has become more of a "weekend gourmet" activity. Young couples are the largest users of take out and home delivery food, which needs only to be warmed in the microwave.[20]

The nutrition movement that began a few years ago appears to have had

[20]Evans Food Group, *Consumer Trends Update,* April 1988.

New Honey Product Introductions
1988, 1989, 1990

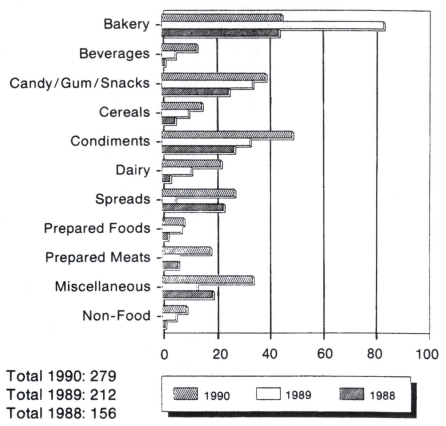

Total 1990: 279
Total 1989: 212
Total 1988: 156

FIGURE 15. *(Courtesy National Honey Board)*

little impact on young people, age 18-29. They are more likely to snack and choose sweets and they consume more soft drinks than any other age group.[21] Television commercials contribute to influencing this age group toward these foods, which illustrates the power of the "tube."

Assuming that the trend toward convenience foods and ease of preparations prevails into the next century, the challenge of the honey industry will be to integrate honey into the food fashion of the period. Whether this will be accomplished using honey as an ingredient, as an additive, fortifying some other food or beverage product, or a completely new concept in packaging, remains to be seen. Merely placing a container of honey on the shelf of the supermarket will not be enough. The beekeeping/honey industry will have to be creative and imaginative.

---

[21]Burros, M., Living, *The New York Times*, Jan. 6, 1988

New Honey Product Introductions - 1990

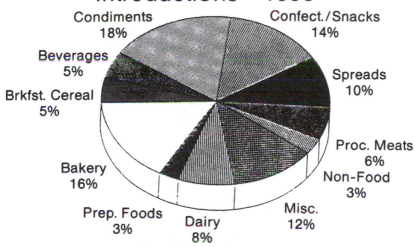

Total number: 279

Figure 16. Honey is also being used in a record number of new products and in some of the most dynamic areas of the food industry, including variety breads and cereals. Honey is being used in new categories such as dairy, snacks and microwave foods. This is important for future growth. *(Courtesy National Honey Board)*

REFERENCES

Associated Marketing, Chicago, Illinois, Survey, July 1987.

Bowsher, C., Report to Congress, Federal Price Support for Honey Should be Phased Out, GAO, RCED-85-107, Aug. 19, 1985.

Bauer, F.W., California Agr. Exp. Sta., Bulletin 776.

Burros, M., Living, New York Times, Jan. 6, 1988.

Diehnelt, W., Honey Acres, Inc., Ashippun, Wisconsin.

Evans, G., President, Sioux Honey Association.

Evans Food Group, Consumer Trends Update, April, 1988.

Federation News Letter, Vol. 15, No. 3-4, March-April, 1958, Page 3.

Foreign Agriculture Service, USDA, AGR. No. CH 8515, 6-21-88.

Gerbore, C., Response to Criticism of Honey Subsidy, Times-Call, Longmont, Colo., April 3, 1988.

Gamber, W., Dutch Gold Honey, Lancaster, Pa.

Graham, J., Editor, *American Bee Journal*, Hamilton, Ill.

Hoff, F., Economic Research Service, Honey, Background for 1985 Farm Legislation, Bulletin #465.

Hallguist, J., Foreign Agricultural Service, USDA, Wash. D.C.

Humann, M., There Is No Substitute for Honey, *American Bee Journal*, Oct., 1987.

Kennedy, Tracey L., Profiles of Potential Markets for U.S. Honey Exports, Preface, 1988.

National Agricultural Statistic Service, USDA, Wash. D.C.

Panko, F., Influence of Demographics on Foods and Feeding in Coming Decade

Payne, T., Thomas Payne and Associates, San Francisco, CA.

Phillips, J., Ag Economist, USDA, Wash. D.C.

Reynolds, G., Pres., Mid U.S. Honey Producers Marketing Assn.

Sioux Honey Assn. Annual Report, 1988.

Smoot, Don, Smoot Honey Co., Power, Mt.

Toth, G., Dr., Cosmetic Use of Hive Products, Facts and Prospects, *American Bee Journal*, June, 1988.

Universal Product Code Council, Inc., Dayton, Ohio.

World Honey Situation, USDA, FAS, Circular FS2-90, Oct. 1990.

CHAPTER 20

WINTERING PRODUCTIVE COLONIES

by B. FURGALA[1]
and D. M. MCCUTCHEON[2]

OUTDOOR WINTERING

The essential requirements, for wintering productive honey bee colonies have been known for almost 130 years. It was Langstroth who wrote: "If the colonies are strong in numbers and stores, have upward ventilation, easy communication from comb to comb, water when needed, and all the hive entrances are sheltered from piercing winds, they have all the conditions essential to wintering successfully in the open air" (Langstroth, 1859).

During the 1940's and 1950's Farrar (1952), while agreeing with the Langstroth requisites, added a significant new dimension to our understanding of the factors affecting wintering honey bee colonies by emphasizing the importance of keeping honey bee colonies "free" of bee diseases.

The former requirements, promoted through the years as means to avoid winter losses (Miller, A.C., 1903; Phillips, 1928; Gooderham, 1950; Dyce and Morse, 1960; Geiger, 1967), were challenged (Miller, C.C., 1903), and still are ignored by some individuals.

Winter losses are usually reported as the percentage of colonies that die before the first extensive examination in late winter or early spring. During the 1920's winter losses in the United States were reported to average 12.5% (Phillips, 1928), with losses in Wisconsin reported to vary from 5 - 30% (Wilson and Milum, 1927). In the 1940's Braun and Geiger (1947) reported average losses of 15% in Manitoba. In the early 1950's Farrar (1952) suggested that winter losses probably averaged 15%. He stressed, with good reason, that the monetary value of the reduced productivity of the colonies which survive poor wintering conditions most likely exceeded the monetary value of the colonies lost during winter and early spring. Wintering surveys conducted in the early 1970's indicate that 20 - 25% of the colonies wintered in Minnesota perish by May 15 (Hyser and Furgala, 1972), and recent surveys continue to show similar losses.

---

[1]Basil Furgala, Professor, Department of Entomology, University of Minnesota, St. Paul, Minnesota, 55108. Author of *Outdoor Wintering* section of this chapter.

[2]D.M. McCutcheon, Supervisor Apiculture program, British Columbia Ministry of Agriculture and Fisheries, Surrey, British Columbia, Canada, retired. Currently - Apiculture Consultant, Phero Tech Inc., Delta, British Columbia, Canada. Author of *Indoor Wintering* section of this chapter.

Obviously, the wintering losses that plagued beekeepers for decades have not declined to acceptable levels. This phenomenon persists despite the urgings of experienced apiculturists and research scientists that beekeepers fulfill the essential requirements when preparing honey bee colonies for winter.

Queen problems (queenlessness, supersedure, reduced fecundity), inadequate food reserves, poor protection, and disease, independently or in concert, continue to decimate many wintering apiaries. Most of the problems associated with winter mortality or with the weak colonies that survive the winter can be avoided if beekeepers satisfy 4 *fundamental principles* in beekeeping management:

1. Every colony must have a young queen of superior genetic stock.
2. Every colony must have an adequate supply of honey and pollen.
3. Every colony must be maintained in a "disease-free" condition.
4. Every colony must be properly established in a well constructed hive and protected from extreme climatic conditions.

If these requirements are fulfilled throughout the year, particularly in late summer and early fall when a colony is being prepared for winter, the colony will not only retain its remarkable ability to adapt to a very broad range of winter conditions, it will also be productive the following season.

Since the basic requirements are the same wherever honey bees are managed, wintering misfortunes must be due to management practices that overlook or conflict with the 4 *fundamental principles noted above.*

THE WINTER CLUSTER

Knowledge of how a colony winters took many decades to assemble. To obtain an overview of how the standards for dependable wintering evolved, a review of how a wintering cluster forms and how it responds to changes in temperature is extremely important.

The honey bees in every normal colony maintain a cluster throughout the year. During the active season it is a loose cluster that parallels, in general, the brood nest. The cluster plays a role in regulating the temperature and humidity in the brood nest. As air temperature falls, the cluster becomes more definitive. It was Phillips and Demuth (1914) who first reported that the winter clustering temperature of honey bees was approximately 57°F (14°C). Their studies indicated that when the temperature of the air immediately surrounding the bees reaches 57°F (14°C) the cluster becomes well defined, and, as the temperature continues to drop, honey bees in the center of the cluster generate heat, while those on the surface serve as insulators. The colder the temperature, the more compact the cluster becomes. The behavior of honey bees in the cluster was described in detail by Wilson and Milum (1927).

A significant contribution was made by Corkins (1930) when he reported that the temperature within the cluster remains constant regardless of the outside temperature. The next year Farrar (1931) cited in Eckert and Shaw, (1960) found that the temperature at the surface of the cluster remains between 43-46°F (6-8°C), even when the temperatures outside the hive are below freezing. The air temperature surrounding the honey bees must be approximately 43-46°F (6-8°C) before all of the bees become a part of the cluster.

When forming a cluster, honey bees on the surface establish an insulating shell which varies in thickness from 1-3 inches. Honey bees will enter the empty cells within the area of the food reserves embodying these cells, forming an integral part of the insulating shell (Fig. 1). As the external temperature rises above 45°F (7°C) the cluster expands. As the temperature drops below 45°F (7°C) the cluster contracts, reducing the surface area from which heat energy is radiated. Since honey bees use their honey reserves most efficiently at 45°F (7°C) (Betts, 1943), they do not consume as much of their stores at lower temperatures as might be expected (Gilbert, 1932).

The honey bees within the cluster are much less compact and generate heat through metabolic processes (Phillips and Demuth, 1914; Farrar, 1963). The heat produced within the cluster is conducted to the surface of the cluster. Sufficient heat is generated to equal the heat radiated from the surface of the cluster at approximately 45°F (7°C) (Farrar, 1952).

The contraction or expansion of the cluster, therefore, is the principal mechanism used by bees to sustain a favorable environment. This action will function as long as the cluster maintains firm contact with its food reserves.

There is a practical limit to which cluster contraction can occur. When this point is reached, continued or increased temperature stress overwhelms the heat-generating capacity of the bees because they lose contact with their food reserves. The internal and external cluster temperatures drop and the cluster perishes (Corkins, 1930). At any given low temperature, smaller clusters are more vulnerable than larger clusters as they must maintain higher inner cluster temperatures than do larger clusters (Farrar, 1963). See Johansson and Johansson (1979) and Winston (1987) for additional information on the winter cluster and thermoregulation of clustering honey bees.

Knowledge of what triggers cluster formation, how the cluster responds to temperature, and how the cluster temperature is regulated by the bees provides the beekeeper with the first ingredient for successful wintering. However, as noted earlier, successful wintering is maximized *only* if the 4 principles of beekeeping management discussed next are fulfilled during the spring, late summer, and early fall manipulations.

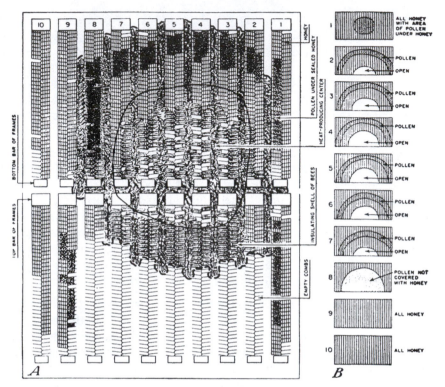

FIGURE 1. A, Diagram of the winter cluster as seen through a vertical section of a two-story hive cut across the middle of the cluster. B, Face view of frames of upper hive body. The numbers indicate the position of the frames. Note how the bees concentrate between combs and in open cells to form a compact insulating shell around a much less compact heat-producing center. The band of pollen covered with honey indicates an accumulation of reserve pollen before the honeyflow.

ESSENTIAL REQUIREMENTS FOR
SUCCESSFUL WINTERING

I. Young Productive Queen

Installing a young queen does not ensure the beekeeper that he/she has a productive queen. Despite this disadvantage beekeepers should replace queens at regular intervals or face a greater disadvantage, namely, having old queens in colonies during fall and winter, a condition that all too often brings about:

1. A supersedure in the fall, too late for the virgin queen to be mated. This results in a drone-layer.

2. A failing old queen in late winter or early spring. In the north, this happens too early for the virgin to mate and too early for the beekeeper

to requeen the colony. A more serious effect is that a void in egg laying occurs during a period when accelerated brood production is requisite to the proper development of the colony.

3. The death of an old queen during the winter, leaving the colony queenless.

Many beekeepers requeen annually. Others allow a queen to lay through two summer periods and one winter period (requeening conducted just prior to the spring nectar flow); or two winter periods and two summer periods (requeening conducted during the late summer flow). Cale (1964) found that it is also possible to requeen colonies after cold weather has set in and brood rearing has ceased.

Most honey producers requeen every year. Unfortunately, too many beekeepers allow colonies to requeen by supersedure. This is not advisable.

How often should a colony be requeened to preclude the failing and/or replacement of an established queen? Much depends on the pressure that has been placed on the egg-laying activities of the queen. In a climate where the season is short and honeyflow is brief, queens may remain prolific for two seasons. On the other hand, where the season is prolonged and/or where heavy egg laying is needed to build up colonies to exploit heavy nectar flows, requeening must be more frequent since the queen will be superseded much sooner. Recent studies in Minnesota (Sugden and Furgala, 1982; Duff and Furgala, 1986; and Duff *et al.*, unpublished) indicate that depending on the source of queens, 13 - 66% are superseded or missing after 16 months.

II. Adequate Reserves of Honey and Pollen

HONEY. The failure of beekeepers to provide wintering colonies with sufficient, properly organized honey reserves is the principal cause of winter mortality (Farrar, 1952; Eckert and Shaw, 1960). Braun and Geiger (1947) reporting from Brandon, Manitoba, Canada, concluded *"Bees do not freeze to death, they starve to death."*

The quantity of honey that is necessary to sustain a normal, healthy colony through the winter months will vary, depending upon latitude, altitude, and local climatic conditions. Strong, healthy colonies in the northern regions consume an average of 42 (Braun and Geiger, 1947), 48 (Boch, 1964), or 50 -55 (Farrar, 1960, 1963) pounds of honey from the time brood rearing ceases in the fall until sufficient nectar is available to support the colony in the spring. Most studies have been concerned with the loss of weight of colonies from late summer to early spring (Braun and Geiger, 1947; Boch, 1964). Sugden and Furgala (1982) comparing 6 commercial stocks of queens found average weight losses of 57 - 66 pounds during the October 15 - April 15 period. Weight loss results do not present the true picture since honey has been converted to bees and brood; the weight may be there in early spring but the food reserves are not. Moreover, since populous colonies consume more honey, a *minimum* of 60 - 70 pounds should be in place in the colony in late summer or early fall (Farrar, 1936; 1952; Arnott and Bland, 1954; Dyce and Morse, 1960; Johansson, T.S.K. and Johansson, M.P., 1969). At the University of Minnesota colonies are wintered in 3 deep Langstroth hive bodies with food reserves of 80 - 90 pounds, and a gross hive weight of 190 - 200 pounds.

In the more temperate regions 40 - 60 pounds of honey are recommended for winter food reserves (Jaycox, 1969).

Since colonies wintering in the southern regions have the opportunity to gather nectar intermittently during mid-winter and regularly during late winter, the food reserve requirements are reduced. Fifteen to 30 pounds (Morse, 1956; Eckert and Shaw, 1960) are usually reported to be sufficient to carry a colony through winter. This presents many southern beekeepers with a false sense of security. Colonies often starve during the winter months when confronted with brief periods of inclement weather.

A normal, healthy colony, with 20 pounds of honey above the minimum requirement is capable of coping with any short-term stressful situation during late winter. If one considers what it costs to replace a dead colony, be it in the cold north or the warm south, the extra honey in place in the fall is a small price to pay for successfully wintering a superior colony. Further, the practice of providing adequate reserves of honey in the fall reduces late winter and early spring labor costs and extra costs associated with the emergency feeding of dry sugar or sugar syrup. Extra honey consumed by a superior wintering colony is generally replaced by the colony with aggressive foraging on spring floral sources. Honey that is not consumed will naturally influence the amount of nectar a colony must collect and "ripen" to provide adequate reserves for the following winter. A colony that is restrained by chronic starvation often gathers only enough nectar to meet its daily needs. Buildup is retarded and surplus honey yields are nominal.

The proper organization of food reserves is as important to a wintering colony as the amount of reserves. There must be an upper food-brood chamber and a lower brood-food chamber(s) as illustrated in Figure 2. The wintering cluster must be able to surround or at least maintain broad contact with honey throughout the winter. The cluster in the wintering unit invariably moves up among the dark brood combs in the upper food-brood chamber(s). It is important that there remain a small honey and pollen-free area to serve as the center of the cluster (Fig. 2), although this is disputed by some. The cluster tends to avoid a food-brood chamber that consists of new, white combs, or even dark combs, if the latter are solid with capped honey.

In the more northern tier(s) of states, an upper (standard Langstroth hive body) food-brood chamber ought to contain approximately 45 pounds of

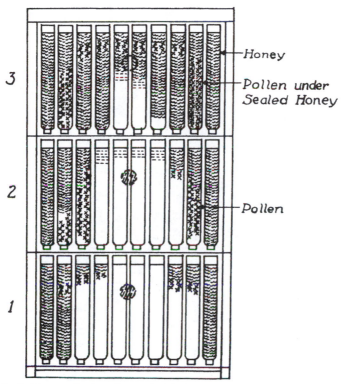

FIGURE 2. Desirable organization of stores in fall: 3-story standard hive. Note: pollen in upper chamber is under sealed honey; honey in ends of center frames does not show in cross section.

honey at the time brood rearing ceases in the fall. This represents seven fully sealed combs of honey, with the remaining combs one-half to two-thirds full. The lower brood-food chamber ought to contain a minimum of 30 pounds of honey. If two brood-food chambers are used, each should contain at least 20 pounds of honey. This arrangement provides reserve honey that the bees can move into the cluster whenever favorable temperatures allow flight and/or movement of bees in late fall, late winter, or early spring.

A colony will starve in mid winter even though there is plenty of honey in the lower brood-food chamber(s), if there is too little honey in the upper food-brood chamber. A colony may starve if the upper food-brood chamber is honey-bound since, as mentioned earlier, the cluster often fails to move up but remains in the lower chamber(s).

The three standard Langstroth hive body unit illustrated in Figure 3 or its equivalent in hive space, with adequate and proper organization of food reserves, will preclude starvation as a significant cause of winter loss in the

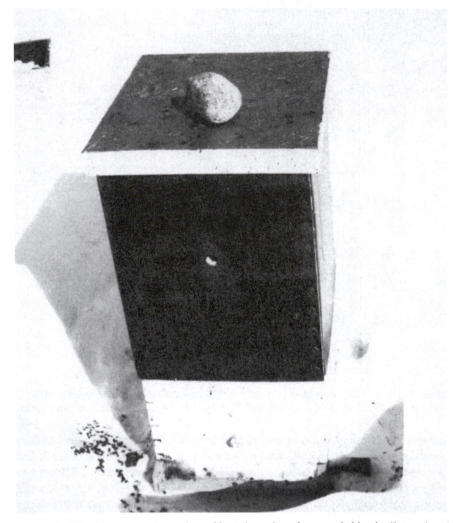

FIGURE 3. Wintering honey bee colony. Note three deep Langstroth hive bodies, reduced bottom entrance, upper entrance and commercial winter carton. Gross weight 180-200 pounds. This unit will provide at least one division by mid-spring.

cold northern states or parts of Canada. Sugden *et al.* (1988) comparing four methods of wintering colonies in Minnesota found that significantly more colonies wintered in three deep hive bodies survived the winter than did colonies wintered in two deep hive bodies. Johansson and Johansson (1971, 1979) examined many factors which affected wintering. They related winter loss after a severe winter in the northeastern area of the United States to: Age and source of queens; food reserves; colony population; wintering location;

and windbreaks. Their results vividly demonstrated the effect of honey stores on successful wintering. Only 18 percent of the colonies with 70 or more pounds of honey stores died compared to 55 percent of those colonies with less than 65 pounds of honey stores. Johansson and Johansson (1984) conclude the most suitable colony for winter is a three hive body unit. Three hive bodies allow sufficient space for an adequate population of bees and adequate stores. Smaller configurations put a stress on either bee population or food reserves. Moeller (1977) reported that colonies which consume more during winter produce more honey the following year. The three chamber unit allows adequate space for bees and brood, with a hive body in place that can be used to make a strong division six to eight weeks before the main nectar flow (Sugden *et al.*, 1988). Further, a three chamber wintering unit can be intensively managed to produce many times the surplus usually obtained from a struggling weak colony.

POLLEN. A colony may remain alive through mid-winter only to dwindle in population during late winter and early spring. Dwindling can be due to diseases (see Chapter 25) and/or insufficient pollen to support normal brood rearing. Lack of the nutrients present in pollen is a cause of spring dwindling in a colony which is otherwise normal and healthy.

Pollen, like honey, must be available in adequate amounts and be readily available to a wintering colony (Farrar, 1934, 1936). When Farrar (1934) varied the amount of pollen in wintering colonies, he found that spring populations in colonies wintered with no pollen were reduced by 78 percent; populations of those colonies wintered with 600 sq. in. of pollen were reduced by only 6 percent. The presence of pollen in the hive is particularly important during winter (in the south) or late winter and early spring (in the north). A wintering colony must replace its fall population with young bees and have a large, active brood nest by the time natural pollen is available in winter, late winter or early spring.

In most northern areas, colonies should enter the wintering period with an equivalent of three to five combs well-filled with pollen. A reserve of approximately 500 sq. in. of cells with pollen is desirable (Farrar, 1934, 1960). Pollen, like honey, must be available within the cluster if it is to be used by the wintering bees to rear a normal amount of brood.

The problem of wintering colonies with an inadequate amount of pollen is more acute in some areas. It is important that each beekeeper determine whether his/her colonies have adequate amounts of pollen and be prepared to remedy any needs by providing one of the various substitutes and/or supplements recommended in the literature (Standifer, 1967; Shesley and Poduska, 1968; Jaycox, 1969, Furgala, 1977). (See Chapter 6).

Excellent results are obtained in the north when patties of pollen substitute or supplement are placed on the top bars of the upper food-brood chamber five to six weeks in advance of the availability of natural pollen. It is important to locate patties so that they are in contact with the cluster (Fig. 4A), and once substitute or supplement feeding is initiated, replacement patties must be continually provided as needed (Fig. 4B) until natural pollen is available (Wolfe, 1950; Jaycox, 1969; Furgala, 1977).

Normal colonies, in the south as well as in the north, begin brood rearing several weeks before an adequate supply of fresh pollen is available. It has been demonstrated that supplemental feeding in late winter and early spring stimulates brood rearing (Rosenthal, 1962; Villumstad, 1964; Langridge and Rufford-Sharpe, 1966; Shesley and Poduska, 1968; Furgala and Hyser, 1974) which in turn leads to increased honey yields (Wolfe, 1950; Walstrom, 1950; Langridge and Rufford-Sharpe, 1966).

Feeding a Colony in the Fall

SUGAR SYRUP. Supplemental feeding in the fall fulfills two requirements: 1. It allows efficient and effective administration of chemotherapeutic agents (see Chapter 25—*Diseases and Enemies of the Honey Bee*); 2. It provides a wintering colony with the recommended amount of winter stores.

Each gallon of heavy syrup (two parts sucrose: one part water, wt/wt or wt/vol) or high fructose corn syrup (HFCS) will increase food reserves by approximately seven pounds. Various methods are used by beekeepers to

FIGURE **4.** A. Pollen substitute (Soybean flour, brewers yeast, dry skim-milk powder + pasteurized honey) fed the first warm days of March (Minnesota, 1973).

FIGURE 4. B. Ten days after feeding pollen substitute. Note second substitute pattie in place.

dispense sucrose syrups: a) Gravity feeders [metal, plastic (Fig. 5), glass], with numerous holes pierced or drilled in the lids; b) Special homemade feeding trays; c) Division board feeders; and, d) In some cases, large drums filled with syrup (open feeding) exposed to active robbing in the apiary. This latter method is not recommended since cattle have died after consuming the syrup.

Southern beekeepers should be particularly aware of the need for supplemental feedings in midwinter since large amounts of food reserves are generally lacking in the fall. A special effort must be made to maintain an equivalent of three to four frames of natural honey in each hive or "honey" elaborated from sugar syrup.

Need for Winter Examination

Despite a beekeeper's most careful preparations (to assure the presence of a good queen, adequate food reserves, good protection, and a disease-free condition in each colony), it is necessary to make midwinter (south) or late winter (north) inspections to answer two significant questions—*Is the colony dead or alive? Should I feed or not feed?* As mentioned earlier, queens in colonies wintered in the north commence egg laying in early to late January (Farrar, 1943; Phillips, 1928) and brood rearing will expand if sufficient pollen is available within the cluster even when the external temperatures are well below 0°F. (-17.8°C.). The availability of food reserves to the cluster becomes critical since the cluster generally will not leave the brood to maintain contact with its food reserves. This problem is of particular signifi-

Figure 5. Seven pound plastic honey pails, covers drilled with 30 - 40, 1/16″ holes. Pails are used as gravity feeders to dispense sugar syrup.

cance in the north during late winter when the brood area may consist of several hundred square inches.

Each colony should be inspected in February, certainly by the first warm days in March. Hives in which the clusters are in contact with frames of honey and where four to five frames of reserve honey are observed are secure for an additional three to four weeks. In Minnesota, Sugden and Furgala (1982), Sugden *et al.* (1988), Duff and Furgala (1986) ensure adequate reserves at all times by wintering in three Langstroth deep hive bodies with 45 pounds of reserves in the top hive body and a gross colony weight of 180+ pounds.

A dead colony must be sealed, or dismantled and moved to a bee-tight unheated facility. If, during the fall inspection, a shortage of pollen is determined to be a limiting factor, a pollen substitute or supplement must be provided (Fig. 4A, B).

Colonies that have sufficient but improperly organized food reserves can be adjusted quickly by centering the cluster and rotating frames of honey to ensure adequate honey in contact with the cluster. Sugar syrup may be fed at this time. Some beekeepers prefer to feed dry sugar over the inner cover as shown in Figure 6. Emergency feeding (syrup or dry) can be avoided by wintering colonies in a three hive body configuration. Pollen combs are often found in the lower hive body, away from the cluster. Pollen substitute or supplement should be provided.

In the temperate regions this initial inspection, with necessary adjustments and/or feedings of honey, sugar syrup, pollen substitute or pollen supplement, will generally assure survival and normal development of the colony until the first sources of natural nectar and pollen are available. Where winters are more severe, one or more follow up inspections (at 2 - 3 week intervals) with additional feedings of pollen substitute or pollen supplement may be necessary.

It must be stressed that late winter and very early spring represent critical periods for wintering honey bee colonies. Consumption of food reserves increases dramatically to satisfy the needs of an expanding brood nest. Holte (1970) monitored a scale colony during the 1969 - 1970 season in St. Paul, Minnesota. Total weight loss for the winter season was 51 pounds. The weight loss during the 86 day period between Nov. 21 - Feb. 15 was 10 pounds. The

FIGURE 6. Wintering honey bee colony feeding on granulated sugar poured over inner cover.

weight losses during the subsequent 28 and 26 day periods were 12 and 18 pounds, respectively. A colony with minimal food reserves can suffer from acute or chronic starvation. This problem can be prevented by examining each colony as early as possible in winter.

III. Keep Colonies "Disease-free"

Honey bee diseases and parasitic mites (Furgala *et al.*, 1989) may take a significant toll of wintering colonies. Careful inspection for the symptoms of American foulbrood, and an examination for honey bee tracheal mites and varroa mites in late summer and early fall, followed by a microscopic examination of adult bee samples in late winter and early spring to detect *Nosema* can reduce winter losses if a reliable chemotherapeutic program is implemented. Diseases and exotic mites are often responsible for winter loss, late winter and early spring dwindling, queenlessness, drone laying queens, and supersedures. (See Chapter 25 for a full discussion of bee diseases).

IV. Proper Protection

Some form of protection has long been considered a requisite for wintering honey bee colonies in the colder regions. During the latter 1800's and the early decades of the 1900's, protection was achieved by confining colonies in cellars and trenches, or by heavily insulating any hives wintered out of doors. Wooden wintering cases or tarpaper wraps, both with several inches of insulating material, were recommended (Miller, C.C., 1903; Phillips and Demuth, 1914; Merrill, 1920, 1923; King, 1923; Wilson and Milum, 1927; Gooderham, 1945). These methods and the recommendation to use heavy insulation were based on the theory that conservation of heat would reduce cluster activity and this in turn would conserve bee energy and reduce food consumption. Proposals to reduce the amount of packing (Miller, A.C., 1901, 1903), based on limited experimentation, were met with scorn (Miller, C.C., 1903). It was believed that fluctuating winter temperatures, accumulation of indigestible material, and lack of windbreaks (Merrill, 1920, 1923; Wilson and Milum, 1927) were responsible for winter losses. Insulation, it was believed, kept bees alive for longer periods, possibly overriding the above mentioned factors (Phillips and Demuth, 1914). Reports that colonies with little or no packing, could be wintered out of doors as readily as those heavily packed, did appear in the journals (Wilson and Milum, 1927; Gilbert, 1932), as did reports that heavily packed colonies consumed as much if not more food reserves as did lightly packed or unprotected colonies (Wilson and Milum, 1927; Braun, 1949). Corkins (1930) presented data from which he concluded that there was no foundation to the theory that subzero temperatures induced honey bees to expend great quantities of energy. He found that a colony consumed more honey when the outside temperature was 28°F, (-2°C) and above, than it did when the average outdoor temperature was

15°F (-9°C) and below. Many researchers and beekeepers continued to be influenced by the theory of conservation of heat because the numerous experiments concerned with wintering presented results that were inconsistent. The 1930's, however, brought about a gradual change; more colonies were being wintered out of doors; lighter packing, wrapping, or no protection at all were being substituted for heavy insulation when colonies were prepared for winter.

During the 1940's, debate continued on the value of heavy insulation versus light insulation or no protection at all. The advantages and disadvantages of an upper entrance to a wintering colony were argued at meetings and symposia (Gates, 1942). The continued shift to light protection was, in part, a delayed response to the early studies (Miller, A.C., 1901, 1903; Wilson and Milum, 1927; Corkins, 1930) which indicated that colonies with little or no protection wintered well in many areas in the north. The persistent writings of Farrar (1943, 1944, 1952), in which he lucidly described the behavior of the winter cluster and convincingly explained the shortcomings of the theory of conservation of energy, probably played the major role in accelerating the *conversion to the less time-consuming methods of preparing colonies for winter.*

The upper entrance became a feature of the wintering colony after it was demonstrated that little if any heat was lost through the opening (Anderson, 1943) and that it served as an emergency exit when the lower entrance was blocked. Recent studies indicate that ventilation above the brood nest in winter gives the best wintering results under a wide range of packing and insulating methods (Cherednikov, 1964; Martinovs, 1972). Other studies (Taranov and Mikhailov, 1960), suggest closed bottom entrances increase CO_2 concentrations, slow air movement, retard condensation of vapors and reduce food consumption. The authors recommend only top entrances for wintering colonies.

The revelation finally prevailed that non-insulated colonies wintered out of doors could successfully survive the cold winters of the northern United States and parts of Canada (Braun and Geiger, 1947; Lysne, 1954; Dyce and Morse, 1960; Boch, 1964). The empirical knowledge gained by beekeepers in trial and error wintering methods encouraged most northern beekeepers to alter their wintering practices (Mraz, 1947, 1962, 1967).

The discoveries demonstrating that the winter cluster does not attempt to heat the inside of the hive laid to rest the concept that heavy packing of wintering colonies conserves a significant amount of heat and energy. In the 1970's, Peer (1978) described an insulating "warm pack" method of wintering honey bee colonies in the Nipawin region of Saskatchewan, Canada. The procedure involved placing four 2-brood chamber colonies on a pallet, wrapping the sides with R11 fiberglass insulation, the top with R22 insulation, and covering the unit with a sheet of 1/4" plywood (Fig. 7A,B,C,D). Appropriate upper entrances were provided. Research in Minnesota (Sugden *et al.*, 1988) recently compared this "Peer warm pack" method with the standard non-insulated commercial winter carton "Minnesota light pack" method (Fig. 8A,B,C,D) and two other configurations ("Peer warm pack" in three hive bodies; "Minnesota light pack" using two hive bodies). Highest winter survival occurred in colonies wintered in three brood chambers. There were no significant differences between the three brood chamber "Peer warm pack" and the three brood chamber "Minnesota light-pack." Significantly more colonies survived the winter in the three brood chamber carton "Minnesota light pack" than in the two brood chamber "Peer warm pack" or a two brood chamber configuration of the "Minnesota light pack." More divisions were made from the three brood chamber configurations. The authors concluded that the convenience of using the "Minnesota light pack" warranted its use, at least in Minnesota.

Most beekeepers in the northern regions continue to use some form of packing or wrapping as a method of providing protection to wintering colonies. Colonies are protected with commercially treated cardboard winter packs (Boch, 1964; Haydak, 1967; Sugden *et al.*, 1988) (see Fig. 7C) or moisture-proof paper (Edmunds, 1961; Mraz, 1962; Jaycox, 1969). The interiors of hives protected in this manner tend to cool off more slowly with declining temperatures than the interiors of unprotected hives, but, conversely, with increases in external temperature they also warm up more slowly (Farrar, 1952). The former condition allows more time for a cluster to shift to a more favorable position, while the latter restricts movement during short-term fluctuations in temperature.

Table 1. Mean winter survival and productivity index. Minnesota locations. 1984-87

Treatment	Winter survival (%)	Mean productivity Index[1] (lbs.)
MN-2[a]	70.2[2]	1510
4 PAK-2[b]	69.0[2]	1545
MN-3[c]	91.0[2]	2185
4 PAK-3[d]	83.3[2]	1944

[1]Productivity index=(No. of colonies surviving winter) x (Mean productivity of parents + mean productivity of divisions)

[2]N = 36

a) Minnesota Light Pack - 2 Deep Brood Chambers
b) Peer Warm Pack - 2 Deep Brood Chambers
c) Minnesota Light Pack - 3 Deep Brood Chambers
d) Peer Warm Pack - 3 Deep Brood Chambers

FIGURE 7A. "Peer warm pack" ready to be pushed together after fall feeding. Note bees on top bars.

FIGURE 7B. Top insulation, cut to size, being layered.

FIGURE 7C. Top cover wrap folded into place.

FIGURE 7D. Ready for winter with upper entrances secured, plywood tied down— WHEW! *Photos by M.A. Sugden.*

A properly prepared colony is one which is provided with some protection (pack or wrap) to temper any drastic temperature fluctuation in late fall and late winter (L'Arrivee, 1961); an upper entrance to ensure ventilation (Anderson, 1943) and a moisture-releaser (insulite board) over the inner cover to allow moisture to be absorbed and released (Edmunds, 1961; Sugden *et al.,* 1988) precluding the accumulation and subsequent freezing of the products of metabolism (Geiger, 1967). Wintering colonies in sunlight, protecting them from the prevailing winds, and locating them where there is good air drainage provide additional tangible and intangible benefits.

Southern beekeepers need not be concerned with the question: *"To wrap or not to wrap."* However, as will be noted later, the effect of external winter temperature fluctuations on the utilization of food reserves by the cluster is a problem often ignored.

WINTERING COLONIES IN THE WARMER REGIONS

In those areas where the average January outdoor temperatures range between 50-68°F. (10-20°C.), 15,000-20,000 bees are sufficient to success-

FIGURE 8A. "Minnesota light pack," note few bees on top bars. They do not have to be grouped on pallets.

FIGURE 8B. Insulate board over inner cover, winter cover being placed over hive.

FIGURE 8C. Outer cover on—Done!

FIGURE 8D. "Peer warm packs" and "Minnesota light packs" wintering near Thief River Falls, MN. *Photos by M.A. Sugden.*

fully carry a colony through the winter season (Morse, 1956; Eckert and Shaw, 1960). The fact that bees are seldom confined to their hive for more than a few days at a time eliminates the need for large quantities of reserve honey and pollen. Additional protection in the form of a tar paper wrap or cardboard sleeve is also unnecessary. However, good queens, adequate food reserves, protection from climatic extremes, and a "disease-free" unit are still requisites for the wintering colony (Eckert, 1954; Eckert and Shaw, 1960). Lower entrances should be reduced, an upper entrance provided, and the colony sheltered from the cool, wintery winds.

Periodic winter inspections should be conducted to determine whether additional food is needed. Adequate reserves, provided in the fall (30-60 pounds), should carry a colony through a prolonged dearth of nectar (Eckert, 1954). A super of honey with pollen is good insurance against starvation and poor buildup in southern California, Texas, Georgia, and Alabama (Anon., 1971). However, brood rearing activity and winter flights do increase food consumption. A winter inspection, with a feeding when necessary, can save a colony which would otherwise die from starvation.

In the less temperate areas or at higher elevations, additional food reserves are recommended. Up to 75 pounds of honey and added protection (tar paper wrap, commercial cardboard sleeve) are generally used to winter a colony.

WINTERING WEAK COLONIES

In the northern regions it is advisable to kill colonies that are not in condition to winter. A weak colony which is otherwise normal can be united with an average colony [Kill the queen of the weak colony; and then place the weak colony under a strong colony, separated by a sheet of newspaper.] Two weak colonies can be united using newspaper between the two colonies. The beekeeper must follow up and organize the food-brood and brood-food chambers after the union is effected. Uniting colonies is not recommended and should be avoided to prevent the possible spread of diseases and exotic pests.

Small, healthy colonies can and do survive winter even under severe climatic conditions, but they seldom expand sufficiently in strength during the next season to provide divisions and/or become superior honey producers or pollinating units. When a beekeeper is faced with the prospect of wintering many weak colonies, he/she should heed the words of Farrar (1944): *"It is good beekeeping practice to take winter losses in the fall."*

A Synopsis

An understanding of how temperature initiates the formation of the winter cluster and continues to affect the activity and behavior of the cluster predicates certain colony standards that must be met to properly and reliably winter honey bee colonies. Winter and early spring losses can be avoided if colonies that do not meet optimum standards for wintering are culled in early fall. *"The problem of management is not how or where, but what kind of colonies are wintered"* (Farrar, 1963).

A normal, unrestricted colony at the close of brood rearing in early fall should consist of up to 30,000 physiologically young bees. This population can consume 10 - 15 pounds of honey (Holte, 1970) and lose 3,000 - 5,000 individuals before egg laying commences in mid winter (Farrar, 1963).

The initiation of brood rearing increases both food consumption (Holte, 1970) and mortality of adult bees (Farrar, 1963). A "disease-free" colony with ample food reserves, will begin to replace the individuals that die. By the time the first spring flowers bloom, a normal productive colony should once again consist of 30,000 adult bees and as many cells of brood in all stages of development. With favorable conditions, the population can increase to a maximum of 45,000 - 60,000 adult honey bees within two months. With this population a colony, given a three to four week nectar flow, should fulfill its maximum productive potential.

OPTIMUM FALL CONDITION OF A WINTERING COLONY
Cold Regions

In those areas where the average temperature of the coldest month is 20°F. (-6.7°C.) or lower, the normal productive colony should meet the following standards at the close of brood rearing in the fall:

1. The wintering unit must have a good productive queen, supported by a worker population that covers 10 - 15 frames.

2. The wintering colony should consist of 3 brood chambers with a minimum of 80 pounds reserve honey (preferably 80 - 100 pounds). The reserves should be properly arranged, with 45 pounds in the top hive body (food-brood chamber) and 15 - 30 pounds in the middle hive body, and 10 - 20 pounds in the bottom hive body (brood-food chamber). Wintering in three hive bodies, (Fig. 2), will significantly reduce winter loss. *There must be room in the fall for both adequate food reserves and a large worker population.*

3. The wintering colony should have approximately 500 sq. in. of pollen distributed as illustrated in Figure 2.

4. Each wintering colony must be "disease-free."

5. Each wintering colony should be provided with adequate protection and other adaptations to reduce the effects of certain stresses.

    i) Supplemental protection in the form of a commercial wintering carton or tar paper wrap with a moisture releaser over the inner cover.

    ii) Shelter from the prevailing winds, with good air drainage and maximum exposure to sunshine.

    iii) Bottom and upper entrances; lower entrance reduced (1/4 x 3 inches), upper entrance an auger hole (1 inch), located just below the handhold of the uppermost hive body.

    iv) Hive stand to provide a dead air space under the hive.

A two-story Langstroth hive (with frames and bees) weighs approximately 70 pounds. A minimum gross weight of 130 pounds will ensure about 60 pounds of food reserves. Winter survival is less reliable and surviving colonies are weaker on average (Sugden *et al.,* 1988. A three-story Langstroth hive weighs approximately 90 - 95 pounds. A minimum gross weight of 180 - 200 pounds will ensure 80 - 100 pounds of food reserves. Winter survival is significantly greater and surviving colonies are stronger (Sugden *et al.,* 1988).

Temperate Regions

In those areas where the average temperature of the coldest month ranges between 25 - 45°F. (-4 to +7°C.), supplemental protection is not necessary. The amount of food reserves can be reduced appropriately, with 30 - 60 pounds generally recommended (Eckert, 1954; Jaycox, 1969). A fall population covering 8 - 10 frames can survive the shorter period of confinement.

A good productive queen, reserve pollen, "disease-free" condition, shelter from prevailing winds, air drainage, upper entrance, and reduced lower entrance remain requisites for efficient wintering.

Warm Regions

In the warm regions where the average temperature of the coldest month exceeds 50°F. (10°C/), food reserves of 15 - 30 pounds have been reported to be adequate (Morse, 1956).

OPTIMUM SPRING CONDITION OF WINTERED COLONY
Cold Regions

The hazards of wintering bees in the cold regions are reduced significantly when increased food reserves are provided in the fall. Starvation and late winter and early spring emergency feedings are avoided if these honey reserves are supplemented with the feeding of pollen substitutes and/or pollen supplements in late winter and early spring. Good productive colonies should be in the following condition at the beginning of dandelion or similar bloom:

1. Good productive queen.
2. 15 - 20 frames of bees (7 - 10 pounds).
3. 8 - 12 frames of brood (all stages).
4. 20 pounds or more of reserve honey.
5. Continuous supply of pollen and/or pollen substitute or supplement.
6. Adequate space for incoming nectar and upward expansion of the brood nest (This can be achieved by reversing hive bodies).

The Relation of Wintering to Productive Management

To provide a colony the optimal requirements for the nonactive season, the beekeeper must apply skillful management during the productive season. To prevent the colony from swarming a strong colony requires more hive space, properly organized for expansion in brood rearing and honey storage. Queen longevity may be reduced in colonies that are provided with pollen or pollen substitutes and/or supplements because they lay more eggs in late winter and spring. Honeyflows may occur from plants formerly thought to be only of minor importance because populous colonies forage more intensely. Increases, made by dividing strong colonies (introducing young queens in the divisions) will be necessary as a swarm-control measure prior to the main nectarflow.

Colony standards have changed since Langstroth's time, but the principles he outlined cover the requisities for wintering and we need only to assign the necessary values. A colony "strong in stores" now has 60 - 90 pounds of honey and 500 sq. in. of pollen. Although Langstroth recognized the need for pollen,

100 years elapsed before the significance of the pollen supply and proper use of pollen substitutes and supplements were understood.

The "upward ventilation" he recommended has been readopted with the common use of notched inner covers or auger-hole upper entrances. Langstroth allowed communication between combs by cutting holes through the center, while today the same objective is accomplished in hives of more than one story by the space between the sets of combs.

The importance of water for spring brood rearing has been appreciated, but it is still left largely to chance. Protection from wind, using natural vegetation surrounding the location, or wrapping the hive, is considered beneficial to the colony. To Langstroth's principles may be added freedom from infectious diseases, infestation by mites, exposure to maximum sunlight, good air drainage, and winter inspection to correct unfavorable cluster positions.

Langstroth, if living today, would be disappointed by the number of neglected colonies. Yet he would be surprised and equally pleased with the strength of colonies made possible through the use of larger hives, more honey, more pollen or pollen substitutes and supplements, productive queens, chemotherapy, and intelligent management. Such colonies produce correspondingly larger and more certain honey crops because many of the elements of chance have been eliminated by intensive management.

INDOOR OVERWINTERING

Wintering bee hives indoors is a successful alternative to outdoor overwintering due to present-day technology. However, wintering inside is not a new idea.

Wintering beehives in cellars is described in great detail in beekeeping publications up to, and including, the 1920's (Gooderham, 1926; Pettit, 1917; Phillips and Demuth, 1918; Sladen, 1920). Areas where cellar wintering was considered desirable included the Canadian Prairie Provinces, Northern Ontario, Quebec, and the North Central United States. (Phillips and Demuth, 1918). The conditions set down for successful cellar wintering were total darkness and even temperatures at which bees would remain quiet. Brood rearing was not expected to begin, nor was it desired until the hives were removed from the cellar.

It appears there were a number of disadvantages to cellar wintering as practiced in the first quarter of the century. There was difficulty in deciding when to move the hives into the cellar. If they were placed inside too early or the weather became too warm following "cellaring" it was difficult to control the cellar temperature.

FIGURE 9. Early cellar wintering was often unreliable and always back-breaking labor.

In the spring, due to restlessness of the bees, hives often had to be removed earlier than would be desirable, since there were usually periods of inclement weather following their removal from the cellar. While bees were "cellared," considerable attention was necessary in providing proper ventilation.

These problems of timing in "cellaring" and removal, and in controlling cellar temperatures and humidity, were factors in lessening the interest in cellar wintering. In addition, there was a considerable amount of heavy work in carrying hives in and out of the cellar (Figure 9). Outdoor wintering methods had become successful and practical, thus cellar wintering became less popular.

From the mid 1920's to the mid 1960's, there were continuing investigations into cellar wintering. Those investigations took place mainly in Manitoba at the Brandon Research Station of Agriculture Canada. Between 1921 and 1939, records were kept on the length of the confinement period. This period averaged 160 days, but varied from 147 to 177 days (Braun, 1940). The lowest outside temperature recorded during the confinement periods was -47°F (-44°C) with -29°F (-34°C) being a common low.

Records at Brandon show they were able to control temperatures and humidity in their cellar (Geiger & Braun, 1955). Relative humidity ranged from 60-70% with temperature at 39°F (4°C). Bees remained quietest at 36 -39°F (2 - 4°C). Weight loss records indicate consumption of stores increased fairly uniformly with the advance of winter. The average winter consumption of stores for the several hundred hives wintered was 33 lbs. (15 kg) per hive for the average 160 days confinement. (Geiger and L'Arrivee, 1965). Investigations on the type of stores giving the best results showed that clover honey and cane sugar were equally good.

INDOOR WINTERING 1960'S - PRESENT
The easy availability of equipment for moving and distributing air, the electronic controls for temperature and humidity and cheap and easy to use insulation materials, have made it possible to winter hives inside more efficiently and economically than earlier in the century. Indoor overwintering has thus become a viable alternative to wintering outdoors.

The first reference to the use of temperature and humidity control equipment for overwintering bees was in 1961. Dr. J.C.M. L'Arrivee installed a high capacity air conditioner at the Agriculture Canada Research Station, Brandon, Manitoba.

The first practical use of such equipment appears to be by J. Kuehl and Max Cook of Loup City, Nebraska, in the late 1960's. These beekeepers wintered over 700 hives in an insulated wooden frame building. The bees were wintered in single brood chamber hives which were made up following the main honey flow and placed in the facility in late November.

Their success was apparent as the word spread to Manitoba, Canada. By 1974, Manitoba commercial beekeepers began overwintering in controlled environment buildings. Since then, many beekeepers in Quebec, the four Western Canadian provinces and the North Central States are overwintering hives indoors.

Physical Requirements
The wintering chamber must be kept completely dark inside (Nelson, 1982; Wrubleski and Bland, 1976). Honey bees are sensitive to noise and vibration. Every reasonable effort should be taken to reduce noise and vibration from motors and fans (Darby, 1988). Optimum temperatures are 39±3°F(4±1°C), but a range of 36 - 48°F (2 - 9°C) is satisfactory. Temperatures of up to 59°F (15°C) are acceptable on a warm spring day if adequate ventilation and airflow are maintained (Fingler and Small, 1982; Konrad, 1970; Nelson, 1979; 1982; Anon., 1980; Wrubleski and Bland, 1976).

The ideal relative humidity for a wintering chamber has not been determined. From 50 - 75% R.H. is considered satisfactory, but levels from 30 - 44% in Saskatchewan produced no apparent ill effects (Fingler & Small, 1982; Nelson, 1979, 1982; Wrubleski and Bland, 1976).

A recirculated airflow of 4.25 cfm (2.0 L/s) per 2.2 lbs. (1 kg) of bees is necessary. An average hive with 5.5 lbs. (2.5 kg) of bees would require an air flow of 10.5 cfm (5.0 L/s) under normal winter conditions (Fingler & Small, 1982). In fall and spring when temperatures may be higher, a flow of up to 10.5 cfm (5.0 L/s) may be required per 2.2 lbs. (1 kg) of bees (Fingler & Small, 1982; Nelson, 1982; Wrubleski & Bland, 1976).

The Chamber

Some beekeepers construct new buildings for wintering hives while others use building facilities available such as the room used for warming supers of honey prior to extracting. This warming room is usually close to the unheated main warehouse area, from which tempered air for ventilation can be drawn.

Size

The suggested space requirement is from 10.6 - 15.8 cubic ft. (.3 - .45 cubic meters) per single brood chamber hive (Fingler and Small, 1982; Nelson and Henn, 1977; Wrubleski and Bland, 1976). For double brood chamber hives, 21 - 26.5 cubic ft. (.6 - .75 cubic meters) per hive is required (Nelson, 1979, 1982; Darby, 1988). The lower loading densities allow for an additional volume of air per hive and thus better air circulation which assists in moderating temperature fluctuations.

Typical wall heights of 13.75 - 15.75 ft. (4.2 to 4.8 m) provide efficient space utilization, and allow for large doors for equipment access. A high hive stacking pattern gains maximum storage capacity for the available space, but is not essential for successful wintering (Darby, 1988).

Space requirements for bee wintering will depend on the method of handling and stacking hives. Most larger producers handle hives on pallets with some type of forklift. The building layout is planned accordingly with respect to stacking patterns and access for equipment.

Construction

The important factor in construction is insulation. The necessary insulation required on the walls is R20 (RSI 3.5) and R28 (RSI 5) on the ceiling. The foundation is insulated on the exterior to at least 2 ft. (60 cm) below grade with 2 in. (5 cm) of polystyrene (Fingler and Small, 1982; Wrubleski and Bland, 1976; Darby, 1988). A polyethylene vapor barrier must be placed on the inside of the wall. Wind-barrier paper under the exterior sheathing is beneficial. A concrete floor is recommended. Doors should be insulated and tight fitting.

A slightly different approach to an inside overwintering facility was taken by the Beaverlodge, Alberta, Agriculture Canada Research Station. The structure is simple and requires a low investment (Nelson & Henn, 1977). Hives are wintered in a building 6 ft. (1.8 m) wide, 6 ft. (1.8 m) high and 23 ft.

(7 m) long (approximately 650 cubic ft. [18.5 m³]). The structure is well insulated and has large double doors for ease in moving hives in and out.

An additional variation of the simple low cost building was designed by a northern Alberta beekeeper, H. Pirker. His building is 8 ft. (2.5 m) wide, 8 ft. (2.5 m) high and 33 ft. (10 m) long; insulated with 3 in. (7.5 cm) of fiberglass insulation. Pirker's system provides flight channels for each hive. These flight channels are 1.5 in. (3.75 cm) inside diameter black PVC plastic pipe, roughened inside to provide sufficient foothold for bees. The flight channels are closed from late fall, when hives are placed in the building, until late winter when they are opened, giving the colonies free flight in good weather. The building has a capacity of 56 hives. Forty-four hives are placed in two tiers facing the south wall. Six hives are placed at each end (Pirker, 1976, 1978).

Light Control

Since complete darkness is necessary, one must take steps to prevent light entry at air intake and discharge ports. Light traps can be designed which will cause minimal restriction for the fan. They are installed on the inside of the building at air exhaust or entry ports and are painted flat black to minimize light reflection. A cross section of a suggested light trap is seen in Fig. 10.

VENTILATION, TEMPERATURE, AND HUMIDITY CONTROL

Ventilation involves drawing in outside air, exhausting an equal amount of air, and distributing the air inside the wintering chamber. The ventilation process controls the carbon dioxide level, humidity and temperature.

Bees generate heat which will assist or completely take care of the heat requirements inside the storage area. Cooling is accomplished by bringing in outside air. Carbon dioxide and humidity levels are controlled by the exhausting of air with high CO_2 and water vapor content and replacing it with outside air. Air is distributed and recirculated within the building by means of fan systems.

Ventilation and Air Circulation

In most wintering buildings a combination of exhaust fans in conjunction with a recirculation fan attached to a polyethylene duct distribution system is used. When air is exhausted, fresh air is automatically drawn in from the air intake. The re-circulation fan is commonly called a jet fan. The polyethylene duct consists of a perforated tube running the length of the chamber perpendicular to the rows of hives. This forces the recirculated air down between the rows.

The perforations or openings are evenly placed and allow air to escape and thus be distributed. These systems are commonly used in greenhouses and poultry setups and ensure uniform conditions throughout the space. Such tubes may be up to 35 in. (90 cm) in diameter with perforations as large as 2.5

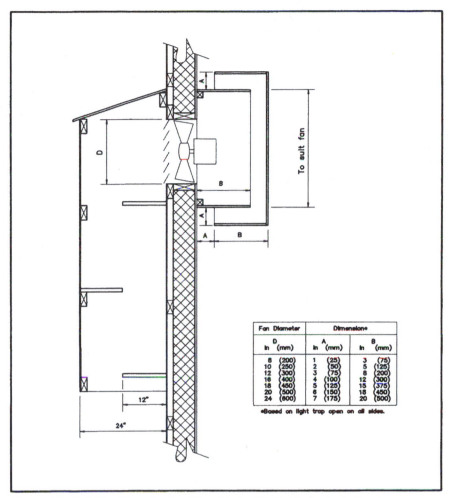

Fan Diameter		Dimensions			
D		A		B	
in	(mm)	in	(mm)	in	(mm)
8	(200)	1	(25)	3	(75)
10	(250)	2	(50)	5	(125)
12	(300)	3	(75)	8	(200)
16	(400)	4	(100)	12	(300)
18	(450)	5	(125)	15	(375)
20	(500)	6	(150)	18	(450)
24	(600)	7	(175)	20	(500)

*Based on light trap open on all sides.

FIGURE 10. Light Trap Cross Section *(From Darby, 1988, Alberta Agriculture).*

in. (6.35 cm). The diameter of the jet fan matches the diameter of the tube. The size and spacing of the perforations can be determined by a local government agricultural engineer extension person. Such persons will have at their disposal, a ventilation handbook for similar situations in poultry, hog and greenhouse setups.

Some wintering chambers have two systems, one being auxiliary in case of need. Others have a 2-speed fan. In jet fan systems, intake air is admitted through an adjustable opening near the jet recirculating fan; thus fresh air will be well mixed with inside air and distributed by the duct. The recommended air inlet size is 10.75 sq. ft. (1 square meter) for each 10,500 cfm (5,000 L/s) airflow. Under ordinary winter conditions, a ventilation air exchange of .5 cfm (0.25 L/s) is sufficient. A very large inlet might be required under very extreme spring temperature conditions when an air exchange of up to 10.5 cfm (5 L/s) per hive may be required (Fingler and Small, 1982; Darby, 1988).

Exhausting of air in large operations is accomplished by a series of 4 exhaust fans. In very cold weather, even though a heater may be operational, one fan operates continuously in some operations to provide humidity control. As the outdoor temperature and thus the indoor temperatures rise, fans 2, 3 and 4 are progressively activated by thermostats. As the temperature increases, the heater will turn off and fan 2 will begin. For each 2°F (1°C) rise in temperature, an additional exhaust fan will become operational (Fingler and Small, 1982). To ensure proper operation of the exhaust fans, thermostats should be located in close proximity at a central location in the room. The air distribution system ensures that conditions are the same in all areas of the room; thus the location of the exhaust fans may be in the most convenient place. Figure 11 shows a schematic ventilation system.

Over-ventilating causes low relative humidity; thus smaller rooms holding 500 hives or less require a much lower exhaust rate, therefore it may be difficult to secure a suitable fan of the size required. One method of overcoming this problem is to use one fan for both steps 1 and 2 in exhausting. This fan would be of a size suitable for step 2, but set to operate 3.3 of each 10 minutes

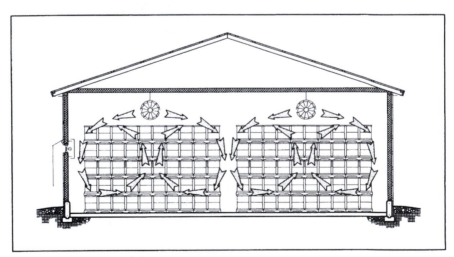

FIGURE 11. Air circulation in a building by polytube duct system. Two ducts are typical for buildings 32 - 50′ wide, ducts 24 - 36″ depending on air flow requirements. Air exchange is controlled by either exhaust fans or proportioning dampers on the air supply. *(From Darby, 1988, Alberta Agriculture)*

for step 1 exhausting. For step 2, the fan would be set to operate at full capability (Fingler and Small, 1982).

The Beaverlodge, Alberta, Research Station ventilation system, in keeping with the simplicity of their system, uses a 155 cfm (70.5 L/s) bathroom fan which pulls air in near the top of one end of the structure and forces it out through two exhaust ports at floor level. Each exhaust port has a slide to control air flow. The fan is controlled thermostatically and by time. Normally the fan operates for two 30-minute periods each day or when the temperature rises above 43°F (6°C).

Temperature Control

Some beekeepers feel the optimum temperature is close to 46°F (8°C), while others feel it is closer to 36°F (2°C), based on what they feel is proper

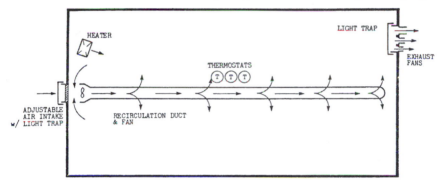

FIGURE 12. Ventilation system for overwintering hives. *(From Fingler and Small, Manitoba Beekeeper 1983)*

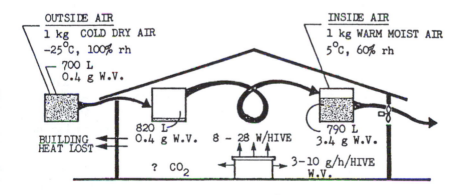

FIGURE 13. Winter ventilation for overwintering hives. *(From Fingler and Small, Manitoba Beekeeper 1983)*

activity within the hive. While it might appear that higher temperatures would reduce food consumption, in fact Quebec and Saskatchewan beekeepers' experience show the opposite is true. The cooler range reduces food consumption (Wrubleski and Bland, 1976). Cooler temperatures also provide an additional safety factor should ventilating and cooling equipment fail.

When the outside temperature is very cold, the heat produced by the bees is used to maintain the required temperature inside the storage area. While it may seem sensible under such conditions to cease operation of the exhaust fans, it is unwise to do so since bee colonies continue to produce water vapor and CO_2. Thus, even under very low temperature conditions, a low level of exhausting is necessary (Fingler and Small, 1982).

The exhausting of warm storage air and replacing it with cold air removes a large amount of heat. To heat outside air at the minimum ventilation rate from a winter temperature of -13°F to 41°F (-25°C to 5°C) requires 13 watts of power per hive. In addition, storage heat loss by conduction through building components equals 4 watts for a total loss of 17 watts/hive (Fingler and Small, 1982). Typically, heat production per hive is from 12 watts to 20 watts (Young and Feddes, 1988). Supplementary heating is therefore needed for the colder part of the winter to keep temperatures constant.

Usually, fan-forced electric heaters are used to supply heat. They are sized on the basis of 10 watts per hive to provide some margin of safety since heat production by bees varies considerably (Fingler and Small, 1982). The heaters are controlled by thermostats to maintain desired temperatures. Nelson indicates 600 - 1000 watts are required to provide heat when needed for 100 colonies (Nelson, 1979). The Beaverlodge Research Station system uses a 500 watt electric baseboard heater, thermostatically controlled to supply heat when necessary for 650 cubic feet (18.5 cubic meters) of space (Nelson and Henn, 1977). At temperatures as low as -40°F (-40°C), a 3 kw heater will provide sufficient heat for 100 hives or more at a high loading density 10 cubic ft. (0.28 m³/hive) in a properly insulated building with air flow of 0.25 cfm/2.2 lb. (0.10 litres/sec/ 1 kg) of bees. In one situation with temperatures as low as -40°F (-40°C), a 3 kw heater provided heat for 100 hives. No heat was required until outside temperatures reached 0°F (-18°C) or less (Barker, 1975). Many beekeepers find supplementary heating is not required in their operation.

Rising temperature problems often occur in late spring before hives are moved outside. With outside temperatures of 59°F (15°C) or more, it is difficult to maintain storage temperatures of 39°F (4°C). Most operations increase airflow rates to keep storage temperatures a few degrees below outdoor temperatures. Higher temperatures last only a few hours; thus this method has proven successful since the colonies can withstand temperatures up to 59°F (15°C) for a short period if a good supply of fresh air is provided.

A practical example of temperature control is by use of the exhaust and fresh air entry system. The main fan can exhaust 12,800 cubic ft. (364 cubic meters) of air per minute. An increase in temperature activates the fan. When the exhaust fan begins, the vacuum created opens a light, flexible plastic flap at the opposite end of the building through which fresh air is drawn. The plastic flap falls and closes the opening when the exhaust fan ceases operation. Additional exhaust fans are pre-set to operate two minutes in each 10 minute period.

The fresh air drawn in comes from the unheated warehouse. Air enters the warehouse from a covered, shaded area outside. Thus, this whole area acts as a cold air sump and provides an ample supply of cold air. Another method is to use a large stone pile with a pipe leading into the center. When cold air is needed, it is pulled from the center of the stone pile.

When an adjacent cool warehouse is not available for cooling air, an excellent method of controlling temperatures is a system of earth tubes through which air is drawn to stabilize temperatures in the wintering room. A 500 hive facility near Dawson Creek, British Columbia, uses 6-inch (40 cm) diameter corrugated plastic pipe buried 39 in. (1 meter) below the soil surface for cooling. Initially, two trenches holding three 100-foot (30 meter) long tubes each were installed to carry a 660 cfm (310 L/s) airflow. More recently 12 additional 6-inch (40 cm) tubes in 4 trenches were added so that most of the air could be buffered before entering the building. The tubes end in a mixing chamber before entering the air intake.

The system operates on the principle that the temperature 39 in. (1 meter) below the surface is slow to react to temperature changes. Thus, when outside air temperature increases, it can be cooled by drawing it through the tubes to cool the inside of the wintering building. Theoretically, air warmer than outside air can be drawn from the tubes during periods when outside temperatures are very cold.

Few overwintering facilities use refrigeration units for cooling during critical temperature periods. Cooling large volumes of fresh air requires a very expensive system and high operation costs. Most refrigeration units installed earlier are no longer used.

Humidity

Tests show that bee colonies are very adaptable to a wide range of relative humidity situations. Three groups of 28 hives each were overwintered at relative humidity ranges of 45 - 60%; 45 - 80%; 60 - 85%. Final results showed little difference in weight of dead bees, stores consumed, spring brood and population counts and in crop results (Proc. Can. Assoc. Prof. Apic., 1978).

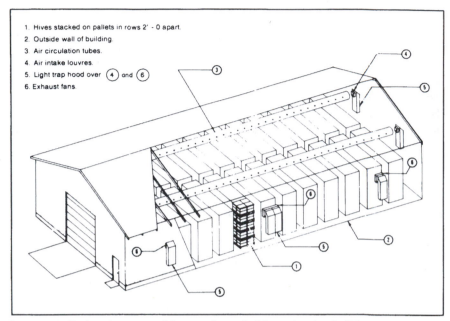

1. Hives stacked on pallets in rows 2' - 0 apart.
2. Outside wall of building.
3. Air circulation tubes.
4. Air intake louvres.
5. Light trap hood over ④ and ⑥
6. Exhaust fans.

FIGURE 14. Design elements and layout of a typical bee wintering building. *(From Darby, Alberta Agriculture 1988)*

Water content of the storage air is contributed from two sources: the outside air, and the bees. Each hive will produce 0.10 - 0.35 oz. (3 - 10 grams) of water vapor per hour (Fingler and Small, 1982). Under normal quiet winter conditions, the production will be 0.10 - 0.17 oz. (3 - 5 grams). Cool air from outside, when heated, increases its water-holding capacity. Thus, slightly warmed outside air can pick up additional moisture produced by the bees. This air is then exhausted, taking with it excess moisture.

In areas of very dry winter climate, cold outside air has a very low relative humidity (15 - 25%). When brought inside this air becomes lower yet in relative humidity since warm air can hold more moisture. This causes problems at times. If stores consist of too much solidly crystallized honey, bees will not be able to reliquefy it for their use (Barker, 1975). Bees will become very distressed under such conditions and large numbers of colonies will succumb to starvation.

Pirker showed that adding moisture to an inside overwintering system by means of a steam type electric humidifier, especially in mid-winter, encourages brood rearing. He states that when outside air contains less than 0.10 oz. (3 grams) of water per 35 cubic ft. (1 cubic meter) of air, moisture must be added inside the building. Inside relative humidity levels of 80% at 52°F

(11°C) in spring encourage a high level of brood rearing (Pirker, 1979, 1980). Figure 12 provides diagrammatic information on ventilation requirements for beehive overwintering buildings.

FALL PREPARATION OF HIVES

The process of selecting hives, re-queening and feeding should be started in August. Use of sugar syrup for overwintering stores appears to give best results. Honey, which crystallizes excessively, appears to cause stress, restlessness and dysentery (Wrubleski and Bland, 1976).

Hive Size and Preparation

The original concept of Max Cook of Loup City, Nebraska, was to prepare nucleus hives in late summer using brood and bees from summer producing colonies. Manitoba beekeepers found that the single brood chamber nucleus hives increased in population in the fall far beyond that experienced by Cook. They also found their more populous hives wintered far better at 39°F (4°C) rather than the 46°F (8°C) used by Cook.

Provision of Stores

Guidelines for provision of stores in the fall are as follows (Nelson and Henn, 1977; Nelson, 1979).

For single brood chambers: stores should total 53 lbs. (24 kg) for a total hive weight of 92 lbs. (42 kg).

For double brood chambers: stores should total 68 lbs. (31 kg) for a total hive weight of 132 lbs. (60 kg).

He also indicates the average consumption of stores for the overwintering duration is as follows:

For single brood chambers: 24 - 29 lbs. (11 - 13 kg)
For double brood chambers: 48 - 53 lbs. (22 - 24 kg)

If honey is used for winter feed, it should not be crystallized. On the Canadian Prairies, care must be exercised to not provide honey from rapeseed (Canola) as it crystallizes and becomes extremely hard. The fall feeding of sucrose syrup is generally recommended to overcome this problem.

Recommendations for feeding sucrose syrup range from 1 part sugar: 1 part water to 2 parts sugar: 1 part water. Fumagillin for Nosema prevention and oxytetracycline hydrochloride for foulbrood prevention are included in the syrup (Konrad, 1970).

Moving Hives Indoors

The usual practice is to finish feeding sucrose syrup as early as possible. In North-Central Alberta, hives are placed inside by late October. In more moderate areas, hives are moved inside during the first half of November (Nelson and Henn, 1977; Wrubleski and Bland, 1976).

Hives are usually placed back to back in 2 rows, with 39 inches (1 meter) spacing between the double rows. Hive entrances are left wide open and lids are usually left on. Hives are usually placed in the wintering room on pallets, four hives per pallet. If double brood chamber hives are used, pallets are stacked three high. To facilitate air circulation, 39 inches (1 m) should be left between hives and outside walls.

THE OVERWINTER PERIOD

Temperatures of 39±3°F (4± 1°C) are recommended as ideal for inside overwintering hives. Temperatures up to 48°F (9°C) caused consumption of additional stores of 2.09 lbs. (.95 kg) and 6 lbs. (2.7) kg per hive in two years of tests. Hives stored at a temperature of 36°F (2°C) show additional consumption of only 0.5 lbs. (.22 kg) and 1.0 lb. (.5 kg) in stores, respectively, for the two years. This information suggests the lowest metabolic rate for inside overwintering bees is accomplished at the 39 - 43°F (4 - 6°C) range (Proc. Can. Assoc. Prof. Apic., 1980). Temperature also has an effect on bee death rate. Results of observations in two rooms each containing 80 hives are as follows (Nelson, 1979):

a. Room temperature 46°F (8°C) for 127 days—November 25 to April 2:
   Total weight dead bees: 213 lbs. (97 kg)
   Average weight loss per hive: 2.66 lbs. (1.21 kg) of bees

b. Room temperature 39°F (4°C) for 127 days—November 25 to April 2:
   Total weight dead bees 106 lbs. (48 kg)
   Average weight loss per hive: 1.32 lbs. (0.6 kg) of bees

Temperatures may rise for a number of reasons. Electrical failure may be one reason, thus, an auxiliary generator is almost a must. Other instances include the covering of the air intake by "hoar frost." Another very likely possibility is warm outside temperatures in early spring. Inside temperatures of up to 79°F (26°C) have been experienced in such situations with little damage evident. Huge numbers of bees come out of the hives, but return when temperatures diminish (Barker, 1976; McCutcheon, 1977).

Brood rearing is reported as early as January 17. Measured brood area ranged from 6 - 20 in² (41-126 cm²) per colony.

A major problem in inside overwintering is Nosema disease, according to the University of Manitoba. (Proc. Can. Assoc. Prof. Apic., 1978). Observations on Nosema in 1977 on two separate occasions show the following:

- 11 of 17 hives sampled were infected with an average 3.58 million Nosema spores per bee (early season).

- 9 of 10 hives sampled were infected with an average of 7.67 million Nosema spores per bee (mid-season).

Crystallized rapeseed (Canola) honey can be a severe problem even when present in small quantities. Reliquefication is difficult and water must be supplied at the entrance to enable colonies to carry out the process. A common water feeder consists of a 2.2 lbs. (1 kg) plastic honey container in which 5 small holes are drilled adjacent to each other and near the lid. The container is filled, capped, inverted and placed at the hive entrance with the holes turned inward.

Nelson (1979) reported on stress factors in overwintered hives. From January 9 to April 4, the increase in weight of bee abdomens was 19% and water content increased 9%. This indicates a build-up of wastes which contributes to stress. Vanderput in Manitoba (personal communication, 1976) relates that by the third week of March bees become restless and temperatures increase in the room. This possibly relates to increasing stress.

SPRING MANAGEMENT

Hives are moved out from mid-March to mid-April, depending on climate and yearly conditions. Dead hives vary from 5 - 23%, mostly in the 5 - 10% range. Other problems during spring contribute to a further loss of 8 - 10% (McCutcheon, 1977; Nelson, 1979).

Following removal from winter quarters, loss of bee population averages 25% in the first 24 hours. These results are evident whether hives are placed in the open or within windbreaks or whether they are moved out during the day or at night. Various combinations of colors and designs on hive fronts and offsetting hive entrances in different directions failed to reduce losses significantly (Proc. Can. Assoc. Prof. Apic. 1977).

Quebec observations show that an average spring population of an inside overwintered hive is 25,000 bees, or 7.0 lbs. (3.15 kg). They also show that protecting hives after removal with black felt paper or by means of black corrugated cardboard does not result in an increased population or honey crop (Proc. Can. Assoc. Prof. Apic., 1975).

Wintering Four-Frame Nuclei

Four-frame nuclei can be overwintered inside successfully if prepared properly. There must be sufficient population to fill the hive space and generous stores are necessary. Inside overwintering will not overcome a small population or a low amount of stores. Results show a 20% loss of hives, with stores consumption averaging 7.26 lbs. (3.3 kg). Hives which succumbed generally were short on stores. A small sample of four-frame nuclei fed syrup and pollen substitutes in February and March produced excellent hives in

May and outproduced outdoor overwintered hives and hives established with a 2.2 lb. (1 kg) package of bees (Apiary Branch Annual Report, 1978, 1981).

That brood is reared in four frame nuclei is evidenced by observations on February 23. Brood amounts varied from a total of 25 in² (160 cm²) of capped brood to 7 in² (44 cm²) of eggs only (Apiary Branch Annual Report, 1977).

SUMMARY

Inside wintering was popular in North America early in the century. The practice fell into disfavor by the 1920's as outside wintering became more successful due to improvement in methods and packing materials.

Outside overwintering and/or the use of package bees were both popular and economic for many decades. More recently, it has become more economic to winter hives even in the colder areas of North America. Added to economics is a perceived need for self-sufficiency. Inside overwintering of bee colonies has become a viable alternative to outside overwintering or to the use of package bees. Considerable information is available on various methods, incorporating the latest technology.

The wintering chamber must be well insulated and dark. A good air distribution system is very important as is an air exhaust-entry system. All are coordinated to operate automatically. Hives are prepared as single or double brood chamber units in late August or early September. Feeding of sucrose syrup is preferred as winter stores. This must be accomplished by early October. Hives are placed in the building in early November.

Losses of hive numbers vary from 5-10% over winter. Stores consumed average 26 lbs. (12 kg) for single brood chamber units and 50 lbs. (23 kg) for double units. It is estimated that inside overwintered hives produce 40% more honey compared to package bees at a cost which almost equals the price of a 2-lb. (.9 kg) package. Many beekeepers have adopted the idea of inside overwintering and have perfected it. These beekeepers reside mainly in the prairie provinces and in Quebec province, Canada. In Quebec, approximately 90% of hives overwintered are placed inside. Thus, the practice has become one of the standard operations in commercial beekeeping there.

METRIC CONVERSION FACTORS

1 mm = .039 in.	1 Pa = .004 in. H_2O	1 W = 3.41 BTU/hr.
1 kg = 2.2 lb.	1 m² = 10.75 ft²	1 RSI = 5.68"R"
1 L/s = 2.12 cfm	1 m³ = 35.2 ft³	

REFERENCES

Outdoor Wintering

Anderson, E.J. (1943). Some research on wintering bees. *Gl. Bee Cult. 71 (12)*:681-683, 715.

Anon. (1971). Starting Right with Bees: A Beginners Handbook on Beekeeping. The A.I. Root Co., Medina, Ohio.

Arnott, J.H. and S.E. Bland (1954) Beekeeping in Saskatchewan. Sask. Dept. Agric., Apiary Div.

Betts, A.D. (1943) Temperature and food consumption of wintering bees. *Bee World 24(8)*:60-62.

Boch, R (1964). Wintering hives in cardboard boxes. *Can. Bee Journ. 75(10)*:7-10.

Braun, E. (1949) Apiculture, Dom. Expt. Farm, Brandon, C.D.A. Expt. Farms Ser. Prog. Rpt. 1937-1947.

Braun, E. and J.E. Geiger. (1947). Winter loss from colonies and related factors. *Gl. Bee Cult. I. 75(11)*:656-659. *II. 75(12)*714-718, 761.

Cale, G.H. Jr. (1964). Beekeeping for Beginners. Journal Printing Co., Carthage, Ill.

Cherednikov, A.V. (1964). Wintering with open upper entrances. *Pchelovodstvo 84(9)*:23.

Corkins, C.L. (1930). The winter activity in the honeybee cluster. Rpt. Ia. St. Apiarist pp. 44-49.

Dyce, E.J. and R.A. Morse. (1960) Wintering honeybees in New York State. N.Y. State Coll. of Agric., Cornell Ext. Bull. 1054.

Duff, S.R. and B. Furgala. (1986). Pollen trapping honey bee colonies in Minnesota. Part I. Effect on amount of pollen trapped, brood reared, winter survival, queen longevity, and adult bee population. *Amer. Bee Jour. 126(10)*:686-689.

Duff, S.R. and B. Furgala. (1989). Unpublished data.

Eckert, J.E. (1954). A handbook for beekeeping in California. Calif. Agric. Expt. Sta. Manual 15.

Eckert, J.E. and F.R. Shaw. (1960) Beekeeping: Successor to "Beekeeping" by Everett F. Phillips. The Macmillan Co., N. Y., New York.

Edmunds, J.W. (1961). Wintering bees in Alberta. *Gl. Bee Cult. 89(11)*:654-657.

Farrar, C.L. (1931). A Measure of Some Factors Affecting the Development of the Honey Bee Colony. Unpubl. Thesis. Mass. Sta. Coll.

Farrar, C.L. (1934). Bees must have pollen. *Gl. Bee Cult. 62(5)*:276-278.

Farrar, C.L. (1936). Influence of pollen reserves on the surviving populations of overwintered colonies. *Am. Bee J. 76(9)*:452-454.

Farrar, C.L. (1943). An interpretation of the problems of wintering the honeybee colony. *Gl. Bee Cult. 71(9)*:513-518.

Farrar, C.L. (1944). Productive management of honeybee colonies in the northern states. U.S.D.A. Circ. No. 702.

Farrar, C.L. (1947). Nosema losses in package bees as related to queen supersedure and honey yields. *J. Econ. Entomol. 40(3)*:333-338.

Farrar, C.L. (1952). Ecological studies on overwintered honey bee colonies. *J. Econ. Entomol. 45(3)*445-449.

Farrar, C.L. (1960). From need to plenty — through the cold of winter *Am. Bee J. 100(8)*:306-308

Farrar, C.L. (1963). The overwintering of productive colonies. *In*: Grout, R. *The Hive and the Honey Bee.* Dadant and Sons, Hamilton, Illinois p. 341.

Furgala, B. (1977). Pollen substitutes and supplements. Agric. Ext. Ser., Univ. Minn., Fact Sheet No. 24.

Furgala, B., S.R. Duff, S. Aboulfaras, D. Ragsdale and R.A. Hyser. (1989). Some effects of the honey bee tracheal mite (*Acarapis woodi*) on non-migratory, wintering honey bee (*Apis mellifera*) colonies in East Central Minnesota. *Amer. Bee Journ. 129(3)*:195-197.

Furgala, B. and R.A. Hyser. (1974). Unpublished results.

Gates, A.H. (1942). Wintering outdoors in the state of Washington. *Gl. Bee Cult.* *70(9)*:554-556

Geiger, J.E. (1967). Winter temperatures and the relative humidity in beehives. *Am. Bee J.* *107(10)*:372-373.

Gilbert, C.H. (1932). Studies of temperature in the bee hive with special reference to radiation of heat from the cluster. Rpt. IA., St. Apiarist, pp. 31-37.

Gooderham, C.B. (1945). Bees and how to keep them. Dept. Agric., Dom. Canada, Publ. 578.

Gooderham, C.B. (1950). Bee Div. Prog. Rpt. 1937-1948, C.D.A. Cent. Expt. Farms Ser., Ottawa.

Haydak, M.H. (1967). Beekeepers still divided on the question of how best to protect their colonies during winter. *Am. Bee J. 107(11)*:418-420.

Holte, F. (1970). Unpublished data.

Hyser, R.A. and B. Furgala. (1972) Unpublished data.

Jaycox, E.R. (1969). Beekeeping in Illinois. Coll. Agric., Univ. of Ill., Circ. 1000.

Johansson, T.S.K. and M.P. Johansson. (1969). Wintering. *Bee World 50(3)*:89-100.

Johansson, T.S.K. and M.P. Johansson. (1971). Winter Losses 1970. *Am. Bee J. 111(1)*:10-12.

Johansson, T.S.K. and M.P. Johansson. (1979). The honey bee colony in winter. *Bee World* *60(4)*:155-170.

Johansson, T.S.K. and M.P. Johansson. (1984). Wintering the honey bee colony: Hives Part I. *Gl. Bee Cult 111(1)*:43-44,47.

Johansson, T.S.K. and M.P. Johansson. (1984). Wintering the honey bee colony: Hives Part II. *Gl. Bee Culture. 111(2)*:90-92.

King, G.E. (1923). A study of factors affecting the outdoor wintering of honeybees. *J. Econ. Entomol. 16*:321-323.

Langridge, D.F. and J. Rufford-Sharpe. (1966). A successful protein supplement for honey bees. *Am. Bee J. 106(9)*:328-329.

Langstroth, L.L. (1859). A Practical Treatise on the Hive and the Honey-bee. 3rd ed. A.O Moore & Co., New York, N.Y.

L'Arrivee, J.C.M. (1961). Wintering bees in western Canada *Gl. Bee Cult. 89(10)*: 616-619,636.

Lysne, J. (1954). Winter management of bees. *Am. Bee J. 94(10)*:383.

Martinovs, A. (1972). How do damp and mould affect the wintering of colonies. *Bitidningen* *72(12)*:358-360.

Merrill, J.H. (1920). Preliminary notes on the value of winter protection for bees. *J. Econ. Entomol. 13*:99-111.

Merrill, J.H. (1923). Value of winter protection for bees. *J. Econ. Entomol. 16*:125-130.

Miller, A.C. (1901). Tarred paper for packing. *Am. Bee J. 41(11)*:718.

Miller, A.C. (1903). Tarred paper for winter protection. *Gl. Bee Cult. 31(12)*:534-535.

Miller, C.C. (1903). *In*: Stray straws. *Gl. Bee Cult. 31(9)*:371.

Moeller, F.E. (1977). Overwintering of honey bee colonies. Prod. Res. Rept. No. 169. USDA, 16 pp.

Morse, R.A. (1956). Florida Beekeeping. St. Plt. Bd. Fla. Vol. 11, Bull. 10.

Mraz, C. (1947). Packing bees for winter. *Gl. Bee Cult. 75(10)*:577-583.

Mraz, C. (1962). Wintering bees in Vermont. *Gl. Bee Cult. 90(11)*:658-660, 669.

Mraz, C. (1967). Learning the hard way. *Gl. Bee Cult. 95(2)*:73-75.

Peer, D.F. (1978). A warm method of wintering honey bee colonies outdoors in cold regions. *Canadian Beekeeping 7(3)*:33-36.

Phillips, E.F. (1928). Beekeeping. The MacMillan Co., N.Y., New York.

Phillips, E.F. (1928). Theories of wintering bees. Rpt. Ia. St. Apiarist, pp. 100-105.

Phillips, E.F. and G.S. Demuth. (1914). The temperature of the honeybee cluster in winter. U.S.D.A. Bull. 93.

Rosenthal, C. (1962). Stimulative feeding of bees on pollen and pollen substitutes.- *Apicultura 15(10)*:32-34.

Shesley, B. and B. Poduska. (1968). Supplemental feeding of honey bees. . . colony strength and pollination results. *Am. Bee J. 108(9)*:357-359.

Standifer, L.N. (1967). Beekeeping in the United States. U.S.D.A. Agric. handbook No. 335.

Sugden, M.A. and B. Furgala. (1982). Evaluation of six commercial honey bee (*Apis mellifera* L.) stocks used in Minnesota. Part I. Wintering ability and queen longevity. *Amer. Bee. Jour. 122(2)*:105-109.

Sugden, M.A., B. Furgala and S.R. Duff. (1988). A comparison of four methods of wintering honey bee colonies outdoors in Minnesota. *Amer. Bee Jour. 128(7)*:484-487.

Taranov, G.F. and K.I. Mikhailov. (1960). The concentration of carbon dioxide in the winter cluster of the honeybee. *Pchelovodstvo. 37(10)*:5-10.

Villumstad, E. (1964). Investigations on feeding soya-bean flour and pollen to bees in 1963. *Birokteren. 80(1)*:9-13.

Walstrom, R.J. (1950). Pollen substitute tests in Nebraska. *Am. Bee J. 90(3)*:118-119.

Wilson, H.F. and V.G. Milum. (1927). Winter protection for the honey bee colony. Wisc. Agric. Expt. St., Res. Bull. 75.

Winston, M.L. (1987). The biology of the honey bee. Harvard Univ. Press, Cambridge, MA 281 pp.

Wolfe, E.A. (1950). The use of pollen supplements. Rpt. Ia. St. Apiarist, pp. 48-50.

Indoor Wintering

Annual Report, Apiary Branch, Cloverdale Office. (1976). B.C. Ministry of Agriculture and Food, Surrey, B.C.

Annual Report, Apiary Branch, Cloverdale Office. (1977). B.C. Ministry of Agriculture and Food, Surrey, B.C.

Annual Report, Apiary Branch, Cloverdale Office. (1978). B.C. Ministry of Agriculture and Food, Surrey, B.C.

Annual Report, Apiary Branch, Cloverdale Office. (1979). B.C. Ministry of Agriculture and Food, Surrey, B.C.

Annual Report, Apiary Branch, Cloverdale Office. (1981). B.C. Ministry of Agriculture and Food, Surrey, B.C.

Barker, R.G. (1975). Indoor Wintering of Honeybee Colonies in Manitoba.

Braun, E. (1940). Two Methods of Wintering Bees for the Prairies Provinces. Publication 689, Canada Department of Agriculture, Experimental Farms, Ottawa.

Darby, D. (1988). Overwintering Buildings for Bees Engineering Branch, Alberta Agriculture, Lethbridge, Alberta. (Unpublished).

Fingler, B., and D. Small (1982). Indoor Wintering in Manitoba, pp. 7-19, Manitoba Beekeeper, Fall Issue. Manitoba Department of Agriculture, Winnipeg.

Geiger, J., and E. Braun (1955). Comparison of Methods for Wintering Honeybees in the Prairie Provinces. Revised Publication 689, Canada Department of Agriculture, Ottawa.

Geiger, J.E. and J.C.M. L'Arrivee (1965). Seventy-five Year Summary of Beekeeping Investigations at the Experimental Farm, Brandon, Manitoba. *Canadian Bee Journal*, June, 1965.

Geiger, J.E. and D.R. Robertson (1967). Wintering Honeybees in Manitoba. Publication 458, Manitoba Department of Agriculture, Winnipeg.

Gooderham, C.B. (1926). Wintering Bees in Canada. Bulletin 75 (new series). Canada Department of Agriculture, Experimental Farm, Ottawa.

Konrad, John. (1970). Inside Wintering of Bees. Centennial Report, Red River Apiarists Association, Winnipeg, Manitoba.

McCutcheon, D.M. (1977). Wintering Honeybees in a Controlled Atmosphere Chamber. *Canadian Beekeeping, 6(8),* Orono, Ontario.

Nelson, D.L. and G.D. Henn (1977). Indoor Wintering Research Highlights. Research Station, Beaverlodge, Alberta. *Canadian Beekeeping, 7(1),* Orono, Ontario.

Nelson, D.L. (1979). Indoor Wintering: Physical Requirements and Problems of Wintering Indoors. (Address to Red River Apiarists Association, Winnipeg, Manitoba.)

Nelson, D.L. (1982). Indoor Wintering: Outline of Basic Requirements, N.R.G. Publication No. 81-1. Research Station, Canada Agriculture, Beaverlodge, Alberta.

Pettit, M. (1917). The Wintering of Bees in Ontario. Bulletin 256, Ontario Department of Agriculture, Toronto, Ontario.

Phillips, E.F. and G.S. Demuth (1918). Wintering Bees in Cellars. Farmers Bulletin 1014, U.S. Department of Agriculture, Washington.

Pirker, H.J. (1976). Here's How Pirker Produces our First Package Bees. *Alberta Bee Culture, 1(2)* Alberta Beekeepers Commission, Edmonton.

Pirker, H.J. (1978). (a) Package bee production in northern Canada. *Can. Beekeep.* 7:17, 20-21.

Pirker, H.J. (1978). (b) Steering factor humidity. *Can. Beekeep.* 7:102-106.

Pirker, H.J. (1980). Brood rearing in the winter: factors and methods. *Can. Beekeep.* 8:69-71.

Research Reports. (1974). Proc., Canadian Association of Professional Apiculturists.

Research Reports. (1975). Proc., Canadian Association of Professional Apiculturists.

Research Reports. (1976). Proc., Canadian Association of Professional Apiculturists.

Research Reports. (1977). Proc., Canadian Association of Professional Apiculturists.

Research Reports. (1978). Proc., Canadian Association of Professional Apiculturists.

Research Reports. (1979). Proc., Canadian Association of Professional Apiculturists.

Research Reports. (1980). Proc., Canadian Association of Professional Apiculturists.

Root, E.R., Root, H.H. (1962). Indoor Wintering, pp. 694-696. *In* ABC and XYZ in Bee Culture. A.I. Root Company, Medina, Ohio.

Sladen, F.W.L. (1920). Wintering Bees in Canada. Bulletin 43 (second series). Canada Department of Agriculture, Experimental Farms, Ottawa.

Wrubleski, E.M. and S.E. Bland (1976). Honeybee Overwintering Management and Facilities. Saskatchewan Department of Agriculture, Regina, Saskatchewan. (unpublished).

Young, B.A. and J.R. Feddes (1988). Indoor and Outdoor Overwintering. Department Animal Science and Agricultural Engineering, University of Alberta, Edmonton, Alberta. (unpublished).

CHAPTER 21

HONEY

by JONATHAN W. WHITE, JR.*

INTRODUCTION

Honey is a most remarkable substance. Revered by the ancients, it can be enjoyed today in exactly the condition in which it was first discovered many thousands of years ago. It is the only sweetening material that requires no manipulation or processing to render it ready to eat. The first known written passages dealing with honey have been dated about 4000 years ago, and honey has been treasured ever since. An extensive essay on the history of honey is available (Crane, 1975). As will become evident, honey is an exceedingly variable and complex material, and we are far from knowing all about it.

TYPES OF HONEY

Honey is classified in several ways: by its source, by the season and/or location of production, its physical state, the means by which it is obtained, and by the form in which it is presented to the consumer.

Source—by this is meant the floral origin of the nectar from which it is derived; this may be from a specific dominating plant or a mixture of plants in bloom depending on the nature of the vegetation in the foraging area of the colony (See Figure 1).

Season—Honeys may be derived from a mixture of many floral nectars, with none predominating. These blends are generally characterized by the time of the year when harvested; depending upon the area, several such periods may occur, with the honey being characterized as "spring honey," "summer blend," or "fall" (autumn) flowers. Honey may also be characterized as "mountain," "desert," etc.

Physical State—Honey may be entirely liquid ("extracted honey"), semi-solid ("honey spread, granulated honey"), or in the comb (see below).

Means of Preparation—Honey may be obtained from the combs by centrifugation ("extracted honey") or by pressing the comb ("pressed honey"). It may be filtered, simply strained, or cleaned only by settling ("filtered honey," "strained honey," "raw honey").

Comb Honey—Such honey may be available in pieces of the original comb, cut and wrapped ("cut comb"), or pieces of comb in a container of liquid honey ("chunk honey") (See Figure 2) A miniature wooden or plastic

---

*Jonathan W. White, Jr., Ph.D., U.S. Department of Agriculture, Agricultural Research Service, retired. Honeydata Corporation, Navasota, Texas.

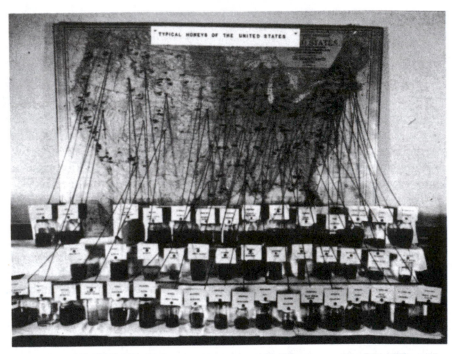

FIGURE 1. This is a display of typical honeys of the United States—a part of the 490 samples collected by Dr. J. W. White. *(USDA photo by M. C. Audsley)*

FIGURE 2. High quality, well prepared chunk comb pack is a popular honey product when it can be found in supermarkets. Note the even cut and the total fill of the jar with comb.

frame filled by the bees with a comb of capped honey is termed "section comb"; the section may be square or round. "Bulk comb honey" is produced in shallow extracting supers fitted with thin super foundation.

THE COMPOSITION OF HONEY

When we consider the factors that contribute to the material we call "honey," it becomes obvious that simply to list the composition of an "average" honey is of limited value. The environment of the area from which the bees collect the nectar (and other sugar-containing materials), influences the types and variety of flora found therein and, hence, the specific composition of the honey. For commercial honey, the extent of blending of lots from different areas and seasons also has a major effect. For example, averages for U.S. honeys from the six major geographic areas show small but notably significant differences among the regions (White *et al.,* 1962).

Average Composition of Honey

Nevertheless, a compilation of average composition data is useful to provide to those wishing to use honey in food manufacture or for other purposes some guide to its effects on their product. A *caveat* must be inserted here: methods of analysis for much of the data in the world literature are to varying extents empirical and do not reflect actual composition. This is especially true in carbohydrate analysis, and these materials make up most of

Table 1. AVERAGE COMPOSITION OF HONEY[a]

Component	Average	Standard Deviation	Range
Moisture	17.2	1.5	12.2 - 22.9
Fructose	38.4	1.8	30.9 - 44.3
Glucose	30.3	3.0	22.9 - 40.7
Sucrose	1.3	0.9	0.2 - 7.6
"Maltose"[b]	7.3	2.1	2.7 - 16.0
Higher Sugars	1.4	1.1	0.1 - 3.8
Free Acid as Gluconic	0.43	0.16	0.13 - 0.92
Lactone as Gluconolactone	0.14	0.07	0.0 - 0.37
Total Acid as Gluconic	0.57	0.20	0.17 - 1.17
Ash	0.169	0.15	0.02 - 1.028
Nitrogen	0.041	0.026	0.00 - 0.133
pH	3.91		3.42 - 6.10
Diastase[c]	20.8	9.8	2.1 - 62.1

[a] Data for 490 samples of U.S. honey (White, Riethof, Kushnir, & Subers, 1962). All values in percentages, except for pH and diastase.

Values for sugars are for 439 of the samples after removal of honeydew outliers (White, 1980).

[b] Reducing disaccharides, calculated as maltose.

[c] Data for 292 of the samples.

the solids of honey. Knowledge of analytical procedures used should be considered in comparing data from different areas and times. In general, more recent (since 1955) data may (but not necessarily will) more closely reflect actual composition. Table 1 shows average values obtained for 490 samples of U.S. honey. These values agree very reasonably with those obtained by Bogdanov and Baumann (1988) who used highly sensitive high-performance liquid chromatography to analyze 118 Swiss honeys.

Variability is a most noteworthy and troublesome attribute of honey composition. Table 2 shows the coefficients of variation (standard deviation/ mean value X 100) for 18 honey constituents for all of the honey types studied in that work and several later projects. All but four are in double digits, emphasizing the complex nature of the factors influencing honey composition. Of the hundreds of nectar-bearing plants which bees visit, relatively few are of commercial importance, though the taste and color characteristics of hundreds have been described (Pellett 1947, Lovell 1956, Crane *et al.*, 1984, Crane and Walker, 1986). Each honey type has a characteristic flavor and color and can thus be identified. They also are fairly consistent in other characteristics— relative amounts of the various sugars, acids, nitrogen compounds, and minerals. Table 3 shows how the average composition of 74 honey types and 4 honeydews compares with the average of all American honeys. In the Table a plus sign means that the honey type is higher than the average in the listed characteristic, a minus sign means that it is lower, and no mark means that it is near the average. An "n" means that insufficient data were available to estimate. These data were obtained in an analytical survey of U.S. honeys (White *et al.*, 1962). In the same work it was shown that area of production has little effect on the composition of "single-source" honeys with cotton

Table 2. VARIABILITY OF HONEY COMPOSITION[a]

Constituent	Samples	CV (%)[b]	Constituent	Samples	CV (%)
Fructose	490	5.4	True Protein	740	42
Glucose	490	9.2	Isomaltose/Maltose	80	68
Total Acidity	490	35	Ash	490	89
Higher Sugars	490	69	Nitrogen	490	63
Sucrose	490	66	Diastase	263	47
"Maltose"	490	29	Polarization[c]	454	30
Fructose/Glucose	490	10	Monosaccharides	439	5.9
Proline	740	40	$\delta^{13}C$	119	3.86
Disaccharides	439	24	Bound Galactose	81	115

[a] White (1987).

[b] Coefficient of variation: standard deviation ÷ average value x 100.

[c] Levorotatory samples only.

honey from three states, alfalfa honey from two areas, and orange honey from two states showing relatively little variation in composition.

Moisture Content

The natural water content ("moisture") of honey in the comb is that remaining from the nectar after it is ripened by the bees. Its concentration is thus a function of the factors involved in ripening, including weather conditions, original moisture of the nectar, its rate of secretion, and strength of the colony. Premature removal from the hive (when not fully capped) will also influence the moisture content of the honey. It may change after removal from the hive as a result of storage conditions either before or after extraction (see hygroscopicity). Moisture content is one of the most important characteristics of honey, profoundly influencing keeping quality, granulation, and body; yet few beekeepers trouble to measure it, relying on guesswork. Moisture measurement is discussed under "refractive index" and "viscosity."

The Sugars of Honey

More than 95% of the solids of honey are carbohydrate in nature, largely simple sugars or monosaccharides. Sugars are classified according to the size and complexity of their molecules. The *monosaccharides* are the building blocks of the more complex types. The dextrose (glucose) and levulose (fructose) in honey are examples. The *disaccharide* sugars are made of two monosaccharides joined together in various ways; many are known. Sucrose (table sugar), lactose (milk sugar) and maltose (malt sugar) are examples. Honey contains a variety of disaccharides. Other more complex sugars (*oligosaccharides*, higher sugars) contain three or more simple sugars; many are present in honey.

Until the middle of this century, honey was believed to be a simple mixture of dextrose, levulose, and sucrose, with an undefined carbohydrate material called "honey dextrin," believed analogous to starch dextrin. Application of powerful new analytical and separation procedures by several groups of investigators in the past 25 years has revealed honey to be a highly complex mixture of sugars; in addition to those named above, Doner (1977) listed in his comprehensive review of the subject maltose (Elser, 1924); isomaltose, nigerose, turanose, maltulose (White and Hoban, 1959); kojibiose (Watanabe and Aso, 1960) $\alpha\beta$-trehalose, gentiobiose, laminaribiose, (Siddiqui and Furgala, 1967) maltotriose, 1-kestose, panose, isomaltosyl glucose, erlose, isomaltosyltriose, theanderose, centose, isopanose, isomaltosyltetraose, isomaltosylpentaose (Siddiqui and Furgala, 1968). Siddiqui found no evidence for the presence of the often-reported raffinose. Recent studies using a highly sensitive specific enzymic procedure (White *et al.*, 1987) support the view that honey contains traces of raffinose. Recently several investigators (Bogdanov & Baumann, 1988; Low & Sporns, 1988; Nikolov *et al.*, 1984) have been able to

Table 3. Characterization of Various Floral Types of Honey[a]

	Color	Granulation	Levulose	Dextrose	Sucrose	Maltose	Higher Sugars	Undetermined	pH	Free Acidity	Lactone	Total Acidity	Lactone/Free Acid	Ash	Nitrogen	Diastase
Alfalfa		+		+	+		−	−					+	−	−	
Aster	+	−			−	+				+		−	−	+		n[1]
Athel Tree	+	+	+	+			−							+	+	n
Bamboo, Japanese			−		+											n
Basswood														−	−	
Bergamot	+		+								+	+				n
Blackberry	+	−		−		+	+		+				−	+		−
Blueberry	+					+			+							n
Blue Curls		+	−	+						−		+				n
Bluevine		−	−													n
Boneset	+	+	−									+		+		
Buckwheat	+	−			+					+		+				+
Canteloupe		+		+							+	+			−	−
Cape vine		−					−								−	
Chinquapin	+	−	−	−		+	+	+	+			−	−			
Clover, crimson	−									−				−	−	
Clover, hubam	−			+			−							−	−	n
Clover, sweet yellow	−	+			+					−		−	+	−	−	n
Coralvine	+	−	−	−			+	+	+	+		+		+	+	n
Cotton		+		+	−		−			+				+		
Cranberry	+	−	−	−			+	+	+					+		
Gallberry		−	+							+					−	
Goldenrod			+		−		−			+	−		−			+
Grape	+	−	−	−		+						+			+	n
Holly	+	−		−		+	+		+							n
Horsemint			+				−			−	+	+	+			
Locust		−	+	−						−		−		−	−	−
Manzanita		+	−	+						−		−			−	n
Marigold			+				−		−		+		+	+	−	+
Mesquite		+	+	+			−					−			−	n
Mexican clover	+	−								+		+				+
Mint	−		+		+			−							−	
Mountain laurel	−	−	−	−	−	+	+	+	+	−	−	−	−			+
Mustard	+	−		−		+				+				+	+	
Orange										−		+	+	−		n

Characterization of Various Floral Types of Honey—*Continued*

	Color	Granulation	Levulose	Dextrose	Sucrose	Maltose	Higher Sugars	Undetermined	pH	Free Acidity	Lactone	Total Acidity	Lactone/Free Acid	Ash	Nitrogen	Diastase
Orange-grapefruit					+								−		−	−
Palmetto		−				+			+	−		−			−	−
Palmetto, saw	+										+	+	+	+	−	−
Pepperbush	+	−						+			+			+		−
Peppermint	+		+						+				−	+		n
Peppervine	+	−	−	−	+										−	−
Poison oak		−		−	+	+	+							+	+	n
Privet	+						−		−	+	+	+				n
Prune	+	+	−	−	+	−			+	−	−	−	−	+	+	n
Raspberry	+	−	−	−			+					+		+	+	−
Rhododendron	−	−	−	−	+			+	+	−	−	−			−	+
Sage		−	+	−												n
Snowbrush	+									+	+					+
Sourwood		−		−	+	+			+		−	−			−	
Spanish needle	+	−	+	−							+	+	+	+	+	+
Spearmint			+											+		n
Sumac	+		−	−			+	+	+	+		+	−	+	+	+
Sunflower	+	−						−			+	+			+	−
Thistle, blue	−	−								−		−		−		n
Thistle, star		−			+		+		−		+	+	+			+
Thyme	+								+					+	+	n
Titi	+						−		+	−	−	−	−	+	−	
Titi, spring	+	−	+	−				+	+	−	−	−	−			n
Trefoil	−											−		−	−	−
Tulip tree	+	−	−	−		+	+	+	+	+	−	+	−	+	+	
Tupelo		−	+	−					−		+	+	−			
Alfalfa honeydew	+	+	−							+	−	+	−	+	+	n
Cedar honeydew	+	−	−	−		−	+	+	+	+	+	−	+			n
Hickory honeydew	+	−	−	−	+		+	+	+	+	−	+	−	+		n
Oak honeydew	+	−	−	−	+			+	+	+		+	−	+	+	n

Near average in all characteristics except diastase, which differs as shown in parentheses: Wild buckwheat (+); clover, alsike; clover, sweet; clover, white; crotalaria (—); cucumber; eucalyptus; fireweed; heartsease (n); palmetto, cabbage; pentstemon (n); purple loosestrife (n); rosinweed (+); vetch; vetch, hairy (—).

[a] White *et. al. (1962).*

[1] "n" means insufficient data were available to allow valid comparison..

measure the amounts of up to 19 of these minor sugars in honey using gas or high-performance liquid chromatography.

Many of these sugars are not found in nectar, but are formed during ripening and storage by the effects of bee enzymes and the acids of honey. The sugars dextrose and levulose predominate (85 - 95% of the total sugars) and give honey its sweetness, energy value, and physical characteristics. Levulose (d-fructose) is largely responsible for the hygroscopicity of honey. It is more soluble than dextrose (d-glucose); the granulation of honey is simply the separation of solid dextrose hydrate crystals from the supersaturated solution which is honey. The diagrams in Figure 3 show the amounts of the three classes of carbohydrates found in honey. In common with all other components, a considerable range is found for carbohydrate composition among honeys from different areas and floral sources. An important attribute of honey which has major influence upon its physical state and properties is the preponderance of levulose over dextrose. The ratio of these two sugars is a characteristic of the floral type of honey. Figure 4 diagrams the amounts of these two sugars in 439 samples of U.S. honey (White *et al.*, 1962, White, 1980). Only two samples in that study contained more dextrose than levulose. This is characteristic of honey from *Trichostema lanceolatum* (blue curls) and a few others.

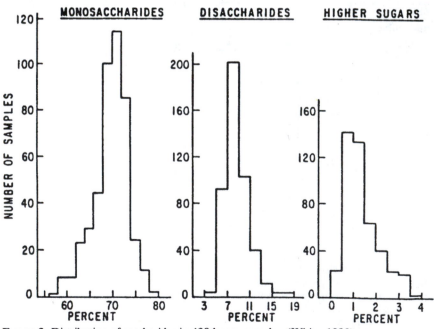

FIGURE 3. Distribution of saccharides in 439 honey samples. (White, 1980)

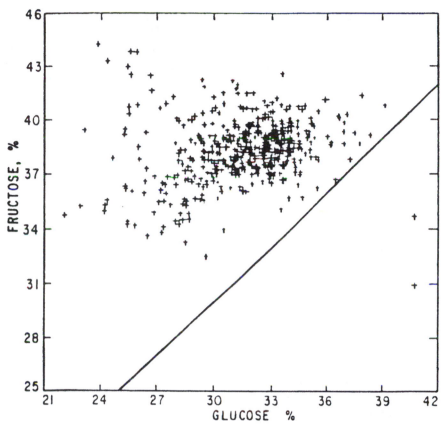

FIGURE 4. Fructose and glucose relationship for 439 honey samples. Line indicates 1 : 1 ratio. (White, 1980)

In two extensive analytical studies of U.S. honey from four crop years, only 19 of 919 samples had more than 5% sucrose; these were mostly citrus (4 samples), and alfalfa or alfalfa-sweetclover (10 samples). These have been known to be high in sucrose probably because they are frequently low in invertase because of heavy nectar flows. The sucrose content of 41 fresh comb honeys measured immediately after removal from the hive averaged 2.87% (White, 1980), significantly higher than for extracted honey. The range was 0.20% to 7.48%.

Acids

Because of its great sweetness, the acidity of honey is not particularly evident to the palate. The acids, however, do contribute to the complex flavor. Years ago it was thought that citric acid predominated, with small amounts of formic, acetic, butyric, malic, and succinic acids (Nelson and Mottern, 1931). More recently it has been found (Stinson *et al.*, 1960) that the predominant acid in honey is gluconic acid, which is derived from dextrose (see glucose oxidase). Other acids identified for the first time in this study were lactic and pyroglutamic.

Reflecting the very early belief that formic acid was the principal acid, its acidity was for many years reported analytically as "% formic acid." It is now more properly expressed in food regulations as "milliequivalents per kilogram honey" or meq/kg. This expression can be converted to the equivalent of any acid by use of the appropriate factor, for example 1 meq/kg is equivalent to 0.0196% gluconic acid (19.6 mg. per 100 g, or 19.6 mg%).

The gluconic acid present in all honey originates largely from the activity of the glucose oxidase which the bee adds at ripening (White *et al.*, 1963), with some contribution from bacterial action during the ripening (Ruiz-Argüeso & Rodriguez-Navarro, 1973). The considerable variation in the amount of this acid in honeys perhaps reflects the time required for the nectar to be completely converted to honey under differing conditions of environment, colony strength, and sugar concentration of the nectar, since the activity of glucose oxidase in full-density honey is negligible (White, 1975, p. 192).

A common expression of the effective acidity level of solutions is the pH value. This is a logarithmic scale representing the concentrations of hydrogen ions in a solution, which is influenced by other factors than the amount of acid present, particularly the mineral content. Values less than pH 7 indicate acidity; honey generally has a pH ranging from about 3.2 to 4.5, averaging about 3.9, which is in the range of a weak vinegar.

Minerals

The ash content (minerals) of honey averages about 0.17 percent of its weight, but varies greatly from 0.02 to over 1.0 percent. Schuette and his students examined the identity of honey minerals rather extensively, as summarized in Table 4. It may be seen from the Table that, in general, dark honeys are richer in minerals than are light-colored honeys. This finding of the Wisconsin group has been confirmed statistically (White *et al.*, 1962). Additional data on minerals in honey from eleven other countries is summarized in a review (White, 1975, pp. 173-175). Analyses for 339 samples of Japanese market honey (both domestic and imported) for potassium, calcium, sodium, phosphorus, magnesium, iron, copper, and zinc are given by Hase *et al.* (1973). Doner and Jackson (1980) have shown that the occasional darkening of tea after adding honey is caused by unusually high (>40 ppm) iron content of some honey.

Honey contains much less sodium than potassium. In a review of a proposal to use the Na/K ratio to detect honey adulteration with high-fructose

Table 4. MINERAL CONSTITUENTS OF HONEY[a]

Mineral	No. samples and color	Range (parts per million)	Average
Potassium	13 light	100-588	205
	18 dark	115-4733	1676
Sodium	13 light	6-35	18
	18 dark	9-400	76
Calcium	14 light	23-68	49
	21 dark	5-266	51
Magnesium	14 light	11-56	19
	21 dark	7-126	35
Iron	10 light	1.20-4.80	2.40
	6 dark	0.70-33.50	9.40
Copper	10 light	0.14-0.70	0.29
	6 dark	0.35-1.04	0.56
Manganese	10 light	0.17-0.44	0.30
	10 dark	0.46-9.53	4.09
Chlorine	10 light	23-75	52
	13 dark	48-201	113
Phosphorus	14 light	23-50	35
	21 dark	27-58	47
Sulfur	10 light	36-108	58
	13 dark	56-126	100
Silicon (as SiO_2)	10 light	7-12	9
	10 dark	5-28	14

[a] Schuette *et al.* (1932, 1937, 1938, 1939).

corn syrup White (1977) examined the available literature on this subject, 24 individual analyses and averages for eight sets totalling 158 samples. Variability rendered the original proposal invalid, but the marked excess of potassium is quite apparent.

Enzymes

Enzymes are complex proteins formed in living cells that bring about the many processes and reactions in living materials. In their presence, even in the test tube, processes are easily carried out that man has not learned to duplicate in their absence.

Invertase—The enzymes in honey are almost totally added by the bee, though some traces of plant enzymes may be present. Echigo *et al.* (1973) studying the slight inversion of sucrose solutions when pollen is present have shown that invertase does not "leak" from pollen grains into the sucrose solution, but the sugar is inverted by a cell-wall invertase, which they consider as a fructoinvertase from examination of its transfructosylation product, as to be expected for a plant enzyme. Of all enzymes added by the bee, the sucrose-splitting enzyme invertase (sucrase) is the most important since by converting nectar sucrose to dextrose and levulose (inverting it) a more stable, higher-solids product can be achieved, which increases the efficiency of the ripening process. Honey invertase has been shown to differ from yeast invertase in that it transfers *glucose* from the sucrose molecule either to water to form free glucose, or to other sugars to form more complex products (White and Maher, 1953). This accounts for some of the less-common sugars present in honey. It is thus a glucoinvertase, or more properly an α-glucosidase; it also splits maltose and other sugars with terminal α-glucose moieties. Yeast invertase, like other plant invertases, is a fructosidase. Unless it is destroyed by heat, the invertase in honey continues its activity after extraction, slowly reducing the sucrose content. The consistently high sucrose content of some honeys (citrus, some clovers, and others) normally results from such heavy nectar flows or such high-sugar nectars that relatively little manipulation by the bees is needed to achieve honey density so that their invertase level is considerably less than most other honeys. Freshly extracted citrus honey is sometimes unacceptable in the international market (maximum sucrose content of 5%) because of higher sucrose. By storing it for a few weeks or months a 24 - 30°C (75 - 86°F), the natural honey invertase will continue its action, eventually reducing the sucrose content to acceptable levels (Deifel *et al.*, 1985).

Glucose Oxidase—This enzyme is also of considerable interest. It is added by the bee (White *et al.*, 1963) to the nectar. During ripening of nectar, the enzyme oxidizes small amounts of glucose to gluconolactone, which equilibrates with gluconic acid, the principal acid of honey. The acidity thus formed contributes to the stability of the ripening nectar against fermentation; in the reaction one molecule of hydrogen peroxide is produced for each molecule of glucose oxidized. This peroxide also helps to stabilize the ripening nectar against spoilage. Unlike invertase, the activity of this enzyme is negligible in full-density honey. It is necessary to dilute honey to measure the activity. Burgett (1974) found glucose oxidase activity in stored honey from nine social *Hymenoptera* from three superfamilies, and suggested that it is a food protective mechanism common to social *Hymenoptera*.

It has been shown (White *et al.*, 1963) that nearly all of the antibacterial activity (other than that due to the high solids content) or "inhibine" of honey is due to the production of hydrogen peroxide by this enzyme during the agar-plate assay for inhibine. Recently Bogdanov (1984) has reported the presence of heat-stable antibiotic activity and suggested that a material "pinocembrine" is responsible.

Diastase—This enzyme, which destroys starch, is also added to nectar by the bee during ripening. In contrast to the preceding two, its function is not known, since nectars are not known to contain starch. Like all enzymes, diastase (amylase) is weakened or destroyed by heat, and since it is relatively easy to measure, it has been used for many years to estimate the extent of heating to which a honey has been exposed. Such information has been required by many countries where heating of honey is believed to reduce or destroy certain ill-defined, health-promoting properties.

There are several difficulties with this approach. First, diastase of fresh honeys varies considerably. Further, it has become recognized within the past 20 years that not only heat as such, but extended storage can attenuate the level of diastase (and other enzymes) in honey (White, Kushnir, & Subers, 1964). In fact, the destructive effect of heat can be approximated mathematically and is as shown in Figure 5. There one can see that 200 days storage at 30°C (86°F) will on average destroy half of the diastase in honey, just as will 1 day at 60°C (140°F) or 4.5 hours at 70°C (158°F). Beekeepers and packers with customers requiring specific minimum diastase levels should be aware of this and handle their honey accordingly.

Other Enzymes—There are minor amounts of other enzymes in honey, less well studied: catalase, which destroys hydrogen peroxide, (Auzinger, 1910, confirmed by Schepartz, 1966) and an acid phosphatase which removes phosphate from organic phosphates (Giri, 1938). No evidence for lactase, protease, lipase, or inulase (Gothe, 1914) has been reported.

Proteins

There are small quantities of proteins in honey, in addition to the enzyme proteins found therein. As with all honey constituents, the concentrations vary greatly among different kinds of honey. The presence of precipitable protein material was the basis of early tests (Langer, 1903, Lund, 1909) for distinguishing natural from artificial honey. Paine *et al.* (1934) noted that about half of the colloidal material in honey was protein in nature; Helvey (1953) reported two proteins of molecular weight 73,000 and 146,000. Later White and Kushnir (1967) found 4 to 7 distinct components in the protein fraction, of which 4 appeared to originate from the bee. Bergner and Diemair (1975) reported 5 proteins in honey, 3 from the bee and 2 from the plant. The complexity of the proteins of honey is further revealed by the newer separation procedures; for example invertase (α-glucosidase) preparations from a bulk honey show up to 18 separable components; that from a single colony fed sugar had only seven (White & Kushnir, 1967); similar complexity is found with other enzymes. Marshall & Williams (1987) have consistently found evidence of at least 19 proteins in honey. White and Kushnir pointed out that 35 - 65% of the nitrogen compounds in honey were non-protein in nature.

Thus, the practice of estimating protein by multiplying nitrogen content by a factor (commonly 6.25) is not accurate for honey. Later, White and Rudyj (1978a) searching for data useful to detect adulteration for honey analyzed 740 samples for "true" protein (isolated by dialysis), measured by a photomet-

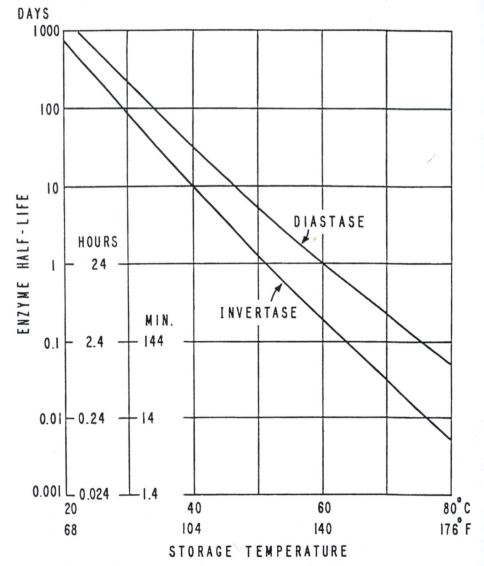

FIGURE 5. Approximate time required at various temperatures for diastase and invertase activities of honey to be reduced to one-half of the original value. (White, 1967)

ric method. They reported that 40 - 80% of the nitrogen in honey was in the protein fraction.

Amino Acids

A large part of this non-protein nitrogen is in the amino acids. With the advent of sensitive and specific analytical methods for these materials, a considerable body of data has become available. Quantitative data for 18 amino acids from five investigators are summarized by White (1975, pp. 178-179); Petrov (1974) provides data from three additional studies. Davies (1975, 1976) proposed that ratios of selected amino acids could be used to characterize the area of production of honey; data from 98 samples from 11 countries were used. Bergner and Hahn (1972) showed by analyzing foragers' honey sac contents that amino acids were added during foraging; the proportions of the various amino acids in honey from sugar-fed bees were strikingly similar to those of honey, although the totals varied with time. Proline is by far the predominant amino acid in honey, averaging for 98 samples (Davies, 1975) about half of the total. Bergner and Hahn's data show that for 13 samples of sugar-fed stores, proline is 89% of the total; for 13 floral honeys, it averaged 75%. They related its concentration to the extent of manipulation by the bees in converting nectar to honey. Searching for ways to detect addition of other sugars to honey, White & Rudyj (1978b) analyzed 740 samples of U.S. honey from two sample collections for proline; the average value was 50.3 mg/100g., range 14.8 - 148. Echigo, Takenaka and Ichimura (1973) examined the relative contributions of pollen and the bee to the amino acids of honey, as well as the possibility that invertase and diastase and flavanoids originate from the pollen. Free amino acids and flavanoids did "leak" from pollen into sugar solutions; diastase or invertase did not. They showed that the proportions of 16 amino acids leaked from pollen into sugar solutions differed from that found in honey and nectar. Asparagine predominated in the pollen extract, lysine in nectar, and proline in both normal honey and that from caged bees fed sugar. It is thus obvious and generally agreed (Davies, 1975) that most of the proline in honey originates from the bee, rather than from nectar or pollen. Analyses of five Indian honeys from *Apis cerana indica* showed an average of only 23% of the total amino acids as proline.

Vitamins

It has been shown that honey contains small and variable amounts of at least six vitamins. Table 5 shows data from four studies. The very low amounts of these materials found in honey have no real nutritional significance; it has been calculated that honey contains only 6% of the niacin and 3% of the thiamine required for the metabolism of its own sugars.

A considerable interest in ascorbic acid in honey has been reviewed by White (1975). Mint honey was shown by Griebel (1938) by a chemical

Table 5. VITAMINS IN HONEY

Samples: number, source	Ribo- flavin	Panto- thenic acid	Niacin	Thia- mine	Pyri- doxin	Ascorbic acid
		(Microgramsper 100 grams honey)				
29 Minnesota[a]	61	105	360[b]	5.5	299	2400
38 U.S. and foreign	63	96	320[b]	6.0	320	2200
21 U.S. 3-7 years old[c]	22	20	124	3.5	7.6	—
19 U.S. 1-2 years old[c]	26	54	108	4.4	10.0	—
4 India[d]	21-54	—	442-978	8-22	—	2000- 3400

[a] Haydak, Palmer, Tanquary, & Vivino (1942).

[b] Corrected from earlier data (Haydak *et al.,* 1943).

[c] Kitzes, Schuette, & Elvehjem (1943).

[d] Kalimi & Sohonie (1965).

method and by bioassay to contain approximately 100 - 300 mg per 100 g. Rahmanian *et al.* (1970) reported honey from the Darmavand area in Iran to contain (by both chemical and bioassay) 118 - 240 mg/100g of ascorbic acid, and suggested its use as a source of this vitamin to alleviate the marginal vitamin C deficiency common to the area. Since ascorbic acid is oxidized by hydrogen peroxide, it is unlikely (Schepartz, 1966) that ascorbic acid in the reduced form would be found in a honey with significant glucose oxidase activity. No data on such activity have been recorded for honeys with high ascorbic acid. It is known, however, (White & Subers, 1963) that glucose oxidase activity as assayed by peroxide accumulation is quite variable among different honeys.

CHARACTERISTIC ATTRIBUTES OF HONEY
Aroma and Flavor
Although aroma and flavor of honey are its most important characteristics from the beekeeper's and consumer's viewpoint, we know relatively little about them. The effect of beekeeping and processing practices upon flavor is often ignored or given too little attention.

The delightful aroma and flavor of fresh honey are remembered with pleasure, yet at times we may be disappointed by the flavor (or off-flavor) of commercial honeys. There are perhaps as many different honey flavors as there are plant nectar sources. Many of these are of only local significance; some kinds of honey not generally acceptable elsewhere may be preferred by those in the specific area of production, since familiarity with them may have been acquired at an early age. There are some kinds of honey that are unpalatable to humans though, of course, quite acceptable to the bees. Beekeepers are usually adept at avoiding the inclusion of these honeys in their marketable product.

The fine flavor and delicate bouquet of honey are particularly vulnerable to heat and improper storage. In addition to the loss of more volatile aromas, excessive heat can change some flavors and induce off-flavors from its effect on the sugars, acids, and protein materials. Honey can be heated to delay granulation and avoid fermentation without danger to flavor if care is given to the duration as well as to the amount of heat. Removal of heat after treatment is absolutely essential to maintaining a quality product. Flavor loss can also be serious during storage, as explained in more detail under "Storage of Honey".

In common with many other natural products, honey has been found to contain many relatively simple volatile organic substances. In the 1960's, considerable attention was given to volatile honey flavor components; one study (Cremer and Riedmann, 1965) separated some 120 substances and identified over half of them. Tan *et al.,* (1989) separated 129 components from honey extracts and identified 81 of them; they suggested that such analyses may be useful for floral source identification.The contributions of these compounds to honey flavor and aroma are difficult to assess. There appears to be a basic "honey flavor" common to most honeys; since nearly every floral type seems to have some of this characteristic flavor, one can speculate that the bee may impart the common aspects of honey flavor, with the volatiles from the specific nectar differentiating the floral type. Maeda *et al.* (1962) ascribed the taste of honey (in general) to the sugars, gluconic acid, and proline. Synthetic honey flavors contain large amounts of several phenylacetic esters; Jacobs (1955) noted that nearly all phenylacetic esters are characterized by a honey taste and aroma. Cremer and Riedmann (1965) speculated on the origin of the phenylethyl alcohol they found in 16 of 22 honeys, proposing that it is produced from the phenylalanine known to occur in honey.

Color

Relatively little is known of the substances responsible for the color of honey. Like most other attributes, color is extremely variable: natural honeys may be nearly colorless, or any of the shades from very light yellow-beige to very dark brown. Fellenberg and Rusiecki (1938) found light-colored honey to contain less water-soluble color than fat-soluble color; with dark honeys the opposite was true. The fat-soluble colors were carotenoid (plant-derived) in nature. It is likely that polyphenols which Browne (1906) found in 25 of 92 honeys he examined, are oxidized by air to dark materials. Echigo *et al.* (1973) consider the color of honey to consist of melanoidin from the amino-carbonyl reaction and flavanoid pigments extracted from pollen by the nectar and honey. An example of the former is the product of a reaction of sugar (especially fructose) with the amino acids. The effect of the acid environment on the sugars (caramelization) and the reaction of the phenolic materials with iron, whether natural or dissolved from containers or processing equipment, also contribute to color. Honey darkens when heated or stored (see Processing and Storage).

Color, in general, is related to flavor in that many light honeys are of mild flavor and dark honeys have a more pronounced taste. There are exceptions to this generalization, however. Color limits are often specified in trading of honey because of this relationship since flavor is so subjective. Several systems are used for the objective measurement of honey color, some designated as official in certain countries. Color is generally not a quality factor in official grading. The various systems for color measurement are reviewed by Rodgers (1975) and Fasler (1975). The Pfund Grader is generally used in international trade; Table 6 shows the U.S. color classes in terms of the Pfund units. The

Table 6. COLOR CLASSIFICATION OF EXTRACTED HONEY

USDA Color Class	Color Range, Pfund Scale (mm)	Absorbance
Water white	8 or less	0.0945
Extra white	> 8, to & including 17	0.189
White	> 17, to & including 34	0.378
Extra light amber	> 34, to & including 50	0.595
Light amber	> 50, to & including 85	1.389
Amber	> 85, to & including 114	3.008
Dark amber	> 114	

[b] Absorbance (optical density) $= \log_{10}$ (100/percent transmittance), at 560 nm for 3.15 cm thickness of caramel-glycerin solutions measured against an equal thickness of glycerin.

FIGURE 6. The USDA Color Comparator is the standard used to classify honey by color and turbidity.

USDA color classifier (Figure 6) and the commercially available Lovibond honey color grader (Figure 7) are officially accepted.

Hydroxymethylfurfural (HMF)

This compound is a decomposition product of simple (monosaccharide) sugars, especially fructose, in the presence of acid. Such decomposition may even begin during the ripening of the nectar in the hive when concentrations of fructose and acid become appropriate. Like all chemical reactions, it is accelerated by heat.

FIGURE 7. The Lovibond Honey Grader is a color comparator that operates similarly to a viewmaster. The honey sample is placed in a cell and compared to a permanent glass color standard.

As discussed in a recent review (White, 1975), the involvement of regulatory officials, food technologists, and honey producers, packers, and dealers with HMF in honey has spanned at least 70 years, and can be resolved into several phases. Earliest was the development and use of a qualitative color test to distinguish artificial from natural honey and to indicate the addition of invert syrup to honey. Being a simple visual color test, it soon became the source of controversy; it was shown that natural honey when sufficiently heated responded to the test. Even though it was later recognized that HMF was responsible for the color, little but controversy enveloped the area until Schou and Abildgaard (1934), Winkler (1955) and Schade *et al.* (1958) developed quantitative methods for measuring HMF in honey. This allowed a more rational approach to quality control by specifying actual HMF limits rather than vague interpretations of color tests. With the use of these methods, several studies of the factors influencing HMF production in honey appeared (Hadorn and Zürcher, 1962, Gonnet, 1963, White, Kushnir and Subers, 1964). It became evident that time and temperature of processing and/or storage were the critical factors, as with enzyme destruction. Figure 8 indicates the relationship between HMF accumulation and temperature of storage. The limits indicated therein are the 4 mg% (40 mg/kg) limit specified in the European Regional Standard of the Codex Alimentarius* and the 20 mg% which has been proposed as an "action level" for the investigation of a sample for invert sugar adulteration (White and Siciliano, 1980). Studies continue on HMF: Pichler *et al.* (1984) have examined the factors controlling HMF production using model solutions simulating honey. Wootton and Ryall (1985) compared several analytical methods and reported that the Codex method seemed more empirical than a high-performance liquid chromatographic method. Neither Winkler's ultraviolet method nor his chemical procedure (essentially identical with the Codex method) qualified for official acceptance by the U.S. Association of Official Analytical Chemists (White 1979a). The two procedures frequently gave differing results on the same samples (White, 1979a, 1979b), and the more accurate chemical method requires a toxic reagent. A new relatively simple method was developed (White, 1979b) and qualified for official use (White *et al.*, 1979). An ultraviolet procedure, it provides results not different from those of the chemical method.

Some indication of the attention given the HMF content of honey is shown by data for 70 honeys imported into Germany and Switzerland

---

*The Codex Alimentarius is the body of food quality regulations used by the Common Market and most other nations except the United States. This standard was superseded in 1987 by a world-wide Codex standard for honey with an HMF limit of 8 mg% (80 mg/kg). It has not yet been widely accepted.

(Hadorn and Kovacs, 1960), 1554 honeys also imported into Germany and Switzerland (Duisberg and Hadorn, 1966), 104 Australian honeys (Chandler *et al.*, 1974), and 481 extracted and 41 comb honeys from the U.S. (White and Siciliano, 1980). In the latter work, HMF in the comb honeys, fresh from the hive, averaged 0.27 mg/100g (range 0.03 - 0.92); for the 481 extracted honeys, which had various heat treatments by their producers, the average was 0.62 mg/100g, range 0.0 - 13.6. The effect of large-scale commercial process- ing (melting from drums, settling, pasteurization, filtration, bottling, casing, warehousing, and storage for a year) has been shown (White and Siciliano, 1980) to increase HMF only by an average of 0.85 mg/100g. In seven smaller processing operations, increases in HMF were in general considerably greater, with an average increase of 1.58 mg/100g, ranging from 0.11 to 3.47.

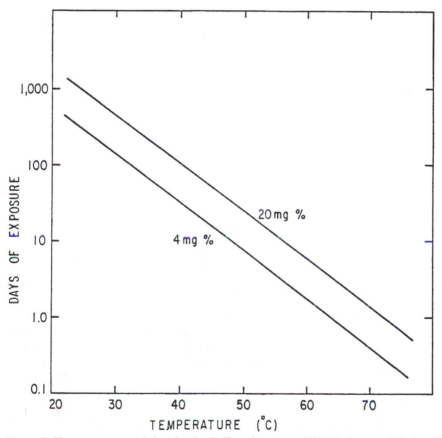

FIGURE 8. Heat exposure needed to develop indicated amount of HMF in honey. (Calculated from data of White, Kushnir, & Subers, 1964)

Toxic Substances

When one considers the enormous number of natural toxic compounds, largely produced by plants, it is remarkable that so little difficulty arises from the transfer in nectar of some of these materials to honey. The likelihood of poisoning from commercial honey is quite remote because generally beekeepers are aware of the times of bloom of the principal sources of toxic nectar such as rhododendron and other Ericaceae and insure that it does not reach market. Further, in some cases, the bees are killed by the toxicants and the colonies are severely weakened. Burnside and Vansell (1936) described 12 instances in the United States of bee kill and identified the sources as yellow jessamine (*Gelsemium sempervirens*), California buckeye (*Aesculus californica*), loco weed (*Austragalis* spp.), Western false hellebore (*Veratrum californicum*), and Southern leatherwood or black ti-ti (*Cyrilla racemiflora*). Toxic honeys have caused problems since at least 401 B.C. when the expedition of Cyrus encountered Rhododendron honey near Trebizond in Asia Minor; instances of poisoning from this source are still reported from this area. Other toxic materials isolated from honey include scopolamine from Egyptian henbane (*Datura metel*) (Örösi-Pál, 1956); atropine from Jimson weed (*Datura stramonium*) and henbane (*Hyoscyamus niger*) (Sviderskaya, 1959), gelsemine from yellow jessamine (*Gelsemium sempervirens*) (Kebler, 1896), aconitine (Saito *et al.*, 1980), and six pyrrolizidine alkaloids from tansy ragwort (*Senecio jacobaea*) (Deinzer *et al.*, 1977). An occurrence of poisoning in New Zealand was investigated in great detail: the pollen of the suspect plant, tu-tu (*Coriaria arborea*), was found harmless and the plant did not secrete nectar, but honeydew gathered from the leaves contained the toxic materials, hyenanchin and tutin (Palmer-Jones, 1947; Paterson, 1947; Hodges and White, 1966). A 1000 square mile area is closed to beekeeping as a result. Toxic honeys have been reviewed in some detail by White (1973, 1981). In the latter review, data are given on toxicants isolated from honey, their toxicity, symptoms of intoxication, and the chemical structures of six toxicants.

PHYSICAL PROPERTIES OF HONEY

Since the sugars make up on the average 98% of the solids in honey, the physical attributes are determined by their identities and concentrations. As with the compositional data, the physical properties are expressed in ranges rather than as constants. Honey is superficially a syrup containing more levulose than dextrose, but it differs in important respects from a simple invert syrup of the same water content. Since most of the properties vary in a regular manner with the water content, many have been used to measure it. A critical examination of the many physical and chemical methods to measure honey moisture is available (White, 1969). Physical attributes that have been used to estimate moisture content are: viscosity, density, specific gravity, and refractive index.

Since moisture content is a most important attribute of honey, affecting as it does keeping quality in terms of fermentation and granulation, the bee-keeper should be aware of it and use adequate means to measure and control it.

Refractive Index

The simplest way to measure moisture is by the use of a properly calibrated hand-held refractometer (Figure 9) with temperature correction. This instrument is sufficiently accurate for all practical purposes, with a standard error of $\pm 0.4\%$ (Pearce and Jegard, 1949). A bench-type refractometer is the instrument of choice for accuracy.

Table 7 gives the relationship between refractive index and moisture content of honey. The refractive index of honey differs systematically from that of a sucrose solution of the same concentration, so a refractometer with a "Brix" or "Sucrose" scale will not provide accurate values for honey. To use such an instrument, a correction (Table 8) must be used.

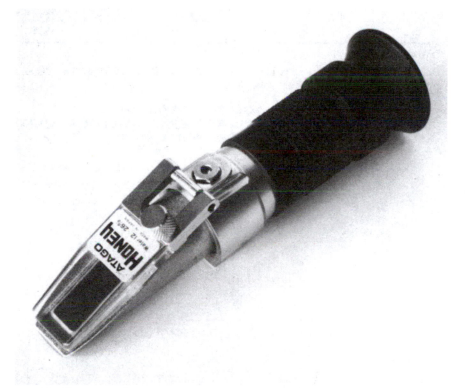

Figure 9. The hand refractometer is probably the best practical way for the beekeeper to monitor the moisture content of his honey.

Table 7. REFRACTIVE INDEX OF HONEY[a]

Water Content (%)	Refractive Index (20° C)[b]	Water Content (%)	Refractive Index (20° C)[b]	Water Content (%)	Refractive Index (20° C)[b]
13.0	1.5044	16.0	1.4966	19.0	1.4890
13.2	1.5038	16.2	1.4961	19.2	1.4885
13.4	1.5033	16.4	1.4956	19.4	1.4880
13.6	1.5028	16.6	1.4951	19.6	1.4875
13.8	1.5023	16.8	1.4946	19.8	1.4870
14.0	1.5018	17.0	1.4940	20.0	1.4865
14.2	1.5012	17.2	1.4935	20.2	1.4860
14.4	1.5007	17.4	1.4930	20.4	1.4855
14.6	1.5002	17.6	1.4925	20.6	1.4850
14.8	1.4997	17.8	1.4920	20.8	1.4845
15.0	1.4992	18.0	1.4915	21.0	1.4840
15.2	1.4987	18.2	1.4910	21.2	1.4835
15.4	1.4982	18.4	1.4905	21.4	1.4830
15.6	1.4976	18.6	1.4900	21.6	1.4825
15.8	1.4971	18.8	1.4895	21.8	1.4820
				22.0	1.4815

[a] Wedmore (1955).

[b] If the reflective index is measured at a temperature above 20° C, add 0.00023 per degree C above 20° C to the reading before using table.

Table 8. HONEY SOLIDS EQUIVALENT OF SUCROSE CONCENTRATION FROM REFRACTOMETER READING IN PERCENT SUCROSE (BRIX) AT 20° C.

% Sucrose (Degrees Brix)	Honey Solids	Honey Moisture
	(%)	(%)
77.00	78.56	21.44
78.00	79.60	20.40
79.00	80.64	19.36
80.00	81.68	18.32
81.00	82.72	17.28
82.00	83.76	16.24
83.00	84.80	15.20
84.00	85.87	14.13

Calculated from data in AOAC Tables 52.012 (1975) and 31:08 (1984).

Viscosity

The viscosity of a material is simply its resistance to flow. It may be called "body" by the beekeeper. A heavy-bodied honey has a high viscosity and flows slowly. Like other physical properties, viscosity depends upon its composition, especially moisture content. Chataway (1932) proposed the determination of moisture of honey by a viscosity measurement. She used the time of fall of a steel ball in a special apparatus and claimed an accuracy equal to direct drying. Oppen and Schuette (1939) have improved Chataway's apparatus. Many beekeepers attempt to use viscosity to estimate content by observing the rise of a large air bubble in a jar of honey. This may be quite misleading, since viscosity is quite sensitive to temperature and protein content of honey. Probably the most extensive data on honey viscosity are those of Munro (1943), shown in Tables 9 and 10. Viscosity is of great practical importance to the beekeeper and honey processor. The high viscosity of honey makes it difficult to empty containers completely and to extract it from the comb. It retards the rate of clarification by settling, and clearing of entrapped air bubbles. As all beekeepers know, the body of honey is reduced by heating; the use of some degree of warming greatly facilitates extraction, straining, settling, flow through pipes and emptying of containers. This is quantified by MacDonald's work in which he examined the effect of temperature on the rates of flow of honey through pipes of different diameters, summarized in Table 11. It should be pointed out that it is better from the standpoint of honey quality to use larger diameter piping at a lower temperature to achieve any given flow rate.

Table 9. EFFECT OF TEMPERATURE ON VISCOSITY OF HONEY

Type	Temperature (° C)	Viscosity (poise)
Sweet Clover *(Melilotus)*	13.7	600.0
(16.1% moisture)	20.6	189.6
	29.0	68.4
	39.4	21.4
	48.1	10.7
	71.1	2.6
Sage *(Eriogonum)*	11.7	729.6
(18.6% moisture)	20.2	184.8
	30.7	55.2
	40.9	19.2
	50.7	9.5
White clover *(Trifolium)*	11.5	600.0
repens) (17.1% moisture)	20.2	122.4
	29.5	44.4
	36.7	19.2
	46.9	10.5
	61.7	3.9

Munro, (1943).

**Table 10. EFFECT OF MOISTURE CONTENT AND FLORAL TYPE
ON VISCOSITY OF HONEY[a]**

Type	Moisture Content (%)	Viscosity (poise)
White clover	13.7	420
(Trifolium repens)	14.2	269
	15.5	138
	17.1	69.0
	18.2	48.1
	19.1	34.9
	20.2	20.4
	21.5	13.6
Sage *(Eriogonum)*	16.5	115
Sweet clover *(Melilotus)*	16.5	87.5
White clover	16.5	94.0

[a]Interpolated from data of Munro (1943). All values at 25° C.

Table 11. RELATIVE FLOW RATE OF HONEY IN PIPES[a]

Pipe inside diameter	Temperature		
	28° C	39° C	50° C
19mm (3/4 in)	149	400	1125
25mm (1 in)	367	973	2353
31mm (1-1/4 in)	729	1895	5000
38mm (1-1/2 in)	1263	2609	6792

[a] Flow rate in pounds per hour through a 4 inch (10 cm) length of pipe with a 4 inch head. Data of MacDonald (1963).

Optical Rotation (Polarization)

Like many other natural materials, honey has the property of rotating the plane of polarization of polarized light. This is due to the sugars, each of which has a specific and characteristic value for this property. It can be measured quite accurately and has been used for many years in the sugar industry to obtain sugar concentration.

It was early used to analyze honey for sugars, but provides only inaccurate information on concentrations of individual sugars because of the complexity of honey. It has long been known that natural honey is levorotatory (*i.e.*, rotates the plane to the left). This is largely due to the excess of levulose (so called because it is levorotatory) over dextrose (dextro, or right-rotatory). Honeydew generally shows dextrorotation to some degree, due in part to differing dextrose and levulose content, but more to the presence of the characteristic sugars of honeydew, melezitose and erlose, which are strongly dextrorotatory. Although the conventional boundary between the two types has been considered to be 0°, White (1980) has proposed that conditions of symmetry argue for a -2° boundary. For 454 levorotatory honeys, he reported an average of -14.70°S, s = 4.37°, CV =29.8%; thus this property reflects the wide variability of the sugar contents of various honeys.

Density and Specific Gravity

The density of a substance is its weight per unit volume; specific gravity (also termed relative density) is the ratio of the weight of a volume of a substance to the weight of the same volume of water, at specified temperatures. These may be determined by weighing known volumes, by the use of a hydrometer, or by a specific gravity balance. These attributes also vary with the temperature of measurement and the moisture content of honey, hence temperature must be specified in tables relating moisture content and any of these properties. The Brix hydrometer is commonly used for measuring sucrose solutions and is calibrated to read percent sucrose directly. When used for honey, the values for solids are too low by the amounts shown in Table 8. A hydrometer calibrated for honey is commercially available.

The variation of density with moisture content is sufficiently large that a low-moisture honey will tend to layer under a higher-moisture honey unless special care is taken to mix them. Honey exposed to moist air will absorb water and form a dilute layer which remains on the surface for a long time due to its lower density (see hygroscopicity).

Hygroscopicity

The hygroscopicity of a material is its ability to remove moisture from the air. It may be expressed as the relative humidity of air which is in equilibrium with the substance, when it neither gains nor loses moisture. This property is important because increase in moisture of honey can lead to fermentation. Moisture retention is said to be improved in foods made with honey.

Since hygroscopicity of honey depends to a great extent upon sugar (especially levulose) concentrations, it varies to some extent among different honeys. Table 12 shows how moisture content of a honey varies with the relative humidity (RH) of the air to which it is exposed. As noted above, the high viscosity of honey permits only slow diffusion of moisture from the absorbing surface into the mass. Martin (1958) showed that a honey of 22.5% moisture exposed in a container of 5.5 cm. diameter for 7 days at 86% relative humidity had 26% water at the surface, but no change was found 2 cm. below; after 24 days the surface moisture was 29.6%, and 23% (only a 0.5% increase) at 2 cm. depth, and even at 95 days no change was found at 6 cm. depth. When exposed to air of R.H. less than the equilibrium value, honey loses moisture at the surface, but much more slowly below the surface. The moisture content of honey (even in the comb) can be reduced by exposure to air of R.H. less than

Table 12. MOISTURE CONTENT OF HONEY AT VARIOUS RELATIVE HUMIDITIES[a]

Relative Humidity (%)	Water Content (%)
50	15.9
55	16.8
60	18.3
65	20.9
70	24.2
75	28.3
80	33.1

[a] Interpolated from the data of Martin (1958) for a clover honey.

the equilibrium value. This may be attained by raising the temperature of the air or by mechanical dehumidification. Using a dehumidifier to reduce R.H. in the drying room to 30%, Killion (1950) reduced the moisture content of honey in 130 supers from 21 to 17% in 23 days at 27°C. Circulation of heated air is also used for the purpose. Townsend (1961) described a drying room holding up to 72 supers in which warmed air at 32 - 35°C (90 - 95°F) is passed through, removing 1 to 3% moisture in 24 hours. The lower the temperature of the air before heating, the more efficient is the process, since its RH is lower after heating.

The surface-to-weight ratio is a controlling factor; honey in the comb has a larger ratio than after it is extracted, so that to reduce moisture after extraction, mechanical means are needed to produce thin layers of honey. Such equipment has been described for use at atmospheric pressure (Platt & Ellis, 1985), or in a vacuum. Paterson and Palmer-Jones (1954) described a vacuum plant that removed 2% water from 2.5 tons of honey in 8 hours. Other equipment for reducing the moisture content of extracted honey is commercially available.

Thermal Properties

The sensitivity of honey to heat damage requires that special attention be given to its thermal properties when designing processing equipment. Little information is available in the literature; equipment in use is empirically designed, or originally intended for another commodity such as milk or sugar.

Helvey (1954) reported that the specific heat of honey at 17.4% moisture is 0.54 ca./g/°C at 20°C (68°F), with a temperature coefficient of 0.02 cal/°C. MacNaughton (Townsend, 1954) obtained somewhat higher values as seen in Table 13.

Detroy (1966) measured the surface conductance (film coefficient) for honey at flow rates of 700 - 900 lb (317 - 408 kg) per hour at two temperatures

Table 13. SPECIFIC HEAT OF HONEY

Moisture Content (%)	Specific Heat cal/g/°C
20.4	0.60
19.8	0.62
18.8	0.64
17.6	0.62
15.8	0.60
14.5	0.56

Data of MacNaughton as reported by Townsend (1954).

of interest in honey processing, preheater (65 - 68°C, 149 - 154°F) and flash heater (85 - 88°C, 185 - 190°F). A diagram of his data is included in a review (White, 1978).

The calorific value was calculated by the U.S. Department of Agriculture using the Atwater system to be 304 Cal/100g (1380 Cal/lb.) (Watt & Merrill, 1963).

Granulation

A supersaturated solution is one that contains more dissolved material than can normally remain in solution. These solutions are more or less unstable and in time will return to the stable saturated condition by crystallizing the excess solute. Many honeys are in this category with respect to their dextrose content and normally equilibrate with crystallization of the excess from solution. The solid material is the monohydrate of dextrose. The tendency and extent of crystallization (granulation, "sugaring") is related to the sugar composition of the honey, its moisture content, and the temperature. Some honeys never crystallize; some will do so within a few days of extraction, or even in the comb in extreme cases. Granulation is characterized by firmness and by the fineness of the crystals or grain. An unheated honey will generally have a fine grain, probably due to the large number of natural crystal nuclei present. A honey that has been heated to prevent fermentation and/or to delay granulation will eventually produce fewer, larger crystals; the crystal-initiating nuclei have been eliminated by the heating. The granulation texture of honey is a major factor in the quality of the semi-solid product marketed as "honey spread" or "creamed honey." It should be soft enough to spread, but not be "runny." Crystal size should be imperceptible to the tongue. This product is susceptible to softening if it is stored at temperatures above 27 - 30°C (81 -86°F).

Attempts have been made to relate the crystallization of honey to its carbohydrate composition. The use of model solutions of dextrose and levulose was examined by Lothrop (1943) and more conclusively by Kelly

(1954). Details and phase diagrams are included in a recent review (White, 1978). Prediction of the tendency of a honey to granulate is of considerable practical interest. Since data from model solutions are not useful, empirical studies of honey in storage are required. Early workers proposed that the levulose-dextrose ratio was the controlling factor, but studies by Austin (1958) and White *et al.* (1962) support the dextrose/water ratio. Table 14 shows the data of White *et al.*

Processing to prevent granulation, or to obtain a finely-granulated product is discussed under "Processing" in this chapter and in Chapter 15.

Table 14. AVERAGE DEXTROSE/WATER RATIOS FOR HONEYS CLASSIFIED BY EXTENT OF GRANULATION[a]

Extent of granulation[b]	No. of samples	Dextrose/water
None	96	1.58
Few scattered crystals	114	1.76
1.5-3 mm layer of crystals	67	1.79
6-12 mm layer of crystals	68	1.83
Few clumps of crystals	19	1.86
1/4 of depth granulated	32	1.99
1/2 of depth granulated	19	1.98
3/4 of depth granulated	16	2.06
Complete soft granulation	18	2.16
Complete hard granulation	28	2.24

[a] White, Riethof, Subers, & Kushnir (1962).

[b] Granulation observed in 0.23 or 0.45 kg jars of heated honey after 6 months undisturbed storage at 23 - 28° C.

MICROORGANISMS IN HONEY AND THEIR EFFECTS
Bacteria

The occurrence and persistence of microorganisms (yeasts and bacteria) in honey has been examined sporadically over the years. While Sackett (1919) and Tysset and Durand (1973) have shown that non-spore-forming intestinal bacteria survive only a few days in full-density honey, only recently has attention been given to the spore-forming bacteria, other than spores of the bee disease organisms which have long been known. In a study to elucidate the role of microorganisms in ripening (and fermentation) of honey, El-Leithy and El-Sibaei (1972) studied their 48 yeast and 87 bacterial isolates from honey and examined those shown in Table 15 for ability to grow under acidity and sugar concentrations normal to honey. Some of the yeasts could do so, but none of the bacteria multiplied above 60% sugar and pH more acid than 4. Ruiz-Argüeso and Rodriguez-Navarro (1973) have consistently isolated from bees and *ripening* honey two gluconic acid-producing bacteria which

Table 15. BACTERIA ISOLATED FROM HONEY[a]

Type	Reference
Bacillus subtilis	A
Bacillus cereus	A , B
Bacillus megaterium	A , B
Bacillus pumilis	A
Bacillus coagulans	A , B
Bacillus alvei	B
Clostridium perfringens	B

[a] Other than those causing disorders of honey bees.

References:
  A. El-Leithy & El-Sibaei (1972).
  B. Kokubo *et al.* (1984).

they characterized as *Gluconobacter*, and proposed that at least part of the gluconic acid present in honey is of bacterial origin. Later (1975) they reported *Lactobacillus viridescens* in the bee and in ripening honey as well as an ethanol-producing bacterium they classified as *Zymomonas mobilis*. All three of these organisms disappeared as the honey became fully ripe. Kokubo *et al.* (1984) examined 71 honeys for *Bacillus* and *Clostridium* spores. Sixty-seven contained spores as listed in Table 15 and others, totaling 12 species.

In the United States, the recognition in the 1970's of infant botulism as a new disease entity prompted examination of potential sources of infection. The causative organism (*Clostridium botulinum*) is widespread in nature, and extensive examination of possible food and environmental sources indicated that honey and corn syrup were the only materials fed to the infants that contained the spores, in up to 10% of market samples (Midura *et al.*, 1979). Although no definite linkage could be shown, it was agreed in the trade and the pediatric community that since these foods are avoidable potential sources, in contrast to some environmental factors, they should not be fed to infants less than one year of age. Older infants are immune. Kokubo *et al.* (1984) found 56 of their samples to contain *Clostridium* spores, with *Cl. perfringens* in 6, but none contained *Cl. botulinum*. Sugiyama *et al.* (1978) reported finding the latter in 18 of 241 honey samples examined. Aureli *et al.* (1985) using the procedure of Midura *et al.* found no spores of *Cl. botulinum* in 107 commercial honey samples, largely of Italian origin. They used a method described by Midura *et al.*; with which Huhtanen *et al.* (1981) found no spores in 80 samples of largely U.S. honey. However, using another method, they found 6 of the samples positive. This subject has been briefly reviewed by Lawrence (1986).

Yeasts and Fermentation

Fermentation of honey is caused by the action of sugar-tolerant yeasts upon the dextrose and levulose, producing alcohol and carbon dioxide. The alcohol may then be converted into acetic acid in the presence of oxygen. Compared with other fermentations, the process is quite slow, and may not become obvious for several months after initiation. Because this type of spoilage has in the past caused considerable losses to beekeepers and packers, it has had much study over the years. Most of the information available was produced by workers in Canada and the northern United States where there were large losses from this problem. Due to the release of carbon dioxide, fermenting honey will exhibit fine gas bubbles with foam at the surface (See Figure 10); if granulated, it shows a lightened color with whitish streaks and will foam excessively upon heating. The extent of damage to quality and especially flavor depends upon the time it is allowed to proceed before being stopped by heating. Most natural honey fermentation in storage occurs after granulation. The reason for this is that the removal of dextrose hydrate (9.09% water) from solution in the honey leaves a higher-moisture liquid phase.

Figure 10. Honey may ferment in the comb before capping if conditions are favorable to spoilage.

Table 16. FERMENTATION LIABILITY OF HONEY[a]

Moisture content	Liability to ferment
Less than 17.1%	Safe, regardless of yeast count
17.1 - 18.0%	Safe if yeast count is <1000/g
18.1 - 19.0%	Safe if yeast count is <10/g
19.1 - 20.0%	Safe if yeast count is <1/g
Above 20.0%	Always in danger

[a] Based on 319 honey samples. Lochhead (1933).

Surface layers exposed to high humidity will absorb moisture and can become liable to fermentation. Table 16 shows the relationship between moisture content, yeast count, and fermentation liability; it can be seen that even a small increase of moisture can considerably elevate the liability to fermentation.

Ordinary yeasts do not grow in honey because of its high sugar concentration. The primary sources of the sugar-tolerant yeasts are the flowers and the soil (Fabian and Quinet, 1928; Lochhead and Heron, 1929). Lochhead and Farrell (1930) have shown that soils in established apiaries contain sugar-tolerant yeasts and the air and equipment in the honey house are always contaminated with them. Combs in the hive, especially those containing honey from the previous season, and wet extracted combs in storage are also abundantly contaminated. One should assume that all honey contains yeasts when planning for its handling. The number of yeast spores in various honeys can vary a million-fold: from 1 in 10 grams to 100,000 per gram; honeys with higher moisture content will contain even more. Uncapped combs will contain more than capped combs in the same super, due to higher moisture in the former. According to Wilson and Marvin (1932) honey yeasts will not grow below 11°C (52°F), hence storage below this should protect from fermentation. Storage above 38°C (100°F) also retards the process, but will severely deteriorate honey quality. Although honey stored in winter cold will not ferment, it is more likely to granulate and thus be more liable to ferment when warm weather arrives. Table 17 lists the yeasts commonly found in honey.

Table 17. YEASTS ISOLATED FROM HONEY

Type	Reference
Nematospora ashbya gossypii	Aoyagi & Oryu, 1968
Saccharomyces bisporus	Aoyagi & Oryu, 1968
Saccharomyces torulosus	Aoyagi & Oryu, 1968
Schizosaccharomyces octosporus	Lochhead & Farrell, 1931
Schwanniomyces occidentilis	Aoyagi & Oryu, 1968
Torula mellis	Fabian & Quinet, 1928
Zygosaccharomyces barkeri	Lochhead & Heron, 1929
Zygosaccharomyces japonicus	Aoyagi & Oryu, 1968
Zygosaccharomyces mellis	Fabian & Quinet, 1928
Zygosaccharomyces mellis acidi	Richter, 1912
Zygosaccharomyces nussbaumeri	Lochhead & Heron, 1929
Zygosaccharomyces priorianus	Fabian & Quinet, 1928
Zygosaccharomyces richteri	Lochhead & Heron, 1929
IN RIPENING HONEY	
Torulopsis magnoliae	Ruiz-Argüeso & Rodriguez-Navarro, 1975
Saccharomyces mellis	Ruiz-Argüeso & Rodriguez-Navarro, 1975
Torulopsis stellata	Ruiz-Argüeso & Rodriguez-Navarro, 1975
Torulopsis apicola	Ruiz-Argüeso & Rodriguez-Navarro, 1975

Table 18 shows minimum conditions needed to destroy yeasts.

To summarize:

1. All unprocessed honey should be considered to contain yeasts.
2. Honey is more liable to fermentation after granulation.
3. Honey over 17% moisture *may*, and over 19% moisture *will* ferment.
4. Storage below 10°C (50°F) will prevent fermentation during such storage but not later.
5. Heating to 63°C (145°F) for 30 minutes, or its equivalent, will destroy honey yeasts and thus prevent fermentation.

Table 18. CONDITIONS REQUIRED TO KILL YEAST IN HONEY[a]

| Temperature | | Time |
Degrees F.	Degrees C.	Minutes
125	51.7	470
130	54.4	170
135	57.2	60
140	60.0	22
145	62.8	7.5
150	65.6	2.8[b]
155	68.3	1.0[b]

[a] Calculated from data of Townsend (1939). "Come-up" time not included.

[b] Extrapolated from logarithmic curve constructed from Townsend's data.

PROCESSING AND STORAGE OF HONEY

Processing

The flavor and desirability of well-ripened honey are at their peak in the comb. Our efforts to convert it to our use must inevitably result in some deterioration, but whether it is significant depends upon how it is treated by the beekeeper and subsequent handlers. The best processing is the least that will meet the objectives. While "raw" honey will always have its devotees, today's markets require a product that will not ferment and will remain in the desired physical state. Sometimes these objectives are met at the expense of honey's most valuable asset, flavor, but with appropriate procedures they can be achieved with preservation of the original full flavor and aroma.

Heat is the only practical means for preventing granulation and fermentation, but can easily deteriorate quality. Not only must time and temperature of application be controlled, but heat must be removed as quickly as possible afterwards. Use of the so-called "flash" method, a continuous process in which honey in a closed system is heated very rapidly, strained or filtered and then rapidly cooled, probably represents the least heat exposure that will meet the desired objective. Treatment in a closed system minimizes losses of volatile aroma during the heating period, and immediate cooling minimizes heat-induced color and flavor changes and allows higher temperatures to be used for a shorter time. Townsend (1961) described equipment that will heat and cool 300 to 600 lb. (135 - 270 kg) of honey per hour, either for a liquid or semi-solid pack. Honey must have a preliminary straining before being heated so that off-odors or off-flavors will not be imparted by the action of hot honey on the extraneous material. Townsend's bulletin (1961) describes comprehensive procedures and equipment for removal of moisture from honey before extraction with details on hot rooms, uncapping, settling, and straining. Details are given on the OAC strainer and a pressure strainer. He describes a cabinet that will melt ten 60-lb (17 kg) cans of honey overnight in which the honey does not exceed 54°C (129°F). A procedure for packing liquid honey and finely granulated honey is given in detail. Processing equipment, such as a continuous-flow mixer for honey spread, and the continuous-flow high-temperature short time pasteurizer mentioned above, are also in the bulletin.

An excellent review is available (Townsend, 1975) describing in detail the various steps in processing honey to delay granulation and to prevent fermentation, and also the preparation of finely-granulated honey. This publication provides specific information on all of the steps needed to produce these types of honey. Information useful to the small producer, and also for the larger packing plant is included. The guiding principle is the use of the least degree of heating that will accomplish the objective. Extraction, straining, filtration, melting for repacking, blending, and storage are all considered, for all sizes of operation. Layouts are diagrammed for processing-packing plants: for a beekeeper's honey house, handling up to 200 kg/hr, and for a continuous-flow operation of the same scale; for a processing-packing plant handling 200 -750 kg/hr; and a layout for a plant handling 750 kg/hr or more.

A detailed discussion of the production of finely-granulated honey (honey spread, Dyce-processed honey) is provided by Dyce's review (1975). The basic steps in the process are the complete liquefaction of the honey, "pasteurization" to eliminate yeasts, adding 5 - 10% of a finely-granulated "starter," thoroughly mixing, and filling into the final containers. The containers are held in a cold room at 14°C (57°F) for up to 6 days, when they are removed to storage not above about 24 - 27°C (75 - 80°F). The original review should be consulted for details. Also, see Chapter 15, *Honey and Wax—Considerations of Processing and Packaging Techniques.*

Storage

Honey is no longer a seasonal commodity; modern merchandising requires year-round availability which increases the importance of proper storage. Honey for export may be stored before and after shipping, and shipping itself may be quite deleterious during long voyages in tropical areas.

Sufficient change in color and flavor quality may take place to bring about refusal of a shipment due to deterioration. Changes that take place during storage have been examined in some detail. Milum (1948) has shown that heat processing does not accelerate the later darkening of honey. His data on the effect of heat on honey color are shown in Table 19. deBoer (1934) pointed out that changes brought about by heating also occur during honey storage; he specifically referred to the production of HMF and a weakening of honey enzymes.

White *et al.* (1962) made an extended study of the effects of storage at ordinary temperatures (23 - 28°C, 73 - 82°F) for periods of 1.5 - 2 years, with statistical evaluation of the results. The original should be consulted for details, but major changes were found in nearly all attributes: in the 2-year period 13% of the dextrose and 5% of the levulose were converted to other carbohydrates, the relatively low sucrose content was increased 73%, other disaccharide sugars increased 68%, and the more complex sugars increased 13%. Diastase decreased at a rate of about 3% per month, and total acidity increased about 0.5% per month. These sugar changes explain the slow liquefaction of granulated honey during long-term storage as seen in Figure 11 and also the softening and partial liquefaction of finely-granulated honey spread in ordinary storage. Such deteriorated spread cannot be reprocessed to its original form because the loss of dextrose has changed the dextrose/water ratio. Effects of storage at 25°C (77°F) upon HMF and invertase are similar (White, Kushnir, & Subers, 1964) and are described elsewhere in this chapter. As indicated in Figure 3, invertase is lost about twice as rapidly as diastase; each of the three samples in the study exceeded the Codex limit for HMF of 40 mg/kg after 1 year storage at 27°C (81°F).

All of these changes can be avoided by storing honey at low temperatures. For this purpose temperature should be 10°C (50°F) for unheated honey;

Table 19. APPROXIMATE RATE OF HONEY DARKENING IN STORAGE[a]

Storage temperature		Darkening rate, mm Pfund/month Original Color		
°F	°C	<34 mm	34 - 50 mm	>50 mm
50	10.0	0.024	0.024	0.024
60	15.6	0.08	0.125	0.10
70	21.1	0.27	0.70	0.40
80	26.7	0.90	4.0	1.50
90	32.2	3.0	7.7	5.0
100	37.8	10.0	14.0	11.0

[a] Calculated from data of Milum (1948).

FIGURE 11. The jar of honey at left has granulated completely. The bubblelike formation toward the top of the jar is the result of fermentation after granulation. The jar in the center represents partial granulation with coarse crystals in the upper half and the lower part solidly granulated. The jar of honey at right shows natural partial liquefaction after complete granulation. Eventually, the sample may completely liquefy.

great advantage will result from avoiding temperatures above 15°C (59°F). Actually, honey kept at very low temperatures for years cannot be distinguished from the freshest honey. However, the interval between 10 - 15°C (50 - 59°F) must be avoided because granulation is promoted with accompanying increased liability to fermentation. Storage above 27°C (81°F) should be avoided, even for relatively short times since deterioration in color, flavor, and enzyme content is particularly rapid.

STANDARDS FOR HONEY
Food and Drug Standards

The U.S. Food and Drug Administration does not have an official definition for honey. The former definition, in force under the original Food and Drug Act of 1906, held that "Honey is the nectar and saccharine exudations of plants, gathered, modified, and stored in the comb by honey bees (*A. mellifera* and *A. dorsata*); is levorototary, contains not more than 25 percent water, not more than 0.25 percent ash, and not more than 8 percent sucrose." The Food, Drug, and Cosmetic Act of 1938, which supersedes the original law, provides for definitions and standards of identity for foods, but it has none for honey. Even though there is no standard of identity, honey is of course covered under the general regulations for foods regarding labelling, absence of adulteration (which includes extraneous matter as well as added sugars). Many of the States have standards for honey.

Department of Agriculture Standards

This department has established voluntary grade standards for extracted and comb honey.* These standards are designed to serve as a convenient basis for sales, for establishing quality control programs, and determining loan values. They also serve as a basis for the inspection of honey by the Federal Inspection Service. There are standards for extracted honey, including crystallized honey, and for comb honey, including section comb, shallow-frame comb, wrapped cut comb, and chunk or bulk comb honey. The grades for extracted honey which had been in effect since 1951 were revised effective April 23, 1985 (Manley, 1985) in response to requests from honey producers and packers to eliminate discrimination against unfiltered honey of otherwise acceptable quality because of reduced clarity when compared to filtered honey. The new standards recognize two "styles" of extracted honey: "filtered" and "strained" and provides grading criteria for each style, with clarity not being scored for strained honey. For each style there are four grades, A, B, C, and Substandard. Table 20 shows the grading criteria for the filtered style. For the strained style, clarity is not considered and no points for it are given. The total score for this style is obtained by adding the values for the two factors actually graded and then multiplying this by 100/90. Color, as shown in Table 6, is not a grade factor. The actual grades are:

U.S. Grade A is the quality of extracted honey that meets the applicable requirements and has a minimum total score of 90 points.

U.S. Grade B is the quality of extracted honey that meets the applicable requirements and has a minimum total score of 80 points.

U.S. Grade C is the quality of honey that meets the applicable requirements and has a minimum total score of 70 points.

Substandard is the quality of extracted honey that fails to meet the requirements of U.S. Grade C.

Other Standards

The Codex Alimentarius Commission of the Food and Agriculture Organization (UN) has issued standards for honey for use in trade by all accepting countries. The United States does not accept these standards, but

---

*Copies of the grade standards for extracted honey (fifth issue, effective April 23, 1985) and comb honey (second issue, effective August, 1933) may be obtained from Processed Products Standardization and Inspection Branch, Fruit and Vegetable Division, Agricultural Marketing Service, U.S. Department of Agriculture, Washington, DC 20250.

Table 20. U. S. STANDARDS FOR EXTRACTED HONEY (FILTERED STYLE)

Factor	Grade A	Grade B	Grade C	SUBSTANDARD
Percent soluble solids (minimum)	81.4	81.4	80.0	Fails grade C
Absence of defects	Practically free-practically none that affect appearance or edibility.	Reasonably free-do not materially affect the appearance or edibility.	Fairly free - do not seriously affect the appearance or edibility.	Fails grade C.
Score points[a]	37 to 40	34 to 36[b]	31 to 33[b]	0 to 30[b]
Flavor & aroma	Good - free from caramelization, smoke, fermentation, chemicals, and other causes.	Reasonably good-practically free from caramelization; free from smoke, fermentation, chemicals, and other causes.	Fairly good - reasonably free from caramelization; free from smoke, fermentation, chemicals, and other causes.	Poor - fails grade C.
Score points[a]	45 to 50	40 to 44[b]	35 to 39[b]	0 to 34[b]
Clarity	Clear - may contain air bubbles that do not materially affect the appearance; may contain a trace of pollen grains or other finely divided particles in suspension that do not affect appearance.	Reasonably clear-may contain air bubbles, pollen grains, or other finely divided particles in suspension that do not materially affect the appearance.	Fairly clear - may contain air bubbles, pollen grains, or other finely divided particles in suspension that do not seriously affect the appearance.	Fails grade C.
Score Points[a]	8 to 10	6 to 7	4 to 5[b]	0 to 3[b].

[a] Numerical ranges are inclusive.

[b] Limiting rule: sample units with score points that fall in this range shall not be graded above the respective grade regardless of the total score.

[c] Partial limiting rule: sample units with score points that fall in this range shall not be graded above U. S. Grade C regardless of the total score.

obviously any exporter must meet the standards to do business with those accepting them. Most of the importing countries subscribe to them. They include considerable detail on composition of the product, as well as the analytical methods required to be used for the measurements. A summary of the compositional requirements is shown below.

Apparent Reducing Sugar:
Blossom honey, when so labelled . . . . Not less than 65%
For honeydew honey and blends with
blossom honey. . . . . Not less than 60%
    Moisture Content: . . . . Not more than 21%
For heather honey . . . . Not more than 23%
    Apparent Sucrose Content: . . . . Not more than 5%
For honeydew honey, and blends with blossom honey, *Robinia,*
Banksia menziesii, and lavender honeys . . . . . Not more than 10%
    Water-insoluble solids: . . . . Not more than 0.1%
For pressed honey . . . . Not more than 0.5%
    Mineral Content (ash): . . . . Not more than 0.6%
For honeydew honey and blends with
blossom honey. . . . . Not more than 1.0%
    Acidity: . . . . Not more than 40 meq/kg
    Diastase activity and HMF content
After processing and blending . . . . Not less than 8 mg/kg
Providing that HMF is . . . . Not more than 40 mg/kg

For honeys with naturally low enzyme
    content e. g. citrus. . . . . Not less than 3
    Providing that HMF is . . . . Not more than 15 mg/kg

The Codex Alimentarius Commission at their 1987 meeting declared the above European Regional Standard to be superseded by a newly-adopted World-wide Codex standard for honey. Such standards are not effective in a country until formally adopted by that country. The Codex procedures state that any member country may indicate its acceptance at three levels: (1) full acceptance, (2) target acceptance (indicating acceptance by a given date with commitment meantime to allow distribution of conforming product), and (3) acceptance with specified deviations or (4) non-acceptance. At the time of writing (1990), no countries have officially accepted the world-wide standard.

The world-wide standard differs in its definitions and descriptive sections and as follows in the composition limits:

Apparent Reducing Sugar:
    Adds Blackboy honey. . . . . Not less than 53%
Moisture content:
    Adds Clover honey. . . . . Not more than 23%
Apparent Sucrose Content:
    Adds citrus, alfalfa, sweet-clover, red gum, acacia, leatherwood,
    Menzies banksia . . . . Not more than 10%
    Redbell, white stringybark
    grand banksia, blackboy . . . . Not more than 15%

Changes:
    Diastase activity after
    blending and processing .... Not less than 3
    HMF content .... Not more than 80 mg/kg
These last two are major changes.

EXHIBITION OF HONEY

Honey shows are popular as a means of reminding the public of the goodness of honey. They are a potential means of improving honey quality by demonstrating the rewards of careful handling and in providing a common ground of competition where the hobbyist and the full-time producer can meet on equal terms.

FIGURE 12. This simple polariscope is useful in judging honey since it shows defects not visible to the naked eye. (White & Maher, 1951)(*USDA photo by M. C. Audsley*)

Showing honey is not difficult, but as in all such competitions, requires a high level of care and attention to detail. All instructions must be followed to the letter; for liquid honey, careful settling and straining are required, as judges note the presence of even a single bubble in the surface. Caps must be clean inside and out and jars should be selected for uniformity and uniformly filled. Dust and fine crystals, perhaps not visible to the unaided eye, may be easily visualized by a judge using the polariscope illustrated in Figures 12 and 13. Body or moisture content is important; beyond a specified moisture content, a sample is disqualified. Generally, additional points are given for lower moisture, with a minimum level beyond which points may be docked, to discourage artificial reduction of moisture content. Flavor is very important, judgment being similar to that in the U. S. Grades. Though many shows do not do

FIGURE 13. Cleanliness and crystallization in honey are readily observed by the light when viewed with a polariscope. *(USDA photo by M. C. Audsley)*

so, means should be taken that entries are properly within the various color classes specified in the entry forms. Judges must have the final decision on the classes into which samples are placed. The score sheet shown below is that of the Eastern Apicultural Society (Connor, 1988).

Honey Judging Score Sheet

Extracted Honey	Points
1. Density	20

    Water content above 18.6% will be disqualified and below 15.5% will be docked points

2. Absence of crystals 10
3. Cleanliness 30
    a. Without lint   7
    b. Without dirt   10
    c. Without wax   7
    d. Without foam   6
4. Flavor 30
Points will be reduced for honey flavor that has been adversely affected by processing.
5. Container appearance 10
TOTAL 100

Creamed Honey
1. Fineness of crystals 30
2. Uniformity and firmness of product 25
3. Cleanliness and freedom from foam 20
4. Flavor 15
    a. Points will be reduced for honey flavor that has been adversely affected by processing.
    b. Disqualified for fermentation.
5. Accuracy of filling and uniformity 10
TOTAL 100

Chunk Honey
1. Neatness and Uniformity of Cut 20
    a. Up-grade for parallel and 4-sided cuts
    b. Down-grade for ragged edges
2. Absence of watery cappings, uncapped cells and pollen 20
3. Cleanliness of product (down-grade for travel stains, foreign matter, wax flakes, foam, crystallization) 20
4. Uniformity of appearance in capping structure color, and thickness of chunks 20
5. Density and flavor of liquid portion
    a. Density (water content above 18.6% will be disqualified, below 15.5% will be docked points) 10
    b. Flavor (points will be reduced for honey flavor that has been adversely affected by processing) 10
    c. Disqualification for fermentation
TOTAL 100

Comb Honey
1. Uniformity of appearance 20
2. Absence of uncapped cells 10

3. Uniformity of color   15
4. Absence of watery cappings   10
5. Cleanliness and absence of travel stains   15
6. Freedom from granulation and pollen   10
7. Uniform weight of each section (except bulk frame)   10
8. Total weight of entry   10
TOTAL   100

Cut Comb Honey
1. Neatness and uniformity of cut, absence of liquid honey   20
2. Absence of watery cappings, uncapped cells and pollen   20
3. Cleanliness of products, absence of travel stains
   crushed wax, and crystallization   20
4. Uniformity of appearance (color of honey, capping
   structure, thickness of comb)   15
5. Uniformity of weight   15
6. Total weight of entry   10
TOTAL   100

Novelty honey (Three identical novel-shaped honey-filled containers or One novelty gift box of packaged honey)
1. General appearance   30
2. Originality   30
3. Quality of honey   25
4. Availability   15
TOTAL   100

ADULTERATION OF HONEY

For at least a century, sugar syrups have been sold as or added to honey in Europe and the United States. Chemical testing for HMF and analysis of sugars and other components had generally dealt successfully with this problem until the introduction of cheap high-fructose corn syrup (HFCS) in the United States in the early 1970's. This highly refined material is similar enough to honey in its sugars that it could not be detected in admixture by existing tests. The application of stable carbon isotope ratio measurement (which Nissenbaum *et al.* [1974] had used in Israel for detection of citrus juice adulteration), allowed detection of this type of falsification (White and Doner, 1978). The basis of this test is the difference in $^{13}C/^{12}C$ ratio between honey plants (those using the Calvin photosynthetic or C_3 cycle) and corn and cane plants (C_4 or Hatch-Slack cycle). For honeys, the values of $\delta^{13}C$ average -25.4‰ while for the C_4 plants it is about -9.7‰. The range found for 119 samples of pure honey was from -22.5 to -27.4‰, which meant that a degree of uncertainty exists when values less negative than -23.5‰ are found.

Another test, the TLC test (Kushnir, 1979) was devised to cover the range between -23.5 and the cut-off of -21.5‰.

After a few years experience with this procedure, it became evident that the test discriminated against citrus honey, which on the basis of a very few samples seemed to be less negative than other types. In order to verify this and also to answer allegations that these less negative values were due to residual HFCS from early stimulatory bee feeding, White and Robinson (1983) measured the isotope ratio values for a number of citrus nectars and citrus honeys from the same season and area. It was found that the average value for citrus nectar ($\delta^{13}C = -23.8‰$) and citrus honey ($\delta^{13}C = -23.8‰$) did not differ significantly and were significantly different from the average values for all honeys and therefore required different limits for judgment.

Use of this test in the USDA honey loan program revealed that certain honey types (mesquite, catsclaw) also differ enough from the average to raise allegations of adulteration. A new approach to isotope ratio testing was developed (White and Winters, 1989) to resolve these difficulties. In this procedure, which is needed only when values for a honey are less negative than -23‰, each honey provides its own standard for comparison, rather than comparing the value for a honey with the average for all honeys. By using the $\delta^{13}C$ values of the protein isolated from a honey as a standard and comparing it with the value for the whole honey, the range of uncertainty is greatly reduced, considerably greater sensitivity is obtained, and the TLC test is no longer needed.

Since this test can indicate small amounts of cane or corn sugars in honey, it is important that care be taken by the beekeeper to avoid including with extracted honey intended for the market any stores from early stimulative feeding of colonies. Simply "passing these sugars through the bee" has no effect on their isotope ratio values, and such honey may be condemned as adulterated when it is only the result of careless beekeeping practice. Sugar-feeding during a honey flow is easily detected by the internal standard isotope ratio test.

HONEYDEW

Honeydew is a sweet liquid excreted by Hemipterous insects, principally plant lice (aphids) and scale insects, feeding on plants. It is frequently gathered and stored by bees and is generally (in the United States) considered inferior to honey in flavor and quality. Europeans regard certain types of honeydews as delicacies, and colonies of ants are maintained in certain forests to encourage and tend aphids producing these honeydews. Honeydew is more likely to be gathered during times of nectar dearth. It may be found on leaves of such trees as oak, beech, tulip poplar, ash, elm, hickory, maple, poplar, linden, and fruit trees as well as fir, cedar, and spruce.

An average composition of honeydew is given in Table 21. It is based on 14 samples, including alfalfa, cedar, hickory, oak, and several unidentified types. These averages compare quite favorably with those obtained for 38 Swiss honeydews by Bogdanov and Baumann (1988) who used a method that allowed measurement of all of the minor sugars known to occur in honey and honeydew. Comparison of these values with those given for honey in Table 1 shows honeydew to be lower in levulose and dextrose, darker in color, and higher in pH, higher sugars, acidity, ash, and nitrogen. These differences have been the basis of a proposal by Kirkwood *et al.* (1960, 1961) to distinguish between honey and honeydew with a calculation relating the pH, ash, and reducing sugar contents.

The sugars of stored honeydew are even more complex than those of honey, since two sets of enzymes, those of the hemipterous insect and of the honey bee, are involved in its production and storage. Several investigators have studied the sugars of honeydew as excreted, using modern methods of analysis. Levulose, dextrose, and melezitose have long been known to be present. Gray and Fraenkel (1953) found erlose (fructomaltose) in several honeydews. Duspiva (1954) found in addition, a series of higher sugars resulting from the stepwise addition of glucose to erlose. This was confirmed by Wolf and Ewart (1955). Bacon and Dickinson (1957) presented evidence that melezitose, common to many honeydews, is not present in the plant, but is produced from plant sap sucrose by an enzyme in the aphid. To the best of our

Table 21. AVERAGE COMPOSITION OF HONEYDEW[a]

Component	Average	Standard Deviation	Range
Moisture (%)	16.3	1.74	12.2 - 18.2
Fructose (%)	31.8	4.2	23.9 - 38.1
Glucose (%)	26.0	3.0	19.2 - 31.9
Sucrose (%)	0.8	0.2	0.4 - 1.1
"Maltose" (%)	8.8	2.5	5.1 - 12.5
Melezitose[b] (%)	2.3	4.6	0.0 - 13.4
Higher Sugars (%)	4.7	1.0	1.3 - 11.5
pH	4.4		3.9 - 4.9
Free Acid (meq/kg)	49.1	10.6	30.3 - 66.0
Lactone (meq/kg)	5.8	3.6	0.4 - 14.1
Total Acid (meq/kg)	54.9	10.8	34.6 - 76.5
Ash (%)	0.74	0.27	0.21 - 1.18
Nitrogen (%)	0.100	0.053	0.047 - 0.223
Diastase[c]	31.9		6.7 - 48.4

[a] Fourteen samples analyzed. White *et al.* (1962).

[b] Eight samples

[c] Four samples

knowledge, melezitose has never been isolated directly from a plant. It is now generally accepted (Maurizio, 1975) that there are at least two types of honeydew, containing erlose or melezitose or occasionally mixtures, depending upon the insect(s) involved. Raffinose is probably also a constituent of some honeydews (Maurizio, 1975). The melezitose type often granulates rapidly, even in the comb, if there is enough of this sugar present. Erlose-type honeydews do not granulate. Another interesting example of the action of insect enzymes upon plant saps to produce toxins is discussed under "toxic honeys."

Honeydews are not considered suitable for winter stores for bees, once thought because of their content of melezitose and "dextrins." Temnov (1958) lays the toxic effects principally to the mineral salts, especially potassium. Maurizio (1975) has written an extensive summary of honeydew composition and production.

The flavor and properties of honeydew resemble those of corn or cane molasses in some cases, which has led to erroneous seizure of market honey based on this and the use of inadequate testing. One useful, if old, test is measurement of polarization (optical rotation). It had long been considered that all honey is levorotatory toward polarized light, and all honeydew is dextrorotatory. In an examination of this, White (1980) has proposed that the dividing line should more properly be at -2°S, rather than 0°.

USES OF HONEY

In the Home

Possibly the greatest single use of honey is at the table, either directly or as an ingredient of a home recipe. Hundreds of cookbooks feature recipes for honey-containing entrees, salads, desserts, beverages, cakes, rolls, breads, and so on (See Figures 14 and 15). As Harriet Grace once wrote, "If you a cook of note would be, use honey in your recipe"; this is just as apt now as it was in 1947.

In Food Manufacture

Honey is a high-energy carbohydrate food. It enjoys a well-deserved reputation as a natural food, much appreciated by those who wish to avoid highly-refined materials and synthetics in their diets. Recognition of this is seen in the marketplace; many products in the stores have "honey" in their names. Unfortunately, many of these may be seen to contain only token amounts of this material, obviously added only to "piggyback" on the reputation that honey enjoys with informed consumers. This can be checked by noting the list of ingredients; they must appear in descending order of their amount in the product. It is not unusual to see "honey" listed sometimes even below the salt. Such use of honey cannot confer the real benefits of honey to the product: the improved flavor, the moisture-retaining qualities, the greater levulose content, and other perhaps undefinable qualities that the consumer looks for when the word "honey" is part of the product name. The National Honey Board has established content standards for use of their emblem on product packaging, but widespread acceptance of this program by the makers of products containing honey is unclear at this time. As long as no legal standards exist for the amount of honey present before use of the name in promotion is permitted, such exploitation of the consumer and the honey industry will continue. While some products can carry only limited amounts of honey without changing their nature, there are many that would benefit from increased levels of honey.

Perhaps the largest single use of honey in the food industry is in baking. Here definite benefits are conferred in "eating quality", keeping quality, texture, flavor; much honey is used in breads, yeast-raised sweet goods, fruit cakes, cookies, pies, and other cakes. Studies some time ago (Johnson *et al.,* 1957) have shown these definite advantages, and provide specifications for the qualities of honey that are important in commercial baking. Bakers need only

FIGURE 14. Honey continues to be a favorite ingredient and topping for baked goods, even in today's fast-paced lifestyle. *(Photo courtesy National Honey Board and Pacific Kitchens, Evans Food Group)*

be careful of flavor and color, and of course need to specify filtered and pasteurized honey.

Honey can and is being included in many over-the-counter pharmaceuti-cals. A study (Rubin *et al.,* 1959) of the stability and palatibility of honey in

FIGURE 15. Warm nonalcoholic drinks made from honey are just a few of the many ways honey can be used in beverages. *(Photo courtesy National Honey Board and Pacific Kitchens, Evans Food Group)*

various drug preparations (riboflavin, ferrous sulfate, sulfonamides, and several cough preparations) concluded that "in all cases where honey proved to be a desirable medical product ingredient, the taste of the preparation was judged by the majority of persons thus testing the preparation to be superior to that of similar solutions made without honey" (Gennaro *et al.,* 1959). The study included the establishment of purchase specifications for honey for this use. Honey is a natural for use in confectionery, and has been so used for many centuries. Honey spreads, of pure honey or containing such materials as cinnamon, butter, and various fruits have been on the market for years. Honey and peanut butter would seem to be a "natural" since it makes such an ideal quick lunch for children. Efforts over many years to commercialize such a mixture have not generally succeeded because the mixture has not yet been produced with sufficient shelf-life for commercial use. It darkens and hardens, and develops off-flavors on storage. This is due to a well-known chemical reaction between sugars and amino acids, the Maillard reaction.

Honey Wines

Wines are, strictly speaking, made from grapes; fermented beverages from honey are more properly called meads. They have been known for thousands of years. An excellent historical and practical review of mead making is that of Morse and Steinkraus (1975). Rather than attempting to brief this article here, it is recommended that anyone seriously interested in making honey wines start by reading it. Detailed instructions are given for producing clover honey mead in 40-gallon lots. Instructions are also given for sparkling mead, sherry mead, and honey vinegar.

*   *   *   *   *   *

Who would ever have thought years ago that this delightful natural product, evolved in the course of countless years to insure the survival of honey bee colonies and consequently, the pollination and survival of so many plants, known to and used by man for millenia, would be revealed by modern analytical techniques to be such a complex and fascinating material? Of one thing we may be certain: we are far from knowing all about it!

REFERENCES

Aoyagi, S. & Oryu, C. (1968). Honeybees and honey. III. Yeasts in honey. *Bull. Fac. Agri. Tamagawa Univ. (7-8)* 203-213.

Aureli, P., Ferrini, A. M., & Negri, S. (1983). *Clostridium botulinum* spores in honey. *Riv. della Soc. Ital. Sci. dell'aliment. 12:*457-460.

Austin, G.H. (1958). Maltose content of Canadian honeys and its probable effects on crystallization. *X Intl. Cong. Entomol. 4:*1001-1006.

Auzinger, A (1910). Über Fermente im Honig und den Wert ihres Nachweises für die Honigbeurteilung. *Z. Unters. Nahr.- u. Genussmittel 17(2):*65-83; Weiter Beiträge zur Kenntis der Fermentreaktion des Honigs. *Ibid 17(7):*353-362.

Bacon, J.S.D.& Dickinson, B. (1957). The origin of melezitose: a biochemical relationship between the lime tree *(Tilia* spp.) and an aphis *(Eucallipteris tiliae* L.). *Biochem. J. 66:* 289-299

Bergner, K.G. & Diemair, S. (1975). Proteins des Bienenhonigs. II. Gelchromatographie, enzymatische Aktivität und Herkunft von Bienenhonig-Proteinen. *Z. Lebensm.-Unter. u. Forsh. 157:*7-13.

Bergner, K.G. & Hahn, H. (1972). Zum Vorkommen und Herkunft der freien Aminosäuren in Honig. *Apidolgie 3(1):*5-34.

Boer, H.E. de. (1934). De invoed van den ouderdom op de samenstelling van honig. Chem. *Weekblad 31:*482-487.

Bogdanov, S. (1984). Characterization of antibacterial substances in honey. *Lebensm. - Wissensch. u.-Technol. 17(2):*74-76

Bogdanov, S. & Baumann, E. (1988). Bestimmung von Honigzuckern mit HPLC. *Mitt. Geb. Lebensm. Hyg. 79:*198-206.

Browne, C.A. (1908). Chemical analysis and composition of American honeys. *U.S. Dept. Agric. Bur. Chem. Bull. 110:* 93 pp.

Burgett, D.W. (1974). Glucose oxidase: a food protective mechanism in social hymenoptera. *Ann. Entomol. Soc. 67(4):*545-546.

Burnside, C.E. & Vansell, G.H. (1936). Plant poisoning of bees. *U.S. Dept. Agric., Bur. Entomol. & Plant Q. Publ. E-398:*12 pp.

Chandler, B.V., Fenwick, D. Orlova, T. & Reynolds, T. (1974). Composition of Australian Honeys. *CSIRO, Australia, Tech. Paper No. 38:* 39 pp.

Chataway, H.D. (1932). Determination of moisture in honey. *Can. J. Res. 6:* 532-547.

Connor, L. (1988). Rules for 1988 EAS competitive shows. *East. Apic. Soc. J. 16(2,3):*19-20.

Crane, E. (1975). History of honey. *In* "Honey: a comprehensive survey" (E. Crane, *ed.*) (London: Heinemann), 439-488.

Crane, E., Walker, P., & Day, R. (1984). Directory of important world honey sources. Internat. *Bee Res. Assn.* London. 384 pp.

Crane, E., & Walker, P. (1986). Honey sources satellite 4. Physical properties, flavour and aroma of some honeys. Internat. Bee Res. Assn. London. 56 pp.

Cremer, E. & Riedemann, M. (1965). Gaschromatographische Untersuchungen zur Frage des Honigaromas. *Mh. Chem. 96(2):*364-368.

Davies, A.M.C. (1975). Amino acid analysis of honeys from eleven countries. *J. Apic. Res. 14(1):*29-39.

Davies, A.M.C. (1976). The application of amino acid analysis to the determination of the geographic origin of honey. *J. Food Technol 11:*515-523.

Deifel, A., Gierschner, K., & Vorwohl, G. (1985). Sucrose and its transglycosylation products in natural honey and honey from sugar-fed bees. *Deutsche Lebensm.-Rundschau 81(11):*356-362.

Deinzer, M.L., Thomson, P.A., Burgett, D.M., & Isaacson, D.L. (1977). Pyrrolizidine alkaloids: their occurrence in honey from tansy ragwort *(Senecio jacobaea* L.). *Science 195:*497-499.

Detroy, B.F. (1966). Determining film coefficient of a viscous liquid. *Trans. ASAE 9(1):*91-93, 97.

Doner, L.W. (1977). The sugars of honey—a review. *J. Sci. Food Agr.* 28:443-456.

Doner, L.W. & Jackson, S.J. (1980). Darkening effect of high iron honey on tea. *Amer. Bee J.* 120:516-517.

Duisberg, H. & Hadorn, H. (1966). Welche Anforderungen sind an Handelshonige zu stellen? Vorschläge auf Grund der Statistischen Auswertung von ca 1600 Honig-Analysen. *Mitt. Geb. Lebensmittelunters. u. Hyg.* 57(5):386-407.

Duspiva, F. (1954). Enzymatische Prozesse bel der Honigtaubildung der Aphiden. *Verhand. Deut. Zool. Gesell.* 440-447.

Dyce, E.J. (1975). Producing finely granulated or creamed honey. *In* "Honey: a comprehensive survey" (E. Crane, ed.) (London: Heinemann), 293-306.

Echigo, T., Takenaka, T., and Ichimira, M. (1973). Effects of chemical constituents in pollen on the process of honey formation. *Bull. Fac. Agr., Tamagawa Univ.* 13:1-9.

El-Leithy, M.A. & El-Sibaei, K.B. (1972). Role of microorganisms isolated from the honeybee (*Apis mellifera*) in ripening and fermentation of honey. Egypt. *J. Microbiol.* 7:89-95.

Elser, E. (1924). Beiträge zur quantitativen Honiguntersuchung. *Arch. Bienenk.* 6:118 only.

Fabian, F.W. & Quinet, R.I. (1928). A study of the cause of honey fermentation. *Mich. Agr. Expt. Sta. Tech. Bull.* 62:41 pp.

Fasler, A. (1975). Honey Standards Legislation. *In* "Honey: a comprehensive survey" (E. Crane, ed.) (London: Heinemann), 329-354.

Fellenberg, T. von & Rusiecki, W. (1938). Bestimmung der Trübung und der Farbe des Honigs. *Mitt. Geb. Lebensmittelunters. u. Hyg.* 29:313-335.

Gennaro, A.R., Sideri, C.N., Rubin, N., & Osol, A. (1959). Use of honey in medicinal preparations. *Amer. Bee J.* 99(12):492-493.

Giri, K.V. (1938). Chemical composition and enzyme content of Indian honey. *Madras Agric. J.* 26:68-72.

Gothe, F. (1914). Die Fermente des Honigs. *Z. Unters. Nahr.-u Genussmittel* 26(6):286-321.

Gray, H.F. & Fraenkel, G. (1953). Fructomaltose, a recently discovered tri-saccharide isolated from honeydew. *Science* 118:304-305.

Griebel, C. (1938). Vitamin C enthaltende Honige. *Z. Unters. Lebensmittel-75*:417-420.

Hadorn, H. & Kovacs, A.S. (1960). Zur Untersuchung und Beurteilung von ausländischen Bienenhonig auf Grund des Hydroxymethylfurfurol- und Diastasegehaltes. *Mitt. Geb. Lebensmittelunters. u. Hyg.* 53(1):6-28.

Hadorn, H. & Zürcher, K. (1962). Über Veränderungen im Bienenhonig bei der grosstechnische Abfüllung. *Mitt. Geb. Lebensmittelunters. u. Hyg.* 53(1):28-34.

Hase, S., Suzuki, O., Odato, M., Miura, A. & Suzuki, S. (1973). [Quality of honey and its analytical methods. III. Qualities of domestic, imported, and commercial honey.] *Shuka. Kenk. Hokoku* (28):72-77.

Haydak, M.H., Palmer, L.S., Tanquary, M.C. & Vivino, A.E. (1942). Vitamin content of honeys. *J. Nutr.* 23:581-588.

Haydak, M.H., Palmer, L.S., Tanquary, M.C. & Vivino, A.E. (1943). The effect of commercial clarification on the vitamin content of honey. *J. Nutr.* 26(3):319-321.

Helvey, T.C. (1953). Colloidal constituents of dark buckwheat honey. *Food Res.* 18(2):-197-205.

Helvey, T.C. (1954). Study on some physical properties of honey. *Food Res.* 19(3):282-292.

Hodges, R. & White, E.P. (1966). Detection and isolation of tutin and hyenanchin in toxic honey. *N.Z.J. Science* 9(1):233-235.

Horwitz, W. (*ed.*) (1975). Official methods of analysis of the Association of Official Analytical Chemists, 12th ed. (Washington, Association of Official Analytical Chemists).

Huhtanen, C.N., Knox, D. & Shimanuki, H. (1981). Incidence and origin of *Clostridium botulinum* spores in honey. *J. Food Prot. 44(11)*:812-814.

Jacobs, M.B. (1955). Flavoring with honey. *Am. Perf. Essent. Oil Rev. 66:(1)*:46-47.

Johnson, J., Nordin, A.P. & Miller, D. (1957). The utilization of honey in baked products. *Bakers Digest 31*:33-34,36,38,40.

Kalimi, M.Y. & Sohonie, K. (1965). Mahabaleshwar honey. III. Vitamin contents (ascorbic acid, thiamine, riboflavin, and niacin) and effect of storage on these vitamins. *J. Nutr. Dietet. 2(1)*:9-11.

Kebler, L.F. (1896). Poisonous honey. *Proc.* Amer. *Pharm. Assoc. 44*:167-174.

Kelley, F.H.C. (1954). Phase equilibria in sugar solutions. IV. Ternary system of water-glucose-fructose. *J. Appl. Chem. Lond. 4*:409-411.

Killion, C.E. (1950). Removing moisture from comb honey. *Amer. Bee J. 90*:14-16.

Kirkwood, K.C., Mitchell, T.J. & Smith, D. (1960). An examination of the occurrence of honeydew in honey. *Analyst 85(1011)*:412-416.

Kirkwood, K.C., Mitchell, T.J. & Ross, I.C. (1961). An examination of the occurrence of honeydew in honey. Part II. *Analyst 86(1020)*:164-165.

Kitzes, G., Schuette, H.A. & Elvehjem, C.A. (1934). The B vitamins in honey *J. Nutr. 26(3)*:241-250.

Kokubo, Y., Jinbo, K., Kaneko, S. & Matsumoto, M. (1984). Prevalence of spore-forming bacteria in commercial honey. *Tokyo Metro. Res. Lab. Pub. Health Ann. Rept. 35*:192-196.

Kushnir, I. (1979). Sensitive thin layer chromatographic detection of high fructose corn syrup and other adulterants in honey. *J. Assn. Off. Anal. Chem. 62*:917-920.

Langer, J. (1903). Fermente im Bienenhonig. *Schweiz. Wschr. Chem. Pharm. 41*:17-18.

Lawrence, W.B. (1986). Infant botulism and its relationship to honey: a review. *Amer. Bee J. 126*:484-486.

Lochhead, A.G. (1933). Factors concerned with the fermentation of honey. *Zentbl. Bakt. Parasit. II Abt. 88*:296-302.

Lochhead, A.G. & Farrell, L. (1930). Soil as a source of infection of honey by sugar-tolerant yeasts. *Can. J. Res. 3(1)*:51-64.

Lochhead, A.G. & Farrell, L. (1931). The types of osmophilic yeasts found in normal honey and their relation to fermentation. *Can. J. Res. 5*:665-672.

Lochhead, A.G. & Heron, D.A. (1929). Microbiological studies of honey. I. Honey fermentation and its cause. II. Infection of honey by sugar-tolerant yeasts. *Can. Dept. Agr., Bull. No. 116*:47 pp.

Lothrop, R.E. (1943). Saturation relations in aqueous solutions of some sugar mixtures with special reference to higher concentrations. *George Washington University: Ph.D. Dissertation.*

Lovell, H.B. (1956). *Honey plants manual.* A.I. Root Co., Medina, OH.

Low, N.H. & Sporns, P. (1988). Analysis and quantitation of minor disaccharides and trisaccharides in honey using capillary gas chromatography. *J. Food Sci. 53(2)*:558-561.

Lund, R. (1909). Albuminate im Naturhonig und Kunsthonig. *Z. Unters. Nahr.- u. Genussmittel 17*:128-130.

MacDonald, J.E. (1963). Honey pumps. *Glean. Bee Cult. 91(2)*:85-87.

Maeda, S., Mukai, A., Kosugi, N. & Okada, Y. (1962). The flavor components of honey. *J. Fd. Sci. Tech. 9(7)*:270-274.

Manley, W.T. (1985). United States grades for extracted honey. *Federal Register 503(78)*:15861--15865.

Marshall, T., & Williams, K.M. (1987). Electrophoresis of honey: Characterization of trace proteins from a complex biological matrix by silver staining. *Anal. Biochem. 167(2)*:301-303.-

Martin, E.C. (1958). Some aspects of hygroscopic properties and fermentation in honey. *Bee Wrld. 39(7)*:165-178.

Maurizio, A. (1975). How bees make honey. *In* "Honey, a comprehensive survey" (E. Crane, *ed.*) Heinemann, London.

Midura, T.F., Snowden, S., Wood, R.M. & Arnon, S.S. (1979). Isolation of *Clostridium botulinum* from honey. *J. Clin. Microbiol. 9(2)*:282-283.

Milum, V.G. (1948). Some factors affecting the color of honey. *J. Econ. Entomol. 41(3)*:495-505.

Morse, R.A. & Steinkraus, K.H. (1975). Wines from the fermentation of honey. Chapter 16, pp. 392-407, from *Honey: a comprehensive survey* ed. E. Crane (1975*a*).

Munro, J.A. (1943). The viscosity and thixotrophy of honey. *J. Econ. Ent. 36(5)*:769-777.

Nelson, E.K. & Mottern, H.H. (1931). Some organic acids in honey. *Ind. Eng. Chem. 23(3)*:335 only.

Nicolov, Z.L., Boskov, Z.M. & Jakovljevic, J.B. (1984). High performance liquid chromatographic separation of oligo saccharides using amine modified silica columns. *Stärke 36(3)*:97-100.

Nissenbaum, A., Lifshitz, A. & Stepek, Y. (1974). Detection of citrus fruit adulteration using the distribution of natural stable isotopes. *Lebensm. -Wiss. u. -Technol. 7(3)*:152-154.

Oppen, F.C. & Schuette, H.A. (1939). Viscometric determination of moisture in honey. *Ind. Eng. Chem., Anal. Ed. 11*:130-133.

Örösi-Pál, A. (1956). A mérgező méz titka nyomában. [On the track of the poisonous honey] *Méhészet 4*:25-27.

Paine, H.S., Gertler, S.I. & Lothrop, R.E. (1934). Colloidal constituents of honey. Influence on properties and commercial value. *Ind. Eng. Chem. 16*:73-81.

Palmer-Jones, T. (1947). A recent outbreak of honey poisoning. Part I. Historical and descriptive. *N.Z.J. Sci. Technol. 29A*:107-114; Part III. The toxicology of the poisonous honey and the antagonism of tutin, mellitoxin, and picrotoxin by barbiturates. *Ibid.* 121-125.

Paterson, C.R. (1947). A recent outbreak of honey poisoning. Part IV. The source of the toxic honey—Field observations. *N.Z.J. Sci. Technol. 29A*:125-129.

Paterson, C.R. & Palmer-Jones, T. (1954). A vacuum plant for removing excess water from honey. *N.Z.J. Sci. Technol. A36(4)*:386-400.

Pearce, J.A. & Jegard, S. (1949). Measuring the solids content of honey and of strawberry jam with a hand refractometer. *Can. J. Res. 27F*:99-103.

Pellett, F.C. (1976). *American honey plants.* 4th Ed. Dadant & Sons, Hamilton, IL.

Petrov, V. (1974). Quantitative determination of amino acids in some Australian honeys. *J. Apic. Res. 13(1)*:61-66.

Pichler, F.J., Vorwohl, G. & Gierschner, K. (1984). Factors controlling the production of HMF in honey. *Apidologie 15*:171-188.

Platt, J.L. Jr. & Ellis, J.R.B. (1985). Removing water from honey at ambient pressure. *U.S. Patent 4,536,973*, Aug. 27, 1985.

Rahmanian, N., Kouhestani, A., Ghavifekr, H., Ter-sarkissian, N., Olzynamarzys, A. & Donoso, G. (1970). High ascorbic acid content in some Iranian honeys. Chemical and biological assays. *Nutr. Metab. 12(3)*:131-135.

Richter, A.A. (1912). Über einen osmophilen Organismus, den Hefepilz *Zygosaccharomyces mellis acidi* sp. n. *Mykol. Zentralbl. 1(3/4)*:67-76.

Rogers, P.E.W. (1975). Honey quality control. *In* "Honey: a comprehensive survey" (E. Crane, *ed.*) (London: Heinemann), 314-325.

Rubin, N., Gennaro, A.R., Sideri, C.N. & Osol, A. (1959). Honey as a vehicle for medicinal preparations. *Amer. J. Pharm. 131*: 246-254.

Ruiz-Argüeso, T. & Rodriguez-Navarro, A. (1973). Gluconic acid-producing bacteria from honey bees and ripening honey. *J. Gen. Microbiol. 76*:211-216.

Ruiz-Argüeso, T. & Rodriguez-Navarro, A. (1975). Microbiology of ripening honey. *Appl. Microbiol. 30*:893-896.

Sackett, W.G. (1919). Honey as a carrier of intestinal diseases. *Col. St. Univ. Agr. Expt. Sta.:18 pp.*

Saito, Y., Mitsura, A., Sasaki, K., Satake, M. & Uchiyama, M. (1980). [Detection of the poisonous substances in honey which caused the intoxication.] Eisei Shikensho Hohoku 98:532-535.

Schade, J.E., Marsh, G.L. & Eckert, J.E. (1958). Diastase activity and hydroxy-methyl-furfural in honey and their usefulness in detecting heat alteration. *Food. Res. 23(5)*:446-463.

Schepartz, A.I. (1966). Honey catalase: occurrence and some kinetic properties. *J. Apic. Res. 5(3)*:167-170.

Schou, S.A. & Abildgaard, J. (1934). [Differentiation between honey and synthetic honey]. *Z. Lebensm.- Untersuch. u. -Forsch. 68*:502-511.

Schuette, H.A. & Remy, K. (1932). Degree of pigmentation and its probable relationship to the mineral constituents of honey. *J. Am. Chem. Soc. 54*:2909-2913.

Schuette, H.A. & Huenink, D.J. (1937). Mineral constituents of honey. II. Phosphorous, calcium, magnesium. *Food Res. 2*:529-538.

Schuette, H.A. & Triller, R.E. (1938). Mineral constituents of honey. III. Sulfur and chlorine. *Food Res. 3(5)*:543-547.

Schuette, H.A. & Woessner, W.W. (1939). Mineral constituents of honey. IV. Sodium and potassium. *Food Res. 4(4)*:349-353.

Siddiqui, I.R. & Furgala, B. (1967). Isolation and characterization of oligosaccharides from honey. Part I. Disaccharides. *J. Apic. Res. 6(3)*:139-145.

Siddiqui, I.R. & Furgala, B. (1968). Isolation and characterization of ologosaccharides from honey. Part II. Trisaccharides. *J. Apic. Res. 7(1)*:51-59.

Stinson, E.E., Subers, M.H., Petty, J. & White, J.W. Jr. (1960). The composition of honey. V. Separation and identification of the organic acids. *Arch Biochem. Biophys. 89(1)*:6-12.

Sugiyama, H., Mills, D.C. & Kuo, L. -J.C. (1978). Number of *Clostridium botulinum* spores in honey. *J. Food Protect. 41(11)*:848-850.

Sviderskaya, Z.I. (1959). [A case of food poisoning from honey]. *Gig. Sanit. 24(5)*:57.

Tan, S.T., Wilkins, A.L., Molan, P.C., Holland, P.T., & Reid, M. (1989). A chemical approach to the determination of floral sources of New Zealand honeys. *J. Apic. Res. 28(4)*:212-222.

Temnov, V.A. (1958). Composition and toxicity of honeydew. *Abstr. XVII Intern. Beek. Congr.*, Rome. 117.

Townsend, G.F. (1954). Private communication.

Townsend, G.F. (1961). Preparation of honey for market. *Ont. Dept. Agr., Publ. No. 544*: 23 pp.

Townsend, G.F. (1975). Processing and storing liquid honey. *In* "Honey: a comprehensive survey" (E. Crane, *ed.*) (London: Heinemann), 269-292.

Tysset, C. & Durand, C. (1973). De la persistance de quelques gaermes a gram negatif non sporules dans les miels du commerce stockes a la temperature ambiante. *Nancy. Univ. Facul. pharm. assoc. diplom. microbiol.* 3-12.

Watanabe, T. & Aso, L. (1960). Studies on honey. II. Isolation of kojibiose, nigerose, maltose, and isomaltose from honey. *Tohoku J. Agr. Res. 11*:105-115.

Watt, B.K. & Merrill, A.L. (1963). Composition of foods. *U.S. Dept. Agr., Agric. Hdbk. No. 8*: 190 pp.

Wedmore, E.B. (1955). The accurate determination of the water content of honeys. I. Introduction and results. *Bee Wrld. 36(11)*:197-206.

White, J.W. Jr. (1967). Measuring honey quality—a rational approach. *Amer. Bee J. 107(10)*:374-375.

White, J.W. Jr. (1969). Moisture in honey: Review of chemical and physical methods. *J. Assn. Off. Anal. Chem. 52*:729-737.

White, J.W. Jr. (1973). Toxic honeys. *In* "Toxicants Occurring Naturally in Foods." Committee on Food Protection, (*Washington, National Academy of Sciences*), 495-507.

White, J.W. Jr. (1975). Composition of honey. *In* "Honey: a comprehensive survey" (E. Crane, *ed.*) (London: Heinemann), 157-206.

White, J.W. Jr. (1977). Sodium-potassium ratios in honey and high-fructose corn syrup. *Bee Wrld. 58(1)*:31-35.

White, J.W. Jr. (1978). Honey. *In* "Advances in Food Research." (C.O. Chichester, E.M. Mrak, & G.F. Stewart, eds.) Vol. 24, 288-374. Academic Press, New York.

White, J.W. Jr. (1979a). Methods for determining carbohydrates, hydroxymethylfurfural, and proline in honey: collaborative study. *J. Assn. Off. Anal. Chem. 62(3)*:515-526.

White, J.W. Jr. (1979b). Spectrophotometric method for hydroxymethylfurfural in honey. *J. Assn. Off. Anal. Chem. 62(3)*:509-514.

White, J.W. Jr. (1980). Detection of honey adulteration by carbohydrate analysis. *J. Assn. Off. Anal. Chem. 63(1)*:11-18.

White, J.W. Jr. (1981). Natural honey toxicants. *Bee Wrld. 62(1)*:23-28.

White, J.W. Jr. (1987). Wiley led the way: a century of federal honey research. *J. Assn. Off. Anal. Chem. 70(2)*:181-189.

White, J.W. Jr. & Doner, L.W. (1978). Mass spectrometric detection of high-fructose corn syrup in honey by use of $^{13}C/^{12}C$ ratio: collaborative study. *J. Assn. Off. Anal. Chem. 61*:746-750.

White, J.W. Jr. & Hoban, N. (1959). Composition of honey. IV. Identification of the disaccharides. *Arch. Biochem. Biophys. 80(2)*:386-392.

White, J.W. Jr., Kushnir, I & Doner, L.W. (1979). Charcoal column/thin layer chromatographic method for high fructose corn sirup in honey and spectrophotometric method for hydroxymethylfurfural in honey:collaborative study. *J. Assn. Off. Anal. Chem. 62(4)*: 921-927.

White, J.W. Jr. & Kushnir, I. (1967). Composition of honey VII. Proteins. *J. Apic. Res.* *6(3):*163-178.

White, J.W. Jr., Kushnir, I. & Subers, M.H. (1964). Effect of storage and processing temperatures on honey quality. *Food Technol. 18(4):*153-156.

White, J.W. Jr. & Maher, J. (1951). Detection of incipient granulation in extracted honey. *Amer. Bee J. 91(9):*376-377.

White, J.W. Jr. & Maher, J. (1953). Transglucosidation by honey invertase. *Arch Biochem. Biophys. 42(2):*360-367.

White, J.W. Jr., Meloy, R.W., Probst, J.L. & Huser, W.F. (1987). Sugars containing galactose occur in honey. *J. Apic. Res. 25(3):*182-185.

White, J.W. Jr., Riethof, M.L., Subers, M.H. & Kushnir, I (1962). Composition of American honeys. *U.S. Dept. Agr., Tech. Bull. 1261:*124 pp.

White, J.W. Jr. & Robinson, F.A. (1983). $^{13}C/^{12}C$ ratios of citrus honeys and their regulatory implications. *J. Assn. Off. Anal. Chem. 66:*1-3

White, J.W. Jr. & Rudyj, O.N. (1978a). The protein content of honey. *J. Apic. Res. 17(4):*234-238

White, J.W. Jr. & Rudyj, O.N. (1978b). Proline content of United States honeys. *J. Apic. Res. 17(2):*89-93.

White, J.W. Jr. & Siciliano, J (1980). Hydroxymethylfurfural and honey adulteration. *J. Assn. Off. Anal. Chem. 63(1):*7-10.

White, J.W. Jr. & Subers, M.H. (1963). Studies on honey inhibine. 2. A chemical assay. *J. Apic. Res. 2(2):*93-100.

White, J.W. Jr., Subers, M.H. & Schepartz, A.I. (1963). The Identification of inhibine, the antibacterial factor in honey, as hydrogen peroxide and its origin in a honey glucose-oxidase system. *Biochem. Biophys. Acta 73:*57-70.

White, J.W. Jr. & Winters, K. (1989). Honey protein as internal standard for stable carbon isotope ratio detection of adulteration of honey. *J. Assn. Off. Anal. Chem. 72(6):*907-911.

Wilson, H.F. & Marvin, G.E. (1932). Relation of temperature to the deterioration of honey in storage. A progress report. *J. Econ. Entomol. 25:*525-528.

Winkler, O., (1955). Beitrag zum Nachweis und zur Bestimmung von Oxymethylfurfural in Honig und Kunsthonig *Z. Lebensmittelunters. u-Forsch. 102(3):*161-167.

Wolf, J.P. & Ewart, W.H. (1955) Carbohydrate composition of honeydew of *Coccus hesperidum L.:*Evidence for the existence of two new oligosaccharides *Arch. Biochem. Biophys. 58:*365-372

Wootton, M. & Ryall, L. (1985). A comparison of Codex Alimentarius Commission and HPLC methods for 5-hydroxymethyl-2-furaldehyde determination in honey. *J. Apic Res. 24 (2):* 120-124.

CHAPTER 22

OTHER PRODUCTS OF THE HIVE

by JUSTIN O. SCHMIDT[1]
and STEPHEN L. BUCHMANN[1]

The major honey bee product, both in terms of familiarity and profit, is honey. Beeswax has also long been considered an important hive product and a secondary source of materials and income. Recently other products of the hive—pollen, propolis, royal jelly, venom, and bee brood—have been gaining importance. Each of these has its own market and potential value that can be a source of income for beekeepers and related industries and can enrich human life. In this chapter, we will discuss the production and value of each of these bee products to the apicultural industry and identify the literature describing their properties and potential for human benefit and health.

Much of the popular literature on other products of the hive has focused on their potential health benefits for humans. The major portion of the professional literature, especially from the medical profession, consists of viewpoints which almost universally dismiss bee products as being of no benefit or "quack-like". This unfortunate polarization suggests that the truth lies somewhere between. Some of the reasons for the general disregard by the medical profession in Western societies for any potential of bee products include: lack of uniformity of most products; lack of knowledge of well designed tests to determine efficacy of the product in question; difficulty of using or prescribing bee products; difficulty of pharmaceutical companies to patent and secure profitability of a bee product; a general dislike of any substance that is not a pure synthesized and well characterized material; and the often ill-informed and potentially unscrupulous nature of promoters and their extravagant literature. These difficulties will be hard to overcome, but with well-designed double blind medical research, education, and responsible behavior on both the part of the bee industry and the medical profession the situation should change. In an effort to provide educational material to help reduce the misinformation, this chapter will emphasize mainly those well-designed experiments that provide useful information. Unfortunately, much of the literature on these subjects is in Eastern European and Asian literature in their national languages and is difficult to obtain and interpret. Moreover, much of this literature is unrefereed, or in journals poorly regarded by the Western scientific community; thus, very little of this information can be cited in this chapter.

[1]Justin O. Schmidt, Ph.D., and Stephen L. Buchmann are Research Entomologists, Carl Hayden Bee Research Center, USDA, ARS, Tucson, AZ 85719

POLLEN

Pollen is the male reproductive cells produced by the anthers of flowering plants for the purpose of transmitting gametes to the stigma of receptive female flowers. Pollination thereby achieved enables the production of seeds. Pollen (Figure 1) is usually transmitted from one flower to another by pollinators such as bees or by the wind.

From a bee's point of view pollen is the most important product of the hive. Since their evolutionary origin over 90 million years ago from sphecid wasps, bees have co-evolved with flowering plants. Bees visit flowers simply to harvest floral rewards for themselves and their larvae. Pollination is an incidental feature of their floral fidelity and the ability of their fuzzy bodies to transport pollen grains.

Pollen supplies all the bee's nutrients for brood rearing as well as for adult growth and development (see chapter on nutrition). Without adequate pollen supplies which are obtained either through foraging or from stores in the form of "bee bread", a colony could not long exist. Historically, pollen was of little commercial or economic consequence to European and North American beekeepers except insofar as having enough available to maintain strong,

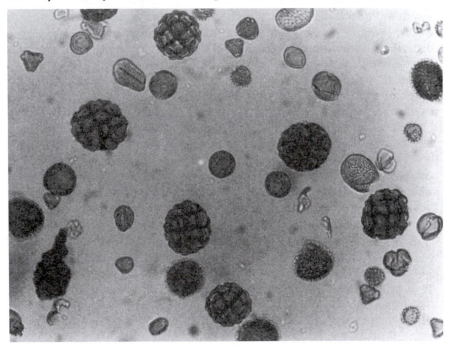

FIGURE 1. Pollen grains representative of the overall yearly diversity of pollen collected by a honey bee colony (*S. L. Buchmann et al., unpublished*).

productive hives. For societies in which honey hunting was an activity, pollen appears to be simply a part of the bounty: honey, brood, stored pollen, wax, etc., all wrapped up in the prized combs which often were eaten whole (usually followed by the spitting out the wax which could be sold for profit or made into useful objects) (Crane, 1983).

In some societies pollen was considered an important part of traditional ceremonies and medicinal products (Durham, 1951) and in the last several decades interest in pollen has dramatically increased, first in eastern Europe, later in western Europe and North America. In North America the interest in pollen for health food and natural supplements increased greatly during the 1970's and 1980's, with even one U.S. president being an avid consumer of pollen containing products (Anon, 1984).

Chemical Composition of Pollen

Unlike an apple, a beef steak, or soybeans, bee-collected pollen is not a uniform, distinct, and easily characterized product. The apple, steak, and soybeans are each derived from one plant or animal species and, thus, their chemical compositions are reasonably uniform and predictable. Bee-collected pollen consists of a blend of pollen grains derived from many, often dozens, of plant species in a given locality. One reason that bees collect a mixture of different pollens is that they are perennial, nutrient-demanding, large colonies that must obtain a year round supply of food; and usually many pollen floral sources are available simultaneously to meet this need. By consuming a mixture of different types of pollen, bees also tend to provide a better nutritional balance and to dilute potentially toxic alkaloids or other toxins. When given a choice bees prefer to consume a mixed pollen diet (Schmidt, 1984) and survive longer when fed mixtures than when fed one-species pollen diets (Schmidt *et al.*, 1987). The problem of chemical analysis of pollen becomes even more difficult because plant pollen sources not only vary by locations throughout the world, but also by season and year in a given locality (O'Rourke and Buchmann, 1990). On a world-wide scale bee-collected pollen is composed of diverse assemblages of pollen grains from thousands of different plants (honey bees have the greatest dietary range of any organism [Schmalzel, 1980]). In this section we will attempt to present a generalized "pollen composition" based on many reported analyses of pollen from individual species and from blends.

The overall chemical composition of pollen is shown in Table 1. Protein is a major component of pollen with an average value of almost 24% (Buchmann, 1986). Carbohydrates constitute about 27% of bee-collected pollen and consist mostly of the simple sugars fructose and glucose. Much of this sugar is added by the foragers in the form of nectar or honey which is used to bind the pollen grains together and allow them to be packed efficiently into the

Table 1. General chemical composition of bee-collected pollen.

Component	No. analyzed	Av. level	Typical ranges	References[1]
Protein	277	23.7%	7.5-35%	1,3-5,7-9,12,13,15-17,20,22,23,25[2]
Lipids	52	4.8%	1-15%	1,3-5,9,20,22
Carbohydrates	47	27%	15-45%	4,9,20,24
Phosphorus	54	.53%	.1-.6%	1,4,9,20,22,
Ash	60	3.12%	1-5%	1,3-5,10,20,22,26
Potassium	56	.58%	.15-1.1%	1,9,10,18,20,26
Calcium	60	.225%	.1-.5%	1,9,10,18,20,22,26
Magnesium	60	.148%	.1-.35%	1,9,10,18,20,22,26
Sodium	30	.044%	.15-.8%	1,9,10,18,26
Iron	51	140μg/g	wide[3]	10,18,20,22,26
Manganese	28	100μg/g	wide[3]	9,10,18,26
Zinc	21	78μg/g	wide[3]	1,9,10,18,26
Copper	27	14μg/g	6-25 μg/g	1,10,18,20
Nickel[4]	23	4.5μg/g	0-? μg/g	2,18
Boron		trace		19
Iodine	?	?	4-10 μg/g	19
Thiamin	8	9.4 μg/g	4-22 μg/g	6,21,22
Niacin	6	157 μg/g	130-210 μg/g	6,22
Riboflavin	8	18.6 μg/g	?	6,21,22
Pyridoxine	2	9 μg/g	?	6
Pantothenic acid	33	28 μg/g	5-50 μg/g	6,11,21,22
Folic acid	8	5.2 μg/g	?	24
Biotin	4	.32 μg/g	.16-.6 μg/g	6,21
Vitamin C	7	350 μg/g	0-740 μg/g	21,22,24
Vitamin A		0		22
Carotenes[5]	4	95 μg/g	50-150 μg/g	22
Vitamin D	4	0		22
Vitamin E	4	14 μg/g	?	22
Vitamin K	4	0		22

[1] References: 1= Bell *et al.*, 1983; 2= Dedic and Koch, 1957; 3= Echigo *et al.*, 1986; 4= Herbert and Shimanuki, 1978; 5= Ibrahim, 1974; 6= Kitzes *et al.*, 1943; 7= Kleinschmidt and Kondos, 1976; 8= Lehnherr *et al.*, 1979; 9= McLellan, 1977; 10= Nation and Robinson, 1971; 11= Pearson, 1942; 12= Rabie *et al.*, 1983; 13= Rayner and Langridge, 1985; 14= Rosenthal, 1967; 15= Schmidt and Johnson, 1984; 16= Schmidt *et al.*, 1984; 17= Schmidt *et al.*, 1987; 18= Solberg and Remedios, 1980; 19= Stanley and Linskens, 1974; 20= Todd and Bretherick, 1942; 21= Togasawa *et al.*, 1967; 22= Vivino and Palmer, 1944; 23= Weaver and Kuiken, 1951; 24= Weygand and Hofmann, 1950; 25= Wille *et al.*, 1985; 26= Youssef *et al.*, 1978.

[2] Based on summation of amino acid contents; to convert to Kjeldahls protein, total amino acid values are divided by 0.8 (Rabie *et al.*, 1983).

[3] The values of these elements vary so widely that no generalizations can be made.

[4] Based on hand-collected pollens; these values should be considered tentative.

[5] Carotenes serve as precursors for Vitamin A and thus fulfill that need in many pollen-feeding organisms.

corbiculae (pollen baskets) on the hind legs. These added sugars are what give bee-collected pollen its sweet taste. Some pollen, especially from grasses, also contains starch which might attain extreme values as high as 18% by weight (Baker and Baker, 1979; Schmidt *et al.*, 1989). Pollen contains on average only 5% fat, most of which is in the form of surface pollenkitt, the outer oily coating on most pollen grains. Much of this lipid is hydrocarbons or waxes (Schmidt, 1985) which are very poorly digested or absorbed by humans (Linscheer and Vergroesen, 1988) and thereby would contribute few calories.

In the popular literature pollen is most noted for its mineral and vitamin composition. Indeed pollen contains substantial quantities of the minerals potassium, calcium, and magnesium, as well as notably high levels of iron, zinc, manganese, and copper. Pollen, like most plant materials, contains low levels of sodium. Bee-collected pollen is extraordinarily rich in most of the B-vitamins, including thiamine, niacin, riboflavin, pyridoxine, pantothenic acid, folic acid, and biotin. It contains highly variable, but not remarkable, levels of vitamin C, in part because this labile vitamin is easily degraded during storage and processing (Herbert *et al.*, 1985).

The Potential of Pollen in Human Nutrition

Pollen is the ideal well-balanced food for bees, but like any other material, is not a "perfect food" for humans and statements or claims implying that pollen is such are not only highly unscientific, but are also unprofessional and potentially damaging to the reputation of the bee industry. Simple evidence that pollen is not a perfect (human) food comes from the fact that pollen contains no, or essentially insignificant, levels of the lipid-soluble vitamins D, K, and E (Table 1).

In contrast to most human food, pollen contains a relatively large amount of indigestible material. This material, mainly cellulose and the carotenoid polymer called sporopollenin, form the hard protective pollen intine and exine wall surrounding the cytoplasm. Because this indigestible cell wall protects the inner nutritious part of the pollen grain, it was commonly believed that humans could not digest and extract the nutrients from pollen. This worry was dispelled when studies by Bell *et al.* (1983) and Schmidt *et al.* (1984), respectively, demonstrated digestibilities of about 55% in rats for pollen from two *Eucalyptus* species and 80% in mice for mesquite pollen. Although some nutrient loss is expected to occur, especially in thick-walled pollen such as the tested *Eucalyptus*, it appears that once in the digestive system, osmotic shock ruptures the grains at the germination pores (Duhoux, 1982) and allows digestion to occur. Direct evidence of digestibility of pollen by humans and monkeys was shown in a study by Wang *et al.* (1987).

Despite confusion created by advertising and the popular literature, pollen has potential as an excellent human food source. To illustrate this point, the

Table 2. Nutritive value per 1000 kilocalories of common foods and pollen.[1]

Item	Protein (g)	Fat (g)	Carbohydrates (g)	Phosphorus (mg)	Potassium (mg)	Sodium (mg)	Calcium (mg)	Iron (mg)	Vit. A (IU)	Thiamin (mg)	Riboflavin (mg)	Niacin (mg)	Vit. C (mg)	Wt./1000 Kcal (lb)
Beef (broiled sirloin)	59.4	82.7	0	468	665	145	26	7.5	143	.17	.46	12.2	0	.57
Chicken (fried breast & leg)	152.8	35.9	6.5	1238	2010	484	60	8.9	484	.28	1.29	57.7	0	1.34
Baked beans	50.1	6.5	155.9	754	1723	3797	443	14.8	1070	.65	.25	4.9	16	1.81
Whole wheat bread	43.2	12.3	196.4	938	1123	2169	407	12.3	trace	1.06	.49	11.5	trace	.91
Apple	3.4	10.3	250.2	171	1900	19	122	5.3	1560	.53	.34	1.9	68	3.80
Raw cabbage	54.1	8.3	224.8	1211	9700	835	2037	16.5	5410	2.11	2.11	12.8	1950	9.17
Tomato	50	8.8	213.8	1225	11000	138	588	22.5	41000	2.75	1.88	31.2	1050	10.64
Pollen	96.3	19.5	109.8	602	2360	179	915	57	14500[2]	3.82	7.56	63.8	142	.90

[1] Data for common foods from Adams (1975) and for pollen from Table 1.
[2] Vitamin A equivalent as based on carotene, with μg carotenes = .375 IU.

established nutrient values for seven common and generally considered highly nutritious foods and pollen are listed in Tables 2 and 3. In Table 2 standard nutrient comparisons are calculated on the basis of overall equal caloric values for the various foods. In terms of protein, pollen ranks above all the listed foods except chicken, and contains over 50% more protein than beef. The fat content of pollen is very low, being about one-half that of chicken and less than one-quarter that of beef. Pollen, like most of the rest of the foods listed, is a good potassium source and is much lower than most foods in sodium, an element whose intake is recommended to be reduced in some people with potential cardiac problems. The calcium levels of pollen is the highest of those listed for all the other foods except cabbage. Cabbage is an interesting food in that to obtain an equal amount of food energy as found in pollen, one would need to consume over ten times as much cabbage as pollen; and cabbage is almost never eaten alone, rather it is often consumed as cole slaw, which contains high levels of fat and correspondingly much lower levels of all other nutrients than cabbage, itself, or as sauerkraut which is mixed with rich pork or sausage.

Pollen is especially noted for its high content of trace mineral elements and vitamins. As listed in Table 2 pollen contains over twice the iron of any of the listed foods and over 7.5 times that of beef, a food generally recommended as a good source of iron. Pollen is extraordinarily rich in carotenes, which are metabolic precursors of vitamin A. In terms of vitamin A equivalents, pollen is several times richer than cabbage, a food considered an excellent source of that vitamin, and is surpassed in Table 2 only by tomatoes. But because of the low caloric nature of tomatoes, almost five times the weight of tomatoes must be eaten to obtain the vitamin A levels of pollen. Pollen is the richest source of the three listed B-vitamins in Table 2 and is, respectively, 1.38, 3.58, and 1.1 times richer in thiamine, riboflavin, and niacin than the next closest food item. Vitamin C levels in bee-collected pollen are listed as being about average for food, but these levels are subject to degradative loss unless the pollen is stored at very low temperatures or is extremely fresh.

Table 3 provides the data for the same seven foods and pollen except the values are listed in terms of amounts per weight of food. This table differs from the previous mainly in that it accounts for the amount of water in the food. For example, tomatoes are mostly water and have only 80 calories of food energy, whereas beef has relatively less water and has over 20 times as much energy per pound as tomatoes. In practical terms, one could easily eat 1,000 calories of beef (9 ounces), but not of tomatoes (12 pounds) and thus measuring nutrients per 1,000 calories as in Table 2 is not practical for most vegetables.

The values listed in the two tables are only those readily available in standard nutritional listings (Adams, 1975). Other trace nutrients in the diet

Table 3. Nutritive value per one pound weight of common foods and pollen.[1]

Item	Energy (Kcal)	Protein (g)	Fat (g)	Carbohydrates (g)	Phosphorus (mg)	Potassium (mg)	Sodium (mg)	Calcium (mg)	Iron (mg)	Vit. A (IU)	Thiamin (mg)	Riboflavin (mg)	Niacin (mg)	Vit. C (mg)
Beef (broiled sirloin)	1750	104.0	145.0	0	820	1166	254	46	13.1	250	.29	.81	21.3	0
Chicken (fried breast & leg)	749	114.5	26.9	4.8	927	1507	362	45	6.6	362	.21	.97	43.2	0
Baked beans	553	27.7	11.8	86.2	417	953	2100	245	8.2	590	.36	.14	2.7	9
Whole wheat bread	1102	47.6	13.6	216.4	1034	1238	2390	449	13.6	trace	1.17	.54	12.7	trace
Apple	236	.9	2.7	65.8	45	499	5	32	1.4	410	.14	.09	.5	18
Raw cabbage	109	5.9	.9	24.5	132	1057	91	222	1.8	590	.23	.23	1.4	213
Tomato	80	4.0	.7	17.1	98	888	11	47	1.8	3280	.22	.15	2.5	84
Pollen	1117	107.6	21.8	122.6	672	2630	200	1022	64	16200[2]	4.27	8.44	71.3	159

[1] Data for common foods from Adams (1975) and for pollen from Table 1.
[2] Vitamin A equivalent as based on carotene, with μg carotenes = .375 IU.

are not commonly listed, though they can be crucial for good health. Pollen is on record as being one of the very richest sources of a variety of these trace nutrients including zinc, copper, manganese, pyridoxine, pantothenic acid, folic acid, and biotin. Recommended Dietary Allowances (RDA) or Estimated Safe and Adequate Daily Dietary Intakes (ESADDI) for these and a few other nutrients essential in the human diet that have been established (Recommended Dietary Allowances, 1989) are listed in Table 4. In that table are also listed the average content of each nutrient in pollen and the amount of pollen that must be consumed to obtain RDA or ESADDI level of that nutrient. Among the elements pollen is remarkably rich in manganese, copper and zinc with one to several ounces providing the entire daily allotment. The same is true for the vitamins pyridoxine, pantothenate, folic acid, and biotin. The vitamin E levels in pollen are marginal, at best, and pollen entirely lacks vitamins D and K. Unfortunately, the levels in pollen of the remainder of the listed factors in the table are unknown.

In summary, pollen is deficient in several of the lipid soluble vitamins, but otherwise has a nutritional composition that surpasses that of virtually any food typically eaten. It remains to be seen whether the nutritional potential of pollen will be achieved and if pollen can become a competitive food item in the human diet, or be developed as a "cottage industry" protein and nutritional supplement for developing nations.

Potential Adverse Reactions from Pollen Consumption

The major adverse reactions reported by people who consume pollen are stomach and gastrointestinal upset. In some studies as many as 12 to 33 percent of the individuals experienced some problems (Feinberg *et al.*, 1940; Maurer and Strauss, 1961). Typical symptoms include stomach pain and diarrhea, although irritation and itching in the mouth and throat area are sometimes experienced. Exceptional reactions might also include headache, general malaise, generalized itching, fatigue, and asthma attacks (Pieroni *et al.*, 1982; Lin *et al.*, 1989). The incidence of gastrointestinal problems is undoubtedly one of the major reasons that most suppliers of pollen products recommend consumption of small amounts of pollen, at least initially.

A second concern about consumption of pollen, especially among the medical profession, is the potential for allergic reactions to orally ingested pollen. This is an intuitively logical, albeit not a well thought out concern, because pollen is probably the major known source of respiratory allergies. The logic runs as follows: since respiratory allergies and anaphylactic reactions are both immunologically based, then allergens such as pollen should be inducers of food anaphylaxis as well. In support of such reasoning there have been a small number of medical reports of individuals experiencing minor to moderate anaphylactic reactions subsequent to the ingestion of pollen (Cohen

Table 4. Recommended Dietary Allowances for minerals and vitamins compared with their levels in pollen.

Nutrient	RDA or ESADDA levels[1]	Pollen levels ($\mu g/g$)[2]	Wt. pollen for needs grams	ounces
Zinc	12mg	78	150	5.5
Copper	1.5-3.0 mg	14	110-120	3-7.6
Manganese	2.0-5.0 mg	100	20-50	.7-1.8
Pyridoxine	1.6 mg	9	180	6.4
Pantothenate	4-7 mg	28	140-250	5.1-8.9
Folate	180 μg	5.2	35	1.2
Biotin	30-100 μg	.32	95-310	3.4-11
Vit. D	5 mg	0	not possible	
Vit. E	8 mg	14	570	20.4
Vit. K	1 μg/kg wt.	0	not possible	
Chromium	50-200 μg	unknown	unknown	
Molybdenum	75-250 μg	unknown	unknown	
Selenium	55 μg	unknown	unknown	
Iodine	150 μg	unknown	unknown	
Flouride	1.5-4.0 mg	unknown	unknown	
Vit. B-12	2 μg	unknown	unknown	

[1] RDA = Recommended Dietary Allowance, ESADDA = Estimated Safe and Adequate Daily Dietary Allowances; values from Recommended Dietary Allowances (1989) for women aged 25-50 years.

[2] Values from Table 1.

et al., 1979; Mansfield and Goldstein, 1981). Superficially this logic is appealing, but on close scrutiny it does not withstand challenge. Respiratory allergies and anaphylactic reactions are strikingly different both in terms of their induction and their treatment. Respiratory allergies are induced by allergens (pollen in this case) coming in contact with moist mucosal membranes, whereas anaphylaxis is essentially always induced by direct contact of allergens (such as penicillin, horse serum, venoms, etc.) with the blood. Treatment of respiratory allergies includes a variety of drugs, but never injection of epinephrine; anaphylaxis is best treated with injection of epinephrine, with allergy drugs being essentially useless. To highlight the differences one need only note that hay fever and respiratory allergies do not cause anaphylactic reactions.

The most convincing evidence that respiratory allergy to pollen and anaphylactic reactions should not be similar is derived from epidemiological data. The allergy literature is full of case histories of strange or unusual examples of allergies, and allergy to ingested pollen would certainly fit into those categories and would be reported (J.O.S. personal observations). Yet, there are remarkably few reports on allergy to ingested pollen: why should this be if respiratory allergy and food allergy to pollen are strongly related and if the profession is keenly alert to examples to demonstrate this? Since there are certainly millions of people ingesting pollen and there are so few cases of allergic reactions to it, the conclusion emerges that pollen as a food is not strongly allergenic. To place pollen as a food allergen in its proper perspective, it is instructive to analyze the incidence of food allergy in the general public. Based on a survey of 722 physicians, the incidence of food allergy in the general U.S. public is $10 \pm 0.7\%$ (Tu *et al.*, 1989). The foods most frequently causing positive reactions on allergy tests with children were: egg (26%), peanuts (24%), milk (23%) and nuts (10%) (Bock and Atkins, 1989). The ultimate food anaphylactic reaction can be death. For eight discovered food-induced anaphylactic deaths the offending foods were: peanuts — 4 (50%); Pecans — 1; shrimp — 1; crabs — 1; and fish — 1 (Squillace *et al.*, 1988). No deaths have ever been reported from consumption of pollen.

After examination of the data, the conclusion emerges that consumption of pollen entails only trivial risks of food allergy, probably much smaller risks than those posed by peanuts, milk, eggs, shellfish, and wheat, and that fears of adverse reactions are based mainly on historical thinking. A more realistic concern is the incidence of gastrointestinal upset subsequent to eating pollen.

Potential for Pollen in Human Health and Well-Being

The best documented healthful benefit of pollen for humans is undoubtedly the treatment of chronic prostatitis. Consumption of pollen preparations have been shown in several studies to reduce the inflammation, discomfort and pathology of patients suffering from benign prostatic inflammation (Ask-Upmark, 1967; Denis, 1966; Hayashi *et al.*, 1986). The exact reasons for the dramatically improved conditions as a result of eating pollen remain unclear. A possible factor in pollen that could be important is zinc. Pollen contains extraordinarily high concentrations of zinc. In pollen zinc functions in enzymatic processes at the growing tip of the pollen germination tube (Ender *et al.*, 1983), and in humans zinc is a key element in prostate gland function. The zinc concentration in that gland is not only several times greater than other body tissues, but appears to be correlated with the ability of the gland to produce an antibacterial factor in semen (Colleen *et al.*, 1975). Men with chronic prostatitis have lower prostate zinc levels than control men (Colleen *et al.*, 1975; Kvist *et al.*, 1988). These observations are only correlations, as specific studies addressing the effect of added dietary zinc on prostatitis have not been reported, but are suggestive that high zinc levels in pollen might be a reason for the therapeutic effects.

Another benefit of pollen is its ability to help protect against the adverse effects of x-ray. Evidence of a radioprotective effect of ingested pollen comes from studies of both animals and humans. Mice irradiated with more than a usual lethal dose of x-rays were fed either diet plus pollen or diet alone. The mice fed diet plus pollen suffered much less damage to their spleens and thymus glands and had much lower death rates than those not fed pollen (Wang *et al.*, 1984). The positive effects of dietary pollen in the medical regimen of cancer patients undergoing radiation treatment has been reported by Hernuss *et al.* (1975). In their study 15 women with cervical carcinoma were given pollen and 10 controls were treated the same except pollen was omitted. Overall the pollen-fed patients subjectively reported feeling less treatment effects and clinically were found to have suffered fewer side effects as measured by changes in blood factor chemistry.

The consumption of pollen or unrefined honey containing traces of local pollen has been widely believed to help reduce the symptoms of hay fever. Although testimonial evidence abounds to support these beliefs, rigorous experimental investigations have revealed, at best, only marginal improvements in patients who consume pollen (Feinberg *et al.*, 1940). Although some positive effects resulting from ingestion of pollen appeared in individuals with simple hay fever without asthma, the overall benefits were so marginal compared to the clear benefits of subcutaneous pollen injections (allergy shots) that oral treatments were not justified (Feinberg *et al.*, 1940). These results indicate that if allergies or asthma are severe enough to merit medical attention, the best treatments are subcutaneous allergy shots; but if they are not that severe, there is no reason for an individual not to enjoy locally produced or raw honey and pollen products — and if the hay fever is helped, that is an added benefit.

Articles periodically appear suggesting that pollen can help reduce or cure some types of cancer. To date there have been no formally documented studies indicating positive effects of pollen in the treatment of cancer, though recent indications suggest that some carotenes may protect against some cancer (Krinsky, 1988) and pollen contains high levels of carotenes. One historically important paper that is sometimes cited as indicating an anticancer effect of pollen is that of Robinson (1948). The author reports in one study an inhibiting effect of ingested pollen on mouse mammary tumors, but the results could not be repeated in his other studies. Moreover, the effects were observed at levels of 1 part pollen to 120,000 parts food! It is inconceivable that such low levels, roughly equivalent to a human consumption of only two bee-collected pollen pellets per day, could have such an effect. Furthermore, since the alleged effects have not been observed in the forty years following that paper, the report is probably best dismissed.

Perhaps the greatest nutritional fame of pollen is based on claims that pollen consumption improves one's physical stamina (plus a variety of other attributes). Undoubtedly pollen in the diet, especially in an otherwise poor or unbalanced diet, would be beneficial, but rigorously controlled tests are largely lacking. Reports such as that of Noyes (1961) involving feeding pollen to football players and recording their weight gains and numbers of absent days and injuries cannot be accepted as valid evidence because the study was reported in an unrefereed and unpublished pamphlet produced by a major pollen producer. Likewise, the often-referred to study of Korchemny in which athletic performance of members of a track team was supposedly enhanced by pollen cannot be accepted; again, because it is an unpublished, unrefereed report issuing from the Pratt Institute of New York. In a refereed, double-blind study of the effects of pollen on the performance of 18 cross country runners, Steben and Boudreaux (1978) reported that neither pollen nor protein extract capsules significantly improved performance or blood levels of potassium, hemoglobin, or hematocrit. Though not statistically significant (in part due to the small sample size of only six individuals per treatment and high variance), their results showed that over the 12-week season the runners given either pollen or protein extract increased their mean speed and gained hemoglobin and hematocrit values. In each case the values for pollen were better than for protein extract. In terms of potassium levels, none of the pollen-treated runners exhibited decreased blood levels, whereas the placebo group had a net loss and the protein extract had an even greater loss. These trends are only suggestive and indicate that larger and more thorough studies are not only justified, but would provide a public service.

Numerous testimonials portray pollen as being of benefit in treating a wide variety of ailments including ulcers, colds, infections, plus a variety of others and improving sexual prowess (Devlin, 1981; Thorsons, 1989). None of these putative health effects have been scientifically tested. The potential of pollen in human health and well being might be greatly improved if indepedent research to address some of these questions were conducted, and if the pollen industry took a lead in funding such research. Although some of the claims for pollen likely have some merit, given the lack of substantiating evidence, the best advice is "user beware".

Uses of Pollen for Animals

A variety of animals are known to feed on pollen and its use in animal diets appears promising. Rats readily eat diets containing pollen and appear to like the flavor of pollen-containing diets over pollen-free diets (Ruiz Abad, 1975). Chickens fed a balanced diet plus 2.5% pollen reportedly showed improved food conversion efficiencies (Costantini and Riciardelli d'Albore, 1971) and the addition of corn pollen to the diets of hens produced egg yolks with a deeper color and higher carotene levels (Tamas *et al.,* 1970). Piglets fed 1 - 3 % corn pollen gained weight more quickly and converted food to weight more efficiently than piglets on the same diet without pollen (Salajan, 1970 as cited in Stanley and Linskens, 1974). Pollen has also been successfully incorporated in the diets of experimental animals including crickets (Sasagawa, 1982),

FIGURE 2. Modified Ontario Agricultural College (OAC) pollen trap showing double 5 mm screen grids and collection tray with pollen.

grasshoppers (Davies and Dadour, 1987), mosquitoes (Eischen and Foster, 1983), thrips (Kirk, 1985), and blister beetles (Leppla *et al.,* 1974), and appears to be an important component in the diet of immature orb-weaving spiders (Smith and Mommsen, 1984). Many birds and bats and other mammals are known to feed on pollen. Use of pollen in diets of these animals in captivity could be beneficial, though not all pollen-feeding birds appear able to digest dietary pollen efficiently (Brice *et al.,* 1989). Pollen might also be a beneficial addition to the food of fish hatcheries, crustacean aquaculture, where the pollen carotenes likely would enhance the red colors of shrimp, prawns, and crayfish plus provide nutrients, and for fish farms.

Pollen Production and Commercial Market Value

Pollen is collected in the form of corbicular pellets (Figure 2) removed from the legs of returning bees. A great variety of pollen traps have been developed for this purpose. Most incorporate two screens with .023 inch diameter wires with 5 per inch that are separated by about 7 mm. Other traps use perforated metal plates with about .188 inch holes. One popular model is the so-called OAC trap which was developed in the Ontario Agricultural College (now called the University of Guelph) and modified by Waller (1980). This pollen trap (Figure 3) features a large collection drawer, drone escape, easy access from the rear of the colony and a large collection surface for bees to enter the colony.

FIGURE 3. Close-up of pollen pellets *(by Marty Cooper [C] 1984 National Geographic Society).*

The design of a pollen trap is crucial both to the effectiveness of collecting pollen and to the welfare of the colony. Factors important to design and manufacture of pollen traps are discussed by Waller (1980) and Shaparew (1985); factors important to pollen handling, cleaning and processing are described by Benson (1984). In brief, the collected pollen should be free of contaminating insect parts, wax moths, debris, mold, etc. and must be kept dry. The trap must not unduly stress the colony by taking too much pollen. If too much pollen is removed from the foragers, severe stress including reduced brood rearing and decrease in honey production can occur. Traps that remove about 60% of the incoming pollen during heavy nectar flows appear about optimal, and can be left in place year round with little adverse effect (Levin and Loper, 1984; Buchmann and Shipman, unpublished). Such rates of pollen harvest can also promote increased pollen foraging activity, thereby increasing the effectiveness of pollination services, without generally adversely affecting the colonies (Levin and Loper, 1984, Webster *et al.,* 1985).

Figures concerning worldwide production of pollen are difficult to obtain, but are large. Major producers are USA, China, USSR, Spain and Europe, Mexico, Argentina, and Australia. In China alone 3,000 - 5,000 metric tons of pollen are harvested yearly (Wang, 1989). The wholesale price paid beekeepers in the U.S. for pollen varies greatly, but is generally in the range of $4 - $15 per kilogram, depending upon the purity, dryness, and other factors of the pollen.

FIGURE 4. Pollen products for human consumption come in a variety of forms including liquids (with honey), capsules, granules and candy bars.

The market for pollen is mainly for human nutritional supplements, feeding to bees, and as an animal food. The use of pollen for race horse care is surprisingly large in North America (R. Brown, personal communication). Pollen is formulated for human consumption into a variety of appealing products including tablets, pollen granules, oral liquids (which usually feature pollen extract in a honey base), candy bars, tonics, etc. (Figure 4). The production of pollen products for human consumption has been growing at a rapid rate, with one company reporting processing 500 tons of pollen to yield gross sales of $13 million in 1987 (Wang, 1989). Prices of pollen products vary enormously and often can yield very high profits for the producers and retailers.

PROPOLIS

Propolis is a sticky plant-derived material used by bees as their ubiquitous caulking, sealing, lining, strengthening, preserving and probably repellent material inside the hive and around the entrance. To beekeepers it is most familiar as the material that sticks frames and other hive parts together. It is also found in a layer as a thin "varnish" over all the inner surfaces of the hive including wax combs. In feral colonies the propolis varnish is spread around

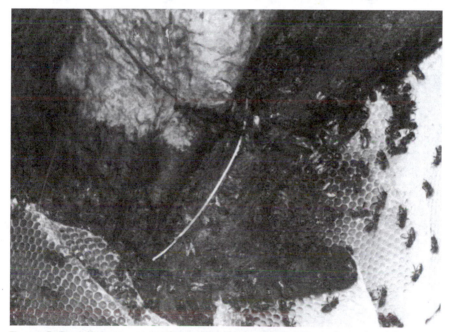

FIGURE 5. Propolis layer surrounding combs of a free-hanging honey bee colony. Propolis is layered over the branch and onto the weathered wood pulp (above the lower left combs) creating a wide dark band between the combs and the lighter unpropolized wood pulp.

the entrances to the nest cavity and along any branches that touch free-hanging combs (Figure 5). Small cracks and holes in the nest cavity are often filled with propolis, damaged combs are repaired with propolis, and objects that cannot be removed from the nest are frequently sealed with propolis. General discussion of the collection and use of propolis by bees is provided by Root (1983) and detailed information with in-depth historical discussion is provided by Haydak (1953).

Propolis is soft and sticky at warm temperatures and can be molded to fill holes and gaps or spread over surfaces. At cool temperatures and as it ages, propolis becomes brittle and hard. It has antimicrobial properties and, thus, is an important part of the chemical arsenal within the hive for combating contamination and pathogen invasion. Although not well investigated, its use as a varnish around colony areas of contact with the external environment (Figure 6) probably serves mainly in defense against ants. This is accomplished both by making the surface slippery, thereby enabling the bees to blow off invading ants (Spangler and Taber, 1970) and by acting as a surface masking agent. A propolis masking agent might reduce the contact chemical signals that otherwise might attract the attention of ants and other marauding potential invaders. This masking would be due to the plant chemical nature of propolis: to potential predators propolis might represent uninteresting or

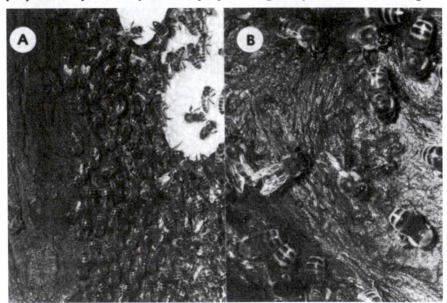

FIGURE 6. (A) Bees adjacent to a comb actively planing a propolized area; (B) close-up view of planing bees. Planing bees adopt a typical head down pose and actively "brush" the surface with their front legs while slowly moving backwards.

repellent plant chemicals rather than signs of a colony. The propolis layer inside the colony might also serve as a moisture barrier to keep humidity high inside, to prevent catastrophic influxes of moisture following heavy rains, and as a barrier against microbial pathogens.

Propolis is a resinous material collected by foragers from a variety of plants, especially the buds of trees. Poplar trees (*Populus* spp.) are a main source of propolis in eastern Europe (Marinescu and Tamas, 1980) and probably wherever members of that large genus are common. Bees will also collect a variety of plant materials, undoubtedly depending in large part on availability, including resins from pine trees (Haydak, 1953; Root, 1983), desert composites (Wollenweber *et al.,* 1987; Wollenweber and Buchmann, unpublished), and Australian grass trees (Ghisalberti *et al.,* 1978). Since propolis is a mixture of locally available plant exudates, it would be expected to differ from one locality to another and from colony to colony. Surprisingly, the composition of propolis samples from diverse sources is remarkably similar (Lindenfelser, 1967; Bunney, 1968; Wollenweber *et al.,* 1987), perhaps reflecting a fundamental similarity of available plant resins, or perhaps indicating that bees somehow recognize and select resins of similar compositions. That bees actively chose resins of similar compositions seems less likely since they are known to collect the synthetic sticky polymer called "tanglefoot" and wet red paint for use as propolis (McGregor, 1952; Lowe, 1980). Even if propolis samples are grossly similar in chemical composition, differences in individual samples mean that propolis—unlike royal jelly and bee venom which are similar from sample to sample—is not a uniform material.

All species of *Apis* as well as many species of stingless bees collect propolis. Stingless bees (*Melipona, Trigona,* etc.) collect large quantities of plant resins, saps, and gums which are incorporated with beeswax to make cerumen. Cerumen is used for constructing entrance tubes to the nest and for attaching and stabilizing cells and various nest structures within the nest. Batumen which is also often made with propolis is used by stingless bees for nest construction material and as a barrier against intruders and the environment (Michener, 1974).

Chemical Composition of Propolis

The chemical composition of propolis varies from sample to sample due to the variety of plant resins, gums, exudates, etc., utilized by the bees and the collection techniques used by beekeepers to obtain propolis from the hive. If hive scrapings are used as a source of propolis, beeswax can be a substantial component of the propolis. Excluding beeswax, which is an additive, propolis consists of a mixture of resins, terpenes and volatile oils, and miscellaneous materials. Resin is a term used for materials that are usually sticky and insoluble, or very poorly soluble, in water or organic solvents; and terpenes and volatile oils are compounds that usually have fragrant odors. Propolis resins appear to be mixtures of natural polymers, that due to their insolubility and inertness, are likely to be important only for their structural properties. Resins are of little interest for humans and they are usually discarded as the residue left after extractions for chemical or biological investigations. Wax, pollen, or minor contaminants of propolis are likewise generally of little commercial or medical interest. Pollen in propolis can, however, yield a long-term record of those plants a honey bee colony has exploited as food (Buchmann, unpublished).

The pharmacologically active constituents of propolis are found in fractions soluble in solvents such as alcohols. Several large classes of compounds have been identified from the diversity of compounds in these fractions. The most important are the flavones, flavonols, and flavanones (collectively called flavonoids) and various phenolics and aromatics (Table 5). Some of the phenolics include cinnamyl alcohol, cinnamic acid, vanillin, benzyl alcohol, benzoic acid, and caffeic and ferulic acids (which are cinnamic acid derivatives with added hydroxyl and/or methoxyl groups). Propolis also contains unusual compounds such as phenolic triglycerides (Popravko *et al.,* 1983), pterostilbene, eugenol, caffeic acid phenethyl ester, a group of caffeic acid pentenyl esters, and xanthorrhoeol, a complex napthanol derivative of Australian grass tree resin.

Flavonoids are a large group of plant-derived natural pigments and active compounds that share many features. All possess two benzene rings connected by a 3-carbon link and all those in propolis have basic structures as shown in Figure 7. The best known flavones and flavanones isolated from propolis are listed in Table 5.

Biological Activities of Propolis

Propolis and some of its constituents exhibit a variety of biological and pharmacological activities (Table 6). The antimicrobial activity of propolis is well documented (Ghisalberti, 1979; Grange and Davey, 1990) and several of its components have been demonstrated to be active. The flavone pinocembrin is active toward a variety of bacteria, fungi, and molds, and it along with galangin, 3-acetyl pinobanksin, and caffeic and ferulic acids is probably responsible for much of the biological activity of propolis. Quercetin is a propolis flavone with both anti-viral activity and capillary strengthening properties. Other flavones and flavanones in propolis exhibit anti-inflammatory activity, topical anesthesia, and spasmolytic activity (inhibition of smooth muscle contractions in the digestive system) (Table 6). Propolis phenolics exhibit various activities: caffeic acid is anti-inflammatory, the phenethyl ester of caffeic acid selectively inhibits melanoma and carcinoma tumor cells, and pterostilbene has anti-diabetic activity. Caffeic acid, luteolin, and quercetin, all

Table 5. Major flavonoids and phenolics isolated from propolis[1]

Common name	Chemical name (IUPAC)
	FLAVONOIDS
Chrysin	5,7-dihydroxyflavone
Tectochrysin	5-hydroxy-7-methoxyflavone
Galangin	3,5,7-trihydroxyflavone
Acacetin	5,7-dihydroxy-4'-methoxyflavone
Isalpinin	3,5-dihydroxy-7-methoxyflavone
—	5-hydroxy-4',7-dimethoxyflavone
Kaempferol	3,4,5,7-tetrahydroxyflavone
Kaempferide	3,5,7-trihydroxy-4'-methoxyflavone
Rhamnocitrin	3,4',5-trihydroxy-7-methoxyflavone
—	3,5-dihydroxy-4',7-dimethoxyflavone
—	5,7-dihydroxy-3,4'-dimethoxyflavone
Pectolinarigenin	5,7-dihydroxy-4',6-dimethoxyflavone
Isorhamnetin	3,4',5,7-tetrahydroxy-3-methoxyflavone
Quercetin	3,3',4',5,7-pentahydroxyflavone
Quercetin-3,3-dimethyl ether	4',5,7-trihydroxy-3,3'-dimethoxyflavone
Pinocembrin	5,7-dihydroxyflavanone
Pinostrobin	5-hydroxy-7-methoxyflavanone
Pinobanksin	3,5,7-trihydroxyflavanone
3-Acetyl pinobanksin	5,7-dihydroxy-3-acetylflavanone
—	5-hydroxy-4',7-dimethoxyflavanone
Sakuranetin	4',5-dihydroxy-7-methoxyflavanone
Isosakuranetin	5,7-dihydroxy-4'-methoxyflavanone
—	3,7-dihydroxy-5-methoxyflavanone
—	2,5-dihydroxy-7-methoxyflavanone
	PHENOLICS
Vanillin	4-hydroxy-3-methoxybenzaldehyde
Isovanillin	3-hydroxy-4-methoxybenzaldehyde
Benzyl alcohol	-hydroxytoluene
—	3,5-dimethoxybenzyl alcohol
Benzoic acid	
Cinnamyl alcohol	3-phenyl-2-propen-1-ol
Cinnamic acid	3-phenyl-2-propenoic acid
Coumaric acid	3-(4-hydroxyphenyl)-prop-2-enoic acid
Caffeic acid	3-(3,4-dihydroxyphenyl)-2-propenoic acid
Ferulic acid	3-(4-hydroxy-3-methoxyphenyl)-2-propenoic acid
Isoferulic acid	3-(3-hydroxy-4-methoxyphenyl)-2-propenoic acid
Eugenol	2-methoxy-4-(2-propenyl)phenol
Cinnamic acid benzyl ester	benzyl 3-phenyl-2-propenoate
Coumaric acid benzyl ester	benzyl 3-(4-hydroxyphenyl)-2-propenoate
Caffeic acid benzyl ester	benzyl 3-(3,4-dihydroxyphenyl)-2-propenoate

Table 5. Major flavonoids and phenolics isolated from propolis[1] *(Continued)*

Common name	Chemical name (IUPAC)
Caffeic acid phenethyl ester	see figure
Caffeic acid 3-methyl-2-butenyl ester	see figure
Caffeic acid isopent-3-enyl ester	3-methyl-3-butenyl 3-(3,4-dihydroxyphenyl)-2-propenoate
Caffeic acid 2-methyl-2-butenyl ester	2-methyl-2-butenyl 3-(3,4-dihydroxyphenyl)-2-propenoate
Ferulic acid 3,3-dimethylallyl ester	3-methyl-2-butenyl-(4-hydroxy-3-methoxyphenyl)-2-propenoate
Ferulic acid isopent-3-enyl ester	3-methyl-3-butenyl 3-(4-hydroxy-3-methoxyphenyl)-2-propenoate
Pterostilbene	see figure
Xanthorrhoeol	see figure

[1] From Ghisalberti (1979, 1978) plus Bankova *et al.,* (1982, 1983, 1987, 1989), Wollenweber *et al.,* (1987) and Grunberger *et al.,* (1988).

compounds having the caffeioylic moiety in common, also exhibit antiviral activity against herpes virus, a virus which is difficult to treat (König and Dustmann, 1985). Flavonoids not so far reported in propolis are known to be capable of aiding in the healing of stomach ulcers, improving pulmonary insufficiency, playing key roles in strengthening capillaries (Table 6) and are essential in acting with vitamin C to prevent scurvy (Roger, 1988).

Propolis has had an ancient history as a curative agent in human health. It was known in the time of Aristotle and discussed in detail by Pliny, the Roman naturalist (Haydak, 1953). Advocates of propolis have provided abundant testimonials to its ability to aid in the treatment of a variety of ailments including colds and sore throats, skin problems, stomach ulcers, burns, hemorrhoids, gum diseases, and wounds (Hill, 1977; Apimondia, 1978; Iannuzzi, 1983; Kosonocka, 1990). Although propolis might have value in treating many of these problems (see Ghisalberti, 1979 and Lindenfelser, 1967 for reviews) there is a paucity of carefully controlled experiments and a plethora of poorly designed or pseudo-scientific reports in the literature. Propolis is clearly antibacterial and antifungal (Lindenfelser, 1967; Ghisalberti, 1979) and is highly active toward *Bacillus larvae,* the pathogenic agent of American foulbrood (Lindenfelser 1968, Mlagan and Salimanovic 1982). Control of pathogens such *Bacillus larvae* is undoubtedly a major role for propolis in the hive (it is spread as a thin coating over most surfaces, including brood cells) though it is, at best, a poor control agent when added to sugar syrup or sprayed on combs for treating the active cases of American foulbrood (Lindenfelser, 1968; Mlagan and Salimanovic, 1982). Propolis and some of its constituents

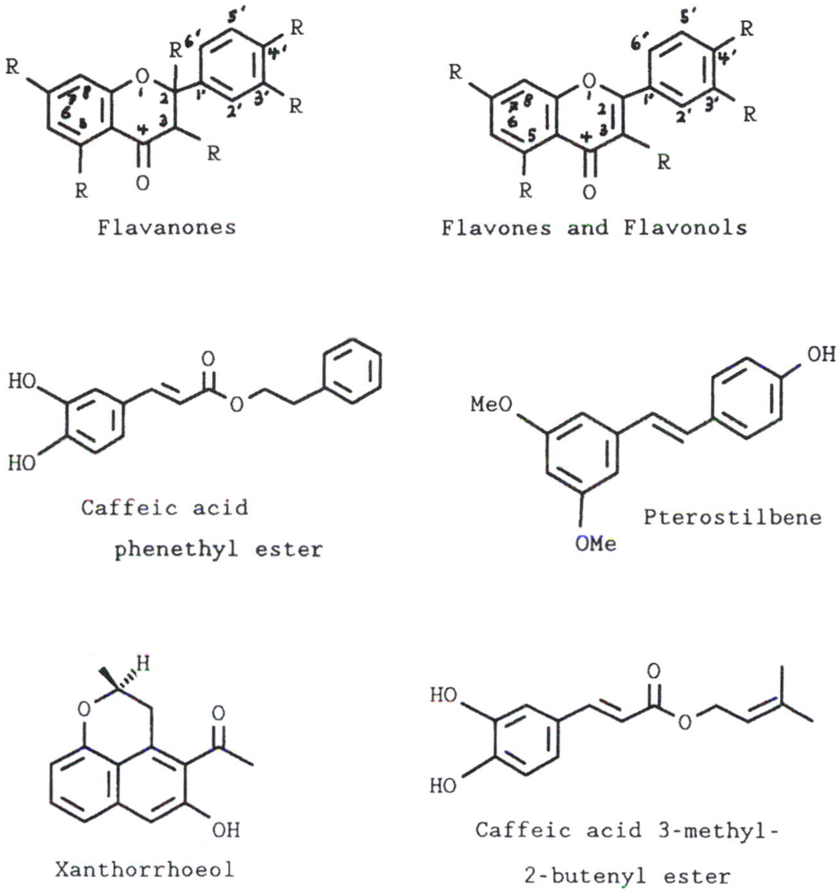

Flavanones

Flavones and Flavonols

Caffeic acid phenethyl ester

Pterostilbene

Xanthorrhoeol

Caffeic acid 3-methyl-2-butenyl ester

FIGURE 7. Chemical structures of flavonoids (general structures) and some phenolic compounds from propolis.

produce anesthesia (Paintz and Metzner, 1979) which in some studies were three times as powerful as cocaine and 52 times that of procaine in the rabbit cornea (see Ghisalberti, 1979). The apparent anesthetic ability of propolis may be the reason that it has received so many favorable reports for treatment of sore throats and gum disorders. Propolis, like many flavonoids, is capable of scavenging free radicals and thereby protecting lipids (Popeskovic *et al.*, 1980) and other compounds (vitamin C) from being oxidized, destroyed or turning rancid (Ghisalberti, 1979). Propolis aids in healing wounds (Schmidt and Wang, unpublished) and is reported also to heal ulcers and gum diseases, reduce inflammations and thromboses, stimulate detoxification enzymes, potentiate or synergize antibiotics, inhibit viruses, and inhibit erythrocyte agglutination (the pathological engorgement of organs with excess red blood cells) (see Ghisalberti, 1979). Flavonoids are well-known biologically active compounds (Farkas *et al.*, 1986; Cody *et al.*, 1988) and are likely responsible for many of these reported activities.

Propolis is not apparently toxic to humans and mammals unless very large quantities are administered (Ghisalberti, 1979). Some of its constituent flavones, *eg.* quercetin, might be mutagenic by the Ames test (Roger, 1988), but mutagenicity *per se* for propolis has not been reported. Hypersensitivity to

Table 6. Activities of known (and related) compounds in propolis[1]

Activity	Active component(s)	References[2]
Anti-bacterial	pinocembrin, galangin caffeic acid, ferulic acid	Vilanueva *et al.*,1970
Anti-fungal	pinocembrin 3-acetyl pinobanksin caffeic acid, p-coumaric acid benzyl ester sakuranetin, pterostilbene	Metzner *et al.*, 1975, 1977 Schneidewind *et al.*, 1975
Anti-mold	pinocembrin	Miyakado *et al.*, 1976
Anti-viral	caffeic acid, lutseolin, quercetin	König and Dustmann, 1985
Tumor cytotoxicity or inhibition	caffeic acid phenethyl ester (methyl caffeate, methyl furuleate)	Grunberger *et al.*, 1988 Inayama *et al.*, 1984
Local anesthetic	pinocembrin, pinostrobin, caffeic esters	Paintz and Metzner, 1979
Anti-inflammatory	caffeic acid acacetin	Bankova *et al.*, 1983
Spasmolytic	quercetin, kaempferide, pectolinarigenin	
Anti-diabetic (un-confirmed)	pterostilbene	
Healing of gastric ulcers	(luteolin, apigenin)	
Helping pulmonary insufficiency	(eriodictyol)	Aviado *et al.*, 1974
Strengthening capillaries	quercetin (3',4'-dihydroxyflavanoids) (flavan-3-ols)	Budavari, 1989 Roger, 1988

[1] Compounds in parentheses are similar to those in propolis.

[2] General source of information is Ghisalberti (1979) with other authors as noted.

propolis is known (Bunney, 1968; Hausen *et al.,* 1987; Hausen and Wollen-weber, 1988) and can present problems for some beekeepers and users. The main symptoms of propolis allergy are contact dermatitis which is caused primarily by the pentenyl esters of caffeic acid and caffeic acid phenethyl ester (Hausen and Wollenweber, 1988). The allergy-inducing mechanisms of these caffeic acids might be similar to that of the structurally related urushiols, the active components of poison ivy. Overall, propolis appears generally not to be harmful, but the shortage of rigorous research into both positive and negative potentials of propolis indicate that common sense and caution should be exercised in the use of propolis.

Production, Commercial Uses, and Economic Value of Propolis

Commercial production of propolis is usually a difficult and time-consuming operation. To obtain the highest grade and purity of propolis, special "inserts" are usually placed in hives. These inserts provide spaces that mimic holes or cracks in the hive, thereby encouraging the bees to fill them with propolis. The resultant propolis is then collected, sorted and packaged. Hive scrapings, though an easier way to obtain propolis, are often contaminated with wood chips, wax, and paint and are of lower commercial quality (Iannuzzi, 1983, 1990a; Wright-Sunflower 1988).

In North America and Europe the main uses of propolis are as natural supplements and herbal medicines. These take the form of tablets in which propolis can be combined with a variety of other ingredients including pollen, royal jelly, and non-hive products; as tinctures in which propolis is extracted with alcohol (usually 70%); and as additives to skin lotions, beauty creams, soaps, shampoos, lipsticks, chewing gums, toothpastes, mouthwashes, and even sunscreens (Iannuzzi, 1990b). Use of propolis tinctures for treatment of sore throats, cuts, and skin rashes is especially popular.

The use of propolis outside North America is much greater than within. In many countries propolis is used for a variety of health purposes in a wide diversity of preparations (Apimondia, 1978). In addition to uses for health, propolis is sometimes used as a varnish, though its use for violins is doubtful or rare (Jolly, 1978). One problem with propolis as a varnish is that it requires a very long time to dry beyond the sticky phase. The potential of propolis as an animal growth stimulant has received little attention, though chickens reportedly gained up to 20% more weight when fed 500 parts per million of propolis in their diets than did controls (see Ghisalberti, 1979).

The price of propolis varies greatly from country to country. In the U. S. and Canada wholesale prices paid for propolis are low, varying from $2 - $6.00 per pound (Iannuzzi, 1990b); in contrast, the price in New Zealand, where propolis is more widely used, is about $26.00 per pound (Jaycox, 1988). Clearly the price of propolis reflects demand and supply in a society, with demand in North America being low. The potential supply of propolis in North America is high; but until new break-throughs in uses or research occur, the demand is not likely to increase dramatically or predictably.

BEE VENOM

Honey bee venom is a well-known pharmacologically active product of the hive. It is synthesized by the venom glands of workers and queens, stored in the venom reservoir, and injected through the sting apparatus during the stinging process (Figure 8). A mature defender or forager contains about 100 -150 µg of venom (Schumacher *et al.,* 1989), and young queens about 700 µg (Schmidt, unpublished). Bee venom is a bitter, hydrolytic blend of proteins with basic pH that is used by the bees for defense. An extensive and growing body of research literature on the chemistry, biochemistry, pharmacological activities on tissue preparations, and allergic properties of bee venom is available (Banks and Shipolini [1986] and in the chapter in this volume on *Allergy to Venomous Insects)* and will not be discussed here.

Humans have been aware of bee venom and its effects, probably since their first encounter with bees, and have often incorporated it into their medical practice. However, only in the past 30 years have efficient means of collecting pure venom been available, thereby providing opportunities for commercial production and use of venom. The potential production and use

Figure 8. Sting apparatus of a worker honey bee. S = sting shaft, V = venom reservoir.

of bee venom has been hindered by a general lack of medical research into its use for other than diagnosing and treating venom hypersensitivity and the typical suspicion and disregard of the medical profession for its use. The main fault for this situation is the reluctance of the medical research community to repeat rigorous, well designed and controlled experiments and clinical trials necessary to determine the potential value of bee venom or its components. Such trials were designed and conducted for the use of insect venoms in the treatment of venom hypersensitivity (Hunt *et al.*, 1978). Remarkably, less evidence of need was available for these trials than exists for tests with bee venom for treatment of arthritis. Like the situation with venom allergy, immense psychological and institutional barriers are present that inhibit needed research that could provide definitive answers (for presentation of this problem, see Mraz [1977], Wells [1977] and Doyle [1983]). As a simple matter of medical ethics, bee venom should be rigorously investigated because it either: 1) has potential to relieve suffering in a large percentage of the population; or 2) it has so little value that its use should be discouraged on the basis of clear evidence, rather than glib opinion. Medicine is based on science; and science only functions in a true sense on the basis of evidence. For this reason, this section will concentrate on scientific evidence relating to the use of bee venom as it affects human welfare.

Medical and Veterinary Research on Bee Venom
    The main sources of information for medicine and veterinary science are chance observations, testimonials, traditional folk medicine, and controlled experiments. Serendipitous observations are probably fundamental in the origin of much traditional folk medicine and today still play important roles in the development of hypotheses to be formulated and tested. Testimonials of good or bad effects as a result of some material or action undoubtedly played a major role in folk medicine and even today are important clues in diagnoses and the discovery of information. Folk medical wisdom is a blend of psychological and physical experiences gathered over generations by concerned and observant individuals. Crude experiments were undoubtedly part of this process. Modern Western medicine differs from traditional medicine in that designed, controlled experimentation is used to test and confirm or reject previous treatments and to provide information from which to design and test further hypotheses.

    The origin of bee venom for treatment of human ailments is unknown, but probably occurred as a result of chance encounters with stinging insects, the subsequent testimonials that resulted from the encounters, and finally societal acceptance after a varying number of repeat experiences and tests. Testimonials and observations indicating the effectiveness of bee venom are common throughout Western and Asian cultures (Beck, 1935; Guyton, 1947; Ryan

1954; Broadman, 1962; Malone, 1979). Testimonials, *per se*, are excellent indicators that some effect is occurring, but cannot provide information on the basis of the effect or whether it has true clinical potential. For example, psychological forces play an important role in a great variety of maladies, and are responsible for the well known placebo effect ("sugar pills" curing a problem). Random factors can also be crucial in distorting and preventing the correct understanding of the medical situation. The importance of such extrinsic statistical effects is clearly illustrated with the example of the belief for almost 50 years that whole body extracts were effective for the treatment of venom hypersensitivity; when, in fact, they were worthless (see discussion in history section of chapter on *Allergy to Venomous Insects* in this volume). It is for these two reasons that testimonials and traditional medicine are not held in high regard by the Western medical profession and its researchers. These are also the reasons that little emphasis will be placed in this discussion on testimonials or research which does not adequately control for the effects of psychological and random effects.

Rheumatoid arthritis is an immunologically medicated inflammatory disease of the joints that often causes intense pain, restricted movement, and disfigurement. It is a disease that has no true cure and often even the best of North American medical treatments are inadequate in achieving meaningful long-term improvement. Typical treatments include: aspirin, ibuprofen, feldane, indomethacin, phenylbutazone, or other non-steroidal anti-inflammatory drugs; penicillamine, gold salts, antimalarials, and prednisone, dexamethasone or other steroidal drugs. All of these drugs, especially the steroids, can have very serious side effects ("rheumatoid arthritis rarely kills the patient; corticosteroids often do" [Calin, 1983]).

Various animal models are useful for investigating the anti-inflammatory effects of treatments or drugs and are often employed for the study of arthritis. In these models, some inflammatory agents, such as turpentine oils, formaldehyde, carrageenin or *Mycobacterium* in mineral oil, is used to inflame a joint and effects of treatments to reduce the inflammation are recorded. Neumann and Stracke (1951) discovered that daily injections of 1 mg/kg of bee venom into rats reduced formaldehyde-induced arthritis in the rat foot pad. In another rat model in which *Mycobacterium* was the inducer of inflammation, intraperitoneal injection of 1 mg of bee venom inhibited the arthritic effect when given daily. If the adrenal glands, the source of corticosteroids, were removed the effect disappeared. Whole venom, but not its individual components, caused a persistent rise in serum corticosterone levels of the animals, leading to the suggestion that bee venom suppresses adjuvant arthritis through its action on the pituitary or adrenal glands which, in turn exert influence by interference with lymphocyte action (Zurier *et al.*, 1973).

Shkenderov (1976) reported that bee venom, its peptides melittin, and apamin, but not venom phospholipase suppress edema, probably through an immunosuppressive effect. The immunosuppressive effect of bee venom in rats was confirmed when high doses of bee venom were injected daily (Eiseman *et al.*, 1982). Chang and Bliven (1979) reported a 35% reduction of carrageenin-induced swelling and attribute the effects to first an alteration of the immune system, then an anti-inflammatory effect via elevation of corticosterols. Whole bee venom and the peptide melittin experimentally lowered the cyclic adenosine monophosphate (cAMP), an intra-cellular mediator, in mouse skin and whole venom and phospholipase caused a dramatic increase in prostaglandin E levels in that tissue (Schmidt, 1978; Schmidt *et al.*, 1978). Bee venom, particularly the peptide apimen, also exerts an anti-complement activity in rats (Gencheva and Shkenderov, 1986). In humans, bee venom inhibits superoxide production by human neutrophils, thereby acting as an anti-inflammatory agent (Somerfield *et al.*, 1984). These experiments suggest avenues for further research to explain possible anti-inflammatory activities of bee venom.

In the rat paw-turpentine model, mast cell degranulating peptide (MCD-peptide, or peptide 401), one of the peptide components of bee venom, exhibited a strong anti-inflammatory effect. This effect appeared due not simply to elevation of the corticosteroid levels, as adrenalectomized animals responded similarly to normal animals. The effect also appeared not to be a side effect of the degranulation of mast cells and consequent mediator release because melittin, which degranulates mast cells, does not suppress the inflammation (Billingham *et al.*, 1973). The peptide was over a hundred times more active in this essay than hydrocortisone. In another study using the same model, 50 μg of MCD-peptide reduced the inflammation better than any of the four tested non-steroidal anti-inflammatory drugs and twice as well as dexamethasone, a popular steroidal anti-inflammatory agent. Melittin, the main peptide in bee venom, did not surpress edema at dosages of 2 mg/kg, nor did it reduce the inflammatory responses to histamine, 5-hydroxytrypamine, or bradykinin, whereas MCD-peptide did; thus, they agreed that the effect was not due to mast cell lysis (Hanson *et al.*, 1974). MCD-peptide still exhibited its anti-inflammatory effects in animals which had nerves to the challenged leg removed, or were adrenalectomized, or were treated with an α-adrenergic blocker. Thus, the effect was not entirely due to elevation of corticosteroids or nervous control. The effective dose for 50 percent reduction (ED_{50}) was .1 mg/kg. The mechanism of action was not discovered, but was hypothesized to be via making the vascular endothelium less responsive to inflammatory stimuli (Hanson *et al.*, 1974).

In dogs the bee venom peptide fractions caused prolonged elevation of plasma cortisol levels, the main steroid produced by the adrenal glands.

Venom phospholipase did not raise the levels, rather exhibited lethal effects (Vick and Shipman, 1972). In monkeys venom or melittin injection also elevates the blood cortisol levels; but when the pituitary gland is removed the venom effect is eliminated. Thus, in these animals, bee venom appears to induce the release of cortisol, a major natural anti-inflammatory steroid, by its effects on the pituitary gland and its release of ACTH, the hormone that acts upon the adrenal glands to induce the ultimate release of cortisol (Vick *et al.,* 1972). Suggestive evidence in support of this hypothesis was provided by Knepel and Gerhards (1987) who reported that melittin induces release of ACTH from pituitary cells by the process of exocytosis.

The above reports indicate that bee venom acts in at least some situations to cause an anti-inflammatory, and therefore, an anti-arthritic effect. However, the results vary depending upon animal assays (in extreme, anti-inflammatory activity appears not to be induced in guinea pigs by venom [Hanson *et al.,* 1974]) and no uniform conclusions can be extrapolated to humans. Although this research can be difficult, discouraging, and the results not uniform across assays (Banks and Shipolini, 1986), the overall results are suggestive that venom is likely to act to reduce arthritic symptoms by enhancing the natural corticosteroid release rates, possibly with additional synergism by other means.

In comparison to the number of testimonials and uncontrolled clinical trials that have indicated that venom treatments improve arthritic conditions, the number of rigorous clinical trials are few. There are, however, some excellent studies that provide good evidence. Kroner *et al.,* (1938) clinically investigated the effects of injection of an undescribed venom preparation in 100 arthritic patients. The injection regimen was conservatively slow and tedious; nevertheless, the authors reported a 73% improvement in their patients. Although the authors had no controls in this investigation, the extremely high improvement rate is suggestive of effects other than psychological and non-related spontaneous remission. Cohen *et al.* (1942) examined the concept that injection of substances that cause "stinging" intradermally near arthritic joints might be of benefit. Their results indicate this to be the case as bee venom, 12.5% aqueous maganesium sulfate, and 10% aqueous glucose (which all cause stinging) all significantly improved arthritic symptoms. In the most carefully designed and thoroughly controlled human clinical trial to date, Steigerwaldt *et al.* (1966) reported an 84% improvement among their 50 venom treated patients and a 55% improvement among their 11 saline-treated control group. The study clearly illustrates the impact of the placebo effect and spontaneous random remission on the results: over half of the control patients, for unclear reasons (the simple fact that they were receiving attention might be important), showed improvement. Nevertheless, the differences between the treated and control groups were significant. When those patients with only "slight" improvement were eliminated, the bee venom effect was even greater — 66% of venom treated experienced moderate to very good improvement as compared to 27% for the controls (Steigerwaldt *et al.,* 1966).

A study that has occasionally been cited as evidence against any value of bee venom in the treatment of arthritis, is that of Hollander (1941). In that study the author injected an undescribed venom preparation of unknown venom concentration into 42 patients and a protein solution into 17 control patients. Because the methods are not detailed, little actual venom appeared to have been injected. No differences in the effects of the venom preparation and the controls were observed. The lack of rigor in this study and the uncertainty of the venom preparation prevent any meaningful conclusions from being drawn (see Broadman [1962] for critique).

A study of severely arthritic dogs focused on actual physical improvement of the animals as a result of venom injections. This study had the advantage that any psychological effects or subjective evaluations were eliminated. Two measures of activity were recorded in both the treated and control-saline-injected dogs in this doubly controlled experiment. The first was the level of plasma cortisol, a major endogenous anti-inflammatory agent of the body, and the second was actual voluntary movement (cage activity) by the dogs. Venom-treated dogs exhibited both prolonged increases in cortisol levels and, after a 7-day lag, significant and meaningful increase in cage activity (Vick *et al.,* 1975). Clearly in this trial, bee venom manifested pronounced anti-arthritic effect.

The lack of more rigorous scientific studies of the potential effect of bee venom in the treatment of arthritis is certainly not due to a lack of interest within society. Numerous newspaper articles, bee journal articles, and literature reports from around the world attest to that. Moreover, the interest cannot be simply attributed to in a recent poll a "small group of fanatic individuals" as evidenced by the astonishing figure that 83% of the American public believed that bee venom is a method of treating arthritis (Price *et al.,* 1983). Of all the suggested treatments, only aspirin scored a more favorable response. Rather than trying to determine the basis for this strong belief in bee venom and suggesting that more studies of bee venom be initiated, the authors simply lump bee venom in with such things as wearing copper bracelets (for which there is no evidence of value) and refer to them as a "quack and/or unproven remedies". Such comments in light of the available information are examples of either irresponsibility or ignorance. Rather than *de facto* restricting the use of bee venom for research and treatment of arthritis (none of the 300 respondents in the survey of Price *et al.,* had used bee venom), a disease that affects 12% of the American population (over 32 million people), the medical profession should be either encouraging research or at least not impeding it. In medical terms bee (and ant) venoms are considered "therapies of potential but unproven benefit" for treatment of arthritis and studies of "venoms are awaited with interest" (Panush and Longley, 1985).

Other Uses of Bee Venom

A small amount of bee venom is used annually for fundamental scientific investigations and for testing and treatment of individuals who are hypersensitive to bee stings. The amount of venom for these uses is very small (Mraz, 1982). Bee venom also serves as a raw material source for enzymes such as phospholipase A_2 and highly active peptides which are widely available through catalogs (*e.g.* Sigma Chemical Co.). Research other than that concerned with inflammation and arthritis is a continuing minor use for bee venom. Examples of the types of research focusing on bee venom include the apparent radioprotective value of bee venom in reducing the lethal effects of x-rays (Shipman and Cole, 1967; Ginsberg *et al.*, 1968).

Production and Value of Bee Venom

Until about 1963 production of quantities of bee venom was inhibited by lack of convenient techniques. The venom collected previous to that time was generally obtained by homogenizing entire sting apparatuses and collecting the extract, or by careful dissection and removal of the venom reservoir which was then either homogenized, or punctured and drained. Any of these variations were, at best, very slow and labor intensive with the consequence that venom was either unavailable commercially, expensive, or of unknown and impure quality. Collection of pure venom either from the sting tip or by careful dissection, isolation of the venom reservoirs, and venom draining into distilled water is still performed to a limited extent and is the best method to yield the purest venom.

The landmark change in bee venom collection occurred with the publication in 1963 of a technique involving simultaneous electrical stimulation of a large portion of the entire population of a bee colony (Benton *et al*, 1963). This technology had followed various electrical methods involving collecting venom from individual or a limited numbers of bees (Marcovic and Molnar, 1955; Palmer, 1961; O'Connor *et al.*, 1963) and, for the first time, allowed the collection of a gram of venom from about 20 colonies in a period of an hour or two. The procedure works best with large colonies and has the disadvantage of making the bees extremely excitable and defensive, with bees stinging people who are within several hundred meters of the affected colony (Morse and Benton, 1964; Mraz, 1983). Although similar electrical methods work to an extent with other stinging insects (Gillaspy and Grant, 1979; Pinnas *et al.*, 1977), this method appears to be a viable method mainly for honey bees.

Virtually all commercial honey bee venom is now collected by means

FIGURE 9. Venom collecting apparatus (*photo courtesy of Charles Mraz*).

similar to those of Benton *et al.*, (1963) (Figure 9). Exact production figures are unavailable, but small. The main venom producer in the U. S. had produced only about 3000 grams of venom over 30 years (Mraz, 1982) and, according to one advisory column, all the venom needed in the U. S. is produced by essentially one beekeeper operating part time (Morse, 1983). The price of bee venom on the U. S. open market varies greatly with a typical 1990 price of about \$100-200/g and much higher for smaller quantities. Many European and Asian producers are also in the market and their prices, as well as those direct from beekeepers, can be considerably lower. Unless bee venom finds accepted usage in treatment of arthritis, its market potential is unlikely to increase.

BEESWAX

Along with royal jelly and venom, beeswax is manufactured by honey bees themselves. Honey, propolis and pollen are botanical in origin and are collected and processed by the bees rather than being produced *de novo*. Beeswax, originally believed to have been collected from flowers or made from pollen, was discovered in 1744 by H. C. Hornbostel to be synthesized by four pairs of wax-secreting epidermal glands on the ventral side of worker abdomens (see Hepburn, 1986 for this and other historical issues). Only four natural resources (pollen, nectar, water and plant resins known as propolis) are necessary to support all the colony's activities. Floral and extrafloral sugars (derived from phloem sap in the case of honeydew) are converted into the energetically expensive commodity we know as beeswax. Beeswax is produced by quiescent bees about 14 days old and worked into intricate complex double-sided hexagonal comb nest architectures (Figure 10) that has allowed all the members of the true honey bee genus *Apis* to become the pre-eminent floral foragers of the insect world. Beeswax was probably a major factor in

FIGURE 10. (A) Free-hanging colony with its massive parallel beeswax combs; (B) close-up of waxen combs illustrating the hexagonal construction of the combs.

their evolution to the pinnacle of the eusocial insects. Beeswax is extremely valuable to bees because much nectar and/or honey is "forfeited", that is lost as potential food and converted into structural material from the colony's annual energy budget to produce the wax combs. For this reason, beeswax is removed, reshaped, molded and used over and over again within the nest. The combs are literally the nursery, walls, storage pantry, home, pharmacy and dance floor for the colony's myriad of occupants. Without their wax-secreting abilities evolved millions of years ago, honey bees would be largely unrecognizable to us.

Ecology of Bees and Wax

After a honey bee swarm has scouted and selected a protected tree or earthen cavity as a new homesite, the workers hang in festoons and begin converting sugars in the "communal" honey stomach into beeswax and forage for additional nectar allowing continued wax synthesis and honey storage. For *Apis mellifera* in the temperate U.S., a typical swarm may consist of 12,000 bees (Fell *et al.*, 1977). On average, each of the swarm bees contains about 35 mg of a 60-65% sugar solution (Combs, 1972). Thus, for a hypothetical swarm of this size, the reserve of concentrated nectar and honey produced in the parental nest and carried with the swarm worker bees amounts to an energy reserve of only about 275 grams of sugar, or ~4500 kj (~1100 kcal). The weight-to-weight efficiency of sugar metabolism and beeswax glandular synthesis ranges from about 0.05 to 0.06 under ideal conditions (Horstmann, 1965; Weiss, 1965; Hepburn, 1968; Buchmann and Schmidt, unpubl.). This 17 or 20:1 conversion ratio is due to the complex biochemistry involved in this energy-demanding process. One gram of beeswax can be worked into about 20 cm^2 two-sided comb surface area (Ribbands, 1953; Root, 1980). It requires about 55 grams of beeswax built into combs to store every kg of ripened and capped honey.

A honey bee colony, with approximately 30,000 or more workers, weighs from 2.4 to 3.6 kg. Their nest, the double-sided hexagonal combs, when empty comprise an area of approximately 2.5 square meters (double sided), weighing about 1.4 kg and containing 100,000 cells (Seeley and Morse, 1976) whose construction requires the metabolism of 25 kg of sugars. A standard Langstroth deep frame can hold 1.3 - 2.7 liters (1.8 - 3.8 kg) of honey, yet the wax necessary to produce these 7,100 cells weighs only 100 g, thus such honey: wax ratios in standard equipment vary from about 17.8 to 19.8:1 (Buchmann, unpubl.). An individual beeswax scale weighs only 1.1 mg (Figure 11) so 910,000 are necessary to equal one kilogram of beeswax. A staggering 9.1 x 10^5 beeswax scales are used in the construction of the 1.4 kg 2.5 m^2 comb present in an average honey bee nest.

Among the 22,000+ described bee species worldwide in the superfamily

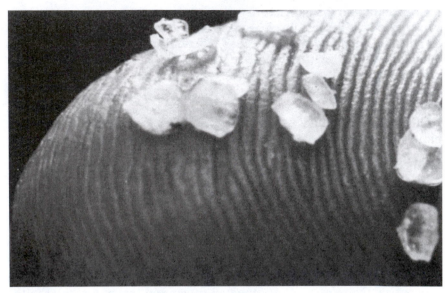

FIGURE 11. Wax scales secreted by worker honey bees.

Apoidea, wax is synthesized and used as a building material by only a few groups of mostly highly social colonial species. These include the worldwide genus of bumblebees (*Bombus* spp.) and their neotropical relatives the orchid bees (tribe Euglossini in the family Apidae) and the extremely diverse and successful 400+ pantropical stingless bees (*e.g. Melipona* and *Trigona* of the tribe Meliponini in the Apidae). These non-*Apis* groups often produce little wax and must adulterate it with plant resins or other materials to construct their brood cells and nest interiors. Bumblebee and Euglossine wax sources have not been exploited by humans. The stingless bees have colonies ranging in size from a few hundred to several thousand individuals with highly developed wax glands that, interestingly, are dorsally-located instead of ventrally as in *Apis*. These bees have been managed since antiquity, especially in the New World tropics by the Mayans and their descendants of the Yucatan peninsula of Mexico. The bees are kept in hollow log hives with end plugs and they are periodically raided for honey and waxen cerumen storage pots. The wax is saved and used in various handicrafts and artworks. Great quantities of beeswax are produced by the giant Asian honey bees (*A. dorsata* and *A. laboriosa*) in producing their open, massive single-comb nests. These bees migrate or abscond and abandon their waxen nests at great energetic cost during seasons of poor forage. Human pressures in the form of honey hunters possibly also contribute to the abandoning of nests. Nest sites are reoccupied year after year, but during the absence of the colony, predators eat the combs so the wax cannot be recycled by the bees. This abandonment of combs by *A dorsata* must be "repayed" by great foraging efforts to allow their subsequent annual reconstruction.

Historical Uses of Beeswax

Beeswax has played a special role in history and human affairs. The earliest honey hunters realized that the wax housed the sweet nutritious worker brood, bee bread and stored honey. It was something to be chewed up and often spat out as an indigestible residue. These earliest hunter-gatherers must have observed that when their smoking torches touched the fragile combs, they melted or caught fire. The likely origin of beeswax candles from such times with the serendipitous discovery of a twig or leaf wick in a pool of molten beeswax drippings are lost to antiquity. One of the earliest uses for beeswax was burning in lamps for illumination.

Actual candles survive from the first century A.D. from France, Denmark and other sites in addition to early paintings depicting candles (see Crane, 1983 for further details). Since very early times beeswax has played important roles in Roman Catholic services where it served as candles and votive offerings.

In the Old World tropics of northern Australia, aborigines some 30,000 to 50,000 years ago used beeswax then, as now, to sculpt waxen figurines for use in religious ceremonies (Crane, 1983). Beeswax, unlike honey, is often preserved in archaeological deposits and forms a rich record available to scholars. It has been put to use to fabricate many decorative objects and for casting metal even before the Bronze Age. Beeswax figurines survive from royal Egyptian tombs dating to 3400 B.P. Throughout much of history beeswax has played a role in commerce and business in the form of wax seals for documents and for many uses as artists' media, binding agents for pigments, or as adhesives. Encaustic (Greek for "burnt in") paintings in which beeswax was mixed with pigments was used in Egypt, and modern artists in southeast Asia use wax in the decorative art process known as batik. One of the most important and widespread uses of beeswax was for cire-perdue, or lost-wax casting (Crane, 1983). This method is extremely ancient, being known to the Summerians, in the Indus Valley and Egypt, and likely China probably prior to 3000 B.P. Many of the world's most famous bronze statues were produced using some variation of the lost-wax process.

Physical and Chemical Properties

Beeswax when first secreted by the wax glands appears as a translucent white ellipsoidal flake (Figure 11). Freshly constructed beeswax combs, prior to their use for food storage or larval growth, are similarly bright white. With storage or the first brood cycle, they become yellow to tan, and if several years old, can be almost brown-black in color. Fresh beeswax is soft to brittle with a slight balsamic taste. Its density is 0.95 to 0.96 with a melting range of 62-5°C.

It is insoluble in water, but quite soluble in organic solvents such as chloroform, benzene or ether.

Beeswax is a complex mixture of lipids and hydrocarbons. Only after 1960 and with the advent of gas-liquid chromatography was it possible to quantitatively and accurately resolve many questions regarding its complicated chemistry. Over 300 individual chemical components have thus far been identified from pure beeswax (Tulloch 1980). Beeswax from *A. mellifera* has been the subject of most chemical studies and consists primarily of monoesters (35%), followed by hydrocarbons (14%), diesters (14%), triesters (3%), hydroxymonoesters (4%), hydroxypolyesters (8%), free fatty acids (12%), acid esters (1%), acid polyesters (2%), free alcohol (1%) and unidentified (6%). However, only four individual components: C^{40} (6%), C^{46} (8%), and C^{48} (6%) monoesters and the C^{24} acid (6%) exceed 5% of the total amount (Table 7). Wax monoesters consist of simple straight-chained alcohols of even carbon numbers from 24 to 36 esterified with straight-chained fatty acids also having even numbers of carbons up to 36 carbons. Some of the fatty acids are hydroxy-fatty acids. Examples of such esters are triacontanyl hexadecanoate and hexacosanyl hexacosanoate. Such esters are admixed with hydrocarbons possessing odd-numbered straight chain lengths from 21 to 33 carbons. Unidentified components, propolis, pollen and pigments make up the remaining 6%.

Table 7. Gross Composition of beeswax[1]

Components	Number of components in fractions		
	Quantity (%)	Major	Minor
Hydrocarbons	14	10	66
Monoesters	35	10	10
Diesters	14	6	24
Triesters	3	5	20
Hydroxy monoesters	4	6	20
Hydroxy polyesters	8	5	20
Acid esters	1	7	20
Acid polyesters	2	5	20
Free acids	12	8	10
Free alcohols	1	5	?
Unidentified	6	7	?
Total	100	74	210

[1] Major components are those that comprise more than 1% of the fractions; for the minor components only estimates are given (after Tulloch, 1980).

Wax from bumblebees has a relatively low melting point range of 30 -40°C and is therefore mixed with pollen for use as a structural building material by bees. Wax from *Bombus rufocinctus* did not contain difunctional components and, therefore, no complex esters (Tulloch, 1970). Recently Blomquist *et al.,* (1985) studied the wax chemistry of two stingless bees *(Trigona buyssoni* and *T. atomaria)* and found this wax to be simpler but considerably different from that of *Apis mellifera.* This wax was 60 - 70% hydrocarbons and 26% monoesters with no diols, hydroxy acids, or their esters.

Modern Production and Uses

There is more demand for beeswax in the United States than can be satisfied domestically. The U.S. imports most of its beeswax from Mexico, Central and South America, the West Indies, Africa and China. Prices paid to beekeepers for beeswax are currently in the range of $1.00 - $1.20 per pound, with one pound of beeswax marketed for each 54 pounds of honey.

The single largest consumer of beeswax in the United States is the cosmetics and related industries. Here it is used in myriad products including facial beauty creams, ointments, lotions and lipsticks. The largest industry using beeswax as a raw material is the candle industry (Figure 12). Until recently, there was a great demand for pure or mostly pure beeswax candles mandated for use during religious services by the Roman Catholic church. The third largest user is the beekeeping industry itself for making into milled hexagonally-stamped beeswax foundation. Minor users are the pharmaceutical and dental industries where beeswax is used in salves, ointments, pill coatings, adhesive waxes, and for impression and base plate wax. Other minor uses of beeswax include uses in waterproofing materials, for floor and furniture polishes, for grinding/polishing optical lenses, and as a minor ingredient in certain adhesives, children's crayons, candy and chewing gum, inks, nursery grafting, musical instruments, ski and ironing wax, and wax for bow strings used in archery.

Prior to invention of the centrifugal honey extractor, the wax was separated from the honey by squeezing/straining and washing with water. In modern times, beekeepers obtain their wax from three primary sources: wax cappings, bits of burr comb scrapings from hive bodies and frames, and old combs which are to be recycled. The best grades of commercial beeswax are light yellow and come from fresh honey cappings. For each ton of extracted honey, only about 20 - 25 pounds of beeswax results. Various methods have been used to separate wax from honey, wax cocoons or brood (*e.g.* straining method, submerged brood chamber method, submerged sac method, solar wax melter, heated wax press, heated centrifugal method). The simplest method for a hobbyist with limited equipment or funds is probably to melt the

The Hive and the Honey Bee

FIGURE 12. Beeswax candles are made in a variety of shapes such as these floating roses enhancing the appeal of a party punch bowl (*candles courtesy of Ann Harman*).

combs in hot water and let the wax rise to the surface and harden. This material can then be strained and remelted/reformed into rectangular molds for shipment or sale to beekeeping supply houses or a commercial wax rendering center. Plans also are readily available for hobbyist beekeepers wishing to let solar power melt and separate their wax from "slum gum" in the form of solar extractors. These solar extractors work any time in the southern and southwestern United States, but only during summer in more northern locations.

ROYAL JELLY

Royal jelly is the hypophyrngeal glandular secretion of young workers that is placed in queen cells as food for larval queens. It is called royal jelly because it is the sole food of queen larvae, in contrast to the food of workers and drones which consist of larval jelly, or brood food. Brood food is initially similar to royal jelly (Haydak and Vivino, 1950; Aesencot and Lensky, 1988), but is modified by day 4 through the addition of pollen and honey (Matskuka *et al.,* 1973; Asencot and Lensky, 1988). Royal jelly is creamy white in appearance and viscous.

Composition

Numerous chemical analyses of royal jelly have been performed over the years. The analyses were initially hampered by the unusual composition of the material, including the extreme acidity (pH 3-4), and later by the complexity of the blend of components. Although there are reports to the contrary, it appears that the composition of royal jelly remains relatively constant over different colonies, bee strains, and time. Some variation can occur as a result of nutritional and age conditions of the secreting worker bees, care of collection/-storage of the royal jelly, and the methods used for analysis. The components with greatest variations in gross composition are probably the sugars, mainly because workers add different amounts of sugars to royal jelly depending upon the age of the queen larva (Asencot and Lensky, 1988). Trace minerals are also likely to vary more than other components because mineral compositions in pollen, the food eaten by workers and metabolically converted into royal jelly, vary dramatically with pollen species (Nation and Robinson, 1971).

Average composition of royal jelly is shown in Table 8. Water constitutes about two-thirds of the fresh weight, with protein and combined sugars the next major components. The proteins consist of six major proteins (Otani *et al.*, 1985) plus smaller peptides. Royal jelly sugar ratios are about the same as found in honey. The unique and chemically most interesting feature of royal jelly is its fatty acids. Unlike fatty acids of most animal and plant materials which consist mainly of triglycerides of fatty acids each having 14 -20 carbon atoms, royal jelly fatty acids are short-chained 8 and 10 carbon free fatty acids that are usually either hydroxy fatty acids or dicarboxylic acids (Table 9). These fatty acids are responsible for most of the noted biological properties of royal jelly. Royal jelly also contains about 1% total ash which is comprised mainly of potassium, but also contains high levels of the trace elements zinc, iron, copper, and manganese. These levels of minerals are generally lower than or the same as those found in pollen, though sodium appears to be slightly elevated (Nation and Robinson, 1971; Schmidt and Schmidt, 1984). The royal jelly mineral levels also appear to be about the same as the levels in adult bees with the exception of of manganese and calcium which are higher in the adults (Nation and Robinson, 1971).

Vitamins are present in royal jelly in varying amounts and in some cases their quantities need further investigation with modern analytical techniques. Levels of the B vitamins in royal jelly are generally high, especially for pantothenic acid (Haydak and Vivino, 1950). Otherwise vitamins in royal jelly are either lacking, or in low to very low levels. In particular, vitamin C levels are low (Melampy and Jones, 1939), and vitamins A and E are absent (or in very low levels) (Evans *et al.*, 1937; Melampy and Jones, 1939). Although these authors used animal nutrition assays, their work was excellent. Vitamins D and K are probably absent in royal jelly; which is not surprising since honey bees do not require any of these lipid-soluble vitamins in their diet (Herbert and Shimanuki, 1978).

Table 8. Typical composition of royal jelly[1]

Component	Quantity
Water	67%
Crude protein	12.5%
Total sugars	11%
Fructose	6.0%
Glucose	4.2%
Sucrose	.3%
Others	.5%
Total fatty acids	5%
Ash	1.0%
K	5500 $\mu g/g$
Mg	700 $\mu g/g$
Na	600 $\mu g/g$
Ca	300 $\mu g/g$
Zn	80 $\mu g/g$
Fe	30 $\mu g/g$
Cu	25 $\mu g/g$
Mn	7 $\mu g/g$
Undetermined	3.5%
Vitamins	
Thiamine	6 $\mu g/g$
Riboflavin	9 $\mu g/g$
Pyridoxine	3 $\mu g/g$
Niacin	50 $\mu g/g$
Pantothenic acid	100 $\mu g/g$
Inositol	100 $\mu g/g$
Biotin	1.5 $\mu g/g$
Folic acid	.2 $\mu g/g$
Vitamin C	4 $\mu g/g$
Vitamin A	~0
Vitamin D	0(?)
Vitamin E	~0
Vitamin K	~0
pH	3.8

[1] Values based on Evans *et al.*, 1937; Melampy and Jones, 1939; Haydak and Palmer, 1942; Haydak and Vivino, 1950; Nation and Robinson, 1971; Lercker *et al.*, 1982; Takenaka, 1984; Howe *et al.*, 1985; Asencot and Lensky, 1988; Karaali *et al.*, 1988.

The sterols in royal jelly, with the exception of 24-methylenecholesterol, are typical of plant sources (Table 9). Twenty-four-methylenecholesterol is an unusual sterol that often appears in very high concentrations in pollen (Svoboda *et al.*, 1983; Nes and Schmidt, 1988); hence, its appearance in royal jelly is not surprising.

Pharmacological and Medical Activities
A wide variety of health and cosmetic properties have been attributed to royal jelly over the years (discussed in Wells, 1976). Nevertheless, no well designed controlled (preferably double blind) medical studies have demonstrated therapeutic effects for royal jelly. This is likely because, other than the fatty acids, there is nothing unusual about the composition of royal jelly. The

Table 9. Typical composition of lipids in royal jelly[1]

Component	Quantity
Hydroxy fatty acids	
3-Hydroxyoctanoic acid	.3%
8-Hydroxyoctanoic acid	5.5%
3-Hydroxydecanoic acid	1.9%
10-Hydroxydecanoic acid	21.6%
(E)-10-Hydroxydec-2-enoic acid	31.8%
3,10-Dihydroxydecanoic acid	1.8%
Dicarboxylic acids	
Octandioic acid	.4%
Decandioic acid	1.4%
Dec-2-endioic acid	2.7%
Simple fatty acids	
Octanoic acid	.1%
Others	
p-Hydroxybenzoic acid	trace
Gluconic acid	24.0%
Undetermined & others	8.4%
Sterols	
24-methylene cholesterol	50 μg/g
β-Stigmasterol	20 μg/g
Δ⁵-Avenasterol	15 μg/g
Cholesterol	10 μg/g
Stigmasterol	2 μg/g
Δ⁷-Avenasterol	.8 μg/g
Testesterone	.012 μg/g

[1] Values based on Takenaka, 1984; Brown *et al.*, 1961; Lecker *et al.*, 1982; Vittek and Slomiany, 1984.

most noteworthy non-lipid component is pantothenic acid which is present in high levels. It is not uncommon to hear that royal jelly has some gonadotropic or otherwise sex-enhancing action, an opinion that is emotionally pleasing since royal jelly is the sole food of one of the most reproductive organisms on earth — the queen honey bee. Any validity to these ideas was thoroughly dispelled over 50 years ago in excellent studies demonstrating that royal jelly had no gonadotropic effects on female rats (Melampy and Stanley, 1940) and contains no nutritionally active levels of vitamin E, the fat soluble vitamin first recognized for its crucial role in reproduction (Evans *et al.*, 1937). Similarly, royal jelly contains the male hormone testosterone, but only at levels of .012 $\mu g/g$ (Vittek and Slomiany, 1984), whereas adult men produce 250,000 to 1 million times as much daily and have almost as high levels circulating in their blood plasma at any given moment as is in royal jelly (Goodman and Gilman, 1975). Obviously any of the hormone contributed by royal jelly would have a negligible effect on humans. Gibberellic acid is a widespread plant growth hormone; it, like the other hypothesized components with sex related activities is not existent in royal jelly (Tucker and Blum, 1972).

Royal jelly is not toxic even when injected in high doses (Hashimoto *et al.*, 1977), or mutagenic (Tamura *et al.*, 1985) and upon intravenous injection is slightly vasodilating. This vasodilating activity is, however, caused by the presence of acetylcholine in royal jelly (Shinoda *et al.*, 1978) and is likely of no meaningful consequence.

The above paragraphs indicate that the potential beneficial properties of royal jelly need to be addressed in responsible ways and through meaningful research. If not, the public and marketplace are not likely to accept royal jelly as a legitimate product and a bad impression could be encouraged (*Bee World,* 1962). The unique fatty acids of royal jelly give it some unusual, and potentially useful properties. Royal jelly and its fatty acids, particularly 10-hydroxydecenoic acid, are strongly antibacterial (McCleskey and Melampy, 1939; Blum *et al.*, 1959; Yatsunami and Echigo, 1985); having an activity toward *Micrococcus pyrogens* one-quarter that of penicillin (Blum *et al.*, 1959). It is also antifungal (Blum *et al.*, 1959), inhibits the germination of pollen grains (Iwanami *et al.*, 1979), and kills some types of tumor cells (Townsend *et al.*, 1960). Unfortunately, almost all of these activities disappear when the fatty acids are neutralized by raising the pH above 5.6 or so (McCleskey and Melampy, 1939; Blum *et al.*, 1959; Townsend *et al.*, 1960). Thus, royal jelly or its eight and ten carbon hydroxy fatty acids have no expected therapeutic potential when injected into the blood, muscle, or peritoneal cavity (all of which will neutralize it to a pH of 7.4), or when consumed (with the possible exception of activities which could take place in the acid environment of the stomach). A report (Cho, 1977) that consumption of royal jelly can reduce plasma levels of cholesterol and triglycerides (conceivably caused by the royal jelly fatty acids) merits further investigation. The most promising antibacterial and textural potential for royal jelly are as a topical cream with both cosmetic and antimicrobial action. There are a number of reports concerning the wound-healing, skin cleansing, and tissue repairing properties of royal jelly (Wells, 1976) and research in these areas needs to be vigorously encouraged. Since both honey and propolis (see earlier section) are also strongly antibiotic, antiviral, and antifungal, cosmetic and clinical testing of combinations of these components would be in order.

Royal jelly has potential as a dietary ingredient in both human foods and for animals. It has been successfully used as a diet for the laboratory rearing of carpenter ant queens (Furukawa, 1982) and is highly digestible for mammals with an approximate digestibility (AD) of 81% in rats, which is better than that for beef (Melampy and Jones, 1939). It is widely used in Asia as an additive to food vials containing pollen extract and honey and has a large following. The products (Figure 13) are attractive and have a pleasant taste (J.O.S., personal observations).

Markets and Market Potential for Royal Jelly
The main markets in North America and Europe for royal jelly are the cosmetic industry which uses royal jelly in moisturizing and skin cream as well as a variety of other products, and the health food market. The antibacterial,

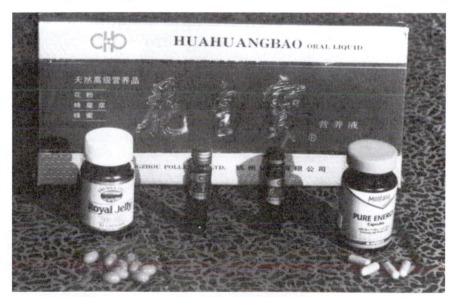

Figure 13. Two royal jelly products for human consumption: "oral liquid" (a combination of honey, pollen and royal jelly) and capsules.

cleansing and textural properties of royal jelly likely account for its popularity in cosmetics. In the health food market royal jelly is often added as a supplement to other ingredients and vitamins which can be taken either as capsules, as parts of beverages, in confectionaries, or mixed with honey as a spread.

Royal jelly has a much larger market in Asia than in North America. In Asia it is commonly found in products including cosmetics, food supplements, and beverages and is used in the medical profession (Inoue and Inoue, 1964).

Production of Royal Jelly

Royal jelly is usually produced in colonies maintained for that purpose. The queen is removed and a frame containing artificial queen cells, each with a 12 - 36 hr old worker larva is inserted. Three days later the frame is removed, the larvae discarded, and the royal jelly collected either with a wooden spoon or a soft suction tube. A good queen cup will yield about 200 - 300 mg of royal jelly. Once collected the royal jelly can be stored in a tight container in the refrigerator for several months, frozen, or freeze dried until used.

In the U.S.A., royal jelly can be purchased for about $6.00 per ounce (28 g) in 2 ounce containers from a variety of sources advertised in the bee journals. The largest producers of royal jelly are the Chinese, Japanese, and Koreans. Annual production levels in China in 1987 were 500 - 800 tons (Wang, 1989). Japan is both a large producer and importer of royal jelly.

BEE BROOD AND ADULTS AS FOOD FOR ANIMALS AND HUMANS

Bee brood is probably the least recognized of all bee products in terms of its use by man as a food or potential material of value. Yet, bee brood was probably the product of second greatest importance, after honey, to early human cultures. Probably most hunter-gatherer societies that had access to any of the honey bee or populous stingless bee species engaged in honey hunting. After finding a honey bee colony and driving off the bees, usually with fire and smoke, the honey hunters typically ate much of the brood comb and honey on the spot. The rest was taken back to the village. Some of the many cultures that are currently proficient honey hunters include the Tongwe of Tanzania (Kakeya, 1976), Shabanese of Zaire (Parent *et al.*, 1978) Kayapo of Brazil (Posey, 1983), Ache of Paraguay (Hill *et al.*, 1984), and the Efe of Zaire (Bailey, 1989). Honey and brood often comprise a considerable portion of the people's yearly calories and protein (Hill *et al.*, 1984; Bailey, 1989) and their frantic craving for honey bee products sends men on long hunting trips in search of colonies (Posey, 1983). In parts of Asia bee brood is commonly sold along with honey in local markets (Figure 14).

In North America there have been only limited investigations into the use

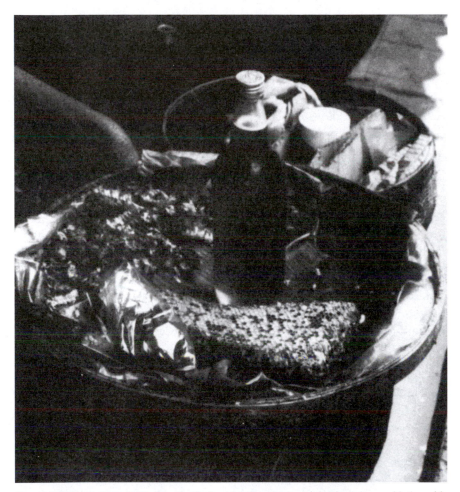

FIGURE 14. Typical market in Thailand where bee combs containing brood are sold alongside honey (*Mike Burgett photo*).

of bee brood as human food. Hocking and Matsumura (1960) pointed out that at that time there were 132 tons of wasted bee brood from colonies destroyed before winter in just the Canadian Prairie Provinces of Manitoba, Saskatchewan, and Alberta. They sought to develop a market for this brood by preparing it in a variety of ways for a volunteer tasting panel. The most acceptable forms were deep fat frying, which imparted a flavor like walnuts, pork cracklings or sunflower seeds, and baked. In the case of their volunteers an initial prejudice proved easier to overcome than expected. Unfortunately, these products were not really suitable for general marketing. Burgett (1990) described "bakuti", a bee brood product produced in Nepal by squeezing brood combs in a woven fabric bag and collecting the juice which is then gently heated and stirred. After 5 minutes a product having a texture similar to scrambled eggs is produced. In taste tests, an American version of bakuti was accepted by 85% of volunteers (Burgett, 1990). In Africa, the people of Shaba are very fond of grilled, salted larvae that they eat as a side dish (Parent *et al.*, 1978). With proper preparation and marketing, it is likely that bee brood could become an interesting high-priced gourmet delicacy.

Bee brood has excellent nutritional properties. It is richer in protein and lower in fat than beef and it has no crunchy cuticle like most insects eaten for food (*e.g.* ants, grasshoppers, termites, beetle grubs). If prepupae and young pupae are used, there is not even any pollen or brood food in the digestive system. The exact nutritional composition of bee brood needs further investigation, but it appears that brood has high quality food value. One particular feature of bee brood that needs clarification is its vitamin D and A content. Preliminary studies (Hocking and Matsumura, 1960) suggested that bee brood was exceedingly rich in both of these vitamins, being 9 - 68 times richer in vitamin D than cod liver oil. These high levels have been used as as basis for statements that eating too much bee brood could cause toxic overdoses of vitamin D. Such "toxic" levels of vitamin D are highly unlikely for several reasons: 1) the assays used to measure the vitamin are subject to false elevated readings due to interfering substances; 2) honey bees have no requirement for either vitamin A or D (Herbert and Shimanuki, 1978) and pollen, their food, contains no vitamins A or D (Schmidt and Schmidt, 1984), so why should they have such high levels; and 3) people like the Efe regularly consume large quantities of brood and no hypervitaminosis from vitamin D has been reported.

The excellent nutritional value of bee brood for non-human animals has been demonstrated. Songbirds, in particular, thrive on bee larvae and pupae (Gary *et al.*, 1961; Guss, 1967; Lanyon and Lanyon, 1969). In a particularly detailed study of hand rearing songbirds from eggs, bee brood turned out to be the crucial element in the diet. The birds not only developed very well when the diet included the brood, but they also ate the brood more avidly than any of the other food provided. The authors (Lanyon and Lanyon, 1969) concluded that bee brood was also superior to any other tested insects because it was not only easier to obtain, but was soft, palatable, and nutritious. Other insects were unsatisfactory for a variety of reasons: mealworms because they were too chitinous; moths because they were too low in nutrition; wax moth larvae because they were too time consuming to rear; and fly maggots because they were too disagreeable to rear.

A potential and very valuable use for bee brood is the rearing of beneficial insects for use in integrated pest management. Many specialized predators such as lady bird beetles (Coccinellidae) and lacewings (Chrysopidae) are exceedingly difficult to rear in laboratory culture. Drone powder made from drone larvae and pupae has been successfully used in rearing lady bird beetle larvae (Matsuka and Takahashi, 1977; Matsuka *et al.*, 1982) including the vedalia beetle which is an extremely important predator of citrus cottony cushion scales. Without drone powder, the vedalia could not be reared in large groups because of cannibalism (Okada and Matsuka, 1973). Lacewings have been also successfully reared for several generations with drone powder (Hasegawa *et al.*, 1983). With the ever increasing resistance of pests to insecticides and ever increasing awareness of the hazards of chemical insecticides, biological control agents are becoming more important to agriculture. Honey bee brood could easily play a crucial role in rearing insectivorous predators and become an important added economic product of the hive.

Production and Marketing of Bee Brood

Perhaps the most serious disadvantage of developing a market for bee brood as a product is the difficulty of non destructively obtaining larvae and pupae. Various methods for extracting brood have been described, but most are labor intensive or destructive. Methods have included the swing - impact method in which a frame is swung so that its top bar impacts a wooden block and larvae are thrown out (Hocking and Matsumura, 1960). This method is inefficient and often damages comb, sealed brood, and frames. Bird breeders have sometimes simply frozen whole frames and removed still frozen pupae and larvae as needed (Guss, 1967), a method that does not lend itself well to large scale bee industry production and marketing. Gary *et al.*, (1961) effectively removed larvae and uncapped pupae with water from a hose. This technique is fast and effective, but injures many larvae and pupae. We have found the best method of removing larvae from the comb is with a stream of water. With this method brood of particular sizes or ages can be selectively removed without injury. The method involves use of a common plastic laboratory "wash bottle" (Figure 15) which allows the operator to direct a fine stream of water into individual cells. With a little practice one can remove all of the larvae (or just the largest larvae) from a frame and within ten minutes collect a hundred grams of larvae. The frame can then be returned unharmed to the hive (Thoenes and Schmidt, 1990).

Because no established markets for bee brood exist, marketing techniques and networks need to be established. The potential demand for bee brood could be easily expanded to include the pet and zoo animal food markets, fish breeding facilities, specialty human gourmet food markets and, insect rearing operations. The pet food market is an obvious local, regional, or national market awaiting creative beekeepers. Birds are expensive, colorful and delicate pets whose owners are very interested in maintaining healthy colorful

FIGURE 15. Rapid, non-destructive method of quickly removing large numbers of larvae from combs: (A) water stream from wash bottles floats out larvae: (B) larvae separated into size classes by pouring through different sized screens.

plumage and alert singing animals. Freeze-dried bee brood or food containing lyophilized bee brood that is formulated into sizes appropriate for the bird should find a good market. This potential market, like the others mentioned, probably requires only creative entrepreneurship, good product research development, and possibly good timing to develop.

Adult Bees

Adult bees are not likely to have market potential without processing and modification. Although adult bees are high in protein and low in fat (Ryan and Jalen, 1982), they have much indigestible chitinous cuticle and have an aversive taste. We are aware of no known cases of humans routinely eating adults, which are considered unpalatable (Hocking and Matsumura, 1960), and are low on the preference scale of animals such as lizards (Schmidt, unpublished). Part of the reason adults are unpalatable is that they contain unpleasant tasting mandibular gland, Nasonov gland, and sting chamber pheromones, and venom. The mandibular gland pheromone 2-heptanone of foraging worker bees imparts a strong "ketonic", almost paint-thinner taste, to the heads. The sting chamber alarm pheromones, including isoamyl acetate and a variety of other compounds, and the venom reservoir impart a disagreeable long-lasting pungent flavor and a bitter corrosive flavor, respectively, to the abdomens of foragers. The thorax has no inherently disagreeable flavor, but contains the bulky, crunchy, indigestible wings and legs. Drones, because they have no pheromone-rich glands or venom are much less aversive than workers (Schmidt, unpublished). Thus, to make adult worker bees suitable as animal food they likely would have to be thoroughly dried to drive off the volatile pheromones and even then might have to be mixed with other ingredients to dilute the effects of the venom. Drones, which are highly palatable to some birds (Schmidt *et al*, 1990), could conceivably be separated and used as food for large birds such as quail and pheasants.

REFERENCES

References for Pollen

Adams, C.F. (1975). Nutritive Value of American Foods in Common Units. USDA Agric. Handb. No. 456. Washington DC: Government Printing Office.

Anon. (1984). Presidential pollen. *Time Mag.* 1984 (Apr. 30): 55.

Ask-Upmark, E. (1967). Prostatitis and its treatment. *Acta Med. Scand*, *181*:355-57.

Baker, H.G. and I. Baker. (1979). Starch in angiosperm pollen grains and its evolutionary significance. *Amer. J. Bot.* 591-600.

Bell, R.R., E.J. Thornber, J.L.L. Seet, M.T. Groves, N.R. Ho and D.T. Bell. (1983). Composition and protein quality of honeybee-collected pollen of *Eucalyptus calophylla. J. Nutr. 113*:2479-84.

Benson, K. (1984). Cleaning and handling pollen. *Amer. Bee J. 124*:301-05.

Bock, S.A. and F.M. Atkins. (1989). Fifteen years of double-blind placebo-controlled food challenges. *J. Allergy Clin. Immunol. 83(1)*:238.

Brice, A.T., K.H. Dahl and C.R. Grau. (1989). Pollen digestibility by hummingbirds and psittacines. *Condor 91*:681-88.

Buchmann, S.L. (1986). Vibratile Pollination in *Solanum* and *Lycopersicon*: A Look at Pollen Chemistry. *In:* Solanaceae Biology and Systematics (W.G. D'Arcy, *ed.*). P. 237-252. New York: Columbia University Press.

Cohen, S.H., J.W. Yunginger, N. Rosenberg and J.N. Fink (1979). Acute allergic reaction after composite pollen ingestion. *J. Allergy Clin. Immunol. 64*:270-74.

Colleen, S., P.-A. Mardh and A. Schytz. (1975). Magnesium and zinc in seminal fluid of healthy males and patients with non-acute prostatitis with and without gonorrhoea. *Scand. J. Urol. Neiphrol. 9*:192-97.

Costantini, F. and G. Riciardelli d'Albore. (1971). Pollen as an additive to the chicken diet. 23rd International Apicultural Congress. p. 539-42.

Crane, E. (1983). The Archaeology of Beekeeping. London: Duckworth.

Davies, P.M. and I.R. Dadour. (1989). A cost of mating by *Requena verticalis* (Orthoptera: Tettigoniidae). *Econ. Entomol. 14*:467-69.

Dedic, G.A. and O.G. Koch. (1957). Zur Kenntinis des Spurrenelementgehaltes von Pollen. *Phyton 9*:65-67.

Denis, L.J. (1966). Chronic prostatitis. *Acta Urol Belg. 34*:49-55.

Devlin, J. (1981). Honeybee Pollen and the New You. Scottsdale, Ariz.: Busiest Bee Publ.

Duhoux, E. (1982). Mechanism of exine rupture of hydrated taxoid type of pollen. *Grana 21*:1-7.

Durham, O.C. (1951). The pollen harvest. *J. Econ. Bot. 5*:211-54.

Echigo, T., T. Takenaka and K. Yatsunami. (1986). Comparative studies on chemical composition of honey, royal jelly and pollen loads. *Bull. Fac. Agric. Tamagawa Univ. no. 26*:1-8.

Eischen, F.A. and W.A. Foster. (1983). Life span and fecundity of adult female *Aedes aegypti* (Diptera: Culicidae) fed aqueous extracts of pollen. *Ann. Entomol. Soc. Amer. 76*:661-63.

Ender, C., M.Q. Li, B. Povh, R. Nobiling, H.-D Reiss and K. Traxel. (1983). Demonstration of polar zinc distribution in pollen tubes of *Lilium longiflorum* with the Heidelberg proton microprobe. *Protoplasma 116*:201-03.

Feinberg, S.M., F.L. Roran, M.R. Lichtenstein, E. Padnos, B.Z. Rappaport, J. Sheldon and M. Zeller. (1940). Oral pollen therapy in ragweed pollinosis. *J. Amer. Med. Assoc. 115*:23-29.

Hayashi, A.U., J. Mitsui, H. Yamakawa *et al.* (1986). Clinical evaluation of Cernilton in benign prostatic hypertrophy. *Hinyokika Kiyo 32*:135-41.

Herbert, E.W. Jr. and H. Shimanuki. (1978). Chemical composition and nutritive value of bee-collected and bee-stored pollen. *Apidologie 9*:33-40.

Herbert, E.W. Jr., J.T. Vanderslice and D.J. Higgs. (1985). Effect of dietary vitamin C levels on the rate of brood production of free-flying and confined colonies of honey bees. *Apidologie. 16*:385-94.

Hernuss, P., E. Müller-Tyl, H. Salzer, H. Sinzinger, L. Wicke, T. Prey and L. Reisinger. (1975) Pollendiät als Adjuvans der Strahlentherapie gynäkologischer Karzinoma. *Strahlentherapie 150*:500-06.

Ibrahim, S.H. (1974). Composition of pollen gathered by honeybees from some major sources. *Agric. Res. Rev., Cairo 52*:121-23.

Kirk, W.D.J. (1985). Pollen-feeding and the host specificity and fecundity of flower thrips (Thysanoptera). *Ecol. Entomol. 10*:281-89.

Kitzes, G., H.A. Schuette and C.A. Elvehjem. (1943). The B vitamins in honey. *J. Nutr.* 26:241-50.

Kleinschmidt, G.J. and A.C. Kondos. (1976). The influence of crude protein levels on colony production. *Australasian Beekeeper* 78:36-39.

Krinsky, N.I. (1988). The evidence for the role of carotenes in preventive health. *Clin Nutr.* 7:107-12.

Kvist, U., S. Kjellberg, L. Björndahl, M. Hammer and G.M. Roomans. (1988). Zinc in sperm chromatin and chromatin stability in fertile men and men in barren unions. *Scand. J. Urol. Nephrol.* 22:1-6.

Lehnherr, V., P. Lavanchy and M. Wille. (1979). Pollensammeln 1978: 5. Eiweiss- und Aminosäuregehalt einiger häufiger Pollenarten. *Schweiz. Bienenzeit.* 102:482-88.

Leppla, N.C., L.N. Standifer and E.H. Erickson, Jr. (1974). Culturing larvae of blister beetles on diets containing different pollens collected by honeybees. *J. Apic. Res.* 13:243-47.

Levin, M.D. and G.M. Loper. (1984). Factors affecting pollen trap efficiency. *Amer. Bee. J.* 124:721-23.

Lin, F.L., T.R. Vaughan, M.L. Vandewalker and R.W. Weber. (1989). Hypereosinophilia, neurologic, and gastrointestinal symptoms after bee-pollen ingestion. *J. Allergy Clin. Immunol.* 83:793-96.

Linscheer, W.G. and A.J. Vergroesen. (1988). Lipids. In: Modern Nutrition in Health and Disease, 7 ed. (M.E. Shils and V.R. Young, eds.). p 72-107. Philadelphia: Lea & Febiger.

Mansfield, L.E. and G.B. Goldstein. (1981). Anaphylactic reaction after ingestion of local bee pollen. *Ann. Allergy* 47:154-56.

Maurer, M.L. and M.B. Strauss. (1961). A new oral treatment for ragweed hay fever. *J. Allergy* 32:343-47.

McLellan, A.R. (1977). Minerals, carbohydrates and amino acids of pollens from some woody and herbaceous plants. *Ann. Bot.* 41:1225-32.

Nation, J.L. and F.A. Robinson. (1971). Concentration of some major and trace elements in honeybees, royal jelly and pollens, determined by atomic absorption spectrophotometry. *J. Apic. Res.* 10:35-43.

Nielson *et al.* (1955). Investigations on the chemical composition of pollen from some plants. *Acta Chemica Scandinavica* 9(1955):1100-1106.

Noyes, C.E. Jr. (1961). The use of cernitin, an extract of organic pollen, to increase body weight and to increase resistance toward infections. Unpublished report of Svenssons Boktryckeri, Bastad, Sweden.

O'Rourke, M.K. and S.L. Buchmann. (1990). Standardized pollen analytical techniques used for various bee collected samples. *Environ. Entomol.* (in Press).

Pearson, P.B. (1942). Pantothenic acid content of pollen. *Proc. Soc. Exp. Biol. Med.* 51:291-92.

Pieroni, R.E., B.E. Phillipson, D.L. Lentz, L. Wittlake and F.C. Gabrielson. (1982). "Miracle" bee pollen: don't let your patients get stung!*J. Med. Assoc. State Alabama* 51:11, 15-16

Rabie, A.L., J.D. Wells and L.K. Dent. (1983). The nitrogen content of pollen protein. *J. Apic. Res.* 22:119-23.

Rayner, C.J. and D.F. Langridge. (1985). Amino acids in bee-collected pollens from Australian indigenous and exotic plants. *Aust. J. Exp. Agric.* 25:722-26.

Recommended Dietary Allowances. (1989). 10th Edition. Washington: *National Academy Press.*

Robinson, W. (1948). Delay in the appearance of palpable mammary tumors in C3H mice following the ingestion of pollenized food. *J. Nat. Cancer Inst. 9:*119-23.

Rosenthal, C. (1967). Chemical composition and importance of pollen. *Apicultura (Bucharest) 20:*11-15. *(Chem. Abst. 69:*3125).

Ruiz Abad, L. (1975). Effect of introducing pollen in the diet of rodents. In: The Hive Products — Food, Health and Beauty (Int. Sym. Apitherapy, Madrid 1974). p 142-46. Bucharest: Apimondia.

Salajan, G. (1970). *Inst. Agron. "Dr. Petru Groza" Luc. Stiit. Ser. Zootech. 26:165*(cited in Stanley and Linskens, 1974, p 113).

Sasagawa, H. (1982). Some experiments on development and nutrition of two spotted cricket, *Gryllus bimaculatus* De Geer with special reference to effects of honeybees' products. *Honeybee Science 3:*135-36.

Schmalzel, R. (1980). The Diet Breadth of *Apis* (Hymenoptera: Apidae). Unp. M. S. Thesis, University of Arizona.

Schmidt, J.O. (1984). Feeding preferences of *Apis mellifera* L. (Hymenoptera: Apidae): individual versus mixed pollen species. *J. Kansas Entomol. Soc. 57:*323-27.

Schmidt, J.O. and P.J. Schmidt. (1984). Pollen digestability and its potential nutritional value. *Gleanings in Bee Culture 112:*320-322.

Schmidt, J.O. (1985). Phagostimulants in pollen. *J. Apic. Res. 24:*107-14.

Schmidt, J.O. and B.E. Johnson. (1984). Pollen feeding preference of *Apis mellifera,* a polylectic bee. *Southwest. Entomol. 9:*41-47.

Schmidt, J.O., S.C. Thoenes and M.D. Levin. (1987). Survival of honey bees, *Apis mellifera* (Hymenoptera: Apidae), fed various pollen sources *Ann. Entomol. Soc. Amer. 80:*176-83.

Schmidt, J.O., S.L. Buchmann and M. Glaiim. (1989). The nutritional value of *Typha latifolia* pollen for bees. *J. Apic. Res. 28:*155-65.

Schmidt, P.J., J.O. Schmidt and C.W. Weber (1984) Mesquite pollen as a dietary protein source for mice. *Nutr. Reports Intl. 30:*513-22.

Shaparew, V. (1985). Pollen trap—design optimization. *Amer. Bee J. 125:*173-75.

Smith, R.B. and T.P. Mommsen. (1984). Pollen feeding in an orb-weaving spider. *Science 226:*1330-32.

Solberg, Y and G. Remedios. (1980). Chemical composition of pure and bee-collected pollen. *Sci. Reports Agric. Univ. Norway 59(18):*1-12.

Squillace, D.L., K.G. Sweeney, R.T. Jones, J.W. Yuninger and R.M. Helm. (1988) Fatal food-induced anaphylaxis. *J. Allergy Clin. Immunol. 81:*239.

Stanley, R.G. and H.F. Linskens. (1974). Pollen: Biology, Biochemistry, Management. Berlin: Springer-Verlag.

Steben, R.E. and P. Boudreaux. (1978). The effects of pollen and protein extracts on selected blood factors and performance of athletes. *J. Sports Med. Phys. Fitness. 18:*221-26.

Tamas, V., G. Salajan and C. Bodea. (1970). Effectul polenului de porumb din hrana gainilor asupra pigmentatiei galbenusului de ou. *Stud. Cerc. Biochem. 13:*423-29.

Thorsons Eds. (1989). The Healing Power of Pollen with Propolis and Royal Jelly. Wellingborough, Northhampshire: Thorsons Publ. Group.

Todd, F.E. and O. Bretherick. (1942). The composition of pollens. *J. Econ. Entomol. 35:*312-17.

Togasawa, Y., T. Katsumata, M. Fukada and T. Motoi. 1967. Biochemical studies on pollen. VII. Vitamins of pollen. *Nippon Nagei Kaishi 41:*184-88.

Tu, L.-C., J.H. Strimas and S.L. Bahna. (1989). Estimated magnitude of food allergy by U.S. Physicians. *Ann. Allergy 63*:261.

Vivino, A.E. and L.S. Palmer. (1944). The chemical composition and nutritional value of pollens collected by bees. *Arch. Biochem. 4:*129-36.

Waller, G.D. (1980). A modification of the O. A. C. pollen trap. *Amer. Bee J. 120*:119-21.

Wang, W. (1989). The development and utilization of the resources of bee-pollen in China. *Proc. Intl. Congr. Apic. (Apimondia) 32*:239.

Wang, W., J. Hu and J. Cheng. (1984). Biological effect of honey bee pollen: I. radioprotective activity on hemotopoitic tissues of irradiated mice. *J. Hangzhou Univ. 11*:231-240.

Wang, W., J. Hu and L. Xu. (1987). Study of the digestibility and absorptibility of unbroken-walled pollen. *Food Science* (Beijing) 1987 *(10)*:1-4.

Weaver, N. and K.A. Kuiken. (1951). Quantitative analysis of the essential amino acids of royal jelly and some pollens. *J. Econ. Entomol. 44*:635-38.

Webster, T.C., R.W. Thorp, D. Briggs, J. Skinner and T. Parisian. (1985). Effects of pollen traps on honey bee (Hymenoptera: Apidae) foraging and brood rearing during almond and prune pollination. *Environ. Entomol. 14*:683-86.

Weygand, F. and H. Hofmann. (1950). Polleninhaltsstoffe, I. Mitteil: Zucker, Folinsäure und Ascorbinsäure. *Chem. Ber. 83*:405-13.

Wille, H., M. Wille, V. Kilchenmann, A. Imdorf and G. Bühlmann. (1985). Pollenernte und Massenwechsel von drei *Apis mellifera*-Völkern auf demselben Bienenstand in zwei aufei-nanderfolgenden Jahren. *Rev. Suisse Zool. 92*:897-914.

Willie, H. *et al.* (1985). Beziehung . . . *Bull. Societe Ent. Suisse 58*:205-214.

Youssef, A.M., R.S. Farag, M.A. Ewies and S.M.A. El-Shakaa. (1978). Chemical studies on pollen collected by honeybees in Giza region, Egypt. *J. Apic. Res. 17*:110-13.

References for Propolis

Apimondia. (1978). A Remarkable Hive Product: Propolis. Apimondia: Bucharest.

Aviado, D.M., L.V. Bacalzo, Jr. and M.A. Belej. (1974). Prevention of acute pulmonary insufficiency by eriodictyol. *J. Pharm. Exp. Therap. 189*:157-66.

Bankova, V.S., S.S. Popov and N.L. Marekov. (1982). High-performance liquid chromato-graphic analysis of flavonoids from propolis. *J. Chromatogr. 242*:135-43.

Bankova. V.S., S.S. Popov and N.L. Marekov. (1983). A study on flavonoids of propolis. *J. Natural Prod. 46*:471-74.

Bankova, V., A. Dyulgerov, S. Popov and N. Marekov. (1987). A GC/MS study of the propolis phenolic constituents. *Z. Naturforsch. 42C:*147-51

Bankova, V.S., S.S. Popov and N.L. Marekov. (1989). Isopentenyl cinnamates from poplar buds and propolis. *Phytochem. 28*:871-73.

Budavari, S. (ed.). (1989). The Merck Index. Merck & Co.: Rahway, NJ.

Bunney, M.H. (1968). Contact dermatitis in beekeepers due to propolis (bee glue). *Br. J. Dermat. 80*:17-23.

Cody, V., E. Middleton, Jr., J.B. Harborne and A. Beretz (eds.) (1988). Plant Flavonoids in Biology and Medicine II Biochemical, Cellular, and Medicinal Properties. Alan R. Liss: New York.

Farkas, L., M. Gabor and F. Kallay (eds.). (1986). Flavonoids and Bioflavonoids, 1985, Elsevier: Amsterdam.

Ghisalberti, E.L. (1979). Propolis: a review. *Bee World 60*:59-84.

Ghisalberti, E.L., P.R. Jefferies, R. Lanteri and J. Mathison. (1978). Constituents of propolis. *Experientia 34*:157-58.

Grange, J.M. and R.W. Davey. (1990). Antibacterial properties of propolis (bee glue). *J. Roy. Soc. Med. 83*:159-60.

Grunberger, D., R. Ganerjee, K. Eisinger, E.M. Oltz, L. Efros, M. Caldwell, V. Estevez and K. Nakanishi. (1988). Preferential cytotoxicity on tumor cells by caffeic acid phenethyl ester isolated from propolis. *Experientia 44*:230-32.

Hausen, B.M. and E. Wollenweber. (1988). Propolis allergy (III). sensitization studies with minor constituents. *Contact Dermatitis 19*:296-303.

Hausen, B.M., E. Wollenweber, H. Senff and B. Post. (1987). Propolis allergy (I). origin, properties, usage and literature review. *Contact Dermatitis 17*:163-70.

Haydak, M.H. (1953). Propolis. Report of the Iowa State Apiarist, pp 74-87. State of Iowa Publ.: Des Moines.

Hill, R. (1977). Propolis the Natural Antibiotic. Thorsons Publishers: Wellingborough, Northamptonshire.

Iannuzzi, J. (1983). Propolis: the most mysterious hive element. *Amer. Bee J. 123*:631-33.

Iannuzzi, J. (1990a). High profits from lowly propolis. *Amer. Bee J. 130*:237-38.

Iannuzzi, J. (1990b). America's propolis king. *Gleanings Bee Cult. 188*:480-81.

Inayama, S., K. Harimaya, H. Hori, T. Ohkura, T. Kawamata, M. Hikichi and T. Yokokura. (1984). Studies on non-sesquiterpenoid constituents of *Gaillardia pulchella*. II. less lipophilic substances, methyl caffeate as an antitumor catecholic. *Chem. Pharm. Bull. 32*:1135-41.

Jaycox, E. (1988). The bee specialist. *Gleanings Bee Cult. 116*:496-99.

Jolly, B.G. (1978). Propolis varnish for violins. *Bee World 59*:157-61.

Konig, B. and J.H. Dustman (1985). Fortschritte . . . *Apidologie 16*:228-30.

Kosonocka, L. (1990). Propolis—snake oil or legitimate medicine? *Amer. Bee J. 130*:451-52.

Lindenfelser, L.A. (1967). Antimicrobial actvity of propolis. *Amer. Bee J. 107*:90-92, 130-31.

Lindenfelser, L.A. (1968). In vivo activity of propolis against *Bacillus larvae*. *J. Invert. Path. 12*:129-31.

Lowe, D.G. (1980). Propolis substitutes. *Bee World 61*:120-21.

Marinescu, I. and M. Tamas (1980). Poplar buds—a source of propolis. *Apiacta 15*:121-26.

McGregor, S.E. (1952). Collection and utilization of propolis and pollen by caged honey bee colonies. *Amer. Bee. J. 92*:20-21.

Metzner, J., H. Bekemeier, E. Schneidewind and R. Schwaiberger. (1975). Bioautographische Erfassung der antimikrobeill wirksamen Inhaltsstoffe von Propolis. *Pharmazie 30*:799-800.

Metzner, J., E.-M. Schneidewind and E. Friedrich. (1977). Zur Wirkung von Propolis und Pinocembrin auf Sprosspilze. *Pharmazie 32*:730.

Michener, C.D. (1974). The Social Behavior of the Bees. Cambridge, MA: Harvard Univ. Press.

Miyakado, M., T. Kato, N. Ohno and T.J. Mabry. (1976). Pinocembrin and (+)-B- endesmol from *Hymenoclea monogyra* and *Baccharis glutinosa*. *Phytochemistry 15*:846.

Mlagan, V. and D. Sulimanovic. (1982). Action of propolis solutions on *Bacillus larvae*. *Apiacta 17*:16-20.

Paintz, M. and J. Metzner. (1979). Zur lokalanästhetischen Wirkung von Propolis und einigen Inhaltsstoffen. *Pharmazie 34*:839-41.

Popeskovic, D., D. Kepcija, M. Dimitrijevic and N. Stojanovic. (1980). The antioxidative properties of propolis and some of its components. *Acta Veterinaria (Beograd). 30*:133-36.

Popravko, S.A., I.V. Sokolov and I.V. Torgov. (1983). New natural phenolic triglycerides. *Chem. Natural Compounds 18*:153-57. (Translation of *Khimiia Prirodnykh Soedinenii 18*:169-73.)

Roger, C.R. (1988). The nutritional incidence of flavonoids: some physiological and metabolic considerations. *Experientia 44*:725-33.

Root, A.I (1983). The ABC and XYZ of Bee Culture. A.I. Root Co.: Medina, Ohio, pp. 538-541.

Schneidewind, E.-M., H. Kala, B. Linzer and J. Metzner. (1975). Zur Kenntnis der Inhaltsstoffe von Propolis. *Pharmazie 30*:803.

Spangler, H.G. and S. Taber, III. (1970). Defensive behavior of honey bees toward ants. *Psyche 77*:184-89.

Vilanueva, V.R., M. Barbier, M. Gonnet and P. Lavie. (1970). Les flavonoides de la propolis. isolement d'une nouvelle substance bacteriostatique: la pinocembrine (dihydroxy 5, 7-flavone). *Ann. Inst. Pasteur, Paris 118*:84-87.

Wollenweber, E., Y. Asakawa, D. Schillo, U. Laelhmann and H. Weigel. (1987). A novel caffeic acid derivative and other constituents of *Populus* bud excretion and propolis (bee-glue). *Z. Naturforsch. 42C*:1030-1034.

Wright-Sunflower, C. (1988). Panning for brown gold. *Gleanings Bee Cult. 116*:414-16.

References for Bee Venom

Banks, B.E.C. and R.A. Shipolini. (1986). Chemistry and pharmacology of honey-bee venom. *In:* Venoms of the Hymenoptera (T. Piek, *ed.*). p. 329-416. London: Academic.

Beck, B.F. (1935). Bee Venom Therapy. New York: Appleton-Century.

Benton, A.W., R.A. Morse and J.D. Stewart. (1963). Venom collection from honey bees. *Science 142*:228-30.

Billingham, M.E.J., J. Morley, J.M. Hanson, R.A. Shipolini and C.A. Vernon. (1973). An anti-inflammatory peptide from bee venom. *Nature 245*:163-64.

Broadman, J. (1962). Bee Venom—The Natural Curative for Arthritis and Rheumatism. New York: Putnam and Sons.

Calin, A. (1983). Diagnosis and Management of Rheumatoid Arthritis. Menlo Park, Calif.: Addison-Wesley.

Chang, Y.-H and M.L. Bliven. (1979). Anti-arthritic effect of bee venom. *Agents Actions 9*:205-11.

Cohen, A., J.B. Pearah, A.W. Dubbs and C.J. Best. (1942). Bee venom in the treatment of chronic arthritis: a comparative study. *Trans. Med. Soc. State Pennsylvania 45*:957-59.

Cole, L.J. and W.H. Shipman (1970). A Novel . . . *Physiology* (1154-1159).

Doyle, L.A. (1983). Bees and arthritis—an interview with Dr. L.A. Doyle, D.O. *Amer. Bee J. 113*:352-55.

Eiseman, J.L., J. von Bredow and A.P. Alvares. (1982). Effect of honeybee (*Apis mellifera*) venom on the course of adjuvant-induced arthritis and depression of drug metabolism in the rat. *Biochem. Pharm. 31*:1139-46.

Forestiera, F. and M. Palmer (1984). Bee Venom in . . . *Apiacta 19*:19-22.

Gencheva, G. and S.V. Shkenderov. (1986). Inhibition of complement activity by certain bee venom components. *Doklady Bolgarskoi Akad. Nauk 39*:137-39.

Gillaspy, J.E. and J.A. Grant. (1979). Mass collection of *Polistes* wasp venom by electrical stimulation. *Southwest. Entomol. 4*:96-101.

Ginsberg, N.J., M. Dauer and K.H. Slotta. (1968). Melittin used as a protective agent against X-irradiation. *Nature 220*:1334.

Guyton, F.E. (1947). Bee sting therapy for arthritis and neuritis. *J. Econ. Entomol. 40*:469-72.

Hanson, J.M., J. Morley and C. Soria-Herrera. (1974). Anti-inflammatory property of 401 (MCD-peptide), a peptide from the venom of the bee *Apis mellifera* (L.). *Brit. J. Pharm. 50*:383-92.

Hollander, J.L. (1941). Bee venom in the treatment of chronic arthritis. *Amer. J. Med. Sci. 201*:796-801.

Hunt, K.J., M.D. Valentine, A.K. Sobotka, A.W. Benton, F.J. Amodio and L.M. Lichtenstein. (1978). A controlled trial of immunotherapy in insect hypersensitivity. *New Engl. J. Med. 299*:157-61.

Knepel, W. and Charles Gerhards. (1987). Stimulation . . . *Prostaglandins 33:(3)*479-491.

Kroner, J., R.M. Lintz, M. Tyndall, L. Andersen and E.E. Nicholls (1938). The treatment of rheumatoid arthritis with an injectable form of bee venom. *Ann. Intern. Med. 11*:1077-83.

Marcovic, O. and L. Molnar. (1955). Prispevok k isolacii a stanoveniu vcelieho jedu. *Chem. Zvesti 8*:80-90.

Malone, F. (1979). Bees Don't Get Arthritis. New York: Dutton.

Morse, R.A. (1983). Research review: bee venom. *Glean. Bee Cult. 111*:234.

Morse, R.A. and A.W. Benton. (1964). Notes on venom collection from honeybees. *Bee World 45*:141-43.

Mraz, C. (1977). Bee venom therapy. *Amer. Bee J. 117*:260.

Mraz, C. (1982). Bee venom for arthritis—an update. *Amer. Bee J. 122*:121-23.

Mraz, C. (1983). Methods of collecting bee venom and its utilization. *Apiacta 18*:33-34, 54.

Neumann, W. and A. Stracke. (1951). Untersuchungen mit Bienengift und Histamin an der Formaldehydarthritis der Ratte. *Arch. Exper. Path. Pharmakol. 213*:8-17.

O'Connor, R., W. Rosenbrook Jr. and R. Erickson. (1963). Hymenoptera: pure venom from bees, wasps, and hornets. *Science 139*:420.

Palmer, D.J. (1961). Extraction of bee venom for research. *Bee World 42*:225-26.

Panush, R.S. and S. Longley. (1985). Therapies of potential but unproven benefit. *In:* Arthritis, Etiology, Diagnosis, Management (P.D. Utsinger, N.J. Zvaifler and G.E. Ehrlich, *eds.*). p. 695.

Pinnas, J.L., R.C. Strunk, T.M. Wang and H.C. Thompson. (1977). Harvester ant sensitivity: in vitro and in vivo studies using whole body extracts and venom. *J. Allergy Clin. Immunol. 59*:10-16.

Price, J.H., K.S. Hillman, M.E. Toral and S. Newell. (1983). The public's perceptions and misperceptions of arthritis. *Arthritis Rheumatism 26*:1023-28.

Ryan, D. (1954). Dr. Carey's bees vanquish arthritis. *Amer. Bee J. 94*:424-25.

Schmidt, D.K. (1978). The Nature of the Response of Prostaglandins and Cyclic AMP to a Bee Sting. PhD Diss. Univ. of Georgia.

Schmidt, D.K., D.B. Destephano and U.E. Brady. (1978). Effect of honey bee venom on prostaglandin levels in mouse skin. *Prostaglandins 16*:233-38.

Schumacher, M.J., J.O. Schmidt and N.B. Egen. (1989). Lethality of "killer" bee stings. *Nature 337*:413.

Shipman, W.H. and L.J. Cole (1967). Increased resistance of mice to X-irradiation after injection of bee venom. *Nature 215*:311-12.

Shkenderov. S. (1976). New pharmacobiochemical data on the anti-inflammatory effect of bee venom. *In:* Animal, Plant, and Microbial Toxins, vol. 2 (A. Ohsada, K. Hayashi and Y. Sawai, *eds.*). p. 319-36. New York: Plenum.

Somerfield, S.D., J.-L. Stach, C. Mraz, F. Gervais and E. Skamene. (1984). Bee venom inhibits superoxide production by human neutrophils. *Inflammation* 8:385-91.

Steigerwaldt, F., H. Mathies and F. Damrau. (1966). Standardized bee venom (SBV) therapy of arthritis. *Indust. Med. Surg.* 35:1045-49.

Vick, J.A. and W.H. Shipman. (1972). Effects of whole bee venom and its fractions (apamin and melittin) on plasma cortisol levels in the dog. *Toxicon* 10:377-80.

Vick, J.A., B. Mehlman, R. Brooks, S.J. Phillips and W. Shipman. (1972). Effect on bee venom and melittin on plasma cortisol in the unanesthetized monkey. *Toxicon* 10:581-86.

Vick, J.A., G.B. Warren and R.B. Brooks. (1975). The effect of treatment with whole bee venom on daily cage activity and plasma cortisol levels in the arthritic dog. *Amer. Bee J.* 115:52-53, 58.

Wells, F.B. (1977). Hive product uses—venom. *Amer. Bee J.* 117:10-22.

Zurier, R.B., H. Mitnick, D. Bloomgarden and G. Weissman. (1973). Effect of bee venom on experimental arthritis. *Ann. Rheumat. Dis.* 32:466-70.

References for Beeswax

Blomquist, G.J., D.W. Roubik and S.L. Buchmann. (1985). Wax chemistry of two stingless bees of the *Trigonisca* group (Apidae: Meliponinae). *Comp. Biochem. Physiol.* 82B:137-42.

Combs, G.F. (1972). The engorgement of swarming worker honey bees. *J. Apic. Res.* 11:121-28.

Crane, E. (1983). The Archaeology of Beekeeping. London: Duckworth.

Crane, E. (1990). Bees and Beekeeping: Science, Practice and World Resources. Ithaca, NY: Comstock Publ.

Hepburn, H.R. (1986). Honeybees and Wax: An Experimental Natural History. Berlin: Springer-Verlag.

Horstmann, H.J. (1965). Einige biochemische Überlegungen zur Bildung von Bienenwachs aus Zucker. *Z. Bienenforsch* 8:125-28.

Ribbands, C.R. (1953). The Behavior and Social Life of Honeybees. London: Bee Research Association.

Root, H.H. (1951). Beeswax: Its Properties, Testing, Production and Applications. Brooklyn, NY: Chemical Publ. Co.

Seeley, T.C. and R.A. Morse. (1976). The nest of the honey bee (*Apis mellifera* L.). *Insectes Soc.* 23:495-512.

Tulloch, A.P. (1970). The composition of beeswax and other waxes secreted by insects. *Lipids* 5:247-58.

Tulloch, A.P. (1980). Beeswax—composition and analysis. *Bee World* 61:47-62.

Weiss, K. (1965). Über den Zuckerverbrauch und die Beanspruchund der Bienen bei der Wachserzeugung. *Z. Bienenforsch* 8:106-24.

References for Royal Jelly

Asencot, M. and Y. Lensky. (1988). The effect of soluble sugars in stored royal jelly on the differentiation of female honeybee (*Apis mellifera* L.) larvae to queens. *Insect Biochem.* 18:127-33.

Bee World. (1962). United States of America vs. Jenasol (civil action No. 1042-58). *Bee World* 43:64-65.

Blum, M.S., A.F. Novak and S. Taber III. (1959). 10-Hydroxy-Δ^2-decenoic acid, an antibiotic found in royal jelly. *Science 130:*452-53.

Brown, W.H., E.E. Felauer and R.J. Freure. (1961). Some new components of royal jelly. *Can. J. Chem. 39:*1086-89.

Cho, Y.T. (1977). Studies on royal jelly and abnormal cholesterol and triglycerides. *Amer. Bee J. 117:*36-38

Evans, H.M., G.A. Emerson and J.E. Eckert. (1937). Alleged vitamin E content in royal jelly. *J. Econ. Entomol. 30:*642-46.

Furukawa, H. (1982). An artificial rearing experiment of black carpenter ant using royal jelly. *Honeybee Sci. 3:*137-38. (Jap. with Eng. Abst.)

Goodman, L.S. and A. Gilman (*eds.*) (1975). The Pharmacological Basis of Therapeutics, 5th edit. New York: MacMillan.

Hashimoto, T., K. Takeuchi, M. Hara and K. Akatsuka. (1977). Pharmacological study on royal jelly (RJ). I. acute and subacute toxicity tests on RJ in mice and rats. *Bull. Meiji Coll. Pharm. No. 7:*1-13. (Abst. in *Apic. Abst. 30:*300-01 [1979]).

Haydak, M.H. and L.S. Palmer. (1942). Royal jelly and bee bread as sources of vitamins B[1], B[2], B[6], C and nicotinic and pantothenic acids. *J. Econ. Entomol. 35:*319-20.

Haydak, M.H. and A.E. Vivino. (1950). The changes in the thiamine, riboflavin, niacin and patothenic acid contents in the food of female honeybees during growth with a note on the vitamin K activity of royal jelly and beebread. *Ann Entomol. Soc. Am. 43:*361-67.

Herbert, E.W. Jr. and H. Shimanuki. (1978). Effect of fat soluble vitamins on the brood rearing capabilities of honey bees fed a synthetic diet. *Ann. Entomol. Soc. Am. 71:*689-91.

Howe, S.R., P.S. Dimick and A.W. Benton. (1985). Composition of freshly harvested and commercial royal jelly. *J. Apis. Res. 24:*52-61.

Inoue, T. and A. Inoue. (1964). The world royal jelly industry; present status and future prospects. *Bee World 45:*59-69.

Iwanami, Y., I. Okada, M. Iwamatsu and T. Iwadare. (1979). Inhibitory effects of royal jelly acid, myrmicacin, and their analogous compounds on pollen germination, pollen tube elongation, and pollen tube mitosis. *Cell Struct. Funct. 4:*135-42.

Karaali, A., F. Meydanoglu and D. Eke. (1988). Studies on composition, freeze-drying and storage of Turkish royal jelly. *J. Apic. Res. 27:*182-85.

Lercker, G., P. Capella, L.S. Conte, F. Ruini and G. Giordani. (1982). Components of royal jelly II. the lipid fraction, hydrocarbons and sterols. *J. Apic. Res. 21:*178-84.

Matsuka, M., N. Watabe and K. Taceuchi. (1973). Analysis of the food of larval drone honeybees. *J. Apic. Res. 12:*3-7.

McCleskey, C.S. and R. M. Melampy. (1939). Bactericidal properties of royal jelly of the honeybee. *J. Econ. Entomol. 32:*581-87.

Melampy, R.M. and D.B. Jones. (1939). Chemical composition and vitamin C content of royal jelly. *Proc. Soc. Rcp. Biol. Med. 41:*382-88.

Melampy R.M. and A.J. Stanley. (1940). Alleged gonadotropic effect of royal jelly. *Science 91:*457-58.

Nation, J.L. and F.A. Robinson. (1971). Concentration of some major and trace elements in honeybees, royal jelly and pollens, determined by atomic absorption spectrophotometry. *J. Apic. Res. 10:*35-43.

Nes, W.D. and J.O. Schmidt. (1988). Isolation of 25(27)-dehydrolanost-8-enol from *Cereus giganteus* and its biosynthetic implications. *Phytochem. 27:*1705-08.

Otani, H., M. Oyama and F. Tokita. (1985). Polyacrylamide gel electrophoretic and immuno-chemical properties of proteins in royal jelly. *Jap. J. Dairy Food Sci. 34:*21-25.

Schmidt, J.O. and P.J. Schmidt. (1984). Pollen digestibility and its potential nutritional value. *Glean. Bee Cult. 112:*320-22.

Shinoda, M., S. Nakajin, T. Oikawa, K. Sato, A. Kamogawa and Y. Akiyama. (1978). Biochemical studies on vasodilative factor in royal jelly. *Yakugaku Zassii 98:*139-45. (Jap. with Eng. Abst).

Svoboda, J.A., E.W. Herbert Jr., W.R. Lusby and M.J. Thompson. (1983). Comparison of sterols of pollens, honeybee workers, and prepupae from field sites. *Arch. Insect Biochem. Physiol. 1:*25-31.

Tamura, T., N. Kuboyama and A. Fujii. (1985). Studies on mutagenicity of royal jelly. *Int. Apic. Congr. Apimondia,* 30th 10-16 Aug. 1985. p.153 (abst.).

Takenaka, T. (1984). Studies on proteins and carboxylic acids in royal jelly. *Bull. Fac. Agr. Tamagawa Univ. No. 24:*101-49. (Jap. with Eng. abst.).

Townsend, G.F., J.F. Morgan, S. Tolnai, B. Hazlett, H.J. Morton and R.W. Shuel. (1960). Studies on the *in vitro* antitumor activity of fatty acids I. 10-hydroxy-2-decenoic acid from royal jelly *Cancer Res. 20:* 503-10.

Tucker, K.W. and M.S. Blum. (1972). No gibberellic acid found in royal jelly. *Ann. Entomol. Soc. Amer. 65:*989-90.

Vittek, J. and B.L. Slomiany. (1984). Testosterone in royal jelly. *Experientia. 40:*104-06.

Wang, W. (1989). The development and utilization of the resources of bee pollen in China. *Proc. Intl. Congr. Apicult. (Apimondia) 32:*239.

Wells, F.B. (1976). Hive product uses — royal jelly. *Amer. Bee J. 116:*560-61,65.

Yatsunami, K. and T. Echigo. (1985). Antibacterial action of royal jelly. *Bull. Fac. Agr. Tamagawa Univ. No. 25:*13-22.

References: Bee Brood and Adults

Bailey, R.C. (1989). The Efe: archers of the African rain forest. *Nat. Geograph. 176:*664-86.

Burgett, M. (1990). Bakuti — a Nepalese culinary preparation of giant honey bee brood. *The Food Insects Newsletter 3(3):*1-2.

Gary, N.E., R.W. Ficken and R.C. Stein. (1961). Honey bee larvae *(Apis mellifera,* L.) for bird food. *Avicult. Mag. 67:*27-32.

Guss, S.B. (1967). Bee larvae as food for caged birds. *Amer. Bee J. 107:*62.

Hasegawa, M., Y. Saeki and Y. Sato. (1983). Artificial rearing of some beneficial insects on drone powder and the possibillity of their application. *Honeybee Sci. 4:*153-56 (Japanese, with English abst.).

Herbert, E.W. Jr. and H. Shimanuki. (1978). Effect of fat soluble vitamins on the brood rearing capabilities of honey bees fed a synthetic diet. *Ann. Entomol. Soc. Amer. 71:*689-91.

Hill, K., K. Hawkes, M. Hurtado and H. Kaplan. (1984). Seasonal variance in the diet of Ache hunter-gatherers in eastern Paraguay. *Human Ecol. 12:*101-35.

Hocking, B. and F. Matsumura. (1960). Bee brood as food. *Bee World 41:*113-20.

Kakeya, M. (1976). Subsistence ecology of the Tongwe, Tanzania. *Kyoto Univ. Afr. Stud. 10:*143-212 (cited in *Apic. Abst. 29:*168-69 [1978]).

Lanyon, W.E. and V.H. Lanyon. (1969). A technique for rearing passerine birds from the egg. *Living Bird 8:*81-93.

Matsuka, M. and S. Takahashi. (1977). Nutritional studies of an aphidophagous coccinellid *Harmonia axyridis* II. significance of minerals for larval growth. *Appl. Ent. Zool. 12:*325-29.

Matsuka, M., M. Watanabe and K. Niijima. (1982). Longevity and oviposition of vedalia beetles on artificial diets. *Environ. Entomol. 11:*816-19.

Okada, I. and M. Matsuka. (1973). Artificial rearing of *Harmonia axyridis* on pulverized drone honey bee brood. *Environ. Entomol. 2:*301-02.

Parent, G., F. Malaisse and C. Verstraeten. (1978). Les miels dans la foret claire du Shaba meridional. *Bull. Rech. Agron. Gemblous 13:*161-76.

Posey, D. A. (1983). Folk apiculture of the Kayapo Indians of Brazil. *Biotropica 15:*154-158.

Ryan, J.K. and P. Jalen (1982). Alkaline extraction of protein from spent honey bees. *J. Food Science 48:*886-88&96.

Schmidt, J.O. and P.J. Schmidt. (1984). Pollen digestibility and its potential nutritional value. *Glean. Bee Cult. 112:*320-22.

Schmidt, J.O., H.G. Spangler and S.C. Thoenes. (1990). Birds as selective predators of drones. *Amer. Bee J. 130:*811.

Thoenes, S.C. and J.O. Schmidt. (1990). A rapid, effective method of non-destructively removing honey bee larvae from combs. *Amer. Bee J. 130:*817.

PRODUCTION OF QUEENS AND PACKAGE BEES

by HARRY H. LAIDLAW, JR.*

QUEEN REARING

INTRODUCTION

Until about the middle of the past century queen rearing was not a part of American beekeeping. Honey bee colonies lived in tree cavities in the wild (feral bees), or in sections of hollow trees that were brought from the forests and arranged upright, or in man-made boxes measuring typically about 3 x 1 x 1 feet that were placed on end, Fig. 1. These "bee gums" and "box hives" received little care and the colonies reared their own queens as needed for supersedure or for replacement of a swarmed queen. Shallow box-like "supers" set on the top of box hives over auger holes in the hive roof, and hives with simple top strips for comb attachment or with crude frames, were little improvement over the bee gums for beekeeping. Colonies in these somewhat advanced hives also were left to themselves to provide needed queens.

Langstroth's invention of the movable-frame hive, that permitted the colony to be intimately observed comb by comb, initiated widespread interest

FIGURE 1. Apiary of "box" beehives.

---

*Harry H. Laidlaw, Jr., Professor Emeritus, University of California, Davis, California. Author of *Contemporary Queen Rearing* and internationally known expert on artificial insemination of queen honey bees.

in honey bees and their behavior and in the production of honey. Shortly thereafter, the importation of Italian queens created a desire to substitute Italian bees for the "black" bees that were everywhere, and this stimulated efforts to rear queens. An increasing demand for Italian queens was met at first by several exceptional beekeepers' experimenting with various ways to entice bees to rear virgin queens. G.M. Doolittle assembled their contributions, plus some of his own into a practical queen rearing system that is still the basis of contemporary queen rearing. Doolittle's system has been modified and refined until it is highly efficient, and is suitable for both commercial queen production and for hobby and sideline needs.

POSITION AND FUNCTION OF THE QUEEN IN THE COLONY

It is a truism among beekeepers that beekeeping is centered on the queen. A colony is not productive unless the mother queen is a vigorous producer of eggs that are the source of a large population of workers. But the queen is more than that. She produces pheromones that her worker bees pass among themselves. These pheromones maintain the cohesion of the colony that in past times was referred to as colony morale.

Fully as significant is the queen's role as custodian of the genetic determinants of the characteristics of her colony. She transmits through her eggs her own genetic contributions to colony characteristics and, additionally, those of her mates by fertilizing her eggs with her mates' sperm that she had acquired at mating and which she carries in her spermatheca. This combination that is a mated queen might be thought of as a biological kit. The mere changing of this kit in a colony will substitute a different colony in the same hive, albeit gradually. By timely queen replacements beekeepers have firm control over the characteristics of their bees.

The quality of queens varies because of hereditary differences and of variations in care during development. Excellent queens are available from commercial queen breeders, and beekeepers can rear excellent queens for themselves. Rearing vigorous, fully-developed queens is not difficult either as a hobby or commercially. The bees rear the queens in both cases and the beekeeper provides the best conditions for the bees' work. It must be mentioned, however, that commercial queen rearing is a business and as such it is far more complex than rearing queens for one's own use.

DETERMINATION OF FEMALE CASTES, AND DEVELOPMENT AND MATURING OF QUEENS

Queens and workers, the female castes of honey bees, develop from fertilized eggs that could have given rise to either caste, but the larva that hatches from the fertilized honey bee egg is neither caste—it is simply female.

The specific female caste in undetermined at this point, though the sex was established when the egg was fertilized. As an egg is laid, one or more sperm enter the egg and the nucleus of one sperm combines with the egg nucleus. Cell divisions begin that under the control of the combined egg and sperm genes form the female pattern. This is complete in 3 days and a female larva hatches. The larva has the genetic constitution to respond to food given it by nurse bees to channel its development toward either a worker or a queen by a route controlled by nurse bees.

The food that newly hatched larvae receive is royal jelly that consists of secretions of the hypopharyngeal glands of the head and of the salivary glands of the head and thorax of worker nurse bees. The compositions of the royal jelly fed to the two castes are similar but have not been shown to be identical. The larva grows rapidly, apparently developing the precursers of both castes, and on the third day following hatching these respond selectively to the larval food to guide the larva toward either the worker or queen caste.

The caste of the worker or queen bee is fixed in a short timespan on the third day from hatching. At that time larvae that are destined to become workers receive some pollen, and the regimen of feeding worker larvae is changed. After this the caste is irrevocable, or at most, partly reversible resulting in intercastes with mixtures of features of both castes.

While female larvae as old as approximately 3 days can develop into adults that are morphologically queens, there is some evidence that divergence of queen and worker characteristics begins soon after the larva hatches from the egg. In any case, the short time period on the third day when caste is determined suggests that physiological characters which diverge extremely widely between the castes, such as longevity, may have responded to feeding variations before the third day. Varying genotypic thresholds of response among features of the two castes seem likely to be present during the development so that intercastes, though physically unrecognizable, may be present among queens reared from larvae that are 3 days old. It seems prudent to rear queens from the youngest larvae it is feasible to use.

The queen larva eats voraciously and grows rapidly the first 6 days. Table 1. The queen cell is provisioned by the nurse bees for only 4 days before it is sealed. These 4 days are crucial and the beekeeper's attention must be focused on them. Virgin queens emerge from the cell on the 11th to the 13th day after larval hatching from the egg, and are sexually mature about 7 days later.

CELL BUILDING
Voluntarily Built Queen Cells
Bees build queen cells when a new queen is needed by the colony. Aging queens are superseded, swarm queens are replaced with young ones, and

Table 1. Honey bee developmental stages

Day	Workers Stages	Workers Moults	Queens Stages	Queens Moults	Drones Stages	Drones Moults
1						
2	Egg		Egg		Egg	
3		(hatching)		(hatching)		(hatching)
4	1st larval	1st moult	1st larval	1st moult	1st larval	1st moult
5	2nd larval	2nd moult	2nd larval	2nd moult	2nd larval	2nd moult
6	3rd larval	3rd moult	3rd larval	3rd moult	3rd larval	3rd moult
7	4th larval	4th moult	4th larval	4th moult	4th larval	4th moult
8	Gorging	(sealing)		4th moult (sealing)		
9			Gorging		Gorging	
10	Pre-pupa		Pre-pupa	5th moult		(sealing)
11		5th moult				
12					Pre-pupa	
13			Pupa			
14						5th moult
15	Pupa					
16			Imago	6th moult (emerging)		
17						
18					Pupa	
19						
20						
21	Imago	6th month (emerging)				
22						
23					Imago	6th moult
24						(emerging)

From Laidlaw, Harry H., Jr. 1979. Modified from Bertholf, L.M. 1925. The moults of the honeybee. *Journal of Economic Entomology 18(2):380-384.*

queens lost by accident or disease are replaced. Successors to superseded and swarmed queens are reared from eggs that were laid in queen cell cups that open downward and which were made by bees at various places on the brood combs, Fig. 2. These young queens are reared in queenright colonies.

Replacements for queens that were lost are reared from young larvae in worker cells at various places on brood combs, Fig. 3. These larvae were originally destined to become worker bees, and the young queens are reared from larvae, not eggs, and in queenless colonies under conditions resembling those prevalent in beekeepers' queen rearing apiaries.

The beekeeper has many choices in queen rearing, some of which are more suitable for particular needs than others. Cutting sealed queen cells from colonies naturally building them, and using them elsewhere, is an easy way to get new queens. For replacing an aging queen, one of any sealed cells that have voluntarily been started can be left along with the old queen or the old queen can be removed leaving only a cell. Alternatively, the old queen may simply be removed so the bees will build queen cells. In this case the young queen will be an emergency queen that is reared from a larva rather than an egg. The queenless part of a divided colony will build queen cells, and a comb with eggs and young larvae can be given to the queenless part of a "divide" so the bees can provide their own queen. Similarly a comb of eggs and young larvae can be inserted into a good but queenless colony for the bees to rear a queen to requeen the colony.

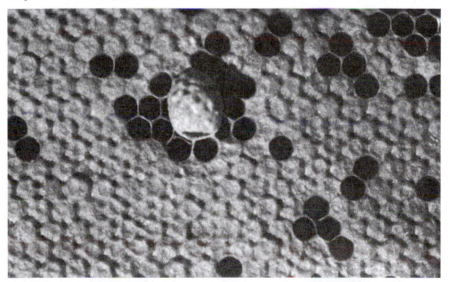

FIGURE 2. Natural queen cell cup on face of the comb. (*From Harry H. Laidlaw, Jr. 1979. Contemporary Queen Rearing.*)

FIGURE 3. Emergency cells drawn over larvae in worker comb. *(From Harry H. Laidlaw, Jr. 1979. Contemporary Queen Rearing.)*

Queens From Larvae in Strips or Plugs of Worker Comb

When queens are needed for several colonies, or over a period of time, new combs with very young larvae can be cut into strips, the cells cut to half their depth with a warm knife, and the cells mashed down leaving every third cell with its larva. The strips can be attached, with the openings of the cells facing downward, to cut lower edges of combs or to cell bars fastened between frame end pieces, and these are put into queenless colonies for cell building. A variation is to cut worker cells that contain newly hatched larvae from new combs with a punch and attach them to comb edges or to cell bars. A similar method is to make a series of grooves on the side of a new comb that has young larvae, and lay the comb with grooved side down, supported by 1 inch thick strips, over the frames of a queenless colony. Cells produced by these methods should be removed from the cell builder in 10 days and individually put into the colonies to be requeened or into mating nuclei because the first virgins to emerge will fight, and they will destroy unemerged virgins in their cells if the cells are left together.

When the cell building colony is strong with bees of all ages, fully developed queens are usually obtained. If the colony is weak or the mix of different age bees is abnormal, small queens can be expected that will be prematurely superseded or will simply disappear.

Grafting Method

The methods mentioned above are useful from time to time. When greater numbers of queens are needed and control of their availability is important, or when it is desired or necessary to control behavior of colonies in the apiaries by selection of stock, it is both easier and more efficient to rear queens by the grafting method, or by other methods that make use of beeswax or plastic queen cell cups. Grafting, the transfer of larvae from worker cells to queen cells cups, is viewed by many beekeepers with such apprehension that they seek to avoid it. In truth, the grafting method of queen rearing is the easiest and most efficient of all methods of controlled queen rearing.

Cell cups. The grafting method of queen rearing requires the use of queen cell cups into which young larvae are transferred. These may be made of beeswax or plastic and can be purchased from bee supply dealers. Naturally occurring beeswax cups are nearly always present in hives as bases of former queen cells from which queens emerged, or as unused bases that are routinely built by bees as part of their comb building regimen, Fig. 2. Strips of new drone comb with the cells cut to half their depth with a warm knife, and every other cell mashed to the midrib are good cups for grafting.

Cups can be made in abundance and easily by dipping a forming stick, Fig. 4, into beeswax that is slightly hotter than the wax melting point. A forming stick made from a section of 3/8″ diameter dowel by slightly tapering the lower 3/4″ of its length and rounding the end is a satisfactory mold. A stick can also be made from a 1/2″ diameter dowel, and the taper extended its entire length from a 3/8″ wide diameter terminal end. The wax should be heated in a thermostatically controlled pan, or in a double jacketed arrangement to keep the wax at the right temperature and to avoid fire hazard.

FIGURE 4. Dipping cell cups with a single forming stick.

The stick is prepared for dipping by soaking it in slightly soapy water for about 15 minutes. The stick is taken from the water and the end touched to a cloth to remove the hanging drop of water. It is then dipped into the wax to a depth of about 1/4″ to 3/8″ and is immediately withdrawn. Dipping into the wax is repeated two or three times. Between dips the stick is held above the wax for a second or two for the wax to cool. After the last dip the stick is dipped into cool water to solidify the wax. With a slight twist the cup can be removed from the stick. Before dipping the next cup, the stick is again momentarily dipped into the soapy water. Three to 5 cups a minute can be dipped by this method with the single stick.

The cups can be fastened by melted wax to cell bars that fit between the end pieces of a frame and are supported by ledges on the inner side of the end pieces or by notches in the end pieces. The cups should be spaced 7/8″ to 1″ apart along the cell bar, Fig. 5. A spacing guide beside the cell bar facilitates the spacing.

When a large quantity of cell cups is needed, 15 dipping sticks fastened together along a bar can be dipped at one time, Fig. 6. After the last dip, the rack of dipping sticks is rested on a support that lets the cell cups touch a cell bar over a container of melted wax, and melted wax is spooned along the bar beside and between the bases of the cups. Alternatively, the face of the cell bar can be touched to the wax, or wax spooned along its upper surface before the cells are rested on it. When the wax has solidified, the cups adhere to the cell bar and can be gently pushed from the forming sticks leaving the cups attached to the cell bar and correctly spaced. Two racks of sticks dipped alternately to allow cooling of cups on one cell bar while the other rack of sticks is dipped is a very rapid way to dip highest quality cups and fasten them to cell bars so they are ready for use.

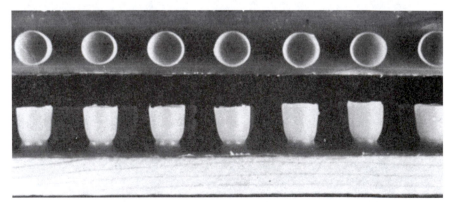

FIGURE 5. Dipped cell cups attached and spaced on cell bars. *(From Harry H. Laidlaw, Jr. 1979. Contemporary Queen Rearing.)*

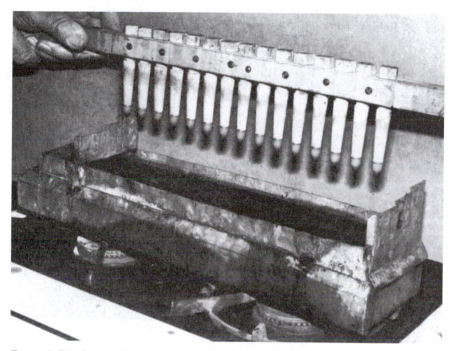

FIGURE 6. Dipping multiple cups at one time. *(From Harry H. Laidlaw, Jr. 1979. Contemporary Queen Rearing.)*

Obtaining larvae for grafting. Larvae 18 to 36 hours old (1 to 1-1/2 times the size of the egg) are the right age for grafting, Fig. 7. They should be abundantly supplied with royal jelly in their worker cells. Finding such larvae is relatively simple when few cells are to be grafted, though this varies with environmental conditions which may cause slowing of oviposition or scant feeding of larvae.

The easiest way to get well fed, proper age larvae for few cells is to put a colony-warmed comb into a breeder hive, and transfer it the next day to a strong queenless colony or above the excluder of a strong queenright colony. Four to 5 days after the comb was given to the breeder colony, there will be well fed larvae of the desired age ready for grafting. When removing the comb for grafting, give it a gentle shake to dislodge most of the bees, and brush off the remaining bees with a soft brush. Hard shaking of the comb will shift the larvae in the cells so they are difficult to pick up with a grafting needle.

In contrast to hobby queen rearing, commercial cell building requires well fed larvae of the proper age in abundance and proximate in the same worker comb so that grafting can be done rapidly and without searching for suitable larvae. An efficient way to obtain such larvae is to confine each breeder queen

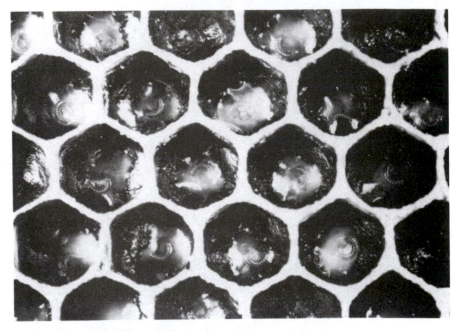

FIGURE 7. Well fed larvae the proper age for grafting.

FIGURE 8. Shallow breeder hive with the breeder queen restricted to three shallow combs. *(From Harry H. Laidlaw, Jr. 1979. Contemporary Queen Rearing.)*

to a single usable comb by partitioning her hive into a queen compartment and a larger larval feeding and brood rearing compartment, Fig. 8. This is done with excluders. A comb is put in the middle of the queen compartment of the breeder hive each day and is transferred the next day to the feeding compartment. Rotation of combs ensures a steady and ready source of graftable larvae.

A second excellent method to get proper larvae is to use the Pritchard insert in breeder hives, Fig. 9. The rotation of the small combs between the queen compartment and the feeding compartment provides abundantly fed day-old larvae every day in a small comb.

Grafting. Young larvae are remarkably hardy, but they dry quickly in dry air, as does their bed of royal jelly. For this reason, grafting should be done in a space protected from air currents. As soon as a bar of cells is grafted, it should be covered with a damp cloth or immediately given to the cell builder colony. For many years it was assumed that cell cups should be "primed" with a drop of royal jelly before larvae are transferred into them, Fig. 10. Experience has shown this to be unnecessary, but it can be helpful to dislodge larvae from the grafting tool when plastic cups are used. Consequently, most contemporary queen rearers "dry graft" larvae.

FIGURE 9. Pritchard insert for obtaining the proper age larvae for grafting. *(From Harry H. Laidlaw, Jr. 1979. Contemporary Queen Rearing.)*

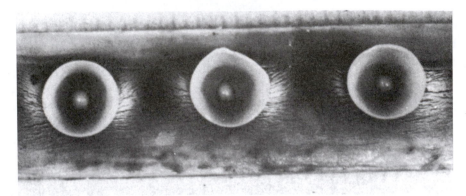

FIGURE 10. Cell cups "primed" with royal jelly.

FIGURE 11. Grafting into queen cups.

Small larvae, not much larger than the egg, blend with their bed of jelly and are difficult for many people to see. A light of about 100 watts directed into the cells of the worker comb and into the cell cups is a necessary aid, Fig. 11. Magnifying glasses or a loupe may be helpful also.

The actual transfer of the larvae from worker comb to cell cups can be made with a finely-bristled artist brush, a "straight" needle with the end fashioned into a small hook-like blade, an "automatic" needle, or needles of other designs. The brush or needle is moved down the side of the comb cell and slipped under the larva leaving, if possible, part of the larva extending over the needle blade to make it easier to dislodge the larva onto the bottom of the cell cup.

Cell Builders

Cell builders may be queenless or queenright. Either must be strong with bees of all ages and particularly strong with nurse bees and have abundant supplies of honey and pollen. When flows of gathered nectar or pollen are interrupted by inclement weather, the cell builders should be supplied with light (50 sugar/50 water) sugar syrup in slow feeders and with pollen stored or dropped into combs, or with pollen and pollen supplement made into patties with sugar syrup. It is, in fact, good practice to supplementally feed cell builders routinely. A shallow super with honey and some pollen placed on the bottom board provides good additional food stores.

Queenless bees accept prepared larvae presented to them and start queen cells more readily than do queenright colonies with unrestricted queens. For this reason beekeepers start queen cells with bees that are partially or completely separated from their queen, such as queenlessness, bees isolated completely but temporarily from their queen, or bees partially separated from their queen by queen excluders.

Queenright cell builders are basic cell builders suitable for commercial queen rearing and also for producing only a few queens. Queen cells can be started and completed in them, or these colonies will finish cells started in other colonies.

Queenright cell builders are strong 2-story colonies with the queen in the lower body which is separated from the upper body by a queen excluder, Fig. 12. Combs of eggs and young larvae are moved from the lower body to the upper one to draw up nurse bees into what is now the cell building area. This is done once or twice each week if cells are built constantly in the colony. At the same time the brood in both bodies is examined for spontaneous queen cells. A well filled comb of pollen is placed between the combs of youngest larvae in the upper body to be easily accessible to the nurses, and is replaced as needed. Collecting pollen combs as they are found during colony examinations and stacking them on a colony in the cell building apiary assures an ample and handy supply of pollen combs at all times.

Queenright cell builders must have young vigorous queens, and their populations maintained by addition of sealed or emerging brood as needed, or by bees shaken from brood nests of other colonies. An efficient way to get bees that are to be added to cell builders is to shake them from the upper body of support colonies which have the queen in the lower body below an excluder and the young brood has been moved above the excluder, exactly as a queenright cell builder is arranged.

A graft of 15 to 30 cells is put into the upper body between the youngest larvae and the pollen comb, and the immediately prior graft of sealed or nearly sealed cells is moved to the opposite side of the comb of young larvae. The "ripe" cells are taken from the colony on the 10th day following grafting by brushing the bees from the cells and distributing the cells to hives or nuclei for the virgins' mating, or they are placed in an incubator overnight.

Free flying queenless cell builders will start and finish queen cells, or

FIGURE 12. Queenright 2-story queen cell builder, or cell finisher.

they may be used to start queen cells that will be finished in queenright cell builders. They are made up by dequeening a colony or by establishing a new strong colony with brood, bees, and pollen and honey, Fig. 13. They need not occupy a standard body—a five-frame nucleus body strong with bees is adequate. Queenless cell builders are maintained by weekly additions of sealed and emerging brood or by bees that are shaken from combs which have young larvae. They are, of course, generously supplied with food by the beekeeper even during a nectar and pollen flow. Thirty to 45 cells can be given to these cell builders at one time.

FIGURE 13. Queenless 1-story queen cell starter or starter/finisher.

Swarm box. When part of a colony of bees is suddenly separated from its queen and is confined, the bees quickly sense the loss of the queen and they are responsive to any possibility to rear a new one. Confined nurse bees secreting royal jelly and having no larvae to feed consistently accept all or nearly all of the viable grafted larvae given to them, and feed the larvae prodigiously. These tendencies are exploited by beekeepers through the use of a swarm box or some variation of it.

The usual swarm box, Fig. 14, has space for five standard frames, with a six-inch deep space below the frames that is screened on both sides. There is no entrance. The box is furnished with one well-filled comb of pollen at the middle, and at each side with a comb of open honey or sugar syrup. A half or

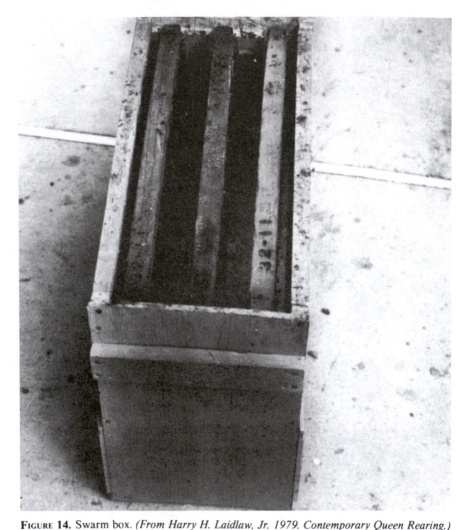

Figure 14. Swarm box. *(From Harry H. Laidlaw, Jr. 1979. Contemporary Queen Rearing.)*

full-depth division-board feeder with thin sugar syrup can be substituted for the feeder combs at one or both sides.

One to two hours before grafting, the swarm box is stocked with 5 to 7 pounds of bees shaken from combs of open brood. Nurse bees are obtained most easily from the upper body of support colonies that are arranged like queenright cell builders. The stocked box is set in a cool dark place and is undisturbed for about two hours after which the bees are jarred down and a frame with a graft of 30 to 45 cells, fastened to three bars, is inserted into the box on each side of the pollen comb.

The box is left undisturbed for about 24 hours. Then the started cells are removed to be distributed to queenright finisher colonies. But this time the swarm box is not jarred. It is taken into the cell finisher yard and opened there and the bees are brushed gently from the cells, or are left on the cells. One to three bars of started cells are given to each queenright finisher. Cells that are started in properly made up and operated starter colonies will produce good queens even when the finishers are somewhat below their best condition. The reverse, however, is less likely.

Modified swarm box. Cells started in swarm boxes are the best that beekeepers can produce. But swarm boxes are troublesome to stock with bees, and their use can draw heavily on the populations of supporting colonies. Beekeepers seek to minimize this cost and retain the excellent cell quality and acceptance of the swarm box by devising economical variations of it.

One of the best such modifications is a cell starter in which the cell building bees are confined above a two-story queenright colony for 24 to 36 hours while cells are being started and undergoing the first day's development. A day prior to grafting, a strong colony is arranged as a queenright cell builder with the queen beneath an excluder in the bottom body along with sealed brood, honey, and empty combs. The young larvae are raised above the excluder into the upper body and a comb of pollen is placed next to the youngest larvae. This arrangement is maintained by moving larvae to the top body and sealed brood, empty combs, and honey-filled combs below at three-day intervals if successive grafts are made. A partially filled shallow body of honey may be put on the bottom board to serve as an auxiliary feeder, and as a place for field bees to congregate when they are not foraging; this placement of space for congregating minimizes swarming.

The queen cells are started on top of the queenright starter colony in a hive body that is specially prepared with two well filled combs of pollen in the middle with a space between them for the grafted cells, and a division board feeder placed next to the outer side of each pollen comb, Fig. 15. There is no entrance. About two hours before newly grafted cells are given to the starter, the prepared cell-starter body is placed beside the starting colony upon a screen made of 8-mesh hardware cloth that is fastened to a hive-body-size rim, Fig. 16, and the feeders are fully or partly filled with 50 sugar/50 water syrup and covered to prevent bees from falling into them.

The top body above the excluder of the starter colony, which has young brood, nurses, and wax builders, is now set upon the prepared cell starter body and the bees are shaken from it into the starter body. The top body is then returned to the starter colony above the excluder and the starter body with its screen is placed on top of it, Fig. 17. The feeders are uncovered before the hive top is put on.

About two hours following stocking the starting body with bees, four bars of grafted cells are put into the space between the two pollen combs. Twenty-four to 36 hours later the started cells are well developed and abundantly provisioned, Fig. 18, and are transferred to queenright finishing colonies. The screen is removed from beneath the starting body and the bees allowed to go

FIGURE 15. Modified swarm box and donor colony. *(From Harry H. Laidlaw, Jr. 1979. Contemporary Queen Rearing.)*

below or they are shaken from the body and the body removed. Alternatively, the cells can be finished in this special starter colony by moving the started cells down into the second body and removing the screen and the prepared body.

When queen cells are 10 days old, Fig. 19, they must be removed and

FIGURE 16. Separatory 8-mesh screen for use with modified swarm box. *(From Harry H. Laidlaw, Jr. 1979. Contemporary Queen Rearing.)*

placed in colonies the queens will head, in mating nuclei, or caged in nursery colonies for special use. If they are left longer in the finisher, any early emerging queens will destroy them, Fig. 20, and the emerged virgins will fight, Fig. 21. Queen cells should *never* be shaken to remove the bees. Bees should

Figure 17. Modified swarm box in operation.

be brushed gently from them because developing queens younger than 11 days from grafting are easily injured.

After cells are sealed they can be removed from the cell builder and incubated in any strong queenless colony or above an excluder of a strong queenright colony. They can also be incubated in an incubator set to 93°F and humidified to approximately 50% relative humidity by a pan of water.

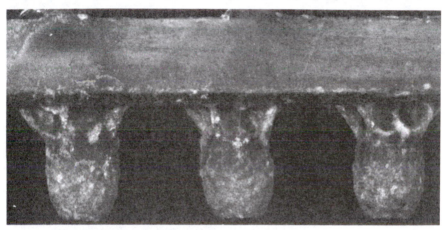

FIGURE 18. Day-old cells from modified swarm box.

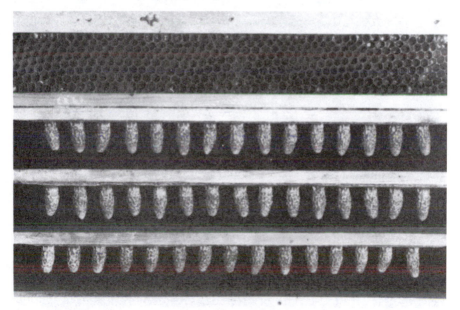

FIGURE 19. "Ripe" 10-day old queen cells from cell finishing colony.

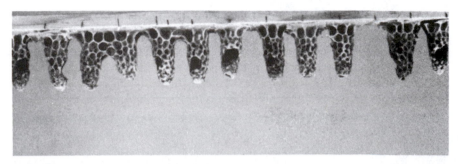

FIGURE 20. Queen cells destroyed by virgin queen.

FIGURE 21. Virgin queens fighting. *(From Harry H. Laidlaw, Jr. 1979. Contemporary Queen Rearing.)*

"Ripe" cells are sometimes *candled* to identify those with non-living queens by tilting them in front of a lamp. Living queens will "rock" in the cells. The unfolded wings of queens in queen cells less than 10 days old can be damaged by rough handling, but queens in older cells are less likely to suffer damage.

Records. Any action involving queens or queen cells requires some kind of record of what was done and when, and a "tickler" of the next action to take. Cell building records *must* have an entry of the date the cells are to be removed from the cell builder. Queens will usually emerge from their cells on the 11th or 12th day after grafting, and their first action is to destroy their emerged or unemerged rivals. Cells should be taken from the cell builder no later than the 10th day after grafting. Other entries such as date grafted, number of cells grafted, breeder number, cell builder or cell starter and cell finisher numbers, are useful additions to the essential "date out" entry. Besides records, weekly examination of cell builders for spontaneous natural queen cells will save grafted queen cells from destruction by virgins from spurious cells.

Arranging cell builders in three rows or groups, Fig. 22, and putting grafted cells into a different group on successive days is a rotation system by which each row receives newly grafted cells and yields 9-day-old cells every third day, or yields 10-day-old cells if started cells were transferred to them. Cells taken out on the 9th day from grafting can be incubated for one or two more days. They can remain for a day, however, toward the side of the cell builder when a new graft is added.

QUEEN MATING

Each virgin queen must be cared for by her own individual colony. She flies from this colony to mate, and subsequently she begins oviposition as the colony's queen. The colony may be any size from full, strong colonies to very small ones, each of which has its place in beekeeping.

Drones

Drones to mate the queens are a primary concern of queen rearers whether this is recognized or not. The quality of a stock can change rapidly if the drones of the mating area are genetically undesirable. Drones should be reared from a predetermined number of selected colonies and in exorbitant numbers. Drone rearing should begin about 45 days before virgins are ready to mate by placing drone combs in the broodnests of drone breeder colonies and adding pollen and honey, or sugar syrup, to the stores of food. Drones need both warmth and proper food to produce and mature spermatozoa and thus should be reared in strong colonies. Drones are ready to mate when about 10 or 12 days old.

FIGURE 22. Queenright cell finishers arranged in rows to simplify record keeping.

Mating Colonies

Full colonies. Virgin queens will mate from full size colonies which have received "ripe" (10 days or older) queen cells or virgin queens to replace the colony queen that was removed, or to requeen a queenless colony.

Colony divisions ("Divides"). Divisions for increase can be established with ripe queen cells or virgin queens, and the virgins will mate from them. This is an economical way to obtain queens for newly established colonies.

Nuclei. When numerous queens are needed, it is necessary to establish special colonies to care for the virgin queens until they mate and begin to lay. These colonies are domiciled in hives of various sizes and shapes called nucleus boxes, or "nucs."

Board divides. The sometimes enormous number of nuclei queen rearers have, and their variety, all too often deter beekeepers from rearing queens for their own use. There is, however, one kind of nucleus all beekeepers have and it is easy to use. It is simply a hive body set on an inner cover. The inner cover has the bee escape hole closed by a screen or thin block, but has a small entrance, Fig. 23, in one end. Two or three combs with emerging or sealed brood and adhering bees, and a frame with honey are moved from the broodnest of a good colony to stock the nucleus. The nucleus is set on the top

FIGURE 23. Board divide for mating queen. *(From Harry H. Laidlaw, Jr. 1979. Contemporary Queen Rearing.)*

body of the donor colony with the nucleus entrance to the rear of that of the colony below, and the nucleus is covered with the hive top. A queen cell is attached to a comb of the nucleus at or soon after makeup and the small entrance is closed with a moving screen that allows the bees to come out and crawl around the entrance. This screen is removed in about 1 or 2 days during which the queen has emerged and the bees have organized into a cohesive colony. The queens produced are convenient for requeening the donor colonies, and are suitable for most queen uses.

FIGURE 24. 5-frame standard mating nucleus.

Standard frame nuclei. Standard full-frame nuclei are the best to use when queen rearing is extended through the summer or into the early fall. They winter well in mild climates. They have five standard combs in a body that is half the width of a standard 10-frame Langstroth hive body, Fig. 24. They are made up with brood, bees, and honey from above the excluder of a queenright colony in the same manner as the board divides. They, too, should be confined by a moving screen over the entrance and lower hive front for 1 or 2 days to protect the virgin and bees as they become a colony, even when they are moved at makeup to another location.

Another popular standard frame nucleus is a standard full depth body and bottom divided lengthwise into two compartments with the entrance of each compartment at opposite ends of the body, Fig. 25.

Shallow frame nuclei. These nuclei are usually shallow 8-frame or 10-frame bodies and bottoms divided crosswise into four nuclei, or are divided both crosswise and lengthwise, Fig. 26. At the close of the queenrearing season, in mild climates, these nuclei can be dequeened, the bottoms removed, and the bodies with combs and bees stacked on nuclei in which one of the four queens has been left. If properly prepared, the resulting colonies will winter and nuclei can be remade from them the following year.

FIGURE 25. Two 5-frame standard mating nuclei in a 10-frame standard body. Robber screen on entrance. *(From Harry H. Laidlaw, Jr. 1979. Contemporary Queen Rearing.)*

The shallow-frame multiple nuclei are particularly convenient to maintain during the queen rearing season by simply transferring brood from a strong nucleus to a weaker one of the same four-plex. This should be done whenever the nucleus failed to yield a laying queen because brood from the queens that are mated from nuclei is needed to maintain bee populations of nuclei.

These nuclei are economical to operate and in this respect are between the standard frame nuclei and the small "baby" nuclei.

Baby nuclei. The proper use of small, or "baby," nuclei can help keep queen production costs comparatively low.

Figure 26. Four-plex shallow body nucleus. *(From Harry H. Laidlaw, Jr. 1979. Contemporary Queen Rearing.)*

Baby nuclei, while similar to each other, are not standard in size or proportion. There are, however, two distinct sizes. The larger is usually as long as the width of an 8-frame shallow body and wide enough to hold three shallow-depth frames, Fig. 27. An auger hole at one end is the entrance. A similar hole at the other end covered by screen wire provides ventilation. The cover may be made of two 1/4" thick "shook" boards half an inch apart. A feeder made of a tin can into which crumpled pigeon wire is inserted is located at the rear of the nucleus box.

The smaller baby nucleus is about 7" long, 6" wide, and 6" deep. Four frames fit crosswise in this nucleus, though usually only two or three are used so that a feeder may occupy the space of one or two frames at the rear of the nucleus box, Fig. 28. An auger hole entrance is at one end and a screen-covered ventilation auger hole is at the other.

Even smaller nuclei, called "pee-wee" nuclei, are sometimes used. In fact, queens can be mated from any domicile that has room for the baby colony. Baby nuclei can be dequeened after they are no longer needed and the frames stacked in special bodies on regular colonies so they can be remade in the Spring with bees and possibly some brood that are divisions of a colony. It is more customary, however, to remake baby nuclei without brood and with bulk bees at the beginning of each queenrearing season. This must be done with care because the small colonies are somewhat unstable.

FIGURE 27. Large "baby" nucleus. *(From Harry H. Laidlaw, Jr. 1979. Contemporary Queen Rearing.)*

FIGURE 28. Medium size "baby" nucleus. Robber screen over entrance hole. Inhabiting colony bees enter the nucleus from beneath the metal strip and the bottom board. *(From Harry H. Laidlaw, Jr. 1979. Contemporary Queen Rearing.)*

To establish the baby nuclei, nucleus boxes with combs are assembled in a "production line" in a protected place. The nuclei entrances are closed and the feeder cans are filled with sugar syrup. A comb is withdrawn, and a ripe queen cell is attached to one of the other combs. One cup of bees from the bees that were shaken from an apiary into a bulk bee box is dumped into each nucleus. The frame is replaced and the nucleus covered. The made-up nuclei are kept in a cool, darkened room for two or three days, after which they are placed on location and the entrances opened.

During the time the bees were confined, the combs were renovated by the bees, the sugar syrup was processed and stored, and the virgin queens emerged so the small colonies became functioning units.

Robbing

Robbing is an ever-present hazard in queenrearing. It is a problem that can be avoided by minimizing exposure of honey to flying bees, and by use of robber screens on the entrances of nuclei and regular hives.

Robber screens are like colony moving-screens in that they allow confined bees to come out of the hive entrance and walk around on the front of the hive. They differ in being open across the top, or in not covering completely the hive entrance but leave a small entrance to one side of the hive, Fig. 25. The bees of the colony use the small entrance, or the one across the top, and robber bees try to enter the hive through the screen because they are attracted to the hive odors that come from the screened main entrance.

Robbing screens for baby nuclei cover the small entrance hole in the front end of the nucleus box, but they protect the same way, Fig. 28.

USING THE QUEENS

A young queen has usually laid a comb well filled with eggs by 14 days after the ripe cell was given to her nucleus, and she is ready to be installed as the new mother of a colony.

Clipping and Marking

When a young queen is used in the queen rearer's own apiary, it is good practice to clip and mark her. In clipping, about half of both wings on one side of the body should be removed. Clipping the right wings in even numbered years and the left wings in odd numbered years is an easy way to register a queen's age, and it is recognized in the apiary.

Color marking the thorax is another way to document a queen's age if different colors are used for different years. An international color marking scheme has been devised where *white* or *grey* represents the years ending in 1 or 6; *yellow* is for years ending in 2 or 7; *red* denotes years ending in 3 or 8; *green* indicates years ending in 4 or 9; and *blue* signifies years ending in 5 or 0.

The paint used should be quick drying and non-offensive to the bees. Paint is easy to apply with a nail that has one end *flat* and the other attached to a knob. The paint bottle is closed with a stopper that has a channel in the center for the daubing nail to slide through. When the nail is withdrawn the paint on the sides is rubbed off leaving only the paint on the flat end. When this is touched to the thorax, Fig. 29, a clean round mark is left. Avoid letting paint run down the side of the thorax. Paint on the thoracic spiracles or on the leg joints will injure the queen and she will be prematurely superseded.

Caging, Shipping, and Storing Queens

Queens must be caged when they are taken from their colony. The standard queen cage in the United States is the Benton 3-hole wooden cage. One hole of the three large holes in the cage is for the candy food, and the other two holes are for the queen and also for her attendants if she is to be shipped. These holes are covered with screen wire. A smaller hole in the candy end of the cage is left exposed when a queen that is being introduced in a colony is to be released by the bees' eating the candy. A like hole in the other end of the cage is the entry hole for the queen and her attendants, Fig. 30. Two-hole cages are used to ship queens without attendants or candy as part of combless packages.

Queen cage candy is made from powdered sugar or Drivert kneaded into

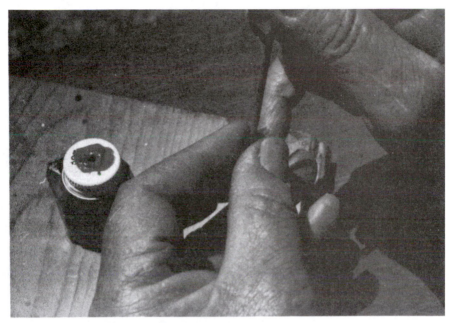

FIGURE 29. Marking laying queen. *(From Harry H. Laidlaw, Jr. 1979. Contemporary Queen Rearing.)*

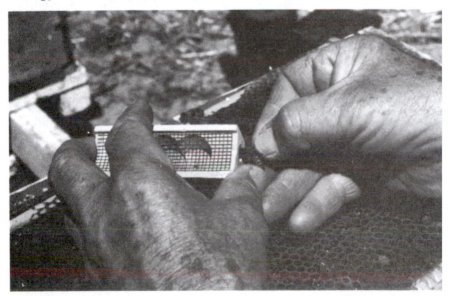

FIGURE 30. Caging laying queen. *(From Harry H. Laidlaw, Jr. 1979. Contemporary Queen Rearing.)*

a firm, but not hard, dough with invert sugar syrup, or with high fructose corn syrup. Honey and powdered sugar make excellent queen cage candy, but since honey may carry bee disease organisms, it is illegal to ship queens with honey-made candy.

Caged queens can be stored in a queenless colony for several weeks to a month or more. The storage colony should be strong and frequently replenished with bees by addition of sealed and emerging brood. Fumagillin in sugar syrup should be fed to the colony as a preventative of Nosema.

Queens are shipped in the 3-hole cages either singly or as bundles or boxes of cages. Single or a few cages with queens ship well in an envelope provided with air holes. Another excellent shipping device is a screened crate that holds about 100 3-hole cages on end in rows. The queens are caged alone in 3-hole cages and pads of queen cage candy are placed between the rows of queen cages. Bees are shaken into the crate as queen attendants.

The nuclei are fed sugar syrup in a feeder, or dry Drivert poured onto the bottom board, immediately after queen removal, and they can be re-celled soon afterward.

Introducing Queens

Installing queens in colonies is known as "introduction," which is an accurate descriptive word because the whole process is one of "enticing" the colony to accept the proffered queen as its own.

The bees and queen recognize each other by their respective pheromones, and both the queen and the bees recognize their mutual colonial relationship. Thus, queen introduction is a matter of developing acquaintance between the queen and the colony. It is habituation, an adjustment wherein hostile reflexive actions diminish with familiarity. It takes time for this, and the protection of the queen in the midst of her new colony is necessary, unless the colony is thrown into complete disorder.

Queens shipped in packages are readily accepted when the packaged bees are installed. Severe disruption of the colony gives similar results. Releasing a queen onto a comb of emerging workers in an incubator or in a hive body above a screen is a sure way to introduce queens, but except for very valuable queens such methods are not practical.

The colony that is to receive a new queen must be queenless, of course, which may be difficult to verify because a supersedure queen or spurious virgin may be overlooked. A light nectar flow from the field, or feeding light sugar syrup, contributes to successful introduction. The queen should preferably be released onto a comb in the broodnest and confined by a "push-in cage." She should be laying in a day or two, and by the 3rd to the 4th day the bees should be quiet on the cage. If so, the cage may be removed. If there is any bee biting the wire of the cage, the queen should be left confined until there is no animosity. Cage biting that continues beyond the 5th day suggests that another queen is loose in the colony.

Queens can be introduced from the 3-hole cage by leaving her confined in the hive for 3 or 4 days, or by letting the bees release the queen by eating the candy in the cage. If the queen is released too quickly after she is put into the hive, she may be killed. The practice of making a hole through the candy from the candy end of the cage should be followed only when the candy is quite hard. Otherwise, the queen may be released too soon.

INSTRUMENTAL INSEMINATION

STATUS AND VALUE OF INSTRUMENTAL INSEMINATION OF HONEY BEE QUEENS

The honey bee virgin queen mates in free flight; never in the hive. This mating independence of queens discouraged beekeepers from attempting to breed bees as other animals are bred, and for this reason control of the mating of virgin queen bees was a tantalizing desire of beekeepers for more than a century before it was attained. Isolation of virgins and drones on islands, prairies, high elevations, or in far northern latitudes was successful and is still used for breeding, but is impractical for most beekeepers and inadequate for genetic and breeding studies. Confinement of virgins and drones in cages or other enclosures for mating was convenient and appealing, but was rarely successful. Confining queens and drones to their hives and releasing them during only certain times and hours was not much better. The obvious solution to the problem was artificial insemination, a method of mating control long practiced on domestic animals, but not on insects.

The history of artificial insemination of honey bee virgin queens (the only insect for which a practical method has been devised) is a long story of interesting attempts that failed until 1927 when Lloyd R. Watson published an account of the research he conducted on artificial insemination of queen bees as a partial requirement for the degree Doctor of Philosophy. Failures still occurred during and after Watson's experiments, but Watson demonstrated that queen bees can be inseminated artificially. He, for the first time, used modern laboratory instruments, and he gave us the term "instrumental insemination."

Watson's success, though partial and erratic, revived hope that practical control of virgin queen mating would become a reality, which it did in 1939 through the researches of Nolan, Laidlaw, and Mackensen. Elements from the contributions of each of these men were combined by Mackensen into a highly efficient method of instrumental insemination. Variations and refinements have been made by others since then, though with only minor improvement in the already high efficiency.

Instrumental insemination of queen bees is not difficult for those familiar with the use of microscopes. Having stated this, it is necessary to add that unless the equipment is suitable, clean, accurate, and functions smoothly, instrumental insemination will be disappointing, if not frustrating. A genuine interest in inseminating queens is most helpful also, but it must be emphasized that success depends on preciseness and strict sanitation.

A necessary auxiliary to instrumental insemination is specialized beekeeping which is often overlooked by learners: rearing and maturing of drones and virgin queens, pre-insemination and post-insemination care of queens, and introduction of inseminated queens are crucial. Satisfactory use of instrumental insemination depends on careful program planning, because instrumental insemination can be good or bad for a beekeeper's stock.

NATURAL MATING

Study of natural mating of queen bees was a logical prelude to artificial insemination attempts. These studies revealed that queens at mating receive semen into the lateral oviducts from which the sperm move to the spermatheca where they are stored for the life of the queen. A flap-like invagination in the lower wall of the vagina, the valvefold, Fig. 31, is retracted by the queen at mating, but prevents flow of semen to the oviducts in instrumental insemination unless it is lowered by the inseminator or otherwise by-passed during injection of semen from the syringe.

It was also found that queens normally mate with several drones on one mating flight or on more than one flight. These observations partially explained failures and incomplete inseminations, and led to modification of early techniques and instruments.

INSTRUMENTAL INSEMINATION
Instruments Needed

Regular biological laboratory equipment required includes a dissection microscope having magnifying power of 10 or 15X (times) to enable the inseminator to accurately guide the movements of the small insemination instruments as they are used. A microscope lamp is essential, and a heat-absorbing disc over the lamp lens is recommended. Small forceps and small scissors are needed. A cylinder of carbon dioxide (CO_2) with valves, gauge, and flexible tubing is also essential laboratory equipment.

Specialized equipment consists of the instrumental insemination instrument or queen manipulator, and the syringe to take up and inject the semen. The most popular insemination instrument is the Mackensen, Fig. 32. A more

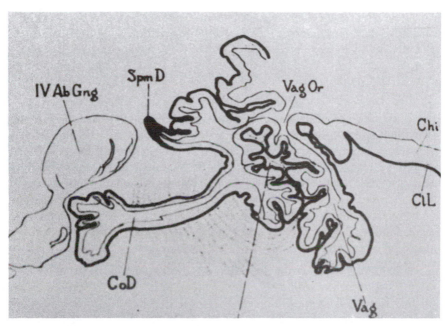

FIGURE 31. Midline histological section through vagina, valvefold, and median oviduct.

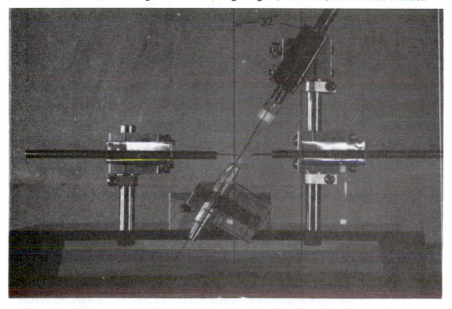

FIGURE 32. Mackensen insemination instrument with syringe and adjustment guide. *(From Harry H. Laidlaw, Jr. 1977. Instrumental Insemination of Honey Bee Queens.)*

advanced and highly engineered instrument, and one based on the Mackensen design, is the West German Schley instrument that is a duplicate of the Ruttner-Schneider-Fresnaye refinement of the Tryasko and Vesely devices. The Schley device is easy to use because it has a syringe holder that is controlled by sliding pieces and screws that give positive and fine control of syringe movement.

Both the Mackensen and the Schley instruments have their own syringe. The Mackensen syringe has a unique design and is easy to use and clean. It is a popular syringe. A third syringe, devised by Harbo, is gaining acceptance. This is simply a glass tube that is moved up and down in or with the syringe holder and in which a similar length glass capillary tube is fastened. The lower end of the inner capillary glass tube has a drawn-glass insemination tip or needle linked to it by small ID flexible tubing, and the upper end is attached by the same kind of flexible tubing and a hypodermic needle base to a Gilmont micrometer syringe.

All three of these syringes need a liquid plunger to move the semen. A *sterile* physiological saline solution is suitable because the semen is only slightly diluted with the fluid plunger while it is taken into the syringe or injected into the queen and is unlikely to be seriously damaged. To this plunger fluid may be added the antibiotic, streptomycin sulfate, in the proportion of 0.25% to the total volume. An antibiotic in the plunger is recommended, but it cannot assure protection of the queen against unsanitary semen handling procedures.

Rearing and Care of Drones

Drones for insemination of queens can be marked at emergence and permitted to fly, or in lieu of marking be caged by some method in their hive. Their colony should be strong and supplied abundantly with pollen and honey or syrup. Drones mature in about 12 days and remain viable for about two or three more weeks. There is much variation in their longevity; and rapid early death is common. It should be recognized that haploid drones are not a true filial generation, but are genetically their mother's genomes and as such their selection and use is a powerful genetic procedure. A drone produces male reproductive cells, the sperm, by mitoses and aborted meioses that clone and multiply the drone's genotype which is the genome received from his mother. Diploid drones are members of the same filial generation as their true sisters, but they are eaten as larvae soon after they hatch from the egg.

Insemination Procedure

The first step in instrumentally inseminating honey bee virgin queens is preparation of a clean laboratory-like work station in a suitable location. The paraphernalia of insemination are put in place and adjusted, and sterile saline solution is taken into the syringe as a plunger; enough fluid must then be extruded to allow a full load of semen to be in the syringe when the plunger is retracted. The plunger is then moved toward the end of the syringe tip leaving space of about 2 mm for an air bubble between the plunger and the semen. The tip may be wiped with clean facial tissue or medical cotton swabs, but should never be wiped by unwashed fingers after drones or queens are handled.

Virgin queens that have been exposed to CO_2 for about 5 minutes the day or two before are brought caged, 3 or 4 at one time, to the work station and a small ball of queen cage candy is put in their cages. Or, their cages are laid over the screen of a cage of worker bees that have been fed sugar syrup by brushing or lightly spraying sugar syrup on the screen of their cage.

Drones are brought caged to the work station, 30 to 50 at one time, and set in a drone flight box, Fig. 33. Only a few are released from their cage into the flight box at one time because they soon tire and then do not ejaculate semen well. The syringe should be filled with semen *before* the queen is opened because the queen's membranes dry quickly when exposed to air which interferes with placement of the syringe in the reproductive tract of the queen.

Semen is obtained from drones that have been induced to evert the copulatory organs and to ejaculate the semen. This can be brought about by pressing the head and thorax of the drone, Fig. 34, which usually stimulates abdominal muscle contraction with pressure on the blood so that the drone's copulatory or intromittent organ, called the penis or phallus, everts to the

FIGURE 33. Drone flight box. *(From Harry H. Laidlaw, Jr. 1977. Instrumental Insemination of Honey Bee Queens.)*

exterior of the abdomen and the seminal vesicles ejaculate the semen. Simply mashing the drones does not initiate this reaction. Eversion normally stops before it is complete, Fig. 34, but ejaculation has taken place, and the semen that passes through the ejaculatory duct is confined in an uneverted part of the penis. Fully mature drones may be so sensitive to handling that they evert fully, which is followed immediately by an explosive ejaculation of semen into the air. This loss of semen can be avoided by piercing the base of the drone's abdomen with a needle before stimulating the drone.

Eversion must be continued before semen is exposed and can be taken into the syringe. This should be done carefully by rolling the thumb and forefinger together toward their tips on the base of the partially everted penis. Pressure should be relaxed as soon as the eversion resumes, so it will stop when the cream-colored semen is exposed adjacent to or on the white extruded mucus, Fig. 35. The semen should be examined carefully at this point for any evidence of fecal material from the queen or the drone. If semisolid or liquid extraneous matter is seen, the drone and semen should be discarded. Likewise, if the semen has touched the operator's fingers or external parts of the drone, it may be contaminated and it should not be used.

The semen and the syringe tip are brought together so they *just touch,* and are then moved slightly apart to form a small cone of semen. Semen is now drawn slowly into the syringe. After the semen is taken into the syringe, it is withdrawn about 2 mm to avoid a seal of dried semen from forming over the tip orifice. A second, and the succeeding drones, are manipulated the same way, except a small drop of semen is extruded from the syringe to touch the

FIGURE 34. Everting and ejaculating the drone. *(From Harry H. Laidlaw, Jr. 1977. Instrumental Insemination of Honey Bee Queens.)*

exposed semen of the penis and thus ensure against taking mucus into the syringe, Fig. 36. Mucus would coagulate in the syringe and the queen's reproductive tract to occlude them. When about 8 cubic millimeters of semen have been taken into the syringe for one queen, or 8 cu mm for each of several queens, the semen is withdrawn about 2 cu mm from the end of the tip and the tip is inserted into saline to take in saline to form a saline "stopper" to prevent drying of the semen, Fig. 37. This saline will be injected into the queen ahead of the semen. The loaded syringe is moved up and aside while the queen is prepared for injection.

A caged virgin queen is now lowered into a container into which carbon dioxide is gently flowing through tubing connected to the container wall near

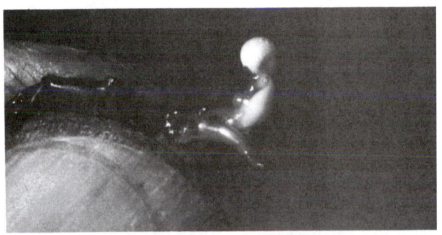

FIGURE 35. Eversion continued with release of semen and mucus to the exterior. *(From Harry H. Laidlaw, Jr. 1977. Instrumental Insemination of Honey Bee Queens.)*

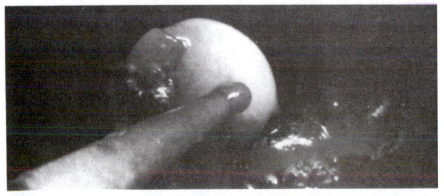

FIGURE 36. Taking semen into the syringe; semen extruded from syringe between successive drones. *(From Harry H. Laidlaw, Jr. 1977. Instrumental Insemination of Honey Bee Queens.)*

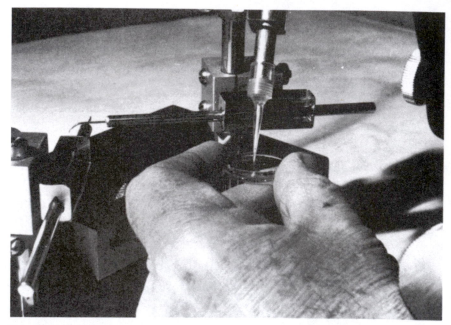

FIGURE 37. Taking minute amount of plunger fluid into end of loaded syringe tip to prevent semen from drying and forming a seal. *(From Harry H. Laidlaw, Jr. 1977. Instrumental Insemination of Honey Bee Queens.)*

FIGURE 38. Anesthetizing a virgin queen.

the bottom, Fig. 38. Soon after the queen becomes motionless, in about 1 or 2 minutes, she is put head first into a backup tube. The base end of a queen-holding tube is then pressed against the open end of the backup tube and the queen is blown gently into the queen-holding tube. The queen-holding tube stopper, which is connected by flexible tubing to a cylinder of CO_2, is inserted into the queen-holding tube until the terminal three abdominal segments of the queen protrude. The tube with queen is then fastened in the queen-tube mount that is slanted about 30° toward the syringe holder, and the flow of CO_2 is adjusted to a *slow stream.*

The opening of the reproductive tract of the queen is situated at the middle of the anterior wall of the sting chamber and is overlaid by the base of the sting. The sting chamber is opened and the sting pulled dorsally with forceps or by specially designed hooks. The dorsal, or sting hook, should be "hoe-shaped" with the "blade," about 0.5 mm wide, perpendicular when in place in the queen, and a section of about 3 mm of the "shaft" behind the blade should be deflected downward about 35°.

The two hooks are inserted into the sting chamber, Fig. 39, and the sting hook is fitted between the sting lancet bases. The ventral hook is positioned

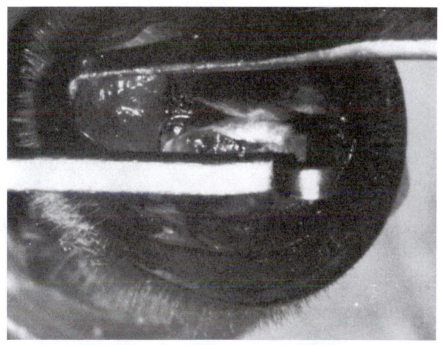

FIGURE 39. Opening the sting chamber with hooks. *(From Harry H. Laidlaw, Jr. 1977. Instrumental Insemination of Honey Bee Queens.)*

against the ventral wall of the sting chamber to pull it ventrally until the sting chamber is sufficiently open, Fig. 40. With the sting hook the sting is now pulled dorsally, Fig. 41, drawing the dorsal edge of the vaginal orifice from under an overlying membranous fold and changing the vaginal orifice from a transverse slit into a dorso-ventral one.

The syringe, in alignment with the queen, is lowered into the sting chamber and is adjusted so the point of the syringe tip is directly over the middle of the vaginal orifice, Fig. 42. A little of the saline solution is discharged from the syringe to lubricate the tip, and the syringe tip is slipped into the vagina about a millimeter, Fig. 43. If properly placed, the syringe tip slides in easily and appears to slide under the ventral membranous fold by-passing the end of the valvefold. The positioning of the syringe is tested by advancing the liquid plunger slightly. If the semen moves easily, injection should continue until the bubble of air between semen and plunger disappears in the queen. If the column of semen begins movement and stops, the syringe is improperly placed, or there is mucus in the syringe. The plunger should be retracted slightly and the syringe re-inserted into the vagina, perhaps after some read-justment of the hooks.

The valvefold may also be by-passed by inserting just the end of the syringe tip into the vaginal orifice, moving it slightly dorsally and then

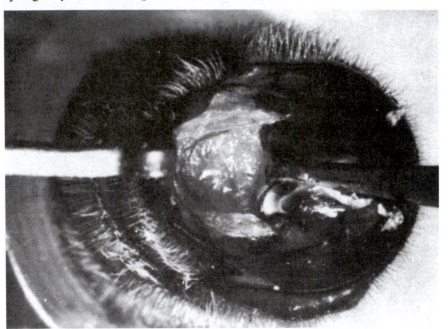

FIGURE 40. Hooks properly positioned.

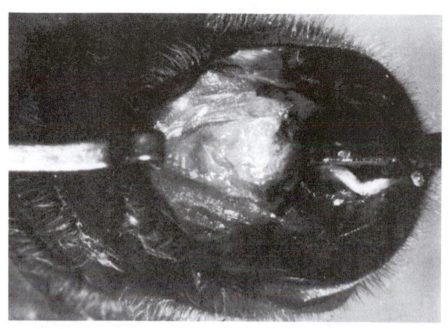

FIGURE 41. Sting pulled dorsally to expose vaginal orifice. *(From Harry H. Laidlaw, Jr. 1977. Instrumental Insemination of Honey Bee Queens.)*

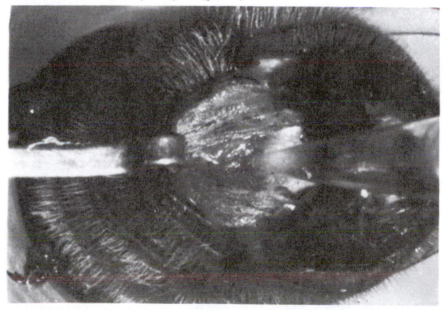

FIGURE 42. Positioning syringe tip for insertion into the reproductive tract of the queen.

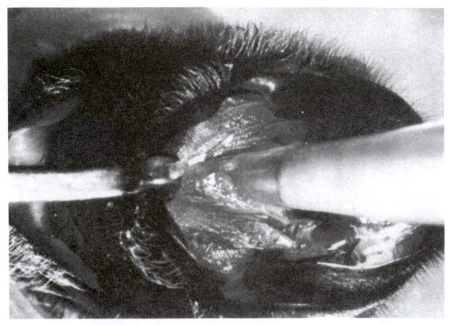

FIGURE 43. Injecting semen into the queen's reproductive tract.

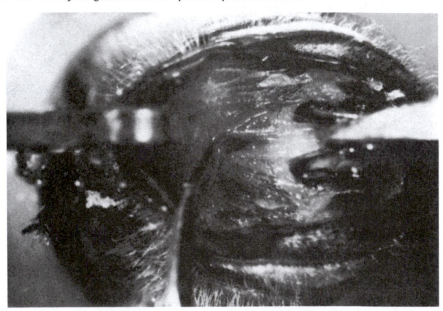

FIGURE 44. Use of probe to lower the valvefold so the syringe can be inserted beyond it. *(From Harry H. Laidlaw, Jr. 1977. Instrumental Insemination of Honey Bee Queens.)*

ventrally before the syringe tip is inserted deeper. The original method of by-passing the valvefold is to push it ventrally with a probe, Fig. 44. This is still preferred by many workers.

Pre- and Post-Insemination Care of Queens

Breeding the honey bee requires fully developed queens free from injuries, as do most genetic studies of honey bees. Virgin queens can be abused and mutilated by worker bees so that they will be handicapped and will be superseded untimely. Caged queens in nurseries may have their feet chewed off unless they are protected by screen wire over the large mesh wire of the cage, that leaves only one row of cage wire exposed for close contact of queen and workers. Caged virgins in nurseries are usually fed by worker bees of the nursery colony, but some may be neglected and starve. It is good practice to supply each nursery cage with a small pellet of queen cage candy.

Virgins emerged in cages in nuclei are less likely to be damaged by the bees of their nuclei than if they are emerged elsewhere and then introduced. Nevertheless, use of the same kind of nursery cages provisioned with candy is added protection for the queen.

The period immediately following insemination is also a time of queen vulnerability, and is especially a critical time for movement of sperm from the oviducts to the spermatheca. Warmth is required for this movement which with other care is best provided by worker bees when the newly inseminated queen is loose among them. Though the number of sperm that reach the spermatheca may be diminished somewhat, it is often necessary or preferable to emerge virgins in cages in nursery colonies and return them caged to the same nursery colony. Such queens have performed equally with queens returned to their nuclei and released following insemination.

Semen Shipment

Instrumental insemination is a way that is available for shipping honey bee genetic material almost anywhere in the world without danger of exporting bee diseases. The semen is shipped in capillary tubes sealed at both ends with Vaseline close to the column of semen. The tubes of semen are kept at ambient temperature and away from sunlight. When shipped, the tubes are laid in saw kerfs in thin pieces of wood and held in place with tape. Semen should survive in the tubes for more than two weeks. Dusting the capillary tubes with streptomycin sulfate before filling them with semen is recommended.

PACKAGE BEES

A.I. Root in 1878 first suggested shipping bees in combless packages, and in 1912 W.D. Achord made the first successful shipment of packaged bees in the U.S. The bees were sent from Alabama to the North. The cages were similar to the ones used currently, but with candy food. The cages were each filled with one to three pounds of bees by shaking the bees through a funnel into the cage. It was necessary to find the queens before bees were shaken from the combs and this procedure has persisted, but has been largely superseded by the use of "shaker boxes."

Package bee production differs from honey production in that the principal commercial product of the package bee industry is bees which must be reared so they are available for shipment in early April. This requirement highlights the need for good management throughout the year and especially immediately after the shipping season closes, and in early spring. Colonies for producing package bees must reach near swarming strength much earlier than is normal for honey-producing colonies. To accomplish this and keep the bees in their hives before they are caged for shipment depends on special beekeeping skills.

Management of Package Bee Producing Colonies

The bee shippers' new year begins with the end of a shipping season. As the season comes to a close, queens in all of the package-yielding colonies are replaced with either young laying queens that were reared toward the end of the package shipping season, or with "ripe" queen cells. The requeened colonies can be used in pollination service and may yield some honey. However, the post-shipping season is mainly a period of developing the colonies into condition for wintering, and for accumulation of colony stores of honey and pollen for next year's early spring buildup.

Trapping pollen and storing it in a freezer, or segregating filled combs of pollen in bodies over excluders on strong colonies will provide convenient sources of pollen when it is needed during late winter and early spring and is a wise precaution even in areas of usual pollen abundance. Trapped pollen can be shaken into combs and sprayed with sugar syrup, or it can be made into patties with sugar syrup for feeding above the frames or in special pollen boards or screens.

When shipping season arrives there is little time to assemble or repair equipment or supplies. For this reason the necessary equipment repairs are made and supplies are acquired during the fall and winter months.

Shipping Cages

Bees are now shipped mainly by specially equipped trucks, and bee shipping cages have become standardized, especially those that are purchased. Beekeepers who make their own cages also tend to adhere to a standard design and construction. The popular 2-pound package cage, Fig. 45, is 8-5/8" high x 12" long x 6" wide. The 3-pound package cage measures 16" long with the other measurements the same as those of the 2-pound cage. The top, bottom, and ends are solid, and the sides are screened. An "H"-shaped or "U"-shaped feeder can support is nailed across the middle of the cage bottom. A hole for a

2-1/2 size can is cut in the center of the top board, and a saw kerf is made in one side of the can hole to receive the metal strip attached to the queen cage. The queens are caged without attendant bees, but their cage is suspended next to the feeder can.

Figure 45. "Standard" bee shipping cage. *(From Harry H. Laidlaw, Jr. 1979. Contemporary Queen Rearing.)*

The feeder cans, Fig. 46, are filled with 50-50 sugar syrup, with liquid sugar diluted with water to bring the solids to 50%, or with high fructose corn syrup similarly diluted. The bees obtain the syrup from the inverted feed can through small holes in the lid or through a large central hole covered with Indian Head Spring Made Decorator line or similar cotton cloth. After the can is sealed with a can sealer and then filled, the cloth patch is fastened to the hole from the outside by a plastic ring pressed into the hole with the cloth but allowing some slack in the cloth. Alternatively, the patch can be fastened to the inside of the can lid with automobile body repair speed spot putty that has been thinned with auto speed putty thinner until it has the consistency of heavy cream or honey. The putty seal must dry for at least 24 hours before use. These cans are filled with syrup and the lids then sealed with a can sealer.

Filling the Package Cages

Each package must have a young laying queen. From necessity, package

FIGURE 46. Package cage feeder cans. *(From Harry H. Laidlaw, Jr. 1979. Contemporary Queen Rearing.)*

bee shippers are also queen rearers because there are no adequate or reliable sources of young laying queens at the time package bees are shipped. A light first "shaking" of bees from the colonies is used to stock queen mating nuclei.

Before the first packages are filled, the queens for them are caged from mating nuclei into 2-hole queen cages without food or attendants, and their cages are fitted into cage-holding frames, Fig. 47, and placed in carrying boxes stocked with bees and food for transport to the apiaries supplying the bees for the packages. After the queens are removed from their nuclei, they may be held a short time in strong, queenless "queen banks" that are fed fumagillin.

Colonies that supply bees for packages are usually over-wintered in two full-depth brood bodies. If the queens are confined to lower bodies by queen excluders, part of the bees of a colony can be driven up into the body above the excluder by removing the hive top and blowing smoke into the colony's entrance, Fig. 48. Drones are excluded, as is the queen, and the bees can be shaken from the combs of the top body through a funnel into the package cages and weighed, Fig. 49. A caged queen is fastened in the package cage so she is near the feed can, Fig. 45.

A similar method is to maintain the queen in two bodies without an excluder between. The bees are smoked up through an excluder placed over the top body and into a "cluster box" that is a shallow body with perpendicu-

FIGURE 47. Two-hole queen cages for shipping queens in packages. *(From Harry H. Laidlaw, Jr. 1979. Contemporary Queen Rearing.)*

FIGURE 48. Bees driven up above excluder in 2-story queenright colony. *(From Harry H. Laidlaw, Jr. 1979. Contemporary Queen Rearing.)*

FIGURE 49. Filling and weighing cage with bees driven above an excluder. *(From Harry H. Laidlaw, Jr. 1979. Contemporary Queen Rearing.)*

FIGURE 50. Cluster box with bees *(From Harry H. Laidlaw, Jr. 1979. Contemporary Queen Rearing.)*

lar cross boards for the bees to cluster on, and a screened top, Fig. 50. The bees are jarred from the cluster box into a shaker box from which the excluder tray has been removed to be poured into package cages, Fig. 51.

Shaking the bees into a modern, light "shaker box," Fig. 52, is one of the easiest and most efficient ways to get package bees from the colonies.

The shaker box is a lightweight contrivance for screening queens and drones from bees as they are shaken from hives and are congregated before being put into packages for shipment. The upper of the three parts is a removable tray made with queen excluders. The middle part is a funnel to slide the screened bees into the lower collecting cage. Queens that may be shaken from the hives are caught in the excluder tray and are returned to their hives. When about 10 or 12 pounds of bees have been accumulated, they are funneled from the collecting cage into a "weighing container" that is set up on a balance, Fig. 53. The bees are dumped from the weighing container into a package cage while a second weighing container of equal weight is being filled.

The empty package cages are fastened together at the warehouse before they are taken to the apiaries to be filled. Emptied shaker boxes are repeatedly exchanged for those with bees so there is a continuous filling of package cages.

FIGURE 51. Jarring bees from cluster box through funnel and into collecting cage. *(From Harry H. Laidlaw, Jr. 1979. Contemporary Queen Rearing.)*

FIGURE 52. Shaker box. *(From Harry H. Laidlaw, Jr. 1979. Contemporary Queen Rearing.)*

FIGURE 53. Dumping bees into weighing container, weighing, and filling package cages. *(From Harry H. Laidlaw, Jr. 1979. Contemporary Queen Rearing.)*

The shaker box was developed by package bee producers through small improvements over more than a decade, and has become an invaluable aid to all commercial beekeepers. The shaker box and weighing container are now indispensable tools of the packaged bee producer.

The queens and feeder cans are put into each cage after it is filled with bees. The bee donor colonies should be fed sugar syrup after filling of the cages is finished.

The filled packages should be kept in a cool dark place until shipment, and should be sprayed lightly with thin sugar syrup or water to quiet the bees.

Installation of Packages

The package bees should be put in a cool darkened place as soon as they are received, and sprayed lightly with thin sugar syrup that contains fumagillin. To be installed, the bees are taken to the apiary, three or four combs are withdrawn from the side of a hive, a division board feeder is filled with syrup, and the bees are mist-sprayed with water and dumped from the package into the space in the hive. The frames are replaced, and the cork removed from the queen cage before it is suspended between the middle combs. The emptied cage may be left in the hive beside the frames so any bees remaining in the cage can join the cluster. Covering the hive temporarily with a cloth before the hive top is returned to the hive prevents bees being crushed between frame bottom bars and the hive bottom board.

A robber screen, Fig. 54, placed on the hive entrance, is effective to protect the new colony.

FIGURE 54. Full-size hive robber screen.

REFERENCES

Harbo, John. R. (1985). Instrumental insemination of queen bees. *Amer. Bee Jour. 125(3)*:197-202; (4):282-287.

Laidlaw, H.H. (1975). Queen rearing. *Amer. Bee Jour. 115(10)*:384-387.

Laidlaw, Harry H., Jr. (1977). Instrumental Insemination of Honey Bee Queens. Dadant and Sons, Inc., Hamilton, Illinois.

Laidlaw, Harry H. Jr. (1979). Contemporary Queen Rearing. Dadant and Sons, Inc., Hamilton, Illinois.

Laidlaw, Harry H., Jr., and J.E. Eckert. (1962). Queen Rearing. University of California Press, Berkeley, Calif.

Mackensen, O., and K.W. Tucker. (1970). Instrumental Insemination of Queen Bees. U.S.D.A. Handbook No. 390. Supt. of Documents, Washington, D.C.

Nolan, W.J. (1932). Breeding the Honey Bee under Controlled Conditions. U.S.D.A. Technical Bulletin No. 326. Supt. of Documents, Washington, D.C.

Ruttner, Friedrich. (1976). The Instrumental Insemination of the Queen Bee. Apimondia Publishing House, Bucharest, Romania.

Ruttner, F. ed. (1983). Queen Rearing. Apimondia Publishing House, Bucharest, Romania.

Schley, Peter. (1983). Praktische Anleitung zur instrumentellen Besamung von Bienenköniginnen. Selbstverlag. Dr. Peter Schley, Heinestrasse 6,6301 Pohlheim, West Germany.

CHAPTER 24

CROP POLLINATION

by ROGER A. HOOPINGARNER[1]
and GORDON D. WALLER[2]

Bees gather nectar and pollen from flowers for food for their own use, and for rearing their offspring. While the bees are visiting flowers to collect nectar and pollen, they pollinate hundreds of different kinds of plants. Estimates of the value of bee-pollinated crops vary widely, but studies indicate that about 90 crops in the United States depend on bees, at least to some extent, for pollination. If we consider the bees' value to be based on the fruit, vegetables and seed resulting directly from pollination, we have a value that is about 150 times the value of the honey and beeswax. Ramifications can, in fact, be carried beyond the crop resulting directly from pollination. For instance, bee-pollinated alfalfa seed is in itself a valuable commodity, but alfalfa seed produces hay, which in turn produces meat and milk. Thus, pollination becomes an essential link in a chain of great significance to agricultural production. When we examine carefully all these contributions, we find that about one-third of our total diet comes directly or indirectly from bee-pollinated crop plants. Pollination of plants by bees also occurs in natural ecosystems. Hundreds of wild flowers, weeds, trees and other non-crop plants are bee pollinated. The whole complex is so involved in nature's web that an accurate estimation of the value is impossible. Without this total contribution by bees, we would certainly live in a very different, less productive and less interesting world.

If less-developed nations are to improve their diets, there is a need for a great world-wide increase, not only in bee-pollinated crops, but also in bees and beekeepers to manage crop pollination. We also need to increase our knowledge and understanding of what a potent force the honey bee is in food production and human nutrition.

Interest in the quality of our environment is causing us to look more carefully at factors upon which food production, health and aesthetic aspects of our environment depend. We are also being alerted to the need to conserve natural resources. Honey bees, and many species of solitary bees are resources of such great significance to human welfare that we must assure their survival for pollination purposes. Problems confront both the long-range survival of solitary bees and the keeping of honey bees. As human populations increase,

[1]Roger A. Hoopingarner, Ph.D., Department of Entomology, Michigan State University, East Lansing, MI 48824-1115.

[2]Gordon D. Waller, Ph.D., Retired, Carl Hayden Bee Research Center, Agricultural Research Service, U.S.D.A., 2000 East Allen Road, Tuscon, AZ 85719.

houses, factories and highways replace open fields and eliminate nectar and pollen plants. Clean cultivation (herbicides) of farm land and large-scale monoculture reduce the availability of wild plants needed to provide a succession or forage for bees throughout the season. Pesticides frequently poison both solitary bees and honey bees. Thus, we have a dilemma, with a reduction of profitable beekeeping and native pollinators on one hand and an increased need for bees for crop pollination on the other. Pollination problems need to be analyzed to develop the appropriate research and extension work. Data is needed on the importance of solitary bees and feral honey bee colonies. We also lack understanding of specific pollination requirements of crops and the biological basis of bees' activities on plants. Without this basic understanding, both beekeeping and crop yields might decrease, with growers simply unaware of pollination problems.

POLLINATION ECOLOGY

Ecology is the study of organisms in relation to each other and to their environment. Ecological relationships between bees and the flowering plants (angiosperms) of the world is an age-old story. Bees likely evolved from a wasp-like ancestor at the same time as the angiosperm plants, about 60 to 100 million years ago. The mutually beneficial relationship that developed between bees and plants has persisted to the present. Flowers provide nectar and pollen for bees and bees provide pollination for plants.

About 250,000 species of flowering plants have been identified. Many of these plants have complex relationships to bees and to other pollinators, including flies, beetles, moths, butterflies, birds and bats (Dowden, 1964; Meeuse, 1961). Cross pollination, the transfer or movement of pollen from one plant to another (outbreeding), provides greater genetic variability in the offspring than does self pollination. Thus, a plant with greater diversity is better able to adapt and compete in new environments and to occupy new ecological niches.

Bees have evolved branched hairs enabling each bee to carry many thousand pollen grains. Honey bees *(Apis* spp.) have pollen baskets, specialized mouthparts, a honey sac for transporting nectar, beeswax comb for storing honey and pollen, and specialized communication, all related to their association with angiosperm plants.

The advantages of cross pollination caused plants to evolve flowers designed to attract bees and mechanisms to prevent self fertilization. Flowers have visible (to humans) colors and also reflect ultraviolet light, which we can't see, but which bees see very well. Nectar guides in some flowers use both color and scent as identifying "labels" which assist bees in their search for nectar. Flowers have also evolved structures, physiological devices, or behavior to lessen or prevent self fertilization. For example, some plants have staminate "male" and pistillate "female" parts on separate flowers on the same plant (monoecious) or on separate plants (dioecious). Other plants increase the chances of cross pollination by having stamens and pistils mature at different times. An example of this is the onion flower where the anthers shed pollen before the stigma of that same flower is receptive to pollen (protandrous).

Among physiological means of avoiding self fertilization is the rather widespread phenomenon of self incompatibility, or relatively low self compatibility. Most apple varieties require the pollen of another clone for successful fertilization. Growers cannot produce a commercial apple crop when they only plant one variety. They must have suitable pollinizer varieties in order to get an adequate yield of fruit. Self incompatibility may be of two types: (a) the pollen tube may not grow to the ovula to deliver the nucleus, or (b) the pollen nucleus may not fuse with the egg nucleus to accomplish successful fertilization. Self incompatibility varies from 100 percent to a slight preference for foreign pollen, and can also vary from one environment to another. Incompatibility of clones (plants of identical genetic material) may be more widespread than agriculturalists have realized. For instance, the early inability to grow red clover seed in New Zealand may have been caused partially by a lack of bumble bees and partially by self incompatibility between plants derived from a single source. The latter may also explain why we could not produce Pellett (kura) clover seed in the United States when this crop was first introduced.

The bee fauna of the world is quite large, possibly 15 to 20 thousand species. There are about 4,000 species and subspecies in the United States and Canada. Of these, possibly 400 are social or semi social, 400 are parasitic and the remainder are solitary. The best-known social bee, the Western honey bee, *Apis mellifera* L., was brought here by European settlers and is not a native of North or South America.

A typical solitary bee's nest may be found in the hollow stem of a plant or in a burrow dug into the ground. Such openings are first lined, usually with cut leaves or other material, by the bee. A pollen ball is then placed in the cell and an egg laid on it. A properly prepared cell is sectioned off individually with pieces of leaf or soil partitions, and the process repeated for another egg. A solitary bee larva eats the pollen and grows into an adult bee without the help of nurse bees (mass provisioning). Some solitary bee species have only one generation per year whereas others produce several. One important aspect of nest-provision behavior of solitary bees is that they are always searching for pollen, and thus will not avoid the anthers (and often the stigma) of the flower. A solitary bee will nearly always pollinate the flower being visited which may not be true of honey bee foragers. Solitary bees may also only gather pollen from a few or just one species of plant. This ensures pollen transfer between the flowers of the same species. However, if they are not foraging the target crop, there is not much you can do to change their foraging preference. Another important aspect of the solitary bees is that they are solitary nest builders. (We will talk more about this later when we discuss in-hive pollen transfer.) Solitary bees are less important to pollination than honey bees because the number of solitary bees has been reduced by insecticides and/or the loss of nesting sites by the elimination of fence rows and woodlots.

Faegri and van der Pijl (1971) have suggested a terminology to more clearly define the ecological relationships between plants and their pollinators (Table 1). The term constancy, or fidelity, is an ecological term which refers to a bee's restriction to one species of flower on one trip or a series of trips. It is a behavioral phenomenon distinctive from monotropy or oligotropy. It is best exemplified by the honey bee which is polytropic, but when an individual honey bee finds a source of nectar or pollen she continues to gather from that one source (constancy). This behavior helps to ensure the arrival of pollen from the same species as the plant being visited.

Table 1. Blossom-Pollinator Relationship

Blossom	Pollinator
Polyphilic - Pollinated by many different species of visitors.	**Polytropic** - Visiting many different species of flowers, *e.g.,* Honey bees.
Oligophilic - Pollinated by a few closely related species of visitors.	**Oligotropic** - Visiting only a few species of plants, *e.g.,* many solitary bees.
Monophilic - Pollinated by one or some closely related species of visitors.	**Monotropic** - Visiting a single or some closely related species of plants, *e.g.* orchid bees.

Foraging Behavior of Bees

The goal of honey bee management is to guide the bees' natural behavior into patterns of activity desired by the beekeeper. Knowledge of bee behavior is essential in management for either honey production or pollination. Honey bees communicate direction, distance and odor of a newly discovered source of food to other bees by dances (von Frisch, 1971). This efficient communication system of honey bees allows them to become rapidly distributed amongst nectar and pollen plants within flight range.

The number of foragers available for pollination of a crop depends on the stage of colony development, time of year and the target crop needing pollination. A colony grows in total population from early spring to mid summer resulting in an increase in the proportion of bees that are potential foragers. The proportion of bees needed for nurse bee duties is lowest when the colony population has reached maximum size, viz., large colonies have proportionally more foragers than small ones. As this phenomenon occurs, the bees-to-brood ratio changes from a figure that is less than one (more brood cells than total number of bees) to a figure near 3:1 (bees:brood) in a 50,000-bee colony.

Foraging honey bees can be classified as: 1) nectar foragers, 2) pollen foragers, 3) bees that collect both on a single trip, or 4) water gatherers. It may not be important in pollination whether the bee is collecting nectar or pollen, as long as the bee touches the stigma. However, for certain flower structures that may not be true (see nectar gatherers under pollination of alfalfa).

The ratio of nectar foragers to pollen foragers is generally about 3:1 (nectar:pollen). This figure changes throughout the season and during each day. In spring a large portion of the foragers collect pollen because brood production is rapidly increasing, and the colony has been depleted of pollen stores during the dearth period of winter. On an hourly basis, a colony nearly always has more pollen foragers in early morning hours than later in the day. Many plants shed their pollen early in the morning when their blossoms first open. Pollen of certain plants can only be collected with higher relative humidity of the early morning, *e.g.,* the pollen of corn or cotton. When a colony is fitted with a pollen trap (which may remove 50 to 70 percent of the incoming pollen) there is an increase in the percentage of pollen collectors (Loper, *et al.,* 1985). Thus, pollen traps can be used when the crop requires pollen gatherers, to insure the fertilization of the flowers, or when pollen is the principal resource attracting bees to a crop.

Honey bees need a diversity of pollen in their diet because some pollen has incomplete proteins (lack certain amino acids). To obtain essential nutrients in proper amounts honey bees collect pollen from a wide cross-section of available plants. This need for a diversity of pollen sources does not work to the benefit of a pollination program, but is contrary to its purposes. In essence, then it is almost impossible to keep all the bees foraging on a particular crop.

Honey bees may start foraging as early as the 10th day after emergence, but most bees do not begin until they are 15 to 20 days old. Once they begin foraging, bees usually continue until they wear out their wings (within about three weeks during the active season), or some other natural (bird, spider or rain), or unnatural (pesticide) factor kills them.

When a bee begins to forage, it will usually do so relatively near the colony. This is important in many pollination situations with unattractive crops because these naive bees usually work the target crop for a day or more prior to expanding their foraging area. A colony may produce up to 2,000 new inexperienced foragers each day. A forager will usually make several trips per day when the weather is favorable. A single foraging trip may last up to an hour or more, with a pollen-gathering trip often taking more time than a nectar gathering trip. Low temperatures severely reduce or prevent foraging, with almost none òccurring below 55° F. and there is a linear rise in the number of foragers up to ca. 90° F. At higher temperatures there is a reduction of foraging that may be attributed to an increased need to cool the hive. Also, as the wind increases, the number of foragers decreases in almost a linear relationship, essentially stopping when the wind speed reaches 20 to 25 mph. Sometimes the wind will cause a shift in foraging to the crop closest to the hive. When it becomes too windy (15-20 mph) most foragers stop visiting apple trees, but will continue to work dandelion blossoms on the orchard "floor". Dandelion, even though its nectar has a lower sugar concentration than apple, becomes a relatively better crop for the bees to forage because there is less wind at ground level. Bees maximize sugar (nectar) income relative to outgoing (flying) cost, and in this case it takes less energy to fly at ground level where the wind speed is lower; therefore the dandelion will produce more net calories.

Beekeepers and growers want bees to pollinate crops efficiently, therefore, they look for methods that attract bees to the crop to be pollinated. Pollination may be increased by rotating colonies between fields at least 2 to 3 miles apart, thus breaking established flight patterns. The new flight patterns will temporarily include greater activity in the crop to be pollinated. Another practice is to place colonies in a crop only after the flowers become attractive to bees. In some crops 10 to 25 percent bloom is recommended. However, if primary blossoms produce the choice fruit, as in apple, watermelon, muskmelon or strawberry, bees should be present in time to pollinate the earliest blossoms.

A few scientists claim that bees can be directed to crops by feeding sugar syrup containing extracts of flowers of that particular crop. This method, to date, has not proven to be practical. Other scientists have felt that bees could be attracted to a crop by spraying it with sugar syrup. Free (1965) showed that more bees visited the crops to collect syrup, but the number of bees working the flowers actually decreased. The problem is to recognize that a bee must contact the flower's sexual parts to be effective in transferring pollen. Honey bees are known to "tag" flowers with scents (pheromones) and it may be possible to attract bees to a crop with the proper chemicals (Free, 1962, 1972; Waller, 1970). A commercial product, Bee-Scent®, has been formulated with some of the components of the natural Nasanov pheromone. It is to be sprayed onto open flowers to attract bees that would visit other flowers. The results, to date, have been variable with no clear picture of success. The most likely candidates would appear to be those crops that are not very attractive that bloom in competition with other flowers. A good example would be early cucumbers that are blooming during the flush of clover blossoms. The cost of these pheromone attractants should be compared to the increased pollination that would occur with the addition of more colonies of bees, maybe with less cost per acre.

Individual foragers tend to restrict successive visits to a relatively small area of a field or orchard. Details of this behavior vary with the crop and can be important, particularly in hybrid seed production or cross-variety pollination. An individual bee may forage down a single row, or within a cultivar, because a distinctive odor or attractive pollen occurs in one parental line but not in the other. Thus, problems in the pollination of hybrid onions, carrots, pickling cucumbers and dwarf tree fruits result. With the impending developments of hybrids of soybean, cotton, muskmelon and other crops, we need more information about honey bee responses to foraging cues.

Honey bees readily fly 2½ miles in all directions from their hive and thus have access to about 12,500 acres. A crop needing pollination is competing for bee visits with all other blooming plants within that area. In searching for and gathering food, bees set up flight patterns which change gradually as available sources or preferences change. It may be helpful to think of the foraging area of a honey bee colony as a large circle. When bees are moved into a field or an orchard their foraging range may be rather small for a day or two. The foraging circle then increases gradually until the bees have become familiar with the available forage within flight range. This changing foraging area has important implications in crop pollination, particularly crops that are not very competitive in attracting bees. Moving bee colonies into the field or orchard after the blossoms have begun to open helps keep foragers from immediately visiting competing crops, thus increasing visitation to the target crop.

Some crops stimulate relatively more foragers because there is an abundance of "attractive" nectar, *e.g.,* citrus. Conversely, other crops will need more colonies because the crop is essentially unattractive to the bees, *e.g.,* onion or cucumber. A good working rule to use, if the relative attractiveness of two competing crops is unknown, is to expect that the crop having more blossoms per unit area will attract most of the foragers.

The Flower

The angiosperms produce seeds in a closed cavity called the ovary and are commonly referred to as flowering plants; flowers are their most distinctive structures. Flowers have developed form, odor, and color to attract bees and other agents of pollination, not to gladden the hearts of people or make poets sing their praises. The beauty of flowers is something for which, to a large extent, we must thank the bees. Bees need flowers to survive and many flowers need bees to reproduce. Beekeepers should learn the basic facts about the structure of flowers, particularly if they make crop pollination part of their business.

The generalized features of a flower are shown in Fig. 1. The outermost part of a flower, which is usually green and originally sheathed the bud, is called the **calyx,** each scale of which is called a **sepal.** Inside the calyx is the **corolla** which is made up of varying numbers and forms of **petals.** Petals are usually shed soon after pollination. The calyx and corolla together constitute the **perianth.** The perianth is the showy envelope of the flower, responsible mostly for protection of inner parts and attraction of pollinators (Guilcher and Noailles, 1966).

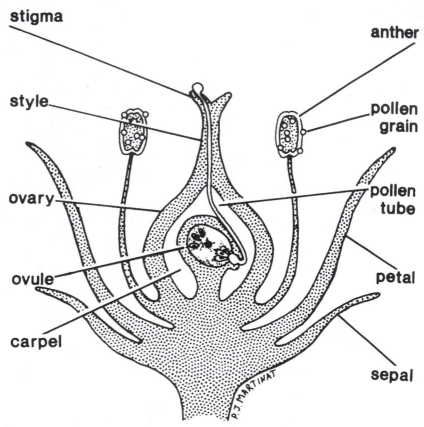

FIGURE 1. Section of a complete flower showing general anatomy and the process of fertilization.

The essential organs of a flower occur within the perianth. Outermost are the **stamens,** the male sex organs, varying in number and arrangement between different species of flowers. Each stamen has a slender stalk, the **filament,** supporting a sac-like **anther** which produces the pollen grains, collectively known as **pollen.** When pollen has matured inside, the anther wall opens and the pollen is discharged in various ways, a process known as dehiscence. Pollen of some plants, such as grasses, is dry and light and is carried by wind currents. Pollen of other plants is sticky and relatively heavy and must be transferred by a bee or other pollinating agent.

Individual pollen grains contain a tube nucleus and one or two sperm nuclei. The tube nucleus produces the pollen tube which grows down the **style** to deliver sperm nuclei to the **ovules.** Fertilization and seed development then follow. Pollen grains from different species of flowers have their own distinctive surface configurations (Fig. 2). It is possible, with the use of a microscope,

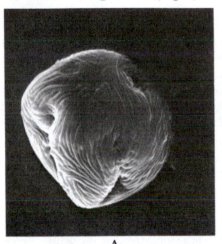

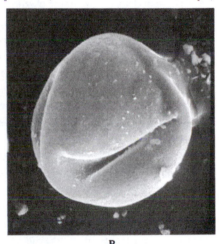

A   B

FIGURE 2. Scanning Electron Microscope picture of pollen grains showing the surface features of the pollen grains. a. tart cherry b. corn (maize) c. bull thistle *(Cirsium vulgare)* d. apple.

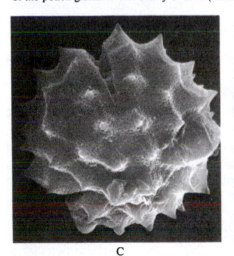

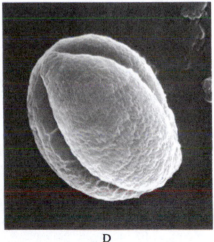

C   D

to identify pollen grains removed from bees or honey and determine the plant species (or even varieties of a species) from which the pollen came. The effective life of pollen varies greatly between species. Pollen from cereals may die within minutes, whereas date palm pollen may live for a year. Apple pollen will usually live for three months under some conditions, but apple pollen from pellets collected by honey bees will seldom germinate. Length of life of pollen is of concern when pollen dispensers are placed on the front of bee hives so that bees will carry suitable live pollen to pollinate a crop. Pollen viability can be tested using artificial media.

The **Pistil**—The innermost part of a complete flower is a pistil or several pistils. A pistil commonly has three parts, the basal **ovary** which is drawn out into an elongated neck, the **style** and a variable but often roughtened and sticky extremity termed the **stigma**. The term **carpel** refers to a single or simple pistil. Pistils are often derived from a union of several carpels. For instance, the edible part of a cucumber is the ripened compound ovary. Slicing a cucumber crosswise will reveal that it is made up of three or four carpels.

The stigma functions to capture pollen grains from visiting bees, or other pollinating agents, and provides a suitable surface on which pollen grains may germinate. A stigma when ready to perform this function is said to be receptive. During the receptive period, the length of which may be quite variable, the stigma is usually moist and sticky.

The ovary is the most important reproductive part of the flower because it contains the ovules. Some flowers produce a single ovule in the pistil whereas other species may produce more than 1,000 ovules. After fertilization, the ovules become the seeds and the ovary and some adhering parts become the fruit. Nature has proliferated fruits of great variety, (see Guilcher and Noailles, 1966 for details on fruit). Examples of true fruit vary from stone fruits to berries, maple keys, legume pods and many others.

Pollination is the transfer of pollen from the anthers of a flower to the stigma of that or another flower.

Fertilization is the union of male and female **gametes** which occurs after pollination. The definition of several other terms will help one to understand various situations involved with pollination.

Self pollination is the transfer of pollen from the anther to the stigma of the same flower automatically, usually through the action of the flower's development. No mixing of different genetic material occurs. This is also called auto pollination.

Cross pollination is the transfer of pollen between plants which do not have identical genetic make up. That is, pollen gametes are not genetically identical to the ovule gametes. For example: cross-variety pollination in apples, cross-inbred-line pollination for hybrid seed production, or cross pollination of plants of slightly different genetic material in nature. Cross pollination has been important in plant survival through the ages by providing a diverse gene pool within plant populations.

Self compatible means being capable of setting fruit or seed with its own pollen. For example, "Montmorency" tart cherries are self compatible. Self-compatible trees are planted in solid blocks with no need for inter-planting with a pollinizer variety. Growers sometimes mistakenly think this term means that the crop does not need bees. The pollen in cherry is heavy and sticky, so for commercial crop production pollen must be moved from flower to flower by bees. The term self fruitful is also used to describe this type of pollination.

Self incompatible varieties do not set commercial crops when self polli-nated. Most apple varieties belong in this category. With a self incompatible crop there is a need to plant other varieties nearby that have compatible pollen so that cross-varietal pollination occurs.

Cross compatible is used when referring to crop plants, or varieties, which set a commercial crop when two or more are properly inter-planted for cross pollination. That is, they are varieties that have compatible pollen.

Cross incompatible varieties are not suitable for inter-planting or cross pollination. Varieties are sometimes incompatible because they bloom at different times (Fig. 5), but cross incompatible usually refers to a physiological incompatibility. In such cases the pollen may not produce a pollen tube (Fig. 3), or the embryo may abort after union of the gametes. Some varieties have sterile (nonviable) pollen. As a rule, physiological incompatibilities exist between varieties with a close genetic relationship. The degree of incompati-bility may be altered by changing environmental conditions.

Monoecious plants are those in which the staminate (male) flowers and the pistillate (female) flowers are separate but occur on the same plant. Cucumber, watermelon and corn (maize) are common examples.

Dioecious plants are those in which the staminate and pistillate flowers occur on different plants as in willow and poplar.

Gynoecious plants—The term gynoecious F_I, hybrids (or varieties) has been applied to certain cultivars of pickling cucumbers. These cultivars have predominantly pistillate (female) flowers, and are used to produce a crop uniform in fruit size in a short time.

Parthenocarpic fruit are produced without fertilization, and so there is no need for pollination in such varieties, *e.g.,* seedless oranges, some varieties of slicing cucumbers and some varieties of pears.

Pollinizer refers to the plant source of pollen used for pollination.

Pollinator is the agent which carries pollen, *e.g.,* honey bee.

Cultivar is a man-made or **culti**vated **var**iety. Plant breeders have been actively re-arranging genetic material of plants to suit present human needs. If carried to extreme the genetic selection process might screen out and discard germ plasm needed in the future. The experience with corn blight destroying large acreages of corn a few years ago exemplifies the need to have access to a broad base of genetic material.

Pollen Transfer Mechanisms

Because of self incompatibility within and between plants, frequently pollen must be moved from one flower to another. Horticultural practices often dictate that pollen must be moved for considerable distances between trees or rows. However, when foraging by individual bees was studied it was learned that individual bees tended to forage within a small area (Free and Spencer-Booth, 1962; Ribbands, 1949). It was only after nectar or pollen were essentially harvested from a bee's foraging area that she moved to a new location. Competition between bees for available nectar or pollen increases bee movement. The tendency for foraging bees to relocate as nectar and pollen becomes scarce is called wandering. It was once believed that if certain crops were to be cross pollinated the foragers had to move directly from one row, or variety, to another. In crops that require cross pollination (*e.g.,* apples or almonds) the growers plant different varieties in alternate rows, or use some other scheme to maximize the main variety and minimize the pollinizer variety. For example, planting a pollinizer every third tree in every third row would allow eight trees of the main variety to be adjacent to (surround) each pollinizer tree. In one analysis of the bees' wandering phenomenon, in apple orchards, it was shown that it was difficult to account for enough cross pollination to set an economic crop (DeGrandi-Hoffman *et al.,* 1986). This can best be seen in Fig. 4 where the ratio of the main crop of apples to pollinizer variety was set at 3:1 (3 rows of main variety to 1 of the pollinizer). Only 18 percent of the blossoms were cross pollinated by wandering bees because much of the movement was between trees of a similar variety. Even excluding the bees that were visiting competitive plants, cross pollination was low considering that all of the foraging bees in this example were treated as wandering bees.

As early as 1889 scientists considered that bees could be exchanging pollen within the hive and then carrying it back out to a different plant (Betts, 1935; Karmo and Vickery, 1954). With the scanning electron microscope (SEM) we can now distinguish pollen from different varieties. Using the ability of the SEM, the level of in-hive pollen transfer was measured. The high proportion of mixed pollen on the bodies of the hive bees demonstrated that even non-foraging worker bees and drones pick up pollen from foragers within the hive (DeGrandi-Hoffman *et al.,* 1986). Free & Williams (1972) showed earlier that foraging bees carried up to 5,000 pollen grains on their bodies when leaving the hive. The amount of in-hive pollen transfer that takes place depends on the amount of foraging that occurs during the day, the crop (pollen type) and the population of a colony. To be effective, in-hive pollen transfer probably has to occur during the same day since most pollen probably has lost its ability to fertilize an ovule when retained overnight within a colony. However, in-hive pollen transfer adds a new dimension to the role that honey

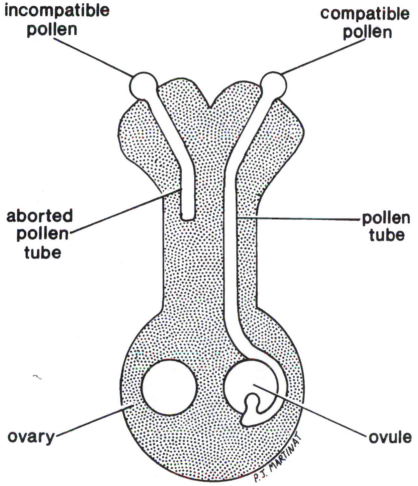

FIGURE 3. Diagram illustrating how compatible pollen will send down a pollen tube to deliver the male gamete to the ovule while incompatible pollen may abort or fail to function in other ways.

bees play compared with solitary bees in the pollination of self-incompatible crops. Thus, a bee that has not previously visited a blossom of a cross-compatible tree or plant may still deposit compatible pollen onto its stigma.

In-hive pollen transfer also opens up new concepts in orchard design and pest management. Theoretically, it should be possible to replace the inter-planting of pollinizers with large, solid blocks of varieties. Planting large blocks of trees within an orchard allows the orchardist to manage insect pests by variety and have a one harvest time within each block. The critical concern is that compatible pollen be within foraging range of each bee colony. Determining the optimum size for these solid blocks depends on the number of colonies within the orchard or area, and the proportion of other varieties. There are too many variables to calculate one optimum size for blocks *viz.*: variety, competition between varieties, weather during bloom which may affect blooming sequence of varieties (Fig. 5), proportion and type of pollinizer variety and number of colonies.

Modern History of Crop Pollination

Credit for recognition of bees as pollinators is generally given to Koelreuter and Sprengel who lived in Germany and published between 1750 and 1800. Charles Darwin also showed that plants benefit from cross pollination. In 1892 the U.S. Department of Agriculture sent Waite to examine an orchard of 22,000 Bartlett pear trees in Virginia that failed to bear fruit. He found that there was only fruit production in orchards having mixed varieties of pears.

APPLE ECOSYSTEM
'Wandering' Forager Distribution

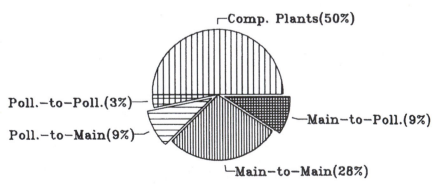

3:1 Ratio of Rows

FIGURE 4. Distribution of wandering foragers in an apple orchard with 3 rows of the main variety to 1 row of the pollinizer variety. Competitive plants are other flowers in bloom at the same time that effectively recruit 50% of the foragers to their blossoms.

FIGURE 5. California almond orchard with one late pollinizer variety not yet blooming and another in an advanced stage of bloom used for early pollination of the main variety.

This led him to the important conclusion that most varieties of pears could not be fertilized with their own pollen; cross-variety pollination was essential for fruit set. This was a very important breakthrough in the story of pollination and people became aware that apples, and some other plants, have a physiological barrier to self fertilization. Waite also showed that bees were necessary for pollen transfer because pear pollen was too heavy and sticky to float in the air. Waite introduced the concept of providing colonies of honey bees for fruit pollination. He recommended that apiaries be at least within three miles of an orchard. Further studies have shown the benefits of having honey bee colonies much closer to (in fact within) the target crop.

Outstanding success with corn breeding in the 1930's showed that hybridization was an important tool for increasing crop yields. Since then plant breeders have made hybrids of various bee-pollinated crops. However, it is with these hybrid crops that some of our most serious pollination problems exist at the present time.

POLLINATION OF CROPS

In the following section we will cover the pollination of certain crops to show examples of important principles. These principles can then be applied to many other crops having similar problems. For information on a particular crop, *e.g.,* details of the crop's flower or its pollination, the reader is referred to the book by McGregor (1976).

Field Crops

Alfalfa (*Medicago sativa* L.)—Alfalfa, or lucerne as it is called in Europe, is the world's most important forage crop. Alfalfa has reached a dominant position because it adapts to widely differing climatic conditions, has high nutritive qualities, and is perennial. Thus, alfalfa seed production is a very important business.

Alfalfa has pollination problems because its flower must be tripped by a bee and pollinated to produce seed. To trip a flower a bee enters the flower and presses its head against the large, upper, standard petal. Pressure from the bee's head trips a mechanism that opens the lower keel petals and releases the flower's male and female parts (Fig. 6). When this happens, the anthers deposit pollen on the bee's body. As a bee moves between alfalfa flowers the process is repeated and pollen is deposited on the stigma each time a bee trips a flower. Thus, cross-plant pollination predominates.

Unfortunately, nectar-gathering honey bees "dislike" the tripping mechanism of alfalfa. They soon learn to insert their tongues between the petals to remove nectar without tripping the flower. Thus, they pollinate a low percentage of flowers visited only for nectar. Pollen-gathering honey bees, by contrast, are efficient pollinators of alfalfa because they must trip the flower to obtain the pollen. Unfortunately, the percentage of honey bees gathering pollen is usually low. In hot, dry weather alfalfa flowers trip more readily (accidental tripping) and nectar-gathering honey bees become effective pollinators. This is why honey bees are major pollinators of alfalfa in arid parts of California, Nevada and Arizona, while solitary bees, such as leafcutter bees (*Megachile rotundata* Fabr.) and alkali bees (*Nomia melanderi* Ckrl.), are more important in Oregon, Washington and Idaho. The alfalfa leafcutter bee and the alkali bee are used extensively for alfalfa pollination because they always trip the alfalfa blossom. Seed growers sometimes supplement the pollination provided by these solitary bees by using colonies of honey bees. In the humid and rainy regions of eastern and central United States alfalfa seed yields are generally low, because of plant diseases and harvest problems. Seed growers there cannot compete with producers in irrigated areas of the West.

Nye and Mackenson (1970) developed inbred lines of honey bees having either high or low alfalfa pollen-collection traits and established that the tendency to gather alfalfa pollen was a heritable trait. This work demonstrated that it is possible to select and breed strains of honey bees suited to the pollination of a specific crop. However, commercial production and use of alfalfa pollen-collecting honey bees has not been economically feasible. Doull (1971) said this type of bee breeding was the selection of honey bees for oligotropy, a behavior pattern exhibited by many species of solitary bees. It is possible that selection for oligotropic behavior might result in reduced brood rearing caused by a nutritional deficiency from the lack of pollen variety. An alternative approach would be to breed a strain of alfalfa that is more easily

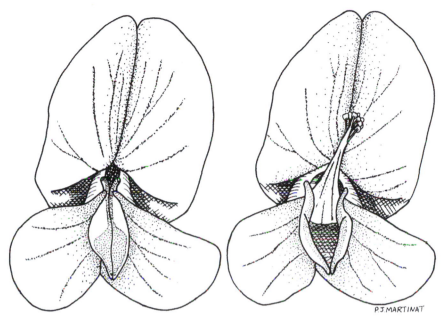

P.J.MARTINAT

FIGURE 6. Alfalfa blossoms. Left untripped, right tripped.

tripped by nectar-gathering honey bees. If this were accomplished, it would alleviate the long-standing problem of obtaining maximum alfalfa seed yields. To make progress along these lines, alfalfa breeders need to be persuaded about the importance of such an effort. Alfalfa lines with increased nectar secretion have been developed to increase the level of honey bee visits that occur in seed production fields.

To produce a 500-pound-per-acre seed crop, about 38 million alfalfa flowers must be tripped and cross pollinated. In areas where honey bees are effective pollinators, four or five colonies per acre are used to accomplish pollination. Most effective use of the bees is accomplished by placing colonies in groups of 10 about one-tenth mile apart; half of the colonies moved into the alfalfa fields about 10 days after the first flowers appear and the remainder one week later. At the five-colony-per-acre density, honey bee colonies probably will not produce a surplus of honey; whereas at two or three colonies per acre they often produce a surplus of honey. Thus, a beekeeper should consider the density of colonies when setting rental fees for pollination.

Some observers have rejected honey bees as significant pollinators of alfalfa because they often do not trip the alfalfa blossom. However, they probably have overlooked certain types of foragers. Within each honey bee colony there may be up to 2,000 inexperienced foragers that enter the field each day. These naive bees do not avoid the tripping mechanism and, thus, become "nectar trippers." Added to these naive bees are those foragers that continue to collect pollen, and the accidental tripping by nectar foragers. The total number of honey bees that are tripping alfalfa blossoms are, in reality, quite a large number. It is not realistic to compare the pollination efficiency of one honey bee with that of one solitary bee without considering the total number of foragers from 3 - 5 honey bee colonies versus the number of solitary bees available.

Cucumber (*Cucumis sativus* L.) — The conventional method of "pickle" production, which is still widely used, is to plant 10,000 to 15,000 plants per acre and handpick the cucumbers as they reach pickling size. At any one time there might be not more than 20,000 female flowers open on each acre. Pollination can be accomplished by relatively few bees, and thus, one colony per three or four acres has been a common recommendation.

The cucumber plant is normally monoecious. Peterson (1960) developed what he called the gynoecious F_1 hybrid cucumber, a variety that is technically monoecious, but has a very high proportion of female flowers. Because some selections of gynoecious cucumbers have nearly 100% female flowers, to assure a pollen supply, standard plants (with both male and female flowers) are inter-planted. From this, a whole new technology for growing cucumbers has evolved that increases the need for effective pollination by bees. This new technology requires planting 50,000 to 100,000 plants per acre of "gynoecious" cucumbers inter-planted with about 10 percent monoecious types to supply pollen. One to three colonies of bees per acre are needed for rapid, uniform pollination. Finally, a single, destructive, machine harvest is done when one or two fruits per vine are of optimum size for pickling. Such high-density plantings of cucumbers creates a challenge to the beekeeper to provide adequate pollen. An individual cucumber flower is open for only one day, and each flower has several hundred ovules (potential seeds). Cucumber pollination is rather sticky and is not actively collected by honey bees. With the large number of ovules, and low pollen adherence to the bees, several visits per flower are needed to produce straight, fully developed fruit. Collison (1976) conducted a definitive study on the actual number of bees needed for pollination of cucumbers. He devised a bee number table for determining adequate pollination of high density pickling cucumbers (Table 2). His results show that the best indicator of cucumber pollination is the number of bees that visit each blossom during a day, not how many colonies are placed in a cucumber field. Collison determined that 15 to 20 visits per blossom per day were needed to produce a profitable crop in Michigan. Table 2 was developed to help growers (or pollination advisors) determine if they have adequate bee visits to cucumber blossoms. Collison's table adjusts for an increase in the

**Table 2. Calculating the Number of Bee Visits
Needed for Cucumber Pollination.**
(See text for instructions on use).

Time of Count	Minimum Number of Bees/30 flowers/ 30 min.*	Time of Count	Minimum Number of Bees/30 flowers/ 30 min.*
8:00 - 9:00	1	1:00 - 2:00	13
9:00 - 10:00	3	2:00 - 3:00	11
10:00 - 11:00	9	3:00 - 4:00	7
11:00 - 12:00	13	4:00 - 5:00	5
12:00 - 1:00	16		

* These numbers will give about 15 visits/flower/day. This is the required number needed for adequate pollination of pickling cucumbers.

number of foragers as the temperature rises during the day, and then a decline in foragers with less light, lower temperature and less available nectar. To monitor whether a cucumber field has enough bees, you watch 10 flowers for 10 minutes and count the number of bees visiting these flowers. This is repeated for a total of 30 minutes of counting. The table is then consulted for the time of day that the observation was taken. If the counts are equal to or greater than the numbers in the table, visits are adequate.

The usual timetable for pollination of high-density cucumber plantings is to move bees into the field when the crop has 15 to 20 percent bloom. Individual cucumber flowers produce rather large quantities of nectar, but even with dense plantings the number of flowers open on any one day is quite low. Thus, cucumbers are not especially attractive to bees, and they often locate more rewarding plants as forage. Pollination continues for about a week, the bees are removed and the crop is machine harvested.

Beekeepers should not expect to produce surplus honey from cucumbers. The rental fee for cucumber pollination should reflect this fact, especially if the pollination period coincides with a major honey flow elsewhere. Some beekeepers report that colonies used in a large acreage of cucumbers winter poorly. The reason is not clear, but it may be because bees gather practically no pollen from this crop. Colonies pollinating small cucumber fields may do the pollination job and still "prosper" because they also have access to a variety of other nectar and pollen sources. Beekeepers should be careful not to claim that good pollination will eliminate the problem of misshapen cucumbers. Inclement weather and excessive moisture also contribute to this condition.

When overhead irrigation is used, most foraging bees are driven from a cucumber field and bees returning to the field do not work the flowers until they are dry (Collison and Martin, 1973). Flowers pollinated less than two hours prior to irrigation fail to produce fruit if water enters the corolla. This may be because water washes pollen from the stigma or interferes with pollen germination or pollen tube growth. The problem might apply to crops exposed to heavy rains or overhead irrigation. A solution would be to overhead irrigate only in the evening or very early morning during bloom.

Sunflower *(Helianthus annuus* L.) — The sunflower is a native of North America, but is now widely distributed throughout the world. In Canada and the northern mid-western states sunflowers are grown extensively for oil which is extracted from the seed, or for confectionery use (seeds to eat). Honey bees are useful pollinators and adequate numbers of bees bring about cross-plant pollination which gives higher seed and oil yield than self pollination (within a single flower head). Older flowers give reduced yields if bad weather delays pollination. For this reason colonies of bees should be located in the fields as soon as flowering begins. There has been considerable effort by plant breeders to produce self-compatible and self-pollinating varieties of sunflowers. However, the transfer of pollen from one variety to another has significantly increased production, sometimes more than 100 percent (Langridge and Goodman, 1974).

For optimum yield the number of bees needed is probably provided by having more than one colony per acre. One study in Australia (Langridge and Goodman, 1974). showed that one bee per plant was the optimum number of bees visiting the crop, while other studies indicated that less than one is sufficient. Sunflowers produce high yields of honey, and self compatibility of the crop has been widely touted, therefore, pollination fees are generally low or nonexistent. With a better understanding of the importance of pollen transfer, more emphasis on providing for sunflower pollination may be forthcoming in this industry.

Onion *(Allium cepa* L.) — Onions are grown worldwide and varieties suited to both warm and cool climates have been developed. Onion seed production, especially for hybrids, has serious problems with low and sporadic yields. Growers of hybrid seed usually plant four to six rows of a male-sterile line alternating with two rows of a male-fertile line. Honey bees collect both nectar and pollen from onion flowers, but in hybrid seed production fields only nectar gatherers move freely between both onion lines and carry on pollination. Studies in New York indicated that only 11 percent of honey bees crossed from male-fertile rows to male-sterile rows. Onion nectar has a characteristic of higher potassium than nectar of most competing plants which reduces relative attractiveness to honey bees (Waller *et al.,* 1972). Thus, nectar-gathering bees are often attracted away from onion to visit other flowering plants in the vicinity.

Growers of hybrid onion seed do not produce the yields common for open-pollinated varieties and inadequate pollination seems to be an important cause. Low viability of pollen, the presence of competing nectar plants, the effect of pesticides on pollinators and on pollen viability, and a whole complex of causes might be involved. The solution appears to be in adjusting production methods so that honey bees will transfer adequate amounts of viable pollen. (Gary *et al.*, 1972) suggests planting larger fields and saturating them with five or more colonies per acre. It is a common practice to introduce honey bees at 20 to 25 percent bloom and distribute them around the field in groups of 6 to 12 colonies.

Cotton *(Gossypium* spp.) — Cotton is an important source of honey, but application of insecticide to cotton in the United States and throughout the world often cause catastrophic losses of bees. McGregor (1958) showed that cross pollination by bees improved yields in some varieties and sets up an earlier crop in others. The development of hybrid cotton has possibilities for increasing cotton yield potential and fiber quality. Cotton pollen is seldom actively collected by bees in considerable quantities (Moffett, 1983). Bees have shown a tendency to work down a single row of a seed production field, thus not transferring pollen between parental lines as needed to produce the hybrid seed. However, with adequate numbers of bees in cotton fields, the pollen transfer process has worked effectively enough to set seed (Moffett, 1983).

An additional side benefit from hybrid cotton seed production is that bees are important, and growers use greater care when pesticides are needed on their fields. Thus, beekeepers can produce a crop of honey as well as receive a pollination fee.

Orchard Crops

Apple *(Malus* spp.) — It is known that apple pollen is carried by wind to some extent, but it has been shown conclusively that wind pollination has little or no significance in fruit production (Free, 1964). Bees, particularly honey bees, provide pollination for this crop. If 5 to 10 percent of full bloom of an apple tree produces fruit, a full commercial crop is realized (Rom, 1970). However, growers must aim for a higher initial set because several fruit drops take place throughout the season. If excessive numbers of fruit are set, some may be removed by using chemical thinners; however, nothing can be done when too few are set, once flowering has passed. If too many apples develop on a tree, they will be smaller than top grade apples, but if pollination is inadequate, a reduced crop of misshapen fruit will result (Fig. 7). Apples have five pistils, each with two ovules, thus, there are 10 potential seeds. Fruit growth and development is stimulated near fertilized, developing seeds. Without adequate pollination the result is low seed numbers and misshapen fruit. In general, fruitlets with the smaller number of seeds are eliminated with a series of early fruit drops (Luckwill *et al.,* 1960). Such self thinning or dropping of insufficiently pollinated fruits are referred to as "June drop".

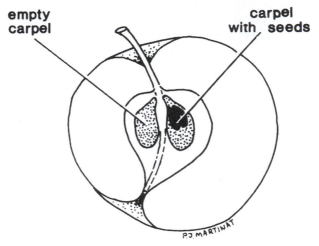

FIGURE 7. Misshapen apple fruit in cross-section.

Growers are generally less aware of factors contributing to adequate pollination than they are to all other cultural practices. Apples are grown mostly in temperate regions where weather during bloom may be unfavorable for bee flight, pollination, pollen tube growth, fertilization and other factors affecting fruit set. Successful growers prepare for this by providing enough pollinizer trees, and renting bees to provide foragers to visit the flowers whenever weather permits. Most apple varieties are self incompatible and certain varieties are also cross incompatible. Varieties do not all bloom at the same time, so a well-designed orchard has enough pollinizer trees that bloom at the same time as the main variety. Two varieties are commonly used, but under some circumstances three varieties must be present because of cross incompatibility or the flowering sequence is not synchronized. Growers should take care in planning an orchard if one variety has sterile pollen. Various plans for inter-planting pollinizer trees are used, as discussed under pollen transfer mechanisms. The ratio of pollinizer trees to main trees often reflects the relative market value of the pollinizer variety.

Apple pollination requires about one colony per acre. The actual need is for enough bees to cover the thousands of blossoms per acre to provide a maximum crop of fully developed fruit. Dwarf trees have a larger number of blossoms per acre and therefore require more bees. The number of colonies of bees needed for good pollination varies because of a number of factors. If there are a large number of feral colonies, bumble bees, and solitary bees, the number of rented colonies could be lower. There are also the factors of the suitability and proportion of pollinizer trees, as well as the weather and the strength of available colonies. Growers generally use less than one colony per acre. There is little doubt that yield and quality suffer from inadequate pollination in many orchards caused either by too few bees or low pollinizer numbers relative to the main variety.

Beekeepers usually distribute groups of four to six hives throughout the crop area (Fig. 8). Pollination would be just as effective in many large orchards if groups of 10 to 20 hives were strategically distributed in sunny, wind-protected spots well within the boundaries of the orchard. It is best to locate colonies near the pollinizer variety, since it is most likely the one with the lowest number of trees. The young bees that exit the colonies will then visit the nearby pollinizer variety, and thus bring its pollen into the hive where in-hive pollen transfer can occur.

Almond *(Amygdalus communis* Batsch*)* — Almond varieties require cross-variety pollination to produce a crop. In order to obtain cross pollination, orchards are often planted with two rows of the main variety alternating with one row of an early-blooming pollinizer and one row of a late-blooming pollinizer variety. The growers' objective is to obtain a maximum set of almonds. In order to obtain this, they seek to provide adequate numbers of

FIGURE 8. Colonies of honey bees on pallets in an almond orchard in California.

pollinizer trees and populous honey bee colonies "properly" distributed throughout the orchard. The distribution of colonies can sometimes be difficult as the almonds are often planted in 1/4 x 1/2 mile blocks, and adequate pollination of the center trees is often not accomplished. In the case where there are large blocks of trees with the colonies of bees located on the edges, more colonies/acre will be needed to ensure adequate pollination of the interior trees (Loper *et al.,* 1985).

Like apples, almonds are self incompatible and pollen from another variety is needed to set fruit. However, in almonds the goal is to set a high percentage of the blossoms; as high as is consistent with continuous year-to-year production. The recommendation is to use two to three colonies per acre. Because the flowering period occurs early in the season (Feb.-Mar.), beekeepers often use special measures to build populous colonies. These measures include fall feeding of sugar and pollen or feeding suitable pollen supplements. Colonies should be moved to the almonds at the start of bloom and remain until 90 percent of the blossoms have dropped.

Tart Cherry *(Prunus cerasus* L.*)* — This pollination example is a crop that is self compatible, but tart cherries need a high percentage fruit set to be commercially successful. Leading commercial varieties are planted in solid blocks of one variety. Pollination does not take place unless bees or other insects are foraging to move the pollen and produce a profitable crop. Blossoms are attractive and honey bees work them well. One strong colony per acre is probably a minimum to ensure adequate yields, as a good cherry crop needs to set 30 to 40 percent of the available blossoms.

Sweet Cherry *(Prunus avium* L.*)* — Most sweet cherry varieties are self incompatible, and also, the three most important cultivars 'Bing'. 'Lambert' and 'Napoleon' ('Royal Ann') are all cross incompatible. Thus, finding the right cultivars to use as pollinizers is crucial. Sweet cherry trees should produce a high percent fruit set and because the trees are self incompatible, the number of colonies used must be higher than for tart cherries. Two colonies per acre are often supplied, though nectar of sweet cherries has a high sugar concentration and therefore attract many solitary and feral bees. In contrast, sweet cherries are one of the earliest fruit crops to bloom in the north (April) and the weather during the blossoming period is often cool and unfavorable for bee flight. Reduced production in some years indicates a need for more bees.

Citrus *(Citrus* spp.*)* — There are several species and many varieties of citrus; most are attractive to bees and colonies on citrus produce delicious honey in commercial quantities. Most citrus do not need pollination as they are self pollinating, or they produce fruit parthenocarpically. Beekeepers are generally anxious to move bees to citrus for the honey crop, so pollination rental is rare. Moffett & Rodney (1971) showed that Fairchild tangerines

(*Citrus* spp.) needed both a pollinizer and bees to produce an adequate crop. They also obtained an increase in yield of lemon (*Citrus limon*) and Orlando tangelo (*C. paradise x C. reticulata*) when comparing bees vs. no bees. Because of the tendency towards parthenocarpy under certain weather conditions, the value of bees for pollination has been difficult to determine with certainty. Also treatment of citrus with insecticides is a continuing concern to beekeepers in all citrus-growing areas of the United States.

Blossom Type and Colony Needs

We can look at the type of blossom (*e.g.*, self incompatible) and the percent fruit set that is desired and often determine the number of bees needed for good pollination. For example, in almonds, which are self incompatible but need a high percent fruit set, the number of colonies is at least double the number needed for apples (also self incompatible, but requiring a low percent fruit set). Almonds also generally receive more than twice the number of colonies as tart cherries because tart cherries need a high fruit set, but are self compatible. We could add another example to this list that maybe does not fit the pattern; highbush blueberries, which has self-compatible blossoms with a high fruit set and needs at least three colonies per acre. Here the problem is the large number of blossoms per acre that require more bees to achieve adequate bee visitation.

Feral (Wild) Colonies and Pollination

In the past there have been estimates that feral colonies made up a large portion of the foraging population, and these bees contributed to the overall pollination of crops (Chang and Hoopingarner, 1990). Growers have been using these bees for their pollination even when they have rented colonies. These feral colonies may be declining, or even cease to exist, if the tracheal and varroa mites destroy them. Evidence from Europe would indicate that varroa kills all unmanaged colonies. Without the feral population the number of rented colonies may have to increase substantially for each crop. For example, if one colony per acre had been recommended for a crop and the feral population was 50 percent, then a doubling of the number of rented colonies (two/acre) would probably be needed. There may be no easy way to determine what is the number of feral colonies near a pollination site. However, beekeepers providing a pollination service should be aware when pollination is inadequate, search for a possible reason for the poor pollination, and recommend to the grower that more colonies are needed to alleviate the situation.

PRACTICAL ASPECTS OF POLLINATION

Quality of Colonies

Overwintered colonies are the first choice for pollination of spring fruit trees. A package colony cannot be started early enough to produce a cycle of brood and have young foragers. Furthermore, a package colony will have lost many bees to natural causes before new bees emerge making the number of available foragers quite low at the time of fruit bloom.

Bees rented for pollination are often described as "strong single-story" or "strong two-story" colonies, etc. Upon examination they can further be described as having 10 frames covered with bees or 6 frames containing brood. McGregor (1976) (p. 61) proposed that remuneration be based on the amount of brood present. The extra value of more populous colonies for pollination has been shown (Waller *et al.*, 1985). A grower should establish guidelines and have the beekeeper agree to either a minimum standard colony or a sliding scale for payment based on the number of frames of brood or frames covered with bees.

Incentive systems for pollination have merit. For example, one designed for onion seed included: 1) use of strong hives with 9 to 12 frames of brood; 2) introduction of 1 or 2 hives per acre when the field is 15 to 20 percent in bloom with additional hives added at 3 or 4 day intervals; 3) colonies placed inside the field as well as around the perimeter; and 4) the beekeeper paid a per hive base price, plus a percentage for all cleaned seed above a predetermined average yield. Such an incentive system should prove profitable for both grower and beekeeper.

Equipment and Costs

It is possible to move bees for pollination without any new or specialized equipment, as long as the operation is relatively small or hand labor is cheap and plentiful. Many pollination situations require the bees to be moved into or removed from the crop in large numbers and on short notice. When that happens, additional labor will have to be used to prevent loss of bees or breaking of the rental agreement. In any case, the pollination fee might not produce the expected profit because of the additional costs incurred. By understanding the pollination of a crop, including the bloom period and colonies needed per acre, the beekeeper-pollinator can recognize the need to move 100 or more colonies each night. The beekeeper may need to purchase labor-saving equipment to accomplish such moves efficiently.

A beekeeper can "grow" into pollination services in much the same way as growing into beekeeping, and that is slowly. You can "test the water" before you jump into a service that you may not enjoy or cannot adequately provide. It is usually not economically sound to purchase specialized equipment, such as loaders, unless such equipment is used for other jobs. Bees can be moved rather inexpensively by using strapping or staples to hold the hive bodies together, even properly tied ropes will hold most colonies. Screens can be used for ventilation or nylon mesh covers used to confine bees to the truck during transit. One simple version of a moving screen is a screen used to prevent robbing (Hoopingarner, 1982). All that is needed is to tack a thin wooden slat over the entrance/exit to confine the bees (Fig. 9). Using a hand truck and a ramp to load and unload a truck are necessary unless two people are available to lift colonies. Keeping bees in standard 10-frame equipment of uniform height is nearly essential if the colonies are to be double stacked on a truck. Migratory (Western style) bottom boards and covers should be used on colonies being moved regularly, as they will take less space on a truck, and can be tied down more securely. Some beekeepers who move colonies regularly have constructed their own pallets to maximize the use of space on a truck.

Assuming that a truck of sufficient size to carry a load of bees is already part of the beekeeping business, then either a forklift, or boom-type loader, is the major additional capital item needed in most pollination operations. Used forklifts are frequently available for purchase, and are suitable for bee work since they may be operated at a lesser demand level than other industries. It would seem wise for a beekeeper to prepare prior to entering the pollination business, by putting colonies on pallets and locating the colonies in apiaries suitable for mechanized loading. For a smaller operation a boom loader would be most desirable, but a ramp and hand truck is less expensive.

Diagram of robbing screen on a colony with flight patterns of hive bees and robbing bees.

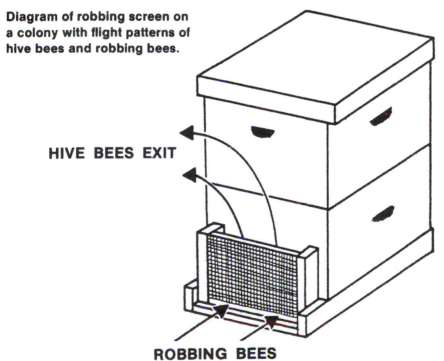

HIVE BEES EXIT

ROBBING BEES

FIGURE 9. Robbing screen on colony. May also be used to move bees by covering exit.

Table 3. POLLINATION AGREEMENT

Date _____ For Season 19 _____

The Beekeeper
    Name _____
    Address _____
    Phone Number _____

The Grower
    Name _____
    Address _____
    Phone Number _____

No. of Colonies Ordered _____

Rental Fee for Grade A Colonies _____

Rental Fee for Grade B Colonies _____

Compensation for Additional Movement
    of Bees or Other Extras _____

Total Rental Fee _____

Name of Crop _____

Location of Crop _____

_____

Distribution Pattern of Colonies shall be _____

_____

The Grower Agrees:
1. To give _____ days notice to bring colonies into the crop.
2. To give _____ days notice to take colonies out of the crop.
3. To pay one-half the agreed total fee when the bees are delivered.
4. To pay in full within _____ days after the delivery date.
5. To pay one percent a month interest on amounts unpaid after the due date.
6. To use no toxic pesticides in the crop during the rental period except with the understanding and consent of the beekeeper, and to warn the beekeeper if neighbors use toxic sprays.
7. To provide an uncontaminated water supply.
8. To assume liability for livestock damage or vandalism.
9. To assume public liability for stinging while the bees are on location in the crop.

The Beekeeper Agrees:
1. To open and demonstrate the strength of colonies randomly as selected by the grower.
2. To leave the bees in the crop for a period necessary for effective pollination estimated to be approximately _____ days and with a maximum period of _____ days, after which time the bees will be removed or a new contract negotiated.
3. To ensure that colonies are properly located and will remain in good condition while pollinating the crop.

Signed: Date _____

_____   _____
Grower   Beekeeper

Contracts

In order that beekeepers maintain the respect and support of grower groups, it is necessary to have high standards of honesty and provide good service. Populous colonies must be moved in and out of the fields and orchards on a timely basis and properly located where the grower wants them. Beekeepers often provide colonies for crop pollination without a written contract. However, such casual arrangements sometimes result in misunderstanding and ill will. To reduce this possibility, provide a written agreement signed by both parties. Of utmost importance is the inclusion of consequences should either party fail to comply with the terms of the contract. There are many items of importance that should be considered in any contract. These include: 1) the rental price and date(s) of payment; 2) the date of expected need, duration, and date and conditions for removal of colonies; 3) number and placement of colonies and availability of access to the area; 4) the type of crop to be pollinated and number of colonies rented per acre of target crop; 5) notification to the beekeeper of pesticide use in the area and remuneration for covering or moving colonies; 6) providing water for the bees; and 7) standard of quality with associated penalties and/or rewards. (See Table 3 for an example of a pollination contract.)

Rental fees for the use of bees in crop pollination vary greatly. In irrigated areas, beekeepers move bees to crops, such as alalfa, for pollination and also to obtain much-needed forage. Because of the dual purpose served by the move, there is a tendency to reduce pollination charges below a legitimate charge for the service. Under these circumstances it would be wise for beekeepers to establish and publish price guidelines through their organizations. Association leadership can often reduce uninformed price cutting. A long-standing guideline used by some beekeeper-pollinators is to use the cost of a package of bees as the basis for setting pollination fees. This is the cost of replacing the colony if lost during pollination, and the cost of bees to a grower if he were to use a package colony for pollinating his crop.

There are a number of variables to consider when determining pollination fees. We have outlined some of these below.

1. Size of the Field — If the crop is one that produces adequate honey and pollen, then having a large field will allow the beekeeper to unload more colonies (complete load) at one field or orchard. However, if the crop is similar to cucumbers, then having a large field causes a severe hardship on the colonies, and the fee for pollination should be increased. Cucumber provides little nectar and pollen for the colonies which makes larger acreage detrimental since the foragers have to fly long distances for pollen necessary to rear brood.

2. Nectar and Pollen Crop — If the target crop produces large amounts of nectar and/or pollen so the colonies prosper during the pollination period, then the fee might be reduced. However, the availability of nectar and the ability of the colony to take advantage of the resource may not always come together. For example, tart cherries often produce an abundance of nectar, though in many years the weather does not allow the bees to collect nectar in surplus.

3. Mechanization — The concern here is not whether the beekeeper uses a forklift to load and unload colonies, but whether the field or orchard will facilitate such equipment. It is of little value to plan to unload colonies with a forklift or boom and then not be able to do so. Bees are often moved into fields at night with slippery, wet conditions of springtime or into irrigated fields in the West. Upsetting colonies in the dark of night encourages the beekeeper to inspect future pollination locations **prior** to moving.

4. Distance to the Crop — The distance that bees should be moved for pollination might be limited by the time it takes to drive and unload bees within one night. Colonies can be screened to prevent flying, or nylon-mesh covers put over the entire load to allow bees to be moved longer distances (Fig. 10). Moving bees more than 350 miles for pollination may not be profitable for the beekeeper. Sometimes beekeepers use the move to a pollination location for other benefits, and longer moves can be justified. For example, many beekeepers move bees from northern states into southern California to pollinate almonds. In addition to earning the pollination fee, these colonies have an early buildup for the northern nectar flows. In this case the pollination fee offsets the moving costs, and the colonies expand their broodnests earlier than if they remained in the North.

Moving Bees to Crops

Moving bees to pollination is hard, tiring work. If done mechanically, a beekeeper can move bees by day or night into or out of the crops in a timely manner. Pollination service provides money early in the season and can give diversity and security to the beekeeping enterprise. Moving colonies between locations can be part of good bee management. Renting bees for crop pollination has become an established business, therefore equipment for rapid and efficient loading and unloading of colonies is commercially available. Growers generally want colonies in and out of crops on schedule, particularly in crops needing insecticidal sprays just before or after the flowering period. Many improvements in the speed and efficiency of moving bees have come about in recent years. The boom-type mechanical lifter (Fig. 11), forklift (Fig. 12), the gantry hive loader (in Australia), bees on mobile platforms (Fig. 13), pallets to lift two (Fig. 14), four or six hives at a time, and strapping to hold hive parts together have all come into common use. Anyone involved in

FIGURE 10. Colonies of bees on truck with nylon mesh over doubles stacked 3-high.

moving bees into and out of crops would be well advised to convert all equipment to Western type lids and bottom boards.

When to Move Bees to the Crop

When the attractiveness of the crop is low, and there is competition from other plants in the area, then the best rule is not to move bees into the field until the plants are 10 to 20 percent in bloom. Moving bees only after there is significant bloom takes advantage of the bees' tendency to slowly expand the foraging area. When there is little or no competition, or the crop is very attractive, then moving bees into the field early will have no adverse effect. There are certain situations where having the bees in the field early is important. For example, in apples the best blossom to have produce a fruit is the king blossom. This is the first (center) blossom in the 5- or 6-flower cluster. Since it is the first blossom to open, it is necessary to have the bees in the orchard prior to bloom. (see the section on apples).

Placement of Colonies

In crops that are self compatible, the placement of colonies should be as evenly spaced as possible. For practical reasons colonies are often grouped in units because the colonies are on pallets or access to the field is limited (Fig. 8). On crops that need cross pollination, grouping of colonies is important since competition between colonies will create more wandering and better distribution of pollen. With increased competition, each bee will be more likely to

FIGURE 11. Boom-type hive loader placing 3-high colonies for watermelon pollination in Arizona. Some colonies fitted with pollen traps.

carry compatible pollen when it visits a flower. It is best to place colonies near the pollinizer trees in those crops requiring cross pollination, especially if the number of pollinizer trees is low. In this case young, naive bees will most likely visit these pollinizer trees on their initial flights and then carry this pollen into the hive where in-hive pollen transfer might occur. Thus, the transfer of the limited (or maybe limiting) pollen will be enhanced. A study of colony placement on almond showed that a high concentration of colonies near the center of the long side of oblong orchards resulted in a uniform concentration of foragers (Loper *et al.,* 1985).

Economics of Pollination

Beekeepers thinking about pollination should not think of colony rental fees simply as added income. You need to give due consideration to real costs. Some of these include: out-of-pocket expenses, extra nightime work, increased pesticide hazard, reduced honey yields, added wear on equipment, greater vulnerability to theft or vandalism, and possible exposure to bee diseases or pests. Thus, there is a cost for each of these items that must be included in the total rental fee. There is a tendency on the part of beekeepers to keep pollination rental fees low to maintain "control" of the area. Obviously, they do themselves a disservice when they attempt to operate below cost. Therefore, fees should be high enough to include a reasonable profit so the operation can provide the necessary pollination service on a continuing basis.

FIGURE 12. A 4-wheel drive forklift hive loader made by this Idaho beekeeper for loading colonies.

When honey prices are high, beekeepers tend to concentrate their efforts on honey production and in so doing neglect the pollination aspect of their business. This may not be in their long-term best interest. It is essential that beekeepers service the pollination business well (consistent with good business practices), and maintain the best possible relationship with growers. Contact between beekeepers and other agricultural groups leads to a better understand-

FIGURE 13. Colonies of bees on a trailer located in blueberries in Michigan.

FIGURE 14. Power-driven hand truck loading two colonies via ramp onto truck in Texas.

ing of each other's problems, thus giving support for necessary legislation, research, extension, disease control, pesticide regulations and public opinion. Interrelationships with growers, through crop pollination, gives beekeepers their greatest opportunity to be recognized as a vital part of agriculture.

Weather

Weather is the single most important factor both to the plant as well as to the bees. Consider the problem of an early, warm spring that advances the blossoming of fruit trees in the North. Unfortunately, honey bees are locked into a brood cycle time of 21 days, plus the time necessary to become forager bees (10-20 days). While the plants can bloom early because of the additional heat units, the colony cannot adequately advance the development of new bees. Although warmer weather may allow the colony to expand the size of its broodnest, the increased temperature does not speed up the development time of the immature bees. Thus, there won't be an increase in early foraging bees to pollinate a crop that blooms ahead of schedule.

Many of the tree fruits have developmental temperatures lower than those at which bees can forage. For example, apples will continue to grow and blossom at 42°F., whereas bees will not break cluster to forage below 55°F. Thus, it is theoretically possible for an apple tree to bloom and drop its blossoms with no bee visits for pollination. Fortunately, such weather extremes rarely occur. Since adequate pollination can occur in a short time period during favorable weather, it is best to think of honey bee colony rentals as insurance against inclement weather. Under adverse weather conditions there might only be a short "window" when bees can forage and pollinate effectively.

Inclement weather can also have an adverse effect on the growth of pollen tubes down the style, and thus reduce fertilization of the ovules. Under extreme heat the pollen may germinate, but the pollen tube will not grow if the stigma has dried. There is evidence that high temperatures might also cause the pollen tube to grow too fast and rupture. Pollen of some cotton varieties is inviable following a period of intensely high temperatures (110-120°F).

Weather may have an important effect on in-hive pollen transfer. Since it takes a period of nearly an hour to complete a foraging trip on many crops, the exchange of pollen within the hive certainly cannot occur before the first forager returns to the hive. When weather allows only short spurts of good flight, in-hive pollen transfers of viable pollen might be insufficient to accomplish cross pollination.

Pollination in Greenhouses and Cages

Honey bee colonies are widely used to pollinate commercial crops grown in confinement and for plant breeding projects. Full-size, queen-right colonies are frequently used in greenhouses and the colony is considered expendable. When the colony population decreases, the beekeeper replaces it with another and charges for the service. Young bees, and bees reared in confinement, forage better than bees which were field bees before they were brought inside. Many older bees fly to the glass or screen and never forage (Moffett and Spangler,1974). Moving a colony to the greenhouses, or cage, in the middle of the day while many older bees are in the field reduces this problem. Under most circumstances many bees will constantly fly against the glass or screen.

Colonies of bees used in confinement often consist of a 3-frame "nuc," or a 5- or 6-frame nuc in large cages. Use of queenless units is generally not satisfactory. If a colony is to survive in confinement over a period of several weeks, pollen cakes or some pollen substitute is needed to support brood rearing. Adequate honey is needed at all times, which may be supplemented with sugar syrup. It is also very important to provide water for the bees. Colonies should be placed in a location where they will not be overheated. Colonies may be placed outside the greenhouse, or cage, but with the entrance to the inside. Some beekeepers use an outside entrance which is opened periodically to allow nectar and pollen gathering and outside cleansing flights. This method should not be used when bees can pick up outside pollen that would contaminate varietal purity.

Mechanical Aids to Pollination

Pollen Inserts — A pollen insert, more properly called a pollen dispenser, is designed to apply viable pollen to foragers as they leave their hive (Fig. 15) (Jaycox, 1979). A pollen dispenser can be used to good advantage in a number of cases. For example, they can be used where pollinizer varieties have not been planted in sufficient quantity or the pollinizer variety has insufficient bloom in a particular year. If the pollinizer variety is out of synchrony with the main variety for a short period, then supplementing compatible pollen in a pollen dispenser will fill in the gap. The pollen dispenser has been successfully used (mostly with apples), but cases of failure have also been reported. The main effort required with pollen dispensers is harvesting or purchasing compatible pollen and storing it properly until it's needed. Diluting the pollen with lycopodium spores and charging the dispensers at regular intervals is also important if bees are to pick it up and move it onto the stigma of receptive flowers.

Other Mechanical Aids — Pollen of tree fruits requiring cross pollination has been commercially available for many years. Helicopters, ground blowers, shot guns, etc., have been suggested and occasionally used to apply metered amounts of compatible pollen directly to fruit trees in bloom. Such pollen is placed in the crop for redistribution by pollinating insects. Tests carried out in California indicated that supplemental artificial pollination did not increase yield of Nonpareil almonds (McGregor, 1976). Pollen distribution by aircraft may find application in the pollination of dioecious, wind-pollinated nut trees. Various mechanical methods have been attempted to trip and pollinate alfalfa with little success, and to aid cranberry pollination. These have been largely discontinued since there is no substitute for repeated bee visits to a population of continually developing flowers.

Pesticides and Pollination

Many crops that require insect pollination also have active pest manage-

POLLEN INSERT

Construction Details

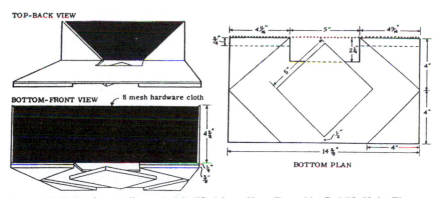

FIGURE 15. Pollen insert (diagram). Modified from Hort. Facts, No. B-4-79, Univ. Ill.

ment programs that include the use of insecticidal sprays, hopefully applied either prior to the bees being moved onto the crop or immediately after their removal. The timing of these sprays is often critical. Under pressure of timely application of insecticide sprays, growers sometimes use methods and materials that endanger bees. Also, on some occasions a grower in one field sprays his crop after the bees had been removed from his field, while there are still colonies of bees in a neighbor's field. Since bees do not recognize fence lines, many bees may be killed under such circumstances. There are also times when a crop is sprayed with long-residual or systemic insecticides just before the bees are moved into the area leaving toxic residues on leaves and petals and/or in the nectar. It is important for the beekeeper-pollinator to recognize some of these potential hazards of managing bees on crops needing pollination. Understanding the growers problems may help the beekeeper anticipate insecticide use and make plans to avoid bee losses. Timely notification of intent to spray by the grower should be part of a pollination contract. Communication established with adjacent farmers prior to moving bees to the crop to be pollinated may also be helpful.

Grower Ownership of Colonies

Occasionally growers producing bee-pollinated seed or fruit crops decide to provide their own bees for pollination. Such ventures rarely succeed. Large-scale commercial beekeeping is very much a specialty and requires a specialist to manage the colonies successfully. Furthermore, when 500 colonies are needed for a single crop, there is the need to manage as many as 750 colonies to have assurance of 500 good colonies for pollination. The beekeeper must manage all these colonies on a year-round basis and not just during the active pollination period. Colonies need to be protected from diseases, pests, starvation and other natural and man-made disasters. A commercial beekeeper is likely to relocate his/her colonies to take advantage of various nectar and pollen sources to buildup and improve the colonies. The beekeeper may also move them for additional pollination fees on other crops. A grower owning bees might need to move his colonies away from the area of operation for no other reason than to avoid insecticide sprays applied to protect the crops. Thus, the seed or fruit grower will be forced to provide full-time management to the bees to ensure that the bees will be in the right condition for crop pollination when needed.

CONCLUSION

It seems that the ever-expanding world populations are changing the land-use picture and seriously affecting beekeeping. Open fields of honey plants continue to be replaced by houses, factories, pavement and hostile environments where profitable beekeeping is difficult. Large-scale monoculture of fruit, corn, soybeans, wheat, etc. maintained weed free by herbicides (necessary for specialized high production) eliminates the continuous source of nectar and pollen that bees require. Such trends require increasing numbers of professionally managed colonies moved to crops during blossom time. Serious problems confront commercial beekeeping, but the long-range survival of the industry is assured because in the job of crop pollination honey bees are indispensable.

ACKNOWLEDGEMENT

We would like to give our thanks to all the pollination biologists who have shared their work with us over the years, especially two great mentors of ours, Dr. E.C. Martin and the late S.E. McGregor.

REFERENCES

Betts, A.D. (1935). The constancy of the pollen-collecting bee. *Bee World 16*:111-113.

Chang, S.Y. and R.A. Hoopingarner. (1990). Relative importance of feral honey bees in apple pollination. VI Intern. Symposium on Pollination. Tillberg, Netherlands.

Collison, C.H. and E.C. Martin. (1973). The effects of overhead irrigation on the pollination of pickling cucumbers. *Pickle Pak Sci. 3*:1-3.

Collison, C.H. (1976). The interrelationships of honey bee activity, foraging behavior, climatic conditions, and flowering in the pollination of pickling cucumbers, *Cucumis sativus* L. Ph.D. Thesis, Michigan State University, East Lansing, MI.

DeGrandi-Hoffman, G., R. Hoopingarner and K. Klomparens. (1986). Influence of honey bee (Hymenoptera: Apidae) in-hive pollen transfer on cross-pollination and fruit set in apple. *Environ. Entomol. 15*:723-725.

Doull, K.M. (1971). An analysis of bee behavior as it relates to pollination. *Am. Bee J. 111*:266, 273, 302-303, 340-341.

Dowden, A.O. (1964). The Secret Life of Flowers. The Odyssey Press, New York.

Faegri, K. and van der Pijl, L. (1971). The Principles of Pollination Ecology. Pergamon Press, Oxford.

Free, J.B. (1962). The attractiveness of geraniol to foraging honeybees. *J. Apic. Res. 1*:52-54.

Free, J.B. (1964). Comparison of the importance of insect and wind pollination of apple trees. *Nature 201*:726-727.

Free, J.B. (1965). Attempts to increase pollination by spraying crops with sugar syrup. *J. Apic. Res. 4*:61-64.

Free, J.B. (1970). Insect Pollination of Crops. Academic Press. New York.

Free, J.B. and Y. Spencer-Booth. (1963). The foraging area of honey-bee colonies in fruit orchards. *J. Hort. Sci. 38*:129-137.

Free, J.B. and I.H. Williams. (1972). The transport of pollen on the body hairs of honeybees (*Apis mellifera* L.) and bumblebees (*Bombus* spp. L.). *J. Appl. Ecol. 9*:609-615.

Free, J.B. and I.H. Williams. (1972). The role of the Nasonov gland pheromone in crop communcation by honeybees (*Apis mellifera* L.) *Behavior 41*:314-318.

Frisch, K. von. (1971). Bees, Their Vision, Chemical Senses and Language. Cornell Univ. Press. Ithaca, NY.

Gary, N.E., P.C. Witherell and J. Marston. (1972). Foraging range and distribution of honey bees used for carrot and onion pollination. *Environ. Entomol. 1*:71-78.

Guilcher, J.M. and R.H. Noailles. (1966). A Fruit is Born. Sterling. New York.

Hoopingarner, R. (1982). The individual hive robbing screen. *Glean. Bee Cult. 110*:92, 109.

Jaycox, E.R. (1979). Pollen inserts for apple pollination. Hort. Facts B-5-79; Making and using pollen inserts. Hort. Facts B-4-79. Univ. Ill. Coop. Extn. Serv.

Karmo, E.A. and V.R. Vickery. (1954). The place of honey bees in orchard pollination. Mimeogr. Circ. Nova Scotia. Dept. Agr. Mktg. No. 67.

Langridge, D.F. and R.D. Goodman. (1974). A study on pollination of sunflowers (*Helianthus annuus*). *Aust. J. Exptl. Agr. Anim. Husb. 14*:201-204.

Loper, G.M., R.W. Thorp and R. Berdel. (1985). Improving honey bee pollination efficiency in almonds. *Calif. Agr. 39*:19-20.

Luckwill, L.C., D.W. Way and J.B. Duggan. The pollination of Fruit Crops. Gibbs & Sons Ltd., Canterbury, England (Reprinted from Sci. Hort. Vols. XIV, 1959-60 and XV 1960-61.)

McGregor, S.E. (1958). Relation of honey bees to pollination of male-sterile cotton. XI Cotton Impr. Conf. Proc. 93-94.

McGregor, S.E. (1976). Insect Pollination of Cultivated Crop Plants. USDA Agr. Handbook No. 496., Supt. Documents, Washington, D.C.

Meeuse, B.J.D. (1961). The Story of Pollination. The Ronald Press. New York.

Moffett, J.O. (1983). Hybrid Cotton. p. 508-514. *In* C.E. Jones and R.J. Little (eds.) Handbook of Experimental Pollination Biology. Sci. and Acad. Ed. Div. of Van Nostrand Reinold Co., New York. 558p.

Moffett, J.O. and D.R. Rodney. (1971). Fairchild tangerines need both honey bees, pollinator trees. *Prog. Agr. in Arizona 23*:6-7.

Moffett, J.O. and H.G. Spangler. (1974). Honey bees flew and visited flowers satisfactorily in polyethylene greenhouses. *Am. Bee J. 114*:62-63.

Nye, W.P. and O. Mackenson. (1970). Selective breeding of honey bees for alfalfa pollen collection: with tests in high and low alfalfa pollen collection regions. *J. Apic. Res. 9*:61-64.

Percival, M.S. (1965). Floral Biology. Pergamon Press. Oxford.

Peterson, C.E. (1960). A gynoecious inbred line of cucumber. *Mich. Agr. Expt. Sta. Quart. Bull. 43*:40-42.

Ribbands, C.R. (1949). The foraging method of individual honeybees. *J. Anim. Ecol. 18*:47-66.

Rom, R.C. (1970). Variety and cultural consideration necessary to assure adequate pollination in apple orchards. *Rept. Pollination Conf. Univ. Ark. 9th, MP 127*:219-225.

Waller, G.D. (1970). Attracting honeybees to alfalfa with citral, geraniol and anise. *J. Apic. Res. 9*:9-12.

Waller, G.D., E.W. Carpenter and D.A. Ziehl. (1972). Potassium in onion nectar and its probable effect on attractiveness of onion flowers to honey bees. *J. Amer. Soc. Hort. Sci. 97*:535-539.

Waller, G.D., N.E. Gary, S.T. Chester, Jr., M.S. Karim, J.H. Martin and B.E. Vaissiere. (1985). Honeybee colony populations and foraging rates on onion seed fields in Arizona. *J. Apic. Res. 24*:93-101.

CHAPTER 25

DISEASES AND PESTS OF HONEY BEES

by H. Shimanuki[1], D. A. Knox[2], B. Furgala[3],
D. M. Caron[4], *and* J. L. Williams[5],

INTRODUCTION

Control of bee diseases has always been recognized as an important part of good beekeeping. In fact, the only Federal bee law (the Honeybee Act) was enacted by Congress in 1922 to restrict the importation of living adult honey bees into the country. In addition to the Honeybee Act, all 50 states have specific laws dealing with bee diseases and beekeeping (Michael *et al.,* 1990). State bee disease laws, were for the most part, enacted to abate American foulbrood. These laws have subsequently been amended to include such things as apiary registration, parasitic mites, and Africanized honey bees.

Beekeeper's Role—Beekeepers, whether they own one or thousands of hives, must develop a bee disease management program based on periodic colony inspections. Beekeepers must learn to recognize signs of bee diseases and be able to differentiate the serious from the not-so-serious and must know the corrective actions for each disease.

Entering an Apiary—The process of disease diagnosis begins as soon as you enter an apiary. You should check such things as flight activity, the presence or absence of dead or moribund bees at colony entrances, and spotting of beehives by feces. Also, hives can be quickly checked for low weights by tilting them or "hefting" them. If you are not inspecting all the hives, choose those hives that appear abnormal. Do not eliminate hives because they are populous or show the greatest activity at the entrance. In some cases, the colony with the largest population may be the one with American foulbrood disease. Furthermore, vigorous activity at a hive entrance may be the result of robber bees removing honey from a diseased colony. Before you open a hive, you should begin to make mental notes of such things

[1]Hachiro Shimanuki, Ph.D., Research Leader, Bee Research Laboratory, USDA, ARS, PSI, Beltsville, Md.

[2]David A. Knox, Entomologist, Bee Research Laboratory, USDA, ARS, PSI, Beltsville, Md.

[3]Basil Furgala, Ph.D., Professor, Department of Entomology, Fisheries and Wildlife, University of Minesota, St. Paul, Minn.

[4]Dewey M. Caron, Ph.D., Professor, Department of Entomology and Applied Ecology, University of Delaware, Newark, Del.

[5]Jon L. Williams, Entomologist, Honey Bee Breeding, Genetics and Physiology Laboratory, Baton Rouge, La.

as—Is there an accumulation of dead bees? Is spotting evident? Are there immature stages of bees at the entrance?

Examining Adult Bees—Most adult bee diseases are difficult to diagnose because the gross symtoms are not unique. The behavior of affected bees and colonies are frequently similar for adult diseases and even for non-communicable disorders such as the effect of pesticides. For instance, inability to fly, unhooked wings, and dysentery are general symptoms associated with many diseases. In most cases, a microscopic examination is required to make a proper diagnosis.

Examining the Brood—In looking for a brood disease select a comb with brood. If no brood is present, look for combs in which brood was previously present. The first thing that you should note is whether there is a uniform egg laying pattern from the center of the comb to the outer edges. Virtually every cell from the center of the comb outward should have eggs, larvae or pupae. The cappings should be uniformly brown or tan and convex (higher in the center than at the margins). The unfinished cappings of healthy brood may appear to have punctures, but since cells are capped from the outer edges to the middle, the holes are always centered and have smooth edges. You should look for the so-called "pepperbox" symptom, a mixture of capped and uncapped brood cells. This is a good symptom to note as it usually signals a disease condition. However, there are times when such conditions result from noninfectious disorders.

The next step is to look for abnormal cappings or cells with brood that appear abnormal. It is generally easier to start the examination by finding sealed brood that have punctured, sunken or discolored cappings. Cells with such cappings should be carefully examined for the decaying remnants of larvae or pupae. Combs should also be examined for scales (the dried remains of larvae or pupae) which are found lying lengthwise on the bottom side of brood cells.

For many years beekeepers distinguished the foulbrood diseases by smell. European foulbrood was the "sour smelling" foulbrood and American foulbrood was the "stinking disease." It wasn't until 1906 that the names American and European foulbrood were given to differentiate these two diseases. One reason for the delay in separating the foulbrood diseases was the lack of agreement on their etiology.

Non-contagious conditions may sometimes produce symptoms that mimic a contagious disease. Neglected brood occurs when there is a shortage of nurse bees and brood dies from either chilling or starvation. Disease-like symptoms also can be the result of a failing queen, laying workers, toxic chemicals, or poisonous plants.

BACTERIAL DISEASES
American Foulbrood Disease

Before parasitic mites were detected in this country, American foulbrood was the most serious bee disease. American foulbrood disease (AFB) can be found on every continent. In the United States, based on statistics submitted by the Apiary Inspectors of America from 44 reporting states, 1.8% of all the colonies inspected had American foulbrood disease in 1984 (Anonymous, 1985).

American foulbrood disease is caused by the bacterium *Bacillus larvae*. Only the spore stage of *B. larvae* can initiate the disease, and the spores can remain viable indefinitely on beekeeping equipment. Worker, drone and queen larvae are susceptible to AFB for up to three days following egg hatch. Drone larvae are less susceptible than queen or worker larvae (Rinderer and Rothenbuhler, 1969). Woodrow (1941a) found that the susceptibility of larvae to American foulbrood disease decreased with increasing age and they become immune 53 hours after egg hatch. Woodrow (1942) concluded that one *B. larvae* spore is sufficient to infect a larva one day after egg hatch. Using Woodrow's data, Bucher (1958) determined the LD_{50}* of *Bacillus larvae* to be 35 spores in one-day-old honey bee larvae.

Spores of *B. larvae* germinate approximately one day after ingestion by the larva. After germination the bacteria multiply in the midgut and penetrate to the body cavity through the gut wall, the larva ultimately dies from a septicemic condition (Bamrick, 1964). American foulbrood may not destroy a colony in the first year. However, if left unchecked, the number of infected individuals may increase, ultimately leading to the death of the colony.

American foulbrood disease is spread from colony to colony by robbing and drifting bees. In addition, beekeepers often spread AFB by inadvertently feeding honey or pollen from diseased colonies or interchanging brood combs among diseased and healthy colonies.

Combs with AFB diseased larvae may exhibit the pepperbox symptom. When the infection is light, the pepperbox symptom may not be apparent. Cappings over diseased brood are dark brown, usually punctured and sunken into the cell (Fig. 1).

Healthy larvae are pearly white in appearance, while diseased individuals show a color change ranging from dull white to brown and finally black with the progression of the disease. When larvae are brown, the symptom of ropiness can be demonstrated (Fig. 2). To do this, macerate the suspected brood with a matchstick or twig and carefully withdraw it from the cell. If

---

*LD_{50}. The dose (number of microorganisms) that will kill 50% of the animals (honey bee larvae) in a test series.

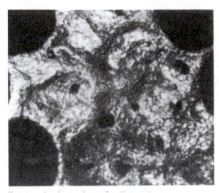

FIGURE 1. American foulbrood—appearance of cappings. *(Photo courtesy of M. V. Smith, Guelph University)*

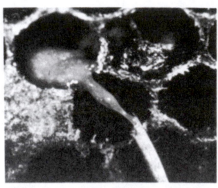

FIGURE 2. American foulbrood—the larval remains "rope out" when a matchstick is inserted and withdrawn. *(Photo courtesy of M. V. Smith, Guelph University)*

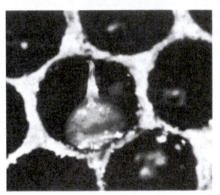

FIGURE 3. American foulbrood—front view of pupa in early stage of breakdown. *(Photo courtesy of M. V. Smith, Guelph University)*

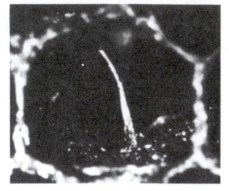

FIGURE 4. American foulbrood—front view of scale showing remains of pupal tongue. *(Photo courtesy of M. V. Smith, Guelph University)*

AFB is present, the brown remains can usually be drawn out like a thread for two or more centimeters. One word of caution—this symptom cannot always be demonstrated in AFB diseased larvae.

The final stage of the disease progression is the formation of the scale. If AFB disease kills the pupa, the resultant scale may have the so-called pupal tongue, which is a threadlike projection that extends away from the scale toward the center of the brood cell (Fig. 3, 4). This too is a valuable symptom, but it may not always be present.

In most cases American foulbrood disease can be diagnosed in the field by experienced beekeepers or by personnel specially trained for disease inspection. However, when a more accurate diagnosis is required, laboratory techniques should be used. Some of the methods available are described in the following section.

Laboratory Diagnostic Techniques

Bacillus larvae is a slender rod-shaped bacterium with slightly rounded ends and a tendency to grow in chains. It varies greatly in length, from about 2.5 to 5 μm and is about 0.5 μm in width. The spore is oval-shaped and approximately twice as long as it is wide, measuring about 1.3 by 0.6 μm. When stained with carbol-fuchsin, the spore walls appear reddish-purple and clear in the center. The spores may form clusters and appear stacked. Approximately 2.5 billion spores are produced in each infected larva. If the larva has been infected for less than 10 days, the vegetative cells are present.

Microscopic Examination—The modified hanging drop technique (Michael, 1957) can be very useful for differentiating American foulbrood from other brood diseases. Residue from a suspected cell is first mixed with water and a drop of this suspension (smear) is placed on a cover glass. The suspension used should always be dilute and only slightly turbid. The smear is dried and fixed under a heat lamp; alternatively, the smear can be air-dried and heat-fixed by passing it rapidly through a Bunsen flame two or three times. The fixed smear is stained with carbol-fuchsin or a suitable spore stain for 10 seconds. Enough stain should be placed on the cover glass to cover the entire smear. The excess stain is then washed off with water. While the smear is still wet, the cover glass is inverted with the smear side down and placed onto a standard microscope slide previously coated with a very thin layer of immersion oil. The slide is gently blotted dry and examined with a microscope using the oil immersion objective. In areas where pockets of water are formed in the oil, the spores of *B. larvae* usually exhibit Brownian movement. This is an extremely valuable diagnostic technique since spores formed by other *Bacillus* species associated with known bee diseases usually remain fixed. Brownian movement can be affected by slide preparation; also debris and other bacteria can exhibit this motion. Therefore, Brownian movement must not be used as the sole criterion for diagnosis. It should be considered together with the characteristic morphology of the spores and the gross larval symptoms.

The Holst Milk Test (Holst 1946)—This is a simple test based on the fact that a high level of proteolytic enzyme is produced by sporulating *B. larvae*. The test is conducted by suspending a suspect scale or a smear of a diseased larva in a tube containing 3 - 4 ml of 1% powdered skim milk in water. The tube is then incubated at 37°C. If *B. larvae* is present, the suspension should clear in 10 - 20 minutes. It should be noted that this test is not always reliable.

Nitrate Reduction—*Bacillus larvae* reduces nitrate to nitrite (Lochhead 1937). This test can be performed on an appropriate medium containing potassium nitrate (1 - 2 mg per liter of medium). After growth has occurred, the addition of a drop of sulfanilic acid-alpha napthol reagent produces a red color if nitrate was reduced to nitrite. This test should not be used alone for diagnosis, but combined with larval gross symptoms, bacterial morphology, and colony growth characteristics.

Catalase Production—For this test, a drop of 3% hydrogen peroxide is placed on an actively growing colony on solid media. Most aerobic bacteria produce catalase which breaks down the peroxide to form water and oxygen. This reaction produces a bubbly foam, but *B. larvae* is usually negative for this reaction (Haynes 1972).

Fluorescent Antibody—This technique requires the preparation of specific antibodies stained with a fluorescent dye. Rabbits are injected with pure cultures of *B. larvae* and the active antiserum collected and stained with a fluorescent dye. This fluorescent antiserum is then mixed with a bacterial smear on a slide and allowed to react. The excess antiserum is washed off the slide and then examined using a fluorescence microscope. *Bacillus larvae* appears as brightly fluorescing bodies on a dark background (Toschkov *et al.,* 1970; Zhavnenko, 1971; Otte, 1973; Peng and Peng, 1979).

Cultivation of *Bacillus larvae*—Thiamine (Vitamin B_1) and some amino acids are required for the growth of *B. larvae.* Routine culture media such as nutrient broth will not support the growth of *B. larvae.* Good vegetative growth occurs on Difco brain heart infusion fortified with 0.1 mg thiamine hydrochloride per liter of medium (BHIT) and adjusted to pH 6.6 with HCl; however, sporulation does not occur. Satisfactory growth and sporulation occurs on the yeast extract, soluble starch, and glucose media recommended by Bailey and Lee (1962). The medium can be liquid, semisolid (0.3% agar), or solid (2% agar).

To culture *B. larvae*, prepare spore suspensions by mixing diseased material (scales) in sterile water (one scale per ml water) in screw-cap tubes. The suspension is heat-shocked at 80°C for 10 minutes (effective time) to kill non-sporeforming bacteria. A sterile cotton swab is then used to spread a portion of the suspension (approx. 0.2 ml) evenly over the surface of BHIT agar plates. The plates are then incubated for 72 hours at 34° C. Individual colonies are small (1 - 2 mm) and opaque; however, if large numbers of viable *B. larvae* spores are inoculated, a solid layer of growth will cover the plate.

Bacillus larvae spores also reproduce in the hemolymph of honey bee larvae, pupae and adults when artifically introduced by injection (Michael, 1960; Wilson and Rothenbuhler, 1968; Wilson, 1970). The injection of pupae with *B. larvae* is a good method of determining spore viability.

Detection of *Bacillus larvae* Spores in Hive Products
    Honey—Occasionally it may be necessary to examine honey for the

presence of *B. larvae.* Due to the high carbohydrate and other natural bacteriostatic substance(s) in honey, the examination of honey requires special techniques. The classical method (Sturtevant, 1932, 1936) is to dilute the honey 1:9 with water, centrifuge the mixture to concentrate the spores in the sediment, and then examine the sediment microscopically for the presence of spores.

Shimanuki and Knox (1988) developed an improved technique to detect *B. larvae* spores in honey. Honey to be examined is heated to 45°C to permit easier handling and to decrease viscosity for more uniform distribution of spores that may be present. Twenty-five ml of honey is placed in a 50-ml beaker and diluted with 10 ml of sterile water. The diluted honey is then transferred into a 1.75-inch (44-mm) dialysis tube. The open end is tied after filling, and the tube is submerged in running water or in a water bath with three to four water changes in an 18-hour period. Following dialysis, the contents of the tube are centrifuged at 2000 g for 20 minutes. The supernatant is carefully removed with a pipet to leave approximately 1 ml of residue. This residue is resuspended in 9 ml of water in a screw-cap vial and heat-shocked at 80°C for 10 minutes to kill non-sporeforming bacteria. Next, 0.5 ml of the suspension is spread onto a BHIT agar plate. The plate is incubated at 37°C for 72 hours and examined for colonies of *B. larvae.* Difficulties can sometimes occur when honey samples contain other sporeforming bacteria that may completely cover the surface of the plate.

Pollen—*Bacillus larvae* spores also can be recovered from bee-collected pollen pellets by physically removing bits of AFB scale. A series of different sized sieves is helpful. If scales are not detected, pass a water-pollen suspension through #2 filter paper, centrifuge the filtrate, and culture the pellet as described above (Gochnauer and Corner, 1974).

Treatment

There are no strains of bees immune to AFB. However, different degrees of resistance to the disease have been shown. The mechanisms responsible for this resistance include: the role of the proventricular valve (Sturtevant and Revell, 1953); removal of diseased larvae (house cleaning behavior) (Woodrow, 1941b); protection of larvae by adults (Thompson and Rothenbuhler, 1957); larval resistance (Rothenbuhler and Thompson, 1956); and different levels of bacterial inhibitors in the brood food provided by different nurse bees (Rose and Briggs, 1969).

Beekeepers are subject to state laws. As such, the options for disease prevention and control may be limited by state regulations. Consequently, beekeepers should first establish the legality of any chemical or other treatment before use.

One of the earliest methods for treating foulbrood disease was the "shaking method." Simply, the method called for holding the adult bees away from the infection source until the potentially contaminated honey carried by the bees was consumed (Howard, 1907).

Burning of hives with AFB disease is required in some states; in others, beekeepers are allowed to retain the hive bodies, bottom boards, and inner and outer covers, and burn only the frames. The parts to be salvaged are first scraped and then scrubbed with a stiff brush and hot soapy water. Some beekeepers will also scorch the inner surface of the hive bodies before reuse. Another treatment option is the use of a lye solution. Here the wooden equipment is completely immersed in a boiling solution containing one pound (454 gm) lye to 10 gallons (38 l) of water for 20 minutes. Since lye solutions are caustic, appropriate safety equipment should be worn.

Haseman and Childers (1944) began a new era when they introduced sulfathiazole for the prevention and control of AFB. Later, Gochnauer (1951) reported that oxytetracycline was also effective in the treatment of the disease. In the U.S., only oxytetracycline HCl (Terramycin®) can be used for the prevention and control of American foulbrood disease, since sodium sulfathiazole is no longer registered for this purpose.

The sterilization of hive equipment has also been found effective in controlling AFB. These techniques include: Colbalt-60 irradiation (Studier, 1958); ethylene oxide (Michael, 1964; Shimanuki, 1967); and high-velocity electron beams (Shimanuki *et al.,* 1984).

Methods of Feeding Terramycin®

Terramycin® (oxytetracycline HCl) is available in several formulations and each carries specific instructions. If you are going to feed Terramycin, you must first read the label carefully to see whether the formulation in hand is registered for use on honey bee colonies. In some cases, the instructions for feeding honey bee colonies may be on the label accompaniment rather than on the primary label. In any case, follow the label instructions. If used incorrectly, residues of Terramycin can contaminate surplus honey.

The methods of feeding Terramycin include dusting, bulk syrup feeding and in an antibiotic extender patty [no label has been approved by the U.S. Food and Drug Administration (FDA) for the latter as of 1990]. Select your feeding method based on your management system, but understand that each method has its advantages and disadvantages.

Dusting—Dusting requires very little additional expense or equipment. All one needs is some confectioners (powdered) sugar, Terramycin and a measuring spoon. In using this method, dust only the ends of the top bars in the brood area. Because sugar is being used as the carrier, the bees consume the dust rapidly. Do not apply the dust on the face of the brood combs as the dust mixture may kill open brood. This method is quick and easy, but it may result in uneven dosages within the hive.

Bulk feeding—Many beekeepers routinely feed sugar syrup to their colonies in the fall and spring. Consequently, Terramycin can be easily incorporated in the syrup. However, if syrup feeding is not a routine management procedure, you will need to purchase feeding containers such as a pail, division board feeder, or equivalent. In addition you will need to mix sugar syrup to serve as a carrier. This method delivers a uniform dosage to the colony and should cause no brood mortality. Two drawbacks with this method are that Terramycin is less stable in syrup than in a dry formulation and the likelihood of contaminating honey is greatest with this method (see Gilliam and Argauer, 1975). For this reason all drug feeding should cease four weeks before the surplus honey flow and all honey should be removed from above the brood nest when drug feeding is terminated.

Extender patty—The use of extender patties has been sanctioned by the FDA; however, no label has been approved. Extender patties have been shown to be effective in the prevention and control of American foulbrood. In comparison to the other methods that use a sugar formulation for feeding Terramycin, the extender patty uses a sugar plus solid vegetable oil as the major ingredients. Since no water is required for this formulation, the Terramycin in the extender patties maintains its activity longer. Consequently, the patties can be prepared before the active bee season. All a beekeeper needs to do is to place a patty on a piece of wax paper laying on the top bars of the brood comb. Two advantages of extender patties are that they last for long periods of time and the likelihood of contamination of the honey crop is the lowest of the three methods (Gilliam and Argauer, 1975).

European Foulbrood Disease

European foulbrood disease (EFB) is worldwide in distribution and in some areas, considered to be as serious a problem as AFB. There is a complex of microflora associated with this disease including the causative bacterium *Melissococcus pluton.* Associated organisms, *Bacillus alvei, B. laterosporus,* and *Enterococcus faecalis,* do not cause the disease, but they do influence the odor and consistency of the dead brood.

European foulbrood disease strikes primarily in mid to late spring (May and June in northern temperate areas), the time when colonies should be building up to maximum populations. Because diseased colonies fail to increase normally, they may not provide the beekeeper with surplus honey. In most cases, the disease symptoms disappear with the onset of a surplus honey flow. Unfortunately, this remedy occurs too late for most colonies to produce a surplus crop. European foulbrood also can be found in the fall, but it is not as common as in the spring. The bacteria overwinter on the sides of the cell wall or in feces and wax debris on the bottom of the hive.

European foulbrood disease is readily transmitted by nurse bees that inadvertently infect the larvae while feeding them. The infectious cycle begins when a larva (less than 48 hours old) consumes food contaminated with *M. pluton.* The organism multiplies in the midgut, destroys the peritrophic membrane and, as the disease progresses, invades the intestinal epithelium (Tarr, 1938). The larva must compete for food with the rapidly multiplying bacteria creating an abnormal demand for larval food. The nurse bees eject those larvae that require more than the usual amount of food (Bailey, 1960); therefore, if populations of nurse bees are sufficient most colonies can overcome the infection without beekeeper assistance.

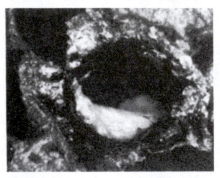

FIGURE 5. European foulbrood—larva twisted in cell. *(Photo courtesy of M. V. Smith, Guelph University)*

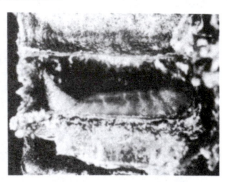

FIGURE 6. European foulbrood—prepupa with head raised. *(Photo courtesy of M. V. Smith, Guelph University)*

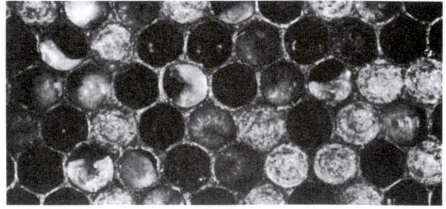

FIGURE 7. European foulbrood—unsealed larvae in various stages of collapse. [From Bull. 100, Brit. Min. of Agric., Fisheries and Food *(Reproduced by permission of the Controller of H. M. Stationary Office)*]

Worker, drone and queen larvae are all susceptible to EFB. Larvae infected with EFB usually die while still in the coiled stage. The larvae first turn yellow and then brown, at which time the tracheal system becomes quite visible. Sometimes diseased larvae apparently die in the upright stage (Fig. 5), but at other times they appear to collapse and seem to be twisted or melted in the bottom of the cell (Fig. 6, 7). At this stage the larval remains can exhibit a slight ropiness (less than 2.5 centimeters), which, when drawn out, appear granular.

If the disease is widespread in a colony, the combs take on a pepperbox appearance (the term used to describe combs with many uncapped cells mixed with normal capped cells). Another useful symptom is the appearance of the cappings: cell cappings of healthy brood are convex, while those over diseased brood are concave and sometimes punctured. The odor of larvae infected with European foulbrood disease varies with the presence of saprophytes. Typically it can be described as sour, hence the German term for the disease, *"sauerbrut."* After some time, the remains of the larva dry out and form a scalelike protrusion in the cell. Characteristically, scales caused by European foulbrood disease are rubbery rather than brittle as are those caused by American foulbrood disease. European foulbrood scales are much easier to remove than are American foulbrood scales.

European foulbrood is positively diagnosed in the laboratory by the identification of the causative organism or associated microflora. *Melissococcus pluton* is generally observed microscopically early in the infection cycle before the appearance of the varied microflora associated with this disease. It is a short, non-sporeforming, lancet-shaped cell measuring 0.5 to 0.7 by $1.0\mu m$ and occurs singly, in pairs, or in chains.

Pinnock and Featherstone (1984) developed an enzyme-linked immunosorbent assay (ELISA) for detecting *M. pluton*. Using this technique, they were able to demonstrate the presence of *M. pluton* even in apparently healthy colonies.

Melissococus pluton can be isolated on the following medium developed by Bailey (1959): 1 gm yeast extract (Difco), 1 gm glucose, 1.35 gm potassium dihydrogen phosphate (KH_2PO_4), 1 gm soluble starch, distilled water to make 100ml, 2 gm agar, adjust ph to 6.6 with KOH and autoclave at 10 lb/in² (116°C) for 20 minutes. It has been found that the addition of cysteine (0.1 gm per 100 ml) improves the multiplication of *M. pluton* (Bailey and Collins, 1982).

It is difficult to isolate *M. pluton* on artificial media because of its growth requirements and competition from other bacteria. Also, once isolated, identification is difficult due to its pleomorphic nature. *Melissococcus pluton* is best isolated when few if any other organisms are present. According to Bailey

(1959), it is best to dry smears of diseased larval midguts on a slide. A water suspension of this material or a suspension prepared from larvae (apparently healthy, infected, or dead), cappings, etc. can be streaked on freshly prepared Bailey's agar media. Alternatively, decimal dilutions of these suspensions can be inoculated into melted Bailey's agar medium (45°C) and poured into petri dishes (Bailey, 1981). The petri dishes are then incubated anaerobically at 34°C. The "Gas Pak" (BBL) Anaerobic System employing a disposable hydrogen and carbon dioxide generator can be used to obtain anaerobic conditions. Small white colonies of *M. pluton* should appear after four days.

Bacillus alvei is the most frequently encountered of the organisms associated with EFB. It is a rod measuring 0.5 to 0.8 μm in width by 2.0 to 5.0 μm in length. Spores measure 0.8 by 1.8 to 2.2 μm. Like *B. larvae*, the spores may be clumped and appear stacked. The sporangium may be observed attached to the spore. Typical strains of *B. alvei* spread luxuriantly on nutrient agar. "Motile colonies" may appear, and free spores may lie side by side in long rows on the agar. The growth of this bacterium produces an unpleasant odor.

Bacillus laterosporus rods measure 0.5 to 0.8 μm by 2.0 to 5.0 μm and the spores 1.0 to 1.3 by 1.2 to 1.5 μm. An important diagnostic feature is the production of a canoe-shaped parasporal body that stains very heavily along one side and the two ends and appears to remain firmly attached to the spore after lysis of the sporangium. The clear portion and the finely outlined wall is the spore. *Bacillus laterosporus* grows moderately well on nutrient agar, becoming dull and opaque. It spreads rapidly if the agar surface is moist. Growth on nutrient agar with 1% glucose added (glucose agar) is thicker and the surface growth may appear wrinkled.

Enterococcus faecalis resembles *M. pluton* and may exhibit Brownian movement in a modified hanging drop. The cells are ovoid (elongated in the direction of chain), 0.5 - 1.0 μm in diameter, and usually are in pairs or short chains. Growth occurs on nutrient agar usually within one day. Colonies are generally less than 2 mm and are smooth and convex with a well defined border. When magnified, the colonies appear light brown in color and granular in structure.

Treatment

When the EFB infection is light, treatment is usually not required. As a rule, the disease disappears during a good, steady nectar flow. However, EFB requires treatment in instances where adult populations of colonies have diminished. In such cases, the colonies do not gather sufficient stores to overwinter and may die.

Differences in susceptibility to EFB have been reported in strains of bees, but no strain is immune to the disease. Requeening is advocated as a treatment. The success of this procedure can be attributed partly to the introduction of a potentially more prolific queen, but more importantly, the break in the brood cycle provides nurse bees with the opportunity to remove affected larvae and polish the cells.

A number of chemotherapeutic agents have been used successfully for the prevention and control of EFB (Katznelson *et al.,* 1952). The most widely used antibiotic today is oxytetracycline HCl (Terramycin®). Various formulations of this antibiotic are available for feeding honey bee colonies. [see "Methods of Feeding Terramycin"]. In the case of EFB, the antibiotic extender patties do not seem to be as effective.

Powdery Scale Disease

Powdery scale disease is seldom a problem and is apparently rare. Powdery scale disease is caused by the bacterium *Bacillus pulvifaciens* which affects only the honey bee larvae. The scale is light brown to yellow and extends from the base to the top of the cell. The remains of the dead larva are powdery rather than brittle as in American foulbrood, or rubbery as in European foulbrood; hence the name powdery scale.

Bacillus pulvifaciens vegetative cells measure 0.3 to 0.6 μm by 1.5 to 3.0 μm. The spores measure 1.0 by 1.3 to 1.5 μm. It can be isolated on nutrient agar, however, growth is heavier on glucose agar. When first isolated, it produces a reddish-brown pigment that can be lost by subculturing. *Bacillus pulvifaciens* closely resembles *B. larvae,* but the spores do not exhibit Brownian movement in the modified hanging drop. Also, *B. pulvifaciens* is distinguished by its ability to grow at 20°C and by its growth on nutrient agar.

Septicemia Disease

Septicemia disease is caused by the bacterium *Pseudomonas aeruginosa.* It is a somewhat rare adult disease and colonies require no treatment. The disease destroys the connective tissues of the thorax, legs, wings, and antennae of adult honey bees. Bees affected with *P. aeruginosa* fall apart when handled and have a putrid odor.

A bacterial smear can easily be prepared after removing a wing from the thorax and dipping the wing base in a drop of water on a microscope slide. The smear can then be stained by the Gram method and examined microscopically. *Pseudomonas aeruginosa* rods measure 0.5 to 0.8 by 1.5 to 3.0 μm; are Gram-negative; and occur singly, in pairs, or in short chains. In culture, this organism is characterized by the excretion of diffusible yellow-green pigments that fluoresce in ultraviolet light (wave length below 260 nm).

Septicemia disease can also be diagnosed by reproducing the disease symptoms in healthy caged bees. This is accomplished by preparing a water extract (macerate the equivalent of one suspect bee per ml of water) and inoculating healthy bees through the thorax or dipping them in the water extract. Bees with septicemia die within 24 hours and exhibit the typical odor and the "break apart" symptom after approximately 48 hours.

Spiroplasmosis

Bee spiroplasma is found in the hemolymph of infected adult honey bees and in nectar. *Spiroplasma* sp. is a helical, motile, cell wall-free prokaryote measuring 0.7 to 1.2 μm in diameter; the length increases with age, ranging from 2 to >10 μm (Clark, 1977a, 1978a). It is likely that most colonies have bees infected with spiroplasmosis, however, there are no gross symptoms and the pathogenicity of this disease has not been demonstrated.

FUNGUS DISEASES

Chalkbrood Disease

Chalkbrood disease (CB) is caused by the fungus, *Ascosphaera apis.* It develops a characteristic spore cyst when opposite thallic strains (+ and -) fuse. Spore cysts measure 47 to 140 μm in diameter. Spore balls enclosed within the cyst range from 9 to 19 μm in diameter, while individual spores are 3.0 to 4.0 μm by 1.4 to 2.0 μm.

The fungus infects larva three to four days after egg hatch and is more commonly found in worker and drone larvae than in queen larvae. The infected larva is soon overgrown with a white cotton-like mycelium which fills the entire brood cell. Finally the white mass dries to form a hard, shrunken mass referred to as a mummy. Mummies of CB are usually white, thus the name chalkbrood. When the + and – mycelia are present in a diseased larva, spore cysts can form, and the resulting mummy appears mottled (black on white) or completely black.

Chalkbrood disease is easily identified by its gross symptoms—mummies in cells of the brood nest, hive entrance, and on the bottom board. Infected individuals can be easily removed from the brood cells by tapping the comb against a solid surface. This easy removal of larval remains also differentiates chalkbrood from other brood diseases.

Ascosphaera apis grows luxuriantly on potato dextrose agar fortified with 4 gm yeast extract per liter. Growth and sporulation also occur on malt agar, but less profusely and with no aerial hyphae; this facilitates subculturing and microscopic examination. Cultures exhibit a fruity odor similar to that of fermenting peaches. The optimum temperature for growth is 30°C. If only one strain (+ or -) is isolated, a fluffy cotton-like growth will eventually cover the plate. When both the + and - thalli are isolated, spore cysts form throughout the culture. The + and - thalli are morphologically identical. They can be distinguished by inoculating isolates on opposing sides of a plate. When opposite thalli grow together, a line of spore cysts forms at the juncture.

No chemotherapeutic agent is available for the control of chalkbrood. The most promising approach for its control is to maintain populous colonies and to use bee stocks that show evidence of resistance.

Stonebrood Disease

Stonebrood is usually caused by *Aspergillus flavus*, occasionally *Aspergillus fumigatus*, and sometimes other *Aspergillus* species. These fungi are common soil inhabitants that are also pathogenic to adult bees and other insects, as well as mammals and birds. The disease is difficult to identify in the early stages of infection. The fungus grows rapidly and forms a characteristic whitish-yellow collar-like ring near the head end of the infected larva. A wet mount prepared from the larva shows mycelium penetrating throughout the insect. After death, the affected larva becomes hardened and difficult to crush, hence the name stonebrood.

Eventually, the fungus erupts from the integument of the insect and forms a false skin (mummy). At this stage the larva may be covered with green powdery fungal spores. The spores of *A. flavus* have a yellow-green color, and *A. fumigatus* spores are grey-green. These spores can become so numerous that they fill the comb cells containing the affected larvae.

Stonebrood can usually be diagnosed from gross symptoms. However positive identification of the fungus requires its cultivation in the laboratory and subsequent examination of its conidial heads. *Aspergillus* spp. can be grown on potato dextrose or Sabouraud dextrose agars.

PROTOZOAN DISEASES

Nosema Disease

Nosema disease, caused by the microsporidian *Nosema apis* Zander, is by far the most widespread of the adult honey bee diseases (Nixon, 1982). In the early 1900's USDA authorities were fearful that the disease might be introduced to the North American continent (Morgenthaler, 1959), only to find that it was already present in 27 states. An interesting historical review, and the results of early studies conducted on the North American continent are presented by White (1919).

The severity of nosema disease has been reported to vary from colony to colony (Doull and Cellier, 1961), apiary to apiary (Doull and Cellier, 1961: Furgala and Hyser, 1969; Foote, 1971), and one geographic area to another (Mussen *et al.*, 1975). Studies in Minnesota, however, suggest that within that state, spring levels of infection in untreated wintered apiaries are relatively high (Furgala and Hyser, 1969) and do not vary significantly from one year to another (Furgala *et al.*, 1973).

Symptoms—The symptoms of nosema disease are often confused with other troublesome conditions that afflict adult honey bees (bee paralysis, starvation, pesticide poisoning, dysentery). The symptoms, when observed,

FIGURE 8. The effect of nosema on the midgut of the bee. *(Photo courtesy of Canada Department of Agriculture)*

include: disjointed wings, distended abdomens and the absence of the stinging reflex (Fantham and Porter, 1912). Honey bees crawling about in the hive and in the grass in front of the hive (unable to fly) have been found to be infected by nosema in some cases, afflicted by other disorders in other cases (Hertig, 1923).

A honey bee ventriculus is normal when straw-brown and the individual circular constrictions are clearly seen. Nosema can definitely be suspected when the ventriculus is white in color, soft in consistency, and swollen, obscuring the constrictions (White, 1919) (See Fig. 8). However, honey bees with this particular symptom are heavily infected and beyond help.

Incidence—The incidence of nosema disease generally varies during the year. Although the highest levels of infection are normally observed in the spring (Doull and Cellier, 1961; Girardeau, 1961; Harder and Kundert, 1951; Furgala and Boch, 1970), a smaller but detectable peak occurs in the fall (Doull and Cellier, 1961; Langridge, 1961; Furgala and Boch, 1970). It is the fall peak that plays a significant role in fueling the spring epizootics among wintered colonies.

Life Cycle—The vegetative stage of *Nosema apis* is not infective (Bailey, 1955b; Goetze *et al.*, 1959). Spores must be swallowed by a honey bee for infection to be initiated. It has been shown that relatively few spores are required to infect workers or drones (Bailey, 1955b, 1972a, 1972b; Kellner and Jacobs, 1978). Spores germinate quickly after entering the ventriculus, and the epithelial cells of the ventriculus are infected when the vegetative stage is introduced by way of the polar filament (Bailey, 1955b; Kramer, 1960; Kudo, 1920). Once inside a cell, the vegetative stage increases in size and multiplies, effecting an apparent concurrent reduction of RNA synthesis in the host cell (Hartwig and Przelecka, 1971; Liu, 1973). Liu (1984) found that although cell membranes remained intact, large vacuoles, aggregated ribosomes, and extensive lysis of the cytoplasm characterized infected cells. In 6 -10 days the host epithelial cell becomes filled with new spores (Burnside and Revell, 1948; White, 1919). Results suggesting that spores formed in a shorter time period (Goetze *et al.*, 1959) were probably based on the presence of a natural infection prior to artificial inoculation.

The honey bee does not secrete digestive enzymes directly into the ventriculus. Epithelial cells are shed into the ventriculus, burst, and release their contents, among which are the digestive enzymes. Infected cells are shed similarly, but release infective spores when they burst. Steche and Held (1981) found that spores free in the lumen migrated to other epithelial cells. If infection of other epithelial cells is not blocked, the digestive function of the epithelium is repressed in a short period (14 - 21 days). Any action that could preclude autoinfection could effect a cure when nosema levels are extremely low. Spores that are voided by infected honey bees remain viable for long periods. They resist refrigeration (Revell, 1960), freezing (Moffett and Wilson, 1971; Cantwell, *et al.*, 1971) lyophilization (Bailey, 1972b), and exposure to microwaves (Cantwell *et al.*, 1971).

Effect on Adult Honey Bees—Nosema infection affects individual honey bee workers in many ways. The life span of infected honey bees is reduced (Furgala and Boch, 1970), particularly under the stress of rearing brood (Poteikina, 1960). Kang *et al.*, (1976) found that the life span of infected workers was less than half (22% - 44%) that of healthy controls. The ability of infected nurse bees to feed brood is greatly reduced (Hassanein, 1951, 1952b). It was long suspected (Hassanein, 1952b) and later confirmed that infection causes ultrastructural changes in the hypopharyngeal glands (Wang and Moeller, 1971). The hypopharyngeal glands of infected nurse bees become atrophied (Wang, 1969) and the behavior of infected bees changes due to rapid physiological aging (Wang and Moeller, 1970). It has been suggested that under unfavorable weather conditions, notably those that cause confinement, active brood rearing may lead to an increase in the incidence and levels of infection in nurse bees (Dreher, 1956; Steche, 1960).

Queens are susceptible to infection, but considerable variation exists in the ability of a queen to continue laying eggs after becoming infected. Some queens are superseded with low levels of infection, others continue to function for longer periods with high levels of infection (Bogdan *et al.*, 1959; Farrar, 1947; Furgala, 1962a).

When queens found dead or being superseded in package colonies were examined, many were infected with nosema (Furgala, 1962b; Doull and Eckert, 1962). Jay and Dixon (1982), in a 5-year survey, detected *Nosema* in 26 - 53% of packages, 0.5 - 18% of queens, and 31 - 47% of the queen attendants examined. Experiments have demonstrated that nosema is transmitted by workers to queens in mating nuclei (Shimanuki *et al.*, 1973), in mailing cages (Lehnert *et al.*, 1973), and in queen banks (Foote, 1971). The cause(s) of queen loss is often difficult to determine. To recover and examine each queen being superseded would require daily colony manipulation (Furgala, 1962b).

Effect on Honey Bee Colonies—Nosema disease affects honey bee colonies in many ways. A) Late winter and early spring dwindling of adult populations are often caused by nosema. B) Loss of queens or queen supersedures are associated with nosema. C) Nosema infection peaks in the spring because infected bees, unable to defecate outside the hive during late fall and early winter, contaminate combs and frames by voiding fecal matter in the hive (Bailey, 1953). D) Wintering bees cleaning and polishing cells to expand the active brood area during late winter and early spring become infected as they pick up spore-laden fecal matter. E) The level of infection and percentage of infected bees increase (Harder and Kundert, 1951) since the adult population is generally confined, particularly in the northern regions. F) Honey bee death rate exceeds honey bee birth rate.

It should be stressed that in contaminated colonies, the percentage of infected hive bees increases whenever brood rearing is reduced (Moeller, 1972; Harder and Kundert, 1951), or hive bees are prevented from flying (Moeller, 1972). This condition is often not recognized unless honey bee samples are examined by microscope. All of these nosema-induced problems lead to reduced nectar collection (Hammer and Karmo, 1947) and depressed honey yields (Kauffeld *et al.*, 1972; Harder and Kundert, 1951; Fritzsch, 1970; Woyke, 1984).

Diagnosis—The behavior and physical appearance of infected honey bees do not change until the bees are close to death. Microscopic examination of suspect bees or their fecal material is the only method that offers positive diagnosis of nosema regardless of the level of infection. *Nosema apis* spores are large oval bodies, 4 to 6 μm in length by 2 to 4 μm in width (Fig. 9), and develop exclusively within the epithelial cells of the ventriculus of the adult honey bee. Samples of bees to be examined can be dried or preserved in alcohol. For quick routine examinations, the abdomens from 10 or more bees are removed, placed in a dish with 1.0 ml water per bee abdomen, and ground with a pestle or the rounded end of a clean test tube. A microscope slide (wet mount) is prepared from the resulting suspension and examined under the

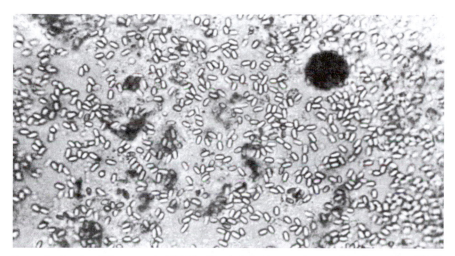

FIGURE 9. Spores of *Nosema apis* (750 X). *(Reproduced by permission of the Controller of H. M. Stationary Office)*

high dry objective (400×) of a compound microscope. Alternatively, individual bees can be examined to obtain an approximate percentage of infected bees in a colony. Also, a quantitative measure of levels of *Nosema* infection can be determined using a hemocytometer as described by Cantwell (1970).

Nosema can also be detected without sacrificing workers or queens by examining their fecal material. The collection of honey bee feces can be facilitated by placing glass plates near a hive entrance. Any fecal material that falls on the plates can be easily scraped off, mixed with water, and the resulting suspension examined with a high dry objective (400 ×) (Wilson and Ellis, 1966). Suspect queens can be held in small petri dishes or in glass tubes and allowed to walk freely. They usually defecate within one hour. Queen feces appear as drops of clear, colorless liquid which are then transferred to a microscope slide with a pipet or capillary tube. Queen fecal material is also examined using a high dry objective (L'Arrivèe and Hrytsak, 1964).

Prevention and Treatment

Nosema is frequently overlooked by the hobbyist, the sideliner, and by some commercial honey producers. Since there are no overt symptoms, such as perforated cappings, rubbery larvae, and distinctive odors, beekeepers must be alert to less obvious symptoms or they may never know that the unthrifty condition of their colony or colonies is due to nosema. Beekeepers need to know the normal course of nosema infection in a colony. Treatment to prevent and control nosema must be properly applied and timed to avoid winter kill, dwindling, queen supersedure, reduced honey yields and contamination of the honey crop with treatment chemicals. All too frequently, beekeepers do not diagnose the cause of an abnormal condition until serious damage occurs and treatment is no longer cost effective.

Some European workers, as mentioned earlier, have suggested that factors which improve brood rearing (colony manipulation and spring feeding) also stimulate the growth of the nosema organism (Goetze *et al.*, 1959) and that losses due to nosema are increased by active brood rearing (Poteikina, 1960; Steche, 1960). However, no one seems to have suggested ways to avoid these stresses if, indeed, they can be avoided. Colony management at regular intervals is essential if diseases are to be detected and held in check. In the northern temperate regions vigorous brood rearing is a must if beekeepers are to manage populous, productive colonies.

Practices Used to Lessen the Effects of Nosema—Beekeepers can combat the ravages of nosema disease with: **Management practices, fumigation of equipment, thermal sterilization of equipment, and chemotherapy.**

Management—(1) Provide each colony in the fall with a good prolific queen and a large population of young bees. (2) Select good apiary sites, with good air drainage, accessible at all times, with protection from prevailing winds, but sufficiently exposed to the sunny periods of winter which may (Langridge, 1961), or may not (Jaycox, 1960) affect wintering. (3) Keep each colony adequately supplied with carbohydrates (honey and bee manipulated syrup), proteins, lipids, minerals, and vitamins, (pollen, pollen substitute or supplement).

Fumigation—Two methods of fumigating beekeeping equipment have been reported in the literature: (1) Fumigation with acetic acid is effective (Bailey, 1957), especially when the bees are transferred as early as possible in the season from contaminated equipment to fumigated equipment [method: a pad of absorbent material soaked with 1/4 pint acetic acid (80%) is placed on the frame top bars in each hive body; hive bodies are stacked, sealed, and undisturbed for one week.] After fumigation hive bodies should be aired for at least two days, preferably one week before use. (2) The use of ethylene oxide (ETO) as a fumigant is a more recent approach to decontaminate equipment (Michael, 1964; Grobov *et al.*, 1970). It has been demonstrated that 100 mg ETO/liter for 24 hours at 100°F (37.8°C) kills spores on combs (Lehnert and Shimanuki, 1973a). Fumigation with ethylene oxide has certain inherent hazards. Beekeepers should contact state or federal agencies before fumigating with ETO.

Thermal Sterilization—Elevated temperatures can be used to decontaminate hive equipment. A temperature of 120°F (49°C) for 24 hours destroys or renders spores nonviable (Cantwell and Shimanuki, 1970). Certain precautions must be taken, the most important being that the combs contain little or no honey or pollen. The temperature must not exceed 120°F (49°C).

It is important to remember that decontamination of equipment by fumigation or thermal methods will be negated if beekeepers hive infected package bees or transfer infected bees on combs to treated equipment. **(decontaminated equipment + infected bees = contaminated equipment + infected bees.)**

Chemotherapy—Of the numerous compounds that have been tested against *Nosema apis,* only one, fumagillin (Fumidil-B®) has proven to be an effective and reliable agent in the prevention and control of nosema disease. Katznelson and Jamieson (1952a, 1952b) first demonstrated the efficacy of fumagillin in suppressing nosema infection in artificially inoculated honey bees. Since that discovery, many studies have been conducted on the chemotherapeutic value of fumagillin (Furgala and Boch, 1970; Fritzsch, 1970; Kulikov, 1961; Hejtmanek, 1970; Moeller, 1967; Szabo and Heikel, 1987b; Wyborn and McCutcheon, 1987). Fumagillin has been found to be effective in suppressing nosema disease in both package colonies and wintered colonies in the northern honey producing areas (Furgala and Boch, 1970; Moeller, 1967; Moffett *et al.,* 1969; Gochnauer and Furgala, 1969). It has also been determined that fumagillin is effective in preventing the transmission of infection to queens in mating nuclei (Shimanuki *et al.,* 1973), colonies (Kauffeld, 1973; Lehnert and Shimanuki, 1973b) and queen cages (Lehnert *et al.,* 1973) in the southern queen and package producing region of the United States.

Fumagillin fed to wintering colonies in the autumn, at a rate of 100 mg fumagillin activity/gallon of 2:1 sugar syrup, with a minimum dose of two gallons per colony, markedly suppresses nosema infection the following spring. Half the recommended dose in half as much syrup is much less effective (Furgala and Gochnauer, 1969; Gochnauer and Furgala, 1969; Girardeau, 1972).

Fumagillin fed to package colonies (when hiving) at a rate of 100 mg activity/gallon in 1:1 sugar syrup, with a minimum dose of one gallon per package colony can repress natural levels of nosema. Two gallons of medicated syrup are necessary when natural levels of nosema in packages are high or when bees are confined for long periods after installation (Furgala and Gochnauer, 1969). Woyke (1984) reported that colonies treated with fumagillin produced significantly more honey.

It is very important that the volume of medicated syrup contains the effective dose of fumagillin when fed for prevention and repression of nosema disease. A study in Georgia has demonstrated that reducing the recommended dose of fumagillin results in inconsistent and ineffective repression of disease (Lehnert and Shimanuki, 1973b).

Wyborn and McCutcheon (1987) and Szabo and Heikel (1987b) recently conducted studies that demonstrated nosema levels in the spring were significantly reduced when a spring treatment with dry fumagillin followed a fall syrup treatment of wintering colonies. These results should be viewed with caution. Reducing spring spore loads of wintering colonies "from 15 million to 5 million spores per bee" with spring dry treatment is mathematically significant, but hardly pathologically significant. It normally requires two fall preventive or control treatments to reduce to insignificant levels nosema spore loads that contaminate wintering bees and hive equipment. It is also important to note that beekeepers must avoid diluting fumagillin treatment dosage by following medicated treatment syrup with nonmedicated syrup. Fumagillin treated syrup diluted with sugar syrup significantly reduces the chemotherapeutic activity of the chemical. Many beekeepers face this problem attempting to provide adequate food stores for wintering. Failure to reduce late winter and spring nosema spore counts with a fall treatment indicates heavy spore loads and need for consecutive fall treatments (with no dilution).

Although dusting fumagillin has been reported (Langridge, 1961), most studies have indicated that fumagillin can be properly distributed throughout the colony only in a syrup. Extender patties are not as reliable in the south as a syrup treatment (Williams, 1973) and dusting and candy patties do not have any appreciable residual effect in the northern regions (Furgala and Gochnauer, 1969).

Amoeba Disease

This disease is caused by a protozoan, *Malpighamoeba mellificae*. The presence of *M. mellificae* has been reported in Europe (Maassen, 1916; Morgenthaler, 1920; Prell, 1926), the Soviet Union (Poltev, 1953), North America (Bulger, 1928), South America (Stejskal, 1958), and New Zealand (Palmer-Jones, 1949).

Adult worker honey bees become infected by ingesting cysts (Bailey, 1963b). The cysts germinate and the amoebae migrate to the malpighian tubules which are in the bee serving to remove waste products from the blood. They feed by pseudopodia between or within the epithelial cells. These extracellular parasites divide by binary fission (Steinhaus, 1963). It takes 2 - 4 weeks for new cysts to form (Fyg, 1932; Jordan, 1936; Hassanein, 1952a). Cysts pass into the intestine, accumulate in the rectum, or are voided. Bailey (1955a) observed amoeba cysts in fecal material that he scraped from hive combs.

Since this protozoan is found in the Malpighian tubules of adult bees, diagnosis can be made only by the removal and microscopic examination of the tubules for the presence of amoeba cysts. The cysts measure 5 to 8 μm in diameter. The Malpighian tubules are long, thread-like projections originating at the junction of the midgut and the hindgut. Infected malpighian tubules are usually swollen and have a shiny appearance. The tubules can be teased away from the digestive tract with a pair of fine tweezers and then placed in a drop of water on a microscope slide. A cover glass is positioned over the tubules while applying uniform pressure to obtain a flat surface for microscopic examination. *Malpighamoeba mellificae* can be discerned using a high dry objective and then changing to the oil immersion objective for more detail.

Many researchers have found that *Malpighamoeba* infections peak in April or May, as do *Nosema* infections (Bondarenko, 1966). This is probably because the cysts deposited on the comb during late fall and winter are ingested in late winter and spring. The amount of inoculum can be heavy in some colonies. Hassanein (1952a) and Giordani (1959) reported that 70-100% of the adult worker honey bees could be infected. Although Schiller (1937) and Johnsen (1951) claimed that honey bee colonies died from *M. mellificae* infection, Prell (1926), Bailey (1968b) contend that the amoeba rarely kills a colony, but often causes a condition they call "spring dwindling" or "disappearing" disease. Further, *M. mellificae* infections are often mixed, especially with *N. apis.* These infections should not be ignored, especially in Europe and South America where honey bee queens become infected.

Prevention and control of amoeba disease is not practiced in North America. In Europe, regular decontamination of equipment with acetic acid is often part of a management system. No chemical has been found that will control amoebae in bees.

Gregarines and Flagellates

Gregarines are the largest protozoans known to be associated with honey bees. They live in the midgut (ventriculus) of adult honey bees attached to the epithelium. There are four gregarines found in honey bees: *Monoica apis, Apigregarina stammeri, Acuta rousseaui,* and *Leidyana apis.* The immature stages (cephalonts) average about 16 by 44 μm. Cephalonts are oval and consist of two distinct body segments; the posterior segment is larger. The mature stages (sporonts) average about 35 by 85μm, with a reduced anterior segment (Fig. 10). There is little indication showing gregarines to be detrimental.

Flagellates (*Crithidia* spp.) have been found either free in the lumen or attached to the epithelium of the hindgut and rectum of adult honey bees (Fyg, 1954). They vary in size (5 to 30 μm), and some appear as pear-like bodies with flagella while others are long thread-like forms or are round without flagella (Lotmar, 1946). There is no evidence that flagellates are harmful to honey bees.

Organisms such as *Spiroplasma* sp., gregarines, and flagellates are not of economic significance. However, when a disease such as nosema is present,

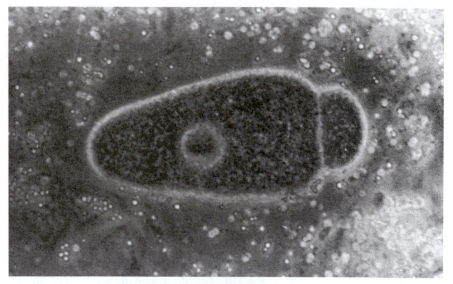

FigURE 10. Gregarine *(Photo by T. B. Clark)*

the added stress of the so-called lesser known diseases may make the problem more serious.

VIRUS DISEASES

Bees are attacked by a number of viruses. Although some insect viruses produce granules or crystals in the host that are visible under ordinary microscopes, most of the viruses attacking bees do not do so. Thus, diagnosis has in the past been restricted to the symptoms of infection.

Sacbrood Disease

The symptoms of sacbrood disease were first described by White (1913, 1917). There are a number of characteristic signs of this disease: (1) Partially uncapped cells scattered among the capped brood or capped cells that remain after surrounding brood has emerged. (2) Partially uncapped or uncapped cells containing individuals with characteristic, darkened heads (Fig. 11, 12). (3) Unlike AFB, suspect individuals are easily removed from the cells intact, but appear sac-like because of an accumulation of what appears to be ecdysial fluid between the propupal and pupal skins. (4) Diseased prepupae fail to pupate and darken prematurely from pearly white to pale yellow, to light brown and finally, dark brown. (5) The dark brown individuals dry, and become wrinkled, brittle scales that are easily removed from the cells.

Beekeepers rarely consider sacbrood a serious threat since the signs are generally observed in a few colonies and only during late winter, spring and early summer. It has been shown in Britain, however, that most of the samples diagnosed as "addled brood" were, in fact, infected with sacbrood (Bailey, 1971b). It has also been shown in surveys that 80% of diseased larvae with no visible pathogens had sacbrood (Bailey, 1971b).

In nature, adult bees normally detect and remove infected larvae very quickly (Bailey *et al.,* 1964). Therefore, when symptoms are observed by the beekeeper, the disease may be too severe for the adult population to cope with the condition. Severity of infection in developing package colonies varies from a few cells per frame to 90% of the brood (Mussen and Furgala, unpublished). Further, the number of colonies infected per apiary varies from 0 to 100%. In view of these facts, it is possible that the disease is more prevalent than is generally believed.

Although White (1917) first suggested that sacbrood was caused by a virus, and others (Steinhaus, 1951; Brcak *et al.,* 1963) later observed spherical virus-like particles in water extracts of infected larvae, Bailey *et al.* (1964) first associated an isometric particle, 28 - 30 nm in diameter, with infectivity, and designated the virus as sacbrood virus (SBV).

Some aspects of the epizootiology of sacbrood disease have been resolved particularly by Bailey and his colleagues in Britain. However, certain features concerning the spread and perennial nature of sacbrood disease are yet to be fully explained. Sacbrood disease has been transmitted experimentally in colonies to larvae as old as four days by inoculating the food in the larval cell (Hitchcock, 1966). The disease develops quickly in young larvae infected in the laboratory. Larvae 12 - 36 hours old that are inoculated with SBV by mouth show symptoms in 48 hours and die shortly thereafter (Lee and Furgala, 1967a). Electron microscopy shows SBV to be extremely abundant (Fig. 13) in the cytoplasm of fat, muscle and tracheal end-cells of larvae with symptoms of sacbrood, and in larvae that had been inoculated but appeared healthy (Lee and Furgala 1967a).

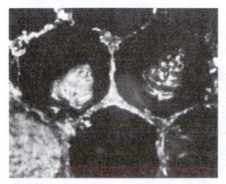 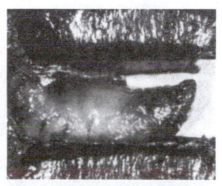

FIGURE 11. Sacbrood—prepupae, front view, cappings removed. *(Photo courtesy of M. V. Smith)*   FIGURE 12. Sacbrood—side view of prepupa stage. *(Photo courtesy of M. V. Smith)*

White (1917) demonstrated that one diseased larva could infect 3000 other larvae. Recent estimates suggest that one larva killed by SBV contains sufficient virus to be lethal to over a million larvae (Bailey, 1969, 1981). Dried smears of larvae that have just succumbed to SBV are highly infective. Why then does the disease disappear each summer and not spread when heavily infected combs (54% infection) are inserted into healthy colonies (Hitchcock 1966)? The explanation, in part, may come from the fact that infectivity of SBV decreases very rapidly (White, 1917; Hitchcock, 1966; Bailey, 1967b). Although dried, blackened larvae still give strong precipitin lines in gel-diffusion tests (Bailey, 1967b), they lose their infectivity after 3 weeks at 64.4°F (18°C). Dried smears of larvae freshly killed by SBV, although still infective after 3 weeks at 18°C, lose their infectivity within 10 months (Bailey, 1965) stored at 64.4°F (18°C). Semi-purified virus stored in royal jelly at 41°F (5°C) remains infective for at least 3 weeks (Mussen and Furgala, unpublished data).

How then does the disease spread in the colony? How does it persist from year to year? Healthy adult bees injected with purified preparations of SBV were shown to contain crystalline arrays of virus-like particles in the cytoplasm of fat body cells five days after injection (Lee and Furgala, 1967b).

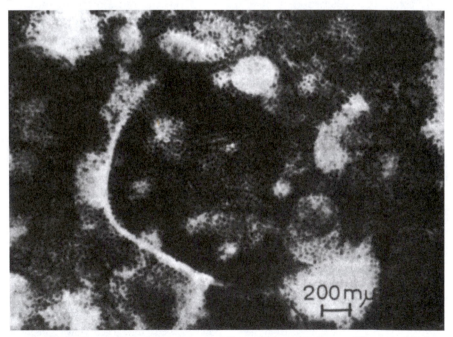

FIGURE 13. SBV particles in infected tissue of diseased honey bee larvae. *(Courtesy, Virology 1965. 25:390, Academic Press, Inc.)*

Curiously, the injected bees appeared normal 12 days after inoculation, suggesting that the results could be interpreted as the replication of SBV in adult bees (Lee and Furgala, 1967b). Bailey (1967a), on the other hand, reported that he was unable to show that SBV multiplied in adult bees and could not infect larvae with extracts of adults injected with SBV. He presented evidence that SBV injections activated a different virus that multiplies in adults and has similar physical characteristics. Further work by Bailey and his colleagues later showed that SBV can multiply (asymptomatically) in adult bees. Multiplication can occur with (Bailey, 1968a) or without (Bailey and Fernando, 1972) an antiserum to protect the bees from lethal doses of SBV.

Electron microscopic observations, serological, and infectivity tests suggest that sacbrood virus can accumulate in the heads of infected adults, with much of the virus accumulating in their hypopharyngeal glands (Bailey, 1969, 1968a). Virus particles are transmitted when adults feed larvae. In adult drones, however, sacbrood virus has been reported to multiply in the brain (Bailey, 1970).

It is apparent from the above discussion that adult honey bees do serve as a reservoir of SBV. Although the virus appears to accumulate in the hypopharyngeal glands, there is no direct evidence that workers transmit the virus to larvae since small colonies composed entirely of young worker bees that were injected with SBV reared only healthy larvae (Bailey, 1970).

Studies on sacbrood disease are needed to determine how the virus is transmitted to larvae in nature, how the virus persists from year to year, and why severe outbreaks occur only during the build-up season.

Bee Paralysis Diseases

Huber (1814) described in his notes of 1809 "History of Some Black Bees" the very symptoms many of the early reference works (Root and Root 1913; Miller 1931) refer to as adult bee paralysis. Adult bees are hairless, appear shiny and greasy, and are afflicted with a trembling motion. Wings are often disjointed, abdomens distended and there is a general inability to fly.

Burnside (1933, 1945) first demonstrated that filtrates from sick "paralyzed" bees, when sprayed on healthy bees, caused symptoms and death. In a significant development Bailey *et al.* (1963) not only repeated Burnside's experiments, but also isolated two bee viruses which caused paralysis and death to adult honey bees.

Chronic Bee Paralysis—This virus (CBPV) was isolated from diseased bees taken from colonies naturally affected by bee paralysis (Bailey *et al.*, 1963). The virus particles were anisometric and varied in size (Bailey *et al.*, 1968). CBPV has been isolated from paralyzed bees in Britain, Austrailia, North America, and Europe (Bailey, 1967b). Similar particles have been reported in the USSR (Alekseenko and Kolomiets, 1967) and elsewhere

(Bailey, 1963a, 1963b). Seventy percent of the samples of "crawling moribund bees," or live bees from colonies producing many crawlers, have been reported as suffering from paralysis (Bailey, 1967b).

Much CBPV has been observed in the liquid contents of paralytic field bees bloated with honey suggesting that the virus has been secreted by the salivary and possibly the hypopharyngeal glands. Electron microscope studies have shown that CBPV multiplies in nerve tissue (Lee and Furgala, 1965). Infectivity and serological tests indicate that CBPV accumulates in the head, with extracts of brains from paralytic bees proving to be very infective (Bailey and Milne, 1969).

While symptomatic bees often suffer from CBPV, the virus commonly occurs in seemingly healthy bees (Bailey *et al.,* 1963; Bailey, 1963b). Also, CBPV has been recovered from pollen pellets carried by apparently healthy bees.

How paralysis spreads from bee to bee and why only isolated colonies in an apiary are decimated by the disease are unknown. The fact that the virus does multiply in pupae suggests that hereditable factors influencing susceptibility are operative. Attempts to document this aspect were successful when Rinderer *et al.* (1975), and Kulincevic and Rothenbuhler (1975) demonstrated that they could select strains that were more susceptible to a "hairless black syndrome" that proved to be CBPV (Rinderer and Green, 1976).

Acute Bee Paralysis—This virus (ABPV) was isolated by Bailey *et al.* (1963) from paralyzed bees. They referred to it as Acute Bee Paralysis Virus because adult bees infected with this virus showed symptoms and died very quickly at 86°F (30°C). Acute Bee Paralysis, according to Bailey, is a "laboratory phenomenon." Although the virus can be detected in apparently healthy bees from many parts of the world (Bailey, 1963b), it does not normally cause overt signs of paralysis and occurs in the same numbers in apparently healthy bees as in paralyzed individuals. It has been suggested that ABPV does not normally cause overt symptoms because bees are unable to acquire a lethal dose in nature (Bailey and Gibbs, 1964). Adult bees injected with ABPV become moribund (Furgala and Lee, 1966) or developed symptoms of paralysis (Bailey *et al.,* 1963) in about 4 days and died within another day or two.

The ABPV particle is isometric, about 28-30nm in diameter and resembles the SBV particle. Although SBV and ABPV share similar physical properties, serological and cross infection studies suggest that they are distinct. Unlike SBV, the ABPV particles remain infective for long periods when preserved at -4°F (-20°C).

When ABPV inoculum is injected into apparently healthy bees, isometric particles resembling the virus are observed in the fat body but not in pieces of mid-intestine, thoracic muscle, brain, or thoracic and abdominal ganglia (Lee and Furgala, 1967b). Electron microscope and infectivity studies by Bailey and Milne (1969) indicate that ABPV accumulates in the hypopharyngeal glands. Fernando (1972) reported that ABPV does not affect hypopharyngeal glands. The particles have also been detected in freshly collected pollen suggesting that bees may secrete ABPV from their glands (Bailey, 1971a).

Filamentous Virus Disease

This disease, which was thought previously to be of rickettsial origin, was found to be caused by a folded nucleocapsid virus within a viral envelope and measure 0.40 by 0.10 μm (Clark 1978b). It can be diagnosed by examining the hemolymph of infected bees using phase-contrast microscopy. The hemolymph of honey bees infected with this virus is milky white and contains many spherical to rod-shaped viral particles of a size close to the limit of resolution for light microscopy.

Other Virus Diseases

Clark (1977b, 1978b) and Bailey and coworkers (see below) isolated and described a series of viruses from honey bees. Some, (black queen virus, and bee virus Y) are found only when *Nosema* is present. According to Bailey (1981) they multiply only in *Nosema*-infected adults. Another virus (bee virus X) is associated with amoeba-infected bees. The pathological importance of Cloudy Wing Particle (Bailey *et al.,* 1980), Slow Paralysis Virus (Bailey, 1976), Kashmir Bee Virus (Bailey and Woods, 1977), Arkansas Bee Virus (Bailey and Woods, 1974), and Egypt Bee Virus (Bailey *et al.,* 1976) is unknown. See Bailey (1981) for more detail regarding these "rare" honey bee viruses.

NONINFECTIOUS DISEASES

Noninfectious disorders can be the result of neglect, lethal genes, pollen or nectar from poisonous plants, toxic chemicals (pesticides), or other causes. Normally bees feed and protect the brood; but if there is a sudden shortage of adult bees, the larvae and pupae suffer and may die of chilling, overheating, or starvation. Most often dead or discolored pupae result from a noninfectious condition. For a good review of noninfectious disorders of honey bees, see Tucker (1978).

Chilled Brood—Chilling usually occurs in early spring when brood nests expand rapidly and there is a shortage of adult bees to cover all the brood. Consequently, chilled brood is found most often on the fringes of the brood area, with healthy brood at the center. However, chilling also can happen during cold weather following any sudden reduction of the worker bee population. Chilled larvae and pupae are often yellowish, tinged with black on segmental margins. They also may be brownish or black, crumbly, pasty, or watery. In extreme cases, brood cells are punctured and uncapped, and pupae are decapitated by the adult bees. It is well to remember that decapitation also could result from the feeding of lesser wax moth larvae.

Overheated Brood—The overheating of brood develops when there is a sudden loss of worker bees available to cool the colony during hot weather. Cappings of brood cells can appear melted, darkened, sunken, and punctured. Larvae dead from overheating become brownish or black and are watery; pupae have a black greasy appearance; and newly emerged adult bees may be wingless.

Starved Brood—Normally when there is shortage of food in a colony, larvae are removed or consumed by the adult bees. However, when there is a sudden shortage of adult bees available to feed the larvae, the larvae starve. Affected larvae are not restricted to the periphery of brood combs. The most striking feature of this syndrome is the appearance of larvae seemingly crawling out of the brood cells to search for food. Starved brood is almost always restricted to the larval stage. However, emerging bees may starve if they were stressed as pupae by chilling or overheating and if there are too few nurse bees to feed them soon after they have chewed through their cappings. These bees usually die with only their heads out of the cells and with their tongues extended.

Overheated Bees—Overheating in worker bees can occur when bees are confined in their hives during hot weather without proper ventilation or access to water. Bees dying from overheating crawl about rapidly while fanning their wings. They are often wet, and their wings appear hazy. In some cases, there may be an abnormally large accumulation of dead bees at the hive entrance.

Lethal Genes—Bees also can die from genetic defects during all stages of development and usually do not exhibit symptoms of known diseases. However, drone brood from laying workers and drone-laying queens often die exhibiting symptoms that resemble EFB, but no known pathogenic agents are present; genetic lethality is the suspected cause of this condition.

PARASITIC BEE MITES

Honey Bee Tracheal Mite—*Acarapis woodi*

The honey bee tracheal mite, *Acarapis woodi,* or acariosis as the disease is known in Europe, afflicts only adult honey bees. The parasite was first described in 1921 in bees in Great Britain (Rennie, 1921). This discovery and concern over the potential impact that this mite would have on beekeeping in the United States led to the enactment of the Honeybee Act of 1922, which restricted the importation of honey bees from countries where this mite was known to exist.

There are three *Acarapis* species associated with adult honey bees: *A. woodi, A. externus,* and *A. dorsalis.* These mites are difficult to detect and differentiate due to their small size and similarity; therefore, they are frequently identified by location on the bee instead of morphological characterristics. However, only *A. woodi* can be positively diagnosed solely on habitat; the position of other species on the host is useful, but not infallible. *Acarapis woodi* lives exclusively in the prothoracic tracheae (Fig. 14); *A. externus,* being external, inhabits the membranous area between the posterior region of the head and thorax or the ventral neck region and the posterior tentorial pits; and *A. dorsalis* is usually found in the dorsal groove between the mesoscutum and mesocutellum and the wing bases. Complete descriptions and illustrations can be found in Delfinado-Baker and Baker (1982).

The *A. woodi* female is 143 - 174 μm in length and the male 125 - 136 μm. The body is oval, widest between the second and third pair of legs, and is whitish or pearly white with shining, smooth cuticle; a few long hairs are present on the body and legs. It has an elongate, beak-like gnathosoma with long, blade-like styles (mouthparts) for feeding.

When over 30 percent of the bees in a colony become parasitized by *A. woodi,* honey production may be reduced and the likelihood of winter survival decreases with a corresponding increase in infestation (Bailey, 1961; Furgala *et al.,* 1989). Individual bees are believed to die because of the disruption to respiration due to the mites clogging the tracheae, the damage caused by the

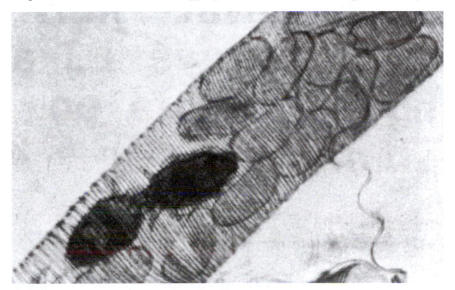

FIGURE 14. *Acarapis woodi* in tracheae. *(American Bee Journal, Vol. 122, No. 7, p. 505)*

mites to the integrity of the tracheae, microorganisms entering the hemolymph (blood) through the damaged tracheae, and from the loss of hemolymph.

The tracheal mite has now been reported on every continent except Australia. Initial detections of *A. woodi* were reported in Brazil in 1974 (Lehnert *et al.*, 1974), in Mexico in 1980 (Wilson and Nunamaker, 1982), and in Texas in 1984 (Riley, 1984). The mites are transmitted bee to bee within a colony by queens, drones and workers. In addition, the movement of package bees and queens, as well as established colonies, has resulted in the dissemination of this mite throughout much of the United States.

One of the first problems that became apparent when the tracheal mite was detected in the United States was the lack of agreement on their economic impact. The literature from Europe did not always agree and beekeepers, research scientists and regulatory officials had differing opinions on the interpretation of the data. However, it soon became evident that the mites were spreading faster than one would have ever predicted and the level of infestation within colonies was higher than expected. It is apparent that the tracheal mite found an extremely susceptible honey bee host in the United States.

There has been considerable speculation on the length of time *A. woodi* has been present in the United States. Bailey (1985) speculated that *A. woodi* may have been in America for as long as the honey bee and was not previously encountered because of limited mite surveys. However, based on the sudden economic impact of the tracheal mites and the periodic surveys that were conducted from 1959 - 1984, Shimanuki and Knox (1989) concluded that *A. woodi* was introduced into the United States about 1983. Apparently, the tracheal mite entered south Texas in honey bee swarms crossing the Rio Grande river from Mexico. The infestation in Florida probably originated with adult bees imported from outside the United States.

Detection Techniques
The population of *A. woodi* in a colony may vary seasonally. During the period of maximum bee population, the percentage of bees with mites is reduced. The likelihood of detecting trachael mites is highest in the fall and winter. No one symptom characterizes this disease; an affected bee could have disjointed wings and be unable to fly, or have a distended abdomen, or both. Absence of these symptoms does not necessarily imply freedom from mites. Positive diagnosis can only be made by microscopic examination of the tracheae; since only *A. woodi* is found in the bee tracheae, this is an important diagnostic feature.

In sampling for *A. woodi,* collect moribund bees that may be crawling near the hive entrance or bees at the entrance as they are leaving or returning to the hive. These bees should be placed in 70% ethyl or methyl alcohol as they are collected. Bees that have been dead for an indeterminate period are less than ideal for trachael mite diagnosis.

There are several techniques for dissecting and examining bees for *A. woodi*; each has its advantages and disadvantages. The classic method is to pin a bee on its back and remove its head and first pair of legs by pushing them off

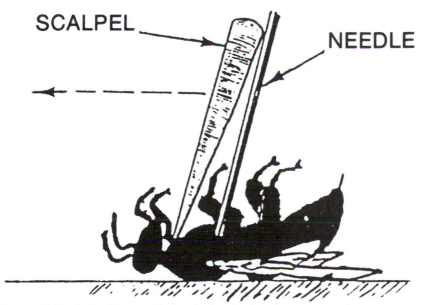

FIGURE 15. Tracheal mites—positioning the bee for dissection.

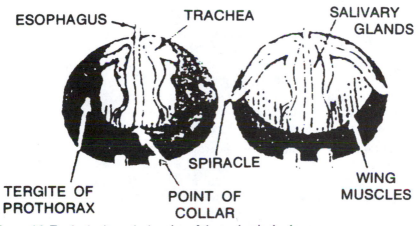

FIGURE 16. Tracheal mites—the location of the trachea in the thorax.

with a scalpel or razor blade in a downward and forward motion (Fig. 15). With the aid of a dissecting microscope, the first ring of the thorax ("collar") is removed with forceps. This exposes the tracheal trunks in the mesothorax (Fig. 16). Using a pair of fine forceps and a probe, the tracheae can be carefully removed, placed in a drop of 85% lactic acid or glycerol on a glass slide, and covered with a cover glass for examination with a compound microscope (40-100 ×).

The tracheae must be examined carefully for the presence of mites. The tracheae of severely infested bees may have brown blotches with crustlike lesions, or they may be black, and are obstructed by numerous mites in different stages of development. The trachea may not always be discolored when mites are present, and cloudy or discolored tracheae do not always contain mites. The tracheae of healthy bees appear cream or white.

A faster examination method is to grasp the bee between your thumb and forefinger and remove the head and first pair of legs as described earlier. Then with a scalpel, razor blade, or fine pair of scissors, a thin transverse section is cut from the anterior face of the thorax in such a way as to obtain a disc. The disc is then placed on a microscope slide in a few drops of 85% lactic acid. The lactic acid makes the material more transparent and helps to facilitate dissection. The tracheae are carefully removed and examined as previously described. This procedure is recommended for a quick examination of a few bees.

Large numbers of bees can be examined by making transverse section discs from the thoraces of 50 or more honey bees. These are placed in 5 percent potassium hydroxide (KOH), and incubated at 37°C for 24 hours. The KOH dissolves the muscle and fat tissue leaving the trachea exposed. The disc-trachea suspension can then be examined under a dissecting microscope for suspicious looking tracheae. These tracheae can be separated from the discs and examined more closely using a compound microscope.

A technique of staining *A. woodi* within the tracheae to make them easier to observe was developed by Peng and Nasr (1985). This method involves boiling transverse section discs in an 8% KOH solution to dissolve and clear the soft tissue, filtering and washing with tap water, staining in a modified methylene blue staining solution, washing with distilled water, and finally rinsing the discs with 70% ethanol. The discs are then examined for stained mites within the tracheae under a dissecting microscope.

Bees that have been dead for some time also can be examined for tracheal mites. Thoracic discs are placed on a slide in a few drops of 10% KOH and the slide is gently heated for one to two minutes. The discs are then covered with a cover glass, crushed lightly, and examined microscopically.

A method of differentiating live from dead mites is described by Eischen *et al.* (1986a). This is the method of choice for evaluating chemicals for the

Figure 17. Sketch of trachea containing mites.

control of tracheal mites. Live bees are anesthetized with carbon dioxide and the abdomens are removed with a scalpel to prevent being stung. The head and first pair of legs of each bee are removed by holding the bee on its back and gently pushing this section off as described earlier. While held in this position, each bee is placed under a dissecting microscope, and the first ring of the thorax is removed with fine forceps exposing the tracheal attachment to the thoracic wall. Tracheae which appear abnormal are removed with forceps and transferred to a glass slide containing a thin film of glycerol. The tracheae are then dissected using a pair of fine needle probes. Mites are considered dead if they do not move; also, dead mites often appear discolored and desiccated. Living mites have a translucent gray or pearl color and move within a few seconds after dissection.

Acarapis woodi also can be detected by using an enzyme-linked immunosorbent assay (ELISA) developed by Ragsdale and Furgala (1987) and further refined by Ragsdale and Kjer (1989).

Another method for detecting *A. woodi* involves homogenizing honey bee thoraces (Colin *et al.,* 1979). They described a technique in which thoraces are homogenized, strained, and centrifuged. The resulting sediment is examined under a microscope for mites. Camazine (1985) also described a method in which the thoraces are broken apart in a blender. The resulting mixture is poured into test tubes where most of the dense material falls to the bottom while the tracheae and air sacs float. This floating layer is examined under a compound microscope for tracheae that may contain mites.

Treatment

For many years chlorobenzilate (Folbex®)* was the material of choice to control acariosis. Since chlorobenzilate is no longer available, beekeepers in Mexico and the United States were forced to look at alternative materials. Eischen *et al.* (1987) evaluated six compounds for control of the tracheal mite, and found Amitraz and bromopropylate to be reasonably effective against adult mites, but not against immature mites.

Menthol is the only material that is currently approved by the Environmental Protection Agency (EPA) for the control of these mites in the United States. Menthol was first tested against *A. woodi* over 20 years ago in Italy (Giavarini and Giordani, 1966). They reported that menthol produced the most effective control of the 39 products tested. It was harmless to bees and did not contaminate the honey. In Texas, and Mexico, Wilson *et al.* (1988) also found the efficacy of menthol to be excellent, with little or no menthol residues detected in the honey from treated colonies.

The variations in the effect of *A. woodi* on individual colonies suggests that breeding bees that are tolerant or resistant to this mite would be feasible (Gary and Page, 1988). The development of such bee stocks that still have all the other desirable traits is the method of choice over the use of chemicals; however, this may take several years. Meanwhile, beekeepers can minimize the impact of tracheal mites by intensive management practices to maintain populous colonies and by using menthol.

Colonies can be treated with menthol when there is no heavy nectar flow and daytime temperatures are expected to reach at least 60° F. The best time being in the spring when the weather is warm, and in the late summer or fall of the year immediately after removing the surplus honey.

Directions for using Menthol: Fifty grams (1.8 ounce) of crystalline menthol should be enclosed in a 7″ x 7″ plastic bag or equally porous material and placed inside a colony for 20-25 days. Menthol placed on the top bars is the preferred method of treatment provided the daytime temperature does not exceed 80 degrees F. During hot weather, the menthol should be placed on the bottom board of the colony. There should be no honey supers on the hive during the treatment, and the menthol should be taken out of a colony at least one month before any anticipated flow. Before using menthol, read and follow the approved label carefully.

Varroa jacobsoni

The parasitic bee mite, *Varroa jacobsoni* Oudemans is one of the most serious pests of the honey bee, *Apis mellifera,* and its introduction into new

---

* The old Folbex (chlorobenzilate) should not be confused with Folbex VA (bromopropylate).

countries is causing much concern to beekeepers throughout the world. *Varroa* was first described on *Apis indica (=cerana)* from Java (Oudemans, 1904). The first reports of *V. jacobsoni* attacking *A. mellifera* (a new host) was in 1962 on a sample sent to the USDA in Beltsville from Hong Kong, and in

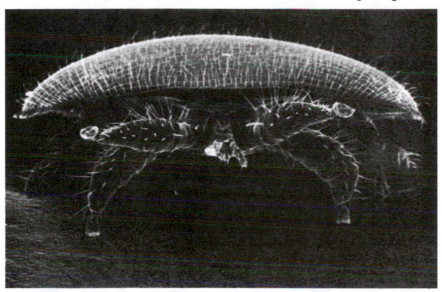

FIGURE 18. (top photo) Anterior (frontal) view of female *Varroa jacobsoni* which possesses a streamlined flattened body. *(Photo courtesy Dr. T.P. Liu, Bee Diseases Research Lab., Beaverlodge, Alberta).* (bottom photo) Immature honey bee with four *Varroa* attached. *(Photo courtesy Apiary Inspectors of America).*

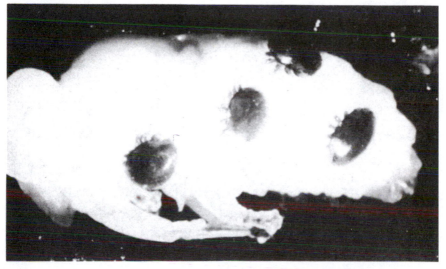

1963 in the Philippines (Delfinado, 1963). The mite has since become established on every continent except Australia and will continue to spread because of the commercial transport of bees and queens; the migratory activities of beekeepers; swarms that may fly long distances, or be carried by ships or aircraft; and drifting bees.

The adult female *Varroa* mite is oval and flat, about 1.1 mm long and 1.5 mm wide, pale to reddish-brown in color, and can be seen easily with the unaided eye (Fig. 18). Male mites are considerably smaller and are pale to lightly tanned (Delfinado-Baker, 1984). Adult bees serve as intermediate hosts when little or no brood is available and as a means of transport. The females attach to the adult bee between the abdominal segments or between body regions (head-thorax-abdomen), making them difficult to detect. These are also places from which they can easily feed on the bee's hemolymph. The adult bee suffers not only the loss of blood, but may be subjected to microbial invasion, leading to a reduced life expectancy (De Jong and De Jong, 1983).

The most severe parasitism occurs on the older larvae and pupae, drone brood being preferred to worker brood (Ritter and Ruttner, 1980). The degree

LIFE CYCLE of <u>Varroa jacobsoni</u>

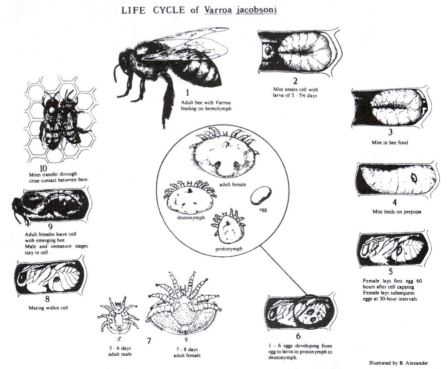

FIGURE 19. Life cycle of *Varroa jacobsoni. (American Bee Journal, Vol. 127, No. 11, p. 755)*

of damage depends on the number of mites parasitizing each bee larva. One or two mites will cause a decrease in vitality of the emerging bee. Higher numbers of *Varroa* per cell result in malformations like shortened abdomens, mis-shapen wings, deformed legs or even in the death of the pupa (De Jong *et al.,* 1982b; Schneider and Drescher, 1987).

Figure 19 summarizes the life cycle of *Varroa jacobsoni.* The adult female mites enter the brood cells shortly before capping and must feed on larval hemolymph before they can lay eggs. Each mite lays 2-6 eggs at approxi-mately 30-hour intervals. The first egg usually develops into a male and the later ones into females. The development proceeds from egg to six-legged larvae, to eight legged protonymphs, to deutonymphs, to sexually mature adult mites in 6 to 10 days (Ritter, 1981). They mate in the capped cells with the males dying soon afterward. All immature mites will die after the emerging bee opens the cell, while the young adult female mites and the mature (gravid) females move on to passing bees. The mite enters another brood cell in 3 to more than 150 days depending on the season and availability of brood.

Detection Techniques

It is important to detect *Varroa* early, before symptoms appear in the bees. When an infestation is discovered late, treatment is relatively unsuccessful and colony losses are high. It is also important to note that the bee louse, *Braula coeca* closely resembles *V. jacobsoni* in size and color. However, *Braula* being an insect, has six legs that extend to the side. *Varroa* being an arachnid has eight legs and these extend forward.

Varroa can be found on the adult bees, on the brood and in hive debris. In looking for *Varroa* remember that the number and location of mites in a colony vary according to time of year, the number being lowest in the spring, increasing during the summer, and highest in the fall. During the spring and summer most mites are found on the brood (especially drone brood). In the late fall and winter most mites are attached to adult worker bees.

Adult Examination—When sampling adult honey bees, 500 to 1000 should be collected. This can be done by brushing honey bees off the comb through a large mouth funnel (made of paper or cardboard, etc.) into a container or by using a modified portable car vacuum cleaner. Individual honey bees can be examined with or without the aid of a hand lens or dissecting microscope. When the mites are moving about on a bee, they are fairly easy to detect; but once they attach themselves between segments, they are difficult to find.

Varroa jacobsoni can be dislodged by shaking bees in 70% ethyl or isopropyl alcohol; the alcohol also kills the bees and preserves them for other purposes such as examination for *A. woodi.* De Jong *et al.* (1982a) found that hand-shaking bees in alcohol for one minute dislodged about 90% of the mites and that mechanical shaking on a rotary shaker for 30 minutes removed 100% of the mites. Following shaking, the mites are collected by passing the bees and alcohol through a wire screen (8 to 12 mesh) to remove the bees and then straining the alcohol through a 30 mesh screen or cotton cloth. The screen or cloth is then examined for mites.

Alternatively, an ether spray can be used to dislodge *Varroa* from adult bees. This technique is a rapid and efficient field detection method and avoids the handling, shipping, and time-consuming procedures associated with shaking adult bees in alcohol or other solvents. The bees (500 - 1,000) are collected in a jar and anesthetized with ether delivered from an aerosol can (this aerosol formulation is easily obtained in auto-parts stores where the product is sold as an aid to start engines). A one- to two-second burst of material is adequate. The bees are then rotated in the jar for about 10 seconds, hence the name ether-roll technique. The mites, the majority of which will have dislodged from their hosts, stick to the inside wall of the jar. To complete the process, the bee sample is deposited on a white surface and spread around. This should cause any remaining mites to fall on the white substrate. The bees should be examined immediately after application of the ether since the mites tend to stick to the bees if left in the jar for more than a few minutes.

Varroa also can be detected without killing the bees. Live adult honey bees are placed in an oven in a wire-based cage over white paper. The bees are heated for 10 - 15 minutes at 46 - 47°C. *Varroa jacobsoni*, if present, can be observed on the white paper (Crane, 1978). In Israel, specially formulated strips of amitraz are ignited to expose a sample of bees in a cage. This method is not harmful to the bees and does not expose hive products to the acaricide.

Brood Examination—To look for mites on brood, pupae (preferably drone) are examined. *Varroa jacobsoni* can be easily recognized against the white surface of worker or drone pupae. Pupae can be removed from their cells with forceps or a hive tool. However, a minimum of 100 pupae should be examined per colony. A quick and easy method of obtaining pupae is to insert a capping scratcher at an angle through the cappings and lift the pupae and cappings upward (Fig. 20) (Szabo, 1989). Another method is to slice off the caps of 4 - 6 in² brood with a long bladed knife. The comb (frame) is then sharply jarred on a hard, flat surface such as a hive top; this dislodges the brood on which the mites are easily observed. Also, be sure to examine the bottom of the cells which once held the pupae.

Inspection of Hive Debris—Debris (wax particles, pollen, dead bees and brood, mites, etc.) normally falls to the hive floor and is removed by house-cleaning bees during warm weather. This material can be collected and examined for the presence of *Varroa jacobsoni*. A magnifying glass or dissecting microscope can be helpful in locating the mites in the debris. The

FIGURE 20. Photo of capping scratcher being used to expose brood for mite examination.

collection of hive debris can be facilitated by placing white construction paper on the hive floor. Stapling the paper under a wood (1/4 inch) and wire (8 to 12 mesh) frame protects both the paper and debris from the bees. The paper is examined for mites that can be easily seen against the white background. Sticky boards or shelf paper, with the adhesive surface exposed, can be used to help hold the debris. However, in Europe, sticky boards or adhesives are not recommended because of problems of bees being killed and messiness (Koeniger, 1990).

Mites also can be separated from hive debris using the flotation method. Hive debris is placed in a pan and covered with 98% ethyl alcohol. The mites float to the surface while the heavier debris sinks (Ritter and Ruttner, 1980).

The acaricides used for treatment of mite infestations also can be applied to colonies to detect *Varroa*. Apistan®is currently approved and available for this purpose. After treatment, the mites drop to the paper and can easily be detected. It is important that the paper have a sticky surface to hold any recovering mites.

Treatment

Infested honey bee colonies will die depending on whether the beekeeper effectively controls *Varroa*. Thus, all colonies in continental Europe with *Varroa* must rely on some control on a regular basis, commonly once a year.

In the United States, at this time, only Apistan is registered for use in controlling *Varroa* on honey bees. All Apistan formulations have been approved by the EPA for the control of *Varroa* in queen mailing cages, packages, and beehives. Contact your state apiarist for current information regarding the use and availability of Apistan and other possible materials to control *Varroa.*

On a worldwide basis, there are literally hundreds of materials that have been used to control this mite. The most common of these chemicals being used today include: amitraz, Apistan (fluvalinate), Apitol (cymiazol HCl), Bayvarol (flumetrin), Folbex VA (bromopropylate), formic acid, lactic acid, (10 - 15%), malathion (1%), and Perizin (coumaphos). No chemical treatment should be applied to honey-producing colonies before and during honey flows. The earliest chemicals can be applied is after the last honey harvest. During this season colonies have large numbers of capped brood cells and all mites in these cells are protected from most chemical treatments. However, in cases of high mite infestations, colonies should be treated immediately. In Europe, usually formic acid which penetrates the cell capping is used when brood is present. Alternatively, Apistan and Bayvarol are constant release methods which are applied over a period of at least 30 days. So, the mites which emerge from the capped brood during treatment are killed (Koeniger and Fuchs, 1989). When mite infestations are moderate or low, it is best to withhold treatment until brood rearing decreases and when no sealed brood is present in the colony (in autumn or winter). However, treatment with acaricides has serious drawbacks, there is always a danger of contaminating honey and beeswax, killing bees and brood, and creating strains of mites resistant to chemicals.

There are few alternatives to using chemicals to control *Varroa*. In Europe, many beekeepers start new colonies each season from treated bees (without *Varroa*). These colonies will be used for honey production in the following year. Colonies which reared brood during the entire season and produced honey are dequeened at the end of the season, the bees treated and used to initiate new units or various other ways (Koeniger, 1990). Direct methods to reduce the level of mites in a colony are mainly the removal of capped drone brood. Drone combs can be placed in a colony and the drone brood removed and destroyed by freezing after it is capped.

A tolerance or resistance to *Varroa* is exhibited by *Apis cerana* that is believed to develop from several factors including a "grooming behavior" that is followed by chewing of the mites (Peng, 1988). This behavior reduces the level of mite infestation in hives of *A. cerana*. What may be more significant, is that in *A. cerana*, the mites reproduce only in drone brood and not in worker or queen brood (Koeniger *et al.*, 1983). It is believed that the developmental time of the worker and queen from egg to emergence is insufficient for mite development.

Studies are also underway to develop *Apis mellifera* stocks that are tolerant or resistant to *Varroa jacobsoni*. This approach, although requiring a long developmental period, should be the method of choice over acaricides.

Tropilaelaps clareae

The distribution of *Tropilaelaps clareae* is still restricted to southeast Asia. These mites also parasitize adult bees and brood and have been reported to infest colonies infested with *Varroa jacobsoni* (Delfinado-Baker and Aggarwal, 1987). Female mites measure about 1 mm long and 0.6 mm wide; the male is slightly smaller. These mites are difficult to detect because of their small size and their brownish coloration that blends in perfectly with brood cappings and comb. *Tropilaelaps clareae* can be found by observing under a magnifying glass or dissecting microscope a brood comb suspected of being infested. In the field, hit the comb on a light colored surface. Dislodged mites may be seen moving on the surface. The mites can be picked up with a fine brush moistened with alcohol.

Studies on the control of *T. clareae* have been based primarily on substances used to control *V. jacobsoni*. Fluvalinate and amitraz have both shown promise in controlling *T. clareae* (Burgett and Kitprasert, 1990). Woyke (1985) reported that a break in brood cycle was all that was necessary to control this mite.

Melittiphis alvearius—This is a little known mite that is associated with honey bees but not considered to be a pest (Gibbins and van Toor, 1990). It is included here because it was found in California during a survey for *Varroa* and because of increasing reports on its distribution (Delfinado-Baker, 1988). There is little chance that *M. alvearius* would be confused with other mites found in honey bee colonies. The adult female is ovate, flattened dorso-ventrally, 0.79 mm long and 0.68 mm wide, brown, and well sclerotized with many stout and spine-like setae.

PESTS AND PREDATORS OF HONEY BEES

A wide variety of pests and predators, besides infectious microorganisms, may attack the honey bee, bee brood, the hive itself, or attempt to remove the honey and pollen the bees have stored. Some pests merely use the hive as a resting place or as a place to build a nest, while other pests attack stored equipment and beeswax comb usually in search of pollen and debris left over from brood rearing.

Insects and Spiders

Insects and spiders are the most common life forms on our planet in numbers of described species. Of 27 orders of insects, 13 have members that are pests of varying degrees of severity to honey bees. Insects may live in or on live or dead honey bees. In the bee colony, they may feed on brood, eat honey and/or pollen or utilize the hive itself as a home. Several insects attack hive parts not in use, especially the beeswax combs. Fortunately most of the insects are only local or occasional pests that do not require control measures.

Insects like bristletails, cockroaches, earwigs and psocids or booklice, live in beehives, but do not do any apparent harm. Nonetheless, cockroaches and earwigs are very offensive when the extractor puts them in extracted honey. Some spiders, dragonflies and a couple of the large predacious bugs capture foraging honey bees, along with other flying insects, but seldom cause noticeable losses to a bee colony. Termites attack the wooden hive parts and are a particularly severe problem in subtropical/tropical areas of the world. Wooden hive parts can be protected by use of wood preservatives or a barrier of grease or oil to reduce damage from termites or ants. Hives placed directly on the ground are especially vulnerable to termite or ant attack.

One insect, the strepsipteran *Stylops melittae* lives on the bee. The female of this unusual insect feeds on blood. She positions herself between abdominal segments of the bee abdomen. Such parasitism is not common and control is not necessary (Adlakha and Sharma, 1976).

The four largest insect orders, the Lepidoptera, Diptera, Coleoptera and the Hymenoptera, have members that are bee pests.

Lepidoptera—The Greater Wax Moth

The greater wax moth, *Galleria mellonella* (L.), is the most serious insect pest of unprotected honey bee combs throughout the warm areas of the world.

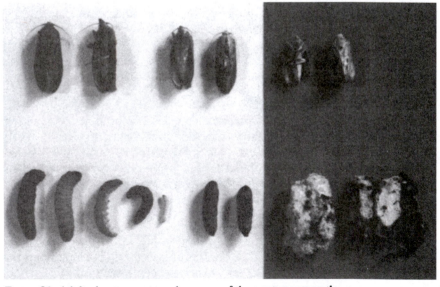

FIGURE 21. Adults, larvae, pupae and cocoons of the greater wax moth.

This moth is thought to have evolved with ancestral honey bees in Asia (Morse, 1975; Paddock, 1918), and it now commonly inhabits the nests of *Apis cerana, Apis dorsata, Apis florea,* and *Apis mellifera* (Ruttner, 1987; Singh, 1962).

The larvae of the greater wax moth (GWM) feed primarily on combs, cast larval skins, pollen and honey. Comb destruction by GWM in the United States is greatest in the southernmost states where the major package bee and queen producing operations are located (Williams, 1976).

During the warmest part of the season, honey bee colonies generally contain large numbers of small GWM larvae. Honey bees constantly remove these larvae and repair damage as soon as it occurs. Thus, GWM larvae are not obvious to the beekeeper. However, severe weakening or death of a colony is commonly followed by rapid growth of these GWM larvae and destruction of all unprotected combs within two to three weeks (Nielsen and Brister, 1979). For example, the larval period at 35°C (95°F) is only about 18 days (Beck, 1960).

In subtropical and tropical areas, honey bee combs are especially prone to attack by GWM because warm conditions favor its constant reproduction and development. Wax moth larvae destroy unprotected combs of colonies whenever worker bee populations decline, especially during dearth periods (Anderson *et al.*, 1983; Singh, 1962). Some beekeepers render the wax from combs from honey supers each season instead of storing them.

In temperate areas, freezing winter temperatures kill all exposed stages of GWM, thus preventing GWM larval populations from reaching destructive levels until August and September (Morse, 1975). In the northern United States, comb losses generally develop from cocooned stages of GWM that overwinter without freezing and infested colonies transported from the south by migratory beekeepers. Losses by the U.S. beekeeping industry attributable to GWM, estimated at over $4 million in 1976 (Williams, 1978), probably reached $8 million in 1988.

Not everyone considers the GWM a pest. It is raised on beeswax-free artificial diets as a standard laboratory insect and commercially as fish bait and live food for certain zoo animals (Beck, 1960; Dadd, 1966; Dutky *et al.*, 1962; Eischen and Dietz, 1990; Marston and Campbell, 1973).

Larvae—Newly hatched wax moth larvae are white, but successive instars are medium to dark gray above with creamy white undersides; the larval head capsule is brown. The larvae feed on honey within an hour of hatching and many burrow into pollen or cell walls. Later, they migrate to the midrib area of the comb. The newly hatched larvae are extremely active for about a day and, weather permitting, some migrate 10 ft. (3 m) or more and infest other hives in an apiary (Nielsen and Brister, 1979). Growing larvae construct silk-lined tunnels through the cell walls over the face of combs. The larval period lasts from about 18 days to 3 months, during which a maximum body length of about 2.5 cm (1″) is reached. The final size of larvae is largely dependent on temperature and the type and amount of food available.

The GWM has evolved various larval defense mechanisms in order to coexist with honey bees. Very young larvae avoid nest-cleaning bees by feeding in or near the midribs of combs and other concealed areas. Older larvae rely on their strong silken tunnels, rapid movement, and a tough collapsible cuticle to resist capture and stinging by nest-cleaning bees. The few GWM larvae able to complete their development in normal bee colonies spin a cocoon and pupate in a protected site.

Dark combs are preferred by larvae because they contain a variety of growth enhancing impurities such as entrapped pollen and cast larval skins. Some newly hatched larvae are able to develop to the adult stage on wax foundation, though their growth rate is slower and the resulting adults are quite small (Milum, 1935).

Cocoons—Cocoons from well-fed larvae are about 1.3 to 2 cm (1/2 to 13/16″) long and 0.5 to 0.7 cm (3/16 to 9/32″) in diameter. Mature larvae usually spin cocoons in oval depressions that they chew on the inside of hive bodies, the inner cover, lid and on frames. The tough silken cocoons are white to gray and may be covered with dark fecal pellets and debris. Large sheets of these cocoons often are cemented together, especially over the inside of the lid and hive bodies. The fully grown larva (prepupa) transforms into a pupa inside the cocoon. The adult then emerges in as little as seven to eight days. In cold climates, the cocooned stage (prepupa or pupa) may last for four months or more if it does not freeze.

Moths—Adult wax moths are small, heavy-bodied moths about 1/2 to 3/4″ (1.3 - 1.9 cm) long. The upper surfaces of the wings vary in color from grayish to brown to dark gray to almost black, whereas the undersides are light gray. The wings are held together against the body in rooflike fashion when the moths are at rest (Milum, 1935; Paddock, 1918). Generally, moths emerge within the nest at night, immediately run outside, and then rest on the hive or nearby objects until their wings expand and harden. Usually within 40 to 60 minutes after emergence, moths fly high into trees where some mating presumably occurs. Females are receptive to mating 24 hours after emergence and start laying eggs shortly after mating (Nielsen and Brister, 1977). A wax moth female produces less than 300 eggs during her life span of 3 to 30 days, but a few may lay as many as 2000 eggs.

Mated females fly to beehives one to three hours after dark, enter, and lay eggs until they leave shortly before daylight. Eggs vary in color from tan to white; most are laid in masses. The female moths preferentially deposit their eggs in small crevices such as the space between pollen and cell walls or adjoining hive parts and frames. Some colonies are highly attractive to the female moths as egg laying sites, while others receive no eggs (Nielsen and Brister, 1977). Eggs hatch in as little as three to five days when incubated near 32°C (90°F), but may require more than 30 days when kept at 10° to 16°C (50° to 60°F).

The attraction of a mate by greater wax moths is based on an interesting combination of short range olfactory and acoustic communication systems. Unlike most moths, the male wax moth, not the female, emits a scent to attract a potential mate (Finn and Payne, 1977). This sex attractant scent or phero- mone is a combination of the chemicals n-nonanal or n-undecanal. It is known for its distinctive odor by those who raise wax moths (Leyrer and Monroe, 1973; Schmidt and Monroe, 1976). Scent released from the male's wing glands is detected by microscopic receptors on the antennae of the female moth (Payne and Finn, 1977).

Both sexes of wax moths produce sound by rapid wing movement, but only the males have specialized acoustic mechanisms. Males release short pulses of ultrasonic sound (75 kHz) by fluttering their wings to flex small sound-emitting organs (tymbals) near the base of each wing (Spangler, 1985). The receptors for ultrasonic sound are present on both sexes as paired abdominal organs (tympana) located in the constriction between the thorax and abdomen, hidden by the wings (Mullen and Tsao, 1971). Female moths produce low-frequency sound (39 Hz) by wing-fanning without the aid of specialized organs. The wings of the male are held rigidly at 45° from the body to receive low-frequency vibrations from the air or substrates such as comb. Apparently males detect low-frequency sound released by female moths by vibration receptors in their legs (Spangler, 1985, 1987, 1988).

Two patterns of behavior are used by the male wax moth to communicate with females, according to whether the male is isolated or in a group. Both types of behavior enable an unmated female to locate a male moth in a dark environment. Wax moths normally are widely separated in the vicinity of bee colonies. A male moth thus isolated often "calls" a female moth from a stationary position on the side of a beehive or other nearby object. The calling male releases a constant low level of scent until a female approaches. Stimu- lated by the female, the male moth flutters his wings briefly and may release pulses of ultrasound (Flint and Merkle, 1983; Spangler, 1985). The female reacts by wing-fanning to release continuous low-frequency sound which, in turn, stimulates the male to abruptly release more pheromone. The female moth follows the gradient of airborne pheromone molecules to the male, at which point courtship and mating take place (Flint and Merkle, 1983; Spangler, 1987).

A different sequence of mate-recruitment behavior ensues when groups of moths are present inside a bee nest or other enclosure. Here the males constantly move about, releasing a low level of scent and periodically wing-fanning to emit pulses of ultrasound. In this situation, a male moth also increases the release of scent in response to low-frequency sound released by an approaching female (Spangler, 1985, 1987, 1988).

Control of the Greater Wax Moth

The presence of numerous developing *Galleria* larvae and damaged combs in a managed honey bee colony usually signals a major problem such as queenlessness, an infectious disease, poisoning, or starvation. The most practical way to prevent comb destruction by GWM in active bee hives is by maintaining strong colonies of a vigorous bee stock. Unfortunately, beekeepers often are unable to routinely inspect colonies in out-apiaries and frequently a certain number of colonies become queenless and, as a result, wax worms are able to damage or destroy large numbers of combs during mid to late season. Italian bees are better at controlling the larvae of wax moth than some other races of *Apis mellifera* (Milum, 1940; Morse, 1978a). The African bee hybrids in South America are more adept at recognizing greater wax moths and preventing their entry into the nest than European honey bees (Eischen *et al.*, 1986b). However, Africanization of European-type honey bees in Venezuela apparently has led to increased GWM problems, a result of additional absconding leaving combs unprotected (Hellmich and Rinderer, 1990).

Prompt treatment of the combs infested with GWM larvae is needed to kill the larvae present and prevent further comb losses. The simplest approach is to set supers of infested combs on strong colonies or place infested combs in a freezer for about 48 hours. Some beekeepers in the Southeast United States place supers of infested combs on colonies of the red imported fire ant (*Solenopsis invicta*) for a day or so.

Parasites and pathogens identified from GWM larvae include a small wasp (*Apanteles galleriae*) (Shimamori, 1987), a bacterium (*Bacillus thuringiensis*), a virus (*Galleria*-specific nuclear polyhedrosis virus), and a protozoan (*Nosema galleriae*). All but *N. galleriae* have been studied with the objective of developing practical biological controls for protection of honey bee combs (Ahmad *et al.*, 1983; Cantwell and Shieh, 1981; Dougherty *et al.*, 1982; Johansen, 1962). Only one, a strain of *B. thuringiensis*, (Certan®) has been made commercially available as a biological insecticide. Unfortunately, Sandoz, Inc., recently stopped production of this product.

Certan® consists of a water dispensable concentrate containing spores of a selected strain of the bacterium *B. thuringiensis* that is highly active against young GWM larvae. Commercial products containing other strains of *B.*

Figure 22. Typical results of uncontrolled wax moth infestations of honey bee combs. Frame comb remains showing cocoons on lower left side of end bar and webbing produced by larvae feeding on the unprotected combs; hive body with cluster of cocoons.

thuringiensis are not as effective as Certan®. Furthermore, these products are not registered for use against the GWM.

As of 1990, controls approved by the U.S. Environmental Protection Agency (EPA) for protection of honey bee combs against insect damage included paradichlorobenzene, *B. thuringiensis,* carbon dioxide fumigation, heat, and exposure to freezing temperatures (Shimanuki, 1981). Only freezing or fumigation with CO_2 are suitable treatments of comb honey (Cantwell *et al.,* 1972).

Paradichlorobenzene—The chemical paradichlorobenzene (PDB) may be used to protect combs in storage by placing ca. 1 tablespoon of crystals on a piece of paper on the top bars of every fourth or fifth super in a stack, beginning at floor level. Not only does PDB kill larvae, pupae and adults (eggs are not affected), but moths are repelled by the chemical. A constant fumigation with PDB is needed to eliminate larvae that hatch in the supers or attempt to migrate to them from other hatching sites.

Cracks between supers and holes should be taped to contain PDB vapor within the stack. Before use, combs should be aired at least 24 hours at warm temperatures to prevent repellency or other adverse effects to bees. Repellency is especially noticeable when package bees are hived on combs harboring considerable PDB residue.

Cold Treatment and Fumigation with Carbon Dioxide—Beekeepers and packers must protect comb honey from wax moth damage from the time it is removed from colonies until it is securely packaged. Feeding damage by a single young larva can destroy the marketability of comb honey through leakage of honey and the unsightly appearance of larval tunnels on the surface of combs. In the United States, the only methods approved for protecting comb honey from wax moth damage are carbon dioxide fumigation and freezing.

Holding comb honey in a home freezer at -15°C (5°F) for 24 hours is sufficient to kill all stages of the greater wax moth (Cantwell and Smith, 1970). Cold-treated comb honey must be packaged in mothproof containers or held in mothproof rooms to prevent reinfestation. Comb honey may be kept in a freezer for months and will maintain its flavor. Freezing prevents crystallization and protects it against the wax moth (Morse, 1978a; Taylor, 1988).

Carbon dioxide fumigation can be used to protect comb honey from wax moth damage. At one time, a carbon dioxide fumigation system was utilized to

Table 1.Control Methods for the Greater Wax Moth in Honey Bee Combs.

Treatment	Paradichlorobenzene	*Bacillus thuringiensis* (Certan®)	Carbon dioxide	Cold
Empty comb	yes	yes	yes	yes
Full comb[1] (food use)	no	no	yes	yes
Full comb[2] (non- food use)	yes	yes	yes	yes
Temperature or dosage level	3 ounces (84 g) or 6 tablespoons per stack of 5 supers	Consult current label	98 percent by volume at 100°F (38°C)[4]	20°F (- 7°C) † 10°F (-13°C) § 5°F (-15°C) ‡
Length of exposure	Keep constant supply while in storage[3]		4 hours	4.5 hours † 3 hours § 2 hours ‡

The methods listed above have been approved by the U.S. Environmental Protection Agency for use with honey bee combs. They may be used ONLY AS SPECIFIED until such time that changes are officially enacted. Consult your state apiary inspector regarding updated control guidelines. Source: USDA Farmers Bulletin 2217, January 1981 revision.

[1] Honey to be used for human consumption.

[2] Honey to be used for bee food (not for human consumption).

[3] Paradichlorobenzene (PDB) is most effective at temperatures above 70°F (21°C). Eggs of the greater wax moth are not destroyed by PDB.

[4] WARNING: Exposure to carbon dioxide (CO_2) gas may be hazardous or even fatal to humans. This gas is heavier than air and should be used in small sealed containers, ventilated areas, or outside buildings.

protect large quantities of gallberry comb honey at the Georgia Sioux Bee Association's packing plant in Waycross, Georgia (Cantwell *et al.,* 1972; Jay *et al.,* 1972). Fumigation of extracted combs with carbon dioxide imparts no residue, but it is generally unsatisfactory due to the cost of materials and the long treatment period.

Other Lepidoptera—In addition to the greater moth, several other moths can be found in live bee colonies or feeding on stored beeswax comb. These include the lesser wax moth, *Achroia grisella,* and the driedfruit moth, *Vitula edmandsae serratilineela.* Lesser wax moths, *A. grisella,* are small, slender-bodied, and silver-gray to buff-colored with a conspicuously orange-yellow head. Adults are a maximum of 10 to 13 mm (3/8 - 1/2") in length. This species has white narrow-bodied larvae that reach a maximum length of about 20 mm (13/16") (Milum, 1935).

The driedfruit moth *V. edmandsae,* ranges through the Rocky Mountain States and western Canada. It infests stored combs primarily and overwinters in the larval stage (Szabo and Heikel, 1987a; Winston *et al.,*1981). The larvae reach a maximum of 15 mm (19/32") long, changing from white to light pink as they mature (Okumura, 1966). Larvae may form a dense mass of webbing over combs without destroying them (Richards, 1984; Wilson and Brewer, 1974).

Two additional species of moth, the Indian Meal Moth *(Plodia interpunctella)* and the Mediterranean flour moth *(Agnagasta kuehniella)* seek pollen and can be serious pests, along with other moths and insects, in pollen traps or pollen harvesting operations (Williams, 1978). All of these pests are controlled by methods used for control of the greater wax moth in stored combs.

The death's head moth *(Acherontia atropos),* so named due to a distinctive skull pattern on the top of the thorax, is a nighttime robber of honey from bee colonies in Europe, Africa and Asia (Brugger, 1946). No control other than strong colonies is indicated.

Diptera—The Diptera or true flies have several species that are pests of living or dead bees. Robber flies are sometimes called "bee-killers." Larger bodied species include honey bees among their prey although none take only honey bees for food.

Several internal fly parasites, a condition termed myiasis, have been recorded from honey bees. Reports include tachinids and sarcophagids, but such reports are rare. Control is not needed. Fruit flies *(Drosophila)* and flesh flies such as sarcophagids are attracted to dead, rotting honey bees and fermenting honey. They probably help hasten decomposition and thus serve a useful role in decomposition and recycling of materials (Knutson, 1978).

One fly, the bee louse *Braula coeca,* is a worldwide ectoparasite of honey

Figure 23. Photo of *Braula coeca* on queen. *(Photo by I. B. Smith, Jr.)*

bees (Smith and Caron, 1984a). The flies ride about on honey bees and take their food at the bee mouthparts. The maggots tunnel beneath wax cappings to eat pollen and debris and may disfigure honey to be sold in the comb. Although numbers of *B. coeca* may congregate on the queen (Fig. 23) in the fall months, the greatest problem with this curious and generally harmless pest is its being mistaken for a mite when spotted on bees (Smith and Caron, 1984b).

Coleoptera—The Coleoptera or beetles comprise 40% of all the described insect species. It is not surprising, therefore, that several are pests of beehives. None are serious pests. Most beetles found in bee colonies are seeking shelter, feed on dead bees or hive debris. This includes carabids (ground beetles), dermestids, scarabaeids and others (Caron, 1978a). In Africa, the small hive beetle *Aethina tumida* has been compared to the wax moth in its potential to cause damage to comb (Lundie, 1940). In Asia, a blister or oil beetle *(Meloe sp.)* is locally a pest of adult bees. The larval stage pierces the adult bee to feed on blood. Control is apparently difficult.

Hymenoptera—Several social hymenopterans are pests of honey bees. Ants like the fire ant, carpenter ant, pharaoh ant and especially tropical ants like army ants may occasionally be serious pests. They steal honey, bee brood or live in the colony where bees can't remove them. Chemical control of ants can be hazardous as the ants may track the insecticide intended for the pest into and onto comb and hive parts. Where the ants are a problem, hive stands with ant guards—barriers of oil or grease—can be effective as long as the barrier remains intact (DeJong, 1978).

Several hornets prey on honey bees. In parts of the middle East and Asia, beekeepers place traps on their colonies to reduce predation by a species of *Vespa,* a large hornet that attacks *en masse,* sometimes killing the entire colony (Caron and Schaefer, 1985). In the U.S. and Europe, yellowjackets (*Vespula* sp.) prey on live or dead adult bees, steal honey and are misidentified as "bees" when they sting humans. Control of these social wasps is difficult as the nest must be found before the population can be eliminated with an insecticide. However, baiting yellowjackets, with or without insecticides, can be quite sucessful (Mayer *et al.,* 1987).

The bee wolf (*Philanthus* sp.), a solitary digger wasp, uses adult bees as food for their larvae in Europe but less so in the United States. The female velvet ant, a wingless wasp, also uses bees to feed her young. Other bees, like bumblebees or the stingless bees of the tropics, can be honey bee pests as they attempt to steal honey from colonies. As with other insect pests, control of these occasional hive pests is difficult.

Spiders—Spiders are predacious arthropods and insects are their frequent prey. Several kinds of spiders eat honey bees, but none prey specifically on them and all include other insects as well as honey bees in their diet. Some spiders, like crab spiders, wait in camouflage to grab foragers at flowers. Other spiders, like the common orb weaver *Argione,* construct a large web to trap flying insects like bees. Black widow spiders *(Latrodectes mactans)* frequently are found under beehives. In areas where they occur, caution is advised against grasping hives and hive stands with bare hands. Since they are only occasional pests, spider control, other than removal of webs in bee flight areas, is not necessary.

Amphibians

Several toads feed on honey bees while frogs, especially the bullfrog, occasionally may eat bees. The giant marine toad *Bufo marinus* is the most serious amphibian pest. In Arizona, the Colorado River Toad eats large numbers of bees during the summer (Waller, 1990). Beekeepers in some tropical areas construct a fence to keep this creature away from hive entrances (Morse, 1978b).

Birds

Generally beekeepers outside of the tropics have few problems with birds. Several woodpeckers attack the beehive especially in colder temperatures

Figure 24. Photo of woodpecker damaged bee hive. *(Photo by J. V. Lindner)*

(Fig. 24). Birds like titmice, some shrikes, swifts and flycatchers, including the kingbird, may predate heavily on honey bees in a very local area. This may present some problems for queen breeders, but otherwise control or concern about bird predation on honey bees is seldom necessary (Ambrose, 1978).

Two groups of birds, the bee-eaters (genus *Merops*) and the honeyguides (genus *Indicator*) of Africa and Asia are more serious bee pests. Bee-eaters eat honey bees and other venomous Hymenoptera almost exclusively and bee-keepers often move apiaries away from their nest locations at certain times of the year to reduce their impact. Honeyguides eat beeswax. At least two species locate bee nests and then lead mammals like honey badgers or baboons to the nest. The larger animal digs out and exposes the bee nest while the bird gets to eat as a reward (Friedmann, 1955). Control, other than moving colonies or controlling the badger, is difficult.

Mammals and Marsupials

Several mammals and one marsupial, the opossum, are occasional bee pests. Some of these animals may require control under some circumstances. Small insectivorous mammals such as shrews, moles and the hedgehog may invade bee colonies, but they apparently do little harm. Rodents like mice, rats and squirrels also nest in colonies or in stored bee equipment where they may damage comb as they build a nest (Fig. 25).

FIGURE 25. Photo of mouse nest in comb. *(Photo by D. M. Caron)*

Mice are the most common and troublesome of these smaller animal pests. Some apiary locations have higher mice populations than others. All of the smaller animals can be kept out of colonies by reducing colony entrances to keep them out, particularly during the winter. To be effective, entrances must be reduced before the mouse or other small animal establishes its nest inside the bee colony. Care is also necessary to keep these small animals out of stored equipment.

Among mammalian carnivores, the jackal, raccoons, coatis, martens, some weasels and the badger are occasional bee pests. In Africa, the honey badger or ratel is a common pest that digs up feral bee colonies in their nocturnal wanderings. They also may invade apiaries and push colonies from hive stands on occasion. Fencing to exclude carnivore visitors or hanging colonies in trees serves to control the occasional carnivore pest.

In North America the skunk is one of a group of nighttime hive visitors which, along with the opossum and raccoon, may cause damage locally. These mammals visit apiaries and prey on the same colonies night after night. Typically, they scratch on the exterior of the colony and then eat the adult bees that come outside in response to the disturbance. With repeated feeding they may weaken colonies, especially during winter. Also bee colonies continually disturbed can become aggressive due to the repeated visits. Control of night-time animal visitors involves trapping of the animal or fencing of the apiary to exclude their visits. Elevation of beehives so the bees can get to the underside of the animal to sting it when their colony is disturbed also can reduce this predation. When laws permit, these animals can be trapped or shot.

A few larger animals like baboons and chimpanzees may bother bees. Bee colonies are usually fenced to exclude large hoofed animals like cattle and horses, so such livestock don't overturn colonies.

Probably the largest animal pest of bee colonies is the bear. The black bear of North America is the greatest threat, but the brown bear of Europe and three bear species of Asia also are bee pests (Caron, 1978b). In a 1978 survey, Lord and Ambrose (1981) stated that 39 of 62 U.S. States and Canadian Provinces reported bee colonies damaged by bears with an average loss of $16,209 (Fig. 26). Nine of these jurisdictions compensated beekeepers for

FIGURE 26. Photo of bear damage. *(Photo by J. V. Lindner)*

bear damage, while 29 of them have bear control programs that involve removal or aversive conditioning of problem bears or which give permits to individuals to destroy individual bears.

Bears are attracted to apiaries where they smash bee equipment to obtain bee brood and honey. The bear returns repeatedly to an apiary, often resulting in destruction of the entire apiary in more isolated areas before their damage is detected. Once a bear begins feeding on the colonies in an apiary, ordinary protective measures are ineffective and colonies need to be moved to avoid further damage.

Beekeepers use electrified fences to protect isolated apiary sites from bears (see review by Caron, 1978c). Fences must be erected before bear damage occurs, be sturdy with three or four strands and they must be continuously charged. Other beekeepers have placed bee colonies on bear-proof platforms, have enclosed colonies in stout wire cages, chained colonies to trees, have used various types of traps or snares, tried aversive conditioning, employed chained dogs, have utilized lion dung, rented electronic noise makers or have tried coating the ground with sulfur or moth ball crystals (paradichlorobenzene) as methods to keep bears away. The most reliable technique is to trap or shoot bears that cause a problem. Occasional bear damage continues to be a significant problem for some beekeepers.

REFERENCES

Adlakha, R.L. and O.P. Sharma. (1976). *Stylops* (Strepsiptera) parasites of honey bees in India. *Amer. Bee Jour. 116*:66.

Ahmad, R.,N. Muzaffar and Q. Ali. (1983). Biological control of the wax moths *Achroia grisella* (F.) and *Galleria mellonella* (L.) (Lepidoptera, Galleridae) by augmentation of the parasite *Apanteles Galleriae* Wilk. (Hymenoptera, Braconidae) in Pakistan. *Apiacta 18*:15-20.

Alekseenko, F.M. and A. Ju. Kolomiets. (1967). Study of the virus paralysis of bees in the Ukraine. XXI Intern. Beekeep. Cong., Prelim. Sci. Meet., p. 492.

Ambrose, T.J. (1978). Birds. *In* Honey Bee Pests, Predators and Diseases R.A. Morse, *ed.* Cornell Univ. Press, Ithaca, New York, pp. 215-226.

Anderson, R.H., B. Buys and M.F. Johannsmeier. (1983). Beekeeping in South Africa. So. Afri. Dept. Agric. Bull. 394.

Anonymous. (1985). 1984 Apiary inspection statistics. Proc. Ann. A.I.A. Conf., San Antonio, Texas.

Bailey, L. (1953). The transmission of nosema disease. *Bee World 34*:171-172.

Bailey, L. (1955a). Control of amoeba disease by the fumigation of combs and by fumagillin. *Bee World 36*:162-163.

Bailey, L. (1955b). The infection of the ventriculus of the adult honey bee by *Nosema apis* (Zander). *Parasitol. 45*:86-94

Bailey, L. (1957). Comb fumigation for nosema disease. *Amer. Bee Jour. 97*:24-26.

Bailey, L. (1959). An improved method for the isolation of *Streptococcus pluton* and observations on its distribution and ecology. *Jour. Insect Pathol. 1*:80-85.

Bailey, L. (1960) The epizoology of European foulbrood of the larval honey bee, *Apis mellifera* L. *Jour. Insect Pathol. 2*:67-83.

Bailey, L. (1961). The natural incidence of *Acarapis woodi* (Rennie) and the winter mortality of honey bee colonies. *Bee World 42*:96-100.

Bailey, L. (1963a). The occurrence of chronic and acute bee paralysis viruses in bees outside Britain. *Jour. Invert. Pathol. 7*:167-169.

Bailey, L. (1963b). Infectious diseases of the honey bee. Land Books, London.

Bailey, L. (1965). Bee diseases and pests. Rpt. Rothamsted Exp. Stn. for 1964. pp. 200-201.

Bailey, L. (1967a). Acute bee paralysis virus in adult honey bees injected with sacbrood virus. *Virol. 32*:368.

Bailey, L. (1967b). The incidence of virus disease in the honey bee. *Ann. Appl. Biol. 60*:43-48.

Bailey, L. (1968a). The multiplication of sacbrood virus in the adult honey bee. *Virol. 36*: 312-313.

Bailey, L. (1968b). The measurement and interrelationships of infections with *Nosema apis* and *Malpighamoeba mellificae* of honey-bee populations. *Jour. Invert. Pathol. 12*:175-179.

Bailey, L. (1969). The multiplication and spread of sacbrood virus of bees. *Ann. Appl. Biol. 63*:483-491.

Bailey, L. (1970). Bee diseases and pests. Rpt. Rothamsted Exp. Stn. for 1969, I. pp. 259-261.

Bailey, L. (1971a). Bee diseases and pests. Rpt. Rothamsted Exp. Stn. for 1970, II. pp. 171-183.

Bailey, L. (1971b). Honey-bee viruses. *Sci. Prog. Oxf. 59*:390-323.

Bailey, L. (1972a). *Nosema apis* in drone honey bees. *Jour. Apic. Res. 11*:171-174.

Bailey, L. (1972b) The preservation of infective microsporidian spores. *Jour. Invert. Pathol. 20*:252-254.

Bailey, L. (1976). Viruses attacking the honey bee. *Adv. Virus Res. 20*:271-304.

Bailey, L. (1981). Honey bee pathology. Academic Press, London.

Bailey, L. (1985). *Acarapis woodi*: a modern appraisal. *Bee World 66*:99-104.

Bailey, L., B.V. Ball, J.M. Carpenter and R.D. Woods. (1980). Small virus-like particles in honey bees associated with chronic paralysis virus and with a previously undescribed disease. *Jour. Gen. Virol. 46*:149-155.

Bailey, L., B.V. Ball and R.D. Woods. (1976). An iridovirus from bees. *Jour. Gen. Virol. 31*:459-461.

Bailey, L. and M.D. Collins. (1982). Reclassification of *Streptococcus pluton (White) in a new genus Melissococcus pluton. Jour. Appl. Bacteriol. 53*:215-217.

Bailey, L. and E.F.W. Fernando. (1972). Effects of sacbrood virus on adult honey bees. *Ann. Appl. Biol. 72*:27-35.

Bailey, L. and A.J. Gibbs. (1964). Acute infection of bees with paralysis virus. *Jour. Insect Pathol. 6*:395-407.

Bailey, L., A.J. Gibbs and R.D. Woods. (1963). Two viruses from adult honey bees (*Apis mellifera L.) Virol. 21*:390-395.

Bailey, L., A.J. Gibbs and R.D. Woods. (1964). Sacbrood virus of the larval honey bee (*Apis mellifera* Linnaeus). *Virol. 23*:425-429.

Bailey, L., A.J. Gibbs and R.D. Woods. (1968). The purification and properties of chronic bee paralysis virus. *Jour. Gen. Virol. 2*:251-260.

Bailey, L., and D.C. Lee (1962). *Bacillus larvae*: its cultivation *in vitro* and its growth *in vivo*. *Jour. Gen. Micobiol. 29*:711-717

Bailey, L. and R.G. Milne. (1969). The multiplication regions and interractions of acute and chronic bee-paralysis viruses in adult honey bees. *Jour. Gen. Virol. 4:*9-14.

Bailey, L. and R.D. Woods. (1974). Three previously undescribed viruses from the honey bee. *Jour. Gen. Virol. 25:*175-186.

Bailey, L. and R.D. Woods. (1977).Two more small RNA viruses from honey bees and further observations on sacbrood and acute bee-paralysis viruses. *Jour. Gen. Virol. 37:*175-182.

Bamrick, J.F. (1964). Resistance to American foulbrood in honey bees. V. Comparative pathogenesis in resistant and susceptible larvae. *Jour. Insect Pathol. 6:*284-304.

Beck, S.D. (1960). Growth and development of the greater wax moth *Galleria mellonella* (L.) (Lepidoptera: Galleriidae). *Trans. Wisc. Acad. Sci. Arts Lett. 49:*137-148.

Bogdan, T., A. Popa and N. Foti. (1959). Research on the possibility of using the coprological test in the diagnosis of nosema diseases of living queens (in Romanian). *Apicultura 32:*6-11.

Bondarenko, O.I. (1966). On the frequency of detection of amoeba in honey bees. *In* Materials of the Annual Scientific Conference of the All-Union Institute of Experimental Veterinary Science, Moscow.

Brcak, J., J. Svoboda and O. Kralik. (1963). Electron microscopic investigation of sacbrood of the honey bee. *Jour. Insect. Pathol. 5:*385-386.

Brugger, A. (1946). The deathhead moth. *Glean. Bee Cult. 74:*602-603, 651.

Bucher, G.E. (1958). General summary and review of utilization of disease to control insects. *Proc. Tenth Internat. Cong. Entomol. 4:*695-701.

Bulger, J.W. (1928). *Malpighamoeba* (Prell) in the adult honey bee found in the United States. *Jour. Econ. Entomol. 21:*376-379.

Burgett, D.M. and C. Kitprasert. (1990). Evaluation of Apistan as a control for *Tropilaelaps clareae* (Acari: Laelapidae), an Asian honey bee brood parasite. *Amer. Bee Jour. 130:*51-53.

Burnside, C.E. (1933). Preliminary observations on paralysis of honey bees. *Jour. Econ. Entomol. 26:*162-168.

Burnside, C.E. (1945). The causes of paralysis of honey bees. *Amer. Bee Jour. 85:*354-355.

Burnside, C.E. and I.L. Revell. (1948). Observations on nosema disease of honey bees. *Jour. Econ. Entomol. 41:*603-607.

Camazine, S. (1985). Tracheal flotation: A rapid method for detection of honey bee acarine disease. *Amer. Bee Jour. 125:*104-105.

Cantwell, G.E. (1970). Standard methods for counting nosema spores. *Amer. Bee Jour. 110:*222-223.

Cantwell, G.E., E.G. Jay, G.C. Pearman, Jr. and J.V. Thompson. (1972). Control of the greater wax moth, *Galleria mellonella* (L.), in comb honey with carbon dioxide. Part I. Laboratory studies. *Amer. Bee Jour. 112:*302-303.

Cantwell, G.E. and T.R. Shieh. (1981). Certan®—A new bacterial insecticide against the greater wax moth, *Galleria mellonella* L. *Amer. Bee Jour. 121:*424-426, 430-431.

Cantwell, G.E. and H. Shimanuki. (1970). The use of heat to control nosema and increase production for the commercial beekeeper. *Amer. Bee Jour. 110:*263.

Cantwell, G.E. and L.J. Smith. (1970). Control of the greater wax moth, *Galleria mellonella,* in honeycomb and comb honey. *Amer. Bee Jour. 110:*141.

Cantwell, G.E., L.J. Smith and T. Lehnert. (1971). Effects of extreme cold and of microwaves on spores of *Nosema apis. Amer. Bee Jour. 111:*188.

Caron, D.M. (1978a). Other insects. *In* Honey bee Pests, Predators and Diseases. R.A. Morse, *ed.* Cornell Univ. Press, Ithaca, New York, pp. 158-187.

Caron, D.M. (1978b). Bears and Beekeeping. *Bee World 59:*18-24.

Caron, D.M. (1978c). Marsupials and mammals. *In* Honey Bee Pests, Predators and Diseases. R.A. Morse, *ed.* Cornell Univ. Press, Ithaca, New York, pp. 227-256.

Caron, D.M. and P.W. Schaefer. (1985). Social wasps as bee pests. *Amer. Bee Jour. 126:*269-271.

Clark, T.B. (1977a). *Spiroplasma* sp., a new pathogen in honey bees. Jour. Invert. Pathol. 29:112-113.

Clark, T.B. (1977b). Another virus in honey bees. *Amer. Bee Jour. 117:*340-341.

Clark, T.B. (1978a). Honey bee spiroplasmosis, a new problem for beekeepers. *Amer. Bee Jour. 118:*18-19, 23.

Clark, T.B. (1978b). A filamentous virus of the honey bee. Jour. Invert. Pathol. 32:332-340.

Colin, M.E., J.P. Faucon, A. Giauffret and C. Sarrazin. (1979). A new technique for the diagnosis of acarine infestation in honeybees. Jour. Apic. Res. 18:222-224.

Crane, E. (1978). The *Varroa* mite. *Bee World 59:*164-167.

Dadd, R.H. (1966). Beeswax in the nutrition of the wax moth, *Galleria mellonella* (L.) *Jour. Insect Physiol. 12:*1479-1492.

Delfinado, M.D. (1963). Mites of the honeybee in Southeast Asia. *Jour. Apic. Res. 2:*113-114.

Delfinado-Baker, M. (1984). The nymphal stages and male of *Varroa jacobsoni* Oudemans—a parasite of honey bees. *Internat. Jour. Acarol. 10:*75-80.

Delfinado-Baker, M. (1988). Incidence of *Melittiphis alvearius* (Berlese), a little known mite of beehives, in the United States. *Amer. Bee Jour. 128:*214.

Delfinado-Baker, M. and K. Aggarwal. (1987). Infestation of *Tropilaelaps clareae* and *Varroa jacobsoni* in *Apis mellifera ligustica* colonies in Papua New Guinea. *Amer. Bee Jour. 127:*443.

Delfinado-Baker, M. and E.W. Baker. (1982). Notes on honey bee mites of the genus *Acarapis* Hirst (Acari:Tarsonemidae). *Internat. Jour. Acarol. 8:*211-226.

DeJong, D. (1978). Insects: Hymenoptera (Ants, Wasps and Bees). *In* Honey Bee Pests, Predators and Diseases. R.A. Morse, *ed.* Cornell Univ. Press, Ithaca, New York, pp. 138-157.

DeJong, D., D. De Andrea Roma and L.S. Goncalves. (1982a). A comparative analysis of shaking solutions for the detection of *Varroa jacobsoni* on adult honeybees. *Apidologie 13:*297-306.

DeJong, D. and P.H. DeJong (1983). Longevity of Africanized honeybees (Hymenoptera: Apidae) infested by *Varroa jacobsoni* (Parasitiformes: Varroidae). *Jour. Econ. Entomol. 76:*766-768.

DeJong, D., P.H. DeJong and L.S. Goncalves. (1982b). Weight loss and other damage to developing worker honeybees from infestation with *Varroa jacobsoni. Jour. Apic. Res. 21:*165-167.

Dougherty, E.M., G.E. Cantwell and M. Kuchinski. (1982). Biological control of the greater wax moth (Lepidoptera: Pyralidae), utilizing in vivo- and in vitro-propagated baculovirus. *Jour. Econ. Entomol. 75:*675-679.

Doull, K.M. and K.M. Cellier. (1961). A survey of the incidence of nosema disease (*Nosema apis* Zander) of the honey bee in South Australia. Jour. Insect Pathol. 3:280-288.

Doull, K.M. and J.E. Eckert. (1962). A survey of the incidence of nosema disease in California. *Jour. Econ. Entomol. 55:*313-317.

Dreher, K. (1956). Research and observations on the nosema problem (in German). *Die Biene 92:*11-14.

Dutky, S.R., J.V. Thompson and G.E. Cantwell. (1962). A technique for mass rearing the greater wax moth (Lepidoptera: Galleriidae). *Proc. Entomol. Soc. Wash. 64:*56-58.

Eischen, F.A. and A. Dietz. (1990). Improved culture technique for mass rearing *Galleria mellonella (Lepidoptera:Pyralidae). Entomol. News 101:*123-128.

Eischen, F.A., J.S. Pettis and A. Dietz. (1986a). Prevention of *Acarapis woodi* infestation in queen bees with amitraz. *Amer. Bee Jour. 126:*498-500.

Eischen, F.A., J.S. Pettis and A. Dietz. (1987). A rapid method of evaluating compounds for the control of *Acarapis woodi* (Rennie). *Amer. Bee Jour. 127:*99-101.

Eischen, F.A., T.E. Rinderer and A. Dietz. (1986b). Nocturnal defensive responses of African-ized and European honey bees to the greater wax moth *(Galleria mellonella* L.). *An. Behav. 34:*1075-1077.

Fantham, H.B. and A. Porter. (1912). Microsporidiosis, a protozoal disease of bees due to *Nosema apis,* and popularly known as Isle of Wight disease. *Ann. Trop. Med. Parasitol. 6:*145-162.

Farrar, C.L. (1947). Nosema losses in package bees as related to queen supersedure and honey yields. *Jour. Econ. Entomol. 40:*333-338.

Fernando, E.F.W. (1972). Sacbrood. Rpt. Rothamsted Exp. Stn. for 1971, I. p. 227.

Finn, W.E. and T.L. Payne. (1977). Attraction of greater wax moth females to male-produced pheromones. *Southwest Entomol. 2:*62-65.

Flint, H.M. and J.R. Merkle. (1983). Mating behavior, sex pheromone responses, and radiation sterilization of the greater wax moth *(Lepidoptera: Pyralidae). Jour. Econ. Entomol. 76:*467-472.

Foote, L. (1971). California nosema survey 1969-1970. *Amer. Bee Jour. 111:*17.

Friedmann, H. (1955). The honey-guides. U.S. Nat. Mus. Bull. 208. Smithsonian Inst., Washington, D.C.

Fritzsch, W. (1970). Testing a treatment for nosema disease (in German). *Arch. F. Exp. Veterinamed. 24:*951-984.

Furgala, B. (1962a). The effect of the intensity of nosema inoculum on queen supersedure in the honey bee, *Apis mellifera* Linnaeus. *Jour. Insect Pathol. 4:*429-432.

Furgala, B. (1962b). Factors affecting queen losses in package bees, *Glean. Bee Cult. 90:*294-295.

Furgala, B. and R. Boch. (1970). The effect of Fumidil-B, Nosemack, and Humatin on *Nosema apis. Jour. Apic. Res. 9:*79-85.

Furgala, B., S.R. Duff, S. Aboulfaraj, D.W. Ragsdale and R.A. Hyser. (1989). Some effects of the honey bee tracheal mite *(Acarapis woodi)* on non-migratory honey bee colonies in east central Minnesota. *Amer. Bee Jour. 129:*195-197.

Furgala, B. and T.A. Gochnauer. (1969). Effect of treatment method with Fumidil-B. *Amer. Bee Jour. 109:*380-381,392.

Furgala, B. and R.A. Hyser. (1969) Minnesota nosema survey—distribution and levels of infection in wintering apiaries. *Amer. Bee Jour. 109:*460-461

Furgala, B., R.A. Hyser and E.C. Mussen (1973). Enzootic levels of nosema disease in untreated and fumagillin treated apiaries in Minnesota. *Amer. Bee Jour. 113:*210,212.

Furgala, B. and P.E. Lee. (1966). Acute bee paralysis virus, a cytoplasmic insect virus. *Virol. 29:*346-348.

Fyg, W. (1932). Observations on amoeba infection ("cyst disease") of the Malpighian tubules of the honey bee (in German). *Schweiz. Bienen-Ztg. 55:*562-572.

Fyg, W. (1954). The occurrence of flagellates in the honey bee rectum (*Apis mellifera* L.) (in German). *Schweiz. Ent. Gesell, Mitt.* 27:423-428.

Gary, N.E. and R.E. Page, Jr. (1988). Factors that affect the infestation of worker honey bees by tracheal mites (*Acarapis woodi*). *In* Africanized Honey Bees and Bee Mites. Needham *et al., eds.* Ellis Horwood Ltd. West Sussex, pp. 506-511

Giavarini, I. and G. Giordani. (1966). Study of acarine disease of honey bee. Final technical Report. Nat. Inst. of Apicult., Bologna, Italy.

Gibbins, B.L. and R.F. van Toor. (1990) Investigation of the parasitic status of *Melittiphis alvearius* (Berlese) on honey bees *Apis mellifera* L. by immunoassay. *Jour. Apic. Res.* 29: (in press).

Gilliam, M. and R.J. Argauer. (1975). Stability of oxytetracycline in diets fed to honeybee colonies for disease control. *Jour. Invert. Pathol.* 26:383-386.

Giordani, G. (1959). Amoeba disease of the honey bee, *Apis mellifera* L., and an attempt at its chemical control. *Jour. Insect Pathol.* 1:245-269.

Girardeau, J.H., Jr. (1961). The effect of fumagillin on a natural infection of *Nosema apis* (Zander). *Bul. Entomol. Soc. Amer.* 7:170.

Girardeau, J.H., Jr. (1972). Fumagillin use in honey bee queen mating hives to suppress *Nosema apis. Environ. Entomol.* 1:519-520.

Gochnauer, T.A. (1951). Drugs fight foulbrood diseases in bees. *Minn. Home Fm. Sci.* 9:15.

Gochnauer, T.A. and J. Corner. (1974). Detection and identification of *Bacillus larvae* in a commercial pollen sample. *Jour. Apic. Res.* 13: 264-267.

Gochnauer, T.A. and B. Furgala. (1969). Compatibility of fumagillin with other chemicals. *Amer. Bee Jour.* 109:309-311.

Goetze, G., F. Eberhardt and B. Zeutzschel. (1959). Research on selfhealing and therapy of nosema disease of the honey bee (in German). Inst. f. Bienenkunde, Friedrich-Wilhelms Univ., Bonn.

Grobov, O.F., A.M. Smirnov and G.D. Volkovskii. (1970). The efficiency of ethylene oxide and methyl bromide against *Nosema apis* spores. (in Russian). *Veterinariya.* 6:66-67.

Hammer, O. and E. Karmo. (1947). Nosema and honey gathering (in German). *Schweiz. Bienen-Ztg.* 70:190-194.

Harder, A. and J. Kundert. (1951). The course and the effects of nosema on the bee establishment of the Confederated Research Station, Wadenswill, in the years 1941-50 (in German). *Schweiz. Bienen-Zatg.* 74:531-544.

Hartwig, A. and A. Przelecka. (1971). Nucleic acids in the intestine of *Apis mellifica* infected with *Nosema apis* and treated with fumagillin DCH: cytochemical and autoradiographic studies. *Jour. Invert. Pathol.* 18:331-336.

Haseman, L. and L.F. Childers. (1944). Controlling American foulbrood with sulfa drugs. *Mo. Agric. Sta. Bull.* 482.

Hassanein, M.H. (1951). The influence of *Nosema apis* on the larval honeybee. *Ann. Appl. Biol.* 38:844-846.

Hassanein, M.H. (1952a). Some studies on amoeba disease. *Bee World* 33:109-112.

Hassanein, M.H. (1952b). The effects of infection with *Nosema apis* on the pharyngeal salivary glands of the worker honey-bee. *Proc. Roy. Entomol. Soc. London, Ser. A.* 27:22-27.

Haynes, W.C. (1972). The catalase test: An aid in the identification of *Bacillus larvae*. *Amer. Bee Jour.* 112:130-131.

Hejtmanek, J. (1970). The effect of fumagillin DCH in the complex of zootechnical measures (in Polish). *Pol'nohospodarstvo* 16:155-162.

Hellmich, R.L. and T.E. Rinderer. (1990). Beekeeping in Venezuela. *In* The African Honey Bee. M. Spivak and M. D. Breed, *eds.* Westview Press, Boulder, CO.

Hertig, M. (1923). The normal and pathological histology of the ventriculus of the honey bee, with special reference to infection with *Nosema apis. Jour. Parasitol. 9*:109-140.

Hitchcock, J.D. (1966). Transmission of sacbrood disease to individual honey bee larvae. *Jour. Econ. Entomol. 59*:1154-1156.

Holst, E.C. (1946). A single field test for American foulbrood. *Amer. Bee Jour. 86*:14,34.

Howard, L.O. (1907). Report of the meeting of inspectors of apiaries San Antonio, Texas, Nov. 12, 1906. *USDA Bur. Entomol. Bull. 61.*

Huber, F. (1814). New Observations Upon Bees. Trans. C.P. Dadant. *Amer. Bee Jour.,* Hamilton, Ill.

Jay, E.G., G.E. Cantwell, G.C. Pearman, Jr. and J.V. Thompson. (1972). Control of the greater wax moth, *Galleria mellonella* (Linnaeus) in comb honey with carbon dioxide. Part II. Field studies. *Amer. Bee Jour. 112*:342,344.

Jay, S.C. and D. Dixon. (1982). Nosema disease in package honey bees, queens, and attendant workers shipped to western Canada. *Jour. Apic. Res. 21*:216-221.

Jaycox, E.R. (1960). Surveys for nosema disease of honey bees in California. *Jour. Econ. Entomol. 53*:95-98.

Johansen, C. (1962). Impregnated foundation for waxmoth control. *Glean. Bee Cult. 90*:682-684.

Johnsen, P. (1951). More about amoeba disease, and an appeal (in Danish). *Tidsskr. Biavl 85*:120-121.

Jordan, R. (1936). The course of amoeba and nosema disease with only one of the pasrasites in the colony (in German). *Deutsche Imker 49*:152-158.

Kang, Y.B., D.S. Kim and D.H. Jang. (1976). Experimental studies on the pathogenicity and developmental stages of *Nosema apis. Korean Jour. Veterin. Res. 16*:11-25.

Katznelson, H., J. Arnott and S.E. Bland. (1952). Preliminary report on the treatment of European foulbrood of honey bees with antibiotics. *Scien. Agric. 32*:180-184.

Katznelson, H. and C.A. Jamieson. (1952a). Antibiotics and other chemotherapeutic agents in the control bee diseases. *Scien. Agric. 32*:219-225.

Katznelson, H. and C.A. Jamieson. (1952b). Control of nosema disease of honey bees with fumagillin. *Science. 115*:70-71.

Kauffeld, N.M. (1973). Queen honey bees in colonies: susceptibility to nosema disease. *Amer. Bee Jour. 113*:12-14.

Kauffeld, N.M., J.L. Williams, T. Lehnert and F.E. Moeller. (1972). Nosema control in package bee production—fumigation with ethylene oxide and feeding with fumagillin. *Amer. Bee Jour. 112*:297-299, 301.

Kellner, N. and F.J. Jacobs. (1978). How long do the spores of *Nosema apis* take to reach the ventriculus of the honey bee? *Vlaams Diegeneeskd Tijdschr. 47*:252-259.

Knutson, L. (1978). Insects: Diptera (flies). *In* Honey Bee Pests, Predators and Diseases. R.A. Morse, *ed.* Cornell Univ. Press, Ithaca, New York, pp. 128-137.

Koeniger, N. (1990). Personal Communication.

Koeniger, N. and S. Fuchs. (1989). Eleven years with Varroa—experiences, retrospects, and prospects. *Bee World 70*:148-159.

Koeniger, N., G. Koeniger, and M. Delfinado-Baker. (1983). Observations on mites of the Asian honeybee species *(Apis cerana, Apis dorsata, Apis florea). Apidologie 14*:197-204.

Kramer, J.P. (1960). Observation on the emergence of the microsporidian sporoplasm. *Jour. Insect Pathol. 2*:433-439.

Kudo, R. (1920). Notes on *Nosema apis* (Zander). *Jour. Parasitol. 7*:85-90.

Kulikov, N.S. (1961). Treating bees with fumagillin for nosema infection (in Russian). *Pchelovodstvo 38*(5):43-44.

Kulincevic, J.M. and W.C. Rothenbuhler. (1975). Selection for resistance and susceptibility to hairless-black syndrome in the honey bee. *Jour. Invert. Pathol. 25*:289-295.

Langridge, D.F. (1961). Nosema disease of the honeybee and some investigations into its control in Victoria, Australia. *Bee World. 42*:36-40.

L'Arrivee, J.C.M. and R. Hrytsak. (1964). Coprological examination for nosematosis in queen bees. *Jour. Insect Pathol. 6*:126-127.

Lee, P.E. and B. Furgala. (1965). Chronic bee paralysis. *Jour. Invert. Pathol. 7*:170-174.

Lee, P.E. and B. Furgala. (1967a). Electron microscopic observations on the localization and development of sacbrood virus. *Jour. Invert. Pathol. 9*:178-187.

Lee, P.E. and B. Furgala. (1967b). Viruslike particles in adult honey bees (*Apis mellifera* Linneaus) following injection with sacbrood virus. *Virol. 32*:11-17.

Lehnert, T., A.S. Michael and M.D. Levin. (1974). Disease survey of South American Africanized bees. *Amer. Bee Jour. 114*:338.

Lehnert, T. and H. Shimanuki. (1973a). The dosage of ethylene oxide necessary to control *Nosema apis* Zander on honey combs at 100°F. *Amer. Bee Jour. 113*:296.

Lehnert, T. and H. Shimanuki. (1973b). Production of nosema-free bees in the South. *Amer. Bee Jour. 113*:381-382.

Lehnert, T., H. Shimanuki and D.A. Knox. (1973). Transmission of nosema disease from infected workers of the honey bee to queens in mailing cages. *Amer. Bee Jour. 113*:313-314.

Leyrer, R.L. and R.F. Monroe. (1973). Isolation and identification of the scent of the moth *Galleria mellonella*, and a reevaluation of its sex pheromone. *Jour. Insect Physiol. 19*:2267-2271.

Liu, T.P. (1973). Effects of Fumidil B on the spores of *Nosema apis* and on lipids of the host cell as revealed by freeze-etching. *Jour. Invert. Pathol. 22*:364-368.

Liu, T.P. (1984). Ultrastructure of the midgut of the worker honey bee *Apis mellifera* heavily infected with *Nosema apis. Jour. Invert. Pathol. 44*:282-291.

Lochhead, A.G. (1937). The nitrate reduction test and its significance in the detection of *Bacillus larvae. Canadian Jour. Res. 15*:79-86.

Lord, W.G. and J.T. Ambrose. (1981). Bear depredation of bee hives in the United States and Canada. *Amer. Bee Jour. 121*:811-815.

Lotmar, R. (1946). Concerning the flagellates and bacteria in the small intestine of the honey bee (in German). *Schweiz. Bienen Ztg., Beih. 2*:49-76.

Lundie, A.E. (1940). The small hive beetle, *Aethina tumida. Sci. Bul 220,* Ent. Series 3, South Africa Dept. Agric. and Forestry.

Maassen, A. (1916). On bee diseases (in German). *Mitt. aus der Kaiserlichen Biol. Anstalt fur Land- und Forstwirtschaft 16*:51-58.

Marston, N. and B. Campbell. (1973). Comparison of nine diets for rearing *Galleria mellonella. Ann Entomol. Soc. Amer. 66*:132-136.

Mayer, D.F., R.D. Akre, A.L. Antonelli, and D.M. Burgett. (1987). Protecting honey bees from yellowjackets. *Amer. Bee Jour. 127*:693.

Michael, A.S. (1957). Droplet method for observation of living unstained bacteria. *Jour. Bacteriology* 74:831-832.

Michael, A.S. (1960). A new technique for studying bee diseases. XVII Internat. Beekeeping Cong., Bologna-Roma, 1958, Official Rept., 2nd Vol., pp. 73-77. Imola Tipografia Galeati, 1960.

Michael, A.S. (1964). Ethylene oxide. A fumigant for control of pests and parasites of the honey bee. *Glean. Bee Cult.* 92:102-104.

Michael, A.S., D.A. Knox and H. Shimanuki. (1990). U.S. and Federal and State laws relating to honey bees, Appendix 3. *In* Honey Bee Pests, Predators, and Diseases, Second Edition. R.A. Morse and R. Nowogrodzki, eds. Cornell Univ. Press, Ithaca, NY, pp. 375-382.

Miller, C.C. (1931). A Thousand Answers to Beekeeping Questions. 6th ed. *Amer. Bee Jour.*, Hamilton, Ill.

Milum, V.G. (1935). Lesser vs. greater wax moth. *Glean. Bee Cult.* 63:662-666.

Milum, V.G. (1940). Moth pests of honeybee combs. *Glean. Bee Cult.* 68:424-428.

Moeller, F. (1967). A study of the incidence of *Nosema* infection in overwintered colonies in Wisconsin. XXI Intern. Beekeep. Congr. Prelim. Sci. Meet. Summary Paper No. 47, p. 489.

Moeller, F. (1972). Effects of emerging bees and of winter flights on nosema disease in honeybee colonies. *Jour. Apic. Res.* 11:117-120.

Moffett, J.O., J.J. Lackett and J.D. Hitchcock. (1969). Compounds tested for control of nosema in honey bees. *Jour. Econ. Entomol.* 62:886-889.

Moffett, J.O. and W.T. Wilson. (1971). The viability and infectivity of frozen nosema spores. *Amer. Bee Jour.* 111:55, 70.

Morgenthaler, O. (1920). Bee diseases in 1919 (in German). *Schweiz. Bienen-Ztg.* 43:146-154.

Morgenthaler, O. (1959). Fifty years of *Nosema apis* Zander (in German). *Sudwest-deutscher Imker* 11:166-172.

Morse, R.A. (1975). Bees and Beekeeping. Cornell Univ. Press, Ithaca, New York.

Morse, R.A. (1978a). Comb Honey Production. Wicwas Press, Ithaca, New York.

Morse, R.A. (1978b). Amphibians (Frogs and toads). *In* Honey Bee Pests, Predators and Diseases. R.A. Morse, *ed.* Cornell Univ. Press, Ithaca, New York, pp. 210-214.

Mullen, M.A. and C.H. Tsao. (1971). Morphology of the tympanic organ of the greater wax moth, *Galleria mellonella. Jour. Georgia Entomol. Soc.* 6:124-132.

Mussen, E.C., B. Furgala and R.A. Hyser. (1975). Enzootic levels of nosema disease in the continental United States (1974). *Amer. Bee Jour.* 115:48-50, 58.

Nielsen, R.A. and D. Brister. (1977). The greater wax moth: adult behavior. *Ann. Entomol. Soc. Amer.* 70:101-103.

Nielsen, R.A. and C.D. Brister. (1979). Greater wax moth: behavior of larvae. *Ann Entomol. Soc. Amer.* 72:811-815.

Nixon, M. (1982). Preliminary world maps of honey bee diseases and parasites. *Bee World* 63:23-42.

Okumura, G.T. (1966). The dried-fruit moth (*Vitula edmandsae serratilineella* Ragonot), pest of dried fruit and honeycombs. *Bull. Calif. Dept. Agric.* 55:180-186.

Otte, E. (1973). A contribution of the laboratory diagnosis of American foulbrood of the honey bee with a particular reference to the immunoflourescence method. *Apidologie* 4:331-339.

Oudemans, A.C. (1904). On a new genus and species of parasitic Acari. *Notes Leyden Mus.* 24:216-222.

Paddock, F.B. (1918). The beemoth or waxworm. *Tex. Agric. Exp. Stn. Bull. Number 231.*

Palmer-Jones, T. (1949). Diseases of bees in New Zealand. *New Zealand Jour. Agric.* 79:483-486.

Payne, T.L. and W.E. Finn. (1977). Pheromone receptor system in the females of the greater wax moth *Galleria mellonella. Jour. Insect Physiol. 23*:879-881.

Peng, Y-S. and M.E. Nasr. (1985). Detection of honey bee tracheal mites *(Acarapis woodi)* by simple staining techniques. *Jour. Invert. Pathol. 46*:325-331.

Peng, Y-S. and K-Y Peng. (1979). A study on the possible utilization of immunodiffusion and immunofluorescence techniques as the diagnostic for American foulbrood of honeybees *(Apis mellifera). Jour. Invert. Pathol. 33*:284-289.

Peng, Y-S. (1988). The resistance mechanism of the Asian honey bee *(Apis cerana)* to the mite *Varroa jacobsoni. In* Africanized Honey Bees and Bee Mites. Needham *et al.* eds. Ellis Horwood Ltd. West Sussex, pp. 426-429.

Pinnock, D.E. and N.E. Featherstone. (1984). Detection and quantification of *Melissococcus pluton* infection in honey bee colonies by means of enzyme-linked immunosorbent assay. *Jour. Apic. Res. 23*:168-170.

Poltev, V.I. (1953). Effect of nosema and amoeba infections on the productivity of honey bee colonies in the Primorsky region (in Russian). *Pchelovodstvo 30(2):*46-48.

Poteikina, E.A. (1960). Physiological state of bees in the fall and longevity of bees infected with nosema (in Russian). *Pchelovodstvo 37(9)*:18-19.

Prell, H. (1926). The amoeba-disease of adult bees: a little-noticed springtime disease. *Bee World 8*:10-13.

Ragsdale, D.W. and B. Furgala. (1987). A serological approach to the detection of *Acarapis woodi* parasitism in honey bees using an enzyme-linked immunosorbent assay. *Apidologie 18*:1-9.

Ragsdale, D.W. and K.M. Kjer. (1989). Diagnosis of tracheal mite *(Acarapis woodi* Rennie) parasitism of honey bees using a monoclonal based enzyme-linked immunosorbent assay. *Amer. Bee Jour. 129*:550-553.

Rennie, J. (1921). Acarine disease: The organism associated with the disease, *Tarsonemus woodi. In* Isle of Wight disease in Hive Bees. *Trans. Roy. Soc. Edin. 52, pt. 4(29)*:737-779.

Revell, I.L. (1960). Longevity of refrigerated nosema spores. *Jour. Econ. Entomol. 53*:1132-1133.

Richards, K.W. (1984). Food preference, growth, and development of larvae of the driedfruit moth, *Vitula edmandsae serratilineella* Ragonot (Lepidoptera: Pyralidae). *Jour. Kan. Entomol. Soc. 57*:28-33.

Riley, D. (1984). APHIS Communication.

Rinderer, T.E. and T.J. Green. (1976). Serological relationship between chronic bee paralysis virus and the virus causing hairless-black syndrome in the honey bee. *Jour Invert. Pathol. 27*:403-405.

Rinderer, T.E. and W.C. Rothenbuhler. (1969). Resistance to American foulbrood in honey bees. X. Comparative morality of queen, worker, and drone larvae. *Jour Invert. Pathol. 13*:81-86.

Rinderer, T.E., W.C. Rothenbuhler and J.M. Kulincevic. (1975). Responses of three genetically different stocks of the honey bee to a virus from bees with hairless-black syndrome. *Jour Invert. Pathol. 25*:297-300.

Ritter, W. (1981). Varroa disease of the honeybee *Apis mellifera. Bee World 62*:141-153.

Ritter, W. and F. Ruttner. (1980). Diagnoseverfahren *(Varroa). Allgemeine deutsche Imkerzeitung (5)*:134-138.

Root, A.I. and E.R. Root. (1913). *The ABC and XYZ of Beekeeping.* The A.I. Root Co., Medina, Ohio.

Rose, R.I. and J.D. Briggs. (1969). Resistance to American foulbrood in honey bees. IX. Effects of honey-bee larval food on the growth and viability of *Bacillus larvae. Jour Invert. Pathol. 13*:74-80.

Rothenbuhler, W.C. and V.C. Thompson. (1956). Resistance to American foulbrood in honey bees. I. Differential survival of larvae of different genetic lines. *Jour. Econ. Entomol. 49*:470-475.

Ruttner, F. (1987). Biogeography and taxonomy of honey bees. Springer-Verlag: Heidelberg.

Schiller, J. (1937). Two lethal cases of pure amoeba infection (in German). *Bienen Vater 69*:382-384.

Schmidt, S.P. and R.E. Monroe. (1976). Biosynthesis of the wax moth sex attractants. *Insect Biochem. 6*:377-380.

Schneider, P. and W. Drescher. (1987). The influence of *Varroa jacobsoni* Oud. on weight, development of weight and hypopharyngeal glands, and longevity of *Apis mellifera* L. *Apidologie 18*:101-110.

Shimamori, K. (1987). On the biology of *Apanteles galleriae*, a parasite of the two species of wax moths (in Japanese with English summary). *Honeybee Sci. 8*:107-112.

Shimanuki, H. (1967). Ethylene oxide and control of American foulbrood: a progress report. *Amer. Bee Jour. 107*:290-291.

Shimanuki, H. (1981). Controlling the greater wax moth—a pest of honeycombs. USDA Farmers' Bull. No. 2217.

Shimanuki, H., E.W. Herbert, Jr. and D.A. Knox (1984). High velocity electron beams for bee disease control. *Amer. Bee Jour. 124*:865-867.

Shimanuki, H. and D.A. Knox. (1988). Improved method for the detection of *Bacillus larvae* spores in honey. *Amer. Bee Jour. 128*:353-354.

Shimanuki, H. and D.A. Knox. (1989). Tracheal mite surveys. *Amer. Bee Jour. 129*:671-672.

Shimanuki, H., T. Lehnert and D.A. Knox. (1973). Transmission of nosema disease from infected honey bee workers to queens in mating nuclei. *Jour. Econ. Entomol. 66*:777-778.

Singh, S. (1962). Beekeeping in India. Indian Council Agric. Res., New Delhi.

Smith, I.B., Jr. and D.M. Caron. (1984a). Distribution of the bee louse *Braula coeca* worldwide and in the U.S. *Amer. Bee Jour. 125*:294-296.

Smith, I.B., Jr. and D.M. Caron. (1984b). Caste preference and colony distribution of the bee louse *Braula coeca* Nitzsch on its host *Apis mellifera* L. *Jour. Apic. Res. 23*:171-176.

Spangler, H.G. (1985). Sound production and communication in the greater wax moth (Lepidoptera: Pyralidae). *Ann. Entomol. Soc. Amer. 78*:54-61.

Spangler, H.G. (1987). Acoustically mediated pheromone release in the greater wax moth (Lepidoptera: Pyralidae). *Jour. Insect Physiol. 33*:465-468.

Spangler, H.G. (1988). Look and Listen. *Glean. Bee Cult. 116*:500-513.

Steche, W. (1960). Etiology and therapy of nosema disease of the honey bee (in German). *Ztschr. F. Bienenforsch. 5*:49-92.

Steche, W. and T. Held. (1981). Scanning electron microscope study of the ontogenosis of *Nosema apis. Apidologie 12*:185-207.

Steinhaus, E.A. (1951). Report on diagnosis of diseased insects 1944-50. *Hilgardia 20*:629-678.

Steinhaus, E.A. (1963). Insect pathology. Academic Press, New York.

Stejskal, M. (1958). Correlations between bee diseases and atmospheric conditions in Venezuela. *Internatl. Beekeeping Cong. 17*:112.

Studier, H. (1958). The sterilization of American foulbrood by irradiation with gamma rays. *Amer. Bee Jour. 98*:192.

Sturtevant, A.P. (1932). Relation of commercial honey to the spread of American foulbrood. *Jour. Agric. Res. 45*:257-285.

Sturtevant, A.P. (1936). Quantitative demonstration of the presence of spores of *Bacillus larvae* in honey contaminated by contact with American foulbrood. *Jour. Agric. Res. 52*:697-704.

Sturtevant, A.P. and I.L. Revell. (1953). Reduction of *Bacillus larvae* spores in liquid food of honey bees by action of the honey stopper, and its relation to the development of American foulbrood. *Jour. Econ. Entomol. 46*:855-860.

Szabo, T. (1989). The capping scratcher: A tool for detection and control of *Varroa jacobsoni. Amer. Bee Jour. 129*:402-403.

Szabo, T. and D.T. Heikel. (1987a). Fumigation with SO_2 to control dried-fruit moth in honeybee combs. *Bee World 68*:37-38.

Szabo, T. and D.T. Heikel. (1987b). Effect of dry fumagillin feeding on spring *Nosema* spores counts in overwintered colonies. *Amer. Bee Jour. 127*:210-211.

Tarr, H.L.A. (1938). Studies on European foulbrood of bees. IV. On the attempted cultivation of *Bacillus pluton*, the susceptibility of individual larvae to inoculation with the organism and its localization within its host. *Ann. Appl. Biol. 25*:815-821.

Taylor, R. (1988). Bee talk. *Glean. Bee Cult. 116*:299-300.

Thompson, V.C. and W.C. Rothenbuhler. (1957). Resistance to American foulbrood in honey bees. II. Differential protection of larvae by adults of different genetic lines. *Jour. Econ. Entomol. 50*:731-737.

Toschkov, A., T. Vallerianov, and A. Tomov. (1970). The fluorescent-immunization method, and the rapid and specific diagnosis of American foulbrood in the bee brood (in German). *Bull. Apicole 13*:13-18.

Tucker, K.W. (1978). Abnormalities and noninfectious diseases. *In* Honey Bee Pests, Predators and Diseases. R.A. Morse, *ed.* Cornell Univ. Press, Ithaca, New York, pp. 257-274.

Waller, G.D. (1990). Personal communication.

Wang, Der-I. (1969). The effects of behavior, amino acids of the hemolymph, and development of hypopharyngeal glands of nosema-diseased worker honey bees, *Apis mellifera* L., on the ability of queens to escape infection by *Nosema apis* Zander. Univ. Wisc. Thesis.

Wang, Der-I. and F.E. Moeller. (1970). The division of labor and queen attendance behavior of nosema-infected worker honey bees. *Jour. Econ. Entomol. 63*:1539-1541.

Wang, Der-I. and F.E. Moeller. (1971). Ultrastructural changes in the hypopharyngeal glands of worker honey bees infected by *Nosema apis. Jour. Invert. Pathol. 17*:308-320.

White, G.F. (1913). Sacbrood, a disease of bees. USDA Bur. Entomol. Circ. 169.

White, G.F. (1917). Sacbrood. USDA Bull. 431.

White, G.F. (1919). Nosema Disease. USDA Bull. 780.

Williams, J.L. (1973). Fumagillin-treated extender patties ineffective for *Nosema* control in nuclei. *Amer. Bee Jour. 113*:58-59.

Williams, J.L. (1976). Status of the greater wax moth, *Galleria mellonella* (L.), in the United States beekeeping industry. *Amer. Bee Jour. 116*:524-526.

Williams, J.L. (1978). Insects: Lepidoptera (Moths). *In* Honey Bee Pests, Predators, and Diseases. R.A. Morse, *ed.* Cornell Univ. Press, Ithaca New York, pp. 105-127.

Wilson, C.A. and L.L. Ellis. (1966). A new technique for the detection of nosema in apiaries. *Amer. Bee Jour. 106*:131.

Wilson, W.T. (1970). Inoculation of the pupal honeybee with spores of *Bacillus larvae. Jour. Apic. Res. 9*:33-37.

Wilson, W.T. and J.W. Brewer. (1974). Beekeeping in the Rocky Mountain Region. Col. St. Univ. Coop. Ext. Serv. WRP-12.

Wilson, W.T. and J.O. Moffett, R.L. Cox, D.L. Maki, H. Richardson and R. Rivera. (1988). Menthol treatment for *Acarapis woodi* control in *Apis mellifera* and the resulting residues in honey. *In* Africanized Honey Bees and Bee Mites. Needham *et al. eds.* Ellis Horwood Ltd. West Sussex, pp. 535-540

Wilson, W.T. and R. Nunamaker. (1982). The infestation of honey bees in Mexico with *Acarapis woodi. Amer. Bee Jour. 122*:503-505,508.

Wilson, W.T. and W.C. Rothenbuhler. (1968). Resistance to American foulbrood in honey bees. VIII. Effects of injecting *Bacillus larvae* spores into adults. *Jour. Invert. Pathol. 12*:418-424.

Winston, M.L., G.G. Grant, K. Slessor and J. Corner. (1981). The moth *Vitula edmandsae:* a pest of honeybee combs. *Bee World. 62*:108-110.

Woodrow, A.W. (1941a). Susceptibility of honeybee larvae to American foulbrood. *Glean. Bee Cult. 69*:148-151,190.

Woodrow, A.W. (1941b). Behavior of honeybees toward brood infected with American foulbrood. *Amer. Bee Jour. 81*:363-366.

Woodrow, A.W. (1942). Susceptibility of honeybee larvae to individual inoculations with spores of *Bacillus larvae. Jour. Econ. Entomol. 35*:892-895.

Woyke, J. (1984). Increases in life-span, unit honey productivity and honey surplus with fumagillin treatment of honeybees. *Jour. Apic. Res. 23*:209-212.

Woyke, J. (1985). *Tropilaelaps clareae,* a serious pest of *Apis mellifera* in the tropics, but not dangerous for apiculture in temperate zones. *Amer. Bee Jour. 125*:497-499.

Wyborn, M.H. and D.M. McCutcheon. (1987). A comparison of dry and wet fumagillin treatments for spring *Nosema* disease suppression of overwintered colonies. *Amer. Bee Jour. 127*:207-209.

Zhavnenko, V.M. (1971). Indirect method of immunofluorescence in the diagnosis of foulbrood (American and European) (in Russian). *Veterinariya 1971(8)*:109-111.

CHAPTER 26

INJURY TO HONEY BEES BY POISONING

by E. Laurence Atkins*

Controlling pests of agricultural crops is as important as good seed, sufficient rain or water of high quality, intelligent use of fertilizers, proper cultivation and control of weeds for economical production of an adequate supply of food and fiber for increasing numbers of people throughout the world. Practically every species of plant has disease, insect or weed pests which significantly influence its growth and production. The demand for an increasing amount of food and fiber has necessitated enlarging the size of farm and ranch production units, utilizing more land for growing crops and accelerating the mechanization of agriculture. The increased area of land in crops and the larger fields of a single crop favors the development of insect, mite and disease problems. This, in turn, often requires that more pesticide applications be made. Agricultural chemicals or pesticides for specific purposes are utilized to alleviate the countless pests of plants. Therefore, individual problems are controlled with chemicals for specific uses such as acaricides, antibiotics, chemosterilants, defoliants, dessicants, fungicides, herbicides, insect growth regulators, insecticides, nematocides and plant growth regulators. Unfortunately, the honey bee is susceptible to many of the pesticides used in such an intensive pest control program. As a result, the honey bee is subjected to an intensive and continuous hazard of chemical poisoning that may overshadow all other problems including bee diseases (see Fig. 1).

Figure 1. Portion of apiary, at edge of citrus grove, killed by pesticide application to blooming citrus tree. Only three colonies out of 120 survived. (*Photo by E.L. Atkins*).

---

* E. Laurence Atkins, B.S., M.S.; R.P.E.; Specialist in Entomology and Apiology, University of California, Riverside.

The increased area for crops has promoted the development of agricultural aviation and the improvement in land machines for the application of pesticides so that more area can be treated with pesticides in a relatively short time (see Fig. 2).

FIGURE 2. Modern fixed wing biplane and helicopter equipped for efficient and fast spraying of agricultural crops. Spraying citrus trees. (*Photo by E.L. Atkins*).

The practice of chemical weed control, the elimination of fences and open ditches has eliminated many of the solitary bees and bumble bees that once nested in such sites and served to pollinate the family orchard or crops on smaller farms and gardens. Most species of solitary bees are more susceptible to pesticides than is the honey bee. Consequently, the honey bee has become more important as a pollinator and is the only pollinator which can be increased easily and which can be moved quickly and in the desired numbers to affect the pollination of cultivated crops and trees. Nevertheless, the beekeeping industry is having a increasingly difficult time in maintaining an adequate number of honey bee colonies in the intensively cultivated areas because of losses from pesticide poisoning.

POISONING FROM PESTICIDES

In collecting nectar and pollen from flowers, honey bees contact pesticides deposited upon the plants. The first sign of poisoning from pesticides is the appearance of large numbers of dead or dying bees at the colony entrances throughout the apiary. Many pesticides are capable of killing larvae (brood) in all stages as well as adult bees. Most often pesticides kill the field bees (foragers) without other serious effects on the colony. In some instances bees die in large numbers after returning to the hive. Many bees are lost in the field and between the treated field and the colony. The colony is weakened but not usually killed. In extreme instances pesticides are carried by the foragers to the hive where they may kill brood and young workers in the colony. The entire population of the colony may die when this happens.

Honey bees may be killed by three distinct actions by pesticides—by direct contact, by stomach poison and by fumigation. Some pesticides kill by only one action while others kill by various combinations of two actions and some kill by all three actions. Contact poisons are absorbed by bees through the integument; stomach poisons are absorbed through the alimentary canal when taken internally through feeding or cleaning activities; and, fumigants are absorbed through the spiracles or respiratory system.

Once a pesticide has gained entrance into the adult bee several modes of action may take place. The pesticide may affect only the alimentary tract. The alimentary tract may be paralyzed or physically altered making it impossible for the adult to nourish itself and hence to function normally. The abdomen is frequently distended. This mode of action results in the adult starving or desiccating causing death. More commonly, however, the mode of action caused by organic pesticides used today affects the nervous system in various ways so that the legs, wings, digestive tract, etc., will not function and the adult is therefore unable to orient itself to find food, to replenish food and water or is unable to utilize food and water and hence indirectly dies from starvation and desiccation. Bees which are without food will become weakened after 3 to 4 hours and will succumb to starvation after 6 to 8 hours.

Typical Pesticide Poisoning Symptoms:

A. Organophosphorous pesticides: EXAMPLES: parathion, methyl parathion, fenitrothion (Accothion), dimethoate (Cygon, DE-FEND), naled (Dibrom), methamidophos (Monitor), methidathion (Supracide), mevinphos (Phosdrin), diazinon, oxydemeton-methyl (Metasystox-R), dichlorvos (DDVP, Vapona), monocrotophos (Azodrin), malathion, phorate (Thimet), phosphamidon (Dimecron), tepp, chlorpyrifos (Dursban, Lorsban), profenofos (Curacron), disulfoton (Di-Syston), etc. **TYPICAL SYMPTOMS:** regurgitation is common (bees are wet); bees are often disoriented, lackadaisical and remain at the hive awaiting paralysis and death. Bees may also exhibit other symptoms such as erratic attempts of cleaning themselves, tumbling about; wings held away from body but usually remaining hooked together; may also have distended abdomen. Queens may cease egg laying. A high percentage of poisoned bees die at the colony.

B. Carbamate Pesticides: EXAMPLES: carbaryl (Sevin), carbofuron (Furadan), carbosulfan (Advantage), aminocarb (Matacil), dimetilan (Dimetilane), mexacarbate, methomyl (Lannate, Nudrin), thiodicarb (Larvin), formetanate hydrochloride (Carzol), etc. **TYPICAL SYMPTOMS:** aggressiveness, erratic movements; then unable to fly; stupefaction as though they had been chilled, followed by paralysis, moribundity and death. Queens often cease egg laying and hive bees often initiate supersedure queen cells before egg laying resumes. Most poisoned bees die at the colony.

C. Pyrethroid, Pyrethrum and Botanical Pesticides: EXAMPLES: Pyrethroids: allethrin (Pynamin), bifenthrin (Brigade, Capture), cyfluthrin (Baythroid), cyhalothrin (Karate), cypermethrin (Ammo, Cymbush), deltamethrin (Decis), esfenvalerate (Asana), fenpropathrin (Danitol), fenvalerate (Pydrin), flucythrinate (Pay-Off), fluvalinate (Mavric, Spur), permethrin (Ambush, Pounce), resmethrin (Synthrin), tetramethrin (Phthalthrin), tralomethrin (Scout), etc.

Botanicals: Pyrethrums: **EXAMPLES:** pyrethrum I, pyrethrum II, Cinerin, I, Cinerin II, jasmolin II, etc. Other Botanicals: azadiractin (neem seed oil, Margosan-O), nicotine, rotenone (cube, derris), ryania, sabadilla, etc. **TYPICAL SYMPTOMS:** direct contact exposure may cause regurgitation from the highly toxic pyrethroids, together with erratic movements; then unable to fly and stupefaction followed by paralysis, moribundity and death within a very short period of time. Some poisoned bees die in the foraging area, others die between the foraging area and the hive, and, the remainder die at the hive. The remainder of bees compounds are nontoxic to honey bees at normal field dosages if application is made when bees are not foraging.

D. Bacterial Pesticides: EXAMPLES: *Bacillus thuringiensis* Berliner. There are several strains and varieties such as *aizawai, israelensis* and *kurstaki, tenebrionis* and many trade names. This bacteria produces a crystalline toxin in the manufacturing fermentation process which is toxic to certain insects. The preparations which contain the endotoxin have been nontoxic to adult honey bees. The strain containing the exotoxin *(thuringiensin)* was toxic as a simple poison to adult bees and bee brood, however, it was nontoxic at dosages which are recommended for field applications.

E. Chlorinated Hydrocarbon Pesticides; EXAMPLES: aldrin, chlordane, DDT, dicofol, dieldrin, endrin, endosulfan, heptachlor, lindane, toxaphene, etc. **TYPICAL SYMPTOMS:** erratic movements, abnormal activities, trembling, dragging the hind legs as if paralyzed and wings held away from the body but usually remaining hooked together. In spite of severe symptoms many bees are able to fly to the field until shortly before morbidity occurs; therefore, a high percentage of poisoned bees die in the field and between the field and colony as well as at the colony.

F. Dinitrophenyl Pesticides: EXAMPLES: binapacryl (Morocide), dinobuton (Dessin), dinocap (Karathane), dinitrocresol, etc. **TYPICAL SYMPTOMS:** similar to those of the chlorinated hydrocarbons, but often accompanied by the regurgitation of substances in the digestive tract as is typical of the organophosphorous pesticides. Most of the affected bees die at the colony. Very few of the dinitro compounds currently being used are toxic to bees at field dosages.

G. Plant Growth Regulators (PGR): Plant growth regulators are chemicals which in minute amounts alter the behavior of plants and/or their produce through physiological (hormonal) action. Some promote plant rooting, prolong or break dormancy, speed up or retard growth, inhibit bloom or fruit set, etc. Following are several specific types, both natural and synthetic and examples such as:

1. Auxins, as indole acetic acid; 2,4-D, isopropyl ester; MCPB, BNOA.
2. Gibberellins, as gibberellic acid.
3. Cytokinins, as zeatin, kinetin, adenine.
4. Ethelene generators, as ethylene thiuram sulfide (ETHISUL; VEGETTA), ethephon (ETHREL).
5. Inhibiters, as benzoic acid, gallic acid, cinnamic acid, maleic hydrazide (MH), diniconazole (Spotless).
6. Retardants, as beta-hydroxy ethyl hydrazine, beta-napthoxy acetic acid, chlorophenoxy propionic acid, 4-CPA, chlormequat-chloride (CYCOCEL), gibberellic acid 1-napthaleneacetic acid, N-metatolylphthalamic acid potassium salt of maleic hydrazide (RETARD; SUCKER STUFF), chlorpropham (SPROUT NIP).

This wide variety of chemical types are used at very low dosages and all of those which have been tested are essentially nontoxic at the dosages utilized to regulate plants.

H. Insect Growth Regulators (IGR): Insect growth regulators are chemicals, both natural and synthetic, which disrupt the action of insect hormones controlling moulting, metamorphosis from pupa to adult, maturation of eggs, sterility of males; modify or disrupt mating behavior; deter feeding; attract to or repel individuals from specific areas; and, infect the arthropods with diseases derived from a variety of sources.

Following are several types of insect and mite (anthropod) growth regulators, both natural and synthetic, and samples such as:

1. Insect growth regulators as FARNESOL, insect hormone; diflubenzuron (DIMILIN); methoprene (ALTOSID) and teflubenzuron (NOMOLT), juvenile hormones.

2. Sex pheromones, communication disruptants, as multi-methylalkenols (STIRRUP), mite pheromone; amlure, insect attractant; ATTRACT'N KILL BORER-Guard, peach tree borer; ATTRACT'N KILL WEEVIL, boll weevil; ATTRACT'N KILL ROOTWORM, corn rootworm; ATTRACT'N KILL HELIOTHIS, cornearworm; ATTRACT'N KILL PBW, pink bollworm; and, ATTRACT'N KILL TOMATO-PINWORM, tomato pinworm.

3. Chemosterilants as beta-asarone (from plant source), hempa, methiotepa and tepa.

4. Insect attractants and repellents as benzyl acetate, melon fly; benzyl benzoate, acaricide and repellent; MEDLURE, attractant for Mediteranean fruit fly; methyl eugenal, food lure for oriental fruit fly.

5. Feeding deterrent as BOLLEX, weevil feeding deterrent.

6. Anthropod diseases such as:
   a. Bacterial insect diseases as *Bacillus popilliae,* milky disease spores, selective for controlling certain beetle larvae.
   b. Fungal diseases for arthropods as *Hirsutella Thompsonii,* citrus red mite.
   c. Granulosis virus diseases for insects as DECYDE, codling moth.
   d. Nuclear polyhedrosis virus diseases for insects as *Heliothis zea* virus (BIOTROL VHZ), corn earworm; *Neodeprion sertifer,* larvicide for European pine sawfly.
   e. Microsporidian (protozoan) disease as *nosema locustae* Canning that significantly reduces grasshoppers and Mormon crickets when

applied in a wheat bran mixture, whereas, honey bees and beneficial organisms are not affected.

This wide variety of chemical and source types are researched extensively
to determine if insect hormones and hormone mimics can disrupt normal
insect development before the insects become adults thus preventing reproduction as a method of insect control. The compounds tested thus far are used
at very low dosages and all of those which have been tested are essentially
nontoxic to adult honey bees at the dosages utilized to regulate arthropods;
however, we have little information concerning their effect on honey bee eggs,
larvae and pupae (brood).

In summary, pesticide poisoning may be suspected especially from heavy
kills at every colony in the apiary. Pesticide kills may also be suspected when
weather has been warm, dry and pleasant for any length of time and the
colonies are located in blooming crop and forage areas. The visual determination of pesticide poisoning of honey bees is essentially a matter of frequent
observations and long experience, just as is the visual determination of adult
bee diseases.

LITTLE DANGER OF HONEY BEING POISONED

When one considers the large quantities of pesticides toxic to honey bees
being applied to orchards, to cultivated truck and field crops, to pastures and
to forage crops, one might suppose that honey would contain some of the
poisons. This is seldom the case. In order to protect human health and welfare,
Federal and State Regulations have been enacted which regulate the sale and
application of a majority of the pesticides that are highly toxic to man or to
domestic animals. Under these regulations all pesticides have to be registered
by federal government before they can be offered for sale and used. Before
they can be registered, data have to be acquired and be presented by the
chemical company showing that each pesticide is effective for the intended use
and is not injurious to plants or animals when applied as directed on the label.
State and county regulations have been enacted in many instances to regulate
the method and timing of the application of the chemical in order to minimize
drift and possible injury caused by the residue. All of these regulations tend to
afford some protection to honey bees and to reduce the hazards of pesticide
contamination of marketable food products.

The anatomy, physiology and behavior of the honey bee is such that when
bees become poisoned they do not behave in a normal manner. If worker bees
are injured physically, they soon lose their sense of orientation and/or their
ability to fly and usually die away from the hive. Data obtained from dead bee
traps attached to the colonies of bees show that 90 percent of bees dying of old
age die away from the hive. This also happens in varying degree when bees are
poisoned by different toxic pesticides at some distance from their hives.

Should bees return to the colony with a load of poisoned nectar, there are further natural provisions against a general contamination of the honey. Foraging bees which become incapacitated at the hive are usually removed from the hive without expelling their load of nectar. When foraging bees eject their load of food, the hive bees process every drop of nectar brought in and thus are exposed to any toxic substance it may contain for a considerable period of time. Hive bees tend to retain the food in their honey stomachs when they become affected by a poison and are removed from the hive by other bees. Guard bees also resist abnormal bees or bees which return with offensive odors and remove them from the hive. When poisons are applied near the hive, some bees may return to their hives with the poisons on their bodies and contaminate some of the house bees. This sometimes results in many dead bees collecting near the entrance, again removing poisons from the hive.

Foraging bees can also collect pollens contaminated with pesticides and carry them back to their hive. If bees are affected seriously when they return to their hives, they are usually removed from the hive without expelling their load of pollen. If contaminated pollen is deposited in the hive from foraging bees, the hive bees may be poisoned in the process of storing the pollen in the brood nest. When nurse bees elaborate such pollens into brood food, they may become poisoned before they have fed many larvae. Such bees usually die on the bottom board or in front of the hive. In a few instances some or all of the larvae are also killed. Populations in hives may become so reduced that the colonies either die or are too small to pollinate blossoms adequately or to gather surplus honey.

The behavior of honey bees prevents almost all contamination of honey and pollen, few instances of pesticide contamination have been detected in nectar, honey and pollen in the hive from normal commercial pesticide applications. More often no detectable residues have been found, although dead bees, pollen and bee brood collected simultaneously did occasionally have variable amounts of pesticides present. In those instances where pesticide contamination has been detected, the amounts found were considered too small to be of significance. A thorough review of the organophosphorous systemic pesticides and their occurrence in the nectar of plants and subsequently in honey and their toxicity to honey bees was given by Anderson & Atkins (1968). Perhaps most encouraging is that we have been unable to find any instance where pesticides have been found in honey packaged by the wholesale or retail trade.

TOXICITY TO HONEY BEES OF PESTICIDES
APPLIED *OUTSIDE* THE HIVE

It was mentioned earlier in this chapter that pesticides or agricultural chemicals include chemicals for the control of specific pests. They are usually named to reflect this pest usage specifically such as acaricides (miticides) for mites, fungicides for fungus, herbicides for weeds, insecticides for insects and nematicides for nematodes; etc. The honey bee is susceptible to many of the more widely used pesticides in our intensive pest control programs for agricultural crop production. This results in the honey bee being subjected to an intensive and continuous hazard of chemical poisoning that overshadows all other problems including bee diseases. It should be pointed out that most pesticides are not hazardous to bees. In the most recent listing of the toxicity of 302 pesticides to honey bees by Atkins, *et al.,* (1981) 29% are highly toxic and 10% are moderately toxic to bees whereas 61% are relatively nontoxic or nontoxic to honey bees. One might question that this statement is misleading in that the quantity of pesticides most commonly used on agricultural crops for pest control comprise mostly the more highly toxic chemicals.

The most significant change in the pesticide picture is that there are 25% fewer pesticides registered for use now than there was 15 years ago (Atkins, *et al.,* 1975). A current survey of this reveals that only 39 percent of the pesticides used are highly toxic or moderately toxic to bees. Anderson & Atkins (1968) in their 1968 review article of Pesticide Usage in Relation to Beekeeping, which examined over 450 articles concluding in May 1967, pointed out that the subject of pesticide toxicity to honey bees has been written about extensively since Shaw's review (1941). This review lists 46 references and begins about the same time that the agricultural industry began using the organic pesticides in commercial amounts. Knowlton (1944 and 1948) reviewed 104 reports on honey bee poisoning. Musgrave & Salkeld (1950) published a fairly complete bibliography of honey bee toxicology research. Hocking (1950) reviewed the literature on the effect of agricultural chemicals on honey bees. In his report he discussed fertilizers, herbicides, fungicides and insecticides, and gave indices for each material. Brown (1951), in his book *Insect Control by Chemicals,* briefly reviewed the pesticide-bee poisoning situation and gave relative toxicity of several of the newer phosphate and chlorinated hydrocarbon materials. Todd & McGregor (1952) reviewed the early history of pesticide poisoning of bees, stating that the problem began in the early 1870's when Paris green was first used for codling moth control on apples. Beran & Neururer (1956), in Germany, surveyed the literature in detail, covering 83 references in two articles. They not only gave toxicity ratings of many pesticides, but also discussed laboratory and field test methods. Three extensive reviews appeared in 1960: Evenius, in Germany; Todd & McGregor, in the U.S.A.; and a paper with 407 references by Pangaud, in France. All added toxicity data for numerous new pesticides. Maurizio (1956) in Switzerland; Johansen (1959); and Palmer-Jones & Forster (1958) in New Zealand, briefly reviewed recent work in their areas. They included toxicity ratings of the new pesticides and suggestions for their safe use. Anderson & Atkins

(1967) and Atkins & Anderson (1967) reviewed and summarized their honey bee-pesticide studies involving thousands of laboratory and hundreds of field tests with several hundred pesticides. Johansen (1966) wrote a very thorough digest on bee poisoning, its effects and prevention. This report covers 114 pesticides and included the length of time the residue of each pesticide remained toxic to foraging honey bees.

More recently a newly developed class of pesticides appeared. Elliott, *et al.,* (1978) in his article containing 145 references traces the development of the early pyrethroids (synthetic pyrethrins) described in 1973. These chemicals have good stability resulting in prolonged residual activity, and high efficacy in controlling many insects. The pyrethroids' much greater activity makes possible significantly lower dosages to affect pest control which results in lower contamination of the environment. Many of the second generation of pyrethroids are approximately another order of magnitude more active so that extremely low dosages are needed to control pests. This recently developed class of pesticides are generally very safe to warm blooded animals.

Nominé (1982) edited a monograph on deltamethrin which was written by 16 authors and contained 257 references. This monograph on deltamethrin, a second generation pyrethroid, gives an excellent summary of this pesticide plus extensive information on pyrethroids in general. It adds to the information covered by Elliott by including structure, physical characteristics, chemical properties, photo degradation, metabolism, mode of action, toxicity to higher vertebrates, agricultural applications; scope of protection for crops, stored products, household insects, control of vectors of endemic diseases of man and animals; residues on food crops and animal products and effect on the environment.

The pyrethroids as a group are very highly toxic to honey bees by direct contact. However, the low dosage needed for adequate pest control and care in applying them during hours when bees are not foraging greatly diminish their hazard to levels which are safer to bees than most of the organophosphate and carbamate pesticides.

An interesting characteristic of the pyrethroids is that the highest registered dosage needed for pest control is approximately equivalent to the direct contact LD_{50} for honey bees (the lethal dose killing 50 percent of the bees), allowing their safe use around bees when they are not foraging. This compares to ethyl parathion which requires a pest control dosage 1.4 to 5.7 times the direct contact bee LD_{50}; and, to carbaryl which requires a dosage of 1.5 to 2.5 times the direct contact bee LD_{50} together with a 5 to 7 day residual period. As in all classes of pesticides there are some of the pyrethroids which have selectively low toxicity to bees while being highly toxic to the pests being controlled.

These reviews describe many of the worldwide research investigations of the effects of pesticides on honey bee adults. There are a few differences of opinion on the bee toxicity ratings of some of these pesticides, but it is remarkable how similar the results are for most of the chemicals. The findings of Atkins, Kellum, and Atkins (1981) are summarized in Table I.

Since the 1975 edition: (1) studies concerning sublethal effects on adult bees have been researched; (2) comparative studies concerning amorphogenic effects on bee brood have been researched.

Studies concerning sublethal and lethal effects on adult bees were conducted by Atkins (1966) in studies of the acute and chronic toxic effects of smog pollution, using the O_3 or ozone component, on longevity or lifespan of adult bees. High levels of ozone were toxic to workers, but not as toxic as the effect of ozone on the plants which bees visit as major sources of California honey. Subsequent research was conducted on other smog components (gasses of fluorine; SO_2, H_2S and $H_2S + CO_2$) were studied using LS_{50} (life-span, when 50% of the bees have died) curves of treated compared to untreated (Atkins, unpublished data from Ann. Project Reports, 1968; 1977).

Atkins (unpublished data) has conducted bee adult acute and chronic feeding tests periodically for many years by adding the pesticide to the honey water syrup food. These tests were to acquire chemical hazard data for individual pesticide manufacturers. The protocols provide data for both acute and chronic dosage lethal and sublethal comparisons measured by mortality and/or adult longevity (life span).

Smirle, *et al.* (1984), citing 17 references, proposed a standardized bioassay to evaluate sublethal pesticide effects on honey bees adults which considers worker bee age and environment (whether in hive or in cages) quantifying worker bee longevity. The protocol provides data for both acute dosage lethal and sublethal comparisons measured by adult longevity.

Haynes (1988) reviewed the recent literature, citing 92 references, concerning sublethal effects of neurotoxic insecticides on insect behavior in general. He outlines the various effects of sublethal exposure and cites effects caused by specific pesticides to honey bee workers.

In the early 1970's it became known that pesticide poisoning was occurring in brood as well as in adult bees; this damage was often undetected by monitoring only dead adult bees following a pesticide application. Comparative studies of the amorphogenic effects on bee brood have been researched by a few bee researchers for over 18 years (Atkins, *et al.*, 1972). Atkins and Kellum (1986), using an inhive individual larval protocol have tested 31 pesticides for normal morphogenesis, amorphogenesis and/or mortality of treated larvae of all ages and gave comparative hazard evaluations between brood and adults and overall predictions of the effect of chemicals on bee colonies when using practical registered field application dosages. Subsequently, additional pesticides have been tested (see TABLE II).

Wittmann (1981) since 1977 has been developing what he calls '*Apis* larvae test' which was used for testing the insect growth regular, diflubenzuron, using a semi-artificial diet in brood frames held in an incubator. He utilized the new *Apis*-larvae-test for comparative testing of diflubenzuron and carbaryl in his dissertation (Wittmann, 1982). Continuing attempts have been tried to determine diets which would successfully allow larvae to be reared in the laboratory (Wittmann & Engels, 1987). He and his colleagues (Wittmann, *et al.*, 1985) compared methods conducted on larvae in brood frame from a free-flying colony to larvae kept in an incubator and fed a semi-artificial diet.

Table I

Relative Toxicity of Pesticides to Honey Bees as Determined by Laboratory Tests in California (1950-1988).

Group I — Highly Toxic Pesticides
(LD_{50}* =0.001-1.99 μg/bee)

Severe losses may be expected if the following pesticides are used when bees are present at treatment time or within a day thereafter, except as indicated by footnotes. *Lethal dose killing 50% of bees. (Arranged by toxicity to honey bees — highest to lowest.)

Pesticide	LD_{50} in μg/bee	Slope Value, Probits
tepp[1][2][3]	0.002	0.68
bifenthrin[2], Brigade®, Capture®	0.016	3.06
avermectin[2]	0.018	3.36
cyfluthrin, Baythroid®	0.029	3.49
bioethanomethrin[2]	0.035	3.95
cyhalothrin[2], Karate®	0.052	3.32
resmethrin[2], Synthrin®	0.054	3.90
cypermethrin[2], Ammo®, Cymbush®	0.060	1.95
d-phenothrin, Sumithrin®	0.067	5.96
decamethrin[2], Decis®	0.067	4.88
flucythrinate[2], Pay-Off®	0.078	3.73
fenpropathrin[2], Danitol®	0.120	2.97
tralomethrin[2], Scout®	0.129	3.07
chlorpyrifos[2], Dursban®, Lorsban®	0.110	10.17
methyl parathion EC[1][2] †	0.111	5.13
dieldrin[1][2]	0.133	2.51
carbofuran[2][5], Furadan®	0.149	6.14
parathion[1][2]	0.175	4.96
fenitrothion, Sumithion®	0.176	5.75
dimethoate[2], Cygon®, DE-FEND®	0.191	5.84
methidathion[2], Supracide®	0.237	8.48
EPN[1][2]	0.237	4.31

Metacide®[1]	0.268	6.50
aldicarb[1 2 5 7], Temik®	0.272	5.00
mexacarbate[2], Zectran®	0.302	4.87
dicrotophos[1 2], Bidrin®	0.305	15.86
mevinphos[1 2 3], Phosdrin®	0.305	7.77
fenthion[2], Baytex®	0.319	6.14
prallethrin, ETOC®	0.337	5.53
fensulfothion[5], Dasanit®	0.337	4.78
methyl parathion[2], Penncap-M®	0.348	5.38
aldrin[2]	0.352	5.06
monocrotophos[1 2], Azodrin®	0.357	8.31
carbosulfan[2], Advantage®	0.362	4.70
diazinon[2], Spectracide®	0.372	8.03
methiocarb, Mesurol®	0.372	3.35
famphur, Famophos®	0.414	5.75
azinphos-methyl[1 2], Guthion®	0.428	7.43
pyrazophos[2], Afugan®	0.431	5.45
naled[2 3], Dibrom®	0.485	16.43
DDVP, dichlorvos[2], Vapona®	0.501	8.61
heptachlor[1 2]	0.526	5.94
lindane[2]	0.562	5.07
BHC[2]	0.562	5.07
malathion[2 4], Cythion®	0.726	7.83
phosmet[2], Imidan®	1.13	3.55
acephate[2], Orthene®	1.20	8.26
methomyl[2 5], Lannate®, Nudrin®	1.29	2.39
propoxur[2], Baygon®	1.34	3.23
methamidophos[2], Monitor®, Tamaron®	1.37	10.61
tetrachlorvinphos[2], Appex®, Gardona®	1.39	13.96
phosphamidon[2], Dimecron®	1.45	12.74
carbaryl[2], Sevin® 80 S	1.54	3.04
pyrazophos[2], Afugan®	1.85	3.48
arsenicals[1]	27.15	1.22

Group II — Moderately Toxic Pesticides
(LD_{50}* = 2.0-10.99 μg/bee)

These can be used in the vicinity of bees if dosage, timing and method of application are correct, but should not be applied directly on bees in the field or at the colonies. *Lethal dose killing 50% of bees. (Arranged by toxicity to honey bees — highest to lowest.)

Pesticide	LD_{50} in μg/bee	Slope Value, Probits
endrin[1 2]	2.02	4.20
aldicarb sulfoxide	2.21	2.33
crotoxyphos	2.26	17.10
propamocarb, Carbamult®	2.36	5.91
Vaporthrin®	2.47	4.29
demeton[1 2], Systox®	2.60	1.85
Pyramat®	2.95	4.07

oxydemeton-methyl[2], Metasystox-R®	3.00	2.32
disulfoton[1] [2] [6], DiSyston®	5.14	1.14
DDT[1] [2] [10]	5.95	4.89
Bacillius thuringiensis, thuringiensin, Di-Beta®	6.06	0.94
thiodicarb[2], Larvin®	6.50	3.48
endosulfan[2], Thiodan®	7.81	3.15
fluvalinate[2], Mavrik®	8.78	2.71
chlordane[2]	8.80	2.34
phosalone, Zolone®	8.94	3.83
phorate[1] [2] [6], Thimet®	10.07	1.34
oxamyl[2], Vydate®	10.32	6.43
formetanate[2], Carzol®	14.27	3.97

Group III — Relatively Nontoxic Pesticides
(LD_{50}* above 11.0 μg/bee)

These can be used around bees with a minimum of injury. *Lethal dose killing 50% of bees. (Arranged alphabetically.)

Acaricides, Diseases, IGRs & Insecticides

allethrin
aldoxycarb, Standak®
amitraz[2], Mitac®
azadirachtin, Margosan-O®
Bacillus thuringiensis, Biotrol®, Dipel®, Thuricide®
B. t., Kurstaki[11], Javelin®, Dipel® 4L
B. t., tenebrionis[11]
chlordimeform, Fundal®, Galecron®
chlorobenzilate, Acaraben®, Folbex®
cryolite, Kryocide®
clofentizine, Apollo®
cymiazole, Apitol®
cyromazine[11], Trigard®
dibromochloropropane, Nemagon®
dicofol, Kelthane®
diflubenzuron, Dimilin®
dinobuton, Dessin®

esfenvalerate[2], Asana®
ethion[2], Ethiol®
methoxychlor, Marlate®
multimethylalkenols, Stirrup®
nicotine[2]
Nosema locustae Canning
oxythioquinox, Morestan®
pirimicarb[2], Pirimor®
polynactins
propargite, Comite®, Omite®
pyrethrum
pyriproxyfen[11]
rotenone[2]
tetradifon, Tedion®
tetraflubenzuron[11], CME
trichlorfon, Dylox®
Z-11-hexadecanol, tomato pinworm pheromone

Fungicides

anilazine, Dyrene®, Kemate®
benomyl, Benlate®
bordeaux mixture[2]
captafol, Difolatan®
captan[2], Orthocide®
copper oxychloride sulfate

fenaminosulf, Lesan®
folpet, Phaltan®
glyodin, Glyoxide®
maneb
nabam, Parzate®
Polyphase™P-100, Troysan®

copper 8-quinolinate
copper sulfate (monohydrated)[2]
cuprous oxide
dazomet[5], Mylone®
diniconazole, Spotless®
dinocap, Karathane®
dithianon, Thynon®
dodine, Cyprex®

prochloraz
prochloraz/carbendazin, Sportac®
sulfur[2]
thiram
triforine, Funginex®
triphenyltin hydroxide, Du-Ter®
ziram, Zerlate®

Herbicides, Defoliants, Desiccants & PGRs

alachlor
amitrole
atrazine, AAtrex®
bentazon, Basagran®
bromacil, Hyvar®
butifos[8], DEF®
chlorbromuron, Maloran®
chloroxuron, Tenoran®
cyanazine, Bladex®
dalapon[12]
DEF®[8]
dicamba, Banvel®
dichlobenil, Casoron®
diquat[9]
diuron, Karmex®
EPTC, Eptam®
ethalfluralin, Sonalan®
etephon, [12], Ethrel®
EXD, Herbisan®
fluometuron, Cotoran®
fluridone, BRAKE®, Sonar®
hydrogen cyanamide[12], Dormex®
imadagylin, Arsenal®
linuron, Lorox®
MCPA[1], Mapica®
metaldehyde[13]

methazole, Probe®
metribuzin, Lexone®, Sencor®
monuron
naptalam, Alanap®, (cloproxydim), Select®
nitrofen, TOK®
norflurazon, Zorial®
paraquat[9]
phenmedipham, Betanal®
picloram[1], Tordon®
prometryn, Caparol®
pronamide, Kerb®
propanil, Stam® F-34
propazine, Milogard®
propham, IPC®, Ban-Hoe®
quinchlorac, FACET®
simizine, Princep®
sodium chlorate[8] [9], KNOCK 'UM OFF®
terbacil, Sinbar®
terbutryn
thiadiazuron[8] [12], DROPP®
tribuphos[8], Folex® 6EC
Uniconazole-P®[12]
2,3,6-TBA[1]
2,4-D[1] [2] [12], 2,4-D®
2,4-DB[1], Butoxon®, Butyrac®
2,4,5-T[1] [2] [12]

---

[1] California state regulation requires permits for most uses of these materials; also for 2,4-D and 2,4,5-T as weed treatments but not as hormone sprays on citrus.

[2] These materials have been laboratory tested and field tested mainly on alfalfa, citrus, cotton, ladino clover, milo, safflower and sweet corn; all others are laboratory tested only.

[3] Mevinphos, naled and tepp have such short residual activity that they kill only bees contacted at treatment time or shortly thereafter. These materials usually are safe to use when bees are not in flight; they are not safe to use around colonies.

[4] Malathion has been used on thousands of acres of blooming alfalfa without serious loss of bees. However, occasional heavy losses have occurred, particularly under high temperature

conditions. If applied to alfalfa in bloom, it should be as a spray and treatment should be made during the night or early morning when bees are not foraging in the field. Undiluted (ULV) technical malathion spray should not be used around bees.

5 Nematocide.

6 Disulfoton and other systemic pesticides used as seed treatments have not caused bee losses.

7 Aldicarb, although highly toxic to bees as a contact poison, is used only in granular form and extensive field usage has not resulted in bee losses.

8 Defoliant.

9 Desiccant.

10 DDT has been withdrawn from use in the U.S.A.

11 Insect growth regulator.

12 Plant growth regulator.

13 Molluscicide.

† The encapsulated methyl parathion formulation, Penncap-M, is highly toxic to foraging bees, young hive bees, and brood. Overall, it is 13 times more toxic to honey bee colonies than the EC formulation (emulsifiable concentrate). Penncap-M is too hazardous to be applied to any area at any time when bees are present in the field or within 1 mile of the area to be treated.

NOTE: For references to chemical and other names of pesticides listed in table, see Atkins, *et al.*, 1975; Meister, 1989.

Table II

SUMMARY: Dosage-mortality data for honey bee larvae treated in the cell with pesticide and type of effect on the larvae; dosage-mortality data of adults are included for comparison.

Abbreviations for formulations: AG = analytical grade; E = emulsion; EC = emulsifiable concentrate; L = liquid; S = solution; SP = soluable powder; T = technical grade; ULV = ultra-low volume; WP = wettable powder. Abbreviations for pesticidal effects: AM = amorphogenic; IGR = insect growth regulator; NI = no information available; NT = nontoxic, or nontoxic at highest level tested; P = poisonous; PS = simple poison.

Pesticide and formulation	Brood LD_{50}* (μg/larva)	Mode of action (brood)	Adult LD_{50}* (μg/bee)
diazinon, 88.4% a.i.[1] T	0.00012	PS	0.372
CGA-112913, 99.3% a.i. T	0.00048	AM;P	NT[2] (100μg)
profenofos, 86.3% a.i. T	0.00787	PS	3.46
flucythrinate, 2.5E	0.029	PS	0.078
chlorpyrifos, 4 E	0.051	PS	0.110
cypermethrin, 3 E	0.066	PS	0.060
UC 84572,2 F (IGR)	0.0698	AM;P	NT (30μg)
permethrin, 3.2EC	0.239	PS	0.159
methidathion, 2 E	0.274	PS	0.237

naled, 8 E	0.284	PS	0.485
dimethoate, 4 EC	0.340	AM;P	0.191
aldicarb, 98.45 a.i. AG	0.356	PS	0.272
oxamyl, 2 L	0.367	AM;P	10.26
mevinphos, 4 EC	0.441	PS	0.305
methomyl, 1.8 L	0.539	PS	1.29
fenvalerate, 2.4 EC	0.672	PS	0.408
malathion, ULV 91 a.i.	0.736	AM;P	0.726
methamidophos, 4 EC	0.766	PS	1.37
aldicarb sulfoxide, 98 a.i. AG	0.854	PS	2.211
methyl parathion, 2 EC	0.919	PS	0.291
aldicarb sulfone, 98 a.i. AG	1.067	NT	399.302
UC 84572, 100 a.i. (IGR)	1.267	AM;P	NT (363μg)
carbaryl, 80 S	1.212	AM;P	1.34
acephate, 75 SP	2.668	PS	1.202
fluvalinate, 2 E	3.047	PS	34.621
captan, 90% technical	3.090	AM;P	215.131
azadirachtin, 0.35 a.i.	4.200	AM;P	NT (181μg)
oxydemeton-methyl, 2 EC	4.440	PS	2.997
thuringiensin, 3 a.i. w/w	16.490	PS	6.063
cyhexatin, 50 WP	25.077	PS	36.70
endosulfan, 2 EC	28.142	PS	21.79
disulfoton, 8 EC	35.296	NT	6.12
oxythioquinox, 25 WP	36.287	NT	66.47
formetanate, 92 SP	67.202	NT	9.21
clofentezine, 50 SC	112.111	NT	NT (79μg)
propargite, 6 E	121.512	NT	67.47
amitraz, 1.5 E	182.607	NT	NT (100μg)
benomyl/ziram, combination	191.9	NT	NI
cryolite, 96 a.i.	2,245.2	NT	NT (218μg)
chlorobenzilate, 4 E	2,600	NT	1,700
ziram, 76 WP	39,000	NT	224.7
benomyl, 50 WP	853,000	NT	NT (121μg)

* LD_{50} is the lethal dose of pesticide in micrograms per bee causing 50 percent mortality.

[1] active ingredient

[2] nontoxic at highest dose (parentheses) tested.

TOXICITY TO HONEY BEES OF PESTICIDES APPLIED *INSIDE* THE HIVE

As mentioned earlier, it became known in the 1970's that honey bee brood was being poisoned by pesticides as well as adults. The research which followed was concerned with pesticides applied *outside* the hive to crops and rangeland for the control of plant pests and pests attacking people such as mosquitoes, etc. which were treated under the supervision of knowledgeable and licensed pest control advisors and applicators. With the general spread throughout the world of the tracheal mite, *Acarapis woodi* (Rennie), and the varroa mite, *Varroa jacobsoni* Oudemans, beekeepers are confronted with actually using pesticides *inside* the hive in attempts to control these honey bee pests.

Many beekeepers are not knowledgeable about handling pesticides and few have acquired the necessary training to properly administer the chemicals safely *inside* a hive containing a colony of bees. Improper dosage and application may control the mites, but could also kill the brood and adult bees. In addition to losses of bees, probable residues of the chemicals in wax, honey and other hive products, Jaycox (1987) listed other potential problems such as resistant mites and changes in consumers' attitudes toward the possibility of contaminated honey, pollen and royal jelly. Many of the chemicals previously and currently being utilized have been found at low levels in honey and in beeswax. Beeswax has shown up to six times as much chemical as honey where contamination has occurred. Jaycox cautions that when residues of mite-control chemicals are found in honey, all beekeepers may be forced to provide analyses showing freedom from contamination in order to sell their honey. Beeswax is used to manufacture wax foundation for use in the hive, making candles and as an additive in many cosmetics. Analyses of honey, pollen and wax are costly and the costs will have to be absorbed by the beekeeper.

Applications of pesticides (acaricides and miticides) *inside* the hive opens up new problems which are usually not encountered from applications made outside the hive. Persons applying pesticides to crops and other outside areas must acquire the necessary training and knowledge to qualify for a pest advisor's and/or applicator's license. Beekeepers should obtain the necessary information about pesticides, dosage levels, frequency of application, proper handling and safety precautions when using pesticides before they attempt inhive do-it-yourself mite control. Beekeepers have endured the continuous problem of misuse of pesticides *outside* the hive. Now they must demonstrate that they can use pesticides responsibly *inside* the hive.

We will return to the studies of Atkins & Kellum (1986) where the comparative toxicological and morphogenic studies of Atkins & Kellum (1986) determines the potential hazard to honey bee brood from pesticides which have contaminated the food in the hive. Each pesticide is tested on worker larvae 1-2, 3-4 and 5-6 days old for mortality and amorphogenic changes.

It is appropriate to define the use of a few terms used in the text and in Table II. When it is stated that a substance acts as a simple poison, or that its mode of action is as a simple poison or toxin, this indicates that it kills bees during some stage of development. Morphogenic means that the bee develops normally and looks normal throughout its existence as an egg through the larval, pupal and adult stages. Amorphogenic means that the normal developmental process has been affected in such a way that abnormalities are visible in the pupa or adult.

Test results indicate that mortality may occur at any stage of larval, prepupal or pupal development or newly emerged adults. Some amorphogenic adults that survived were light in color, and of light weight, and often had deformed wings or no wings; they died soon after eclosion. Some chemicals were less toxic, some equally toxic, and, some more toxic to brood than to adults.

This information is being utilized to determine choice of chemicals, dosage, timing of application, and substitution strategies to decrease the hazards of pesticides which contaminate brood food in the hive.

In more recent years Atkins and coworkers in honey bee brood morphogenic testing have emphasized testing acaricides which can be used in hive to control mite infestations caused by tracheal mites *(acariosis)* and caused by varroa mites *(varroasis)*. Such testing indicates which chemicals and their dosage levels that are safe to both brood and adult bees in the hive, Therefore, only those chemicals which have been shown in these tests to be safe within the hive need to be tested for inhive mite control. This should eliminate many of the catch-as-catch-can screening which has been tried the last few years. Several of these acaricides have already been tested and their hazards to both brood and adult bees are listed in Table II.

Studies by Rivera, *et al.,* (1987) found contamination in beeswax, honey, and also in honey bees from colonies being experimentally treated with menthol to control tracheal mites and under conditions being widely used by some beekeepers. Menthol contamination increased exponentially as the quantity of menthol volatilized within the hive was increased.

Other studies (Rivera & Wilson, 1989) have shown that menthol volatilized inside hives containing honey bees infested with mites accumulated in beeswax up to 14 times more than in honey. They also indicated that heating the menthol contaminated product at 65°C (149°F) dissipated most all of the menthol in 2 days in honey; 3 days in beeswax.

Methods have been developed for rapid determination of acaricides in honey (Smith, 1989). The author did not name the acaricides which were found in honey and does not say whether the honey samples were fortified (acaricides added to uncontaminated honey) to provide standard recovery quantities to compare to the unknown samples being tested.

THE STUDY OF PESTICIDE EFFECTS ON HONEY BEES

The study of pesticide effect on honey bees is vital because agriculture must have pesticides to control many agricultural pests; agriculture also requires honey bees for pollinating over 50 of the 250 crops grown in the U.S.A. to produce high quality commercial quantities of seeds and fruits. Since this is so, agriculture must learn how to use pesticides so that pests are controlled and honey bees and other beneficial insects survive.

The ideal pesticide is selectively nontoxic to honey bees while it is controlling a specific pest. Since it is seldom possible to develop a product wholly meeting the ideal, the best compromise must be looked for. To find this, the relative toxicity of various chemicals to honey bees have been comparatively investigated.

Laboratory examinations and field examinations must both be conducted separately if the toxic effect of each pesticide on honey bees is to be discovered. To learn how to apply a pesticide to the best advantage, detailed qualitative and quantitative study of its effects is necessary. This is best done under the controlled conditions in the laboratory. Then, large-scale field studies based on the preliminary date obtained in laboratory tests must be conducted to determine the practical effect from actual use. These chemicals are applied in many forms. The most important forms are as baits, dips, dusts, fumigants, granules, side dressings, and sprays. Comparisons are made in field tests to determine toxic differences that occur under different treatment conditions such as day and night applications, airplane and ground machine treatments, high-volume, low volume and ultra-low volume sprays, treatments over or near covered colonies, dust and spray applications, etc.

Recently interest has been renewed in the principles of *economic entomology* under the current term of *plant pest management* and/or *insect pest management.* In the most recent revision list giving the toxicity of pesticides to honey bees (Atkins *et al.*, 1981) determined from laboratory tests is provided information which enables one to predict the expected hazard of a pesticide to honey bees in field applications. This information is included in Tables I and II along with the list of pesticides most commonly applied for pest control.

Through experience and observations of field applications and field tests a useful rule of thumb method of determining the anticipated toxicity hazard of a pesticide to honey bees in the field is available by utilizing the laboratory data. In most instances the LD_{50} value* of a pesticide in micrograms (μg) per bee can be directly converted to the equivalent number of pounds of chemical per acre when applied as a spray or dust to the aerial portions of plants (μg/bee x 1.12 is equivalent to the number of kg/hectare). Remember that the LD_{50} value is the amount of pesticide which will kill 50 percent of the bees contacted. Then, for example, since the LD_{50} of parathion is 0.175 μg/bee, we

---

*LD_{50} is the lethal dosage in micrograms per bee of a pesticide causing 50 percent mortality of the honey bees contacting the pesticide.

would expect that 0.17 lb/acre of parathion would kill 50 percent of the bees foraging in a treated area at the time of treatment or shortly afterwards.

The slope value (probits) of a pesticide may also be utilized to determine the anticipated increase or decrease of honey bee hazard in relation to the LD_{50} value. In general, a pesticide with a slope value of 4 probits or higher can often be made safer to honey bees by lowering the dosage only slightly. Conversely, by increasing the dosage only slightly the pesticide can become highly hazardous to bees. This information is particularly useful when the LD_{50} in μg/bee is approximately equal to the normal dosage in lbs/acre needed in the field to control pest populations. For example, consider a pesticide, which is normally applied at dosages of 0.5 to 1.5 lbs/acre to control pest insects, having a LD_{50} of 1.0 μg/bee. Furthermore, suppose that the slope value of this pesticide is 2.0 probits. Then, if this chemical is applied at 0.5 lb/acre, we would expect a 28 percent kill of bees in the field; at 1 lb/acre, we would expect a 50 percent kill; and, at 1.5 lbs/acre, we would expect a 64 percent kill.

Then, for example, suppose the slope value of this pesticide is 16 probits. Under these conditions, if this chemical is applied at 0.5 lb/acre, we would expect no kill of bees in the field; at 1 lb/acre, we would expect a 50 percent kill; and, at 1.5 lbs/acre, we would expect 100 percent kill. These examples illustrate the basic principle that the property of a pesticide being toxic or nontoxic is determined by its dosage. Pesticides normally considered toxic always have a dosage below which they cause no harmful effect.

For additional examples of anticipated bee mortalities at other selected slope values and field dosages of a pesticide having an LD_{50} value of 1.0 μg/bee, refer to Table III.

Table III Examples of Anticipated Honey Bee Mortality
When a Pesticide with a LD_{50} Value of 1.0 is Applied
at Selected Slope Values and Increasing and Decreasing Dosages

Slope Value	Percent mortality at following dosage (lb/acre):									
	0.1	0.25	0.5	0.75	1.0	1.25	1.5	1.75	3.0	10.0
	Below LD_{50}*				LD_{50}*	Above LD_{50}*				
2	3	12	28	42	50	57	64	68	82	97
4	—	1	12	32	50	66	72	82	96	—
6	—	—	2	17	50	76	91	97	—	—
16	—	—	—	3	50	93	—	—	—	—

* LD_{50} is the lethal dosage in micrograms per bee of a pesticide causing 50 percent mortality of the honey bees contacting the pesticide.

Any pesticide having known LD_{50} and slope values can be similarly perused by substituting the LD_{50} value of the chemical into the LD_{50} (the 1.0 lb/acre or center) column of this table and multiplying the LD_{50} value by the other factors (0.1, 0.25, 0.5, 0.75, 1.0, 1.25, 1.5, 1.75, 3.0, or 10.0) to obtain the proper range of field dosages per acre. Then, using the slope value closest to the known slope value for the particular pesticide, the anticipated percent mortalities will be valid for that chemical.

We wish to emphasize that there are a few exceptions to the above rule of thumb method—those pesticides which are less hazardous as well as more hazardous than one can anticipate from the laboratory data. Pesticides which have exceptionally long residual properties are most likely to be more hazardous to bees in the field than indicated by the calculated mortalities obtained in the laboratory.

It is our desire, that by presenting these data and these methods, decisions can be made (to select a pesticide, determine the dosage, and apply the chemical in the safest way and at the most appropriate time of day) maximizing the control of the pest species, while minimizing the adverse effects upon honey bees and other beneficial nontarget species in the treated areas.

THINGS TO CONSIDER WHEN USING HAZARDOUS PESTICIDES AROUND HONEY BEES

Many pesticides are toxic to honey bees and other beneficial insects and mites. The farmer, the beekeeper, the pest control advisor and the pesticide applicator must cooperate closely to keep losses of these beneficials minimal. For protection of beneficial insects and mites, use only the safest of the recommended pesticides.

Some states have laws and regulations, which require that before a pesticide application toxic to bees can be made, the owner of the bees within 1 mile of the treated area must be notified and the beekeeper is given 48 hours to move or otherwise protect his bees. This regulation is inconvenient for the grower in many instances. It seldom protects the bees since the beekeeper does not have any practical method for protecting his colonies from toxic pesticides nor is he able to move his bees to another location. The commercial beekeeper seldom has the manpower, equipment and time to move large numbers of colonies. He seldom has an alternate location nearby to move to. Often he moves his colonies to a different location to avoid poisoning only to discover that he has chosen an area requiring an application of pesticides. Therefore, this procedure ends up as the beekeeper's version of the game of musical chairs. Another problem directly related to this is that a beekeeper is usually under contract with a grower to provide pollination at the particular location he has his colonies. Therefore, if he moves his colonies to avoid poisoning, he is violating his pollination contract. It is imperative that other steps must be utilized to reduce poisoning of bees.

During daylight, bees usually will leave the hive to forage when temperatures are above 55-60°F. Foraging does not occur at night although the bees may cluster on the outside of the hive if the temperature is above 70°F. A strong wind may cause clustered bees to enter the hive at temperatures above 70°F. Bees in crowded hives are more likely to cluster than those in uncrowded hives. Clustering can usually be alleviated if an additional empty super is added to the hive. The time and intensity of bee visitation in a given crop depends on abundance and attractiveness of the bloom. For example, alfalfa, cotton and safflower crops in bloom may be attractive to bees all day, while cucumbers and melons are usually attractive from midmorning through early afternoon, and milo and sweet corn are attractive only in the morning.

Location of bees is important. Colonies located in the field and treated over may sustain higher losses than colonies at the edge or outside the field that are not treated over. Bee losses are usually not significant to colonies one-fourth mile or more away from treatments unless the treated crop is the only attractive field in the area. In this case injury may occur to colonies up to several miles away. The farther the colonies are from the treated area the less critical is the treatment time. Colonies moved into the field 2 or 3 days after treatment usually escape damage.

Time of application and the location of colonies are important and depend on bloom period and attractiveness of the honey bee forage crop. Treatments made when bees are foraging in the field are usually the most hazardous. Treatments made over colonies in hot weather when bees are clustering on the outside of the hives may cause severe losses. Treatments made during the early morning before bees are foraging are more safe; treatments made during the night are the safest.

Treating a nonblooming crop with a hazardous pesticide when cover crops, weeds or wild flowers are in bloom in the field or close by may cause heavy bee kills. Destroy cover crops and weeds in bloom by mowing, discing or herbicides when bees are in the crop area. Cover crops and weeds may be contaminated from pesticide applications. Their blooms may also reduce the number of bees foraging crop blossoms. Beekeepers should be encouraged to move colonies out of crops being pollinated when 90% petal fall occurs or as soon as pollination is accomplished to prevent further bee losses from pesticides. Each pollination project should be covered with a written pollination contract which includes these agreements.

Pesticide drift to neighboring fields that are attractive to bees may cause losses (see Fig. 3). Treating large areas and/or repeating applications may cause greater bee kills. With few exceptions, pesticides applied as dusts are more hazardous to honey bees than those applied as sprays. Pesticide spray formulations vary significantly in their toxicity to honey bees. Wettable powder formulations are often more hazardous to bees than either emulsifiable or water soluble concentrate formulations. Water soluble formulations are often safer than emulsifiable formulations. Pesticides toxic to bees which are microencapsulated are often more hazardous than when the pesticide is nonencapsulated. Fine sprays are less toxic than coarse sprays. Sprays of undiluted technical pesticide (ULV) may be more toxic than diluted sprays. Pesticide applications by aircraft over bees in flight are more hazardous than applications by ground equipment. Granular applications are usually the safest method of treatment. Baits on material such as apple pomace may be attractive to bees and therefore hazardous.

Covering colonies with tarpaulins of burlap or dark plastic material for 2 to 4 hours during and after treatment in early morning may give added protection (see Fig. 4). Colonies may be covered for two days if burlap tarpaulins are kept dampened with water. Adequate colony protection often can be obtained by laying water-soaked burlap bags across the entrance to each colony during application. Covering may be important when pesticides are applied by aircraft. In some instances when bees will be exposed to highly toxic pesticides having long residual properties, it is prudent to move the colonies away from the area to be treated.

Kinds and amounts of pesticides are important. Use the proper dosage and formulation of the safest pesticides on bees that will give adequate pest control. Combinations of toxic pesticides are usually no more toxic to bees than a single pesticide since each are often used at their full dosage. Read the label

FIGURE 3. Modern fixed wing airplane spraying seed alfalfa in the early morning. Note honey bee colonies infield next to cotton trailers utilized as bee landmarks to reduce drifting. *(Photo by David E. Sbur)*

FIGURE 4. Five colonies covered with black plastic tarpaulin to confine bees to hives and protect them from exposure to pesticide application. Five uncovered colonies to right; airplane spraying in background. *(Photo by E. L. Atkins)*

and follow approved local, state and federal recommendations. When using pesticides hazardous to bees, notify the beekeeper so that he may protect his colonies or so he can administer prompt intensive care following poisoning to alleviate further damage.

In general, the organic pesticides are less hazardous to honey bees than the inorganic arsenicals. Most organic pesticides have a shorter residual period than the inorganic since they begin to deteriorate as soon as they are applied. The inorganic pesticides such as the arsenicals, cryolite and sulfur persist until they are rubbed off by abrasion or fall off after loss of adhesion or until plant growth reduces the accessibility of the pesticide to visiting bees. Although the organic pesticides are used in large quantities over large areas and on an ever-increasing number of crops, they usually can be applied with safety if the facts and precautions presented herein are taken into consideration and utilized.

HISTORY OF HONEY BEE COLONY LOSSES
FROM PESTICIDE APPLICATIONS

Table IV which gives the honey bee losses in California from official statistics for 1962 through 1981 show an average of approximately 80,000 colonies of bees were killed by pesticides through 1971. In addition, another 80,000 colonies were killed by "other causes" which includes death of queens, winter kill and disease. These causes account for about 20% of the total number of colonies lost by other causes. We have recently found that these losses are from pesticides affecting the brood and results in the so-called summer dwindling.

Certain chlorinated hydrocarbon pesticides, particularly DDT and toxaphene, largely replaced the aresenicals from about 1945 through 1965. These chemicals were usually less toxic to honey bees and demonstrated that beekeeping was feasible in the vicinity of cotton fields, for example, when these compounds were used in place of the arsenicals or other more highly toxic compounds. However, the limitations of the amount of residue of hydrocarbons on hay and forage crops, as well as on vegetables and fruit, necessitated the substitution of carbaryl or of certain organophosphorous chemicals to be applied on such crops, and even on other crops grown adjacent to them. In 1965 the use of DDT was drastically reduced and was subsequently banned from most crops. This change resulted in a marked increase in the loss of bees (see Table IV for colonies lost due to pesticides, particularly for 1966-1971).

Table IV

**Honey Bee Colony Losses from Pesticides & Other Causes
In Relation to the Total Number of Colonies in California.[1]**

Year	Colonies of Bees	Pesticides	Other Losses	Total
1962	—	82,000	—	—
1963	—	41,000	—	—
1964	—	41,000	—	—
1965	—	49,000	—	—
1966	—	55,000	85,000	140,000
1967	559,000	76,000	86,000	162,000
1968	565,000	83,000	84,000	167,000
1969	537,000	82,000	117,000	199,000
1970	521,000	89,000	70,000	159,000
1971	511,000	76,000	32,000	108,000
1972	500,000	40,000	30,000	70,000
1973	500,000	36,000	31,000	67,000
1974	500,000	54,000	33,000	87,000
1975	500,000	31,000	28,000	59,000
1976	525,000	46,000	39,000	85,000
1977	525,000	51,000	31,000	82,000
1978	504,000	60,000	20,000	80,000
1979	504,000	54,000	34,000	88,000
1980	504,000	51,000	38,000	89,000
1981	500,000	56,000	32,000	88,000

[1] Estimates from the California Crop & Livestock Reporting Service from Bee & Honey Inquiries to a significant sample of beekeepers annually. They are reported by the California Department of Food and Agriculture in the annual Apiary Inspectors' Newsletter. (Compiled by: E. L. Atkins)

In the early years many insecticides were applied as dusts; this is reflected in 1962 poisoned colonies figures. This bee kill problem was somewhat reduced by replacing the more hazardous dusts with the safer spray formulations.

Most pesticide applications were being applied from late morning throughout most of the daylight hours prior to 1963, resulting in many colony kills. At about this time treatments were moved up to early morning applications (from daylight to 7 a.m.). This practice greatly reduced the kill of foraging bees directly contacted by insecticide applications.

Prior to 1967 the more highly toxic organophosphate and carbamate chemicals were used somewhat sparingly. Beginning in 1968 their use increased and, as shown by the dead colony figures, this increased the number of colony losses.

In 1972 California began applying approximately 85% of the pesticide applications to agriculture crops by airplane at night. This reduced the bee colony kill over 50% by providing more time for the pesticide to become less toxic to bees by dissipation. Interestingly, the number of colonies lost to "other causes" also dropped 50%. The success of this advance depended upon the agricultural airplane industry developing adequate lighting systems allowing safe night flying.

Night applications are effective in reducing bee kills with most of the commonly used pesticides. It does not reduce bee kills to tolerable levels with all of the long residual pesticides such as carbaryl and carbofuran; the long residual formulation of ultra low volume insecticides such as malathion; or the long residual formulation of microencapsulated insecticides such as methyl parathion (Penncap-M®). Generally, because of their very high bee hazard, these last two insecticide formulations are not used on agricultural crops in California.

These statistics demonstrate that California has significantly reduced the loss of colonies caused by insecticides since 1962 (Atkins, *et al.*, 1978).

APPEARANCE OF PESTICIDE INJURY IN THE FIELD AND ORCHARD

When pesticides are applied to fruit trees or to cultivated crops in bloom, even though the applications are made at night or early in the morning when bees are not foraging, by mid morning the trees or fields will usually be quiet without the sound of any bees or other pollinating insects working the blossoms. The bees that were accustomed to foraging the treated areas will not be present. If the pesticide used is toxic to bees for a considerable period of time, many of the bees probably have been killed. If large numbers of bees are killed, many dead or dying bees may be seen on the ground in the field and on the ground between the treated field and the apiary. The pesticide treatment suppresses pollination activity by killing off all or some of the foragers and by disorganizing the surviving field force. Some bees may be found hanging to blossoms or lying on the ground beneath the trees or plants. The bees may occasionally be driven away by the change of odor in the field, but the pesticides seldom have sufficient repellency to prevent workers from foraging. When the field force is killed by a pesticide, it may be one day up to several days before other bees will be attracted to the field, the time interval depends on the residual characteristics of the pesticides and the severity of the kill. Fields treated with pesticides having toxic residues which persist for less than 4 days will accumulate nectar and in some instances pollen and hence become more attractive to the pollinators. However, if all of the pollinators have been killed by the treatment and there is not an abundance of pollinators in the area, there is often a delay in repopulating the treated area. Occasionally, when the foragers are killed in a treated field and the surrounding untreated area remains highly attractive to bees, no bees will relocate in the area where the foragers were killed. In this instance, if pollination is critical for a crop, new colonies of bees should be brought in to pollinate the treated area. Otherwise, scout bees will have to rediscover the blossoms and convey the information back to their hives. If these scout bees are killed, as well as those already oriented to the treated fields, the time interval for the field or orchard to be repopulated with bees will be longer. When the hive bees as well as the field bees are decimated, normal colonies may have to be moved into the treated field to provide a sufficient force of pollinators.

The set of seed or fruit can be correlated directly with the effort of the pollinators when plants are in the proper physiological condition to set seed or fruit and weather conditions are favorable. A period of 10 days to 2 weeks without adequate pollination activity may cause an economic loss in production or a delay in the time of harvest, depending on the particular crop. The crown set of melons, for example, may be lost if these blossoms are not adequately pollinated, and not only the quantity and quality of the yield will be affected, but the time of harvesting will be delayed.

As mentioned above, under certain conditions when only part of the foraging force is killed by a pesticide having a short residual period, such as mevinphos, naled or tepp, the number of foragers visiting the blossoms will increase for a time equivalent to the period of time that foraging activity was suppressed by the pesticide application. This intensification of foraging activity apparently is due to the greater attractiveness of the blossoms caused by the accumulation of nectar and/or available pollen, causing more foragers to be attracted to the area from nearby colonies of bees.

In one pesticide study the effects of weekly applications of carbaryl for pink bollworm, *Pectinphora gossypiella* (Saunders), supression treatments in Imperial County, California, were studied from September to November, 1966. Thirty test colonies provided with Todd Dead Bee Hive Entrance Traps (see Fig. 5) were strategically located 0 to 2 miles from isolated cotton areas and bee kills were monitored daily for nine weeks. Up to 20,000 bees per colony were lost (see Fig. 6). Most colonies were weakened and a few colonies

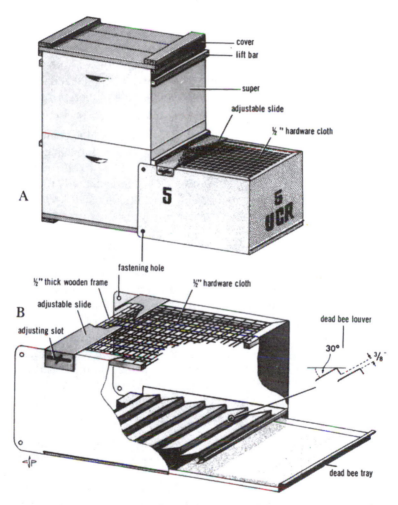

FIGURE 5. UCR modified TODD DEAD BEE HIVE ENTRANCE TRAP used in field research, especially to retain bees dying at the hive from natural causes, diseases, pesticides or other poisons. *(Photo by E. L. Atkins)*

died out after the treatments ended. Colony evaluation in January revealed no evidence of residual carbaryl in frames or colonies; there was evidence of chemical thinning; most colonies were short of honey stores (more so than colonies not in the treated areas); pollen was in short supply or absent (however, this was also true of colonies in untreated areas). All colonies still alive in January were developing normally at the time of colony evaluation conducted in the middle of April, 1967.

Distance did provide protection from carbaryl sprays. A safety mechanism operated to lessen the possibility of killing out a colony where frequent heavy kills of foraging bees occurred. When a heavy or disastrous field force kill occurred, all remaining bees were required to cover and tend brood; and, therefore, until new bees emerged, there were few or no bees available to forage and hence to be killed when the next scheduled (seven days) carbaryl treatment was applied. Some of the weakest colonies in this study missed two weekly treatments.

This study also showed that distance provided significant safety to honey bees foraging cotton from colonies located at various distances from the treated fields. Colonies which were located 1/4 to 1/2 mile from the cotton had 35 percent lower bee kill; colonies which were located 1/2 to 1 mile from the cotton had 60 percent lower bee kill; and colonies located 1 to 2 miles from the cotton had 90 percent lower bee kill from pesticide treatments than the colonies located at the edge of the field.

EVIDENCE OF POISONING AT THE HIVE

Symptoms of pesticide poisoning are somewhat similar to symptoms caused by plant poisoning or by adult bee diseases. The most positive indication of serious pesticide poisoning is the appearance of massive numbers of dying, or dead bees at the colony entrances throughout the apiary (see Fig. 7). The quantities are far larger than those found with the most serious instances of adult bee diseases and plant poisoning. Pesticide kills may also be suspected when the weather has been warm, dry and pleasant for any length of time and the colonies are located in blooming crop and forage areas. A knowledge of local pesticide programs is important in avoiding poisoning.

Some pesticides cause slightly affected bees to become excessively aggressive. Affected bees usually are restless and are found crawling out of the hive, often unable to fly except for short distances. Many are trembling or crawling and tumbling about aimlessly. Many others remain in the hive badly affected and completely disoriented or are being carried out of the hive by other hive workers. The legs of affected workers may be dragged along as if paralyzed and the rear wings may be unhooked from the front wings and held at

FIGURE 6. Tray from Todd Dead Bee Hive Entrance Trap containing 24-hour collection of 3,847 bees (approximately 1 quart) killed by pesticide application to seed alfalfa with colonies in the field. *(Photo by E. L. Atkins)*

FIGURE 7. Disastrous colony kill caused by improper application of highly toxic pesticide dust. *(Photo by E. L. Atkins)*

abnormal angles. The pesticides in each class of chemicals often affect bees in the same typical manner.

EVIDENCE OF POISONING WITHIN THE HIVE

The overall appearance of injury to a colony will vary with the mode of action and degree of poisoning caused by a specific pesticide. A 3-story colony that loses its foragers may have a super partially or entirely filled with honey, but with the bees covering only part of the combs in the two lower chambers. An examination of the combs will indicate brood in all stages. The queen may discontinue egg-laying slightly over the following week, but within two weeks the colony population will have returned to normal, replenished by emerging bees. Flight activity will gradually increase, but the colony will have lost most of its productivity in the meantime.

If a colony not only loses its foragers, but also many of the hive bees from a heavy poisoning, within a day or two the population will be drastically reduced to the queen, a few nurse bees, and recently emerged bees. An examination of the brood combs will reveal many cells, from which larvae have been removed, and many sick or dead larvae not yet removed from their cells. After a few days, many pupae may also be found dead in their cells, the cappings of some cells having been partially removed. Some or all of the larvae may die of starvation or neglect and others may die from either chilling or overheating because there are not enough hive bees remaining to tend the brood or to maintain the brood nest temperature. In very severe cases, only a handful of young bees will remain with the queen and sealed brood. Nurse bees may be killed if they feed on contaminated pollen or honey. The queen usually survives because she receives most of her food from nurse bees which fed on uncontaminated food.

Occasionally, in a severely poisoned apiary small emergency swarms, frequently of less than a quart of bees in volume, will issue from the affected hive.

While numerous accounts of losses of bees have been reported from different states and from various parts of the world, the reports of colony loss in California will illustrate the extent of losses from pesticides (see Tab. IV). In addition to the 15 percent of colonies killed outright, the remaining colonies are weakened to such an extent that they are no longer productive or effective pollinators, nor can they be used for divides to increase colony numbers. This type of economic loss probably greatly exceeds the loss from colonies killed outright by pesticides. The monetary loss to beekeepers is large, but the value of the seed and fruit loss through inadequate pollination is 50 to 100 times greater.

The use of pesticides around apiaries and around bee houses to control insect pests can be hazardous to honey bees. Ants are often a problem around colonies, particularly in permanent apiaries. It has become common practice to use a pesticide to control ants in the apiary. Many pesticides are good ant killers and most are equally toxic to bees. Chlordane applied as a spray is very effective for this use because it has a residual period of several weeks. Chlordane is also highly toxic to bees and should be used with extreme caution. Even with extreme caution heavy kills of bees can occur under certain circumstances. One of our urgent needs is research to discover and develop a specific pesticide which will kill ants and be completely safe to use around bee hives.

Another pesticide problem is controlling ants, wax moths and other insects in and around the honey house and hive equipment storage building. No pesticides should be used inside these facilities since most of these chemicals are absorbed by beeswax. Chlordane, used for ants, will be absorbed into wax foundation, drawn comb and brood comb. When this happens, the absorbed chlordane may kill bees in colonies on which the equipment is placed for several weeks or months.

More recently, there have been numerous reports of beekeepers using various aerosol space spraying devices in their bee houses. These devices dispense pesticides such as lindane, DDVP or resmethrin; all have been shown to contaminate beeswax and honey. Also, NO-PEST STRIPS containing Vapona (DDVP, dichlorvos) have become popular for controlling insects in buildings. They are highly effective in controlling the insects and are usually safe to people; however, they give off fumes which are readily absorbed into beeswax and honey and will all kill bees placed in contaminated equipment for several months. In some instances the equipment can be salvaged only by rendering out the beeswax and starting over. In other instances the wax must be destroyed because of excessive pesticide contamination.

TREATMENT OF COLONIES POISONED BY PESTICIDES
When a colony has been poisoned by a pesticide, if the colony loss represents only its foragers, bees do not carry in any poisoned nectar or pollen, and the colony has plenty of good honey and pollen, nothing needs to be done. If the poison applied remains residual for a long time, the colony should be moved to a safer location. It will not return to productivity until its population has been restored by emerging bees. If the loss occurs at the beginning of a nectar flow, there will be little if any surplus honey by the end of the season, and the bees may have to be fed their winter stores.

If further applications of toxic pesticides are not to be applied, and there is no evidence of poisoned honey or pollen, colonies need not be moved if conditions are favorable for their development.

If the brood and nurse bees also are affected, poisoned pollen may have been gathered and, not only should the apiary be moved to a safe location, but the combs containing the contaminated pollen should be removed. Experience has shown that as long as brood combs contain poisoned pollen, whether from pesticides or from a plant source, normal development of the colony will be retarded if the colony survives at all. In many instances, package bees and swarms have died when placed on combs from colonies which previously had been poisoned. Even when such colonies survive, their development frequently is so retarded that they may be of little value for several weeks or months.

Combs containing brood as well as pollen can be concentrated in a limited number of hives until the bees emerge, and then the combs treated to remove the pollen. The combs can be soaked in water for 24 hours and then the pollen washed from the cells and the combs dried. Any pollen that may remain in the cells will become hard and will be removed by the bees without further injury.

The addition of one or more pounds of bees to a colony that has been weakened by the loss of its foragers, but which has no poisoned pollen, will enable a quicker recovery to producing strength. Occasionally, weakened colonies can be united, and will produce more than if no effort is made to bring them all back to producing strength. In all cases, surplus combs should be removed from badly weakened colonies and protected from robbing bees and the wax moth until they are needed, and the size of the entrance opening should be regulated to strength of the colony and weather conditions.

Additional procedures may be undertaken to restore poisoned colonies to productivity, if the beekeeper considers them necessary. Colonies may be moved to an area where intensive care can be provided. Intensive care may be provided in numerous ways. Colonies can be moved to a pesticide-free field in full nectar flow with plenty of pollen available. Supplemental feeding of either natural pollen, pollen substitute or a combination of natural pollen and pollen substitute is given to the bees either in dry form, in patties or mixed with honey syrup, sugar syrup or drivert sugar. Using various combinations of these procedures, we have returned a living but drastically poisoned colony to full recovery and productivity after only two complete brood cycles.

PESTICIDE CONTROL MEASURES

The federal and state regulations governing the sale and use of pesticides are designed to protect the producer as well as the consumer from the misuse of poisonous materials. Generally, these are governed by health considerations for man, domestic animals, and wildlife. In many instances the regulations are for the benefit of the majority of interests in any community in which food products are produced or sold. Beekeeping is given a considerable measure of protection by these regulations, but the use of pesticides still remains an ever-present hazard to the beekeeping industry.

The numerous research agencies at the state, federal, and commercial levels generally take into consideration the need for conserving honey bees as well as beneficial and other nontarget insects and animals when they make recommendations for the application of pesticides for the control of injurious pests and diseases. However, this is not always the case and the fact that hives of bees are transportable is given too much weight in determining the type of poison and the method and timing of its application since it is seldom possible for the beekeeper to move all the bees in danger in 48 hours and he has no other practical method of adequately protecting his bees. If bees are of little or no benefit to the crops to be treated, little or no consideration may be given to the need of using materials or methods which will cause the least injury to the bees. However, if crops that require bees for full production or crops which benefit from bees visiting the crop are grown in the near vicinity, and the chemicals cannot be confined to the fields treated, then the chemicals that are the least toxic to the bees are frequently required, and the timing and method of application are also recommended.

The formation of regional, county, and state committees, composed of the representatives of all parties interested in the end results of pest control, would be helpful in formulating policies and regulations governing the sale and use of toxic compounds.

Beekeepers can reduce their losses by becoming familiar with the chemical control programs of the regions in which they operate their colonies. It is also helpful when beekeepers register the number of hives and their locations with the county regulatory officials who have authority over the use of pesticides, and request notification in advance of the use of chemical compounds that are likely to be highly injurious to their bees. If the beekeeper knows in advance that an apiary may be poisoned, he may wish to contact the grower and compromise the pesticide used, the timing of the application or the method of application. With notification, the beekeeper is aware that his bees may be poisoned and he can help them recover.

In California, bee laws and regulations, which are enforced by County Agricultural Commissioners, provide the basis for an effective apiary inspection program to help beekeepers protect honey bees from diseases, pesticides and theft. In addition, beekeeping in some localities is governed by city and county ordinances so beekeepers should also consult local authorities.

In California all apiaries must be registered each November 1 with the Agricultural Commissioner of the county in which bee colonies are located. Registration consists of listing the location of each apiary and the number of colonies at each location. Newly acquired colonies and colonies brought into the state must be registered within 30 days of establishment. If the beekeeper changes the location of an apiary, he must inform the County Agricultural Commissioner within 5 days of the change of location. He must identify each apiary, in a conspicuous place near the entrance side of the apiary, using a sign stating in black letters not less than 1 inch in height on a background of contrasting color the owner's name, address and phone number (or information outlining any alternate means of contacting the beekeeper). If the beekeeper wishes to be notified of pesticide applications within a 1 mile radius of his apiary, he must submit a written request to the County Agricultural Commissioner requesting notifications. In some counties, the beekeeper may make a choice of the pesticides or groups of pesticides for which he wishes to be notified. If the beekeeper has not complied with these requirements, he may not be able to recover losses caused by pesticide applications.

Also in California, regulations require each registered and/or licensed pest control operator using pesticides highly toxic (Group I), or moderately toxic (Group II) to bees to obtain approval of the Agricultural Commissioner prior to the start of each operation—unless stated otherwise in the "exceptions." In addition, the use of any pesticide requiring a permit must be reported to the District Inspector 12 hours prior to each application by air, and, all those used in the vicinity of schools, dwellings, hospitals, recreational areas and livestock enclosures. The state regulations governing agricultural pest control operations require that all owners of apiaries located within 1 mile of the area to be treated are to be notified, at the apiary owner's expense, of the intent to use a pesticide, or combinations of pesticides, known to be harmful to bees; providing that the beekeeper has made a request in writing to the Agricultural Commissioner for such notification. The apiary owner must be allowed reasonable time, not to exceed 48 hours, to move or otherwise protect his bees. The apiary locations of record are available by calling the county agricultural commissioner's office.

COMPENSATION FOR HONEY BEES KILLED BY PESTICIDES

During the years of 1967 through 1978 the Agricultural Act of 1970 authorized the federal government to indemnify beekeepers who through no fault of their own suffered provable losses of honey bees as a result of the use of registered and approved pesticides on nearby or adjacent properties. The beekeeper filed evidence of his losses with his county office of the Agricultural Stabilization & Conservation Service. If his claim was accepted, he received indemnity payments accordingly: $20 for each colony destroyed; $15 for each colony severely damaged; $5 for each colony moderately damaged; $7.50 for each queen nucleus destroyed; $5 for each queen nucleus severely damaged. As an alternative he could ask that his payment be determined by the County

Committee on the basis of evidence submitted by him relating the following: cost of bees obtained to replace those lost; loss of sales of honey and beeswax; loss of pollination fees; and, loss of sales of queen bees and package bees. This act was subsequently cancelled.

This is recalled here because the requirements for compensation for honey bee colony losses caused by pesticides were carefully outlined in the Agricultural Act of 1970. They are excellent guidelines which the beekeeper should use if he suffers extensive bee losses.

The beekeeper had to prove such losses by submitting evidence such as: official reports of losses filed with the state or local authorities; daybooks or regularly kept business records in which the losses of bees were recorded at the time of loss; signed statements of certain disinterested persons having personal knowledge of the beekeeper's loss of bees; photographs, provided they were accurately identified as to date, location, and accurately showed the conditions that existed; reports filed by state or local apiary inspectors; tax returns substantiating losses of honey bees; etc. In addition, the beekeeper had to show proof of use of pesticides and submit evidence that he took reasonable steps to protect his bees. The proof of use of pesticides included: chemical tests, if any, performed on the dead bees, pollen and honey; chemical residue tests of plant parts in the treated area; records, signed statements or official reports of pesticide applications; statements of farmers who applied the pesticide(s) or contracted for its application within the area; records, signed statements or official reports of local, state or federal government agencies or colleges or universities having knowledge of the uses of pesticides in the area; etc. Proofs of reasonable care consisted of: signed statements by the beekeeper as to whether or not he received advance notice that the pesticides were going to be applied near or adjacent to his apiary; a statement describing the steps he took to protect his bees if he received advance notice, or why there was no suitable action he could take; a statement by the beekeeper describing what steps he took after his bees were exposed to pesticides to improve the condition of his colonies, or why there were no suitable steps he could take; etc.

The users and applicators of highly toxic pesticides are generally held liable for any damages caused to the property or interests of others from the misuse or careless application of the pesticides. Drifting poisons constitute a major hazard of all dust applications and, to a lesser extent, when sprays are applied. In case of losses to crops or to bees from chemicals applied for pest control, the injured parties should immediately notify the county and state regulatory authorities. The purpose of this notification is to establish the type of poison applied and the method used, as well as the conditions at the time of application, so that preventive measures can be taken to forestall similar losses in the future.

If pesticides are substantially confined to the fields treated and are used only in minimum quantity when necessary under suitable atmospheric and plant conditions, with notice issued to all interested parties who might be adversely affected, the loss of bees will be kept at a minimum. Compounds which are least toxic to beneficial insects should be used if they give control of the injurious pests. Applications made at night, early morning, or late in the afternoon, when pollinators are not active, result in less damage than if made during the hours of major flight of bees. Above all, certain highly toxic pesticides should not be applied to fruit trees, legumes, or other bee-attractive crops while they are in bloom.

Many farmers use land rigs to apply pesticides at night when the air is cool and calm and pollinators are not working. Night applications are made by airplanes to fields that are relatively free of surrounding hazards to such flights. Lesser amounts of the pesticides can be used, and with greater efficiency, when drift is cut to a minimum by using properly maintained equipment and optimum weather conditions.

Night applications are specially desirable since during the night most pest insects leave the cover of the crowns and lower foliage after a day of high temperature to feed. Therefore, they are more exposed to treatments and are actively feeding; whereas, the honey bee pollinators are safely away from direct exposure to the pesticide application.

USE OF REPELLENTS

A bee repellent used on a crop or orchard must be strong enough to overcome the natural attractiveness to prevent bees from foraging plants treated with a toxic pesticide, yet it must not injure any part of the plant or harm the operator applying the poison. Carbolic acid, creosote, lime sulfur, naphthalene, nicotine sulfate, and more recently, benzaldehyde and propionic anhydride, commonly mentioned as bee repellents, have only limited use because the efficiency of each depends upon its volatility. Additionally, some of the above listed repellents cause an herbicidal effect on some plants.

We think that a bee repellent for use on crops which, when combined with a toxic pesticide, will repel bees significantly for the daylight hours of one day, and together with the hours of darkness, would reduce bee losses from pesticides a minimum of 50%. Such a bee repellent would be worthy of development, if its cost was economical. The residues of most pesticides remain toxic for 2 to 5 days. But since the number of bees killed is reduced by 1/2 each day following application, only the kill of bees the first day is critical.

Several investigators have experimented with thousands of chemicals in search of suitable repellents to bees. Few of the compounds tried were found to be very effective, although some did decrease the number of bee visits. The author has been involved in testing chemicals for repellency to bees in both laboratory and field tests. Eventually three pesticides which have low hazards to bees were field-tested in large plots. In 1970 a laboratory screening program was initiated which continued through 1977 detecting 50 repellents while testing 509 chemicals (Atkins *et al.*, 1975). Field testing of repellents discovered in laboratory tests began in 1972. Most of these chemicals were exotic and would have to be commercially manufactured, tested and registered for use. This process takes 5 to 10 years. Therefore, efforts were directed to already existing and available selective pesticides and repellents known to have low hazard to bees while repelling or otherwise modifying the behavior of foraging bees (Atkins *et al.*, 1977).

In routine field tests to determine toxicity to bees the pyrethroid permethrin (Ambush®, Pounce®) was discovered in 1976 to exhibit true bee repellence (Atkins *et al.*, 1978; 1981). Later the same year it was discovered that a night application of demeton (Systox®) when added to a toxic pesticide modified bee foraging by supressing bee visitation for two days reducing the bee hazard 75% from the toxic pesticide (Atkins *et al.*, 1977; 1978). In 1979 disulfoton (Di-Syston®) was found to be effective both as a night and early morning application, when added to a combination of two toxic pesticides. It modified bee foraging by suppressing bee visitation up to two days and reducing the bee hazard 67 to 78 percent from the toxic pesticides (Atkins & Kellum, 1979). These series of tests indicated that both demeton and disulfoton could be utilized as bee repellents to significantly reduce pesticide kills from applications made at night (darkness to 04:00) or during the early morning (05:00 to 07:00) before bees began foraging.

The plots used in these tests were sprayed using commercial airplanes applying 5 to 10 gallon spray per acre; generally plots were 16 acres and were separated from one another by more than 1300 ft. laterally and longitudinally; seed alfalfa was in full bloom; field was stocked with three to four well established commercial bee colonies for pollination; and, the colonies were located in one group in the center of each plot. Each treatment was applied directly over unprotected bee colonies. Each test series consisted of the toxic pesticide(s) applied alone, the repellent applied alone, the repellent added to the pesticide(s), and an untreated check plot (Atkins & Kellum, 1979, 1980; and Atkins, 1981).

Subsequent tests showed that demeton, disulfoton and permethrin (applied at the rate of 0.25, 0.5 and 0.15, respectively) are all effective as repellents for 1 to 2 days for protecting bees from toxic pesticides on cotton and safflower as well as for seed alfalfa (Atkins & Kellum, 1981). We have not found a repellent which will repel honey bees from citrus in bloom.

Demeton and disulfoton are both effective organophosphate systemic insecticides and acaricides (miticides) having a low hazard for bees. Neither are true bee repellents, but are rather bee behavior modifiers in that bees actually forage the treated in a normal manner but at significantly reduced levels and with shorter intervals per flower. Demeton and disulfoton are both essentially nontoxic at field dosages.

Permethrin, a pyrethroid insecticide, is effective for controlling many agricultural pest insects. In addition, permethrin is also the most effective honey bee repellent we have found for field use. Permethrin is highly toxic to bees directly contacted or when confined in cages on foliage and/or blossom residues. However, if permethrin is applied at night and early morning hours when bees are not foraging, permethrin is a true repellent preventing foraging bees from alighting on residue-contaminated plants as long as the residue is present at toxic levels.

In 1980 permethrin was applied early morning at four different dosages in a successful attempt to establish a dosage repellence curve (Atkins & Kellum, 1980). The data from these four repellent plots showed that permethrin, because of its repellence to bees, is nonhazardous to honey bee pollinators, if it is applied when bees are not in the crop foraging.

The length of repellence from permethrin can now be customized to the length of time pesticides are hazardous by adding the appropriate dosage to the toxic spray. Data obtained from Table V enables one to construct a linear regression showing the dosage-time repellency relationships of permethrin. Therefore, if it is desirable to repel bees from a pesticide residue for 2.5 days, one can add permethrin as a bee repellent at a dosage of 0.1 lbs ai per acre; if one wants repellency for 4.5 days, for example, permethrin can be added at 0.2 lb ai per acre; etc. Other dosages and their resulting lengths of repellence can be obtained from the data in Table V (Atkins, 1981).

In seed alfalfa pollination the most important problem with permethrin has been that when using 0.2 lb ai per acre it has repelled bees for more than 4 days. Repelling bees from alfalfa blossoms and thus preventing pollination for this length of time may seriously reduce seed set. If one wished to use a bee-toxic pesticide with an extended residual interval where bees are present for pollination, then consideration should first be given to using a less persistent pesticide or a combination of pesticides which will control the pest populations equally well. The appropriate bee repelling dosage could be added to this application to further reduce the hazard to bee pollinators.

Extensive field testing and field usage will be necessary to determine the limitations of using permethrin, demeton, disulfoton and other repellents as bee repellents on seed alfalfa, cotton and safflower as well as on other agricultural crops. The results of this recent research are cause for optimism that other repellents will be discovered and developed.

Table V. Dosage/Time* Repellency of Permethrin For Worker Honey Bees in Seed Alfalfa

With A Permethrin (lbs. ai/A)	Dosage of:[1] (Kg ai/ha)	Honey Bee Repellency In Field Is:	
		Number Of: Hours	Days
0.05[2]	0.056[2]	48	2.0
0.075	0.084	54	2.25
0.1[2]	0.112[2]	60	2.5
0.125	0.140	72	3.0
0.15[2]	0.168[2]	84	3.5
0.175	0.196	96	4.0
0.2[2]	0.224[2]	108	4.5
0.2125	0.238	120	5.0
0.225	0.252	132	5.5
0.25	0.280	150	6.25
0.3	0.336	200	8.33
0.35	0.392	270	11.25
0.4	0.448	365	15.25

* Dosage-Time repellency curve = the dosage (indicated in this table), in either pounds per acre or kilograms per hectare, which when applied to a field will repel honey bees for a given number of hours or days as indicated.

[1] Dosage/Time Repellency linear regression:
Correlation Coefficient = 0.99759
Slope, probits = 4.07291
Intercept, probits = -3.39237

[2] Dosages applied in field tests.

METHODS OF REDUCING HAZARDS FROM POISONING

Beekeepers who have experienced heavy losses from pesticidal poisoning are wary about moving their colonies into seed or fruit areas where little consideration is given to the protection of pollinators in the pesticide control programs. The rentals for colonies in such areas are generally higher, reflecting the greater hazard each beekeeper takes in accepting pollination contracts. When the grower fully understands the relationship of an adequate number of bees to an economic production, he is willing to pay a sufficient rental to get the necessary number of colonies of bees. He will then take an interest in his investment to the extent of not only properly timing the applications of his pesticides and in using the least toxic substance that will control the pests, but also will interest his neighbor growers in adopting similar programs.

Growers as well as beekeepers are interested in suitable legislation that will protect their crops and bees from the misuse of toxic substances by other growers or commercial agencies. It is desirable, therefore, for grower committees at local, county, state and federal levels to meet with the legislative and regulatory officials to secure suitable legislation and regulation for the use of toxic substances for the mutual benefit of all interested parties. This frequently involves the need for research and extension agencies to gather evidence and to disseminate and evaluate information about pesticides and their use. The Pure Food and Drug regulations set tolerances that generally form a basis for the recommended uses of a majority of the toxic pesticides.

Highly toxic materials when applied in commercial quantities should be sold and used under a permit or prescription system that would give the regulatory agencies a definite record of their intended use, as well as the power to regulate such uses for the benefit of all of the interests in the areas in which they are used. The 48-hour notice to the beekeeper or seed grower of the intended use of a highly toxic pesticide to corn, milo, cotton, safflower or sugar beets in the vicinity of fruit and seed acreage is helpful, but is not an adequate answer to the poisoning problem. Frequently, the colonies cannot be moved because of the condition of the soil, the number of the colonies involved, or for other reasons. Only under definitely established emergency conditions and supervision should the highly toxic pesticides and herbicides be applied to plants in bloom, and never under conditions where the drift will prove injurious to other interests. The least toxic chemicals should be used and only when needed, and then in minimum quantities necessary and by such means that will confine the material to the fields treated and allow the maximum time for reducing injurious residues.

Beekeepers should become familiar with the pest control programs that might be used seasonally on the crops within a 2-mile radius of their apiaries. They should register and identify their apiary locations and become acquainted with the neighboring farmers, farm advisors, and regulatory officials in diversified agricultural areas, as well as in the fruit and seed areas in which they keep their colonies of bees. They should know the condition of their colonies and notify the regulatory authorities as soon as any pesticide injury of the colonies is apparent.

The temporary covering or closing of hives during the application of pesticides is costly and not always practical, especially in warm weather. Bees require water and adequate ventilation at all times and excessive heat and dryness are definite hazards to the welfare of the colony.

POISONING OF BEES BY PLANTS

The poisoning of bees from plant products is usually confined to limited areas and, therefore, is much less a problem than pesticidal poisoning. Nevertheless, poisonous plants do cause serious losses under certain conditions in limited areas. Fortunately, the honey produced from such plants is seldom injurious to man. Two of the exceptions: the honey from mountain laurel, which sometimes causes acute illness soon after being eaten, and a honeydew in New Zealand.

The mountain laurel, *Kalmia latifolia,* is found from southern Maine to Florida and Louisana on rocky hillsides and acid swamps. The plants contain a poison andromedotoxin which poisons and sometimes it occurs in honey. After eating a spoonful of such honey, people may feel numbness and may lose consciousness for several hours. No aftereffects have been reported (Lovell, 1956). Small amounts of honey may be produced in these restricted areas which has to be discarded by the beekeepers. In New Zealand there is an area where bees collect honeydew from a leafhopper, *Scolypopa australis* feeding on Tutu, *Coriaria aborea.* The honeydew is toxic to bees and to guinea pigs. There is one instance of the toxin, tutin, being toxic to man dating back to 1901. The area is closed to beekeeping (Palmer-Jones & White, 1949). In New Zealand, nectar from the karaka tree, *Corynocarpus laevigata,* is very toxic to foraging bees and the toxicity disappears from the honey in about 9 to 26 weeks. The nectar is also toxic to guinea pigs, leading to paralysis, tremors and slight convulsions. No cases of poisoning have been recorded from people eating the honey since apparently the bees use the honey before the main honey flow (Palmer-Jones, 1968; Palmer-Jones & Line, 1962 and 1969).

Of the innumerable plants visited by honey bees, comparatively few produce nectar or pollen poisonous to bees or to their brood. The effect of the injurious plants varies with environmental conditions, and the severity of injury is affected by the amount of nectar and pollen gathered from other plants in the same area. The following are the most important plants poisonous to bees or suspected as such in the U.S.A.: California buckeye, *Aesculus californica* (Spach); black nightshade, *Solanum nigrum;* death camas, *Zygadenus venenosus;* dodder, *Cuscuta* spp.; summer titi or leatherwood, *Cyrilla racemiflora;* spotted locoweed, *Astragalus diphysus* Gray = *A. lentiginosus* v. *diphysus* (Gray) Jones; Snow-on-the-mountain, *Euphorbia marginata* Pursh; *Datura* spp.; black nightshade, *Solanum nigrum;* mountain laurel, *Kalmia latifolia* L.; seaside arrowgrass, *Triglochin maritima;* whorled milkweed, *Asclepias subverticillata;* and, western false hellebore, *Veratrum californicum;* summer titi, *Crilla racemiflora;* yellow jessamine, *Gelsemium sempervirens;* tansy ragwort, *Senecio jacobaea;* and, rhododendrons, *Rhododendron* spp., and other members of the health family, *Ericaceae,* such as *Pieres* spp., *Andromeda* and *Leucothoe.*

Records from other countries list some of the more important plants poisonous to bees or suspected as such: In Russia, hellebore, *Veratrum album,* pollen is toxic to bees (Perepelova, 1949); both adult bees and brood were killed by henbane, *Hyoscyamus niger,* which contains a deadly toxin, especially to fowls (Shaginyan, 1956). In Hungary, honey from nightshade,

Datura metel, contains a poison (White, 1981). In Denmark, nectar and/or pollen of the horse chestnut, *Aesculus hippocastanum,* is toxic to the brood causing deformed bees (Johnsen, 1952). In Scotland, ornamental rhododendrons, various *Rhododendron* spp., and hybrids, contain toxins toxic to bees (MacGregor, 1960). Toxic honey from France, *Rhododendron* spp. (Hannoteaux, 1981).; from Nepel, *R. arboreum* and/or *R. campanulatum,* where visitors suffered poisoning symptoms after eating local honey (Kerkvliet, 1981). In Japan, oral administration to animals of poisonous honey which originated from azaelea, *Tripetaleia paniculata*, caused nausea, vomiting and diarrhea (Tsuchiya *et al.,* 1977). From Canada, the honey from *Astragalis miser* v. *serotinus* was toxic to the honey bee (Majak *et al.,* 1980.

Injury Caused by Poisonous Plants

The substances in poisonous plants which are toxic to bees are specific in action and may be in both pollen and nectar or may be confined to the nectar or simply to the pollen. Symptoms of plant poisoning are sometimes difficult to recognize or to be substantiated by chemical or microscopical diagnosis. The presence of symptoms usually is limited to the blooming period of the plant if the nectar is poisonous and, if the colony survives, the symptoms may disappear with the bloom. However, if the toxic substance is in the pollen, the symptoms may linger as long as the supply of pollen remains in the combs.

When only the adult bees are affected, piles of them may be found dead in front of the hive entrance, and there may not be enough adults to care for the brood or cover the brood combs. The field bees may die away from the hive and newly emerged bees may leave the hive and crawl upon the ground, or lie there stupified or dead. Newly emerged bees may have crumpled wings, or fail to shed the last pupal case from the abdomen. When poisoned by California buckeye, some of the field bees appear black and shiny from loss of hair and may tremble as in an advanced stage of paralysis. For additional information concerning paralysis, see Chapter 25, "Diseases and Enemies of the Honey Bee."

Eggs may be killed by the toxins before the queen deposits them in the cells. Brood affected by plant poisons may die any time between the hatching of the egg and the emergence of the adult. The dead brood generally lacks the brown or black colors associated with American or European foulbrood. In buckeye poisoning, larvae die soon after hatching and are removed quickly.

In one instance, attributed to locoweed poisoning, many individuals died in the late pupal stage, and bees about to emerge dried or mummified in their cells. Queens affected by buckeye poisoning and much of the injury to the colony seems to stem from their behavior. Affected queens produce eggs that do not hatch or the larvae die soon after hatching. The queens also may become incapable of laying or may lay only drone eggs. They frequently recover their egg-laying ability, partially or completely, after the colonies are removed from the buckeye territory or when other plants furnish a pollen source. Affected colonies often try to supersede their queens, but usually fail and colony mortality may be high. Some hybrid strains of bees are more resistant to certain plant poisons than purer strains.

In cases of plant poisoning reported from Florida and Georgia, larvae in all stages, accumulate in the cells, and appear purple. This is caused by summer titi, *Cyrilla racemiflora*, which grows as a shrub or small tree up to 30 feet high and ranges from Florida to Texas and north to Missouri and Virginia. The honey is said to cause the 'disease' known as "purple brood." The brood in affected colonies dies and turns blue to purple in color. Some feel that the problem is caused by bees foraging on the juice from rotten berries; others believe that the honey poisons brood only if it is not properly ripened (Lovell, 1956; Sanford, 1989). Some beekeepers located where this condition occurs move their colonies or feed them syrup to reduce the effect on their colonies.

In British Columbia, Canada, it has been reported that bees are poisoned while foraging flowering timber milkvetch, *Astragalus miser* v. *serotinus.* The nectar of this plant contains miserotoxin, a nitropropanol glycoside. Although there was no sign of brood poisoning, the dead adult bees at the hive entrance were newly emerged and/or young workers which were affected by the poisonous nectar (Majak *et al.,* 1980). Other *Astragalus* spp. of locoweed are also known to be toxic to bees.

In New Zealand the passion vine hopper, *Scolypea autralis*, produces honeydew when it feeds on the tutu tree, *Cariara arborea*, whose sap contains tutin and hyenanchin toxins. The honey produced by bees collecting the contaminated honeydew is toxic to bees. The New Zealand Ministry of Agriculture and Fisheries has the power to order removal of bee hives from certain areas where toxic honey is produced from honeydew containing these two toxins (Clinch, P.G. and J.C. Turner, 1975; Hammond, 1976).

Honey bees forage for both pollen and nectar on the 29 native species of rhododendrons in North America. An examination of brood, emerging, and older adults revealed no ill effects from collection, storage and consumption of the *Rhododendron* spp. studied. It is the opinion of many investigators that uncapped, unripe honey may be poisonous to man; but, that capped honey which is properly ripened is no longer harmful. This is also said to be true for mountain laurel and yellow jasmine honey, both of which are poisonous when uncapped, but found to be edible when capped (Olszowy, 1977).

Bee Toxic Plant Toxins
Many of the honey bee toxic plant toxins isolated from honey, nectar, or

flowers have been identified and chemical analyses protocols have been determined. White (1981) has listed the plants, their country of origin, and the toxicant(s) each contain. He lists the following:

Timber milkvetch,
   Astragalus miser v. serotinus
    Canada
		miserotoxin

Nightshade,
   Datura metel
    Hungary
	honey	scopolamine

Yellow, jessamine
   Gelsemium sempervirens
    U.S.A.
	honey
	flowers
		gelsemine

Mountain laurel,
   Kalmia latifolia
    U.S.A.
	honey	acetylandromedol
		(grayanotoxin I,
		andromedotoxin,
		rhodotoxin, asebotoxin)

Tansy ragwort,
   Senecio jacobaea
    U.S.A.
	honey	pyrrolizidine alkaloids
		(senecionine,
		seneciphylline, jacoline,
		jacobine, jacozine

Tutu, *Cariara aborea* New Zealand	honeydew honey	tutin; hyenanchin
Unknown	honey	andromedol (grayanotoxin III, diacetyl andromedotoxin
Unknown	honey	androandromedol
Unknown	honey	desacetypieristoxin B

Detection of Plant Poisoned Bees

There is no specific rule for differentiating between plant and pesticidal poisoning. Whenever symptoms of poisoning occur, a careful examination should be made of the brood, the amount of pollen present, the colony strength, and the accumulation of dead and deformed bees. In plant poisoning, other than buckeye, investigation may reveal large numbers of dead bees beneath or around the plants, in front of the hives, and all over the ground for some distance away from the hives. The effects usually are more gradual and for a longer period of time than in pesticidal poisoning and generally occur in the same area at similar periods each year, but not necessarily with equal severity.

Treatment

Familiarity with the nectar and pollen plants within flight range of each apiary is a definite aid in formulating practices to prevent plant poisoning. The seasonal succession of natural plants frequently is changed by extensive grass or brush fires. Wherever the injury causes colony death or reduced colony strength below a producing level, bees should be kept away from suspected territory during the blooming period of the plant in question. With California buckeye, about 15 million acres seem to be involved, and thus an important area of land is rendered unavailable for beekeepers for about 6 weeks in early summer. In seasons when other plants within this area provide a substantial source of pollen and nectar, the damage is less severe.

When toxic plant substances occur in the pollen, the removal of pollen-clogged combs from affected colonies is definitely helpful. Requeening and strengthening by the addition of brood and bees from normal colonies will enable those injured to make a more rapid recovery. Stimulative feeding is also helpful. Invariably, however, prevention is more effective than cure.

CHEMICAL CONTROL OF WEEDS

The chemical control of weeds by the use of various sprays, while seldom physically injurious to bees, has become so general as to cause a definite reduction in the pollen and nectar plants in many farming areas. The use of herbicide, 2,4-dichlorophenoxyacetic acid more commonly known as 2, 4-D, one of the first in general use, resulted in the elimination of many pollen and nectar plants along highways, ditchbanks, pastures, grain fields, and waste places. Such plants as the mustards, chickweed, dandelion, thistles, sweet clover, willows, and many others of importance as food sources to pollinating insects are susceptible to one or more applications of 2, 4-D. Other herbicides are presently available for killing weeds which are not susceptible to 2, 4-D (see herbicide list in Table I). Because this practice is generally less expensive than cultivation or hand labor, one can expect the use of weed control by chemical means to become more prevalent than at present. It may change the value of beekeeping locations and cause beekeepers to engage in migratory beekeeping practices in order to build numerically strong colonies for pollination or for honey production. The use of additional bee-collected pollen, pollen supplements, drivert sugar, and isomerized corn syrup has assumed a more important role in colony management.

Beekeepers can reduce some of these adverse effects by encouraging the planting of pollen- and nectar-producing plants and trees along streets, highways, watersheds, irrigation canals, and streams. In certain areas of California the Department of Highways mows around nectar and pollen plants on the highways until after the blooming period has ended. This practice should be encouraged and should be extended to other areas.

INDUSTRIAL FUMES AND SMOG POISONING

Before the use of filters, the fumes of smelters caused heavy deposits of arsenic on the plants and on and in the soil for many miles around their locations. In recent years, however, the installation of filters and precipitation processes, necessitated by the concerns to reduce air pollution, have taken large quantities of the previously injurious chemicals out of the fumes coming from the smelters. There is doubt that flourine has caused poisoning of bees attributed to it at the levels found in smelter fumes and on plants surrounding these areas (Alstad *et al.,* 1982). Certain plants take up destructive chemicals from the soil and poison bees that collect pollen and nectar. The high content of arsenicals in some soils contributes to the deposit of arsenicals on the surfaces of plants by wind action and may prove injurious to bees.

Because air pollutants (smog) have moved into many agricultural areas, beekeepers wanted to know whether bees were affected (Atkins, 1966). Caged honey bee workers were continuously exposed for four days to various concentrations of ozone, a smog component. At 0.25 part per million (ppm) of ozone little effect was detectable; at 0.5 ppm the bees were disturbed but soon settled down and no significant effect was shown. However, at 1 and 5 ppm the bees buzzed continuously and erratically, lost their appetite and lived only one-sixth as long as similar bees in filtered air.

In another set of tests, four-frame observation hives exposed continuously to 0.5 ppm ozone for four days showed an accelerated worker mortality of 9.5 percent during exposure and a 9.7 percent reduction of worker mortality for four days post exposure as compared to colonies in filtered air. In other words, at 0.5 ppm worker bees were killed at an accelerated rate, but following exposure the surviving bees lived an equally longer period of time. At the time that these tests were conducted, an ozone level of 0.25 ppm was common in the area for four to six hours per day in the late afternoon and evening. Other plant research has shown that crops foraged by bees—citrus and alfalfa, for example—are damaged by ozone well below 0.5 ppm. Thus, bees are more ozone-tolerant than are their forage plants.

Other smog component tests were conducted in 1968 to study the effect of fluorine gas on caged worker honey bees (Atkins *et al.,* 1970). Control bees receiving filtered air were compared to bees continuously receiving four to five parts per billion (ppb) fluorine. Duplicate sets of five cages containing approximately 26 bees per cage were placed in each greenhouse which was maintained at 75°F and 40% RH. These tests demonstrated that a continuous exposure to fluorine gas at 4-5 ppb is somewhat toxic to worker bees. The toxicity was expressed as an early acceleration of bee morality which reached its highest differential at four weeks and then tapered off. The last survivor in the fluorine groups outlived the last check bee by two days. The significance of these data indicate that life-span of worker bees exposed to fluorine is shortened 13 percent at the point that workers are performing as foragers and pollinators, two of the most important divisions of labor in a bee colony. On the other hand, many of the plants tested for adverse effects from fluorine were more seriously affected than were the bees.

In 1977 a study utilizing controlled gas fumigation of caged honey bee workers was made in greenhouses of the California Statewide Air Pollution Research Center whereby the effects were determined on the lifespan and mortalities of bees subjected to continuous exposure to various concentrations of SO_2 on lettuce, alfalfa, sugar beets, cotton, silver and big-leaf maple, Thompson seedless grapes and two species of pine trees growing in containers (Atkins *et al.,* 1977).

Approximately 25 bees were placed in 5- by 5- by 5-inch cages provisioned with a 15 ml. feeder containing honey-water (1:1) syrup for food. Four cage replicates were used for each dosage in the gas chamber and in the carbon-filtered air untreated check test series. Temperature and relative humidity were measured. The tests were repeated a minimum of three times at each dosage. The continuous exposure dosages utilized were SO_2 at 300 ppb at mean temperatures of (1) 62°F and (2) 81°F. Bee mortality readings were recorded daily near mid-day. The data for each trial was calculated by linear regression analyses to obtain the $LD_{10, 50}$ and $_{90}$ values (lethal dose giving 10, 50 and 90% mortality) expressed as a time-mortality curve giving the number of days of lifespan (LS). It is interesting to note that a continuous exposure to 300 ppb of SO_2 at a mean temperature of 62°F shortened bee lifespan 7.5 days (23%) at LS_{50} (lifespan time when 50% of bees have died) more than the lifespan of the check bees; tests at a mean temperature of 81°F shortened bee lifespan 26.5 days (81.3%) at LS_{50} more than the lifespan of the check bees. The effect of the 19° increase in temperature over the 62°F. level reduced bee lifespan 19 days (75.7%) with the same dosage of SO_2.

These data showed that SO_2 at a continuous exposure to 300 ppb was detrimental to honey bee workers; and, the higher mean temperature of 81°F was more highly toxic than the same dosage at 62°F. Therefore, both dosage and temperature are important considerations with this pollutant.

The significance of these data under natural conditions would need to be considered as to levels and duration of the levels of SO_2; and, as to the mean temperature prevailing during the pollution incident. Another consideration, of course, is whether plants attractive to bees can withstand these levels of contamination. In other words if no plants can survive in the area, then honey bees will indirectly not survive because of the lack of food to forage upon.

Another series of tests were conducted in 1977 studying hydrogen sulfide gas as an atmospheric air pollutant to determine the environmental effects on the lifespan and mortality of exposed worker honey bees. Utilized were concentrations of hydrogen sulfide (H_2S) at 30, 100 and 300 parts per billion (ppb); a combination of H_2S and carbon dioxide (CO_2) at 300 ppb and 50 parts per million (ppm), respectively.

The tests were made in greenhouses of the California Statewide Air Pollution Research Center whereby the effects were determined on the lifespan and mortalities of bees subjected to continuous exposure to various concentrations of H_2S on black lettuce, Hayden alfalfa, sugar beets, cotton, silver and big-leaf maple, Thompson seedless grapes and two species of pine trees growing in containers (Atkins *et al.*, 1977).

Bees were obtained from the Departmental colonies by shaking bees from hive frames into a stock bee cage. In the laboratory, only young workers were aspirated into 5- by 5- by 5-inch cages provisioned with a 15 ml glass vial containing 1:1 honey-water, v/v, syrup for food. Four to five cages were used as replicates for each dosage of gas fumigation and for the carbon-filtered air untreated check for each test series. The ambient temperature and relative humidity were recorded. Mortality readings were recorded daily near mid-day.

The data obtained from the various treatments were corrected by Abbott's formula against the untreated checks for each series to determine the net effect caused by the treatment (Abbott, 1925). The tests were repeated a minimum of three times at each dosage.

The data from each dosage was calculated as linear regression analyses obtaining the $LD_{10, \, 50}$ and $_{90}$ values (lethal dose giving 10, 50 and 90% mortality) expressed as the number of days of lifespan (LS). The data were also expressed as a time-mortality curve (LT_{50}, lethal time 50% mortality). It is interesting to note that as each continuous dosage increased, the slope (hazard potential) increased progressively. These data indicate that worker honey bee lifespan (LS_{50}, lifespan when 50% of bees have died) is lengthened (is beneficial) 10.9 days (33.5%) when continuously exposed to 30 ppb H_2S, than is the lifespan of unexposed check bees; that when bees are continuously exposed more than 13 days to 100 ppb H_2S, the lifespan (LS_{50}) is shortened (is detrimental) 10.4 days (32%) than is the LS_{50} of the check bees; and, that when bees are continuously exposed more than 14 days to 300 ppb of H_2S, the lifespan (LS_{50}) is shortened 16.5 days (50.8%) than is the LS_{50} of the check bees. When bees are continuously exposed more than 15 days to a combination of 300 ppb H_2S + 50 ppm CO_2, the lifespan, LS_{50}, is shortened 0.7 day (4.4%) more than with 300 ppb H_2S alone. The mean temperature and/or relative humidity did not exert a direct effect on the hazard to bees.

These data show that continuous exposure to H_2S gas as an air pollutant significantly shortens the lifespan of honey bee workers at concentrations of

100 ppb upwards. The LD_{50} is 123 ppb at 19 days of continuous exposure. The CO_2 in the combination of 300 ppb + 50 ppm CO_2 is more hazardous than 300 ppb of H_2S alone.

The significance of these data in relation to natural conditions has to be considered as to the presence and duration of the levels of H_2S. In other words, if no plants can survive in the area, then honey bees will indirectly not be present because of the absence of food plants.

ABNORMAL CONDITIONS

Abnormal environmental conditions during the blooming period of certain plants may produce effects that often are confused with plant or pesticidal poisoning. For example, many varieties of eucalyptus in California bloom during the winter, and at times thousands of dead bees are found on the ground under the trees. Many beekeepers attribute these deaths to injurious effects of the nectar, but it seems more probable that the bees are paralyzed by cold. On two separate occasions, hundreds of dead bees were found under black locust trees in bloom. Portions of the intestinal tracts of the bees had been forced between the segments of their abdominal walls as if by some internal explosion. In warm weather, favorable to flight, bees produce a white honey of excellent quality from the bloom of the black locust with no injury to bees. Frequently, cold, damp and windy weather occurs during the blooming period, and the injury undoubtedly is associated with some abnormal condition of this sort.

Not infrequently colonies expand their brood during periods in early spring beyond their ability to keep brood warm when the advent of cold weather forces the cluster to contract. The exposed brood is frequently killed by exposure. Careful observation will reveal the cause as exposure rather than from pesticidal or plant poisoning.

ACID RAIN

With reference to all of the discussions concerning the detrimental effects of acid rain on the environment, we have been asked what effect acid rain has on honey bees. A perusal of the literature covering the last 25 years has not revealed any references dealing directly to the effect of acid rain on honey bees.

At the lower elevations in California the soil is alkaline and the water ranges from neutral to alkaline (neutral is pH 7.0 and alkaline is above and acid is below pH 7). The surface water from the mountains is always alkaline by the time it is delivered for domestic consumption and for irrigation use.

Bradford *et al.* (1981) monitored the changes in acidity of water in 114 California lakes. Their tests revealed that there had been no essential change of acidity of water in Sierra lakes during the period from 1965 through 1980. The water in all of the 114 lakes tested had lower acidity than "normal" rainwater.

The research of Johansen and Eves (1972) showed that 10 of 11 proprietary nutrient and agricultural spray acidifiers tested reduced the pH of alkaline spray waters of eastern Washington to below the neutral level. These acidifiers enhanced the control effectiveness on pest insects of some, but not all, insecticides studied. The acidified sprays were not more hazardous to alfalfa leafcutter bees, *Megachile rotundata* (F.); alkali bees, *Nomia melanderi* Cockerell; or, of the beneficial predators studied. However, acidification markedly increased the hazard of one insecticide, trichlorfon, to honey bees, especially at the lowest pH levels.

Atkins and Sbur (1988) conducted studies to determine whether the characteristics of two pesticides applied at night were: toxic to honey bees; repelled bees foraging blooming alfalfa; or were phytotoxic to the alfalfa when applied alone; when combined with a spray acidifier; or, at different pH levels of the finished sprays. Some of the sprays were buffered to the desired pH level using potassium hydroxide (KOH). In general, the finished sprays were applied having pH levels of approximately 7.0 and between 5.72 and 3.20. Caged bees directly exposed to the pesticide sprays at application were all killed, whereas the caged bees exposed to the spray containing acidifier alone (pH 3.20) were not killed. The sprays of each pesticide killed approximately the same number of bee adults regardless of the pH level. One of the pesticides killed up to two times as many bees as the other pesticide. It was concluded that the range of pH levels in this study had no direct effect on honey bees.

The phytotoxicity studies showed that there was a slight yellowing of the leaflet tip margin in all plots at 192 hours posttreatment, including the water only plot (pH 8.6). All of the other sprays caused somewhat more extensive phytotoxic symptoms. The sprays containing the pesticide and the acidifier combination were slightly more phytotoxic than the other spray combinations; however, none was serious. The foliage affected was middle-aged leaflets which were fully expanded. Older leaflets and actively expanding leaflets were not affected. All of the sprays, including water alone, suppressed foraging for approximately two hours (until the foliage had dried following application). Foraging began to resume thereafter and by two days visitation had recovered to pretreatment levels. In some instances bee visitation reached 160 percent of pretreatment levels. This phenomenon is believed to be caused by an accumulation of nectar and/or pollen following suppression of visitation caused by spraying. There were no indications that either of the pesticide sprays, the spray combinations, or, the pH levels caused or demonstrated bee visitation repellence either directly by toxicity to bees, or, indirectly by phytotoxic action of the sprays upon the foliage or floral parts of blooming alfalfa plants.

It is the author's opinion that, where acid rain has actually significantly lowered the pH environment of a plant-growing area, honey bees would not be directly affected since bees are less affected than plants. However, as studies with components of various smogs have shown, if the plants can no longer survive in the acidic area, or, if plant growth and/or flowering are reduced or eliminated, then honey bees could be indirectly affected by having no source of nectar and/or pollen available and could not survive in the affected area.

REFERENCES

Abbott, W.S. (1925). A method of computing the effectiveness of an insecticide. *J. Econ. Entomol.* 18(2):265-267

Alstad, D.N., C.F. Edmunds, Jr., and L.H. Weinstein. (1982). Effects sof air pollutants on insect populations. *Ann. Rev. Entomol.* 27:369-384.

Anderson, L.D. and E.L. Atkins. (1967). Toxicity of pesticides to honey bees in the field. Apimondia, XXI Intern. Apicult. Congr., 194-199.

Anderson, L.D. and E.L. Atkins. (1968). Pesticide usage in relation to beekeeping. *Ann. Rev. Entomol.* 13:213-238.

Anderson, L.D., E.L. Atkins, H. Nakaihara and E.A. Greywood. (1971). Toxicity of pesticides and other agricultural chemicals to honey bees. Univ. of Calif. Agr. Extn. AXT-251, 8 pp. Rev.

Atkins, E.L. (1966). Well, Almost—Bees Breathe Smog With Ease. *Amer. Bee Jour.* 106(10):374.

Atkins, E.L. (1977). Pesticides, repellents and honey bees. pp. 8 - 11 *in:* Proc. Alfalfa Seed Production Symposium. 49 pp., Mar. 1, 3, 1977. Univ. of Calif., Coop. Ext. Service and Calif. Alfalfa Seed Production Research Program.

Atkins, E.L. (1979). Pesticides, repellents and honey bees. pp. 16 - 21 *in:* Proc. Alfalfa Seed Production Symposium. 48 pp., Mar. 6, 8, 1979. Univ. of Calif., Coop. Ext. Service and Calif. Alfalfa Seed Production Research Program.

Atkins, E.L. (1981). Repellents reduce insecticidal kills of honeybees. pp. 305-310. *In:* Proc. XXVIII International Congress of Apiculture, APIMONDIA Publishing House, Bucharest, Romania, 568 pp. English Editor: C. Meletinov.

Atkins, E.L. and L.D. Anderson. (1967). Toxicity of pesticides to honey bees in the laboratory. Apimondia, XXI Intern. Apicult. Contr., 188-194.

Atkins, E.L. and D. Kellum. (1979). Effect of pesticides on apiculture. 1979 Ann. Rpt. Project 1499, Univ. of Calif. Riverside: 631-657.

Atkins, E.L. and D. Kellum. (1980). Effect of pesticides on apiculture. 1980 Ann. Rpt. Project 1499, Univ. of Calif. Riverside: 658-701.

Atkins, E.L. and D. Kellum. (1981). Effect of pesticides on apiculture. 1981 Ann. Rpt. Project 1499, Univ. of Calif. Riverside: 702-732.

Atkins, E.L. and D. Kellum. (1986). Comparative morphogenic and toxicity studies on the effect of pesticides on honeybee brood. *J. Apicultural Research.* 25(4):242-255.

Atkins, E.L. and D.E. Sbur. (1988). Effect of pesticides on apiculture; maximizing the effectiveness of honey bees as pollinators. 1988 Ann. Rpt. Project 1499, Univ. of Calif., 31 pages.

Atkins, E.L., L.D. Anderson and E.A. Greywood. (1970). Research on the effect of pesticides on honey bees 1968-69: Part I. *Am. Bee J. 110*:387-389.

Atkins, E.L., E.A. Greywood and R.L. Macdonald. (1972). Effect of pesticides on apiculture. 1972 Ann. Rpt. Project 1499, Univ. of Calif. Riverside: 417-418, 421.

Atkins, E.L., E.A. Greywood and R.L. Macdonald. (1975). Toxicity of pesticides and other agricultural chemicals to honey bees: Laboratory Studies. Univ. of Calif., Div. of Agric. Sci., LEAFLET 2287, reprinted, 38 pp.

Atkins, E.L., D. Kellum and K.W. Atkins. (1977). Effect of pesticides on apiculture. 1977 Ann. Rpt. Project 1499, Univ. of Calif. Riverside: 589-593.

Atkins, E.L., D. Kellum and K.W. Atkins. (1978). Integrated pest management strategies for protecting honey bees from pesticides. *Amer. Bee J. 118(8)*:542-3, 547-8.

Atkins, E.L., D. Kellum and K.W. Atkins. (1981). Reducing pesticide hazards to honey bees: mortality prediction and integrated management strategies. Univ. of Calif., Div. of Agric. Sci., LEAFLET 2883, revised, 22 pp.

Atkins, E.L., D. Kellum and K.J. Neuman. (1977). Repellent additives to reduce pesticide hazards to honey bees. *Amer. Bee J. 117(7)*:438-9, 457.

Atkins, E.L., R.L. Macdonald and E.A. Greywood-Hale. (1975). Repellent additives to reduce pesticide hazards to honey bees: Field Tests. *Environ. Entomol. 4(2)*:207-210.

Atkins, E.L., R.L. Macdonald, T.P. McGovern, M. Berosa and E.A. Greywood-Hale. (1975). Repellent additives to reduce pesticide hazards to honey bees: Laboratory Tests. *J. Apicultural Res. 14(2)*:85-97.

Beran, F. and J. Neururer. (1956). Bundesanstalt für Plflanzenschutz, Wein II, Austria (Toxicity to bees of plant protection substances). *Pflanzenschutz Ber. 15*:97-147.

Bradford, G.R., A.L. Page and J.R. Straughan. (1981). Are Sierra lakes becoming acid? *Calif. Agric. 35(5/6)*:6-7.

Brown, A.W.A. (1951). Insect Control By Chemicals. John Wiley & Sons, Inc., New York.

Clinch, P.G. and J.C. Turner. (1975). Estimation of tutin and hyenanchin in honey: 3. Toxicity of honey samples from test hives, 1968-74. *New Zealand J. of Sci. 18(3)*:323-328.

Eliott, M., N.F. Janes and C. Potter. (1978). The future of pyrethroids in insect control. *Ann. Rev. Entomol. 23*:443-469.

Evenius, J. (1960). Bienezucht und pflanzenschutz (Beekeeping & plant protection). Biene und Bienezucht, 338-346.

Hammond, J. (1976). Honey poisoning. *J. New Zealand Dietetic Assn. 30(1)*:27-28.

Hannoteaux, L. (1981). Rhododendron—un mellifère soupçommé. Rev. Française d'Apiculture No. 399, 351-353.

Haynes, K.F. (1988). Sublethal effects of neurotoxic insecticides on insect behavior. *Ann. Rev. Entomol. 33*:149-168.

Hocking, B. (1950). The honeybee and agricultural chemicals. *Bee World 31*:49-53.

Jaycox, E.R. (1987). The Bee Specialist: Active Beekeeping. *Gleanings in Bee Culture, 115(12)*:704-705.

Johansen, C. (1959). The bee-poisoning hazard. *Wash. State Hort. Assn. Proc. 55*:12-14.

Johansen, C. (1966). Digest on bee poisoning, its effect and prevention. *Bee World 47*:9-25.

Johansen, C. and J. Eves. (1972). Acidified sprays, pollinator safety, and integrated pest control on alfalfa grown for seed. *J. Econ. Entomol. 65(2)*:456-551.

Johnsen, P. (1952). Deformed bees and horse chestnut poisoning. *Nord. Bitidskr. 4*:44-47.

Kerkvliet, J.D. (1981). Analysis of a toxic rhododendron honey. *J. Api. Res. 20(4)*:249-253.

Knowlton, G.F. (1944). Poisoning of honey bees. *Utah Agr. Expt. Sta. Mimeo. Ser. 310*:1-11.

Knowlton, G.F. (1948). Some relationships of the newer insecticides to honey bees. Utah State Agr. Coll. Extn. MS-725.

Knowlton, G.F., A.P. Sturtevant and C.J. Sorenson. (1950). Adult honey bee losses in Utah as related to arsenic poisoning. Utah Agr. Expt. Sta. Bull. No. 340.

Lovell, H.B. (1956). Honey Plant Manual. The A.I. Root Co., Medina, Ohio.

MacGregor, J.L. (1960). Poisoning of bees by *Rhododendron* nectars. *Scot. Beekpr. 36*:52-54.

Majak, W., R. Neufeld and J. Corner. (1980). Toxicity of *Astragalus miser* v. *serotinus* to the honeybee. *J. Apic. Res. 19(3)*:196-199.

Maurizio, A. (1960). Bestimmung det letalen Dosis einiger Flourverbindungen für Methodik der Giftwertbestimmung in Bienenversuchen. *Intern. Congr. Plant Protect., 4th, 1957, 2*:1709-1713.

McGregor, S.E. (1976). Insect pollination of cultivated crop plants. USDA, Agric. Res. Serv. Agric., Handbook No. 496. 411 pp.

Meister, R.T. (1989). Farm Chemicals Handbook '89. (666 pages) Meister Pub. Co., Willoughby, OH 44094, USA.

Musgrave, A.J. and E.H. Salkeld. (1950). A bibliography of honey bee toxicology. *Canad. Entomol. 82*:177-179.

Nominé, G. (editor) *et al.* (1982). Deltamethrin Monograph. Roussel-Uclaf, 412 pp.

Olszowy, D.R. (1977). Of bees, rhododendrons and honey. *Amer. Bee J. 117(8)*:498-500.

Palmer-Jones, T. (1968). Nectar from karaka trees poisonous to honey bees. *N.Z. Jour. Agr. 117*:77.

Palmer-Jones, T. and I. Forster. (1958). Effect on honey bees of DDT applied from the air as a spray to lucerne; notes on lucerne pollination. N.Z. Jour. Agr. Res. 1:627-632.

Palmer-Jones, T. and L.J.S. Line. (1962). Poisoning of honeybees by nectar from the karaka tree (*corynocarpus laevigata* J.R. *et* G. Forst). *N.Z. Jour. Agr. Res. 5*:433-436.

Palmer-Jones, T. and L.J.S. Line. (1969). Poisoning of bees from karaka trees, *N.Z. Beekpr. 30*:8-10.

Palmer-Jones, T. and E.P White. (1949). A recent outbreak of honey poisoning. Part VII: observations on the toxicity and toxin of the tutu (*Coriaria aborea* Lindsay). *N.Z. Jour. Sci. & Tech. 31A*:46-56.

Pangaud, C. (1960). Reveu bibliographique des intoxications de l'abeille. *Bull. Apicult. Inform. 3*:109-180.

Perepelova. L.I. (1949). Effect of hellebore pollen on bees. *Works. vet. Sect. Lenin Acad. Agr. Sci. Session. 27*:55-65.

Rivera, R. and W.T. Wilson. (1989). Presence and removal of menthol in honey and beeswax. Proc. Amer. Bee Res. Conf., *Amer. Bee J. 129(12)*:821.

Rivera, R., J.O. Moffett and R.L. Cox. (1987). Menthol residues in honey, beeswax, and honey bees. Proc. Amer. Bee Res. Conf., *Amer. Bee J. 127(12)*:850.

Sanford, T. (1989). Southeast exposure "Florida's Favorites". *Gleanings in Bee Culture 117(1)*:18-19.

Shaginyan, E.G. (1956). Poisoning of bees by the alkaloids of henbane. *Pchelovodstvo 33*:45-46.

Shaw, F.R. (1941). Bee poisoning: review of the more important literature. *Jour. Econ. Entomol. 34*:16-21.

Smirle, M.J., W.L. Winston and K.L. Woodward. (1984). Development of a sensitive bioassay for evaluating sublethal pesticide effects on the honey bee (Hymenoptera: Apidae). *J. Econ. Entomol* 77(1):63-67.

Smith, R.K. (1989). A rapid acaricide detection for honey analysis. Proc. Amer. Bee Res. Conf., *Amer. Bee J.* 129(12):823.

Todd, F.E. and S.E. McGregor. (1952). Insecticides and bees. U.S. Dept. Agr. Insects, The Yearbook of Agriculture, 131-135.

Todd, F.E. and S.E. McGregor. (1960). The use of honey bees in the production of crops. *Ann. Rev. Entomol* 5:265-278.

Tsuchiya, H. (and 5 others). (1977). Studies on a poisonous honey originated from azalea, *Tripetaleia paniculata.* Kanagawa-kan Eisei Kenkyusho Kenkyu Hokoku No. 7, 19-28, Yokohama, Japan.

White, J.W. Jr., (1981). Natural honey toxicants. *Bee World 62(1)*:23-28.

Wittmann, D. (1981). Determination of LC$_{50}$ of Dimilin 25 WP on honey bee brood using new "*Apis* larvae test." *Zeitschrift für Angewandte Entomologie 92(2)*:165-172.

Wittmann, D. (1982). Development of methods and experiments for testing the effects of insecticides on honeybee larvae. Dissertation, Univ. of Tübingen, German Federal Republic, vi + 96 pp.

Wittmann, D. and W. Engels. (1987). On which diet can worker honeybees be reared *in vitro*? *Apidologie 18(3)*:279-288.

Wittmann, D., R. Kuhn, R. Hertland, and W. Engels. (1985). Application of new *Apis*-larvae-test for determination of CL$_{50}$ and DL$_{50}$ of pesticides on brood frames and *in vitro*. *Mitteilungen Dtsch. Ges. Allg. Angew. Ent.* 4:336-338.

CHAPTER 27

ALLERGY TO VENOMOUS INSECTS

by JUSTIN O. SCHMIDT*

O n first contact with a beekeeper, members of the general public frequently make statements to the effect that they, or a close relative/ friend are "highly allergic to bee (or insect) stings and are likely to die if stung again." Besides being unfortunate, incorrect, and misleading, this public perception of the risk from bees is the cause for much unnecessary personal limitation and suffering.

How has this situation arisen? It is probably a misunderstanding based on the natural human fear and dislike of pain, the fear of uncertainty, and the wide publicity of serious allergic reactions that have resulted from stings. An important role of beekeepers, entomologists, and physicians should be educating of the public about the realistic dangers of sting allergies and, importantly, helping to dispel unfounded and unrealistic fears. The latter task is by far the more difficult of the two and has received proportionately little attention. The object of this chapter is to present detailed descriptions of the entomological and medical aspects of allergies to stings and to provide information and guidance necessary for rationally dealing with both minor and serious allergies, as well as perceived allergies.

STINGING INSECTS

Stinging insects are members of the aculeate Hymenoptera, a large group consisting of several thousand species of ants, bees, and hunting wasps. Species of stinging insects that are capable of inflicting painful or damaging stings to humans are small in number and restricted mainly to the social Hymenoptera plus a few exceptional solitary taxa. Among this small group of venomous insects, only a very few species are of major concern to humans. These include the yellowjackets of the genus *Vespula* (sometimes split into two genera, *Vespula* and *Paravespula*), the aerial yellowjackets of the genus *Dolichovespula* (which also includes the baldfaced hornet), the (true) hornets of the genus *Vespa*, the paper wasps of the genus *Polistes*, the honey bee *Apis mellifera*, and several species of ants including the fire ants (mainly *Solenopsis invicta*) and the harvester ants (*Pogonomyrmex* spp.). The first four are all members of the family Vespidae and are often simply referred to as vespids. Honey bees are in the family Apidae, which also includes the bumble bees (*Bombus* spp.). Ants are in another family, the Formicidae. A more inclusive list of Hymenoptera known to have painful stings or to cause allergic reactions to their stings is

*Justin O. Schmidt, Ph.D., Entomologist, Carl Hayden Bee Research Center, U.S. Dept. of Agriculture, Agric. Res. Serv., 2000 E. Allen Road, Tucson, AZ 85719

provided in Figure 1. More detailed biological information on the groups included in Figure 1 is available in a number of reviews (Spradbery, 1973; Edwards, 1980; Michener, 1974; Cole, 1968; Hölldobler and Wilson, 1990).

Among the general public there is an unfortunate tendency to blame any allergic reactions to stinging insects on honey bees. This is because the honey bee is familiar, in name at least, to almost everyone, and because most people do not like insects. Consequently, many people cannot, or do not, distinguish among honey bees, yellowjackets and other general flying insects and use the term "honey bee" simply because it is the first name that comes to mind. This public generalization places unfair blame on the honey bee and beekeepers, and greatly increases the difficulty of properly diagnosing and treating the allergic situation. If the actual species causing the problem cannot be definitively identified, precautions, advice and treatment often must be generic for stinging insects, rather than specific for the offending species.

The stinging insects responsible for most of the venom allergy in the United States are yellowjackets plus the baldfaced hornet, honey bees, fire

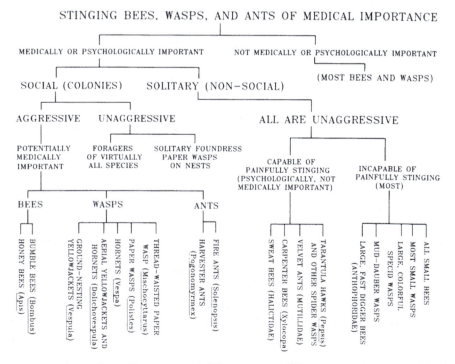

FIGURE 1. Categories and importance of different groups of Hymenoptera that are capable of stinging humans.

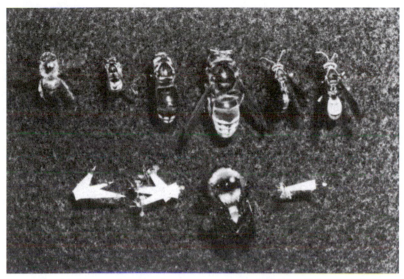

Figure 2. Stinging insects of medical importance (shown at life size: From top left) honey bee (*Apis mellifera*); yellowjacket (*Vespula* sp.); baldfaced hornet (*Dolichovespula maculata*); European hornet (*Vespa crabro*); two species of paper wasps (*Polistes* spp.); (bottom from left) Fire ants (*Solenopsis* sp.); harvester ants (*Pogonomyrmex* sp.); bumble bee (*Bombus* sp.); sweatbee (Halictidae: *Lasioglossum* sp.).

ants, and paper wasps (Figure 2). Minor offending species include the harvester ants, the introduced European hornet, bumble bees, and sweat bees. In Europe sting reactions are most frequently caused by yellowjackets (often called simply "wasps"), honey bees, paper wasps, and hornets. In Asia the incidence of venom allergy is less well studied, but the species causing most of the allergic problems are yellowjackets, honey bees, paper wasps, and hornets with a variety of potential minor species such as other honey bee species (*Apis cerana, A. dorsata, A. florea, S. andreniformis,* etc.) and a host of ant species. In the United States yellowjackets are generally considered the most important causes of sting allergy, with honey bees in close second place, and the remaining species of considerably less importance (Hoffman *et. al.,* 1980; Lockey *et al.,* 1988, 1989).

A recent and very serious cause of allergy in the southern United States are the fire ants, mainly *Solenopsis invicta,* that were introduced from South America during the first half of the 20th Century. The relative importance of the different groups varies with geographical area. Yellowjackets and other vespid wasps overall are the most serious offenders in the United States and cause over twice as many allergic reactions as honey bees (Lockey *et al.,* 1988). Moreover, these vespids cause more serious reactions, with the incidence of serious life-threatening reactions being induced three times more frequently by wasps than bees (Lockey *et al.,* 1988). Vespids are a particularly serious problem in the northeastern and northwestern United States. Fire ants are a very serious source of allergic reactions in the states of South Carolina to Florida and along the Gulf Coast to Texas. Honey bees are the most important stinging insects in the Pacific Southwest (Hoffman *et al.,* 1980). The Sonoran desert areas of Arizona are unique in the United States because yellowjackets do not exist there and, sting allergy is often induced by paper wasps, harvester ants, and the biting bugs called "kissing bugs" (*Triatomma*) as well as by honey bees.

VENOM COMPONENTS OF BEES AND
OTHER STINGING INSECTS

The venoms of most stinging insects consist of proteins, peptides, and a variety of smaller molecules (Table 1). The pharmacological and biochemical activities of the various stinging insect venoms are remarkably convergent. Most venoms induce immediate pain, contain phospholipases, hyaluronidase, and other enzymatic activities, and are capable of destroying red blood cells (Schmidt *et al.,* 1986). Most hymenopterous venoms also contain low molecular weight peptides that are highly basic and have isoelectric points ranging from pH 9 to 12 (Habermann and Jentsch, 1967; Haux, 1967; Haux *et al.,* 1969; King *et al.,* 1976; Nakajima, 1986; Hirai *et al.,* 1979; Bernheimer *et al.,* 1980, 1982; Argiolas and Pisano, 1985).

In contrast to the pharmacological similarities of insect venoms, their biochemical structures differ remarkably among the various taxa. The phospholipases derived from honey bees, yellowjackets, and fire ants have very different molecular weights: 16,000 for bees (Shipolini *et al.,* 1974), 30,000 to 37,000 for yellowjackets (Hoffman and Wood, 1984; Hoffman 1985a; King *et al.,* 1985; Takasaki and Fukumoto, 1989), and 28,000 in the form of two roughly equal subunits for fire ants (Hoffman *et al.,* 1988a). Their specific activities are also strikingly different as indicated by the assignments of A_2, A_1, or B activity (Table 2). These activities refer to the site of hydrolysis of the phospholipids and how many sites can be hydrolyzed by the enzyme (designations for the "A" activity of the ants is not known and can be either A_1 or A_2). The amino acid compositions and sequences of the phospholipases are also different as determined by direct analysis (Shipolini *et al.,* 1974; King *et al.,* 1984; Hoffman *et al.,* 1988a) and lack of cross-reactivity by antibodies (Nair *et al.,* 1976; King *et al.,* 1978, 1985; Hoffman *et al.,* 1988b).

The venoms of honey bees, yellowjackets, paper wasps, true hornets, and harvester ants all contain small, highly basic pain-inducing peptides. These algogenic peptides exhibit almost no structural similarities among the venoms of the different insect families. The honey bee peptide, melittin, contains 26

Table 1. Biochemical composition of venoms of insects that frequently sting humans.[a]

	Honey bees	Yellow-jackets	Paper wasps	Hornets	Fire ants	Harvester ants	Bull ants	Bumble bees
ENZYMES								
Phospholipase A_2	10-12%	0	0	0	+[b]	+[b]	+[b]	+
A_1	0	+	+	+				
B	1%	+	+	+	+	+	0	+
Hyaluronidase	1-2%	+	+	+	+/−	+	+	+
Acid phosphatase	1%	0	0		+	+		+
Alkaline phosphatase	+	0	0			+		
Lipase	0	+	+			+		
Esterase	+	+	+			+		
Protease	0	0	0	0(?)	0	0		
PEPTIDES								
Melittin	40-50%	0	0	0		0		0
Apamin	3%	(0)	(0)	(0)				
MCD-peptide	2%	(0)	(0)	(0)				
Secapin	.5%							
Tertiapin	.1%							
Bombolitins	0							+
Kinins	0	+	+	+		+		
Mastoparans	0	+	+	+				
Chemotactic peptides	0	+		+				
Antigen 5	0	+	+	+				
Vespid neurotoxins	0	+		+				
Barbatolysin	0	0	0	0		+		
SMALL MOLECULES								
Histamine	.7-1.6%	+	+	+		+	+	
Dopamine	.1-1.0%	+	+	+				
Norepinephrine	.1-.2%	+	0			+		
Acetylcholine	0	0	0	+			0	+
Putrescine	0							+
Serotonin	0	+	+	+		0	0	
Tyramine	0	+	+	+				
Leukotrienes	.003%	+	+					
Alkaloids	0	0	0	0	95%	0		

a Data from: Piek, 1986; Schmidt, 1982, 1985; Schmidt *et al.*, 1986; Hoffman, 1982, 1985c; Hoffman *et al.*, 1988a; King *et al.*, 1984; Benton, 1967; Piek *et al.*, 1989; Takasaki and Fukumoto, 1989; Czarnetzki *et al.*, 1990.

b Specificity of A_1 or A_2 unknown.

amino acids, seven which form a very hydrophilic end of the molecule and 19 which form a highly hydrophobic end. In water, melittin readily forms into groups of four peptides (tetramers) held by non-covalent bonds (Habermann and Reiz, 1965; Banks and Shipolini, 1986). In vespid wasp venoms the algogenic peptides analogous to melittin are called kinins. Although kinins vary in structure among the various species of wasps, they all contain from 15 to 22 amino acids. Unlike melittin, kinins do not form tetramers in solution or have hydrophilic or hydrophobic ends (Nakajima, 1986). In harvester ants a presumed major algogenic peptide is called barbatolysin, a peptide of 32 amino acids of strikingly different composition from those in the bee or wasp peptides (Bernheimer *et al.,* 1980). Bumble bee venom contains no melittin (Hoffman, 1982). Instead it contains five peptides called bombolitins which are similar to honey bee melittin in some activities, but are structurally different (Argiolas and Pisano, 1985).

In addition to proteins and peptides, most insect venoms contain small active organic molecules including various biogenic amines and alkaloids. Fire ant venoms are unusual in that their compositions are about 95% piperidine alkaloids and only a small percent protein (Blum *et al.,* 1958; Baer *et al.,* 1979; Hoffman *et al.,* 1988a). The alkaloids cause most of the local sting reactions (Blum *et al.,* 1958; Jones and Blum, 1983), whereas the proteins cause allergic reactions (Baer *et al.,* 1979; Hoffman *et al.,* 1988a). All insect venoms investigated other than fire ants (and likely their near relatives) contain one or more of histamine, acetylcholine, serotonin (5-hydroxy-tryptamine), and catacholamines (dopamine, norepinephrine, etc.). These small molecules are normally neurotransmitters or hormones in the human body and, when injected into the skin, cause various reactions such as pain, constriction or dilation of blood vessels, and swelling. Their exact functions in venoms are unknown, but they might be of use in inducing local reactions and synergizing the activities of other venom components.

The allergens in bee, wasp, and ant venoms are all proteins (Table 2). These proteins are not generally believed to cause immediate sting pain (melittin, a minor allergen is an exception [Prince *et al.,* 1985]), but cause, or enhance, toxicity; and all are capable of inducing hypersensitive reactions in some individuals. An evident feature in Table 2 is that all of these allergens except hyaluronidase and possibly acid phosphatase are restricted to the members of only one specific family of stinging insects. The special case of hyaluronidase will be discussed under the subject of cross-reactivity of venom allergens.

From an evolutionary standpoint the pharmacological similarities of hymenopterous venoms are not surprising. There appear to be only a few truly effective ways to repel attackers (pain, tissue damage, swelling, lethality, etc.)

Table 2. Known allergens in insect venoms[a]

Allergen	Mol. Wt.	Percent of Venom			allergenic importance
		Honey bee	Vespid wasps	Fire ants	
Phospholipase A_2	15,800	10-12	—	—	major
Phospholipase A_1B	31-37,000	—	10-25	—	major
Phospholipase A_2B	28,200	—	—	$10\text{-}12^c$	major
Hyaluronidase	$40\text{-}46,000^b$	1-2	1.5-5	+/–	maj/moder
Acid phosphatase	$98,000^d$	1	—	<1	mod/major
Antigen 5	22-25,000	—	15-40	—	major
Allergen C	102,000	<1	—	—	moderate
V mac 1	97,000	—	1	—	moderate
V mac 3	39,000	—	<1	—	minor
Solenopsis I	35,000	—	—	$.5\text{-}1^c$	major
Solenopsis III	26,000	—	—	3^c	major
Solenopsis IV	14,000	—	—	$1\text{-}1.5^c$	major
Melittin	2,800	40-60	—	—	minor

[a] Data from: Aukrust *et al.,* 1982; Hoffman, 1985a,c; Hoffman and Wood, 1984; Hoffman and Shipman, 1976; Hoffman *et al.,* 1987,1988a,b; Kemeny *et al.,*1983a,1984; King *et al.,* 1976,1978,1983,1984; Takasaki and Fukumoto, 1989.

[b] 40,000 for honey bee, 46,000 for yellowjacket.

[c] Percent of venom protein.

[d] For honey bee.

with a venom. Consequently, selection pressure has caused the convergence of many venoms toward these activities. Selection pressure would not be expected to cause the evolution of identical biochemical structures for the venom components that are responsible for the activities, only to cause the final activities to be similar. Thus, biochemical differences in the insect venoms are to be expected because of the independent evolution of each species with its own set of molecules. The fact that venoms are pharmacologically similar, yet biochemically different, has been the source of major confusion in the scientific and medical literature.

Further discussions of individual venom components from various stinging Hymenoptera will not be presented here. In-depth analyses are available in reviews (Bettini, 1978; Schmidt, 1982; Piek, 1986).

ALLERGIC REACTIONS TO INSECT STINGS

Allergy is a general term that describes a variety of human symptoms and reactions to a diversity of materials including pollen, animal dander, foods, drugs, dust mites (house dust), stinging insects and others. Stinging insect allergy refers to sting-induced systemic reactions of the body that occur at body locations distant from the sting site. Allergic reactions do not include the immediate pain caused by the sting itself or to the burning, redness, itching and swelling that might occur around the sting site. Such reactions including very large local swelling are referred to as "local reactions."

In theory any stinging insect species can cause allergic reactions in humans. This is because an insect sting introduces venom—which essentially is a blend of foreign proteins—into the body where it contacts the immune system and can induce production of allergy-causing antibodies. In practice, however, only very few stinging species are responsible for most of the allergic problems. This is because sting allergy usually requires at least two stings separated by a period of time; and, with extremely rare exceptions, only social species defending their colonies ever sting humans. Solitary (*i.e.* nonsocial) bees and wasps do not defend nests and will only sting if actually held or trapped against the skin. Thus, the opportunity for several separate stinging occasions by solitary Hymenoptera is very rare.

An allergic reaction typically occurs after the second or subsequent stinging event by the same or a closely related species. The first sting, (or stings), induces the production of the allergy causing antibody, immunoglobulin E (IgE), by the body resulting in the sensitization of the individual to the venom. Later when the now hypersensitive individual is stung again, the venom causes an IgE-mediated allergic reaction.

Normal and allergic reactions to stings can vary enormously from individual to individual (Table 3). Normal reactions are those that virtually everybody experiences and are characterized mainly by pain and burning that typically are intense for a few minutes and then decrease over time. After the intense pain decreases a redness and swelling are often observed and these can last several hours to a day or more. Normal reactions, though painful and frightening to some people, usually accomplish the goals of the insect and cause the stung person to retreat and stop disturbing the insect or its nest.

Like normal (non-allergic) reactions, large local reactions are nothing to be feared. Though they are thought to be immunologically based reactions (Hoffman, 1978b; Abrecht *et al.,* 1980; Green *et al.,* 1980; Mauriello *et al.,* 1984; Graft *et al.,* 1984), they rarely progress to systemic reactions (Abrecht *et al.,* 1980; Mauriello *et al.,* 1984; Graft *et al.,* 1984; Reisman, 1988; Golden, 1989). Moreover, the frequency of individuals who experience large local reactions later having systemic reactions is no greater than that of people not experiencing large locals (Abrecht *et al.,* 1980; Mauriello *et al.,* 1984; Graft *et al.,* 1984). Large local reactions are characterized by local swelling around the sting site that begins several hours to a day after the sting. The swelling might continue to increase for a few days and sometimes becomes enormous. Huge swellings, including those extending as much as from the finger tips up the arm past the elbow are not rare, occurring in as much as 17% of the general

Table 3. Normal and allergic reactions to insect stings

I. Normal, non-allergic reactions at the time of the sting
    Pain, sometimes sharp and piercing
    Burning, or itching burn
    Redness (erythema) around the sting site
    A white area (wheal) immediately
        surrounding the sting puncture mark
    Swelling (edema)
    Tenderness to touch

II. Normal, non-allergic reactions hours or days after the sting
    Itching
    Residual redness
    A small brown or red damage spot at the puncture site
    Swelling at the sting site

III. Large local reactions
    Massive swelling (angioedema) around the sting site
        extending over an area of 10 cm or more and
        frequently increasing in size for 24 to 72 hours,
        sometimes lasting up to a week in duration

IV. Cutaneous allergic reactions
    Urticaria (hives, nettle rash) anywhere on the skin
    Angiodema (massive swelling) remote from the sting
        site
    Generalized pruritis (itching) of the skin
    Generalized erythema (redness) of the skin remote
        from the sting site

V. Non life-threatening systemic allergic reactions
    Allergic rhinitis or conjunctivitis
    Minor respiratory symptoms
    Abdominal cramps
    Severe gastrointestinal upset
    Weakness
    Fear or other subjective feelings

VI. Life-threatening systemic allergic reactions
    Shock
    Unconsciousness
    Hypotension or fainting
    Respiratory distress (difficulty in breathing)
    Laryngeal blockage (massive swelling in the throat)

population (Golden *et al.,* 1982b). Large local reactions appear to occur more frequently as a result of bee stings than yellowjacket stings (Graft *et al.,* 1984), for reasons not understood. Many experienced beekeepers have large local reactions the first time they are stung in the spring of the year and after that have little or no reaction.

Large local reactions can be a problem because they 1) are a painful nuisance and often restrict movement of the affected part for days, 2) cause concern in the reactive individual and those around him or her, and 3) sometimes cause patients and/or doctors to demand or prescribe unnecessary and expensive treatments. Large local reactions do not require medical testing or treatment. Skin tests, radioallergosorbent tests (RAST, or blood tests), and/or immunotherapy are not necessary nor recommended (Albrecht *et al.,* 1980; Parker *et al.,* 1982; Mauriello *et al.,* 1984; Graft *et al.,* 1984; Reisman, 1988). The only examples where large local reactions without other symptoms might require medical treatment are where they occur in the neck or inside the mouth and acutely threaten to block breathing or air flow. In those cases, prompt medical care is indicated. Self treatment, if desired to alleviate the symptoms of large local reactions, consists of administering aspirin (or other anti-inflammatory agents) and antihistamines to help reduce the swelling. These treatments are especially helpful for individuals known to be large local reactors who take the medications as soon as the first sting occurs.

The next category (Table 3) of reactions to insect stings, the "cutaneous allergic reactions" has been the subject of much medical debate. Some schools of thought recommend treatment including immunotherapy (allergy shots) (Norman and van Metre, 1990) and others believe immunotherapy is unnecessary (Reisman *et al.,* 1985; Kampelmacher and van der Zwan 1987; Engel *et al.,* 1988). The current trend appears to favor lack of aggressive treatment such as immunotherapy. Cutaneous allergic reactions all have one feature in common—they affect only the skin. They can be very unpleasant and even frightening to those who fear, or have been led to fear, that such reactions will progress to more serious reactions. This line of reasoning (plus the fact that these reactions are clearly mediated by the allergy-inducing antibody IgE) is the basis for the belief that cutaneous reactions often should be treated with immunotherapy. There is no evidence that cutaneous allergic reactions frequently progress to serious life-threatening systemic reactions (Kampelmacher and van der Zwan, 1987). An instructive way to view cutaneous allergic reactions is to realize that skin reactions cannot threaten one's life. Nobody has ever died of skin reactions to stings. Therefore, management to save one's life, which is an ultimate goal of immunotherapy, is not necessary for cutaneous reactions.

Category V in Table 3 is a miscellaneous category of allergic reactions that sometimes follow a sting and are neither skin reactions, nor life-threatening. These reactions can be extremely unpleasant and/or frightening to the victim. Their management depends upon the individual situation. A full understanding of sting allergy, an evaluation of one's own life style, and consultation with an allergist experienced in venom allergy can be helpful in arriving at a final and appropriate plan of management.

Category VI in Table 3, the life-threatening systemic allergic reactions, includes the reactions most discussed and feared. Fortunately, they are also relatively rare. These reactions affect the body's life support systems, the circulatory and respiratory systems, and any serious direct attack on either of these systems can ultimately lead to death. These reactions include shock, extreme hypotension which might result in fainting and loss of consciousness, and serious difficulty in breathing either due to constricted lungs or bronchia or to swelling in the throat area that blocks air flow. Any of these reactions must be considered serious, and appropriate first aid and treatment are recommended.

CAUSES AND RISK FACTORS IN STING ALLERGY DEATHS

Allergic reaction to venom allergens is the ultimate cause of death in sting-related deaths; however, the proximate, or immediate, causes of death can reveal some interesting information of value for risk analysis. Sting related deaths are not randomly distributed throughout the general population, but are clustered in certain groups. For example, children and young people are at very low risk of death, while those over the age of 40 are the most susceptible (Table 4). This is in marked contrast to the situation with respect to snake and spider bites, where the young are at the greatest risk by a large margin (Table 4). The reasons for these differences reflect the nature of the immediate causes of death. In the case of snake and spider bites, death inevitably occurs as a result of toxin overload; and since the ability of the body to handle toxins (or drugs, alcohol, etc.) is related to the body size, it is of little surprise that young children are particularly susceptible to snake bites. Older people are also more susceptible simply because their bodies are no longer as strong and able to deal with severe challenges as they were when younger.

In sting-induced deaths, the venom toxins, themselves, are of no direct consequence in causing death. Hence, adult (or large) size is of no intrinsic benefit. What are important are: 1) how long the body has had to experience previous stings and develop sensitization, 2) how well and how normally the immune system functions, and 3) how the rest of the body reacts during an acute anaphylactic episode. Increased age adversely affects all of these factors. Although clear evidence is difficult to obtain, it is apparent that the development of hypersensitivity in those predisposed to develop it is related to the

Table 4. Ages at death of victims of Hymenoptera stings and snake/spider bites.

Age at death (years)	% of Deaths	U.S. population in age bracket[a]	Relative risk (age 20-29 = 1.00)
	INSECT STINGS (n = 335 Deaths)[b]		
0 - 9	4.8	14.8	1.3
10 - 19	2.1	18.1	.5
20 - 29	3.6	15	1.00
30 - 39	14.6	15	4.0
40 - 49	23.0	10.4	9.0
50 - 59	23.9	8	12.4
60 - 69	17.3	7	10.3
70 +	10.7	10	5.0
	SNAKE/SPIDER BITES (n = 217 Deaths)[c]		
0 - 9	30.4	14.8	5.5
10 - 19	8.3	18.1	1.3
20 - 29	5.5	15	1.00
30 - 39	6.9	15	1.3
40 - 49	12.0	10.4	3.1
50 - 59	16.6	8	5.6
60 - 69	7.8	7	3.0
70 +	12.5	10	4.0

[a] Vital Statistics of the United States (1986).

[b] Combined totals from Parrish, 1963; Sommerville *et al.,* 1975; Ennik, 1980; Mosbech, 1983.

[c] Combined total from Parish, 1963; and Ennik, 1980.

number of stinging incidences. For typical non-beekeepers, the occurrence of stings is infrequent and more or less random over time; thus the longer one lives, the greater the number of opportunities to be stung and develop allergy. As the human body ages, it becomes less smooth and perfect in its operation. This is true of the immune system as well. With increasing age comes decreased function and increased incidence of malfunction. In oversimplified terms, venom-induced allergy can be viewed as a malfunctioning of the immune system and its system of regulation.

Perhaps the most important reason for increased death rate in individuals over 40 years of age relates to how their bodies react to the acute stress of a hypersensitive reaction. During the anaphylactic reaction, large quantities of highly active substances are released into the blood stream. These, in turn, can cause movement of fluid from the blood stream to adjoining tissues, swelling and increased fluid production in the respiratory system, contraction of smooth muscles, drop in blood pressure, and other actions. A normal, healthy, near perfect body can usually handle these stresses and recover. But an individual already burdened with, for example, coronary disease is at severe risk of having the circulatory system collapse under the added demands of anaphylactic trauma (Rubenstein, 1982; Barach *et al.,* 1984; Freye and Ehrlich, 1989). Moreover, as the body ages, it becomes less tolerant of hypotension and loses some of its ability to transfer oxygen from the air through the lungs to the rest of the body. Thus, during periods of anaphylactically induced low blood pressure, the trauma to the brain from hypoxia is increased by the already lowered abilities of the oxygen transport system.

Table 5 is a fact sheet listing some of the data relating to insect sting allergy as reported by the insect Committee of the American Academy of Allergy and Immunology (Lockey *et al.,* 1988). This data was collected by 84 allergists practicing throughout most of North America and necessarily reflects some collection technique influences. For example, the data was collected only from patients who sought an allergist's assistance and does not reflect the numbers, or frequencies of, reactions for which either no assistance or assistance from nonallergists was sought. Thus, it probably reflects a cross-sampling of the most concerned citizens and most serious reactions. The findings in this report were typical of earlier reports (Insect Allergy Committee, 1965) and represent a wealth of information derived from 2,866 patients who experienced at least one systemic (categories III - VI of Table 3) reaction. Of those, 3.1% were beekeepers and 24% had experienced a systemic reaction, yet were skin test negative. Also, 24.8% of the individuals who experienced serious life-threatening reactions (category VI in Table 3) had no previous systemic reaction. These figures indicate that skin tests do not always detect allergy and that very serious reactions frequently occur without prior warning (Jensen, 1962; Barnard, 1973; Mosbech, 1983).

The majority of patients listed in Table 5 experienced only one previous systemic reaction before entering the survey. Most of the rest had only two previous reactions and less than 10% experienced more than two reactions. These data suggest that the number of individuals who have multiple systemic reactions is low (but in the case of beekeeping households this probably is not the case due to the frequency of repeated stinging events). If we focus just on the truly serious systemic reactions, the life-threatening reactions, we find that reaction symptoms generally occur very rapidly after the sting. Seventy percent occur within 10 minutes and almost 98% occur within one hour. The frequency of life-threatening reactions is much greater in older age groups than younger age groups. Children under 17 years of age reported only 12.7% of reactions as serious, as opposed to 25.9% for the 17-39 group, and 38% for the 40 and over group. These figures are highly significantly different between each adjoining age group (x^2 test, p <.001).

A final figure in Table 5 that is of interest, especially with the arrival of Africanized bees in the United States, is the relationship between number of

Table 5. Frequencies and characteristics of allergic reactions to stinging insects.[2]

Total surveyed individuals with systemic reactions:		2866
Percent beekeepers:		3.1%
Percent who experienced a reaction, yet were skin test negative		24%
Percent who experienced serious life-threatening reactions, yet had no previous systemic reactions:		24.8%
Number of previous systemic reactions:		
1 Only	70.3%	
2	20.3%	
>2	9.4%	
Time to onset of reaction in only those patients who experience serious life-threatening reactions:		
0 - 10 min	70.0%	
11 - 20 min	16.3%	
21 - 40 min	8.5%	27.8%
41 - 60 min	3.0%	
1 - 6 hr	1.5%	2.2%
>6 hr	.7%	
Percent of systemic reactions that are serious life-threatening reactions by age group:		
<17 yrs	12.7%	
17 - 39 yrs	25.9%	
>39 yrs	38.0%	

Number of stings as a factor in seriousness of reactions:

Stings	Life-threatening (%)	Non systemic Reactions (%)
1	16.4	32.9
>10	33.3	15.7

[2] Data from Lockey *et al., 1988.*

stings received at one time and the type and severity of the reaction. General statements and implications of the literature have indicated that the number of stings is not an important factor in the reaction outcome (see Lockey *et al.,* 1988, for an example). This might be true if all systemic, or the moderate and serious reactions are lumped together, but for the truly serious life-threatening reactions this is not the case. For example, more than twice as many patients stung by more than 10 insects experienced serious life-threatening reactions than did those individuals who were stung only once (33.3 versus 16.4%, P <.01, x^2 test). Also, less than half as many stung greater than 10 times experienced only local reactions as compared to those stung only once (15.7 versus 32.9%; P <.05, x^2 test).

An understanding of the risk factors accompanying potential anaphylaxis can help reduce the risk of serious anaphylactic injury. People who have known coronary problems, for example, should not take undue risks around stinging insects. Also middle-aged and older people should not take heroic actions when unnecessary to "help" a child who is being stung. Obviously, good judgment is needed in such situations; but if the child is in no imminent threat, that is can run away and is old enough to have the sense to escape to safety, then the older adult should remain in safety (perhaps yelling instructions and encouragement). The chances of the child "going into anaphylactic shock" are much less than those of a rescuing and stung adult. Also, if the child does have a serious anaphylactic reaction, the adult can rescue him or her; the reverse situation is not necessarily true. Moreover, risk factors indicate that in situations, such as the elimination of yellowjacket or wasp nests, where no clear advantage accrues to the older person doing the task, the younger person should consider doing the job (besides, teenagers are often quite capable and willing to perform such undertakings).

MASS ENVENOMATION

On rare occasions humans and animals receive hundreds or thousands of bee stings (Figure 3). These mass envenomating attacks result in the direct poisoning of the body and the effects are due to the venom itself, rather than the body's immune system reacting to the venom. Mass envenomations are somewhat analogous to snake bites (the stinging insects can be viewed as a snake divided into a thousand little parts, each carrying a small amount of venom) and can be extremely serious or fatal (Table 6). If all, or most of a bee's venom is injected during stinging and the subsequent continued venom delivery by the embedded sting, and humans are about equally sensitive to venom as other mammals, then an average sized person in good health should survive 1,140 bee stings half of the time (Table 6). Children and babies being smaller will be in serious danger of lethal envenomation with as few as 100 -400 stings. Older people are at higher risk of being massively stung as can be seen in the table (median age of 16 individuals is 65 years) because they have less alert senses, less vigorous bodies, and are less able to escape an attack. Most adults stung by a thousand bees survived (7 of 8) and the highest number of survived stings is 2,243 (Murray, 1964) which suggests that the human LD$_{50}$ of around 1,100 stings/adult is about right. Nevertheless, many fewer stings, in the order of 100 - 300 have been known to kill people and any stinging event involving this number of stings should be considered as serious.

Mass envenomations most frequently occur as a result of honey bee attacks, but yellowjackets, and especially hornets are also responsible for these envenomations. In the case of yellowjacket and hornet mass envenomation and, unlike the usual mass attack by honey bees, serious kidney failure frequently occurs (James and Walker, 1952; Scragg and Szent-Ivany, 1965;

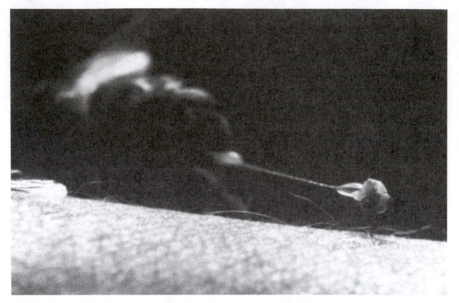

FIGURE 3. A bee leaving the scene after successfully stinging its victim. Note the glistening, transparent venom reservoir attached to the upper left part of the sting apparatus.

Hoh *et al.,* 1966; Shilkin *et al.,* 1972; Chugh *et al.,* 1976; Gädeke *et al.,* 1977; Bousquet *et al.,* 1984a). Any incident of vespid attack with 50 stings should be closely followed for signs of kidney failure (blood in urine or severe reduction of urine flow).

Proper treatment of mass envenomations by either bees or wasps is different from treatment of allergic reactions. The first problem can be logistic—how to stop the attack and rescue the victim, who often is injured, trapped, unconscious, very young, or old. Rescue sometimes is virtually impossible for the unequipped person because the stinging insects usually attack any person in the area. Once the person is removed from the attack area, medical attention should be obtained as soon as possible. Treatment depends on the particular situation, but is best handled in a hospital where blood analyses, life support systems, and dialysis equipment are available, should they be necessary.

In the temperate northern hemisphere so few mass envenomations occur that no specific treatment protocols are in place. The situation may change in the southern part of the United States after arrival of the Africanized bees which are known to be more aggressive than bees from temperate climates and are responsible for some deaths in Latin America. How to deal with mass envenomations by Africanized bees is a matter of some concern and many questions remain. Africanized honey bees are smaller than European honey

Table 6. Projected number of bee stings to cause human deaths and examples of toxic envenomation by bees.

Median lethal dose (LD$_{50}$) for bee venom:	2.8 mg venom/kg body wt[a]
Weight of venom per bee:	147 μg/bee[a]
Number of stings for LD$_{50}$/kg body wt.:	19 stings/kg[b]

Predicted number of stings to provide a median lethal dose of venom for:

Body weight	Number of stings for LD$_{50}$
Child of 10 kg (=22 lbs)	190
Child of 30 kg (=66 lbs)	570
Adult of 60 kg (=132 lbs)	1140
Adult of 90 kg (=198 lbs)	1710

CASE EXAMPLES

Number of stings	Body size[c]	(Age)	Factor times projected LD$_{50}$	Fate	Reference
100–200+	50 kg	(66)	0.1-0.2	died	Anon, 1989
300+	60 ? kg	(64)	.26	lived	Jex-Blake, 1942
400	50 ? kg	(75)	.4	lived	Jex-Blake, 1942
500	50 ? kg	(25?)	.52	lived	Jex-Blake, 1942
600+	60 ? kg	(25)	.52	lived	Koszalka, 1949
1000+	60 ? kg	(77)	.88	lived	Mejia *et al.*, 1986
1000+	60 ? kg	(47)	.88	lived	Mejia *et al.*, 1986
1000+	60 ? kg	(80)	.88	lived	Mejia *et al.*, 1986
1000+	60 ? kg	(62)	.88	lived	Mejia *et al.*, 1986
1000+	60 ? kg	(74)	.88	died	Mejia *et al.*, 1986
1000+	60 ? kg	(31)	.88	lived	Humblet *et al.*, 1982
1000+	60 ? kg	(?)	.88	lived	Borham and Roubik, 1987
1000+	60 ? kg	(78)	.88	lived	Patrick *et al.*, 1987
1200+	60 ? kg	(85)	1.05	died	Beck 1935, p85
2243	60 ? kg	(30)	1.97	lived	Murray, 1964
5-6000	85 kg	(77)	3.1-3.7	died	Moret *et al.*, 1983
≦8000	60 ? kg	(25?)	≦7.0	died	Miami Herald, 1986; C. Angspurgen, unpubl.

[a] Data from Schumacher *et al.*, 1989.
[b] Assuming all venom is injected
[c] If not given, assumption is 50 kg for women and 60 kg for men.

bees and contain less venom than European bees. Their venoms are, however, of approximately equal lethality, both having LD$_{50}$ values of about 2.8 mg/kg (Schumacher *et al.*, 1989). Although the venoms of European and Africanized bees are not identical (Nelson *et al.*, 1989; Schumacher *et al.*, 1990), preliminary evidence indicates that they are immunologically cross-reactive and European venom can probably be used for skin testing and immunologically treating persons sensitive to Africanized bee venom (Schmidt *et al.*, unpublished). One concern with Africanized bees is whether their tendency to sting *en masse* will result in cases of venom allergy in people who can tolerate one or a few stings, but who will have an allergic reaction following dozens or hundreds of stings. Examples of allergic reactions that occur only after numerous stings apparently overwhelmed the protective immune system are known (Haydak, 1956; Ewan, 1985). Another serious problem concerning Africanized bees is the current lack of an antivenin capable of neutralizing the venom.

CAUSES AND CLINICAL DIAGNOSIS OF VENOM ALLERGY

Typical allergic reactions to stings are classified as "immediate hypersensitive", or type 1, allergic reactions. These are often called anaphylactic reactions because the original descriptions were based on animals that went into anaphylactic shock (Wasserman and Marquardt, 1988) and anaphylactic shock is one of the more profound of the immediate hypersensitivity reactions. As mentioned earlier, allergic reactions are generally believed to require at least one "sensitizing" sting prior to the sting that induces the allergic response. In individuals who develop hypersensitivity to stings, the initial sensitizing sting(s) cause an immune response in the body that results in the production of immunoglobulin E (IgE), the allergy-inducing antibody. Unlike other human antibodies which remain soluble in the blood or other body fluid, IgE is unusual in that most of these antibodies become attached to the membranes of mast cells and basophils. Mast cells are abundant in the skin, digestive system lining and similar areas that are most frequently exposed to factors from outside the body.

Basophils are cells that comprise about 0.5 to 1 percent of the white blood cells. Both cells contain releasible granules and an abundance of chemical mediators. When properly stimulated these cells degranulate and release or generate histamine, heparin, leukotrienes (the slow-reacting-substances of anaphylaxis), prostaglandins, and a host of other active compounds (Marom and Casale, 1983; Wasserman and Marquardt, 1988). These compounds are part of the body's normal defense against injury or invaders and operate by increasing the blood and fluid supply to the area and by helping to attract macrophages and other white blood cells to the area. In excess, however, these compounds can cause allergic attacks by acting systemically to induce swelling, itching, constriction of the respiratory system, cardiac hypotension, and the other symptoms observed in allergic attacks. IgE is the key antibody in causing allergic reactions because when it is contacted by the antigen (venom), it induces the degranulation and mediator release by the mast cells and basophils.

The normal non-allergic human immunological reaction to a bee sting consists of the production of mainly immunoglobin G (IgG), the most abundant of the antibody classes in the body. IgG, as the primary blood borne antibody, quickly binds with and inactivates foreign antigens such as venom proteins. IgG production is a main reason why the reactions of most beekeepers to stings decrease over time. In allergic individuals, including allergic beekeepers, the body typically reacts to sting venom proteins by producing too much IgE and not enough IgG. The balance between these two antibodies appears to be a factor in the cause of allergy (Urbanek *et al.*, 1983; Mosbech *et al.*, 1986; Bousquet *et al.*, 1989), but no simple predictive relationship seems to exist (Kemeny *et al.*, 1980; Przybilla *et al.*, 1989). Excessive IgE, not simply its presence (many beekeepers and others have low levels of IgE), is often correlated with allergic reactions.

A variety of tests have been developed to diagnose and analyze venom allergy. Two of these have become popular methods for routine use with patients. The simplest of these is the skin test. In brief, skin tests usually start with a scratch or prick test which involves placing small amounts of venom in saline on the skin and a small scratch or prick is made. A positive reaction consists of the development of a white raised wheal and a red flare around the site. For more precision, the prick or scratch tests are followed by the skin test, itself, which consists of injecting just under the skin surface 20-50 μl of venom-saline solutions (Schwartz *et al.*, 1981). These results are recorded essentially the same as for the scratch or prick tests. Venom solutions used are very low, usually starting about 10^{-10} g venom/ml saline and are increased by factors of ten over a period of time. A positive skin test is generally considered one in which a reaction is observed at a level of 10^{-6} g/ml, or 1 μg/ml (Schwartz *et al.*, 1981; Golden, 1987), though some researchers and practitioners consider reactions at 0.1 μg/ml to be a better indicator. This is because at 1 μg/ml many patients give a positive reaction (even though they are not allergic to the venom) due to the general toxic and irritating nature of the venom (Parker *et al.*, 1982; Georgitis and Reisman, 1985; Day, 1986; Lantner and Reisman, 1989). Interestingly, the degree of sensitivity also does not appear to depend upon the concentration of venom that causes the positive skin test (Reisman *et al.*, 1981; Blaauw and Smithuis, 1985; Golden, 1987; Lantner and Reisman, 1989) or the RAST levels (Reisman and DeMasi, 1989).

The second clinical test for venom hypersensitivity is the RAST (radioallergosorbent test). The RAST involves testing a sample of blood for the presence and levels of IgE antibodies to the venom. RAST has the advantage over skin tests in that no direct venom challenge or long times in the office are required. A further advantage of the RAST is that allergen challenge and the possible consequent anaphylaxis are avoided. The disadvantages of the RAST over the skin test are that it is a more difficult test to conduct and diagnostic value of the test is generally lower (Golden *et al.*, 1989a; Knicker, 1989).

Skin tests and RAST tests have two main functions. They give indications of venom sensitivity and, perhaps more importantly, they provide information as to the taxonomic identity of the stinging insect(s) to which the person is sensitive. Confirmation of hypersensitivity is a necessity for venom-allergy treatments and knowledge of the offending species is very helpful for design of the specific treatment and for advising the individual of avoidance strategies.

Complicating and unfortunate factors in the diagnosis of insect sting allergy are the very low rate of predictability based on either skin tests or RAST. Quite frequently individuals with systemic reactions have negative skin tests (Insect Allergy Committee, 1965; Santrach *et al.*, 1980; Clayton *et al.*, 1985; Lockey *et al.*, 1989; Lantner and Reisman, 1989; Golden *et al.*, 1989a). The incidence of systemic reactions in skin test positive individuals is only about 10 - 15% (Golden *et al.*, 1989a). Moreover, even in those individuals who had a previous systemic reaction and currently have a positive skin test, the systemic reaction rate on re-sting is only 20 - 60% (Hunt *et al.*, 1978; Settipane and Chafee, 1979; Parker *et al.*, 1982; Blaauw and Smithuis, 1985; Kampelmacher and vanderZwan, 1987; Lantner and Reisman, 1989) and decreases with increasing time interval (Settipane and Chafee, 1979; Savliwala and Reisman, 1987). Similar trends and conclusions can be observed from data derived from RAST results (Busse and Yunginger, 1978; Santrach *et al.*, 1980; Lantner and Reisman, 1989; Golden *et al.*, 1989a). Patients with a history of serious systemic reactions have been reported who experience a subsequent series of 3 - 6 stinging events, yet have only one randomly distributed systemic reaction during that series (Reisman and Lantner, 1989). In practice, a history of actual systemic reaction is of paramount importance in decision-making processes; other information from skin tests and RAST is secondary. A further major reason most physicians require both a history of systemic reactions and a positive skin test (or RAST) is that experience indicates that these combined factors are more indicative of future systemic reactions than just a history alone (Hunt *et al.*, 1978; Lichtenstein *et al.*, 1979; Valentine and Golden, 1988; Golden *et al.*, 1989). A negative test suggests IgE antibodies were not involved in the previous reactions or else the sensitivity was lost and therefore no treatment such as immunotherapy is needed. The skin test is generally the first diagnostic test used for routine clinical applications because its diagnosis rate is considered better than that of the RAST and it is easier to conduct.

From the above discussion it becomes evident that there is no current *a priori* method for determining which individuals will become hypersensitive and which will not. Skin tests and, especially, RAST tests have been conducted on populations of people for the purpose of determining the frequency of allergic antibodies to venoms and have yielded incidences from about 1 to

20% (Reisman and Georgitis, 1984; Müller *et al.*, 1977; Herbert and Salkie, 1982; Stuckey *et al.*, 1982; Jarisch *et al.*, 1982; Golden *et al.*, 1982b; Schwartz *et al.*, 1988; Zora *et al.*, 1988; Golden *et al.*, 1989a). Unfortunately, these survey tests tell only that IgE antibodies exist and have no predictive value as to the actual incidence of future allergic reactions. It is clear that much more than the mere presence of venom-specific IgE is needed to cause a hypersensitive reaction (Blaauw and Smithuis, 1985; Kemeny *et al.*, 1983b). The majority of individuals who have venom IgE antibodies have never experienced an allergic reaction (Golden *et al.*, 1989a).

STINGS AND STATISTICS: DEATH RATES AND STINGING INSECTS

Mortality rate tables can reveal much about human values, behaviors and fears. Table 7 is a partial listing of the death rates in the United States due to various diseases, accidents and other causes. By far the greatest killers are cardiovascular disease, cancer, smoking and alcohol. These major causes of death are followed by accidents caused by motor vehicles, suicide and murder, with each of the remaining causes accounting for 1% or less of the deaths. When combined into broad categories the causes of death listed in Table 7 are: diseases, 69.2%; anthropogenically caused deaths (smoking, alcohol, suicide, murder), 14.3%; accidents, 2.95%; and miscellaneous (all the rest), 1.0%.

Mortality statistics can be viewed several ways. An instructive approach is to compare individual causes of death and thereby determine the relative importance of each. When this is done, the 41 deaths per year due to all insect stings are dwarfed by the figures for almost all other causes. For example, the death rates due to smoking are 150,000 per year, or more than 3,000 times greater than due to insect stings. Motor vehicle accidents also kill over 1000 times more people than insect stings. Even such unlikely deaths as freezing, hunger and thirst, and horse (and other animal) riding cause 25, 5, and 3 times as many respective deaths as stings. Accidents such as motor vehicle accidents, drowning, and electrocutions cause, respectively, 1100, 110, and 20 times as many deaths as insect stings. The environmental pollutant radon gas is responsible for 317 more deaths than insect stings and the allergic or immunologically caused diseases of asthma and penicillin allergy account for 95 and 7 times as many deaths as stinging insects. Finally, even dog (and other animal) bites, lightning and sports collisions cause more deaths than insect stings. Overall death rates due to insect stings are so low that most of the causes of death that are lower include such unusual and curious listings as "overexertion". The point of these comparisons is to place the importance of deaths due to insects stings in true perspective. Compared to other causes of death, insect stings are a trivial contribution, causing less than 0.002% of deaths. And from the perspective of deaths caused by honey bees, this level is even lower,

Table 7. Death rates in the U.S. from various diseases, accidents, and unusual causes

Cause	Number of Deaths/year	Death rate per 1,000,000/year	% of total deaths	Reference
All	2,086,440	8739	100	a
Cardiovascular disease	977,700	4096	46.9	a
Cancer	461,400	1933	22.1	a
Smoking	150,000	750	7.8	b
Alcohol	100,000	500	5.2	b
Motor vehicle accidents	45,901	192	2.2	a
Suicide	29,453	123	1.4	a
Murder	19,628	83	.9	a
Radon gas	13,000	54	.62	c
Pedestrian-vehicle	7,641	32	.37	a
Drowning	4,407	18.4	.21	a
Home fires	3,964	16.6	.19	a
Asthma	3,880	16.2	.18	a
Poisonings	3,612	15.1	.17	a
Fire arm accidents	1,649	6.9	.079	a
Freezing	1,010	4.2	.048	a
Electrical accidents	802	3.4	.039	a
Slips & falls while walking	404	1.7	.019	a
Penicillin allergy	300	1.5	.016	d
Hunger, thirst, etc.	195	.82	.0039	a
Horse & other animal riding	108	.45	.0051	a
Animal bites (dogs, etc.)	101	.42	.0048	a
Lightning	85	.36	.0041	a
Sports collisions	42	.18	.0021	a
Insect stings (all)	41	.17	.0019	a
Honey bee stings	17	.07	.0008	e
Overexertion	28	.12	.0014	a

[a] Vital Statistics of the United States (1986).

[b] Upton, 1982

[c] Kerr, 1988

[d] Idsoe *et al.*, 1968

[e] Author's estimate.

0.0008%. Numbers as small as this are hard to grasp; described another way, heart attacks and other cardiovascular diseases cause almost 60,000 times as many deaths as honey bee stings and may be indirectly responsible for many sting-related deaths (Rubenstein, 1982; Barnard, 1973).

In light of these statistics, why then, is there so much fear among the public of stinging insects, and why are so many people and doctors so preoccupied with venom allergy? There are, of course, no simple answers to these questions; but I submit it is partly a result of natural human fear instincts. Modern man with his vehicles and other conveniences, and with his rich diet and long lifespan is only a very recent development in the long human history. During most of the evolution and history of humans, lifespans were short and deaths were frequent due to diseases and other causes in nature. Death due to large predatory animals and to venomous animals was a real part of life. Under these conditions, our species would naturally develop a fear of animals, including stinging insects, that directly attack our bodies. Since tobacco and alcohol were not a part of human life (or, presumably, at least were not until very recent times in terms of evolution) there were, at most, only minor fears due to indirect and subtle causes of death such as cancer and heart attacks due to smoking. Because long lifespans were rare, cancer and cardiovascular disease had less opportunity to develop and cause fears. Hence fear instincts to these modern-day killers were not developed in our ancestors. Perhaps even more important to understanding our usually irrational fear of stinging insects is the nature of the threat. As mentioned above, stinging insects directly assault our bodies. Heart disease, smoking and alcohol damage, etc. are not immediate and direct assaults on our bodies, but rather are the delayed result of apparently benign and often pleasurable immediate activities. Is it any wonder then, that we do not fear the far greater killers of smoking, alcohol, and fatty diets as much as honey bee stings?

Statistics such as those in Table 7 can be used to investigate if deaths from various causes can be prevented and at what cost. Such analyses can indicate areas where preventative action would be most profitable. Of those listed in the table, deaths due to smoking and alcohol are clearly preventable. Human behaviors to eliminate smoking and heavy alcohol consumption would eliminate these 250,000 annual deaths. Reduction of alcohol consumption, more care, and more vehicle safety designs could dramatically reduce motor vehicle and pedestrian-vehicle accident rates. Reduction of excess fat in the diet would greatly reduce cardiovascular disease and cancer. These preventative measures would cost nothing or relatively little, yet would save many lives. Indeed, public health officials in the United States recognize this fact and have mounted large and successful campaigns to reduce smoking, alcohol consumption, and excessively fatty diets among the populace. On the other hand, public health officials and organizations have mounted no campaigns with regard to stinging insects because these threats to health and life are recognized as so minor.

The purpose of presenting statistics such as those in Table 7 is only to put the importance of threats from stinging insects in perspective. Clearly stinging insects are not a serious threat to the vast majority of people and fear of these insects and the restrictive lifestyles that this fear can cause are not justified or warranted.

STINGS AND STATISTICS:
VENOM ALLERGY AND DEATH RATES

A question of great theoretical and practical interest is the frequency of allergy among the population to insect stings. Published estimates of the insect sting hypersensitivity within the general public vary greatly, ranging from about 0.5% to 4% (Chafee, 1970; Settipane and Boyd, 1970; Abrishami *et al.,* 1971; Herbert and Salkie, 1982; Stuckey *et al.,* 1982, Charpin *et al,* 1988; Golden *et al.,* 1989a). These differences in incidence result from the variety of different populations studied, analytical methods used, and criteria used for assigning hypersensitivity. Different authors cite different statistics, usually based on their own particular studies and expectations. A Gallup poll in the United Kingdom surveyed the general public and reported an incidence of being stung of 10% with only 0.73% of these stings being severe enough to warrant further investigation. This translates into a 0.073% rate in the United Kingdom of sting allergy needing medical attention (Riches, 1988). A reasonable and generally medically accepted estimate of the incidence of clinical hypersensitivity among the European and North American populace is in the range of 1 to 2%.

Rather than focusing on the incidence of insect venom hypersensitivity, or of the presence of IgE antibodies as measured by skin tests or RAST, attention should be paid to the real threat to hypersensitive people. That is, what percentage of hypersensitive individuals are at risk of dying as a result of another sting? Also, what are the risks of dying from a sting based on the type of hypersensitive reaction experienced by the individual? If we assume 2% of the public is hypersensitive to venom, that equates to roughly 5 million people in the United States. Each year approximately 41 individuals die of venom allergy, or only 1 individual per 120,000 hypersensitive individuals dies per year. The percentage of individuals among the assumed 2% hypersensitive group who are life-threateningly hypersensitive (category VI on table 3) is difficult to estimate. If, for argument's sake, a conservative value of 10% is assumed, then the risk of dying among these truly hypersensitive people is still only 1 in 12,000 per year. As can be seen, this risk is really very minor and similar analyses have prompted comments that the risk of dying in a traffic accident while traveling to the hospital after a sting are probably much greater than the risks from the sting, itself (Rubenstein, 1980a).

HISTORICAL PERSPECTIVES OF VENOM ALLERGY
AND TREATMENTS

Historical fancy recorded the first death due to an insect sting as that of the great Egyptian King Menes, who, as the story goes, was stung and died about 2641 B.C. while on a warship near Britain. As delightful as this story is, the king most likely was actually killed by a hippopotamus while navigating the waters of the Nile river (Cohen, 1989). Thus, the first actual recorded death resulting from an insect sting may be difficult to determine.

The medical discovery of anaphylactic death and the coining of the term anaphylaxis occurred in 1902. The discoverers, Portier and Richet, experimentally demonstrated that traumatic, rapid death could be induced by first immunizing a dog with small amounts of sea anemone venom and then challenging it with another dose of that venom (cf. May, 1985). Shortly thereafter Waterhouse (1914) described a patient who suffered serious cardiovascular and respiratory compromise after a bee sting and, thus, identified and documented the first case of insect sting anaphylaxis.

The first use of bee venom both for skin testing and for treating hypersensitivity was performed by Braun (1925). In that report he also demonstrated the first use of epinephrine (adrenalin) as the treatment of choice for arresting the reactions of insect sting anaphylaxis, and developed and used a protocol involving venom (extract of the last 1/8 of the bee body) for desensitizing the patient. The desensitizing protocol he developed consisted of starting with injections of tiny amounts of venom, followed over the course of several days with injections of ever increasing amounts of venom, and finally, when desensitization was judged adequate, with sting challenge. This magificently perceptive first study embodied essentially all of the procedure used today in accepted treatment of venom allergy.

The period from 1930 until almost 1980 could be called the dark ages of the medicine of venom allergy treatment. In 1930 Benson and Semenov published a major study in which they desensitized patients allergic to both bee venom and bee body proteins. This report of dual sensitivities was unfortunate because it set a precedent for a belief that the extract of the entire bee body (whole body extract) was the proper material for immunotherapy. The conclusion of dual sensitivity of allergic patients to both bee venom and bee body was reasonable given the data (Benson, 1939), but the equally plausible, and, as it turned out, correct conclusion of independent allergies to venom and body proteins was not pursued further. Thus, this period was known as the "whole-body-extract" period of venom immunotherapy during which time virtually all the diagnoses and treatments of venom allergy used whole body extracts.

During most of the reign of whole body extract treatments for venom hypersensitivity, there was one lone voice of dissent. That was by Dr. Mary Loveless who reasoned that it was venom that was injected during the stinging act, and not general body proteins. Accordingly, from about 1950 until about 1980 she successfully treated allergic patients with pure venoms derived from both honey bees and yellowjackets (Loveless and Fackler, 1956; Loveless,

1977). Interestingly, her work was largely ignored; possibly because other physicians were reluctant to handle live insects and there was no commercially available source of pure venom, and possibly because she never performed a rigorous double blind test of venom versus whole body extracts for treatments of hypersensitivity.

Use of whole body extracts for treatments flourished for such a long period of time partly because many testimonials and statistics indicated success (Insect Allergy Report, 1965). That the "placebo" effect actually explained most of the results was not widely recognized or accepted (but see Loveless, 1957). Deaths of individuals undergoing whole body extract treatments occurred (Ordman, 1968) including the nagging report of eight patients who were treated with whole body extracts and yet died of stings (Torsney, 1973). These reports led to the suspicion that whole body extracts were ineffective treatments. This knowledge, presumably coupled with the demonstration that whole bee bodies without venom sacs contained no antigens in common with venom (Schulman *et al.,* 1966), led to a well designed double blind test of venom versus body extract for immunotherapy treatment. The results clearly showed that venom treatment reduced systemic reactions and that whole body extract treatment did not differ from no treatment (Hunt *et al.,* 1978). Except for a few skirmishes in the literature, the period of whole body extract treatment ended with this report. Since gaining wide acceptance in the early 1980's, the use of venom for immunotherapy has continued to the present. In the future, it is hoped that venom will be replaced with specific genetically engineered proteins (or blends of proteins designed to suit each individual case) which will be more effective than venom and have fewer side effects.

STING KITS AND IMMUNOTHERAPY

Over the years, a variety of treatments have been performed in hope of preventing or arresting anaphylactic or other sting hypersensitivity reactions. Two main basic tools are used in the treatment of sting-induced allergic reactions: epinephrine and immunotherapy. Home remedies, usually the same ones used in attempts to reduce the immediate sting pain, are occasionally used by people in hope of preventing allergic reactions. There is, however, no medical evidence that any of these have the slightest value; to the contrary there is good scientific and medical reasoning to indicate that none should have any effect—none are capable of interfering with the immunological mechanisms involved in a reaction. There are other valuable procedures and life support systems used in the emergency room or under physician's direction to help an individual undergoing acute reactions, but these could not be readily performed outside medical settings and are not practical for routine use. These procedures will not be discussed here, but see Lindzon and Silvers (1988) for emergency room treatments of anaphylaxis.

Epinephrine is the only treatment that is capable of arresting ongoing systemic reactions (Barach *et al.*, 1984) and is clearly the treatment of choice. Other treatments such as antihistimines and intravenous fluids can be secondarily helpful, but they cannot block the effects of leukotrienes which are much more serious mediators of anaphylaxis than histamine. Steroids, though of benefit for type 3, or serum sickness allergic reactions, are not indicated for type 1 acute anaphylactic reactions. Epinephrine is a hormone naturally produced by the adrenal glands and is usually supplied in saline in a 1:1000 dilution. The usual treatment for allergic reactions is 0.3 ml of this solution injected intramuscularly into the thigh or upper arm. Occasionally several injections are required. Epinephrine is conveniently packaged in sting kits (Anakit produced by Hollister Stier, and EpiPen manufactured by Center Laboratories) that can be carried in a purse, shirt pocket, or attached to the belt (Figure 4). These sting kits are easy to use and have been used to stop innumerable reactions and perhaps save lives. The one caveat with use of epinephrine concerns its reduced effectiveness and perhaps danger in patients using beta-adrenergic blocking drugs (beta-blockers) such as propanolol, a drug used to treat cardiac arrhythmias, angina pectoris, and hypertension (Barach *et al.*, 1984, Bousquet *et al.*, 1987b). Anyone who is exquisitely sensitive (*e.g.* becomes unconscious after a sting) should certainly let his or her physician know of that fact and inquire if any of their medications include beta-blockers. Some physicians suggest avoiding beta-blockers for their anaphylactic patients (Bousquet *et al.*, 1987b; Valentine and Golden, 1988), or recommend not providing immunotherapy unless beta-blockers can be replaced with other treatments (Norman and van Metre, 1990).

An alternative means for delivering epinephrine is via inhaler. These devices are commonly used by asthma patients and deliver medication directly to the bronchial tubes and lungs. Although most asthma patients use other, less harsh materials in their inhalers, epinephrine inhalers can be readily obtained as over-the-counter preparations. In cases of respiratory involvement, use of epinephrine via inhaler seems the logical best route of delivery and some evidence supports this concept (Ganderton, 1979; Frankland, 1976; Heilborn *et al.*, 1986; Riches, 1988). The use of inhaled epinephrine has not gained much medical popularity, perhaps partly because it is probably not as effective for controlling allergic reactions not involving the respiratory system, and partly because it is more difficult to administer in medical settings. Inhalers are, however, easy for patients to use, avoid the fears of hypodermic needles, and are readily available. Their value for treatment of allergic reactions merits further medical study, particularly, clinical testing involving actual cases of anaphylaxis.

In all respects immunotherapy, sometimes called venom allergy shots, is

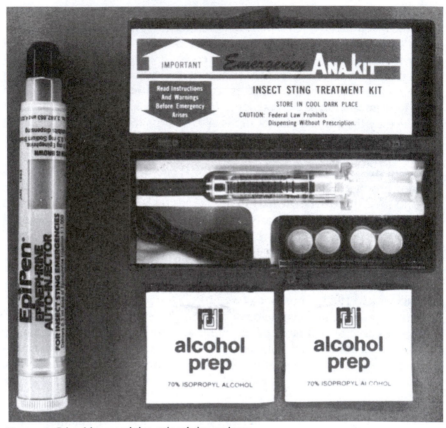

FIGURE 4. Sting kits containing epinephrine syringes.

the ultimate preventative treatment of venom hypersensitivity—it is the last
treament option available, it is the most complicated, it is the most inconvenient and labor intensive, and it the most expensive treatment. Nevertheless,
venom immunotherapy does eliminate future incidence of systemic reactions
in 85 - 95+% of treated individuals and generally reduces the severity of
reactions in those individuals who do have reactions (Golden *et al.,* 1981a;
1982a, 1986; Peppe *et al.,* 1983; Reisman *et al.,* 1985; Bousquet *et al.,* 1987;
Lantner and Reisman, 1989; Livingston and Reisman, 1990) and is the
preferred treatment for life-threatening reactions. Nevertheless, there are some
treatment failures, that is systemic reactions of individuals on maintenance
immunotherapy, including serious life-threatening reactions. Bousquet *et al.,*
(1988, 1989) present some examples and discuss in detail treatment failures.

    Venom immunotherapy treatment basically consists of a series of periodic
injections of venom or venom extract given while in a physician's office or a

clinic (see Bousquet *et al.,* 1987b; Graft, 1987 for details). The venom injections start with very small doses, usually around 0.0001 to 0.01 μg and are progressively increased to 50 μg (Reisman *et al.,* 1985; Livingston and Reisman, 1990), 100 μg (the most common top dose; *e.g.* Golden *et al.,* 1981a), or even 200 or 300 μg for individuals who do not exhibit the desired responses at lesser amounts (Yunginger *et al.,* 1979; Hoffman *et al.,* 1981; Golden *et al.,* 1982a; Bousquet *et al.,* 1988). These injections can be given according to a wide variety of time tables, ranging from 3.5 hr to 35 days before reaching the usual maintenance dose (the top dose) (Golden *et al.,* 1980; Nataf *et al.,* 1984; Birnbaum *et al.,* 1990a). There are no clear indications that any particular protocol is better than the others (Golden *et al.,* 1980; Thurnheer *et al.,* 1983; Bousquet *et al.,* 1987a; Birnbaum *et al.* 1990a). The consensus in Europe appears to favor the rapid (Rush) treatments, whereas in North America the slower regimens are more common.

As mentioned above, immunotherapy does not always protect an individual from future systemic reactions after a sting. Ideally it would be helpful to be able to predict treatment successes and failures and then concentrate further efforts on those patients who will be failures. Various criteria have been investigated for their ability to predict treatment success. These have included a rise above a certain level of the patients venom-specific IgG (Golden *et al.,* 1982a, 1986; Bousquet *et al.,* 1987a; but see Yunginger *et al.,* 1979), the lowering of the specific IgE levels (Reisman and Lantner, 1989), the ratio of IgG to IgE (Hoffman *et al.,* 1981), and the loss of skin test and/or RAST sensitivity (Thurnheer *et al.,* 1983; Reisman and Lantern, 1989; but see Kampelmacher and van der Zwan, 1977; Bousquet *et al.,* 1989). None of these evaluations provides certain prediction, and there are many exceptions to any generalization; however, generally positive indicators include a high level of IgE, a low level of IgE, a large ratio of IgG/IgE, and the development of a negative skin test.

Many questions and problems relating to immunotherapy still exist. The incidences ranging from 9.6 to 50% of large local reactions (Golden *et al.,* 1980; Ramirez *et al.,* 1981; Thurnheer *et al.,* 1983) as well as an incidence 1.6 - 30% systemic reactions during immunotherapy are troubling (Golden *et al.,* 1980; Thurnheer *et al.,* 1983; Muller *et al.,* 1990; Birnbaum *et al.,* 1990a; Lockey *et al.,* 1990b). Particularly distressing in this regard is the apparent drop out rate as a result of these reactions in the truly life-threatening sensitive individuals (Golden *et al.,* 1980; Kemeny *et al.,* 1983b; Collins and Engler 1988; Bousquet *et al.,* 1988, 1989; Lantner and Reisman, 1989; Lockey *et al.,* 1990b)—those who presumably need the treatment the most. Other questions concern who exactly to treat with immunotherapy (see Table 8), how often maintenance injections are needed (monthly, every six weeks, every two

Table 8. Indications for venom immunotherapy.

Type of reaction	Age group	Skin test	RAST	Venom immunotherapy
severe systemic cardiovascular or respiratory (category VI of Table 4)	all	+ – –	+/– + –	yes many authors recommend no
Moderate reactions: rhinitis, conjunctivitis, vomiting, diarrhea (category V of Table 4)	all	+ –	+/– +	generally yes some authors recommend
Mild cutaneous reactions: angio-oedema, hives, etc. (category IV of Table 4)	adults children	+ +/–	+/– +/–	probably no; depends on patient and situation no; no testing needed
Large local reactions (category III of Table 4)	all	should not be tested		no
Local reactions (categories I and II of Table 4)	all	should not be tested		no

SPECIAL SITUATIONS

Serious cardiac condition and/or beta-blocker medications	all	+	+/–	only for serious cardiovascular or respiratory reactions and then after careful evaluation
Delayed reactions: vasculitis, nephrosis, neuritis, serum sickness, encephalopathy	all	testing only for research		no
Any systemic reaction that occurred more than seven years previously	all	testing not necessary		no (Saliwala and Reisman, 1987)

months: Golden *et al.*, 1981b; Gadde *et al.*, 1985; Bousquet *et al.*, 1987b; Goldberg and Reisman, 1988), and when to stop giving immunotherapy (*i.e.* when it is "safe" to stop because protection is adequate).

Criteria for terminating immunotherapy have captured much of the recent attention and debate in the literature and are likely to continue to be modified in the future. Part of the reason for this attention is likely that patient dropout rate during immunotherapy is enormous (Kemeny *et al.*, 1983b; Golden *et al.*, 1986; Reisman and Lantner, 1989; Lockey *et al.*, 1990) and it would be advantageous if a physician could provide a treatment termination date for the patient before asking him or her to make a decision about the treatment. The older literature generally recommended immunotherapy forever (Lichtenstein *et al.*, 1979; Blaauw and Smithuis, 1985), but recently various studies have provided information that is suggestive that treatments can be terminated earlier based upon the elevation of the patient's venom-specific IgG levels (Urbanek *et al.*, 1983; Bousquet *et al.*, 1987a; but see Bousquet *et al.*, 1989), the lowering of the IgE levels (Urbanek *et al.,* 1983, Randolph and Reisman, 1986), loss of skin test and/or RAST sensitivity (Graft *et al.,* 1984; Reisman and Lantner, 1989; Bousquet *et al.,* 1989), and termination purely on the basis of time—5 years (Schuberth, *et al.,* 1988; Golden *et al.,* 1989b; Stone *et al.,* 1990), 3 years (Heibling *et al.,* 1990), 2 years (Reisman and Lantner, 1989), or possibly even 1 year (Keating *et al.,* 1990). Unfortunately, there still is no parameter that predicts the evolution of an individual patient's sensitivity (Bousquet *et al.,* 1989) and severe cases (often involving loss of consciousness) are not only the most difficult, but also the most dangerous (Lantner and Reisman, 1989).

TREATMENT OF ALLERGY TO VENOMOUS INSECTS

Before any treatment is planned or initiated, an invividual should first collect information and do some serious evaluation of his/her own situation and psyche. The old adage "know thyself" could never be more true than when it comes to dealing with insect sting allergies. The first action to take is to determine if, in fact, hypersensitivity exists. Although this should be easy enough to determine, a great deal of emotion is often involved, both on the part of the concerned individual and those surrounding that person. Self diagnosis is where to start. The afflicted individual should make a cold, rational evaluation based on the symptoms and criteria listed in Table 3. If the symptoms are only those of categories I - III, then no systemic hypersensitivity exists and the person should not unduly worry or seek medical attention. In the case of exceedingly large local reactions, the person might want to try to avoid undue future stings and possibly take a small dose of anti-inflammatory and/or antihistamine medication, an hour before working bees or immediately upon being stung.

If a stung individual has a cutaneous reaction only (category IV in Table 3), then the professional recommendations become cloudy. Many physicians recommend sting kits and/or immunotherapy. Many also recommend against immunotherapy, especially for children (Schuberth *et al.,* 1982; Golden and Valentine, 1988), but also adults (Reisman *et al.,* 1985). These cutaneous hypersensitive reactions are situations in which the afflicted individual should "take charge" of the situation him/herself. The first point to remember is that nobody has ever died or been seriously physically impaired as a result of a cutaneous reaction. The second point to consider is that, contrary to some well-ingrained beliefs, the frequency of experiencing a serious life-threatening reaction on the next sting is no greater for the individuals who have cutaneous systemic reactions than for those who have no systemic or large local reaction (Abrecht *et al.,* 1980; Mauriello *et al.,* 1984; Graft *et al.,* 1984).

Past experience has indicated that individuals who have reactions, such as cutaneous reactions, tend to have repeat reactions of the same type in the future (Yunginger *et al.*, 1979; Lichtenstein *et al.*, 1979; Golden, 1987). Thus, fear and emotion should not play a role in decision making. If the individual feels comfortable knowing that another cutaneous reaction might occur on another sting, but that probability indicates that nothing more serious is likely to occur, then (s)he might well decide to live with that knowledge and avoid undue exposure to future stings. An option for action in the case of cutaneous reactions is to see a physician and obtain a prescription (except in North Carolina where it is an "over-the-counter" item) for an epinephrine "sting kit." These kits are quite inexpensive and are extremely effective for arresting the progression of symptoms. If nothing else, these kits, when properly used, will reduce the unpleasantness of the reaction as well as provide psychological assurance.

The last option sometimes suggested for treatment of individuals who have cutaneous reactions is to initiate immunotherapy. This option is not necessary based on medical grounds, but might be considered for some terrified individuals. Immunotherapy will in these cases definitely reduce the serious morbidity (fear) of such patients (Golden, 1987; Golden and Valentine, 1981) and therefore can be justified. If the morbidity can be reduced by less dramatic means, then immunotherapy probably is not needed.

Individuals who experience moderately severe systemic reactions (category V in Table 3) present a difficult decision in terms of treatment. Current wisdom indicates that such people should be treated with immunotherapy. But information to the contrary comes in the form of knowledge that, as in cutaneous reactions, nobody has ever died of these symtoms of moderate systemic reactions. Unfortunately, there are not enough data to indicate whether or not these reactions are likely to progress to life-threatening reactions. The individual with moderate systemic reactions, but not serious life-threatening reactions, might keep in mind the mortality statistics in Table 7, that deaths are truly rare, and that immunotherapy can be expensive and a major inconvenience; but beyond that he/she will need to evaluate the options and take a hard look before deciding. Certainly individuals in this category would be advised to obtain a sting kit, learn to use it, and use it promptly if stung again.

Individuals who experience serious life-threatening reactions are generally considered candidates for immunotherapy. This is because these reactions threaten the body's life-support systems and can actually cause death. And, although the percentage of life-threatening reactions that actually result in death is very tiny, the risk is there. If there are any individuals who are serious candidates for immunotherapy, it is these people. A sting kit is mandatory for such individuals and likely is of more value than any other single item or action (including immunotherapy) that is available. Use of the sting kit should become automatic and without delay after a sting for these people, and, indeed, data indicates that the majority of individuals do properly use them (Lockey *et al.*, 1988).

Table 8 also lists some special situations concerning individuals with particular allergic situations. People who have severe cardiac conditions and/or are on beta-blocker drugs should not consider undergoing immunotherapy unless they previously experienced loss of consciousness, serious respiratory problems, or other category VI of Table 3 symptoms. If they experienced those serious symptoms, then immunotherapy should only be considered after consultation with a trained and experienced allergist. Individuals who have delayed reactions such as vasculitis (blood vessel inflammation), etc., are not candidates for immunotherapy (Reisman, 1988). The same advice is recommended for individuals who have not been stung again within seven years of a previous reaction. The recommendation not to test these individuals or administer immunotherapy is based on the self-limiting nature (spontaneous disappearance of allergy) of venom hypersensitivity, plus the observation that reactions in such individuals are infrequent (Savliwala and Reisman, 1987).

CROSS-REACTIVITY AMONG VENOMS

An important question relating to venom allergy concerns the ability of different insect venoms to induce similar allergic reactions in an individual. Such an ability is termed venom cross-reactivity and is the result of the allergens from the venom of one species (such as a yellowjacket) being antigenically similar enough to those of another species (such as the baldfaced hornet) to cause an allergic reaction in an individual who has not been sensitized to that venom. Understanding cross-reactivity among venoms of stinging species is important if, for no other reason, than because so many people have misinformation about the subject. Moreover, this subject is not simply of academic interest—it can profoundly affect the life of a venom-sensitive person. For example, if a person is hypersensitive to fire ant stings, is he also sensitive to honey bee stings? If the answer is yes, then beekeeping as an activity might be difficult; if the answer is no, then there is no need to worry about being a beekeeper or to take special precautions to avoid bee stings (but precaution to avoid fire ants while working bees might be necessary).

The frequency of cross-reactivity to different venoms depends upon a variety of factors such as the species of insects involved, the taxonomic relatedness of the species, the particular venom antigen(s) that cause(s) the reaction, and the diagnostic test performed. As based on the presence of IgE in the blood (by RAST), the overall frequency of individuals who are sensitive to only one of honey bee, yellowjacket, or paper wasp venoms (as opposed to two or three of these venoms) varies from 55.5 to 78.8%, depending upon region within the United States (Figure 5). Throughout the northeast of the country as exemplified by the state of New York, the most frequent single sensitivity is to yellowjacket venom (36.3% of hypersensitive individuals) followed by honey bee venom (with 18.3%) (Figure 5A). In contrast, in southern California where yellowjackets are rare, only 7.9% of the single sensitivities are to yellowjackets and 70.4% are to honey bees (Figure 5B).

In the southeastern United States, though a major region for beekeeping, single sensitivities to honey bees occur at a frequency of only 7.4%, with single sensitivity to paper wasp venom occurring in 25.9% of hypersensitive individuals (Figure 5C). Overall for the entire country the single sensitivity figures are 33.5% for honey bees, 25.7% for yellowjackets, and 4.1% for paper wasps (Figure 5D). These figures reflect, in part, the relative abundance of the species in the different areas, but, importantly also reflect the frequencies of cross-reactivity or multiple reactivities among the venoms. When frequencies of single venom sensitivities are measured by skin test instead of by RAST, similar results are obtained (Figure 5E). The main differences are that RAST appears to be more specific (single venom sensitivity of 63.3% versus only 46.7% for skin test) and tends to indicate a higher frequency of sensitivity to honey bees than does skin test.

The numbers in Figure 5 indicate the sensitivity to only one group of venoms is the situation for a slight majority of hypersensitive people. These data indicate that concerned hypersensitive persons should see a physician to have their sensitivities to other venoms determined. If the sensitivity is indeed to any one venom group, then lifestyles can be altered to target avoidance of just that one species, and not stinging insects in general.

The venoms of fire ants and harvester ants are biochemically rather different from those of honey bees and vespid wasps (Schmidt and Blum, 1978; Schmidt et al., 1986; Hoffman et al., 1988a). Their venoms also appear to cross-react little with those of honey bees and vespids (Pinnas et al., 1977; Watkins et al., 1988; Hoffman et al., 1988a; Schmidt and Pinnas, unpublished). Thus, unless an individual becomes independently sensitive to both ants and the flying stinging insects (multiple sensitivity in contrast to cross-reactivity), there is little need to worry about a reaction from a sting of the other group.

The more closely related two groups of stinging insects are, the more likely they are to possess venoms that cross-react. This is particularly true in the case of species in the same genus. For example harvester ants in the genus *Pogonomyrmex* exhibit virtually total cross-reactivity (Schmidt et al., 1984) as do fire ants (*Solenopsis*) (Hoffman et al., 1988a) paper wasps (Findlay et al., 1977; Hoffman and McDonald, 1982b; Grant et al., 1983), yellowjackets in

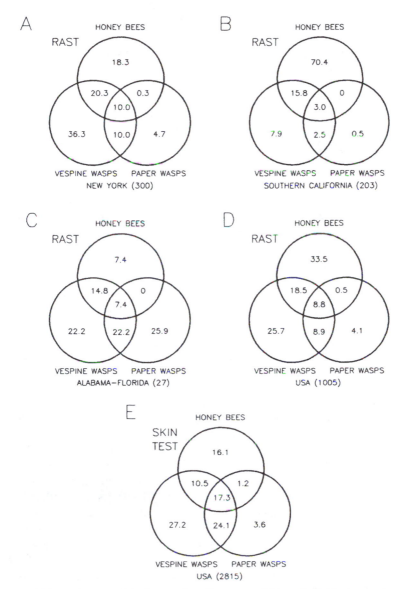

FIGURE 5. Venn diagrams of the frequencies of hypersensitivities of individuals to various wasp and bee venoms. The numbers in the diagrams are percentages of patients, those in parentheses are the total number of individuals surveyed. The numbers in the intersecting areas are the percentages of individuals with sensitivities to two or three venoms. Sensitivities based on RAST measurement of IgE for: A) New York state; B) southern California; c) Alabama-Florida; D) the entire U.S.A. E) is data derived from skin tests for the entire U.S.A. Data for A -D from Hoffman *et al.*, 1980; for E from Lockey *et al.*, 1989.

the *vulgaris* species group (often called *Paravespula*) (Hoffman and McDonald, 1982a; Hoffman *et al.*, 1980; King *et al.*, 1983), and the aerial hornets of the genus *Dolichovespula* (King *et al.*, 1978). The degree of cross-reactivity decreases as the taxonomic relatedness decreases. For example, although the yellowjackets are all in the genus *Vespula*, the two species *V. squamosa* and *V. sulphurea* form a sister group as do the four or five species examined within the *vulgaris* group. The venoms within these two species groupings likewise are more closely cross-reactive than between the groups (Wicher *et al.*, 1980; Hoffman and McDonald, 1982a).

A similar situation is observed between species groups within the *Polistes* (Reisman *et al.*, 1982; Hoffman and McDonald, 1982b; Hoffman *et al.*, 1990; King *et al.*, 1984). The only example of sister genera that have been examined for cross-reactivity are the yellowjacket and hornet genera of *Vespula, Dolichovespula,* and *Vespa*. Among these genera there is extensive cross-reactivity (King *et al.*, 1978; Hoffman, 1981c, 1985c; Hoffman *et al.*, 1987; Bousquet *et al.*, 1987). At the next level of taxonomic relatedness, the subfamily level, the paper wasps in the subfamily Polistinae can be compared to the Vespinae just described. Again, cross-reactivity between subfamilies occurs though not to the degree observed between the vespine genera (Reisman *et al.*, 1982; Grant *et al.*, 1983; King *et al.*, 1984). Honey bees and bumble bees are usually also considered to be separate subfamilies within the family Apidae. What little data is available for these two groups indicates a fair degree of cross-reactivity exists, but not to all known antigens (Hoffman, 1982).

Comparisons between venoms separated at the highest taxonomic levels, the family level, reveal far less cross-reactivity than between more closely related groups. For example sweat bees (family Halictidae) and honey bees (Apidae) exhibit no cross-reactivity (Pence *et al.*, 1985, unpublished); likewise for the carpenter bees (Anthophoridae: *Xylocopa*) and the honey bees (Pence and Schmidt, unpublished). As mentioned previously, there also appears to be little or no cross-reactivity between the Formicidae (ants) and either honey bees or vespid venoms. Although generally not cross reactive (Charavejasarn *et al.*, 1975), the picture for the two families of greatest concern, the Vespidae and the Apidae, is less clear. Cross-reactivity between paper wasp and bee venoms is about 20% (Grant *et al.*, 1983). As indicated in Figure 4, 45.7% of all honey bees venom allergic individuals in the United States are also sensitive to yellowjacket venoms, and 44.1% of individuals hypersensitive to jellowjacket venoms are also sensitive to honey bee venom. Some of this cross-sensitivity appears to be multiple-sensitivity rather than cross-reactivity, as well as minor reactivity in one venom to a major allergen in another venom (Reisman *et al.*, 1987). The cross-reactivity between honey bees and yellowjackets appears to be based mainly on the cross-reaction of their hyaluronidase antigens with little, or no, cross reactivity among phospholipases or other antigens (Hoffman 1986; Wypych *et al.,* 1989). Recent data suggests that the venom hyaluronidase molecules, in general, are highly phylogenetically conserved, or at least immunologically related and might well cause some allergic cross-reactivity between fire ant venoms and the venoms of bees and wasps (Hoffman and Wood, 1984, Hoffman *et al.,* 1988b). The phospholipases and other venom antigens within the vespid, bee and ant families appear not to strongly cross-react (Hoffman, 1986; Hoffman *et al.,* 1988b).

ATOPY AND VENOM ALLERGY

Atopy generally consists of allergy to airborne pollen (hay fever) or food, asthma, or allergic dermatitis. The rates of atopy in the general adult public is measured by elevated IgE (RAST) are about 22% for men and 16% for women and are higher in young than older individuals (Haddi *et al.,* 1989). Because atopy is typically an allergic reaction caused by IgE antibodies, as is venom allergy, it is natural to suspect that the presence of atopy in an individual predisposes him or her to venom hypersensitivity. This question has been addressed around the world, including North America, Europe, and Africa, and, although a universal consensus has not emerged, a variety of studies indicate that atopy and venom allergy are not related.

Settipane and Boyd (1970) studied almost 5000 Boy Scouts and reported a venom hypersensitivity rate of about 0.4%. One of their associates, Chafee (1970), investigated 3705 patients who were referred to allergists for asthma or hay fever and discovered that this group also exhibited a 0.4% hypersensitivity to insect stings. In a joint study of 2964 Boy Scouts using more detailed criteria for hypersensitivity, they reported a 0.8% sting allergy rate for non atopics and a 0.9% rate for the atopic group (Settipane *et al.,* 1972). The overall atopy rate in these boys was 15% and, interestingly, the parents diagnosed hay fever 94% correctly, but venom hypersensitivity only 20% correctly.

In another approach to the question of a possible relationship between atopy and venom hypersensitivity 587 venom allergic patients were analyzed for incidence of atopy. Of these venom hypersensitive patients 22% were atopic (Settipane *et al.,* 1980), about the same rate as in the population at large, and the same as individuals having food or pollen allergy reported in a study of 400 venom allergic patients (Brown and Bernton, 1970). Among a similar group of 458 venom hypersensitive patients only 13.1% had personal atopy (Huber *et al.,* 1983). In Africa, a 22% atopy rate was reported among 386 patients sensitive to bee stings (Ordman, 1968). The atopy rate among a group of 158 patients with life-threatening reactions was also 20% (Lantner and Reisman, 1989). Kleisbauer *et al.* (1987) and Birnbaum *et al.,* (1990b) likewise reported no difference between the atopy rate of venom hypersensitive and non-hypersensitive patients in France when skin testing for atopy was conducted, findings in agreement with a British study (Kampelmacher and van der Zwan, 1987). However, when basing the statistics only on adult venom reactions Kleisbauer *et al.* (1987) reported higher incidence of atopy in reactors than nonreactors. The American Academy of Allergy and Immunology Committee on insects conducted an extensive survey of patients who were seen by allergists for insect sting problems. They report an overall atopy rate among the 2,866 cases of 31% and note that, although this incidence is slightly higher than for the general public, the slight elevation is expected and likely a result of subject selection, "since all reporting physicians were allergists/immunologists and subjects were selected from their patients" (Lockey *et al.,* 1988).

For the general public, there appears to be no serious evidence that atopy predisposes one to venom allergy. Settipane *et al.* (1978, 1980) and Huber *et al.* (1983) did, however, note one caveat to their studies. Those individuals who were both atopic and venom hypersensitive often had more severe respiratory symptoms during a systemic reaction than individuals who were not atopics. This is not surprising as atopic individuals have an already compromised respiratory function, and any aggravation to their respiratory system would have an additive effect. Nevertheless, no mortality studies indicate an increased incidence of atopics in the populations of venom hypersensitivity fatalities.

BEEKEEPERS AND VENOM ALLERGY

Beekeepers and their families are a special group in terms of venom allergy. Unlike the general public, beekeepers experience repeated stings, and stings that are almost always from honey bees. Family members of beekeepers are often exposed to bee stings at a much reduced rate relative to the beekeeper, yet higher than most of the public. In addition, beekeeper family members often are subject to "odors" of bees and allergens derived from bee body parts and debris.

Rates of systemic reactions to bee stings among beekeepers and family members have been reported to range from 8.9 to 43% (Bell and Hahlbohm, 1983; Bousquet *et al.*, 1982) but these figures can be misleading. For example, in a study by Yunginger *et al.* (1978) only 1% (1 of 91) of active beekeepers were hypersensitive to bee stings; whereas 74% (26 of 35) of family members were hypersensitive. Among family members who had "no contact" with bees, that is, received only 1 or 2 stings per year, the systemic reaction rate was 89% (25 of 28). Combining the data for both beekeepers and family members revealed that among individuals who were stung daily or weekly, there were no incidences of hypersensitivity; among those stung monthly, the rate was 8% (2 of 25); and among those stung only once or twice yearly, it was 93% (25 of 27 (Yunginger *et al.*, 1978). These data indicated that the incidence of bee sting allergy was inversely correlated with the number of stings received by the individual. Similar results were reported in a European study of 250 beekeepers. Although 43% of these beekeepers had experienced hypersensitive reactions, only 7.5% were anaphylactic, most of which were not serious, and the even smaller figure of 1.2% were serious reactions (Bousquet *et al.*, 1982) Again beekeepers who receive repeated stings during the beekeeping season experienced fewer and less severe reactions than those receiving few stings. None of the 79 beekeepers who received over 200 stings annually had a systemic reaction, 9% of those stung annually by 50 to 200 bees had systemic reactions, and 20% of those stung 25 to 50 times annually experienced systemic reactions. The highest incidence of systemic reactions experienced was 45% for the group receiving only 0 to 25 stings annually (Bousquet *et al.*, 1984a).

The trends recorded for sting reactions correlated reasonably well with the levels of "blocking" IgG antibodies and "allergic" IgE antibodies in the blood of beekeepers. Beekeepers who experienced frequent stings typically had high levels of IgG and low, or mixed, levels of IgE (Light *et al.*, 1975; Mueller *et al.*, 1977, 1978; Yunginger *et al.*, 1978; Bousquet *et al.*, 1982, 1984a; Blaauw and Smithuis, 1985). Beekeepers who were not sensitive also typically had negative skin tests, high IgG, and low or mixed IgE (Light *et al.*, 1975; Yunginer *et al.*, 1978; Miyachi *et al.*, 1979; Bousquet *et al.*, 1982, 1984a). However, the data for IgE levels are so scattered and overlapping among sensitive and nonsensitive beekeepers as to preclude their value as a diagnostic tool for predicting sensitivity (Yunginger *et al.*, 1978; Bousquet *et al.*, 1982, 1984a). The best "predictive index" appears to be the number of stings received yearly—over 50 correlates with no reactions, and lesser numbers indicate increased risk of reaction (Bousquet *et al.*, 1984a).

Unlike within the general public, there appears to be a relationship between bee venom hypersensitivity and atopy. Atopic beekeepers and family members had significantly higher levels of IgE antibodies than nonatopic individuals. This was true for both sting sensitive and nonsensitive individuals (Yunginger *et al.*, 1978; Bousquet *et al.*, 1982). Venom-sensitive beekeepers also report a much higher incidence of atopy than nonsensitive beekeepers, though their atopy was not clinically confirmed (Bousquet *et al.*, 1982) and requires more thorough study. Despite these correlations, beekeeping, in itself, does not appear to predispose one to atopy: the incidence of atopy among beekeepers in one report was only 8.8% (Light *et al.*, 1975) and random surveys of the beekeeping population have not been conducted.

Several unusual sting-related observations have been reported among beekeepers. Bousquet *et al.* (1982) reported that 2% of their patients experienced late (24 - 48 hours after the sting) joint inflammation, and several cases of individuals who had never experienced a sting reaction until they underwent surgery or other medical treatments that were independent of their beekeeping activities. Whether these reactions were just the result of random statistical variation or because of medical stress cannot be determined. The blood of beekeepers was found to have very minor differences from nonbeekeepers, but these differences were not related to the number of stings received and were considered not worthy of concern (Yunginger *et al.*, 1978). These negative indications from the only systematic study that had attempted to discover other medical effects on individuals who had long-term exposure to bee stings are suggestive that there are no long-term nonimmunological side effects of beekeeping.

A frequent series of reactions to subsequent stings (and one experienced by the author) upon beginning beekeeping is increased severity of reaction after subsequent stings, followed by a peaking of effects and gradual, usually total, loss of symptoms. This "bell-shaped"curve of reactions often takes the form of large local reactions that increase in size for the first 5 to 10 stinging incidents, then decrease in size, until the reactions finally disappear altogether. Often during this process and just when the person starts to despair because of the severity of the reactions (can extend from fingers to the elbow), subsequent stings induce smaller reactions. The local reactions might return again if many stings, for example 25, are received in one area such as a hand.

Several overall observations and conclusions can be made relative to venom reactions and beekeeping. Reactions typically occur in individuals during their first year of beekeeping (Bousquet *et al.*, 1982). Reactions in individuals in subsequent years of beekeeping typically occur after the first sting of the spring season, before IgG antibodies build up (Bousquet *et al.*, 1984a). These early season reactions are often mild, consisting of cutaneous or minor respiratory problems. The longer a beekeeper has been working bees and the more stings received, the less frequent the sting reactions and the less severe are those reactions that do occur (Light *et al.*, 1975; Mueller *et al.*, 1977; Yunginger *et al.*, 1978; Bousquet *et al.*, 1982, 1984a). Long-term beekeepers very rarely, if ever, have serious life-threatening systemic reactions; though retired and inactive beekeepers can have elevated systemic reaction rates (Light *et al.*, 1975). Nonbeekeeping family members of beekeepers more frequently experience sting reactions than the beekeepers, themselves. Finally, those people who are interested in beekeeping and experience side effects to stings often give up beekeeping (Light *et al.*, 1975; Schmidt, unpublished observations).

ALLERGY TO BEE BODY PARTS

Allergy to bee "odor"or dust is distinct from venom allergy and is a form of respiratory allergy that resembles hay fever or pollen-induced asthma. It is caused by identical immunological mechanisms as typical inhalant atopy and differs only in the respect that the inducing allergens are derived from the proteins of bee body parts. Allergy to bee bodies is often considered a form of occupational atopy because it is usually observed in beekeepers, members of beekeeping families, and workers in honey and bee processing plants.

Allergy to bee body proteins was first clearly demonstrated in 1932 in a thorough study conducted by Ellis and Ahrens (1932). They described two cases, a researcher who was a hobbyist beekeeper, and a beekeeper who developed asthma when around bees or bee odor. By skin tests using local pollens including those species collected by the bees, they showed that the reaction was not caused by the pollen carried by the bees. Extracts of bees, bee heads and thoraces (minus the abdomen with its venom), bee larvae, royal jelly, and even bumble bees caused strong positive skin tests. That the allergic skin reaction could be induced in a nonsensitive person by passive transfer (Prausnitz-Küstner reaction) demonstrated that the blood of the hypersensitive patient actually contained the allergic antibodies. Finally Ellis and Ahrens (1932) desensitized one patient with extract of bee head-thoraces and achieved total protection from respiratory asthma upon challenge by contact with bees. That patient returned to working bees and conducting his research.

Several other workers in the 1930's (Benson and Semenov, 1930; Benson, 1939) discovered allergy to bee body parts, but unfortunately they confused allergy to bee body proteins with allergy to bee venom. Light *et al.* (1975) demonstrated in their 34 beekeepers that the levels of whole body IgE did not differ between venom hypersensitive and nonhypersensitive groups; hence confirming that allergic antibodies to bee venom and body proteins were different and did not cross react. Yunginger *et al.* (1978) reported that two of their surveyed beekeepers developed asthma when they were cleaning bee equipment. Upon questioning they also discovered that 14 of 22 atopic beekeepers reported increased hay fever or asthma upon exposure to bee dust. These findings suggest that those individuals who have a tendency toward atopy, are more prone than nonatopics to develop allergy to bee body materials.

Although allergy to bee body proteins and bee venom are independent processes, individuals can develop sensitivity to one, or both types of allergens. A son of a beekeeper was found to have sensitivities to both types of bee proteins; and in another case a honey plant worker developed inhalant allergy to bee body materials, yet did not react to bee venom (Reisman *et al.,* 1983). A further patient who worked in a honey packing plant developed severe inhalant allergy to bee parts and to filtrate of air in the plant, but not to honey, pollen, or bee stings. RAST-inhibition using whole body extracts showed that the reaction was to the body proteins. Two of the other 15 workers in the plant mentioned they also experienced inhalant allergy symptoms (Ostrom *et al.,* 1986).

PRECAUTIONS TO AVOID STINGS

Numerous printed sources provide advice on how to avoid stinging insects and being stung. Most rely on "common sense", and are often overly cautious and even incorrect. The most useful advice to provide hypersensitive patients is to learn the basic biology of stinging insects and when and why they might sting. The stinging insects of concern are all social insects and sting only in defense of their colonies. Foraging bees and wasps do not readily sting when away from their colonies. I and many of my colleagues have never been spontaneously stung by any foraging insect on a flower or vegetation, even when intentionally provoked.

Most stinging events occur when a person becomes inadvertently close to and threatening to a wasp or bee colony. Avoidance of such confrontations is simple. Yellowjacket colonies usually nest in the ground near rotting logs or vegetation, and along streambanks. They usually are located at the edges of yards at the base of shrubbery or trees, or near or under buildings and structures. They are most abundant and potentially dangerous in mid summer to late fall. Baldfaced hornets, aerial yellowjackets and paper wasps tend to nest under eaves of building or in dense vegetation. Honey bees tend to nest in large hollow trees, in buildings, in holes in the ground (in hot dry parts of the world), or in managed apiaries. For individuals who are life-threateningly sensitive to insect stings, they should learn which species are the threat and avoid areas and times when they are likely to encounter a nest. Simple precautions such as not mowing the lawn, trimming shrubbery, painting buildings or clearing yard debris—especially during the summer and fall—are ways to avoid yellowjackets and wasps. In most areas, these activities can be safely performed during the winter or early spring, and it is best to do yard "clean up" at this time, or else have someone else do that work. If yellowjacket and wasp nests are discovered, they should be removed by someone who is not hypersensitive, preferably a responsible adolescent or young person, or simply avoided. Lawn mowers are probably the single most common means of disturbing a yellowjacket colony and are particularly risky because they not only make so much noise that the operator cannot hear the buzzing of attacking wasps, but they also present all the proper stimuli for attack— warmth, vibration, movement, and odor and carbon dioxide emission. Feral honey bees are more difficult to avoid predictably, but the risks of contact can be reduced by keeping ones eyes and ears alert for the flight patterns and sounds of bees leaving a colony. Honey bees, like yellowjackets tend to nest near streams or other water sources (Schmidt *et al.*, unpublished) and when in these areas one should be particularly alert.

To reduce the chances of being stung by attacking wasps or bees a few simple behaviors will help. Attacking wasps and bees are attracted to moving, large, dark, warm objects (Free 1961, Maschwitz 1964). Motion is probably the most important of these cues. Thus, to avoid being stung, motion should be slow and deliberate. Swatting, flailing, and other panic motions are sure to attract the attention of defenders and greatly increase the chances of receiving what was to be avoided, stings. If the insects are just "buzzing" around the nest entrance, the best escape procedure is to lower one's head and torso and walk away. If the insects are actively attacking, then retreat *without flailing* (holding breath for the first few seconds can help eliminate the cues of warm moist, odor and carbon dioxide-laden breath). If someone else has provoked a colony for any reason, avoid that area.

Although rare events, stings by foragers distant from their colonies can occur. To minimize the probability of such stings, one can wear footwear to avoid stepping on a bee on a flower, watch where he/she is about to sit (such as edges of swimming pools where bees and wasps are collecting water), and keep one's hair either covered, tied, or short. Once a stinging insect gets caught in hair, the chances of being stung are great. Also, to crush and kill and insect trapped in hair requires such a hard blow to the area that one risks giving himself a headache. Entanglement of stinging insects in loose hair is fairly common, and if it occurs by a single insect and away from a known or likely nest, a good means of removal is to insert a comb between the scalp and insect and comb it out.

Individual forager yellowjackets can present a serious hazard in late summer and fall. Many species are scavengers and frequent garbage cans, soft drink cups, ripe fruit, and especially picnics. Foragers can sometimes become such a nuisance at outdoor activities and picnics, as occurred in the Sierra Nevada mountains of California during the summer of 1989, that they even attracted the attention of national newspapers and magazines (Harris, 1989). The problem with these yellowjackets is that they will land on food being eaten, or in beverage containers and can be inadvertently consumed, resulting in a nasty sting in the mouth. They also land and crawl on surfaces, causing stings when accidentally sat upon, and generally buzz around anybody in the area, thereby making themselves a nuisance as well as a risk. To minimize these problems, outdoor activities can be reduced during peak yellowjacket periods and sweet and meaty food avoided at picnics (yellowjackets do not eat cheese, peanut butter, bread, diet beverages, salads, water, etc.) If a yellow-jacket or other flying insect enters a car window, all the windows should be opened to allow the insect to leave, and the car pulled over to the side of the road until it leaves or can be crushed.

Fire ants present different avoidance problems than wasps and bees. Although they cannot fly, they can be omnipresent outdoors in parts of the South and Southeastern United States. The best way to avoid fire ant stings is to be always attentive as to where one is walking, standing, or sitting, to wear shoes, and to avoid grassy and disturbed land where fire ants abound.

Beyond what is mentioned above, no other precautions or alterations of life style are suggested unless the individual is exquisitely life-threateningly hypersensitive (such as having experienced loss of consciousness from a previous sting). In that case utmost care needs to be taken in addition to receiving immunotherapy. Golf courses, outdoor hunting, fishing, camping and similar recreation should be avoided during the stinging insect season, and some alteration of clothing when in such situations is in order. Stinging insects attack large, dark, rough textured objects more readily than their opposites. Thus black, brown, blue, and red (stinging insects cannot readily see red and it looks black to them) clothes, and wool and loose weave should be avoided. White, beige, yellow, light green, and any other light color is recommended. There is only anecdotal evidence that brightly colored or "floral patterned" clothes or perfumes attract stinging insects, so there is little need to avoid these.

ECONOMIC CONSIDERATIONS OF VENOM ALLERGY

Direct as well as indirect economic considerations may affect decisions and activities of individuals allergic to insect venom. Direct considerations include the actual costs for medical advice and treatment, the time and cost to get to the medical facilities, and the possible salary loss during the treatment time. Indirect costs include the nuisance and time expenditure necessary to deal with the hypersensitivity, the potential cost of work time loss, or alteration of work habits resulting from the affliction, and the potential alteration of lifestyles and general psychological apprehension and worry engendered in the sensitive individual and those around that person.

The direct costs for various medical treatment options are the easiest costs to determine. To assign a value to the indirect costs is much more difficult, and will vary greatly from individual situation to situation. The possible enormous indirect costs in terms of personal morbidity, *i.e.* the uncontrolled fear and anxiety resulting from knowledge of the hypersensitivity, are extremely difficult to measure. These morbidity costs—not saving lives—are the main actual justification for administering immunotherapy to venom hypersensitive patients (Golden and Valentine 1981; Golden 1987).

The costs of diagnosis of sting hypersensitivity are relatively low. Usually only one or two visits to an allergist, plus the requisite skin tests, and possibly

RAST tests are all that is required. If hypersensitivity is diagnosed, the costs can range from nothing to extremely high, depending upon the actions taken. If the hypersensitivity is in a child and is nonlife-threatening, or if it mainly involves the skin, minimal treatment is all that is necessary. This typically involves advice, plus the optional addition of an epinephrine sting kit. Sting kits are quite reasonable in price, usually costing between $15 and $30 in the United States. If the hypersensitivity is more severe, treatment costs can increase dramatically, depending upon what action the physician recommends and the patient accepts. Usually immunotherapy is recommended, but the patient does not need to automatically accept this suggestion. The patient is perhaps the most important part of venom hypersensitivity treatment and needs to be aware and make decisions. Unless the hypersensitivity is life-threatening, that is falls into category VI of Table 3, there are no medical indications that immunotherapy can prevent future loss of life. In these situations, patient awareness and possession of an epinephrine sting kit are all that are required for safety. If severe morbidity is involved as well, then immunotherapy might be a reasonable option as a means of alleviating this problem.

Immunotherapy can be expensive. Immunotherapy for insect venom hypersensitivity often involves repeated visits to an allergy clinic for routine injections over a very long period of time and requires the use of expensive venoms for injection. If the hypersensitivity is just for one species such as the honey bee, then only one venom is needed and the cost is reduced. If, however, multiple, or cross-sensitivity is detected, then two, three, or more venoms might be needed. Treatment recommendations typically include 12 treatment visits over a 15-week period, followed by monthly treatments thereafter. This means 20 separate treatment occasions the first year (Graft, 1987) at an estimated cost of $500 or more and 12 visits per year thereafter at costs of $300 or more per year (Lockey, 1980). Some immunotherapy regimens are faster, taking as little as 3.5 hours (Bousquet *et al.,* 1987b; Birnbaum *et al.,* 1990a) for initial treatment and only monthly treatments thereafter. These "Rush" procedures reduce the number of visits and costs for the initial year only.

Immunotherapy with its accompanying inconveniences and expenses is probably a good investment for some venom hypersensitive individuals. Those individuals experiencing life-threatening reactions (category VI of Table 3), particularly those who lose consciousness as a result of a sting, definitely need immunotherapy treatment. Individuals who also have cardiac insufficiency or compromise, particularly if over 40 years of age, such that stress to the heart can cause insufficient blood flow to the brain or infarction are definite candidates for immunotherapy. Moreover, although use of epinephrine sting kits is not known to have caused any deaths, epinephrine administration can be dangerous for individuals with severe cardiac problems (Barach *et al.,* 1984; Freye and Ehrlich, 1989). Thus, immunotherapy is recommended under careful attention to prevent a future hypersensitive reaction in the case of a sting, or the necessity to use one or more epinephrine injections to stop a hypersensitive reaction.

PSYCHOLOGICAL ASPECTS

Much of the pathology of insect venom hypersensitivity is psychological. Fear can be a medical problem, especially if irrational fear. Insects, in general, are an aversion to many people and stinging insects, in particular, can engender great fear or phobias in some individuals. The irrational swatting, flailing and similar movements by many people when a flying or stinging insect approaches them are evidence of general aversions and phobias of these insects (Crane, 1976; Olkowski and Olkowski, 1976; Byrne *et al.,* 1984). Such feelings are biologically natural, as no organism seeks pain such as that delivered by a sting. However, excessive preoccupation with the prospect of the pain or a hypersensitive reaction is not natural and can lead to morbidity and a consequent reduction in quality of life.

Fear and morbidity of insect stings and sting reactions can be reduced through a variety of treatments including counseling and immunotherapy. But the simplest, easiest, and least expensive solution to morbidity is often obtained by gaining knowledge of the situation and obtaining the support of those around the afflicted individual. Once knowledge of the habits and biology of stinging insects is achieved, their unpredictability can be reduced in the mind of the affected person. Most public libraries have good, readable books on beekeeping and stinging insects. These plus encyclopedias can provide fascinating insights into the lives of stinging insects and often include beautiful pictures of them. If one can gain an appreciation of the beauty and intricacy of these insects, often they can be tolerated and enjoyed, at least in a vicarious way. If the inherent morbid fear of stinging insects, themselves, can be eliminated, then reason and medical information can provide the needed support for reducing the fear of "dying from the next sting." As presented earlier, the chances of dying from an insect sting are less than of falling down stairs, crossing the street, or playing athletics; yet most people do not fear climbing stairs, crossing streets, or participating in sports. So, why is there any need to fear insect sting reactions, a far less threat to life? Even among life-threateningly hypersensitive individuals, the chances of dying as a result of a sting are less than that of drowning, burning in a fire, or of succumbing to asthma. Again, there is no reason to have irrational fear of sting-related death, even for these people. As with many common activities, such as crossing the street, normal caution and awareness are in order when dealing with chances of encountering stinging insects. In the last analysis the statement by Ruben-stein (1980a) that the chances of dying in an automobile accident while en route to a hospital following an insect sting may be greater than the chances of dying due to the sting may be a telling analysis.

THE BEEKEEPER'S ROLE

Beekeepers, because of their knowledge and expertise with honey bees (and often other stinging insects), are in a unique position to provide a supportive and stabilizing influence in situations relating to stings. As most beekeepers know, they are also frequently asked by friends and strangers about bee stings, venom allergy, and what to do about them. The questioners are usually coming with either a great deal of misinformation, apprehension, or both. An informed beekeeper can do much to correct this misinformation and to allay any fears. This can be done by providing realistic information, and, above all, by projecting confidence and strength. If the beekeeper is not worried or frightened by stings and the potential for sting reactions, then that confidence can positively affect the inquiring person.

Beekeepers and physicians are the first line of contact for the public when questions about stings arise. Physicians usually have very little knowledge of the insects and their actual threats, have been trained to look for exceptions and to treat them, are overly cautious because of the prospects of lawsuits, and make their living by treating patients. Beekeepers are usually motivated by a desire to make honey, to enjoy interesting activities involving bees, to learn about bees, and to pass on their enthusiasm to others. Because beekeepers generally like their bees, they are excellent ambassadors to provide positive information about stinging insects. Physicians typically only see negative sides of stinging insects and generally convey negative information. Positive infor-mation is not only helpful to allay and reduce irrational fears and behaviors, it also benefits the honey industry as a whole by maintaining a friendly attitude toward bees and their products.

The beekeeper, by understanding sting allergy can provide a great public service. By knowing symptoms and conditions relating to a person's health and sting reactions, a beekeeper is in an excellent position to provide sound advice of actions to take. The beekeeper can help determine if the afflicted individual is at risk and how important it is to see a physician. Thus, informed beekeepers can often act as indirect and valuable referral services.

EPILOGUE: OTHER EVIDENCE AND PERSPECTIVES

Strong arguments can be made against the trends of current medical treatment for sting reactions. Arguments against current testing and treatment are based on the following five points: 1) the "disease" has a mortality rate near zero; 2) there is no ability to predict who will and who will not suffer a fatal outcome of a sting; 3) for the majority of actual sting deaths, current treatment would not have prevented the death; 4) fear reduction (not saving of lives) is the main espoused justification of treatment; and 5) treatment is a major inconvenience and expense.

An ultimate goal of medicine is saving lives. How necessary is medical treatment for diseases whose mortality rate approaches zero, especially when those at risk have not been identified? Medicine has no abiltiy to predict who will and who will not suffer a fatal sting reaction. Medicine can predict the probability of an individual experiencing an allergic reaction based on past history plus skin tests and RAST measurements; but these predictions relate to having a *reaction*, not to dying. Instead of concentrating on prevention of death, which should be the main concern, medical focus has shifted to the reaction, itself. This shift of medical emphasis is because reactions can be statistically reduced, whereas reduction of death is uncertain. Is it responsible medicine to focus so much attention on nonlethal reactions (Rubenstein, 1980b)?

A common public perception of venom immunotherapy is that it is administered to save lives, a perception which, by necessity, must have its origins in the medical profession (*e.g.* "Each year thousands of patients have severe, systemic reactions to stings resulting in death to a considerable number of these individuals" [Hutcheson and Slavin, 1990]). Yet the majority of sting-induced deaths occur without prior warning (Jensen, 1962; Barnard, 1973; Mosbech, 1983; Golden *et al.*, 1989a); that is, as the first reaction experienced by the individual. Since the individual had never experienced a previous reaction, he/she would have no way to know to take special precautions or to have immunotherapy treatment. Thus, even if immunotherapy could save lives*, it would not have been possible to help these individuals or to save their lives. This reduces the already tiny number of lives that could be potentially saved to an even more diminishingly small number.

The actual justification of immunotherapy treatments for venom sensitivity is reduction of fear in the patients (Golden and Valentine, 1981; Golden,

---

*Although direct evidence that venom immunotherapy actually saves lives would be virtually impossible to obtain, indirect (and not discovered by the medical community) evidence indicates that lives probably have been saved by venom immunotherapy. This evidence comes in the discovery that over a period of 11 years eight individuals who received significant whole body extract therapy died of insect stings (Torsney, 1973). There are no reports over the 10-year period of venom immunotherapy (ca. 1980 - 1990) of a sting-related death of anyone undergoing therapy. Thus, assuming that equal numbers of individuals were undergoing immunotherapy during both periods, we predict that about eight people were statistically saved by venom immunotherapy during that period (if venom immunotherapy were ineffective we would have expected eight deaths to have occurred).

1987) and not protection of lives. This justification is based on such statements as: "Fear of fatal reactions and the consequent change in life-style is more widespread because 0.4 to 0.8 per cent of the United States population has survived a systemic reaction to a sting" (Hunt *et al.*, 1978); "80% of the parents believed that their child would be *at risk of death* (author emphasis) after a subsequent sting" (Chipps *et al.*, 1980); "it is rarely possible to dissuade such a patient from the notion that the next sting may have severe consequences" (Lichtenstein *et al.*, 1980); and "many patients describe an overwhelming fear of being stung and have modified their lifestyles to avoid the outdoors and many activities previously enjoyed" (Golden and Valentine, 1981). Immunotherapy, especially followed by an uneventful sting challenge, can reduce this fear (Chipps *et al.*, 1980). But such reports alone are not sufficient to justify immunotherapy as the best treatment: far better evidence would be provided by a classical experiment in which one group received immunotherapy and the other was provided sound factual and epidemiological information along with counseling and psychological support. After the treatments were completed, independent surveys of the opinions and fears then could be conducted to determine which procedure was more effective in reducing the fear.

Fear concerning insect sting reactions is clearly present, but measures can be taken to prevent the accentuation of the fear and to reduce any fear that does exist. Fear of insect sting reactions appears to be a learned reaction. In a study by Mickalide *et al.*, (1985) "two-thirds of the parents rated it (their child's sting allergy) as a moderately or very serious health problem," yet "most children believed that they could control being stung" and "62% were slightly or not at all worried about its (a sting reaction) occurrence." A conclusion that emerges from these statements is that children have a low innate fear of sting reactions and that if positive input were provided, they likely would retain that low fear (rather than becoming like their parents). Indeed, as stated by Rubenstein (1982) "physicians have helped patients to master other frightening but usually benign syndromes including hyperventilation . . . and acute anxiety attacks" — so why cannot they help patients to master fear of insect stings? Serious attempts to use other approaches instead of (or in addition to) immunotherapy for controlling fear of insect stings is clearly owed to patients.

The last contraindication for immunotherapy as a primary treatment of venom hypersensitivity is its extreme cost (Yunginger, 1981) and inconvenience. This "hassle" is so great, that statements have been recorded such as: "Every one of them indicated that cost and inconvenience of treatment were major deterrents and that they preferred to take their chances with avoidance techniques and adrenalin kits" (Golden *et al.*, 1986) and "None of our subjects with systemic reactions had sought medical attention or advice from their physicians. In fact, they all refused our advice to enter venom immunotherapy" (Golden *et al.*, 1989a). Indeed, with a cost of $300 to $1000 or more a year (Lockey, 1980; Rubenstein, 1980a)—plus the hassle—plus the dubious need for immunotherapy—is it any surprise that so few people readily endure venom immunotherapy? Rather than education to inform all of these people to get treated, maybe what really is needed is education to "strive to liberate patients from unrealistic fears of highly improbable events (and not focus their lives on reactions some call 'life threatening') in a disease in which death is very rare" (Rubenstein, 1982).

REFERENCES

Abrecht, I., G. Eichler, U. Müller and R. Hoigne. (1980). On the significance of severe local reactions to Hymenoptera stings. *Clin. Allergy 10*:675-682.

Abrishami, M.A., G.K. Boyd and G.A. Settipane. (1971). Prevalence of bee sting allergy in 2,010 Girl Scouts. *Acta. Allergol. 26*:117-20.

Anon. (1989). Woman dies after hitting beehive with truck. *Speedy Bee (August)*:13.

Argiolas, A. and J.J. Pisano. (1985). Bombolitins, a new class of mast cell degranulating peptides from the venom of the bumblebee, *Megabombus pennsylvanicus. J. Biol. Chem. 260*:1437-44.

Aukrust, L., E. Einarsson, S. Öhman and S.G.O. Johansson. (1982). Crossed radioimmunoelectrophoretic studies of bee venom allergens. *Allergy 37*: 265-71.

Baer, H., T.-Y. Liu, M.C. Anderson, M. Blum, W.H. Schmid and F.J. James. (1979). Protein components of fire ant venom (*Solenopsis invicta*). *Toxicon 17*: 397-405.

Banks, B.E.C. and R.A. Shipolini. (1986). Chemistry and Pharmacology of Honey-bee Venom. *In*: Venoms of the Hymenoptera, pp. 330-416 (T. Piek, *ed.*). Academic Press:London.

Barach, E.M., R.M. Nowak, T.G. Lee and M.C. Tomlanovich. (1984) Epinephrine for treatment of anaphylactic shock. JAMA 251:2118-22.

Barnard, J.H. (1973). Studies of 400 Hymenoptera sting deaths in the United States. J. Allergy Clin. *Immunol. 52*:259-64.

Beck, B.F. (1935). Bee Venom Therapy. Appleton-Century:New York.

Bell, T.D. and D.F. Hahlbohm. (1983). Hymenoptera allergy:clustering in beekeeping households. *Ann. Allergy 50*:356.

Benson, R.L. (1939). Diagnosis of hypersensitiveness to the bee and to the mosquito. Arch. *Intern. Med. 64*:1306-27.

Benson, R.L. and H. Semenov. (1930). Allergy in its relation to bee sting. *J. Allergy 1*:105-16.

Benton, A.W. (1967) Esterases and phosphatases of honeybee venom. *J. Apic. Res. 6*:91-94.

Bernheimer, A.W., L.S. Avigad and J.O. Schmidt. (1980). A hemolytic polypeptide from the venom of the red harvester ant, *Pogonomyrmex barbatus. Toxicon 18*:271-78.

Bernheimer, A.W., L.S. Avigad, J.O. Schmidt and J.S. Ishay. (1982). Proteins in venoms of two wasps, *Polistes comanchus navajoe* and *Vespa orientalis. Comp. Biochem. Physiol. 71C*:203-07.

Bettini, S. (1978). Arthropod Venoms, Vol. 48. Handbook of Experimental Pharmacology. Springer-Verlag:Berlin.

Birnbaum, J., C. Charpin and D. Vervloet. (1990a). Rush immunotherapy to Hymenoptera venoms:comparisons of 3 protocols. *J. Allergy Clin. Immunol. 85*:212.

Birnbaum, J.C. Charpin, P. Bongrand and D. Vervloet. (1990b) General adverse reactions during rush venom immunotherapy (RIT) and levels of specific Ige and IgG antibodies. *J. Allergy Clin. Immunol. 85:*210.

Blaauw, P.J. and L.O.M.J. Smithuis. (1985). The evaluation of the common diagnostic methods of hypersensitivity for bee and yellowjacket venom by means of an in-hospital insect sting. *J. Allergy Clin. Immunol. 75:*556-62.

Blum, M.S., J.R. Walker, P.S. Callahan and A.F. Novak. (1958). Chemical, insecticidal and antibiotic properties of fire ant venom. *Science 128:*306-07.

Boreham, M.M. and D.W. Roubik. (1987). Population change and control of Africanized honey bees (Hymenoptera:Apidae) in the Panama Canal area. Bull. *Entomol Soc. Amer. 33:*34-39.

Bousquet, J., Y. Coulomb, M. Robinet-Levy and F.B. Michel. (1982). Clinical and immunological surveys in bee keepers. *Clin Allergy 12:*331-42.

Bousquet, J., J-L Menardo, R. Aznar, M. Robinet-Levy and F.-B. Michel. (1984a) Clinical and immunologic survey in beekeepers in relation to their sensitization. *J. Allergy Clin. Immunol. 73:*332-40.

Bousquet, J., G. Huchard and F.-B. Michel. (1984b). Toxic reactions induced by Hymenoptera venom. *Ann. Allergy 52:*371-74.

Bousquet, J., A. Fontez, R. Aznar, M. Robinet-Levy and F.-B. Michel. (1987a). Combination of passive and active immunization in honeybee venom immunotherapy. *J. Allergy Clin Immunol. 79:*947-54.

Bousquet, J., U.R. Müller, S. Dreborg, J. Jarisch, H.-J Malling, H. Mosbech, R. Urbanek and L. Youlten. (1987b) Immunotherapy with Hymenoptera venoms. *Allergy 42:*401-13.

Bousquet, J., J.-L. Menardo, G. Velasquez and F.-B. Michel (1988). Systemic reactions during maintenance immunotherapy with honey bee venom. *Ann. Allergy 61:*63-68.

Bousquet, J., J. Knani, G. Velasquez, J.L. Menardo, L. Guillous and F.B. Michel. (1989) Evolution of sensitivity to Hymenoptera venom in 200 allergic patients followed for up to 3 years. *J. Allergy Clin. Immunol. 84:*944-50.

Braun, L.I.B. (1925). Notes on desensitization of a patient hypersensitive to bee stings. *S. Afr. Med. Rec. 23:*408-09.

Brown, H. and H.S. Bernton. (1970) Allergy to the Hymenoptera V. clinical study of 400 patients. *Arch. Intern. Med. 125:*665-69.

Busse, W.W. and J.W. Yunginger. (1978). The use of the radioallergosorbent test in the diagnosis of Hymenoptera anaphylaxis. *Clin. Allergy 8:*471-77.

Byrne, D.H., E.H. Carpenter, E.M. Thoms and S.T. Cotty. (1984). Public attitudes toward urban arthropods. Bull. Entomol. *Soc.Amer. 30(2):* 40-43.

Chafee, F.H. (1970). The prevalence of bee sting allergy in an allergic population. *Acta Allerg. 25:* 292-93

Charavejasarn, C.C., R.E. Reisman and C.E. Arbesman. (1975). Reactions of anti-bee venom mouse reagins and other antibodies with related antigens. Int. *Arch. Allergy Appl. Immunol. 48:* 691-97.

Charpin, D., D. Vervloet, M. Tafforeau, E. Haddi, T. Djime, H. Deyme, G. Kulling, A. Lanteaume, J.P. Kleisbauer and J. Charpin. (1988). Allergic aux hypenopteres: frequence et degre d' information du public. *La Presse Med. 17:* 1309-11.

Chipps, B.E., M.D. Valentine, A. Kagey-Sobotka, K.C. Schuberth and L.M. Lichtenstein. (1980). Diagnosis and treatment of anaphylactic reactions to Hymenoptera stings in children. *J. Pediatr. 97:* 177-84.

Chugh, B.K.S., B.K. Sharma and P.C. Singhal. (1976). Acute renal failure following hornet stings. *J. Trop. Med. Hyg. 79:* 42-44.

Clayton, W.F., J.W. Georgitis and R.E. Reisman. (1985). Insect sting anaphylaxis in patients without detectable serum venom-specific IgE. *Clin. Allergy 15:* 329-33.

Cohen, S.G. (1989). The pharaoh and the wasp. *Allergy Proc. 10:* 149-51.

Cole, A.C., Jr. (1968). Pogonomyrmex Harvester Ants. Univ of Tennessee Press: Knoxville.

Collins, L.C. and R.J.M. Engler (1988). Enhanced yellowjacket (YJ) specific IgE (VS-E) following natural sting associated anaphylaxis (ANP) despite maintenance venom immunotherapy (VIT). J. Allergy Clin. *Immunol. 81:* 169.

Crane, E. (1976). The range of human attitudes to bees. *Bee World 57:* 14-18.

Czarnetzki, B.M., T. Thiele and T. Rosenbach. (1990). Evidence for leukotrienes in animal venoms. *J. Allergy Clin. Immunol. 85:* 505-09.

Day, J.H. (1986). A comparison of venom concentrations of 0.1 µg/ml and 1.0 µg/ml as indicator of sensitivity to honey bee stings. *J. Allergy Clin. Immunol. 77:* 143.

Edwards, R. (1980). Social Wasps. Rentokil: E. Grinstead, W. Sussex.

Ellis, R.V. and H.G. Ahrens (1932). Hypersensitiveness to air borne bee allergen. *J. Allergy 3:*247-52.

Engel, T., J.H. Heinig and E.R. Weeke. (1988). Prognosis of patients reacting with urticaria to insect sting. *Allergy 43:*289-93.

Ennik, F. (1980). Deaths from bites and stings of venomous animals. *West. J. Med. 133:*463-68.

Ewan, P.W. (1985). Allergy to insect stings: a review. *J. Roy. Soc. Med. 78:*234-39.

Findlay, S.R., J.E. Gillaspy, R. Lord, L.S. Weiner and J.A. Grant. (1977). *Polistes* wasp hypersensitivity; diagnosis by venom-induced release of histamine in vitro. *J. Allergy Clin. Immunol. 60:*230-235.

Frankland, A.W. (1976). Bee sting allergy. *Bee World 57:*145-50.

Free, J.B. (1961). The stimuli releasing the stinging response of honeybees. *Anim. Behav. 9:*193-96.

Freye, H.B. and B. Ehrlich. (1989). Acute myocardial infarction following Hymenoptera envenomation. *Allergy Proc. 10:*119-26.

Gadde, J., A.K. Sobotka, M.D. Valentine, L.M. Lichtenstein and D.B.K. Golden. (1985). Intervals of six and eight weeks in maintenance venom immunotherapy (MVIT). *Ann. Allergy 54:*348.

Gädeke, R., H. Helwig, M. Otto, F. Schindera and B. Weineck. (1977). Tödliche Vergiftungskrankheit eines Kindes nach massenhaften Wespenstichen. *Med. Klin. 72:*1487-92.

Ganderton, M.A. (1979). Anaphylactic reactions to wasp and bee stings. *Brit. Med. J. 1:*1216-17.

Georgitis, J.W. and R.E. Reisman. (1985). Venom skin tests in insect-allergic and insect-nonallergic populations. *J. Allergy Clin. Immunol. 76:*803-07.

Goldberg, A. and R.E. Reisman. (1988). Prolonged interval maintenance venom immunotherapy. *Ann. Allergy 61:*177-79.

Golden, D.B.K. (1987). Diagnosis and prevalence of stinging insect allergy *Clin Rev. Allergy 5:*119-36.

Golden, D.B.K. (1989). Epidemiology of allergy to insect venoms and stings. *Allergy Proc. 10:*103-07.

Golden, D.B.K. and M.D. Valentine. (1981). Insect sting allergy. *in:* Clinical Immunology Update, pp. 169-96 (E.C. Franklin, *ed.*) Elsevier: NY.

Golden, D.B.K., M.D. Valentine, A. Kagey-Sobotka and L.M. Lichtenstein. (1980). Regimens of Hymenoptera venom immunotherapy. *Ann. Intern. Med.* 92:620-24.

Golden, D.B.K., A. Kagey-Sobotka, M.D. Valentine and L.M. Lichtenstein. (1981a). Dose dependence of Hymenoptera venom immunotherapy. *J. Allergy Clin. Immunol.* 67:370-74.

Golden, D.B.K., A. Kagey-Sobotka, M.D. Valentine and L.M. Lichtenstein. (1981b). Prolonged maintenance interval in Hymenoptera venom immunotherapy. *J. Allergy Clin. Immunol.* 67:482-84.

Golden, D.B.K., D.A. Meyers, A. Kagey-Sobotka, M.D. Valentine and L.M. Lichtenstein. (1982a). Clinical relevance of the venom-specific immunoglobin G antibody level during immunotherapy. *J. Allergy Clin. Immunol.* 69:489-93.

Golden, D.B.K., M.D. Valentine, A. Kagey-Sobotka and L.M. Lichtenstein. (1982b). Prevalence of Hymenoptera venom allergy. *J. Allergy Clin. Immunol.* 69:124.

Golden, D.B.K., K. Johnson, B.I. Addison, M.D. Valentine, A. Kagey-Sobotka and L.M. Lichtenstein (1986). Clinical and immunologic observation in patients who stop venom immunotherapy. *J. Allergy Clin. Immunol.* 77:435-42.

Golden, D.B.K., D.G. Marsh, A. Kagey-Sobotka, L. Freidhoff, M. Szklo, M.D. Valentine and L.M. Lichtenstein. (1989a). Epidemiology of insect venom senstivity. *JAMA* 262:240-44.

Golden, D.B.K., K.A. Kwiterovich, A. Kagey-Sobotka, M.D. Valentine and L.M. Lichtenstein. (1989b). Discontinuing venom immunotherapy (VIT): determinants of clinical reactivity. *J. Allergy Clin. Immunol.* 83:273.

Graft, D.F. (1987). Venom immunotherapy for stinging insect allergy. *Clin. Rev. Allergy* 5:149-59.

Graft, D.F., K.C. Schuberth, A. Kagey-Sobotka, K.A. Kwiterovich, Y. Niv, L.M. Lichtenstein and M.D. Valentine. (1984). A prospective study of the natural history of large local reactions after Hymenoptera stings in children. *J. Pediatr.* 104:664-68.

Grant, J.A., R. Rahr, D.O. Thueson, M.A. Lett-Brown, J.A. Hokanson and J.W. Yunginger. (1983). Diagnosis of *Polistes* wasp hypersensitivity. *J. Allergy Clin. Immunol.* 72:399-406.

Green, A.W., R.E. Reisman and C.E. Arbesman. (1980). Clinical and immunologic studies of patients with large local reactions following insect stings. *J. Allergy Clin. Immunol.* 66:186-89.

Habermann, E. and J. Jentsch. (1967). Sequenzanalyse des Melittins aus den tryptischen und peptischen Spaltstücken. *Hoppe-Seyler's Z. Physiol. Chem* 348:37-50.

Habermann, E. and K.G. Reiz. (1965). Zur Biolchmie der Bienengiftpeptide Melittin and Apamin. *Biochem. Z.* 343:192-203.

Haddi, E., C. Segalen, M. Tafforeau, G. Kulling, D. Charpin, A. Lanteaume, and D. Vervloet. (1989). Distribution of atopy in an adult general population sample. *J. Allergy Clin. Immunol.* 83:210.

Harris, T. (1989). Explosion of yellowjackets rumbles along the West Coast. Sacramento Bee: September 6, page B-1.

Haux, P. (1969). Die Aminosäurensequenz von MCD-Peptid, einem spezifisch Mastzellen-degranulierended Peptid aus Bienengift. *Hoppe-Seyler's Z. Physiol. chem* 350:536-46.

Haux, P., H. Sawerthal and E. Habermann. (1967). Sequenzanalyse des bienengift-neutrotoxins (Apamin) aus seinen tryptishchen and chymotrptischen Spaltstücke. *Hoppe-Seyler's Z. Physiol. Chem.* 348:737-38.

Haydak, M.H. (1956). Bee stings. *Amer. Bee J.* 96:200-01.

Heibling, A., E. Berchtold and U. Müller. (1990). Allergy to honey bee stings: results of sting challenge (CH) one year after discontinuation of venom immunotherapy (VIT). *J. Allergy Clin. Immunol. 85*: 210.

Heilborn, H., P. Hjemdahl, M. Daleskog and U. Adamsson. (1986). Comparison of subcutaneous injection and high-dose inhalation of epinephrine—implications for self-treatment to prevent anaphylaxis. *J. Allergy Clin. Immunol. 78*:1174-79.

Herbert, F.A. and M.L. Salkie. (1982). Sensitivity to Hymenoptera in adult males. *Ann. Allergy 48*: 12-13.

Hirai, Y., T. Yashuhara, H. Yoshida, T. Nakajima, M. Fujino and C. Kitada. (1979). A new mast cell degranulating peptide "mastoparan" in the venom of *Vespula lewisii. Chem. Pharm. Bull. 27*:1942-44.

Hoffman, D.R. (1978). Honey bee venom allergy: immunological studies of systemic and large local reactions. *Ann. Allergy 41*:278-82.

Hoffman, D.R. (1981). Allergens in Hymenoptera venom. VI. Cross reactivity of human IgE antibodies to the three vespid venoms and between vespid and paper wasp venoms. *Ann. Allergy 46*:304-09.

Hoffman, D.R. (1982). Allergenic cross-reactivity between honey bee and bumblebee venoms. *J. Allergy Clin. Immunol. 69*:139.

Hoffman, D.R. (1985a). Allergens in Hymenoptera venom XIII: isolation and purification of protein components from three species of vespid venoms. *J. Allergy Clin. Immunol. 75*:599-605.

Hoffman, D.R. (1985b) Allergens in Hymenoptera venom XIV: IgE binding activities of venom proteins from three species of vespids. *J. Allergy Clin. Immunol. 75*:606-09.

Hoffman, D.R. (1985c). Allergens in Hymenoptera venom XV: the immunologic basis of vespid venom cross-reactivity. *J. Allergy Clin. Immunol. 75*:611-13.

Hoffman, D.R. (1986). Allergens in Hymenoptera venom XVI: studies of the structures and cross-reactivities of vespid venom phospholipases. *J. Allergy Clin. Immunol. 78*:337-43.

Hoffman, D.R. and C.A. McDonald. (1982a). Allergens in Hymenoptera venom VIII. immunologic comparison of venoms from six species of *Vespula* (yellow jackets). Ann. Allergy 48:78-81.

Hoffman, D.R. and C.A. McDonald. (1982b). Allergens in Hymenoptera venom IX. species specificity to *Polistes* (paper wasps) venoms. *Ann. Allergy 48*:82-86.

Hoffman, D.R. and W.H. Shipman. (1976). Allergens in bee venom I. Separation and identification of the major allergens. *J. Allergy Clin. Immunol. 58*:551-62.

Hoffman, D.R. and C.L. Wood. (1984). Allergens in Hymenoptera venom XI. isolation of protein allergens from *Vespula maculifrons* (yellow jacket) venom. *J. Allergy Clin. Immunol. 74*:93-103.

Hoffman, D.R., J.S. Miller and J.L. Sutton. (1980). Hymenoptera venom allergy: a geographic study. *Ann. Allergy 45*:276-79.

Hoffman, D.R., S.A. Gillman, L.H. Cummins, P.P. Kozak, Jr. and A. Oswald. (1981). Correlation of IgG and IgE antibody levels to honey bee venom allergens with protection to sting challenge. *Ann. Allergy 46*:17-23.

Hoffman, D.R., R.S. Jacobson and R. Zerboni. (1987). Allergens in Hymenoptera venom XIX. allergy to *Vespa crabro*, the European hornet. *Int. Arch. Allergy Appl. Immunol. 84*:25-31.

Hoffman, D.R., D.E. Dove and R.S. Jacobson. (1988a). Allergens in Hymenoptera venom XX. isolation of four allergens from imported fire ant (*Solenopsis invicta*) venom. *J. Allergy Clin. Immunol. 82*:818-27.

Hoffman, D.R., D.E. Dove, J.E. Moffitt and C.T. Stafford. (1988b). Allergens in Hymenoptera venom XXI. cross-reactivity and multiple reactivity between fire ant venom and bee and wasp venoms. *J. Allergy Clin. Immunol.* 82:828-34.

Hoffman, D.R., R. Jacobson, and M. Blanca. (1990). Allergy to venom of *Polistes dominulus*, a paper wasp introduced from Europe. *J. Allergy Clin. Immunol.* 85:211.

Hoh, T., C.L. Soong and C.T. Cheng. (1966). Fatal haemolysis from wasp and hornet sting. *Singapore Med. J.* 7:122-26.

Hölldobler, B. and E.O. Wilson. (1990). The Ants. Harvard Univ. Press; Cambridge, MA.

Huber, P., R. Hoigne, P. Schmid, M. Dozzi and U. Müller. (1983). Atopy and generalized allergic reactions to Hymenoptera stings. Monogr. *Allergy 18*:147-49.

Humblet, Y., J. Sonnet, C. van Ypersele de Strihou. (1982). Bee stings and acute tubular necrosis. *Nephron 31*:187-88.

Hunt, K.J., M.D. Valentine, A.K. Sobotka, A.W. Benton, F.J. Amodio and L.M. Lichtenstein. (1978). a controlled trial of immunotherapy in insect hypersensivity. *New Eng. J. Med. 299*:157-61.

Hutcheson, P.S. and R.G. Slavin. (1990). Lack of preventive measures given to patients with stinging insect anaphylaxis in hospital emergency rooms. *Ann. Allergy 64*:306-07.

Idsøe, O., T. Guthe, R.R. Willcox and A.L. de Weck. (1968). Nature and extent of penicillin side-reactions, with particular reference to fatalities from anaphylactic shock. *Bull WHO 38*:159-88.

Insect Allergy Committee.(1965). Insect-sting allergy. JAMA 193:115-20.

James, E.S. and W.G. Walker. (1952). ACTH in the treatment of multiple wasp stings. *Can. Med. Assoc. J.* 67:50-51.

Jarisch, R., J. Zajc and A. Buzath. (1982). The risk of sensitization of nonallergic persons to bee venom. *Arch. Dermat. Res. 273*:173-74.

Jensen. O.M. (1962). Sudden death due to stings from bees and wasps. *Acta Path. Microbiol. Scand. 54*:9-29.

Jex-Blake, A.J. (1942). Bee stings. *E. Afr. Med. J. 19*:74-86.

Jones, T.H. and M.S. Blum. (1983). Arthropod alkaloids: distribution functions, and chemistry, pp. 33-84. *In:*Alkaloids, Chemical and Biological Perspectives, Vol. 1 (S.W. Pelletier, *ed*). John Wiley: New York.

Kampelmacher, J. and J.C. van der Zwan. (1987). Provocation test with a living insect as a diagnostic tool in systemic reactions to bee and wasp venom: a prospective study with emphasis on the clinical aspects. *Clin. Allergy 17*:317-27.

Keating, M.U., A. Kagey-Sobotka, R.G. Hamilton and J.W. Yunginger. (1990). Long-term follow-up to patients who discontinue venom immunotherapy: clinical and immunologic findings. *J. Allergy Clin. Immunol.* 85:210.

Kemeny, D.M., M.H. Lessof and A.K. Trull. (1980). IgE and IgG antibodies to bee venom as measured by a modification of the RAST method. *Clin. Allergy. 10*:413-21.

Kemeny, D.M., M.G. Harries, L.J.F. Youlten, M. Mackenzie-Mills and M.H. Lessof. (1983a). Antibodies to purified bee venom proteins and peptides I. development of a highly specified RAST for bee venom antigens and its application to bee sting allergy. *J. Allergy Clin. Immunol. 71*:505-14.

Kemeny, D.M., M. Mackenzie-Mills, M.G. Harries, L.J.F. Youlten and M.H. Lessof. (1983b). Antibodies to purified bee venom proteins and peptides II. a detained study of changes in IgE and IgG antibodies to individual bee venom antigens. *J. Allergy Clin. Immunol. 72*:376-85.

Kemeny, D.M. N. Dalton, A.J. Lawrence, F.L. Pearce and C.A. Vernon. (1984). The purification and characterisation of hyaluronidase from the venom of the honey bee. *Apis mellifera. Eur. J. Biochem. 139*:217-23.

Kerr, R.A. (1988. Indoor radon: the deadliest pollutant. *Science 240:*606-08.

King, T.P., A.K. Sobotka, L. Kochoumian and L.M. Lichtenstein. (1976). Allergens of honey bee venom. *Arch. Biochem. Biophys. 172*:661-71.

King, T.P., A.K. Sobotka, A. Alagon., L. Kochoumian and L.M. Lichtenstein. (1978). Protein allergens of white-faced hornet, yellow hornet, and yellow jacket venoms. *Biochemistry 17*:5165-74.

King, T.P., A.C. Alagon, J. Kuan, A.K. Sobotka, and L.M. Lichtenstein. (1983). Immuno-chemical studies of yellowjacket venom proteins. *Molecular Immunol. 20*:297-308.

King, T.P., L. Kochoumian and A. Joslyn (1984). Wasp venom proteins: Phospholipase Al and B1. Arch. Biochem. *Biophys. 230*:1-12.

King, T.P., A. Joslyn and, L. Kochoumian (1985). Antigenic cross-reactivity of venom proteins from hornet, wasps, and yellowjackets. *J. Allergy Clin. Immunol. 75*:621-28.

Kleisbauer, J.P., D. Vervloet, D. Charpin, A. Lanteaume and J. Charpin. (1987). Atopy and systemic reactions to Hymenoptera stings. *J. Allergy Clin. Immunol. 79*:232

Knicker, W.T. (1989). Is the choice of allergy skin testing versus *in vitro* determination of specific IgE no longer a scientific issue? *Ann. Allergy 62*:373-74.

Koszalka, M.F. (1949). Multiple bee stings with hemoglobinuria and recovery. *Bull. U.S. Army Med. Dept. 9*:212-17.

Lantner, R. and R.E. Reisman. (1989). Clinical and immunologic features and subsequent course of patients with severe insect-sting anaphylaxis. *J. Allergy Clin. Immunol. 84*:900-06.

Lichtenstein, L.M., M.D. Valentine and A.K. Sobotka. (1979). Insect allergy: the state of the art. *J. Allergy Clin. Immunol. 64*:5-12.

Lichtenstein, L.M., A. Kagey-Sobotka, D.B.K. Golden and M.D. Valentine. (1980). *JAMA 244*:1683-84.

Light, W.C., R.E. Reisman, J.I. Wypych and C.E. Arbesman. (1975). Clinical and immunolog-ical studies of beekeepers. *Clin. Allergy 5*:389-95.

Lindzon, R.D. and W.S. Silvers. (1988). Anaphylaxis. *In:* Emergency Medicine, concepts and clinical practice, Vol. I (P. Rosen, F.J. Baker II, R.M. Barkin, G.R. Braen, R.H. Dailey and R.C. Levy, eds.) pp. 203-31. C.V. Mosby: St. Louis.

Livingston, A. and R. Reisman (1990). Venom immunotherapy (VIT); ten years experience with administration of single venoms and fifty microgram maintenance doses. *J. Allergy Clin. Immunol. 85*:210.

Lockey, R.F. (1980). Cost of testing and treating with Hymenoptera venom extracts. *J. Allergy Clin. Immunol. 65*:398-400.

Lockey, R.F., P.C. Turkeltaub, I.A. Baird-Warren, C.A. Olive, E.S. Olive, B.C. Peppe and S.C. Bukantz (1988). The Hymenoptera venom study I, 1979-1982: demographics and history-sting data. *J. Allergy Clin. Immunol. 82*:370-81.

Lockey, R.F., P.C. Turkeltaub, C.A. Olive, I.A. Baird-Warren, E.S. Olive and S.C. Bukantz. (1989). The Hymenoptera venom study II. Skin test results and safety of venom skin testing. *J. Allergy Clin. Immunol. 84*:967-74.

Lockey, R.F., P.C. Turkeltaub, J.M. Hubbard, E.S. Olive and S.C. Bukantz. (1990a). The Hymenoptera venom study (HVS). V. 1979-1982, efficacy. *J. Allergy Clin. Immunol. 85*:209.

Lockey, R.F., P.C. Turkeltaub, E.S. Olive, J.M. Hubbard, I.A. Baird-Warren and S.C. Bukantz. (1990b). The Hymenoptera venom study III: Safety of venom immunotherapy. *J. Allergy Clin. Immunol.* *86*:775-80.

Loveless, M.H. (1957). Repository immunization in pollen allergy. *J. Immunol.* *79*:68-79.

Loveless, M.H. (1977). Triple stings by captive wasps to appraise and to booster immunity in venom allergy. *Ann. Allergy 38*:299.

Loveless, M.H. and W.R. Fackler. (1956). Wasp venom allergy and immunity. *Ann. Allergy 14*:347-66.

Marom, Z. and T.B. Casale. (1983). Mast cells and their mediators. *Ann. Allergy 50*:367-70.

Maschwitz, U. (1964). Gefahrenalrarmstoffe und Gefahrenalarmierung bei sozialen Hymenoptera. *Z. Vergl. Physiol. 47*:596-655.

Mauriello, P.M., S.H. Barde, J.W. Georgitis and R.E. Reisman. (1984). Natural history of large local reactions from stinging insects. *J. Allergy Clin. Immunol. 74*:494-98.

May, C.D. (1985). The ancestry of allergy: being an account of the original experimental induction of hypersensitivity recognizing the contribution of Paul Portier. *J. Allergy Clin. Immunol. 75*:485-95.

Mejia, G., M. Arbelaez, J.E. Henao, A.A. Sus and J.L. Arango. (1986). Acute renal failure due to multiple stings by Africanized bees. *Ann. Intern. Med. 104*:210-11.

Miami Herald. (1986). "Killer bee" attack leaves student dead. Miami Herald 8 August 1986:18A.

Michener, C.D. (1974). The Social Biology of Bees. Harvard Univ. Press:Cambridge, MA.

Mickalide, A.D., M.D. Valentine, M.R. Dear, K.C. Schuberth and K.A. Kwiterovich. (1985). Insect sting allergy in children: what is the real cost of the disease? *Int. Arch. Allergy Appl. Immunol. 77*:206-09.

Miyachi, S., M.H. Lessof and D.M. Kemeny. (1979). Evaluation of bee sting allergy by skin tests and serum antibody assays. *Int. Arch. Allergy Appl. Immunol. 60*:148-53.

Moret, C., C. Enzel, M. Leclercq, L. Bosson and J. Lecompte. (1983). Un cas d'envenimation mortelle par piqures multiples d'abeilles (*Apis mellifera* L.). *Rev. Med. Liege. 38*:815-22.

Mosbech, H. (1983). Death caused by wasp and bee stings in Denmark 1960-80. *Allergy 38*:195-200.

Mosbech, H., J. Christensen, A. Dirksen and M. Søborg. (1986). Insect allergy. Predictive value of diagnostic tests. A three-year follow-up study. *Clin. Allergy 16*:433-40.

Mueller, U., J. Spiess and A. Roth. (1977). Serological investigations in Hymenoptera sting allergy: IgE and haemagglutinating antibodies against bee venom in patients with bee sting allergy, bee keepers and non-allergic blood donors. *Clin. Allergy 7*:147-54.

Mueller, U., S.G.O. Johansson and C. Streit. (1978). Hymenoptera sting hypersensitivity: IgE, IgG and haemagglutinating antibodies to bee venom constituents in relation to exposure and clinical reaction to bee stings. *Clin. Allergy 8*:267-72.

Mueller, U., A. Heibling and E. Berchtold. (1990). Are there differences between hypersensitivity to honey bee venom (BV) and to *Vespula* venom (VV).*J. Allergy Clin. Immunol. 85*:211.

Murray, J.A. (1964). A case of multiple bee stings. *Central Afr. J. Med. 10*:249-51.

Nair, B.C., C. Nair, S. Denne, J. Wypych, C.E. Arbesman and W.B. Elliott. (1976). Immunologic comparison of phospholipases A present in Hymenoptera insect venoms. *J. Allergy Clin. Immunol. 58*:101-109.

Nakajima, T. (1986). Pharmacological Biochemistry of Vespid Venoms. *In*: Venoms of the Hymenoptera, pp. 309-27 (T. Piek, *ed.*) Academic Press: London.

Nataf, P., M.T. Guinnepain and D. Herman. (1984). Rush venom immunotherapy: a 3-day programme for Hymenoptera sting allergy. *Clin. Allergy 14*:269-75.

Nelson, D.R., A.M. Collins, R.L. Hellmich, R.T. Jones, R.M. Helm, D.L. Squillace and J.W. Yunginger. (1990). Biochemical and immunochemical comparison of Africanized and European honeybee venoms. *J. Allergy Clin. Immunol. 85*:80-85.

Norman, P.S. and T.E. van Metre, Jr. (1990). The safety of allergenic immunotherapy. *J. Allergy Clin. Immunol. 85*:522-25.

Olkowski, H. and W. Olkowski. (1976). Entomophobia in the urban ecosystem, some observations and suggestions. Bull. Entomol. *Soc. Amer. 22*:313-17.

Ordman, D. (1968). Bee stings in South Africa. *S. Afr. Med. J. 42*:1194-98.

Ostrom, N.K., M.C. Swanson, M.K. Agarwal and J.W. Yunginger. (1986). Occupational allergy to honeybee-body dust in a honey-processing plant. *J. Allergy Clin. Immunol. 77*:736-40.

Parker, J.L., P.J. Santrach, M.J.E. Dahlberg and J.W. Yunginger. (1982). Evaluation of Hymenoptera-sting sensitivity with deliberate sting challenges: inadequacy; of present diagnostic methods. *J. Allergy Clin. Immunol. 69*:200-07.

Parrish, H.M. (1963). Analysis of 460 fatalities from venomous animals in the United States. *Amer. J. Med. Sci. 245*:129-41.

Patrick, A., L. Roberts, P. Poon-King and V. Jeelal. (1987). Acute renal failure due to multiple stings by Africanized bees, report of the first case in Trinidad. *West Ind. Med. J. 36*:43-44.

Pence, H., T. Wilson, J. Schmidt. and A. White. (1985). Evaluation of sweat bee allergy with sweat bee venom. *J. Allergy Clin. Immunol. 75*:209.

Peppe, B.C., R.F. Lockey, J. Madden, I. Baird and P. Turkeltaub. (1983). Hymenoptera venom study (HVS), treatment results, 9/30/82. *J. Allergy Clin. Immunol. 71*:120.

Piek, T. (ed.) (1986). Venoms of the Hymenoptera. Academic Press; London.

Piek, T., J.O. Schmidt, J.M. de Jong and P. Mantel. (1989). Kinins in ant venoms—a comparison with venom of related Hymenoptera. *Comp. Biochem. Physiol. 92C*:117-24.

Pinnas, J.L., R.C. Strunk, T.M. Wang and H.C. Thompson. (1977). Harvester ant sensitivity: in vitro and in vivo studies using whole body extracts and venom. *J. Allergy Clin. Immunol. 59*:10-16.

Prince, R.C., D.E. Gunson and A. Scarpa. (1985). Sting like a bee! the ionophoric properties of melittin. *Trends Biochem. Sci. 10*:99.

Przybilla, B., J. Ring and B. Griesshammer. (1989). Diagnostic findings in Hymenoptera venom (HV) allergy: interrelationship of history, skin test (ST) and RAST. *J. Allergy Clin. Immunol. 83*:229.

Ramirez, D.A., S. Londono and R. Evans. (1981). Adverse reactions to venom immunotherapy. *Ann. Allergy 47*:435-39.

Randolph, C.C. and R.E. Reisman. (1986). Evaluation of decline in serum venom-specific IgE as a criterion for stopping venom immunotherapy. *J. Allergy Clin. Immunol. 77*:823-27.

Reisman, R.E. (1988). Insect Allergy. *In:* Allergy Principles and Practice, Vol. II, 3rd edit., pp. 1345-64 (E. Middleton, Jr., C.E. Reed, E.F. Ellis, N.F. Adkinson, Jr. and J.W. Yunginger, *eds).* C.V. Mosby: St. Louis.

Reisman, R.E. and J.M. DeMasi. (1989). Relationship of serum venom-specific IgE titers to clinical aspects of stinging insect allergy. *Int. Arch Allergy Appl. Immunol. 89*:67-70.

Reisman, R.E. and J.W. Georgitis. (1984). Frequency of positive venom skin tests in insect-allergic and nonallergic populations. *J. Allergy Clin. Immunol. 73*:187.

Reisman, R.E. and R. Lantner. (1989). Further observations of stopping venom immunotherapy: comparison of patients stopped because of a fall in serum venom-specific IgE to insignificant levels with patients stopped prematurely by self-choice. *J. Allergy Clin. Immunol 83:*1049-54.

Reisman, R.E., M. Lazell and J. Doerr. (1981). Insect venom allergy: a prospective case study showing lack of correlation between immunologic reactivity and clinical sensitivity. *J. Allergy Clin. Immunol. 68:*406-08.

Reisman, R.E., J.I. Wypych, W.R. Mueller and J.A. Grant. (1982). Comparison of the allergenicity and antigenicity of *Polistes* venom and other vespid venoms. *J. Allergy Clin. Immunol. 70:*281-87.

Reisman, R.E., R. Hale and J.I. Wypych. (1983). Allergy to honeybee body components: distinction from bee venom sensitivity. *J. Allergy Clin. Immunol. 71:*18-20.

Reisman, R.E., K.J. Dvorin, C.C. Randolph and J.W. Georgitis. (1985). Stinging insect allergy: natural history and modification with venom immunotherapy. *A. Allergy Clin. Immunol. 76:*735-40.

Reisman, R.E., J.I. Wypych and M.I. Lazell. (1987). Further studies of patients with both honeybee- and yellow-jacket-venom-specific IgE. Int. Arch. *Allergy Appl. Immunol. 82:*190-94.

Riches, H.R.C. (1989). Bee venom hypersensitivity update. *Bee World 70:*12-18.

Rubenstein, H.S. (1980a). Allergists who alarm the public, a problem in medical ethics. *JAMA 243:*793-94.

Rubenstein, H.S. (1980b). Bee venom immunotherapy. *JAMA 244:*1672.

Rubenstein, H.S. (1982). Bee-sting diseases: who is at risk? what is the treatment? *Lancet 1:*496-99.

Santrach, P.J., L.G. Peterson and J.W. Yunginger. (1980). Comparison of diagnostic tests for Hymenoptera sting allergy. *Ann. Allergy 45:*130-36.

Savliwala, M.N. and R.E. Reisman. (1987). Studies of the natural history of stinging-insect allergy: Long-term follow-up of patients without immunotherapy. *J. Allergy Clin. Immunol. 80:*741-45.

Schmidt, J.O. (1982). Biochemistry of insect venoms. *Annu. Rev. Entomol. 27:*339-68.

Schmidt, J.O. (1985). Proteolytic activities of Hymenoptera venoms. *Toxicon 23:*38.

Schmidt, J.O. and M.S. Blum. (1978). A harvester ant venom:chemistry and pharmacology. *Science 200:*1064-66.

Schmidt, J.O., M.S. Blum and W.L. Overal. (1984). Hemolytic activities of stinging insect venoms. *Arch. Insect Biochem. Physiol. 1:*155-60.

Schmidt, J.O., M.S. Blum and W.L. Overal. (1986). Comparative enzymology of venoms from stinging Hymenoptera. *Toxicon 24:*907-21.

Schuberth, K.C., L.M. Lichtenstein, A. Kagey-Sobotka, M. Szklo, K.A. Kwiterovich and M.D. Valentine. (1982). An epidemiologic study of insect allergy in children. I. characteristics of the disease. *J. Pediatr. 100:*546-51.

Schuberth, K.C., L.M. Lichtenstein, A. Kagey-Sobotka, K.A. Kwiterovich and M.D. Valentine. (1988). Starting and stopping venom immunotherapy (VIT) in children with insect allergy. *J. Allergy Clin. Immunol. 81:*200.

Schulman, S., F. Bigelsen, R. Lang and C. Arbesman. (1966). The allergic response to stinging insects: biochemical and immunologic studies on bee venom and other bee body preparations. *J. Immunol. 96:*29-38.

Schumacher, M.J., J.O. Schmidt and W.B. Egen. (1989). Lethality of "killer" bee stings. *Nature* *337*: 413.

Schumacher, M.J., J.O. Schmidt, N.B. Egen and J.E. Lowry. (1990). Quantity, analysis and lethality of European and Africanized honey bee venoms. *Amer. J. Trop. Med. Hyg.* *43*:79-86.

Schwartz, H.J., R.F. Lockey, A.L. Sheffer, J. Parrino, W.W. Busse and J.W. Yunginger. (1981). A multicenter study on skin-test reactivity of human volunteers to venom as compared with whole body Hymenoptera antigens. *J. Allergy Clin. Immunol. 67*:81-85.

Schwartz, H.J., C. Sutheimer, M.B. Gauerke and J.W. Yunginger. (1988) Hymenoptera venom-specific IgE antibodies in *post-mortem* sera from victims of sudden, unexpected death. *Clin. Allergy. 18*:461-68.

Scragg, R.F.R. and J.J.H. Szent-Ivany. (1965). Fatalities caused by multiple hornet stings in the territory of Papua and New Guinea. *J. Med. Ent. 2*:309-13.

Settipane, G.A. and G.K. Boyd. (1970). Prevalence of bee sting allergy in 4992 boy scouts. *Acta Allergol. 25*:286-91.

Settipane, G.A. and F.H. Chafee. (1979). Natural history of allergy to Hymenoptera. *Clin. Allergy 9*:385-390.

Settipane, G.A., G.J. Newstead and G.K. Boyd. (1972). Frequency of Hymenoptera allergy in an atopic and normal population. *J. Allergy Clin. Immunol. 50*:146-50.

Settipane, G.A., D.E. Klein and G.K. Boyd. (1978). Relationship of atopy and anaphylactic sensitization: a bee sting allergy model. *Clin. Allergy 8*:259-65.

Settipane, G.A., F.H. Chafee, D.E. Klein, G.K. Boyd, J.H. Sturam and H.B. Freye. (1980). Anaphylactic reactions to Hymenoptera stings in asthmatic patients. *Clin. Allergy 10*:659-65.

Shipolini, R.A., G.L. Callewaert, R.C. Cottrell and C.A. Vernon. (1974). The amino-acid sequence and carbohydrate content of phospholipase A2 from bee vonom. *Eur. J. Biochem. 48*:465-76.

Shilkin, K.B., B.T.M. Chen and O.T. Khoo. (1972) Rhabdomyolysis caused by hornet venom. *Brit. Med. J. 1*:156-57.

Somerville, R., D. Till, M. Leclercq and J. Lecompte. (1975). Les morts par pique d'Hymenopteres aculeates en angleterre et au pays de Galler (statistiques pour la periode (1959-1971). *Rev. Med. Liege 30*:76-78.

Spradbery, J.P. (1973). Wasps. Univ. of Washington Press: Seattle.

Stone, B.D., P.S. Hutcheson and R.G. Slavin. (1990). Ten years of observations of a midwest venom referral center. *Ann. Allergy 64*:70.

Stuckey, M., T. Cobain, M. Sears, J. Cheney and R.L. Dawkins. (1982). Bee venom hypersensitivity in Busselton. *Lancet 2*:41.

Takasaki, C. and M. Fukumoto. (1989). Phospholipases B from Japanese yellow hornet (*Vespa xanthoptera*) venom. *Toxicon 27*:449-58.

Thurnheer, U., U. Müller, R. Stoller, A. Lanner and R. Hoigne. (1983). Venom immunotherapy in Hymenoptera sting allergy. *Allergy 38*:465-75.

Torsney, P.J. (1973). Treatment failure: insect desensitization. *J. Allergy Clin. Immunol. 52*:303-06.

Upton, A.C. (1982). The biological effects of low-level ionizing radiation. *Sci. Amer. 246(2)*:41-49.

Urbanek, R., U. Krauss, J. Ziupa and G. Smedegard. (1983). Venom-specific IgE and IgG antibodies as a measure of the degree of protection in insect-sting-sensitive patients. *Clin. Allergy 13*:229-34.

Valentine, M.D. and D.B.K. Golden. (1988). Insect venom allergy. *In*: Immunological Diseases, Vol. II, 4th edit. pp. 1173-84 (M. Samter, D.W. Talmage, M.M. Frank, K.F. Austen and H.N. Claman, *eds.*). Little, Brown and Co.: Boston.

Vital Statistics of the U.S., (1985). Vol. II, Mortality, Part A. Dept. Health Human Services Pub. No. (PHS) 88-1122. U.S. Gov. Printing Off.: Washington.

Wasserman, S.I. and D.L. Marquardt. (1988). Anaphylaxis. *In:*Allergy, Principles and Practice, Vol. II, 3rd Edit., pp. 1365-76. (E. Middleton, Jr., C.E. Reed, E.F. Ellis, N.F. Adkinson, Jr. and J.W. Yunginger, *eds.*) C.V. Mosby: St. Louis.

Waterhouse, A.T. (1914). Bee-stings and anaphylaxis. *Lancet 2*:946.

Watkins, D.J., M.A. Reed and B.T. Butcher. (1988). Lack of cross-reactivity between imported fire ant venom (IFAV) and other Hymenoptera venoms. *J. Allergy Clin. Immunol. 81*:204.

Wicher, K., R.E. Reisman, J. Wypych, W. Elliott, R. Steger, R.S. Mathews and C.E. Arbesman. (1980). Comparison of the venom immunogenicity of various species of yellowjackets (genus *Vespula*) *J. Allergy Clin. Immunol. 66*:244-49.

Wypych, J.I., C.J. Abeyounis and R.E. Reisman. (1989). Analysis of differing patterns of cross-reactivity of honeybee and yellowjacket venom-specific IgE: use of purified venom fractions. *Int. Arch. Allergy Appl. Immunol. 89*:60-66.

Yunginger, J.W. (1981). Advances in the diagnosis and treatment of stinging insect allergy. *Pediatr. 67*:325-28.

Yunginger, J.W., R.T. Jones, K.M. Leiferman, B.R. Paull, P.W. Welsh and G.J. Gleich. (1978). Immunological and biochemical studies in beekeepers and their family members. *J. Allergy Clin. Immunol. 61*:93-101.

Yunginger, J.W., B.R. Paull, R.T. Jones and P.J. Santrach. (1979). Rush venom immunotherapy program for honeybee sting sensitivity. *J. Allergy Clin. Immunol. 63*:340-47.

Zora, J.A., M.C. Swanson and J.W. Yunginger. (1988). A study of the prevalence and clinical significance of venom-specific IgE. *J. Allergy Clin. Immunol. 81*:77-82.

Index

Benson, R.L., and H. Semenov, 1249
Bentley, B., 402
Benton, A.W., 1213
Benton, A.W. *et al,* 958, 959
Beran F. and J. Neururer, 1161
Bergner, K.G. and S. Diemair, 881
Bergner, K.G. and H. Hahn, 883
Bermant, G. and N.E. Gary, 318
Bernheimer, A.W. *et al,* 1212
Bertsch, A., 416
Betts, A.D., 831, 1054
Beutler, R., 172, 326, 402, 411, 415
Beutler, R. and E. Opfinger, 202
Bevan, E. 13
Bicchi, C. *et al,* 407
Bieberdorf, F.W. *et al,* 199
Bieleski, R.L. and R.J. Redgewell, 406, 409
Billingham, M.E.J., 955
Birnbaum, J.C. *et al,* 1237, 1245, 1253
Bischoff, H., 24, 25
Blaauw, P.J. and L.O.M.J. Smithuis, 1227, 1228,
    1229, 1238, 1247
Blomquist, G.J. *et al,* 965
Blum, M.S., 373-400, 382, 388
Blum, M.S. and H.M. Fales, 388
Blum, M.S. *et al,* 380, 385, 388, 392, 970, 1214
Boch, R., 292, 833, 843, 844
Boch, R. and D.A. Shearer, 85, 380, 386, 388
Boch, R.D. *et al,* 200, 225, 380, 381, 384, 388
Bodenheimer, F.S., 52, 53
Bogdan, T. *et al,* 1099
Bogdanov, S., 880
Bogdanov, S. and E. Baumann, 872, 873, 914
Bond, D.A., and J.L. Fyfe, 410
Bondarenko, O.I., 1105
Bonnier, G., 314, 411, 413
Boreham, M.M. and D.W. Roubik, 51, 52, 247,
    1225
Bosi, G., 406
Bousquet, J. *et al,* 1224, 1227, 1235, 1236, 1237
Boyer, W.D. 423
Bozina, K.D., 84
Bradford, G.R. *et al,* 1203
Braun, E., 842, 851
Braun, E. and J.E. Geiger, 829, 833, 843
Braun, L.I.B., 1233
Break, J. *et al,* 1107
Breed, M.D. *et al,* 309
Brewer, J.W. *et al,* 405
Brice, A.T. *et al,* 941
Brisco, N.A. *et al,* 782
Broadman, J., 954, 957
Brother Adam, 53, 55, 56, 57, 58, 59, 60
Brouwers, E.V.M., 203, 226
Brouwers, E.V.M. *et al,* 81
Brown, A.W.A., 1161

Brown, H. and H.S. Bernton, 1245
Brown, R., 943
Brown, W.H., 403
Brown, W.H. *et al,* 225, 969
Browne, C.A., 885
Brugger, A., 1133
Bucher, G.E., 1085
Buchmann, S.L., 927-988, 929, 946, 961
Buchmann, S.L. and J.O. Schmidt, 961
Buco, S.M. *et al,* 247
Budel, A., 14
Bulanov, N.A., 416
Bulger, J.W., 1104
Bunney, M.H., 945, 951
Burgett, D.M., 350, 880
Burgett, D.M. and C. Kitprasert, 1125
Burgett, D.M. and R.A. Morse, 90
Burgett, M. 973, 974
Burns, G., 421
Burnside, C.E., 1109
Burnside, C.E. and I.L. Revell, 1099
Burnside, C.E. and G.H. Vansell, 890
Burres, M., 826
Busse, W.W. and J.W. Yunginger, 1228
Butler, C.G., 8, 23, 25, 27, 92, 338, 344, 381, 395
Butler, C.G. and D.H. Calam, 386
Butler, C.G. and R.K. Callow, 384
Butler, C.G. and J.B. Free, 306, 312
Butler, C.G. and J. Simpson, 384, 388
Butler, C.G. *et al,* 385, 391, 402, 415
Buttel-Reepen, H., 23, 27, 52
Buzzotta, V.R. *et al,* 782
Byrne, D.H. *et al,* 1254

Calderone, N.W. and R.E. Page, Jr., 87, 249, 251,
    273, 278, 279
Cale, G.H. Jr., 61, 833
Cale, G.H. and J.W. Gowan, 61
Cale, G.H. *et al,* 655
Calin, A., 954
Callow, R.K. and N.C. Johnston, 383
Callow, R.K. *et al,* 383
Camareno, J.E. and I. Pecho, 425
Camazine, S., 1117
Cantwell, G.E., 1101
Cantwell, G.E. and T.R. Shieh, 1130
Cantwell, G.E. and H. Shimanuki, 1102
Cantwell, G.E. and L.J. Smith, 1132
Cantwell, G.E. *et al,* 1099, 1131, 1133
Carlisle, E. and M. Ryle, 418
Carlson, D.A., 247
Caron, D.M. 90, 1083-1151, 1134, 1138, 1139,
Caron, D.M. and P.W. Schaefer, 1135
Carpenter, F.M. and H.R. Hermann, 23, 24
Casteel, G.B., 301, 302, 322, 325
Chafee, F.H., 1232, 1245

Nominé, G., 1162
Norman, P.S. and T.E. VanMetre, Jr., 1218, 1235
Noyes, C.E., Jr., 939
Nye, W.P. and O. Mackenson, 249, 259, 273, 425, 1058
O'Connor, R. *et al*, 958
O'Rourke, M.K. and S.L. Buchmann, 929
Oertel, E., 84, 347, 359, 410, 418, 437
Okada, I. and M. Matsuka, 975
Okumura, G.T., 1133
Olkowski, H. and W. Olkowski, 1254
Onions, G.W., 240
Ono, M. *et al*, 188
Ontario Agricultural College, 940, 941
Oppen, F.C. and H.A. Schuette, 893
Ordman, D., 1245
Örösi-Pál, A., 890
Örösi-Pál, Z., 303, 350
Ostrom, N.K. *et al*, 1250
Otani, H. *et al*, 967
Otis, G.W., 84, 91, 94, 359
Otis, G.W. *et al*, 92
Otte, E., 1088
Oudemans, A.C., 1119
Owens, C.D., 88, 188
Owens, C.D. *et al*, 723

Paddock, F.B., 1127, 1128
Page, R.E., 242
Page, R.E. and N.E. Gary, 253, 259
Page, R.E. and R.A. Metcalf, 242
Page, R.E. *et al*, 245, 250, 258, 259, 392, 395
Page, R.E. Jr. and H.H. Laidlaw, Jr., 235-268, 224, 258, 259
Pain, J., 210
Pain, J. and B. Roger, 380, 385
Pain, J. *et al*, 225, 383, 384, 385
Paine, H.S. *et al*, 881
Paintz, M. and J. Metzner, 949
Pais, M.S.S. and H.J. Chaves des Neves, 406
Palmer, D.J., 958
Palmer-Jones, T., 890, 1104, 1195
Palmer-Jones, T. and I. Forster, 1161
Palmer-Jones, T. and L.J.S. Line, 1195
Palmer-Jones, T. and E.P. White, 1195
Panco, F., 828
Pangaud, C. 1161
Pankiw, P, and J.L. Bolton, 412
Panush, R.S. and S. Longley, 958
Parent, G.*et al*, 972, 974
Park, O.W., 58, 315, 316, 325, 326, 327, 328, 331, 332, 333, 334, 335, 416
Parker, J.L. *et al*, 1218, 1227, 1228
Parker, R.L., 26, 325, 326
Parks, H.B., 332
Parrish, H.M., 1220

Paterson, C.R. 890
Paterson, C.R. and T. Palmer-Jones, 896
Patrick, A. *et al*, 1225
Payne, T., 828
Payne, T.L. and W.E. Finn, 1129
Pearson, P.B., 930
Pedersen, M.W., 413, 417
Pedersen, M.W. *et al*, 409, 410
Peer, D.F., 347, 352, 843
Pellett, F.C., 246, 420, 437, 872
Pellett, M., 525
Pence, H. and J. Schmidt, 1244
Pence, H. *et al*, 1244
Peng, Y-S, 1124
Peng, Y.S. and S.C. Jay, 226
Peng, Y-S and M.E. Nasr, 1116
Peng, Y.S., and K-Y Peng, 1088
Peng, Y.S. *et al*, 61, 173
Peppe, B.C. *et al*, 1236
Percival, M.S., 402, 405, 406, 411, 413, 423, 424, 425
Perepelova, L.I. 316, 344, 1195
Peterson, C.E., 1060
Petit, M., 850
Petrov, V., 883
Pettis, J.S. *et al*, 44
Phillips, E.F., 58, 60, 829, 839
Phillips, E.F. and G.S. Demuth, 88, 345, 830, 831, 842, 850
Phillips, J., 828
Pichler, F.J. *et al*, 888
Pickett, J.A. *et al*, 386, 388, 390
Piek, T., 1213, 1215
Piek, T. *et al*, 1213
Pieroni, R.E. *et al*, 935
Pinnas, J.L. *et al*, 958, 1242
Pinnock, D.E. and N.E. Featherstone, 1093
Pirker, H.J., 854, 861
Plass, F., 417
Platt, J.L. and J.R.B. Ellis, 896
Pleasants, J.M. and S.J. Chaplin, 412
Polhemus, M.S. *et al*, 242, 243, 254, 255
Poltev, V.E., 1104
Popeskovic, D. *et al*, 949
Popravko, S.A. *et al*, 946
Porlier, B., 171
Porter, J.W., 415
Posey, D.A., 972
Post, D.C. *et al*, 392
Poteikina, E.A., 1099, 1102
Prell, H., 1105
Price, J.H. *et al*, 957
Prince, R.C. *et al*, 1214
Proc. Canadian, Association of Professional Apiculturists, 859, 862, 863
Przybilla, B. *et al*, 1227

Sasagawa, H. *et al.*, 180
Savliwala, M.N. and R.E. Reisman, 1228, 1238, 1240
Schade, J.E. *et al.*, 888
Schaffer, W.M. and M.V. Schaffer, 411
Schepartz, A.I., 881, 884
Schick, W., 300
Schiller, J., 1105
Schmalzel, R., 929
Schmidt, D.K., 955
Schmidt, D.K. *et al.*, 955
Schmidt, J.O., 200, 423, 1209-1269, 1213, 1215, 1248
Schmidt, J.O. and M.S. Blum, 1242
Schmidt, J.O. and S.L. Buchmann, 198, 927-988, 929, 931
Schmidt, J.O. and P.J. Schmidt, 967, 974
Schmidt, J.O. and S.C. Thoenes, 394
Schmidt, J.O. and W. Wang, 949
Schmidt, J.O. *et al.*, 929, 930, 931, 975, 1213, 1225, 1242, 1251
Schneider, F., 300
Schneider, P. and W. Drescher, 1121
Schnepf, E., 404, 408
Schöntag, A., 417
Schou, S.A. and J. Abildgaard, 888
Schuberth, K.C. *et al.*, 1239
Schulman, S. *et al.*, 1234
Schumacher, M.J. *et al.*, 952, 1225
Schwartz, H.F., 7
Schwartz, H.J. *et al.*, 1227, 1229
Scragg, R.F.R. and J.J.H. Szent-Ivany, 1223
Scullen, H.A., 416
Seeley, T.D., 23, 27, 51, 73, 74, 84, 90, 92, 174, 184, 185, 251, 515
Seeley, T.D. and R.A. Levien, 515
Seeley, T.D. and R.A. Morse, 33, 74, 94, 961
Seeley, T.D. and P.K. Visscher, 91, 92
Seeley, T.D. *et al.*, 341
Settipane, G.A. and G.K. Boyd, 1232, 1245
Settipane, G.A. and F.H. Chafee, 1228
Settipane, G.A. *et al.*, 1245, 1246
Severson, D.W., 341
Severson, D.W. and E.H. Erickson, Jr., 406
Severson, D.W. and J.E. Parry, 423, 425
Severson, D.W. *et al.*, 247, 260, 416
Shaginyan, E.G., 1195
Shaparew, V., 942
Shaw, F. *et al.*, 418
Shaw, F.R., 1161
Shearer, D.A. and R. Boch, 85, 380, 386
Sheppard, W.S. and M.D. Huettal, 32
Sherriff, B., 659
Shesley, B. and B. Poduska, 837, 838
Shilkin, K.B. *et al.*, 1224
Shimamori, K., 1130

Shimanuki, H., 1083-1151, 1090, 1130
Shimanuki, H. and D.A. Knox, 1089, 1114
Shimanuki, H. *et al.*, 1090, 1100, 1103
Shinoda, M.S. *et al.*, 970
Shipman, W.H. and J. Cole, 958
Shipolini, R.A. *et al.*, 1212
Shkenderov, S., 955
Shuel, R.W., 401-436, 409, 411, 412, 413, 414, 415, 416, 417, 418,
Shuel, R.W. and S.E. Dixon, 225
Shuel, R.W. and J.A. Shivas, 417
Shuel, R.W. and W. Tsao, 403, 412
Shuel, R.W. *et al.*, 225
Siddiqui, I.R. and B. Furgala, 873
Simpson, J., 88, 90, 341, 381
Simpson, J. *et al.*, 85
Singh, S., 316, 325, 1127
Sklanowska, K. 417
Sklanowska, K. and S. Pluta, 425
Skorikov, A.S., 52, 53
Sladen, F.W.L., 850
Slessor, K.N. *et al.*, 356, 384
Smirle, M.J. *et al.*, 1163
Smith, F.G., 418
Smith, R.H. and W.C. Johnson, 418
Smith, D.A., 8
Smith, D.R. *et al.*, 247, 253
Smith, F.G., 37, 40, 41
Smith, I.B. Jr., and D.M. Caron, 1134
Smith, M.V., 1086, 1092, 1107
Smith, R.B. and T.P. Mommsen, 941
Smith, R.K., 32, 247, 1171
Smoot, D., 828
Snodgrass, R.E. and E.H. Erickson, 103-170
Solberg, Y. and G. Remedios, 930
Sommerfield, S.D. *et al.*, 955
Sommerville, R. *et al.*, 1220
Southwick, E.E., 172, 188, 191, 411, 412
Southwick, E.E. and G. Heldmaier, 188
Southwick, E.E. and R.F.A. Moritz, 186, 187, 309, 312
Southwick, E.E. and J.N. Mugaas, 88
Southwick, E.E. *et al.*, 172, 193, 406
Southwick, L. Jr., and E.E. Southwick, 43
Spangler, H.G., 179, 274, 1129, 1130
Spangler, H.G. and S. Taber, 944
Spitzner, M.J.E., 282
Spivak, M.T. *et al.*, 247, 253
Spradbery, J.P., 1210
Sprengel, C.K., 9
Standifer, L.N., 206, 423, 837,
Standifer, L.N. and J.P. Mills, 211
Standifer, L.N. *et al.*, 203, 216, 217, 218
Stanley, R.G. and H.F. Linskens, 421, 422, 423, 425, 426, 930
Starratt, A.N. and R. Boch, 209

possible control, 43
preference of nesting sites, 48, 51
swarming rates, 51
After swarms, 341, 347
Agastache, 522, 523
Agastache foeniculum, 522
Agastache rugosa, 522
Agave, 411
Age polyethism, 250
Aging, 180, 250, 251, *also see* Fig. 5, p. 251
Agnagasta kuehniella, 1133
Ah Mucan Cab, 7
Air sacs, 161, *also see* Fig. 19, p. 151
Aizawai, 1157
Alaska, 497
Alfalfa, 249, 259, 317, 410, 412, 413, 414, 426,
    499, 501, 503, 504, 506, 511, 521, 877,
    1175, 1192
  pollination, 1057-1060, *also see* Fig. 6, p. 1059
Alimentary canal, The, **156-158**, 112, 124, *also
    see* Fig. 17, p. 147
  poisoning, 1155
  proventiculus, 156
  honey stomach, 156, 172
  mouth, 156
  sucking pump, 156
Alkaloids, 929
Alleles, 236, 239, 253, 255, 258, 259, *also see*
    genetics
Allergens, *also see* proteins
  in insect venoms, 1214, *also see* Table 2, p.
    1215
Allergic reactions, 935, 937
Allergy,
  definition, 1215
  medical costs of, 1252
  relationships, 1246
  respiratory, 1249
  to bee body parts, 1249
  to bee protein, 1249
  to pollen, 1245
Allergy to Venomous Insects, 1209-1269
Allergy, venom, *also see* anaphylaxis
  analysis, 1227
  atopy, 1245, 1246
  causes, 1226
  diagnosis, 1227
  historical perspectives of, 1232
  radioallergosorbent test (RAST), 1227
  shots, 1235, 1237, *also see* immunotherapy
  skin test, 1228, 1233, 1237, *also see* Table 8, p.
    1238
  treatment, 1233, 1234, *also see* immunotherapy
Allium cepa L., 1062, *also see* onion
Allomones, 374, 375, 381
Almond

pollination of, 1065, 1066, *also see* Fig. 8, p.
    1065
Alsike, 413, 418
Altitude studies, 31, 39, 44
Ambush, 1191
American Bee Journal, 438, 568, 630, 794, 828,
    1113
American Beekeeping Federation, 808
American Honey Queen Program, 808
Amino Acid, 198, 199, 406, 407, 422, 883, *also
    see* acids
  essential, 198, 199
  free, 199
  proline, 883
Amitraz, 566, 1118
Amphibians, 1135
Amygdalus communis Batch, 1065, *also see* almond
Anakit, 1235
Anaphylactic reactions
  to pollen ingestion, 935, 937
Anaphylactic shock, 1226
  death, 1233
Anaphylaxis, *also see* sting reactions
  history of, 1233
  prevention, 1234, *also see* Fig. 4, p. 1236
  risk factors, 1222, 1223
  treatment, 1235
Anatolia, 53, *also see A.m. anatoliaca*
Anatomy of the Honey Bee, The, 103-170
Andromeda, 1195
Angiosperm, 420
Animal,
  diets and pollen, 940
Anise hyssop, 522
Antenna, 108, 112, *also see* Plate 6, p. 117
  as sensory organs, 116, 164, 177
Antenna cleaner, The, **132-135**
Anther, 1050, *also see* Fig. 1, p. 1050
Antibody, 1249
  defense against stings, 1226
  levels for termination of immunotherapy, 1238,
    1239
Antigens,
  hyaluronidase, 1244
  phospholipases, 1245
Ants,
  army, 1134
  carpenter, 1134
  control of, 1135
  fire (*solenopsis*), 1130, 1209, 1212, 1241, 1242,
    1252, *also see* Fig. 1, p. 1210
  harvester, 1209, 1211, 1242
  pesticides, 1185,
  pharoah, 1134
  tropical, 1134
Anus, 108, 147, 156, *also see* Fig. 2, p. 107, Fig.

discovery of drone as male, 8
discovery of queen as female, 8
discovery of worker as female, 8
establishment in the new world, 8
first drawings under a microscope, 8
first facts of mating, 9
first movable comb hives, 10
before 1600, 1
1600-1851, 8
Beekeeping in,
    Africa, 19
    Americas, 19
    Asia, 18
    Australia, 20
    Europe, 17, 18
    USSR, 18
Beekeeping Management, 601-656
    controlling sun exposure, 644-645
    drifting, 650-652, *also see* drifting, behavior of
        bees
    fall preparation of hives, 861, *also see* hives
    feeding, **635-642,** 833
    for the beginner and hobbyist, **575-600,**
        757-782
    inspections, 839, 840, 841, 1084
    moving bees, **646-649,** 654
    of package bees, 1034
    provision of water source, 643
    record keeping, 654, 655, **750-754**
    removing bees from supers, **660-663**
    robbing, **652-654,** *also see* behavior of bees
    seasonal, **613-623**
    seasonal supplemental feeding, 217, *also see*
        feeding
    site selection, 602, 604, 643, 644
    spring, 863
    stock selection, 631
    swarm prevention, 90, 601, 849
    two-queen system, **632-635**
    wintering principles, 830, 847, *also see*
        wintering
Beekeeping, old and new world,
    commercial vs. sideline, 17
    diseases, 17
    mechanization, 17
    yields, 16
Beekeeping operations in the U.S.,
    numbers, 783
Beekeeping techniques
    feeding pollen substitutes, 216, *also see* Fig. 3,
        p. 216
    feeding syrup, 222, 223, 224, 838
    inspection of combs, 553
    installation of hives, 339, 588
    providing water, 215
Bees, *see Apis*, drones, queens, workers

adaptation to environment, 36, 87, 187
adaptibility of, 28, 42, 88, 246
Africanized, *see* Africanized bees
aging characteristics of, 180
alkali, 1058
as food for humans and animals, 975
as pollinators, 16, 43, **1044-1082**
as social insects, 235, 1045
cell builders, 1001
classification of, 35, 53
color perception of, 164, 651, 652
dark (*A.m.mellifera*), 26, *also see Apis*
development timing of, 83
evolution of, 24
fanning, 311, 312-313, *also see* Fig. 17, p. 311
geographic types, 27
guard, 306, 307, 311, 375, 380, 653, 1160, *also*
    see Fig. 15, p. 307
leafcutter, 1058
life history of, **73-102**
mechanical, 294, *also see* Fig. 6, p. 294
morphological character of, 24
morphological differences of, 25
morphometric identification of, 44
nervous system of, *see* nervous system
nurse, 303, 304, 356, 357, 380, 1003, 1004,
    1045, 1046
planing, 944, *also see* Fig. 6, p. 944
queenless, 1001
recruit, 295
respiratory system of, *see* respiratory system
robber, 310, 311, 653, 1069, *also see* Fig. 9, p.
    1069
scenting, 385, 387
scout, 292, 652, 1180
sensitivity to noise, 852
solitary, 1058
stingless, 945
sweat, 1211
symbol perception of, 652
transportation of, 729
wandering, 1054, *also see* Fig. 4, p. 1056
Bee-Scent, 1048
Bees, solitary
    susceptibility to pesticides, 1155
Bee traits
    body color, 31, 34, 39, 53, 55, 56, 57, 58, 59,
        246
    hair, 35, 55, 56, 57, 110, 135, 150, 163, 191,
        322, 323, 324, *also see* Fig. 24, p. 324,
        1044
    size, 53, 59, 247
    tongue length, 33, 34, 55, 57, 59, **120-122,** 124,
        also see Fig. 3, p. 34, 322
    wing venation, 35, 39, 55
Beeswax, **960-966,** *also see* wax

1308   The Hive and the Honey Bee

Mortality statistics
from bee stings, 1229, *also see* Tables 6, p.
1225; 7, p. 1230
Mosaics and Gynandromorphs, **241-242**
Moth
control, 566, 1130, 1132, 1133
death's head, 1133
dried fruit, 1133
greater wax, 1126, *also see* pests of bees, Fig.
21, p. 1126, Table 1, p. 1132
lesser wax, 1133
Indian meal, 1133
Mother-daughter mating, 255
Movement of bees, **646-650**
and commercial pollination, 729, 1073, *also see*
Figs. 10, p. 1073; 11, p. 1074
seasonal, 729, 746, 861
Mt. Meru, 39
Musa paradisica, 411
Muscle action, 181-182
Muscles,
basalar, 144
compressor, 124, 125, 148, *also see* Fig. 15, p.
145
depressor, 128, 142, 181
dilator, 124, 125, 148, *also see* Fig. 15, p. 145
elevator, 181, *also see* Fig. 3, p. 182
flexor, 124
for retracting *proboscis,* 124
intersegmental, 149
mesothoracic, 142
of the *trochanter,* 128
pretarsal, 130
protractor, 148
retractor, 148, 155
subalar, 144
thoracic, 137, 191
wing, 138, 142, 181, 182
Mutation, 238, 254, *also see* genetics

Narcosis, 407, *also see* diseases
Nasanov gland pheromone, **385-387**
Nassanoff, *see* Nasanov, glands, scent
National Honey Board, **808-810,** 825
Near East group, **52-55,** *also see* races of bees,
geographic distribution
Neck foramen, 113, 124, 125
Nectar, **401-420**
artificial, 283, *also see* Fig. 1, p. 283
as carbohydrate, 92, 172, 203
availability, 602, 745
composition, 405-407, 1062
concentration of, 404-405
conversion into honey, 328
energy investment in, 404
enzyme addition, 92

evaporation, 93
external factors affecting, 412
flavor, 288
foragers, 1047
humidity effect on, 415
insecticides in, 407, 745
internal factors affecting, 411
measurement, 408
physiology of secretion, 408
pollination and fertilization, effect of, 411
production, **409-420**
quantity and quality, 602, 658
ripening process, 329-332, *also see* Fig. 27, p.
330
soil factors, effect of, 416-417
storage, 620, *also see* Fig. 1, p. 705
sucrose/hexose ratio in, 406
sugar content, 203, 328
sugar yield of plants, 419, *also see* Fig. 2, p. 419
to bind pollen grains, 929
transfer, 328, *also see* Fig. 27, p. 330
weight, 327
yield, 409
Nectar collecting, **325-328,** 1047, 1100, *also see*
Fig. 25, p. 327
Nectar collecting behavior, 249-250
Nectaries, *also see* Fig. 1, p. 403, Fig. 1, p. 404
anatomy of, 403
cross section of, 404, *also see* Fig. 1, p. 404
extrafloral, 402
extranuptial, 402
floral, 402
locations of, 403
metabolic activity in, 404
nuptial, 402
stomates, 408, *also see* Fig. 2, p. 405
ultra structure of, 409
vascular supply to, 404
Nectar producing plants, *also see* Table 2, p. 419
citrus family (Rutaceae), 419
composites (Compositae), 419
eucalyptus (Myrtaceae), 419
fig wort family (Scrophulariaceae) 419
heath family (Ericaceae), 419
legumes (Leguminosae), 419
Melaleuca quinquenervia, 745
mint family (Labiatae), 419
mustard family (Cruciferae), 419
rose family (Rosaceae), 419
Shinus terebinthifolius, 745
Neodeprion sertifer, 1158
Nephrocytes, 160
Nerve cells, 272
Nerve masses, *see* ganglion
Nerves, also see Fig. 20, p. 151
for orientation, 179

Self pollination, 1052, 1062
Selling, *also see* beekeeping business practice, marketing
retail, **760-769**
wholesale, 769, 772
Semen, 1025, *also see* Fig. 35, p. 1027
shipment of, 1033
Seminal vesicle, 165
Semio chemicals, 374, 382
Senecio jacobaea, 1195
Sense organs, 162, *also see* eyes, brain
Sensory and Nervous System, **162-165**
Sensory nerves, 162, 163
Sensory receptors, 177, *also see* hairs, antennae
Sensory Physiology and Nerve Function, **177-179**
Sepal, 1050, *also see* Fig. 1, p. 1050
Setose membrane, 379, *also see* Fig. 1, p. 377
Sex, *also see* genetics
determinants, 180, 239, 240
habrobracon, 239, 240
mosaics, 241, *also see* gynandromorphs
Shaft, 151, 153, 155, *also see* Plate 21, p. 157
Shaker box, 1041, *also see* Fig. 52, p. 1040
Shipping stress, 607
Sicilian, *see* races of bees, *A.m. sicula*
Sioux Honey Association, 787, 807, *also see* Fig. 4, p. 787
Sitosterol, 207, *also see* sterols
Skeps, *see* hives, early
Slumgum, 688, 689
Smog pollution, *also see* pollutants
effect on bees, 1163
Smoke concentrate, 562-563
Smoking hives, 269, 308-309, 561-563, 577, 588, 595, 608-609, *also see* Fig. 2, p. 577
Social breathing, 186
Sodium content of pollen, 931
Soil,
affecting nectar production, 412, 416-418, 497
maps to locate nectar sources, 746
Solanum nigrum, 1195
Solar
power to melt beeswax, 966
wax melters, 687, *also see* Fig. 22, p. 687
Solar radiation, 603, 644
affecting nectar production, 413, 418, *also see* Fig. 3, p. 414
Solenopsis invicta, 1130, 1209, 1211, 1242, *also see* Fig. 2, p. 1211; ants
Solitary bee, 1045, 1046
Sophora microphylla, 407
Sorbitol, 406, 409, *also see* sugar
Sound,
perception, 179, 852
production, 179
Sourwood, *see* Fig. 9, p. 508

South America
spread of Africanized honey bee, 42, 246, 358
Specialization in activity, 277
Specialty items, 759, 773, 790, 796
Species diversity preference, 319
Species of bees, 1045, *also see* races of bees
Sperm, 238, 1022
released from queen's egg, 81, 105, *also see* Plate 1B, p. 104
Spermatheca, 39, 105, 167, 180, 242, 348, 990, 1022, *also see* Fig. 22, p. 168
Spermatozoa, 39, 104, 165, 167, 348, *also see* Plate 1, p. 104
Spermatozoon, 238
Spiders, 1135
Spin-float honey/wax separator, 669
Spinneret, 107, 108, *also see* Fig. 2, p. 107
Spiracles, 107, 108, 161, 1155, *also see* Figs. 4, p. 111; 19, p. 151
Spiroplasma spp, 1105
Spore, 420
Sporonts, 1105
Sporulation, 1088
Spring conditions of a wintered colony, 849, 850
Spring dwindling, 59
Stage: nest architecture, **73-79**
Stages of honey bee castes, *also see* Fig. 4, p. 82; 4a, p. 83
adult, 81
egg, 81
larva, 81
life span, 84
prepupal, 83
pupa, 81
timing of development, 83
Stamens, 1050
Standards,
Codex Alimentarius, 888
for exhibits of honey, 909
for honey, **905-909**
for HMF, 888, *also see* Fig. 8, p. 889
health, 677
hygiene, 685
Starch content of pollen, 931
Starter seed, 702
Steam chests, 688
Sternal plates, 144
Sternum, 127, 149, 150, 155, *also see* Fig. 15, p. 145
Steroids
for treatment of venom allergy, 1235
Sterols, 206, *also see* lipids; Table 1, p. 208
effect on brood production, 208
in royal jelly, 969, *also see* Table 9, p. 969
isolation, 206, 207
Stigma, 1052, *also see* Fig. 1, p. 1050